STUDENT'S SOLUTIONS MANUAL

JEFFERY A. COLE
Anoka-Ramsey Community College

BEGINNING AND INTERMEDIATE ALGEBRA

FOURTH EDITION

Margaret L. Lial
American River College

John Hornsby
University of New Orleans

Terry McGinnis

PEARSON

Addison
Wesley

Boston San Francisco New York
London Toronto Sydney Tokyo Singapore Madrid
Mexico City Munich Paris Cape Town Hong Kong Montreal

Reproduced by Pearson Addison-Wesley from electronic files supplied by the author.

Copyright © 2008 Pearson Education, Inc.
Publishing as Pearson Addison-Wesley, 75 Arlington Street, Boston, MA 02116.

ISBN-13: 978-0-321-44954-2
ISBN-10: 0-321-44954-1

1 2 3 4 5 6 BB 10 09 08 07

Preface

This *Student's Solutions Manual* contains solutions to selected exercises in the text *Beginning and Intermediate Algebra, Fourth Edition* by Margaret L. Lial, John Hornsby, and Terry McGinnis. It contains solutions to the odd-numbered exercises in each section, all Relating Concepts exercises, as well as solutions to all the exercises in the review sections, the chapter tests, and the cumulative review sections.

This manual is a text supplement and should be read along *with* the text. You should read all exercise solutions in this manual because many concept explanations are given and then used in subsequent solutions. All concepts necessary to solve a particular problem are not reviewed for every exercise. If you are having difficulty with a previously covered concept, refer back to the section where it was covered for more complete help.

A significant number of today's students are involved in various outside activities, and find it difficult, if not impossible, to attend all class sessions; this manual should help meet the needs of these students. In addition, it is my hope that this manual's solutions will enhance the understanding of all readers of the material and provide insights to solving other exercises.

I appreciate feedback concerning errors, solution correctness or style, and manual style. Any comments may be sent directly to me at the address below, at jeff.cole@anokaramsey.edu, or in care of the publisher, Pearson Addison-Wesley.

I would like to thank Ken Grace, of Anoka-Ramsey Community College, and Jeannine Grace, for typesetting the manuscript and providing assistance with many features of the manual; Marv Riedesel and Mary Johnson, for their careful accuracy checking and valuable suggestions; Jim McLaughlin, for his help with the entire art package; and the authors and Maureen O'Connor and Lauren Morse, of Pearson Addison-Wesley, for entrusting me with this project.

Jeffery A. Cole
Anoka-Ramsey Community College
11200 Mississippi Blvd. NW
Coon Rapids, MN 55433

Table of Contents

1 The Real Number System...1

    1.1   Fractions • 1

    1.2   Exponents, Order of Operations, and Inequality • 5

    1.3   Variables, Expressions, and Equations • 7

    1.4   Real Numbers and the Number Line • 10

    1.5   Adding and Subtracting Real Numbers • 12

    1.6   Multiplying and Dividing Real Numbers • 15

          Summary Exercises on Operations with Real Numbers • 18

    1.7   Properties of Real Numbers • 19

    1.8   Simplifying Expressions • 22

          Chapter 1 Review Exercises • 24

          Chapter 1 Test • 30

2 Linear Equations and Inequalities in One Variable..33

    2.1   The Addition Property of Equality • 33

    2.2   The Multiplication Property of Equality • 35

    2.3   More on Solving Linear Equations • 38

          Summary Exercises on Solving Linear Equations • 41

    2.4   An Introduction to Applications of Linear Equations • 42

    2.5   Formulas and Applications from Geometry • 47

    2.6   Ratios and Proportions • 50

    2.7   Further Applications of Linear Equations • 54

    2.8   Solving Linear Inequalities • 58

          Chapter 2 Review Exercises • 62

          Chapter 2 Test • 69

          Cumulative Review Exercises Chapters 1–2 • 71

3 Linear Equations in Two Variables...75

    3.1   Reading Graphs; Linear Equations in Two Variables • 75

    3.2   Graphing Linear Equations in Two Variables • 79

    3.3   The Slope of a Line • 84

    3.4   Equations of a Line • 87

          Chapter 3 Review Exercises • 92

          Chapter 3 Test • 97

          Cumulative Review Exercises Chapters 1–3 • 98

4 Exponents and Polynomials...101

    4.1   The Product Rule and Power Rules for Exponents • 101

    4.2   Integer Exponents and the Quotient Rule • 103

          Summary Exercises on the Rules for Exponents • 105

    4.3   An Application of Exponents: Scientific Notation • 107

    4.4   Adding and Subtracting Polynomials; Graphing Simple Polynomials • 110

    4.5   Multiplying Polynomials • 113

    4.6   Special Products • 117

    4.7   Dividing Polynomials • 120

          Chapter 4 Review Exercises • 124

          Chapter 4 Test • 131

          Cumulative Review Exercises Chapters 1–4 • 133

5 Factoring and Applications..137

    5.1   The Greatest Common Factor; Factoring by Grouping • 137

    5.2   Factoring Trinomials • 140

    5.3   More on Factoring Trinomials • 143

    5.4   Special Factoring Techniques • 148

          Summary Exercises on Factoring • 151

    5.5   Solving Quadratic Equations by Factoring • 153

    5.6   Applications of Quadratic Equations • 159

          Chapter 5 Review Exercises • 163

          Chapter 5 Test • 169

          Cumulative Review Exercises Chapters 1–5 • 171

6   Rational Expressions and Applications..**175**
   6.1   The Fundamental Property of Rational Expressions • 175
   6.2   Multiplying and Dividing Rational Expressions • 178
   6.3   Least Common Denominators • 182
   6.4   Adding and Subtracting Rational Expressions • 185
   6.5   Complex Fractions • 190
   6.6   Solving Equations with Rational Expressions • 194
          Summary Exercises on Rational Expressions and Equations • 200
   6.7   Applications of Rational Expressions • 202
          Chapter 6 Review Exercises • 206
          Chapter 6 Test • 215
          Cumulative Review Exercises Chapters 1–6 • 217

7   Equations of Lines; Functions..**223**
   7.1   Review of Graphs and Slopes of Lines • 223
   7.2   Review of Equations of Lines; Linear Models • 229
          Summary Exercises on Slopes and Equations of Lines • 233
   7.3   Functions • 235
   7.4   Operations on Functions and Composition • 238
   7.5   Variation • 241
          Chapter 7 Review Exercises • 244
          Chapter 7 Test • 250
          Cumulative Review Exercises Chapters 1–7 • 254

8   Systems of Linear Equations..**259**
   8.1   Solving Systems of Linear Equations by Graphing • 259
   8.2   Solving Systems of Linear Equations by Substitution • 263
   8.3   Solving Systems of Linear Equations by Elimination • 268
          Summary Exercises on Solving Systems of Linear Equations • 272
   8.4   Systems of Linear Equations in Three Variables • 275
   8.5   Applications of Systems of Linear Equations • 281
   8.6   Solving Systems of Linear Equations by Matrix Methods • 287
          Chapter 8 Review Exercises • 292
          Chapter 8 Test • 300
          Cumulative Review Exercises Chapters 1–8 • 305

9   Inequalities and Absolute Value..**311**
   9.1   Set Operations and Compound Inequalities • 311
   9.2   Absolute Value Equations and Inequalities • 314
          Summary Exercises on Solving Linear and Absolute Value
          Equations and Inequalities • 318
   9.3   Linear Inequalities in Two Variables • 319
          Chapter 9 Review Exercises • 323
          Chapter 9 Test • 327
          Cumulative Review Exercises Chapters 1–9 • 328

10 Roots, Radicals, and Root Functions..**333**
   10.1   Radical Expressions and Graphs • 333
   10.2   Rational Exponents • 335
   10.3   Simplifying Radical Expressions • 338
   10.4   Adding and Subtracting Radical Expressions • 340
   10.5   Multiplying and Dividing Radical Expressions • 342
          Summary Exercises on Operations with Radicals and Rational Exponents • 346
   10.6   Solving Equations with Radicals • 348
   10.7   Complex Numbers • 352
          Chapter 10 Review Exercises • 354
          Chapter 10 Test • 362
          Cumulative Review Exercises Chapters 1–10 • 364

11 Quadratic Equations, Inequalities, and Functions..**371**

11.1 Solving Quadratic Equations by the Square Root Property • 371
11.2 Solving Quadratic Equations by Completing the Square • 373
11.3 Solving Quadratic Equations by the Quadratic Formula • 377
11.4 Equations Quadratic in Form • 382
      Summary Exercises on Solving Quadratic Equations • 389
11.5 Formulas and Further Applications • 391
11.6 Graphs of Quadratic Functions • 395
11.7 More About Parabolas and Their Applications • 398
11.8 Quadratic and Rational Inequalities • 403
      Chapter 11 Review Exercises • 409
      Chapter 11 Test • 423
      Cumulative Review Exercises Chapters 1–11 • 427

12 Inverse, Exponential, and Logarithmic Functions..**433**

12.1 Inverse Functions • 433
12.2 Exponential Functions • 436
12.3 Logarithmic Functions • 438
12.4 Properties of Logarithms • 441
12.5 Common and Natural Logarithms • 442
12.6 Exponential and Logarithmic Equations; Further Applications • 444
      Chapter 12 Review Exercises • 448
      Chapter 12 Test • 455
      Cumulative Review Exercises Chapters 1–12 • 458

13 Nonlinear Functions, Conic Sections, and Nonlinear Systems.............................**463**

13.1 Additional Graphs of Functions • 463
13.2 The Circle and the Ellipse • 464
13.3 The Hyperbola and Functions Defined by Radicals • 468
13.4 Nonlinear Systems of Equations • 472
13.5 Second-Degree Inequalities and Systems of Inequalities • 477
      Chapter 13 Review Exercises • 482
      Chapter 13 Test • 489
      Cumulative Review Exercises Chapters 1–13 • 492

14 Sequences and Series...**499**

14.1 Sequences and Series • 499
14.2 Arithmetic Sequences • 501
14.3 Geometric Sequences • 503
14.4 The Binomial Theorem • 506
      Chapter 14 Review Exercises • 508
      Chapter 14 Test • 512
      Cumulative Review Exercises Chapters 1–14 • 514

Appendix B Review of Decimals and Percents • 521
Appendix C Sets • 523
Appendix D Mean, Median, and Mode • 525
Appendix E The Metric System and Conversions • 527
Appendix F Review of Exponents, Polynomials, and Factoring • 529
Appendix G Synthetic Division • 531
Appendix H Determinants and Cramer's Rule • 533

CHAPTER 1   THE REAL NUMBER SYSTEM

1.1   Fractions

1. True; the number above the fraction bar is called the numerator and the number below the fraction bar is called the denominator.

3. False; the fraction $\dfrac{17}{51}$ can be written in lowest terms as $\dfrac{1}{3}$ since $\dfrac{17}{51} = \dfrac{17 \cdot 1}{17 \cdot 3} = \dfrac{1}{3}$.

5. False; *product* refers to multiplication, so the product of 8 and 2 is 16. The *sum* of 8 and 2 is 10.

7. Since 19 has only itself and 1 as factors, it is a prime number.

9. $64 = 2 \cdot 32$
$= 2 \cdot 2 \cdot 16$
$= 2 \cdot 2 \cdot 2 \cdot 8$
$= 2 \cdot 2 \cdot 2 \cdot 2 \cdot 4$
$= 2 \cdot 2 \cdot 2 \cdot 2 \cdot 2 \cdot 2$

Since 64 has factors other than itself and 1, it is a composite number.

11. $3458 = 2 \cdot 1729$
$= 2 \cdot 7 \cdot 247$
$= 2 \cdot 7 \cdot 13 \cdot 19$

Since 3458 has factors other than itself and 1, it is a composite number.

13. As stated in the text, the number 1 is neither prime nor composite, by agreement.

15. $30 = 2 \cdot 15$
$= 2 \cdot 3 \cdot 5$

Since 30 has factors other than itself and 1, it is a composite number.

17. $500 = 2 \cdot 250$
$= 2 \cdot 2 \cdot 125$
$= 2 \cdot 2 \cdot 5 \cdot 25$
$= 2 \cdot 2 \cdot 5 \cdot 5 \cdot 5,$

so 500 is a composite number.

19. $124 = 2 \cdot 62$
$= 2 \cdot 2 \cdot 31,$

so 124 is a composite number.

21. Since 29 has only itself and 1 as factors, it is a prime number.

23. $\dfrac{8}{16} = \dfrac{1 \cdot 8}{2 \cdot 8} = \dfrac{1}{2}$

25. $\dfrac{15}{18} = \dfrac{3 \cdot 5}{3 \cdot 6} = \dfrac{5}{6}$

27. $\dfrac{18}{90} = \dfrac{1 \cdot 18}{5 \cdot 18} = \dfrac{1}{5}$

29. $\dfrac{144}{120} = \dfrac{6 \cdot 24}{5 \cdot 24} = \dfrac{6}{5}$

31. $\dfrac{16}{24} = \dfrac{2 \cdot 8}{3 \cdot 8} = \dfrac{2}{3}$

Therefore, **C** is correct.

33. $\dfrac{4}{5} \cdot \dfrac{6}{7} = \dfrac{4 \cdot 6}{5 \cdot 7} = \dfrac{24}{35}$

35. $\dfrac{1}{10} \cdot \dfrac{12}{5} = \dfrac{1 \cdot 12}{10 \cdot 5} = \dfrac{1 \cdot 2 \cdot 6}{2 \cdot 5 \cdot 5} = \dfrac{6}{25}$

37. $\dfrac{15}{4} \cdot \dfrac{8}{25} = \dfrac{15 \cdot 8}{4 \cdot 25}$
$= \dfrac{3 \cdot 5 \cdot 4 \cdot 2}{4 \cdot 5 \cdot 5}$
$= \dfrac{3 \cdot 2}{5}$
$= \dfrac{6}{5}, \text{ or } 1\dfrac{1}{5}$

39. $3\dfrac{1}{4} \cdot 1\dfrac{2}{3}$

Change both mixed numbers to improper fractions.

$3\dfrac{1}{4} = 3 + \dfrac{1}{4} = \dfrac{12}{4} + \dfrac{1}{4} = \dfrac{13}{4}$
$1\dfrac{2}{3} = 1 + \dfrac{2}{3} = \dfrac{3}{3} + \dfrac{2}{3} = \dfrac{5}{3}$

$3\dfrac{1}{4} \cdot 1\dfrac{2}{3} = \dfrac{13}{4} \cdot \dfrac{5}{3}$
$= \dfrac{13 \cdot 5}{4 \cdot 3}$
$= \dfrac{65}{12}, \text{ or } 5\dfrac{5}{12}$

41. $2\dfrac{3}{8} \cdot 3\dfrac{1}{5}$

Change both mixed numbers to improper fractions.

$2\dfrac{3}{8} = 2 + \dfrac{3}{8} = \dfrac{16}{8} + \dfrac{3}{8} = \dfrac{19}{8}$
$3\dfrac{1}{5} = 3 + \dfrac{1}{5} = \dfrac{15}{5} + \dfrac{1}{5} = \dfrac{16}{5}$

$2\dfrac{3}{8} \cdot 3\dfrac{1}{5} = \dfrac{19}{8} \cdot \dfrac{16}{5}$
$= \dfrac{19 \cdot 16}{8 \cdot 5}$
$= \dfrac{19 \cdot 2 \cdot 8}{8 \cdot 5}$
$= \dfrac{38}{5}, \text{ or } 7\dfrac{3}{5}$

43. $\dfrac{5}{4} \div \dfrac{3}{8} = \dfrac{5}{4} \cdot \dfrac{8}{3}$ *Multiply by the reciprocal of the second fraction.*

$$= \dfrac{5 \cdot 8}{4 \cdot 3}$$

$$= \dfrac{5 \cdot 4 \cdot 2}{4 \cdot 3}$$

$$= \dfrac{5 \cdot 2}{3}$$

$$= \dfrac{10}{3}, \text{ or } 3\dfrac{1}{3}$$

45. $\dfrac{32}{5} \div \dfrac{8}{15} = \dfrac{32}{5} \cdot \dfrac{15}{8}$ *Multiply by the reciprocal of the second fraction.*

$$= \dfrac{32 \cdot 15}{5 \cdot 8}$$

$$= \dfrac{8 \cdot 4 \cdot 3 \cdot 5}{1 \cdot 5 \cdot 8}$$

$$= \dfrac{4 \cdot 3}{1} = 12$$

47. $\dfrac{3}{4} \div 12 = \dfrac{3}{4} \cdot \dfrac{1}{12}$ *Multiply by the reciprocal of 12.*

$$= \dfrac{3 \cdot 1}{4 \cdot 12}$$

$$= \dfrac{3 \cdot 1}{4 \cdot 3 \cdot 4}$$

$$= \dfrac{1}{4 \cdot 4} = \dfrac{1}{16}$$

49. $2\dfrac{1}{2} \div 1\dfrac{5}{7}$

Change both mixed numbers to improper fractions.

$$2\dfrac{1}{2} = 2 + \dfrac{1}{2} = \dfrac{4}{2} + \dfrac{1}{2} = \dfrac{5}{2}$$

$$1\dfrac{5}{7} = 1 + \dfrac{5}{7} = \dfrac{7}{7} + \dfrac{5}{7} = \dfrac{12}{7}$$

$$2\dfrac{1}{2} \div 1\dfrac{5}{7} = \dfrac{5}{2} \div \dfrac{12}{7}$$

$$= \dfrac{5}{2} \cdot \dfrac{7}{12}$$ *Multiply by the reciprocal of the second fraction.*

$$= \dfrac{5 \cdot 7}{2 \cdot 12}$$

$$= \dfrac{35}{24}, \text{ or } 1\dfrac{11}{24}$$

51. $2\dfrac{5}{8} \div 1\dfrac{15}{32}$

Change both mixed numbers to improper fractions.

$$2\dfrac{5}{8} = 2 + \dfrac{5}{8} = \dfrac{16}{8} + \dfrac{5}{8} = \dfrac{21}{8}$$

$$1\dfrac{15}{32} = 1 + \dfrac{15}{32} = \dfrac{32}{32} + \dfrac{15}{32} = \dfrac{47}{32}$$

$$2\dfrac{5}{8} \div 1\dfrac{15}{32} = \dfrac{21}{8} \div \dfrac{47}{32}$$

$$= \dfrac{21}{8} \cdot \dfrac{32}{47}$$

$$= \dfrac{21 \cdot 32}{8 \cdot 47}$$

$$= \dfrac{21 \cdot 8 \cdot 4}{8 \cdot 47}$$

$$= \dfrac{21 \cdot 4}{47}$$

$$= \dfrac{84}{47}, \text{ or } 1\dfrac{37}{47}$$

53. To multiply two fractions, multiply their numerators to get the numerator of the product and multiply their denominators to get the denominator of the product. For example,

$$\dfrac{2}{3} \cdot \dfrac{8}{5} = \dfrac{2 \cdot 8}{3 \cdot 5} = \dfrac{16}{15}.$$

To divide two fractions, replace the divisor with its reciprocal and then multiply. For example,

$$\dfrac{2}{5} \div \dfrac{7}{9} = \dfrac{2}{5} \cdot \dfrac{9}{7} = \dfrac{2 \cdot 9}{5 \cdot 7} = \dfrac{18}{35}.$$

55. $\dfrac{7}{12} + \dfrac{1}{12} = \dfrac{7 + 1}{12}$

$$= \dfrac{8}{12}$$

$$= \dfrac{2 \cdot 4}{3 \cdot 4} = \dfrac{2}{3}$$

57. $\dfrac{5}{9} + \dfrac{1}{3}$

Since $9 = 3 \cdot 3$, and 3 is prime, the LCD (least common denominator) is $3 \cdot 3 = 9$.

$$\dfrac{1}{3} = \dfrac{1}{3} \cdot \dfrac{3}{3} = \dfrac{3}{9}$$

Now add the two fractions with the same denominator.

$$\dfrac{5}{9} + \dfrac{1}{3} = \dfrac{5}{9} + \dfrac{3}{9} = \dfrac{8}{9}$$

59. $3\dfrac{1}{8} + 2\dfrac{1}{4}$

$$3\dfrac{1}{8} = 3 + \dfrac{1}{8} = \dfrac{24}{8} + \dfrac{1}{8} = \dfrac{25}{8}$$

$$2\dfrac{1}{4} = 2 + \dfrac{1}{4} = \dfrac{8}{4} + \dfrac{1}{4} = \dfrac{9}{4}$$

$$3\dfrac{1}{8} + 2\dfrac{1}{4} = \dfrac{25}{8} + \dfrac{9}{4}$$

Since $8 = 2 \cdot 2 \cdot 2$ and $4 = 2 \cdot 2$, the LCD is $2 \cdot 2 \cdot 2$ or 8.

$$\begin{aligned} 3\dfrac{1}{8} + 2\dfrac{1}{4} &= \dfrac{25}{8} + \dfrac{9 \cdot 2}{4 \cdot 2} \\ &= \dfrac{25}{8} + \dfrac{18}{8} \\ &= \dfrac{43}{8}, \ \text{ or } \ 5\dfrac{3}{8} \end{aligned}$$

61. $3\dfrac{1}{4} + 1\dfrac{4}{5}$

$$3\dfrac{1}{4} = 3 + \dfrac{1}{4} = \dfrac{12}{4} + \dfrac{1}{4} = \dfrac{13}{4}$$

$$1\dfrac{4}{5} = 1 + \dfrac{4}{5} = \dfrac{5}{5} + \dfrac{4}{5} = \dfrac{9}{5}$$

Since $4 = 2 \cdot 2$, and 5 is prime, the LCD is $2 \cdot 2 \cdot 5 = 20$.

$$\begin{aligned} 3\dfrac{1}{4} + 1\dfrac{4}{5} &= \dfrac{13 \cdot 5}{4 \cdot 5} + \dfrac{9 \cdot 4}{5 \cdot 4} \\ &= \dfrac{65}{20} + \dfrac{36}{20} \\ &= \dfrac{101}{20}, \ \text{ or } \ 5\dfrac{1}{20} \end{aligned}$$

63. $\dfrac{13}{15} - \dfrac{3}{15} = \dfrac{13 - 3}{15}$

$$\begin{aligned} &= \dfrac{10}{15} \\ &= \dfrac{2 \cdot 5}{3 \cdot 5} = \dfrac{2}{3} \end{aligned}$$

65. $\dfrac{7}{12} - \dfrac{1}{9}$

Since $12 = 2 \cdot 2 \cdot 3$ and $9 = 3 \cdot 3$, the LCD is $2 \cdot 2 \cdot 3 \cdot 3 = 36$.

$$\dfrac{7}{12} = \dfrac{7}{12} \cdot \dfrac{3}{3} = \dfrac{21}{36} \text{ and } \dfrac{1}{9} \cdot \dfrac{4}{4} = \dfrac{4}{36}$$

Now subtract fractions with the same denominator.

$$\dfrac{7}{12} - \dfrac{1}{9} = \dfrac{21}{36} - \dfrac{4}{36} = \dfrac{17}{36}$$

67. $4\dfrac{3}{4} - 1\dfrac{2}{5}$

$$4\dfrac{3}{4} = 4 + \dfrac{3}{4} = \dfrac{16}{4} + \dfrac{3}{4} = \dfrac{19}{4}$$

$$1\dfrac{2}{5} = 1 + \dfrac{2}{5} = \dfrac{5}{5} + \dfrac{2}{5} = \dfrac{7}{5}$$

Since $4 = 2 \cdot 2$, and 5 is prime, the LCD is $2 \cdot 2 \cdot 5 = 20$.

$$\begin{aligned} 4\dfrac{3}{4} - 1\dfrac{2}{5} &= \dfrac{19 \cdot 5}{4 \cdot 5} - \dfrac{7 \cdot 4}{5 \cdot 4} \\ &= \dfrac{95}{20} - \dfrac{28}{20} \\ &= \dfrac{67}{20}, \ \text{ or } \ 3\dfrac{7}{20} \end{aligned}$$

69. $6\dfrac{1}{4} - 5\dfrac{1}{3}$

$$6\dfrac{1}{4} = 6 + \dfrac{1}{4} = \dfrac{24}{4} + \dfrac{1}{4} = \dfrac{25}{4}$$

$$5\dfrac{1}{3} = 5 + \dfrac{1}{3} = \dfrac{15}{3} + \dfrac{1}{3} = \dfrac{16}{3}$$

Since $4 = 2 \cdot 2$, and 3 is prime, the LCD is $2 \cdot 2 \cdot 3 = 12$.

$$\begin{aligned} 6\dfrac{1}{4} - 5\dfrac{1}{3} &= \dfrac{25}{4} - \dfrac{16}{3} \\ &= \dfrac{25 \cdot 3}{4 \cdot 3} - \dfrac{16 \cdot 4}{3 \cdot 4} \\ &= \dfrac{75}{12} - \dfrac{64}{12} \\ &= \dfrac{11}{12} \end{aligned}$$

71. Multiply the number of cups of water per serving by the number of servings.

$$\begin{aligned} \dfrac{3}{4} \cdot 8 &= \dfrac{3}{4} \cdot \dfrac{8}{1} \\ &= \dfrac{3 \cdot 8}{4 \cdot 1} \\ &= \dfrac{3 \cdot 2 \cdot 4}{4 \cdot 1} \\ &= \dfrac{3 \cdot 2}{1} = 6 \text{ cups} \end{aligned}$$

For 8 microwave servings, 6 cups of water will be needed.

73. The difference in length is found by subtracting.

$$\begin{aligned} 3\dfrac{1}{4} - 2\dfrac{1}{8} &= \dfrac{13}{4} - \dfrac{17}{8} \\ &= \dfrac{13 \cdot 2}{4 \cdot 2} - \dfrac{17}{8} \quad LCD = 8 \\ &= \dfrac{26}{8} - \dfrac{17}{8} \\ &= \dfrac{9}{8}, \ \text{ or } \ 1\dfrac{1}{8} \end{aligned}$$

The difference is $1\dfrac{1}{8}$ inches.

75. The difference between the two measures is found by subtracting, using 16 as the LCD.

$$\frac{3}{4} - \frac{3}{16} = \frac{3 \cdot 4}{4 \cdot 4} - \frac{3}{16}$$
$$= \frac{12}{16} - \frac{3}{16}$$
$$= \frac{12 - 3}{16} = \frac{9}{16}$$

The difference is $\frac{9}{16}$ inch.

77. The perimeter is the sum of the measures of the 5 sides.

$$196 + 98\frac{3}{4} + 146\frac{1}{2} + 100\frac{7}{8} + 76\frac{5}{8}$$
$$= 196 + 98\frac{6}{8} + 146\frac{4}{8} + 100\frac{7}{8} + 76\frac{5}{8}$$
$$= 196 + 98 + 146 + 100 + 76 + \frac{6 + 4 + 7 + 5}{8}$$
$$= 616 + \frac{22}{8} \quad \left(\frac{22}{8} = 2\frac{6}{8} = 2\frac{3}{4}\right)$$
$$= 618\frac{3}{4} \text{ feet}$$

The perimeter is $618\frac{3}{4}$ feet.

79. Divide the total board length by 3.

$$15\frac{5}{8} \div 3 = \frac{125}{8} \div \frac{3}{1}$$
$$= \frac{125}{8} \cdot \frac{1}{3}$$
$$= \frac{125 \cdot 1}{8 \cdot 3}$$
$$= \frac{125}{24}, \text{ or } 5\frac{5}{24}$$

The length of each of the three pieces must be $5\frac{5}{24}$ inches.

81. To find the number of cakes the caterer can make, divide $15\frac{1}{2}$ by $1\frac{3}{4}$.

$$15\frac{1}{2} \div 1\frac{3}{4} = \frac{31}{2} \div \frac{7}{4}$$
$$= \frac{31}{2} \cdot \frac{4}{7}$$
$$= \frac{31 \cdot 2 \cdot 2}{2 \cdot 7}$$
$$= \frac{62}{7}, \text{ or } 8\frac{6}{7}$$

There is not quite enough sugar for 9 cakes. The caterer can make 8 cakes with some sugar left over.

83. Multiply the amount of fabric it takes to make one costume by the number of costumes.

$$2\frac{3}{8} \cdot 7 = \frac{19}{8} \cdot \frac{7}{1}$$
$$= \frac{19 \cdot 7}{8 \cdot 1}$$
$$= \frac{133}{8}, \text{ or } 16\frac{5}{8} \text{ yd}$$

For 7 costumes, $16\frac{5}{8}$ yards of fabric would be needed.

85. Subtract the heights to find the difference.

$$10\frac{1}{2} - 7\frac{1}{8} = \frac{21}{2} - \frac{57}{8}$$
$$= \frac{21 \cdot 4}{2 \cdot 4} - \frac{57}{8} \quad LCD = 8$$
$$= \frac{84}{8} - \frac{57}{8}$$
$$= \frac{27}{8}, \text{ or } 3\frac{3}{8}$$

The difference in heights is $3\frac{3}{8}$ inches.

87. The sum of the fractions representing the U.S. foreign-born population from Latin America, Asia, or Europe is

$$\frac{27}{50} + \frac{1}{4} + \frac{7}{50} = \frac{27 \cdot 2}{50 \cdot 2} + \frac{1 \cdot 25}{4 \cdot 25} + \frac{7 \cdot 2}{50 \cdot 2}$$
$$= \frac{54 + 25 + 14}{100}$$
$$= \frac{93}{100}.$$

So the fraction representing the U.S. foreign-born population from other regions is

$$1 - \frac{93}{100} = \frac{100}{100} - \frac{93}{100}$$
$$= \frac{7}{100}.$$

89. Multiply the fraction representing the U.S. foreign-born population from Europe, $\frac{7}{50}$, by the total number of foreign-born people in the U.S., approximately 34 million.

$$\frac{7}{50} \cdot 34 = \frac{7}{50} \cdot \frac{34}{1} = \frac{7 \cdot 2 \cdot 17}{2 \cdot 25} = \frac{119}{25}, \text{ or } 4\frac{19}{25}$$

There were approximately $4\frac{19}{25}$ million foreign-born people in the U.S. in 2004 who were born in Europe.

91. Observe that there are 24 dots in the entire figure, 6 dots in the triangle, 12 dots in the rectangle, and 2 dots in the overlapping region.

(a) $\frac{12}{24} = \frac{1}{2}$ of all the dots are in the rectangle.

(b) $\frac{6}{24} = \frac{1}{4}$ of all the dots are in the triangle.

(c) $\frac{2}{6} = \frac{1}{3}$ of the dots in the triangle are in the overlapping region.

(d) $\frac{2}{12} = \frac{1}{6}$ of the dots in the rectangle are in the overlapping region.

1.2 Exponents, Order of Operations, and Inequality

1. False; $4 + 3(8 - 2) = 4 + 3 \cdot 6 = 4 + 18 = 22$. The common error leading to 42 is adding 4 to 3 and then multiplying by 6. One must follow the rules for order of operations.

3. False; the correct interpretation is $4 = 16 - 12$.

5. $7^2 = 7 \cdot 7 = 49$

7. $12^2 = 12 \cdot 12 = 144$

9. $4^3 = 4 \cdot 4 \cdot 4 = 64$

11. $10^3 = 10 \cdot 10 \cdot 10 = 1000$

13. $3^4 = 3 \cdot 3 \cdot 3 \cdot 3 = 81$

15. $4^5 = 4 \cdot 4 \cdot 4 \cdot 4 \cdot 4 = 1024$

17. $\left(\frac{2}{3}\right)^4 = \frac{2}{3} \cdot \frac{2}{3} \cdot \frac{2}{3} \cdot \frac{2}{3} = \frac{16}{81}$

19. $(0.4)^3 = (0.4)(0.4)(0.4) = 0.064$

21. To evaluate 6^3, multiply the base, 6, by itself 3 times. The exponent, 3, indicates the number of times to multiply the base by itself.

23. $64 \div 4 \cdot 2 = (64 \div 4) \cdot 2$
$= 16 \cdot 2$
$= 32$

25. $13 + 9 \cdot 5 = 13 + 45$ *Multiply.*
$= 58$ *Add.*

27. $25.2 - 12.6 \div 4.2 = 25.2 - 3$ *Divide.*
$= 22.2$ *Subtract.*

29. $\frac{1}{4} \cdot \frac{2}{3} + \frac{2}{5} \cdot \frac{11}{3} = \frac{1}{6} + \frac{22}{15}$ *Multiply.*

$= \frac{5}{30} + \frac{44}{30}$ *LCD = 30*

$= \frac{49}{30}$, or $1\frac{19}{30}$ *Add.*

31. $9 \cdot 4 - 8 \cdot 3 = 36 - 24$ *Multiply.*
$= 12$ *Subtract.*

33. $20 - 4 \cdot 3 + 5 = 20 - 12 + 5$ *Multiply.*
$= 8 + 5$ *Subtract.*
$= 13$ *Add.*

35. $10 + 40 \div 5 \cdot 2 = 10 + 8 \cdot 2$ *Divide.*
$= 10 + 16$ *Multiply.*
$= 26$ *Add.*

37. $18 - 2(3 + 4) = 18 - 2(7)$ *Add inside parentheses.*
$= 18 - 14$ *Multiply.*
$= 4$ *Subtract.*

39. $3(4 + 2) + 8 \cdot 3 = 3 \cdot 6 + 8 \cdot 3$ *Add.*
$= 18 + 24$ *Multiply.*
$= 42$ *Add.*

41. $18 - 4^2 + 3$
$= 18 - 16 + 3$ *Use the exponent.*
$= 2 + 3$ *Subtract.*
$= 5$ *Add.*

43. $5[3 + 4(2^2)]$
$= 5[3 + 4(4)]$ *Use the exponent.*
$= 5(3 + 16)$ *Multiply.*
$= 5(19)$ *Add.*
$= 95$ *Multiply.*

45. $3^2[(11 + 3) - 4]$
$= 3^2[14 - 4]$ *Add inside parentheses.*
$= 3^2[10]$ *Subtract.*
$= 9[10]$ *Use the exponent.*
$= 90$ *Multiply.*

47. Simplify the numerator and denominator separately; then divide.

$$\frac{6(3^2 - 1) + 8}{8 - 2^2} = \frac{6(9 - 1) + 8}{8 - 4}$$

$$= \frac{6(8) + 8}{4}$$

$$= \frac{48 + 8}{4}$$

$$= \frac{56}{4} = 14$$

49. $\dfrac{4(6 + 2) + 8(8 - 3)}{6(4 - 2) - 2^2} = \dfrac{4(8) + 8(5)}{6(2) - 2^2}$

$$= \frac{4(8) + 8(5)}{6(2) - 4}$$

$$= \frac{32 + 40}{12 - 4}$$

$$= \frac{72}{8} = 9$$

51. Begin by squaring 2. Then subtract 1 to get a result of $4 - 1 = 3$ within the parentheses. Next, raise 3 to the third power to get $3^3 = 27$. Multiply this result by 3 to obtain 81. Finally, add this result to 4 to get the final answer, 85.

$$
\begin{aligned}
4 + 3\left(2^2 - 1\right)^3 &= 4 + 3(4 - 1)^3 \\
&= 4 + 3(3)^3 \\
&= 4 + 3(27) \\
&= 4 + 81 = 85
\end{aligned}
$$

53.
$$
\begin{aligned}
9 \cdot 3 - 11 &\leq 16 \\
27 - 11 &\leq 16 \\
16 &\leq 16
\end{aligned}
$$

The statement is true since $16 = 16$ is true.

55.
$$
\begin{aligned}
5 \cdot 11 + 2 \cdot 3 &\leq 60 \\
55 + 6 &\leq 60 \\
61 &\leq 60
\end{aligned}
$$

The statement is false since 61 *is greater than* 60.

57.
$$
\begin{aligned}
0 &\geq 12 \cdot 3 - 6 \cdot 6 \\
0 &\geq 36 - 36 \\
0 &\geq 0
\end{aligned}
$$

The statement is true since $0 = 0$ is true.

59.
$$
\begin{aligned}
45 &\geq 2[2 + 3(2 + 5)] \\
45 &\geq 2[2 + 3(7)] \\
45 &\geq 2[2 + 21] \\
45 &\geq 2[23] \\
45 &\geq 46
\end{aligned}
$$

The statement is false since 45 *is less than* 46.

61.
$$
\begin{aligned}
[3 \cdot 4 + 5(2)] \cdot 3 &> 72 \\
[12 + 10] \cdot 3 &> 72 \\
[22] \cdot 3 &> 72 \\
66 &> 72
\end{aligned}
$$

The statement is false since 66 *is less than* 72.

63.
$$
\begin{aligned}
\frac{3 + 5(4 - 1)}{2 \cdot 4 + 1} &\geq 3 \\
\frac{3 + 5(3)}{8 + 1} &\geq 3 \\
\frac{3 + 15}{9} &\geq 3 \\
\frac{18}{9} &\geq 3 \\
2 &\geq 3
\end{aligned}
$$

The statement is false since 2 *is less than* 3.

65.
$$
\begin{aligned}
3 &\geq \frac{2(5 + 1) - 3(1 + 1)}{5(8 - 6) - 4 \cdot 2} \\
3 &\geq \frac{2(6) - 3(2)}{5(2) - 8} \\
3 &\geq \frac{12 - 6}{10 - 8} \\
3 &\geq \frac{6}{2} \\
3 &\geq 3
\end{aligned}
$$

The statement is true since $3 = 3$ is true.

67. "Fifteen is equal to five plus ten" is written

$$15 = 5 + 10.$$

69. "Nine is greater than five minus four" is written

$$9 > 5 - 4.$$

71. "Sixteen is not equal to nineteen" is written

$$16 \neq 19.$$

73. "One-half is less than or equal to two-fourths" is written

$$\frac{1}{2} \leq \frac{2}{4}.$$

75. "$7 < 19$" means "seven is less than nineteen." The statement is true.

77. "$3 \neq 6$" means "three is not equal to six." The statement is true.

79. "$8 \geq 11$" means "eight is greater than or equal to eleven." The statement is false.

81. Answers will vary. One example is

$$5 + 3 \geq 2 \cdot 2.$$

The statement is true since $8 > 4$.

83. $5 < 30$ becomes $30 > 5$ when the inequality symbol is reversed.

85. $2.5 \geq 1.3$ becomes $1.3 \leq 2.5$ when the inequality symbol is reversed.

87. In comparing age, "is younger than" expresses the idea of "is less than."

89. **(a)** Substitute "40" for "age" in the expression for women.

$$14.7 - 40 \cdot 0.13$$

(b) $14.7 - 40 \cdot 0.13 = 14.7 - 5.2$ *Multiply.*
$$ = 9.5 \qquad \text{\emph{Subtract.}}$$

(c) 85% of $9.5 = 0.85(9.5) = 8.075$

Walking at 5 mph is associated with 8.0 METs, which is the table value closest to 8.075.

91. Answers will vary.

93. Look for the bars that are lower than the preceding bars. The corresponding years are 1998, 1999, and 2000.

95. $3 \cdot 6 + 4 \cdot 2 = 60$

Listed below are some possibilities. We'll use trial and error until we get the desired result.

$$(3 \cdot 6) + 4 \cdot 2 = 18 + 8 = 26 \neq 60$$
$$(3 \cdot 6 + 4) \cdot 2 = 22 \cdot 2 = 44 \neq 60$$
$$3 \cdot (6 + 4 \cdot 2) = 3 \cdot 14 = 42 \neq 60$$
$$3 \cdot (6 + 4) \cdot 2 = 3 \cdot 10 \cdot 2 = 30 \cdot 2 = 60$$

97. $10 - 7 - 3 = 6$

$$10 - (7 - 3) = 10 - 4 = 6$$

99. $8 + 2^2 = 100$

$$(8 + 2)^2 = 10^2 = 10 \cdot 10 = 100$$

1.3 Variables, Expressions, and Equations

1. If $x = 3$, then the value of $x + 7$ is $3 + 7$, or 10.

3. The sum of 12 and x is represented by the expression $12 + x$. If $x = 9$, then the value of $12 + x$ is $12 + 9$, or 21.

5. This question is equivalent to asking "is a number ever equal to four more than itself?" Since that never occurs, the answer is no.

7. $2x^3 = 2 \cdot x \cdot x \cdot x$, while $2x \cdot 2x \cdot 2x = (2x)^3$. The last expression is equal to $8x^3$.

9. The exponent 2 applies only to its base, which is x. (The expression $(4x)^2$ would require multiplying 4 by $x = 3$ first.)

11. (Answers will vary.) Two such pairs are $x = 0$, $y = 6$ and $x = 1$, $y = 4$. To find a pair, choose one number, substitute it for a variable, and then calculate the value for the other variable.

In part (a) of Exercises 13–26, replace x with 4. In part (b), replace x with 6. Then use the order of operations.

13. **(a)** $x + 9 = 4 + 9$
$$= 13$$

(b) $x + 9 = 6 + 9$
$$= 15$$

15. **(a)** $5x = 5(4) = 20$

(b) $5x = 5(6) = 30$

17. **(a)** $4x^2 = 4 \cdot 4^2$
$$= 4 \cdot 16$$
$$= 64$$

(b) $4x^2 = 4 \cdot 6^2$
$$= 4 \cdot 36$$
$$= 144$$

19. **(a)** $\dfrac{x+1}{3} = \dfrac{4+1}{3}$
$$= \dfrac{5}{3}$$

(b) $\dfrac{x+1}{3} = \dfrac{6+1}{3}$
$$= \dfrac{7}{3}$$

21. **(a)** $\dfrac{3x-5}{2x} = \dfrac{3 \cdot 4 - 5}{2 \cdot 4}$
$$= \dfrac{12 - 5}{8}$$
$$= \dfrac{7}{8}$$

(b) $\dfrac{3x-5}{2x} = \dfrac{3 \cdot 6 - 5}{2 \cdot 6}$
$$= \dfrac{18 - 5}{12}$$
$$= \dfrac{13}{12}$$

23. **(a)** $3x^2 + x = 3 \cdot 4^2 + 4$
$$= 3 \cdot 16 + 4$$
$$= 48 + 4 = 52$$

(b) $3x^2 + x = 3 \cdot 6^2 + 6$
$$= 3 \cdot 36 + 6$$
$$= 108 + 6 = 114$$

25. **(a)** $6.459x = 6.459 \cdot 4$
$$= 25.836$$

(b) $6.459x = 6.459 \cdot 6$
$$= 38.754$$

In part (a) of Exercises 27–42, replace x with 2 and y with 1. In part (b), replace x with 1 and y with 5.

27. **(a)** $8x + 3y + 5 = 8(2) + 3(1) + 5$
$$= 16 + 3 + 5$$
$$= 19 + 5$$
$$= 24$$

(b) $8x + 3y + 5 = 8(1) + 3(5) + 5$
$$= 8 + 15 + 5$$
$$= 23 + 5$$
$$= 28$$

29. **(a)** $3(x + 2y) = 3(2 + 2 \cdot 1)$
$$= 3(2 + 2)$$
$$= 3(4)$$
$$= 12$$

(b) $3(x + 2y) = 3(1 + 2 \cdot 5)$
$$= 3(1 + 10)$$
$$= 3(11)$$
$$= 33$$

31. **(a)** $x + \dfrac{4}{y} = 2 + \dfrac{4}{1}$
$$= 2 + 4$$
$$= 6$$

(b) $x + \dfrac{4}{y} = 1 + \dfrac{4}{5}$
$$= \dfrac{5}{5} + \dfrac{4}{5}$$
$$= \dfrac{9}{5}$$

33. **(a)** $\dfrac{x}{2} + \dfrac{y}{3} = \dfrac{2}{2} + \dfrac{1}{3}$
$$= \dfrac{6}{6} + \dfrac{2}{6}$$
$$= \dfrac{8}{6} = \dfrac{4}{3}$$

(b) $\dfrac{x}{2} + \dfrac{y}{3} = \dfrac{1}{2} + \dfrac{5}{3}$
$$= \dfrac{3}{6} + \dfrac{10}{6}$$
$$= \dfrac{13}{6}$$

35. **(a)** $\dfrac{2x + 4y - 6}{5y + 2} = \dfrac{2(2) + 4(1) - 6}{5(1) + 2}$
$$= \dfrac{4 + 4 - 6}{5 + 2}$$
$$= \dfrac{8 - 6}{7}$$
$$= \dfrac{2}{7}$$

(b) $\dfrac{2x + 4y - 6}{5y + 2} = \dfrac{2(1) + 4(5) - 6}{5(5) + 2}$
$$= \dfrac{2 + 20 - 6}{25 + 2}$$
$$= \dfrac{22 - 6}{27}$$
$$= \dfrac{16}{27}$$

37. **(a)** $2y^2 + 5x = 2 \cdot 1^2 + 5 \cdot 2$
$$= 2 \cdot 1 + 5 \cdot 2$$
$$= 2 + 10$$
$$= 12$$

(b) $2y^2 + 5x = 2 \cdot 5^2 + 5 \cdot 1$
$$= 2 \cdot 25 + 5 \cdot 1$$
$$= 50 + 5$$
$$= 55$$

39. **(a)** $\dfrac{3x + y^2}{2x + 3y} = \dfrac{3(2) + 1^2}{2(2) + 3(1)}$
$$= \dfrac{3(2) + 1}{4 + 3}$$
$$= \dfrac{6 + 1}{7}$$
$$= \dfrac{7}{7}$$
$$= 1$$

(b) $\dfrac{3x + y^2}{2x + 3y} = \dfrac{3(1) + 5^2}{2(1) + 3(5)}$
$$= \dfrac{3(1) + 25}{2 + 15}$$
$$= \dfrac{3 + 25}{17}$$
$$= \dfrac{28}{17}$$

41. **(a)** $0.841x^2 + 0.32y^2$
$$= 0.841 \cdot 2^2 + 0.32 \cdot 1^2$$
$$= 0.841 \cdot 4 + 0.32 \cdot 1$$
$$= 3.364 + 0.32$$
$$= 3.684$$

(b) $0.841x^2 + 0.32y^2$
$$= 0.841 \cdot 1^2 + 0.32 \cdot 5^2$$
$$= 0.841 \cdot 1 + 0.32 \cdot 25$$
$$= 0.841 + 8$$
$$= 8.841$$

43. "Twelve times a number" translates as $12 \cdot x$ or $12x$.

45. "Added to" indicates addition. "Seven added to a number" translates as $x + 7$.

47. "Two subtracted from a number" translates as $x - 2$.

49. "A number subtracted from seven" translates as $7 - x$.

51. "The difference between a number and 6" translates as $x - 6$.

53. "12 divided by a number" translates as $\dfrac{12}{x}$.

55. "The product of 6 and four less than a number" translates as $6(x - 4)$.

57. "Please excuse me, but I would like to point out that one *solves* an equation, but *simplifies* an expression. You might change 'Solve' to 'Simplify'."

59. $5m + 2 = 7$; 1

$$5(1) + 2 = 7 \ ? \quad \textit{Let m = 1.}$$
$$5 + 2 = 7 \ ?$$
$$7 = 7 \quad \textit{True}$$

Because substituting 1 for m results in a true statement, 1 is a solution of the equation.

61. $2y + 3(y - 2) = 14$; 3

$$2 \cdot 3 + 3(3 - 2) = 14 \ ? \quad \textit{Let y = 3.}$$
$$2 \cdot 3 + 3 \cdot 1 = 14 \ ?$$
$$6 + 3 = 14 \ ?$$
$$9 = 14 \quad \textit{False}$$

Because substituting 3 for y results in a false statement, 3 is not a solution of the equation.

63. $6p + 4p + 9 = 11$; $\frac{1}{5}$

$$6\left(\frac{1}{5}\right) + 4\left(\frac{1}{5}\right) + 9 = 11 \ ? \quad \textit{Let p = }\frac{1}{5}.$$
$$\frac{6}{5} + \frac{4}{5} + 9 = 11 \ ?$$
$$\frac{10}{5} + 9 = 11 \ ?$$
$$2 + 9 = 11 \ ?$$
$$11 = 11 \quad \textit{True}$$

The true result shows that $\frac{1}{5}$ is a solution of the equation.

65. $3r^2 - 2 = 46$; 4

$$3(4)^2 - 2 = 46 \ ? \quad \textit{Let r = 4.}$$
$$3 \cdot 16 - 2 = 46 \ ?$$
$$48 - 2 = 46 \ ?$$
$$46 = 46 \quad \textit{True}$$

The true result shows that 4 is a solution of the equation.

67. $\dfrac{z + 4}{2 - z} = \dfrac{13}{5}$; $\dfrac{1}{3}$

$$\frac{\frac{1}{3} + 4}{2 - \frac{1}{3}} = \frac{13}{5} \ ? \quad \textit{Let z = }\frac{1}{3}.$$
$$\frac{\frac{1}{3} + \frac{12}{3}}{\frac{6}{3} - \frac{1}{3}} = \frac{13}{5} \ ?$$

$$\frac{\frac{13}{3}}{\frac{5}{3}} = \frac{13}{5} \ ?$$
$$\frac{13}{3} \cdot \frac{3}{5} = \frac{13}{5} \ ?$$
$$\frac{13}{5} = \frac{13}{5} \quad \textit{True}$$

The true result shows that $\frac{1}{3}$ is a solution of the equation.

69. "The sum of a number and 8 is 18" translates as

$$x + 8 = 18.$$

Try each number from the given set, $\{2, 4, 6, 8, 10\}$, in turn.

$$x + 8 = 18 \quad \textit{Given equation}$$
$$2 + 8 = 18 \quad \textit{False}$$
$$4 + 8 = 18 \quad \textit{False}$$
$$6 + 8 = 18 \quad \textit{False}$$
$$8 + 8 = 18 \quad \textit{False}$$
$$10 + 8 = 18 \quad \textit{True}$$

The only solution is 10.

71. "Sixteen minus three-fourths of a number is 13" translates as

$$16 - \frac{3}{4}x = 13.$$

Try each number from the given set, $\{2, 4, 6, 8, 10\}$, in turn.

$$16 - \tfrac{3}{4}x = 13 \quad \textit{Given equation}$$
$$16 - \tfrac{3}{4}(2) = 13 \quad \textit{False}$$
$$16 - \tfrac{3}{4}(4) = 13 \quad \textit{True}$$
$$16 - \tfrac{3}{4}(6) = 13 \quad \textit{False}$$
$$16 - \tfrac{3}{4}(8) = 13 \quad \textit{False}$$
$$16 - \tfrac{3}{4}(10) = 13 \quad \textit{False}$$

The only solution is 4.

73. "One more than twice a number is 5" translates as

$$2x + 1 = 5.$$

Try each number from the given set. The only resulting true equation is

$$2 \cdot 2 + 1 = 5,$$

So the only solution is 2.

75. "Three times a number is equal to 8 more than twice the number" translates as

$$3x = 2x + 8.$$

Try each number from the given set.

$$
\begin{array}{ll}
3x = 2x + 8 & \textit{Given equation} \\
3(2) = 2(2) + 8 & \textit{False} \\
3(4) = 2(4) + 8 & \textit{False} \\
3(6) = 2(6) + 8 & \textit{False} \\
3(8) = 2(8) + 8 & \textit{True} \\
3(10) = 2(10) + 8 & \textit{False}
\end{array}
$$

The only solution is 8.

77. There is no equals sign, so $3x + 2(x - 4)$ is an expression.

79. There is an equals sign, so $7t + 2(t + 1) = 4$ is an equation.

81. There is an equals sign, so $x + y = 3$ is an equation.

83. $y = 0.212x - 347$
$\quad = 0.212(1943) - 347$
$\quad = 64.916 \approx 64.9$

The life expectancy of an American born in 1943 is about 64.9 years.

85. $y = 0.212x - 347$
$\quad = 0.212(1980) - 347$
$\quad = 72.76 \approx 72.8$

The life expectancy of an American born in 1980 is about 72.8 years.

1.4   Real Numbers and the Number Line

1. Use the integer 2,845,000 since "increased by 2,845,000" indicates a positive number.

3. Use the integer -2809 since "a decrease of 2809" indicates a negative number.

5. Use the rational numbers -2.4 and 5.2 since "declined 2.4%" and "rose 5.2%" indicate a negative number and a positive number, respectively.

7. Use the rational number 52.59 since "closed up 52.59" indicates a positive number.

9. The only integer between 3.5 and 4.5 is 4.

11. There is only one whole number that is not positive and that is less than 1: the number 0.

13. An irrational number that is between $\sqrt{11}$ and $\sqrt{13}$ is $\sqrt{12}$. There are others.

15. True; every natural number is positive.

17. True; every integer is a rational number. For example, 5 can be written as $\frac{5}{1}$.

19. $\left\{ -9, -\sqrt{7}, -1\frac{1}{4}, -\frac{3}{5}, 0, \sqrt{5}, 3, 5.9, 7 \right\}$

(a) The natural numbers in the given set are 3 and 7, since they are in the natural number set $\{1, 2, 3, \dots\}$.

(b) The set of whole numbers includes the natural numbers and 0. The whole numbers in the given set are 0, 3, and 7.

(c) The integers are the set of numbers $\{\dots, -3, -2, -1, 0, 1, 2, 3, \dots\}$. The integers in the given set are -9, 0, 3, and 7.

(d) Rational numbers are the numbers which can be expressed as the quotient of two integers, with denominators not equal to 0.

We can write numbers from the given set in this form as follows:

$$-9 = \frac{-9}{1}, -1\frac{1}{4} = \frac{-5}{4}, -\frac{3}{5} = \frac{-3}{5}, 0 = \frac{0}{1},$$
$$3 = \frac{3}{1}, 5.9 = \frac{59}{10}, \text{ and } 7 = \frac{7}{1}.$$

Thus, the rational numbers in the given set are $-9, -1\frac{1}{4}, -\frac{3}{5}, 0, 3, 5.9$, and 7.

(e) Irrational numbers are real numbers that are not rational. $-\sqrt{7}$ and $\sqrt{5}$ can be represented by points on the number line but cannot be written as a quotient of integers. Thus, the irrational numbers in the given set are $-\sqrt{7}$ and $\sqrt{5}$.

(f) Real numbers are all numbers that can be represented on the number line. All the numbers in the given set are real.

21. The *natural numbers* are the numbers with which we count. An example is 1. The *whole numbers* are the natural numbers with 0 also included. An example is 0. The *integers* are the whole numbers and their negatives. An example is -1. The *rational numbers* are the numbers that can be represented by a quotient of integers with denominator not 0, such as $\frac{1}{2}$. The *irrational numbers*, such as $\sqrt{2}$, cannot be represented as a quotient of integers. The *real numbers* include all positive numbers, negative numbers, and zero. All the numbers listed in this section are reals.

23. Graph 0, 3, -5, and -6.

Place a dot on the number line at the point that corresponds to each number. The order of the numbers from smallest to largest is $-6, -5, 0, 3$.

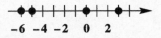

25. Graph $-2, -6, -4, 3,$ and $4.$

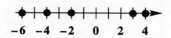

27. Graph $\frac{1}{4}, 2\frac{1}{2}, -3\frac{4}{5}, -4,$ and $-1\frac{5}{8}.$

$$-3\frac{4}{5} \quad -1\frac{5}{8} \quad \frac{1}{4} \quad 2\frac{1}{2}$$

29. (a) $|-7| = 7$ **(A)**

The distance between -7 and 0 on the number line is 7 units.

(b) $-(-7) = 7$ **(A)**

The opposite of -7 is $7.$

(c) $-|-7| = -(7) = -7$ **(B)**

(d) $-|-(-7)| = -|7|$ *Work inside absolute value symbols first*

$$= -(7)$$
$$= -7 \quad \textbf{(B)}$$

31. (a) The opposite of -4 is found by changing the sign of $-4.$ The opposite of -4 is $4.$

(b) The absolute value of -4 is the distance between 0 and -4 on the number line.

$$|-4| = 4$$

The absolute value of -4 is $4.$

33. (a) The opposite of 6 is $-6.$

(b) The distance between 0 and 6 on the number line is 6 units, so the absolute value of 6 is $6.$

35. Since -6 is a negative number, its absolute value is the additive inverse of -6; that is,

$$|-6| = -(-6) = 6.$$

37. $-\left|-\frac{2}{3}\right| = -\left[-\left(-\frac{2}{3}\right)\right] = -\left[\frac{2}{3}\right] = -\frac{2}{3}$

39. $|6 - 3| = |3| = 3$

41. $-12, -4$

Since -12 is to the left of -4 on the number line, -12 is the lesser number.

43. $-8, -7$

Since -8 is located to the left of -7 on the number line, -8 is the lesser number.

45. $3, |-4|$

Since $|-4| = 4$, 3 is the lesser of the two numbers.

47. $|-3.5|, |-4.5|$

Since $|-3.5| = 3.5$ and $|-4.5| = 4.5$, $|-3.5|$ or 3.5 is the lesser number.

49. $-|-6|, -|-4|$

Since $-|-6| = -6$ and $-|-4| = -4$, $-|-6|$ is to the left of $-|-4|$ on the number line, so $-|-6|$ or -6 is the lesser of the two numbers.

51. $|5 - 3|, |6 - 2|$

Since $|5 - 3| = |2| = 2$ and $|6 - 2| = |4| = 4$, $|5 - 3|$ or 2 is the lesser of the two numbers.

53. $-5 < -2$

Since -5 is to the *left* of -2 on the number line, -5 is *less than* -2, and the statement $-5 < -2$ is true.

55. $-4 \leq -(-5)$

Since $-(-5) = 5$ and $-4 < 5$, $-4 \leq -(-5)$ is true.

57. $|-6| < |-9|$

Since $|-6| = 6$ and $|-9| = 9$, $|-6| < |-9|$ is true.

59. $-|8| > |-9|$

Since $-|8| = -8$ and $|-9| = -(-9) = 9$, $-|8| < |-9|$, so $-|8| > |-9|$ is false.

61. $-|-5| \geq -|-9|$

Since $-|-5| = -5$, $-|-9| = -9$, and $-5 > -9$, $-|-5| \geq -|-9|$ is true.

63. $|6 - 5| \geq |6 - 2|$

Since $|6 - 5| = |1| = 1$ and $|6 - 2| = |4| = 4$, $|6 - 5| < |6 - 2|$, so $|6 - 5| \geq |6 - 2|$ is false.

65. The number that represents the greatest percentage increase is 25.9, which corresponds to petroleum refineries from 2002 to 2003.

67. The number with the smallest absolute value in the table is 1.4, so the least change corresponds to construction machinery manufacturing from 2002 to 2003.

69. Three examples of positive real numbers that are not integers are $\frac{1}{2}, \frac{5}{8},$ and $1\frac{3}{4}.$ Other examples are $0.7, 4\frac{2}{3},$ and $5.1.$

71. Three examples of real numbers that are not whole numbers are $-3\frac{1}{2}, -\frac{2}{3},$ and $\frac{3}{7}.$ Other examples are $-4.3, -\sqrt{2},$ and $\sqrt{7}.$

73. Three examples of real numbers that are not rational numbers are $\sqrt{5}, \pi,$ and $-\sqrt{3}.$ All irrational numbers are real numbers that are not rational.

75. The statement "Absolute value is always positive." is not true. The absolute value of 0 is 0, and 0 is not positive. A more accurate way of describing absolute value is to say that *absolute value is never negative,* or *absolute value is always nonnegative.*

1.5   Adding and Subtracting Real Numbers

1. The sum of two negative numbers will always be a *negative* number. In the illustration, we have $-2 + (-3) = -5$.

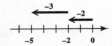

3. If I am adding a positive number and a negative number, and the negative number has the larger absolute value, the sum will be a *negative* number. In the illustration, the absolute value of -4 is larger than the absolute value of 2, so the sum is a negative number; that is, $-4 + 2 = -2$.

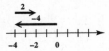

5. To add two numbers with the same sign, add their absolute values and keep the same sign for the sum. For example, $3 + 4 = 7$ and $-3 + (-4) = -7$. To add two numbers with different signs, subtract the lesser absolute value from the larger absolute value, and use the sign of the number with the larger absolute value. For example, $6 + (-4) = 2$ and $(-6) + 4 = -2$.

7. $-6 + (-2)$

The sum of two negative numbers is negative.
$$-6 + (-2) = -8$$

9. $-3 + (-9)$

Because the numbers have the same sign, add their absolute values:
$$3 + 9 = 12.$$

Because both numbers are negative, their sum is negative:
$$-3 + (-9) = -12.$$

11. $5 + (-3)$

To add $5 + (-3)$, find the difference between the absolute values of the numbers.
$$|5| = 5 \text{ and } |-3| = 3$$
$$5 - 3 = 2$$

Since $|5| > |-3|$, the sum will be positive:
$$5 + (-3) = 2.$$

13. $6 + (-8)$

Since the numbers have different signs, find the difference between their absolute values:
$$8 - 6 = 2.$$

Because -8 has the larger absolute value, the sum is negative:
$$6 + (-8) = -2.$$

15. $-3.5 + 12.4$

Since the numbers have different signs, find the difference between their absolute values:
$$12.4 - 3.5 = 8.9.$$

Since 12.4 has the larger absolute value, the answer is positive:
$$-3.5 + 12.4 = 8.9.$$

17. $4 + [13 + (-5)]$

Perform the operation inside the brackets first, then add.
$$4 + [13 + (-5)] = 4 + 8 = 12$$

19. $8 + [-2 + (-1)] = 8 + [-3] = 5$

21. $-2 + [5 + (-1)] = -2 + [4] = 2$

23. $-6 + [6 + (-9)] = -6 + [-3] = -9$

25. $[(-9) + (-3)] + 12 = [-12] + 12 = 0$

27. $-\dfrac{1}{6} + \dfrac{2}{3} = -\dfrac{1}{6} + \dfrac{4}{6} = \dfrac{3}{6} = \dfrac{1}{2}$

29. Since $8 = 2 \cdot 2 \cdot 2$ and $12 = 2 \cdot 2 \cdot 3$, the LCD is $2 \cdot 2 \cdot 2 \cdot 3 = 24$.

$$\frac{5}{8} + \left(-\frac{17}{12}\right) = \frac{5 \cdot 3}{8 \cdot 3} + \left(-\frac{17 \cdot 2}{12 \cdot 2}\right)$$
$$= \frac{15}{24} + \left(-\frac{34}{24}\right)$$
$$= -\frac{19}{24}$$

31. $2\dfrac{1}{2} + \left(-3\dfrac{1}{4}\right) = \dfrac{5}{2} + \left(-\dfrac{13}{4}\right)$
$$= \frac{10}{4} + \left(-\frac{13}{4}\right)$$
$$= -\frac{3}{4}$$

33. $-6.1 + [3.2 + (-4.8)] = -6.1 + [-1.6]$
$$= -7.7$$

35. $[-3 + (-4)] + [5 + (-6)] = [-7] + [-1]$
$$= -8$$

37. $[-4 + (-3)] + [8 + (-1)] = [-7] + [7]$
$$= 0$$

39. $[-4 + (-6)] + [(-3) + (-8)] + [12 + (-11)]$
$$= ([-10] + [-11]) + [1]$$
$$= (-21) + 1$$
$$= -20$$

In Exercises 41–64, use the definition of subtraction to find the differences.

41. $3 - 6 = 3 + (-6) = -3$

43. $5 - 9 = 5 + (-9) = -4$

45. $-6 - 2 = -6 + (-2) = -8$

47. $-9 - 5 = -9 + (-5) = -14$

49. $6 - (-3) = 6 + (3) = 9$

51. $-6 - (-2) = -6 + (2) = -4$

53. $2 - (3 - 5) = 2 - [3 + (-5)]$
$$= 2 - [-2]$$
$$= 2 + (2)$$
$$= 4$$

55. $\dfrac{1}{2} - \left(-\dfrac{1}{4}\right) = \dfrac{1}{2} + \dfrac{1}{4}$
$$= \dfrac{2}{4} + \dfrac{1}{4} = \dfrac{3}{4}$$

57. $-\dfrac{3}{4} - \dfrac{5}{8} = -\dfrac{3}{4} + \left(-\dfrac{5}{8}\right)$
$$= -\dfrac{6}{8} + \left(-\dfrac{5}{8}\right)$$
$$= -\dfrac{11}{8}, \text{ or } -1\dfrac{3}{8}$$

59. $\dfrac{5}{8} - \left(-\dfrac{1}{2} - \dfrac{3}{4}\right)$
$$= \dfrac{5}{8} - \left[-\dfrac{1}{2} + \left(-\dfrac{3}{4}\right)\right]$$
$$= \dfrac{5}{8} - \left[-\dfrac{2}{4} + \left(-\dfrac{3}{4}\right)\right]$$
$$= \dfrac{5}{8} - \left(-\dfrac{5}{4}\right)$$
$$= \dfrac{5}{8} + \dfrac{5}{4}$$
$$= \dfrac{5}{8} + \dfrac{10}{8}$$
$$= \dfrac{15}{8}, \text{ or } 1\dfrac{7}{8}$$

61. $3.4 - (-8.2) = 3.4 + 8.2$
$$= 11.6$$

63. $-6.4 - 3.5 = -6.4 + (-3.5)$
$$= -9.9$$

65. $(4 - 6) + 12 = [4 + (-6)] + 12$
$$= [-2] + 12$$
$$= 10$$

67. $(8 - 1) - 12 = [8 + (-1)] + (-12)$
$$= [7] + (-12)$$
$$= -5$$

69. $6 - (-8 + 3) = 6 - (-5)$
$$= 6 + 5$$
$$= 11$$

71. $2 + (-4 - 8) = 2 + [-4 + (-8)]$
$$= 2 + [-12]$$
$$= -10$$

73. $|-5 - 6| + |9 + 2| = |-5 + (-6)| + |11|$
$$= |-11| + |11|$$
$$= -(-11) + 11$$
$$= 11 + 11$$
$$= 22$$

75. $|-8 - 2| - |-9 - 3| = |-8 + (-2)| - |-9 + (-3)|$
$$= |-10| - |-12|$$
$$= -(-10) - [-(-12)]$$
$$= 10 - [12]$$
$$= -2$$

77. $-9 + [(3 - 2) - (-4 + 2)] = -9 + [1 - (-2)]$
$$= -9 + [1 + 2]$$
$$= -9 + 3$$
$$= -6$$

79. $-3 + [(-5 - 8) - (-6 + 2)]$
$$= -3 + [(-5 + (-8)) - (-4)]$$
$$= -3 + [-13 + 4]$$
$$= -3 + [-9]$$
$$= -12$$

81. $-9.1237 + [(-4.8099 - 3.2516) + 11.27903]$
$$= -9.1237 + [(-4.8099 + (-3.2516)) + 11.27903]$$
$$= -9.1237 + [-8.0615 + 11.27903]$$
$$= -9.1237 + 3.21753$$
$$= -5.90617$$

83. "The sum of -5 and 12 and 6" is written $-5 + 12 + 6$.
$$-5 + 12 + 6 = [-5 + 12] + 6$$
$$= 7 + 6 = 13$$

85. "14 added to the sum of -19 and -4" is written $[-19 + (-4)] + 14$.

$$[-19 + (-4)] + 14 = (-23) + 14$$
$$= -9$$

87. "The sum of -4 and -10, increased by 12," is written $[-4 + (-10)] + 12$.

$$[-4 + (-10)] + 12 = -14 + 12$$
$$= -2$$

89. "$\frac{2}{7}$ more than the sum of $\frac{5}{7}$ and $-\frac{9}{7}$" is written $\left[\frac{5}{7} + \left(-\frac{9}{7}\right)\right] + \frac{2}{7}$.

$$\left[\frac{5}{7} + \left(-\frac{9}{7}\right)\right] + \frac{2}{7} = -\frac{4}{7} + \frac{2}{7}$$
$$= -\frac{2}{7}$$

91. "The difference between 4 and -8" is written $4 - (-8)$.

$$4 - (-8) = 4 + 8 = 12$$

93. "8 less than -2" is written $-2 - 8$.

$$-2 - 8 = -2 + (-8) = -10$$

95. "The sum of 9 and -4, decreased by 7" is written $[9 + (-4)] - 7$.

$$[9 + (-4)] - 7 = 5 + (-7) = -2$$

97. "12 less than the difference between 8 and -5" is written $[8 - (-5)] - 12$.

$$[8 - (-5)] - 12 = [8 + (5)] - 12$$
$$= 13 - 12$$
$$= 13 + (-12)$$
$$= 1$$

99. The outlay for 2002 is $370.6 billion and the outlay for 2003 is $367.0 billion. Thus, the *change in outlay* is

$$367.0 - 370.6 = 367.0 + (-370.6)$$
$$= -3.6$$

billion dollars (a decrease since it is negative).

101. The outlay for 2004 is $374.8 billion and the outlay for 2005 is $403.0 billion. Thus, the *change in outlay* is

$$403.0 - 374.8 = 403.0 + (-374.8)$$
$$= 28.2$$

billion dollars (an increase).

103. $17,400 - (-32,995) = 17,400 + 32,995$
$$= 50,395$$

The difference between the height of Mt. Foraker and the depth of the Philippine Trench is 50,395 feet.

105. $-23,376 - (-24,721) = -23,376 + 24,721$
$$= 1345$$

The Cayman Trench is 1345 feet deeper than the Java Trench.

107. $14,246 - 14,110 = 14,246 + (-14,110)$
$$= 136$$

Mt. Wilson is 136 feet higher than Pikes Peak.

109. $[-5 + (-4)] + (-3) = -9 + (-3)$
$$= -12$$

The total number of seats that New York, Pennsylvania, and Ohio are projected to lose is twelve, which can be represented by the signed number -12.

111. To find the low temperature, start with 44 and subtract 100.

$$44 - 100 = 44 + (-100)$$
$$= -56$$

The temperature fell to $-56°$F.

113. $33°$F lower than $-36°$F can be represented as

$$-36 - 33 = -36 + (-33)$$
$$= -69.$$

The record low in Utah is $-69°$F.

115. $0 + (-130) + (-54) = -130 + (-54)$
$$= -184$$

Their new altitude is 184 meters below the surface, which can be represented by the signed number -184.

117. (a) $10.8 - (-0.5) = 10.8 + 0.5$
$$= 11.3$$

The difference is 11.3%.

(b) Americans spent more money than they earned, which means they had to dip into savings or increase borrowing.

119. $1879 + 869 - 579 = 2748 - 579$ *Add.*
$$= 2169 \qquad \textit{Subtract.}$$

The average was $2169.

121. Add the scores of the four turns to get the final score.

$$-19 + 28 + (-5) + 13 = 9 + (-5) + 13$$
$$= 4 + 13$$
$$= 17$$

His final score for the four turns was 17.

123. Sum of checks:

$$\$35.84 + \$26.14 + \$3.12 = \$61.98 + \$3.12$$
$$= \$65.10$$

Sum of deposits:

$$\$85.00 + \$120.76 = \$205.76$$

Final balance = Beginning balance − checks + deposits
$$= \$904.89 - \$65.10 + \$205.76$$
$$= \$839.79 + \$205.76$$
$$= \$1045.55$$

Her account balance at the end of August was $1045.55.

125.

-870.00	*amount owed*
$+185.90$	*2 return credits* ($35.90 + $150.00)
-684.10	
-102.50	*3 purchases* ($82.50 + $10.00 + $10.00)
-786.60	
$+500.00$	*payment*
-286.60	
-37.23	*finance charge*
-323.83	

She still owes $323.83.

127. The expression $x - y$ would have to be *positive* since subtracting a negative number from a positive number is the same as adding a positive number to a positive number, which is a positive number.

129. $x + |y|$

Since $|y|$ is positive, $x + |y|$ is the sum of two positive numbers, which is *positive*.

1.6 Multiplying and Dividing Real Numbers

1. The product or the quotient of two numbers with the same sign is greater than 0 , since the product or quotient of two positive numbers is positive and the product or quotient of two negative numbers is positive.

3. If three negative numbers are multiplied together, the product is less than 0 , since a negative number times a negative number is a positive number, and that positive number times a negative number is a negative number.

5. If a negative number is squared and the result is added to a positive number, the final answer is greater than 0 , since a negative number squared is a positive number, and a positive number added to another positive number is a positive number.

7. If three positive numbers, five negative numbers, and zero are multiplied together, the product is equal to 0 . Since one of the numbers is zero, the product is zero (regardless of what the other numbers are).

9. The quotient formed by any nonzero number divided by 0 is undefined , and the quotient formed by 0 divided by any nonzero number is 0 . Examples include $\frac{1}{0}$, which is undefined, and $\frac{0}{1}$, which equals 0.

11. $3(-4) = -(3 \cdot 4) = -12$

Note that the product of a positive number and a negative number is negative.

13. $-3(-4) = 3 \cdot 4 = 12$

Note that the product of two negative numbers is positive.

15. $-10(-12) = 10 \cdot 12 = 120$

17. $3(-11) = -(3 \cdot 11) = -33$

19. $-6.8(0.35) = -(6.8 \cdot 0.35) = -2.38$

21.
$$-\frac{3}{8} \cdot \left(-\frac{10}{9}\right) = \frac{3}{8}\left(\frac{10}{9}\right)$$
$$= \frac{3 \cdot 10}{8 \cdot 9}$$
$$= \frac{3 \cdot (2 \cdot 5)}{(4 \cdot 2) \cdot (3 \cdot 3)}$$
$$= \frac{3 \cdot 2 \cdot 5}{4 \cdot 2 \cdot 3 \cdot 3}$$
$$= \frac{5}{4 \cdot 3} = \frac{5}{12}$$

23.
$$\frac{2}{15}\left(-1\frac{1}{4}\right) = \frac{2}{15}\left(-\frac{5}{4}\right)$$
$$= -\frac{2 \cdot 5}{15 \cdot 4}$$
$$= -\frac{2 \cdot 5}{3 \cdot 5 \cdot 2 \cdot 2}$$
$$= -\frac{1}{3 \cdot 2} = -\frac{1}{6}$$

25. $-8\left(-\frac{3}{4}\right) = 8\left(\frac{3}{4}\right) = \frac{24}{4} = 6$

27. Using only positive integer factors, 32 can be written as $1 \cdot 32$, $2 \cdot 16$, or $4 \cdot 8$. Including the negative integer factors, we see that the integer factors of 32 are $-32, -16, -8, -4, -2, -1, 1, 2, 4, 8, 16,$ and 32.

29. The integer factors of 40 are $-40, -20, -10, -8, -5, -4, -2, -1, 1, 2, 4, 5, 8, 10, 20,$ and 40.

31. The integer factors of 31 are $-31, -1, 1,$ and 31.

33. $\dfrac{15}{5} = \dfrac{5 \cdot 3}{5} = \dfrac{3}{1} = 3$

35. $\dfrac{-30}{6} = -\dfrac{2 \cdot 3 \cdot 5}{2 \cdot 3} = -5$

Note that the quotient of two numbers having different signs is negative.

37. $\dfrac{-28}{-4} = \dfrac{4 \cdot 7}{4} = 7$

Note that the quotient of two numbers having the same sign is positive.

39. $\dfrac{96}{-16} = -\dfrac{6 \cdot 16}{16} = -6$

41. Dividing by a fraction (in this case, $-\frac{1}{8}$) is the same as multiplying by the reciprocal of the fraction (in this case, $-\frac{8}{1}$).

$$\left(-\dfrac{4}{3}\right) \div \left(-\dfrac{1}{8}\right) = \left(-\dfrac{4}{3}\right) \cdot \left(-\dfrac{8}{1}\right)$$
$$= \dfrac{4 \cdot 8}{3 \cdot 1}$$
$$= \dfrac{32}{3}, \text{ or } 10\dfrac{2}{3}$$

43. $\dfrac{-8.8}{2.2} = -\dfrac{4(2.2)}{2.2} = -4$

45. $\dfrac{0}{-2} = 0$, because 0 divided by any nonzero number is 0.

47. $\dfrac{12.4}{0}$ is *undefined* because we cannot divide by 0.

In Exercises 49–62, use the order of operations.

49. $7 - 3 \cdot 6 = 7 - 18$
$= -11$

51. $-10 - (-4)(2) = -10 - (-8)$
$= -10 + 8$
$= -2$

53. $-7(3 - 8) = -7[3 + (-8)]$
$= -7(-5) = 35$

55. $(12 - 14)(1 - 4) = (-2)(-3)$
$= 6$

57. $(7 - 10)(10 - 4) = (-3)(6)$
$= -18$

59. $(-2 - 8)(-6) + 7 = (-10)(-6) + 7$
$= 60 + 7$
$= 67$

61. $3(-5) + |3 - 10| = -15 + |-7|$
$= -15 + 7$
$= -8$

63. $\dfrac{-5(-6)}{9 - (-1)} = \dfrac{30}{10}$
$= \dfrac{3 \cdot 10}{10} = 3$

65. $\dfrac{-21(3)}{-3 - 6} = \dfrac{-63}{-3 + (-6)}$
$= \dfrac{-63}{-9} = 7$

67. $\dfrac{-10(2) + 6(2)}{-3 - (-1)} = \dfrac{-20 + 12}{-3 + 1}$
$= \dfrac{-8}{-2} = 4$

69. $\dfrac{3^2 - 4^2}{7(-8 + 9)} = \dfrac{9 - 16}{7(1)} = \dfrac{-7}{7} = -1$

71. $\dfrac{8(-1) - |(-4)(-3)|}{-6 - (-1)} = \dfrac{-8 - |12|}{-6 + 1}$
$= \dfrac{-8 - 12}{-5}$
$= \dfrac{-20}{-5} = 4$

73. $\dfrac{-13(-4) - (-8)(-2)}{(-10)(2) - 4(-2)}$
$= \dfrac{52 - 16}{-20 - (-8)}$
$= \dfrac{36}{-20 + 8}$
$= \dfrac{36}{-12} = -3$

75. If x is negative, $4x$ will be the product of a positive and a negative number, which is negative. If y is negative, $8y$ will likewise be negative. Then $4x + 8y$ will be the sum of two negative numbers, which is negative.

In Exercises 77–88, replace x with 6, y with -4, and a with 3. Then use the order of operations to evaluate the expression.

77. $5x - 2y + 3a = 5(6) - 2(-4) + 3(3)$
$= 30 - (-8) + 9$
$= 30 + 8 + 9$
$= 38 + 9$
$= 47$

79. $(2x + y)(3a) = [2(6) + (-4)][3(3)]$
$$= [12 + (-4)](9)$$
$$= (8)(9)$$
$$= 72$$

81. $\left(\dfrac{1}{3}x - \dfrac{4}{5}y\right)\left(-\dfrac{1}{5}a\right)$

$$= \left[\dfrac{1}{3}(6) - \dfrac{4}{5}(-4)\right]\left[-\dfrac{1}{5}(3)\right]$$

$$= \left[2 - \left(-\dfrac{16}{5}\right)\right]\left(-\dfrac{3}{5}\right)$$

$$= \left(2 + \dfrac{16}{5}\right)\left(-\dfrac{3}{5}\right)$$

$$= \left(\dfrac{10}{5} + \dfrac{16}{5}\right)\left(-\dfrac{3}{5}\right)$$

$$= \left(\dfrac{26}{5}\right)\left(-\dfrac{3}{5}\right)$$

$$= -\dfrac{78}{25}$$

83. $(-5 + x)(-3 + y)(3 - a)$
$$= (-5 + 6)[-3 + (-4)][3 - 3]$$
$$= (1)(-7)(0)$$
$$= 0$$

85. $-2y^2 + 3a = -2(-4)^2 + 3(3)$
$$= -2(16) + 9$$
$$= -32 + 9$$
$$= -23$$

87. $\dfrac{2y^2 - x}{a + 10} = \dfrac{2(-4)^2 - (6)}{3 + 10}$

$$= \dfrac{2(16) - 6}{13}$$

$$= \dfrac{32 - 6}{13}$$

$$= \dfrac{26}{13}$$

$$= 2$$

89. "The product of -9 and 2, added to 9" is written $9 + (-9)(2)$.
$$9 + (-9)(2) = 9 + (-18)$$
$$= -9$$

91. "Twice the product of -1 and 6, subtracted from -4" is written $-4 - 2[(-1)(6)]$.
$$-4 - 2[(-1)(6)] = -4 - 2(-6)$$
$$= -4 - (-12)$$
$$= -4 + 12 = 8$$

93. "Nine subtracted from the product of 1.5 and -3.2 is written $(1.5)(-3.2) - 9$.
$$(1.5)(-3.2) - 9 = -4.8 - 9$$
$$= -4.8 + (-9)$$
$$= -13.8$$

95. "The product of 12 and the difference between 9 and -8" is written $12[9 - (-8)]$.
$$12[9 - (-8)] = 12[9 + 8]$$
$$= 12(17) = 204$$

97. "The quotient of -12 and the sum of -5 and -1" is written
$$\dfrac{-12}{-5 + (-1)},$$
and
$$\dfrac{-12}{-5 + (-1)} = \dfrac{-12}{-6} = 2.$$

99. "The sum of 15 and -3, divided by the product of 4 and -3" is written
$$\dfrac{15 + (-3)}{4(-3)},$$
and
$$\dfrac{15 + (-3)}{4(-3)} = \dfrac{12}{-12} = -1.$$

101. "Two-thirds of the difference between 8 and -1" is written
$$\tfrac{2}{3}[8 - (-1)],$$
and
$$\tfrac{2}{3}[8 - (-1)] = \tfrac{2}{3}[8 + (1)] = \tfrac{2}{3}[9] = 6.$$

103. "20% of the product of -5 and 6" is written
$$0.20(-5 \cdot 6),$$
and
$$0.20(-5 \cdot 6) = 0.20(-30) = -6.$$

105. "The sum of $\frac{1}{2}$ and $\frac{5}{8}$, times the difference between $\frac{3}{5}$ and $\frac{1}{3}$" is written
$$\left(\tfrac{1}{2} + \tfrac{5}{8}\right)\left(\tfrac{3}{5} - \tfrac{1}{3}\right),$$
and
$$\left(\tfrac{1}{2} + \tfrac{5}{8}\right)\left(\tfrac{3}{5} - \tfrac{1}{3}\right) = \left(\tfrac{4}{8} + \tfrac{5}{8}\right)\left(\tfrac{9}{15} - \tfrac{5}{15}\right)$$
$$= \tfrac{9}{8}\left(\tfrac{4}{15}\right)$$
$$= \dfrac{3 \cdot 3 \cdot 4}{2 \cdot 4 \cdot 3 \cdot 5} = \dfrac{3}{10}.$$

107. "The product of $-\frac{1}{2}$ and $\frac{3}{4}$, divided by $-\frac{2}{3}$" is written $\dfrac{-\frac{1}{2}\left(\frac{3}{4}\right)}{-\frac{2}{3}}$.

$$\frac{-\frac{1}{2}\left(\frac{3}{4}\right)}{-\frac{2}{3}} = \frac{-\frac{3}{8}}{-\frac{2}{3}}$$
$$= -\frac{3}{8} \cdot \left(-\frac{3}{2}\right)$$
$$= \frac{9}{16}$$

109. "The quotient of a number and 3 is -3" is written
$$\frac{x}{3} = -3.$$

The solution is -9, since
$$\frac{-9}{3} = -3.$$

111. "6 less than a number is 4" is written
$$x - 6 = 4.$$

The solution is 10, since
$$10 - 6 = 4.$$

113. "When 5 is added to a number, the result is -5" is written
$$x + 5 = -5.$$

The solution is -10, since
$$-10 + 5 = -5.$$

115. Add the numbers and divide by 5.
$$\frac{(23 + 18 + 13) + [(-4) + (-8)]}{5}$$
$$= \frac{54 - 12}{5}$$
$$= \frac{42}{5}, \text{ or } 8\frac{2}{5}$$

117. Add the integers from -10 to 14.
$$(-10) + (-9) + \cdots + 14 = 50$$

[the 3 dots indicate that the pattern continues]

There are 25 integers from -10 to 14 (10 negative, zero, and 14 positive). Thus, the average is $\frac{50}{25} = 2$.

119. The average of a group of numbers is the sum of all the numbers divided by the number of numbers. If the average is 0, then the sum of all the numbers must be 0 since the only way to make a quotient 0 is to have its numerator equal to 0.

121. (a) 3,473,986 is divisible by 2 because its last digit, 6, is divisible by 2.

(b) 4,336,879 is not divisible by 2 because its last digit, 9, is not divisible by 2.

123. (a) 6,221,464 is divisible by 4 because the number formed by its last two digits, 64, is divisible by 4.

(b) 2,876,335 is not divisible by 4 because the number formed by its last two digits, 35, is not divisible by 4.

125. (a) 1,524,822 is divisible by 2 because its last digit, 2, is divisible by 2. It is also divisible by 3 because the sum of its digits,
$$1 + 5 + 2 + 4 + 8 + 2 + 2 = 24,$$

is divisible by 3.

Because 1,524,822 is divisible by *both* 2 and 3, it is divisible by 6.

(b) 2,873,590 is divisible by 2 because its last digit, 0, is divisible by 2. However, it is not divisible by 3 because the sum of its digits,
$$2 + 8 + 7 + 3 + 5 + 9 + 0 = 34,$$

is not divisible by 3.

Because 2,873,590 is not divisible by *both* 2 and 3, it is not divisible by 6.

127. (a) 4,114,107 is divisible by 9 because the sum of its digits,
$$4 + 1 + 1 + 4 + 1 + 0 + 7 = 18,$$

is divisible by 9.

(b) 2,287,321 is not divisible by 9 because the sum of its digits,
$$2 + 2 + 8 + 7 + 3 + 2 + 1 = 25,$$

is not divisible by 9.

Summary Exercises on Operations with Real Numbers

1.
$$14 - 3 \cdot 10 = 14 - 30$$
$$= 14 + (-30)$$
$$= -16$$

3.
$$(3 - 8)(-2) - 10 = (-5)(-2) - 10$$
$$= 10 - 10$$
$$= 0$$

5.
$$7 - (-3)(2 - 10) = 7 - (-3)(-8)$$
$$= 7 - (24)$$
$$= -17$$

7.
$$(-4)(7) - (-5)(2) = (-28) - (-10)$$
$$= -28 + (10)$$
$$= -18$$

9. $40 - (-2)[8 - 9] = 40 - (-2)[-1]$
$= 40 - (2)$
$= 38$

11. $\dfrac{-3 - (-9 + 1)}{-7 - (-6)} = \dfrac{-3 - (-8)}{-7 + 6}$
$= \dfrac{-3 + 8}{-1}$
$= \dfrac{5}{-1} = -5$

13. $\dfrac{6^2 - 8}{-2(2) + 4(-1)} = \dfrac{36 - 8}{-4 + (-4)}$
$= \dfrac{28}{-8}$
$= -\dfrac{4 \cdot 7}{2 \cdot 4} = -\dfrac{7}{2}, \ \text{ or } \ -3\dfrac{1}{2}$

15. $\dfrac{9(-6) - 3(8)}{4(-7) + (-2)(-11)} = \dfrac{-54 - 24}{-28 + 22}$
$= \dfrac{-78}{-6} = 13$

17. $\dfrac{(2 + 4)^2}{(5 - 3)^2} = \dfrac{(6)^2}{(2)^2}$
$= \dfrac{36}{4} = 9$

19. $\dfrac{-9(-6) + (-2)(27)}{3(8 - 9)} = \dfrac{(54) + (-54)}{3(-1)}$
$= \dfrac{0}{-3} = 0$

21. $\dfrac{6(-10 + 3)}{15(-2) - 3(-9)} = \dfrac{6(-7)}{(-30) - (-27)}$
$= \dfrac{-42}{-30 + 27}$
$= \dfrac{-42}{-3} = 14$

23. $\dfrac{(-10)^2 + 10^2}{-10(5)} = \dfrac{100 + 100}{-50}$
$= \dfrac{200}{-50} = -4$

25. $\dfrac{1}{2} \div \left(-\dfrac{1}{2}\right) = \dfrac{1}{2} \cdot \left(-\dfrac{2}{1}\right)$
$= -\dfrac{2}{2} = -1$

27. $\left[\dfrac{5}{8} - \left(-\dfrac{1}{16}\right)\right] + \dfrac{3}{8} = \left[\dfrac{10}{16} + \dfrac{1}{16}\right] + \dfrac{6}{16}$
$= \left[\dfrac{11}{16}\right] + \dfrac{6}{16}$
$= \dfrac{17}{16}, \ \text{ or } \ 1\dfrac{1}{16}$

29. $-0.9(-3.7) = 0.9(3.7)$
$= 3.33$

31. $-3^2 - 2^2 = -\left(3^2\right) - \left(2^2\right)$
$= -9 - 4$
$= -13$

33. $40 - (-2)[-5 - 3] = 40 - (-2)[-8]$
$= 40 - (16)$
$= 24$

In Exercises 34–42, replace x with -2, y with 3, and a with 4. Then use the order of operations to evaluate the expression.

35. $(x + 6)^3 - y^3 = (-2 + 6)^3 - 3^3$
$= (4)^3 - 27$
$= 64 - 27$
$= 37$

37. $\left(\dfrac{1}{2}x + \dfrac{2}{3}y\right)\left(-\dfrac{1}{4}a\right) = \left(\dfrac{1}{2}(-2) + \dfrac{2}{3}(3)\right)\left(-\dfrac{1}{4}(4)\right)$
$= (-1 + 2)(-1)$
$= (1)(-1)$
$= -1$

39. $\dfrac{x^2 - y^2}{x^2 + y^2} = \dfrac{(-2)^2 - 3^2}{(-2)^2 + 3^2}$
$= \dfrac{4 - 9}{4 + 9}$
$= \dfrac{-5}{13} = -\dfrac{5}{13}$

41. $\left(\dfrac{x}{y}\right)^3 = \left(\dfrac{-2}{3}\right)^3 = \left(-\dfrac{2}{3}\right)\left(-\dfrac{2}{3}\right)\left(-\dfrac{2}{3}\right)$
$= -\dfrac{8}{27}$

1.7 Properties of Real Numbers

1. $-12 + 6 = 6 + \underline{(-12)}$
by the *commutative property of addition*.

3. $-6 \cdot 3 = \underline{3} \cdot (-6)$
by the *commutative property of multiplication*.

5. $(4 + 7) + 8 = 4 + (\underline{7} + 8)$
by the *associative property of addition*.

7. $8 \cdot (3 \cdot 6) = (\underline{8} \cdot 3) \cdot 6$
by the *associative property of multiplication*.

9. **(a) B**, since 0 is the identity element for addition.

(b) F, since 1 is the identity element for multiplication.

(c) C, since $-a$ is the additive inverse of a.

(d) I, since $\dfrac{1}{a}$ is the multiplicative inverse, or reciprocal, of any nonzero number a.

(e) B, since 0 is the only number that is equal to its negative; that is, $0 = -0$.

(f) D and **F**, since -1 has reciprocal $\dfrac{1}{(-1)} = -1$ and 1 has a reciprocal $\dfrac{1}{(1)} = 1$; that is, -1 and 1 are their own multiplicative inverses.

(g) B, since the multiplicative inverse of a number a is $\dfrac{1}{a}$ and the only number that we *cannot* divide by is 0.

(h) A

(i) G, since we can consider $(5 \cdot 4)$ to be one number, $(5 \cdot 4) \cdot 3$ is the same as $3 \cdot (5 \cdot 4)$ by the commutative property.

(j) H

11. $7 + 18 = 18 + 7$

The order of the two numbers has been changed, so this is an example of the commutative property of addition: $a + b = b + a$.

13. $5 \cdot (13 \cdot 7) = (5 \cdot 13) \cdot 7$

The numbers are in the same order but grouped differently, so this is an example of the associative property of multiplication: $(ab)c = a(bc)$.

15. $-6 + (12 + 7) = (-6 + 12) + 7$

The numbers are in the same order but grouped differently, so this is an example of the associative property of addition: $(a + b) + c = a + (b + c)$.

17. $-6 + 6 = 0$

The sum of the two numbers is 0, so they are additive inverses (or opposites) of each other. This is an example of the additive inverse property: $a + (-a) = 0$.

19. $\dfrac{2}{3}\left(\dfrac{3}{2}\right) = 1$

The product of the two numbers is 1, so they are multiplicative inverses (or reciprocals) of each other. This is an example of the multiplicative inverse property: $a \cdot \dfrac{1}{a} = 1 \, (a \neq 0)$.

21. $2.34 + 0 = 2.34$

The sum of a number and 0 is the original number. This is an example of the identity property of addition: $a + 0 = a$.

23. $(4 + 17) + 3 = 3 + (4 + 17)$

The order of the numbers has been changed, but not the grouping, so this is an example of the commutative property of addition: $a + b = b + a$.

25. $6(x + y) = 6x + 6y$

The number 6 outside the parentheses is "distributed" over the x and y. This is an example of the distributive property.

27. $-\dfrac{5}{9} = -\dfrac{5}{9} \cdot \dfrac{3}{3} = -\dfrac{15}{27}$

$\frac{3}{3}$ is a form of the number 1. We use it to rewrite $-\frac{5}{9}$ as $-\frac{15}{27}$. This is an example of the identity property of multiplication.

29. $5(2x) + 5(3y) = 5(2x + 3y)$

This is an example of the distributive property. The number 5 is "distributed " over $2x$ and $3y$.

31. ADDITION:

(i) The identity property of addition states that adding zero to any number leaves it unchanged.

(ii) The inverse property of addition states that the sum of a number and its opposite is zero.

MULTIPLICATION:

(i) The identity property of multiplication states that multiplying any number by 1 leaves the number unchanged.

(ii) The inverse property of multiplication states that the product of any nonzero number and its reciprocal is 1.

33. Jack recognized the identity property of addition.

35. $\begin{aligned} 97 + 13 + 3 + 37 &= (97 + 3) + (13 + 37) \\ &= 100 + 50 \\ &= 150 \end{aligned}$

37. $\begin{aligned} 1999 + 2 + 1 + 8 &= (1999 + 1) + (2 + 8) \\ &= 2000 + 10 \\ &= 2010 \end{aligned}$

39. $\begin{aligned} 159 + 12 + 141 + 88 &= (159 + 141) + (12 + 88) \\ &= 300 + 100 \\ &= 400 \end{aligned}$

41. $\begin{aligned} &843 + 627 + (-43) + (-27) \\ &= [843 + (-43)] + [627 + (-27)] \\ &= 800 + 600 \\ &= 1400 \end{aligned}$

43. $\begin{aligned} &6t + 8 - 6t + 3 \\ &= 6t + 8 + (-6t) + 3 \qquad \textit{Definition of} \\ &\qquad\qquad\qquad\qquad\qquad\;\; \textit{subtraction} \\ &= (6t + 8) + (-6t) + 3 \quad \textit{Order of} \\ &\qquad\qquad\qquad\qquad\qquad\;\; \textit{operations} \end{aligned}$

$= (8 + 6t) + (-6t) + 3$ *Commutative property*

$= 8 + [6t + (-6t)] + 3$ *Associative property*

$= 8 + 0 + 3$ *Inverse property*

$= (8 + 0) + 3$ *Order of operations*

$= 8 + 3$ *Identity property*

$= 11$ *Add.*

45. $\dfrac{2}{3}x - 11 + 11 - \dfrac{2}{3}x$

$= \dfrac{2}{3}x + (-11) + 11 + \left(-\dfrac{2}{3}x\right)$

Definition of subtraction

$= \left[\dfrac{2}{3}x + (-11)\right] + 11 + \left(-\dfrac{2}{3}x\right)$

Order of operations

$= \dfrac{2}{3}x + (-11 + 11) + \left(-\dfrac{2}{3}x\right)$

Associative property

$= \dfrac{2}{3}x + 0 + \left(-\dfrac{2}{3}x\right)$ *Inverse property*

$= \left(\dfrac{2}{3}x + 0\right) + \left(-\dfrac{2}{3}x\right)$

Order of operations

$= \dfrac{2}{3}x + \left(-\dfrac{2}{3}x\right)$ *Identity property*

$= 0$ *Inverse property*

47. $\left(\dfrac{9}{7}\right)(-0.38)\left(\dfrac{7}{9}\right)$

$= \left[\left(\dfrac{9}{7}\right)(-0.38)\right]\left(\dfrac{7}{9}\right)$ *Order of operations*

$= \left[(-0.38)\left(\dfrac{9}{7}\right)\right]\left(\dfrac{7}{9}\right)$ *Commutative property*

$= (-0.38)\left[\left(\dfrac{9}{7}\right)\left(\dfrac{7}{9}\right)\right]$ *Associative property*

$= (-0.38)(1)$ *Inverse property*

$= -0.38$ *Identity property*

49. $t + (-t) + \frac{1}{2}(2)$

$= t + (-t) + 1$ *Inverse property*

$= [t + (-t)] + 1$ *Order of operations*

$= 0 + 1$ *Inverse property*

$= 1$ *Identity property*

51. $25 - (6 - 2) = 25 - (4)$
$= 21$
$(25 - 6) - 2 = 19 - 2$
$= 17$

Since $21 \neq 17$, this example shows that subtraction is not associative.

53. $-3(4 - 6)$

When distributing a negative number over a quantity, be careful not to "lose" a negative sign. The problem should be worked in the following way.

$-3(4 - 6) = -3(4) - 3(-6)$
$= -12 + 18$
$= 6$

55. $5(9 + 8) = 5 \cdot 9 + 5 \cdot 8$
$= 45 + 40$
$= 85$

57. $4(t + 3) = 4 \cdot t + 4 \cdot 3$
$= 4t + 12$

59. $-8(r + 3) = -8(r) + (-8)(3)$
$= -8r + (-24)$
$= -8r - 24$

61. $-5(y - 4) = -5(y) + (-5)(-4)$
$= -5y + 20$

63. $-\frac{4}{3}(12y + 15z)$

$= -\frac{4}{3}(12y) + \left(-\frac{4}{3}\right)(15z)$

$= \left[\left(-\frac{4}{3}\right) \cdot 12\right]y + \left[\left(-\frac{4}{3}\right) \cdot 15\right]z$

$= -16y + (-20)z$

$= -16y - 20z$

65. $8z + 8w = 8(z + w)$

67. $7(2v) + 7(5r) = 7(2v + 5r)$

69. $8(3r + 4s - 5y)$
$= 8(3r) + 8(4s) + 8(-5y)$
Distributive property
$= (8 \cdot 3)r + (8 \cdot 4)s + [8(-5)]y$
Associative property
$= 24r + 32s - 40y$ *Multiply.*

71. $-3(8x + 3y + 4z)$
$$= -3(8x) + (-3)(3y) + (-3)(4z)$$
  Distributive property
$$= (-3 \cdot 8)x + (-3 \cdot 3)y + (-3 \cdot 4)z$$
  Associative property
$$= -24x - 9y - 12z \quad Multiply.$$

73. $5x + 15 = 5x + 5 \cdot 3$
$$= 5(x + 3)$$

75. $-(4t + 3m)$

$= -1(4t + 3m)$ *Identity property*

$= -1(4t) + (-1)(3m)$ *Distributive property*

$= (-1 \cdot 4)t + (-1 \cdot 3)m$ *Associative property*

$= -4t - 3m$ *Multiply.*

77. $-(-5c - 4d)$

$= -1(-5c - 4d)$ *Identity property*

$= -1(-5c) + (-1)(-4d)$ *Distributive property*

$= (-1 \cdot -5)c + (-1 \cdot -4)d$ *Associative property*

$= 5c + 4d$ *Multiply.*

79. $-(-q + 5r - 8s)$
$$= -1(-q + 5r - 8s)$$
$$= -1(-q) + (-1)(5r) + (-1)(-8s)$$
$$= (-1 \cdot -1)q + (-1 \cdot 5)r + (-1 \cdot -8)s$$
$$= q - 5r + 8s$$

81. Answers will vary. For example, "putting on your socks" and "putting on your shoes" are everyday operations that are not commutative.

83. $-3[5 + (-5)] = -3(0) = 0$

84. $-3[5 + (-5)] = -3(5) + (-3)(-5)$

85. $-3(5) = -15$

86. We must interpret $(-3)(-5)$ as 15, since it is the additive inverse of -15.

87. **(a)** The left side of the statement is
$$-2(5 \cdot 7) = -2(35) = -70.$$

The right side of the statement is
$$(-2 \cdot 5) \cdot (-2 \cdot 7) = (-10) \cdot (-14) = 140.$$

So the statement is *false*.

(b) The original statement looks like the following true statement:
$$-2(5 + 7) = (-2)(5) + (-2)(7)$$

So it was probably the *distributive property* that was erroneously applied.

1.8   Simplifying Expressions

1. $6x - 2x = (6 - 2)x = 4x$
The correct response is **C**.

3. The numerical coefficient of $5x^3 y^7$ is 5.
The correct response is **A**.

5. $4r + 19 - 8 = 4r + 11$

7. $5 + 2(x - 3y) = 5 + 2(x) + 2(-3y)$
$$= 5 + 2x - 6y$$

9. $-2 - (5 - 3p) = -2 - 1(5 - 3p)$
$$= -2 - 1(5) - 1(-3p)$$
$$= -2 - 5 + 3p$$
$$= -7 + 3p$$

11. $6 + (4 - 3x) - 8 = 6 + 4 - 3x - 8$
$$= 10 - 3x - 8$$
$$= 10 - 8 - 3x$$
$$= 2 - 3x$$

13. The numerical coefficient of the term $-12k$ is -12.

15. The numerical coefficient of the term $5m^2$ is 5.

17. Because xw can be written as $1 \cdot xw$, the numerical coefficient of the term xw is 1.

19. Since $-x = -1x$, the numerical coefficient of the term $-x$ is -1.

21. Since $\frac{x}{5} = \frac{1}{5}x$, the numerical coefficient of the term $\frac{x}{5}$ is $\frac{1}{5}$.

23. $8r$ and $-13r$ are *like* terms since they have the same variable with the same exponent (which is understood to be 1).

25. $5z^4$ and $9z^3$ are *unlike* terms. Although both have the variable z, the exponents are not the same.

27. All numerical terms (constants) are considered like terms, so 4, 9, and -24 are *like* terms.

29. x and y are *unlike* terms because they do not have the same variable.

31. The student made a sign error when applying the distributive property.
$$7x - 2(3 - 2x) = 7x - 2(3) - 2(-2x)$$
$$= 7x - 6 + 4x$$
$$= 11x - 6$$

The correct answer is $11x - 6$.

33. $9y + 8y = (9 + 8)y$
$\qquad = 17y$

35. $-4a - 2a = (-4 - 2)a$
$\qquad = -6a$

37. $12b + b = 12b + 1b$
$\qquad = (12 + 1)b$
$\qquad = 13b$

39. $2k + 9 + 5k + 6 = 2k + 5k + 9 + 6$
$\qquad = (2 + 5)k + 15$
$\qquad = 7k + 15$

41. $-5y + 3 - 1 + 5 + y - 7$
$\quad = (-5y + 1y) + (3 + 5) + (-1 - 7)$
$\quad = (-5 + 1)y + (8) + (-8)$
$\quad = -4y + 8 - 8$
$\quad = -4y$

43. $-2x + 3 + 4x - 17 + 20$
$\quad = (-2x + 4x) + (3 - 17 + 20)$
$\quad = (-2 + 4)x + 6$
$\quad = 2x + 6$

45. $16 - 5m - 4m - 2 + 2m$
$\quad = (16 - 2) + (-5m - 4m + 2m)$
$\quad = 14 + (-5 - 4 + 2)m$
$\quad = 14 - 7m$

47. $-10 + x + 4x - 7 - 4x$
$\quad = (-10 - 7) + (1x + 4x - 4x)$
$\quad = -17 + (1 + 4 - 4)x$
$\quad = -17 + 1x$
$\quad = -17 + x$

49. $1 + 7x + 11x - 1 + 5x$
$\quad = (1 - 1) + (7x + 11x + 5x)$
$\quad = 0 + (7 + 11 + 5)x$
$\quad = 23x$

51. $-\dfrac{4}{3} + 2t + \dfrac{1}{3}t - 8 - \dfrac{8}{3}t$
$\quad = \left(2t + \dfrac{1}{3}t - \dfrac{8}{3}t\right) + \left(-\dfrac{4}{3} - 8\right)$
$\quad = \left(2 + \dfrac{1}{3} - \dfrac{8}{3}\right)t + \left(-\dfrac{4}{3} - 8\right)$
$\quad = \left(\dfrac{6}{3} + \dfrac{1}{3} - \dfrac{8}{3}\right)t + \left(-\dfrac{4}{3} - \dfrac{24}{3}\right)$
$\quad = -\dfrac{1}{3}t - \dfrac{28}{3}$

53. $6y^2 + 11y^2 - 8y^2 = (6 + 11 - 8)y^2$
$\qquad = 9y^2$

55. $2p^2 + 3p^2 - 8p^3 - 6p^3$
$\quad = (2p^2 + 3p^2) + (-8p^3 - 6p^3)$
$\quad = (2 + 3)p^2 + (-8 - 6)p^3$
$\quad = 5p^2 - 14p^3 \text{ or } -14p^3 + 5p^2$

57. $2(4x + 6) + 3 = 2(4x) + 2(6) + 3$
$\qquad = 8x + 12 + 3$
$\qquad = 8x + 15$

59. $100[0.05(x + 3)]$
$\quad = [100(0.05)](x + 3) \qquad \text{\textit{Associative property}}$
$\quad = 5(x + 3)$
$\quad = 5(x) + 5(3) \qquad \text{\textit{Distributive property}}$
$\quad = 5x + 15$

61. $-4(y - 7) - 6$
$\quad = -4(y) + (-4)(-7) - 6 \qquad \text{\textit{Distributive property}}$
$\quad = -4y + 28 - 6$
$\quad = -4y + 22$

63. $-\dfrac{4}{3}(y - 12) - \dfrac{1}{6}y$
$\quad = -\dfrac{4}{3}y - \dfrac{4}{3}(-12) - \dfrac{1}{6}y$
$\quad = -\dfrac{4}{3}y + 16 - \dfrac{1}{6}y$
$\quad = -\dfrac{4}{3}y - \dfrac{1}{6}y + 16$
$\quad = \left(-\dfrac{8}{6} - \dfrac{1}{6}\right)y + 16$
$\quad = -\dfrac{3}{2}y + 16 \qquad \left[-\dfrac{9}{6} = -\dfrac{3}{2}\right]$

65. $-5(5y - 9) + 3(3y + 6)$
$\quad = -5(5y) + (-5)(-9) + 3(3y) + 3(6)$
$\qquad\qquad\qquad \text{\textit{Distributive property}}$
$\quad = -25y + 45 + 9y + 18$
$\quad = (-25y + 9y) + (45 + 18)$
$\quad = (-25 + 9)y + 63$
$\quad = -16y + 63$

67. $-3(2r - 3) + 2(5r + 3)$
$\quad = -3(2r) + (-3)(-3) + 2(5r) + 2(3)$
$\qquad\qquad\qquad \text{\textit{Distributive property}}$
$\quad = -6r + 9 + 10r + 6$
$\quad = (-6r + 10r) + (9 + 6)$
$\quad = (-6 + 10)r + 15$
$\quad = 4r + 15$

69. $8(2k - 1) - (4k - 3)$

$= 8(2k - 1) - 1(4k - 3)$

               Replace $-$ with -1.

$= 8(2k) + 8(-1) + (-1)(4k) + (-1)(-3)$

$= 16k - 8 - 4k + 3$

$= 12k - 5$

71. $-2(-3k + 2) - (5k - 6) - 3k - 5$

$= -2(-3k) + (-2)(2) - 1(5k - 6) - 3k - 5$

$= 6k - 4 + (-1)(5k) + (-1)(-6) - 3k - 5$

$= 6k - 4 - 5k + 6 - 3k - 5$

$= -2k - 3$

73. $-4(-3k + 3) - (6k - 4) - 2k + 1$

$= -4(-3k + 3) - 1(6k - 4) - 2k + 1$

$= 12k - 12 - 6k + 4 - 2k + 1$

              Distributive property

$= (12k - 6k - 2k) + (-12 + 4 + 1)$

              Group like terms.

$= 4k - 7$ *Combine like terms.*

75. $-7.5(2y + 4) - 2.9(3y - 6)$

$= -7.5(2y) - 7.5(4) - 2.9(3y) - 2.9(-6)$

              Distributive property

$= -15y - 30 - 8.7y + 17.4$ *Multiply.*

$= -23.7y - 12.6$ *Combine like terms.*

77. "Five times a number, added to the sum of the number and three" is written $(x + 3) + 5x$.

$(x + 3) + 5x = x + 3 + 5x$

$= (x + 5x) + 3$

$= 6x + 3$

79. "A number multiplied by -7, subtracted from the sum of 13 and six times the number" is written $(13 + 6x) - (-7x)$.

$(13 + 6x) - (-7x) = 13 + 6x + 7x$

$= 13 + 13x$

81. "Six times a number added to -4, subtracted from twice the sum of three times the number and 4" is written $2(3x + 4) - (-4 + 6x)$.

$2(3x + 4) - (-4 + 6x)$

$= 2(3x + 4) - 1(-4 + 6x)$

$= 6x + 8 + 4 - 6x$

$= 6x + (-6x) + 8 + 4$

$= 0 + 12 = 12$

83. $9x - (x + 2)$

Wording will vary. One example is "the difference between 9 times a number and the sum of the number and 2." Another example is "the sum of a number and 2 subtracted from 9 times a number."

85. For widgets, the fixed cost is $1000 and the variable cost is $5 per widget, so the cost to produce x widgets is

$$1000 + 5x \text{ (dollars)}.$$

86. For gadgets, the fixed cost is $750 and the variable cost is $3 per gadget, so the cost to produce y gadgets is

$$750 + 3y \text{ (dollars)}.$$

87. The total cost to make x widgets and y gadgets is

$$1000 + 5x + 750 + 3y \text{ (dollars)}.$$

88. $1000 + 5x + 750 + 3y$

$= (1000 + 750) + 5x + 3y$

$= 1750 + 5x + 3y,$

so the total cost to make x widgets and y gadgets is

$$1750 + 5x + 3y \text{ (dollars)}.$$

Chapter 1 Review Exercises

1. $\dfrac{8}{5} \div \dfrac{32}{15} = \dfrac{8}{5} \cdot \dfrac{15}{32}$

$= \dfrac{8 \cdot (3 \cdot 5)}{5 \cdot (8 \cdot 4)}$

$= \dfrac{8 \cdot 3 \cdot 5}{5 \cdot 8 \cdot 4}$

$= \dfrac{3}{4}$

2. $\dfrac{3}{8} + 3\dfrac{1}{2} - \dfrac{3}{16} = \dfrac{3}{8} + \dfrac{7}{2} - \dfrac{3}{16}$

$= \dfrac{3 \cdot 2}{8 \cdot 2} + \dfrac{7 \cdot 8}{2 \cdot 8} - \dfrac{3}{16}$ *LCD = 16*

$= \dfrac{6}{16} + \dfrac{56}{16} - \dfrac{3}{16}$

$= \dfrac{62}{16} - \dfrac{3}{16}$

$= \dfrac{59}{16}, \text{ or } 3\dfrac{11}{16}$

3. $\dfrac{3}{8} + \dfrac{2}{5} = \dfrac{3 \cdot 5}{8 \cdot 5} + \dfrac{2 \cdot 8}{5 \cdot 8}$ *LCD = 40*

$= \dfrac{15}{40} + \dfrac{16}{40}$

$= \dfrac{31}{40}$

Since the entire pie chart represents $\frac{40}{40}$, this leaves $\frac{9}{40}$ unaccounted for. Thus, $\frac{9}{40}$ of the group did not have an opinion.

4. $\frac{3}{8}$ of the 400 people responded "yes."

$$\frac{3}{8} \cdot 400 = \frac{3}{8} \cdot \frac{400}{1}$$
$$= \frac{3 \cdot (8 \cdot 50)}{8 \cdot 1}$$
$$= \frac{3 \cdot 8 \cdot 50}{8 \cdot 1}$$
$$= 150$$

150 people responded "yes."

5. $5^4 = 5 \cdot 5 \cdot 5 \cdot 5 = 625$

6. $\left(\frac{3}{5}\right)^3 = \frac{3}{5} \cdot \frac{3}{5} \cdot \frac{3}{5} = \frac{27}{125}$

7. $(0.02)^2 = (0.02)(0.02)$
$= 0.0004$

8. $(0.1)^3 = (0.1)(0.1)(0.1)$
$= 0.001$

9. $8 \cdot 5 - 13 = 40 - 13 = 27$

10. $7\left[3 + 6\left(3^2\right)\right] = 7[3 + 6(9)]$
$= 7(3 + 54)$
$= 7(57)$
$= 399$

11. $\dfrac{9(4^2 - 3)}{4 \cdot 5 - 17} = \dfrac{9(16 - 3)}{20 - 17}$
$$= \frac{9(13)}{3}$$
$$= \frac{3 \cdot 3 \cdot 13}{3} = 39$$

12. $\dfrac{6(5 - 4) + 2(4 - 2)}{3^2 - (4 + 3)} = \dfrac{6(1) + 2(2)}{9 - (4 + 3)}$
$$= \frac{6 + 4}{9 - 7}$$
$$= \frac{10}{2} = 5$$

13. $12 \cdot 3 - 6 \cdot 6 = 36 - 36 = 0$

Since $0 = 0$ is true, so is $0 \le 0$, and therefore, the statement "$12 \cdot 3 - 6 \cdot 6 \le 0$" is true.

14. $3[5(2) - 3] = 3(10 - 3) = 3(7) = 21$

Therefore, the statement "$3[5(2) - 3] > 20$" is true.

15. $4^2 - 8 = 16 - 8 = 8$

Since $9 \le 8$ is false, the statement "$9 \le 4^2 - 8$" is false.

16. "Thirteen is less than seventeen" is written $13 < 17$.

17. "Five plus two is not equal to ten" is written $5 + 2 \ne 10$.

18. **(a)** The years in which there were *fewer than* 600 people naturalized are 1995(488), 1997(598), 1998(463), 2002(574), 2003(463), and 2004(537).

(b) The years in which there were *at least* 700 people naturalized are 1996(1045), 1999(840), and 2000(899).

(c) The five years having the largest numbers of naturalizations are the three years in part (b) along with 1997(598) and 2001(606). The *total* number of naturalizations in those five years is $1045 + 840 + 899 + 598 + 606 = 3988$, which represents 3988 thousand.

In Exercises 19–22, replace x with 6 and y with 3.

19. $2x + 6y = 2(6) + 6(3)$
$= 12 + 18 = 30$

20. $4(3x - y) = 4[3(6) - 3]$
$= 4(18 - 3)$
$= 4(15) = 60$

21. $\dfrac{x}{3} + 4y = \dfrac{6}{3} + 4(3)$
$$= 2 + 12 = 14$$

22. $\dfrac{x^2 + 3}{3y - x} = \dfrac{6^2 + 3}{3(3) - 6}$
$$= \frac{36 + 3}{9 - 6}$$
$$= \frac{39}{3} = 13$$

23. "Six added to a number" translates as $x + 6$.

24. "A number subtracted from eight" translates as $8 - x$.

25. "Nine subtracted from six times a number" translates as $6x - 9$.

26. "Three-fifths of a number added to 12" translates as $12 + \frac{3}{5}x$.

27. $5x + 3(x + 2) = 22; \ 2$
$5x + 3(x + 2) = 5(2) + 3(2 + 2)$ *Let x = 2.*
$\qquad\qquad = 5(2) + 3(4)$
$\qquad\qquad = 10 + 12 = 22$

Since the left side and the right side are equal, 2 is a solution of the given equation.

28. $\dfrac{t+5}{3t} = 1; \; 6$

$$\dfrac{t+5}{3t} = \dfrac{6+5}{3(6)} \quad \textit{Let t = 6.}$$

$$= \dfrac{11}{18}$$

Since the left side, $\frac{11}{18}$, is not equal to the right side, 1, 6 is not a solution of the equation.

29. "Six less than twice a number is 10" is written

$$2x - 6 = 10.$$

Letting x equal 0, 2, 4, 6, and 10 results in a false statement, so those values are not solutions.

Since $2(8) - 6 = 16 - 6 = 10$, the solution is 8.

30. "The product of a number and 4 is 8" is written

$$4x = 8.$$

Since $4(2) = 8$, the solution is 2.

31. $-4, -\frac{1}{2}, 0, 2.5, 5$

Graph these numbers on a number line. They are already arranged in order from smallest to largest.

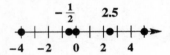

32. $-2, |-3|, -3, |-1|$

Recall that $|-3| = 3$ and $|-1| = 1$. From smallest to largest, the numbers are $-3, -2, |-1|, |-3|$.

33. Since $\frac{4}{3}$ is the quotient of two integers, it is a *rational number*. Since all rational numbers are also real numbers, $\frac{4}{3}$ is a *real number*.

34. Since the decimal representation of $\sqrt{6}$ does not terminate nor repeat, it is an *irrational number*. Since all irrational numbers are also real numbers, $\sqrt{6}$ is a *real number*.

35. $-10, 5$

Since any negative number is less than any positive number, -10 is the lesser number.

36. $-8, -9$

Since -9 is to the left of -8 on the number line, -9 is the lesser number.

37. $-\dfrac{2}{3}, -\dfrac{3}{4}$

To compare these fractions, use a common denominator.

$$-\dfrac{2}{3} = -\dfrac{8}{12}, \quad -\dfrac{3}{4} = -\dfrac{9}{12}$$

Since $-\frac{9}{12}$ is to the left of $-\frac{8}{12}$ on the number line, $-\frac{3}{4}$ is the lesser number.

38. $0, -|23|$

Since $-|23| = -23$ and $-23 < 0$, $-|23|$ is the lesser number.

39. $12 > -13$

This statement is true since 12 is to the right of -13 on the number line.

40. $0 > -5$

This statement is true since 0 is to the right of -5 on the number line.

41. $-9 < -7$

This statement is true since -9 is to the left of -7 on the number line.

42. $-13 \geq -13$

This is a true statement since $-13 = -13$.

43. **(a)** The opposite of the number -9 is its negative; that is, $-(-9) = 9$.

(b) Since $-9 < 0$, the absolute value of the number -9 is $|-9| = -(-9) = 9$.

44. 0

(a) $-0 = 0$

(b) $|0| = 0$

45. 6

(a) $-(6) = -6$

(b) $|6| = 6$

46. $-\frac{5}{7}$

(a) $-\left(-\frac{5}{7}\right) = \frac{5}{7}$

(b) $\left|-\frac{5}{7}\right| = -\left(-\frac{5}{7}\right) = \frac{5}{7}$

47. $|-12| = -(-12) = 12$

48. $-|3| = -3$

49. $-|-19| = -[-(-19)] = -19$

50. $-|9 - 2| = -|7| = -7$

51. $-10 + 4 = -6$

52. $14 + (-18) = -4$

53. $-8 + (-9) = -17$

54. $\dfrac{4}{9} + \left(-\dfrac{5}{4}\right) = \dfrac{4 \cdot 4}{9 \cdot 4} + \left(-\dfrac{5 \cdot 9}{4 \cdot 9}\right)$ *LCD = 36*

$\qquad = \dfrac{16}{36} + \left(-\dfrac{45}{36}\right)$

$\qquad = -\dfrac{29}{36}$

55. $-13.5 + (-8.3) = -21.8$

56. $(-10 + 7) + (-11) = (-3) + (-11)$

$\qquad\qquad\qquad\qquad\quad = -14$

57. $[-6 + (-8) + 8] + [9 + (-13)]$

$\qquad = \{[-6 + (-8)] + 8\} + (-4)$

$\qquad = [(-14) + 8] + (-4)$

$\qquad = (-6) + (-4) = -10$

58. $(-4 + 7) + (-11 + 3) + (-15 + 1)$

$\qquad = (3) + (-8) + (-14)$

$\qquad = [3 + (-8)] + (-14)$

$\qquad = (-5) + (-14) = -19$

59. $-7 - 4 = -7 + (-4) = -11$

60. $-12 - (-11) = -12 + (11) = -1$

61. $5 - (-2) = 5 + (2) = 7$

62. $-\dfrac{3}{7} - \dfrac{4}{5} = -\dfrac{3 \cdot 5}{7 \cdot 5} - \dfrac{4 \cdot 7}{5 \cdot 7}$

$\qquad = -\dfrac{15}{35} - \dfrac{28}{35}$ *LCD = 35*

$\qquad = -\dfrac{15}{35} + \left(-\dfrac{28}{35}\right)$

$\qquad = -\dfrac{43}{35}, \ \text{ or } \ -1\dfrac{8}{35}$

63. $2.56 - (-7.75) = 2.56 + (7.75)$

$\qquad\qquad\qquad\qquad = 10.31$

64. $(-10 - 4) - (-2) = [-10 + (-4)] + 2$

$\qquad\qquad\qquad\quad = (-14) + (2)$

$\qquad\qquad\qquad\quad = -12$

65. $(-3 + 4) - (-1) = (-3 + 4) + 1$

$\qquad\qquad\qquad\quad = 1 + 1$

$\qquad\qquad\qquad\quad = 2$

66. $-(-5 + 6) - 2 = -(1) + (-2)$

$\qquad\qquad\qquad = -1 + (-2)$

$\qquad\qquad\qquad = -3$

67. "19 added to the sum of -31 and 12" is written

$\qquad (-31 + 12) + 19 = (-19) + 19$

$\qquad\qquad\qquad\qquad\quad = 0.$

68. "13 more than the sum of -4 and -8" is written

$\qquad [-4 + (-8)] + 13 = -12 + 13$

$\qquad\qquad\qquad\qquad\quad = 1.$

69. "The difference between -4 and -6" is written

$\qquad -4 - (-6) = -4 + 6$

$\qquad\qquad\qquad\quad = 2.$

70. "Five less than the sum of 4 and -8" is written

$\qquad [4 + (-8)] - 5 = (-4) + (-5)$

$\qquad\qquad\qquad\qquad = -9.$

71. $x + (-2) = -4$

Because

$\qquad (-2) + (-2) = -4,$

the solution is -2.

72. $12 + x = 11$

Because

$\qquad 12 + (-1) = 11,$

the solution is -1.

73. $-23.75 + 50.00 = 26.25$

He now has a positive balance of $26.25.

74. $-26 + 16 = -10$

The high temperature was $-10°$F.

75. $-28 + 13 - 14 = (-28 + 13) - 14$

$\qquad\qquad\qquad = (-28 + 13) + (-14)$

$\qquad\qquad\qquad = -15 + (-14)$

$\qquad\qquad\qquad = -29$

His present financial status is $-$29.

76. $-3 - 7 = -3 + (-7)$

$\qquad\qquad = -10$

The new temperature is $-10°$.

77. $8 - 12 + 42 = [8 + (-12)] + 42$

$\qquad\qquad\qquad = -4 + 42$

$\qquad\qquad\qquad = 38$

The total net yardage is 38.

78. To get the closing value for the previous day, we can add the amount it was down to the amount at which it closed.

$\qquad 26.73 + 10{,}892.32 = 10{,}919.05$

79. $(-12)(-3) = 36$

80. $15(-7) = -(15 \cdot 7)$

$\qquad\qquad = -105$

81. $-\dfrac{4}{3}\left(-\dfrac{3}{8}\right) = \dfrac{4}{3} \cdot \dfrac{3}{8}$

$\qquad\qquad = \dfrac{4 \cdot 3}{3 \cdot 4 \cdot 2}$

$\qquad\qquad = \dfrac{1}{2}$

82. $(-4.8)(-2.1) = 10.08$

83. $5(8 - 12) = 5[8 + (-12)]$
$= 5(-4) = -20$

84. $(5 - 7)(8 - 3) = [5 + (-7)][8 + (-3)]$
$= (-2)(5) = -10$

85. $2(-6) - (-4)(-3) = -12 - (12)$
$= -12 + (-12)$
$= -24$

86. $3(-10) - 5 = -30 + (-5) = -35$

87. $\dfrac{-36}{-9} = \dfrac{4 \cdot 9}{9} = 4$

88. $\dfrac{220}{-11} = -\dfrac{20 \cdot 11}{11} = -20$

89. $-\dfrac{1}{2} \div \dfrac{2}{3} = -\dfrac{1}{2} \cdot \dfrac{3}{2} = -\dfrac{3}{4}$

90. $-33.9 \div (-3) = \dfrac{-33.9}{-3} = 11.3$

91. $\dfrac{-5(3) - 1}{8 - 4(-2)} = \dfrac{-15 + (-1)}{8 - (-8)}$
$= \dfrac{-16}{8 + 8}$
$= \dfrac{-16}{16} = -1$

92. $\dfrac{5(-2) - 3(4)}{-2[3 - (-2)] - 1} = \dfrac{-10 - 12}{-2(3 + 2) - 1}$
$= \dfrac{-10 + (-12)}{-2(5) - 1}$
$= \dfrac{-22}{-10 + (-1)}$
$= \dfrac{-22}{-11} = 2$

93. $\dfrac{10^2 - 5^2}{8^2 + 3^2 - (-2)} = \dfrac{100 - 25}{64 + 9 + 2}$
$= \dfrac{75}{75} = 1$

94. $\dfrac{(0.6)^2 + (0.8)^2}{(-1.2)^2 - (-0.56)} = \dfrac{0.36 + 0.64}{1.44 + 0.56}$
$= \dfrac{1.00}{2.00} = 0.5$

In Exercises 95–98, replace x with -5, y with 4, and z with -3.

95. $6x - 4z = 6(-5) - 4(-3)$
$= -30 - (-12)$
$= -30 + 12 = -18$

96. $5x + y - z = 5(-5) + (4) - (-3)$
$= (-25 + 4) + 3$
$= -21 + 3 = -18$

97. $5x^2 = 5(-5)^2$
$= 5(25)$
$= 125$

98. $z^2(3x - 8y) = (-3)^2[3(-5) - 8(4)]$
$= 9(-15 - 32)$
$= 9[-15 + (-32)]$
$= 9(-47) = -423$

99. "Nine less than the product of -4 and 5" is written
$$-4(5) - 9 = -20 + (-9)$$
$$= -29.$$

100. "Five-sixths of the sum of 12 and -6" is written
$$\tfrac{5}{6}[12 + (-6)] = \tfrac{5}{6}(6)$$
$$= 5.$$

101. "The quotient of 12 and the sum of 8 and -4" is written
$$\dfrac{12}{8 + (-4)} = \dfrac{12}{4} = 3.$$

102. "The product of -20 and 12, divided by the difference between 15 and -15" is written
$$\dfrac{-20(12)}{15 - (-15)} = \dfrac{-240}{15 + 15}$$
$$= \dfrac{-240}{30} = -8.$$

103. "8 times a number is -24" is written
$$8x = -24.$$

If $x = -3$,
$$8x = 8(-3) = -24.$$

The solution is -3.

104. "The quotient of a number and 3 is -2" is written
$$\dfrac{x}{3} = -2.$$

If $x = -6$,
$$\dfrac{x}{3} = \dfrac{-6}{3} = -2.$$

The solution is -6.

105. Find the average of the eight numbers.

$$\frac{26 + 38 + 40 + 20 + 4 + 14 + 96 + 18}{8}$$

$$= \frac{256}{8} = \frac{8 \cdot 32}{8} = 32$$

106. Find the average of the six numbers.

$$\frac{-12 + 28 + (-36) + 0 + 12 + (-10)}{6}$$

$$= \frac{-18}{6} = -3$$

107. $6 + 0 = 6$

This is an example of an identity property.

108. $5 \cdot 1 = 5$

This is an example of an identity property.

109. $-\frac{2}{3}\left(-\frac{3}{2}\right) = 1$

This is an example of an inverse property.

110. $17 + (-17) = 0$

This is an example of an inverse property.

111. $5 + (-9 + 2) = [5 + (-9)] + 2$

This is an example of an associative property.

112. $w(xy) = (wx)y$

This is an example of an associative property.

113. $3x + 3y = 3(x + y)$

This is an example of the distributive property.

114. $(1 + 2) + 3 = 3 + (1 + 2)$

This is an example of a commutative property.

115. $7y + 14 = 7y + 7 \cdot 2$
$\qquad = 7(y + 2)$

116. $-12(4 - t) = -12(4) - (-12)(t)$
$\qquad\qquad = -48 + 12t$

117. $3(2s) + 3(5y) = 3(2s + 5y)$

118. $-(-4r + 5s) = -1(-4r + 5s)$
$\qquad\qquad = (-1)(-4r) + (-1)(5s)$
$\qquad\qquad = 4r - 5s$

119. $25 - (5 - 2) = 25 - 3 = 22$
$(25 - 5) - 2 = 20 - 2 = 18$

Because different groupings lead to different results, we conclude that, in general, subtraction is not associative.

120. $180 \div (15 \div 5) = 180 \div 3 = 60$
$(180 \div 15) \div 5 = 12 \div 5 = 2.4$

Because different groupings lead to different results, we conclude that, in general, division is not associative.

121. $2m + 9m = (2 + 9)m$ *Distributive property*
$\qquad\qquad = 11m$

122. $15p^2 - 7p^2 + 8p^2$
$\qquad = (15 - 7 + 8)p^2$ *Distributive property*
$\qquad = 16p^2$

123. $5p^2 - 4p + 6p + 11p^2$
$\qquad = (5 + 11)p^2 + (-4 + 6)p$
$\qquad\qquad$ *Distributive property*
$\qquad = 16p^2 + 2p$

124. $-2(3k - 5) + 2(k + 1)$
$\qquad = -6k + 10 + 2k + 2$
$\qquad\qquad$ *Distributive property*
$\qquad = -4k + 12$

125. $7(2m + 3) - 2(8m - 4)$
$\qquad = 14m + 21 - 16m + 8$
$\qquad\qquad$ *Distributive property*
$\qquad = (14 - 16)m + 29$
$\qquad = -2m + 29$

126. $-(2k + 8) - (3k - 7)$
$\qquad = -1(2k + 8) - 1(3k - 7)$
$\qquad\qquad$ *Replace $-$ with -1.*
$\qquad = -2k - 8 - 3k + 7$
$\qquad\qquad$ *Distributive property*
$\qquad = -5k - 1$

127. "Seven times a number, subtracted from the product of -2 and three times the number" is written

$$-2(3x) - 7x = -6x - 7x = -13x.$$

128. "A number multiplied by 8, added to the sum of 5 and four times the number" is written

$$(5 + 4x) + 8x = 5 + (4x + 8x) = 5 + 12x.$$

129. **[1.5]** $[(-2) + 7 - (-5)] + [-4 - (-10)]$
$\qquad = \{[(-2) + 7] - (-5)\} + (-4 + 10)$
$\qquad = (5 + 5) + 6$
$\qquad = 10 + 6 = 16$

130. **[1.6]** $\left(-\frac{5}{6}\right)^2 = \left(-\frac{5}{6}\right)\left(-\frac{5}{6}\right)$
$\qquad\qquad = \frac{25}{36}$

131. [1.6] $\dfrac{6(-4) + 2(-12)}{5(-3) + (-3)} = \dfrac{-24 + (-24)}{-15 + (-3)}$

$$= \dfrac{-48}{-18} = \dfrac{8 \cdot 6}{3 \cdot 6}$$

$$= \dfrac{8}{3}, \text{ or } 2\dfrac{2}{3}$$

132. [1.5] $\dfrac{3}{8} - \dfrac{5}{12} = \dfrac{3 \cdot 3}{8 \cdot 3} - \dfrac{5 \cdot 2}{12 \cdot 2}$

$$= \dfrac{9}{24} - \dfrac{10}{24}$$

$$= \dfrac{9}{24} + \left(-\dfrac{10}{24}\right)$$

$$= -\dfrac{1}{24}$$

133. [1.6] $\dfrac{8^2 + 6^2}{7^2 + 1^2} = \dfrac{64 + 36}{49 + 1}$

$$= \dfrac{100}{50} = 2$$

134. [1.6] $-16(-3.5) - 7.2(-3)$

$$= 56 - [(7.2)(-3)]$$

$$= 56 - (-21.6)$$

$$= 56 + 21.6$$

$$= 77.6$$

135. [1.5] $2\dfrac{5}{6} - 4\dfrac{1}{3} = \dfrac{17}{6} - \dfrac{13}{3}$

$$= \dfrac{17}{6} - \dfrac{13 \cdot 2}{3 \cdot 2}$$

$$= \dfrac{17}{6} - \dfrac{26}{6}$$

$$= \dfrac{17}{6} + \left(-\dfrac{26}{6}\right)$$

$$= -\dfrac{9}{6} = -\dfrac{3}{2}, \text{ or } -1\dfrac{1}{2}$$

136. [1.5] $-8 + [(-4 + 17) - (-3 - 3)]$

$$= -8 + \{(13) - [-3 + (-3)]\}$$

$$= -8 + [13 - (-6)]$$

$$= -8 + (13 + 6)$$

$$= -8 + 19 = 11$$

137. [1.6] $-\dfrac{12}{5} \div \dfrac{9}{7} = -\dfrac{12}{5} \cdot \dfrac{7}{9}$

$$= -\dfrac{12 \cdot 7}{5 \cdot 9}$$

$$= -\dfrac{3 \cdot 4 \cdot 7}{5 \cdot 3 \cdot 3}$$

$$= -\dfrac{28}{15}, \text{ or } -1\dfrac{13}{15}$$

138. [1.6] $(-8 - 3) - 5(2 - 9)$

$$= [-8 + (-3)] - 5[2 + (-9)]$$

$$= -11 - 5(-7)$$

$$= -11 - (-35)$$

$$= -11 + 35 = 24$$

139. [1.8] $5x^2 - 12y^2 + 3x^2 - 9y^2$

$$= (5x^2 + 3x^2) + (-12y^2 - 9y^2)$$

$$= (5 + 3)x^2 + (-12 - 9)y^2$$

$$= 8x^2 - 21y^2$$

140. [1.8] $-4(2t + 1) - 8(-3t + 4)$

$$= -4(2t) - 4(1) - 8(-3t) - 8(4)$$

$$= -8t - 4 + 24t - 32$$

$$= 16t - 36$$

141. [1.6] Dividing 0 *by* a nonzero number gives a quotient of 0. However, dividing a number *by* 0 is undefined.

142. [1.6] It is not correct, because it does not consider the operation involved. Multiplying two negative numbers gives a positive number, but adding two negative numbers gives a negative number.

143. [1.6] "The product of 5 and the sum of a number and 7" is translated as

$$5(x + 7) = 5(x) + 5(7)$$

$$= 5x + 35.$$

144. [1.5] $118 - 165 = 118 + (-165)$

$$= -47$$

The lowest temperature ever recorded in Iowa was $-47°$F.

Chapter 1 Test

1. $\dfrac{63}{99} = \dfrac{7 \cdot 9}{11 \cdot 9} = \dfrac{7}{11}$

2. The denominators are 8, 12, and 15; or equivalently, $2^3, 2^2 \cdot 3$, and $3 \cdot 5$. So the LCD is $2^3 \cdot 3 \cdot 5 = 120$.

$$\dfrac{5}{8} + \dfrac{11}{12} + \dfrac{7}{15}$$

$$= \dfrac{5 \cdot 15}{8 \cdot 15} + \dfrac{11 \cdot 10}{12 \cdot 10} + \dfrac{7 \cdot 8}{15 \cdot 8}$$

$$= \dfrac{75}{120} + \dfrac{110}{120} + \dfrac{56}{120}$$

$$= \dfrac{241}{120}, \text{ or } 2\dfrac{1}{120}$$

3. $\dfrac{19}{15} \div \dfrac{6}{5} = \dfrac{19}{15} \cdot \dfrac{5}{6} = \dfrac{19 \cdot 5}{3 \cdot 5 \cdot 6} = \dfrac{19}{18}, \text{ or } 1\dfrac{1}{18}$

4. (a) The number of passengers that used air travel is $\frac{2}{5}$ of 1230 million.

$$\frac{2}{5} \cdot 1230 = 492$$

So 492 million passengers used air travel.

(b) Since $\frac{3}{10}$ of the passengers used the bus, $\frac{7}{10}$ did not.

$$\frac{7}{10} \cdot 1230 = 861$$

So 861 million passengers did not use the bus.

5. $4[-20 + 7(-2)] = 4[-20 + (-14)]$
$$= 4(-34) = -136$$

Since $-136 \leq 135$, the statement "$4[-20 + 7(-2)] \leq 135$" is true.

6. $-1, -3, |-4|, |-1|$

Recall that $|-4| = 4$ and $|-1| = 1$. From smallest to largest, the numbers are $-3, -1, |-1|, |-4|$.

7. The number $-\frac{2}{3}$ can be written as a quotient of two integers with denominator not 0, so it is a *rational number*. Since all rational numbers are real numbers, it is also a *real number*.

8. If -8 and -1 are both graphed on a number line, we see that the point for -8 is to the *left* of the point for -1. This indicates that -8 is *less than* -1.

9. "The quotient of -6 and the sum of 2 and -8" is written $\dfrac{-6}{2 + (-8)}$,

and $\dfrac{-6}{2 + (-8)} = \dfrac{-6}{-6} = 1.$

10. $-2 - (5 - 17) + (-6)$
$$= -2 - [5 + (-17)] + (-6)$$
$$= -2 - (-12) + (-6)$$
$$= (-2 + 12) + (-6)$$
$$= 10 + (-6) = 4$$

11. $-5\dfrac{1}{2} + 2\dfrac{2}{3} = -\dfrac{11}{2} + \dfrac{8}{3}$
$$= -\dfrac{11 \cdot 3}{2 \cdot 3} + \dfrac{8 \cdot 2}{3 \cdot 2}$$
$$= -\dfrac{33}{6} + \dfrac{16}{6}$$
$$= -\dfrac{17}{6}, \text{ or } -2\dfrac{5}{6}$$

12. $-6 - [-7 + (2 - 3)]$
$$= -6 - [-7 + (-1)]$$
$$= -6 - (-8)$$
$$= -6 + 8 = 2$$

13. $4^2 + (-8) - (2^3 - 6)$
$$= 16 + (-8) - (8 - 6)$$
$$= [16 + (-8)] - 2$$
$$= 8 - 2 = 6$$

14. $(-5)(-12) + 4(-4) + (-8)^2$
$$= (-5)(-12) + 4(-4) + 64$$
$$= [60 + (-16)] + 64$$
$$= 44 + 64 = 108$$

15. $\dfrac{-7 - (-6 + 2)}{-5 - (-4)} = \dfrac{-7 - (-4)}{-5 + 4}$
$$= \dfrac{-7 + 4}{-1}$$
$$= \dfrac{-3}{-1} = 3$$

16. $\dfrac{30(-1 - 2)}{-9[3 - (-2)] - 12(-2)}$
$$= \dfrac{30(-3)}{-9(5) - (-24)}$$
$$= \dfrac{-90}{-45 + 24}$$
$$= \dfrac{-90}{-21}$$
$$= \dfrac{30 \cdot 3}{7 \cdot 3} = \dfrac{30}{7}, \text{ or } 4\dfrac{2}{7}$$

17. $-x + 3 = -3$

If $x = 6$,
$$-6 + 3 = -3.$$

Therefore, the solution is 6.

18. $-3x = -12$

If $x = 4$,
$$-3x = -3(4) = -12.$$

Therefore, the solution is 4.

19. $3x - 4y^2$
$$= 3(-2) - 4(4^2) \qquad \textit{Let } x = -2, y = 4.$$
$$= 3(-2) - 4(16)$$
$$= -6 - 64 = -70$$

20.
$$\frac{5x + 7y}{3(x + y)}$$
$$= \frac{5(-2) + 7(4)}{3(-2 + 4)} \quad \textit{Let } x = -2, \, y = 4.$$
$$= \frac{-10 + 28}{3(2)}$$
$$= \frac{18}{6} = 3$$

21. The difference between the highest and lowest elevations is

$6960 - (-40) = 6960 + 40 = 7000$ meters.

22. 4 saves (3 points per save)

+ 3 wins (3 points per win)

+ 2 losses (-2 points per loss)

+ 1 blown save (-2 points per blown save)

$= 4(3) + 3(3) + 2(-2) + 1(-2)$
$= 12 + 9 - 4 - 2$
$= 15$ points

He has a total of 15 points.

23. **(a)** The change in enrollment from 1980 to 1985 was $15.52 - 17.38 = -1.86$ million students. Expressed as an integer, this number is $-1,860,000$.

(b) The change in enrollment from 1985 to 1990 was $14.27 - 15.52 = -1.25$ million students. Expressed as an integer, this number is $-1,250,000$.

(c) The change in enrollment from 1995 to 2000 was $17.34 - 15.75 = 1.59$ million students. Expressed as an integer, this number is $1,590,000$.

(d) The change in enrollment from 2000 to 2003 was $18.17 - 17.34 = 0.83$ million students. Expressed as an integer, this number is $830,000$.

24. Commutative property

$(5 + 2) + 8 = 8 + (5 + 2)$

illustrates a commutative property because the order of the numbers is changed, but not the grouping. The correct response is **B**.

25. Associative property

$-5 + (3 + 2) = (-5 + 3) + 2$

illustrates an associative property because the grouping of the numbers is changed, but not the order. The correct response is **D**.

26. Inverse property

$$-\frac{5}{3}\left(-\frac{3}{5}\right) = 1$$

illustrates an inverse property. The correct response is **E**.

27. Identity property

$3x + 0 = 3x$

illustrates an identity property. The correct response is **A**.

28. Distributive property

$-3(x + y) = -3x + (-3y)$

illustrates the distributive property. The correct response is **C**.

29. $3(x + 1) = 3 \cdot x + 3 \cdot 1$
$\qquad\qquad\; = 3x + 3$

The distributive property is used to rewrite $3(x + 1)$ as $3x + 3$.

30. **(a)** $-6[5 + (-2)] = -6(3) = -18$

(b) $-6[5 + (-2)] = -6(5) + (-6)(-2)$
$\qquad\qquad\qquad\quad = -30 + 12 = -18$

(c) The distributive property assures us that the answers must be the same, because $a(b + c) = ab + ac$ for all a, b, c.

31. $8x + 4x - 6x + x + 14x$
$= (8 + 4 - 6 + 1 + 14)x$
$= 21x$

32. $5(2x - 1) - (x - 12) + 2(3x - 5)$
$= 5(2x - 1) - 1(x - 12) + 2(3x - 5)$
$= 10x - 5 - x + 12 + 6x - 10$
$= (10 - 1 + 6)x + (-5 + 12 - 10)$
$= 15x - 3$

CHAPTER 2   LINEAR EQUATIONS AND INEQUALITIES IN ONE VARIABLE

2.1 The Addition Property of Equality

1. Equations that have exactly the same solution sets are **equivalent equations**.

A.
$$x + 2 = 6$$
$$x + 2 - 2 = 6 - 2 \quad \textit{Subtract 2.}$$
$$x = 4$$

So $x + 2 = 6$ and $x = 4$ *are* equivalent equations.

B.
$$10 - x = 5$$
$$10 - x - 10 = 5 - 10 \quad \textit{Subtract 10.}$$
$$-x = -5$$
$$-1(-x) = -1(-5) \quad \textit{Multiply by } -1.$$
$$x = 5$$

So $10 - x = 5$ and $x = -5$ *are not* equivalent equations.

C. Subtract 3 from both sides to get $x = 6$, so $x + 3 = 9$ and $x = 6$ *are* equivalent equations.

D. Subtract 4 from both sides to get $x = 4$. The second equation is $x = -4$, so $4 + x = 8$ and $x = -4$ *are not* equivalent equations.

3. The addition property of equality says that the same number (or expression) added to each side of an equation results in an equivalent equation. *Example*: $-x$ can be added to each side of $2x + 3 = x - 5$ to get the equivalent equation $x + 3 = -5$.

For Exercises 5–43, all solutions should be checked by substituting into the original equation. Checks will be shown here for only a few of the exercises.

5.
$$x - 4 = 8$$
$$x - 4 + 4 = 8 + 4$$
$$x = 12$$

Check this solution by replacing x with 12 in the original equation.

$$x - 4 = 8$$
$$12 - 4 = 8 \ ? \quad \textit{Let x = 12.}$$
$$8 = 8 \quad \textit{True}$$

Because the final statement is true, $\{12\}$ is the solution set.

7.
$$x - 12 = 19$$
$$x - 12 + 12 = 19 + 12$$
$$x = 31$$

Check $x = 31$:

$$31 - 12 = 19 \ ? \quad \textit{Let x = 31.}$$
$$19 = 19 \quad \textit{True}$$

Thus, $\{31\}$ is the solution set.

9.
$$x - 5 = -8$$
$$x - 5 + 5 = -8 + 5$$
$$x = -3$$

Checking yields a true statement, so $\{-3\}$ is the solution set.

11.
$$r + 9 = 13$$
$$r + 9 - 9 = 13 - 9$$
$$r = 4$$

Checking yields a true statement, so $\{4\}$ is the solution set.

13.
$$x + 26 = 17$$
$$x + 26 - 26 = 17 - 26$$
$$x = -9$$

Checking yields a true statement, so $\{-9\}$ is the solution set.

15.
$$7 + r = -3$$
$$r + 7 = -3$$
$$r + 7 - 7 = -3 - 7$$
$$r = -10$$

The solution set is $\{-10\}$.

17.
$$2 = p + 15$$
$$2 - 15 = p + 15 - 15$$
$$-13 = p$$

The solution set is $\{-13\}$.

19.
$$-2 = x - 12$$
$$-2 + 12 = x - 12 + 12$$
$$10 = x$$

The solution set is $\{10\}$.

21.
$$x - 8.4 = -2.1$$
$$x - 8.4 + 8.4 = -2.1 + 8.4$$
$$x = 6.3$$

The solution set is $\{6.3\}$.

23.
$$t + 12.3 = -4.6$$
$$t + 12.3 - 12.3 = -4.6 - 12.3$$
$$t = -16.9$$

The solution set is $\{-16.9\}$.

25.
$$3x + 7 = 2x + 4$$
$$3x + 7 - 2x = 2x + 4 - 2x$$
$$x + 7 = 4$$
$$x + 7 - 7 = 4 - 7$$
$$x = -3$$

The solution set is $\{-3\}$.

27.
$$8t + 6 = 7t + 6$$
$$8t + 6 - 7t = 7t + 6 - 7t$$
$$t + 6 = 6$$
$$t + 6 - 6 = 6 - 6$$
$$t = 0$$

The solution set is $\{0\}$.

29.
$$-4x + 7 = -5x + 9$$
$$-4x + 7 + 5x = -5x + 9 + 5x$$
$$x + 7 = 9$$
$$x + 7 - 7 = 9 - 7$$
$$x = 2$$

The solution set is $\{2\}$.

31.
$$\frac{2}{5}w - 6 = \frac{7}{5}w$$
$$\frac{2}{5}w - 6 - \frac{2}{5}w = \frac{7}{5}w - \frac{2}{5}w \quad \textit{Subtract } \frac{2}{5}w.$$
$$-6 = \frac{5}{5}w$$
$$-6 = w$$

The solution set is $\{-6\}$.

33.
$$5.6x + 2 = 4.6x$$
$$5.6x + 2 - 4.6x = 4.6x - 4.6x$$
$$1.0x + 2 = 0$$
$$x + 2 - 2 = 0 - 2$$
$$x = -2$$

The solution set is $\{-2\}$.

35.
$$3p = 2p$$
$$3p - 2p = 2p - 2p$$
$$p = 0$$

The solution set is $\{0\}$.

37.
$$1.2y - 4 = 0.2y - 4$$
$$1.2y - 4 - 0.2y = 0.2y - 4 - 0.2y$$
$$1.0y - 4 = -4$$
$$y - 4 + 4 = -4 + 4$$
$$y = 0$$

The solution set is $\{0\}$.

39.
$$\frac{1}{2}x + 2 = -\frac{1}{2}x$$
$$\frac{1}{2}x + \frac{1}{2}x + 2 = -\frac{1}{2}x + \frac{1}{2}x$$
$$x + 2 = 0$$
$$x + 2 - 2 = 0 - 2$$
$$x = -2$$

The solution set is $\{-2\}$.

41.
$$3x + 7 - 2x = 0$$
$$x + 7 = 0$$
$$x + 7 - 7 = 0 - 7$$
$$x = -7$$

The solution set is $\{-7\}$.

43.
$$4x + 30 - 3x = 0$$
$$x + 30 = 0$$
$$x + 30 - 30 = 0 - 30$$
$$x = -30$$

The solution set is $\{-30\}$.

45. A sample answer might be, "A linear equation in one variable is an equation that can be written using only one variable term with the variable to the first power."

47.
$$5t + 3 + 2t - 6t = 4 + 12$$
$$(5 + 2 - 6)t + 3 = 16$$
$$t + 3 - 3 = 16 - 3$$
$$t = 13$$

Check $t = 13$: $16 = 16$ *True*

The solution set is $\{13\}$.

49.
$$6x + 5 + 7x + 3 = 12x + 4$$
$$13x + 8 = 12x + 4$$
$$13x + 8 - 12x = 12x + 4 - 12x$$
$$x + 8 = 4$$
$$x + 8 - 8 = 4 - 8$$
$$x = -4$$

Check $x = -4$: $-44 = -44$ *True*

The solution set is $\{-4\}$.

51.
$$5.2q - 4.6 - 7.1q = -0.9q - 4.6$$
$$-1.9q - 4.6 = -0.9q - 4.6$$
$$-1.9q - 4.6 + 0.9q = -0.9q - 4.6 + 0.9q$$
$$-1.0q - 4.6 = -4.6$$
$$-1.0q - 4.6 + 4.6 = -4.6 + 4.6$$
$$-q = 0$$
$$q = 0$$

Check $q = 0$: $-4.6 = -4.6$ *True*

The solution set is $\{0\}$.

53.

$$\frac{5}{7}x + \frac{1}{3} = \frac{2}{5} - \frac{2}{7}x + \frac{2}{5}$$

$$\frac{5}{7}x + \frac{1}{3} = \frac{4}{5} - \frac{2}{7}x$$

$$\frac{5}{7}x + \frac{2}{7}x + \frac{1}{3} = \frac{4}{5} - \frac{2}{7}x + \frac{2}{7}x \quad \text{Add } \frac{2}{7}x.$$

$$\frac{7}{7}x + \frac{1}{3} = \frac{4}{5} \qquad \begin{array}{l}\textit{Combine} \\ \textit{like terms.}\end{array}$$

$$1x + \frac{1}{3} - \frac{1}{3} = \frac{4}{5} - \frac{1}{3} \qquad \textit{Subtract } \frac{1}{3}.$$

$$x = \frac{12}{15} - \frac{5}{15} \qquad \textit{LCD} = 15$$

$$x = \frac{7}{15}$$

Check $x = \frac{7}{15}$: $\frac{2}{3} = \frac{2}{3}$ *True*

The solution set is $\left\{\frac{7}{15}\right\}$.

55. $(5y + 6) - (3 + 4y) = 10$

$$5y + 6 - 3 - 4y = 10 \qquad \begin{array}{l}\textit{Distributive} \\ \textit{property}\end{array}$$

$$y + 3 = 10 \qquad \textit{Combine terms.}$$

$$y + 3 - 3 = 10 - 3 \qquad \textit{Subtract 3.}$$

$$y = 7$$

Check $y = 7$: $10 = 10$ *True*

The solution set is $\{7\}$.

57. $2(p + 5) - (9 + p) = -3$

$$2p + 10 - 9 - p = -3$$

$$p + 1 = -3$$

$$p + 1 - 1 = -3 - 1$$

$$p = -4$$

Check $p = -4$: $-3 = -3$ *True*

The solution set is $\{-4\}$.

59. $-6(2b + 1) + (13b - 7) = 0$

$$-12b - 6 + 13b - 7 = 0$$

$$b - 13 = 0$$

$$b - 13 + 13 = 0 + 13$$

$$b = 13$$

Check $b = 13$: $0 = 0$ *True*

The solution set is $\{13\}$.

61.

$$10(-2x + 1) = -19(x + 1)$$

$$-20x + 10 = -19x - 19$$

$$-20x + 10 + 19x = -19x - 19 + 19x$$

$$-x + 10 = -19$$

$$-x + 10 - 10 = -19 - 10$$

$$-x = -29$$

$$x = 29$$

Check $x = 29$: $-570 = -570$ *True*

The solution set is $\{29\}$.

63.

$$-2(8p + 2) - 3(2 - 7p) - 2(4 + 2p) = 0$$

$$-16p - 4 - 6 + 21p - 8 - 4p = 0$$

$$p - 18 = 0$$

$$p - 18 + 18 = 0 + 18$$

$$p = 18$$

Check $p = 18$: $0 = 0$ *True*

The solution set is $\{18\}$.

65. $4(7x - 1) + 3(2 - 5x) - 4(3x + 5) = -6$

$$28x - 4 + 6 - 15x - 12x - 20 = -6$$

$$x - 18 = -6$$

$$x - 18 + 18 = -6 + 18$$

$$x = 12$$

Check $x = 12$: $-6 = -6$ *True*

The solution set is $\{12\}$.

67. Answers will vary. One example is $x - 6 = -8$.

69. "Three times a number is 17 more than twice the number."

$$3x = 2x + 17$$

$$3x - 2x = 2x + 17 - 2x$$

$$x = 17$$

The number is 17 and $\{17\}$ is the solution set.

71. "If six times a number is subtracted from seven times the number, the result is -9."

$$7x - 6x = -9$$

$$x = -9$$

The number is -9 and $\{-9\}$ is the solution set.

73. $\dfrac{2}{3}\left(\dfrac{3}{2}\right) = \dfrac{2 \cdot 3}{3 \cdot 2} = 1$

75. $-\dfrac{5}{4}\left(-\dfrac{4}{5}x\right) = -\dfrac{5}{4}\left(-\dfrac{4}{5}\right)x$

$$= \frac{5 \cdot 4}{4 \cdot 5}x$$

$$= 1x = x$$

77. $9\left(\dfrac{r}{9}\right) = 9\left(\dfrac{1}{9}r\right)$

$$= 9\left(\frac{1}{9}\right)r$$

$$= 1r = r$$

2.2 The Multiplication Property of Equality

1. The multiplication property of equality says that the same nonzero number (or expression) multiplied on each side of the equation results in an equivalent equation. *Example*: Multiplying each side of $7x = 4$ by $\frac{1}{7}$ gives the equivalent equation $x = \frac{4}{7}$.

3. Choice **C** doesn't require the use of the multiplicative property of equality. After the equation is simplified, the variable x is alone on the left side.

$$5x - 4x = 7$$
$$x = 7$$

5. To get x alone on the left side, divide by 4, the coefficient of x.

7. $\frac{2}{3}x = 8$

    To get just x on the left side, multiply both sides of the equation by the reciprocal of $\frac{2}{3}$, which is $\frac{3}{2}$.

9. $\frac{x}{10} = 3$

    This equation is equivalent to $\frac{1}{10}x = 3$. To get just x on the left side, multiply both sides of the equation by the reciprocal of $\frac{1}{10}$, which is 10.

11. $-\frac{9}{2}x = -4$

    To get just x on the left side, multiply both sides of the equation by the reciprocal of $-\frac{9}{2}$, which is $-\frac{2}{9}$.

13. $-x = 0.36$

    This equation is equivalent to $-1x = 0.36$. To get just x on the left side, multiply both sides of the equation by the reciprocal of -1, which is -1.

15. $6x = 5$

    To get just x on the left side, divide both sides of the equation by the coefficient of x, which is 6.

17. $-4x = 13$

    To get just x on the left side, divide both sides of the equation by the coefficient of x, which is -4.

19. $0.12x = 48$

    To get just x on the left side, divide both sides of the equation by the coefficient of x, which is 0.12.

21. $-x = 23$

    This equation is equivalent to $-1x = 23$. To get just x on the left side, divide both sides of the equation by the coefficient of x, which is -1.

23. $5x = 30$
$$\frac{5x}{5} = \frac{30}{5} \quad \textit{Divide by 5.}$$
$$1x = 6$$
$$x = 6$$

    Check $x = 6$: $30 = 30$ *True*

    The solution set is $\{6\}$.

25. $2m = 15$
$$\frac{2m}{2} = \frac{15}{2} \quad \textit{Divide by 2.}$$
$$m = \frac{15}{2}$$

    Check $m = \frac{15}{2}$: $15 = 15$ *True*

    The solution set is $\left\{\frac{15}{2}\right\}$.

27. $3a = -15$
$$\frac{3a}{3} = \frac{-15}{3} \quad \textit{Divide by 3.}$$
$$a = -5$$

    Check $a = -5$: $-15 = -15$ *True*

    The solution set is $\{-5\}$.

29. $-3x = 12$
$$\frac{-3x}{-3} = \frac{12}{-3} \quad \textit{Divide by -3.}$$
$$x = -4$$

    Check $x = -4$: $12 = 12$ *True*

    The solution set is $\{-4\}$.

31. $10t = -36$
$$\frac{10t}{10} = \frac{-36}{10} \quad \textit{Divide by 10.}$$
$$t = -\frac{36}{10} = -\frac{18}{5} \quad \textit{Lowest terms}$$

    Check $t = -\frac{18}{5}$: $-36 = -36$ *True*

    The solution set is $\left\{-\frac{18}{5}\right\}$.

33. $-6x = -72$
$$\frac{-6x}{-6} = \frac{-72}{-6} \quad \textit{Divide by -6.}$$
$$x = 12$$

    Check $x = 12$: $-72 = -72$ *True*

    The solution set is $\{12\}$.

35. $2r = 0$
$$\frac{2r}{2} = \frac{0}{2} \quad \textit{Divide by 2.}$$
$$r = 0$$

    Check $r = 0$: $0 = 0$ *True*

    The solution set is $\{0\}$.

37. $0.2t = 8$
$$\frac{0.2t}{0.2} = \frac{8}{0.2}$$
$$t = 40$$

    Check $t = 40$: $8 = 8$ *True*

    The solution set is $\{40\}$.

39. $\quad -2.1m = 25.62$

$$\frac{-2.1m}{-2.1} = \frac{25.62}{-2.1}$$

$$m = -12.2$$

Check $m = -12.2$: $25.62 = 25.62$ *True*

The solution set is $\{-12.2\}$.

41. $\quad \frac{1}{4}x = -12$

$$4 \cdot \frac{1}{4}x = 4(-12) \quad \textit{Multiply by 4.}$$

$$1x = -48$$

$$x = -48$$

Check $x = -48$: $-12 = -12$ *True*

The solution set is $\{-48\}$.

43. $\quad \frac{z}{6} = 12$

$$\frac{1}{6}z = 12$$

$$6 \cdot \frac{1}{6}z = 6 \cdot 12$$

$$z = 72$$

Check $z = 72$: $12 = 12$ *True*

The solution set is $\{72\}$.

45. $\quad \frac{x}{7} = -5$

$$\frac{1}{7}x = -5$$

$$7\left(\frac{1}{7}x\right) = 7(-5)$$

$$x = -35$$

Check $x = -35$: $-5 = -5$ *True*

The solution set is $\{-35\}$.

47. $\quad \frac{2}{7}p = 4$

$$\frac{7}{2}\left(\frac{2}{7}p\right) = \frac{7}{2}(4) \quad \begin{array}{c}\textit{Multiply by}\\ \textit{the reciprocal}\\ \textit{of } \frac{2}{7}.\end{array}$$

$$p = 14$$

Check $p = 14$: $4 = 4$ *True*

The solution set is $\{14\}$.

49. $\quad -\frac{5}{6}t = -15$

$$-\frac{6}{5}\left(-\frac{5}{6}t\right) = -\frac{6}{5}(-15) \quad \begin{array}{c}\textit{Multiply by}\\ \textit{the reciprocal}\\ \textit{of } -\frac{5}{6}.\end{array}$$

$$t = 18$$

Check $t = 18$: $-15 = -15$ *True*

The solution set is $\{18\}$.

51. $\quad -\frac{7}{9}c = \frac{3}{5}$

$$-\frac{9}{7}\left(-\frac{7}{9}c\right) = -\frac{9}{7} \cdot \frac{3}{5} \quad \begin{array}{c}\textit{Multiply by}\\ \textit{the reciprocal}\\ \textit{of } -\frac{7}{9}.\end{array}$$

$$c = -\frac{27}{35}$$

Check $c = -\frac{27}{35}$: $\frac{3}{5} = \frac{3}{5}$ *True*

The solution set is $\left\{-\frac{27}{35}\right\}$.

53. $\quad -x = 12$

$$-1 \cdot (-x) = -1 \cdot 12 \quad \textit{Multiply by -1.}$$

$$x = -12$$

Check $x = -12$: $12 = 12$ *True*

The solution set is $\{-12\}$.

55. $\quad -x = -\frac{3}{4}$

$$-1 \cdot (-x) = -1 \cdot \left(-\frac{3}{4}\right)$$

$$x = \frac{3}{4}$$

Check $x = \frac{3}{4}$: $-\frac{3}{4} = -\frac{3}{4}$ *True*

The solution set is $\left\{\frac{3}{4}\right\}$.

57. $\quad -0.3x = 9$

$$\frac{-0.3x}{-0.3} = \frac{9}{-0.3} \quad \textit{Divide by -0.3.}$$

$$x = -30$$

Check $x = -30$: $9 = 9$ *True*

The solution set is $\{-30\}$.

59. $\quad 4x + 3x = 21$

$$7x = 21$$

$$\frac{7x}{7} = \frac{21}{7}$$

$$x = 3$$

Check $x = 3$: $21 = 21$ *True*

The solution set is $\{3\}$.

61. $\quad 3r - 5r = 10$

$$-2r = 10$$

$$\frac{-2r}{-2} = \frac{10}{-2}$$

$$r = -5$$

Check $r = -5$: $10 = 10$ *True*

The solution set is $\{-5\}$.

63. $\quad 5m + 6m - 2m = 63$

$$9m = 63$$

$$\frac{9m}{9} = \frac{63}{9}$$

$$m = 7$$

Check $m = 7$: $63 = 63$ *True*

The solution set is $\{7\}$.

65.
$$-6x + 4x - 7x = 0$$
$$-9x = 0$$
$$\frac{-9x}{-9} = \frac{0}{-9}$$
$$x = 0$$

Check $x = 0$: $0 = 0$ *True*

The solution set is $\{0\}$.

67.
$$9w - 5w + w = -3$$
$$5w = -3$$
$$\frac{5w}{5} = \frac{-3}{5}$$
$$w = -\frac{3}{5}$$

Check $w = -\frac{3}{5}$: $-3 = -3$ *True*

The solution set is $\left\{-\frac{3}{5}\right\}$.

69.
$$\frac{1}{3}x - \frac{1}{4}x + \frac{1}{12}x = 3$$
$$\left(\frac{1}{3} - \frac{1}{4} + \frac{1}{12}\right)x = 3 \quad \text{\textit{Distributive property}}$$
$$\left(\frac{4}{12} - \frac{3}{12} + \frac{1}{12}\right)x = 3 \quad \text{\textit{LCD = 12}}$$
$$\frac{1}{6}x = 3 \quad \text{\textit{Lowest terms}}$$
$$6\left(\frac{1}{6}x\right) = 6(3) \quad \text{\textit{Multiply by 6.}}$$
$$x = 18$$

Check $x = 18$: $6 - 4.5 + 1.5 = 3$ *True*

The solution set is $\{18\}$.

71. Answers will vary. One example is
$$\frac{3}{2}x = -6.$$

73. "When a number is multiplied by 4, the result is 6."
$$4x = 6$$
$$\frac{4x}{4} = \frac{6}{4}$$
$$x = \frac{3}{2}$$

The number is $\frac{3}{2}$ and $\left\{\frac{3}{2}\right\}$ is the solution set.

75. "When a number is divided by -5, the result is 2."
$$\frac{x}{-5} = 2$$
$$(-5)\left(-\frac{1}{5}x\right) = (-5)(2)$$
$$x = -10$$

The number is -10 and $\{-10\}$ is the solution set.

77.
$$-(3m + 5)$$
$$= -1(3m + 5)$$
$$= -1(3m) - 1(5)$$
$$= -3m - 5$$

79.
$$4(-5 + 2p) - 3(p - 4)$$
$$= 4(-5) + 4(2p) - 3(p) - 3(-4)$$
$$= -20 + 8p - 3p + 12$$
$$= -20 + 12 + 8p - 3p$$
$$= -8 + 5p$$

81.
$$4x + 5 + 2x = 7x$$
$$6x + 5 = 7x \quad \text{\textit{Combine terms.}}$$
$$5 = x \quad \text{\textit{Subtract 6x.}}$$

Check $x = 5$: $20 + 5 + 10 = 35$ *True*

The solution set is $\{5\}$.

2.3 More on Solving Linear Equations

1. *Step 1*: Clear parentheses and combine like terms, as needed.

Step 2: Use the addition property to get all variable terms on one side of the equation and all numbers on the other. Then combine like terms.

Step 3: Use the multiplication property to get the equation in the form $x = $ a number.

Step 4: Check the solution. Examples will vary.

3. Equations **A**, **B**, and **C** each have $\{$all real numbers$\}$ for their solution set. However, equation **D** gives
$$3x = 2x$$
$$3x - 2x = 2x - 2x$$
$$x = 0.$$

The only solution of this equation is 0, so the correct choice is **D**.

5.
$$3x + 8 = 5x + 10$$
$$-2x + 8 = 10 \quad \text{\textit{Subtract 5x.}}$$
$$-2x = 2 \quad \text{\textit{Subtract 8.}}$$
$$x = -1 \quad \text{\textit{Divide by} -2.}$$

Check $x = -1$: $5 = 5$ *True*

The solution set is $\{-1\}$.

7.
$$12h - 5 = 11h + 5 - h$$
$$12h - 5 = 10h + 5 \quad \text{\textit{Combine terms.}}$$
$$2h - 5 = 5 \quad \text{\textit{Subtract 10h.}}$$
$$2h = 10 \quad \text{\textit{Add 5.}}$$
$$h = 5 \quad \text{\textit{Divide by 2.}}$$

Check $h = 5$: $55 = 55$ *True*

The solution set is $\{5\}$.

9. $3(4x + 2) + 5x = 30 - x$

$$12x + 6 + 5x = 30 - x \qquad \textit{Distributive property}$$
$$17x + 6 = 30 - x \qquad \textit{Combine terms.}$$
$$18x + 6 = 30 \qquad \textit{Add 1x.}$$
$$18x = 24 \qquad \textit{Subtract 6.}$$
$$x = \frac{24}{18} = \frac{4}{3} \qquad \textit{Divide by 18.}$$

Check $x = \frac{4}{3}$: $\frac{86}{3} = \frac{86}{3}$ *True*

The solution set is $\left\{ \frac{4}{3} \right\}$.

11. $-2p + 7 = 3 - (5p + 1)$

$$-2p + 7 = 3 - 5p - 1 \qquad \textit{Distributive property}$$
$$-2p + 7 = -5p + 2 \qquad \textit{Combine terms.}$$
$$3p + 7 = 2 \qquad \textit{Add 5p.}$$
$$3p = -5 \qquad \textit{Subtract 7.}$$
$$p = -\frac{5}{3}$$

Check $p = -\frac{5}{3}$: $\frac{31}{3} = \frac{31}{3}$ *True*

The solution set is $\left\{ -\frac{5}{3} \right\}$.

13. $6(3w + 5) = 2(10w + 10)$

$$18w + 30 = 20w + 20$$
$$18w = 20w - 10 \qquad \textit{Subtract 30.}$$
$$-2w = -10 \qquad \textit{Subtract 20w.}$$
$$w = 5 \qquad \textit{Divide by -2.}$$

Check $w = 5$: $120 = 120$ *True*

The solution set is $\{5\}$.

15. $6(4x - 1) = 12(2x + 3)$

$$24x - 6 = 24x + 36$$
$$-6 = 36 \qquad \textit{Subtract 24x.}$$

The variable has "disappeared," and the resulting equation is false. Therefore, the equation has no solution set, symbolized by $\emptyset$.

17. $3(2x - 4) = 6(x - 2)$

$$6x - 12 = 6x - 12$$
$$-12 = -12 \qquad \textit{Subtract 6x.}$$
$$0 = 0 \qquad \textit{Add 12.}$$

The variable has "disappeared." Since the resulting statement is a *true* one, *any* real number is a solution. We indicate the solution set as {all real numbers}.

19. $7r - 5r + 2 = 5r - r$

$$2r + 2 = 4r \qquad \textit{Combine terms.}$$
$$2 = 2r \qquad \textit{Subtract 2r.}$$
$$1 = r \qquad \textit{Divide by 2.}$$

Check $r = 1$: $4 = 4$ *True*

The solution set is $\{1\}$.

21. $11x - 5(x + 2) = 6x + 5$

$$11x - 5x - 10 = 6x + 5$$
$$6x - 10 = 6x + 5$$
$$-10 = 5 \qquad \textit{Subtract 6x.}$$

The variable has "disappeared," and the resulting equation is false. Therefore, the equation has no solution set, symbolized by $\emptyset$.

23. $\frac{3}{5}t - \frac{1}{10}t = t - \frac{5}{2}$

The least common denominator of all the fractions in the equation is 10.

$$10\left(\frac{3}{5}t - \frac{1}{10}t\right) = 10\left(t - \frac{5}{2}\right)$$
$$\textit{Multiply both sides by 10.}$$
$$10\left(\frac{3}{5}t\right) + 10\left(-\frac{1}{10}t\right) = 10t + 10\left(-\frac{5}{2}\right)$$
$$\textit{Distributive property}$$
$$6t - t = 10t - 25$$
$$5t = 10t - 25$$
$$-5t = -25 \qquad \textit{Subtract 10t.}$$
$$\frac{-5t}{-5} = \frac{-25}{-5} \qquad \textit{Divide by -5.}$$
$$t = 5$$

Check $t = 5$: $\frac{5}{2} = \frac{5}{2}$ *True*

The solution set is $\{5\}$.

25. $-\frac{1}{4}(x - 12) + \frac{1}{2}(x + 2) = x + 4$

The LCD of all the fractions is 4.

$$4\left[-\frac{1}{4}(x - 12) + \frac{1}{2}(x + 2)\right] = 4(x + 4)$$
$$\textit{Multiply by 4.}$$
$$4\left(-\frac{1}{4}\right)(x - 12) + 4\left(\frac{1}{2}\right)(x + 2) = 4x + 16$$
$$\textit{Distributive property}$$
$$(-1)(x - 12) + 2(x + 2) = 4x + 16$$
$$\textit{Multiply.}$$
$$-x + 12 + 2x + 4 = 4x + 16$$
$$\textit{Distributive property}$$
$$x + 16 = 4x + 16$$
$$-3x + 16 = 16$$
$$-3x = 0$$
$$\frac{-3x}{-3} = \frac{0}{-3} \qquad \textit{Divide by -3.}$$
$$x = 0$$

Check $x = 0$: $4 = 4$ *True*

The solution set is $\{0\}$.

27. $\frac{2}{3}k - \left(k - \frac{1}{2}\right) = \frac{1}{6}(k - 51)$

The least common denominator of all the fractions in the equation is 6, so multiply both sides by 6 and solve for k.

$$6\left[\frac{2}{3}k - \left(k - \frac{1}{2}\right)\right] = 6\left[\frac{1}{6}(k - 51)\right]$$
$$6\left(\frac{2}{3}k\right) - 6\left(k - \frac{1}{2}\right) = 6\left[\frac{1}{6}(k - 51)\right]$$
Distributive property
$$4k - 6k + 3 = 1(k - 51)$$
$$-2k + 3 = k - 51$$
$$-3k + 3 = -51$$
$$-3k = -54$$
$$k = 18$$

Check $k = 18$: $-\frac{11}{2} = -\frac{11}{2}$ *True*

The solution set is $\{18\}$.

29. $0.2(60) + 0.05x = 0.10(60 + x)$

To eliminate the decimal in 0.2 and 0.10, we need to multiply the equation by 10. But to eliminate the decimal in 0.05, we need to multiply by 100, so we choose 100.

$$100[0.2(60) + 0.05x] = 100[0.10(60 + x)]$$
Multiply by 100.
$$100[0.2(60)] + 100(0.05x) = 100[0.10(60 + x)]$$
Distributive property
$$20(60) + 5x = 10(60 + x)$$
Multiply.
$$1200 + 5x = 600 + 10x$$
$$1200 - 5x = 600$$
$$-5x = -600$$
$$x = \frac{-600}{-5} = 120$$

Check $x = 120$: $18 = 18$ *True*

The solution set is $\{120\}$.

31. $1.00x + 0.05(12 - x) = 0.10(63)$

To clear the equation of decimals, we multiply both sides by 100.

$$100[1.00x + 0.05(12 - x)] = 100[0.10(63)]$$
$$100(1.00x) + 100[0.05(12 - x)] = (100)(0.10)(63)$$
$$100x + 5(12 - x) = 10(63)$$
$$100x + 60 - 5x = 630$$
$$95x + 60 = 630$$
$$95x = 570$$
$$x = \frac{570}{95} = 6$$

Check $x = 6$: $6.3 = 6.3$ *True*

The solution set is $\{6\}$.

33. $0.6(10,000) + 0.8x = 0.72(10,000 + x)$
$$100[0.6(10,000)] + 100(0.8x) =$$
$$100[0.72(10,000 + x)]$$
Multiply by both sides by 100.
$$60(10,000) + 80x = 72(10,000 + x)$$
$$600,000 + 80x = 720,000 + 72x$$
$$600,000 + 8x = 720,000$$
$$8x = 120,000$$
$$x = \frac{120,000}{8} = 15,000$$

Check $x = 15,000$: $18,000 = 18,000$ *True*

The solution set is $\{15,000\}$.

35. $10(2x - 1) = 8(2x + 1) + 14$
$$20x - 10 = 16x + 8 + 14$$
$$20x - 10 = 16x + 22$$
$$4x - 10 = 22$$
$$4x = 32$$
$$x = 8$$

Check $x = 8$: $150 = 150$ *True*

The solution set is $\{8\}$.

37. $-(4x + 2) - (-3x - 5) = 3$
$$-1(4x + 2) - 1(-3x - 5) = 3$$
$$-4x - 2 + 3x + 5 = 3$$
$$-x + 3 = 3$$
$$-x = 0$$
$$x = 0$$

Check $x = 0$: $3 = 3$ *True*

The solution set is $\{0\}$.

39. $\frac{1}{2}(x + 2) + \frac{3}{4}(x + 4) = x + 5$

To clear fractions, multiply both sides by the LCD, which is 4.

$$4\left[\frac{1}{2}(x + 2) + \frac{3}{4}(x + 4)\right] = 4(x + 5)$$
$$4\left(\frac{1}{2}\right)(x + 2) + 4\left(\frac{3}{4}\right)(x + 4) = 4x + 20$$
$$2(x + 2) + 3(x + 4) = 4x + 20$$
$$2x + 4 + 3x + 12 = 4x + 20$$
$$5x + 16 = 4x + 20$$
$$x + 16 = 20$$
$$x = 4$$

Check $x = 4$: $9 = 9$ *True*

The solution set is $\{4\}$.

41.
$$0.1(x + 80) + 0.2x = 14$$
To eliminate the decimals, multiply both sides by 10.
$$10[0.1(x + 80) + 0.2x] = 10(14)$$
$$1(x + 80) + 2x = 140$$
$$x + 80 + 2x = 140$$
$$3x + 80 = 140$$
$$3x = 60$$
$$x = 20$$

Check $x = 20$: $14 = 14$ *True*

The solution set is $\{20\}$.

43.
$$4(x + 8) = 2(2x + 6) + 20$$
$$4x + 32 = 4x + 12 + 20$$
$$4x + 32 = 4x + 32$$
$$4x = 4x$$
$$0 = 0$$

Since $0 = 0$ is a *true* statement, the solution set is {all real numbers}.

45.
$$9(v + 1) - 3v = 2(3v + 1) - 8$$
$$9v + 9 - 3v = 6v + 2 - 8$$
$$6v + 9 = 6v - 6$$
$$9 = -6$$

Because $9 = -6$ is a *false* statement, the equation has no solution set, symbolized by $\emptyset$.

47. The sum of q and the other number is 11. To find the other number, you would subtract q from 11, so an expression for the other number is $11 - q$.

49. The product of k and the other number is 9. To find the other number, you would divide 9 by k, so an expression for the other number is $\dfrac{9}{k}$.

51. An expression for the total number of yards is $x + 7$.

53. If a baseball player gets 65 hits in one season, and h of the hits are in one game, then $65 - h$ of the hits came in the rest of the games.

55. If Monica is a years old now, then 12 years from now, she will be $a + 12$ years old. Two years ago, she was $a - 2$ years old.

57. Since the value of each quarter is 25 cents, the value of r quarters is $25r$ cents.

59. Since each bill is worth 5 dollars, the number of bills is $\dfrac{t}{5}$.

61. Since each adult ticket costs b dollars, the cost of 3 adult tickets is $3b$. Since each child's ticket costs d dollars, the cost of 2 children's tickets is $2d$. Therefore, the total cost is $3b + 2d$ (dollars).

63. "A number added to -6" is written
$$-6 + x.$$

65. "A number decreased by 9" is written
$$x - 9.$$

67. "The quotient of -6 and a nonzero number" is written
$$\dfrac{-6}{x}.$$

69. "The product of 12 and the difference between a number and 9" is written
$$12(x - 9).$$

Summary Exercises on Solving Linear Equations

1.
$$a + 2 = -3$$
$$a = -5 \quad \textit{Subtract 2.}$$

Check $a = -5$: $-3 = -3$ *True*

The solution set is $\{-5\}$.

3.
$$12.5k = -63.75$$
$$k = \frac{-63.75}{12.5} \quad \textit{Divide by 12.5.}$$
$$= -5.1$$

Check $k = -5.1$: $-63.75 = -63.75$ *True*

The solution set is $\{-5.1\}$.

5.
$$\tfrac{4}{5}x = -20$$
$$x = \left(\tfrac{5}{4}\right)(-20) \quad \textit{Multiply by } \tfrac{5}{4}.$$
$$= -25$$

Check $x = -25$: $-20 = -20$ *True*

The solution set is $\{-25\}$.

7.
$$5x - 9 = 4(x - 3)$$
$$5x - 9 = 4x - 12 \quad \begin{array}{l}\textit{Distributive}\\\textit{property}\end{array}$$
$$x - 9 = -12 \quad \textit{Subtract 4x.}$$
$$x = -3 \quad \textit{Add 9.}$$

Check $x = -3$: $-24 = -24$ *True*

The solution set is $\{-3\}$.

9.
$$-3(m - 4) + 2(5 + 2m) = 29$$
$$-3m + 12 + 10 + 4m = 29$$
$$m + 22 = 29$$
$$m = 7$$

Check $m = 7$: $29 = 29$ *True*

The solution set is $\{7\}$.

11.
$$0.08x + 0.06(x + 9) = 1.24$$
To eliminate the decimals, multiply both sides by 100.
$$100[0.08x + 0.06(x + 9)] = 100(1.24)$$
$$8x + 6(x + 9) = 124$$
$$8x + 6x + 54 = 124$$
$$14x + 54 = 124$$
$$14x = 70$$
$$x = 5$$

Check $x = 5$: $0.4 + 0.84 = 1.24$ *True*

The solution set is $\{5\}$.

13.
$$4x + 2(3 - 2x) = 6$$
$$4x + 6 - 4x = 6$$
$$6 = 6$$

Since $6 = 6$ is a *true* statement, the solution set is $\{$all real numbers$\}$.

15.
$$-x = 6$$
$$x = -6 \text{ Multiply by } -1.$$

Check $x = -6$: $6 = 6$ *True*

The solution set is $\{-6\}$.

17.
$$7m - (2m - 9) = 39$$
$$7m - 2m + 9 = 39$$
$$5m + 9 = 39$$
$$5m = 30$$
$$m = 6$$

Check $m = 6$: $39 = 39$ *True*

The solution set is $\{6\}$.

19.
$$-2t + 5t - 9 = 3(t - 4) - 5$$
$$-2t + 5t - 9 = 3t - 12 - 5$$
$$3t - 9 = 3t - 17$$
$$-9 = -17$$

Because $-9 = -17$ is a *false* statement, the equation has no solution set, symbolized by $\emptyset$.

21.
$$0.2(50) + 0.8r = 0.4(50 + r)$$
To eliminate the decimals, multiply both sides by 10.
$$10[0.2(50) + 0.8r] = 10[0.4(50 + r)]$$
$$2(50) + 8r = 4(50 + r)$$
$$100 + 8r = 200 + 4r$$
$$100 + 4r = 200$$
$$4r = 100$$
$$r = 25$$

Check $r = 25$: $10 + 20 = 30$ *True*

The solution set is $\{25\}$.

23.
$$2(3 + 7x) - (1 + 15x) = 2$$
$$6 + 14x - 1 - 15x = 2$$
$$-x + 5 = 2$$
$$-x = -3$$
$$x = 3$$

Check $x = 3$: $48 - 46 = 2$ *True*

The solution set is $\{3\}$.

25.
$$2(4 + 3r) = 3(r + 1) + 11$$
$$8 + 6r = 3r + 3 + 11$$
$$8 + 6r = 3r + 14$$
$$8 + 3r = 14$$
$$3r = 6$$
$$r = 2$$

Check $r = 2$: $20 = 20$ *True*

The solution set is $\{2\}$.

27.
$$\frac{1}{4}x - 4 = \frac{3}{2}x + \frac{3}{4}x$$
To clear fractions, multiply both sides by the LCD, which is 4.
$$4\left(\frac{1}{4}x - 4\right) = 4\left(\frac{3}{2}x + \frac{3}{4}x\right)$$
$$x - 16 = 6x + 3x$$
$$x - 16 = 9x$$
$$-16 = 8x$$
$$x = -2$$

Check $x = -2$: $-4.5 = -3 - 1.5$ *True*

The solution set is $\{-2\}$.

29.
$$\frac{3}{4}(a - 2) - \frac{1}{3}(5 - 2a) = -2$$
To clear fractions, multiply both sides by the LCD, which is 12.
$$12\left[\frac{3}{4}(a - 2) - \frac{1}{3}(5 - 2a)\right] = 12(-2)$$
$$9(a - 2) - 4(5 - 2a) = -24$$
$$9a - 18 - 20 + 8a = -24$$
$$17a - 38 = -24$$
$$17a = 14$$
$$a = \frac{14}{17}$$

Check $a = \frac{14}{17}$: $-\frac{15}{17} - \frac{19}{17} = -2$ *True*

The solution set is $\left\{\frac{14}{17}\right\}$.

2.4   An Introduction to Applications of Linear Equations

1.   Choice **D**, $6\frac{1}{2}$, is *not* a reasonable answer in an applied problem that requires finding the number of cars on a dealer's lot, since you cannot have $\frac{1}{2}$ of a car. The number of cars must be a whole number.

3.   Choice **A**, -10, is *not* a reasonable answer since distance cannot be negative.

The applied problems in this section should be solved by using the six-step method shown in the text. These steps will only be listed in a few of the solutions, but all of the solutions are based on this method.

5.   *Step 2*

Let $x =$ the unknown number. Then $5x + 2$ represents "2 is added to five times a number," and $4x + 5$ represents "5 more than four times a number."

Step 3 $5x + 2 = 4x + 5$

Step 4 $5x + 2 = 4x + 5$
$$x + 2 = 5$$
$$x = 3$$

Step 5
The number is 3.

Step 6
Check that 3 is the correct answer by substituting this result into the words of the original problem. 2 added to five times a number is $2 + 5(3) = 17$ and 5 more than four times the number is $5 + 4(3) = 17$. The values are equal, so the number 3 is the correct answer.

7.   *Step 2*

Let $x =$ the unknown number. Then $x - 2$ is two subtracted from the number, $3(x - 2)$ is triple the difference, and $x + 6$ is six more than the number.

Step 3 $3(x - 2) = x + 6$

Step 4 $3x - 6 = x + 6$
$$2x - 6 = 6$$
$$2x = 12$$
$$x = 6$$

Step 5
The number is 6.

Step 6
Check that 6 is the correct answer by substituting this result into the words of the original problem. Two subtracted from the number is $6 - 2 = 4$. Triple this difference is $3(4) = 12$, which is equal to 6 more than the number, since $6 + 6 = 12$.

9.   *Step 2*

Let $x =$ the unknown number. Then $3x$ is three times the number, $x + 7$ is 7 more than the number, $2x$ is twice the number, and $-11 - 2x$ is the difference between -11 and twice the number.

Step 3 $3x + (x + 7) = -11 - 2x$

Step 4 $4x + 7 = -11 - 2x$
$$6x + 7 = -11$$
$$6x = -18$$
$$x = -3$$

Step 5
The number is -3.

Step 6
Check that -3 is the correct answer by substituting this result into the words of the original problem. The sum of three times a number and 7 more than the number is $3(-3) + (-3 + 7) = -5$ and the difference between -11 and twice the number is $-11 - 2(-3) = -5$. The values are equal, so the number -3 is the correct answer.

11.   Let $x =$ the number of drive-in movie screens in New York.
Then $x + 11 =$ the number of drive-in movie screens in California.

Since the total number of screens was 107, we can write the equation
$$x + (x + 11) = 107.$$
Solve this equation.
$$2x + 11 = 107$$
$$2x = 96$$
$$x = 48$$

Since $x = 48$, $x + 11 = 48 + 11 = 59$.

There were 48 drive-in movie screens in New York and 59 in California. Since $48 + 59 = 107$, this answer checks.

13.   Let $x =$ the number of Democrats.
Then $x + 11 =$ the number of Republicans.

The number of Democrats	plus	the number of Republicans
$\downarrow$	$\downarrow$	$\downarrow$
x	$+$	$(x + 11)$

equals	the number of members of the Senate.
$\downarrow$	$\downarrow$
$=$	99

Solve the equation.
$$x + (x + 11) = 99$$
$$2x + 11 = 99$$
$$2x = 88$$
$$x = 44$$

There were 44 Democrats and $44 + 11 = 55$ Republicans.

15. Let x = revenue from ticket sales for
Bruce Springsteen and
the E Street Band.

Then $x - 35.4$ = revenue from ticket sales for
Céline Dion.

Since the total revenue from ticket sales was
$196.4 (all numbers in millions), we can write
the equation
$$x + (x - 35.4) = 196.4.$$
Solve this equation.
$$2x - 35.4 = 196.4$$
$$2x = 231.8$$
$$x = 115.9$$

Since $x = 115.9$, $x - 35.4 = 80.5$.

Bruce Springsteen and the E Street Band took in
$115.9 million and Céline Dion took in $80.5
million. Since 80.5 is 35.4 less than 115.9 and
$115.9 + 80.5 = 196.4$, this answer checks.

17. Let x = the number of games the Suns lost.
Then $3x + 2$ = the number of games the
Suns won.

Since the total number of games played was
82, we can write the equation
$$x + (3x + 2) = 82.$$
Solve this equation.
$$4x + 2 = 82$$
$$4x = 80$$
$$x = 20$$

Since $x = 20$, $3x + 2 = 62$.

The Suns won 62 games and lost 20 games. Since
$62 + 20 = 82$, this answer checks.

19. Let x = the value of the 1945 nickel.
Then $\frac{8}{7}x$ = the value of the 1950 nickel.

The total value of the two coins is $15.00, so
$$x + \frac{8}{7}x = 15.$$
Solve this equation. First multiply both sides by
7 to clear fractions.
$$7\left(x + \frac{8}{7}x\right) = 7(15)$$
$$7x + 8x = 105$$
$$15x = 105$$
$$x = 7 \qquad \textit{Divide by 15.}$$

Since $x = 7$, $\frac{8}{7}x = \frac{8}{7}(7) = 8$.

The value of the 1945 Philadelphia nickel is $7.00
and the value of the 1950 Denver nickel is $8.00.

21. Let x = the number of kg of onions.
Then $6.6x$ = the number of kg of grilled steak.
The total weight of these two ingredients was
617.6 kg, so
$$x + 6.6x = 617.6.$$
Solve this equation.
$$1x + 6.6x = 617.6$$
$$7.6x = 617.6$$
$$x = \frac{617.6}{7.6} \approx 81.3$$

Since $x = \frac{617.6}{7.6}$, $6.6x = 6.6\left(\frac{617.6}{7.6}\right) \approx 536.3$.

To the nearest tenth of a kilogram, 81.3 kg of
onions and 536.3 kg of grilled steak were used to
make the taco.

23. Let x = the number of CDs sold.
Then $\frac{8}{5}x$ = the number of DVDs sold.

The total number of CDs and DVDs sold was 273,
so
$$x + \frac{8}{5}x = 273.$$

Solve this equation.
$$1x + \frac{8}{5}x = 273$$
$$\frac{13}{5}x = 273$$
$$\frac{5}{13}\left(\frac{13}{5}x\right) = \frac{5}{13}(273)$$
$$x = \frac{5}{\cancel{13}} \cdot \frac{\overset{21}{\cancel{273}}}{1} = 105$$

Since $x = 105$, $\frac{8}{5}x = \frac{8}{5}(105) = 168$.

There were 168 DVDs sold.

25. Let x = the number of ounces of rye flour.
Then $4x$ = the number of ounces of whole wheat
flour.

The total number of ounces would be 32, so
$$x + 4x = 32.$$

Solve this equation.
$$5x = 32$$
$$x = \frac{32}{5} = 6.4$$

Since $x = 6.4$, $4x = 4(6.4) = 25.6$. To make a
loaf of bread weighing 32 oz, use 6.4 oz of rye
flour and 25.6 oz of whole wheat flour.

27. Let x = the number of tickets booked on
United Airlines.

Then $x + 7$ = the number of tickets booked on
American Airlines, and

$2x + 4$ = the number of tickets booked on
Southwest Airlines.

The total number of tickets booked was 55, so

$$x + (x + 7) + (2x + 4) = 55.$$

Solve this equation.

$$4x + 11 = 55$$
$$4x = 44$$
$$x = \frac{44}{4} = 11$$

Since $x = 11$, $x + 7 = 11 + 7 = 18$, and $2x + 4 = 2(11) + 4 = 26$. He booked 18 tickets on American, 11 tickets on United, and 26 tickets on Southwest.

29.   Let x = the length of the shortest piece.
Then $x + 5$ = the length of the middle piece, and $x + 9$ = the length of the longest piece.

The total length is 59 inches, so

$$x + (x + 5) + (x + 9) = 59.$$

Solve this equation.

$$3x + 14 = 59$$
$$3x = 45$$
$$x = 15$$

Since $x = 15$, $x + 5 = 20$, and $x + 9 = 24$.

The shortest piece should be 15 inches, the middle piece should be 20 inches, and the longest piece should be 24 inches. The answer checks since

$$15 + 20 + 24 = 59.$$

31.   Let x = the distance of Mercury from the sun.
Then $x + 31.2$ = the distance of Venus from the sun, and
$x + 57$ = the distance of Earth from the sun.

Since the total of the distances from these three planets is 196.2 (all distances in millions of miles), we can write the equation

$$x + (x + 31.2) + (x + 57) = 196.2.$$

Solve this equation.

$$3x + 88.2 = 196.2$$
$$3x = 108$$
$$x = 36$$

Mercury is 36 million miles from the sun, Venus is $36 + 31.2 = 67.2$ million miles from the sun, and Earth is $36 + 57 = 93$ million miles from the sun. The answer checks since

$$36 + 67.2 + 93 = 196.2.$$

33.   Let x = the measure of angles A and B.
Then $x + 60$ = the measure of angle C.

The sum of the measures of the angles of any triangle is 180°, so

$$x + x + (x + 60) = 180.$$

Solve this equation.

$$3x + 60 = 180$$
$$3x = 120$$
$$x = 40$$

Angles A and B have measures of 40 degrees, and angle C has a measure of $40 + 60 = 100$ degrees. The answer checks since

$$40 + 40 + 100 = 180.$$

35.   Subtract one of the numbers (m) from the sum (k) to express the other number. The other number is $k - m$.

37.   An angle cannot have its supplement equal to its complement. The sum of an angle and its supplement equals 180°, while the sum of an angle and its complement equals 90°. If we try to solve the equation

$$90 - x = 180 - x,$$

we will get

$$90 - x + x = 180 - x + x$$
$$90 = 180 \quad \textit{False}$$

so this equation has no solution.

39.   If x represents an integer, the next smaller consecutive integer is 1 less than x, that is, $x - 1$.

41.   Let x = the measure of the angle.
Then $90 - x$ = the measure of its complement.

The "complement is four times its measure" can be written as

$$90 - x = 4x.$$

Solve this equation.

$$90 = 5x$$
$$x = \frac{90}{5} = 18$$

The measure of the angle is 18°. The complement is $90° - 18° = 72°$, which is four times 18°.

43.   Let x = the measure of the angle. Then $90 - x$ = the measure of its complement, and $180 - x$ = the measure of its supplement.

Its supplement	measures	39°
$\downarrow$	$\downarrow$	$\downarrow$
$180 - x$	$=$	39

more than	twice its complement.
$\downarrow$	$\downarrow$
$+$	$2(90 - x)$

continued

Solve the equation.

$$180 - x = 39 + 2(90 - x)$$
$$180 - x = 39 + 180 - 2x$$
$$180 - x = 219 - 2x$$
$$x + 180 = 219$$
$$x = 39$$

The measure of the angle is $39°$. The complement is $90° - 39° = 51°$. Now $39°$ more than twice its complement is $39° + 2(51°) = 141°$, which is the supplement of $39°$ since $180° - 39° = 141°$.

45.   Let $x =$ the measure of the angle. Then
$180 - x =$ the measure of its supplement, and
$90 - x =$ the measure of its complement.

	The difference	
between the measure of its supplement and	↓	three times the measure of its complement
↓	↓	↓
$(180 - x)$	$-$	$3(90 - x)$

	is	$10°$.
↓		↓
$=$		10

Solve the equation.

$$(180 - x) - 3(90 - x) = 10$$
$$180 - x - 270 + 3x = 10$$
$$2x - 90 = 10$$
$$2x = 100$$
$$x = 50$$

The measure of the angle is $50°$. The supplement is $180° - 50° = 130°$ and the complement is $90° - 50° = 40°$. The answer checks since $130° - 3(40°) = 10°$.

47.   Let $x =$ the number on the first locker.
Then $x + 1 =$ the number on the next locker.

Since the numbers have a sum of 137, we can write the equation

$$x + (x + 1) = 137.$$

Solve the equation.

$$2x + 1 = 137$$
$$2x = 136$$
$$x = \frac{136}{2} = 68$$

Since $x = 68$, $x + 1 = 69$.

The lockers have numbers 68 and 69. Since $68 + 69 = 137$, this answer checks.

49.   Let $x =$ the lesser even integer.
Then $x + 2 =$ the greater even integer.

The lesser added to three times the greater gives a sum of 46 can be written as

$$x + 3(x + 2) = 46.$$
$$x + 3x + 6 = 46$$
$$4x + 6 = 46$$
$$4x = 40$$
$$x = 10$$

Since $x = 10$, $x + 2 = 12$.

The integers are 10 and 12. This answer checks since $10 + 3(12) = 46$.

51.   Because the two pages are back-to-back, they must have page numbers that are consecutive integers.

Let $x =$ the lesser page number.
Then $x + 1 =$ the greater page number.

$$x + (x + 1) = 203$$
$$2x + 1 = 203$$
$$2x = 202$$
$$x = 101$$

Since $x = 101$, $x + 1 = 102$.

The page numbers are 101 and 102. This answer checks since the sum is 203.

53.   Let $x =$ the lesser integer.
Then $x + 1 =$ the greater integer.

$$x + 3(x + 1) = 43$$
$$x + 3x + 3 = 43$$
$$4x + 3 = 43$$
$$4x = 40$$
$$x = 10$$

Since $x = 10$, $x + 1 = 11$.

The integers are 10 and 11. This answer checks since $10 + 3(11) = 43$.

55.   Let $x =$ the first even integer.
Then $x + 2 =$ the second even integer, and
$x + 4 =$ the third even integer.

$$x + (x + 2) + (x + 4) = 60$$
$$3x + 6 = 60$$
$$3x = 54$$
$$x = 18$$

Since $x = 18$, $x + 2 = 20$, and $x + 4 = 22$.

The first even integer is 18. This answer checks since $18 + 20 + 22 = 60$.

57. Let $x =$ the first odd integer.
Then $x + 2 =$ the second odd integer, and
$x + 4 =$ the third odd integer.

$$2[(x + 4) - 6] = [x + 2(x + 2)] - 23$$
$$2(x - 2) = x + 2x + 4 - 23$$
$$2x - 4 = 3x - 19$$
$$-4 = x - 19$$
$$15 = x$$

Since $x = 15$, $x + 2 = 17$, and $x + 4 = 19$.

The integers are 15, 17, and 19.

59. Let $x =$ the amount of Head Start funding in the first year (in billions of dollars).
Then $x + 0.13 =$ the amount of funding in the second year, and
$(x + 0.13) + 0.10 = x + 0.23 =$ the amount of funding in the third year.

The total funding was 19.98 billion dollars, so

$$x + (x + 0.13) + (x + 0.23) = 19.98.$$

Solve this equation.

$$3x + 0.36 = 19.98$$
$$3x = 19.62$$
$$x = \frac{19.62}{3} = 6.54$$

Since $x = 6.54$, $x + 0.13 = 6.67$, and $x + 0.23 = 6.77$.

The Head Start funding was 6.54 billion dollars in the first year, 6.67 billion dollars in the second year, and 6.77 billion dollars in the third year.

61. $L = 6$ and $W = 4$, so $LW = 6 \cdot 4 = 24$.

63. $L = 8$ and $W = 2$, so
$2L + 2W = 2(8) + 2(2) = 16 + 4 = 20$.

65. $B = 27$ and $h = 6$, so $\frac{1}{2}Bh = \frac{1}{2}(27)(6) = 81$.

2.5 Formulas and Applications from Geometry

1. **(a)** The perimeter of a plane geometric figure is the distance around the figure. It can be found by adding up the lengths of all the sides. Perimeter is a one-dimensional (linear) measurement, so it is given in linear units (inches, centimeters, feet, etc.).

(b) The area of a plane geometric figure is the measure of the surface covered or enclosed by the figure. Area is a two-dimensional measurement, so it is given in square units (square centimeters, square feet, etc.).

3. You would need to be given 4 values in a formula with 5 variables to find the value of any one variable.

5. Carpeting for a bedroom covers the surface of the bedroom floor, so *area* would be used.

7. To measure fencing for a yard, use *perimeter* since you would need to measure the lengths of the sides of the yard.

9. Tile for a bathroom covers the surface of the bathroom floor, so *area* would be used.

11. To determine the cost for replacing a linoleum floor with a wood floor, use *area* since you need to know the measure of the surface covered by the wood.

In Exercises 13–32, substitute the given values into the formula and then solve for the remaining variable.

13. $P = 2L + 2W$; $L = 8$, $W = 5$

$$P = 2L + 2W$$
$$= 2(8) + 2(5)$$
$$= 16 + 10$$
$$P = 26$$

15. $A = \frac{1}{2}bh$; $b = 8$, $h = 16$

$$A = \frac{1}{2}bh$$
$$= \frac{1}{2}(8)(16)$$
$$A = 64$$

17. $P = a + b + c$; $P = 12$, $a = 3$, $c = 5$

$$P = a + b + c$$
$$12 = 3 + b + 5$$
$$12 = b + 8$$
$$4 = b$$

19. $d = rt$; $d = 252$, $r = 45$

$$d = rt$$
$$252 = 45t$$
$$\frac{252}{45} = \frac{45t}{45}$$
$$5.6 = t$$

21. $I = prt$; $p = 7500$, $r = 0.035$, $t = 6$

$$I = prt$$
$$= (7500)(0.035)(6)$$
$$I = 1575$$

23. $A = \frac{1}{2}h(b + B)$; $A = 91$, $h = 7$, $b = 12$

$$A = \frac{1}{2}h(b + B)$$
$$91 = \frac{1}{2}(7)(12 + B)$$
$$182 = (7)(12 + B)$$
$$12 + B = \frac{1}{7}(182)$$
$$B = 26 - 12 = 14$$

25. $C = 2\pi r$; $C = 16.328$, $\pi = 3.14$

$$C = 2\pi r$$
$$16.328 = 2(3.14)r$$
$$16.328 = 6.28r$$
$$2.6 = r$$

27. $C = 2\pi r$; $C = 20\pi$

$$C = 2\pi r$$
$$20\pi = 2\pi r$$
$$10 = r \qquad \text{Divide by } 2\pi.$$

29. $A = \pi r^2$; $r = 4$, $\pi = 3.14$

$$A = \pi r^2$$
$$= 3.14(4)^2$$
$$= 3.14(16)$$
$$A = 50.24$$

31. $S = 2\pi rh$; $S = 120\pi$, $h = 10$

$$S = 2\pi rh$$
$$120\pi = 2\pi r(10)$$
$$120\pi = 20\pi r$$
$$6 = r \qquad \text{Divide by } 20\pi.$$

In Exercises 33–38, substitute the given values into the formula and then evaluate V.

33. $V = LWH$; $L = 10, W = 5, H = 3$

$$V = LWH$$
$$= (10)(5)(3)$$
$$V = 150$$

35. $V = \frac{1}{3}Bh$; $B = 12$, $h = 13$

$$V = \frac{1}{3}Bh$$
$$= \frac{1}{3}(12)(13)$$
$$V = 52$$

37. $V = \frac{4}{3}\pi r^3$; $r = 12$, $\pi = 3.14$

$$V = \frac{4}{3}\pi r^3$$
$$= \frac{4}{3}(3.14)(12)^3$$
$$= \frac{4}{3}(3.14)(1728)$$
$$V = 7234.56$$

39.
$$P = 2L + 2W$$
$$54 = 2(W + 9) + 2W \qquad \text{Let } L = W + 9.$$
$$54 = 2W + 18 + 2W$$
$$54 = 4W + 18$$
$$36 = 4W$$
$$9 = W$$

The width is 9 inches and the length is $9 + 9 = 18$ inches.

41.
$$P = 2l + 2w$$
$$36 = 2(3w + 2) + 2w \qquad \text{Let } l = 3w + 2.$$
$$36 = 6w + 4 + 2w$$
$$36 = 8w + 4$$
$$32 = 8w$$
$$4 = w$$

The width is 4 meters and the length is $3(4) + 2 = 14$ meters.

43. Let s = the length of the shortest side, in inches;
$s + 2$ = the length of the medium side, and,
$s + 3$ = the length of the longest side.

The perimeter is 20 inches, so

$$s + (s + 2) + (s + 3) = 20.$$
$$3s + 5 = 20$$
$$3s = 15$$
$$s = 5$$

Since $s = 5$, $s + 2 = 7$, and $s + 3 = 8$. The lengths of the sides are 5, 7, and 8 inches. The perimeter is $5 + 7 + 8 = 20$, as required.

45. Let s = the length of the two sides that have equal length, in meters; and $2s - 4$ = the length of the third side.

The perimeter is 24 meters, so

$$s + s + (2s - 4) = 24.$$
$$4s - 4 = 24$$
$$4s = 28$$
$$s = 7$$

Since $s = 7$, $2s - 4 = 10$. The lengths of the sides are 7, 7, and 10 meters. The perimeter is $7 + 7 + 10 = 24$, as required.

47. The diameter of the circle is 443 feet, so its radius is $\frac{443}{2} = 221.5$ ft. Use the area of a circle formula to find the enclosed area.

$$A = \pi r^2$$
$$= \pi(221.5)^2$$
$$\approx 154{,}133.6 \text{ ft}^2,$$

or about $154{,}000 \text{ ft}^2$. (If 3.14 is used for π, the value is 154,055.465.)

49. The page is a rectangle with length 1.5 m and width 1.2 m, so use the formulas for the perimeter and area of a rectangle.

$$P = 2L + 2W$$
$$= 2(1.5) + 2(1.2)$$
$$= 3 + 2.4$$
$$P = 5.4 \text{ meters}$$

$$A = LW$$
$$= (1.5)(1.2)$$
$$A = 1.8 \text{ square meters}$$

51. Use the formula for the area of a triangle with $A = 70$ and $b = 14$.

$$A = \tfrac{1}{2}bh$$
$$70 = \tfrac{1}{2}(14)h$$
$$70 = 7h$$
$$10 = h$$

The height of the sign is 10 feet.

53. Use the formula for the area of a trapezoid with $B = 115.80$, $b = 171.00$, and $h = 165.97$.

$$A = \tfrac{1}{2}(B + b)h$$
$$= \tfrac{1}{2}(115.80 + 171.00)(165.97)$$
$$= \tfrac{1}{2}(286.80)(165.97)$$
$$= 23{,}800.098$$

To the nearest hundredth of a square foot, the combined area of the two lots is 23,800.10 square feet.

55. The girth is $4 \cdot 18 = 72$ inches. Since the length plus the girth is 108, we have

$$L + G = 108$$
$$L + 72 = 108$$
$$L = 36 \text{ in.}$$

The volume of the box is

$$V = LWH$$
$$= (36)(18)(18)$$
$$= 11{,}664 \text{ in.}^3$$

57. The two angles are supplementary, so the sum of their measures is $180°$.

$$(x + 1) + (4x - 56) = 180$$
$$5x - 55 = 180$$
$$5x = 235$$
$$x = 47$$

Since $x = 47$, $x + 1 = 47 + 1 = 48$, and $4x - 56 = 4(47) - 56 = 132$.

The measures of the angles are $48°$ and $132°$.

59. The two angles are vertical angles, which have equal measures. Set their measures equal to each other and solve for x.

$$5x - 129 = 2x - 21$$
$$3x - 129 = -21$$
$$3x = 108$$
$$x = 36$$

Since $x = 36$, $5x - 129 = 5(36) - 129 = 51$, and $2x - 21 = 2(36) - 21 = 51$.

The measure of each angle is $51°$.

61. The angles are vertical angles, so their measures are equal.

$$12x - 3 = 10x + 15$$
$$2x - 3 = 15$$
$$2x = 18$$
$$x = 9$$

Since $x = 9$, $12x - 3 = 12(9) - 3 = 105$, and $10x + 15 = 10(9) + 15 = 105$.

The measure of each angle is $105°$.

63. $d = rt$ for t

$$d = rt$$
$$\frac{d}{r} = \frac{rt}{r} \quad \text{Divide by } r.$$
$$\frac{d}{r} = t \quad \text{or} \quad t = \frac{d}{r}$$

65. $A = bh$ for b

$$A = bh$$
$$\frac{A}{h} = \frac{bh}{h} \quad \text{Divide by } h.$$
$$\frac{A}{h} = b \quad \text{or} \quad b = \frac{A}{h}$$

67. $C = \pi d$ for d

$$C = \pi d$$
$$\frac{C}{\pi} = \frac{\pi d}{\pi} \quad \text{Divide by } \pi.$$
$$\frac{C}{\pi} = d \quad \text{or} \quad d = \frac{C}{\pi}$$

69. $V = LWH$ for H

$$V = LWH$$
$$\frac{V}{LW} = \frac{LWH}{LW} \quad \text{Divide by } LW.$$
$$\frac{V}{LW} = H \quad \text{or} \quad H = \frac{V}{LW}$$

71. $I = prt$ for r

$$I = prt$$
$$\frac{I}{pt} = \frac{prt}{pt} \quad \text{Divide by } pt.$$
$$\frac{I}{pt} = r \quad \text{or} \quad r = \frac{I}{pt}$$

73. $A = \tfrac{1}{2}bh$ for h

$$2A = 2\left(\tfrac{1}{2}bh\right) \quad \text{Multiply by 2.}$$
$$2A = bh$$
$$\frac{2A}{b} = \frac{bh}{b} \quad \text{Divide by } b.$$
$$\frac{2A}{b} = h \quad \text{or} \quad h = \frac{2A}{b}$$

75. $V = \frac{1}{3}\pi r^2 h$ for h

$$3V = 3\left(\frac{1}{3}\right)\pi r^2 h \qquad \textit{Multiply by 3.}$$

$$3V = \pi r^2 h$$

$$\frac{3V}{\pi r^2} = \frac{\pi r^2 h}{\pi r^2} \qquad \textit{Divide by } \pi r^2.$$

$$\frac{3V}{\pi r^2} = h \quad \text{or} \quad h = \frac{3V}{\pi r^2}$$

77. $P = a + b + c$ for b

$$P - a - c = a + b + c - a - c$$
$$\qquad\qquad\qquad\quad \textit{Subtract a and c.}$$
$$P - a - c = b \quad \text{or} \quad b = P - a - c$$

79. $P = 2L + 2W$ for W

$$P - 2L = 2L + 2W - 2L \quad \textit{Subtract 2L.}$$

$$P - 2L = 2W$$

$$\frac{P - 2L}{2} = \frac{2W}{2} \qquad \textit{Divide by 2.}$$

$$\frac{P - 2L}{2} = W \quad \text{or} \quad W = \frac{P - 2L}{2}$$

81. $y = mx + b$ for m

$$y - b = mx + b - b \quad \textit{Subtract b.}$$

$$y - b = mx$$

$$\frac{y - b}{x} = \frac{mx}{x} \qquad \textit{Divide by x.}$$

$$\frac{y - b}{x} = m \quad \text{or} \quad m = \frac{y - b}{x}$$

83. $Ax + By = C$ for y

$$By = C - Ax \qquad \textit{Subtract Ax.}$$

$$\frac{By}{B} = \frac{C - Ax}{B} \qquad \textit{Divide by B.}$$

$$y = \frac{C - Ax}{B}$$

85. $M = C(1 + r)$ for r

$$M = C + Cr \quad \textit{Distributive Property}$$
$$M - C = Cr \qquad \textit{Subtract C.}$$
$$\frac{M - C}{C} = \frac{Cr}{C} \qquad \textit{Divide by C.}$$
$$\frac{M - C}{C} = r \quad \text{or} \quad r = \frac{M - C}{C}$$

Alternative solution:

$$M = C(1 + r)$$
$$\frac{M}{C} = 1 + r \qquad \textit{Divide by C.}$$
$$\frac{M}{C} - 1 = r \qquad \textit{Subtract 1.}$$

87. $\dfrac{x}{12} = \dfrac{12}{72}$

$$12\left(\frac{1}{12}x\right) = 12\left(\frac{1}{6}\right) \quad \textit{Multiply by 12.}$$
$$x = 2$$

The solution set is $\{2\}$.

89.
$$0.06x = 300$$
$$x = \frac{300}{0.06} = 5000$$

The solution set is $\{5000\}$.

91.
$$\frac{3}{4}x = 21$$
$$\frac{4}{3}\left(\frac{3}{4}x\right) = \frac{4}{3}(21)$$
$$x = 28$$

The solution set is $\{28\}$.

93.
$$-3x = \frac{1}{4}$$
$$\frac{-3x}{-3} = \frac{\frac{1}{4}}{-3}$$
$$x = \frac{1}{4}\left(-\frac{1}{3}\right) = -\frac{1}{12}$$

The solution set is $\left\{-\frac{1}{12}\right\}$.

2.6 Ratios and Proportions

1. **(a)** 75 to 100 is $\dfrac{75}{100} = \dfrac{3}{4}$ or 3 to 4.
The answer is **C**.

(b) 5 to 4 or $\dfrac{5}{4} = \dfrac{5 \cdot 3}{4 \cdot 3} = \dfrac{15}{12}$ or 15 to 12.
The answer is **D**.

(c) $\dfrac{1}{2} = \dfrac{1 \cdot 50}{2 \cdot 50} = \dfrac{50}{100}$ or 50 to 100
The answer is **B**.

(d) 4 to 5 or $\dfrac{4}{5} = \dfrac{4 \cdot 20}{5 \cdot 20} = \dfrac{80}{100}$ or 80 to 100.
The answer is **A**.

3. The ratio of 40 miles to 30 miles is

$$\frac{40 \text{ miles}}{30 \text{ miles}} = \frac{40}{30} = \frac{4}{3}.$$

5. The ratio of 120 people to 90 people is

$$\frac{120 \text{ people}}{90 \text{ people}} = \frac{4 \cdot 30}{3 \cdot 30} = \frac{4}{3}.$$

7. To find the ratio of 20 yards to 8 feet, first convert 20 yards to feet.

$$20 \text{ yards} = 20 \text{ yards} \cdot \frac{3 \text{ feet}}{1 \text{ yard}} = 60 \text{ feet}$$

The ratio of 20 yards to 8 feet is then

$$\frac{60 \text{ feet}}{8 \text{ feet}} = \frac{60}{8} = \frac{15 \cdot 4}{2 \cdot 4} = \frac{15}{2}.$$

9. Convert 2 hours to minutes.

$$2 \text{ hours} = 2 \text{ hours} \cdot \frac{60 \text{ minutes}}{1 \text{ hour}}$$
$$= 120 \text{ minutes}$$

The ratio of 24 minutes to 2 hours is then

$$\frac{24 \text{ minutes}}{120 \text{ minutes}} = \frac{24}{120} = \frac{1 \cdot 24}{5 \cdot 24} = \frac{1}{5}.$$

11. 2 yards $= 2 \cdot 3 = 6$ feet
6 feet $= 6 \cdot 12 = 72$ inches
The ratio of 60 inches to 2 yards is then

$$\frac{60 \text{ inches}}{72 \text{ inches}} = \frac{5 \cdot 12}{6 \cdot 12} = \frac{5}{6}.$$

In Exercises 13–21, to find the best buy, divide the price by the number of units to get the unit cost. Each result was found by using a calculator and rounding the answer to three decimal places. The *best buy* (based on price per unit) is the smallest unit cost.

13.

Size	Unit Cost (dollars per lb)
4 lb	$\frac{\$1.78}{4} = \$0.445$
10 lb	$\frac{\$4.39}{10} = \$0.439$ (∗)

The 10 lb size is the best buy.

15.

Size	Unit Cost (dollars per oz)
16 oz	$\frac{\$2.44}{16} = \$0.153$
32 oz	$\frac{\$2.98}{32} = \$0.093$ (∗)
48 oz	$\frac{\$4.95}{48} = \$0.103$

The 32 oz size is the best buy.

17.

Size	Unit Cost (dollars per oz)
16 oz	$\frac{\$1.54}{16} = \$0.096$
24 oz	$\frac{\$2.08}{24} = \$0.087$
64 oz	$\frac{\$3.63}{64} = \$0.057$
128 oz	$\frac{\$5.65}{128} = \$0.044$ (∗)

The 128 oz size is the best buy.

19.

Size	Unit Cost (dollars per oz)
14 oz	$\frac{\$1.39}{14} = \$0.099$
24 oz	$\frac{\$1.55}{24} = \$0.065$
36 oz	$\frac{\$1.78}{36} = \$0.049$ (∗)
64 oz	$\frac{\$3.99}{64} = \$0.062$

The 36 oz size is the best buy.

21.

Size	Unit Cost (dollars per oz)
87 oz	$\frac{\$7.88}{87} = \$0.091$
131 oz	$\frac{\$10.98}{131} = \$0.084$
263 oz	$\frac{\$19.96}{263} = \$0.076$ (∗)

The 263 oz size is the best buy.

23. $\frac{5}{35} = \frac{8}{56}$

Check to see whether the cross products are equal.

$$5 \cdot 56 = 280$$
$$35 \cdot 8 = 280$$

The cross products are *equal*, so the proportion is *true*.

25. $\frac{120}{82} = \frac{7}{10}$

Compare the cross products.

$$120 \cdot 10 = 1200$$
$$82 \cdot 7 = 574$$

The cross products are *different*, so the proportion is *false*.

27. $\frac{\frac{1}{2}}{5} = \frac{1}{10}$

Compare the cross products.

$$\tfrac{1}{2} \cdot 10 = 5$$
$$5 \cdot 1 = 5$$

The cross products are *equal*, so the proportion is *true*.

29. $\frac{k}{4} = \frac{175}{20}$
$20k = 4(175)$ *Cross products are equal.*
$20k = 700$
$\frac{20k}{20} = \frac{700}{20}$ *Divide by 20.*
$k = 35$

The solution set is $\{35\}$.

31. $\frac{49}{56} = \frac{z}{8}$
$56z = 49(8)$ *Cross products are equal.*
$56z = 392$
$\frac{56z}{56} = \frac{392}{56}$ *Divide by 56.*
$z = 7$

The solution set is $\{7\}$.

33. $\frac{a}{24} = \frac{15}{16}$
$16a = 24(15)$ *Cross products are equal.*
$16a = 360$
$\frac{16a}{16} = \frac{360}{16}$ *Divide by 16.*
$a = \frac{45 \cdot 8}{2 \cdot 8} = \frac{45}{2}$

The solution set is $\left\{\frac{45}{2}\right\}$.

35. $\dfrac{z}{2} = \dfrac{z+1}{3}$

$3z = 2(z+1)$ *Cross products are equal.*

$3z = 2z + 2$ *Distributive property*

$z = 2$ *Subtract 2z.*

The solution set is $\{2\}$.

37. $\dfrac{3y-2}{5} = \dfrac{6y-5}{11}$

$11(3y-2) = 5(6y-5)$ *Cross products are equal.*

$33y - 22 = 30y - 25$ *Dist. prop.*

$3y - 22 = -25$ *Subtract 30y.*

$3y = -3$ *Add 22.*

$y = -1$ *Divide by 3.*

The solution set is $\{-1\}$.

39. $\dfrac{5k+1}{6} = \dfrac{3k-2}{3}$

$3(5k+1) = 6(3k-2)$ *Cross products*

$15k + 3 = 18k - 12$ *Dist. prop.*

$-3k + 3 = -12$ *Subtract 18k.*

$-3k = -15$ *Subtract 3.*

$k = 5$ *Divide by −3.*

The solution set is $\{5\}$.

41. $\dfrac{2p+7}{3} = \dfrac{p-1}{4}$

$4(2p+7) = 3(p-1)$ *Cross products*

$8p + 28 = 3p - 3$ *Dist. prop.*

$5p + 28 = -3$ *Subtract 3p.*

$5p = -31$ *Subtract 28.*

$p = -\dfrac{31}{5}$ *Divide by 5.*

The solution set is $\left\{-\dfrac{31}{5}\right\}$.

43. $\dfrac{0.5x+2}{3} = \dfrac{2.25x+27}{9}$

$9(0.5x+2) = 3(2.25x+27)$ *Cross products*

$4.5x + 18 = 6.75x + 81$ *Dist. prop.*

$18 = 2.25x + 81$ *Subtract 4.5x.*

$-63 = 2.25x$ *Subtract 81.*

$-28 = x$ *Divide by 2.25.*

The solution set is $\{-28\}$.

45. Let $x =$ the cost of 24 candy bars.
Set up a proportion.

$$\frac{x}{24} = \frac{\$20.00}{16}$$

$$16x = 24(20)$$

$$16x = 480$$

$$x = 30$$

The cost of 24 candy bars is $30.00.

47. Let $x =$ the cost of 5 quarts of oil.
Set up a proportion.

$$\frac{x}{5} = \frac{\$14.00}{8}$$

$$8x = 5(14)$$

$$8x = 70$$

$$x = 8.75$$

The cost of five quarts of oil is $8.75.

49. Let $x =$ the cost of five pairs of jeans.

$$\frac{9 \text{ pairs}}{\$121.50} = \frac{5 \text{ pairs}}{x}$$

$$9x = 5(121.50)$$

$$9x = 607.5$$

$$\frac{9x}{9} = \frac{607.5}{9}$$

$$x = 67.5$$

The cost of five pairs is $67.50.

51. Let $x =$ the cost for filling a 15-gallon tank.
Set up a proportion.

$$\frac{x \text{ dollars}}{\$19.56} = \frac{15 \text{ gallons}}{6 \text{ gallons}}$$

$$6x = 15(19.56)$$

$$6x = 293.4$$

$$x = 48.90$$

It would cost $48.90 to completely fill a 15-gallon tank.

53. Let $x =$ the distance between Memphis and Philadelphia on the map (in feet).

Set up a proportion with one ratio involving map distances and the other involving actual distances.

$$\frac{x \text{ feet}}{2.4 \text{ feet}} = \frac{1000 \text{ miles}}{600 \text{ miles}}$$

$$\frac{x}{2.4} = \frac{1000}{600}$$

$$600x = (2.4)(1000)$$

$$600x = 2400$$

$$x = 4$$

The distance on the map between Memphis and Philadelphia would be 4 feet.

55. Let $x =$ the number of inches between St. Louis and Des Moines on the map.
Set up a proportion.

$$\frac{8.5 \text{ inches}}{x \text{ inches}} = \frac{1040 \text{ miles}}{333 \text{ miles}}$$

$$1040x = 8.5(333)$$

$$1040x = 2830.5$$

$$x \approx 2.72$$

St. Louis and Des Moines are about 2.7 inches apart on the map.

57. Let $x =$ the number of inches between Moscow and Berlin on the globe.

Set up a proportion.

$$\frac{12.4 \text{ inches}}{x \text{ inches}} = \frac{10{,}080 \text{ km}}{1610 \text{ km}}$$
$$10{,}080x = 12.4(1610)$$
$$10{,}080x = 19{,}964$$
$$x \approx 1.98$$

Moscow and Berlin are about 2.0 inches apart on the globe.

59. Let $x =$ the number of cups of cleaner.

Set up a proportion with one ratio involving the number of cups of cleaner and the other involving the number of gallons of water.

$$\frac{x \text{ cups}}{\frac{1}{4} \text{ cup}} = \frac{10\frac{1}{2} \text{ gallons}}{1 \text{ gallon}}$$
$$x \cdot 1 = \tfrac{1}{4}\left(10\tfrac{1}{2}\right)$$
$$x = \tfrac{1}{4}\left(\tfrac{21}{2}\right) = \tfrac{21}{8}$$

The amount of cleaner needed is $2\frac{5}{8}$ cups.

61. Let $x =$ the number of U.S. dollars Ashley exchanged.

Set up a proportion.

$$\frac{\$1.2128}{x \text{ dollars}} = \frac{1 \text{ euro}}{300 \text{ euros}}$$
$$x \cdot 1 = 1.2128(300)$$
$$x = 363.84$$

She exchanged $363.84.

63. Let $x =$ the number of fish in Grand Bay.

Set up a proportion with one ratio involving the sample and the other involving the total number of fish.

$$\frac{7 \text{ fish}}{700 \text{ fish}} = \frac{500 \text{ fish}}{x \text{ fish}}$$
$$7x = (700)(500)$$
$$7x = 350{,}000$$
$$x = 50{,}000$$

We estimate that there are 50,000 fish in Grand Bay.

65.
$$\frac{x}{12} = \frac{3}{9}$$
$$9x = 12 \cdot 3 = 36$$
$$x = 4$$

Other possibilities for the proportion are:

$$\frac{12}{x} = \frac{9}{3}, \quad \frac{x}{12} = \frac{5}{15}, \quad \frac{12}{x} = \frac{15}{5}$$

67.
$$\frac{x}{3} = \frac{2}{6} \qquad\qquad \frac{y}{\frac{4}{3}} = \frac{6}{2}$$
$$6x = 3 \cdot 2 = 6 \qquad 2y = 6\left(\tfrac{4}{3}\right) = 8$$
$$x = 1 \qquad\qquad\quad y = 4$$

69. (a)

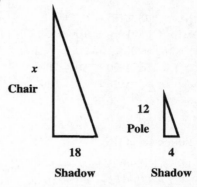

(b) These two triangles are similar, so their sides are proportional.

$$\frac{x}{12} = \frac{18}{4}$$
$$4x = 18(12)$$
$$4x = 216$$
$$x = 54$$

The chair is 54 feet tall.

71. Let $x =$ the 1996 price of groceries.

$$\frac{1990 \text{ price}}{1990 \text{ index}} = \frac{1996 \text{ price}}{1996 \text{ index}}$$
$$\frac{120}{130.7} = \frac{x}{156.9}$$
$$130.7x = 120(156.9)$$
$$x = \frac{120(156.9)}{130.7} \approx 144.06$$

The 1996 price would be about $144.

73. Let $x =$ the 2002 price of groceries.

$$\frac{1990 \text{ price}}{1990 \text{ index}} = \frac{2002 \text{ price}}{2002 \text{ index}}$$
$$\frac{120}{130.7} = \frac{x}{179.9}$$
$$130.7x = 120(179.9)$$
$$x = \frac{120(179.9)}{130.7} \approx 165.17$$

The 2002 price would be about $165.

75.
$$I = prt$$
$$\$1280 = (\$8000)(r)(4)$$
$$r = \frac{1280}{8000(4)} = 0.04$$

The rate of interest is 4%.

77.
$$I = prt$$
$$\$5700 = (\$19,000)(0.03)(t)$$
$$t = \frac{5700}{19,000(0.03)} = 10$$

The money earned interest for 10 years.

79.
$$0.15x + 0.30(3) = 0.20(3 + x)$$
$$100[0.15x + 0.30(3)] = 100[0.20(3 + x)]$$
$$15x + 30(3) = 20(3 + x)$$
$$15x + 90 = 60 + 20x$$
$$90 = 60 + 5x$$
$$30 = 5x$$
$$6 = x$$

The solution set is $\{6\}$.

81.
$$0.92x + 0.98(12 - x) = 0.96(12)$$
$$100[0.92x + 0.98(12 - x)] = 100[0.96(12)]$$
$$92x + 98(12 - x) = 96(12)$$
$$92x + 1176 - 98x = 1152$$
$$1176 - 6x = 1152$$
$$-6x = -24$$
$$x = 4$$

The solution set is $\{4\}$.

2.7 Further Applications of Linear Equations

1. The amount of pure alcohol in 150 liters of a 30% alcohol solution is

150	×	0.30	= 45 liters.
↑		↑	↑
Amount of solution		Rate of concentration	Amount of pure alcohol

3. If $25,000 is invested at 3% simple interest for one year, the amount of interest earned is

$25,000	×	0.03	×	1	= $750.
↑		↑		↑	↑
Principal		Interest Rate		Time	Interest earned

5. The monetary value of 35 half-dollars is

35	×	$0.50	= $17.50.
↑		↑	↑
Number of coins		Denomination	Monetary value

7. 42.1% is about 40%. The population of 1,819,000 is about 1.8 million, and 40% of 1.8 million is 720,000. So choice **A** is the correct answer.

9. **(a)** 13% of 4 million $= 0.13(4,000,000)$
$= 520,000$ white cars.

(b) 24% of 4 million $= 0.24(4,000,000)$
$= 960,000$ silver cars.

(c) 6% of 4 million $= 0.06(4,000,000)$
$= 240,000$ blue cars.

11. The problem can be stated as follows: "8,149,000 is what percent of 147,401,000?"

Substitute $a = 8,149,000$ and $b = 147,401,000$ into the percent proportion; then find p.

$$\frac{p}{100} = \frac{a}{b}$$
$$\frac{p}{100} = \frac{8,149,000}{147,401,000}$$
$$p = \frac{8,149,000(100)}{147,401,000} \approx 5.53$$

The percent of unemployment, to the nearest tenth, is 5.5%.

13. The concentration of the new solution could not be more than the strength of the stronger of the original solutions, so the correct answer is **D**, since 32% is stronger than both 20% and 30%.

15. *Step 2*
Let $x =$ the number of liters of 25% solution to be used.

Step 3
Use the box diagram in the textbook to write the equation.

25% acid solution	40% acid solution	30% acid solution
↓	↓	↓
$0.25x$ +	$0.40(80)$ =	$0.30(x + 80)$

Step 4
Multiply by 100 to clear decimals.

$$25x + 40(80) = 30(x + 80)$$
$$25x + 3200 = 30x + 2400$$
$$25x + 800 = 30x$$
$$800 = 5x$$
$$160 = x$$

Step 5
160 liters of 25% acid solution must be added.

Step 6
25% of 160 liters plus 40% of 80 liters is 40 liters plus 32 liters, or 72 liters, of pure acid; which is equal to 30% of $(160 + 80)$ liters.
$[0.30(240) = 72]$

17. Let $x =$ the number of liters of 5% drug solution.

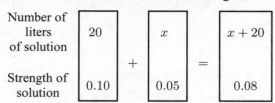

Number of liters of solution	20		x		$x + 20$
		$+$		$=$	
Strength of solution	0.10		0.05		0.08

pure drug solution in 10% solution   plus   pure drug solution in 5% solution
$$\downarrow \qquad\qquad \downarrow \qquad\qquad \downarrow$$
$$0.10(20) \qquad + \qquad 0.05x$$

is   pure drug solution in 8% solution.
$$\downarrow \qquad\qquad \downarrow$$
$$= \qquad 0.08(x + 20)$$

Solve the equation.

$$0.10(20) + 0.05x = 0.08(x + 20)$$
$$10(20) + 5x = 8(x + 20)$$
$$200 + 5x = 8x + 160$$
$$200 = 3x + 160$$
$$40 = 3x$$
$$x = \frac{40}{3} = 13\frac{1}{3}$$

The pharmacist needs $13\frac{1}{3}$ liters of 5% drug solution.

Check $x = 13\frac{1}{3}$:

LS and RS refer to the left side and right side of the original equation.

LS: $\;0.10(20) + 0.05\left(13\frac{1}{3}\right) = 2\frac{2}{3}$
RS: $\qquad 0.08\left(13\frac{1}{3} + 20\right) = 2\frac{2}{3}$

19. Let $x =$ the number of liters of 60% acid solution.
Then $20 - x =$ the number of liters of 75% acid solution.

Number of liters of solution	x		$20 - x$		20
		$+$		$=$	
Concentration of acid	0.60		0.75		0.72

Pure acid in 60% solution   plus   pure acid in 75% solution
$$\downarrow \qquad\quad \downarrow \qquad\qquad \downarrow$$
$$0.60x \qquad + \qquad 0.75(20 - x)$$

is   pure acid in 72% solution.
$$\downarrow \qquad\qquad \downarrow$$
$$= \qquad 0.72(20)$$

Solve the equation.

$$0.60x + 0.75(20 - x) = 0.72(20)$$
$$60x + 75(20 - x) = 72(20)$$
$$60x + 1500 - 75x = 1440$$
$$1500 - 15x = 1440$$
$$-15x = -60$$
$$x = 4$$

4 liters of 60% acid solution must used.

Check $x = 4$:
LS: $\;0.60(4) + 0.75(20 - 4) = 14.4$
RS: $\qquad\qquad\qquad 0.72(20) = 14.4$

21. Let $x =$ the number of gallons of 12% indicator solution.
Then $10 - x =$ the number of gallons of 20% indicator solution.

Gallons of solution	x		$10 - x$		10
		$+$		$=$	
Strength of solution	0.12		0.20		0.14

Indicator in 12% solution   indicator in 20% solution   indicator in 14% solution
$$\downarrow \qquad\qquad \downarrow \qquad\qquad \downarrow$$
$$0.12x \quad + \quad 0.20(10 - x) \;=\; 0.14(10)$$

$$12x + 20(10 - x) = 14(10)$$
$$12x + 200 - 20x = 140$$
$$200 - 8x = 140$$
$$-8x = -60$$
$$x = \frac{-60}{-8}$$
$$x = \frac{15}{2} \text{ or } 7\frac{1}{2}$$

$7\frac{1}{2}$ gallons of 12% indicator solution must be used.

Check $x = 7\frac{1}{2}$:
LS: $\quad 0.12\left(7\frac{1}{2}\right) + 0.20\left(10 - 7\frac{1}{2}\right) = 1.4$
RS: $\qquad\qquad\qquad\qquad 0.14(10) = 1.4$

23. Let $x =$ the amount of water to be added.
Then $20 + x =$ the amount of 2% solution.

There is no minoxidil in water.

Number of milliliters of solution	x		20		$20 + x$
		$+$		$=$	
Concentration of solution	0		0.04		0.02

Pure minoxidil in x milliliters of water   plus   pure minoxidil in 4% solution   is   pure minoxidil in 2% solution
$$\downarrow \qquad\quad \downarrow \qquad\qquad \downarrow \qquad \downarrow \qquad\qquad \downarrow$$
$$0(x) \qquad + \qquad 0.04(20) \;=\; 0.02(20 + x)$$

continued

Solve the equation.

$$0x + 0.04(20) = 0.02(20 + x)$$
$$4(20) = 2(20 + x)$$
$$80 = 40 + 2x$$
$$40 = 2x$$
$$20 = x$$

20 milliliters of water should be used.

Check $x = 20$:

LS: $0(20) + 0.04(20) = 0.8$
RS: $0.02(20 + 20) = 0.8$

This answer should make common sense; that is, equal amounts of 0% and 4% solutions should produce a 2% solution.

25. Let $x =$ the amount invested at 5% (in dollars).
Then $x - 1200 =$ the amount invested at 4% (in dollars).

Amount invested (in dollars)	Rate of interest	Interest for one year
x	0.05	$0.05x$
$x - 1200$	0.04	$0.04(x - 1200)$

Since the total annual interest was $141, the equation is

$$0.05x + 0.04(x - 1200) = 141.$$
$$5x + 4(x - 1200) = 100(141)$$
$$5x + 4x - 4800 = 14{,}100$$
$$9x - 4800 = 14{,}100$$
$$9x = 18{,}900$$
$$x = 2100$$

Since $x = 2100$, $x - 1200 = 900$.
Eduardo invested $2100 at 5% and $900 at 4%.

27. Let $x =$ the amount invested at 6%.
Then $3x + 6000 =$ the amount invested at 5%.

$$0.06x + 0.05(3x + 6000) = 825$$
$$6x + 5(3x + 6000) = 100(825)$$
$$6x + 15x + 30{,}000 = 82{,}500$$
$$21x + 30{,}000 = 82{,}500$$
$$21x = 52{,}500$$
$$x = 2500$$

Since $x = 2500$, $3x + 6000 = 13{,}500$.
The artist invested $2500 at 6% and $13,500 at 5%.

29. Let $x =$ the number of nickels.
Then $x + 2 =$ the number of dimes.

The value of nickels	plus	the value of dimes		is	$1.70
↓	↓	↓		↓	↓
$0.05x$	$+$	$0.10(x + 2)$	$=$		1.70

$$5x + 10(x + 2) = 100(1.70)$$
$$5x + 10x + 20 = 170$$
$$15x + 20 = 170$$
$$15x = 150$$
$$x = 10$$

The collector has 10 nickels.

31. Let $x =$ the number of 39-cent stamps.
Then $45 - x =$ the number of 24-cent stamps.

The value of the 39-cent stamps is $0.39x$ and the value of the 24-cent stamps is $0.24(45 - x)$. The total value is $15.00, so

$$0.39x + 0.24(45 - x) = 15.00.$$
$$39x + 24(45 - x) = 100(15)$$
$$39x + 1080 - 24x = 1500$$
$$15x + 1080 = 1500$$
$$15x = 420$$
$$x = 28$$

Since $x = 28$, $45 - x = 17$. She bought 28 39-cent stamps valued at $10.92 and 17 24-cent stamps valued at $4.08, for a total value of $15.00.

33. Let $x =$ the number of pounds of Colombian Decaf beans.
Then $2x =$ the number of pounds of Arabian Mocha beans.

Number of Pounds	Cost per Pound	Total Value (in $)
x	$8.00	$8x$
$2x$	$8.50	$8.5(2x)$

The total value is $87.50, so

$$8x + 8.5(2x) = 87.50.$$
$$8x + 17x = 87.50$$
$$25x = 87.50$$
$$x = 3.5$$

Since $x = 3.5$, $2x = 7$. She can buy 3.5 pounds of Colombian Decaf and 7 pounds of Arabian Mocha.

35. To estimate the average speed of the trip, round 405 to 400 and 8.2 to 8.

Use $r = \dfrac{d}{t}$ with $d = 400$ and $t = 8$.

$$r = \frac{d}{t} = \frac{400}{8} = 50$$

The best estimate is **A**, 50 miles per hour.

37. Use the formula $d = rt$ with $r = 53$ and $t = 10$.

$$d = rt$$
$$= (53)(10)$$
$$= 530$$

The distance between Memphis and Chicago is 530 miles.

39. Use $d = rt$ with $d = 500$ and $r = 157.603$.

$$d = rt$$
$$500 = 157.603t$$
$$t = \frac{500}{157.603} \approx 3.173$$

His time was about 3.173 hours.

41. $r = \dfrac{d}{t} = \dfrac{100 \text{ meters}}{12.37 \text{ seconds}} \approx 8.084$

Her rate was about 8.08 meters per second.

43. $r = \dfrac{d}{t} = \dfrac{400 \text{ meters}}{47.63 \text{ seconds}} \approx 8.398$

His rate was about 8.40 meters per second.

45. Let t = the number of hours until John and Pat meet.
The distance John travels and the distance Pat travels total 440 miles.

John's distance	and	Pat's distance	equal	total distance.
↓	↓	↓	↓	↓
$60t$	$+$	$28t$	$=$	440

$$88t = 440$$
$$t = 5$$

It will take 5 hours for them to meet.

47. Let t = the number of hours until the trains are 315 kilometers apart.

Distance of northbound train	plus	distance of southbound train	is	total distance
↓	↓	↓	↓	↓
$85t$	$+$	$95t$	$=$	315

$$180t = 315$$
$$t = \frac{315}{180} = \frac{7}{4}$$

It will take $1\frac{3}{4}$ hours for the trains to be 315 kilometers apart.

49. Let t = the number of hours Marco and Celeste traveled.

Make a chart using the formula $d = rt$.

	r	t	d
Marco	10	t	$10t$
Celeste	12	t	$12t$

Marco's distance	minus	Celeste's distance	is	15.
↓	↓	↓	↓	↓
$12t$	$-$	$10t$	$=$	15

$$12t - 10t = 15$$
$$2t = 15$$
$$t = \frac{15}{2} \text{ or } 7\frac{1}{2}$$

They will be 15 miles apart in $7\frac{1}{2}$ hours.

51. Let x = the rate of the westbound plane.
Then $x - 150$ = the rate of the eastbound plane.

Using the formula $d = rt$ and the chart in the text, we see that

$$d_{\text{west}} + d_{\text{east}} = d_{\text{total}}$$
$$x(3) + (x - 150)(3) = 2250$$
$$3x + 3x - 450 = 2250$$
$$6x = 2700$$
$$x = 450$$

Since $x = 450$, $x - 150 = 300$.

The speed of the westbound plane is 450 mph and the speed of the eastbound plane is 300 mph.

53. Let x = the rate of the slower car.
Then $x + 20$ = the rate of the faster car.

Use the formula $d = rt$ and the fact that each car travels for 4 hours.

$$d_{\text{faster}} + d_{\text{slower}} = d_{\text{total}}$$
$$(x + 20)(4) + (x)(4) = 400$$
$$4x + 80 + 4x = 400$$
$$8x = 320$$
$$x = 40$$

The speed of the slower car is 40 mph and the speed of the faster car is 60 mph.

55. $-2x + 6$
$= -2(3) + 6$ *Let x = 3.*
$= -6 + 6$
$= 0$

57. $-2x + 6$
$= -2(0) + 6$ *Let x = 0.*
$= 0 + 6$
$= 6$

59. $x + 6 = 0$
$-6 + 6 = 0$ *Let x = −6.*
$0 = 0$ *True*

Since a true statement results, -6 is a solution of the given equation.

61. $-6 + x = 0$
$-6 + (-6) = 0$ *Let x = −6.*
$-12 = 0$ *False*

Since a false statement results, -6 is not a solution of the given equation.

2.8   Solving Linear Inequalities

1.   Use a parenthesis if the inequality symbol is $>$ or $<$. Use a square bracket if the inequality symbol is $\geq$ or $\leq$. Examples:

A parenthesis would be used for the inequalities $x < 2$ and $x > 3$. A square bracket would be used for the inequalities $x \leq 2$ and $x \geq 3$. Note that a parenthesis is *always* used with the symbols $-\infty$ and ∞.

3.   The set of numbers graphed corresponds to the inequality $x > -4$.

5.   The set of numbers graphed corresponds to the inequality $x \leq 4$.

7.   The statement $k \leq 4$ says that k can represent any number less than or equal to 4. The interval is written as $(-\infty, 4]$. To graph the inequality, place a square bracket at 4 (to show that 4 is part of the graph) and draw an arrow extending to the left.

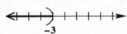

9.   The statement $x < -3$ says that x can represent any number less than -3. The interval is written as $(-\infty, -3)$. To graph the inequality, place a parenthesis at -3 (to show that -3 is *not* part of the graph) and draw an arrow extending to the left.

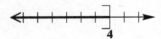

11.   The statement $t > 4$ says that t can represent any number greater than 4. The interval is written $(4, \infty)$. To graph the inequality, place a parenthesis at 4 (to show that 4 is *not* part of the graph) and draw an arrow extending to the right.

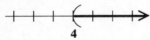

13.   The statement $8 \leq x \leq 10$ says that x can represent any number between 8 and 10, including 8 and 10. To graph the inequality, place brackets at 8 and 10 (to show that 8 and 10 are part of the graph) and draw a line segment between the brackets. The interval is written as $[8, 10]$.

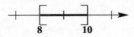

15.   The statement $0 < y \leq 10$ says that y can represent any number between 0 and 10, excluding 0 and including 10. To graph the inequality, place a parenthesis at 0 and a bracket at 10 and draw a line segment between them. The interval is written as $(0, 10]$.

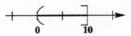

17.
$$4 > -x > -3$$
$$-1(4) < -1(-x) < -1(-3) \quad \begin{array}{l}\textit{Multiply by } -1;\\ \textit{reverse the}\\ \textit{symbols.}\end{array}$$
$$-4 < x < 3$$

Graph the solution set $(-4, 3)$.

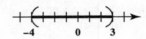

19.   It is wrong to write $3 < x < -2$ because it would imply that $3 < -2$, a false statement. Also, note that $3 < x < -2$ would require x to be a number which is both less than -2 and greater than 3 at the same time, which is impossible.

21.
$$z - 8 \geq -7$$
$$z - 8 + 8 \geq -7 + 8 \quad \textit{Add 8.}$$
$$z \geq 1$$

Graph the solution set $[1, \infty)$.

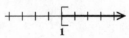

23.
$$2k + 3 \geq k + 8$$
$$2k + 3 - k \geq k + 8 - k \quad \textit{Subtract k.}$$
$$k + 3 \geq 8$$
$$k + 3 - 3 \geq 8 - 3 \qquad \textit{Subtract 3.}$$
$$k \geq 5$$

Graph the solution set $[5, \infty)$.

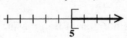

25.
$$3n + 5 < 2n - 6$$
$$3n - 2n + 5 < 2n - 2n - 6 \quad \textit{Subtract 2n.}$$
$$n + 5 < -6$$
$$n + 5 - 5 < -6 - 5 \qquad \textit{Subtract 5.}$$
$$n < -11$$

Graph the solution set $(-\infty, -11)$.

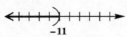

27.   The inequality symbol must be reversed when one is multiplying or dividing by a negative number.

29. $3x < 18$

$\dfrac{3x}{3} < \dfrac{18}{3}$ *Divide by 3.*

$x < 6$

Graph the solution set $(-\infty, 6)$.

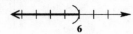

31. $2y \geq -20$

$\dfrac{2y}{2} \geq \dfrac{-20}{2}$ *Divide by 2.*

$y \geq -10$

Graph the solution set $[-10, \infty)$.

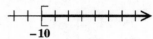

33. $-8t > 24$

$\dfrac{-8t}{-8} < \dfrac{24}{-8}$ *Divide by -8;*
reverse the symbol
from $>$ to $<$.

$t < -3$

Graph the solution set $(-\infty, -3)$.

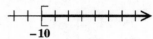

35. $-x \geq 0$

$-1x \geq 0$

$\dfrac{-1x}{-1} \leq \dfrac{0}{-1}$ *Divide by -1;*
reverse the symbol
from $\geq$ to $\leq$.

$x \leq 0$

Graph the solution set $(-\infty, 0]$.

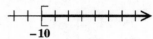

37. $-\frac{3}{4}r < -15$

$\left(-\frac{4}{3}\right)\left(-\frac{3}{4}r\right) > \left(-\frac{4}{3}\right)(-15)$

Multiply by $-\frac{4}{3}$ (the reciprocal of $-\frac{3}{4}$);
reverse the symbol from $<$ to $>$.

$r > 20$

Graph the solution set $(20, \infty)$.

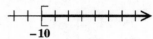

39. $-0.02x \leq 0.06$

$\dfrac{-0.02x}{-0.02} \geq \dfrac{0.06}{-0.02}$ *Divide by -0.02;*
reverse the symbol
from $\leq$ to $\geq$.

$x \geq -3$

Graph the solution set $[-3, \infty)$.

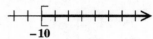

41. $5r + 1 \geq 3r - 9$

$2r + 1 \geq -9$ *Subtract $3r$.*

$2r \geq -10$ *Subtract 1.*

$r \geq -5$ *Divide by 2.*

Graph the solution set $[-5, \infty)$.

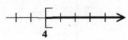

43. $6x + 3 + x < 2 + 4x + 4$

$7x + 3 < 4x + 6$ *Combine like terms.*

$3x + 3 < 6$ *Subtract $4x$.*

$3x < 3$ *Subtract 3.*

$x < 1$ *Divide by 3.*

Graph the solution set $(-\infty, 1)$.

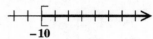

45. $-x + 4 + 7x \leq -2 + 3x + 6$

$6x + 4 \leq 4 + 3x$

$3x + 4 \leq 4$

$3x \leq 0$

$x \leq 0$

Graph the solution set $(-\infty, 0]$.

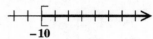

47. $5(x + 3) - 6x \leq 3(2x + 1) - 4x$

$5x + 15 - 6x \leq 6x + 3 - 4x$

$-x + 15 \leq 2x + 3$

$-3x + 15 \leq 3$

$-3x \leq -12$

$\dfrac{-3x}{-3} \geq \dfrac{-12}{-3}$ *Divide by -3;*
reverse the symbol.

$x \geq 4$

Graph the solution set $[4, \infty)$.

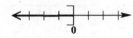

49. $\frac{2}{3}(p + 3) > \frac{5}{6}(p - 4)$

$6\left(\frac{2}{3}\right)(p + 3) > 6\left(\frac{5}{6}\right)(p - 4)$

Multiply by 6, the LCD.

$4(p + 3) > 5(p - 4)$

$4p + 12 > 5p - 20$

$-p + 12 > -20$

$-p > -32$

$\dfrac{-p}{-1} < \dfrac{-32}{-1}$ *Divide by -1;*
reverse the symbol.

$p < 32$

Graph the solution set $(-\infty, 32)$.

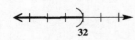

51. $-\frac{1}{4}(p+6) + \frac{3}{2}(2p-5) < 10$

Multiply each term by 4 to clear the fractions.

$-1(p+6) + 6(2p-5) < 40$

$-p - 6 + 12p - 30 < 40$

$11p - 36 < 40$

$11p < 76$

$p < \frac{76}{11}$

Check that the solution set is the interval $\left(-\infty, \frac{76}{11}\right)$.

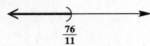

53. $4x - (6x+1) \le 8x + 2(x-3)$

$4x - 6x - 1 \le 8x + 2x - 6$

$-2x - 1 \le 10x - 6$

$-12x - 1 \le -6$

$-12x \le -5$

$\frac{-12x}{-12} \ge \frac{-5}{-12}$ *Divide by −12;*
                                          reverse the symbol.

$x \ge \frac{5}{12}$

Graph the solution set $\left[\frac{5}{12}, \infty\right)$.

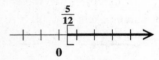

55. $5(2k+3) - 2(k-8) > 3(2k+4) + k - 2$

$10k + 15 - 2k + 16 > 6k + 12 + k - 2$

$8k + 31 > 7k + 10$

$k + 31 > 10$

$k > -21$

Graph the solution set $(-21, \infty)$.

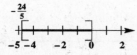

57. The graph corresponds to the inequality $-1 < x < 2$, excluding both -1 and 2.

59. The graph corresponds to the inequality $-1 < x \le 2$, excluding -1 but including 2.

61. $-5 \le 2x - 3 \le 9$

$-5 + 3 \le 2x - 3 + 3 \le 9 + 3$ *Add 3 to each*
                                       part.

$-2 \le 2x \le 12$

$\frac{-2}{2} \le \frac{2x}{2} \le \frac{12}{2}$ *Divide each*
                                                     part by 2.

$-1 \le x \le 6$

Graph the solution set $[-1, 6]$.

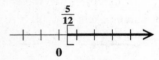

63. $5 < 1 - 6m < 12$

$5 - 1 < 1 - 6m - 1 < 12 - 1$ *Subtract 1 from*
                                   each part.

$4 < -6m < 11$

$\frac{4}{-6} > \frac{-6m}{-6} > \frac{11}{-6}$ *Divide each*
                                                    part by −6;
                                                    reverse both symbols.

$-\frac{2}{3} > m > -\frac{11}{6}$

or $-\frac{11}{6} < m < -\frac{2}{3}$ *Equivalent inequality*

Graph the solution set $\left(-\frac{11}{6}, -\frac{2}{3}\right)$.

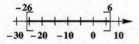

65. $10 < 7p + 3 < 24$

$7 < 7p < 21$ *Subtract 3.*

$1 < p < 3$ *Divide by 7.*

Graph the solution set $(1, 3)$.

67. $-12 \le \frac{1}{2}z + 1 \le 4$

$2(-12) \le 2\left(\frac{1}{2}z + 1\right) \le 2(4)$ *Multiply each*
                                                         part by 2.

$-24 \le z + 2 \le 8$

$-26 \le z \le 6$ *Subtract 2.*

Note: We could have started this solution by
subtracting 1 from each part.

Graph the solution set $[-26, 6]$.

69. $1 \le 3 + \frac{2}{3}p \le 7$

$3(1) \le 3\left(3 + \frac{2}{3}p\right) \le 3(7)$ *Multiply by 3.*

$3 \le 9 + 2p \le 21$

$-6 \le 2p \le 12$ *Subtract 9.*

$-3 \le p \le 6$ *Divide by 2.*

Graph the solution set $[-3, 6]$.

71. $-7 \le \frac{5}{4}r - 1 \le -1$

$-6 \le \frac{5}{4}r \le 0$ *Add 1.*

$\frac{4}{5}(-6) \le \frac{4}{5}\left(\frac{5}{4}r\right) \le \frac{4}{5}(0)$ *Multiply by $\frac{4}{5}$.*

$-\frac{24}{5} \le r \le 0$

Graph the solution set $\left[-\frac{24}{5}, 0\right]$.

73. $3x + 2 = 14$

$\qquad 3x = 12$

$\qquad\ \ x = 4$

Solution set: $\{4\}$

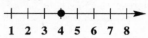

74. $3x + 2 > 14$

$\qquad 3x > 12$

$\qquad\ \ x > 4$

Solution set: $(4, \infty)$

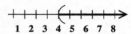

75. $3x + 2 < 14$

$\qquad 3x < 12$

$\qquad\ \ x < 4$

Solution set: $(-\infty, 4)$

76. If you were to graph all the solutions from Exercises 73–75 on the same number line, the graph would be the complete number line, that is, all real numbers.

77. Let $x =$ the score on the third test.

The average of the three tests	is at least	80.
↓	↓	↓
$\dfrac{76 + 81 + x}{3}$	$\geq$	80

$$\frac{157 + x}{3} \geq 80$$

$$3\left(\frac{157 + x}{3}\right) \geq 3(80)$$

$$157 + x \geq 240$$

$$x \geq 83$$

In order to average at least 80, Inkie's score on her third test must be 83 or greater.

79. Let $n =$ the number.

"When 2 is added to the difference between six times a number and 5, the result is greater than 13 added to five times the number" translates to

$$(6n - 5) + 2 > 5n + 13.$$

Solve the inequality.

$$6n - 5 + 2 > 5n + 13$$

$$6n - 3 > 5n + 13$$

$$n - 3 > 13 \qquad \textit{Subtract } 5n.$$

$$n > 16 \qquad \textit{Add } 3.$$

All numbers greater than 16 satisfy the given condition.

81. The Fahrenheit temperature must correspond to a Celsius temperature that is greater than or equal to -25 degrees.

$$C = \tfrac{5}{9}(F - 32) \geq -25$$

$$\tfrac{9}{5}\left[\tfrac{5}{9}(F - 32)\right] \geq \tfrac{9}{5}(-25)$$

$$F - 32 \geq -45$$

$$F \geq -13$$

The temperature in Minneapolis on a certain winter day is never less than $-13°$ Fahrenheit.

83. $P = 2L + 2W;\ P \geq 400$

From the figure, we have $L = 4x + 3$ and $W = x + 37$. Thus, we have the inequality

$$2(4x + 3) + 2(x + 37) \geq 400.$$

Solve this inequality.

$$8x + 6 + 2x + 74 \geq 400$$

$$10x + 80 \geq 400$$

$$10x \geq 320$$

$$x \geq 32$$

The rectangle will have a perimeter of at least 400 if the value of x is 32 or greater.

85.

$$2 + 0.30x \leq 5.60$$

$$10(2 + 0.30x) \leq 10(5.60)$$

$$20 + 3x \leq 56$$

$$3x \leq 36$$

$$x \leq 12$$

Jorge can use the phone for a maximum of 12 minutes after the first three minutes. This means that the maximum *total* time he can use the phone is 15 minutes.

87. Let $x =$ the number of gallons.

The amount she spends can be represented by $\$4.50 + \$3.20x$. This must be less than or equal to $\$38.10$.

$$4.5 + 3.2x \leq 38.10$$

$$3.2x \leq 33.6 \qquad \textit{Subtract 4.5.}$$

$$\frac{3.2x}{3.2} \leq \frac{33.6}{3.2} \qquad \textit{Divide by 3.2.}$$

$$x \leq 10.5$$

She can purchase 10.5 gallons of gasoline.

89. "Revenue from the sales of the DVDs is $5 per DVD less sales costs of $100" translates to

$$R = 5x - 100,$$

where x represents the number of DVDs to be produced.

91. $P = R - C$
$= (5x - 100) - (125 + 4x)$
$= 5x - 100 - 125 - 4x$
$= x - 225$

We can use this expression for P to solve the inequality.

$$P > 0$$
$$x - 225 > 0$$
$$x > 225$$

In part (a) of Exercises 93–98, replace x with -2. In part (b), replace x with 4. Then use the order of operations.

93. **(a)** $y = 5x + 3$
$y = 5(-2) + 3$
$y = -10 + 3$
$y = -7$

(b) $y = 5x + 3$
$y = 5(4) + 3$
$y = 20 + 3$
$y = 23$

95. **(a)** $6x - 2 = y$
$6(-2) - 2 = y$
$-12 - 2 = y$
$-14 = y$

(b) $6x - 2 = y$
$6(4) - 2 = y$
$24 - 2 = y$
$22 = y$

97. **(a)** $2x - 5y = 10$
$2(-2) - 5y = 10$
$-4 - 5y = 10$
$-5y = 14$
$y = -\frac{14}{5}$

(b) $2x - 5y = 10$
$2(4) - 5y = 10$
$8 - 5y = 10$
$-5y = 2$
$y = -\frac{2}{5}$

Chapter 2 Review Exercises

1. $x - 5 = 1$
$x = 6$ *Add 5.*

The solution set is $\{6\}$.

2. $x + 8 = -4$
$x = -12$ *Subtract 8.*

The solution set is $\{-12\}$.

3. $3k + 1 = 2k + 8$
$k + 1 = 8$ *Subtract 2k.*
$k = 7$ *Subtract 1.*

The solution set is $\{7\}$.

4. $5k = 4k + \frac{2}{3}$
$k = \frac{2}{3}$ *Subtract 4k.*

The solution set is $\left\{\frac{2}{3}\right\}$.

5. $(4r - 2) - (3r + 1) = 8$
$(4r - 2) - 1(3r + 1) = 8$ *Replace – with –1.*
$4r - 2 - 3r - 1 = 8$ *Dist. prop.*
$r - 3 = 8$
$r = 11$ *Add 3.*

The solution set is $\{11\}$.

6. $3(2x - 5) = 2 + 5x$
$6x - 15 = 2 + 5x$ *Dist. prop.*
$x - 15 = 2$ *Subtract 5y.*
$x = 17$ *Add 15.*

The solution set is $\{17\}$.

7. $7k = 35$
$k = 5$ *Divide by 7.*

The solution set is $\{5\}$.

8. $12r = -48$
$r = -4$ *Divide by 12.*

The solution set is $\{-4\}$.

9. $2p - 7p + 8p = 15$
$3p = 15$
$p = 5$ *Divide by 3.*

The solution set is $\{5\}$.

10. $\frac{m}{12} = -1$
$m = -12$ *Multiply by 12.*

The solution set is $\{-12\}$.

11. $\frac{5}{8}k = 8$
$\frac{8}{5}\left(\frac{5}{8}k\right) = \frac{8}{5}(8)$ *Multiply by $\frac{8}{5}$.*
$k = \frac{64}{5}$

The solution set is $\left\{\frac{64}{5}\right\}$.

12. $12m + 11 = 59$
$12m = 48$ *Subtract 11.*
$m = 4$ *Divide by 12.*

The solution set is $\{4\}$.

13. $3(2x + 6) - 5(x + 8) = x - 22$
$6x + 18 - 5x - 40 = x - 22$
$x - 22 = x - 22$

This is a true statement, so the solution set is {all real numbers}.

14. $5x + 9 - (2x - 3) = 2x - 7$
$5x + 9 - 2x + 3 = 2x - 7$
$3x + 12 = 2x - 7$
$x + 12 = -7$
$x = -19$

The solution set is $\{-19\}$.

15. $\frac{1}{2}r - \frac{r}{3} = \frac{r}{6}$
$6\left(\frac{1}{2}r\right) - 6\left(\frac{r}{3}\right) = 6\left(\frac{r}{6}\right)$ *Multiply by 6.*
$3r - 2r = r$
$r = r$

This is a true statement, so the solution set is {all real numbers}.

16.
$$0.1(x + 80) + 0.2x = 14$$
$$10[0.1(x + 80) + 0.2x] = 10(14) \quad \textit{Multiply by 10.}$$
$$(x + 80) + 2x = 140 \quad \textit{Dist. prop.}$$
$$3x + 80 = 140$$
$$3x = 60$$
$$x = 20$$

The solution set is $\{20\}$.

17.
$$3x - (-2x + 6) = 4(x - 4) + x$$
$$3x + 2x - 6 = 4x - 16 + x$$
$$5x - 6 = 5x - 16$$
$$-6 = -16$$

This statement is false, so there is no solution set, symbolized by $\emptyset$.

18.
$$2(y - 3) - 4(y + 12) = -2(y + 27)$$
$$2y - 6 - 4y - 48 = -2y - 54$$
$$-2y - 54 = -2y - 54$$

This is a true statement, so the solution set is {all real numbers}.

19. *Step 2*
Let $x =$ the number of Republicans.
Then $x - 13 =$ the number of Democrats.

Step 3 $x + (x - 13) = 101$

Step 4
$$2x - 13 = 101$$
$$2x = 114$$
$$x = 57$$

Step 5
Since $x = 57$, $x - 13 = 44$.
There were 44 Democrats and 57 Republicans.

Step 6
There are 13 fewer Democrats than Republicans and the total is 101.

20. *Step 2*
Let $x =$ the land area of Rhode Island.
Then $x + 5213 =$ land area of Hawaii.

Step 3
The areas total 7637 square miles, so
$$x + (x + 5213) = 7637.$$

Step 4
$$2x + 5213 = 7637$$
$$2x = 2424$$
$$x = 1212$$

Step 5
Since $x = 1212$, $x + 5213 = 6425$. The land area of Rhode Island is 1212 square miles and that of Hawaii is 6425 square miles.

Step 6
The land area of Hawaii is 5213 square miles greater than the land area of Rhode Island and the total is 7637 square miles.

21. *Step 2*
Let $x =$ the height of Rhaiadr Falls.
Then $\frac{11}{8}x =$ the height of Kegon Falls.

Step 3
The sum of the heights is 570 feet, so
$$x + \frac{11}{8}x = 570.$$

Step 4
$$\frac{8}{8}x + \frac{11}{8}x = 570$$
$$\frac{19}{8}x = 570$$
$$\frac{8}{19}\left(\frac{19}{8}x\right) = \frac{8}{19}(570)$$
$$x = 240$$

Step 5
Since $x = 240$, $\frac{11}{8}x = \frac{11}{8}(240) = 330$. The height of Rhaiadr Falls is 240 feet and that of Kegon Falls is 330 feet.

Step 6
The height of Kegon Falls is $\frac{11}{8}$ the height of Rhaiadr Falls and the sum is 570.

22. *Step 2*
Let $x =$ the height of Twin Falls.
Then $\frac{5}{2}x =$ the height of Seven Falls.

Step 3
The sum of the heights is 420 feet, so
$$x + \frac{5}{2}x = 420.$$

Step 4
$$2\left(x + \frac{5}{2}x\right) = 2(420)$$
$$2x + 5x = 840$$
$$7x = 840$$
$$x = 120$$

Step 5
Since $x = 120$, $\frac{5}{2}x = \frac{5}{2}(120) = 300$. The height of Twin Falls is 120 feet and that of Seven Falls is 300 feet.

Step 6
The height of Seven Falls is $\frac{5}{2}$ the height of Twin Falls and the sum is 420.

23. *Step 2*
Let $x =$ the measure of the angle.
Then $90 - x =$ the measure of its complement and $180 - x =$ the measure of its supplement.

Step 3 $180 - x = 10(90 - x)$

Step 4
$$180 - x = 900 - 10x$$
$$9x + 180 = 900$$
$$9x = 720$$
$$x = 80$$

Step 5
The measure of the angle is 80°.
Its complement measures $90° - 80° = 10°$, and its supplement measures $180° - 80° = 100°$.

Step 6
The measure of the supplement is 10 times the measure of the complement.

24. *Step 2*
Let x = lesser odd integer.
Then $x + 2$ = greater odd integer.

Step 3 $x + 2(x + 2) = (x + 2) + 24$

Step 4 $x + 2x + 4 = x + 26$
$$3x + 4 = x + 26$$
$$2x + 4 = 26$$
$$2x = 22$$
$$x = 11$$

Step 5
Since $x = 11$, $x + 2 = 13$. The consecutive odd numbers are 11 and 13.

Step 6
The lesser plus twice the greater is $11 + 2(13) = 37$, which is 24 more than the greater.

In Exercises 25–28, substitute the given values into the given formula and then solve for the remaining variable.

25. $A = \frac{1}{2}bh$; $A = 44$, $b = 8$
$$A = \tfrac{1}{2}bh$$
$$44 = \tfrac{1}{2}(8)h$$
$$44 = 4h$$
$$11 = h$$

26. $A = \frac{1}{2}h(b + B)$; $h = 8$, $b = 3$, $B = 4$
$$A = \tfrac{1}{2}h(b + B)$$
$$A = \tfrac{1}{2}(8)(3 + 4)$$
$$= \tfrac{1}{2}(8)(7)$$
$$= (4)(7)$$
$$A = 28$$

27. $C = 2\pi r$; $C = 29.83$, $\pi = 3.14$
$$C = 2\pi r$$
$$29.83 = 2(3.14)r$$
$$29.83 = 6.28r$$
$$\frac{29.83}{6.28} = \frac{6.28r}{6.28}$$
$$4.75 = r$$

28. $V = \frac{4}{3}\pi r^3$; $r = 6$, $\pi = 3.14$
$$V = \tfrac{4}{3}\pi r^3$$
$$= \tfrac{4}{3}(3.14)(6)^3$$
$$= \tfrac{4}{3}(3.14)(216)$$
$$= \tfrac{4}{3}(678.24)$$
$$V = 904.32$$

29. $A = bh$ for h
$$\frac{A}{b} = \frac{bh}{b} \quad \textit{Divide by } b.$$
$$\frac{A}{b} = h \quad \text{or} \quad h = \frac{A}{b}$$

30. $A = \frac{1}{2}h(b + B)$ for h
$$2A = 2\left[\tfrac{1}{2}h(b + B)\right] \quad \textit{Multiply by 2.}$$
$$2A = h(b + B)$$
$$\frac{2A}{(b + B)} = \frac{h(b + B)}{(b + B)} \quad \textit{Divide by } b + B.$$
$$\frac{2A}{b + B} = h \quad \text{or} \quad h = \frac{2A}{b + B}$$

31.   Because the two angles are supplementary,
$$(8x - 1) + (3x - 6) = 180.$$
$$11x - 7 = 180$$
$$11x = 187$$
$$x = 17$$

Since $x = 17$, $8x - 1 = 135$, and $3x - 6 = 45$.

The measures of the two angles are 135° and 45°.

32.   The angles are vertical angles, so their measures are equal.
$$3x + 10 = 4x - 20$$
$$10 = x - 20$$
$$30 = x$$

Since $x = 30$, $3x + 10 = 100$, and $4x - 20 = 100$.

Each angle has a measure of 100°.

33.   Let W = the width of the rectangle.
Then $W + 12$ = the length of the rectangle.

The perimeter of the rectangle is 16 times the width can be written as
$$2L + 2W = 16W$$
since the perimeter is $2L + 2W$.

Because $L = W + 12$, we have
$$2(W + 12) + 2W = 16W.$$
$$2W + 24 + 2W = 16W$$
$$4W + 24 = 16W$$
$$-12W + 24 = 0$$
$$-12W = -24$$
$$W = 2$$

The width is 2 cm and the length is
$2 + 12 = 14$ cm.

34. First, use the formula for the circumference of a circle to find the value of r.

$$C = 2\pi r$$
$$62.5 = 2(3.14)(r) \quad \textit{Let } C = 62.5,$$
$$\qquad\qquad\qquad\quad \pi = 3.14.$$
$$62.5 = 6.28r$$
$$\frac{62.5}{6.28} = \frac{6.28r}{6.28}$$
$$9.95 \approx r$$

The radius of the turntable is approximately 9.95 feet. The diameter is twice the radius, so the diameter is approximately 19.9 feet.

Now use the formula for the area of a circle.

$$A = \pi r^2$$
$$= (3.14)(9.95)^2 \quad \textit{Let } \pi = 3.14, r = 9.95.$$
$$= (3.14)(99.0025)$$
$$A \approx 311$$

The area of the turntable is approximately 311 square feet.

35. The sum of the three marked angles in the triangle is 180°.

$$45° + (x + 12.2)° + (3x + 2.8)° = 180°$$
$$4x + 60 = 180$$
$$4x = 120$$
$$x = 30$$

Since $x = 30$, $(x + 12.2)° = 42.2°$, and $(3x + 2.8)° = 92.8°$.

36. Knowing the values of h and b is not enough information to find the value of A. We would also need to know the value of B. Note that B and b are different variables. In general, to find the numerical value of one variable in a formula, we need to know the values of all the other variables.

37. The ratio of 60 centimeters to 40 centimeters is

$$\frac{60 \text{ cm}}{40 \text{ cm}} = \frac{3 \cdot 20}{2 \cdot 20} = \frac{3}{2}.$$

38. To find the ratio of 5 days to 2 weeks, first convert 2 weeks to days.

$$2 \text{ weeks} = 2 \cdot 7 = 14 \text{ days}$$

Thus, the ratio of 5 days to 2 weeks is $\frac{5}{14}$.

39. To find the ratio of 90 inches to 10 feet, first convert 10 feet to inches.

$$10 \text{ feet} = 10 \cdot 12 = 120 \text{ inches}$$

Thus, the ratio of 90 inches to 10 feet is

$$\frac{90}{120} = \frac{3 \cdot 30}{4 \cdot 30} = \frac{3}{4}.$$

40. To find the ratio of 3 months to 3 years, first convert 3 years to months.

$$3 \text{ years} = 3 \cdot 12 = 36 \text{ months}$$

Thus, the ratio of 3 months to 3 years is

$$\frac{3}{36} = \frac{1 \cdot 3}{12 \cdot 3} = \frac{1}{12}.$$

41.
$$\frac{p}{21} = \frac{5}{30}$$
$$30p = 105 \quad \textit{Cross products are equal.}$$
$$\frac{30p}{30} = \frac{105}{30} \quad \textit{Divide by 30.}$$
$$p = \frac{105}{30} = \frac{7 \cdot 15}{2 \cdot 15} = \frac{7}{2}$$

The solution set is $\left\{ \frac{7}{2} \right\}$.

42.
$$\frac{5 + x}{3} = \frac{2 - x}{6}$$
$$6(5 + x) = 3(2 - x) \quad \begin{array}{l}\textit{Cross products}\\ \textit{are equal.}\end{array}$$
$$30 + 6x = 6 - 3x \quad \begin{array}{l}\textit{Distributive}\\ \textit{property}\end{array}$$
$$30 + 9x = 6 \quad \textit{Add 3x.}$$
$$9x = -24 \quad \textit{Subtract 30.}$$
$$x = \frac{-24}{9} = -\frac{8}{3}$$

The solution set is $\left\{ -\frac{8}{3} \right\}$.

43. Let $x =$ the number of pounds of fertilizer needed to cover 500 square feet.

$$\frac{x \text{ pounds}}{2 \text{ pounds}} = \frac{500 \text{ square feet}}{150 \text{ square feet}}$$
$$150x = 2(500)$$
$$x = \frac{1000}{150} = \frac{20 \cdot 50}{3 \cdot 50}$$
$$= \frac{20}{3} = 6\frac{2}{3}$$

$6\frac{2}{3}$ pounds of fertilizer will cover 500 square feet.

44. Let $x =$ the tax on a \$36.00 item.
Set up a proportion with one ratio involving sales tax and the other involving the costs of the items.

$$\frac{x \text{ dollars}}{\$2.04} = \frac{\$36}{\$24}$$
$$24x = (2.04)(36) = 73.44$$
$$x = \frac{73.44}{24} = 3.06$$

The sales tax on a \$36.00 item is \$3.06.

45. Let x = the actual distance between the second pair of cities (in kilometers).

Set up a proportion with one ratio involving map distances and the other involving actual distances.

$$\frac{x \text{ kilometers}}{150 \text{ kilometers}} = \frac{80 \text{ centimeters}}{32 \text{ centimeters}}$$
$$32x = (150)(80) = 12,000$$
$$x = \frac{12,000}{32} = 375$$

The cities are 375 kilometers apart.

46. Let x = the number of bronze medals earned by China.

$$\frac{x \text{ bronze medals}}{63 \text{ medals}} = \frac{2 \text{ bronze medals}}{9 \text{ medals}}$$
$$9x = 2(63) = 126$$
$$x = 14$$

At the 2004 Olympics, 14 bronze medals were earned by China.

In Exercises 47–48, to find the best buy, divide the price by the number of units to get the unit cost. Each result was found by using a calculator and rounding the answer to three decimal places. The *best buy* (based on price per unit) is the smallest unit cost.

47.

Size	Unit Cost (dollars per oz)
15 oz	$\frac{\$2.69}{15} = \$0.179$
20 oz	$\frac{\$3.29}{20} = \$0.165$
25.5 oz	$\frac{\$3.49}{25.5} = \$0.137$ (∗)

The 25.5 oz size is the best buy.

48.

Size	Unit Cost (dollars per oz)
32 oz	$\frac{\$1.95}{32} = \$0.061$
48 oz	$\frac{\$2.89}{48} = \$0.060$
64 oz	$\frac{\$3.29}{64} = \$0.051$ (∗)

The 64 oz size is the best buy.

49. $160 million is what percent of $290 million?

$$\frac{160}{290} \approx 0.5517$$

Approximately 55.2% of the cost of the new stadium was borrowed.

50. Let x = the number of liters of the 60% solution to be used.
Then $x + 15$ = the number of liters of the 20% solution.

Liters of solution	15		x		$x + 15$
Strength of solution	0.10	+	0.60	=	0.20

$$
\begin{array}{ccccc}
\text{Drug amount} & & \text{Drug amount} & & \text{Drug amount} \\
\text{in 10\% solution} & \text{plus} & \text{in 60\% solution} & \text{is} & \text{in 20\% solution} \\
\downarrow & \downarrow & \downarrow & \downarrow & \downarrow \\
0.10(15) & + & 0.60(x) & = & 0.20(x + 15)
\end{array}
$$

Multiply by 10 to clear decimals.

$$1(15) + 6x = 2(x + 15)$$
$$15 + 6x = 2x + 30$$
$$15 + 4x = 30$$
$$4x = 15$$
$$x = \frac{15}{4} = 3.75$$

3.75 liters of 60% solution are needed.

51. Let x = the amount invested at 5%.
Then $10,000 - x$ = the amount invested at 6%.

$$
\begin{array}{ccccc}
\text{Interest} & & \text{Interest} & & \\
\text{at 5\%} & \text{plus} & \text{at 6\%} & & \text{equals \$550.} \\
\downarrow & \downarrow & \downarrow & \downarrow & \downarrow \\
0.05x & + & 0.06(10,000 - x) & = & 550
\end{array}
$$

$$5x + 6(10,000 - x) = 100(550)$$
$$5x + 60,000 - 6x = 55,000$$
$$-x = -5000$$
$$x = 5000$$

Todd invested $5000 at 5% and $10,000 - 5000 = \$5000$ at 6%.

52. Use the formula $d = rt$ or $r = \frac{d}{t}$.

$$r = \frac{d}{t} = \frac{3150}{384} \approx 8.203$$

Rounded to the nearest tenth, the *Yorkshire's* average speed was 8.2 mph.

53. Use the formula $d = rt$ or $t = \frac{d}{r}$.

$$t = \frac{d}{r} = \frac{819}{63} = 13$$

Honey drove for 13 hours.

54. Let t = the number of hours until the planes are 1925 miles apart.

Use $d = rt$.

$$
\begin{array}{ccccc}
\text{The distance} & & \text{the distance} & & \text{the distance} \\
\text{one plane} & \text{plus} & \text{the other plane} & \text{is} & \text{between} \\
\text{flies north} & & \text{flies south} & & \text{the planes.} \\
\downarrow & \downarrow & \downarrow & \downarrow & \downarrow \\
350t & + & 420t & = & 1925
\end{array}
$$

$$770t = 1925$$
$$t = \frac{1925}{770} = \frac{5}{2} = 2\frac{1}{2}$$

The planes will be 1925 miles apart in $2\frac{1}{2}$ hours.

55. The statement $x \geq -4$ can be written as $[-4, \infty)$.

56. The statement $x < 7$ can be written as $(-\infty, 7)$.

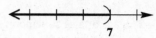

57. The statement $-5 \leq x < 6$ can be written as $[-5, 6)$.

58. By examining the choices, we see that $-4x \leq 36$ is the only inequality that has a negative coefficient of x. Thus, **B** is the only inequality that requires a reversal of the inequality symbol when it is solved.

59. $x + 6 \geq 3$
$\qquad x \geq -3 \quad$ *Subtract 6.*

Graph the solution set $[-3, \infty)$.

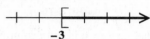

60. $5x < 4x + 2$
$\qquad x < 2 \qquad$ *Subtract 4x.*

Graph the solution set $(-\infty, 2)$.

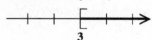

61. $-6x \leq -18$
$\dfrac{-6x}{-6} \geq \dfrac{-18}{-6} \quad$ *Divide by -6;*
$\qquad \qquad \qquad \quad$ *reverse the symbol.*
$\qquad x \geq 3$

Graph the solution set $[3, \infty)$.

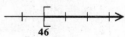

62. $8(x - 5) - (2 + 7x) \geq 4$
$\quad 8x - 40 - 2 - 7x \geq 4$
$\qquad \qquad \quad x - 42 \geq 4$
$\qquad \qquad \qquad \quad x \geq 46$

Graph the solution set $[46, \infty)$.

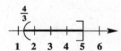

63. $4x - 3x > 10 - 4x + 7x$
$\qquad \quad x > 10 + 3x$
$\quad -2x > 10$
$\dfrac{-2x}{-2} < \dfrac{10}{-2} \quad$ *Divide by -2;*
$\qquad \qquad \qquad$ *reverse the symbol.*
$\qquad x < -5$

Graph the solution set $(-\infty, -5)$.

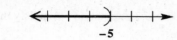

64. $3(2x + 5) + 4(8 + 3x) < 5(3x + 2) + 2x$
$\quad 6x + 15 + 32 + 12x < 15x + 10 + 2x$
$\qquad \qquad 18x + 47 < 17x + 10$
$\qquad \qquad \quad x + 47 < 10$
$\qquad \qquad \qquad \quad x < -37$

Graph the solution set $(-\infty, -37)$.

65. $-3 \leq 2x + 1 \leq 4$
$-4 \leq 2x \leq 3 \qquad$ *Subtract 1.*
$-2 \leq x \leq \frac{3}{2} \qquad$ *Divide by 2.*

Graph the solution set $\left[-2, \frac{3}{2}\right]$.

66. $9 < 3x + 5 \leq 20$
$4 < 3x \leq 15 \qquad$ *Subtract 5.*
$\frac{4}{3} < x \leq 5 \qquad$ *Divide by 3.*

Graph the solution set $\left(\frac{4}{3}, 5\right]$.

67. Let $x = $ the score on the third test.

The average of the three tests	is at least	90.
↓	↓	↓
$\dfrac{94 + 88 + x}{3}$	$\geq$	90

$\qquad \qquad \dfrac{182 + x}{3} \geq 90$

$\qquad \quad 3\left(\dfrac{182 + x}{3}\right) \geq 3(90)$

$\qquad \qquad \quad 182 + x \geq 270$

$\qquad \qquad \qquad \quad x \geq 88$

In order to average at least 90, Carlotta's score on her third test must be 88 or more.

68. Let n = the number.

"If nine times a number is added to 6, the result is at most 3" can be written as

$$9n + 6 \leq 3.$$

Solve the inequality.

$$9n \leq -3 \quad \textit{Subtract 6.}$$
$$n \leq \tfrac{-3}{9} \quad \textit{Divide by 9.}$$

All numbers less or equal to $-\tfrac{1}{3}$ satisfy the given condition.

69. **[2.6]** $\dfrac{x}{7} = \dfrac{x-5}{2}$

$$2x = 7(x - 5) \quad \textit{Cross products are equal.}$$
$$2x = 7x - 35$$
$$-5x = -35$$
$$x = 7$$

The solution set is $\{7\}$.

70. **[2.5]** $I = prt$ for r

$$\frac{I}{pt} = \frac{prt}{pt} \quad \textit{Divide by pt.}$$

$$\frac{I}{pt} = r \quad \text{or} \quad r = \frac{I}{pt}$$

71. **[2.8]** $-2x > -4$

$$\frac{-2x}{-2} < \frac{-4}{-2} \quad \begin{array}{l}\textit{Divide by –2;} \\ \textit{reverse the symbol.}\end{array}$$
$$x < 2$$

The solution set is $(-\infty, 2)$.

72. **[2.3]** $2k - 5 = 4k + 13$

$$-2k - 5 = 13 \quad \textit{Subtract 4k.}$$
$$-2k = 18 \quad \textit{Add 5.}$$
$$k = -9 \quad \textit{Divide by –2.}$$

The solution set is $\{-9\}$.

73. **[2.2]** $0.05x + 0.02x = 4.9$

To clear decimals, multiply both sides by 100.

$$100(0.05x + 0.02x) = 100(4.9)$$
$$5x + 2x = 490$$
$$7x = 490$$
$$x = 70$$

The solution set is $\{70\}$.

74. **[2.3]** $2 - 3(x - 5) = 4 + x$

$$2 - 3x + 15 = 4 + x$$
$$17 - 3x = 4 + x$$
$$17 - 4x = 4$$
$$-4x = -13$$
$$x = \frac{-13}{-4} = \frac{13}{4}$$

The solution set is $\left\{ \frac{13}{4} \right\}$.

75. **[2.3]** $9x - (7x + 2) = 3x + (2 - x)$

$$9x - 7x - 2 = 3x + 2 - x$$
$$2x - 2 = 2x + 2$$
$$-2 = 2$$

Because $-2 = 2$ is a false statement, the given equation has no solution, symbolized by $\emptyset$.

76. **[2.3]** $\tfrac{1}{3}s + \tfrac{1}{2}s + 7 = \tfrac{5}{6}s + 5 + 2$

$$\tfrac{1}{3}s + \tfrac{1}{2}s = \tfrac{5}{6}s \quad \textit{Subtract 7.}$$

The least common denominator is 6.

$$6\left(\tfrac{1}{3}s + \tfrac{1}{2}s\right) = 6\left(\tfrac{5}{6}s\right)$$
$$2s + 3s = 5s$$
$$5s = 5s$$

Because $5s = 5s$ is a true statement, the solution set is {all real numbers}.

77. **[2.3]** Let $x = 6$ in the equation.

$$3 - (8 + 4x) = 2x + 7$$
$$3 - [8 + 4(6)] = 2(6) + 7$$
$$-29 = 19$$

This is false, so $x = 6$ is not a solution of the equation.

Solve the equation.

$$3 - (8 + 4x) = 2x + 7$$
$$3 - 8 - 4x = 2x + 7$$
$$-5 - 4x = 2x + 7$$
$$-6x = 12$$
$$x = -2$$

The solution set is $\{-2\}$. The student probably got the incorrect answer by writing

$$3 - (8 + 4x) = 3 - 8 + 4x$$

and then solving the equation, which *does* have solution set $\{6\}$.

78. **[2.6]** Let x = the number of calories a 175-pound athlete can consume.

Set up a proportion with one ratio involving calories and the other involving pounds.

$$\frac{x \text{ calories}}{50 \text{ calories}} = \frac{175 \text{ pounds}}{2.2 \text{ pounds}}$$
$$2.2x = 50(175)$$
$$x = \frac{8750}{2.2} \approx 3977.3$$

To the nearest hundred calories, a 175-pound athlete in a vigorous training program can consume 4000 calories per day.

79. **[2.4]** Let $x =$ the length of the Brooklyn Bridge. Then $x + 2604 =$ the length of the Golden Gate Bridge.

$$x + (x + 2604) = 5796$$
$$2x + 2604 = 5796$$
$$2x = 3192$$
$$x = 1596$$

Since $x = 1596$, $x + 2604 = 4200$.

The length of the Brooklyn Bridge is 1596 feet and that of the Golden Gate Bridge is 4200 feet.

80. **[2.6]** The unit costs are rounded to four decimal places.

Size	Unit Cost (dollars per oz)
32 oz	$\frac{\$1.19}{32} = \$0.0372$
48 oz	$\frac{\$1.79}{48} = \$0.0373$
64 oz	$\frac{\$1.99}{64} = \$0.0311$ $(*)$

The 64 ounce size is the best buy.

81. **[2.6]** Let $x =$ the number of quarts of oil needed for 192 quarts of gasoline.
Set up a proportion with one ratio involving oil and the other involving gasoline.

$$\frac{x \text{ quarts}}{1 \text{ quart}} = \frac{192 \text{ quarts}}{24 \text{ quarts}}$$
$$x \cdot 24 = 1 \cdot 192 \qquad \textit{Cross products}$$
$$x = 8 \qquad \textit{Divide by 24.}$$

The amount of oil needed is 8 quarts.

82. **[2.7]** Let $x =$ the speed of the slower train. Then $x + 30 =$ the speed of the faster train.

	r	t	d
Slower train	x	3	$3x$
Faster train	$x + 30$	3	$3(x + 30)$

The sum of the distances traveled by the two trains is 390 miles, so

$$3x + 3(x + 30) = 390.$$
$$3x + 3x + 90 = 390$$
$$6x + 90 = 390$$
$$6x = 300$$
$$x = 50$$

Since $x = 50$, $x + 30 = 80$.

The speed of the slower train is 50 miles per hour and the speed of the faster train is 80 miles per hour.

83. **[2.5]** Let $x =$ the length of the first side. Then $2x =$ the length of the second side.

Use the formula for the perimeter of a triangle, $P = a + b + c$, with perimeter 96 and third side 30.

$$x + 2x + 30 = 96$$
$$3x + 30 = 96$$
$$3x = 66$$
$$x = 22$$

The sides have lengths 22 meters, 44 meters, and 30 meters. The length of the longest side is 44 meters.

84. **[2.8]** Let $s =$ the length of a side of the square. The formula for the perimeter of a square is $P = 4s$.

The perimeter   cannot be greater than   200.
$$\downarrow \qquad\qquad\qquad \downarrow \qquad\qquad\qquad \downarrow$$
$$4s \qquad\qquad\qquad \leq \qquad\qquad\qquad 200$$

$$4s \leq 200$$
$$s \leq 50$$

The length of a side is 50 meters or less.

Chapter 2 Test

1.
$$5x + 9 = 7x + 21$$
$$-2x + 9 = 21 \qquad \textit{Subtract 7x}$$
$$-2x = 12 \qquad \textit{Subtract 9}$$
$$x = -6 \qquad \textit{Divide by –2}$$

The solution set is $\{-6\}$.

2.
$$-\tfrac{4}{7}x = -12$$
$$\left(-\tfrac{7}{4}\right)\left(-\tfrac{4}{7}x\right) = \left(-\tfrac{7}{4}\right)(-12)$$
$$x = 21$$

The solution set is $\{21\}$.

3.
$$7 - (x - 4) = -3x + 2(x + 1)$$
$$7 - x + 4 = -3x + 2x + 2$$
$$-x + 11 = -x + 2$$

Because the last statement is false, the equation has no solution set, symbolized by $\emptyset$.

4. $0.6(x + 20) + 0.8(x - 10) = 46$
To clear decimals, multiply both sides by 10.

$$10[0.6(x + 20) + 0.8(x - 10)] = 10(46)$$
$$6(x + 20) + 8(x - 10) = 460$$
$$6x + 120 + 8x - 80 = 460$$
$$14x + 40 = 460$$
$$14x = 420$$
$$x = 30$$

The solution set is $\{30\}$.

5.
$$-8(2x + 4) = -4(4x + 8)$$
$$-16x - 32 = -16x - 32$$

Because the last statement is true, the solution set is $\{$all real numbers$\}$.

6. Let x = the number of games the Cardinals lost. Then $2x - 24$ = the number of games the Cardinals won.

The total number of games played was 162.

$$x + (2x - 24) = 162$$
$$3x - 24 = 162$$
$$3x = 186$$
$$x = 62$$

Since $x = 62$, $2x - 24 = 100$.
The Cardinals won 100 games and lost 62 games.

7. Let x = the area of Kauai (in square miles). Then $x + 177$ = the area of Maui (in square miles), and $(x + 177) + 3293 = x + 3470$ = the area of Hawaii.

$$x + (x + 177) + (x + 3470) = 5300$$
$$3x + 3647 = 5300$$
$$3x = 1653$$
$$x = 551$$

Since $x = 551$, $x + 177 = 728$, and $x + 3470 = 4021$.

The area of Hawaii is 4021 square miles, the area of Maui is 728 square miles, and the area of Kauai is 551 square miles.

8. Let x = the measure of the angle. Then $90 - x$ = the measure of its complement, and $180 - x$ = the measure of its supplement.

$$180 - x = 3(90 - x) + 10$$
$$180 - x = 270 - 3x + 10$$
$$180 - x = 280 - 3x$$
$$180 + 2x = 280$$
$$2x = 100$$
$$x = 50$$

The measure of the angle is 50°. The measure of its supplement, 130°, is 10° more than three times its complement, 40°.

9. **(a)** Solve $P = 2L + 2W$ for W.

$$P - 2L = 2W$$
$$\frac{P - 2L}{2} = W \quad \text{or} \quad W = \frac{P - 2L}{2}$$

(b) Substitute 116 for P and 40 for L in the formula obtained in part (a).

$$W = \frac{P - 2L}{2}$$
$$= \frac{116 - 2(40)}{2}$$
$$= \frac{116 - 80}{2} = \frac{36}{2} = 18$$

10. The angles are vertical angles, so their measures are equal.

$$3x + 15 = 4x - 5$$
$$15 = x - 5$$
$$20 = x$$

Since $x = 20$, $3x + 15 = 75$ and $4x - 5 = 75$.

Both angles have measure 75°.

11.
$$\frac{z}{8} = \frac{12}{16}$$
$$16z = 8(12) \quad \textit{Cross products are equal}$$
$$16z = 96$$
$$\frac{16z}{16} = \frac{96}{16} \quad \textit{Divide by 16}$$
$$z = 6$$

The solution set is $\{6\}$.

12.
$$\frac{x + 5}{3} = \frac{x - 3}{4}$$
$$4(x + 5) = 3(x - 3)$$
$$4x + 20 = 3x - 9$$
$$x + 20 = -9$$
$$x = -29$$

The solution set is $\{-29\}$.

13.

Size	Unit Cost (dollars per slice)
8 slices	$\frac{\$2.19}{8} = \$0.27375$ $(*)$
12 slices	$\frac{\$3.30}{12} = \$0.275$

The better buy is 8 slices for $2.19.

14. Let x = the actual distance between Seattle and Cincinnati.

$$\frac{x \text{ miles}}{1050 \text{ miles}} = \frac{92 \text{ inches}}{42 \text{ inches}}$$
$$42x = 92(1050) = 96{,}600$$
$$x = \frac{96{,}600}{42} = 2300$$

The actual distance between Seattle and Cincinnati is 2300 miles.

15. Let x = the amount invested at 3%. Then $x + 6000$ = the amount invested at 4.5%.

Amount invested (in dollars)	Rate of interest	Interest for one year
x	0.03	$0.03x$
$x + 6000$	0.045	$0.045(x + 6000)$

$$0.03x + 0.045(x + 6000) = 870$$
$$1000(0.03x) + 1000[0.045(x + 6000)] = 1000(870)$$
$$30x + 45(x + 6000) = 870,000$$
$$30x + 45x + 270,000 = 870,000$$
$$75x + 270,000 = 870,000$$
$$75x = 600,000$$
$$x = 8000$$

Since $x = 8000$, $x + 6000 = 14,000$.

Keith invested \$8000 at 3% and \$14,000 at 4.5%.

16. Use the formula $d = rt$ and let t be the number of hours they traveled.

	r	t	d
First car	50	t	$50t$
Second car	65	t	$65t$

First car's   and   second car's
distance   distance
   ↓   ↓   ↓
  $50t$ $+$ $65t$

is   total
       distance.
↓   ↓
$=$ 460

$$50t + 65t = 460$$
$$115t = 460$$
$$t = 4$$

The two cars will be 460 miles apart in 4 hours.

17. $-4x + 2(x - 3) \geq 4x - (3 + 5x) - 7$
$$-4x + 2x - 6 \geq 4x - 3 - 5x - 7$$
$$-2x - 6 \geq -x - 10$$
$$-x - 6 \geq -10$$
$$-x \geq -4$$
$$\frac{-1x}{-1} \leq \frac{-4}{-1} \quad \textit{Divide by –1;}$$
$$\qquad\qquad \textit{reverse the symbol}$$
$$x \leq 4$$

Graph the solution set $(-\infty, 4]$.

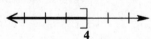

18. $-10 < 3x - 4 \leq 14$
$$-6 < 3x \leq 18 \qquad \textit{Add 4.}$$
$$-2 < x \leq 6 \qquad \textit{Divide by 3.}$$

Graph the solution set $(-2, 6]$.

19.   Let $x =$ the score on the third test.

The average of the three tests	is at least	80.
↓	↓	↓
$\dfrac{76 + 81 + x}{3}$	$\geq$	80

$$\frac{157 + x}{3} \geq 80$$
$$3\left(\frac{157 + x}{3}\right) \geq 3(80)$$
$$157 + x \geq 240$$
$$x \geq 83$$

In order to average at least 80, Twylene's score on her third test must be 83 or more.

20.   When an inequality is multiplied or divided by a negative number, the direction of the inequality symbol must be reversed.

Cumulative Review Exercises (Chapters 1–2)

1. $\dfrac{108}{144} = \dfrac{3 \cdot 36}{4 \cdot 36} = \dfrac{3}{4}$

2. $\dfrac{5}{6} + \dfrac{1}{4} - \dfrac{7}{15} = \dfrac{50}{60} + \dfrac{15}{60} - \dfrac{28}{60}$
$$= \frac{65 - 28}{60}$$
$$= \frac{37}{60}$$

3. $\dfrac{9}{8} \cdot \dfrac{16}{3} \div \dfrac{5}{8} = \dfrac{9}{8} \cdot \dfrac{16}{3} \cdot \dfrac{8}{5}$
$$= \frac{3 \cdot 3 \cdot 16 \cdot 8}{8 \cdot 3 \cdot 5}$$
$$= \frac{48}{5}$$

4.   "The difference between half a number and 18" is written
$$\tfrac{1}{2}x - 18.$$

5.   "The quotient of 6 and 12 more than a number is 2" is written
$$\frac{6}{x + 12} = 2.$$

6. $\dfrac{8(7) - 5(6 + 2)}{3 \cdot 5 + 1} \geq 1$
$$\frac{8(7) - 5(8)}{3 \cdot 5 + 1} \geq 1$$
$$\frac{56 - 40}{15 + 1} \geq 1$$
$$\tfrac{16}{16} \geq 1$$
$$1 \geq 1$$

The statement is *true*.

7. $9 - (-4) + (-2) = (9 + 4) + (-2)$
$$= 13 - 2$$
$$= 11$$

8. $\dfrac{-4(9)(-2)}{-3^2} = \dfrac{-36(-2)}{-1 \cdot 3^2}$
$$= \dfrac{72}{-9}$$
$$= -8$$

9. $(-7 - 1)(-4) + (-4) = (-8)(-4) + (-4)$
$$= 32 + (-4)$$
$$= 28$$

10. $\dfrac{3x^2 - y^3}{-4z} = \dfrac{3(-2)^2 - (-4)^3}{-4(3)}$ *Let $x = -2$,*
                  $y = -4, z = 3$.
$$= \dfrac{3(4) - (-64)}{-12}$$
$$= \dfrac{12 + 64}{-12}$$
$$= \dfrac{76}{-12}$$
$$= -\dfrac{19}{3}$$

11. $7(k + m) = 7k + 7m$

The multiplication of 7 is distributed over the sum, which illustrates the distributive property.

12. $3 + (5 + 2) = 3 + (2 + 5)$

The order of the numbers added in the parentheses is changed, which illustrates the commutative property.

13. $-4(k + 2) + 3(2k - 1)$
$$= (-4)(k) + (-4)(2) + (3)(2k) + (3)(-1)$$
$$= -4k - 8 + 6k - 3$$
$$= -4k + 6k - 8 - 3$$
$$= 2k - 11$$

14. $2r - 6 = 8r$
$$-6 = 6r$$
$$-1 = r$$

Check $r = -1$: $-8 = -8$ *True*

The solution set is $\{-1\}$.

15. $4 - 5(a + 2) = 3(a + 1) - 1$
$$4 - 5a - 10 = 3a + 3 - 1$$
$$-5a - 6 = 3a + 2$$
$$-8a - 6 = 2$$
$$-8a = 8$$
$$a = -1$$

Check $a = -1$: $-1 = -1$ *True*

The solution set is $\{-1\}$.

16. $\dfrac{2}{3}x + \dfrac{3}{4}x = -17$
$$12\left(\dfrac{2}{3}x + \dfrac{3}{4}x\right) = 12(-17) \quad LCD = 12$$
$$8x + 9x = -204$$
$$17x = -204$$
$$x = -12$$

Check $x = -12$: $-17 = -17$ *True*

The solution set is $\{-12\}$.

17. $\dfrac{2x + 3}{5} = \dfrac{x - 4}{2}$
$$(2x + 3)(2) = (5)(x - 4)$$
$$4x + 6 = 5x - 20$$
$$6 = x - 20$$
$$26 = x$$

Check $x = 26$: $11 = 11$ *True*

The solution set is $\{26\}$.

18. $3x + 4y = 24$ for y
$$4y = 24 - 3x$$
$$y = \dfrac{24 - 3x}{4}$$

19. $A = P(1 + ni)$ for n
$$A = P + Pni$$
$$A - P = Pni$$
$$\dfrac{A - P}{Pi} = n$$

20. $6(r - 1) + 2(3r - 5) \le -4$
$$6r - 6 + 6r - 10 \le -4$$
$$12r - 16 \le -4$$
$$12r \le 12$$
$$r \le 1$$

Graph the solution set $(-\infty, 1]$.

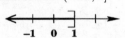

21. $-18 \le -9z < 9$
$$2 \ge z > -1 \qquad \begin{array}{l}\textit{Divide by } -9; \\ \textit{reverse the symbols.}\end{array}$$

or $-1 < z \le 2$

Graph the solution set $(-1, 2]$.

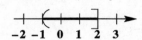

22. Let x = the length of the middle-sized piece.
Then $3x$ = the length of the longest piece, and
$x - 5$ = the length of the shortest piece.

$$x + 3x + (x - 5) = 40$$
$$5x - 5 = 40$$
$$5x = 45$$
$$x = 9$$

The length of the middle-sized piece is 9
centimeters, of the longest piece is 27 centimeters,
and of the shortest piece is 4 centimeters.

23. Let r = the radius and use 3.14 for π.
Using the formula for circumference, $C = 2\pi r$,
and $C = 78$, we have

$$2\pi r = 78.$$
$$r = \frac{78}{2\pi} \approx 12.4204$$

To the nearest hundredth, the radius is 12.42 cm.

24. $$\frac{x \text{ cups}}{1\frac{1}{4} \text{ cups}} = \frac{20 \text{ people}}{6 \text{ people}}$$
$$6x = \left(1\frac{1}{4}\right)(20)$$
$$6x = 25$$
$$x = \frac{25}{6}, \quad \text{or} \quad 4\frac{1}{6} \text{ cups}$$

$4\frac{1}{6}$ cups of cheese will be needed to serve 20
people.

25. Let x = speed of slower car.
Then $x + 20$ = speed of faster car.

Use the formula $d = rt$.

$$d_{\text{slower}} + d_{\text{faster}} = d_{\text{total}}$$
$$(x)(4) + (x + 20)(4) = 400$$
$$4x + 4x + 80 = 400$$
$$8x + 80 = 400$$
$$8x = 320$$
$$x = 40$$

The speeds are 40 mph and 60 mph.

CHAPTER 3 LINEAR EQUATIONS IN TWO VARIABLES

3.1 Reading Graphs; Linear Equations in Two Variables

1. Locate the two tallest bars. Follow the bar down to see which state it is and move across from the top of the bar to the vertical scale to estimate their production. Ohio (OH) and Iowa (IA) are the top two egg-producing states. Ohio produced about 8 billion eggs and Iowa produced about 10 billion eggs.

3. Locate the shortest bar. It's above North Carolina (NC). Follow the top of the bar across to the vertical scale to estimate North Carolina's production, which is about 2.5 billion eggs.

5. The line segments between 1998 and 1999, 1999 and 2000, and 2003 and 2004 fall, so these are the pairs of consecutive years in which the unemployment rate decreased.

7. For 2003, the unemployment rate was about 6%. For 2004, the unemployment rate was about 5.5%. The decline is $6\% - 5.5\% = 0.5\%$.

9. The symbol (x, y) *does* represent an ordered pair, while the symbols $[x, y]$ and $\{x, y\}$ *do not* represent ordered pairs. (Note that only parentheses are used to write ordered pairs.)

11. All points having coordinates in the form

(negative, positive)

are in quadrant II, so the point whose graph has coordinates $(-4, 2)$ is in quadrant II.

13. All ordered pairs that are solutions of the equation $y = 3$ have y-coordinates equal to 3, so the ordered pair $(4, 3)$ is a solution of the equation $y = 3$.

15. A linear equation in one variable can be written in the form $Ax + B = C$, where $A \neq 0$. Examples are $2x + 5 = 0$, $3x + 6 = 2$, and $x = -5$. A linear equation in two variables can be written in the form $Ax + By = C$, where A and B cannot both equal 0. Examples are $2x + 3y = 8$, $3x = 5y$, and $x - y = 0$.

17. $x + y = 8; \; (0, 8)$

To determine whether $(0, 8)$ is a solution of the given equation, substitute 0 for x and 8 for y.

$$x + y = 8$$
$$0 + 8 = 8 \quad ? \quad \textit{Let x = 0, y = 8.}$$
$$8 = 8 \quad \quad \textit{True}$$

The result is true, so $(0, 8)$ is a solution of the given equation $x + y = 8$.

19. $2x + y = 5; \; (3, -1)$

Substitute 3 for x and -1 for y.

$$2x + y = 5$$
$$2(3) + (-1) = 5 \quad ? \quad \textit{Let x = 3, y = -1.}$$
$$6 - 1 = 5 \quad ?$$
$$5 = 5 \quad \quad \textit{True}$$

The result is true, so $(3, -1)$ is a solution of $2x + y = 5$.

21. $5x - 3y = 15; \; (5, 2)$

Substitute 5 for x and 2 for y.

$$5x - 3y = 15$$
$$5(5) - 3(2) = 15 \quad ? \quad \textit{Let x = 5, y = 2.}$$
$$25 - 6 = 15 \quad ?$$
$$19 = 15 \quad \quad \textit{False}$$

The result is false, so $(5, 2)$ is not a solution of $5x - 3y = 15$.

23. $x = -4y; \; (-8, 2)$

Substitute -8 for x and 2 for y.

$$x = -4y$$
$$-8 = -4(2) \quad ? \quad \textit{Let x = -8, y = 2.}$$
$$-8 = -8 \quad \quad \textit{True}$$

The result is true, so $(-8, 2)$ is a solution of $x = -4y$.

25. $y = 2; \; (4, 2)$

Since x does not appear in the equation, we just substitute 2 for y.

$$y = 2$$
$$2 = 2 \quad \textit{Let y = 2; true}$$

The result is true, so $(4, 2)$ is a solution of $y = 2$.

27. $x - 6 = 0; \; (4, 2)$

Since y does not appear in the equation, we just substitute 4 for x.

$$x - 6 = 0$$
$$4 - 6 = 0 \quad ? \quad \textit{Let x = 4.}$$
$$-2 = 0 \quad \quad \textit{False}$$

The result is false, so $(4, 2)$ is not a solution of $x - 6 = 0$.

29. No, the ordered pair $(3, 4)$ represents the point 3 units to the right of the origin and 4 units up from the x-axis. The ordered pair $(4, 3)$ represents the point 4 units to the right of the origin and 3 units up from the x-axis.

31. $y = 2x + 7;\ (5,\ \ )$

In this ordered pair, $x = 5$. Find the corresponding value of y by replacing x with 5 in the given equation.

$$y = 2x + 7$$
$$y = 2(5) + 7 \quad \textit{Let } x = 5.$$
$$y = 10 + 7$$
$$y = 17$$

The ordered pair is $(5, 17)$.

33. $y = 2x + 7;\ (\ \ , -3)$

In this ordered pair, $y = -3$. Find the corresponding value of x by replacing y with -3 in the given equation.

$$y = 2x + 7$$
$$-3 = 2x + 7 \quad \textit{Let } y = -3.$$
$$-10 = 2x$$
$$-5 = x$$

The ordered pair is $(-5, -3)$.

35. $y = -4x - 4;\ (\ \ , 0)$

$$y = -4x - 4$$
$$0 = -4x - 4 \quad \textit{Let } y = 0.$$
$$4 = -4x$$
$$-1 = x$$

The ordered pair is $(-1, 0)$.

37. $y = -4x - 4;\ (\ \ , 24)$

$$y = -4x - 4$$
$$24 = -4x - 4 \quad \textit{Let } y = 24.$$
$$28 = -4x$$
$$-7 = x$$

The ordered pair is $(-7, 24)$.

39. Substitute 0 for x in the equation $y = mx + b$.

$$y = mx + b$$
$$y = m \cdot 0 + b$$
$$y = b$$

For the equation $y = mx + b$, the y-value corresponding to $x = 0$ for *any* value of m is b.

41. $4x + 3y = 24$

If $x = 0$,

$$4(0) + 3y = 24$$
$$0 + 3y = 24$$
$$y = 8.$$

If $y = 0$,

$$4x + 3(0) = 24$$
$$4x + 0 = 24$$
$$x = 6.$$

If $y = 4$,

$$4x + 3(4) = 24$$
$$4x + 12 = 24$$
$$4x = 12$$
$$x = 3.$$

The completed table of values is shown below.

x	y	ordered pair
0	8	$(0, 8)$
6	0	$(6, 0)$
3	4	$(3, 4)$

43. $4x - 9y = -36$

If $y = 0$,

$$4x - 9(0) = -36$$
$$4x - 0 = -36$$
$$4x = -36$$
$$x = -9.$$

If $x = 0$,

$$4(0) - 9y = -36$$
$$0 - 9y = -36$$
$$-9y = -36$$
$$y = 4.$$

If $y = 8$,

$$4x - 9(8) = -36$$
$$4x - 72 = -36$$
$$4x = 36$$
$$x = 9.$$

The completed table of values is shown below.

x	y	ordered pair
-9	0	$(-9, 0)$
0	4	$(0, 4)$
9	8	$(9, 8)$

45. $x = 12$

No matter which value of y is chosen, the value of x will always be 12. Each ordered pair can be completed by placing 12 in the first position.

x	y	ordered pair
12	3	$(12, 3)$
12	8	$(12, 8)$
12	0	$(12, 0)$

47. $y = -10$

No matter which value of x is chosen, the value of y will always be -10. Each ordered pair can be completed by placing -10 in the second position.

x	y	ordered pair
4	-10	$(4, -10)$
0	-10	$(0, -10)$
-4	-10	$(-4, -10)$

49. The given equation, $y + 2 = 0$, may be written $y = -2$. For any value of x, the value of y will always be -2.

x	y	ordered pair
9	-2	$(9, -2)$
2	-2	$(2, -2)$
0	-2	$(0, -2)$

For Exercises 51–60, the ordered pairs are plotted on the graph following the solution for Exercise 60.

51. To plot $(6, 2)$, start at the origin, go 6 units to the right, and then go up 2 units.

53. To plot $(-4, 2)$, start at the origin, go 4 units to the left, and then go up 2 units.

55. To plot $\left(-\frac{4}{5}, -1\right)$, start at the origin, go $\frac{4}{5}$ unit to the left, and then go down 1 unit.

57. To plot $(0, 4)$, start at the origin and go up 4 units. The point lies on the y-axis.

59. To plot $(4, 0)$, start at the origin and go right 4 units. The point lies on the x-axis.

61. The point with coordinates (x, y) is in quadrant III if x is *negative* and y is *negative*.

63. The point with coordinates (x, y) is in quadrant IV if x is *positive* and y is *negative*.

65. $x - 2y = 6$

x	y
0	
	0
2	
	-1

Substitute the given values to complete the ordered pairs.

$$x - 2y = 6$$
$$0 - 2y = 6 \qquad \textit{Let x = 0.}$$
$$-2y = 6$$
$$y = -3$$

$$x - 2y = 6$$
$$x - 2(0) = 6 \qquad \textit{Let y = 0.}$$
$$x - 0 = 6$$
$$x = 6$$

$$x - 2y = 6$$
$$2 - 2y = 6 \qquad \textit{Let x = 2.}$$
$$-2y = 4$$
$$y = -2$$

$$x - 2y = 6$$
$$x - 2(-1) = 6 \qquad \textit{Let y = -1.}$$
$$x + 2 = 6$$
$$x = 4$$

The completed table of values follows.

x	y
0	-3
6	0
2	-2
4	-1

Plot the points $(0, -3)$, $(6, 0)$, $(2, -2)$, and $(4, -1)$ on a coordinate system.

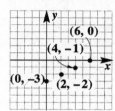

67. $3x - 4y = 12$

x	y
0	
	0
-4	
	-4

Substitute the given values to complete the ordered pairs.

$$3x - 4y = 12$$
$$3(0) - 4y = 12 \qquad \textit{Let x = 0.}$$
$$0 - 4y = 12$$
$$-4y = 12$$
$$y = -3$$

$$3x - 4y = 12$$
$$3x - 4(0) = 12 \qquad \textit{Let y = 0.}$$
$$3x - 0 = 12$$
$$3x = 12$$
$$x = 4$$

$$3x - 4y = 12$$
$$3(-4) - 4y = 12 \qquad \textit{Let x = -4.}$$
$$-12 - 4y = 12$$
$$-4y = 24$$
$$y = -6$$

$$3x - 4y = 12$$
$$3x - 4(-4) = 12 \qquad \textit{Let y = -4.}$$
$$3x + 16 = 12$$
$$3x = -4$$
$$x = -\frac{4}{3}$$

The completed table is as follows.

x	y
0	-3
4	0
-4	-6
$-\frac{4}{3}$	-4

continued

Plot the points $(0, -3)$, $(4, 0)$, $(-4, -6)$, and $\left(-\frac{4}{3}, -4\right)$ on a coordinate system.

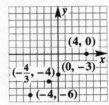

69. The given equation, $y + 4 = 0$, can be written as $y = -4$. So regardless of the value of x, the value of y is -4.

x	y
0	-4
5	-4
-2	-4
-3	-4

Plot the points $(0, -4)$, $(5, -4)$, $(-2, -4)$, and $(-3, -4)$ on a coordinate system.

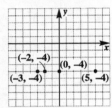

71. The points in each graph appear to lie on a straight line.

73. **(a)** When $x = 5$, $y = 45$. The ordered pair is $(5, 45)$.

(b) When $y = 50$, $x = 6$. The ordered pair is $(6, 50)$.

75. **(a)** We can write the results from the table as ordered pairs (x, y).

$(1996, 53.3)$, $(1997, 52.8)$, $(1998, 52.1)$, $(1999, 51.6)$, $(2000, 51.2)$, $(2001, 50.9)$

(b) x represents the year and y represents the graduation rate.

$(2002, 51.0)$ means that in 2002, the graduation rate for 4-year college students within 5 years was 51.0%.

(c)

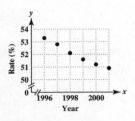

(d) The points appear to be approximated by a straight line. Graduation rates for 4-year college students within 5 years are decreasing.

77. **(a)** Substitute $x = 20, 40, 60, 80$ in the equation $y = -0.85x + 187$.

$y = -0.85(20) + 187 = -17 + 187 = 170$
$y = -0.85(40) + 187 = -34 + 187 = 153$
$y = -0.85(60) + 187 = -51 + 187 = 136$
$y = -0.85(80) + 187 = -68 + 187 = 119$

The completed table follows.

Age	Heartbeats (per minute)
20	170
40	153
60	136
80	119

(b) In the ordered pairs (x, y), x represents age and y represents heartbeats. $(20, 170)$, $(40, 153)$, $(60, 136)$, $(80, 119)$

(c)

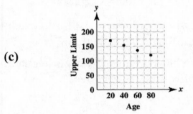

Yes, the points lie in a linear pattern.

79. $3x + 6 = 0$
$\qquad 3x = -6$
$\qquad\ x = \frac{-6}{3} = -2$

Check $x = -2$: $-6 + 6 = 0$

The solution set is $\{-2\}$.

81. $9 - m = -4$
$\qquad -m = -13$
$\qquad\ \ m = 13$

Check $m = 13$: $9 - 13 = -4$

The solution set is $\{13\}$.

83. $\qquad x = -y + 6$
$\ x + y = 6$ \qquad *Add y.*
$\qquad\ \ y = -x + 6$ \quad *Subtract x.*

85. $2x + 3y = 12$
$\qquad\ \ 3y = 12 - 2x$ \quad *Subtract 2x.*
$\qquad\ \ \ y = \dfrac{12 - 2x}{3}$ \quad *Divide by 3.*

3.2 Graphing Linear Equations in Two Variables

1. $x + y = 5$

$(0, \quad), (\quad, 0), (2, \quad)$

If $x = 0,$ If $y = 0,$
$0 + y = 5$ $x + 0 = 5$
$y = 5.$ $x = 5.$

If $x = 2,$
$2 + y = 5$
$y = 3.$

The ordered pairs are $(0, 5)$, $(5, 0)$, and $(2, 3)$. Plot the corresponding points and draw a line through them.

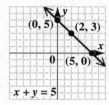

3. $y = \frac{2}{3}x + 1$

$(0, \quad), (3, \quad), (-3, \quad)$

If $x = 0,$ If $x = 3,$
$y = \frac{2}{3}(0) + 1$ $y = \frac{2}{3}(3) + 1$
$y = 0 + 1$ $y = 2 + 1$
$y = 1.$ $y = 3.$

If $x = -3,$
$y = \frac{2}{3}(-3) + 1$
$y = -2 + 1$
$y = -1.$

The ordered pairs are $(0, 1)$, $(3, 3)$, and $(-3, -1)$. Plot the corresponding points and draw a line through them.

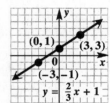

5. $3x = -y - 6$

$(0, \quad), (\quad, 0), \left(-\frac{1}{3}, \quad\right)$

If $x = 0,$ If $y = 0,$
$3(0) = -y - 6$ $3x = -0 - 6$
$0 = -y - 6$ $3x = -6$
$y = -6.$ $x = -2.$

If $x = -\frac{1}{3},$
$3\left(-\frac{1}{3}\right) = -y - 6$
$-1 = -y - 6$
$y - 1 = -6$
$y = -5.$

The ordered pairs are $(0, -6)$, $(-2, 0)$, and $\left(-\frac{1}{3}, -5\right)$. Plot the corresponding points and draw a line through them.

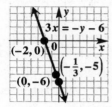

7. To determine which equation has x-intercept $(4, 0)$, set y equal to 0 and see which equation is equivalent to $x = 4$. Choice **C** is correct since

$$2x - 5y = 8$$
$$2x - 5(0) = 8$$
$$2x = 8$$
$$x = 4$$

9. If the graph of the equation passes through the origin, then substituting 0 for x and 0 for y will result in a true statement. Choice **D** is correct since

$$x + 4y = 0$$
$$(0) + 4(0) = 0$$
$$0 = 0$$

is a true statement.

11. Choice **B** is correct since the graph of $y = -3$ is a horizontal line.

13. $3x = y - 9$

If $x = 0,$ If $y = 0,$
$3(0) = y - 9$ $3x = 0 - 9$
$0 = y - 9$ $3x = -9$
$9 = y.$ $x = -3.$

The graph of this equation is a line with x-intercept $(-3, 0)$ and y-intercept $(0, 9)$.

15. $3y = -6$
$y = -2$ *Divide by 3.*

The graph of this equation is a horizontal line with y-intercept $(0, -2)$.

17. To find the x-intercept, let $y = 0$.

$$2x - 3y = 24$$
$$2x - 3(0) = 24$$
$$2x - 0 = 24$$
$$2x = 24$$
$$x = 12$$

The x-intercept is $(12, 0)$.

To find the y-intercept, let $x = 0$.

$$2x - 3y = 24$$
$$2(0) - 3y = 24$$
$$0 - 3y = 24$$
$$-3y = 24$$
$$y = -8$$

The y-intercept is $(0, -8)$.

19. To find the x-intercept, let $y = 0$.

$$x + 6y = 0$$
$$x + 6(0) = 0$$
$$x + 0 = 0$$
$$x = 0$$

The x-intercept is $(0, 0)$. Since we have found the point with x equal to 0, this is also the y-intercept.

21. To find the x-intercept, let $y = 0$.

$$5x - 2y = 20$$
$$5x - 2(0) = 20$$
$$5x = 20$$
$$x = 4$$

The x-intercept is $(4, 0)$.

To find the y-intercept, let $x = 0$.

$$5x - 2y = 20$$
$$5(0) - 2y = 20$$
$$-2y = 20$$
$$y = -10$$

The y-intercept is $(0, -10)$.

23. To find the x-intercept, let $y = 0$.

$$y = -2x + 4$$
$$0 = -2x + 4$$
$$2x = 4$$
$$x = 2$$

The x-intercept is $(2, 0)$.

To find the y-intercept, let $x = 0$.

$$y = -2x + 4$$
$$y = -2(0) + 4$$
$$y = 0 + 4$$
$$y = 4$$

The y-intercept is $(0, 4)$.

25. To find the x-intercept, let $y = 0$.

$$y = \tfrac{1}{3}x - 2$$
$$0 = \tfrac{1}{3}x - 2$$
$$2 = \tfrac{1}{3}x$$
$$6 = x$$

The x-intercept is $(6, 0)$.

To find the y-intercept, let $x = 0$.

$$y = \tfrac{1}{3}x - 2$$
$$y = \tfrac{1}{3}(0) - 2$$
$$y = 0 - 2$$
$$y = -2$$

The y-intercept is $(0, -2)$.

27. $x - 4 = 0$ is equivalent to $x = 4$. This is an equation of a vertical line. Its x-intercept is $(4, 0)$ and there is no y-intercept.

29. $y = 2.5$ is an equation of a horizontal line. Its y-intercept is $(0, 2.5)$ and there is no x-intercept.

31. The equation of the x-axis is $y = 0$. The equation of the y-axis is $x = 0$.

33. Begin by finding the intercepts.

$$x = y + 2$$
$$x = 0 + 2 \quad \textit{Let y = 0.}$$
$$x = 2$$

$$x = y + 2$$
$$0 = y + 2 \quad \textit{Let x = 0.}$$
$$-2 = y$$

The x-intercept is $(2, 0)$ and the y-intercept is $(0, -2)$. To find a third point, choose $y = 1$.

$$x = y + 2$$
$$x = 1 + 2 \quad \textit{Let y = 1.}$$
$$x = 3$$

This gives the ordered pair $(3, 1)$. Plot $(2, 0)$, $(0, -2)$, and $(3, 1)$ and draw a line through them.

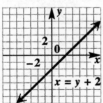

35. Find the intercepts.

$$x - y = 4$$
$$x - 0 = 4 \quad \textit{Let y = 0.}$$
$$x = 4$$

$x - y = 4$
$0 - y = 4$ *Let x = 0.*
$\quad y = -4$

The x-intercept is $(4, 0)$ and the y-intercept is $(0, -4)$. To find a third point, choose $y = 1$.

$$x - y = 4$$
$$x - 1 = 4 \quad \text{\textit{Let y = 1.}}$$
$$x = 5$$

This gives the ordered pair $(5, 1)$. Plot $(4, 0)$, $(0, -4)$, and $(5, 1)$ and draw a line through them.

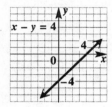

37. Find the intercepts.

$2x + y = 6$
$2x + 0 = 6$ *Let y = 0.*
$\quad 2x = 6$
$\quad x = 3$

$2x + y = 6$
$2(0) + y = 6$ *Let x = 0.*
$\quad 0 + y = 6$
$\quad y = 6$

The x-intercept is $(3, 0)$ and the y-intercept is $(0, 6)$. To find a third point, choose $x = 1$.

$$2x + y = 6$$
$$2(1) + y = 6 \quad \text{\textit{Let x = 1.}}$$
$$2 + y = 6$$
$$y = 4$$

This gives the ordered pair $(1, 4)$. Plot $(3, 0)$, $(0, 6)$, and $(1, 4)$ and draw a line through them.

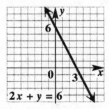

39. Find the intercepts.

$y = 2x - 5$
$0 = 2x - 5$ *Let y = 0.*
$5 = 2x$
$\frac{5}{2} = x$

$y = 2x - 5$
$y = 2(0) - 5$ *Let x = 0.*
$y = 0 - 5$
$y = -5$

The x-intercept is $\left(\frac{5}{2}, 0\right)$ and the y-intercept is $(0, -5)$. To find a third point, choose $x = 1$.

$$y = 2x - 5$$
$$y = 2(1) - 5 \quad \text{\textit{Let x = 1.}}$$
$$y = 2 - 5$$
$$y = -3$$

This gives the ordered pair $(1, -3)$. Plot $\left(\frac{5}{2}, 0\right)$, $(0, -5)$, and $(1, -3)$ and draw a line through them.

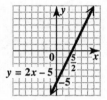

41. Find the intercepts.

$3x + 7y = 14$
$3x + 7(0) = 14$ *Let y = 0.*
$3x + 0 = 14$
$3x = 14$
$x = \frac{14}{3}$

$3x + 7y = 14$
$3(0) + 7y = 14$ *Let x = 0.*
$0 + 7y = 14$
$7y = 14$
$y = 2$

The x-intercept is $\left(\frac{14}{3}, 0\right)$ and the y-intercept is $(0, 2)$. To find a third point, choose $x = 2$.

$$3x + 7y = 14$$
$$3(2) + 7y = 14 \quad \text{\textit{Let x = 2.}}$$
$$6 + 7y = 14$$
$$7y = 8$$
$$y = \frac{8}{7}$$

This gives the ordered pair $\left(2, \frac{8}{7}\right)$. Plot $\left(\frac{14}{3}, 0\right)$, $(0, 2)$, and $\left(2, \frac{8}{7}\right)$. Writing $\frac{14}{3}$ as the mixed number $4\frac{2}{3}$ and $\frac{8}{7}$ as $1\frac{1}{7}$ will be helpful for plotting. Draw a line through these three points.

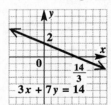

43. $y = -\frac{3}{4}x + 3$

x-intercept $(y = 0)$ **y-intercept $(x = 0)$**

$0 = -\frac{3}{4}x + 3$ $y = -\frac{3}{4}(0) + 3$

$\frac{3}{4}x = 3$ $y = 0 + 3$

$x = \frac{4}{3}(3) = 4$ **$(4, 0)$** $y = 3$ **$(0, 3)$**

third point $(x = 2)$

$y = -\frac{3}{4}(2) + 3$

$y = -\frac{3}{2} + 3$

$y = \frac{3}{2}$ **$\left(2, \frac{3}{2}\right)$**

Plot $(4, 0)$, $(0, 3)$, and $\left(2, \frac{3}{2}\right)$ and draw a line through these points.

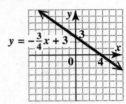

45. $y - 2x = 0$

If $y = 0$, $x = 0$. Both intercepts are the origin, $(0, 0)$. Find two additional points.

$y - 2x = 0$

$y - 2(1) = 0$ *Let x = 1.*

$y - 2 = 0$

$y = 2$

$y - 2x = 0$

$y - 2(-3) = 0$ *Let x = -3.*

$y + 6 = 0$

$y = -6$

Plot $(0, 0)$, $(1, 2)$, and $(-3, -6)$ and draw a line through them.

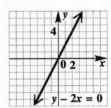

47. $y = -6x$

Find three points on the line.

If $x = 0$, $y = -6(0) = 0$.

If $x = 1$, $y = -6(1) = -6$.

If $x = -1$, $y = -6(-1) = 6$.

Plot $(0, 0)$, $(1, -6)$, and $(-1, 6)$ and draw a line through these points.

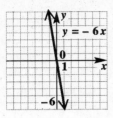

49. $y = -1$

For any value of x, the value of y is -1. Three ordered pairs are $(-4, -1)$, $(0, -1)$, and $(3, -1)$. Plot these points and draw a line through them. The graph is a horizontal line.

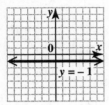

51. $x + 2 = 0$

$x = -2$

For any value of y, the value of x is -2. Three ordered pairs are $(-2, 3)$, $(-2, 0)$, and $(-2, -4)$. Plot these points and draw a line through them. The graph is a vertical line.

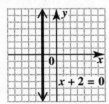

53. $-3y = 15$

$y = \frac{15}{-3} = -5$

For any value of x, the value of y is -5. Three ordered pairs are $(-2, -5)$, $(0, -5)$, and $(1, -5)$. Plot these points and draw a line through them. The graph is a horizontal line.

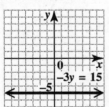

55. $x + 2 = 8$

$x = 6$

For any value of y, the value of x is 6. Three ordered pairs are $(6, -2)$, $(6, 0)$, and $(6, 1)$. Plot these points and draw a line through them. The graph is a vertical line.

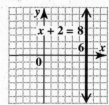

57. Find two ordered pairs that satisfy the equation, usually the intercepts. Plot the corresponding points on a coordinate system. Draw a straight line through the two points. As a check, find a third ordered pair and verify that it lies on the line you drew.

59. **(a)** See the solution for Exercise 3.1.77. The ordered pairs are $(20, 170)$, $(40, 153)$, $(60, 136)$, and $(80, 119)$.

(b)

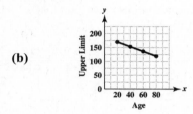

(c) Follow the vertical line between the tick marks 20 and 40. It appears that the graph of the line intersects this vertical line at about 160.

(d) $y = -0.85x + 187$
$y = -0.85(30) + 187$
$y = -25.5 + 187$
$y = 161.5 \approx 162$

(e) They are quite close.

61. No. To go beyond the given data at either end assumes that the graph continues in the same way, which may not be true.

63. **(a)** $y = 5.5x - 220$

$y = 5.5(62) - 220 = 121$ *Let x = 62.*
$y = 5.5(66) - 220 = 143$ *Let x = 66.*
$y = 5.5(72) - 220 = 176$ *Let x = 72.*

The approximate weights of men whose heights are 62 in., 66 in., and 72 in. are 121 lb, 143 lb, and 176 lb, respectively.

(b) Plot the points $(62, 121)$, $(66, 143)$, and $(72, 176)$ and connect them with a smooth line.

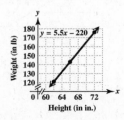

(c) Locate 155 on the vertical scale, then move across to the line, then down to the horizontal scale. From the graph, the height of a man who weighs 155 lb is about 68 in. Now substitute 155 for y in the equation.

$$y = 5.5x - 220$$
$$155 = 5.5x - 220$$
$$375 = 5.5x$$
$$x = \tfrac{375}{5.5} \approx 68.18$$

From the equation, the height is 68 in. to the nearest inch.

65. **(a)** $y = 1.983x + 53.66$

For 1997, let $x = 7$.
$y = 1.983(7) + 53.66 = 67.541$

For 2000, let $x = 10$.
$y = 1.983(10) + 53.66 = 73.49$

For 2002, let $x = 12$.
$y = 1.983(12) + 53.66 = 77.456$

The approximate sales for 1997, 2000, and 2002 are $67.5 billion, $73.5 billion, and $77.5 billion, respectively.

(b) Locate 7, 10, and 12 on the horizontal scale, then find the corresponding value on the vertical scale.

The approximate sales for 1997, 2000, and 2002 are $67 billion, $74 billion, and $78 billion, respectively.

(c) The values are very close.

67. **(a)** Plot the points $(40, 2000)$ and $(30, 2500)$ and connect them with a smooth line.

(b) See the graph in part (a). Locate 20 on the horizontal scale, then move up to the line and across to read the value on the vertical scale. The demand will be 3000 units if the price drops to $20.

(c) Locate 3500 on the vertical scale, then move across to the line and down to read the value on the horizontal scale. The price is $10 if the demand is 3500 units.

69. $5(2x - 1) - 4(2x + 1) - 7 = 0$
$$10x - 5 - 8x - 4 - 7 = 0$$
$$2x - 16 = 0$$
$$2x = 16$$
$$x = \frac{16}{2} = 8$$

This is the same as the x-intercept on the calculator screen.

71. $-\frac{2}{7}x + 2x - \frac{1}{2}x - \frac{17}{2} = 0$

Multiply both sides by the LCD, 14, to clear fractions.

$$14\left(-\frac{2}{7}x + 2x - \frac{1}{2}x - \frac{17}{2}\right) = 14(0)$$
$$-4x + 28x - 7x - 119 = 0$$
$$17x - 119 = 0$$
$$17x = 119$$
$$x = \frac{119}{17} = 7$$

This is the same as the x-intercept on the calculator screen.

73. $\frac{4 - 2}{8 - 5} = \frac{2}{3}$

75. $\frac{-2 - (-4)}{3 - (-1)} = \frac{-2 + 4}{3 + 1} = \frac{2}{4} = \frac{1}{2}$

77. Solve $d = rt$ for t.
$$\frac{d}{r} = \frac{rt}{r} \qquad \text{\textit{Divide by r.}}$$
$$\frac{d}{r} = t$$

79. Solve $P = a + b + c$ for c.
$$P - a = b + c \qquad \text{\textit{Subtract a.}}$$
$$P - a - b = c \qquad \text{\textit{Subtract b.}}$$

3.3 The Slope of a Line

1. Rise is the vertical change between two different points on a line.

Run is the horizontal change between two different points on a line.

3. The indicated points have coordinates $(-1, -4)$ and $(1, 4)$.

$$\frac{\text{rise}}{\text{run}} = \frac{\text{vertical change}}{\text{horizontal change}}$$
$$= \frac{4 - (-4)}{1 - (-1)} = \frac{8}{2} = 4$$

5. The indicated points have coordinates $(-3, 2)$ and $(5, -2)$.

$$\frac{\text{rise}}{\text{run}} = \frac{\text{vertical change}}{\text{horizontal change}}$$
$$= \frac{-2 - 2}{5 - (-3)} = \frac{-4}{8} = -\frac{1}{2}$$

7. The indicated points have coordinates $(-2, -4)$ and $(4, -4)$.

$$\frac{\text{rise}}{\text{run}} = \frac{\text{vertical change}}{\text{horizontal change}}$$
$$= \frac{-4 - (-4)}{4 - (-2)} = \frac{-4 + 4}{4 + 2} = \frac{0}{6} = 0$$

9. **(a)** The indicated points have coordinates $(-1, 2)$ and $(2, 0)$.

$$\frac{\text{rise}}{\text{run}} = \frac{\text{vertical change}}{\text{horizontal change}}$$
$$= \frac{0 - 2}{2 - (-1)} = \frac{-2}{3} = -\frac{2}{3} \quad \text{(choice \textbf{C})}$$

(b) The indicated points have coordinates $(-1, 2)$ and $(-4, 0)$.

$$\frac{\text{rise}}{\text{run}} = \frac{\text{vertical change}}{\text{horizontal change}} = \frac{0 - 2}{-4 - (-1)}$$
$$= \frac{-2}{-4 + 1} = \frac{-2}{-3} = \frac{2}{3} \quad \text{(choice \textbf{A})}$$

(c) The indicated points have coordinates $(-1, 2)$ and $(1, -1)$.

$$\frac{\text{rise}}{\text{run}} = \frac{\text{vertical change}}{\text{horizontal change}}$$
$$= \frac{-1 - 2}{1 - (-1)} = \frac{-3}{2} = -\frac{3}{2} \quad \text{(choice \textbf{D})}$$

(d) The indicated points have coordinates $(-1, 2)$ and $(-3, -1)$.

$$\frac{\text{rise}}{\text{run}} = \frac{\text{vertical change}}{\text{horizontal change}}$$
$$= \frac{-1 - 2}{-3 - (-1)} = \frac{-3}{-2} = \frac{3}{2} \quad \text{(choice \textbf{B})}$$

11. Negative slope

Sketches will vary. The line must fall from left to right. One such line is shown in the following graph.

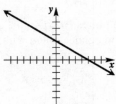

13. Undefined slope

Sketches will vary. The line must be vertical. One such line is shown in the following graph.

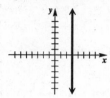

15. The slope of a line is the ratio (or quotient) of the rise, the change in y, and the run, the change in x.

17. **(a)** Because the line *falls* from left to right, its slope is *negative*.

(b) Because the line intersects the y-axis *at the* origin, the y-value of its y-intercept is *zero*.

19. **(a)** Because the line *rises* from left to right, its slope is *positive*.

(b) Because the line intersects the y-axis *below* the origin, the y-value of its y-intercept is *negative*.

21. **(a)** The line is *horizontal*, so its slope is *zero*.

(b) The line intersects the y-axis *below* the origin, so the y-value of its y-intercept is *negative*.

23. The slope (or grade) of the hill is the ratio of the rise to the run, or the ratio of the vertical change to the horizontal change. Since the rise is 32 and the run is 108, the slope is

$$\frac{32}{108} = \frac{8 \cdot 4}{27 \cdot 4} = \frac{8}{27}.$$

25. Because he found the difference $3 - 5 = -2$ in the numerator, he should have subtracted in the same order in the denominator to get $-1 - 2 = -3$. The correct slope is $\frac{-2}{-3} = \frac{2}{3}$. Note that the student's slope and the correct slope are opposites of one another.

27. Use the slope formula with $(1, -2) = (x_1, y_1)$ and $(-3, -7) = (x_2, y_2)$.

$$\text{slope } m = \frac{\text{change in } y}{\text{change in } x} = \frac{y_2 - y_1}{x_2 - x_1}$$

$$= \frac{-7 - (-2)}{-3 - 1} = \frac{-5}{-4} = \frac{5}{4}$$

29. Use the slope formula with $(0, 3) = (x_1, y_1)$ and $(-2, 0) = (x_2, y_2)$.

$$\text{slope } m = \frac{\text{change in } y}{\text{change in } x} = \frac{y_2 - y_1}{x_2 - x_1}$$

$$= \frac{0 - 3}{-2 - 0} = \frac{-3}{-2} = \frac{3}{2}$$

31. Use the slope formula with $(4, 3) = (x_1, y_1)$ and $(-6, 3) = (x_2, y_2)$.

$$\text{slope } m = \frac{\text{change in } y}{\text{change in } x} = \frac{y_2 - y_1}{x_2 - x_1}$$

$$= \frac{3 - 3}{-6 - 4} = \frac{0}{-10} = 0$$

33. Use the slope formula with $(-2, 4) = (x_1, y_1)$ and $(-3, 7) = (x_2, y_2)$.

$$\text{slope } m = \frac{\text{change in } y}{\text{change in } x} = \frac{y_2 - y_1}{x_2 - x_1}$$

$$= \frac{7 - 4}{-3 - (-2)} = \frac{3}{-1} = -3$$

35. Use the slope formula with $(-12, 3) = (x_1, y_1)$ and $(-12, -7) = (x_2, y_2)$.

$$\text{slope } m = \frac{\text{change in } y}{\text{change in } x} = \frac{y_2 - y_1}{x_2 - x_1}$$

$$= \frac{-7 - 3}{-12 - (-12)} = \frac{-10}{0},$$

which is *undefined*.

37. Use the slope formula with $(4.8, 2.5) = (x_1, y_1)$ and $(3.6, 2.2) = (x_2, y_2)$.

$$\text{slope } m = \frac{\text{change in } y}{\text{change in } x} = \frac{y_2 - y_1}{x_2 - x_1}$$

$$= \frac{2.2 - 2.5}{3.6 - 4.8} = \frac{-0.3}{-1.2} = \frac{1}{4}$$

39. Use the slope formula with $\left(-\frac{7}{5}, \frac{3}{10}\right) = (x_1, y_1)$ and $\left(\frac{1}{5}, -\frac{1}{2}\right) = (x_2, y_2)$.

$$\text{slope } m = \frac{\text{change in } y}{\text{change in } x} = \frac{y_2 - y_1}{x_2 - x_1}$$

$$= \frac{-\frac{1}{2} - \frac{3}{10}}{\frac{1}{5} - \left(-\frac{7}{5}\right)} = \frac{-\frac{5}{10} - \frac{3}{10}}{\frac{1}{5} + \frac{7}{5}} = \frac{-\frac{8}{10}}{\frac{8}{5}}$$

$$= \left(-\frac{8}{10}\right)\left(\frac{5}{8}\right) = -\frac{5}{10} = -\frac{1}{2}$$

41. $y = 5x + 12$

Since the equation is already solved for y, the slope is given by the coefficient of x, which is 5. Thus, the slope of the line is 5.

43. Solve the equation for y.

$$4y = x + 1$$
$$y = \tfrac{1}{4}x + \tfrac{1}{4} \quad \text{Divide by 4.}$$

The slope of the line is given by the coefficient of x, so the slope is $\frac{1}{4}$.

45. Solve the equation for y.

$$3x - 2y = 3$$
$$-2y = -3x + 3 \quad \textit{Subtract 3x.}$$
$$y = \tfrac{3}{2}x - \tfrac{3}{2} \quad \textit{Divide by −2.}$$

The slope of the line is given by the coefficient of x, so the slope is $\tfrac{3}{2}$.

47. Solve the equation for y.

$$-3x + 2y = 5$$
$$2y = 3x + 5 \quad \textit{Add 3x.}$$
$$y = \tfrac{3}{2}x + \tfrac{5}{2} \quad \textit{Divide by 2.}$$

The slope of the line is given by the coefficient of x, so the slope is $\tfrac{3}{2}$.

49. $y = -5$

This is an equation of a horizontal line. Its slope is 0. (This equation may be rewritten in the form $y = 0x - 5$, where the coefficient of x gives the slope.)

51. $x = 6$

This is an equation of a vertical line. Its slope is *undefined*.

53. Solve the equation for y.

$$3x + y = 7$$
$$y = -3x + 7 \quad \textit{Subtract 3x.}$$

The slope of the given line is -3, so the slope of a line whose graph is parallel to the graph of the given line is also -3.

The slope of a line whose graph is perpendicular to the graph of the given line is the negative reciprocal of -3, that is, $\tfrac{1}{3}$.

55. If two lines are both vertical or both horizontal, they are *parallel*. Choice **A** is correct.

57. Find the slope of each line by solving the equations for y.

$$2x + 5y = 4$$
$$5y = -2x + 4 \quad \textit{Subtract 2x.}$$
$$y = -\tfrac{2}{5}x + \tfrac{4}{5} \quad \textit{Divide by 5.}$$

The slope of the first line is $-\tfrac{2}{5}$.

$$4x + 10y = 1$$
$$10y = -4x + 1 \quad \textit{Subtract 4x.}$$
$$y = -\tfrac{4}{10}x + \tfrac{1}{10} \quad \textit{Divide by 10.}$$
$$y = -\tfrac{2}{5}x + \tfrac{1}{10} \quad \textit{Lowest terms}$$

The slope of the second line is $-\tfrac{2}{5}$.

The slopes are equal, so the lines are *parallel*.

59. Find the slope of each line by solving the equations for y.

$$8x - 9y = 6$$
$$-9y = -8x + 6 \quad \textit{Subtract 8x.}$$
$$y = \tfrac{8}{9}x - \tfrac{2}{3} \quad \textit{Divide by −9.}$$

The slope of the first line is $\tfrac{8}{9}$.

$$8x + 6y = -5$$
$$6y = -8x - 5 \quad \textit{Subtract 8x.}$$
$$y = -\tfrac{4}{3}x - \tfrac{5}{6} \quad \textit{Divide by 6.}$$

The slope of the second line is $-\tfrac{4}{3}$.

The slopes are not equal, so the lines are not parallel. The slopes are not negative reciprocals (the negative reciprocal of $\tfrac{8}{9}$ is $-\tfrac{9}{8}$), so the lines are not perpendicular. Thus, the lines are *neither parallel nor perpendicular*.

61. Find the slope of each line by solving the equations for y.

$$3x - 2y = 6$$
$$-2y = -3x + 6 \quad \textit{Subtract 3x.}$$
$$y = \tfrac{3}{2}x - 3 \quad \textit{Divide by −2.}$$

The slope of the first line is $\tfrac{3}{2}$.

$$2x + 3y = 3$$
$$3y = -2x + 3 \quad \textit{Subtract 2x.}$$
$$y = -\tfrac{2}{3}x + 1 \quad \textit{Divide by 3.}$$

The slope of the second line is $-\tfrac{2}{3}$.

The product of the slopes is

$$\tfrac{3}{2}\left(-\tfrac{2}{3}\right) = -1,$$

so the lines are *perpendicular*.

63. Find the slope of each line by solving the equations for y.

$$5x - y = 1$$
$$-y = -5x + 1 \quad \textit{Subtract 5x.}$$
$$y = 5x - 1 \quad \textit{Divide by −1.}$$

The slope of the first line is 5.

$$x - 5y = -10$$
$$-5y = -x - 10 \quad \textit{Subtract x.}$$
$$y = \tfrac{1}{5}x + 2 \quad \textit{Divide by −5.}$$

The slope of the second line is $\tfrac{1}{5}$.

The slopes are not equal, so the lines are not parallel. The slopes are not negative reciprocals (the negative reciprocal of 5 is $-\tfrac{1}{5}$), so the lines are not perpendicular. Thus, the lines are *neither parallel nor perpendicular*.

65. We use the points with coordinates $(1990, 11{,}338)$ and $(2005, 14{,}818)$.

$$m = \frac{14{,}818 \text{ thousand } - 11{,}338 \text{ thousand}}{2005 - 1990}$$

$$= \frac{3480 \text{ thousand}}{15}$$

$$= 232 \text{ thousand } \quad \text{or} \quad 232{,}000.$$

66. The slope of the line in Figure A is *positive*. This means that during the period represented, enrollment *increased* in grades 9–12.

67. The increase is approximately 232,000 students per year.

68. We use the points with coordinates $(1990, 20)$ and $(2002, 4.9)$.

$$m = \frac{4.9 - 20}{2002 - 1990} = \frac{-15.1}{12}$$

$$\approx -1.26 \text{ students per computer per year.}$$

69. The slope of the line in Figure B is *negative*. This means that the number of students per computer *decreased* during the period represented.

70. The decrease is 1.26 students per computer per year.

71. The *y*-values change 0.1 billion (or 100,000,000) square feet each year. Since the change each year is the same, the graph is a straight line.

73. Two of the ordered pairs have coordinates $(-12, -0.8)$ and $(0, 4)$. Therefore, the slope is

$$m = \frac{4 - (-0.8)}{0 - (-12)} = \frac{4.8}{12} = 0.4.$$

75. From the table, when $X = 0$, $Y_1 = 4$. Therefore, the *y*-intercept is $(0, 4)$.

77. $\quad 2x + 5y = 15$

$\qquad\quad 5y = -2x + 15$

$\qquad\quad\ y = -\frac{2}{5}x + 3$

79. $\qquad 10x = 30 + 3y$

$\quad 10x - 30 = 3y$

$\quad \frac{10}{3}x - 10 = y$

81. $\quad y - (-8) = 2(x - 4)$

$\qquad\ y + 8 = 2x - 8$

$\qquad\quad\ y = 2x - 16$

83. $\quad y - \left(-\frac{3}{5}\right) = -\frac{1}{2}[x - (-3)]$

$\qquad\quad y + \frac{3}{5} = -\frac{1}{2}(x + 3)$

$\qquad\ y + \frac{6}{10} = -\frac{1}{2}x - \frac{3}{2}$

$\qquad\qquad y = -\frac{1}{2}x - \frac{15}{10} - \frac{6}{10}$

$\qquad\qquad y = -\frac{1}{2}x - \frac{21}{10}$

3.4   Equations of a Line

1. The point-slope form of the equation of a line with slope -2 passing through the point $(4, 1)$ is

$$y - 1 = -2(x - 4).$$

Choice **E** is correct.

3. A line that passes through the points $(0, 0)$ [its *y*-intercept] and $(4, 1)$ has slope

$$m = \frac{\text{rise}}{\text{run}} = \frac{1 - 0}{4 - 0} = \frac{1}{4}.$$

Its slope-intercept form is

$$y = \tfrac{1}{4}x + 0 \quad \text{or} \quad y = \tfrac{1}{4}x.$$

Choice **B** is correct.

5. **(a)** $y = x + 3 = 1x + 3$

The graph of this equation is a line with slope 1 and *y*-intercept $(0, 3)$. The only graph which has a *positive* slope and intersects the *y*-axis *above* the origin is **C**.

(b) $y = -x + 3 = -1x + 3$

The graph of this equation is a line with slope -1 and *y*-intercept $(0, 3)$. The only graph which has a *negative* slope and intersects the *y*-axis *above* the origin is **B**.

(c) $y = x - 3 = 1x - 3$

The graph of this equation is a line with slope 1 and *y*-intercept $(0, -3)$. The only graph which has a *positive* slope and intersects the *y*-axis *below* the origin is **A**.

(d) $y = -x - 3 = -1x - 3$

The graph of this equation is a line with slope -1 and *y*-intercept $(0, -3)$. The only graph which has a *negative* slope and intersects the *y*-axis *below* the origin is **D**.

7. The rise is 3 and the run is 1, so the slope is given by

$$m = \frac{\text{rise}}{\text{run}} = \frac{3}{1} = 3.$$

The *y*-intercept is $(0, -3)$, so $b = -3$. The equation of the line, written in slope-intercept form, is

$$y = 3x - 3.$$

9. Since the line falls from left to right, the "rise" is negative. For this line, the rise is -3 and the run is 3, so the slope is

$$m = \frac{\text{rise}}{\text{run}} = \frac{-3}{3} = -1.$$

The y-intercept is $(0, 3)$, so $b = 3$. The slope-intercept form of the equation of the line is

$$y = -1x + 3$$
$$y = -x + 3.$$

11. $m = 4$, $(0, -3)$

Since the y-intercept is $(0, -3)$, we have $b = -3$. Use the slope-intercept form.

$$y = mx + b$$
$$y = 4x + (-3)$$
$$y = 4x - 3$$

13. $m = 0$, $(0, 3)$

Since the y-intercept is $(0, 3)$, we have $b = 3$. Use the slope-intercept form.

$$y = mx + b$$
$$y = 0x + 3$$
$$y = 3$$

15. Undefined slope, $(0, -2)$

Since the slope is undefined, the line is vertical and has equation $x = k$. Because the line goes through a point $(x, y) = (0, -2)$, we must have $x = 0$.

17. $y = 3x + 2$

The slope is $3 = \frac{3}{1} = \frac{\text{change in } y}{\text{change in } x}$, and the y-intercept is $(0, 2)$. Graph that point and count up 3 units and right 1 unit to get to the point $(1, 5)$. Draw a line through the points.

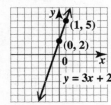

19. $y = -\frac{1}{3}x + 4$

The slope is $-\frac{1}{3} = \frac{-1}{3} = \frac{\text{change in } y}{\text{change in } x}$, and the y-intercept is $(0, 4)$. Graph that point and count down 1 unit and right 3 units to get to the point $(3, 3)$. Draw a line through the points.

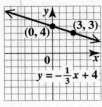

21. $2x + y = -5$, so $y = -2x - 5$

The slope is $-2 = \frac{-2}{1} = \frac{\text{change in } y}{\text{change in } x}$, and the y-intercept is $(0, -5)$. Graph that point and count down 2 units and right 1 unit to get to the point $(1, -7)$. Draw a line through the points.

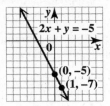

23. Solve the equation for y.

$$4x - 5y = 20 \qquad \textit{Given equation}$$
$$-5y = -4x + 20 \quad \textit{Subtract 4x.}$$
$$y = \tfrac{4}{5}x - 4 \qquad \textit{Divide by } -5.$$

The slope is the coefficient of x, $\frac{4}{5}$.

The slope is $\frac{4}{5} = \frac{\text{change in } y}{\text{change in } x}$, and the y-intercept is $(0, -4)$. Graph that point and count up 4 units and right 5 units to get to the point $(5, 0)$. Draw a line through the points.

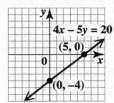

25. $(0, 1)$, $m = 4$

First, locate the point $(0, 1)$, which is the y-intercept of the line to be graphed. Write the slope as

$$m = \frac{\text{rise}}{\text{run}} = \frac{4}{1}.$$

Locate another point by counting 4 units up and then 1 unit to the right. Draw a line through this new point, $(1, 5)$, and the given point, $(0, 1)$.

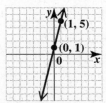

27. $(1, -5)$, $m = -\frac{2}{5}$

First, locate the point $(1, -5)$. Write the slope as

$$m = \frac{\text{rise}}{\text{run}} = \frac{-2}{5}.$$

Locate another point by counting 2 units down (because of the negative sign) and then 5 units to the right. Draw a line through this new point, $(6, -7)$, and the given point, $(1, -5)$.

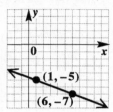

29. $(-1, 4)$, $m = \frac{2}{5}$

First, locate the point $(-1, 4)$. The slope is

$$m = \frac{\text{rise}}{\text{run}} = \frac{2}{5}.$$

Locate another point by counting 2 units up and then 5 units to the right. Draw a line through this new point, $(4, 6)$, and the given point, $(-1, 4)$.

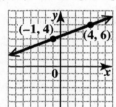

31. $(-2, 3)$, $m = 0$

First, locate the point $(-2, 3)$. Since the slope is 0, the line will be horizontal. Draw a horizontal line through the point $(-2, 3)$.

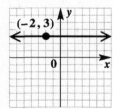

33. $(2, 4)$, undefined slope

First, locate the point $(2, 4)$. Since the slope is undefined, the line will be vertical. Draw the vertical line through the point $(2, 4)$.

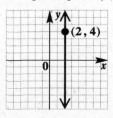

35. The common name given to a line with slope 0 (a horizontal line) and whose y-intercept is the origin is the x-axis.

37. $(-1, 3)$, $m = -4$

The given point is $(-1, 3)$, so $x_1 = -1$ and $y_1 = 3$. Also, $m = -4$. Substitute these values into the point-slope form. Then solve for y to obtain the slope-intercept form.

$$\begin{aligned}
y - y_1 &= m(x - x_1) \\
y - 3 &= -4[x - (-1)] \\
y - 3 &= -4(x + 1) \\
y - 3 &= -4x - 4 \quad \textit{Dist. property} \\
y &= -4x - 1 \quad \textit{Add 3.}
\end{aligned}$$

39. $(-2, 5)$, $m = \frac{2}{3}$

Use the values $x_1 = -2$, $y_1 = 5$, and $m = \frac{2}{3}$ in the point-slope form.

$$\begin{aligned}
y - y_1 &= m(x - x_1) \\
y - 5 &= \frac{2}{3}[x - (-2)] \\
y - 5 &= \frac{2}{3}(x + 2) \\
y - 5 &= \frac{2}{3}x + \frac{4}{3} \\
y &= \frac{2}{3}x + \frac{19}{3} \quad \textit{Add 5} = \frac{15}{3}.
\end{aligned}$$

41. $(-4, 1)$, $m = \frac{3}{4}$

Use the values $x_1 = -4$, $y_1 = 1$, and $m = \frac{3}{4}$ in the point-slope form.

$$\begin{aligned}
y - y_1 &= m(x - x_1) \\
y - 1 &= \frac{3}{4}[x - (-4)] \\
y - 1 &= \frac{3}{4}(x + 4) \\
y - 1 &= \frac{3}{4}x + 3 \\
y &= \frac{3}{4}x + 4
\end{aligned}$$

43. $(2, 1)$, $m = \frac{5}{2}$

Use the values $x_1 = 2$, $y_1 = 1$, and $m = \frac{5}{2}$ in the point-slope form.

$$\begin{aligned}
y - y_1 &= m(x - x_1) \\
y - 1 &= \frac{5}{2}(x - 2) \\
y - 1 &= \frac{5}{2}x - 5 \\
y &= \frac{5}{2}x - 4
\end{aligned}$$

45. An equation of the line which passes through the origin and a second point whose x- and y-coordinates are equal is $y = x$. This equation may be written in other forms, such as $x - y = 0$ or $y - x = 0$.

47. $(4, 10)$ and $(6, 12)$

First, find the slope of the line.

$$m = \frac{12 - 10}{6 - 4} = \frac{2}{2} = 1$$

Now use the point $(4, 10)$ for (x_1, y_1) and $m = 1$ in the point-slope form.

$$y - y_1 = m(x - x_1)$$
$$y - 10 = 1(x - 4)$$
$$y - 10 = x - 4$$
$$y = x + 6$$

The same result would be found by using $(6, 12)$ for (x_1, y_1).

49. $(-2, -1)$ and $(3, -4)$

$$m = \frac{-4 - (-1)}{3 - (-2)} = \frac{-3}{5} = -\frac{3}{5}$$

Use the point-slope form with $(x_1, y_1) = (3, -4)$ and $m = -\frac{3}{5}$.

$$y - y_1 = m(x - x_1)$$
$$y - (-4) = -\frac{3}{5}(x - 3)$$
$$y + 4 = -\frac{3}{5}(x - 3)$$
$$y + 4 = -\frac{3}{5}x + \frac{9}{5}$$
$$y = -\frac{3}{5}x - \frac{11}{5}$$

51. $(-4, 0)$ and $(0, 2)$

$$m = \frac{2 - 0}{0 - (-4)} = \frac{2}{4} = \frac{1}{2}$$

Use the point-slope form with $(x_1, y_1) = (-4, 0)$ and $m = \frac{1}{2}$.

$$y - y_1 = m(x - x_1)$$
$$y - 0 = \frac{1}{2}[x - (-4)]$$
$$y = \frac{1}{2}(x + 4)$$
$$y = \frac{1}{2}x + 2$$

53. $\left(-\frac{2}{3}, \frac{8}{3}\right)$ and $\left(\frac{1}{3}, \frac{7}{3}\right)$

$$m = \frac{\frac{7}{3} - \frac{8}{3}}{\frac{1}{3} - \left(-\frac{2}{3}\right)} = \frac{-\frac{1}{3}}{\frac{3}{3}} = \frac{-\frac{1}{3}}{1} = -\frac{1}{3}$$

Use the point-slope form with $(x_1, y_1) = \left(\frac{1}{3}, \frac{7}{3}\right)$ and $m = -\frac{1}{3}$.

$$y - y_1 = m(x - x_1)$$
$$y - \frac{7}{3} = -\frac{1}{3}\left(x - \frac{1}{3}\right)$$
$$y - \frac{7}{3} = -\frac{1}{3}x + \frac{1}{9}$$
$$y = -\frac{1}{3}x + \frac{22}{9}$$

55. When $C = 0$, $F = 32$. This gives the ordered pair $(0, 32)$. When $C = 100$, $F = 212$. This gives the ordered pair $(100, 212)$.

56. Use the two points $(0, 32)$ and $(100, 212)$.

$$m = \frac{212 - 32}{100 - 0} = \frac{180}{100} = \frac{9}{5}$$

57. Write the point-slope form as

$$F - F_1 = m(C - C_1).$$

We may use either the point $(0, 32)$ or the point $(100, 212)$ with the slope $\frac{9}{5}$, which we found in Exercise 56. Using $F_1 = 32$, $C_1 = 0$, and $m = \frac{9}{5}$, we obtain

$$F - 32 = \frac{9}{5}(C - 0).$$

Using $F_1 = 212$, $C_1 = 100$, and $m = \frac{9}{5}$, we obtain

$$F - 212 = \frac{9}{5}(C - 100).$$

58. We want to write an equation in the form

$$F = mC + b.$$

From Exercise 56, we have $m = \frac{9}{5}$. The point $(0, 32)$ is the y-intercept of the graph of the desired equation, so we have $b = 32$. Thus, an equation for F in terms of C is

$$F = \frac{9}{5}C + 32.$$

59. The expression for F in terms of C obtained in Exercise 58 is

$$F = \frac{9}{5}C + 32.$$

To obtain an expression for C in terms of F, solve this equation for C.

$$F - 32 = \frac{9}{5}C \qquad \textit{Subtract 32.}$$

$$\frac{5}{9}(F - 32) = \frac{5}{9} \cdot \frac{9}{5}C \qquad \textit{Multiply by } \frac{5}{9}, \textit{ the reciprocal of } \frac{9}{5}.$$

$$\frac{5}{9}(F - 32) = C \quad \text{or} \quad C = \frac{5}{9}F - \frac{160}{9}$$

60. $F = \frac{9}{5}C + 32$

If $C = 30$,

$$F = \frac{9}{5}(30) + 32$$
$$= 54 + 32 = 86.$$

Thus, when $C = 30$, $F = 86$.

61. $C = \frac{5}{9}(F - 32)$

When $F = 50$,

$$C = \frac{5}{9}(50 - 32)$$
$$= \frac{5}{9}(18) = 10.$$

Thus, when $F = 50$, $C = 10$.

62. Let $F = C$ in the equation obtained in Exercise 58.

$$F = \tfrac{9}{5}C + 32$$
$$C = \tfrac{9}{5}C + 32 \qquad \textit{Let } F = C.$$
$$5C = 5\left(\tfrac{9}{5}C + 32\right) \qquad \textit{Multiply by 5.}$$
$$5C = 9C + 160$$
$$-4C = 160 \qquad \textit{Subtract 9C.}$$
$$C = -40 \qquad \textit{Divide by } -4.$$

(The same result may be found by using either form of the equation obtained in Exercise 59.) The Celsius and Fahrenheit temperatures are equal $(F = C)$ at -40 degrees.

63. Solve the equation for y.

$$3x = 4y + 5$$
$$-4y = -3x + 5$$
$$y = \tfrac{3}{4}x - \tfrac{5}{4}$$

The slope is $\tfrac{3}{4}$. A line parallel to this line has the same slope. Now use the point-slope form with $m = \tfrac{3}{4}$ and $(x_1, y_1) = (2, -3)$.

$$y - y_1 = m(x - x_1)$$
$$y - (-3) = \tfrac{3}{4}(x - 2)$$
$$y + 3 = \tfrac{3}{4}x - \tfrac{3}{2}$$
$$y = \tfrac{3}{4}x - \tfrac{3}{2} - \tfrac{6}{2}$$
$$y = \tfrac{3}{4}x - \tfrac{9}{2}$$

65. Solve the equation for y.

$$x - 2y = 7$$
$$-2y = -x + 7$$
$$y = \tfrac{1}{2}x - \tfrac{7}{2}$$

The slope is $\tfrac{1}{2}$. A line perpendicular to this line has slope -2 $\left(\text{the negative reciprocal of } \tfrac{1}{2}\right)$. Now use the slope-intercept form with $m = -2$ and y-intercept $(0, -3)$.

$$y = mx + b$$
$$y = -2x - 3$$

67. **(a)** The fixed cost is $400.

(b) The variable cost is $0.25.

(c) Substitute $m = 0.25$ and $b = 400$ into $y = mx + b$ to get the cost equation

$$y = 0.25x + 400.$$

(d) Let $x = 100$ in the cost equation.

$$y = 0.25(100) + 400$$
$$y = 25 + 400$$
$$y = 425$$

The cost to produce 100 snow cones will be $425.

(e) Let $y = 775$ in the cost equation.

$$775 = 0.25x + 400$$
$$375 = 0.25x \qquad \textit{Subtract 400.}$$
$$x = \tfrac{375}{0.25} = 1500 \qquad \textit{Divide by 0.25.}$$

If the total cost is $775, 1500 snow cones will be produced.

69. **(a)** x represents the year and y represents the cost in the ordered pairs (x, y). The ordered pairs are $(1, 13{,}785)$, $(3, 15{,}518)$, $(5, 17{,}377)$, $(7, 18{,}950)$, and $(9, 21{,}235)$.

(b)

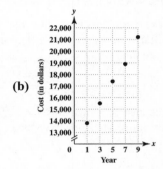

Yes, the points lie approximately on a straight line.

(c) Find the slope using $(x_1, y_1) = (3, 15{,}518)$ and $(x_2, y_2) = (9, 21{,}235)$.

$$m = \frac{y_2 - y_1}{x_2 - x_1} = \frac{21{,}235 - 15{,}518}{9 - 3} = \frac{5717}{6}$$
$$= 952.8\overline{3} = 952.8 \text{ (to the nearest tenth)}$$

Now use the point-slope form with $m = 952.8$ and $(x_1, y_1) = (3, 15{,}518)$.

$$y - y_1 = m(x - x_1)$$
$$y - 15{,}518 = 952.8(x - 3)$$
$$y - 15{,}518 = 952.8x - 2858.4$$
$$y = 952.8x + 12{,}659.6$$

Note that if $(9, 21{,}235)$ is used, the equation of the line will be $y = 952.8x + 12{,}659.8$.

(d) For 2008, $x = 2008 - 1996 = 12$.

$$y = 952.8x + 12{,}659.6$$
$$y = 952.8(12) + 12{,}659.6$$
$$y = 24{,}093.2 \approx 24{,}093$$

In 2008, the estimate of the average annual cost at private 4-year colleges is $24,093.

71. From the calculator screens, we see that the points with coordinates $(-1, 9)$ and $(4, -6)$ are on the line. First, find the slope.

$$m = \frac{-6 - 9}{4 - (-1)} = \frac{-15}{5} = -3$$

Use the point-slope form with $(x_1, y_1) = (-1, 9)$ and $m = -3$.

$$y - 9 = -3[x - (-1)]$$
$$y - 9 = -3(x + 1)$$
$$y - 9 = -3x - 3$$
$$y = -3x + 6$$

73. We can choose any two points from the table. The two points we will use are $(0, 1)$ and $(4, 4)$. First, find the slope of the line.

$$m = \frac{4 - 1}{4 - 0} = \frac{3}{4}$$

Since the y-intercept is $(0, 1)$, we have $b = 1$ for the slope-intercept form. Thus, we have the equation

$$Y_1 = \frac{3}{4}X + 1.$$

75. $2 \cdot 2 \cdot 2 \cdot 2 \cdot 2 \cdot 2 = 64$

77. $5 \cdot 5 \cdot 5 \cdot 5 = 625$

79. $\frac{2}{3} \cdot \frac{2}{3} \cdot \frac{2}{3} = \frac{8}{27}$

Chapter 3 Review Exercises

1. **(a)** Locate 2000 on the horizontal scale and follow the line up to the line graph. Then move across to read the value on the vertical scale. The cost for a gallon of gas in 2000 was about $1.50.

(b) Using $1.50 for 2000 and $1.90 for 2004, we get an increase of $1.90 − $1.50 = $0.40.

(c) The graph falls from 2000 to 2001 and also from 2001 to 2002, so the price of a gallon of gas decreased from 2000 to 2001 and 2001 to 2002.

(d) The steepest rise of the line occurs between 1999 and 2000, so the greatest increase in gasoline prices occurred from 1999 to 2000. Using $1.50 for 2000 and $1.15 for 1999, we get an increase of $1.50 − $1.15 = $0.35.

2. $y = 3x + 2;\ (-1, \quad)\ (0, \quad)\ (\quad, 5)$

$$y = 3x + 2$$
$$y = 3(-1) + 2 \quad \textit{Let x = -1.}$$
$$y = -3 + 2$$
$$y = -1$$

$$y = 3x + 2$$
$$y = 3(0) + 2 \quad \textit{Let x = 0.}$$
$$y = 0 + 2$$
$$y = 2$$

$$y = 3x + 2$$
$$5 = 3x + 2 \quad \textit{Let y = 5.}$$
$$3 = 3x$$
$$1 = x$$

The ordered pairs are $(-1, -1)$, $(0, 2)$, and $(1, 5)$.

3. $4x + 3y = 6;\ (0, \quad)\ (\quad, 0)\ (-2, \quad)$

$$4x + 3y = 6$$
$$4(0) + 3y = 6 \quad \textit{Let x = 0.}$$
$$3y = 6$$
$$y = 2$$

$$4x + 3y = 6$$
$$4x + 3(0) = 6 \quad \textit{Let y = 0.}$$
$$4x = 6$$
$$x = \frac{6}{4} = \frac{3}{2}$$

$$4x + 3y = 6$$
$$4(-2) + 3y = 6 \quad \textit{Let x = -2.}$$
$$-8 + 3y = 6$$
$$3y = 14$$
$$y = \frac{14}{3}$$

The ordered pairs are $(0, 2)$, $\left(\frac{3}{2}, 0\right)$, and $\left(-2, \frac{14}{3}\right)$.

4. $x = 3y;\ (0, \quad)\ (8, \quad)\ (\quad, -3)$

$$x = 3y$$
$$0 = 3y \quad \textit{Let x = 0.}$$
$$0 = y$$

$$x = 3y$$
$$8 = 3y \quad \textit{Let x = 8.}$$
$$\frac{8}{3} = y$$

$$x = 3y$$
$$x = 3(-3) \quad \textit{Let y = -3.}$$
$$x = -9$$

The ordered pairs are $(0, 0)$, $\left(8, \frac{8}{3}\right)$, and $(-9, -3)$.

5. $x - 7 = 0;\ (\quad, -3)\ (\quad, 0)\ (\quad, 5)$
The given equation may be written $x = 7$. For any value of y, the value of x will always be 7. The ordered pairs are $(7, -3)$, $(7, 0)$, and $(7, 5)$.

6. $x + y = 7;\ (2, 5)$
Substitute 2 for x and 5 for y in the given equation.

$$x + y = 7$$
$$2 + 5 = 7 \quad ?$$
$$7 = 7 \quad \textit{True}$$

Yes, $(2, 5)$ is a solution of the given equation.

7. $2x + y = 5; \quad (-1, 3)$
    Substitute -1 for x and 3 for y in the given equation.

    $$2x + y = 5$$
    $$2(-1) + 3 = 5 \quad ?$$
    $$-2 + 3 = 5 \quad ?$$
    $$1 = 5 \quad \textit{False}$$

    No, $(-1, 3)$ is not a solution of the given equation.

8. $3x - y = 4; \quad \left(\frac{1}{3}, -3\right)$
    Substitute $\frac{1}{3}$ for x and -3 for y in the given equation.

    $$3x - y = 4$$
    $$3\left(\frac{1}{3}\right) - (-3) = 4 \quad ?$$
    $$1 + 3 = 4 \quad ?$$
    $$4 = 4 \quad \textit{True}$$

    Yes, $\left(\frac{1}{3}, -3\right)$ is a solution of the given equation.

Graph for Exercises 9–12

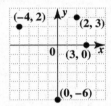

9. To plot $(2, 3)$, start at the origin, go 2 units to the right, and then go up 3 units. (See above graph.) The point lies in quadrant I.

10. To plot $(-4, 2)$, start at the origin, go 4 units to the left, and then go up 2 units. The point lies in quadrant II.

11. To plot $(3, 0)$, start at the origin, go 3 units to the right. The point lies on the x-axis (not in any quadrant).

12. To plot $(0, -6)$, start at the origin, go down 6 units. The point lies on the y-axis (not in any quadrant).

13. The product of two numbers is positive whenever the two numbers have the same sign. If $xy > 0$, either $x > 0$ and $y > 0$, so that (x, y) lies in quadrant I, or $x < 0$ and $y < 0$, so that (x, y) lies in quadrant III.

14. To find the x-intercept, let $y = 0$.

    $$y = 2x + 5$$
    $$0 = 2x + 5 \quad \textit{Let y = 0.}$$
    $$-2x = 5$$
    $$x = -\frac{5}{2}$$

    The x-intercept is $\left(-\frac{5}{2}, 0\right)$.

    To find the y-intercept, let $x = 0$.

    $$y = 2x + 5$$
    $$y = 2(0) + 5 \quad \textit{Let x = 0.}$$
    $$y = 5$$

    The y-intercept is $(0, 5)$.

    To find a third point, choose $x = -1$.

    $$y = 2x + 5$$
    $$y = 2(-1) + 5 \quad \textit{Let x = -1.}$$
    $$y = 3$$

    This gives the ordered pair $(-1, 3)$. Plot $\left(-\frac{5}{2}, 0\right)$, $(0, 5)$, and $(-1, 3)$ and draw a line through them.

    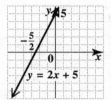

15. To find the x-intercept, let $y = 0$.

    $$3x + 2y = 8$$
    $$3x + 2(0) = 8 \quad \textit{Let y = 0.}$$
    $$3x = 8$$
    $$x = \frac{8}{3}$$

    The x-intercept is $\left(\frac{8}{3}, 0\right)$.

    To find the y-intercept, let $x = 0$.

    $$3x + 2y = 8$$
    $$3(0) + 2y = 8 \quad \textit{Let x = 0.}$$
    $$2y = 8$$
    $$y = 4$$

    The y-intercept is $(0, 4)$.

    To find a third point, choose $x = 1$.

    $$3x + 2y = 8$$
    $$3(1) + 2y = 8 \quad \textit{Let x = 1.}$$
    $$2y = 5$$
    $$y = \frac{5}{2}$$

    This gives the ordered pair $\left(1, \frac{5}{2}\right)$.
    Plot $\left(\frac{8}{3}, 0\right)$, $(0, 4)$, and $\left(1, \frac{5}{2}\right)$ and draw a line through them.

    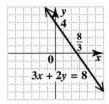

16. $x + 2y = -4$

Find the intercepts.

If $y = 0$, $x = -4$, so the x-intercept is $(-4, 0)$.

If $x = 0$, $y = -2$, so the y-intercept is $(0, -2)$.

To find a third point, choose $x = 2$.

$$x + 2y = -4$$
$$2 + 2y = -4 \quad \textit{Let x = 2.}$$
$$2y = -6$$
$$y = -3$$

This gives the ordered pair $(2, -3)$. Plot $(-4, 0)$, $(0, -2)$, and $(2, -3)$ and draw a line through them.

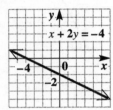

17. Let $(2, 3) = (x_1, y_1)$ and $(-4, 6) = (x_2, y_2)$.

$$\text{slope } m = \frac{\text{change in } y}{\text{change in } x} = \frac{y_2 - y_1}{x_2 - x_1}$$
$$= \frac{6 - 3}{-4 - 2} = \frac{3}{-6} = -\frac{1}{2}$$

18. Let $(2, 5) = (x_1, y_1)$ and $(2, 8) = (x_2, y_2)$.

$$\text{slope } m = \frac{\text{change in } y}{\text{change in } x} = \frac{y_2 - y_1}{x_2 - x_1}$$
$$= \frac{8 - 5}{2 - 2} = \frac{3}{0},$$

which is *undefined*.

19. $y = 3x - 4$

The equation is already solved for y, so the slope of the line is given by the coefficient of x. Thus, the slope is 3.

20. $y = 5$ is an equation of a horizontal line. Its slope is 0.

21. The indicated points have coordinates $(0, -2)$ and $(2, 1)$. Use the definition of slope with $(0, -2) = (x_1, y_1)$ and $(2, 1) = (x_2, y_2)$.

$$m = \frac{\text{change in } y}{\text{change in } x} = \frac{y_2 - y_1}{x_2 - x_1}$$
$$= \frac{1 - (-2)}{2 - 0} = \frac{3}{2}$$

22. The indicated points have coordinates $(0, 1)$ and $(3, 0)$. Use the definition of slope with $(0, 1) = (x_1, y_1)$ and $(3, 0) = (x_2, y_2)$.

$$m = \frac{\text{change in } y}{\text{change in } x} = \frac{y_2 - y_1}{x_2 - x_1}$$
$$= \frac{0 - 1}{3 - 0} = \frac{-1}{3} = -\frac{1}{3}$$

23. From the table, we choose the two points $(0, 1)$ and $(2, 4)$. Therefore,

$$\text{slope } m = \frac{\text{change in } y}{\text{change in } x} = \frac{y_2 - y_1}{x_2 - x_1}$$
$$= \frac{4 - 1}{2 - 0} = \frac{3}{2}.$$

24. **(a)** Because parallel lines have equal slopes and the slope of the graph of $y = 2x + 3$ is 2, the slope of a line parallel to it will also be 2.

(b) Because perpendicular lines have slopes which are negative reciprocals of each other and the slope of the graph of $y = -3x + 3$ is -3, the slope of a line perpendicular to it will be

$$-\frac{1}{-3} = \frac{1}{3}.$$

25. Find the slope of each line by solving the equations for y.

$$3x + 2y = 6$$
$$2y = -3x + 6 \quad \textit{Subtract 3x.}$$
$$y = -\frac{3}{2}x + 3 \quad \textit{Divide by 2.}$$

The slope of the first line is $-\frac{3}{2}$.

$$6x + 4y = 8$$
$$4y = -6x + 8 \quad \textit{Subtract 6x.}$$
$$y = -\frac{6}{4}x + 2 \quad \textit{Divide by 4.}$$
$$y = -\frac{3}{2}x + 2 \quad \textit{Lowest terms}$$

The slope of the second line is $-\frac{3}{2}$. The slopes are equal so the lines are *parallel*.

26. Find the slope of each line by solving the equations for y.

$$x - 3y = 1$$
$$-3y = -x + 1 \quad \textit{Subtract x.}$$
$$y = \frac{1}{3}x - \frac{1}{3} \quad \textit{Divide by -3.}$$

The slope of the first line is $\frac{1}{3}$.

$$3x + y = 4$$
$$y = -3x + 4 \quad \textit{Subtract 3x.}$$

The slope of the second line is -3.

The product of the slopes is

$$\tfrac{1}{3}(-3) = -1,$$

so the lines are *perpendicular*.

27. Find the slope of each line by solving the equations for y.

$$x - 2y = 8$$
$$-2y = -x + 8 \quad \textit{Subtract x.}$$
$$y = \tfrac{1}{2}x - 4 \quad \textit{Divide by –2.}$$

The slope of the first line is $\tfrac{1}{2}$.

$$x + 2y = 8$$
$$2y = -x + 8 \quad \textit{Subtract x.}$$
$$y = -\tfrac{1}{2}x + 4 \quad \textit{Divide by 2.}$$

The slope of the second line is $-\tfrac{1}{2}$.

The slopes are not equal and their product is

$$\left(\tfrac{1}{2}\right)\left(-\tfrac{1}{2}\right) = -\tfrac{1}{4} \neq -1,$$

so the lines are *neither* parallel nor perpendicular.

28. $m = -1, b = \tfrac{2}{3}$

Use the slope-intercept form, $y = mx + b$.

$$y = -1 \cdot x + \tfrac{2}{3} \quad \text{or} \quad y = -x + \tfrac{2}{3}$$

29. Through $(2, 3)$ and $(-4, 6)$

$$m = \frac{6 - 3}{-4 - 2} = \frac{3}{-6} = -\frac{1}{2}$$

Use $(2, 3)$ and $m = -\tfrac{1}{2}$ in the point-slope form.

$$y - y_1 = m(x - x_1)$$
$$y - 3 = -\tfrac{1}{2}(x - 2)$$
$$y - 3 = -\tfrac{1}{2}x + 1$$
$$y = -\tfrac{1}{2}x + 4$$

30. Through $(4, -3)$, $m = 1$

Use the point-slope form.

$$y - y_1 = m(x - x_1)$$
$$y - (-3) = 1(x - 4)$$
$$y + 3 = x - 4$$
$$y = x - 7$$

31. Through $(-1, 4)$, $m = \tfrac{2}{3}$

Use the point-slope form.

$$y - y_1 = m(x - x_1)$$
$$y - 4 = \tfrac{2}{3}[x - (-1)]$$
$$y - 4 = \tfrac{2}{3}(x + 1)$$
$$y - 4 = \tfrac{2}{3}x + \tfrac{2}{3}$$
$$y = \tfrac{2}{3}x + \tfrac{14}{3}$$

32. Through $(1, -1)$, $m = -\tfrac{3}{4}$

Use the point-slope form.

$$y - (-1) = -\tfrac{3}{4}(x - 1)$$
$$y + 1 = -\tfrac{3}{4}x + \tfrac{3}{4}$$
$$y = -\tfrac{3}{4}x - \tfrac{1}{4}$$

33. $m = -\tfrac{1}{4}, b = \tfrac{3}{2}$

Use the slope-intercept form, $y = mx + b$.

$$y = -\tfrac{1}{4}x + \tfrac{3}{2}$$

34. Slope 0, through $(-4, 1)$

Horizontal lines have 0 slope and equations of the form $y = k$. In this case, k must equal 1 since the line goes through $(-4, 1)$, so the equation is

$$y = 0x + 1 \quad \text{or} \quad y = 1.$$

35. Through $\left(\tfrac{1}{3}, -\tfrac{5}{4}\right)$ with undefined slope

Vertical lines have undefined slope and equations of the form $x = k$. In this case, k must equal $\tfrac{1}{3}$ since the line goes through $\left(\tfrac{1}{3}, -\tfrac{5}{4}\right)$, so the equation is

$$x = \tfrac{1}{3}.$$

It is not possible to express this equation as $y = mx + b$.

36. **[3.3]** Vertical lines have undefined slopes. The answer is **A**.

37. **[3.2]** Two graphs pass through $(0, -3)$. **C** and **D** are the answers.

38. **[3.2]** Three graphs pass through the point $(-3, 0)$. **A**, **B**, and **D** are the answers.

39. **[3.3]** Lines that fall from left to right have negative slope. The answer is **D**.

40. **[3.2]** $y = -3$ is a horizontal line passing through $(0, -3)$. **C** is the answer.

41. **[3.3]** **B** is the only graph that has a positive slope, so it is the only one we need to investigate. **B** passes through the points $(0, 3)$ and $(-3, 0)$. Find the slope.

$$m = \frac{0 - 3}{-3 - 0} = \frac{-3}{-3} = 1$$

B is the answer.

42. **[3.3]** $y = -2x - 5$

The equation is in the slope-intercept form, so the slope is -2 and the y-intercept is $(0, -5)$. To find the x-intercept, let $y = 0$.

$$0 = -2x - 5 \quad \textit{Let y = 0.}$$
$$2x = -5$$
$$x = -\tfrac{5}{2}$$

The x-intercept is $\left(-\tfrac{5}{2}, 0\right)$.

Graph the line using the intercepts.

continued

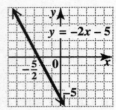

43. **[3.3]** $x + 3y = 0$

Solve the equation for y.

$$x + 3y = 0$$
$$3y = -x \quad \textit{Subtract x.}$$
$$y = -\tfrac{1}{3}x \quad \textit{Divide by 3.}$$

From this slope-intercept form, we see that the slope is $-\tfrac{1}{3}$ and the y-intercept is $(0, 0)$, which is also the x-intercept. To find another point, let $y = 1$.

$$x + 3(1) = 0 \quad \textit{Let y = 1.}$$
$$x = -3$$

So the point $(-3, 1)$ is on the graph. Graph the line through $(0, 0)$ and $(-3, 1)$.

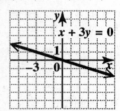

44. **[3.3]** $y - 5 = 0$ or $y = 5$

This is a horizontal line passing through the point $(0, 5)$, which is the y-intercept.

There is no x-intercept.
Horizontal lines have slopes of 0.

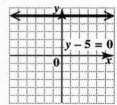

45. **[3.4]** $m = -\tfrac{1}{4}$, $b = -\tfrac{5}{4}$

Substitute the values $m = -\tfrac{1}{4}$ and $b = -\tfrac{5}{4}$ into the slope-intercept form.

$$y = mx + b$$
$$y = -\tfrac{1}{4}x - \tfrac{5}{4}$$

46. **[3.4]** Through $(8, 6)$, $m = -3$
Use the point-slope form with $(x_1, y_1) = (8, 6)$ and $m = -3$.

$$y - y_1 = m(x - x_1)$$
$$y - 6 = -3(x - 8)$$
$$y - 6 = -3x + 24$$
$$y = -3x + 30$$

47. **[3.4]** Through $(3, -5)$ and $(-4, -1)$

First, find the slope of the line.

$$m = \frac{-1 - (-5)}{-4 - 3} = \frac{4}{-7} = -\frac{4}{7}$$

Now use either point and the slope in the point-slope form. If we use $(3, -5)$, we get the following.

$$y - y_1 = m(x - x_1)$$
$$y - (-5) = -\tfrac{4}{7}(x - 3)$$
$$y + 5 = -\tfrac{4}{7}x + \tfrac{12}{7}$$
$$y = -\tfrac{4}{7}x - \tfrac{23}{7} \quad \textit{Subtract 5 = } \tfrac{35}{7}.$$

48. In 1997, 44.2% of four-year college students in public institutions earned a degree within five years. In 2002, it was 41.2%. The total decrease in percent was

$$44.2\% - 41.2\% = 3.0\%.$$

49. Because the graph falls from left to right, the slope is negative.

50. Let x represent the year and y represent the percent in the ordered pairs (x, y). The ordered pairs are $(1997, 44.2)$ and $(2002, 41.2)$.

51. Find the slope using $(x_1, y_1) = (1997, 44.2)$ and $(x_2, y_2) = (2002, 41.2)$.

$$m = \frac{\text{change in } y}{\text{change in } x} = \frac{y_2 - y_1}{x_2 - x_1}$$
$$= \frac{41.2 - 44.2}{2002 - 1997} = \frac{-3}{5} = -0.6$$

Now use the point-slope form with $m = -0.6$ and $(x_1, y_1) = (1997, 44.2)$.

$$y - y_1 = m(x - x_1)$$
$$y - 44.2 = -0.6(x - 1997)$$
$$y - 44.2 = -0.6x + 1198.2$$
$$y = -0.6x + 1242.4$$

52. From the slope-intercept form, $y = mx + b$, the slope of the line is -0.6. Yes, the slope agrees with the answer in Exercise 49 because the slope is negative.

53. Substitute 1998, 1999, 2000, and 2001 for x in
$y = -0.6x + 1242.4$ to complete the table.

Year (x)	Percent (y)
1998	43.6
1999	43.0
2000	42.4
2001	41.8

54. For 2003, let $x = 2003$.

$$y = -0.6(2003) + 1242.4$$
$$= -1201.8 + 1242.4$$
$$= 40.6$$

The prediction is 40.6% for 2003. We cannot be
sure that this prediction is accurate since the
equation is based on data only from 1997 through
2002.

Chapter 3 Test

1. $3x + 5y = -30;\ (0,\ \ ),\ (\ \ ,0),\ (\ \ ,-3)$

$$\begin{aligned} 3x + 5y &= -30 \\ 3(0) + 5y &= -30 \quad \textit{Let x = 0.} \\ 5y &= -30 \\ y &= -6 \end{aligned}$$

$$\begin{aligned} 3x + 5y &= -30 \\ 3x + 5(0) &= -30 \quad \textit{Let y = 0.} \\ 3x &= -30 \\ x &= -10 \end{aligned}$$

$$\begin{aligned} 3x + 5y &= -30 \\ 3x + 5(-3) &= -30 \quad \textit{Let y = -3.} \\ 3x - 15 &= -30 \\ 3x &= -15 \\ x &= -5 \end{aligned}$$

The ordered pairs are $(0, -6)$, $(-10, 0)$, and
$(-5, -3)$.

2.
$$\begin{aligned} 4x - 7y &= 9 \\ 4(4) - 7(-1) &= 9 \quad ? \quad \textit{Let x = 4, y = -1.} \\ 16 + 7 &= 9 \quad ? \\ 23 &= 9 \quad ? \quad \textit{No} \end{aligned}$$

So $(4, -1)$ is not a solution of $4x - 7y = 9$.

3. To find the x-intercept, let $y = 0$ and solve for x.
To find the y-intercept, let $x = 0$ and solve for y.

4. $3x + y = 6$

If $y = 0$, $x = 2$, so the x-intercept is $(2, 0)$.
If $x = 0$, $y = 6$, so the y-intercept is $(0, 6)$.

A third point, such as $(1, 3)$, can be used as a
check. Draw a line through $(0, 6)$, $(1, 3)$, and
$(2, 0)$.

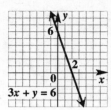

5. $y - 2x = 0$

Solving for y gives us the slope-intercept form of
the line, $y = 2x$. We see that the y-intercept is
$(0, 0)$ and so the x-intercept is also $(0, 0)$. The
slope is 2 and can be written as

$$m = \frac{\text{rise}}{\text{run}} = \frac{2}{1}.$$

Starting at the origin and moving to the right 1
unit and then up 2 units gives us the point $(1, 2)$.
Draw a line through $(0, 0)$ and $(1, 2)$.

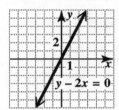

6. $x + 3 = 0$ can also be written as $x = -3$. Its
graph is a vertical line with x-intercept $(-3, 0)$.
There is no y-intercept.

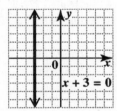

7. $y = 1$ is the graph of a horizontal line with
y-intercept $(0, 1)$. There is no x-intercept.

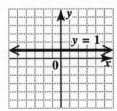

8. $x - y = 4$

If $y = 0$, $x = 4$, so the x-intercept is $(4, 0)$.
If $x = 0$, $y = -4$, so the y-intercept is $(0, -4)$.

A third point, such as $(2, -2)$, can be used as a
check. Draw a line through $(0, -4)$, $(2, -2)$, and
$(4, 0)$.

continued

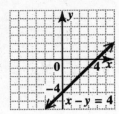

9. Through $(-4, 6)$ and $(-1, -2)$

Use the definition of slope with $(x_1, y_1) = (-4, 6)$ and $(x_2, y_2) = (-1, -2)$.

$$\text{slope } m = \frac{\text{change in } y}{\text{change in } x} = \frac{y_2 - y_1}{x_2 - x_1}$$
$$= \frac{-2 - 6}{-1 - (-4)} = \frac{-8}{3} = -\frac{8}{3}$$

10. $2x + y = 10$

To find the slope, solve the given equation for y.

$$2x + y = 10$$
$$y = -2x + 10$$

The equation is now written in $y = mx + b$ form, so the slope is given by the coefficient of x, which is -2.

11. $x + 12 = 0$ can also be written as $x = -12$. Its graph is a vertical line with x-intercept $(-12, 0)$. The slope is undefined.

12. The indicated points are $(0, -4)$ and $(2, 1)$. Use the definition of slope with $(x_1, y_1) = (0, -4)$ and $(x_2, y_2) = (2, 1)$.

$$\text{slope } m = \frac{\text{change in } y}{\text{change in } x} = \frac{y_2 - y_1}{x_2 - x_1}$$
$$= \frac{1 - (-4)}{2 - 0} = \frac{5}{2}$$

13. $y - 4 = 6$ can also be written as $y = 10$. Its graph is a horizontal line with y-intercept $(0, 10)$. Its slope is 0, as is the slope of any line parallel to it.

14. Through $(-1, 4)$; $m = 2$

Let $x_1 = -1$, $y_1 = 4$, and $m = 2$ in the point-slope form.

$$y - y_1 = m(x - x_1)$$
$$y - 4 = 2[x - (-1)]$$
$$y - 4 = 2(x + 1)$$
$$y - 4 = 2x + 2$$
$$y = 2x + 6$$

15. The indicated points are $(0, -4)$ and $(2, 1)$. The slope of the line through those points is

$$m = \frac{1 - (-4)}{2 - 0} = \frac{5}{2}.$$

The y-intercept is $(0, -4)$, so the slope-intercept form is

$$y = \frac{5}{2}x - 4.$$

16. Through $(2, -6)$ and $(1, 3)$

The slope of the line through these points is

$$m = \frac{3 - (-6)}{1 - 2} = \frac{9}{-1} = -9.$$

Use the point-slope form of a line with $(1, 3) = (x_1, y_1)$ and $m = -9$.

$$y - y_1 = m(x - x_1)$$
$$y - 3 = -9(x - 1)$$
$$y - 3 = -9x + 9$$
$$y = -9x + 12$$

17. The slope of the line is positive since food and drink sales are increasing.

18. Two ordered pairs are $(0, 43)$ and $(30, 376)$. Use these points to find the slope.

$$m = \frac{\text{change in } y}{\text{change in } x} = \frac{y_2 - y_1}{x_2 - x_1}$$
$$= \frac{376 - 43}{30 - 0} = \frac{333}{30} = 11.1$$

The slope is 11.1.

19. For 1990, $x = 1990 - 1970 = 20$.

$$y = 11.1x + 43$$
$$y = 11.1(20) + 43$$
$$y = 265$$

For 1995, $x = 1995 - 1970 = 25$.

$$y = 11.1x + 43$$
$$y = 11.1(25) + 43$$
$$y = 320.5$$

The approximate food and drink sales for 1990 and 1995 were $265 billion and $320.5 billion.

20. $(30, 376)$; $x = 30$ represents $1970 + 30 = 2000$. In 2000, food and drink sales were $376 billion.

Cumulative Review Exercises (Chapters 1–3)

1. $10\frac{5}{8} - 3\frac{1}{10} = \frac{85}{8} - \frac{31}{10}$

$$= \frac{425}{40} - \frac{124}{40}$$

$$= \frac{301}{40}, \text{ or } 7\frac{21}{40}$$

2. $\dfrac{3}{4} \div \dfrac{1}{8} = \dfrac{3}{4} \cdot \dfrac{8}{1} = \dfrac{3 \cdot 2 \cdot 4}{4 \cdot 1} = 3 \cdot 2 = 6$

3. $5 - (-4) + (-2) = 9 + (-2) = 7$

4. $\dfrac{(-3)^2 - (-4)(2^4)}{5(2) - (-2)^3}$

$= \dfrac{9 - (-4)(16)}{10 - (-8)}$ *Do exponents first.*

$= \dfrac{9 - (-64)}{10 - (-8)}$ *Multiply.*

$= \dfrac{9 + 64}{10 + 8} = \dfrac{73}{18}$, or $4\dfrac{1}{18}$

5. $\dfrac{4(3 - 9)}{2 - 6} \geq 6$?

$\dfrac{4(-6)}{-4} \geq 6$?

$\dfrac{-24}{-4} \geq 6$?

$6 \geq 6$

The statement is *true* since $6 = 6$.

6. $xz^3 - 5y^2 = (-2)(-1)^3 - 5(-3)^2$

Let x = −2, y = −3, z = −1.

$= (-2)(-1) + (-5)(9)$

$= 2 + (-45)$

$= -43$

7. $3(-2 + x) = 3 \cdot (-2) + 3(x)$

$= -6 + 3x$

illustrates the *distributive property*.

8. $-4p - 6 + 3p + 8 = (-4p + 3p) + (-6 + 8)$

$= -p + 2$

9. $V = \frac{1}{3}\pi r^2 h$

$3V = \pi r^2 h$ *Multiply by 3.*

$\dfrac{3V}{\pi r^2} = h$ *Divide by πr^2.*

10. $6 - 3(1 + a) = 2(a + 5) - 2$

$6 - 3 - 3a = 2a + 10 - 2$

$3 - 3a = 2a + 8$

$-5a = 5$

$a = -1$

The solution set is $\{-1\}$.

11. $-(m - 3) = 5 - 2m$

$-m + 3 = 5 - 2m$ *Distributive property*

$m + 3 = 5$ *Add 2m.*

$m = 2$ *Subtract 3.*

The solution set is $\{2\}$.

12. $\dfrac{x - 2}{3} = \dfrac{2x + 1}{5}$

$(x - 2)(5) = (3)(2x + 1)$ *Cross products*

$5x - 10 = 6x + 3$

$-10 = x + 3$

$-13 = x$

The solution set is $\{-13\}$.

13. $-2.5x < 6.5$

$\dfrac{-2.5x}{-2.5} > \dfrac{6.5}{-2.5}$

Divide by −2.5; reverse the symbol.

$x > -2.6$

Thus, the solution set is $(-2.6, \infty)$.

-2.6

14. $4(x + 3) - 5x < 12$

$4x + 12 - 5x < 12$ *Distributive property*

$-x + 12 < 12$ *Combine like terms.*

$-x < 0$ *Subtract 12.*

$x > 0$ *Divide by −1; reverse the symbol.*

Thus, the solution set is $(0, \infty)$.

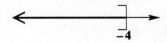

0

15. $\frac{2}{3}x - \frac{1}{6}x \leq -2$

$6\left(\frac{2}{3}x - \frac{1}{6}x\right) \leq 6(-2)$ *Multiply by 6.*

$6\left(\frac{2}{3}x\right) - 6\left(\frac{1}{6}x\right) \leq 6(-2)$ *Dist. property*

$4x - x \leq -12$

$3x \leq -12$

$x \leq -4$

Thus, the solution set is $(-\infty, -4]$.

-4

16. Let x = average annual earnings for a person with a high school diploma.

Then $x + 18{,}436$ = average annual earnings for a person with a bachelor's degree.

$x + (x + 18{,}436) = 61{,}590$

$2x + 18{,}436 = 61{,}590$

$2x = 43{,}154$

$x = 21{,}577$

A person with a high school diploma can expect to earn \$21,577/year while a person with a bachelor's degree can expect to earn

$\$21{,}577 + \$18{,}436 = \$40{,}013/\text{year}.$

17. $C = 2\pi r$ *Circumference formula*
$80 = 2\pi r$ *Let C = 80.*
$\dfrac{80}{2\pi} = r$ *Divide by 2π.*
$r \approx 13$

The radius is about 13 miles.

18. (a) $y = -0.5075x + 95.4179$

Let $x = 12$.
$y = -0.5075(12) + 95.4179 = 89.3279 \approx 89.33$

Let $x = 28$.
$y = -0.5075(28) + 95.4179 = 81.2079 \approx 81.21$

Let $x = 36$.
$y = -0.5075(36) + 95.4179 = 77.1479 \approx 77.15$

The completed table follows.

x	y
12	89.33
28	81.21
36	77.15

(b) $(20, 85.27)$; $x = 20$ represents
$1960 + 20 = 1980$. In 1980, the winning time was
85.27 sec.

19. (a) Multiply 14% (or 0.14) by the total of
$50,000.

$$0.14(50,000) = 7000$$

$7000 is expected to go toward home purchase.

(b) Multiply 20% (or 0.20) by the total of
$50,000.

$$0.20(50,000) = 10,000$$

$10,000 is expected to go toward retirement.

(c) Since the sector for paying off debt or funding
children's education is about three times larger
than the sector for retirement, 3($10,000) or about
$30,000 is expected to go toward paying off debt
or funding children's education.

20.   To find the x-intercept, let $y = 0$.

$$-3x + 4y = 12$$
$$-3x + 4(0) = 12$$
$$-3x = 12$$
$$x = -4$$

The x-intercept is $(-4, 0)$.

To find the y-intercept, let $x = 0$.

$$-3x + 4y = 12$$
$$-3(0) + 4y = 12$$
$$4y = 12$$
$$y = 3$$

The y-intercept is $(0, 3)$.

21.   To find the slope of the line, solve the equation for
y.

$$-3x + 4y = 12$$
$$4y = 3x + 12$$
$$y = \tfrac{3}{4}x + 3$$

The slope is the coefficient of x, $\tfrac{3}{4}$.

22.   To find a third point, let $x = 4$.

$$-3x + 4y = 12$$
$$-3(4) + 4y = 12$$
$$4y = 12 + 12 = 24$$
$$y = 6$$

Plot the points $(-4, 0)$, $(0, 3)$, and $(4, 6)$ and draw
a line through them.

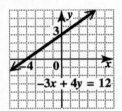

23. $x + 5y = -6$
$5y = -x - 6$
$y = -\tfrac{1}{5}x - \tfrac{6}{5}$

The slope of the first line is $-\tfrac{1}{5}$.

The slope of the second line, $y = 5x - 8$, is 5.
Since $-\tfrac{1}{5}$ is the negative reciprocal of 5, the lines
are *perpendicular*.

24.   Through $(2, -5)$ with slope 3.

Use the point-slope form of a line.

$$y - y_1 = m(x - x_1)$$
$$y - (-5) = 3(x - 2)$$
$$y + 5 = 3x - 6$$
$$y = 3x - 11$$

25.   Through $(0, 4)$ and $(2, 4)$

$$\text{slope } m = \frac{\text{change in } y}{\text{change in } x} = \frac{4 - 4}{2 - 0} = \frac{0}{2} = 0$$

Since the slope is 0, the line is horizontal.
Horizontal lines have equations of the form $y = k$.
An equation of the line is $y = 4$.

CHAPTER 4 EXPONENTS AND POLYNOMIALS

4.1 The Product Rule and Power Rules for Exponents

1. $3^3 = 3 \cdot 3 \cdot 3 = 27$, so the statement $3^3 = 9$ is *false*.

3. $(a^2)^3 = a^{2(3)} = a^6$, so the statement $(a^2)^3 = a^5$ is *false*.

5. $\underbrace{w \cdot w \cdot w \cdot w \cdot w \cdot w}_{6w\text{'s}} = w^6$

7. $\left(\frac{1}{2}\right)\left(\frac{1}{2}\right)\left(\frac{1}{2}\right)\left(\frac{1}{2}\right)\left(\frac{1}{2}\right)\left(\frac{1}{2}\right) = \left(\frac{1}{2}\right)^6$

9. $(-4)(-4)(-4) = (-4)^3$

11. $(-7x)(-7x)(-7x)(-7x) = (-7x)^4$

13. In $(-3)^4$, -3 is the base.
$$(-3)^4 = (-3)(-3)(-3)(-3) = 81$$
In -3^4, 3 is the base.
$$-3^4 = -(3 \cdot 3 \cdot 3 \cdot 3) = -81$$

15. In the exponential expression 3^5, the base is 3 and the exponent is 5.
$$3^5 = 3 \cdot 3 \cdot 3 \cdot 3 \cdot 3 = 243$$

17. In the expression $(-3)^5$, the base is -3 and the exponent is 5.
$$(-3)^5 = (-3)(-3)(-3)(-3)(-3) = -243$$

19. In the expression $(-6x)^4$, the base is $-6x$ and the exponent is 4.

21. In the expression $-6x^4$, -6 is not part of the base. The base is x and the exponent is 4.

23. The product rule does not apply to $5^2 + 5^3$ because the expression is a sum, not a product. The product rule would apply if we had $5^2 \cdot 5^3$.
$$5^2 + 5^3 = 25 + 125 = 150$$

25. $5^2 \cdot 5^6 = 5^{2+6} = 5^8$

27. $4^2 \cdot 4^7 \cdot 4^3 = 4^{2+7+3} = 4^{12}$

29. $(-7)^3(-7)^6 = (-7)^{3+6} = (-7)^9$

31. $t^3 \cdot t^8 \cdot t^{13} = t^{3+8+13} = t^{24}$

33. $(-8r^4)(7r^3) = -8 \cdot 7 \cdot r^4 \cdot r^3$
$$= -56r^{4+3}$$
$$= -56r^7$$

35. $(-6p^5)(-7p^5) = (-6)(-7)p^5 \cdot p^5$
$$= 42p^{5+5}$$
$$= 42p^{10}$$

37. $(5x^2)(-2x^3)(3x^4)$
$$= (5)(-2)(3)x^2 \cdot x^3 \cdot x^4$$
$$= (-10)(3)x^{2+3+4}$$
$$= -30x^9$$

39. $3^8 + 3^9$ is a sum, so the product rule does not apply.

41. $5^8 \cdot 3^9$ is a product with different bases, so the product rule does not apply.

43. $(4^3)^2 = 4^{3 \cdot 2}$ *Power rule (a)*
$$= 4^6$$

45. $(t^4)^5 = t^{4 \cdot 5} = t^{20}$ *Power rule (a)*

47. $(7r)^3 = 7^3 r^3$ *Power rule (b)*

49. $(5xy)^5 = 5^5 x^5 y^5$ *Power rule (b)*

51. $(-5^2)^6 = (-1 \cdot 5^2)^6$
$$= (-1)^6 \cdot (5^2)^6 \quad \textit{Power rule (b)}$$
$$= 1 \cdot 5^{2 \cdot 6} \qquad \textit{Power rule (a)}$$
$$= 1 \cdot 5^{12} = 5^{12}$$

53. $(-8^3)^5 = (-1 \cdot 8^3)^5$
$$= (-1)^5 \cdot (8^3)^5 \quad \textit{Power rule (b)}$$
$$= -1 \cdot 8^{3 \cdot 5} \qquad \textit{Power rule (a)}$$
$$= -8^{15}$$

55. $8(qr)^3 = 8q^3 r^3$ *Power rule (b)*

57. $\left(\frac{9}{5}\right)^8 = \frac{9^8}{5^8}$ *Power rule (c)*

59. $\left(\frac{1}{2}\right)^3 = \frac{1^3}{2^3} = \frac{1}{2^3}$ *Power rule (c)*

61. $\left(\frac{a}{b}\right)^3 (b \neq 0) = \frac{a^3}{b^3}$ *Power rule (c)*

63. $\left(\frac{5}{2}\right)^3 \cdot \left(\frac{5}{2}\right)^2 = \left(\frac{5}{2}\right)^{3+2}$ *Product rule*
$$= \left(\frac{5}{2}\right)^5$$
$$= \frac{5^5}{2^5} \qquad \textit{Power rule (c)}$$

65. $\left(\dfrac{9}{8}\right)^3 \cdot 9^2 = \dfrac{9^3}{8^3} \cdot \dfrac{9^2}{1}$ *Power rule (c)*

$\qquad = \dfrac{9^3 \cdot 9^2}{8^3 \cdot 1}$ *Multiply fractions*

$\qquad = \dfrac{9^{3+2}}{8^3}$ *Product rule*

$\qquad = \dfrac{9^5}{8^3}$

67. $(2x)^9(2x)^3 = (2x)^{9+3}$ *Product rule*

$\qquad = (2x)^{12}$

$\qquad = 2^{12}x^{12}$ *Product rule (b)*

69. $(-6p)^4(-6p)$

$\qquad = (-6p)^4(-6p)^1$

$\qquad = (-6p)^5$ *Product rule*

$\qquad = (-1)^5 6^5 p^5$ *Power rule (b)*

$\qquad = -6^5 p^5$

71. $(6x^2y^3)^5 = 6^5(x^2)^5(y^3)^5$ *Power rule (b)*

$\qquad = 6^5 x^{2\cdot5} y^{3\cdot5}$ *Power rule (a)*

$\qquad = 6^5 x^{10} y^{15}$

73. $(x^2)^3(x^3)^5 = x^6 \cdot x^{15}$ *Power rule (a)*

$\qquad = x^{21}$ *Product rule*

75. $(2w^2x^3y)^2(x^4y)^5$

$\qquad = [2^2(w^2)^2(x^3)^2y^2][(x^4)^5y^5]$ *Power rule (b)*

$\qquad = (2^2 w^4 x^6 y^2)(x^{20} y^5)$ *Power rule (a)*

$\qquad = 2^2 w^4 (x^6 x^{20})(y^2 y^5)$

Commutative and associative properties

$\qquad = 4w^4 x^{26} y^7$

77. $(-r^4 s)^2(-r^2 s^3)^5$

$\qquad = [(-1)r^4 s]^2 [(-1)r^2 s^3]^5$

$\qquad = [(-1)^2(r^4)^2 s^2][(-1)^5(r^2)^5(s^3)^5]$

$\qquad$ *Power rule (b)*

$\qquad = [(-1)^2 r^8 s^2][(-1)^5 r^{10} s^{15}]$ *Power rule (a)*

$\qquad = (-1)^7 r^{18} s^{17}$ *Product rule*

$\qquad = -r^{18} s^{17}$

79. $\left(\dfrac{5a^2 b^5}{c^6}\right)^3$ $(c \neq 0)$

$\qquad = \dfrac{(5a^2 b^5)^3}{(c^6)^3}$ *Power rule (c)*

$\qquad = \dfrac{5^3(a^2)^3(b^5)^3}{(c^6)^3}$ *Power rule (b)*

$\qquad = \dfrac{125 a^6 b^{15}}{c^{18}}$ *Power rule (a)*

81. To simplify $(10^2)^3$ as 1000^6 is not correct. Using power rule (a) to simplify $(10^2)^3$, we obtain

$\qquad (10^2)^3 = 10^{2\cdot3}$

$\qquad\qquad = 10^6$

$\qquad\qquad = 10\cdot10\cdot10\cdot10\cdot10\cdot10$

$\qquad\qquad = 1{,}000{,}000.$

83. Use the formula for the area of a rectangle, $A = LW$, with $L = 4x^3$ and $W = 3x^2$.

$\qquad A = (4x^3)(3x^2)$

$\qquad\quad = 4\cdot3\cdot x^3 \cdot x^2$

$\qquad\quad = 12x^5$

85. Use the formula for the area of a parallelogram, $A = bh$, with $b = 2p^5$ and $h = 3p^2$.

$\qquad A = (2p^5)(3p^2)$

$\qquad\quad = 2\cdot3\cdot p^5 \cdot p^2$

$\qquad\quad = 6p^7$

87. Use the formula for the volume of a cube, $V = e^3$, with $e = 5x^2$.

$\qquad V = (5x^2)^3$

$\qquad\quad = 5^3(x^2)^3$

$\qquad\quad = 125x^6$

89. If a is a number greater than 1, $a^4 > a^3$, $-(-a)^3$ is positive, $-a^3$ is negative, $(-a)^4$ is positive, and $-a^4$ is negative. Therefore, in order from least to greatest, we have

$\qquad -a^4, \ -a^3, \ -(-a)^3, \ (-a)^4.$

Another way to determine the order is to choose a number greater than 1 and substitute it for a in each expression, and then arrange the terms from least to greatest.

91. Use the formula $A = P(1 + r)^n$ with $P = \$250$, $r = 0.04$, and $n = 5$.

$\qquad A = 250(1 + 0.04)^5$

$\qquad\quad = 250(1.04)^5$

$\qquad\quad \approx 304.16$

The amount of money in the account will be $304.16.

93. Use the formula $A = P(1 + r)^n$ with $P = \$1500$, $r = 0.035$, and $n = 6$.

$\qquad A = 1500(1 + 0.035)^6$

$\qquad\quad = 1500(1.035)^6$

$\qquad\quad \approx 1843.88$

The amount of money in the account will be $1843.88.

95. The reciprocal of 5 is $\frac{1}{5}$.

97. The reciprocal of $-\frac{1}{4}$ is $\frac{1}{-\frac{1}{4}} = 1\left(-\frac{4}{1}\right) = -4$.

99. $8 - (-2) = 8 + 2 = 10$

101. "Subtract -5 from -2" translates and simplifies as:
$$-2 - (-5) = -2 + 5$$
$$= 3$$

4.2 Integer Exponents and the Quotient Rule

1. By definition, $a^0 = 1 \; (a \neq 0)$, so $9^0 = 1$.

3. $(-4)^0 = 1$ *Definition of zero exponent*

5. $-9^0 = -(9^0) = -(1) = -1$

7. $-(-4)^0 = -1 \cdot (-4)^0 = -1 \cdot 1 = -1$

9. $(-2)^0 - 2^0 = 1 - 1 = 0$

11. $\dfrac{0^{10}}{10^0} = \dfrac{0}{1} = 0$

13. **(a)** $x^0 = 1$ (Choice **B**)

(b) $-x^0 = -1 \cdot x^0 = -1 \cdot 1 = -1$ (Choice **C**)

(c) $3x^0 = 3 \cdot x^0 = 3 \cdot 1 = 3$ (Choice **D**)

(d) $(3x)^0 = 1$ (Choice **B**)

(e) $-3x^0 = -3 \cdot x^0 = -3 \cdot 1 = -3$ (Choice **E**)

(f) $(-3x)^0 = 1$ (Choice **B**)

15. $7^0 + 9^0 = 1 + 1 = 2$

17. $4^{-3} = \dfrac{1}{4^3}$ *Definition of negative exponent*
$$= \tfrac{1}{64}$$

19. When we evaluate a fraction raised to a negative exponent, we can use a shortcut. Note that
$$\left(\frac{a}{b}\right)^{-n} = \frac{1}{\left(\frac{a}{b}\right)^n} = \frac{1}{\frac{a^n}{b^n}} = \frac{b^n}{a^n} = \left(\frac{b}{a}\right)^n.$$

In words, a fraction raised to the negative of a number is equal to its reciprocal raised to the number. We will use the simple phrase "$\frac{a}{b}$ and $\frac{b}{a}$ are reciprocals" to indicate our use of this evaluation shortcut.
$$\left(\frac{1}{2}\right)^{-4} = 2^4 = 16 \quad \text{\small $\frac{1}{2}$ and 2 are reciprocals.}$$

21. $\left(\dfrac{6}{7}\right)^{-2} = \left(\dfrac{7}{6}\right)^2$ *$\frac{6}{7}$ and $\frac{7}{6}$ are reciprocals.*
$$= \frac{7^2}{6^2} \quad \textit{Power rule (c)}$$
$$= \frac{49}{36}$$

23. $(-3)^{-4} = \dfrac{1}{(-3)^4}$
$$= \tfrac{1}{81}$$

25. $5^{-1} + 3^{-1} = \dfrac{1}{5} + \dfrac{1}{3}$
$$= \frac{3}{15} + \frac{5}{15} = \frac{8}{15}$$

27. $3^{-2} - 2^{-1} = \dfrac{1}{3^2} - \dfrac{1}{2^1}$
$$= \tfrac{1}{9} - \tfrac{1}{2}$$
$$= \frac{2}{18} - \frac{9}{18} = -\frac{7}{18}$$

29. $\dfrac{5^8}{5^5} = 5^{8-5} = 5^3 = 125$ *Quotient rule*

31. $\dfrac{3^{-2}}{5^{-3}} = \dfrac{5^3}{3^2}$ *Changing from negative to positive exponents*
$$= \tfrac{125}{9}$$

33. $\dfrac{5}{5^{-1}} = \dfrac{5^1}{5^{-1}} = 5^1 \cdot 5^1$ *Changing from negative to positive exponents*
$$= 5^{1+1} = 5^2 = 25$$

35. $\dfrac{x^{12}}{x^{-3}} = x^{12} \cdot x^3$ *Changing from negative to positive exponents*
$$= x^{12+3} = x^{15}$$

37. $\dfrac{1}{6^{-3}} = 6^3 = 216$ *Changing from negative to positive exponents*

39. $\dfrac{2}{r^{-4}} = 2r^4$ *Changing from negative to positive exponents*

41. $\dfrac{4^{-3}}{5^{-2}} = \dfrac{5^2}{4^3} = \dfrac{25}{64}$ *Changing from negative to positive exponents*

43. $p^5 q^{-8} = \dfrac{p^5}{q^8}$ *Changing from negative to positive exponents*

45. $\dfrac{r^5}{r^{-4}} = r^5 \cdot r^4 = r^{5+4} = r^9$

Or we can use the quotient rule:
$$\frac{r^5}{r^{-4}} = r^{5-(-4)} = r^{5+4} = r^9$$

47. $\dfrac{x^{-3}y}{4z^{-2}} = \dfrac{yz^2}{4x^3}$

49. Treat the expression in parentheses as a single variable; that is, treat $(a+b)$ as you would treat x.
$$\frac{(a+b)^{-3}}{(a+b)^{-4}} = (a+b)^{-3-(-4)}$$
$$= (a+b)^{-3+4}$$
$$= (a+b)^1 = a + b$$

continued

Another Method:

$$\frac{(a+b)^{-3}}{(a+b)^{-4}} = \frac{(a+b)^4}{(a+b)^3}$$
$$= (a+b)^{4-3}$$
$$= (a+b)^1 = a+b$$

51. $\dfrac{(x+2y)^{-3}}{(x+2y)^{-5}} = (x+2y)^{-3-(-5)}$

$$= (x+2y)^{-3+5}$$
$$= (x+2y)^2$$

53. In simplest form, $\frac{25}{25} = 1$.

54. $\dfrac{25}{25} = \dfrac{5^2}{5^2}$

55. $\dfrac{5^2}{5^2} = 5^{2-2} = 5^0$

56. $1 = 5^0$; this supports the definition of 0 as an exponent.

57. $\dfrac{(7^4)^3}{7^9} = \dfrac{7^{4\cdot3}}{7^9}$ *Power rule (a)*

$$= \frac{7^{12}}{7^9}$$
$$= 7^{12-9} \quad \textit{Quotient rule}$$
$$= 7^3 = 343$$

59. $x^{-3} \cdot x^5 \cdot x^{-4}$

$$= x^{-3+5+(-4)} \quad \textit{Product rule}$$
$$= x^{-2}$$
$$= \frac{1}{x^2} \qquad \begin{array}{l}\textit{Definition of}\\ \textit{negative exponent}\end{array}$$

61. $\dfrac{(3x)^{-2}}{(4x)^{-3}} = \dfrac{(4x)^3}{(3x)^2}$ $\begin{array}{l}\textit{Changing from}\\ \textit{negative to}\\ \textit{positive exponents}\end{array}$

$$= \frac{4^3 x^3}{3^2 x^2} \qquad \textit{Power rule (b)}$$
$$= \frac{4^3 x^{3-2}}{3^2} \qquad \textit{Quotient rule}$$
$$= \frac{4^3 x}{3^2} = \frac{64x}{9}$$

63. $\left(\dfrac{x^{-1}y}{z^2}\right)^{-2} = \dfrac{(x^{-1}y)^{-2}}{(z^2)^{-2}}$ *Power rule (c)*

$$= \frac{(x^{-1})^{-2}y^{-2}}{(z^2)^{-2}} \quad \textit{Power rule (b)}$$
$$= \frac{x^2 y^{-2}}{z^{-4}} \qquad \textit{Power rule (a)}$$
$$= \frac{x^2 z^4}{y^2} \qquad \begin{array}{l}\textit{Definition of}\\ \textit{negative exponent}\end{array}$$

65. $(6x)^4(6x)^{-3} = (6x)^{4+(-3)}$ *Product rule*
$$= (6x)^1 = 6x$$

67. $\dfrac{(m^7 n)^{-2}}{m^{-4}n^3} = \dfrac{(m^7)^{-2}n^{-2}}{m^{-4}n^3}$

$$= \frac{m^{7(-2)}n^{-2}}{m^{-4}n^3}$$
$$= \frac{m^{-14}n^{-2}}{m^{-4}n^3}$$
$$= m^{-14-(-4)}n^{-2-3}$$
$$= m^{-10}n^{-5}$$
$$= \frac{1}{m^{10}n^5}$$

69. $\dfrac{(x^{-1}y^2 z)^{-2}}{(x^{-3}y^3 z)^{-1}} = \dfrac{(x^{-1})^{-2}(y^2)^{-2}z^{-2}}{(x^{-3})^{-1}(y^3)^{-1}z^{-1}}$

$$= \frac{x^2 y^{-4}z^{-2}}{x^3 y^{-3}z^{-1}}$$
$$= \frac{x^2 y^3 z^1}{x^3 y^4 z^2}$$
$$= \frac{1}{xyz}$$

71. $\left(\dfrac{xy^{-2}}{x^2 y}\right)^{-3} = \dfrac{x^{-3}(y^{-2})^{-3}}{(x^2)^{-3}y^{-3}}$

$$= \frac{x^{-3}y^6}{x^{-6}y^{-3}}$$
$$= \frac{x^6 y^6 y^3}{x^3}$$
$$= x^3 y^9$$

73. $\dfrac{(4a^2 b^3)^{-2}(2ab^{-1})^3}{(a^3 b)^{-4}}$

$$= \frac{(a^3 b)^4 (2ab^{-1})^3}{(4a^2 b^3)^2}$$
$$= \frac{(a^3)^4 b^4 2^3 a^3 (b^{-1})^3}{4^2 (a^2)^2 (b^3)^2}$$
$$= \frac{a^{12}b^4 8a^3 b^{-3}}{16a^4 b^6}$$
$$= \frac{8a^{15}b^1}{16a^4 b^6}$$
$$= \frac{a^{11}}{2b^5}$$

75. $\dfrac{(2y^{-1}z^2)^2(3y^{-2}z^{-3})^3}{(y^3z^2)^{-1}}$

$= \dfrac{2^2y^{-2}z^4 3^3y^{-6}z^{-9}}{y^{-3}z^{-2}}$

$= \dfrac{4 \cdot 27 y^{-8}z^{-5}}{y^{-3}z^{-2}}$

$= 108y^{-5}z^{-3}$

$= \dfrac{108}{y^5z^3}$

77. $\dfrac{(9^{-1}z^{-2}x)^{-1}(4z^2x^4)^{-2}}{(5z^{-2}x^{-3})^2}$

$= \dfrac{9^1z^2x^{-1}4^{-2}z^{-4}x^{-8}}{5^2z^{-4}x^{-6}}$

$= \dfrac{9z^{-2}x^{-9}}{25 \cdot 4^2 z^{-4}x^{-6}}$

$= \dfrac{9z^2x^{-3}}{400}$

$= \dfrac{9z^2}{400x^3}$

79.   The student attempted to use the quotient rule with unequal bases. The correct way to simplify this expression is

$$\frac{16^3}{2^2} = \frac{(2^4)^3}{2^2} = \frac{2^{12}}{2^2} = 2^{10} = 1024.$$

81. $10 = 10^1$; move the decimal point to the *right* 1 place when *multiplying*.

$$10(6427) = 64{,}270$$

83. $1000(1.23) = 1230$ $(1000 = 10^3)$

85. $10 = 10^1$; move the decimal point to the *left* 1 place when *dividing*.

$$34 \div 10 = 3.4$$

87. $237 \div 1000 = 0.237$ $(1000 = 10^3)$

Summary Exercises on the Rules for Exponents

1. $\left(\dfrac{6x^2}{5}\right)^{12} = \dfrac{(6x^2)^{12}}{5^{12}}$

$= \dfrac{6^{12}(x^2)^{12}}{5^{12}}$

$= \dfrac{6^{12}x^{24}}{5^{12}}$

3. $(10x^2y^4)^2(10xy^2)^3$

$= 10^2(x^2)^2(y^4)^2 \cdot 10^3x^3(y^2)^3$

$= 10^2x^4y^8 10^3x^3y^6$

$= 10^5x^7y^{14}$

5. $\left(\dfrac{9wx^3}{y^4}\right)^3 = \dfrac{(9wx^3)^3}{(y^4)^3}$

$= \dfrac{9^3w^3(x^3)^3}{y^{12}}$

$= \dfrac{729w^3x^9}{y^{12}}$

7. $\dfrac{c^{11}(c^2)^4}{(c^3)^3(c^2)^{-6}} = \dfrac{c^{11}c^8}{c^9c^{-12}}$

$= \dfrac{c^{19}}{c^{-3}}$

$= c^{19-(-3)} = c^{22}$

9. $5^{-1} + 6^{-1} = \dfrac{1}{5^1} + \dfrac{1}{6^1}$

$= \dfrac{6}{30} + \dfrac{5}{30}$

$= \dfrac{11}{30}$

11. $\dfrac{(2xy^{-1})^3}{2^3x^{-3}y^2} = \dfrac{2^3x^3y^{-3}}{2^3x^{-3}y^2}$

$= x^6y^{-5}$

$= \dfrac{x^6}{y^5}$

13. $(z^4)^{-3}(z^{-2})^{-5}$

$= z^{-12}z^{10} = z^{-2} = \dfrac{1}{z^2}$

15. $\dfrac{(3^{-1}x^{-3}y)^{-1}(2x^2y^{-3})^2}{(5x^{-2}y^2)^{-2}}$

$= \dfrac{(5x^{-2}y^2)^2(2x^2y^{-3})^2}{(3^{-1}x^{-3}y)^1}$

$= \dfrac{5^2x^{-4}y^4 2^2x^4y^{-6}}{3^{-1}x^{-3}y}$

$= \dfrac{25x^0y^{-2} \cdot 4}{3^{-1}x^{-3}y}$

$= 100x^3y^{-3} \cdot 3$

$= \dfrac{300x^3}{y^3}$

17. $\left(\dfrac{-2x^{-2}}{2x^2}\right)^{-2} = \left(\dfrac{2x^2}{-2x^{-2}}\right)^2$

$= \left(\dfrac{x^4}{-1}\right)^2 = \dfrac{(x^4)^2}{(-1)^2}$

$= \dfrac{x^8}{1} = x^8$

19. $\dfrac{(a^{-2}b^3)^{-4}}{(a^{-3}b^2)^{-2}(ab)^{-4}}$

$= \dfrac{a^8 b^{-12}}{a^6 b^{-4} a^{-4} b^{-4}}$

$= \dfrac{a^8 b^{-12}}{a^2 b^{-8}}$

$= a^6 b^{-4}$

$= \dfrac{a^6}{b^4}$

21. $5^{-2} + 6^{-2} = \dfrac{1}{5^2} + \dfrac{1}{6^2}$

$= \dfrac{1}{25} + \dfrac{1}{36}$

$= \dfrac{36}{25 \cdot 36} + \dfrac{25}{25 \cdot 36}$

$= \dfrac{36 + 25}{900} = \dfrac{61}{900}$

23. $\left(\dfrac{7a^2 b^3}{2}\right)^3 = \dfrac{(7a^2 b^3)^3}{2^3}$

$= \dfrac{7^3 a^6 b^9}{8} = \dfrac{343 a^6 b^9}{8}$

25. $-(-12)^0 = -1$

27. $\dfrac{(2xy^{-3})^{-2}}{(3x^{-2}y^4)^{-3}}$

$= \dfrac{(3x^{-2}y^4)^3}{(2xy^{-3})^2}$

$= \dfrac{3^3 x^{-6} y^{12}}{2^2 x^2 y^{-6}}$

$= \dfrac{27 x^{-8} y^{18}}{4}$

$= \dfrac{27 y^{18}}{4x^8}$

29. $(6x^{-5}z^3)^{-3}$

$= 6^{-3} x^{15} z^{-9}$

$= \dfrac{x^{15}}{6^3 z^9}$

$= \dfrac{x^{15}}{216 z^9}$

31. $\dfrac{(xy)^{-3}(xy)^5}{(xy)^{-4}} = (xy)^{-3+5-(-4)}$

$= (xy)^6 = x^6 y^6$

33. $\dfrac{(7^{-1}x^{-3})^{-2}(x^4)^{-6}}{7^{-1}x^{-3}}$

$= \dfrac{7^2 x^6 x^{-24}}{7^{-1} x^{-3}}$

$= 7^{2-(-1)} x^{6-24-(-3)}$

$= 7^3 x^{-15} = \dfrac{343}{x^{15}}$

35. $(5p^{-2}q)^{-3}(5pq^3)^4$

$= 5^{-3} p^6 q^{-3} 5^4 p^4 q^{12}$

$= 5 p^{10} q^9$

37. $\left[\dfrac{4r^{-6}s^{-2}t}{2r^8 s^{-4}t^2}\right]^{-1}$

$= \left[\dfrac{2s^2}{r^{14}t}\right]^{-1}$

$= \dfrac{r^{14}t}{2s^2}$

39. $\dfrac{(8pq^{-2})^4}{(8p^{-2}q^{-3})^3}$

$= \dfrac{8^4 p^4 q^{-8}}{8^3 p^{-6} q^{-9}}$

$= 8^{4-3} p^{4-(-6)} q^{-8-(-9)}$

$= 8 p^{10} q$

41. $-(-3^0)^0 = -1$

43. (a) $2^0 + 2^0 = 1 + 1 = 2$ **(D)**

(b) $2^1 \cdot 2^0 = 2 \cdot 1 = 2$ **(D)**

(c) $2^0 - 2^{-1} = 1 - \dfrac{1}{2} = \dfrac{1}{2}$ **(E)**

(d) $2^1 - 2^0 = 2 - 1 = 1$ **(B)**

(e) $2^0 \cdot 2^{-2} = 1 \cdot \dfrac{1}{2^2} = 1 \cdot \dfrac{1}{4} = \dfrac{1}{4}$ **(J)**

(f) $2^1 \cdot 2^1 = 2 \cdot 2 = 4$ **(F)**

(g) $2^{-2} - 2^{-1} = \dfrac{1}{2^2} - \dfrac{1}{2^1}$

$= \dfrac{1}{4} - \dfrac{1}{2} = \dfrac{1}{4} - \dfrac{2}{4} = -\dfrac{1}{4}$ **(I)**

(h) $2^0 \cdot 2^0 = 1 \cdot 1 = 1$ **(B)**

(i) $2^{-2} \div 2^{-1} = \dfrac{1}{2^2} \div \dfrac{1}{2^1}$

$= \dfrac{1}{4} \div \dfrac{1}{2} = \dfrac{1}{4} \cdot \dfrac{2}{1} = \dfrac{1}{2}$ **(E)**

(j) $2^0 \div 2^{-2} = 1 \div \dfrac{1}{2^2}$

$= 1 \div \dfrac{1}{4} = 1 \cdot \dfrac{4}{1} = 4$ **(F)**

4.3 An Application of Exponents: Scientific Notation

1. **(a)** Move the decimal point to the left 4 places due to the exponent, -4.

    $$4.6 \times 10^{-4} = 0.000\,46$$

    Choice **C** is correct.

    (b) $4.6 \times 10^4 = 46{,}000$

    Choice **A** is correct.

    (c) Move the decimal point to the right 5 places due to the exponent, 5.

    $$4.6 \times 10^5 = 460{,}000$$

    Choice **B** is correct.

    (d) $4.6 \times 10^{-5} = 0.000\,046$

    Choice **D** is correct.

3. 4.56×10^3 is written in scientific notation because 4.56 is between 1 and 10, and 10^3 is a power of 10.

5. $5{,}600{,}000$ is not in scientific notation. It can be written in scientific notation as 5.6×10^6.

7. 0.8×10^2 is not in scientific notation because $|0.8| = 0.8$ is not greater than or equal to 1 and less than 10. It can be written in scientific notation as 8×10^1.

9. 0.004 is not in scientific notation because $|0.004| = 0.004$ is not between 1 and 10. It can be written in scientific notation as 4×10^{-3}.

11. A number is written in scientific notation if it is written as a product of two numbers, the first of which has absolute value less than 10 and greater than or equal to 1 and the second of which is a power of 10. Some examples are 2.3×10^{-4} and 6.02×10^{23}.

13. $5{,}876{,}000{,}000$

    Move the decimal point to the right of the first nonzero digit and count the number of places the decimal point was moved.

    $$5{,}87\,6{,}0\,0\,0{,}0\,0\,0 \quad \textit{9 places}$$

    Because moving the decimal point to the *left* made the number *smaller*, we must multiply by a *positive* power of 10 so that the product 5.876×10^n will equal the larger number. Thus, $n = 9$, and

    $$5{,}876{,}000{,}000 = 5.876 \times 10^9.$$

15. $82{,}350$

    Move the decimal point left 4 places so it is to the right of the first nonzero digit.

    $$8.2\,3\,5\,0 \quad \textit{4 places}$$

    Since the number got smaller, multiply by a positive power of 10.

    $$82{,}350 = 8.2350 \times 10^4 = 8.235 \times 10^4$$

    (Note that the final zero need not be written.)

17. $0.000\,007$

    Move the decimal point to the right of the first nonzero digit.

    $$0.0\,0\,0\,0\,0\,7. \quad \textit{6 places}$$

    Since moving the decimal point to the *right* made the number *larger*, we must multiply by a *negative* power of 10 so that the product 7×10^n will equal the smaller number. Thus, $n = -6$, and

    $$0.000\,007 = 7 \times 10^{-6}.$$

19. $0.002\,03$

    To move the decimal point to the right of the first nonzero digit, we move it 3 places. Since 2.03 is larger than 0.00203, the exponent on 10 must be negative.

    $$0.002\,03 = 2.03 \times 10^{-3}$$

21. 7.5×10^5

    Since the exponent is positive, make 7.5 larger by moving the decimal point 5 places to the right.

    $$7.5 \times 10^5 = 750{,}000$$

23. 5.677×10^{12}

    Since the exponent is positive, make 5.677 larger by moving the decimal point 12 places to the right. We need to add 9 zeros.

    $$5.677 \times 10^{12} = 5{,}677{,}000{,}000{,}000$$

25. 1×10^{12}

    Since the exponent is positive, make 1 larger by moving the decimal point 12 places to the right. We need to add 12 zeros.

    $$1 \times 10^{12} = 1{,}000{,}000{,}000{,}000$$

27. 6.21×10^0

    Because the exponent is 0, the decimal point should not be moved.

    $$6.21 \times 10^0 = 6.21$$

    We know this result is correct because $10^0 = 1$.

29. 7.8×10^{-4}

Since the exponent is negative, make 7.8 smaller by moving the decimal point 4 places to the left.

$$7.8 \times 10^{-4} = 0.000\,78$$

31. 5.134×10^{-9}

Since the exponent is negative, make 5.134 smaller by moving the decimal point 9 places to the left.

$$5.134 \times 10^{-9} = 0.000\,000\,005\,134$$

33. **(a)** $\left(2 \times 10^8\right)\left(3 \times 10^3\right)$

$$= (2 \times 3)\left(10^8 \times 10^3\right) \qquad \begin{array}{l} \textit{Commutative} \\ \textit{and associative} \\ \textit{properties} \\ \textit{Product rule} \\ \textit{for exponents} \end{array}$$

$$= 6 \times 10^{11}$$

(b) $6 \times 10^{11} = 600{,}000{,}000{,}000$

35. **(a)** $\left(5 \times 10^4\right)\left(3 \times 10^2\right)$

$$= (5 \times 3)\left(10^4 \times 10^2\right)$$
$$= 15 \times 10^6$$
$$= 1.5 \times 10^7$$

(b) $15 \times 10^6 = 15{,}000{,}000$

37. **(a)** $\left(3 \times 10^{-4}\right)\left(2 \times 10^8\right)$

$$= (3 \times 2)\left(10^{-4} \times 10^8\right)$$
$$= 6 \times 10^4$$

(b) $6 \times 10^4 = 60{,}000$

39. **(a)** $\left(6 \times 10^3\right)\left(4 \times 10^{-2}\right)$

$$= (6 \times 4)\left(10^3 \times 10^{-2}\right)$$
$$= 24 \times 10^1$$
$$= 2.4 \times 10^2$$

(b) $2.4 \times 10^2 = 240$

41. **(a)** $\left(9 \times 10^4\right)\left(7 \times 10^{-7}\right)$

$$= (9 \times 7)\left(10^4 \times 10^{-7}\right)$$
$$= 63 \times 10^{-3}$$
$$= 6.3 \times 10^{-2}$$

(b) $6.3 \times 10^{-2} = 0.063$

43. **(a)** $\dfrac{9 \times 10^{-5}}{3 \times 10^{-1}} = \dfrac{9}{3} \times \dfrac{10^{-5}}{10^{-1}}$

$$= 3 \times 10^{-5-(-1)}$$
$$= 3 \times 10^{-4}$$

(b) $3 \times 10^{-4} = 0.0003$

44. **(a)** $\dfrac{12 \times 10^{-4}}{4 \times 10^{-3}} = \dfrac{12}{4} \times \dfrac{10^{-4}}{10^{-3}}$

$$= 3 \times 10^{-4-(-3)}$$
$$= 3 \times 10^{-1}$$

(b) $3 \times 10^{-1} = 0.3$

45. **(a)** $\dfrac{8 \times 10^3}{2 \times 10^2} = \dfrac{8}{2} \times \dfrac{10^3}{10^2}$

$$= 4 \times 10^1$$

(b) $4 \times 10^1 = 40$

46. **(a)** $\dfrac{15 \times 10^4}{3 \times 10^3} = \dfrac{15}{3} \times \dfrac{10^4}{10^3}$

$$= 5 \times 10^1$$

(b) $5 \times 10^1 = 50$

47. **(a)** $\dfrac{2.6 \times 10^{-3}}{2 \times 10^2} = \dfrac{2.6}{2} \times \dfrac{10^{-3}}{10^2}$

$$= 1.3 \times 10^{-5}$$

(b) $1.3 \times 10^{-5} = 0.000\,013$

48. **(a)** $\dfrac{9.5 \times 10^{-1}}{5 \times 10^3} = \dfrac{9.5}{5} \times \dfrac{10^{-1}}{10^3}$

$$= 1.9 \times 10^{-1-3}$$
$$= 1.9 \times 10^{-4}$$

(b) $1.9 \times 10^{-4} = 0.000\,19$

49. **(a)** $\dfrac{4 \times 10^5}{8 \times 10^2} = \dfrac{4}{8} \times \dfrac{10^5}{10^2}$

$$= 0.5 \times 10^{5-2}$$
$$= 0.5 \times 10^3$$
$$= 5 \times 10^2$$

(b) $5 \times 10^2 = 500$

51. To work in scientific mode on the TI-83, press MODE and then change from "Normal" to "Sci."

$$0.000\,000\,47 = 4.7 \times 10^{-7}$$

Prediction: $4.7\,\text{E}^-7$

53. $(8\,\text{E}5)/(4\,\text{E}^-2) = \dfrac{8 \times 10^5}{4 \times 10^{-2}}$

$$= \tfrac{8}{4} \times 10^{5-(-2)}$$
$$= 2 \times 10^7$$

Prediction: $2\,\text{E}7$

55. $(2\,\text{E}6)*(2\,\text{E}^-3)/(4\,\text{E}2)$

$$= \dfrac{\left(2 \times 10^6\right)\left(2 \times 10^{-3}\right)}{4 \times 10^2}$$

$$= \dfrac{(2 \times 2) \times \left(10^6 \times 10^{-3}\right)}{4 \times 10^2}$$

$$= \dfrac{4 \times 10^3}{4 \times 10^2}$$

$$= \tfrac{4}{4} \times 10^{3-2}$$

$$= 1 \times 10^1$$

Prediction: $1\,\text{E}1$

57. $\dfrac{650,000,000(0.000\,003\,2)}{0.000\,02}$

$= \dfrac{(6.5 \times 10^8)(3.2 \times 10^{-6})}{2 \times 10^{-5}}$

$= \dfrac{20.8 \times 10^2}{2 \times 10^{-5}}$

$= \dfrac{20.8}{2} \times \dfrac{10^2}{10^{-5}}$

$= 10.4 \times 10^7$

$= 1.04 \times 10^8$

59. $\dfrac{0.000\,000\,72(0.000\,23)}{0.000\,000\,018}$

$= \dfrac{(7.2 \times 10^{-7})(2.3 \times 10^{-4})}{1.8 \times 10^{-8}}$

$= \dfrac{16.56 \times 10^{-11}}{1.8 \times 10^{-8}}$

$= 9.2 \times 10^{-3}$

61. $\dfrac{0.000\,001\,6(240,000,000)}{0.000\,02(0.0032)}$

$= \dfrac{(1.6 \times 10^{-6})(2.4 \times 10^8)}{(2 \times 10^{-5})(3.2 \times 10^{-3})}$

$= \dfrac{3.84 \times 10^2}{6.4 \times 10^{-8}}$

$= 0.6 \times 10^{10}$

$= 6 \times 10^9$

63. $10 \text{ billion} = 10,000,000,000$

$\phantom{10 \text{ billion}} = 1 \times 10^{10}$

65. $2 \times 10^9 = 2,000,000,000$

67. $1,341,000,000 = 1.341 \times 10^9$

69. $\$25,240,000,000 = \$2.524 \times 10^{10}$

71. $1200(2.3 \times 10^{-4}) = (1.2 \times 10^3)(2.3 \times 10^{-4})$

$\phantom{1200(2.3 \times 10^{-4})} = (1.2 \times 2.3) \times (10^3 \times 10^{-4})$

$\phantom{1200(2.3 \times 10^{-4})} = 2.76 \times 10^{3+(-4)}$

$\phantom{1200(2.3 \times 10^{-4})} = 2.76 \times 10^{-1}$

$\phantom{1200(2.3 \times 10^{-4})} = 0.276$

There is about 0.276 lb of copper in 1200 such people.

73. Use the formula $d = rt$.

distance = rate × time

$6.68 \times 10^7 = (1.86 \times 10^5) \times \text{time}$

$\dfrac{6.68 \times 10^7}{1.86 \times 10^5} = \text{time}$

$\text{time} = \dfrac{6.68}{1.86} \times \dfrac{10^7}{10^5}$

$\phantom{\text{time}} \approx 3.59 \times 10^2$

It takes about 359 seconds for light to travel from the sun to Venus.

75. To get the average ticket price, divide the total gross by the attendance.

$\dfrac{\$7.69 \times 10^8}{1.15 \times 10^7} = \dfrac{\$7.69}{1.15} \times \dfrac{10^8}{10^7}$

$\phantom{\dfrac{\$7.69 \times 10^8}{1.15 \times 10^7}} \approx \$6.687 \times 10^1$

The average ticket price was about \$66.87.

77. To find the debt per person, divide the debt by the population.

$\dfrac{\$9 \times 10^{12}}{300 \text{ million}} = \dfrac{\$9 \times 10^{12}}{300,000,000}$

$\phantom{\dfrac{\$9 \times 10^{12}}{300 \text{ million}}} = \dfrac{\$9 \times 10^{12}}{3 \times 10^8}$

$\phantom{\dfrac{\$9 \times 10^{12}}{300 \text{ million}}} = \$3 \times 10^4$

The debt will be about \$30,000 for every person.

79. To find the number of miles, multiply the number of light-years by the number of miles per light-year.

$(25,000)(6,000,000,000,000)$

$= (2.5 \times 10^4)(6 \times 10^{12})$

$= (2.5 \times 6)(10^4 \times 10^{12})$

$= 15 \times 10^{16}$

$= 1.5 \times 10^{17}$

The nebula is about 1.5×10^{17} miles from Earth.

81. $\$6730(290,000,000)$

$= (\$6.73 \times 10^3)(2.9 \times 10^8)$

$= (\$6.73 \times 2.9)(10^3 \times 10^8)$

$= \$19.517 \times 10^{11}$

$= \$1.9517 \times 10^{12}$

In 2003, the U.S. government collected about \$1,951,700,000,000 in taxes.

83. To find the amount each person would have to contribute, divide the amount by the population.

$\dfrac{\$1,000,000,000,000}{281.4 \text{ million}} = \dfrac{\$1 \times 10^{12}}{281,400,000}$

$\phantom{\dfrac{\$1,000,000,000,000}{281.4 \text{ million}}} = \dfrac{\$1 \times 10^{12}}{2.814 \times 10^8}$

$\phantom{\dfrac{\$1,000,000,000,000}{281.4 \text{ million}}} = \dfrac{\$1}{2.814} \times \dfrac{10^{12}}{10^8}$

$\phantom{\dfrac{\$1,000,000,000,000}{281.4 \text{ million}}} \approx \$0.3554 \times 10^4$

$\phantom{\dfrac{\$1,000,000,000,000}{281.4 \text{ million}}} = \$3.554 \times 10^3$

Each person would have to contribute about \$3554.

85. $-3(2x + 4) + 4(2x - 6)$

$= -6x - 12 + 8x - 24$

$= 2x - 36$

87. $2x^2 - 3x + 9$
$$= 2(3)^2 - 3(3) + 9 \quad \text{Let } x = 3.$$
$$= 2(9) - 9 + 9$$
$$= 18$$

89. $4x^3 - 5x^2 + 2x - 6$
$$= 4(3)^3 - 5(3)^2 + 2(3) - 6 \quad \text{Let } x = 3.$$
$$= 4(27) - 5(9) + 6 - 6$$
$$= 108 - 45$$
$$= 63$$

4.4   Adding and Subtracting Polynomials; Graphing Simple Polynomials

1. In the term $7x^5$, the coefficient is _7_ and the exponent is _5_.

3. The degree of the term $-4x^8$ is _8_, the exponent.

5. When $x^2 + 10$ is evaluated for $x = 4$, the result is
$$4^2 + 10 = 16 + 10 = \underline{26}.$$

7. $3xy + 2xy - 5xy = (3 + 2 - 5)xy$
$$= (0)xy$$
$$= \underline{0}$$

9. The polynomial $6x^4$ has one term. The coefficient of this term is 6.

11. The polynomial t^4 has one term. Since $t^4 = 1 \cdot t^4$, the coefficient of this term is 1.

13. The polynomial $-19r^2 - r$ has two terms. The coefficient of r^2 is -19 and the coefficient of r is -1.

15. The polynomial $x + 8x^2 + 5x^3$ has three terms. The coefficient of x is 1, the coefficient of x^2 is 8, and the coefficient of x^3 is 5.

In Exercises 17–28, use the distributive property to add like terms.

17. $-3m^5 + 5m^5 = (-3 + 5)m^5 = 2m^5$

19. $2r^5 + \left(-3r^5\right) = [2 + (-3)]r^5$
$$= -1r^5 = -r^5$$

21. The polynomial $0.2m^5 - 0.5m^2$ cannot be simplified. The two terms are unlike because the exponents on the variables are different, so they cannot be combined.

23. $-3x^5 + 2x^5 - 4x^5 = (-3 + 2 - 4)x^5$
$$= -5x^5$$

25. $-4p^7 + 8p^7 + 5p^9 = (-4 + 8)p^7 + 5p^9$
$$= 4p^7 + 5p^9$$

In descending powers of the variable, this polynomial is written $5p^9 + 4p^7$.

27. $-4xy^2 + 3xy^2 - 2xy^2 + xy^2$
$$= (-4 + 3 - 2 + 1)xy^2$$
$$= -2xy^2$$

29. $6x^4 - 9x$

This polynomial has no like terms, so it is already simplified. It is already written in descending powers of the variable x. The highest degree of any nonzero term is 4, so the degree of the polynomial is 4. There are two terms, so this is a *binomial*.

31. $5m^4 - 3m^2 + 6m^4 - 7m^3$
$$= \left(5m^4 + 6m^4\right) + \left(-7m^3\right) + \left(-3m^2\right)$$
$$= 11m^4 - 7m^3 - 3m^2$$

The resulting polynomial is a *trinomial* of degree 4.

33. $\frac{5}{3}x^4 - \frac{2}{3}x^4 = \left(\frac{5}{3} - \frac{2}{3}\right)x^4$
$$= \frac{3}{3}x^4 = x^4$$

The resulting polynomial is a *monomial* of degree 4.

35. $0.8x^4 - 0.3x^4 - 0.5x^4 + 7$
$$= (0.8 - 0.3 - 0.5)x^4 + 7$$
$$= 0x^4 + 7 = 7$$

Since 7 can be written as $7x^0$, the degree of the polynomial is 0. The simplified polynomial has one term, so it is a *monomial*.

37. **(a)** $2x^2 - 3x - 5$
$$= 2(2)^2 - 3(2) - 5 \quad \text{Let } x = 2.$$
$$= 2(4) - 6 - 5$$
$$= 8 - 6 - 5$$
$$= 2 - 5$$
$$= -3$$

(b) $2x^2 - 3x - 5$
$$= 2(-1)^2 - 3(-1) - 5 \quad \text{Let } x = -1.$$
$$= 2(1) + 3 - 5$$
$$= 2 + 3 - 5$$
$$= 5 - 5$$
$$= 0$$

39. **(a)** $-3x^2 + 14x - 2$
$$= -3(2)^2 + 14(2) - 2 \quad \text{Let } x = 2.$$
$$= -3(4) + 28 - 2$$
$$= -12 + 28 - 2$$
$$= 16 - 2$$
$$= 14$$

(b) $-3x^2 + 14x - 2$

$= -3(-1)^2 + 14(-1) - 2$ *Let x = –1.*

$= -3(1) - 14 - 2$

$= -3 - 14 - 2$

$= -17 - 2$

$= -19$

41. (a) $2x^5 - 4x^4 + 5x^3 - x^2$

$= 2(2)^5 - 4(2)^4 + 5(2)^3 - (2)^2$ *Let x = 2.*

$= 2(32) - 4(16) + 5(8) - 4$

$= 64 - 64 + 40 - 4$

$= 36$

(b) $2x^5 - 4x^4 + 5x^3 - x^2$

$= 2(-1)^5 - 4(-1)^4 + 5(-1)^3 - (-1)^2$

Let x = –1.

$= 2(-1) - 4(1) + 5(-1) - 1$

$= -2 - 4 - 5 - 1$

$= -12$

43. Add, column by column.

$$\begin{array}{r} 2x^2 - 4x \\ 3x^2 + 2x \\ \hline 5x^2 - 2x \end{array}$$

45. Add, column by column.

$$\begin{array}{r} 3m^2 + 5m + 6 \\ 2m^2 - 2m - 4 \\ \hline 5m^2 + 3m + 2 \end{array}$$

47. Add.

$$\begin{array}{r} \frac{2}{3}x^2 + \frac{1}{5}x + \frac{1}{6} \\ \frac{1}{2}x^2 - \frac{1}{3}x + \frac{2}{3} \\ \hline \end{array}$$

Rewrite the fractions so that the fractions in each column have a common denominator; then add column by column.

$$\begin{array}{r} \frac{4}{6}x^2 + \frac{3}{15}x + \frac{1}{6} \\ \frac{3}{6}x^2 - \frac{5}{15}x + \frac{4}{6} \\ \hline \frac{7}{6}x^2 - \frac{2}{15}x + \frac{5}{6} \end{array}$$

49. $\left(9m^3 - 5m^2 + 4m - 8\right) + \left(-3m^3 + 6m^2 - 6\right)$

$= \left(9m^3 - 3m^3\right) + \left(-5m^2 + 6m^2\right) + 4m$

$\quad + (-8 - 6)$

$= 6m^3 + m^2 + 4m - 14$

51. Subtract.

$$\begin{array}{r} 5y^3 - 3y^2 \\ 2y^3 + 8y^2 \\ \hline \end{array}$$

Change all signs in the second row, and then add.

$$\begin{array}{r} 5y^3 - 3y^2 \\ -2y^3 - 8y^2 \\ \hline 3y^3 - 11y^2 \end{array}$$

53. Subtract.

$$\begin{array}{r} 12x^4 - x^2 + x \\ 8x^4 + 3x^2 - 3x \\ \hline \end{array}$$

Change all signs in the second row, and then add.

$$\begin{array}{r} 12x^4 - x^2 + x \\ -8x^4 - 3x^2 + 3x \\ \hline 4x^4 - 4x^2 + 4x \end{array}$$

55. Subtract.

$$\begin{array}{r} 12m^3 - 8m^2 + 6m + 7 \\ -3m^3 + 5m^2 - 2m - 4 \\ \hline \end{array}$$

Change all signs in the second row, and then add.

$$\begin{array}{r} 12m^3 - 8m^2 + 6m + 7 \\ 3m^3 - 5m^2 + 2m + 4 \\ \hline 15m^3 - 13m^2 + 8m + 11 \end{array}$$

57. Vertical addition and subtraction of polynomials is preferable because like terms are arranged in columns.

59. $\left(8m^2 - 7m\right) - \left(3m^2 + 7m - 6\right)$

$= \left(8m^2 - 7m\right) + \left(-3m^2 - 7m + 6\right)$

$= (8 - 3)m^2 + (-7 - 7)m + 6$

$= 5m^2 - 14m + 6$

61. $\left(16x^3 - x^2 + 3x\right) + \left(-12x^3 + 3x^2 + 2x\right)$

$= 16x^3 - x^2 + 3x - 12x^3 + 3x^2 + 2x$

$= (16 - 12)x^3 + (-1 + 3)x^2 + (3 + 2)x$

$= 4x^3 + 2x^2 + 5x$

63. $\left(7y^4 + 3y^2 + 2y\right) - \left(18y^4 - 5y^2 + y\right)$

$= \left(7y^4 + 3y^2 + 2y\right) + \left(-18y^4 + 5y^2 - y\right)$

$= (7 - 18)y^4 + (3 + 5)y^2 + (2 - 1)y$

$= -11y^4 + 8y^2 + y$

65. $\left(9a^4 - 3a^2 + 2\right) + \left(4a^4 - 4a^2 + 2\right)$

$\quad + \left(-12a^4 + 6a^2 - 3\right)$

$= \left(9a^4 + 4a^4 - 12a^4\right)$

$\quad + \left(-3a^2 - 4a^2 + 6a^2\right) + (2 + 2 - 3)$

$= a^4 - a^2 + 1$

67. $\left[(8m^2 + 4m - 7) - (2m^2 - 5m + 2)\right]$
$\qquad - (m^2 + m + 1)$
$= (8m^2 + 4m - 7) + (-2m^2 + 5m - 2)$
$\qquad + (-m^2 - m - 1)$
$= (8 - 2 - 1)m^2 + (4 + 5 - 1)m$
$\qquad + (-7 - 2 - 1)$
$= 5m^2 + 8m - 10$

69. $\left[(3x^2 - 2x + 7) - (4x^2 + 2x - 3)\right]$
$\qquad - \left[(9x^2 + 4x - 6) + (-4x^2 + 4x + 4)\right]$
$= \left[(3 - 4)x^2 + (-2 - 2)x + (7 + 3)\right]$
$\qquad - \left[(9 - 4)x^2 + (4 + 4)x + (-6 + 4)\right]$
$= (-x^2 - 4x + 10) - (5x^2 + 8x - 2)$
$= -x^2 - 4x + 10 - 5x^2 - 8x + 2$
$= -6x^2 - 12x + 12$

71. The coefficients of the x^2 terms are -4, $-(-2)$,
and -8. The sum of these numbers is
$$-4 + 2 - 8 = -10.$$

73. $(6b + 3c) + (-2b - 8c)$
$= (6b - 2b) + (3c - 8c)$
$= 4b - 5c$

75. $(4x + 2xy - 3) - (-2x + 3xy + 4)$
$= (4x + 2xy - 3) + (2x - 3xy - 4)$
$= (4x + 2x) + (2xy - 3xy) + (-3 - 4)$
$= 6x - xy - 7$

77. $\left(5x^2y - 2xy + 9xy^2\right)$
$\qquad - \left(8x^2y + 13xy + 12xy^2\right)$
$= \left(5x^2y - 2xy + 9xy^2\right)$
$\qquad + \left(-8x^2y - 13xy - 12xy^2\right)$
$= \left(5x^2y - 8x^2y\right) + (-2xy - 13xy)$
$\qquad + \left(9xy^2 - 12xy^2\right)$
$= -3x^2y - 15xy - 3xy^2$

79. Use the formula for the perimeter of a rectangle,
$P = 2L + 2W$, with length $L = 4x^2 + 3x + 1$
and width $W = x + 2$.
$$P = 2L + 2W$$
$$= 2(4x^2 + 3x + 1) + 2(x + 2)$$
$$= 8x^2 + 6x + 2 + 2x + 4$$
$$= 8x^2 + 8x + 6$$

81. Use the formula for the perimeter of a square,
$P = 4s$, with $s = \frac{1}{2}x^2 + 2x$.
$$P = 4s$$
$$= 4\left(\tfrac{1}{2}x^2 + 2x\right)$$
$$= 4\left(\tfrac{1}{2}x^2\right) + 4(2x)$$
$$= 2x^2 + 8x$$

83. **(a)** Use the formula for the perimeter of a triangle,
$P = a + b + c$, with $a = 2y - 3t$, $b = 5y + 3t$,
and $c = 16y + 5t$.
$$P = (2y - 3t) + (5y + 3t) + (16y + 5t)$$
$$= (2y + 5y + 16y) + (-3t + 3t + 5t)$$
$$= 23y + 5t$$

The perimeter of the triangle is $23y + 5t$.

(b) Use the fact that the sum of the angles of any
triangle is $180°$.
$$(7x - 3)° + (5x + 2)° + (2x - 1)° = 180°$$
$$(7x + 5x + 2x) + (-3 + 2 - 1) = 180$$
$$14x - 2 = 180$$
$$14x = 182$$
$$x = \tfrac{182}{14} = 13$$

If $x = 13$,
$$7x - 3 = 7(13) - 3 = 88,$$
$$5x + 2 = 5(13) + 2 = 67,$$
$$\text{and} \quad 2x - 1 = 2(13) - 1 = 25.$$

The measures of the angles are $25°$, $67°$, and $88°$.

85. $(3x^2 - 2) - (9x^2 - 6x + 5)$
$= (3x^2 - 2) + (-9x^2 + 6x - 5)$
$= (3x^2 - 9x^2) + 6x + (-2 - 5)$
$= -6x^2 + 6x - 7$

87. First find the two sums and then find their
difference.
$$\left[(5x^2 + 2x - 3) + (x^2 - 8x + 2)\right]$$
$$- \left[(7x^2 - 3x + 6) + (-x^2 + 4x - 6)\right]$$
$$= (6x^2 - 6x - 1) - (6x^2 + x)$$
$$= (6x^2 - 6x - 1) + (-6x^2 - x)$$
$$= (6x^2 - 6x^2) + (-6x - x) + (-1)$$
$$= -7x - 1$$

89. $y = x^2 - 4$

x	$y = x^2 - 4$
-2	$(-2)^2 - 4 = 4 - 4 = 0$
-1	$(-1)^2 - 4 = 1 - 4 = -3$
0	$(0)^2 - 4 = 0 - 4 = -4$
1	$(1)^2 - 4 = 1 - 4 = -3$
2	$(2)^2 - 4 = 4 - 4 = 0$

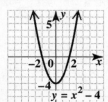

91. $y = 2x^2 - 1$

x	$y = 2x^2 - 1$
-2	$2(-2)^2 - 1 = 2 \cdot 4 - 1 = 7$
-1	$2(-1)^2 - 1 = 2 \cdot 1 - 1 = 1$
0	$2(0)^2 - 1 = 2 \cdot 0 - 1 = -1$
1	$2(1)^2 - 1 = 2 \cdot 1 - 1 = 1$
2	$2(2)^2 - 1 = 2 \cdot 4 - 1 = 7$

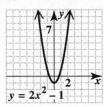

93. $y = -x^2 + 4$

x	$y = -x^2 + 4$
-2	$-(-2)^2 + 4 = -4 + 4 = 0$
-1	$-(-1)^2 + 4 = -1 + 4 = 3$
0	$-(0)^2 + 4 = -0 + 4 = 4$
1	$-(1)^2 + 4 = -1 + 4 = 3$
2	$-(2)^2 + 4 = -4 + 4 = 0$

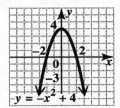

95. $y = (x + 3)^2$

x	$y = (x + 3)^2$
-5	$(-5 + 3)^2 = (-2)^2 = 4$
-4	$(-4 + 3)^2 = (-1)^2 = 1$
-3	$(-3 + 3)^2 = (0)^2 = 0$
-2	$(-2 + 3)^2 = (1)^2 = 1$
-1	$(-1 + 3)^2 = (2)^2 = 4$

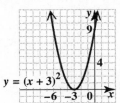

97. If $x = 9$, then $7x = 7(9) = 63$. If a dog is $\underline{9}$ in dog years, then it is $\underline{63}$ in human years.

98. If $x = 6$,
$$2x + 15 = 2(6) + 15 = 12 + 15 = 27.$$
If the saw is rented for $\underline{6}$ days, the cost is $\underline{\$27}$.

99. If $x = 2.5$,
$$-16x^2 + 60x + 80$$
$$= -16(2.5)^2 + 60(2.5) + 80$$
$$= -16(6.25) + 150 + 80$$
$$= -100 + 150 + 80$$
$$= 130.$$
If $\underline{2.5}$ seconds have elapsed, the height of the object is $\underline{130}$ feet.

100. Using the hint that any power of 1 is equal to 1, we add the coefficients and the constant:
$$1331 - 9390 + 68,849 = 60,790$$
So the number of Iowa families receiving food stamps in 1997 was approximately 60,790.

101. $(2a)(-5ab) = (2)(-5)(a \cdot a)b$
$$= -10a^2 b$$

103. $\left(-m^2\right)\left(m^5\right) = -m^{2+5} = -m^7$

105. $5(x + 4) = 5x + 5(4) = 5x + 20$

107. $4(2a + 6b) = 4(2a) + 4(6b)$
$$= 8a + 24b$$

4.5 Multiplying Polynomials

1. **(a)** $5x^3\left(6x^5\right)$
$$= 5 \cdot 6x^{3+5}$$
$$= 30x^8$$
Choice **B** is correct.

(b) $-5x^5\left(6x^3\right)$
$$= -5 \cdot 6x^{5+3}$$
$$= -30x^8$$
Choice **D** is correct.

(c) $(5x^5)^3 = (5)^3(x^5)^3$
$$= 125x^{5 \cdot 3}$$
$$= 125x^{15}$$
Choice **A** is correct.

(d) $(-6x^3)^3 = (-6)^3(x^3)^3$
$$= -216x^{3 \cdot 3}$$
$$= -216x^9$$
Choice **C** is correct.

3. $5y^4\left(3y^7\right) = 5(3)y^{4+7}$
$$= 15y^{11}$$

5. $-15a^4\left(-2a^5\right) = -15(-2)a^{4+5}$
$$= 30a^9$$

7. $5p\left(3q^2\right) = 5(3)pq^2$
$$= 15pq^2$$

9. $-6m^3(3n^2) = -6(3)m^3n^2$
$\qquad = -18m^3n^2$

11. $2m(3m + 2) = 2m(3m) + 2m(2)$
$\qquad = 6m^2 + 4m$

13. $3p(-2p^3 + 4p^2) = 3p(-2p^3) + 3p(4p^2)$
$\qquad = -6p^4 + 12p^3$

15. $-8z(2z + 3z^2 + 3z^3)$
$\qquad = -8z(2z) + (-8z)(3z^2) + (-8z)(3z^3)$
$\qquad = -16z^2 - 24z^3 - 24z^4$

17. $2y^3(3 + 2y + 5y^4)$
$\qquad = 2y^3(3) + 2y^3(2y) + 2y^3(5y^4)$
$\qquad = 6y^3 + 4y^4 + 10y^7$

19. $-4r^3(-7r^2 + 8r - 9)$
$\qquad = -4r^3(-7r^2) + (-4r^3)(8r) + (-4r^3)(-9)$
$\qquad = 28r^5 - 32r^4 + 36r^3$

21. $3a^2(2a^2 - 4ab + 5b^2)$
$\qquad = 3a^2(2a^2) + 3a^2(-4ab) + 3a^2(5b^2)$
$\qquad = 6a^4 - 12a^3b + 15a^2b^2$

23. $7m^3n^2(3m^2 + 2mn - n^3)$
$\qquad = 7m^3n^2(3m^2) + 7m^3n^2(2mn) + 7m^3n^2(-n^3)$
$\qquad = 21m^5n^2 + 14m^4n^3 - 7m^3n^5$

In Exercises 25–36, we can multiply the polynomials horizontally or vertically. The following solutions illustrate these two methods.

25. $(6x + 1)(2x^2 + 4x + 1)$
$\qquad = 6x(2x^2) + 6x(4x) + 6x(1)$
$\qquad\quad + 1(2x^2) + 1(4x) + 1(1)$
$\qquad = 12x^3 + 24x^2 + 6x + 2x^2 + 4x + 1$
$\qquad = 12x^3 + 26x^2 + 10x + 1$

27. $(9a + 2)(9a^2 + a + 1)$

Multiply vertically.

$$
\begin{array}{rrrr}
9a^2 & + & a & + 1 \\
& & 9a & + 2 \\
\hline
18a^2 & + & 2a & + 2 \\
81a^3 + & 9a^2 & + 9a & \\
\hline
81a^3 + & 27a^2 & + 11a & + 2 \\
\end{array}
$$

29. $(4m + 3)(5m^3 - 4m^2 + m - 5)$

Multiply vertically.

$$
\begin{array}{rrrrr}
5m^3 & - 4m^2 & + m & - 5 \\
& & 4m & + 3 \\
\hline
15m^3 & - 12m^2 & + 3m & - 15 \\
20m^4 - 16m^3 & + 4m^2 & - 20m & \\
\hline
20m^4 - & m^3 & - 8m^2 & - 17m & - 15 \\
\end{array}
$$

31. $(2x - 1)(3x^5 - 2x^3 + x^2 - 2x + 3)$

Multiply vertically.

$$
\begin{array}{rrrrrr}
3x^5 & - 2x^3 & + x^2 & - 2x & + 3 \\
& & & 2x & - 1 \\
\hline
-3x^5 & + 2x^3 & - x^2 & + 2x & - 3 \\
6x^6 & - 4x^4 + 2x^3 & - 4x^2 & + 6x & \\
\hline
6x^6 - 3x^5 & - 4x^4 + 4x^3 & - 5x^2 & + 8x & - 3 \\
\end{array}
$$

33. $(5x^2 + 2x + 1)(x^2 - 3x + 5)$

Multiply vertically.

$$
\begin{array}{rrrrr}
5x^2 & + 2x & + 1 \\
x^2 & - 3x & + 5 \\
\hline
25x^2 & + 10x & + 5 \\
-15x^3 & - 6x^2 & - 3x \\
5x^4 + 2x^3 & + x^2 \\
\hline
5x^4 - 13x^3 & + 20x^2 & + 7x & + 5 \\
\end{array}
$$

35. $(6x^4 - 4x^2 + 8x)(\tfrac{1}{2}x + 3)$
$\qquad = 6x^4(\tfrac{1}{2}x) + -4x^2(\tfrac{1}{2}x) + 8x(\tfrac{1}{2}x)$
$\qquad\quad + 6x^4(3) + -4x^2(3) + 8x(3)$
$\qquad = 3x^5 - 2x^3 + 4x^2 + 18x^4 - 12x^2 + 24x$
$\qquad = 3x^5 + 18x^4 - 2x^3 - 8x^2 + 24x$

37. $(m + 7)(m + 5)$

$\qquad\ \ \ \textbf{F} \qquad\ \textbf{O} \qquad\ \textbf{I} \qquad\ \textbf{L}$
$\qquad = m(m) + m(5) + 7(m) + 7(5)$
$\qquad = m^2 + 5m + 7m + 35$
$\qquad = m^2 + 12m + 35$

39. $(x + 5)(x - 5)$

$\qquad\ \ \ \textbf{F} \qquad\ \textbf{O} \qquad\ \textbf{I} \qquad\ \textbf{L}$
$\qquad = x(x) + x(-5) + 5(x) + 5(-5)$
$\qquad = x^2 - 5x + 5x - 25$
$\qquad = x^2 - 25$

41. $(2x + 3)(6x - 4)$

$\qquad\ \ \ \textbf{F} \qquad\ \textbf{O} \qquad\ \textbf{I} \qquad\ \textbf{L}$
$\qquad = 2x(6x) + 2x(-4) + 3(6x) + 3(-4)$
$\qquad = 12x^2 - 8x + 18x - 12$
$\qquad = 12x^2 + 10x - 12$

43. $(3x - 2)(3x - 2)$

$$\quad\quad\;\; \textbf{F}\quad\quad\;\; \textbf{O}\quad\quad\;\; \textbf{I}\quad\quad\;\;\;\; \textbf{L}$$
$$= 3x(3x) + 3x(-2) + (-2)(3x) + (-2)(-2)$$
$$= 9x^2 - 6x - 6x + 4$$
$$= 9x^2 - 12x + 4$$

45. $(5a + 1)(2a + 7)$

$$\quad\quad\;\; \textbf{F}\quad\quad\;\; \textbf{O}\quad\quad\; \textbf{I}\quad\quad\;\; \textbf{L}$$
$$= 5a(2a) + 5a(7) + 1(2a) + 1(7)$$
$$= 10a^2 + 35a + 2a + 7$$
$$= 10a^2 + 37a + 7$$

47. $(6 - 5m)(2 + 3m)$

$$\quad\; \textbf{F}\quad\quad\;\; \textbf{O}\quad\quad\quad\;\; \textbf{I}\quad\quad\quad\;\; \textbf{L}$$
$$= 6(2) + 6(3m) + (-5m)(2) + (-5m)(3m)$$
$$= 12 + 18m - 10m - 15m^2$$
$$= 12 + 8m - 15m^2$$

49. $(5 - 3x)(4 + x)$

$$\quad\;\; \textbf{F}\quad\;\; \textbf{O}\quad\quad\;\; \textbf{I}\quad\quad\quad \textbf{L}$$
$$= 5(4) + 5(x) + (-3x)(4) + (-3x)(x)$$
$$= 20 + 5x - 12x - 3x^2$$
$$= 20 - 7x - 3x^2$$

51. $(4x + 3)(2y - 1)$

$$\quad\;\; \textbf{F}\quad\quad\;\; \textbf{O}\quad\quad\; \textbf{I}\quad\quad\;\; \textbf{L}$$
$$= 4x(2y) + 4x(-1) + 3(2y) + 3(-1)$$
$$= 8xy - 4x + 6y - 3$$

53. $(3x + 2y)(5x - 3y)$

$$\quad\;\; \textbf{F}\quad\quad\;\; \textbf{O}\quad\quad\;\; \textbf{I}\quad\quad\;\; \textbf{L}$$
$$= 3x(5x) + 3x(-3y) + 2y(5x) + 2y(-3y)$$
$$= 15x^2 - 9xy + 10xy - 6y^2$$
$$= 15x^2 + xy - 6y^2$$

55. $3y^3(2y + 3)(y - 5)$

$$(2y + 3)(y - 5)$$
$$\quad\;\; \textbf{F}\quad\quad\; \textbf{O}\quad\;\; \textbf{I}\quad\;\; \textbf{L}$$
$$= 2y(y) + 2y(-5) + 3(y) + 3(-5)$$
$$= 2y^2 - 10y + 3y - 15$$
$$= 2y^2 - 7y - 15$$

Now multiply this result by $3y^3$.

$$3y^3(2y^2 - 7y - 15)$$
$$= 3y^3(2y^2) + 3y^3(-7y) + 3y^3(-15)$$
$$= 6y^5 - 21y^4 - 45y^3$$

57. $-8r^3(5r^2 + 2)(5r^2 - 2)$

$$(5r^2 + 2)(5r^2 - 2)$$
$$\quad\quad\; \textbf{F}\quad\quad\;\; \textbf{O}\quad\quad\; \textbf{I}\quad\quad\;\; \textbf{L}$$
$$= 5r^2(5r^2) + 5r^2(-2) + 2(5r^2) + 2(-2)$$
$$= 25r^4 - 10r^2 + 10r^2 - 4$$
$$= 25r^4 - 4$$

Now multiply this result by $-8r^3$.

$$-8r^3(25r^4 - 4)$$
$$= -8r^3(25r^4) + (-8r^3)(-4)$$
$$= -200r^7 + 32r^3$$

59. **(a)** Use the formula for the area of a rectangle, $A = LW$, with $L = 3y + 7$ and $W = y + 1$.

$$A = (3y + 7)(y + 1)$$
$$= (3y + 7)(y) + (3y + 7)(1)$$
$$= 3y^2 + 7y + 3y + 7$$
$$= 3y^2 + 10y + 7$$

(b) Use the formula for the perimeter of a rectangle, $P = 2L + 2W$, with $L = 3y + 7$ and $W = y + 1$.

$$P = 2(3y + 7) + 2(y + 1)$$
$$= 2(3y) + 2(7) + 2(y) + 2(1)$$
$$= 6y + 14 + 2y + 2$$
$$= 8y + 16$$

61. $\left(3p + \frac{5}{4}q\right)\left(2p - \frac{5}{3}q\right)$ *Use FOIL.*
$$= 3p(2p) + 3p\left(-\frac{5}{3}q\right) + \frac{5}{4}q(2p) + \frac{5}{4}q\left(-\frac{5}{3}q\right)$$
$$= 6p^2 - 5pq + \frac{5}{2}pq - \frac{25}{12}q^2$$
$$= 6p^2 + \left(-\frac{10}{2} + \frac{5}{2}\right)pq - \frac{25}{12}q^2$$
$$= 6p^2 - \frac{5}{2}pq - \frac{25}{12}q^2$$

63. $(x + 7)^2 = (x + 7)(x + 7)$ *Use FOIL.*
$$= x^2 + 7x + 7x + 49$$
$$= x^2 + 14x + 49$$

65. $(a - 4)(a + 4)$ *Use FOIL.*
$$= a^2 + 4a - 4a - 16$$
$$= a^2 - 16$$

67. $(2p - 5)^2 = (2p - 5)(2p - 5)$ *Use FOIL.*
$$= 4p^2 - 10p - 10p + 25$$
$$= 4p^2 - 20p + 25$$

69. $(5k + 3q)^2 = (5k + 3q)(5k + 3q)$ *Use FOIL.*
$$= 25k^2 + 15kq + 15kq + 9q^2$$
$$= 25k^2 + 30kq + 9q^2$$

71. Recall that a^3 means $(a)(a)(a)$, so $(m-5)^3 = (m-5)(m-5)(m-5)$. We'll start by finding

$$(m-5)(m-5) = m^2 - 5m - 5m + 25$$
$$= m^2 - 10m + 25.$$

Now multiply that result by $m - 5$.

$$
\begin{array}{r}
m^2 - 10m + 25 \\
m - 5 \\
\hline
-5m^2 + 50m - 125 \\
m^3 - 10m^2 + 25m \\
\hline
m^3 - 15m^2 + 75m - 125
\end{array}
$$

73. $(2a+1)^3 = (2a+1)(2a+1)(2a+1)$
$(2a+1)(2a+1) = 4a^2 + 2a + 2a + 1$
$= 4a^2 + 4a + 1$

Now multiply vertically.

$$
\begin{array}{r}
4a^2 + 4a + 1 \\
2a + 1 \\
\hline
4a^2 + 4a + 1 \\
8a^3 + 8a^2 + 2a \\
\hline
8a^3 + 12a^2 + 6a + 1
\end{array}
$$

75. $7(4m-3)(2m+1)$
$= 7(8m^2 + 4m - 6m - 3)$ *FOIL*
$= 7(8m^2 - 2m - 3)$
$= 56m^2 - 14m - 21$

77. $-3a(3a+1)(a-4)$
$= -3a(3a^2 - 12a + a - 4)$ *FOIL*
$= -3a(3a^2 - 11a - 4)$
$= -9a^3 + 33a^2 + 12a$

79. $(3r-2s)^4 = (3r-2s)^2(3r-2s)^2$
First we find $(3r-2s)^2$.
$(3r-2s)^2 = (3r-2s)(3r-2s)$
$= 9r^2 - 6rs - 6rs + 4s^2$
$= 9r^2 - 12rs + 4s^2$

Now multiply this result by itself.

$$
\begin{array}{r}
9r^2 - 12rs + 4s^2 \\
9r^2 - 12rs + 4s^2 \\
\hline
36r^2s^2 - 48rs^3 + 16s^4 \\
-108r^3s + 144r^2s^2 - 48rs^3 \\
81r^4 - 108r^3s + 36r^2s^2 \\
\hline
81r^4 - 216r^3s + 216r^2s^2 - 96rs^3 + 16s^4
\end{array}
$$

81. $3p^3(2p^2+5p)(p^3+2p+1)$
$= [3p^3(2p^2) + 3p^3(5p)](p^3 + 2p + 1)$
     Distributive property
$= (6p^5 + 15p^4)(p^3 + 2p + 1)$

Now multiply vertically.

$$
\begin{array}{r}
p^3 + 2p + 1 \\
6p^5 + 15p^4 \\
\hline
15p^7 + 30p^5 + 15p^4 \\
6p^8 + 12p^6 + 6p^5 \\
\hline
6p^8 + 15p^7 + 12p^6 + 36p^5 + 15p^4
\end{array}
$$

83. $-2x^5(3x^2 + 2x - 5)(4x + 2)$
$= [-2x^5(3x^2) + (-2x^5)(2x)$
    $+ (-2x^5)(-5)](4x + 2)$
     Distributive property
$= (-6x^7 - 4x^6 + 10x^5)(4x + 2)$

Now multiply vertically.

$$
\begin{array}{r}
-6x^7 - 4x^6 + 10x^5 \\
4x + 2 \\
\hline
-12x^7 - 8x^6 + 20x^5 \\
-24x^8 - 16x^7 + 40x^6 \\
\hline
-24x^8 - 28x^7 + 32x^6 + 20x^5
\end{array}
$$

85. The area A of the shaded region is the difference between the area of the larger square, which has sides of length $x + 7$, and the area of the smaller square, which has sides of length x.

$$
\begin{aligned}
A &= (x+7)^2 - (x)^2 \\
&= (x+7)(x+7) - x^2 \\
&= (x^2 + 7x + 7x + 49) - x^2 \\
&= x^2 + 14x + 49 - x^2 \\
&= 14x + 49
\end{aligned}
$$

87. The area A of the shaded region is the difference between the area of the circle, which has radius x, and the area of the square, which has sides of length 3.

$$
\begin{aligned}
A &= \pi r^2 - s^2 \\
&= \pi(x)^2 - (3)^2 \\
&= \pi x^2 - 9
\end{aligned}
$$

89. Use the formula for the area of a rectangle, $A = LW$, with $L = 3x + 6$ and $W = 10$.

$$
\begin{aligned}
A &= (3x + 6)(10) \\
&= 3x(10) + 6(10) \\
&= 30x + 60
\end{aligned}
$$

90. If the area A is 600 square yards, we have the equation

$$
\begin{aligned}
30x + 60 &= 600. \\
30x &= 540 \\
x &= \tfrac{540}{30} = 18
\end{aligned}
$$

91. If $x = 18$,
$$3x + 6 = 3(18) + 6$$
$$= 54 + 6$$
$$= 60.$$

The dimensions of the rectangle are 10 yards by 60 yards.

92. To find the cost of covering the lawn with sod, we use area. Since the area is 600 square yards and the cost of sod is \$3.50 per square yard, the total cost is given by
$$600(3.50) = 2100.$$

The total cost is \$2100.

93. Use the formula for the perimeter of a rectangle, $P = 2L + 2W$, with $L = 60$ and $W = 10$.
$$P = 2(60) + 2(10)$$
$$= 120 + 20$$
$$= 140$$

The perimeter is 140 yards.

94. To find the cost of constructing a fence around the lawn, we use perimeter. If the cost is \$9.00 per yard to fence the lawn and 140 yards must be fenced, the total cost to fence the yard is given by
$$9.00(140) = 1260.$$

The total cost is \$1260.

95. **(a)** From Exercise 89, the area of the lawn is given by the polynomial $30x + 60$. If it costs k dollars per square yard to sod the lawn, the cost to sod the entire lawn will be
$$(30x + 60)(k) = 30xk + 60k \text{ (dollars)}$$

(b) The total cost to fence the lawn is given by multiplying the perimeter by r. First, use the formula for the perimeter of a rectangle, $P = 2L + 2W$, with $L = 3x + 6$ and $W = 10$.
$$P = 2(3x + 6) + 2(10)$$
$$= 6x + 12 + 20$$
$$= 6x + 32$$

Then the total cost is given by
$$(6x + 32)(r) = 6xr + 32r.$$

The total cost is $6xr + 32r$ (dollars).

97. $(3m)^2 = 3^2 m^2 = 9m^2$

99. $(-2r)^2 = (-2)^2 r^2 = 4r^2$

101. $(4x^2)^2 = 4^2 (x^2)^2 = 16x^4$

4.6 Special Products

1. $(2x + 3)^2$

(a) The square of the first term is
$$(2x)^2 = (2x)(2x) = 4x^2.$$

(b) Twice the product of the two terms is
$$2(2x)(3) = 12x.$$

(c) The square of the last term is
$$3^2 = 9.$$

(d) The final product is the trinomial
$$4x^2 + 12x + 9.$$

In Exercises 3–20, use one of the following formulas for the square of a binomial:
$$(x + y)^2 = x^2 + 2xy + y^2$$
$$(x - y)^2 = x^2 - 2xy + y^2$$

3. $(m + 2)^2 = m^2 + 2(m)(2) + 2^2$
$$= m^2 + 4m + 4$$

5. $(r - 3)^2 = r^2 - 2(r)(3) + 3^2$
$$= r^2 - 6r + 9$$

7. $(x + 2y)^2 = x^2 + 2(x)(2y) + (2y)^2$
$$= x^2 + 4xy + 4y^2$$

9. $(5p + 2q)^2 = (5p)^2 + 2(5p)(2q) + (2q)^2$
$$= 25p^2 + 20pq + 4q^2$$

11. $(4a + 5b)^2 = (4a)^2 + 2(4a)(5b) + (5b)^2$
$$= 16a^2 + 40ab + 25b^2$$

13. $(7t + s)^2 = (7t)^2 + 2(7t)(s) + s^2$
$$= 49t^2 + 14ts + s^2$$

15. $\left(6m - \frac{4}{5}n\right)^2$
$$= (6m)^2 - 2(6m)\left(\frac{4}{5}n\right) + \left(\frac{4}{5}n\right)^2$$
$$= 36m^2 - \frac{48}{5}mn + \frac{16}{25}n^2$$

17. $t(3t - 1)^2$
$$(3t - 1)^2 = (3t)^2 - 2(3t)(1) + 1^2$$
$$= 9t^2 - 6t + 1$$

Now multiply by t.
$$t\left(9t^2 - 6t + 1\right) = 9t^3 - 6t^2 + t$$

19. $-(4r - 2)^2$
$$(4r - 2)^2 = (4r)^2 - 2(4r)(2) + 2^2$$
$$= 16r^2 - 16r + 4$$

Now multiply by -1.
$$-1\left(16r^2 - 16r + 4\right) = -16r^2 + 16r - 4$$

21. **(a)** $7x(7x) = 49x^2$

    (b) $7x(-3y) + 3y(7x)$
$$= -21xy + 21xy = 0$$

    (c) $3y(-3y) = -9y^2$

    (d) $49x^2 - 9y^2$

The sum found in part (b) is omitted because it is 0. Adding 0, the identity element for addition, would not change the answer.

23. The square of a binomial leads to a polynomial with *three* terms. The product of the sum and difference of two terms leads to a polynomial with *two* terms.

In Exercises 24–41, use the formula for the product of the sum and difference of two terms.
$$(x + y)(x - y) = x^2 - y^2$$

25. $(k + 5)(k - 5) = k^2 - 5^2$
$$= k^2 - 25$$

27. $(4 - 3t)(4 + 3t) = 4^2 - (3t)^2$
$$= 16 - 9t^2$$

29. $(5x + 2)(5x - 2) = (5x)^2 - 2^2$
$$= 25x^2 - 4$$

31. $(6a - p)(6a + p) = (6a)^2 - p^2$
$$= 36a^2 - p^2$$

33. $(10x + 3y)(10x - 3y) = (10x)^2 - (3y)^2$
$$= 100x^2 - 9y^2$$

35. $\left(2x^2 - 5\right)\left(2x^2 + 5\right) = (2x^2)^2 - 5^2$
$$= 4x^4 - 25$$

37. $\left(\frac{3}{4} - x\right)\left(\frac{3}{4} + x\right) = \left(\frac{3}{4}\right)^2 - x^2$
$$= \frac{9}{16} - x^2$$

39. $\left(9y + \frac{2}{3}\right)\left(9y - \frac{2}{3}\right) = (9y)^2 - \left(\frac{2}{3}\right)^2$
$$= 81y^2 - \frac{4}{9}$$

41. $q(5q - 1)(5q + 1)$

$(5q - 1)(5q + 1) = (5q)^2 - 1^2$
$$= 25q^2 - 1$$

Now multiply by q.

$q\left(25q^2 - 1\right) = 25q^3 - q$

43. $(x + 1)^3$

$= (x + 1)^2(x + 1)$ $a^3 = a^2 \cdot a$

$= \left(x^2 + 2x + 1\right)(x + 1)$ *Square the binomial.*

$= x^3 + 2x^2 + x + x^2 + 2x + 1$ *Multiply polynomials.*

$= x^3 + 3x^2 + 3x + 1$ *Combine like terms.*

45. $(t - 3)^3$

$= (t - 3)^2(t - 3)$ $a^3 = a^2 \cdot a$

$= \left(t^2 - 6t + 9\right)(t - 3)$ *Square the binomial.*

$= t^3 - 6t^2 + 9t - 3t^2 + 18t - 27$ *Multiply polynomials.*

$= t^3 - 9t^2 + 27t - 27$ *Combine like terms.*

47. $(r + 5)^3$

$= (r + 5)^2(r + 5)$ $a^3 = a^2 \cdot a$

$= \left(r^2 + 10r + 25\right)(r + 5)$ *Square the binomial.*

$= r^3 + 10r^2 + 25r + 5r^2 + 50r + 125$
     Multiply polynomials.

$= r^3 + 15r^2 + 75r + 125$ *Combine like terms.*

49. $(2a + 1)^3$

$= (2a + 1)^2(2a + 1)$ $a^3 = a^2 \cdot a$

$= \left(4a^2 + 4a + 1\right)(2a + 1)$ *Square the binomial.*

$= 8a^3 + 8a^2 + 2a + 4a^2 + 4a + 1$ *Multiply polynomials.*

$= 8a^3 + 12a^2 + 6a + 1$ *Combine like terms.*

51. $(4x - 1)^4$

$= (4x - 1)^2(4x - 1)^2$ $a^4 = a^2 \cdot a^2$

$= \left(16x^2 - 8x + 1\right)\left(16x^2 - 8x + 1\right)$
     Square each binomial.

$= 256x^4 - 128x^3 + 16x^2 - 128x^3 + 64x^2$
$$- 8x + 16x^2 - 8x + 1$$
     Multiply polynomials.

$= 256x^4 - 256x^3 + 96x^2 - 16x + 1$
     Combine like terms.

53. $(3r - 2t)^4$

$= (3r - 2t)^2(3r - 2t)^2$ $a^4 = a^2 \cdot a^2$

$= \left(9r^2 - 12rt + 4t^2\right)\left(9r^2 - 12rt + 4t^2\right)$
     Square each binomial.

$= 81r^4 - 108r^3t + 36r^2t^2 - 108r^3t$
$$+ 144r^2t^2 - 48rt^3 + 36r^2t^2 - 48rt^3 + 16t^4$$
     Multiply polynomials.

$= 81r^4 - 216r^3t + 216r^2t^2 - 96rt^3 + 16t^4$
     Combine like terms.

55. The large square has sides of length $a + b$, so its area is $(a + b)^2$.

56. The red square has sides of length a, so its area is a^2.

57. Each blue rectangle has length a and width b, so each has an area of ab. Thus, the sum of the areas of the blue rectangles is

$$ab + ab = 2ab.$$

58. The yellow square has sides of length b, so its area is b^2.

59. Sum $= a^2 + 2ab + b^2$

60. The area of the largest square equals the sum of the areas of the two smaller squares and the two rectangles. Therefore, $(a + b)^2$ must equal $a^2 + 2ab + b^2$.

61. $35^2 = (35)(35)$

$$\begin{array}{r} 35 \\ 35 \\ \hline 175 \\ 105 \\ \hline 1225 \end{array}$$

62. $(a + b)^2 = a^2 + 2ab + b^2$
$(30 + 5)^2 = 30^2 + 2(30)(5) + 5^2$

63. $30^2 + 2(30)(5) + 5^2$
$= 900 + 60(5) + 25$
$= 900 + 300 + 25$
$= 1225$

64. The answers are equal.

65. $101 \times 99 = (100 + 1)(100 - 1)$
$= 100^2 - 1^2$
$= 10{,}000 - 1$
$= 9999$

67. $201 \times 199 = (200 + 1)(200 - 1)$
$= 200^2 - 1^2$
$= 40{,}000 - 1$
$= 39{,}999$

69. $20\frac{1}{2} \times 19\frac{1}{2} = \left(20 + \frac{1}{2}\right)\left(20 - \frac{1}{2}\right)$
$= 20^2 - \left(\frac{1}{2}\right)^2$
$= 400 - \frac{1}{4}$
$= 399\frac{3}{4}$

71. Use the formula for area the of a triangle, $A = \frac{1}{2}bh$, with $b = m + 2n$ and $h = m - 2n$.

$$A = \tfrac{1}{2}(m + 2n)(m - 2n)$$
$$= \tfrac{1}{2}[m^2 - (2n)^2]$$
$$= \tfrac{1}{2}(m^2 - 4n^2)$$
$$= \tfrac{1}{2}m^2 - 2n^2$$

73. Use the formula for the area of a parallelogram, $A = bh$, with $b = 3a + 2$ and $h = 3a - 2$.

$$A = (3a + 2)(3a - 2)$$
$$= (3a)^2 - 2^2$$
$$= 9a^2 - 4$$

75. Use the formula for the area of a circle, $A = \pi r^2$, with $r = x + 2$.

$$A = \pi(x + 2)^2$$
$$= \pi(x^2 + 4x + 4)$$
$$= \pi x^2 + 4\pi x + 4\pi$$

77. Use the formula for the volume of a cube, $V = e^3$, with $e = x + 2$.

$$V = (x + 2)^3$$
$$= (x + 2)^2(x + 2)$$
$$= (x^2 + 4x + 4)(x + 2)$$
$$= x^3 + 4x^2 + 4x + 2x^2 + 8x + 8$$
$$= x^3 + 6x^2 + 12x + 8$$

79. $\dfrac{1}{2p}\left(4p^2 + 2p + 8\right)$

$$= \frac{1}{2p}\left(4p^2\right) + \frac{1}{2p}(2p) + \frac{1}{2p}(8)$$
$$= 2p + 1 + \frac{4}{p}$$

81. $\dfrac{1}{3m}\left(m^3 + 9m^2 - 6m\right)$

$$= \frac{1}{3m}\left(m^3\right) + \frac{1}{3m}\left(9m^2\right) + \frac{1}{3m}(-6m)$$
$$= \frac{m^2}{3} + 3m - 2$$

83. $-3k\left(8k^2 - 12k + 2\right)$
$= -3k\left(8k^2\right) + (-3k)(-12k) + (-3k)(2)$
$= -24k^3 + 36k^2 - 6k$

85. $(-2k + 1)\left(8k^2 + 9k + 3\right)$
$= -2k\left(8k^2\right) + (-2k)(9k) + (-2k)(3)$
$\quad + 1\left(8k^2\right) + 1(9k) + 1(3)$
$= -16k^3 - 18k^2 - 6k + 8k^2 + 9k + 3$
$= -16k^3 - 10k^2 + 3k + 3$

87. Subtract.

$$\begin{array}{rrrrr} -4x^3 & + \ 2x^2 & - \ 3x & + \ 7 \\ -4x^3 & - \ 8x^2 & + \ x & - \ 4 \\ \hline \end{array}$$

Change all signs in the second row, and then add.

$$\begin{array}{rrrr} -4x^3 & + \ 2x^2 & - \ 3x & + \ 7 \\ 4x^3 & + \ 8x^2 & - \ x & + \ 4 \\ \hline & 10x^2 & - \ 4x & + \ 11 \end{array}$$

4.7 Dividing Polynomials

1. In the statement $\dfrac{6x^2 + 8}{2} = 3x^2 + 4$, $\underline{6x^2 + 8}$ is the dividend, $\underline{2}$ is the divisor, and $\underline{3x^2 + 4}$ is the quotient.

3. To check the division shown in Exercise 1, multiply $\underline{3x^2 + 4}$ by $\underline{2}$ (or $\underline{2}$ by $\underline{3x^2 + 4}$) and show that the product is $\underline{6x^2 + 8}$.

5. In this section, we are dividing a polynomial by a monomial. The problem
$$\frac{16m^3 - 12m^2}{4m}$$
is an example of such a division. However, in the problem
$$\frac{4m}{16m^3 - 12m^2},$$
we are dividing a monomial by a binomial. Therefore, the methods of this section do not apply.

7. $\dfrac{60x^4 - 20x^2 + 10x}{2x}$

$= \dfrac{60x^4}{2x} - \dfrac{20x^2}{2x} + \dfrac{10x}{2x}$

$= \dfrac{60}{2}x^{4-1} - \dfrac{20}{2}x^{2-1} + \dfrac{10}{2}$

$= 30x^3 - 10x + 5$

9. $\dfrac{20m^5 - 10m^4 + 5m^2}{5m^2}$

$= \dfrac{20m^5}{5m^2} - \dfrac{10m^4}{5m^2} + \dfrac{5m^2}{5m^2}$

$= \dfrac{20}{5}m^{5-2} - \dfrac{10}{5}m^{4-2} + \dfrac{5}{5}$

$= 4m^3 - 2m^2 + 1$

11. $\dfrac{8t^5 - 4t^3 + 4t^2}{2t}$

$= \dfrac{8t^5}{2t} - \dfrac{4t^3}{2t} + \dfrac{4t^2}{2t}$

$= 4t^4 - 2t^2 + 2t$

13. $\dfrac{4a^5 - 4a^2 + 8}{4a}$

$= \dfrac{4a^5}{4a} - \dfrac{4a^2}{4a} + \dfrac{8}{4a}$

$= a^4 - a + \dfrac{2}{a}$

15. $\dfrac{12x^5 - 9x^4 + 6x^3}{3x^2}$

$= \dfrac{12x^5}{3x^2} - \dfrac{9x^4}{3x^2} + \dfrac{6x^3}{3x^2}$

$= 4x^3 - 3x^2 + 2x$

17. $\dfrac{3x^2 + 15x^3 - 27x^4}{3x^2}$

$= \dfrac{3x^2}{3x^2} + \dfrac{15x^3}{3x^2} - \dfrac{27x^4}{3x^2}$

$= 1 + 5x - 9x^2 \quad \text{or} \quad -9x^2 + 5x + 1$

19. $\dfrac{36x + 24x^2 + 6x^3}{3x^2}$

$= \dfrac{36x}{3x^2} + \dfrac{24x^2}{3x^2} + \dfrac{6x^3}{3x^2}$

$= \dfrac{12}{x} + 8 + 2x \quad \text{or} \quad 2x + 8 + \dfrac{12}{x}$

21. $\dfrac{4x^4 + 3x^3 + 2x}{3x^2}$

$= \dfrac{4x^4}{3x^2} + \dfrac{3x^3}{3x^2} + \dfrac{2x}{3x^2}$

$= \dfrac{4x^2}{3} + x + \dfrac{2}{3x}$

23. $\dfrac{-27r^4 + 36r^3 - 6r^2 - 26r + 2}{-3r}$

$= \dfrac{-27r^4}{-3r} + \dfrac{36r^3}{-3r} - \dfrac{6r^2}{-3r} - \dfrac{26r}{-3r} + \dfrac{2}{-3r}$

$= 9r^3 - 12r^2 + 2r + \dfrac{26}{3} - \dfrac{2}{3r}$

25. $\dfrac{2m^5 - 6m^4 + 8m^2}{-2m^3}$

$= \dfrac{2m^5}{-2m^3} - \dfrac{6m^4}{-2m^3} + \dfrac{8m^2}{-2m^3}$

$= -m^2 + 3m - \dfrac{4}{m}$

27. $\left(20a^4 - 15a^5 + 25a^3\right) \div \left(5a^4\right)$

$= \dfrac{20a^4 - 15a^5 + 25a^3}{5a^4}$

$= \dfrac{20a^4}{5a^4} - \dfrac{15a^5}{5a^4} + \dfrac{25a^3}{5a^4}$

$= 4 - 3a + \dfrac{5}{a} \quad \text{or} \quad -3a + 4 + \dfrac{5}{a}$

29. $\left(120x^{11} - 60x^{10} + 140x^9 - 100x^8\right) \div \left(10x^{12}\right)$

$= \dfrac{120x^{11} - 60x^{10} + 140x^9 - 100x^8}{10x^{12}}$

$= \dfrac{120x^{11}}{10x^{12}} - \dfrac{60x^{10}}{10x^{12}} + \dfrac{140x^9}{10x^{12}} - \dfrac{100x^8}{10x^{12}}$

$= \dfrac{12}{x} - \dfrac{6}{x^2} + \dfrac{14}{x^3} - \dfrac{10}{x^4}$

31. $\left(120x^5y^4 - 80x^2y^3 + 40x^2y^4 - 20x^5y^3\right)$

$\div \left(20xy^2\right)$

$= \dfrac{120x^5y^4 - 80x^2y^3 + 40x^2y^4 - 20x^5y^3}{20xy^2}$

$= \dfrac{120x^5y^4}{20xy^2} - \dfrac{80x^2y^3}{20xy^2} + \dfrac{40x^2y^4}{20xy^2} - \dfrac{20x^5y^3}{20xy^2}$

$= 6x^4y^2 - 4xy + 2xy^2 - x^4y$

33. Use the formula for the area of a rectangle, $A = LW$, with $A = 15x^3 + 12x^2 - 9x + 3$ and $W = 3$.

$$15x^3 + 12x^2 - 9x + 3 = L(3)$$

$$\dfrac{15x^3 + 12x^2 - 9x + 3}{3} = L$$

$$\dfrac{15x^3}{3} + \dfrac{12x^2}{3} - \dfrac{9x}{3} + \dfrac{3}{3} = L$$

$$5x^3 + 4x^2 - 3x + 1 = L$$

35.
$$\begin{array}{r} 1423 \\ 2\overline{\smash{\big)}\,2846} \end{array}$$

36. $1423 = \left(1 \times 10^3\right) + \left(4 \times 10^2\right)$
$+ \left(2 \times 10^1\right) + \left(3 \times 10^0\right)$

37. $\dfrac{2x^3 + 8x^2 + 4x + 6}{2}$

$= \dfrac{2x^3}{2} + \dfrac{8x^2}{2} + \dfrac{4x}{2} + \dfrac{6}{2}$

$= x^3 + 4x^2 + 2x + 3$

38. They are similar in that the coefficients of powers of 10 are equal to the coefficients of the powers of x. They are different in that one is a constant while the other is a polynomial. They are equal if $x = 10$ (the base of our decimal system).

39. $\dfrac{x^2 - x - 6}{x - 3}$

$$\begin{array}{r} x + 2 \\ x - 3 \overline{\smash{\big)}\, x^2 - x - 6} \\ \underline{x^2 - 3x} \\ 2x - 6 \\ \underline{2x - 6} \\ 0 \end{array}$$

The remainder is 0. The answer is the quotient, $x + 2$.

41. $\dfrac{2y^2 + 9y - 35}{y + 7}$

$$\begin{array}{r} 2y - 5 \\ y + 7 \overline{\smash{\big)}\, 2y^2 + 9y - 35} \\ \underline{2y^2 + 14y} \\ -5y - 35 \\ \underline{-5y - 35} \\ 0 \end{array}$$

The remainder is 0. The answer is the quotient, $2y - 5$.

43. $\dfrac{p^2 + 2p + 20}{p + 6}$

$$\begin{array}{r} p - 4 \\ p + 6 \overline{\smash{\big)}\, p^2 + 2p + 20} \\ \underline{p^2 + 6p} \\ -4p + 20 \\ \underline{-4p - 24} \\ 44 \end{array}$$

The remainder is 44. Write the remainder as the numerator of a fraction that has the divisor $p + 6$ as its denominator. The answer is

$$p - 4 + \dfrac{44}{p + 6}.$$

45. $\left(r^2 - 8r + 15\right) \div (r - 3)$

$$\begin{array}{r} r - 5 \\ r - 3 \overline{\smash{\big)}\, r^2 - 8r + 15} \\ \underline{r^2 - 3r} \\ -5r + 15 \\ \underline{-5r + 15} \\ 0 \end{array}$$

The remainder is 0. The answer is the quotient, $r - 5$.

47. $\dfrac{12m^2 - 20m + 3}{2m - 3}$

$$\begin{array}{r} 6m - 1 \\ 2m - 3 \overline{\smash{\big)}\, 12m^2 - 20m + 3} \\ \underline{12m^2 - 18m} \\ -2m + 3 \\ \underline{-2m + 3} \\ 0 \end{array}$$

The remainder is 0. The answer is the quotient, $6m - 1$.

49. $\dfrac{4a^2 - 22a + 32}{2a + 3}$

$$
\begin{array}{r}
2a \quad - 14 \\
2a + 3 \overline{)\,4a^2\ - 22a\ + 32} \\
4a^2\ +\ 6a \\
\hline
-28a\ + 32 \\
-28a\ - 42 \\
\hline
74
\end{array}
$$

The remainder is 74. The answer is

$$2a - 14 + \dfrac{74}{2a + 3}.$$

51. $\dfrac{8x^3 - 10x^2 - x + 3}{2x + 1}$

$$
\begin{array}{r}
4x^2\ - 7x\ + 3 \\
2x + 1 \overline{)\,8x^3\ - 10x^2\ -\ x\ + 3} \\
8x^3\ +\ 4x^2 \\
\hline
-14x^2\ -\ x \\
-14x^2\ - 7x \\
\hline
6x\ + 3 \\
6x\ + 3 \\
\hline
0
\end{array}
$$

The remainder is 0. The answer is the quotient,

$$4x^2 - 7x + 3.$$

53. $\dfrac{8k^4 - 12k^3 - 2k^2 + 7k - 6}{2k - 3}$

$$
\begin{array}{r}
4k^3 \qquad\quad -\ k\ + 2 \\
2k - 3 \overline{)\,8k^4\ - 12k^3\ - 2k^2\ + 7k\ - 6} \\
8k^4\ - 12k^3 \\
\hline
-2k^2\ + 7k \\
-2k^2\ + 3k \\
\hline
+\ 4k\ - 6 \\
+\ 4k\ - 6 \\
\hline
0
\end{array}
$$

The remainder is 0. The answer is the quotient,

$$4k^3 - k + 2.$$

55. $\dfrac{5y^4 + 5y^3 + 2y^2 - y - 3}{y + 1}$

$$
\begin{array}{r}
5y^3 \qquad\quad +\ 2y\ - 3 \\
y + 1 \overline{)\,5y^4\ + 5y^3\ + 2y^2\ -\ y\ - 3} \\
5y^4\ + 5y^3 \\
\hline
2y^2\ -\ y \\
2y^2\ + 2y \\
\hline
-3y\ - 3 \\
-3y\ - 3 \\
\hline
0
\end{array}
$$

The remainder is 0. The answer is the quotient,

$$5y^3 + 2y - 3.$$

57. $\dfrac{3k^3 - 4k^2 - 6k + 10}{k - 2}$

$$
\begin{array}{r}
3k^2\ + 2k\ -\ 2 \\
k - 2 \overline{)\,3k^3\ - 4k^2\ - 6k\ + 10} \\
3k^3\ - 6k^2 \\
\hline
2k^2\ - 6k \\
2k^2\ - 4k \\
\hline
-2k\ + 10 \\
-2k\ +\ 4 \\
\hline
6
\end{array}
$$

The remainder is 6.

The quotient is $3k^2 + 2k - 2.$

The answer is $3k^2 + 2k - 2 + \dfrac{6}{k - 2}.$

59. $\dfrac{6p^4 - 16p^3 + 15p^2 - 5p + 10}{3p + 1}$

$$
\begin{array}{r}
2p^3\ -\ 6p^2\ + 7p\ -\ 4 \\
3p + 1 \overline{)\,6p^4\ - 16p^3\ + 15p^2\ - 5p\ + 10} \\
6p^4\ +\ 2p^3 \\
\hline
-18p^3\ + 15p^2 \\
-18p^3\ -\ 6p^2 \\
\hline
21p^2\ -\ 5p \\
21p^2\ +\ 7p \\
\hline
-12p\ + 10 \\
-12p\ -\ 4 \\
\hline
14
\end{array}
$$

The remainder is 14.

The quotient is $2p^3 - 6p^2 + 7p - 4.$

The answer is $2p^3 - 6p^2 + 7p - 4 + \dfrac{14}{3p + 1}.$

61. $(x^3 - 2x^2 - 9) \div (x - 3)$

Use 0 as the coefficient of the missing x-term.

$$
\begin{array}{r}
x^2\ +\ x\ + 3 \\
x - 3 \overline{)\,x^3\ - 2x^2\ + 0x\ - 9} \\
x^3\ - 3x^2 \\
\hline
x^2\ + 0x \\
x^2\ - 3x \\
\hline
3x\ - 9 \\
3x\ - 9 \\
\hline
0
\end{array}
$$

$$(x^3 - 2x^2 - 9) \div (x - 3) = x^2 + x + 3$$

63. $(2x^3 + x + 2) \div (x + 1)$

Use 0 as the coefficient of the missing x^2-term.

$$
\begin{array}{r}
2x^2 - 2x + 3 \\
x + 1 \overline{\smash{\big)}\, 2x^3 + 0x^2 + x + 2} \\
\underline{2x^3 + 2x^2} \\
-2x^2 + x \\
\underline{-2x^2 - 2x} \\
3x + 2 \\
\underline{3x + 3} \\
-1
\end{array}
$$

$$\left(2x^3 + x + 2\right) \div (x + 1) = 2x^2 - 2x + 3 + \frac{-1}{x + 1}$$

65. $\dfrac{5 - 2r^2 + r^4}{r^2 - 1}$

Use 0 as the coefficient for the missing terms. Rearrange terms of the dividend in descending powers.

$$
\begin{array}{r}
r^2 - 1 \\
r^2 + 0r - 1 \overline{\smash{\big)}\, r^4 + 0r^3 - 2r^2 + 0r + 5} \\
\underline{r^4 + 0r^3 - r^2} \\
-r^2 + 0r + 5 \\
\underline{-r^2 + 0r + 1} \\
4
\end{array}
$$

The remainder is 4.

The quotient is $r^2 - 1$.

The answer is $r^2 - 1 + \dfrac{4}{r^2 - 1}$.

67. $\dfrac{y^3 + 1}{y + 1}$

$$
\begin{array}{r}
y^2 - y + 1 \\
y + 1 \overline{\smash{\big)}\, y^3 + 0y^2 + 0y + 1} \\
\underline{y^3 + y^2} \\
-y^2 + 0y \\
\underline{-y^2 - y} \\
y + 1 \\
\underline{y + 1} \\
0
\end{array}
$$

The remainder is 0. The answer is the quotient, $y^2 - y + 1$.

69. $\dfrac{a^4 - 1}{a^2 - 1}$

$$
\begin{array}{r}
a^2 + 1 \\
a^2 + 0a - 1 \overline{\smash{\big)}\, a^4 + 0a^3 + 0a^2 + 0a - 1} \\
\underline{a^4 + 0a^3 - a^2} \\
a^2 + 0a - 1 \\
\underline{a^2 + 0a - 1} \\
0
\end{array}
$$

The remainder is 0. The answer is the quotient, $a^2 + 1$.

71. $\dfrac{x^4 - 4x^3 + 5x^2 - 3x + 2}{x^2 + 3}$

$$
\begin{array}{r}
x^2 - 4x + 2 \\
x^2 + 0x + 3 \overline{\smash{\big)}\, x^4 - 4x^3 + 5x^2 - 3x + 2} \\
\underline{x^4 + 0x^3 + 3x^2} \\
-4x^3 + 2x^2 - 3x \\
\underline{-4x^3 + 0x^2 - 12x} \\
2x^2 + 9x + 2 \\
\underline{2x^2 + 0x + 6} \\
9x - 4
\end{array}
$$

$$\frac{x^4 - 4x^3 + 5x^2 - 3x + 2}{x^2 + 3}$$
$$= x^2 - 4x + 2 + \frac{9x - 4}{x^2 + 3}$$

73. $\dfrac{2x^5 + 9x^4 + 8x^3 + 10x^2 + 14x + 5}{2x^2 + 3x + 1}$

$$
\begin{array}{r}
x^3 + 3x^2 - x + 5 \\
2x^2 + 3x + 1 \overline{\smash{\big)}\, 2x^5 + 9x^4 + 8x^3 + 10x^2 + 14x + 5} \\
\underline{2x^5 + 3x^4 + x^3} \\
6x^4 + 7x^3 + 10x^2 \\
\underline{6x^4 + 9x^3 + 3x^2} \\
-2x^3 + 7x^2 + 14x \\
\underline{-2x^3 - 3x^2 - x} \\
10x^2 + 15x + 5 \\
\underline{10x^2 + 15x + 5} \\
0
\end{array}
$$

The remainder is 0. The answer is the quotient, $x^3 + 3x^2 - x + 5$.

75. $\left(3a^2 - 11a + 17\right) \div (2a + 6)$

$$
\begin{array}{r}
\frac{3}{2}a \quad - 10 \\
2a + 6 \overline{\smash{\big)}\ 3a^2 - 11a + 17} \\
\underline{3a^2 + 9a} \\
-20a + 17 \\
\underline{-20a - 60} \\
77
\end{array}
$$

The remainder is 77.

The quotient is $\frac{3}{2}a - 10$.

The answer is $\dfrac{3}{2}a - 10 + \dfrac{77}{2a + 6}$.

77. Use $A = LW$ with

$$A = 5x^3 + 7x^2 - 13x - 6$$

and $W = 5x + 2$.

$$5x^3 + 7x^2 - 13x - 6 = L(5x + 2)$$
$$\frac{5x^3 + 7x^2 - 13x - 6}{5x + 2} = L$$

$$
\begin{array}{r}
x^2 + \quad x - 3 \\
5x + 2 \overline{\smash{\big)}\ 5x^3 + 7x^2 - 13x - 6} \\
\underline{5x^3 + 2x^2} \\
5x^2 - 13x \\
\underline{5x^2 + 2x} \\
-15x - 6 \\
\underline{-15x - 6} \\
0
\end{array}
$$

The length L is $(x^2 + x - 3)$ units.

79. Use the distance formula, $d = rt$, with $d = (5x^3 - 6x^2 + 3x + 14)$ miles and $r = (x + 1)$ miles per hour.

$$\left(5x^3 - 6x^2 + 3x + 14\right) = (x + 1)t$$
$$\frac{\left(5x^3 - 6x^2 + 3x + 14\right)}{(x + 1)} = t$$

$$
\begin{array}{r}
5x^2 - 11x + 14 \\
x + 1 \overline{\smash{\big)}\ 5x^3 - 6x^2 + 3x + 14} \\
\underline{5x^3 + 5x^2} \\
-11x^2 + 3x \\
\underline{-11x^2 - 11x} \\
14x + 14 \\
\underline{14x + 14} \\
0
\end{array}
$$

The time t is $(5x^2 - 11x + 14)$ hours.

81. We can write 18 as

$$1 \cdot 18, \quad 2 \cdot 9, \quad \text{or} \quad 3 \cdot 6,$$

so the positive integer factors of 18 are

$$1, 2, 3, 6, 9, \quad \text{and} \quad 18.$$

83. We can write 48 as

$$1 \cdot 48, \, 2 \cdot 24, \, 3 \cdot 16, \, 4 \cdot 12, \quad \text{or} \quad 6 \cdot 8,$$

so the positive integer factors of 48 are

$$1, 2, 3, 4, 6, 8, 12, 16, 24, \quad \text{and} \quad 48.$$

Chapter 4 Review Exercises

1. $4^3 \cdot 4^8 = 4^{3+8} = 4^{11}$

2. $(-5)^6(-5)^5 = (-5)^{6+5} = (-5)^{11}$

3. $\left(-8x^4\right)\left(9x^3\right) = -8(9)\left(x^4\right)\left(x^3\right)$
$$= -72x^{4+3} = -72x^7$$

4. $\left(2x^2\right)\left(5x^3\right)\left(x^9\right) = 2(5)\left(x^2\right)\left(x^3\right)\left(x^9\right)$
$$= 10x^{2+3+9} = 10x^{14}$$

5. $(19x)^5 = 19^5x^5$

6. $(-4y)^7 = (-4)^7y^7$

7. $5(pt)^4 = 5p^4t^4$

8. $\left(\dfrac{7}{5}\right)^6 = \dfrac{7^6}{5^6}$

9. $(3x^2y^3)^3 = 3^3(x^2)^3(y^3)^3$
$$= 3^3x^{2\cdot3}y^{3\cdot3}$$
$$= 3^3x^6y^9$$

10. $(t^4)^8(t^2)^5 = t^{4\cdot8}t^{2\cdot5}$
$$= t^{32}t^{10}$$
$$= t^{32+10}$$
$$= t^{42}$$

11. $(6x^2z^4)^2(x^3yz^2)^4$
$$= 6^2(x^2)^2(z^4)^2(x^3)^4(y)^4(z^2)^4$$
$$= 6^2x^4z^8x^{12}y^4z^8$$
$$= 6^2x^{4+12}y^4z^{8+8}$$
$$= 6^2x^{16}y^4z^{16}$$

12. $\left(\dfrac{2m^3n}{p^2}\right)^3 = \dfrac{2^3(m^3)^3n^3}{(p^2)^3}$
$$= \dfrac{2^3m^9n^3}{p^6}$$

13. The product rule does not apply to $7^2 + 7^4$ because you are adding powers of 7, not multiplying them.

14. $6^0 + (-6)^0 = 1 + 1 = 2$

15. $-(-23)^0 = -1 \cdot (-23)^0 = -1 \cdot 1 = -1$

16. $-10^0 = -(10^0) = -(1) = -1$

17. $-7^{-2} = -\dfrac{1}{7^2} = -\dfrac{1}{49}$

18. $\left(\dfrac{5}{8}\right)^{-2} = \left(\dfrac{8}{5}\right)^2 = \dfrac{64}{25}$

19. $(5^{-2})^{-4} = 5^{(-2)(-4)}$ *Power rule*
$= 5^8$

20. $9^3 \cdot 9^{-5} = 9^{3+(-5)} = 9^{-2} = \dfrac{1}{9^2} = \dfrac{1}{81}$

21. $2^{-1} + 4^{-1} = \dfrac{1}{2^1} + \dfrac{1}{4^1}$
$= \tfrac{1}{2} + \tfrac{1}{4}$
$= \tfrac{2}{4} + \tfrac{1}{4} = \tfrac{3}{4}$

22. $\dfrac{6^{-5}}{6^{-3}} = \dfrac{6^3}{6^5} = \dfrac{1}{6^2} = \dfrac{1}{36}$

23. $\dfrac{x^{-7}}{x^{-9}} = \dfrac{x^9}{x^7} = x^{9-7} = x^2$ *Quotient rule*

24. $\dfrac{y^4 \cdot y^{-2}}{y^{-5}} = \dfrac{y^4 \cdot y^5}{y^2} = \dfrac{y^9}{y^2} = y^7$

25. $(3r^{-2})^{-4} = (3)^{-4}(r^{-2})^{-4}$
$= (3^{-4})(r^{-2(-4)})$
$= \dfrac{1}{3^4} r^8$
$= \dfrac{r^8}{81}$

26. $(3p)^4(3p^{-7}) = (3^4 p^4)(3p^{-7})$
$= 3^{4+1} \cdot p^{4+(-7)}$
$= 3^5 p^{-3}$
$= \dfrac{3^5}{p^3}$

27. $\dfrac{ab^{-3}}{a^4 b^2} = \dfrac{a}{a^4 b^2 b^3} = \dfrac{1}{a^3 b^5}$

28. $\dfrac{(6r^{-1})^2(2r^{-4})}{r^{-5}(r^2)^{-3}} = \dfrac{(6^2 r^{-2})(2r^{-4})}{r^{-5} r^{-6}}$
$= \dfrac{72 r^{-6}}{r^{-11}}$
$= \dfrac{72 r^{11}}{r^6}$
$= 72 r^5$

29. $48{,}000{,}000 = 4.8 \times 10^7$
Move the decimal point left 7 places so it is to the right of the first nonzero digit. 48,000,000 is *greater* than 4.8, so the power is *positive*.

30. $28{,}988{,}000{,}000 = 2.8988 \times 10^{10}$
Move the decimal point left 10 places so it is to the right of the first nonzero digit. 28,988,000,000 is *greater* than 2.8988, so the power is *positive*.

31. $0.000\,000\,082\,4 = 8.24 \times 10^{-8}$
Move the decimal point right 8 places so it is to the right of the first nonzero digit. 0.000 000 082 4 is *less* than 8.24, so the power is *negative*.

32. $2.4 \times 10^4 = 24{,}000$
Move the decimal point 4 places to the right.

33. $7.83 \times 10^7 = 78{,}300{,}000$
Move the decimal point 7 places to the right.

34. $8.97 \times 10^{-7} = 0.000\,000\,897$
Move the decimal point 7 places to the left.

35. $(2 \times 10^{-3})(4 \times 10^5)$
$= (2 \times 4)(10^{-3} \times 10^5)$
$= 8 \times 10^{-3+5} = 8 \times 10^2 = 800$

36. $\dfrac{8 \times 10^4}{2 \times 10^{-2}} = \dfrac{8}{2} \times \dfrac{10^4}{10^{-2}} = 4 \times 10^{4-(-2)}$
$= 4 \times 10^6 = 4{,}000{,}000$

37. $\dfrac{12 \times 10^{-5} \times 5 \times 10^4}{4 \times 10^3 \times 6 \times 10^{-2}}$
$= \dfrac{12 \times 5}{4 \times 6} \times \dfrac{10^{-5} \times 10^4}{10^3 \times 10^{-2}}$
$= \dfrac{60}{24} \times \dfrac{10^{-1}}{10^1}$
$= \tfrac{5}{2} \times 10^{-1-1}$
$= 2.5 \times 10^{-2}$
$= 0.025$

38. $2 \times 10^{-6} = 0.000\,002$

39. $1.6 \times 10^{-12} = 0.000\,000\,000\,001\,6$

40. There are 41 zeros, so the number is 4.2×10^{42}.

41. $97{,}000 = 9.7 \times 10^4$; $5000 = 5 \times 10^3$

42. There are 100 zeros, so the number is 1×10^{100}.

43. $1000 = 1 \times 10^3$; $2000 = 2 \times 10^3$;
$50{,}000 = 5 \times 10^4$; $100{,}000 = 1 \times 10^5$

44. $9m^2 + 11m^2 + 2m^2 = (9 + 11 + 2)m^2$
$= 22m^2$

The degree is 2.

To determine if the polynomial is a monomial, binomial, or trinomial, count the number of terms in the final expression.

There is one term, so this is a *monomial*.

45. $-4p + p^3 - p^2 + 8p + 2$
$$= p^3 - p^2 + (-4 + 8)p + 2$$
$$= p^3 - p^2 + 4p + 2$$

The degree is 3.

To determine if the polynomial is a monomial, binomial, or trinomial, count the number of terms in the final expression. Since there are four terms, it is none of these.

46. $12a^5 - 9a^4 + 8a^3 + 2a^2 - a + 3$ cannot be simplified further and is already written in descending powers of the variable.

The degree is 5.

This polynomial has 6 terms, so it is none of the names listed.

47. $-7y^5 - 8y^4 - y^5 + y^4 + 9y$
$$= -7y^5 - 1y^5 - 8y^4 + 1y^4 + 9y$$
$$= (-7 - 1)y^5 + (-8 + 1)y^4 + 9y$$
$$= -8y^5 - 7y^4 + 9y$$

The degree is 5.

There are three terms, so the polynomial is a *trinomial*.

48. $\left(12r^4 - 7r^3 + 2r^2\right) - \left(5r^4 - 3r^3 + 2r^2 - 1\right)$
$$= \left(12r^4 - 7r^3 + 2r^2\right)$$
$$+ \left(-5r^4 + 3r^3 - 2r^2 + 1\right)$$

Change signs in the second polynomial and add.
$$= 12r^4 - 7r^3 + 2r^2 - 5r^4 + 3r^3 - 2r^2 + 1$$
$$= 7r^4 - 4r^3 + 1$$

The degree is 4.

The polynomial is a *trinomial*.

49. $\left(5x^3y^2 - 3xy^5 + 12x^2\right)$
$$- \left(-9x^2 - 8x^3y^2 + 2xy^5\right)$$
$$= \left(5x^3y^2 - 3xy^5 + 12x^2\right)$$
$$+ \left(9x^2 + 8x^3y^2 - 2xy^5\right)$$
$$= \left(5x^3y^2 + 8x^3y^2\right) + \left(-3xy^5 - 2xy^5\right)$$
$$+ \left(12x^2 + 9x^2\right)$$
$$= 13x^3y^2 - 5xy^5 + 21x^2$$

50. Add.
$$\begin{array}{r} -2a^3 + 5a^2 \\ 3a^3 - a^2 \\ \hline a^3 + 4a^2 \end{array}$$

51. Subtract.
$$\begin{array}{r} 6y^2 - 8y + 2 \\ 5y^2 + 2y - 7 \end{array}$$

Change all signs in the second row and then add.
$$\begin{array}{r} 6y^2 - 8y + 2 \\ -5y^2 - 2y + 7 \\ \hline y^2 - 10y + 9 \end{array}$$

52. Subtract.
$$\begin{array}{r} -12k^4 - 8k^2 + 7k \\ k^4 + 7k^2 - 11k \end{array}$$

Change all signs in the second row and then add.
$$\begin{array}{r} -12k^4 - 8k^2 + 7k \\ -k^4 - 7k^2 + 11k \\ \hline -13k^4 - 15k^2 + 18k \end{array}$$

53. $y = -x^2 + 5$

$x = -2 : y = -(-2)^2 + 5 = -4 + 5 = 1$
$x = -1 : y = -(-1)^2 + 5 = -1 + 5 = 4$
$x = 0 : y = -(0)^2 + 5 = 0 + 5 = 5$
$x = 1 : y = -(1)^2 + 5 = -1 + 5 = 4$
$x = 2 : y = -(2)^2 + 5 = -4 + 5 = 1$

x	-2	-1	0	1	2
y	1	4	5	4	1

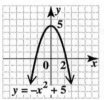

54. $y = 3x^2 - 2$

$x = -2 : y = 3(-2)^2 - 2 = 3 \cdot 4 - 2 = 10$
$x = -1 : y = 3(-1)^2 - 2 = 3 \cdot 1 - 2 = 1$
$x = 0 : y = 3(0)^2 - 2 = 3 \cdot 0 - 2 = -2$
$x = 1 : y = 3(1)^2 - 2 = 3 \cdot 1 - 2 = 1$
$x = 2 : y = 3(2)^2 - 2 = 3 \cdot 4 - 2 = 10$

x	-2	-1	0	1	2
y	10	1	-2	1	10

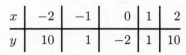
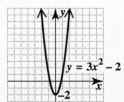

55. $(a+2)(a^2-4a+1)$

Multiply vertically.

$$
\begin{array}{r}
a^2 - 4a + 1 \\
a + 2 \\
\hline
2a^2 - 8a + 2 \\
a^3 - 4a^2 + a \\
\hline
a^3 - 2a^2 - 7a + 2
\end{array}
$$

56. $(3r-2)(2r^2+4r-3)$

Multiply vertically.

$$
\begin{array}{r}
2r^2 + 4r - 3 \\
3r - 2 \\
\hline
-4r^2 - 8r + 6 \\
6r^3 + 12r^2 - 9r \\
\hline
6r^3 + 8r^2 - 17r + 6
\end{array}
$$

57. $\left(5p^2+3p\right)\left(p^3-p^2+5\right)$

$\qquad = 5p^2\left(p^3\right) + 5p^2\left(-p^2\right) + 5p^2(5)$

$\qquad\quad + 3p\left(p^3\right) + 3p\left(-p^2\right) + 3p(5)$

$\qquad = 5p^5 - 5p^4 + 25p^2 + 3p^4 - 3p^3 + 15p$

$\qquad = 5p^5 - 2p^4 - 3p^3 + 25p^2 + 15p$

58. $(m-9)(m+2)$

$\qquad\quad$ **F** $\qquad$ **O** $\qquad$ **I** $\qquad$ **L**

$\qquad = m(m) + m(2) + (-9)(m) + (-9)(2)$

$\qquad = m^2 + 2m - 9m - 18$

$\qquad = m^2 - 7m - 18$

59. $(3k-6)(2k+1)$

$\qquad\quad$ **F** $\qquad$ **O** $\qquad$ **I** $\qquad$ **L**

$\qquad = 3k(2k) + 3k(1) + (-6)(2k) + (-6)(1)$

$\qquad = 6k^2 + 3k - 12k - 6$

$\qquad = 6k^2 - 9k - 6$

60. $(a+3b)(2a-b)$

$\qquad\quad$ **F** $\qquad$ **O** $\qquad$ **I** $\qquad$ **L**

$\qquad = a(2a) + a(-b) + 3b(2a) + 3b(-b)$

$\qquad = 2a^2 - ab + 6ab - 3b^2$

$\qquad = 2a^2 + 5ab - 3b^2$

61. $(6k+5q)(2k-7q)$

$\qquad\quad$ **F** $\qquad$ **O** $\qquad$ **I** $\qquad$ **L**

$\qquad = 6k(2k) + 6k(-7q) + 5q(2k) + 5q(-7q)$

$\qquad = 12k^2 - 42kq + 10kq - 35q^2$

$\qquad = 12k^2 - 32kq - 35q^2$

62. $(s-1)^3 = (s-1)^2(s-1)$

$\qquad\qquad = \left(s^2 - 2s + 1\right)(s-1)$

Now, use vertical multiplication.

$$
\begin{array}{r}
s^2 - 2s + 1 \\
s - 1 \\
\hline
-s^2 + 2s - 1 \\
s^3 - 2s^2 + s \\
\hline
s^3 - 3s^2 + 3s - 1
\end{array}
$$

$(s-1)^3 = s^3 - 3s^2 + 3s - 1$

63. $(a+4)^2 = a^2 + 2(a)(4) + 4^2$

$\qquad\qquad = a^2 + 8a + 16$

64. $(2r+5t)^2 = (2r)^2 + 2(2r)(5t) + (5t)^2$

$\qquad\qquad = 4r^2 + 20rt + 25t^2$

65. $(6m-5)(6m+5) = (6m)^2 - 5^2$

$\qquad\qquad\qquad = 36m^2 - 25$

66. $(5a+6b)(5a-6b) = (5a)^2 - (6b)^2$

$\qquad\qquad\qquad = 25a^2 - 36b^2$

67. $(r+2)^3 = (r+2)^2(r+2)$

$\qquad\qquad = \left(r^2 + 4r + 4\right)(r+2)$

$\qquad\qquad = r^3 + 4r^2 + 4r + 2r^2 + 8r + 8$

$\qquad\qquad = r^3 + 6r^2 + 12r + 8$

68. $t(5t-3)^2 = t\left(25t^2 - 30t + 9\right)$

$\qquad\qquad = 25t^3 - 30t^2 + 9t$

69. Answers will vary. One example is given here.

(a) $(x+y)^2 \neq x^2 + y^2$

Let $x = 2$ and $y = 3$.

$$(x+y)^2 = (2+3)^2 = 5^2 = 25$$
$$x^2 + y^2 = 2^2 + 3^2 = 4 + 9 = 13$$

Since $25 \neq 13$,

$$(x+y)^2 \neq x^2 + y^2.$$

(b) $(x+y)^3 \neq x^3 + y^3$

Let $x = 2$ and $y = 3$.

$$(x+y)^3 = (2+3)^3 = 5^3 = 125$$
$$x^3 + y^3 = 2^3 + 3^3 = 8 + 27 = 35$$

Since $125 \neq 35$,

$$(x+y)^3 \neq x^3 + y^3.$$

70. To find the third power of a binomial, such as $(a+b)^3$, first square the binomial and then multiply that result by the binomial:

$$(a+b)^3 = (a+b)^2(a+b)$$
$$= \left(a^2 + 2ab + b^2\right)(a+b)$$
$$= \left(a^3 + 2a^2b + ab^2\right)$$
$$+ \left(a^2b + 2ab^2 + b^3\right)$$
$$= a^3 + 3a^2b + 3ab^2 + b^3$$

71. If we chose to let $x = 0$ and $y = 1$, we would get the true equation $1 = 1$ for both (a) and (b). These results would not be sufficient to illustrate the truth, in general, of the inequalities. The next step in working the exercise would be to use two other values instead of $x = 0$ and $y = 1$.

72. Use the formula for the volume of a cube, $V = e^3$, with $e = x^2 + 2$ centimeters.

$$\begin{aligned} V &= (x^2 + 2)^3 \\ &= (x^2 + 2)^2(x^2 + 2) \\ &= (x^4 + 4x^2 + 4)(x^2 + 2) \end{aligned}$$

Now use vertical multiplication.

$$\begin{array}{r} x^4 + 4x^2 + 4 \\ x^2 + 2 \\ \hline 2x^4 + 8x^2 + 8 \\ x^6 + 4x^4 + 4x^2 \\ \hline x^6 + 6x^4 + 12x^2 + 8 \end{array}$$

The volume of the cube is

$$x^6 + 6x^4 + 12x^2 + 8$$

cubic centimeters.

73. Use the formula for the volume of a sphere, $V = \frac{4}{3}\pi r^3$, with $r = x + 1$ inches.

$$\begin{aligned} V &= \frac{4}{3}\pi(x + 1)^3 \\ &= \frac{4}{3}\pi(x + 1)^2(x + 1) \\ &= \frac{4}{3}\pi(x^2 + 2x + 1)(x + 1) \end{aligned}$$

Now use vertical multiplication.

$$\begin{array}{r} x^2 + 2x + 1 \\ x + 1 \\ \hline x^2 + 2x + 1 \\ x^3 + 2x^2 + x \\ \hline x^3 + 3x^2 + 3x + 1 \end{array}$$

$$\begin{aligned} V &= \frac{4}{3}\pi(x^3 + 3x^2 + 3x + 1) \\ &= \frac{4}{3}\pi x^3 + 4\pi x^2 + 4\pi x + \frac{4}{3}\pi \end{aligned}$$

The volume of the sphere is

$$\frac{4}{3}\pi x^3 + 4\pi x^2 + 4\pi x + \frac{4}{3}\pi$$

cubic inches.

74. $\dfrac{-15y^4}{9y^2} = \dfrac{-15y^{4-2}}{9} = \dfrac{-5y^2}{3}$

75. $\begin{aligned} \dfrac{6y^4 - 12y^2 + 18y}{6y} &= \dfrac{6y^4}{6y} - \dfrac{12y^2}{6y} + \dfrac{18y}{6y} \\ &= y^3 - 2y + 3 \end{aligned}$

76. $(-10m^4n^2 + 5m^3n^2 + 6m^2n^4) \div (5m^2n)$

$$\begin{aligned} &= \frac{-10m^4n^2 + 5m^3n^2 + 6m^2n^4}{5m^2n} \\ &= \frac{-10m^4n^2}{5m^2n} + \frac{5m^3n^2}{5m^2n} + \frac{6m^2n^4}{5m^2n} \\ &= -2m^2n + mn + \frac{6n^3}{5} \end{aligned}$$

77. Let P be the polynomial that when multiplied by $6m^2n$ gives the product

$$12m^3n^2 + 18m^6n^3 - 24m^2n^2.$$

$$\begin{aligned} (P)(6m^2n) &= 12m^3n^2 + 18m^6n^3 - 24m^2n^2 \\ P &= \frac{12m^3n^2 + 18m^6n^3 - 24m^2n^2}{6m^2n} \\ &= \frac{12m^3n^2}{6m^2n} + \frac{18m^6n^3}{6m^2n} - \frac{24m^2n^2}{6m^2n} \\ &= 2mn + 3m^4n^2 - 4n \end{aligned}$$

78. The error made was not dividing both terms in the numerator by 6. The correct method is as follows:

$$\frac{6x^2 - 12x}{6} = \frac{6x^2}{6} - \frac{12x}{6} = x^2 - 2x.$$

79. $\dfrac{2r^2 + 3r - 14}{r - 2}$

$$\begin{array}{r} 2r + 7 \\ r - 2 \overline{\smash{\big)}\, 2r^2 + 3r - 14} \\ \underline{2r^2 - 4r} \\ 7r - 14 \\ \underline{7r - 14} \\ 0 \end{array}$$

The remainder is 0.

The answer is the quotient, $2r + 7$.

80. $\dfrac{10a^3 + 9a^2 - 14a + 9}{5a - 3}$

$$\begin{array}{r} 2a^2 + 3a - 1 \\ 5a - 3 \overline{\smash{\big)}\, 10a^3 + 9a^2 - 14a + 9} \\ \underline{10a^3 - 6a^2} \\ 15a^2 - 14a \\ \underline{15a^2 - 9a} \\ -5a + 9 \\ \underline{-5a + 3} \\ 6 \end{array}$$

The answer is

$$2a^2 + 3a - 1 + \frac{6}{5a - 3}.$$

81. $\dfrac{x^4 - 5x^2 + 3x^3 - 3x + 4}{x^2 - 1}$

Write the dividend in descending powers and use 0 as the coefficient of the missing term.

$$
\begin{array}{r}
x^2 + 3x - 4 \\
x^2 + 0x - 1 \enclose{longdiv}{x^4 + 3x^3 - 5x^2 - 3x + 4} \\
\underline{x^4 + 0x^3 - x^2} \\
3x^3 - 4x^2 - 3x \\
\underline{3x^3 + 0x^2 - 3x} \\
-4x^2 + 4 \\
\underline{-4x^2 + 4} \\
0
\end{array}
$$

The remainder is 0. The answer is the quotient,

$$x^2 + 3x - 4.$$

82. $\dfrac{m^4 + 4m^3 - 12m - 5m^2 + 6}{m^2 - 3}$

Write the dividend in descending powers and use 0 as the coefficient of the missing term.

$$
\begin{array}{r}
m^2 + 4m - 2 \\
m^2 + 0m - 3 \enclose{longdiv}{m^4 + 4m^3 - 5m^2 - 12m + 6} \\
\underline{m^4 + 0m^3 - 3m^2} \\
4m^3 - 2m^2 - 12m \\
\underline{4m^3 + 0m^2 - 12m} \\
-2m^2 + 0m + 6 \\
\underline{-2m^2 + 0m + 6} \\
0
\end{array}
$$

The remainder is 0. The answer is the quotient,

$$m^2 + 4m - 2.$$

83. $\dfrac{16x^2 - 25}{4x + 5}$

$$
\begin{array}{r}
4x - 5 \\
4x + 5 \enclose{longdiv}{16x^2 + 0x - 25} \\
\underline{16x^2 + 20x} \\
-20x - 25 \\
\underline{-20x - 25} \\
0
\end{array}
$$

The remainder is 0.
The answer is the quotient, $4x - 5$.

84. $\dfrac{25y^2 - 100}{5y + 10}$

$$
\begin{array}{r}
5y - 10 \\
5y + 10 \enclose{longdiv}{25y^2 + 0y - 100} \\
\underline{25y^2 + 50y} \\
-50y - 100 \\
\underline{-50y - 100} \\
0
\end{array}
$$

The remainder is 0.
The answer is the quotient, $5y - 10$.

85. $\dfrac{y^3 - 8}{y - 2}$

$$
\begin{array}{r}
y^2 + 2y + 4 \\
y - 2 \enclose{longdiv}{y^3 + 0y^2 + 0y - 8} \\
\underline{y^3 - 2y^2} \\
2y^2 + 0y \\
\underline{2y^2 - 4y} \\
4y - 8 \\
\underline{4y - 8} \\
0
\end{array}
$$

The remainder is 0.
The answer is the quotient, $y^2 + 2y + 4$.

86. $\dfrac{1000x^6 + 1}{10x^2 + 1}$

$$
\begin{array}{r}
100x^4 - 10x^2 + 1 \\
10x^2 + 1 \enclose{longdiv}{1000x^6 + 0x^4 + 0x^2 + 1} \\
\underline{1000x^6 + 100x^4} \\
-100x^4 \\
\underline{-100x^4 - 10x^2} \\
10x^2 + 1 \\
\underline{10x^2 + 1} \\
0
\end{array}
$$

The remainder is 0.
The answer is the quotient, $100x^4 - 10x^2 + 1$.

87. $\dfrac{6y^4 - 15y^3 + 14y^2 - 5y - 1}{3y^2 + 1}$

$$
\begin{array}{r}
2y^2 - 5y + 4 \\
3y^2 + 1 \enclose{longdiv}{6y^4 - 15y^3 + 14y^2 - 5y - 1} \\
\underline{6y^4 + 2y^2} \\
-15y^3 + 12y^2 - 5y \\
\underline{-15y^3 - 5y} \\
12y^2 - 1 \\
\underline{12y^2 + 4} \\
-5
\end{array}
$$

The answer is $2y^2 - 5y + 4 + \dfrac{-5}{3y^2 + 1}$.

88. $\dfrac{4x^5 - 8x^4 - 3x^3 + 22x^2 - 15}{4x^2 - 3}$

$$
\begin{array}{r}
x^3 - 2x^2 + 4 \\
4x^2 - 3 \enclose{longdiv}{4x^5 - 8x^4 - 3x^3 + 22x^2 + 0x - 15} \\
\underline{4x^5 - 3x^3} \\
-8x^4 + 22x^2 \\
\underline{-8x^4 + 6x^2} \\
16x^2 - 15 \\
\underline{16x^2 - 12} \\
-3
\end{array}
$$

The answer is $x^3 - 2x^2 + 4 + \dfrac{-3}{4x^2 - 3}$.

89. **[4.2]** $5^0 + 7^0 = 1 + 1 = 2$

90. **[4.1]** $\left(\dfrac{6r^2p}{5}\right)^3 = \dfrac{6^3 r^{2\cdot3} p^3}{5^3}$

$= \dfrac{6^3 r^6 p^3}{5^3}$

91. **[4.6]** $(12a+1)(12a-1) = (12a)^2 - 1^2$

$= 144a^2 - 1$

92. **[4.2]** $2^{-4} = \dfrac{1}{2^4} = \dfrac{1}{16}$

93. **[4.2]** $(8^{-3})^4 = 8^{(-3)(4)}$

$= 8^{-12}$

$= \dfrac{1}{8^{12}}$

94. **[4.7]** $\dfrac{2p^3 - 6p^2 + 5p}{2p^2}$

$= \dfrac{2p^3}{2p^2} - \dfrac{6p^2}{2p^2} + \dfrac{5p}{2p^2}$

$= p - 3 + \dfrac{5}{2p}$

95. **[4.2]** $\dfrac{(2m^{-5})(3m^2)^{-1}}{m^{-2}(m^{-1})^2}$

$= \dfrac{(2m^{-5})(3^{-1}m^{-2})}{m^{-2}(m^{-2})}$

$= \dfrac{2}{3} \cdot \dfrac{m^{-5+(-2)}}{m^{-2+(-2)}}$

$= \dfrac{2}{3} \cdot \dfrac{m^{-7}}{m^{-4}}$

$= \dfrac{2}{3} \cdot \dfrac{m^4}{m^7}$

$= \dfrac{2}{3m^3}$

96. **[4.5]** $(3k-6)(2k^2 + 4k + 1)$

Multiply vertically.

$$
\begin{array}{rrrr}
2k^2 & + \quad 4k & + \quad 1 & \\
 & 3k & - \quad 6 & \\
\hline
-12k^2 & - \quad 24k & - \quad 6 & \\
6k^3 \quad + \quad 12k^2 & + \quad 3k & & \\
\hline
6k^3 & - \quad 21k & - \quad 6 &
\end{array}
$$

97. **[4.2]** $\dfrac{r^9 \cdot r^{-5}}{r^{-2} \cdot r^{-7}} = \dfrac{r^9 r^2 r^7}{r^5}$

$= \dfrac{r^{9+2+7}}{r^5}$

$= \dfrac{r^{18}}{r^5} = r^{13}$

98. **[4.6]** $(2r+5s)^2$

$= (2r)^2 + 2(2r)(5s) + (5s)^2$

$= 4r^2 + 20rs + 25s^2$

99. **[4.4]** $(-5y^2 + 3y - 11) + (4y^2 - 7y + 15)$

$= -5y^2 + 4y^2 + 3y - 7y - 11 + 15$

$= -y^2 - 4y + 4$

100. **[4.5]** $(2r+5)(5r-2)$

$\qquad \mathbf{F \qquad O \qquad I \qquad L}$

$= 2r(5r) + 2r(-2) + 5(5r) + 5(-2)$

$= 10r^2 - 4r + 25r - 10$

$= 10r^2 + 21r - 10$

101. **[4.7]** $\dfrac{2y^3 + 17y^2 + 37y + 7}{2y + 7}$

$$
\begin{array}{r}
y^2 \ + \ 5y \ + 1 \\
2y+7\overline{\smash{\big)}\,2y^3 \ + 17y^2 \ + 37y \ +7} \\
\underline{2y^3 \ + \ 7y^2 } \\
10y^2 \ + 37y \\
\underline{10y^2 \ + 35y } \\
2y \ + 7 \\
\underline{2y \ + 7} \\
0
\end{array}
$$

The remainder is 0.
The answer is the quotient, $y^2 + 5y + 1$.

102. **[4.7]** $(25x^2y^3 - 8xy^2 + 15x^3y) \div (10x^2y^3)$

$= \dfrac{25x^2y^3 - 8xy^2 + 15x^3y}{10x^2y^3}$

$= \dfrac{25x^2y^3}{10x^2y^3} - \dfrac{8xy^2}{10x^2y^3} + \dfrac{15x^3y}{10x^2y^3}$

$= \dfrac{5}{2} - \dfrac{4}{5xy} + \dfrac{3x}{2y^2}$

103. **[4.4]** $(6p^2 - p - 8) - (-4p^2 + 2p - 3)$

$= (6p^2 - p - 8) + (4p^2 - 2p + 3)$

$= 10p^2 - 3p - 5$

104. **[4.7]** $\dfrac{3x^3 - 2x + 5}{x - 3}$

Use 0 as the coefficient for the missing term.

$$
\begin{array}{r}
3x^2 \ + 9x \ + 25 \\
x-3\overline{\smash{\big)}\,3x^3 \ + 0x^2 \ - 2x \ + 5} \\
\underline{3x^3 \ - 9x^2 } \\
9x^2 \ - 2x \\
\underline{9x^2 \ - 27x } \\
25x \ + 5 \\
\underline{25x \ - 75} \\
80
\end{array}
$$

$\dfrac{3x^3 - 2x + 5}{x - 3} = 3x^2 + 9x + 25 + \dfrac{80}{x - 3}$

105. [4.6] $(-7+2k)^2$
$$= (-7)^2 + 2(-7)(2k) + (2k)^2$$
$$= 49 - 28k + 4k^2$$

106. [4.2] $\left(\dfrac{x}{y^{-3}}\right)^{-4} = \dfrac{x^{-4}}{(y^{-3})^{-4}}$
$$= \dfrac{x^{-4}}{y^{12}}$$
$$= \dfrac{1}{x^4 y^{12}}$$

107. [4.5] **(a)** Use the formula for the perimeter of a rectangle, $P = 2L + 2W$, with $L = 2x - 3$ and $W = x + 2$.
$$P = 2(2x - 3) + 2(x + 2)$$
$$= 4x - 6 + 2x + 4$$
$$= 6x - 2$$

The perimeter of the rectangle is
$$(6x - 2) \text{ units.}$$

(b) Use the formula for the area of a rectangle, $A = LW$, with $L = 2x - 3$ and $W = x + 2$.
$$A = (2x - 3)(x + 2)$$
$$= (2x)(x) + (2x)(2) + (-3)(x) + (-3)(2)$$
$$= 2x^2 + 4x - 3x - 6$$
$$= 2x^2 + x - 6$$

The area of the rectangle is
$$\left(2x^2 + x - 6\right) \text{ square units.}$$

108. [4.6] **(a)** Use the formula for the perimeter of a square, $P = 4s$, with $s = 5x^4 + 2x^2$.
$$P = 4\left(5x^4 + 2x^2\right)$$
$$= 20x^4 + 8x^2$$

The perimeter of the square is
$$\left(20x^4 + 8x^2\right) \text{ units.}$$

(b) Use the formula for the area of a square, $A = s^2$, with $s = 5x^4 + 2x^2$.
$$A = \left(5x^4 + 2x^2\right)^2$$
$$= (5x^4)^2 + 2\left(5x^4\right)\left(2x^2\right) + (2x^2)^2$$
$$= 25x^8 + 20x^6 + 4x^4$$

The area of the square is
$$\left(25x^8 + 20x^6 + 4x^4\right) \text{ square units.}$$

Chapter 4 Test

1. $5^{-4} = \dfrac{1}{5^4} = \dfrac{1}{625}$

2. $(-3)^0 + 4^0 = 1 + 1 = 2$

3. $4^{-1} + 3^{-1} = \dfrac{1}{4^1} + \dfrac{1}{3^1} = \dfrac{3}{12} + \dfrac{4}{12} = \dfrac{7}{12}$

4. $\dfrac{(3x^2 y)^2 (xy^3)^2}{(xy)^3} = \dfrac{3^2 (x^2)^2 y^2 x^2 (y^3)^2}{x^3 y^3}$
$$= \dfrac{9x^4 y^2 x^2 y^6}{x^3 y^3}$$
$$= \dfrac{9x^6 y^8}{x^3 y^3}$$
$$= 9x^3 y^5$$

5. $\dfrac{8^{-1} \cdot 8^4}{8^{-2}} = \dfrac{8^{(-1)+4}}{8^{-2}} = \dfrac{8^3}{8^{-2}} = 8^{3-(-2)} = 8^5$

6. $\dfrac{(x^{-3})^{-2}(x^{-1}y)^2}{(xy^{-2})^2} = \dfrac{(x^{-3})^{-2}(x^{-1})^2(y)^2}{(x)^2(y^{-2})^2}$
$$= \dfrac{x^6 x^{-2} y^2}{x^2 y^{-4}}$$
$$= \dfrac{x^4 y^2}{x^2 y^{-4}}$$
$$= x^{4-2} y^{2-(-4)}$$
$$= x^2 y^6$$

7. **(a)** $3^{-4} = \dfrac{1}{3^4} = \dfrac{1}{81}$, which is *positive*.
A negative exponent indicates a reciprocal, not a negative number.

(b) $(-3)^4 = 81$, which is *positive*.

(c) $-3^4 = -1 \cdot 3^4 = -81$, which is *negative*.

(d) $3^0 = 1$, which is *positive*.

(e) $(-3)^0 - 3^0 = 1 - 1 = 0$ (*zero*)

(f) $(-3)^{-3} = \dfrac{1}{(-3)^3} = \dfrac{1}{-27}$, which is *negative*.

8. **(a)** $45{,}000{,}000{,}000 = 4.5 \times 10^{10}$
Move the decimal point left 10 places so it is to the right of the first nonzero digit. $45{,}000{,}000{,}000$ is *greater* than 4.5, so the power is *positive*.

(b) $3.6 \times 10^{-6} = 0.000\,003\,6$
Move the decimal point 6 places to the left.

(c) $\dfrac{9.5 \times 10^{-1}}{5 \times 10^3} = \dfrac{9.5}{5} \times \dfrac{10^{-1}}{10^3}$
$$= 1.9 \times 10^{-1-3}$$
$$= 1.9 \times 10^{-4}$$
$$= 0.00019$$

9. **(a)** $1000 = 1 \times 10^3$
$5{,}890{,}000{,}000{,}000 = 5.89 \times 10^{12}$

(b) $\left(1 \times 10^3 \text{ light-years}\right)\left(5.89 \times 10^{12} \ \dfrac{\text{miles}}{\text{light-year}}\right)$

$= 5.89 \times 10^{3+12} \text{ miles}$

$= 5.89 \times 10^{15} \text{ miles}$

10. $5x^2 + 8x - 12x^2 = 5x^2 - 12x^2 + 8x$

$\qquad\qquad\qquad = -7x^2 + 8x$

degree 2; binomial (2 terms)

11. $13n^3 - n^2 + n^4 + 3n^4 - 9n^2$

$= n^4 + 3n^4 + 13n^3 - n^2 - 9n^2$

$= 4n^4 + 13n^3 - 10n^2$

degree 4; trinomial (3 terms)

12. $y = 2x^2 - 4$

$x = -2 : y = 2(-2)^2 - 4 = 2 \cdot 4 - 4 = 4$

$x = -1 : y = 2(-1)^2 - 4 = 2 \cdot 1 - 4 = -2$

$x = 0 : y = 2(0)^2 - 4 = 2 \cdot 0 - 4 = -4$

$x = 1 : y = 2(1)^2 - 4 = 2 \cdot 1 - 4 = -2$

$x = 2 : y = 2(2)^2 - 4 = 2 \cdot 4 - 4 = 4$

x	-2	-1	0	1	2
y	4	-2	-4	-2	4

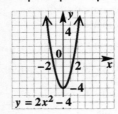

$y = 2x^2 - 4$

13. $\left(2y^2 - 8y + 8\right) + \left(-3y^2 + 2y + 3\right)$

$\quad - \left(y^2 + 3y - 6\right)$

$= \left(2y^2 - 8y + 8\right) + \left(-3y^2 + 2y + 3\right)$

$\quad + \left(-y^2 - 3y + 6\right)$

$= \left(2y^2 - 3y^2 - y^2\right) + \left(-8y + 2y - 3y\right)$

$\quad + (8 + 3 + 6)$

$= -2y^2 - 9y + 17$

14. $\left(-9a^3b^2 + 13ab^5 + 5a^2b^2\right)$

$\quad - \left(6ab^5 + 12a^3b^2 + 10a^2b^2\right)$

$= \left(-9a^3b^2 + 13ab^5 + 5a^2b^2\right)$

$\quad + \left(-6ab^5 - 12a^3b^2 - 10a^2b^2\right)$

$= \left(-9a^3b^2 - 12a^3b^2\right) + \left(13ab^5 - 6ab^5\right)$

$\quad + \left(5a^2b^2 - 10a^2b^2\right)$

$= -21a^3b^2 + 7ab^5 - 5a^2b^2$

15. Subtract.

$$
\begin{array}{rrrrr}
9t^3 & - & 4t^2 & + 2t & + 2 \\
9t^3 & + & 8t^2 & - 3t & - 6 \\
\end{array}
$$

Change all signs in the second row and then add.

$$
\begin{array}{rrrr}
9t^3 & - \ \ 4t^2 & + 2t & + 2 \\
-9t^3 & - \ \ 8t^2 & + 3t & + 6 \\
\hline
& -12t^2 & + 5t & + 8 \\
\end{array}
$$

16. $3x^2\left(-9x^3 + 6x^2 - 2x + 1\right)$

$= 3x^2\left(-9x^3\right) + 3x^2\left(6x^2\right) + 3x^2(-2x) + 3x^2(1)$

$= -27x^5 + 18x^4 - 6x^3 + 3x^2$

17.
$$\qquad\ \ \mathbf{F \quad O \quad I \quad L}$$
$(t - 8)(t + 3) = t^2 + 3t - 8t - 24$

$\qquad\qquad\quad = t^2 - 5t - 24$

18. $(4x + 3y)(2x - y)$
$$\qquad \mathbf{F \quad O \quad I \quad L}$$
$= 8x^2 - 4xy + 6xy - 3y^2$

$= 8x^2 + 2xy - 3y^2$

19. $(5x - 2y)^2 = (5x)^2 - 2(5x)(2y) + (2y)^2$

$\qquad\qquad\ = 25x^2 - 20xy + 4y^2$

20. $(10v + 3w)(10v - 3w) = (10v)^2 - (3w)^2$

$\qquad\qquad\qquad\qquad\ = 100v^2 - 9w^2$

21. $(2r - 3)\left(r^2 + 2r - 5\right)$

Multiply vertically.

$$
\begin{array}{rrrr}
r^2 & + \ \ 2r & - 5 & \\
& 2r & - 3 & \\
\hline
-3r^2 & - \ \ 6r & + 15 & \\
2r^3 & + \ \ 4r^2 & - 10r & \\
\hline
2r^3 & + \quad r^2 & - 16r & + 15 \\
\end{array}
$$

22. Use the formula for the perimeter of a square, $P = 4s$, with $s = 3x + 9$.

$P = 4s$

$\quad = 4(3x + 9)$

$\quad = 4(3x) + 4(9)$

$\quad = 12x + 36$

The perimeter of the square is $(12x + 26)$ units.

Use the formula for the area of a square, $A = s^2$, with $s = 3x + 9$.

$A = s^2$

$\quad = (3x + 9)^2$

$\quad = (3x)^2 + 2(3x)(9) + 9^2$

$\quad = 9x^2 + 54x + 81$

The area of the square is $(9x^2 + 54x + 81)$ square units.

23. $\dfrac{8y^3 - 6y^2 + 4y + 10}{2y}$

$= \dfrac{8y^3}{2y} - \dfrac{6y^2}{2y} + \dfrac{4y}{2y} + \dfrac{10}{2y}$

$= 4y^2 - 3y + 2 + \dfrac{5}{y}$

24. $\left(-9x^2y^3 + 6x^4y^3 + 12xy^3\right) \div (3xy)$

$= \dfrac{-9x^2y^3 + 6x^4y^3 + 12xy^3}{3xy}$

$= \dfrac{-9x^2y^3}{3xy} + \dfrac{6x^4y^3}{3xy} + \dfrac{12xy^3}{3xy}$

$= -3xy^2 + 2x^3y^2 + 4y^2$

25. $\left(3x^3 - x + 4\right) \div (x - 2)$

$$
\require{enclose}
\begin{array}{r}
3x^2 + 6x + 11 \\
x-2 \enclose{longdiv}{3x^3 + 0x^2 - x + 4 } \\
\underline{3x^3 - 6x^2 } \\
6x^2 - x \\
\underline{6x^2 - 12x } \\
11x + 4 \\
\underline{11x - 22 } \\
26
\end{array}
$$

The answer is

$$3x^2 + 6x + 11 + \dfrac{26}{x - 2}.$$

Cumulative Review Exercises
(Chapters 1–4)

1. $\dfrac{28}{16} = \dfrac{7 \cdot 4}{4 \cdot 4} = \dfrac{7}{4}$

2. $\dfrac{55}{11} = \dfrac{5 \cdot 11}{1 \cdot 11} = \dfrac{5}{1} = 5$

3. $\frac{2}{3} + \frac{1}{8} = \frac{16}{24} + \frac{3}{24} = \frac{19}{24}$

4. $\frac{7}{4} - \frac{9}{5} = \frac{35}{20} - \frac{36}{20} = -\frac{1}{20}$

5. Each shed requires $1\frac{1}{4}$ cubic yards of concrete, so the total amount of concrete needed for 25 sheds would be

$$25 \times 1\tfrac{1}{4} = 25 \times \tfrac{5}{4}$$
$$= \tfrac{125}{4}$$
$$= 31\tfrac{1}{4} \text{ cubic yards.}$$

6. Use the formula for simple interest, $I = Prt$, with $P = \$34,000$, $r = 5.4\%$, and $t = 1$.

$$I = Prt$$
$$= (34,000)(0.054)(1)$$
$$= 1836$$

She earned $1836 in interest.

7. The positive integer factors of 45 are

$$1, 3, 5, 9, 15, \text{ and } 45.$$

8. $\dfrac{4x - 2y}{x + y} = \dfrac{4(-2) - 2(4)}{(-2) + 4}$ *Let $x = -2$,*
$y = 4$.

$= \dfrac{-8 - 8}{2} = \dfrac{-16}{2} = -8$

9. $\dfrac{(-13 + 15) - (3 + 2)}{6 - 12} = \dfrac{2 - 5}{-6} = \dfrac{-3}{-6} = \dfrac{1}{2}$

10. $-7 - 3[2 + (5 - 8)] = -7 - 3[2 + (-3)]$
$= -7 - 3[-1]$
$= -7 + 3 = -4$

11. $(9 + 2) + 3 = 9 + (2 + 3)$

The numbers are in the same order but grouped differently, so this is an example of the associative property of addition.

12. $6(4 + 2) = 6(4) + 6(2)$

The number 6 outside the parentheses is "distributed" over the 4 and the 2. This is an example of the distributive property.

13. $-3\left(2x^2 - 8x + 9\right) - \left(4x^2 + 3x + 2\right)$
$= -6x^2 + 24x - 27 - 4x^2 - 3x - 2$
$= -10x^2 + 21x - 29$

14. $2 - 3(t - 5) = 4 + t$
$2 - 3t + 15 = 4 + t$
$-3t + 17 = 4 + t$
$-4t + 17 = 4$
$-4t = -13$
$t = \frac{-13}{-4} = \frac{13}{4}$

The solution set is $\left\{\frac{13}{4}\right\}$.

15. $2(5h + 1) = 10h + 4$
$10h + 2 = 10h + 4$
$2 = 4$ *False*

The false statement indicates that the equation has no solution, symbolized by $\emptyset$.

16. Solve $d = rt$ for r.

$\dfrac{d}{t} = \dfrac{rt}{t}$ *Divide by t.*

$\dfrac{d}{t} = r$

17. $\dfrac{x}{5} = \dfrac{x - 2}{7}$
$7x = 5(x - 2)$ *Cross products are equal.*
$7x = 5x - 10$
$2x = -10$
$x = -5$

The solution set is $\{-5\}$.

18. $\frac{1}{3}p - \frac{1}{6}p = -2$

To clear fractions, multiply both sides of the equation by the least common denominator, which is 6.

$$6\left(\tfrac{1}{3}p - \tfrac{1}{6}p\right) = (6)(-2)$$
$$6\left(\tfrac{1}{3}p\right) - 6\left(\tfrac{1}{6}p\right) = -12$$
$$2p - p = -12$$
$$p = -12$$

The solution set is $\{-12\}$.

19. $0.05x + 0.15(50 - x) = 5.50$

To clear decimals, multiply both sides of the equation by 100.

$$100[0.05x + 0.15(50 - x)] = 100(5.50)$$
$$100(0.05x) + 100[0.15(50 - x)] = 100(5.50)$$
$$5x + 15(50 - x) = 550$$
$$5x + 750 - 15x = 550$$
$$-10x + 750 = 550$$
$$-10x = -200$$
$$x = 20$$

The solution set is $\{20\}$.

20. $4 - (3x + 12) = (2x - 9) - (5x - 1)$
$$4 - 3x - 12 = 2x - 9 - 5x + 1$$
$$-3x - 8 = -3x - 8 \qquad \textit{True}$$

The true statement indicates that the solution set is $\{$all real numbers$\}$.

21. Let $x =$ the number of calories burned in thermoregulation.

Then $5\frac{3}{8}x = \frac{43}{8}x =$ the number of calories burned in exertion.

$$x + \tfrac{43}{8}x = 11,200$$
$$8\left(x + \tfrac{43}{8}x\right) = 8(11,200)$$
$$8x + 43x = 89,600$$
$$51x = 89,600$$
$$x = \tfrac{89,600}{51} \approx 1757$$

A husky burns approximately 1757 calories for thermoregulation and $\frac{43}{8}\left(\frac{89,600}{51}\right) \approx 9443$ calories for exertion.

22. Let $x =$ the unknown number.
$$3(8 - x) = 3x$$
$$24 - 3x = 3x$$
$$24 = 6x$$
$$x = 4$$

The unknown number is 4.

23. Let $x =$ one side of the triangle.
Then $2x =$ the other (unknown) side.

The third side is 17 feet. The perimeter of the triangle cannot be more than 50 feet. This is equivalent to stating that the sum of the lengths of the sides must be less than or equal to 50 feet. Write this statement as an inequality and solve.

$$x + 2x + 17 \leq 50$$
$$3x + 17 \leq 50$$
$$3x \leq 33$$
$$x \leq 11$$

One side cannot be more than 11 feet. The other side cannot be more than $2 \cdot 11 = 22$ feet.

24. $-8x \leq -80$
$$\frac{-8x}{-8} \geq \frac{-80}{-8} \qquad \textit{Divide by -8;}$$
$$\textit{reverse the symbol.}$$
$$x \geq 10$$

The solution set is $[10, \infty)$.

25. $-2(x + 4) > 3x + 6$
$$-2x - 8 > 3x + 6$$
$$-2x > 3x + 14$$
$$-5x > 14$$
$$\frac{-5x}{-5} < \frac{14}{-5} \qquad \textit{Divide by -5;}$$
$$\textit{reverse the symbol.}$$
$$x < -\tfrac{14}{5}$$

The solution set is $\left(-\infty, -\frac{14}{5}\right)$.

26. $-3 \leq 2x + 5 < 9$
$$-8 \leq 2x < 4 \qquad \textit{Subtract 5.}$$
$$\frac{-8}{2} \leq \frac{2x}{2} < \frac{4}{2} \qquad \textit{Divide by 2.}$$
$$-4 \leq x < 2$$

The solution set is $[-4, 2)$.

27. We recognize $y = -3x + 6$ as the equation of a line with y-intercept 6 and slope -3.

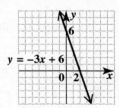

28. **(a)** Use the definition of slope with $(x_1, y_1) = (-1, 5)$ and $(x_2, y_2) = (2, 8)$.

$$m = \frac{\text{change in } y}{\text{change in } x} = \frac{y_2 - y_1}{x_2 - x_1}$$
$$= \frac{8 - 5}{2 - (-1)} = \frac{3}{3} = 1$$

(b) Use the point-slope form of the equation of a line with $m = 1$ and $(x_1, y_1) = (-1, 5)$.

$$y - y_1 = m(x - x_1)$$
$$y - 5 = 1[x - (-1)]$$
$$y - 5 = x + 1$$
$$y = x + 6$$

29. $4^{-1} + 3^0 = \dfrac{1}{4^1} + 1 = 1\dfrac{1}{4}$ or $\dfrac{5}{4}$

30. $2^{-4} \cdot 2^5 = 2^{-4+5} = 2^1 = 2$

31. $\dfrac{8^{-5} \cdot 8^7}{8^2} = \dfrac{8^{-5+7}}{8^2} = \dfrac{8^2}{8^2} = 1$

32. $\dfrac{(a^{-3}b^2)^2}{(2a^{-4}b^{-3})^{-1}} = \dfrac{(a^{-3})^2(b^2)^2}{2^{-1}(a^{-4})^{-1}(b^{-3})^{-1}}$

$$= \dfrac{a^{-6}b^4}{2^{-1}a^4b^3}$$

$$= \dfrac{2b^4}{a^6a^4b^3} = \dfrac{2b}{a^{10}}$$

33. $34{,}500 = 3.45 \times 10^4$

Move the decimal point left 4 places so it is to the right of the first nonzero digit. 34,500 is *greater* than 3.45, so the power is *positive*.

34. 5.36×10^{-7}

Move the decimal point 7 places to the left.
$5.36 \times 10^{-7} = 0.000\,000\,536$

35. $\left(3.6 \times 10^1 \text{ sec}\right)\left(3.0 \times 10^5 \dfrac{\text{km}}{\text{sec}}\right)$

$$= 3.6 \times 3.0 \times 10^{1+5} \text{ km}$$
$$= 10.8 \times 10^6 \text{ km}$$

Venus is about 10,800,000 km from the sun.

36. $y = (x + 4)^2$

$x = -6 : y = (-6 + 4)^2 = (-2)^2 = 4$
$x = -5 : y = (-5 + 4)^2 = (-1)^2 = 1$
$x = -4 : y = (-4 + 4)^2 = (0)^2 = 0$
$x = -3 : y = (-3 + 4)^2 = (1)^2 = 1$
$x = -2 : y = (-2 + 4)^2 = (2)^2 = 4$

x	-6	-5	-4	-3	-2
y	4	1	0	1	4

37. $(7x^3 - 12x^2 - 3x + 8) + (6x^2 + 4)$
$\quad - (-4x^3 + 8x^2 - 2x - 2)$
$\quad = (7x^3 - 12x^2 - 3x + 8) + (6x^2 + 4)$
$\qquad + (4x^3 - 8x^2 + 2x + 2)$
$\quad = (7 + 4)x^3 + (-12 + 6 - 8)x^2$
$\qquad + (-3 + 2)x + (8 + 4 + 2)$
$\quad = 11x^3 - 14x^2 - x + 14$

38. $6x^5(3x^2 - 9x + 10)$
$\quad = 6x^5(3x^2) + 6x^5(-9x) + 6x^5(10)$
$\quad = 18x^7 - 54x^6 + 60x^5$

39. $(7x + 4)(9x + 3)$
$\quad = 63x^2 + 21x + 36x + 12 \quad FOIL$
$\quad = 63x^2 + 57x + 12$

40. $(5x + 8)^2 = (5x)^2 + 2(5x)(8) + (8)^2$
$\quad = 25x^2 + 80x + 64$

41. $\dfrac{14x^3 - 21x^2 + 7x}{7x}$

$$= \dfrac{14x^3}{7x} - \dfrac{21x^2}{7x} + \dfrac{7x}{7x}$$
$$= 2x^2 - 3x + 1$$

42. $\dfrac{y^3 - 3y^2 + 8y - 6}{y - 1}$

$$
\begin{array}{r}
y^2 \phantom{{}- 2y} - 2y + 6 \\
y - 1 \overline{\smash{\big)}\ y^3 - 3y^2 + 8y - 6} \\
\underline{y^3 - y^2 \phantom{{}+ 8y - 6}} \\
-2y^2 + 8y \phantom{{}- 6} \\
\underline{-2y^2 + 2y \phantom{{}- 6}} \\
6y - 6 \\
\underline{6y - 6} \\
0
\end{array}
$$

The remainder is 0. The answer is the quotient,

$$y^2 - 2y + 6.$$

CHAPTER 5   FACTORING AND APPLICATIONS

5.1   The Greatest Common Factor; Factoring by Grouping

1. Find the prime factored form of each number.

$$40 = 2 \cdot 2 \cdot 2 \cdot 5$$
$$20 = 2 \cdot 2 \cdot 5$$
$$4 = 2 \cdot 2$$

The least number of times 2 appears in all the factored forms is 2. There is no 5 in the prime factored form of 4, so the

$$\text{GCF} = 2^2 = 4.$$

3. Find the prime factored form of each number.

$$18 = 2 \cdot 3 \cdot 3$$
$$24 = 2 \cdot 2 \cdot 2 \cdot 3$$
$$36 = 2 \cdot 2 \cdot 3 \cdot 3$$
$$48 = 2 \cdot 2 \cdot 2 \cdot 2 \cdot 3$$

The least number of times the primes 2 and 3 appear in all four factored forms is once, so

$$\text{GCF} = 2 \cdot 3 = 6.$$

5. 6, 8, 9

Find the prime factored form of each number.

$$6 = 2 \cdot 3$$
$$8 = 2 \cdot 2 \cdot 2$$
$$9 = 3 \cdot 3$$

There are no primes common to all three numbers, so the GCF is 1.

7. First, verify that you have factored completely. Then multiply the factors. The product should be the original polynomial.

9. Write each term in prime factored form.

$$16y = 2^4 \cdot y$$
$$24 = 2^3 \cdot 3$$

There is no y in the second term, so y will not appear in the GCF. Thus, the GCF of $16y$ and 24 is

$$2^3 = 8.$$

11.
$$30x^3 = 2 \cdot 3 \cdot 5 \cdot x^3$$
$$40x^6 = 2^3 \cdot 5 \cdot x^6$$
$$50x^7 = 2 \cdot 5^2 \cdot x^7$$

The GCF of the coefficients, 30, 40, and 50, is $2^1 \cdot 5^1 = 10$. The smallest exponent on the variable x is 3. Thus the GCF of the given terms is $10x^3$.

13.
$$12m^3n^2 = 2^2 \cdot 3 \cdot m^3 \cdot n^2$$
$$18m^5n^4 = 2 \cdot 3^2 \cdot m^5 \cdot n^4$$
$$36m^8n^3 = 2^2 \cdot 3^2 \cdot m^8 \cdot n^3$$

The GCF is $2 \cdot 3 \cdot m^3 \cdot n^2 = 6m^3n^2$.

15.
$$-x^4y^3 = -1 \cdot x^4 \cdot y^3$$
$$-xy^2 = -1 \cdot x \cdot y^2$$

The GCF is xy^2, or $-xy^2$.

17. $2k^2(5k)$ is written as a product of $2k^2$ and $5k$ and hence, it is *factored.*

19. $2k^2 + (5k + 1)$ is written as a sum of $2k^2$ and $(5k + 1)$ and hence, it is *not factored.*

21. Yes, $-xy$ is a common factor of $-x^4y^3$ and $-xy^2$. When $-xy$ is multiplied by x^3y^2, the result is $-x^4y^3$.

23. $9m^4 = 3m^2(3m^2)$

Factor out $3m^2$ from $9m^4$ to obtain $3m^2$.

25. $-8z^9 = -4z^5(2z^4)$

Factor out $-4z^5$ from $-8z^9$ to obtain $2z^4$.

27. $6m^4n^5 = 3m^3n(2mn^4)$

Factor out $3m^3n$ from $6m^4n^5$ to obtain $2mn^4$.

29. $12y + 24 = 12 \cdot y + 12 \cdot 2$
$$\qquad\qquad = 12(y + 2)$$

31. $10a^2 - 20a = 10a(a) - 10a(2)$
$$\qquad\qquad\quad = 10a(a - 2)$$

33. $8x^2y + 12x^3y^2 = 4x^2y(2) + 4x^2y(3xy)$
$$\qquad\qquad\qquad = 4x^2y(2 + 3xy)$$

35. $27m^3 - 9m$
The GCF is $9m$.

$$27m^3 - 9m = 9m(3m^2) + 9m(-1)$$
$$\qquad\qquad = 9m(3m^2 - 1)$$

37. $16z^4 + 24z^2$
The GCF is $8z^2$.

$$16z^4 + 24z^2 = 8z^2(2z^2) + 8z^2(3)$$
$$\qquad\qquad = 8z^2(2z^2 + 3)$$

39. $\frac{1}{4}d^2 - \frac{3}{4}d$
As in Example 3(e), we factor out $\frac{1}{4}d$.

$$\frac{1}{4}d^2 - \frac{3}{4}d = \frac{1}{4}d(d) + \frac{1}{4}d(-3)$$
$$\qquad\qquad = \frac{1}{4}d(d - 3)$$

41. $12x^3 + 6x^2$
The GCF is $6x^2$.

$$12x^3 + 6x^2 = 6x^2(2x) + 6x^2(1)$$
$$\qquad\qquad = 6x^2(2x + 1)$$

43. $65y^{10} + 35y^6$
The GCF is $5y^6$.

$$65y^{10} + 35y^6 = \left(5y^6\right)\left(13y^4\right) + \left(5y^6\right)(7)$$
$$= 5y^6\left(13y^4 + 7\right)$$

45. $11w^3 - 100$

The two terms of this expression have no common factor (except 1).

47. $8m^2n^3 + 24m^2n^2$
The GCF is $8m^2n^2$.

$$8m^2n^3 + 24m^2n^2$$
$$= \left(8m^2n^2\right)(n) + \left(8m^2n^2\right)(3)$$
$$= 8m^2n^2(n + 3)$$

49. $13y^8 + 26y^4 - 39y^2$
The GCF is $13y^2$.

$$13y^8 + 26y^4 - 39y^2$$
$$= 13y^2\left(y^6\right) + 13y^2\left(2y^2\right) + 13y^2(-3)$$
$$= 13y^2\left(y^6 + 2y^2 - 3\right)$$

51. $36p^6q + 45p^5q^4 + 81p^3q^2$
The GCF is $9p^3q$.

$$36p^6q + 45p^5q^4 + 81p^3q^2$$
$$= 9p^3q\left(4p^3\right) + 9p^3q\left(5p^2q^3\right) + 9p^3q(9q)$$
$$= 9p^3q\left(4p^3 + 5p^2q^3 + 9q\right)$$

53. $a^5 + 2a^3b^2 - 3a^5b^2 + 4a^4b^3$
The GCF is a^3.

$$a^5 + 2a^3b^2 - 3a^5b^2 + 4a^4b^3$$
$$= a^3\left(a^2\right) + a^3\left(2b^2\right) + a^3\left(-3a^2b^2\right)$$
$$+ a^3\left(4ab^3\right)$$
$$= a^3\left(a^2 + 2b^2 - 3a^2b^2 + 4ab^3\right)$$

55. The GCF of the terms of $c(x + 2) - d(x + 2)$ is the binomial $x + 2$.

$$c(x + 2) - d(x + 2)$$
$$= (x + 2)(c) + (x + 2)(-d)$$
$$= (x + 2)(c - d)$$

57. The GCF of the terms of
$m(m + 2n) + n(m + 2n)$ is the binomial
$m + 2n$.

$$m(m + 2n) + n(m + 2n)$$
$$= (m + 2n)(m) + (m + 2n)(n)$$
$$= (m + 2n)(m + n)$$

59. $8(7t + 4) + x(7t + 4)$

This expression is the *sum* of two terms, $8(7t + 4)$ and $x(7t + 4)$, so it is not in factored form. We can factor out $7t + 4$.

$$8(7t + 4) + x(7t + 4)$$
$$= (7t + 4)(8) + (7t + 4)(x)$$
$$= (7t + 4)(8 + x)$$

61. $(8 + x)(7t + 4)$

This expression is the *product* of two factors, $8 + x$ and $7t + 4$, so it is in factored form.

63. $18x^2(y + 4) + 7(y + 4)$

This expression is the *sum* of two terms, $18x^2(y + 4)$ and $7(y + 4)$, so it is not in factored form. We can factor out $y + 4$.

$$18x^2(y + 4) + 7(y + 4)$$
$$= (y + 4)\left(18x^2\right) + (y + 4)(7)$$
$$= (y + 4)\left(18x^2 + 7\right)$$

65. It is not possible to factor the expression in Exercise 64 because the two terms, $12k^3(s - 3)$ and $7(s + 3)$, do not have a common factor.

67. $p^2 + 4p + pq + 4q$

The first two terms have a common factor of p, and the last two terms have a common factor of q. Thus,

$$p^2 + 4p + pq + 4q$$
$$= \left(p^2 + 4p\right) + (pq + 4q)$$
$$= p(p + 4) + q(p + 4).$$

Now we have two terms which have a common binomial factor of $p + 4$. Thus,

$$p^2 + 4p + pq + 4q$$
$$= p(p + 4) + q(p + 4)$$
$$= (p + 4)(p + q).$$

69. $a^2 - 2a + ab - 2b$

$$= \left(a^2 - 2a\right) + (ab - 2b) \quad \textit{Group the terms.}$$

$$= a(a - 2) + b(a - 2) \quad \textit{Factor each group.}$$

$$= (a - 2)(a + b) \quad \textit{Factor out } a - 2.$$

71. $7z^2 + 14z - az - 2a$

$$= \left(7z^2 + 14z\right) + (-az - 2a) \quad \textit{Group the terms.}$$

$$= 7z(z + 2) - a(z + 2) \quad \textit{Factor each group.}$$

$$= (z + 2)(7z - a) \quad \textit{Factor out } z + 2.$$

73. $18r^2 + 12ry - 3xr - 2xy$

$$= \left(18r^2 + 12ry\right) + (-3xr - 2xy) \quad \textit{Group the terms.}$$

$$= 6r(3r + 2y) - x(3r + 2y) \quad \textit{Factor each group.}$$

$$= (3r + 2y)(6r - x) \quad \textit{Factor out } 3r + 2y.$$

75. $3a^3 + 3ab^2 + 2a^2b + 2b^3$

$$= \left(3a^3 + 3ab^2\right) + \left(2a^2b + 2b^3\right) \quad \textit{Group the terms.}$$

$$= 3a\left(a^2 + b^2\right) + 2b\left(a^2 + b^2\right) \quad \textit{Factor each group.}$$

$$= \left(a^2 + b^2\right)(3a + 2b) \quad \textit{Factor out } a^2 + b^2.$$

77. $1 - a + ab - b$

$$= (1 - a) + (ab - b) \quad \textit{Group the terms.}$$

$$= 1(1 - a) - b(-a + 1) \quad \textit{Factor each group.}$$

$$= 1(1 - a) - b(1 - a)$$

$$= (1 - a)(1 - b) \quad \textit{Factor out } 1 - a.$$

79. $16m^3 - 4m^2p^2 - 4mp + p^3$

$$= \left(16m^3 - 4m^2p^2\right) + \left(-4mp + p^3\right)$$

$$= 4m^2\left(4m - p^2\right) - p\left(4m - p^2\right)$$

$$= \left(4m - p^2\right)\left(4m^2 - p\right)$$

81. $5m - 6p - 2mp + 15$

We need to rearrange these terms to get two groups that each have a common factor. We could group $5m$ with either $-2mp$ or 15.

$$5m + 15 - 2mp - 6p \quad \textit{Rearrange.}$$

$$= (5m + 15) + (-2mp - 6p) \quad \textit{Group the terms.}$$

$$= 5(m + 3) - 2p(m + 3) \quad \textit{Factor each group.}$$

$$= (m + 3)(5 - 2p) \quad \textit{Factor out } m + 3.$$

83. $18r^2 - 2ty + 12ry - 3rt$

We'll rearrange the terms so that $18r^2$ is grouped with another term containing r.

$$18r^2 + 12ry - 3rt - 2ty$$

$$\textit{Rearrange.}$$

$$= \left(18r^2 + 12ry\right) + (-3rt - 2ty)$$

$$\textit{Group the terms.}$$

$$= 6r(3r + 2y) - t(3r + 2y)$$

$$\textit{Factor each group.}$$

$$= (3r + 2y)(6r - t)$$

$$\textit{Factor out } 3r + 2y.$$

85. $a^5 - 3 + 2a^5b - 6b$

$$= a^5 + 2a^5b - 3 - 6b \quad \textit{Rearrange.}$$

$$= \left(a^5 + 2a^5b\right) + (-3 - 6b) \quad \textit{Group the terms.}$$

$$= a^5(1 + 2b) - 3(1 + 2b) \quad \textit{Factor each group.}$$

$$= (1 + 2b)\left(a^5 - 3\right) \quad \textit{Factor out } 1 + 2b.$$

87.   In order to rewrite

$$2xy + 12 - 3y - 8x$$

as

$$2xy - 8x - 3y + 12,$$

we must change the order of the terms. The property that allows us to do this is the commutative property of addition.

88.   After we group both pairs of terms in the rearranged polynomial, we have

$$(2xy - 8x) + (-3y + 12).$$

The greatest common factor for the first pair of terms is $2x$. The GCF for the second pair is -3. Factoring each group gives us

$$2x(y - 4) - 3(y - 4).$$

89.   The expression obtained in Exercise 88 is the *difference* between two terms, $2x(y - 4)$ and $3(y - 4)$, so it is *not* in factored form.

90. $2x(y - 4) - 3(y - 4)$

$$= (y - 4)(2x - 3)$$

or $\quad = (2x - 3)(y - 4)$

Yes, this is the same result as the one shown in Example 6(b), even though the terms were grouped in a different way.

91.   **(a)** To determine whether the result $(a - 1)(b - 1)$ is correct, multiply the factors by using FOIL.

$$(a - 1)(b - 1) = ab - a - b + 1$$

Rearranging the terms of this product, we obtain

$$ab - a - b + 1 = 1 - a + ab - b,$$

which is the polynomial given in Exercise 77, so the student's answer is correct.

(b) Both answers are acceptable because in each case the product of the factors is the given polynomial. We can also see the two factored forms are equivalent in the following way:

$$1 - a = -1(a - 1)$$
$$1 - b = -1(b - 1).$$

continued

Thus,

$$(1-a)(1-b)$$
$$= [-1(a-1)] \cdot [-1(b-1)]$$
$$= (-1)(-1)(a-1)(b-1)$$
$$= 1(a-1)(b-1)$$
$$= (a-1)(b-1).$$

93. $(x+6)(x-9) = x^2 - 9x + 6x - 54$
$$= x^2 - 3x - 54$$

95. $(x+2)(x+7) = x^2 + 7x + 2x + 14$
$$= x^2 + 9x + 14$$

97. $2x^2(x^2 + 3x + 5) = 2x^4 + 6x^3 + 10x^2$

5.2 Factoring Trinomials

1.   Product: 48   Sum: -19

Factors of 48	Sums of Factors
1, 48	$1 + 48 = 49$
$-1, -48$	$-1 + (-48) = -49$
2, 24	$2 + 24 = 26$
$-2, -24$	$-2 + (-24) = -26$
3, 16	$3 + 16 = 19$
$-3, -16$	$-3 + (-16) = -19 \leftarrow$
4, 12	$4 + 12 = 16$
$-4, -12$	$-4 + (-12) = -16$
6, 8	$6 + 8 = 14$
$-6, -8$	$-6 + (-8) = -14$

The pair of integers whose product is 48 and whose sum is -19 is -3 and -16.

3.   Product: -24 Sum: -5

Factors of -24	Sums of Factors
1, -24	$1 + (-24) = -23$
$-1, 24$	$-1 + 24 = 23$
2, -12	$2 + (-12) = -10$
$-2, 12$	$-2 + 12 = 10$
3, -8	$3 + (-8) = -5 \leftarrow$
$-3, 8$	$-3 + 8 = 5$
4, -6	$4 + (-6) = -2$
$-4, 6$	$-4 + 6 = 2$

The pair of integers whose product is -24 and whose sum is -5 is 3 and -8.

5.   If the coefficient of the last term of the trinomial is negative, then a and b must have different signs, one positive and one negative.

7.   A *prime polynomial* is one that cannot be factored using only integers in the factors.

9. $x^2 - 12x + 32$

Multiply each of the given pairs of factors to determine which one gives the required product.

A. $(x-8)(x+4) = x^2 - 4x - 32$

B. $(x+8)(x-4) = x^2 + 4x - 32$

C. $(x-8)(x-4) = x^2 - 12x + 32$

D. $(x+8)(x+4) = x^2 + 12x + 32$

Choice **C** is the correct factored form.

11. $p^2 + 11p + 30 = (p+5)(\quad)$

Look for an integer whose product with 5 is 30 and whose sum with 5 is 11. That integer is 6.

$$p^2 + 11p + 30 = (p+5)(p+6)$$

13. $x^2 + 15x + 44 = (x+4)(\quad)$

Look for an integer whose product with 4 is 44 and whose sum with 4 is 15. That integer is 11.

$$x^2 + 15x + 44 = (x+4)(x+11)$$

15. $x^2 - 9x + 8 = (x-1)(\quad)$

Look for an integer whose product with -1 is 8 and whose sum with -1 is -9. That integer is -8.

$$x^2 - 9x + 8 = (x-1)(x-8)$$

17. $y^2 - 2y - 15 = (y+3)(\quad)$

Look for an integer whose product with 3 is -15 and whose sum with 3 is -2. That integer is -5.

$$y^2 - 2y - 15 = (y+3)(y-5)$$

19. $x^2 + 9x - 22 = (x-2)(\quad)$

Look for an integer whose product with -2 is -22 and whose sum with -2 is 9. That integer is 11.

$$x^2 + 9x - 22 = (x-2)(x+11)$$

21. $y^2 - 7y - 18 = (y+2)(\quad)$

Look for an integer whose product with 2 is -18 and whose sum with 2 is -7. That integer is -9.

$$y^2 - 7y - 18 = (y+2)(y-9)$$

23. $y^2 + 9y + 8$

Look for two integers whose product is 8 and whose sum is 9. Both integers must be positive because b and c are both positive.

Factors of 8	Sums of Factors
1, 8	$9 \leftarrow$
2, 4	6

Thus,

$$y^2 + 9y + 8 = (y+8)(y+1).$$

25. $b^2 + 8b + 15$

Look for two integers whose product is 15 and whose sum is 8. Both integers must be positive because b and c are both positive.

Factors of 15	Sums of Factors
1, 15	16
3, 5	8 ←

Thus,

$$b^2 + 8b + 15 = (b + 3)(b + 5).$$

27. $m^2 + m - 20$

Look for two integers whose product is -20 and whose sum is 1. Since c is negative, one integer must be positive and one must be negative.

Factors of -20	Sums of Factors
$-1, 20$	19
$1, -20$	-19
$-2, 10$	8
$2, -10$	-8
$-4, 5$	1 ←
$4, -5$	-1

Thus,

$$m^2 + m - 20 = (m - 4)(m + 5).$$

29. $y^2 - 8y + 15$

Find two integers whose product is 15 and whose sum is -8. Since c is positive and b is negative, both integers must be negative.

Factors of 15	Sums of Factors
$-1, -15$	-16
$-3, -5$	-8 ←

Thus,

$$y^2 - 8y + 15 = (y - 5)(y - 3).$$

31. $x^2 + 4x + 5$

Look for two integers whose product is 5 and whose sum is 4. Both integers must be positive since b and c are both positive.

Product	Sum
$5 \cdot 1 = 5$	$5 + 1 = 6$

There is no other pair of positive integers whose product is 5. Since there is no pair of integers whose product is 5 and whose sum is 4, $x^2 + 4x + 5$ is a *prime* polynomial.

33. $z^2 - 15z + 56$

Find two integers whose product is 56 and whose sum is -15. Since c is positive and b is negative, both integers must be negative.

Factors of 56	Sums of Factors
$-1, -56$	-57
$-2, -28$	-30
$-4, -14$	-18
$-7, -8$	-15

Thus,

$$z^2 - 15z + 56 = (z - 7)(z - 8).$$

35. $r^2 - r - 30$

Look for two integers whose product is -30 and whose sum is -1. Because c is negative, one integer must be positive and the other must be negative.

Factors of -30	Sums of Factors
$-1, 30$	29
$1, -30$	-29
$-2, 15$	13
$2, -15$	-13
$-3, 10$	7
$3, -10$	-7
$-5, 6$	1
$5, -6$	-1 ←

Thus,

$$r^2 - r - 30 = (r + 5)(r - 6).$$

37. $a^2 - 8a - 48$

Find two integers whose product is -48 and whose sum is -8. Since c is negative, one integer must be positive and one must be negative.

Factors of -48	Sums of Factors
$-1, 48$	47
$1, -48$	-47
$-2, 24$	22
$2, -24$	-22
$-3, 16$	13
$3, -16$	-13
$-4, 12$	8
$4, -12$	-8 ←
$-6, 8$	2
$6, -8$	-2

Thus,

$$a^2 - 8a - 48 = (a + 4)(a - 12).$$

39. $x^2 + 3x - 39$

Look for two integers whose product is -39 and whose sum is 3. Because c is negative, one integer must be positive and one must be negative.

Factors of -39	Sums of Factors
$-1, 39$	38
$1, -39$	-38
$-3, 13$	10
$3, -13$	-10

This list does not produce the required integers, and there are no other possibilities to try. Therefore, $x^2 + 3x - 39$ is *prime*.

41. Factor $8 + 6x + x^2$ directly to get $(2 + x)(4 + x)$. Alternatively, use the commutative property to write the trinomial as $x^2 + 6x + 8$ and factor to get $(x + 2)(x + 4)$, an equivalent answer.

43. $r^2 + 3ra + 2a^2$

Look for two expressions whose product is $2a^2$ and whose sum is $3a$. They are $2a$ and a, so

$$r^2 + 3ra + 2a^2 = (r + 2a)(r + a).$$

45. $t^2 - tz - 6z^2$

Look for two expressions whose product is $-6z^2$ and whose sum is $-z$. They are $2z$ and $-3z$, so

$$t^2 - tz - 6z^2 = (t + 2z)(t - 3z).$$

47. $x^2 + 4xy + 3y^2$

Look for two expressions whose product is $3y^2$ and whose sum is $4y$. The expressions are $3y$ and y, so

$$x^2 + 4xy + 3y^2 = (x + 3y)(x + y).$$

49. $v^2 - 11vw + 30w^2$

Factors of $30w^2$	Sums of Factors
$-30w, -w$	$-31w$
$-15w, -2w$	$-17w$
$-10w, -3w$	$-13w$
$-5w, -6w$	$-11w$

The completely factored form is

$$v^2 - 11vw + 30w^2 = (v - 5w)(v - 6w).$$

51. $4x^2 + 12x - 40$

First, factor out the GCF, 4.

$$4x^2 + 12x - 40 = 4(x^2 + 3x - 10)$$

Now factor $x^2 + 3x - 10$.

Factors of -10	Sums of Factors
$-1, 10$	9
$1, -10$	-9
$2, -5$	-3
$-2, 5$	$3 \leftarrow$

Thus,

$$x^2 + 3x - 10 = (x - 2)(x + 5).$$

The completely factored form is

$$4x^2 + 12x - 40 = 4(x - 2)(x + 5).$$

53. $2t^3 + 8t^2 + 6t$

First, factor out the GCF, $2t$.

$$2t^3 + 8t^2 + 6t = 2t(t^2 + 4t + 3)$$

Then factor $t^2 + 4t + 3$.

$$t^2 + 4t + 3 = (t + 1)(t + 3)$$

The completely factored form is

$$2t^3 + 8t^2 + 6t = 2t(t + 1)(t + 3).$$

55. $2x^6 + 8x^5 - 42x^4$

First, factor out the GCF, $2x^4$.

$$2x^6 + 8x^5 - 42x^4 = 2x^4(x^2 + 4x - 21)$$

Now factor $x^2 + 4x - 21$.

Factors of -21	Sums of Factors
$1, -21$	-20
$-1, 21$	20
$3, -7$	-4
$-3, 7$	$4 \leftarrow$

Thus,

$$x^2 + 4x - 21 = (x - 3)(x + 7).$$

The completely factored form is

$$2x^6 + 8x^5 - 42x^4 = 2x^4(x - 3)(x + 7).$$

57. $5m^5 + 25m^4 - 40m^2$

Factor out the GCF, $5m^2$.

$$5m^5 + 25m^4 - 40m^2 = 5m^2(m^3 + 5m^2 - 8)$$

59. $m^3n - 10m^2n^2 + 24mn^3$

First, factor out the GCF, mn.

$$m^3n - 10m^2n^2 + 24mn^3$$
$$= mn(m^2 - 10mn + 24n^2)$$

The expressions $-6n$ and $-4n$ have a product of $24n^2$ and a sum of $-10n$. The completely factored form is

$$m^3n - 10m^2n^2 + 24mn^3$$
$$= mn(m - 6n)(m - 4n).$$

61. $(2x + 4)(x - 3)$

$$\begin{array}{cccc} \textbf{F} & \textbf{O} & \textbf{I} & \textbf{L} \end{array}$$
$$= 2x(x) + 2x(-3) + 4(x) + 4(-3)$$
$$= 2x^2 - 6x + 4x - 12$$
$$= 2x^2 - 2x - 12$$

It is incorrect to completely factor $2x^2 - 2x - 12$ as $(2x + 4)(x - 3)$ because $2x + 4$ can be factored further as $2(x + 2)$. The first step should be to factor out the GCF, 2. The correct factorization is

$$2x^2 - 2x - 12 = 2(x^2 - x - 6)$$
$$= 2(x + 2)(x - 3).$$

63. $a^5 + 3a^4b - 4a^3b^2$

The GCF is a^3, so

$$a^5 + 3a^4b - 4a^3b^2$$
$$= a^3(a^2 + 3ab - 4b^2).$$

Now factor $a^2 + 3ab - 4b^2$. The expressions $4b$ and $-b$ have a product of $-4b^2$ and a sum of $3b$. The completely factored form is

$$a^5 + 3a^4b - 4a^3b^2 = a^3(a + 4b)(a - b).$$

65. $y^3z + y^2z^2 - 6yz^3$

The GCF is yz, so

$$y^3z + y^2z^2 - 6yz^3 = yz(y^2 + yz - 6z^2).$$

Now factor $y^2 + yz - 6z^2$. The expressions $3z$ and $-2z$ have a product of $-6z^2$ and a sum of z. The completely factored form is

$$y^3z + y^2z^2 - 6yz^3 = yz(y + 3z)(y - 2z).$$

67. $z^{10} - 4z^9y - 21z^8y^2$
$$= z^8(z^2 - 4zy - 21y^2) \quad GCF \text{ is } z^8.$$
$$= z^8(z - 7y)(z + 3y)$$

69. The GCF is $(a + b)$, so

$$(a + b)x^2 + (a + b)x - 12(a + b)$$
$$= (a + b)(x^2 + x - 12).$$

Now factor $x^2 + x - 12$.

$$x^2 + x - 12 = (x + 4)(x - 3)$$

The completely factored form is

$$(a + b)x^2 + (a + b)x - 12(a + b)$$
$$= (a + b)(x + 4)(x - 3).$$

71. The GCF is $(2p + q)$, so
$$(2p + q)r^2 - 12(2p + q)r + 27(2p + q)$$
$$= (2p + q)(r^2 - 12r + 27).$$

Now factor $r^2 - 12r + 27$.

$$r^2 - 12r + 27 = (r - 9)(r - 3)$$

The completely factored form is

$$(2p + q)r^2 - 12(2p + q)r + 27(2p + q)$$
$$= (2p + q)(r - 9)(r - 3).$$

73. Multiply the factors using FOIL to determine the polynomial.

$$(a + 9)(a + 4)$$

$$\begin{array}{cccc} \textbf{F} & \textbf{O} & \textbf{I} & \textbf{L} \end{array}$$
$$= a(a) + a(4) + 9(a) + 9(4)$$
$$= a^2 + 4a + 9a + 36$$
$$= a^2 + 13a + 36$$

75. $(2y - 7)(y + 4) = 2y^2 + 8y - 7y - 28$
$$= 2y^2 + y - 28$$

77. $(5z + 2)(3z - 2) = 15z^2 - 10z + 6z - 4$
$$= 15z^2 - 4z - 4$$

79. $(4p + 1)(2p - 3) = 8p^2 - 12p + 2p - 3$
$$= 8p^2 - 10p - 3$$

5.3 More on Factoring Trinomials

1. $10t^2 + 5t + 4t + 2$
$$= (10t^2 + 5t) + (4t + 2) \quad \textit{Group terms.}$$
$$= 5t(2t + 1) + 2(2t + 1) \quad \textit{Factor each group.}$$
$$= (2t + 1)(5t + 2) \quad \textit{Factor out } 2t + 1.$$

3. $15z^2 - 10z - 9z + 6$
$$= (15z^2 - 10z) + (-9z + 6) \quad \textit{Group terms.}$$
$$= 5z(3z - 2) - 3(3z - 2) \quad \textit{Factor each group.}$$
$$= (3z - 2)(5z - 3) \quad \textit{Factor out } 3z - 2.$$

5. $8s^2 - 4st + 6st - 3t^2$
$$= (8s^2 - 4st) + (6st - 3t^2) \quad \textit{Group terms.}$$
$$= 4s(2s - t) + 3t(2s - t) \quad \textit{Factor each group.}$$
$$= (2s - t)(4s + 3t) \quad \textit{Factor out } 2s - t.$$

7. $2m^2 + 11m + 12$

(a) Find two integers whose product is $2 \cdot 12 = \underline{24}$ and whose sum is $\underline{11}$.

(b) The required integers are $\underline{3}$ and $\underline{8}$. (Order is irrelevant.)

(c) Write the middle term, $11m$, as $\underline{3m} + \underline{8m}$.

(d) Rewrite the given trinomial as
$\underline{2m^2 + 3m + 8m + 12}$.

(e) $(2m^2 + 3m) + (8m + 12)$ *Group terms.*
$\quad = m(2m + 3) + 4(2m + 3)$ *Factor each group.*
$\quad = (2m + 3)(m + 4)$ *Factor out $2m + 3$.*

(f) $(2m + 3)(m + 4)$

$$\begin{array}{cccc} \mathbf{F} & \mathbf{O} & \mathbf{I} & \mathbf{L} \end{array}$$
$$= 2m(m) + 2m(4) + 3(m) + 3(4)$$
$$= 2m^2 + 8m + 3m + 12$$
$$= 2m^2 + 11m + 12$$

9. To factor $12y^2 + 5y - 2$, we must find two integers with a product of $12(-2) = -24$ and a sum of 5. The only pair of integers satisfying those conditions is 8 and -3, choice **B**.

11. $2x^2 - x - 1$

Multiply the factors in the choices together to see which ones give the correct product. Since

$$(2x - 1)(x + 1) = 2x^2 + x - 1$$

and $(2x + 1)(x - 1) = 2x^2 - x - 1$,
the correct factored form is choice **B**,
$(2x + 1)(x - 1)$.

13. $4y^2 + 17y - 15$

Multiply the factors in the choices together to see which ones give the correct product. Since

$$(y + 5)(4y - 3) = 4y^2 + 17y - 15$$

and $(2y - 5)(2y + 3) = 4y^2 - 4y - 15$,
the correct factored form is choice **A**,
$(y + 5)(4y - 3)$.

15. $6a^2 + 7ab - 20b^2 = (3a - 4b)(\qquad)$

The first term in the missing expression must be $2a$ since

$$(3a)(2a) = 6a^2.$$

The second term in the missing expression must be $5b$ since

$$(-4b)(5b) = -20b^2.$$

Checking our answer by multiplying, we see that
$(3a - 4b)(2a + 5b) = 6a^2 + 7ab - 20b^2$,
as desired.

17. $2x^2 + 6x - 8 = 2(x^2 + 3x - 4)$

To factor $x^2 + 3x - 4$, we look for two integers whose product is -4 and whose sum is 3. The integers are 4 and -1. Thus,

$$2x^2 + 6x - 8 = 2(x + 4)(x - 1).$$

19. Since 2 is not a factor of $12x^2 + 7x - 12$, it cannot be a factor of any factor of $12x^2 + 7x - 12$. Since 2 is a factor of $2x - 6$, this means that $2x - 6$ cannot be a factor of $12x^2 + 7x - 12$.

Note: In Exercises 21–77, either the trial and error method (which uses FOIL in reverse) or the grouping method can be used to factor each polynomial.

21. $3a^2 + 10a + 7$

Factor by the grouping method. Look for two integers whose product is $3(7) = 21$ and whose sum is 10. The integers are 3 and 7. Use these integers to rewrite the middle term, $10a$, as $3a + 7a$, and then factor the resulting four-term polynomial by grouping.

$3a^2 + 10a + 7$
$= 3a^2 + 3a + 7a + 7$ *$10a = 3a + 7a$*
$= (3a^2 + 3a) + (7a + 7)$ *Group the terms.*
$= 3a(a + 1) + 7(a + 1)$ *Factor each group.*
$= (a + 1)(3a + 7)$ *Factor out $a + 1$.*

23. $2y^2 + 7y + 6$

Factor by the grouping method. Look for two integers whose product is $2(6) = 12$ and whose sum is 7. The integers are 3 and 4.

$2y^2 + 7y + 6$
$= 2y^2 + 3y + 4y + 6$ *$7y = 3y + 4y$*
$= (2y^2 + 3y) + (4y + 6)$ *Group terms.*
$= y(2y + 3) + 2(2y + 3)$ *Factor each group.*
$= (2y + 3)(y + 2)$ *Factor out $2y + 3$.*

25. $15m^2 + m - 2$

Factor by the grouping method. Look for two integers whose product is $15(-2) = -30$ and whose sum is 1. The integers are 6 and -5.

$15m^2 + m - 2$
$= 15m^2 + 6m - 5m - 2$ *$m = 6m - 5m$*
$= (15m^2 + 6m) + (-5m - 2)$ *Group the terms.*

$$= 3m(5m + 2) - 1(5m + 2)$$ *Factor each group.*

$$= (5m + 2)(3m - 1)$$ *Factor out $5m + 2$.*

27. $12s^2 + 11s - 5$

Factor by trial and error.

Possible factors of $12s^2$ are s and $12s$, $2s$ and $6s$, or $3s$ and $4s$.

Factors of -5 are -1 and 5 or -5 and 1.

$(2s - 1)(6s + 5) = 12s^2 + 4s - 5$ *Incorrect*
$(2s + 1)(6s - 5) = 12s^2 - 4s - 5$ *Incorrect*
$(3s - 1)(4s + 5) = 12s^2 + 11s - 5$ *Correct*

29. $10m^2 - 23m + 12$

Factor by the grouping method. Look for two integers whose product is $10(12) = 120$ and whose sum is -23. The integers are -8 and -15.

$10m^2 - 23m + 12$

$= 10m^2 - 8m - 15m + 12$ $\quad -23m = -8m - 15m$

$= (10m^2 - 8m) + (-15m + 12)$ *Group the terms.*

$= 2m(5m - 4) - 3(5m - 4)$ *Factor each group.*

$= (5m - 4)(2m - 3)$ *Factor out $5m - 4$.*

31. $8w^2 - 14w + 3$

Factor by trial and error. Possible factors of $8w^2$ are w and $8w$ or $2w$ and $4w$.
Factors of 3 are -1 and -3 (since $b = -14$ is negative).

$(4w - 3)(2w - 1) = 8w^2 - 10w + 3$ *Incorrect*
$(4w - 1)(2w - 3) = 8w^2 - 14w + 3$ *Correct*

33. $20y^2 - 39y - 11$

Factor by the grouping method. Look for two integers whose product is $20(-11) = -220$ and whose sum is -39. The integers are -44 and 5.

$20y^2 - 39y - 11$

$= 20y^2 - 44y + 5y - 11$ $\quad -39y = -44y + 5y$

$= (20y^2 - 44y) + (5y - 11)$ *Group the terms.*

$= 4y(5y - 11) + 1(5y - 11)$ *Factor each group.*

$= (5y - 11)(4y + 1)$ *Factor out $5y - 11$.*

35. $3x^2 - 15x + 16$

Factor by the grouping method. Look for two integers whose product is $3(16) = 48$ and whose sum is -15. The negative factors and their sums are:

$$-1 + (-48) = -49$$
$$-2 + (-24) = -26$$
$$-3 + (-16) = -19$$
$$-4 + (-12) = -16$$
$$-6 + (-8) = -14$$

So there are no integers satisfying the conditions and the polynomial is *prime*.

37. First, factor out the greatest common factor, 2.

$$20x^2 + 22x + 6 = 2(10x^2 + 11x + 3)$$

Now factor $10x^2 + 11x + 3$ by trial and error to obtain

$$10x^2 + 11x + 3 = (5x + 3)(2x + 1).$$

The complete factorization is

$$20x^2 + 22x + 6 = 2(5x + 3)(2x + 1).$$

39. Factor out the GCF, 3.

$$24x^2 - 42x + 9 = 3(8x^2 - 14x + 3)$$

Use the grouping method to factor $8x^2 - 14x + 3$. Look for two integers whose product is $8(3) = 24$ and whose sum is -14. The integers are -12 and -2.

$24x^2 - 42x + 9$

$= 3(8x^2 - 12x - 2x + 3)$ $\quad -14x = -12x - 2x$

$= 3[(8x^2 - 12x) + (-2x + 3)]$ *Group the terms.*

$= 3[4x(2x - 3) - 1(2x - 3)]$ *Factor each group.*

$= 3(2x - 3)(4x - 1)$ *Factor out $2x - 3$.*

41. $40m^2q + mq - 6q$

First, factor out the greatest common factor, q.

$$40m^2q + mq - 6q = q(40m^2 + m - 6)$$

Now factor $40m^2 + m - 6$ by trial and error to obtain

$$40m^2 + m - 6 = (5m + 2)(8m - 3).$$

The complete factorization is

$$40m^2q + mq - 6q = q(5m + 2)(8m - 3).$$

43. Factor out the GCF, $3n^2$.

$$15n^4 - 39n^3 + 18n^2 = 3n^2(5n^2 - 13n + 6)$$

Factor $5n^2 - 13n + 6$ by the trial and error method. Possible factors of $5n^2$ are $5n$ and n.

Possible factors of 6 are -6 and -1, -3 and -2, -2 and -3, or -1 and -6.

$(5n - 6)(n - 1) = 5n^2 - 11n + 6$ *Incorrect*
$(5n - 3)(n - 2) = 5n^2 - 13n + 6$ *Correct*

The completely factored form is
$$15n^4 - 39n^3 + 18n^2 = 3n^2(5n - 3)(n - 2).$$

45. Factor out the GCF, y^2.
$$15x^2y^2 - 7xy^2 - 4y^2 = y^2(15x^2 - 7x - 4)$$

Factor $15x^2 - 7x - 4$ by the grouping method. Look for two integers whose product is $15(-4) = -60$ and whose sum is -7. The integers are -12 and 5.

$$15x^2y^2 - 7xy^2 - 4y^2$$
$$= y^2(15x^2 - 12x + 5x - 4)$$
$$= y^2[3x(5x - 4) + 1(5x - 4)]$$
$$= y^2(5x - 4)(3x + 1)$$

47. $5a^2 - 7ab - 6b^2$

Factor by the grouping method. Look for two integers whose product is $5(-6) = -30$ and whose sum is -7. The integers are -10 and 3.

$$5a^2 - 7ab - 6b^2$$
$$= 5a^2 - 10ab + 3ab - 6b^2$$
$$= (5a^2 - 10ab) + (3ab - 6b^2)$$
$$= 5a(a - 2b) + 3b(a - 2b)$$
$$= (a - 2b)(5a + 3b)$$

49. $12s^2 + 11st - 5t^2$

Factor by the grouping method. Look for two integers whose product is $12(-5) = -60$ and whose sum is 11. The integers are 15 and -4.

$$12s^2 + 11st - 5t^2$$
$$= 12s^2 + 15st - 4st - 5t^2$$
$$= (12s^2 + 15st) + (-4st - 5t^2)$$
$$= 3s(4s + 5t) - t(4s + 5t)$$
$$= (4s + 5t)(3s - t)$$

51. Factor out the GCF, m^4n.

$$6m^6n + 7m^5n^2 + 2m^4n^3$$
$$= m^4n(6m^2 + 7mn + 2n^2)$$

Now factor $6m^2 + 7mn + 2n^2$ by trial and error.

Possible factors of $6m^2$ are $6m$ and m or $3m$ and $2m$. Possible factors of $2n^2$ are $2n$ and n.

$(3m + 2n)(2m + n) = 6m^2 + 7mn + 2n^2$
                              Correct

The completely factored form is
$$6m^6n + 7m^5n^2 + 2m^4n^3$$
$$= m^4n(3m + 2n)(2m + n).$$

53. $5 - 6x + x^2$
$$= 5 - 5x - x + x^2 \qquad -6x = -5x - x$$
$$= (5 - 5x) + (-x + x^2) \qquad \text{Group the terms.}$$
$$= 5(1 - x) - x(1 - x) \qquad \text{Factor each group.}$$
$$= (1 - x)(5 - x) \qquad \text{Factor out } 1 - x.$$

55. $16 + 16x + 3x^2$

Factor by the grouping method. Find two integers whose product is $(16)(3) = 48$ and whose sum is 16. The numbers are 4 and 12.

$$16 + 16x + 3x^2$$
$$= 16 + 4x + 12x + 3x^2$$
$$= (16 + 4x) + (12x + 3x^2)$$
$$= 4(4 + x) + 3x(4 + x)$$
$$= (4 + x)(4 + 3x)$$

57. $-10x^3 + 5x^2 + 140x$

First, factor out $-5x$; then complete the factoring by trial and error, using FOIL to test various possibilities until the correct one is found.

$$-10x^3 + 5x^2 + 140x$$
$$= -5x(2x^2 - x - 28)$$
$$= -5x(2x + 7)(x - 4)$$

59. $12x^2 - 47x - 4$

Factor by the grouping method. Find two integers whose product is $12(-4) = -48$ and whose sum is -47. The numbers are 1 and -48.

$$12x^2 - 47x - 4$$
$$= 12x^2 + x - 48x - 4$$
$$= (12x^2 + x) + (-48x - 4)$$
$$= x(12x + 1) - 4(12x + 1)$$
$$= (12x + 1)(x - 4)$$

61. $24y^2 - 41xy - 14x^2$

Factor by using trial and error. Only positive factors of 24 should be considered: 1 and 24, 2 and 12, 3 and 8, or 4 and 6. The factors of -14 include 7 and -2. The correct factorization is

$$24y^2 - 41xy - 14x^2 = (24y + 7x)(y - 2x).$$

63. $36x^4 - 64x^2y + 15y^2$

Factor by using trial and error. For $15y^2$, consider only $-y$ and $-15y$, and $-3y$ and $-5y$.

$$36x^4 - 64x^2y + 15y^2 = (18x^2 - 5y)(2x^2 - 3y)$$

65. $48a^2 - 94ab - 4b^2$

Factor out the GCF, 2.

$$48a^2 - 94ab - 4b^2 = 2(24a^2 - 47ab - 2b^2)$$

Now factor $24a^2 - 47ab - 2b^2$ by the grouping method. Look for two integers whose product is $24(-2) = -48$ and whose sum is -47. The integers are 1 and -48.

$$\begin{aligned} 24a^2 &- 47ab - 2b^2 \\ &= 24a^2 + ab - 48ab - 2b^2 \\ &= (24a^2 + ab) + (-48ab - 2b^2) \\ &= a(24a + b) - 2b(24a + b) \\ &= (24a + b)(a - 2b) \end{aligned}$$

The completely factored form is

$$48a^2 - 94ab - 4b^2 = 2(24a + b)(a - 2b).$$

67. $10x^4y^5 + 39x^3y^5 - 4x^2y^5$

Factor out the GCF, x^2y^5.

$$\begin{aligned} 10x^4y^5 &+ 39x^3y^5 - 4x^2y^5 \\ &= x^2y^5(10x^2 + 39x - 4) \end{aligned}$$

Now factor $10x^2 + 39x - 4$ by the grouping method. Look for two integers whose product is $10(-4) = -40$ and whose sum is 39. The integers are -1 and 40.

$$\begin{aligned} 10x^2 &+ 39x - 4 \\ &= 10x^2 - x + 40x - 4 \\ &= (10x^2 - x) + (40x - 4) \\ &= x(10x - 1) + 4(10x - 1) \\ &= (10x - 1)(x + 4) \end{aligned}$$

The completely factored form is

$$\begin{aligned} 10x^4y^5 &+ 39x^3y^5 - 4x^2y^5 \\ &= x^2y^5(10x - 1)(x + 4). \end{aligned}$$

69. $36a^3b^2 - 104a^2b^2 - 12ab^2$

Factor out the GCF, $4ab^2$.

$$\begin{aligned} 36a^3b^2 &- 104a^2b^2 - 12ab^2 \\ &= 4ab^2(9a^2 - 26a - 3) \end{aligned}$$

Now factor $9a^2 - 26a - 3$ by the grouping method. Look for two integers whose product is $9(-3) = -27$ and whose sum is -26. The integers are 1 and -27.

$$\begin{aligned} 9a^2 &- 26a - 3 \\ &= 9a^2 + a - 27a - 3 \\ &= (9a^2 + a) + (-27a - 3) \\ &= a(9a + 1) - 3(9a + 1) \\ &= (9a + 1)(a - 3) \end{aligned}$$

The completely factored form is

$$\begin{aligned} 36a^3b^2 &- 104a^2b^2 - 12ab^2 \\ &= 4ab^2(9a + 1)(a - 3). \end{aligned}$$

71. $24x^2 - 46x + 15$

Factor by the grouping method. Look for two integers whose product is $24(15) = 360$ and whose sum is -46. The integers are -10 and -36.

$$\begin{aligned} 24x^2 &- 46x + 15 \\ &= 24x^2 - 10x - 36x + 15 \\ &= (24x^2 - 10x) + (-36x + 15) \\ &= 2x(12x - 5) - 3(12x - 5) \\ &= (12x - 5)(2x - 3) \end{aligned}$$

73. $24x^4 + 55x^2 - 24$

Factor by the grouping method. Look for two integers whose product is $24(-24) = -576$ and whose sum is 55. The integers are -9 and 64. Note the x^2-term.

$$\begin{aligned} 24x^4 &+ 55x^2 - 24 \\ &= 24x^4 - 9x^2 + 64x^2 - 24 \\ &= (24x^4 - 9x^2) + (64x^2 - 24) \\ &= 3x^2(8x^2 - 3) + 8(8x^2 - 3) \\ &= (8x^2 - 3)(3x^2 + 8) \end{aligned}$$

75. $24x^2 + 38xy + 15y^2$

Factor by the grouping method. Look for two integers whose product is $24(15) = 360$ and whose sum is 38. The integers are 18 and 20. Note the xy-term.

$$\begin{aligned} 24x^2 &+ 38xy + 15y^2 \\ &= 24x^2 + 18xy + 20xy + 15y^2 \\ &= (24x^2 + 18xy) + (20xy + 15y^2) \\ &= 6x(4x + 3y) + 5y(4x + 3y) \\ &= (4x + 3y)(6x + 5y) \end{aligned}$$

77. $24x^2z^4 - 113xz^2 - 35$

Factor by the grouping method. Look for two integers whose product is $24(-35) = -840$ and whose sum is -113. The integers are -120 and 7. Note the xz^2-term.

$$24x^2z^4 - 113xz^2 - 35$$
$$= 24x^2z^4 - 120xz^2 + 7xz^2 - 35$$
$$= \left(24x^2z^4 - 120xz^2\right) + \left(7xz^2 - 35\right)$$
$$= 24xz^2\left(xz^2 - 5\right) + 7\left(xz^2 - 5\right)$$
$$= \left(xz^2 - 5\right)\left(24xz^2 + 7\right)$$

79. $-x^2 - 4x + 21 = -1\left(x^2 + 4x - 21\right)$
$$= -1(x + 7)(x - 3)$$

81. $-3x^2 - x + 4 = -1\left(3x^2 + x - 4\right)$
$$= -1(3x + 4)(x - 1)$$

83. $-2a^2 - 5ab - 2b^2$
$$= -1\left(2a^2 + 5ab + 2b^2\right)$$
Factor out −1.
$$= -1\left(2a^2 + 4ab + ab + 2b^2\right)$$
5ab = 4ab + ab
$$= -1\left[\left(2a^2 + 4ab\right) + \left(ab + 2b^2\right)\right]$$
Group the terms.
$$= -1[2a(a + 2b) + b(a + 2b)]$$
Factor each group.
$$= -1(a + 2b)(2a + b)$$
Factor out a + 2b.

85. Yes, $(x + 7)(3 - x)$ is equivalent to $-1(x + 7)(x - 3)$ because $-1(x - 3) = -x + 3 = 3 - x$.

87. First, factor out the GCF, $(m + 1)^3$; then factor the resulting trinomial by trial and error.

$$25q^2(m + 1)^3 - 5q(m + 1)^3 - 2(m + 1)^3$$
$$= (m + 1)^3\left(25q^2 - 5q - 2\right)$$
$$= (m + 1)^3(5q - 2)(5q + 1)$$

89. $15x^2(r + 3)^3 - 34xy(r + 3)^3 - 16y^2(r + 3)^3$
$$= (r + 3)^3\left(15x^2 - 34xy - 16y^2\right)$$
$$= (r + 3)^3(5x + 2y)(3x - 8y)$$

91. $5x^2 + kx - 1$

Look for two integers whose product is $5(-1) = -5$ and whose sum is k.

Factors of −5	Sums of Factors
−5, 1	−4
5, −1	4

Thus, there are two possible integer values for k: −4 and 4.

93. $2m^2 + km + 5$

Look for two integers whose product is $2(5) = 10$ and whose sum is k.

Factors of 10	Sums of Factors
−10, −1	−11
10, 1	11
−5, −2	−7
5, 2	7

Thus, there are four possible integer values for k: $-11, -7, 7,$ and 11.

95. $(7p + 3)(7p - 3) = (7p)^2 - 3^2$
$$= 49p^2 - 9$$

97. $\left(r^2 + \frac{1}{2}\right)\left(r^2 - \frac{1}{2}\right) = (r^2)^2 - \left(\frac{1}{2}\right)^2$
$$= r^4 - \frac{1}{4}$$

99. $(3t + 4)^2 = (3t)^2 + 2(3t)(4) + 4^2$
$$= 9t^2 + 24t + 16$$

5.4 Special Factoring Techniques

1.

$1^2 = \underline{1}$	$2^2 = \underline{4}$	$3^2 = \underline{9}$
$4^2 = \underline{16}$	$5^2 = \underline{25}$	$6^2 = \underline{36}$
$7^2 = \underline{49}$	$8^2 = \underline{64}$	$9^2 = \underline{81}$
$10^2 = \underline{100}$	$11^2 = \underline{121}$	$12^2 = \underline{144}$
$13^2 = \underline{169}$	$14^2 = \underline{196}$	$15^2 = \underline{225}$
$16^2 = \underline{256}$	$17^2 = \underline{289}$	$18^2 = \underline{324}$
$19^2 = \underline{361}$	$20^2 = \underline{400}$	

3.

$1^3 = \underline{1}$	$2^3 = \underline{8}$	$3^3 = \underline{27}$
$4^3 = \underline{64}$	$5^3 = \underline{125}$	$6^3 = \underline{216}$
$7^3 = \underline{343}$	$8^3 = \underline{512}$	$9^3 = \underline{729}$
$10^3 = \underline{1000}$		

5. **(a)** $64x^6y^{12} = (8x^3y^6)^2$, so $64x^6y^{12}$ is a perfect square.

$64x^6y^{12} = (4x^2y^4)^3$, so $64x^6y^{12}$ is also a perfect cube.

Therefore, the answer is "both of these."

(b) $125t^6 = (5t^2)^3$, so $125t^6$ is a perfect cube. Since 125 is not a perfect square, $125t^6$ is not a perfect square.

(c) $49x^{12} = (7x^6)^2$, so $49x^{12}$ is a perfect square. Since 49 is not a perfect cube, $49x^{12}$ is not a perfect cube.

(d) $81r^{10} = (9r^5)^2$, so $81r^{10}$ is a perfect square. It is not a perfect cube.

7. $y^2 - 25$

To factor this binomial, use the rule for factoring a difference of squares.

$$a^2 - b^2 = (a + b)(a - b)$$
$$\downarrow \quad \downarrow \qquad \downarrow \; \downarrow \; \downarrow \; \downarrow$$
$$y^2 - 25 = y^2 - 5^2 = (y + 5)(y - 5)$$

9. $p^2 - \frac{1}{9} = p^2 - \left(\frac{1}{3}\right)^2$
$$= \left(p + \frac{1}{3}\right)\left(p - \frac{1}{3}\right)$$

11. $m^2 + 64$

This binomial is the *sum* of squares and the terms have no common factor. Unlike the *difference* of squares, it cannot be factored. It is a *prime* polynomial.

13. $9r^2 - 4 = (3r)^2 - 2^2$
$$= (3r + 2)(3r - 2)$$

15. $36m^2 - \frac{16}{25} = (6m)^2 - \left(\frac{4}{5}\right)^2$
$$= \left(6m + \frac{4}{5}\right)\left(6m - \frac{4}{5}\right)$$

17. $36x^2 - 16$

First factor out the GCF, 4; then use the rule for factoring the difference of squares.

$$36x^2 - 16 = 4\left(9x^2 - 4\right)$$
$$= 4\left[(3x)^2 - 2^2\right]$$
$$= 4(3x + 2)(3x - 2)$$

19. $196p^2 - 225 = (14p)^2 - 15^2$
$$= (14p + 15)(14p - 15)$$

21. $16r^2 - 25a^2 = (4r)^2 - (5a)^2$
$$= (4r + 5a)(4r - 5a)$$

23. $100x^2 + 49$

This binomial is the *sum* of squares and the terms have no common factor. Unlike the *difference* of squares, it cannot be factored. It is a *prime* polynomial.

25. $p^4 - 49 = (p^2)^2 - 7^2$
$$= \left(p^2 + 7\right)\left(p^2 - 7\right)$$

27. $x^4 - 1$

To factor this binomial completely, factor the difference of squares twice.

$$x^4 - 1 = (x^2)^2 - 1^2$$
$$= \left(x^2 + 1\right)\left(x^2 - 1\right)$$
$$= \left(x^2 + 1\right)\left(x^2 - 1^2\right)$$
$$= \left(x^2 + 1\right)(x + 1)(x - 1)$$

29. $p^4 - 256$

To factor this binomial completely, factor the difference of squares twice.

$$p^4 - 256 = (p^2)^2 - 16^2$$
$$= \left(p^2 + 16\right)\left(p^2 - 16\right)$$
$$= \left(p^2 + 16\right)\left(p^2 - 4^2\right)$$
$$= \left(p^2 + 16\right)(p + 4)(p - 4)$$

31.   The student's answer is not a complete factorization because $x^2 - 9$ can be factored further. The correct complete factorization is

$$x^4 - 81 = \left(x^2 + 9\right)(x + 3)(x - 3).$$

In Exercises 33–50, use the rules for factoring perfect square trinomials:

$$a^2 + 2ab + b^2 = (a + b)^2$$
$$a^2 - 2ab + b^2 = (a - b)^2.$$

33. $w^2 + 2w + 1$

The first and last terms are perfect squares, w^2 and 1^2. This trinomial is a perfect square, since the middle term is twice the product of w and 1, or

$$2 \cdot w \cdot 1 = 2w.$$

Therefore,
$$w^2 + 2w + 1 = (w + 1)^2.$$

35. $x^2 - 8x + 16$

The first and last terms are perfect squares, x^2 and $(-4)^2$. This trinomial is a perfect square, since the middle term is twice the product of x and -4, or
$$2 \cdot x \cdot (-4) = -8x.$$

Therefore,
$$x^2 - 8x + 16 = (x - 4)^2.$$

37. $t^2 + t + \frac{1}{4}$

t^2 is a perfect square, and $\frac{1}{4}$ is a perfect square since $\frac{1}{2} \cdot \frac{1}{2} = \frac{1}{4}$. The middle term is twice the product of t and $\frac{1}{2}$, or

$$t = 2(t)\left(\tfrac{1}{2}\right).$$

Therefore,
$$t^2 + t + \tfrac{1}{4} = \left(t + \tfrac{1}{2}\right)^2.$$

39. $x^2 - 1.0x + 0.25$

The first and last terms are perfect squares, x^2 and $(-0.5)^2$. The trinomial is a perfect square, since the middle term is
$$2 \cdot x \cdot (-0.5) = -1.0x.$$

Therefore,
$$x^2 - 1.0x + 0.25 = (x - 0.5)^2.$$

41. $2x^2 + 24x + 72$

First, factor out the GCF, 2.

$$2x^2 + 24x + 72 = 2(x^2 + 12x + 36)$$

Now factor $x^2 + 12x + 36$ as a perfect square trinomial.

$$x^2 + 12x + 36 = (x + 6)^2$$

The final factored form is

$$2x^2 + 24x + 72 = 2(x + 6)^2.$$

43. $16x^2 - 40x + 25$

The first and last terms are perfect squares, $(4x)^2$ and $(-5)^2$. The middle term is

$$2(4x)(-5) = -40x.$$

Therefore,

$$16x^2 - 40x + 25 = (4x - 5)^2.$$

45. $49x^2 - 28xy + 4y^2$

The first and last terms are perfect squares, $(7x)^2$ and $(-2y)^2$. The middle term is

$$2(7x)(-2y) = -28xy.$$

Therefore,

$$49x^2 - 28xy + 4y^2 = (7x - 2y)^2.$$

47. $64x^2 + 48xy + 9y^2$
$$= (8x)^2 + 2(8x)(3y) + (3y)^2$$
$$= (8x + 3y)^2$$

49. $50h^2 - 40hy + 8y^2$
$$= 2(25h^2 - 20hy + 4y^2)$$
$$= 2\left[(5h)^2 - 2(5h)(2y) + (2y)^2\right]$$
$$= 2(5h - 2y)^2$$

51. $4k^3 - 4k^2 + 9k$

First, factor out the GCF, k.

$$4k^3 - 4k^2 + 9k = k(4k^2 - 4k + 9)$$

Since $4k^2 - 4k + 9$ cannot be factored, $k(4k^2 - 4k + 9)$ is the final factored form.

53. $25z^4 + 5z^3 + z^2$

First, factor out the GCF, z^2.

$$25z^4 + 5z^3 + z^2 = z^2(25z^2 + 5z + 1)$$

Since $25z^2 + 5z + 1$ cannot be factored, $z^2(25z^2 + 5z + 1)$ is the final factored form.

55.   Find b so that

$$x^2 + bx + 25 = (x + 5)^2.$$

Since $(x + 5)^2 = x^2 + 10x + 25$, $b = 10$.

57.   Find a so that

$$ay^2 - 12y + 4 = (3y - 2)^2.$$

Since $(3y - 2)^2 = 9y^2 - 12y + 4$, $a = 9$.

59. $a^3 - 1$

Let $x = a$ and $y = 1$ in the pattern for the difference of cubes.

$$x^3 - y^3 = (x - y)(x^2 + xy + y^2)$$
$$a^3 - 1 = a^3 - 1^3 = (a - 1)(a^2 + a \cdot 1 + 1^2)$$
$$= (a - 1)(a^2 + a + 1)$$

61. $m^3 + 8$

Let $x = m$ and $y = 2$ in the pattern for the sum of cubes.

$$x^3 + y^3 = (x + y)(x^2 - xy + y^2)$$
$$m^3 + 8 = m^3 + 2^3 = (m + 2)(m^2 - m \cdot 2 + 2^2)$$
$$= (m + 2)(m^2 - 2m + 4)$$

63.   Factor $27x^3 - 64$ as the difference of cubes.

$$27x^3 - 64 = (3x)^3 - 4^3$$
$$= (3x - 4)\left[(3x)^2 + 3x \cdot 4 + 4^2\right]$$
$$= (3x - 4)(9x^2 + 12x + 16)$$

65. $6p^3 + 6 = 6(p^3 + 1)$ *GCF = 6*
$$= 6(p^3 + 1^3)$$ *Sum of cubes*
$$= 6(p + 1)(p^2 - p \cdot 1 + 1^2)$$
$$= 6(p + 1)(p^2 - p + 1)$$

67. $5x^3 + 40 = 5(x^3 + 8)$ *GCF = 5*
$$= 5(x^3 + 2^3)$$ *Sum of cubes*
$$= 5(x + 2)(x^2 - x \cdot 2 + 2^2)$$
$$= 5(x + 2)(x^2 - 2x + 4)$$

69. $2x^3 - 16y^3$
$$= 2(x^3 - 8y^3)$$ *GCF = 2*
$$= 2\left[x^3 - (2y)^3\right]$$ *Difference of cubes*
$$= 2(x - 2y)\left[x^2 + x \cdot 2y + (2y)^2\right]$$
$$= 2(x - 2y)(x^2 + 2xy + 4y^2)$$

71.   Factor $8p^3 + 729q^3$ as the sum of cubes.

$$8p^3 + 729q^3$$
$$= (2p)^3 + (9q)^3$$
$$= (2p + 9q)\left[(2p)^2 - 2p \cdot 9q + (9q)^2\right]$$
$$= (2p + 9q)(4p^2 - 18pq + 81q^2)$$

73. Factor $27a^3 + 64b^3$ as the sum of cubes.

$$27a^3 + 64b^3$$
$$= (3a)^3 + (4b)^3$$
$$= (3a + 4b)\left[(3a)^2 - 3a \cdot 4b + (4b)^2\right]$$
$$= (3a + 4b)\left(9a^2 - 12ab + 16b^2\right)$$

75. Factor $125t^3 + 8s^3$ as the sum of cubes.

$$125t^3 + 8s^3$$
$$= (5t)^3 + (2s)^3$$
$$= (5t + 2s)\left[(5t)^2 - 5t \cdot 2s + (2s)^2\right]$$
$$= (5t + 2s)\left(25t^2 - 10ts + 4s^2\right)$$

77. Factor $8x^3 - 125y^6$ as the difference of cubes.

$$8x^3 - 125y^6$$
$$= (2x)^3 - (5y^2)^3$$
$$= (2x - 5y^2)\left[(2x)^2 + 2x(5y^2) + (5y^2)^2\right]$$
$$= (2x - 5y^2)\left(4x^2 + 10xy^2 + 25y^4\right)$$

79. Factor $27m^6 + 8n^3$ as the sum of cubes.

$$27m^6 + 8n^3$$
$$= (3m^2)^3 + (2n)^3$$
$$= (3m^2 + 2n)\left[(3m^2)^2 - 3m^2(2n) + (2n)^2\right]$$
$$= (3m^2 + 2n)\left(9m^4 - 6m^2n + 4n^2\right)$$

81. Factor $x^9 + y^9$ as the sum of cubes.

$$x^9 + y^9$$
$$= (x^3)^3 + (y^3)^3$$
$$= (x^3 + y^3)\left[(x^3)^2 - x^3(y^3) + (y^3)^2\right]$$
$$= (x + y)(x^2 - xy + y^2)(x^6 - x^3y^3 + y^6)$$

83. $(m + n)^2 - (m - n)^2$

Factor as the difference of squares. Substitute into the rule using $x = m + n$ and $y = m - n$.

$$(m + n)^2 - (m - n)^2$$
$$= [(m + n) + (m - n)] \cdot [(m + n) - (m - n)]$$
$$= (2m)(2n)$$
$$= 4mn$$

85. $m^2 - p^2 + 2m + 2p$

This expression can be factored by grouping.

$$= (m + p)(m - p) + 2(m + p) \qquad \text{\textit{Factor each group.}}$$
$$= (m + p)(m - p + 2) \qquad \text{\textit{Factor out } } m + p.$$

87. $m - 4 = 0$
$$m = 4 \quad \textit{Add 4.}$$

The solution set is $\{4\}$.

89. $4z - 9 = 0$
$$4z = 9 \quad \textit{Add 9.}$$
$$z = \tfrac{9}{4} \quad \textit{Divide by 4.}$$

The solution set is $\left\{\tfrac{9}{4}\right\}$.

91. $9x - 6 = 0$
$$9x = 6 \quad \textit{Add 6.}$$
$$x = \tfrac{6}{9} = \tfrac{2}{3} \quad \textit{Divide by 9.}$$

The solution set is $\left\{\tfrac{2}{3}\right\}$.

Summary Exercises on Factoring

1. $a^2 - 4a - 12 = (a - 6)(a + 2)$

3. $6y^2 - 6y - 12 = 6\left(y^2 - y - 2\right)$
$$= 6(y - 2)(y + 1)$$

5. $6a + 12b + 18c$
$$= 6(a + 2b + 3c)$$

7. $p^2 - 17p + 66 = (p - 11)(p - 6)$

9. $10z^2 - 7z - 6$

Use the grouping method.
Look for two integers whose product is $10(-6) = -60$ and whose sum is -7. The integers are -12 and 5.

$$10z^2 - 7z - 6$$
$$= 10z^2 - 12z + 5z - 6$$
$$= 2z(5z - 6) + 1(5z - 6) \qquad \text{\textit{Factor each group.}}$$
$$= (5z - 6)(2z + 1) \qquad \text{\textit{Factor out } } 5z - 6.$$

11. $17x^3y^2 + 51xy = 17xy(x^2y + 3)$

13. $8a^5 - 8a^4 - 48a^3$
$$= 8a^3\left(a^2 - a - 6\right)$$
$$= 8a^3(a - 3)(a + 2)$$

15. $z^2 - 3za - 10a^2 = (z - 5a)(z + 2a)$

17. $x^2 - 4x - 5x + 20$
$$= x(x - 4) - 5(x - 4) \qquad \text{\textit{Factor each group.}}$$
$$= (x - 4)(x - 5) \qquad \text{\textit{Factor out } } x - 4.$$

19. $6n^2 - 19n + 10 = (3n - 2)(2n - 5)$

21. $16x + 20 = 4(4x + 5)$

23. $6y^2 - 5y - 4$

Factor by grouping. Find two integers whose product is $6(-4) = -24$ and whose sum is -5. The integers are -8 and 3.

$$6y^2 - 5y - 4$$
$$= 6y^2 - 8y + 3y - 4$$
$$= 2y(3y - 4) + 1(3y - 4)$$
$$= (3y - 4)(2y + 1)$$

25. $6z^2 + 31z + 5 = (6z + 1)(z + 5)$

27. $4k^2 - 12k + 9$
$= (2k)^2 - 2 \cdot 2k \cdot 3 + 3^2$
$= (2k - 3)^2$ *Perfect square trinomial*

29. $54m^2 - 24z^2$
$= 6(9m^2 - 4z^2)$
$= 6[(3m)^2 - (2z)^2]$
$= 6(3m + 2z)(3m - 2z)$

31. $3k^2 + 4k - 4 = (3k - 2)(k + 2)$

33. $14k^3 + 7k^2 - 70k$
$= 7k(2k^2 + k - 10)$
$= 7k(2k + 5)(k - 2)$

35. $y^4 - 16$
$= (y^2)^2 - 4^2$
$= (y^2 + 4)(y^2 - 4)$ *Difference of squares*
$= (y^2 + 4)(y + 2)(y - 2)$ *Difference of squares*

37. $8m - 16m^2 = 8m(1 - 2m)$

39. Factor $z^3 - 8$ as the difference of cubes.
$$z^3 - 8 = z^3 - 2^3$$
$$= (z - 2)(z^2 + z \cdot 2 + 2^2)$$
$$= (z - 2)(z^2 + 2z + 4)$$

41. $k^2 + 9$ cannot be factored because it is the sum of squares with no GCF. The expression is prime.

43. $32m^9 + 16m^5 + 24m^3$
$= 8m^3(4m^6 + 2m^2 + 3)$

45. $16r^2 + 24rm + 9m^2$
$= (4r)^2 + 2 \cdot 4r \cdot 3m + (3m)^2$
$= (4r + 3m)^2$ *Perfect square trinomial*

47. $15h^2 + 11hg - 14g^2$

Factor by grouping. Look for two integers whose product is $15(-14) = -210$ and whose sum is 11. The integers are 21 and -10.

$15h^2 + 11hg - 14g^2$
$= 15h^2 + 21hg - 10hg - 14g^2$
$= 3h(5h + 7g) - 2g(5h + 7g)$ *Factor each group.*
$= (5h + 7g)(3h - 2g)$ *Factor out $5h + 7g$.*

49. $k^2 - 11k + 30 = (k - 5)(k - 6)$

51. $3k^3 - 12k^2 - 15k$
$= 3k(k^2 - 4k - 5)$
$= 3k(k - 5)(k + 1)$

53. $1000p^3 + 27$
$= (10p)^3 + 3^3$ *Sum of Cubes*
$= (10p + 3)[(10p)^2 - 10p \cdot 3 + 3^2]$
$= (10p + 3)(100p^2 - 30p + 9)$

55. $6 + 3m + 2p + mp$
$= (6 + 3m) + (2p + mp)$
$= 3(2 + m) + p(2 + m)$
$= (2 + m)(3 + p)$

57. $16z^2 - 8z + 1$
$= (4z)^2 - 2 \cdot 4z \cdot 1 + 1^2$
$= (4z - 1)^2$ *Perfect square trinomial*

59. $108m^2 - 36m + 3$
$= 3(36m^2 - 12m + 1)$
$= 3(6m - 1)^2$ *Perfect square trinomial*

61. $x^2 - xy + y^2$ is *prime*. The middle term would have to be $+2xy$ or $-2xy$ in order to make this a perfect square trinomial.

63. $32z^3 + 56z^2 - 16z$
$= 8z(4z^2 + 7z - 2)$
$= 8z(4z - 1)(z + 2)$

65. $20 + 5m + 12n + 3mn$
$= 5(4 + m) + 3n(4 + m)$
$= (4 + m)(5 + 3n)$

67. $6a^2 + 10a - 4$
$= 2(3a^2 + 5a - 2)$
$= 2(3a - 1)(a + 2)$

69. $a^3 - b^3 + 2a - 2b$

Factor by grouping. The first two terms form a difference of cubes.

$a^3 - b^3 + 2a - 2b$
$= (a^3 - b^3) + (2a - 2b)$
$= (a - b)(a^2 + ab + b^2) + 2(a - b)$
$= (a - b)(a^2 + ab + b^2 + 2)$ *Factor out $(a - b)$.*

71. $64m^2 - 80mn + 25n^2$
$= (8m)^2 - 2(8m)(5n) + (5n)^2$
$= (8m - 5n)^2$ *Perfect square trinomial*

73. $8k^2 - 2kh - 3h^2$
$= 8k^2 - 6kh + 4kh - 3h^2$
$= 2k(4k - 3h) + h(4k - 3h)$ *Factor each group.*
$= (4k - 3h)(2k + h)$ *Factor out $4k - 3h$.*

75. $2x^3 + 128$
$= 2(x^3 + 64)$ *GCF = 2*
$= 2(x^3 + 4^3)$ *Sum of Cubes*
$= 2(x + 4)(x^2 - x \cdot 4 + 4^2)$
$= 2(x + 4)(x^2 - 4x + 16)$

77. $10y^2 - 7yz - 6z^2$
$= 10y^2 - 12yz + 5yz - 6z^2$
$= 2y(5y - 6z) + z(5y - 6z)$ *Factor each group.*
$= (5y - 6z)(2y + z)$ *Factor out $5y - 6z$.*

79. $8a^2 + 23ab - 3b^2$
$= 8a^2 + 24ab - ab - 3b^2$
$= 8a(a + 3b) - b(a + 3b)$ *Factor each group.*
$= (a + 3b)(8a - b)$ *Factor out $a + 3b$.*

81. $x^6 - 1 = (x^3)^2 - 1^2$
$= (x^3 + 1)(x^3 - 1)$
or $(x^3 - 1)(x^3 + 1)$

82. From Exercise 81, we have
$$x^6 - 1 = (x^3 - 1)(x^3 + 1).$$
Use the rules for the difference and sum of cubes to factor further.
Since
$$x^3 - 1 = (x - 1)(x^2 + x + 1)$$
and
$$x^3 + 1 = (x + 1)(x^2 - x + 1),$$
we obtain the factorization
$$x^6 - 1 = (x - 1)(x^2 + x + 1)$$
$$\cdot (x + 1)(x^2 - x + 1).$$

83. $x^6 - 1 = (x^2)^3 - 1^3$
$= (x^2 - 1)\left[(x^2)^2 + x^2 \cdot 1 + 1^2\right]$
$= (x^2 - 1)(x^4 + x^2 + 1)$

84. From Exercise 83, we have
$$x^6 - 1 = (x^2 - 1)(x^4 + x^2 + 1).$$
Use the rule for the difference of two squares to factor the binomial.
$$x^2 - 1 = (x - 1)(x + 1)$$
Thus, we obtain the factorization
$$x^6 - 1 = (x - 1)(x + 1)(x^4 + x^2 + 1).$$

85. The result in Exercise 82 is the completely factored form.

86. Multiply the trinomials from the factored form in Exercise 82 vertically.

$$
\begin{array}{rrr}
x^2 & +x & +1 \\
x^2 & -x & +1 \\
\hline
x^2 & +x & +1 \\
-x^3 & -x^2 & -x \\
x^4 & +x^3 & +x^2 \\
\hline
x^4 & +x^2 & +1
\end{array}
$$

87. In general, if I must choose between factoring first using the method for difference of squares or the method for difference of cubes, I should choose the *difference of squares* method to eventually obtain the completely factored form.

88. $x^6 - 729 = (x^3)^2 - 27^2$
$= (x^3 - 27)(x^3 + 27)$
$= (x^3 - 3^3)(x^3 + 3^3)$
$= (x - 3)(x^2 + 3x + 9)$
$\quad \cdot (x + 3)(x^2 - 3x + 9)$

5.5 Solving Quadratic Equations by Factoring

1. A quadratic equation is an equation that can be put in the form $\underline{ax^2 + bx + c} = 0$. $(a \neq 0)$

3. If a quadratic equation is in standard form, to solve the equation we should begin by attempting to *factor* the polynomial.

5. If a quadratic equation $ax^2 + bx + c = 0$ has $c = 0$, then $\underline{0}$ *must* be a solution because $\underline{x}$ is a factor of the polynomial. Note that if $c = 0$, the equation becomes $ax^2 + bx = 0$, which can be factored as $x(ax + b) = 0$, and has solutions $x = 0$ and $x = -b/a$.

7. $(x + 3)(2x - 7)(x - 4) = 0$

Set each factor equal to zero and solve the resulting linear equations.

$$x + 3 = 0 \quad \text{or} \quad 2x - 7 = 0 \quad \text{or} \quad x - 4 = 0$$
$$2x = 7$$
$$x = -3 \quad \text{or} \quad x = \tfrac{7}{2} \quad \text{or} \quad x = 4$$

The solution set is $\left\{-3, \tfrac{7}{2}, 4\right\}$.

9. We can consider the factored form as $2 \cdot x(3x - 4)$, the product of three factors, $2, x,$ and $3x - 4$. Applying the zero-factor property yields three equations,

$$2 = 0 \quad \text{or} \quad x = 0 \quad \text{or} \quad 3x - 4 = 0.$$

Since "$2 = 0$" is impossible and thus has no solution, we end up with the two solutions, $x = 0$ and $x = \tfrac{4}{3}$. We conclude that multiplying a polynomial by a constant does not affect the solutions of the corresponding equation.

For all equations in this section, answers should be checked by substituting into the original equation. These checks will be shown here for only a few of the exercises.

11. $(x + 5)(x - 2) = 0$

By the zero-factor property, the only way that the product of these two factors can be zero is if at least one of the factors is zero.

$$x + 5 = 0 \quad \text{or} \quad x - 2 = 0$$

Solve each of these linear equations.

$$x = -5 \quad \text{or} \quad x = 2$$

Check $x = -5$: $0(-7) = 0$ *True*
Check $x = 2$: $7(0) = 0$ *True*

The solution set is $\{-5, 2\}$.

13. $(2m - 7)(m - 3) = 0$

Set each factor equal to zero and solve the resulting linear equations.

$$2m - 7 = 0 \quad \text{or} \quad m - 3 = 0$$
$$2m = 7 \quad \text{or} \quad m = 3$$
$$m = \tfrac{7}{2}$$

The solution set is $\left\{3, \tfrac{7}{2}\right\}$.

15. $t(6t + 5) = 0$

Set each factor equal to zero and solve the resulting linear equations.

$$t = 0 \quad \text{or} \quad 6t + 5 = 0$$
$$6t = -5$$
$$t = -\tfrac{5}{6}$$

The solution set is $\left\{-\tfrac{5}{6}, 0\right\}$.

17. $2x(3x - 4) = 0$

Set each factor equal to zero and solve the resulting linear equations.

$$2x = 0 \quad \text{or} \quad 3x - 4 = 0$$
$$x = 0 \quad \text{or} \quad 3x = 4$$
$$x = \tfrac{4}{3}$$

The solution set is $\left\{0, \tfrac{4}{3}\right\}$.

19. $\left(x + \tfrac{1}{2}\right)\left(2x - \tfrac{1}{3}\right) = 0$

Set each factor equal to zero and solve the resulting linear equations.

$$x + \tfrac{1}{2} = 0 \quad \text{or} \quad 2x - \tfrac{1}{3} = 0$$
$$x = -\tfrac{1}{2} \quad \text{or} \quad 2x = \tfrac{1}{3}$$
$$x = \tfrac{1}{6}$$

The solution set is $\left\{-\tfrac{1}{2}, \tfrac{1}{6}\right\}$.

21. $(0.5z - 1)(2.5z + 2) = 0$

Set each factor equal to zero and solve the resulting linear equations.

$$0.5z - 1 = 0 \quad \text{or} \quad 2.5z + 2 = 0$$
$$0.5z = 1 \quad \text{or} \quad 2.5z = -2$$
$$z = \tfrac{1}{0.5} \quad \text{or} \quad z = -\tfrac{2}{2.5}$$
$$z = 2 \quad \text{or} \quad z = -0.8$$

The solution set is $\{-0.8, 2\}$.

23. $(x - 9)(x - 9) = 0$

Set each factor equal to zero and solve the resulting linear equations.

$$x - 9 = 0 \quad \text{or} \quad x - 9 = 0$$
$$x = 9 \quad \text{or} \quad x = 9$$

9 is called a *double solution* for $(x - 9)^2 = 0$ because it occurs twice when the equation is solved.

The solution set is $\{9\}$.

25. We can consider the factored form as $3 \cdot x(5x - 4)$, the product of three factors, $3, x,$ and $5x - 4$. Applying the zero-factor property yields three equations,

$$3 = 0 \quad \text{or} \quad x = 0 \quad \text{or} \quad 5x - 4 = 0.$$

Since "$3 = 0$" is impossible, it has no solution, so we end up with the two solutions, $x = 0$ and $x = \tfrac{4}{5}$.

The solution set is $\left\{0, \tfrac{4}{5}\right\}$.

We conclude that multiplying a polynomial by a constant does not affect the solutions of the corresponding equation.

27. $y^2 + 3y + 2 = 0$

Factor the polynomial.

$$(y + 2)(y + 1) = 0$$

Set each factor equal to 0.

$$y + 2 = 0 \quad \text{or} \quad y + 1 = 0$$

Solve each equation.

$$y = -2 \quad \text{or} \quad y = -1$$

Check these solutions by substituting -2 for y and then -1 for y in the original equation.

$$y^2 + 3y + 2 = 0$$
$$(-2)^2 + 3(-2) + 2 = 0 \text{ ? } \textit{Let } y = -2.$$
$$4 - 6 + 2 = 0 \text{ ?}$$
$$-2 + 2 = 0 \quad \textit{True}$$
$$y^2 + 3y + 2 = 0$$
$$(-1)^2 + 3(-1) + 2 = 0 \text{ ? } \textit{Let } y = -1.$$
$$1 - 3 + 2 = 0 \text{ ?}$$
$$-2 + 2 = 0 \quad \textit{True}$$

The solution set is $\{-2, -1\}$.

29. $y^2 - 3y + 2 = 0$

Factor the polynomial.

$$(y - 1)(y - 2) = 0$$

Set each factor equal to 0.

$$y - 1 = 0 \quad \text{or} \quad y - 2 = 0$$

Solve each equation.

$$y = 1 \quad \text{or} \quad y = 2$$

The solution set is $\{1, 2\}$.

31. $x^2 = 24 - 5x$

Write the equation in standard form.

$$x^2 + 5x - 24 = 0$$

Factor the polynomial.

$$(x + 8)(x - 3) = 0$$

Set each factor equal to 0.

$$x + 8 = 0 \quad \text{or} \quad x - 3 = 0$$

Solve each equation.

$$x = -8 \quad \text{or} \quad x = 3$$

The solution set is $\{-8, 3\}$.

33. $x^2 = 3 + 2x$

Write the equation in standard form.

$$x^2 - 2x - 3 = 0$$
$$(x + 1)(x - 3) = 0$$
$$x + 1 = 0 \quad \text{or} \quad x - 3 = 0$$
$$x = -1 \quad \text{or} \quad x = 3$$

The solution set is $\{-1, 3\}$.

35. $z^2 + 3z = -2$

Write the equation in standard form.

$$z^2 + 3z + 2 = 0$$

Factor the polynomial.

$$(z + 2)(z + 1) = 0$$

Set each factor equal to 0.

$$z + 2 = 0 \quad \text{or} \quad z + 1 = 0$$
$$z = -2 \quad \text{or} \quad z = -1$$

The solution set is $\{-2, -1\}$.

37. $m^2 + 8m + 16 = 0$

Factor $m^2 + 8m + 16$ as a perfect square trinomial.

$$(m + 4)^2 = 0$$

Set the factor $m + 4$ equal to 0 and solve.

$$m + 4 = 0$$
$$m = -4$$

The solution set is $\{-4\}$.

39. $3x^2 + 5x - 2 = 0$

Factor the polynomial.

$$(3x - 1)(x + 2) = 0$$

Set each factor equal to 0.

$$3x - 1 = 0 \quad \text{or} \quad x + 2 = 0$$
$$3x = 1 \quad \text{or} \quad x = -2$$
$$x = \tfrac{1}{3}$$

The solution set is $\left\{-2, \tfrac{1}{3}\right\}$.

41.
$$12p^2 = 8 - 10p$$
$$6p^2 = 4 - 5p \quad \textit{Divide by 2.}$$
$$6p^2 + 5p - 4 = 0$$
$$(3p + 4)(2p - 1) = 0$$
$$3p + 4 = 0 \quad \text{or} \quad 2p - 1 = 0$$
$$3p = -4 \quad \text{or} \quad 2p = 1$$
$$p = -\tfrac{4}{3} \quad \text{or} \quad p = \tfrac{1}{2}$$

The solution set is $\left\{-\tfrac{4}{3}, \tfrac{1}{2}\right\}$.

43.
$$9s^2 + 12s = -4$$
$$9s^2 + 12s + 4 = 0$$
$$(3s + 2)^2 = 0$$

Set the factor $3s + 2$ equal to 0 and solve.

$$3s + 2 = 0$$
$$3s = -2$$
$$s = -\frac{2}{3}$$

The solution set is $\left\{-\frac{2}{3}\right\}$.

45.
$$y^2 - 9 = 0$$
$$(y + 3)(y - 3) = 0$$

$$y + 3 = 0 \quad \text{or} \quad y - 3 = 0$$
$$y = -3 \quad \text{or} \quad y = 3$$

The solution set is $\{-3, 3\}$.

47.
$$16k^2 - 49 = 0$$
$$(4k + 7)(4k - 7) = 0$$

$$4k + 7 = 0 \quad \text{or} \quad 4k - 7 = 0$$
$$4k = -7 \quad \text{or} \quad 4k = 7$$
$$k = -\frac{7}{4} \quad \text{or} \quad k = \frac{7}{4}$$

The solution set is $\left\{-\frac{7}{4}, \frac{7}{4}\right\}$.

49.
$$n^2 = 121$$
$$n^2 - 121 = 0$$
$$(n + 11)(n - 11) = 0$$

$$n + 11 = 0 \quad \text{or} \quad n - 11 = 0$$
$$n = -11 \quad \text{or} \quad n = 11$$

The solution set is $\{-11, 11\}$.

51.
$$x^2 = 7x$$
$$x^2 - 7x = 0$$
$$x(x - 7) = 0$$

$$x = 0 \quad \text{or} \quad x - 7 = 0$$
$$x = 0 \quad \text{or} \quad x = 7$$

Check $x = 0$: $0 = 0$ *True*
Check $x = 7$: $49 = 49$ *True*

The solution set is $\{0, 7\}$.

53.
$$6r^2 = 3r$$
$$6r^2 - 3r = 0$$
$$3r(2r - 1) = 0$$

$$3r = 0 \quad \text{or} \quad 2r - 1 = 0$$
$$r = 0 \quad \text{or} \quad 2r = 1$$
$$r = \frac{1}{2}$$

The solution set is $\left\{0, \frac{1}{2}\right\}$.

55.
$$x(x - 7) = -10$$
$$x^2 - 7x = -10$$
$$x^2 - 7x + 10 = 0$$
$$(x - 2)(x - 5) = 0$$

$$x - 2 = 0 \quad \text{or} \quad x - 5 = 0$$
$$x = 2 \quad \text{or} \quad x = 5$$

The solution set is $\{2, 5\}$.

57.
$$3z(2z + 7) = 12$$
$$z(2z + 7) = 4 \quad \textit{Divide by 3.}$$
$$2z^2 + 7z = 4$$
$$2z^2 + 7z - 4 = 0$$
$$(2z - 1)(z + 4) = 0$$

$$2z - 1 = 0 \quad \text{or} \quad z + 4 = 0$$
$$2z = 1 \quad \text{or} \quad z = -4$$
$$z = \frac{1}{2}$$

The solution set is $\left\{-4, \frac{1}{2}\right\}$.

59.
$$2y(y + 13) = 136$$
$$y(y + 13) = 68 \quad \textit{Divide by 2.}$$
$$y^2 + 13y = 68 \quad \textit{Multiply.}$$
$$y^2 + 13y - 68 = 0 \quad \textit{Standard form}$$
$$(y + 17)(y - 4) = 0 \quad \textit{Factor.}$$

$$x + 17 = 0 \quad \text{or} \quad y - 4 = 0 \quad \textit{Zero-factor property}$$
$$y = -17 \text{ or} \quad y = 4 \quad \textit{Solve.}$$

Check $y = -17$:

$$2y(y + 13) = 136$$
$$2(-17)(-17 + 13) = 136 \quad ? \quad \textit{Let y = -17.}$$
$$(-34)(-4) = 136 \quad \textit{True}$$

Check $y = 4$:

$$2(4)(4 + 13) = 136 \quad ? \quad \textit{Let y = 4.}$$
$$8(17) = 136 \quad \textit{True}$$

The solution set is $\{-17, 4\}$.

61. $(2r + 5)\left(3r^2 - 16r + 5\right) = 0$

Begin by factoring $3r^2 - 16r + 5$.

$(2r + 5)(3r - 1)(r - 5) = 0$

Set each of the three factors equal to 0 and solve the resulting equations.

$$2r + 5 = 0 \quad \text{or} \quad 3r - 1 = 0 \quad \text{or} \quad r - 5 = 0$$
$$2r = -5 \qquad\qquad 3r = 1$$
$$r = -\frac{5}{2} \quad \text{or} \quad r = \frac{1}{3} \quad \text{or} \quad r = 5$$

The solution set is $\left\{-\frac{5}{2}, \frac{1}{3}, 5\right\}$.

63. $(2x + 7)(x^2 + 2x - 3) = 0$

$\quad (2x + 7)(x + 3)(x - 1) = 0$

$2x + 7 = 0 \quad$ or $\quad x + 3 = 0 \quad$ or $\quad x - 1 = 0$

$\quad 2x = -7$

$\quad x = -\frac{7}{2} \quad$ or $\qquad x = -3 \quad$ or $\qquad x = 1$

The solution set is $\left\{-\frac{7}{2}, -3, 1\right\}$.

65. $9y^3 - 49y = 0$

To factor the polynomial, begin by factoring out the greatest common factor.

$$y(9y^2 - 49) = 0$$

Now factor $9y^2 - 49$ as the difference of squares.

$$y(3y + 7)(3y - 7) = 0$$

Set each of the three factors equal to 0 and solve.

$y = 0 \quad$ or $\quad 3y + 7 = 0 \quad$ or $\quad 3y - 7 = 0$

$\qquad\qquad\qquad 3y = -7 \qquad\qquad 3y = 7$

$y = 0 \quad$ or $\qquad y = -\frac{7}{3} \quad$ or $\qquad y = \frac{7}{3}$

The solution set is $\left\{-\frac{7}{3}, 0, \frac{7}{3}\right\}$.

67. $r^3 - 2r^2 - 8r = 0$

$\quad r(r^2 - 2r - 8) = 0 \quad$ *Factor out r.*

$\quad r(r - 4)(r + 2) = 0 \quad$ *Factor.*

Set each factor equal to zero and solve.

$r = 0 \quad$ or $\quad r - 4 = 0 \quad$ or $\quad r + 2 = 0$

$r = 0 \quad$ or $\qquad r = 4 \quad$ or $\qquad r = -2$

The solution set is $\{-2, 0, 4\}$.

69. $a^3 + a^2 - 20a = 0$

$\quad a(a^2 + a - 20) = 0 \quad$ *Factor out a.*

$\quad a(a + 5)(a - 4) = 0 \quad$ *Factor.*

Set each factor equal to zero and solve.

$a = 0 \quad$ or $\quad a + 5 = 0 \quad$ or $\quad a - 4 = 0$

$a = 0 \quad$ or $\qquad a = -5 \quad$ or $\qquad a = 4$

The solution set is $\{-5, 0, 4\}$.

71. $r^4 = 2r^3 + 15r^2$

Rewrite with all terms on the left side.

$\quad r^4 - 2r^3 - 15r^2 = 0$

$\quad r^2(r^2 - 2r - 15) = 0 \quad$ *Factor out r^2.*

$\quad r^2(r - 5)(r + 3) = 0 \quad$ *Factor.*

Set each factor equal to zero and solve.

$r^2 = 0 \quad$ or $\quad r - 5 = 0 \quad$ or $\quad r + 3 = 0$

$r = 0 \quad$ or $\qquad r = 5 \quad$ or $\qquad r = -3$

The solution set is $\{-3, 0, 5\}$.

73. $\qquad 3x(x + 1) = (2x + 3)(x + 1)$

$\qquad 3x^2 + 3x = 2x^2 + 5x + 3$

$\qquad x^2 - 2x - 3 = 0$

$\qquad (x + 1)(x - 3) = 0$

$x + 1 = 0 \qquad$ or $\qquad x - 3 = 0$

$\quad x = -1 \qquad$ or $\qquad\qquad x = 3$

The solution set is $\{-1, 3\}$.

Alternatively, we could begin by moving all the terms to the left side and then factoring out $x + 1$.

$$3x(x + 1) - (2x + 3)(x + 1) = 0$$
$$(x + 1)[3x - (2x + 3)] = 0$$
$$(x + 1)(x - 3) = 0$$

The rest of the solution is the same.

75. $\qquad x^2 + (x + 1)^2 = (x + 2)^2$

$\quad x^2 + x^2 + 2x + 1 = x^2 + 4x + 4$

$\qquad\quad x^2 - 2x - 3 = 0$

$\qquad (x + 1)(x - 3) = 0$

$x + 1 = 0 \quad$ or $\quad x - 3 = 0$

$\quad x = -1 \quad$ or $\qquad x = 3$

The solution set is $\{-1, 3\}$.

77. $(2x)^2 = (2x + 4)^2 - (x + 5)^2$

$\quad 4x^2 = 4x^2 + 16x + 16 - (x^2 + 10x + 25)$

$\quad 4x^2 = 4x^2 + 16x + 16 - x^2 - 10x - 25$

$\quad 4x^2 = 3x^2 + 6x - 9$

$\quad x^2 - 6x + 9 = 0$

$\quad (x - 3)(x - 3) = 0$

Set the factor $x - 3$ equal to 0 and solve.

$$x - 3 = 0$$
$$x = 3$$

The solution set is $\{3\}$.

79. $6p^2(p + 1) = 4(p + 1) - 5p(p + 1)$

$\quad 6p^2(p + 1) + 5p(p + 1) - 4(p + 1) = 0$

$\qquad\qquad$ *Rewrite with all terms on the left.*

$\qquad (p + 1)(6p^2 + 5p - 4) = 0$

$\qquad\qquad$ *Factor out $p + 1$.*

$\qquad (p + 1)(2p - 1)(3p + 4) = 0$

$\qquad\qquad$ *Factor.*

Set each factor equal to zero and solve.

$p + 1 = 0 \quad$ or $\quad 2p - 1 = 0 \quad$ or $\quad 3p + 4 = 0$

$\qquad\qquad\qquad\qquad 2p = 1 \qquad\qquad 3p = -4$

$\quad p = -1 \quad$ or $\qquad p = \frac{1}{2} \quad$ or $\qquad p = -\frac{4}{3}$

The solution set is $\left\{-\frac{4}{3}, -1, \frac{1}{2}\right\}$.

81.
$$(k+3)^2 - (2k-1)^2 = 0$$
$$(k^2 + 6k + 9) - (4k^2 - 4k + 1) = 0$$
Square the binomials.
$$k^2 + 6k + 9 - 4k^2 + 4k - 1 = 0$$
$$-3k^2 + 10k + 8 = 0$$
Combine like terms.
$$-1(3k^2 - 10k - 8) = 0$$
Factor out −1.
$$-1(3k + 2)(k - 4) = 0$$
Factor.

Set each factor equal to zero and solve.

$$-1 = 0 \quad \text{or} \quad 3k + 2 = 0 \quad \text{or} \quad k - 4 = 0$$
$$3k = -2$$
$$k = -\tfrac{2}{3} \quad \text{or} \quad k = 4$$

The solution set is $\left\{-\tfrac{2}{3}, 4\right\}$.

Alternatively we could begin by factoring the left side as the difference of two squares.

$$(k+3)^2 - (2k-1)^2 = 0$$
$$[(k+3) + (2k-1)][(k+3) - (2k-1)] = 0$$
$$[3k + 2][-k + 4] = 0$$

The same solution set is obtained.

83. (a) $d = 16t^2$

$$t = 2: \quad d = 16(2)^2 = 16(4) = 64$$
$$t = 3: \quad d = 16(3)^2 = 16(9) = 144$$
$$d = 256: \quad 256 = 16t^2; \; 16 = t^2; \; t = 4$$
$$d = 576: \quad 576 = 16t^2; \; 36 = t^2; \; t = 6$$

t in seconds	0	1	2	3	4	6
d in feet	0	16	64	144	256	576

(b) When $t = 0$, $d = 0$, no time has elapsed, so the object hasn't fallen (been released) yet.

(c) Time cannot be negative.

85. From the calculator screens, we see that the solution set of

$$x^2 + 0.4x - 0.05 = 0$$

is $\{-0.5, 0.1\}$. Substituting -0.5 for x gives us

$$(-0.5)^2 + 0.4(-0.5) - 0.05 = 0 \, ?$$
$$0.25 - 0.20 - 0.05 = 0 \, ?$$
$$0.25 - 0.25 = 0 \quad \textit{True}$$

This shows that -0.5 is a solution of the given quadratic equation.

A similar check shows that 0.1 is also a solution.

The solution set is $\{-0.5, 0.1\}$.

87. From the calculator screens, we see that the solution set of

$$2x^2 + 7.2x + 5.5 = 0$$

is $\{-2.5, -1.1\}$. Verify as in Exercise 85.

89. Let $x = $ the number of counties in California.
Then $x + 9 = $ the number of counties in Florida.
Together, the two states have 125 counties, so

$$x + (x + 9) = 125.$$
$$2x + 9 = 125$$
$$2x = 116$$
$$x = 58$$

Since $x = 58$, $x + 9 = 67$, so California has 58 counties and Florida has 67 counties.

91. Let $x = $ the width of the rectangle.
Then $x + 3 = $ the length of the rectangle.
The perimeter is 34 meters, so

$$2(x) + 2(x + 3) = 34$$
$$2x + 2x + 6 = 34$$
$$4x + 6 = 34$$
$$4x = 28$$
$$x = 7$$

The width of the rectangle is 7 meters and the length is $7 + 3 = 10$ meters.

93. Let $x = $ the first consecutive integer. Then $x + 1 = $ the second consecutive integer.
"Twice the sum of two consecutive integers is 28 more than the second integer" is translated as

$$2[x + (x + 1)] = 28 + (x + 1).$$

Solve this equation.

$$2(2x + 1) = 28 + x + 1$$
$$4x + 2 = x + 29$$
$$3x + 2 = 29$$
$$3x = 27$$
$$x = 9$$

Since $x = 9$, $x + 1 = 10$, so the integers are 9 and 10.

95. Use the formula for the area of a triangle.

$$A = \tfrac{1}{2}bh$$
$$48 = \tfrac{1}{2}(12)h \quad \textit{Let A = 48, b = 12.}$$
$$48 = 6h$$
$$8 = h$$

The height is 8 inches.

5.6   Applications of Quadratic Equations

1.   Read; variable; equation; Solve; answer; Check, original

3. $A = bh$; $A = 45$, $b = 2x + 1$, $h = x + 1$

Step 3 $A = bh$
$$45 = (2x + 1)(x + 1)$$

Step 4
Solve the equation.
$$45 = 2x^2 + 3x + 1$$
$$0 = 2x^2 + 3x - 44$$
$$0 = (2x + 11)(x - 4)$$

$2x + 11 = 0$ or $x - 4 = 0$
$2x = -11$
$x = -\frac{11}{2}$ or $x = 4$

Step 5
Substitute these values for x in the expressions $2x + 1$ and $x + 1$ to find the values of b and h.
$$b = 2x + 1 = 2\left(-\frac{11}{2}\right) + 1$$
$$= -11 + 1 = -10$$
or $b = 2x + 1 = 2(4) + 1$
$$= 8 + 1 = 9$$

We must discard the first solution because the base of a parallelogram cannot have a negative length. Since $x = -\frac{11}{2}$ will not give a realistic answer for the base, we only need to substitute 4 for x to compute the height.
$$h = x + 1 = 4 + 1 = 5$$

The base is 9 units and the height is 5 units.

Step 6
$bh = 9 \cdot 5 = 45$, the desired value of A.

5. $A = LW$; $A = 80$, $L = x + 8$, $W = x - 8$

Step 3 $A = LW$
$$80 = (x + 8)(x - 8)$$

Step 4
Solve the equation.
$$80 = x^2 - 64$$
$$0 = x^2 - 144$$
$$0 = (x + 12)(x - 12)$$

$x + 12 = 0$ or $x - 12 = 0$
$x = -12$ or $x = 12$

Step 5
The solution cannot be $x = -12$ since, when substituted, $-12 + 8$ and $-12 - 8$ are negative numbers and length and width cannot be negative. Thus, $x = 12$ and
$$L = x + 8 = 12 + 8 = 20$$
$$W = x - 8 = 12 - 8 = 4.$$

The length is 20 units, and the width is 4 units.

Step 6
$LW = 20 \cdot 4 = 80$, the desired value of A.

7.   *Step 2*
Let $x =$ the width of the case.
Then $x + 2 =$ the length of the case.

Step 3
$$A = LW$$
$$168 = (x + 2)x   \text{Substitute.}$$

Step 4
$$168 = x^2 + 2x   \text{Multiply.}$$
$$x^2 + 2x - 168 = 0   \text{Standard form}$$
$$(x + 14)(x - 12) = 0   \text{Factor.}$$

$x + 14 = 0$ or $x - 12 = 0$ *Zero-factor property*
$x = -14$ or $x = 12$ *Solve.*

Step 5
Because a width cannot be negative, $x = 12$ and $x + 2 = 14$. The width is 12 cm and the length is 14 cm.

Step 6
The length is 2 cm more than the width and the area is $14(12) = 168$ cm^2, as required.

9.   Let $x =$ the width of the aquarium.
Then $x + 3 =$ the height of the aquarium.

Use the formula for the volume of a rectangular box.
$$V = LWH$$
$$2730 = 21x(x + 3)$$
$$130 = x(x + 3)   \text{Divide by 21.}$$
$$130 = x^2 + 3x$$
$$0 = x^2 + 3x - 130$$
$$0 = (x + 13)(x - 10)$$

$x + 13 = 0$ or $x - 10 = 0$
$x = -13$ or $x = 10$

We discard -13 because the width cannot be negative. The width is 10 inches. The height is $10 + 3 = 13$ inches.

11. Let $h =$ the height of the triangle.
Then $2h + 2 =$ the base of the triangle.

The area of the triangle is 30 square inches.

$$A = \tfrac{1}{2}bh$$
$$30 = \tfrac{1}{2}(2h + 2) \cdot h$$
$$60 = (2h + 2)h$$
$$60 = 2h^2 + 2h$$
$$0 = 2h^2 + 2h - 60$$
$$0 = 2(h^2 + h - 30)$$
$$0 = 2(h + 6)(h - 5)$$

$$h + 6 = 0 \quad \text{or} \quad h - 5 = 0$$
$$h = -6 \quad \text{or} \quad h = 5$$

The solution $h = -6$ must be discarded since a triangle cannot have a negative height. Thus,

$$h = 5 \text{ and } 2h + 2 = 2(5) + 2 = 12.$$

The height is 5 inches, and the base is 12 inches.

13. *Step 2*
Let $x =$ the width of the monitor.
Then $x + 3 =$ the length of the monitor.
The area is $x(x + 3)$.

Step 3
If the length were doubled $[2(x + 3)]$ and if the width were decreased by 1 in. $[x - 1]$, the area would be increased by 150 in.^{2} $[x(x + 3) + 150]$. Write an equation.

$$LW = A$$
$$[2(x + 3)](x - 1) = x(x + 3) + 150$$

Step 4

$$(2x + 6)(x - 1) = x^2 + 3x + 150$$
$$2x^2 + 4x - 6 = x^2 + 3x + 150$$
$$x^2 + x - 156 = 0$$
$$(x + 13)(x - 12) = 0$$

$$x + 13 = 0 \quad \text{or} \quad x - 12 = 0$$
$$x = -13 \quad \text{or} \quad x = 12$$

Step 5
Reject -13, so the width is 12 inches and the length is $12 + 3 = 15$ inches.

Step 6
The length is 3 inches more than the width. The area of the monitor is $15(12) = 180$ in.^{2}. Doubling the length and decreasing the width by 1 inch gives us an area of $30(11) = 330$ in.^{2}, which is 150 in.^{2} more than the area of the original monitor, as required.

15. Let $x =$ the length of a side of the square painting.
Then $x - 2 =$ the length of a side of the square mirror.

Since the formula for the area of a square is $A = s^2$, the area of the painting is x^2, and the area of the mirror is $(x - 2)^2$. The difference between their areas is 32, so

$$x^2 - (x - 2)^2 = 32$$
$$x^2 - (x^2 - 4x + 4) = 32$$
$$x^2 - x^2 + 4x - 4 = 32$$
$$4x - 4 = 32$$
$$4x = 36$$
$$x = 9.$$

The length of a side of the painting is 9 feet. The length of a side of the mirror is $9 - 2 = 7$ feet.

Check: $9^2 - 7^2 = 81 - 49 = 32$

17. Let $x =$ the first volume number.
Then $x + 1 =$ the second volume number.
The product of the numbers is 420.

$$x(x + 1) = 420$$
$$x^2 + x - 420 = 0$$
$$(x - 20)(x + 21) = 0$$

$$x - 20 = 0 \quad \text{or} \quad x + 21 = 0$$
$$x = 20 \quad \text{or} \quad x = -21$$

The volume number cannot be negative, so we reject -21. The volume numbers are 20 and $x + 1 = 20 + 1 = 21$.

19. Let $x =$ the first integer.
Then $x + 1 =$ the second integer, and $x + 2 =$ the third integer.

The product of the		more	10 times	
second and third	is	2	than	the first
$\downarrow$	$\downarrow$	$\downarrow$	$\downarrow$	$\downarrow$
$(x + 1)(x + 2)$	$=$	2	$+$	$10x$

$$x^2 + 3x + 2 = 2 + 10x$$
$$x^2 - 7x = 0$$
$$x(x - 7) = 0$$

$$x = 0 \quad \text{or} \quad x - 7 = 0$$
$$x = 7$$

If $x = 0$, then $x + 1 = 1$, and $x + 2 = 2$.
If $x = 7$, then $x + 1 = 8$, and $x + 2 = 9$.
So there are two sets of consecutive integers that satisfy the condition: 0, 1, 2 and 7, 8, 9.

Check 0, 1, 2: $1(2) \stackrel{?}{=} 2 + 10(0)$
$$2 = 2 + 0 \qquad \textit{True}$$

Check 7, 8, 9: $8(9) \stackrel{?}{=} 2 + 10(7)$
$$72 = 2 + 70 \qquad \textit{True}$$

21. Let $x =$ the first odd integer.
Then $x + 2 =$ the second odd integer and
$x + 4 =$ the third odd integer.

$$3[x + (x + 2) + (x + 4)] = x(x + 2) + 18$$
$$3(3x + 6) = x^2 + 2x + 18$$
$$9x + 18 = x^2 + 2x + 18$$
$$0 = x^2 - 7x$$
$$0 = x(x - 7)$$
$$x = 0 \quad \text{or} \quad x - 7 = 0$$
$$x = 7$$

We must discard 0 because it is even and the problem requires the integers to be odd. If $x = 7$, $x + 2 = 9$, and $x + 4 = 11$. The three integers are 7, 9, and 11.

23. Let $x =$ the first even integer. Then $x + 2$ and $x + 4$ are the next two even integers.

$$x^2 + (x + 2)^2 = (x + 4)^2$$
$$x^2 + x^2 + 4x + 4 = x^2 + 8x + 16$$
$$x^2 - 4x - 12 = 0$$
$$(x - 6)(x + 2) = 0$$
$$x - 6 = 0 \quad \text{or} \quad x + 2 = 0$$
$$x = 6 \quad \text{or} \quad x = -2$$

If $x = 6$, $x + 2 = 8$, and $x + 4 = 10$.
If $x = -2$, $x + 2 = 0$, and $x + 4 = 2$.

The three integers are 6, 8, and 10 or -2, 0, and 2.

25. Let $x =$ the length of the longer leg of the right triangle.
Then $x + 1 =$ the length of the hypotenuse and $x - 7 =$ the length of the shorter leg.

Refer to the figure in the text. Use the Pythagorean formula with $a = x, b = x - 7$, and $c = x + 1$.

$$a^2 + b^2 = c^2$$
$$x^2 + (x - 7)^2 = (x + 1)^2$$
$$x^2 + (x^2 - 14x + 49) = x^2 + 2x + 1$$
$$2x^2 - 14x + 49 = x^2 + 2x + 1$$
$$x^2 - 16x + 48 = 0$$
$$(x - 12)(x - 4) = 0$$
$$x - 12 = 0 \quad \text{or} \quad x - 4 = 0$$
$$x = 12 \quad \text{or} \quad x = 4$$

Discard 4 because if the length of the longer leg is 4 centimeters, by the conditions of the problem, the length of the shorter leg would be $4 - 7 = -3$ centimeters, which is impossible. The length of the longer leg is 12 centimeters.

Check: $12^2 + 5^2 = 13^2$; $169 = 169$ *True*

27. Let $x =$ Alan's distance from home. Then $x + 1 =$ the distance between Tram and Alan.

Refer to the diagram in the textbook.
Use the Pythagorean formula.

$$a^2 + b^2 = c^2$$
$$x^2 + 5^2 = (x + 1)^2$$
$$x^2 + 25 = x^2 + 2x + 1$$
$$24 = 2x$$
$$12 = x$$

Alan is 12 miles from home.

Check: $12^2 + 5^2 = 13^2$; $169 = 169$ *True*

29. Let $x =$ the length of the ladder. Then
$x - 4 =$ the distance from the bottom of the ladder to the building and
$x - 2 =$ the distance on the side of the building to the top of the ladder.

Substitute into the Pythagorean formula.

$$a^2 + b^2 = c^2$$
$$(x - 2)^2 + (x - 4)^2 = x^2$$
$$x^2 - 4x + 4 + x^2 - 8x + 16 = x^2$$
$$x^2 - 12x + 20 = 0$$
$$(x - 10)(x - 2) = 0$$
$$x - 10 = 0 \quad \text{or} \quad x - 2 = 0$$
$$x = 10 \quad \text{or} \quad x = 2$$

The solution cannot be 2 because then a negative distance results. Thus, $x = 10$ and the top of the ladder reaches $x - 2 = 10 - 2 = 8$ feet up the side of the building.

Check: $8^2 + 6^2 = 10^2$; $100 = 100$ *True*

31. **(a)** Let $h = 64$ in the given formula and solve for t.

$$h = -16t^2 + 32t + 48$$
$$64 = -16t^2 + 32t + 48$$
$$16t^2 - 32t + 16 = 0$$
$$16(t^2 - 2t + 1) = 0$$
$$16(t - 1)^2 = 0$$
$$t - 1 = 0$$
$$t = 1$$

The height of the object will be 64 feet after 1 second.

(b) To find the time when the height is 60 feet, let $h = 60$ in the given equation and solve for t.

$$h = -16t^2 + 32t + 48$$
$$60 = -16t^2 + 32t + 48$$
$$16t^2 - 32t + 12 = 0$$
$$4(4t^2 - 8t + 3) = 0$$
$$4(2t - 1)(2t - 3) = 0$$

$$2t - 1 = 0 \quad \text{or} \quad 2t - 3 = 0$$
$$2t = 1 \quad \text{or} \quad 2t = 3$$
$$t = \tfrac{1}{2} \quad \text{or} \quad t = \tfrac{3}{2}$$

The height of the object is 60 feet after $\tfrac{1}{2}$ second (on the way up) and after $\tfrac{3}{2}$ or $1\tfrac{1}{2}$ seconds (on the way down).

(c) To find the time when the object hits the ground, let $h = 0$ and solve for t.

$$h = -16t^2 + 32t + 48$$
$$0 = -16t^2 + 32t + 48$$
$$16t^2 - 32t - 48 = 0$$
$$16(t^2 - 2t - 3) = 0$$
$$16(t + 1)(t - 3) = 0$$

$$t + 1 = 0 \quad \text{or} \quad t - 3 = 0$$
$$t = -1 \quad \text{or} \quad t = 3$$

We discard -1 because time cannot be negative. The object will hit the ground after 3 seconds.

(d) The negative solution, -1, does not make sense, since t represents time, which cannot be negative.

33. **(a)** $x = 6$ in 1996.
$$y = 0.740x^2 + 2.56x + 3.46$$
$$y = 0.740(6)^2 + 2.56(6) + 3.46$$
$$y = 45.46 \approx 45.5$$

In 1996, the model predicts there were about 45.5 million cellular phone subscribers. The result using the model is a little more than 44 million, the actual number for 1996.

(b) $x = 2004 - 1990 = 14$
$x = 14$ corresponds to 2004.

(c) $y = 0.740(14)^2 + 2.56(14) + 3.46$
$\quad y = 184.34 \approx 184.3$

In 2004, the model predicts there were 184.3 million cellular phone subscribers.

The result is a little more than 182 million, the actual number for 2004.

(d) $x = 2006 - 1990 = 16$
$\quad y = 0.740(16)^2 + 2.56(16) + 3.46$
$\quad y = 233.86 \approx 233.9$

In 2006, the model predicts there will be 233.9 million cellular phone subscribers.

34. $378.7 - 271.3 = 107.4$

The trade deficit increased \$107.4 billion from 1999 to 2000.

107.4 is what percent of 271.3?

$$\frac{a}{b} = \frac{p}{100}$$
$$\frac{107.4}{271.3} = \frac{p}{100}$$
$$271.3p = 10{,}740$$
$$p \approx 39.59$$

This is about a 40% increase.

35. $y = 40.8x + 66.9$

In 1997, $x = 2$.

$$y = 40.8(2) + 66.9 = 148.5$$

In 1999, $x = 4$.

$$y = 40.8(4) + 66.9 = 230.1$$

In 2000, $x = 5$.

$$y = 40.8(5) + 66.9 = 270.9$$

The deficits in billions of dollars for 1997, 1999, and 2000 are 148.5, 230.1, and 270.9, respectively.

36. The answers using the linear equation are not at all close to the actual data.

37. $y = 18.5x^2 - 33.4x + 104$

In 1997, $x = 2$.

$$y = 18.5(2)^2 - 33.4(2) + 104 = 111.2$$

In 1999, $x = 4$.

$$y = 18.5(4)^2 - 33.4(4) + 104 = 266.4$$

In 2000, $x = 5$.

$$y = 18.5(5)^2 - 33.4(5) + 104 = 399.5$$

The trade deficit in billions of dollars for 1997, 1999, and 2000 is 111.2, 266.4, and 399.5, respectively.

38. The answers in Exercise 37 are fairly close to the actual data. The quadratic equation models the data better.

39. The x-coordinates are 0, 1, 2, 3, 4, and 5. The y-coordinates are the deficits given in the table. The ordered pairs are: $(0, 97.5)$, $(1, 104.3)$, $(2, 104.7)$, $(3, 164.3)$, $(4, 271.3)$, $(5, 378.7)$.

40.

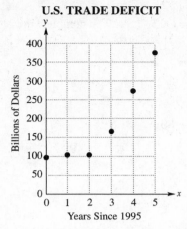

U.S. TRADE DEFICIT

No, the ordered pairs do not lie in a linear pattern.

41. In 2002, $x = 7$.

$$y = 18.5(7)^2 - 33.4(7) + 104 = 776.7$$

In 2002, the model predicts the trade deficit to be $776.7 billion.

42. **(a)** $776.7 - 417.9 = 358.8$

The actual deficit is quite a bit less than the prediction.

(b) No, the equation is based on data for the years 1995–2000 and is valid only for those years. Data for later years might not follow the same pattern.

43. $\dfrac{50}{72} = \dfrac{25 \cdot 2}{36 \cdot 2} = \dfrac{25}{36}$

45. $\dfrac{26}{13} = \dfrac{2 \cdot 13}{1 \cdot 13} = \dfrac{2}{1}$ or 2

47. $\dfrac{48}{-27} = \dfrac{16 \cdot 3}{-9 \cdot 3} = \dfrac{16}{-9}$ or $-\dfrac{16}{9}$

Chapter 5 Review Exercises

1. $7t + 14 = 7 \cdot t + 7 \cdot 2 = 7(t + 2)$

2. $60z^3 + 30z = 30z \cdot 2z^2 + 30z \cdot 1$
$$= 30z(2z^2 + 1)$$

3. $2xy - 8y + 3x - 12$
$= (2xy - 8y) + (3x - 12)$ *Group terms.*
$= 2y(x - 4) + 3(x - 4)$ *Factor each group.*
$= (x - 4)(2y + 3)$ *Factor out $x - 4$.*

4. $6y^2 + 9y + 4xy + 6x$
$= (6y^2 + 9y) + (4xy + 6x)$
$= 3y(2y + 3) + 2x(2y + 3)$
$= (2y + 3)(3y + 2x)$

5. $x^2 + 5x + 6$

Find two integers whose product is 6 and whose sum is 5. The integers are 3 and 2. Thus,

$$x^2 + 5x + 6 = (x + 3)(x + 2).$$

6. $y^2 - 13y + 40$

Find two integers whose product is 40 and whose sum is -13.

Factors of 40	Sums of Factors
$-1, -40$	-41
$-2, -20$	-22
$-4, -10$	-14
$-5, -8$	-13

The integers are -5 and -8, so

$$y^2 - 13y + 40 = (y - 5)(y - 8).$$

7. $q^2 + 6q - 27$

Find two integers whose product is -27 and whose sum is 6. The integers are -3 and 9, so

$$q^2 + 6q - 27 = (q - 3)(q + 9).$$

8. $r^2 - r - 56$

Find two integers whose product is -56 and whose sum is -1. The integers are 7 and -8, so

$$r^2 - r - 56 = (r + 7)(r - 8).$$

9. $r^2 - 4rs - 96s^2$

Find two expressions whose product is $-96s^2$ and whose sum is $-4s$. The expressions are $8s$ and $-12s$, so

$$r^2 - 4rs - 96s^2 = (r + 8s)(r - 12s).$$

10. $p^2 + 2pq - 120q^2$

Find two expressions whose product is $-120q^2$ and whose sum is $2q$. The expressions are $12q$ and $-10q$, so

$$p^2 + 2pq - 120q^2 = (p + 12q)(p - 10q).$$

11. $8p^3 - 24p^2 - 80p$

First, factor out the GCF, $8p$.

$$8p^3 - 24p^2 - 80p = 8p(p^2 - 3p - 10)$$

Now factor $p^2 - 3p - 10$.

$$p^2 - 3p - 10 = (p + 2)(p - 5)$$

The completely factored form is

$$8p^3 - 24p^2 - 80p = 8p(p + 2)(p - 5).$$

12. $3x^4 + 30x^3 + 48x^2$
$= 3x^2(x^2 + 10x + 16)$
$= 3x^2(x + 2)(x + 8)$

13. $p^7 - p^6q - 2p^5q^2 = p^5(p^2 - pq - 2q^2)$
$$= p^5(p + q)(p - 2q)$$

14. $3r^5 - 6r^4s - 45r^3s^2$
$= 3r^3(r^2 - 2rs - 15s^2)$
$= 3r^3(r + 3s)(r - 5s)$

15. $9x^4y - 9x^3y - 54x^2y$
$$= 9x^2y(x^2 - x - 6)$$
$$= 9x^2y(x + 2)(x - 3)$$

16. $2x^7 + 2x^6y - 12x^5y^2$
$$= 2x^5(x^2 + xy - 6y^2)$$
$$= 2x^5(x - 2y)(x + 3y)$$

17. To begin factoring $6r^2 - 5r - 6$, the possible first terms of the two binomial factors are r and $6r$, or $2r$ and $3r$, if we consider only positive integer coefficients.

18. When factoring $2z^3 + 9z^2 - 5z$, the first step is to factor out the GCF, z.

In Exercises 19–27, either the trial and error method or the grouping method can be used to factor each polynomial.

19. Factor $2k^2 - 5k + 2$ by trial and error.
$$2k^2 - 5k + 2 = (2k - 1)(k - 2)$$

20. Factor $3r^2 + 11r - 4$ by grouping. Look for two integers whose product is $3(-4) = -12$ and whose sum is 11. The integers are 12 and -1.
$$3r^2 + 11r - 4 = 3r^2 + 12r - r - 4$$
$$= (3r^2 + 12r) + (-r - 4)$$
$$= 3r(r + 4) - 1(r + 4)$$
$$= (r + 4)(3r - 1)$$

21. Factor $6r^2 - 5r - 6$ by grouping. Find two integers whose product is $6(-6) = -36$ and whose sum is -5. The integers are -9 and 4.
$$6r^2 - 5r - 6 = 6r^2 - 9r + 4r - 6$$
$$= (6r^2 - 9r) + (4r - 6)$$
$$= 3r(2r - 3) + 2(2r - 3)$$
$$= (2r - 3)(3r + 2)$$

22. Factor $10z^2 - 3z - 1$ by trial and error.
$$10z^2 - 3z - 1 = (5z + 1)(2z - 1)$$

23. Factor $8v^2 + 17v - 21$ by grouping. Look for two integers whose product is $8(-21) = -168$ and whose sum is 17. The integers are 24 and -7.
$$8v^2 + 17v - 21 = 8v^2 + 24v - 7v - 21$$
$$= (8v^2 + 24v) + (-7v - 21)$$
$$= 8v(v + 3) - 7(v + 3)$$
$$= (v + 3)(8v - 7)$$

24. $24x^5 - 20x^4 + 4x^3$

Factor out the GCF, $4x^3$. Then complete the factoring by trial and error.
$$24x^5 - 20x^4 + 4x^3$$
$$= 4x^3(6x^2 - 5x + 1)$$
$$= 4x^3(3x - 1)(2x - 1)$$

25. $-6x^2 + 3x + 30 = -3(2x^2 - x - 10)$
$$= -3(2x - 5)(x + 2)$$

26. $10r^3s + 17r^2s^2 + 6rs^3$
$$= rs(10r^2 + 17rs + 6s^2)$$
$$= rs(5r + 6s)(2r + s)$$

27. $48x^4y + 4x^3y^2 - 4x^2y^3$
$$= 4x^2y(12x^2 + xy - y^2)$$
$$= 4x^2y(3x + y)(4x - y)$$

28. The student stopped too soon. He needs to factor out the common factor $4x - 1$ to get $(4x - 1)(4x - 5)$ as the correct answer.

29. Only choice **B**, $4x^2y^2 - 25z^2$, is the difference of squares. In **A**, 32 is not a perfect square. In **C**, we have a sum, not a difference. In **D**, y^3 is not a square. The correct choice is **B**.

30. Only choice **D**, $x^2 - 20x + 100$, is a perfect square trinomial because $x^2 = x \cdot x$, $100 = 10 \cdot 10$, and $-20x = -2(x)(10)$.

In Exercises 31–34, use the rule for factoring a difference of squares.

31. $n^2 - 49 = n^2 - 7^2 = (n + 7)(n - 7)$

32. $25b^2 - 121 = (5b)^2 - 11^2$
$$= (5b + 11)(5b - 11)$$

33. $49y^2 - 25w^2 = (7y)^2 - (5w)^2$
$$= (7y + 5w)(7y - 5w)$$

34. $144p^2 - 36q^2 = 36(4p^2 - q^2)$
$$= 36[(2p)^2 - q^2]$$
$$= 36(2p + q)(2p - q)$$

35. $x^2 + 100$

This polynomial is *prime* because it is the sum of squares and the two terms have no common factor.

In Exercises 36–37, use the rules for factoring a perfect square trinomial.

36. $r^2 - 12r + 36 = r^2 - 2(6)(r) + 6^2$
$$= (r - 6)^2$$

37. $9t^2 - 42t + 49 = (3t)^2 - 2(3t)(7) + 7^2$
$$= (3t - 7)^2$$

In Exercises 38–39, use the rule for factoring a sum of cubes.

38. $m^3 + 1000$

$= m^3 + 10^3$

$= (m + 10)(m^2 - 10 \cdot m + 10^2)$

$= (m + 10)(m^2 - 10m + 100)$

39. $125k^3 + 64x^3$

$= (5k)^3 + (4x)^3$

$= (5k + 4x)[(5k)^2 - 5k \cdot 4x + (4x)^2]$

$= (5k + 4x)(25k^2 - 20kx + 16x^2)$

In Exercises 40–41, use the rule for factoring a difference of cubes.

40. $343x^3 - 64$

$= (7x)^3 - 4^3$

$= (7x - 4)[(7x)^2 + 7x \cdot 4 + 4^2]$

$= (7x - 4)(49x^2 + 28x + 16)$

41. $1000 - 27x^6$

$= 10^3 - (3x^2)^3$

$= (10 - 3x^2)[10^2 + 10(3x^2) + (3x^2)^2]$

$= (10 - 3x^2)(100 + 30x^2 + 9x^4)$

42. $x^6 - y^6$

$= (x^3)^2 - (y^3)^2$ *Difference of squares*

$= (x^3 + y^3)(x^3 - y^3)$

Now factor as the sum and difference of cubes.

$= (x + y)(x^2 - xy + y^2)(x - y)(x^2 + xy + y^2)$

In Exercises 43–56, all solutions should be checked by substituting in the original equations. The checks will not be shown here.

43. $(4t + 3)(t - 1) = 0$

$4t + 3 = 0$ or $t - 1 = 0$

$4t = -3$ or $t = 1$

$t = -\frac{3}{4}$

The solution set is $\left\{-\frac{3}{4}, 1\right\}$.

44. $(x + 7)(x - 4)(x + 3) = 0$

$x + 7 = 0$ or $x - 4 = 0$ or $x + 3 = 0$

$x = -7$ or $x = 4$ or $x = -3$

The solution set is $\{-7, -3, 4\}$.

45. $x(2x - 5) = 0$

$x = 0$ or $2x - 5 = 0$

$2x = 5$

$x = \frac{5}{2}$

The solution set is $\left\{0, \frac{5}{2}\right\}$.

46. $z^2 + 4z + 3 = 0$

$(z + 3)(z + 1) = 0$

$z + 3 = 0$ or $z + 1 = 0$

$z = -3$ or $z = -1$

The solution set is $\{-3, -1\}$.

47. $m^2 - 5m + 4 = 0$

$(m - 1)(m - 4) = 0$

$m - 1 = 0$ or $m - 4 = 0$

$m = 1$ or $m = 4$

The solution set is $\{1, 4\}$.

48. $x^2 = -15 + 8x$

$x^2 - 8x + 15 = 0$

$(x - 3)(x - 5) = 0$

$x - 3 = 0$ or $x - 5 = 0$

$x = 3$ or $x = 5$

The solution set is $\{3, 5\}$.

49. $3z^2 - 11z - 20 = 0$

$(3z + 4)(z - 5) = 0$

$3z + 4 = 0$ or $z - 5 = 0$

$3z = -4$ or $z = 5$

$z = -\frac{4}{3}$

The solution set is $\left\{-\frac{4}{3}, 5\right\}$.

50. $81t^2 - 64 = 0$

$(9t + 8)(9t - 8) = 0$

$9t + 8 = 0$ or $9t - 8 = 0$

$9t = -8$ or $9t = 8$

$t = -\frac{8}{9}$ or $t = \frac{8}{9}$

The solution set is $\left\{-\frac{8}{9}, \frac{8}{9}\right\}$.

51. $y^2 = 8y$

$y^2 - 8y = 0$

$y(y - 8) = 0$

$y = 0$ or $y - 8 = 0$

$y = 8$

The solution set is $\{0, 8\}$.

52. $n(n - 5) = 6$

$n^2 - 5n = 6$

$n^2 - 5n - 6 = 0$

$(n + 1)(n - 6) = 0$

$n + 1 = 0$ or $n - 6 = 0$

$n = -1$ or $n = 6$

The solution set is $\{-1, 6\}$.

53.
$$t^2 - 14t + 49 = 0$$
$$(t - 7)^2 = 0$$
$$t - 7 = 0$$
$$t = 7$$

The solution set is $\{7\}$.

54.
$$t^2 = 12(t - 3)$$
$$t^2 = 12t - 36$$
$$t^2 - 12t + 36 = 0$$
$$(t - 6)^2 = 0$$
$$t - 6 = 0$$
$$t = 6$$

The solution set is $\{6\}$.

55.
$$(5z + 2)(z^2 + 3z + 2) = 0$$
$$(5z + 2)(z + 2)(z + 1) = 0$$

$5z + 2 = 0$ or $z + 2 = 0$ or $z + 1 = 0$
$5z = -2$
$z = -\frac{2}{5}$ or $z = -2$ or $z = -1$

The solution set is $\left\{-2, -1, -\frac{2}{5}\right\}$.

56.
$$x^2 = 9$$
$$x^2 - 9 = 0$$
$$(x + 3)(x - 3) = 0$$

$x + 3 = 0$ or $x - 3 = 0$
$x = -3$ or $x = 3$

The solution set is $\{-3, 3\}$.

57.   Let x = the width of the rug.
Then $x + 6$ = the length of the rug.
$$A = LW$$
$$40 = (x + 6)x$$
$$40 = x^2 + 6x$$
$$0 = x^2 + 6x - 40$$
$$0 = (x + 10)(x - 4)$$

$x + 10 = 0$ or $x - 4 = 0$
$x = -10$ or $x = 4$

Reject -10 since the width cannot be negative. The width of the rug is 4 feet and the length is $4 + 6$ or 10 feet.

58.   From the figure, we have $L = 20$, $W = x$, and $H = x + 4$.
$$S = 2WH + 2WL + 2LH$$
$$650 = 2x(x + 4) + 2x(20) + 2(20)(x + 4)$$
$$650 = 2x^2 + 8x + 40x + 40(x + 4)$$
$$650 = 2x^2 + 48x + 40x + 160$$
$$0 = 2x^2 + 88x - 490$$
$$0 = 2(x^2 + 44x - 245)$$
$$0 = 2(x + 49)(x - 5)$$

$x + 49 = 0$ or $x - 5 = 0$
$x = -49$ or $x = 5$

Reject -49 because the width cannot be negative. The width of the chest is 5 feet.

59.   Let x = the width of the rectangle.
Then $3x$ = the length. Use $A = LW$.

The width increased by 3	times	the same length	would be	an area of 30.
↓	↓	↓	↓	↓
$(x + 3)$	$\cdot$	$3x$	$=$	30

Solve the equation.
$$(x + 3)(3x) = 30$$
$$3x^2 + 9x = 30$$
$$3x^2 + 9x - 30 = 0$$
$$3(x^2 + 3x - 10) = 0$$
$$3(x + 5)(x - 2) = 0$$

$x + 5 = 0$ or $x - 2 = 0$
$x = -5$ or $x = 2$

Reject -5. The width of the original rectangle is 2 meters and the length is $3(2)$ or 6 meters.

60.   Let x = the length of the box.
Then $x - 1$ = the height of the box.
$$V = LWH$$
$$120 = x(4)(x - 1)$$
$$120 = 4x^2 - 4x$$
$$4x^2 - 4x - 120 = 0$$
$$4(x^2 - x - 30) = 0$$
$$4(x - 6)(x + 5) = 0$$

$x - 6 = 0$ or $x + 5 = 0$
$x = 6$ or $x = -5$

Reject -5. The length of the box is 6 meters and the height is $6 - 1 = 5$ meters.

61.   Let x = the first integer.
Then $x + 1$ = the next integer.

The product of the integers is 29 more than their sum, so
$$x(x + 1) = 29 + [x + (x + 1)].$$

Solve this equation.
$$x^2 + x = 29 + 2x + 1$$
$$x^2 - x - 30 = 0$$
$$(x - 6)(x + 5) = 0$$

$x - 6 = 0$ or $x + 5 = 0$
$x = 6$ or $x = -5$

If $x = 6$, $x + 1 = 6 + 1 = 7$.
If $x = -5$, $x + 1 = -5 + 1 = -4$.

The consecutive integers are 6 and 7 or -5 and -4.

62. Let $x =$ the distance traveled west.
Then $x - 14 =$ the distance traveled south,
and $(x - 14) + 16 = x + 2 =$ the distance
between the cars.

These three distances form a right triangle with x
and $x - 14$ representing the lengths of the legs and
$x + 2$ representing the length of the hypotenuse.
Use the Pythagorean formula.

$$a^2 + b^2 = c^2$$
$$x^2 + (x - 14)^2 = (x + 2)^2$$
$$x^2 + x^2 - 28x + 196 = x^2 + 4x + 4$$
$$x^2 - 32x + 192 = 0$$
$$(x - 8)(x - 24) = 0$$

$$x - 8 = 0 \quad \text{or} \quad x - 24 = 0$$
$$x = 8 \quad \text{or} \quad x = 24$$

If $x = 8$, then $x - 14 = -6$, which is not possible
because a distance cannot be negative.

If $x = 24$, then $x - 14 = 10$ and $x + 2 = 26$.
The cars were 26 miles apart.

63. **(a)** In 2005, $x = 3$.

$$y = 0.07x^2 + 0.89x + 1.21$$
$$y = 0.07(3)^2 + 0.89(3) + 1.21$$
$$y = 4.51$$

In 2005, the model predicts $4.51 billion in annual
revenue for eBay.

(b) Yes, the prediction seems reliable.
If eBay revenues in the last half of 2005 are
comparable to those for the first half of the year,
annual revenue in 2005 would be about $4.24
billion.

64. $h = 128t - 16t^2$
$h = 128(1) - 16(1)^2$ *Let t = 1.*
$= 128 - 16$
$= 112$

After 1 second, the height is 112 feet.

65. $h = 128t - 16t^2$
$h = 128(2) - 16(2)^2$ *Let t = 2.*
$= 256 - 16(4)$
$= 256 - 64$
$= 192$

After 2 seconds, the height is 192 feet.

66. $h = 128t - 16t^2$
$h = 128(4) - 16(4)^2$ *Let t = 4.*
$= 512 - 256$
$= 256$

After 4 seconds, the height is 256 feet.

67. The object hits the ground when $h = 0$.

$$h = 128t - 16t^2$$
$$0 = 128t - 16t^2 \quad \text{Let h = 0.}$$
$$0 = 16t(8 - t)$$

$$16t = 0 \quad \text{or} \quad 8 - t = 0$$
$$t = 0 \quad \text{or} \quad 8 = t$$

The solution $t = 0$ represents the time before the
object is projected upward. The object returns to
the ground after 8 seconds.

68. **[5.1]** **D** is not factored completely.
$3(7t + 4) + x(7t + 4) = (7t + 4)(3 + x)$

69. **[5.1]** The factor $2x + 8$ has a common factor of 2.
The completely factored form is
$2(x + 4)(3x - 4)$.

70. **[5.2]** $z^2 - 11zx + 10x^2 = (z - x)(z - 10x)$

71. **[5.3]** $3k^2 + 11k + 10$

Two integers with product $3(10) = 30$ and sum 11
are 5 and 6.

$$3k^2 + 11k + 10$$
$$= 3k^2 + 5k + 6k + 10$$
$$= (3k^2 + 5k) + (6k + 10)$$
$$= k(3k + 5) + 2(3k + 5)$$
$$= (3k + 5)(k + 2)$$

72. **[5.1]**
$15m^2 + 20m - 12mp - 16p$
$= 5m(3m + 4) - 4p(3m + 4)$ *Factor by grouping.*
$= (3m + 4)(5m - 4p)$

73. **[5.4]**
$y^4 - 625$
$= (y^2)^2 - 25^2$
$= (y^2 + 25)(y^2 - 25)$ *Difference of squares*
$= (y^2 + 25)(y + 5)(y - 5)$ *Difference of squares*

74. **[5.3]**
$6m^3 - 21m^2 - 45m$
$= 3m(2m^2 - 7m - 15)$
$= 3m[(2m^2 - 10m) + (3m - 15)]$ *Factor by grouping.*
$= 3m[2m(m - 5) + 3(m - 5)]$
$= 3m(m - 5)(2m + 3)$

75. **[5.1]** $24ab^3c^2 - 56a^2bc^3 + 72a^2b^2c$
$$= 8abc(3b^2c - 7ac^2 + 9ab)$$

76. **[5.3]** $25a^2 + 15ab + 9b^2$ is a *prime* polynomial.

77. **[5.1]** $12x^2yz^3 + 12xy^2z - 30x^3y^2z^4$
$$= 6xyz(2xz^2 + 2y - 5x^2yz^3)$$

78. **[5.2]** $2a^5 - 8a^4 - 24a^3$
$$= 2a^3(a^2 - 4a - 12)$$
$$= 2a^3(a - 6)(a + 2)$$

79. **[5.1]** $12r^2 + 18rq - 10r - 15q$
$$= 6r(2r + 3q) - 5(2r + 3q) \quad \text{\textit{Factor by grouping.}}$$
$$= (2r + 3q)(6r - 5)$$

80. **[5.4]** $1000a^3 + 27$
$$= (10a)^3 + 3^3$$
$$= (10a + 3)[(10a)^2 - 10a \cdot 3 + 3^2]$$
$$= (10a + 3)(100a^2 - 30a + 9)$$

81. **[5.4]** $49t^2 + 56t + 16$
$$= (7t)^2 + 2(7t)(4) + 4^2$$
$$= (7t + 4)^2$$

82. **[5.5]** $t(t - 7) = 0$
$$t = 0 \quad \text{or} \quad t - 7 = 0$$
$$t = 7$$

The solution set is $\{0, 7\}$.

83. **[5.5]** $x^2 + 3x = 10$
$$x^2 + 3x - 10 = 0$$
$$(x + 5)(x - 2) = 0$$
$$x + 5 = 0 \quad \text{or} \quad x - 2 = 0$$
$$x = -5 \quad \text{or} \quad x = 2$$

The solution set is $\{-5, 2\}$.

84. **[5.5]** $25x^2 + 20x + 4 = 0$
$$(5x)^2 + 2(5x)(2) + 2^2 = 0$$
$$(5x + 2)^2 = 0$$
$$5x + 2 = 0$$
$$5x = -2$$
$$x = -\tfrac{2}{5}$$

The solution set is $\left\{-\tfrac{2}{5}\right\}$.

85. **[5.6]** $y = 5.5x^2 + 4.1x + 445$

(a) If $x = 5$ (for 2005), then $y = 603$, so the prediction is 603,000 vehicles.

(b) The estimate may be unreliable because the conditions that prevailed in the years 2001–2004 may have changed, causing either a greater increase or a decrease in the number of alternative-fueled vehicles.

86. **[5.6]** Let $x =$ the length of the shorter leg. Then $2x + 6 =$ the length of the longer leg and $(2x + 6) + 3 = 2x + 9 =$ the length of the hypotenuse.

Use the Pythagorean formula, $a^2 + b^2 = c^2$.
$$x^2 + (2x + 6)^2 = (2x + 9)^2$$
$$x^2 + 4x^2 + 24x + 36 = 4x^2 + 36x + 81$$
$$x^2 - 12x - 45 = 0$$
$$(x - 15)(x + 3) = 0$$
$$x - 15 = 0 \quad \text{or} \quad x + 3 = 0$$
$$x = 15 \quad \text{or} \quad x = -3$$

Reject -3 because a length cannot be negative. The sides of the lot are 15 meters, $2(15) + 6 = 36$ meters, and $36 + 3 = 39$ meters.

87. **[5.6]** Let $x =$ the width of the base. Then $x + 2 =$ the length of the base.

The area of the base, B, is given by LW, so
$$B = x(x + 2).$$

Use the formula for the volume of a pyramid,
$$V = \tfrac{1}{3} \cdot B \cdot h.$$
$$48 = \tfrac{1}{3}x(x + 2)(6)$$
$$48 = 2x(x + 2)$$
$$24 = x^2 + 2x$$
$$x^2 + 2x - 24 = 0$$
$$(x + 6)(x - 4) = 0$$
$$x + 6 = 0 \quad \text{or} \quad x - 4 = 0$$
$$x = -6 \quad \text{or} \quad x = 4$$

Reject -6. The width of the base is 4 meters and the length is $4 + 2$ or 6 meters.

88. **[5.6]** Let $x =$ the first integer. Then $x + 1$ and $x + 2$ are the next two integers.

The product of the first two of three consecutive integers is equal to 23 plus the third.
$$x(x + 1) = 23 + (x + 2)$$
$$x^2 + x = 23 + x + 2$$
$$x^2 - 25 = 0$$
$$(x + 5)(x - 5) = 0$$
$$x + 5 = 0 \quad \text{or} \quad x - 5 = 0$$
$$x = -5 \quad \text{or} \quad x = 5$$

If $x = -5$, then $x + 1 = -4$ and $x + 2 = -3$.
If $x = 5$, then $x + 1 = 6$ and $x + 2 = 7$.
The integers are -5, -4, and -3, or 5, 6, and 7.

89. **[5.6]** $d = 16t^2$
$$t = 4: \ d = 16(4)^2 = 256$$

In 4 seconds, the object would fall 256 feet.

90. **[5.6]** $d = 16t^2$

$$t = 8: \quad d = 16(8)^2 = 1024$$

In 8 seconds, the object would fall 1024 feet.

91. **[5.6]** Let x = the width of the house.
Then $x + 7$ = the length of the house.

Use $A = LW$ with 170 for A, $x + 7$ for L, and x for W.

$$170 = (x + 7)(x)$$
$$170 = x^2 + 7x$$
$$0 = x^2 + 7x - 170$$
$$0 = (x + 17)(x - 10)$$

$$x + 17 = 0 \quad \text{or} \quad x - 10 = 0$$
$$x = -17 \quad \text{or} \quad x = 10$$

Discard -17 because the width cannot be negative. If $x = 10$, $x + 7 = 10 + 7 = 17$.

The width is 10 meters and the length is 17 meters.

92. **[5.6]** Let b = the base of the sail.
Then $b + 4$ = the height of the sail.

Use the formula for the area of a triangle.

$$A = \tfrac{1}{2}bh$$
$$30 = \tfrac{1}{2}(b)(b + 4) \quad \textit{Let A = 30.}$$
$$60 = b^2 + 4b$$
$$0 = b^2 + 4b - 60$$
$$0 = (b + 10)(b - 6)$$

$$b + 10 = 0 \quad \text{or} \quad b - 6 = 0$$
$$b = -10 \quad \text{or} \quad b = 6$$

Discard -10 since the base of a triangle cannot be negative. The base of the triangular sail is 6 meters.

Chapter 5 Test

1. $2x^2 - 2x - 24 = 2\left(x^2 - x - 12\right)$
$$= 2(x + 3)(x - 4)$$

The correct completely factored form is choice **D**. Note that the factored forms **A**, $(2x + 6)(x - 4)$, and **B**, $(x + 3)(2x - 8)$, also can be multiplied to give a product of $2x^2 - 2x - 24$, but neither of these is completely factored because $2x + 6$ and $2x - 8$ both contain a common factor of 2.

2. $12x^2 - 30x = 6x(2x - 5)$

3. $2m^3n^2 + 3m^3n - 5m^2n^2$
$$= m^2n(2mn + 3m - 5n)$$

4. $2ax - 2bx + ay - by$
$$= 2x(a - b) + y(a - b)$$
$$= (a - b)(2x + y)$$

5. $x^2 - 5x - 24$
Find two integers whose product is -24 and whose sum is -5. The integers are 3 and -8.

$$x^2 - 5x - 24 = (x + 3)(x - 8)$$

6. Factor $2x^2 + x - 3$ by trial and error.

$$2x^2 + x - 3 = (2x + 3)(x - 1)$$

7. Factor $10z^2 - 17z + 3$ by trial and error.

$$10z^2 - 17z + 3 = (2z - 3)(5z - 1)$$

8. $t^2 + 2t + 3$

We cannot find two integers whose product is 3 and whose sum is 2. This polynomial is prime.

9. $x^2 + 36$

This polynomial is *prime* because the sum of squares cannot be factored and the two terms have no common factor.

10. $12 - 6a + 2b - ab$
$$= (12 - 6a) + (2b - ab)$$
$$= 6(2 - a) + b(2 - a)$$
$$= (2 - a)(6 + b)$$

11. $9y^2 - 64 = (3y)^2 - 8^2$
$$= (3y + 8)(3y - 8)$$

12. $4x^2 - 28xy + 49y^2$
$$= (2x)^2 - 2(2x)(7y) + (7y)^2$$
$$= (2x - 7y)^2$$

13. $-2x^2 - 4x - 2$
$$= -2\left(x^2 + 2x + 1\right)$$
$$= -2\left(x^2 + 2 \cdot x \cdot 1 + 1^2\right)$$
$$= -2(x + 1)^2$$

14. $6t^4 + 3t^3 - 108t^2$
$$= 3t^2\left(2t^2 + t - 36\right)$$
$$= 3t^2(2t + 9)(t - 4)$$

15. $r^3 - 125 = r^3 - 5^3$
$$= (r - 5)\left(r^2 + 5 \cdot r + 5^2\right)$$
$$= (r - 5)\left(r^2 + 5r + 25\right)$$

16. $8k^3 + 64 = 8\left(k^3 + 8\right)$
$$= 8\left(k^3 + 2^3\right)$$
$$= 8(k + 2)\left(k^2 - 2 \cdot k + 2^2\right)$$
$$= 8(k + 2)\left(k^2 - 2k + 4\right)$$

17. $x^4 - 81 = (x^2)^2 - 9^2$
$$= \left(x^2 + 9\right)\left(x^2 - 9\right)$$
$$= \left(x^2 + 9\right)(x + 3)(x - 3)$$

18. $81x^4 - 16y^4 = (9x^2)^2 - (4y^2)^2$
$$= (9x^2 + 4y^2)(9x^2 - 4y^2)$$
$$= (9x^2 + 4y^2)[(3x)^2 - (2y)^2]$$
$$= (9x^2 + 4y^2)[(3x + 2y)(3x - 2y)]$$
$$= (3x + 2y)(3x - 2y)(9x^2 + 4y^2)$$

19. $9x^6y^4 + 12x^3y^2 + 4$
$$= (3x^3y^2)^2 + 2(3x^3y^2)(2) + (2)^2$$
$$= (3x^3y^2 + 2)^2$$

20. $(p + 3)(p + 3)$ is not the correct factored form of $p^2 + 9$ because
$$(p + 3)(p + 3) = p^2 + 6p + 9$$
$$\neq p^2 + 9.$$

The binomial $p^2 + 9$ is a *prime* polynomial.

21. $2r^2 - 13r + 6 = 0$
$(2r - 1)(r - 6) = 0$

$2r - 1 = 0$ or $r - 6 = 0$
$2r = 1$ or $r = 6$
$r = \frac{1}{2}$

The solution set is $\left\{\frac{1}{2}, 6\right\}$.

22. $25x^2 - 4 = 0$
$(5x + 2)(5x - 2) = 0$

$5x + 2 = 0$ or $5x - 2 = 0$
$5x = -2$ or $5x = 2$
$x = -\frac{2}{5}$ or $x = \frac{2}{5}$

The solution set is $\left\{-\frac{2}{5}, \frac{2}{5}\right\}$.

23. $x(x - 20) = -100$
$x^2 - 20x = -100$
$x^2 - 20x + 100 = 0$
$(x - 10)^2 = 0$
$x - 10 = 0$
$x = 10$

The solution set is $\{10\}$.

24. $t^3 = 9t$
$t^3 - 9t = 0$
$t(t^2 - 9) = 0$
$t(t + 3)(t - 3) = 0$

$t = 0$ or $t + 3 = 0$ or $t - 3 = 0$
$t = 0$ or $t = -3$ or $t = 3$

The solution set is $\{-3, 0, 3\}$.

25. Let x = the width of the flower bed.
Then $2x - 3$ = the length of the flower bed.

Use the formula $A = LW$.

$$x(2x - 3) = 54$$
$$2x^2 - 3x = 54$$
$$2x^2 - 3x - 54 = 0$$
$$(2x + 9)(x - 6) = 0$$

$2x + 9 = 0$ or $x - 6 = 0$
$2x = -9$ or $x = 6$
$x = -\frac{9}{2}$

Reject $-\frac{9}{2}$. If $x = 6$, $2x - 3 = 2(6) - 3 = 9$.
The dimensions of the flower bed are 6 feet by 9 feet.

26. Let x = the first integer.
Then $x + 1$ = the second integer.

The square of the sum of the two integers is 11 more than the first integer.

$$[x + (x + 1)]^2 = x + 11$$
$$(2x + 1)^2 = x + 11$$
$$4x^2 + 4x + 1 = x + 11$$
$$4x^2 + 3x - 10 = 0$$
$$(4x - 5)(x + 2) = 0$$

$4x - 5 = 0$ or $x + 2 = 0$
$4x = 5$ or $x = -2$
$x = \frac{5}{4}$

Reject $\frac{5}{4}$ because it is not an integer. If $x = -2$, $x + 1 = -1$. The integers are -2 and -1.

27. Let x = the length of the stud.
Then $3x - 7$ = the length of the brace.

The figure shows that a right triangle is formed with the brace as the hypotenuse. Use the Pythagorean formula, $a^2 + b^2 = c^2$.

$$x^2 + 15^2 = (3x - 7)^2$$
$$x^2 + 225 = 9x^2 - 42x + 49$$
$$0 = 8x^2 - 42x - 176$$
$$0 = 2(4x^2 - 21x - 88)$$
$$0 = 2(4x + 11)(x - 8)$$

$4x + 11 = 0$ or $x - 8 = 0$
$4x = -11$ or $x = 8$
$x = -\frac{11}{4}$

Reject $-\frac{11}{4}$. If $x = 8$, $3x - 7 = 24 - 7 = 17$, so the brace should be 17 feet long.

28. For 2000, $x = 2000 - 1984 = 16$.

$$y = 1.06x^2 - 4.77x + 47.9$$
$$y = 1.06(16)^2 - 4.77(16) + 47.9 = 242.94$$

In 2000, the model estimates that the number of cable TV channels is 243.

Cumulative Review Exercises
(Chapters 1–5)

1. $3x + 2(x - 4) = 4(x - 2)$
$$3x + 2x - 8 = 4x - 8$$
$$5x - 8 = 4x - 8$$
$$x - 8 = -8$$
$$x = 0$$

The solution set is $\{0\}$.

2. $0.3x + 0.9x = 0.06$

Multiply both sides by 100 to clear decimals.

$$100(0.3x + 0.9x) = 100(0.06)$$
$$30x + 90x = 6$$
$$120x = 6$$
$$x = \frac{6}{120} = \frac{1}{20} = 0.05$$

The solution set is $\{0.05\}$.

3. $\frac{2}{3}m - \frac{1}{2}(m - 4) = 3$

To clear fractions, multiply both sides by the least common denominator, which is 6.

$$6\left[\frac{2}{3}m - \frac{1}{2}(m - 4)\right] = 6(3)$$
$$4m - 3(m - 4) = 18$$
$$4m - 3m + 12 = 18$$
$$m + 12 = 18$$
$$m = 6$$

The solution set is $\{6\}$.

4.
$$A = P + Prt$$
$$A = P(1 + rt) \quad \textit{Factor out P.}$$
$$\frac{A}{1 + rt} = \frac{P(1 + rt)}{1 + rt} \quad \textit{Divide by 1 + rt.}$$
$$\frac{A}{1 + rt} = P \quad \text{or} \quad P = \frac{A}{1 + rt}$$

5. The angles are supplementary, so the sum of the angles is $180°$.

$$(2x + 16) + (x + 23) = 180$$
$$3x + 39 = 180$$
$$3x = 141$$
$$x = 47$$

Since $x = 47$, $2x + 16 = 2(47) + 16 = 110$ and $x + 23 = 47 + 23 = 70$.
The angles are $110°$ and $70°$.

6. Let x = number of bronze medals.
Then $x + 5$ = number of gold medals,
and $(x + 5) + 1 = x + 6$ = the number of silver medals.

The total number of medals was 29.

$$x + (x + 5) + (x + 6) = 29$$
$$3x + 11 = 29$$
$$3x = 18$$
$$x = 6$$

Since $x = 6$, $x + 5 = 11$, and $x + 6 = 12$. Germany won 11 gold medals, 12 silver medals, and 6 bronze medals.

7. A 36% increase is 136% of the original number. 136% of what number is 69.4 million?

$$\frac{\text{part}}{\text{whole}} = \frac{p}{100}$$
$$\frac{69.4}{b} = \frac{136}{100}$$
$$136b = 6940$$
$$b \approx 51.03 \approx 51.0$$

About 51.0 million people in the United States used broadband connections to access the Internet from home in December 2004.

8. 69% of 500 is what number?

$$\frac{\text{part}}{\text{whole}} = \frac{p}{100}$$
$$\frac{a}{500} = \frac{69}{100}$$
$$100a = 69(500)$$
$$a = 345$$

42% of 500 is what number?

$$\frac{\text{part}}{\text{whole}} = \frac{p}{100}$$
$$\frac{a}{500} = \frac{42}{100}$$
$$100a = 500(42)$$
$$a = 210$$

What percent of 500 is 190?

$$\frac{\text{part}}{\text{whole}} = \frac{p}{100}$$
$$\frac{190}{500} = \frac{p}{100}$$
$$500p = 190(100)$$
$$p = 38$$

continued

What percent of 500 is 75?

$$\frac{\text{part}}{\text{whole}} = \frac{p}{100}$$

$$\frac{75}{500} = \frac{p}{100}$$

$$500p = 75(100)$$

$$p = 15$$

Item	Percent	Number
Toilet paper	69%	345
Zipper	42%	210
Frozen foods	38%	190
Self-stick note pads	15%	75

9. The point with coordinates (a, b) is in

   (a) quadrant II if a is *negative* and b is *positive*.

   (b) quadrant III if a is *negative* and b is *negative*.

10. The equation $y = 12x + 3$ is in slope-intercept form, so the y-intercept is $(0, 3)$.

Let $y = 0$ to find the x-intercept.

$$0 = 12x + 3$$
$$-3 = 12x$$
$$-\frac{1}{4} = x$$

The x-intercept is $\left(-\frac{1}{4}, 0\right)$.

11. The equation $y = 12x + 3$ is in slope-intercept form, so the slope is the coefficient of x, that is, 12.

12.

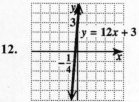

$y = 12x + 3$

13. **(a)** $(1995, 9889)$, $(2002, 10{,}569)$

$$m = \frac{y_2 - y_1}{x_2 - x_1}$$

$$= \frac{10{,}569 - 9889}{2002 - 1995}$$

$$= \frac{680}{7} = 97 \text{ (to the nearest whole number)}$$

A slope of 97 means that the number of radio stations increased by about 97 stations per year.

(b) The graph of the line intersects the vertical grid line over 2000 about halfway between 10,250 and 10,500. The average of these numbers is

$$\frac{10{,}250 + 10{,}500}{2} = \frac{20{,}750}{2} = 10{,}375.$$

The ordered pair is $(2000, 10{,}375)$.

14. Slope $-\frac{1}{2}$, y-intercept $(0, 5)$

The slope-intercept form of a line with slope m and y-intercept $(0, b)$ is $y = mx + b$. In this case, the equation is

$$y = -\frac{1}{2}x + 5.$$

15. Through $\left(\frac{1}{4}, -\frac{2}{3}\right)$, slope 0

The equation of a line with 0 slope is $y = k$. In this case, we have $y = -\frac{2}{3}$. Note that the value $\frac{1}{4}$ is not used to find the equation.

16. $2^{-3} \cdot 2^5 = 2^{-3+5} = 2^2 = 4$

17. $\left(\frac{3}{4}\right)^{-2} = \left(\frac{4}{3}\right)^2 = \frac{16}{9}$

18. $\left(\frac{4^{-3} \cdot 4^4}{4^5}\right)^{-1} = \left(\frac{4^5}{4^{-3} \cdot 4^4}\right)^1$

$$= \frac{4^5}{4^1} = 4^4 = 256$$

19. $\dfrac{(p^2)^3 p^{-4}}{(p^{-3})^{-1} p} = \dfrac{p^{2 \cdot 3} p^{-4}}{p^{(-3)(-1)} p}$

$$= \frac{p^6 p^{-4}}{p^3 p^1}$$

$$= \frac{p^{6-4}}{p^{3+1}}$$

$$= \frac{p^2}{p^4} = \frac{1}{p^2}$$

20. $\dfrac{(m^{-2})^3 m}{m^5 m^{-4}} = \dfrac{m^{-2(3)} m^1}{m^{5+(-4)}}$

$$= \frac{m^{-6+1}}{m^1}$$

$$= \frac{m^{-5}}{m^1} = \frac{1}{m^6}$$

21. $\left(2k^2 + 4k\right) - \left(5k^2 - 2\right) - \left(k^2 + 8k - 6\right)$

$$= \left(2k^2 + 4k\right) + \left(-5k^2 + 2\right)$$
$$\quad + \left(-k^2 - 8k + 6\right)$$
$$= 2k^2 + 4k - 5k^2 + 2 - k^2 - 8k + 6$$
$$= -4k^2 - 4k + 8$$

22. $(9x + 6)(5x - 3)$

$$\text{F} \qquad \text{O} \qquad \text{I} \qquad \text{L}$$
$$= 9x(5x) + 9x(-3) + 6(5x) + 6(-3)$$
$$= 45x^2 - 27x + 30x - 18$$
$$= 45x^2 + 3x - 18$$

23. $(3p + 2)^2 = (3p)^2 + 2 \cdot 3p \cdot 2 + 2^2$

$$= 9p^2 + 12p + 4$$

24. $\dfrac{8x^4 + 12x^3 - 6x^2 + 20x}{2x}$

$= \dfrac{8x^4}{2x} + \dfrac{12x^3}{2x} - \dfrac{6x^2}{2x} + \dfrac{20x}{2x}$

$= 4x^3 + 6x^2 - 3x + 10$

25. $55{,}000 = 5.5 \times 10^4$

Move the decimal point left 4 places so it is to the right of the first nonzero digit. 55,000 is *greater* than 5.5, so the power is *positive*.

$2{,}000{,}000 = 2.0 \times 10^6$

Move the decimal point left 6 places so it is to the right of the first nonzero digit. 2,000,000 is *greater* than 2, so the power is *positive*.

26. Factor $2a^2 + 7a - 4$ by trial and error.

$2a^2 + 7a - 4 = (a + 4)(2a - 1)$

27. $10m^2 + 19m + 6$

To factor by grouping, find two integers whose product is $10(6) = 60$ and whose sum is 19. The integers are 15 and 4.

$10m^2 + 19m + 6 = 10m^2 + 15m + 4m + 6$
$= 5m(2m + 3) + 2(2m + 3)$
$= (2m + 3)(5m + 2)$

28. Factor $8t^2 + 10tv + 3v^2$ by trial and error.

$8t^2 + 10tv + 3v^2 = (4t + 3v)(2t + v)$

29. $4p^2 - 12p + 9 = (2p - 3)(2p - 3)$
$= (2p - 3)^2$

30. $25r^2 - 81t^2 = (5r)^2 - (9t)^2$
$= (5r + 9t)(5r - 9t)$

31. $2pq + 6p^3q + 8p^2q$
$= 2pq\left(1 + 3p^2 + 4p\right)$
$= 2pq\left(3p^2 + 4p + 1\right)$
$= 2pq(3p + 1)(p + 1)$

32. $6m^2 + m - 2 = 0$
$(3m + 2)(2m - 1) = 0$

$3m + 2 = 0 \quad$ or $\quad 2m - 1 = 0$
$3m = -2 \quad$ or $\quad 2m = 1$
$m = -\dfrac{2}{3} \quad$ or $\quad m = \dfrac{1}{2}$

The solution set is $\left\{-\dfrac{2}{3}, \dfrac{1}{2}\right\}$.

33. $\qquad 8x^2 = 64x$
$8x^2 - 64x = 0$
$8x(x - 8) = 0$

$8x = 0 \quad$ or $\quad x - 8 = 0$
$x = 0 \quad$ or $\qquad x = 8$

The solution set is $\{0, 8\}$.

34. Let $x =$ the length of the shorter leg. Then $x + 7 =$ the length of the longer leg, and $2x + 3 =$ the length of the hypotenuse.

Use the Pythagorean formula.

$x^2 + (x + 7)^2 = (2x + 3)^2$
$x^2 + \left(x^2 + 14x + 49\right) = 4x^2 + 12x + 9$
$2x^2 + 14x + 49 = 4x^2 + 12x + 9$
$0 = 2x^2 - 2x - 40$
$0 = 2\left(x^2 - x - 20\right)$
$0 = (x - 5)(x + 4)$

$x - 5 = 0 \quad$ or $\quad x + 4 = 0$
$x = 5 \quad$ or $\qquad x = -4$

Reject -4 because the length of a leg cannot be negative. Since $x = 5$, $x + 7 = 12$, and $2x + 3 = 2(5) + 3 = 13$. The length of the sides are 5 meters, 12 meters, and 13 meters.

CHAPTER 6 RATIONAL EXPRESSIONS AND APPLICATIONS

6.1 The Fundamental Property of Rational Expressions

1. A rational expression is a quotient of two polynomials, such as $\dfrac{x^2 + 3x - 6}{x + 4}$. One can think of this as an algebraic fraction.

3. **(a)** $\dfrac{3x + 1}{5x} = \dfrac{3(2) + 1}{5(2)}$ *Let x = 2.*

$= \dfrac{7}{10}$

(b) $\dfrac{3x + 1}{5x} = \dfrac{3(-3) + 1}{5(-3)}$ *Let x = -3.*

$= \dfrac{-8}{-15} = \dfrac{8}{15}$

5. **(a)** $\dfrac{x^2 - 4}{2x + 1} = \dfrac{(2)^2 - 4}{2(2) + 1}$ *Let x = 2.*

$= \dfrac{0}{5} = 0$

(b) $\dfrac{x^2 - 4}{2x + 1} = \dfrac{(-3)^2 - 4}{2(-3) + 1}$ *Let x = -3.*

$= \dfrac{5}{-5} = -1$

7. **(a)** $\dfrac{(-2x)^3}{3x + 9} = \dfrac{(-2 \cdot 2)^3}{3 \cdot 2 + 9}$ *Let x = 2.*

$= \dfrac{-64}{15} = -\dfrac{64}{15}$

(b) $\dfrac{(-2x)^3}{3x + 9} = \dfrac{[-2(-3)]^3}{3(-3) + 9}$ *Let x = -3.*

$= \dfrac{216}{0}$

Since substituting -3 for x makes the denominator zero, the given rational expression is undefined when $x = -3$.

9. **(a)** $\dfrac{7 - 3x}{3x^2 - 7x + 2}$

$= \dfrac{7 - 3(2)}{3(2)^2 - 7(2) + 2}$ *Let x = 2.*

$= \dfrac{7 - 6}{12 - 14 + 2} = \dfrac{1}{0}$

Since substituting 2 for x makes the denominator zero, the given rational expression is undefined when $x = 2$.

(b) $\dfrac{7 - 3x}{3x^2 - 7x + 2}$

$= \dfrac{7 - 3(-3)}{3(-3)^2 - 7(-3) + 2}$ *Let x = -3.*

$= \dfrac{7 + 9}{27 + 21 + 2} = \dfrac{16}{50} = \dfrac{8}{25}$

11. Division by 0 is undefined, so if the denominator of a rational expression equals 0, the expression is undefined.

13. $\dfrac{12}{5y}$

The denominator $5y$ will be zero when $y = 0$, so the given expression is undefined for $y = 0$. We write the answer as $y \neq 0$.

15. $\dfrac{x + 1}{x - 6}$

To find the values for which this expression is undefined, set the denominator equal to zero and solve for x.

$$x - 6 = 0$$
$$x = 6$$

Because $x = 6$ will make the denominator zero, the given expression is undefined for 6. We write the answer as $x \neq 6$.

17. $\dfrac{4x^2}{3x + 5}$

To find the values for which this expression is undefined, set the denominator equal to zero and solve for x.

$$3x + 5 = 0$$
$$3x = -5$$
$$x = -\tfrac{5}{3}$$

Because $x = -\tfrac{5}{3}$ will make the denominator zero, the given expression is undefined for $-\tfrac{5}{3}$. We write the answer as $x \neq -\tfrac{5}{3}$.

19. $\dfrac{5m + 2}{m^2 + m - 6}$

To find the numbers that make the denominator 0, we must solve

$$m^2 + m - 6 = 0$$
$$(m + 3)(m - 2) = 0$$

$m + 3 = 0 \qquad$ or $\qquad m - 2 = 0$
$\quad\; m = -3 \qquad$ or $\qquad\;\; m = 2$

The given expression is undefined for $m = -3$ and for $m = 2$. We write the answer as $m \neq -3$ and $m \neq 2$.

21. $\dfrac{x^2 + 3x}{4}$ is never undefined since the

denominator is never zero.

23. $\dfrac{3x - 1}{x^2 + 2}$

This denominator cannot equal zero for any value of x because x^2 is always greater than or equal to zero, and adding 2 makes the sum greater than zero. Thus, the given rational expression is never undefined.

25. (a) $\dfrac{x^2 + 4x}{x + 4}$

The two terms in the numerator are x^2 and $4x$. The two terms in the denominator are x and 4.

(b) To express the rational expression in lowest terms, factor the numerator and denominator and divide both by the common factor $x + 4$ to get x.

$$\frac{x^2 + 4x}{x + 4} = \frac{x(x + 4)}{x + 4} = \frac{x}{1} = x$$

27. $\dfrac{18r^3}{6r} = \dfrac{3r^2(6r)}{1(6r)}$ *Factor.*

$\qquad = 3r^2$ *Fundamental property*

29. $\dfrac{4(y - 2)}{10(y - 2)} = \dfrac{2 \cdot 2(y - 2)}{5 \cdot 2(y - 2)}$ *Factor.*

$\qquad = \dfrac{2}{5}$ *Fundamental property*

31. $\dfrac{(x + 1)(x - 1)}{(x + 1)^2} = \dfrac{(x + 1)(x - 1)}{(x + 1)(x + 1)}$

$\qquad = \dfrac{x - 1}{x + 1}$ *Fundamental property*

33. $\dfrac{7m + 14}{5m + 10} = \dfrac{7(m + 2)}{5(m + 2)}$ *Factor.*

$\qquad = \dfrac{7}{5}$ *Fundamental property*

35. $\dfrac{6m - 18}{7m - 21} = \dfrac{6(m - 3)}{7(m - 3)}$ *Factor.*

$\qquad = \dfrac{6}{7}$ *Fundamental property*

37. $\dfrac{m^2 - n^2}{m + n} = \dfrac{(m + n)(m - n)}{m + n}$

$\qquad = m - n$

39. $\dfrac{2t + 6}{t^2 - 9} = \dfrac{2(t + 3)}{(t + 3)(t - 3)}$

$\qquad = \dfrac{2}{t - 3}$

41. $\dfrac{12m^2 - 3}{8m - 4} = \dfrac{3(4m^2 - 1)}{4(2m - 1)}$

$\qquad = \dfrac{3(2m + 1)(2m - 1)}{4(2m - 1)}$

$\qquad = \dfrac{3(2m + 1)}{4}$

43. $\dfrac{3m^2 - 3m}{5m - 5} = \dfrac{3m(m - 1)}{5(m - 1)}$

$\qquad = \dfrac{3m}{5}$

45. $\dfrac{9r^2 - 4s^2}{9r + 6s} = \dfrac{(3r + 2s)(3r - 2s)}{3(3r + 2s)}$

$\qquad = \dfrac{3r - 2s}{3}$

47. $\dfrac{5k^2 - 13k - 6}{5k + 2} = \dfrac{(5k + 2)(k - 3)}{5k + 2}$

$\qquad = k - 3$

49. $\dfrac{x^2 + 2x - 15}{x^2 + 6x + 5} = \dfrac{(x + 5)(x - 3)}{(x + 5)(x + 1)}$

$\qquad = \dfrac{x - 3}{x + 1}$

51. $\dfrac{2x^2 - 3x - 5}{2x^2 - 7x + 5} = \dfrac{(2x - 5)(x + 1)}{(2x - 5)(x - 1)}$

$\qquad = \dfrac{x + 1}{x - 1}$

53. Factor the numerator and denominator by grouping.

$\dfrac{zw + 4z - 3w - 12}{zw + 4z + 5w + 20}$

$\qquad = \dfrac{z(w + 4) - 3(w + 4)}{z(w + 4) + 5(w + 4)}$

$\qquad = \dfrac{(w + 4)(z - 3)}{(w + 4)(z + 5)}$

$\qquad = \dfrac{z - 3}{z + 5}$

55. $\dfrac{pr + qr + ps + qs}{pr + qr - ps - qs}$

$\qquad = \dfrac{r(p + q) + s(p + q)}{r(p + q) - s(p + q)}$

$\qquad = \dfrac{(p + q)(r + s)}{(p + q)(r - s)}$

$\qquad = \dfrac{r + s}{r - s}$

57. $\dfrac{m^2 - n^2 - 4m - 4n}{2m - 2n - 8}$

$= \dfrac{(m+n)(m-n) - 4(m+n)}{2(m-n-4)}$ *Factor by grouping.*

$= \dfrac{(m+n)(m-n-4)}{2(m-n-4)}$

$= \dfrac{m+n}{2}$ *Fundamental property*

59. The numerator is the sum of cubes.

$\dfrac{1+p^3}{1+p} = \dfrac{1^3 + p^3}{1+p}$

$\qquad = \dfrac{(1+p)(1-p+p^2)}{1+p}$

$\qquad = 1 - p + p^2$

61. The numerator is the difference of cubes and the denominator is the difference of squares.

$\dfrac{b^3 - a^3}{a^2 - b^2}$

$= \dfrac{(b-a)(b^2 + ba + a^2)}{(a-b)(a+b)}$

$= (-1) \cdot \dfrac{(b^2 + ba + a^2)}{(a+b)} \qquad \dfrac{(b-a)}{(a-b)} = -1$

$= -\dfrac{b^2 + ba + a^2}{a+b}$

63. The numerator is the sum of cubes. The denominator has a common factor of z.

$\dfrac{z^3 + 27}{z^3 - 3z^2 + 9z} = \dfrac{z^3 + 3^3}{z(z^2 - 3z + 9)}$

$\qquad = \dfrac{(z+3)(z^2 - 3z + 9)}{z(z^2 - 3z + 9)}$

$\qquad = \dfrac{z+3}{z}$

65. A. $\dfrac{2x+3}{2x-3} \neq -1$

B. $\dfrac{2x-3}{3-2x} = \dfrac{-1(3-2x)}{3-2x} = -1$

C. $\dfrac{2x+3}{3+2x} = 1 \neq -1$

D. $\dfrac{2x+3}{-2x-3} = \dfrac{2x+3}{-1(2x+3)} = -1$

B and **D** are equal to -1.

67. $\dfrac{6-t}{t-6} = \dfrac{-1(t-6)}{1(t-6)} = \dfrac{-1}{1} = -1$

Note that $6 - t$ and $t - 6$ are opposites, so we know that their quotient will be -1.

69. $\dfrac{m^2 - 1}{1 - m} = \dfrac{(m+1)(m-1)}{-1(m-1)}$

$\qquad = \dfrac{m+1}{-1}$

$\qquad = -(m+1) \quad \text{or} \quad -m - 1$

71. $\dfrac{q^2 - 4q}{4q - q^2} = \dfrac{q(q-4)}{q(4-q)}$

$\qquad = \dfrac{q-4}{4-q} = -1$

$q - 4$ and $4 - q$ are opposites.

73. In the expression $\dfrac{p+6}{p-6}$, neither numerator nor denominator can be factored. It is already in lowest terms. Note: $(p+6)$ and $(p-6)$ are not opposites.

75. A. $\dfrac{3-x}{x-4} = \dfrac{-1(3-x)}{-1(x-4)} = \dfrac{-3+x}{-x+4} = \dfrac{x-3}{4-x}$

C. $-\dfrac{3-x}{4-x} = \dfrac{-1(3-x)}{4-x} = \dfrac{-3+x}{4-x} = \dfrac{x-3}{4-x}$

D. $-\dfrac{x-3}{x-4} = \dfrac{x-3}{-1(x-4)} = \dfrac{x-3}{-x+4} = \dfrac{x-3}{4-x}$

Since **A**, **C**, and **D** are equivalent to $\dfrac{x-3}{4-x}$, **B** is the one that is not.

There are many possible answers for Exercises 77–82.

77. To write four equivalent expressions for $-\dfrac{x+4}{x-3}$, we will follow the outline in Example 7. Applying the negative sign to the numerator we have

$\dfrac{-(x+4)}{x-3}.$

Distributing the negative sign gives us

$\dfrac{-x-4}{x-3}.$

Applying the negative sign to the denominator yields

$\dfrac{x+4}{-(x-3)}.$

Again, we distribute to get

$\dfrac{x+4}{-x+3}.$

79. $-\dfrac{2x-3}{x+3}$ is equivalent to each of the following:

$$\dfrac{-(2x-3)}{x+3}, \quad \dfrac{-2x+3}{x+3},$$

$$\dfrac{2x-3}{-(x+3)}, \quad \dfrac{2x-3}{-x-3}$$

81. $-\dfrac{3x-1}{5x-6}$ is equivalent to each of the following:

$$\dfrac{-(3x-1)}{5x-6}, \quad \dfrac{-3x+1}{5x-6},$$

$$\dfrac{3x-1}{-(5x-6)}, \quad \dfrac{3x-1}{-5x+6}$$

83. $L \cdot W = A$

$$W = \dfrac{A}{L}$$

$$W = \dfrac{x^4 + 10x^2 + 21}{x^2 + 7}$$

$$= \dfrac{(x^2 + 7)(x^2 + 3)}{x^2 + 7}$$

$$= x^2 + 3$$

Note: If it is not apparent that we can factor A as $x^4 + 10x^2 + 21 = (x^2 + 7)(x^2 + 3)$, we may use "long division" to find the quotient $\dfrac{A}{L}$. Remember to insert zeros for the coefficients of the missing terms in the dividend and divisor.

$$
\begin{array}{r}
x^2 \qquad\quad + 3 \\
x^2 + 0x + 7\,\overline{\big)\,x^4 + 0x^3 + 10x^2 + 0x + 21} \\
\underline{x^4 + 0x^3 + 7x^2} \qquad\qquad\\
3x^2 + 0x + 21 \\
\underline{3x^2 + 0x + 21} \\
0
\end{array}
$$

The width of the rectangle is $x^2 + 3$.

85. Let $w = \dfrac{x^2}{2(1-x)}$.

(a) $w = \dfrac{(0.1)^2}{2(1-0.1)}$ *Let x = 0.1.*

$$= \dfrac{0.01}{2(0.9)} \approx 0.006$$

To the nearest tenth, w is 0.

(b) $w = \dfrac{(0.8)^2}{2(1-0.8)}$ *Let x = 0.8.*

$$= \dfrac{0.64}{2(0.2)} = 1.6$$

(c) $w = \dfrac{(0.9)^2}{2(1-0.9)}$ *Let x = 0.9.*

$$= \dfrac{0.81}{2(0.1)} = 4.05 \approx 4.1$$

(d) Based on the answers in (a), (b), and (c), we see that as the traffic intensity increases, the waiting time also increases.

87. $\dfrac{2}{3} \cdot \dfrac{5}{6} = \dfrac{2 \cdot 5}{3 \cdot 2 \cdot 3} = \dfrac{5}{3 \cdot 3} = \dfrac{5}{9}$

89. $\dfrac{6}{15} \cdot \dfrac{25}{3} = \dfrac{2 \cdot 3 \cdot 5 \cdot 5}{3 \cdot 5 \cdot 3} = \dfrac{2 \cdot 5}{3} = \dfrac{10}{3}$

91. $\dfrac{10}{3} \div \dfrac{5}{6} = \dfrac{10}{3} \cdot \dfrac{6}{5}$

$$= \dfrac{2 \cdot 5 \cdot 2 \cdot 3}{3 \cdot 5}$$

$$= \dfrac{2 \cdot 2}{1} = 4$$

6.2   Multiplying and Dividing Rational Expressions

1. **(a)** $\dfrac{5x^3}{10x^4} \cdot \dfrac{10x^7}{2x} = \dfrac{5 \cdot 10 \cdot x^3 \cdot x^7}{10 \cdot 2 \cdot x^4 \cdot x}$

$$= \dfrac{5x^{10}}{2x^5}$$

$$= \dfrac{5x^5}{2} \textbf{ (B)}$$

(b) $\dfrac{10x^4}{5x^3} \cdot \dfrac{10x^7}{2x} = \dfrac{10 \cdot 10 \cdot x^4 \cdot x^7}{5 \cdot 2 \cdot x^3 \cdot x}$

$$= \dfrac{10x^{11}}{1x^4}$$

$$= 10x^7 \textbf{ (D)}$$

(c) $\dfrac{5x^3}{10x^4} \cdot \dfrac{2x}{10x^7} = \dfrac{5 \cdot 2 \cdot x^3 \cdot x}{10 \cdot 10 \cdot x^4 \cdot x^7}$

$$= \dfrac{1x^4}{10x^{11}}$$

$$= \dfrac{1}{10x^7} \textbf{ (C)}$$

(d) $\dfrac{10x^4}{5x^3} \cdot \dfrac{2x}{10x^7} = \dfrac{10 \cdot 2 \cdot x^4 \cdot x}{5 \cdot 10 \cdot x^3 \cdot x^7}$

$$= \dfrac{2x^5}{5x^{10}}$$

$$= \dfrac{2}{5x^5} \textbf{ (A)}$$

3. $\dfrac{15a^2}{14} \cdot \dfrac{7}{5a} = \dfrac{3 \cdot 5 \cdot a \cdot a \cdot 7}{2 \cdot 7 \cdot 5 \cdot a}$ *Multiply and factor.*

$$= \dfrac{3 \cdot a(5 \cdot 7 \cdot a)}{2(5 \cdot 7 \cdot a)}$$

$$= \dfrac{3a}{2} \qquad \textit{Lowest terms}$$

5. $\dfrac{12x^4}{18x^3} \cdot \dfrac{-8x^5}{4x^2} = \dfrac{-96x^9}{72x^5}$ *Multiply numerators; multiply denominators.*

$\qquad = \dfrac{-4x^4(24x^5)}{3(24x^5)}$ *Group common factors.*

$\qquad = -\dfrac{4x^4}{3}$ *Lowest terms*

7. $\dfrac{2(c+d)}{3} \cdot \dfrac{18}{6(c+d)^2}$

$\qquad = \dfrac{3 \cdot 3 \cdot 2 \cdot 2(c+d)}{3 \cdot 3 \cdot 2(c+d)(c+d)}$ *Multiply and factor.*

$\qquad = \dfrac{2}{c+d}$ *Lowest terms*

9. $\dfrac{(x-y)^2}{2} \cdot \dfrac{24}{3(x-y)}$

$\qquad = \dfrac{6 \cdot 4(x-y)(x-y)}{6(x-y)}$

$\qquad = 4(x-y)$

11. $\dfrac{t-4}{8} \cdot \dfrac{4t^2}{t-4}$

$\qquad = \dfrac{4t^2(t-4)}{2 \cdot 4(t-4)}$

$\qquad = \dfrac{t^2}{2}$

13. $\dfrac{3x}{x+3} \cdot \dfrac{(x+3)^2}{6x^2}$

$\qquad = \dfrac{3x(x+3)(x+3)}{2 \cdot 3 \cdot x \cdot x(x+3)}$

$\qquad = \dfrac{x+3}{2x}$

15. $\dfrac{9z^4}{3z^5} \div \dfrac{3z^2}{5z^3} = \dfrac{9z^4}{3z^5} \cdot \dfrac{5z^3}{3z^2}$

$\qquad = \dfrac{9 \cdot 5z^7}{3 \cdot 3z^7}$

$\qquad = 5$

17. $\dfrac{4t^4}{2t^5} \div \dfrac{(2t)^3}{-6} = \dfrac{4t^4}{2t^5} \cdot \dfrac{-6}{(2t)^3}$

$\qquad = \dfrac{4t^4}{2t^5} \cdot \dfrac{-6}{8t^3}$

$\qquad = \dfrac{-24t^4}{16t^8}$

$\qquad = \dfrac{-3(8t^4)}{2t^4(8t^4)}$

$\qquad = \dfrac{-3}{2t^4} = -\dfrac{3}{2t^4}$

19. $\dfrac{3}{2y-6} \div \dfrac{6}{y-3} = \dfrac{3}{2y-6} \cdot \dfrac{y-3}{6}$

$\qquad = \dfrac{3}{2(y-3)} \cdot \dfrac{y-3}{6}$

$\qquad = \dfrac{3(y-3)}{2 \cdot 2 \cdot 3(y-3)}$

$\qquad = \dfrac{1}{2 \cdot 2} = \dfrac{1}{4}$

21. $\dfrac{7t+7}{-6} \div \dfrac{4t+4}{15}$

$\qquad = \dfrac{7t+7}{-6} \cdot \dfrac{15}{4t+4}$

$\qquad = \dfrac{7(t+1)}{-2 \cdot 3} \cdot \dfrac{3 \cdot 5}{4(t+1)}$

$\qquad = \dfrac{3 \cdot 5 \cdot 7(t+1)}{-2 \cdot 3 \cdot 4(t+1)} = -\dfrac{35}{8}$

23. $\dfrac{2x}{x-1} \div \dfrac{x^2}{x+2}$

$\qquad = \dfrac{2x}{x-1} \cdot \dfrac{x+2}{x^2}$

$\qquad = \dfrac{2x(x+2)}{x \cdot x(x-1)} = \dfrac{2(x+2)}{x(x-1)}$

25. $\dfrac{(x-3)^2}{6x} \div \dfrac{x-3}{x^2}$

$\qquad = \dfrac{(x-3)^2}{6x} \cdot \dfrac{x^2}{x-3}$

$\qquad = \dfrac{x \cdot x(x-3)(x-3)}{6x(x-3)} = \dfrac{x(x-3)}{6}$

27. Suppose I want to multiply $\dfrac{a^2-1}{6} \cdot \dfrac{9}{2a+2}$. I start by factoring where possible:

$$\dfrac{(a+1)(a-1)}{2 \cdot 3} \cdot \dfrac{3 \cdot 3}{2(a+1)}.$$

Next, I divide out common factors in the numerator and denominator to get $\dfrac{a-1}{2} \cdot \dfrac{3}{2}$. Finally, I multiply numerator times numerator and denominator times denominator to get the final product, $\dfrac{3(a-1)}{4}$.

29. $\dfrac{5x-15}{3x+9} \cdot \dfrac{4x+12}{6x-18}$

$\qquad = \dfrac{5(x-3)}{3(x+3)} \cdot \dfrac{4(x+3)}{6(x-3)}$

$\qquad = \dfrac{5 \cdot 4 \cdot (x-3)(x+3)}{3 \cdot 6 \cdot (x-3)(x+3)}$

$\qquad = \dfrac{10}{9}$

31. $\dfrac{2-t}{8} \div \dfrac{t-2}{6} = \dfrac{2-t}{8} \cdot \dfrac{6}{t-2}$ *Multiply by reciprocal.*

$= \dfrac{6(2-t)}{8(t-2)}$ *Multiply numerators; multiply denominators.*

$= \dfrac{6(-1)}{8} \qquad \dfrac{2-t}{t-2} = -1$

$= -\dfrac{3}{4}$ *Lowest terms*

33. $\dfrac{27-3z}{4} \cdot \dfrac{12}{2z-18}$

$= \dfrac{3(9-z)}{4} \cdot \dfrac{3 \cdot 4}{2(z-9)}$

$= \dfrac{3 \cdot 3 \cdot 4(9-z)}{4 \cdot 2(z-9)}$

$= \dfrac{3 \cdot 3 \cdot (-1)}{2}$

$= -\dfrac{9}{2}$

35. $\dfrac{p^2 + 4p - 5}{p^2 + 7p + 10} \div \dfrac{p-1}{p+4}$

$= \dfrac{p^2 + 4p - 5}{p^2 + 7p + 10} \cdot \dfrac{p+4}{p-1}$

$= \dfrac{(p+5)(p-1) \cdot (p+4)}{(p+5)(p+2) \cdot (p-1)}$

$= \dfrac{p+4}{p+2}$

37. $\dfrac{m^2 - 4}{16 - 8m} \div \dfrac{m+2}{8}$

$= \dfrac{(m+2)(m-2)}{8(2-m)} \cdot \dfrac{8}{m+2}$

$= \dfrac{8(m+2)(m-2)}{8(m+2)(2-m)}$

$= -1$

39. $\dfrac{2x^2 - 7x + 3}{x-3} \cdot \dfrac{x+2}{x-1}$

$= \dfrac{(2x-1)(x-3)}{x-3} \cdot \dfrac{x+2}{x-1}$

$= \dfrac{(2x-1)(x+2)}{x-1}$

41. $\dfrac{2k^2 - k - 1}{2k^2 + 5k + 3} \div \dfrac{4k^2 - 1}{2k^2 + k - 3}$

$= \dfrac{2k^2 - k - 1}{2k^2 + 5k + 3} \cdot \dfrac{2k^2 + k - 3}{4k^2 - 1}$

$= \dfrac{(2k+1)(k-1)(2k+3)(k-1)}{(2k+3)(k+1)(2k+1)(2k-1)}$

$= \dfrac{(k-1)(k-1)}{(k+1)(2k-1)}$

$= \dfrac{(k-1)^2}{(k+1)(2k-1)}$

43. $\dfrac{2k^2 + 3k - 2}{6k^2 - 7k + 2} \cdot \dfrac{4k^2 - 5k + 1}{k^2 + k - 2}$

$= \dfrac{(2k-1)(k+2)}{(3k-2)(2k-1)} \cdot \dfrac{(4k-1)(k-1)}{(k+2)(k-1)}$

$= \dfrac{(2k-1)(k+2)(4k-1)(k-1)}{(3k-2)(2k-1)(k+2)(k-1)}$

$= \dfrac{4k-1}{3k-2}$

45. $\dfrac{m^2 + 2mp - 3p^2}{m^2 - 3mp + 2p^2} \div \dfrac{m^2 + 4mp + 3p^2}{m^2 + 2mp - 8p^2}$

$= \dfrac{m^2 + 2mp - 3p^2}{m^2 - 3mp + 2p^2} \cdot \dfrac{m^2 + 2mp - 8p^2}{m^2 + 4mp + 3p^2}$

$= \dfrac{(m+3p)(m-p)(m+4p)(m-2p)}{(m-2p)(m-p)(m+3p)(m+p)}$

$= \dfrac{m+4p}{m+p}$

47. $\dfrac{m^2 + 3m + 2}{m^2 + 5m + 4} \cdot \dfrac{m^2 + 10m + 24}{m^2 + 5m + 6}$

$= \dfrac{(m+2)(m+1)}{(m+4)(m+1)} \cdot \dfrac{(m+6)(m+4)}{(m+3)(m+2)}$

$= \dfrac{m+6}{m+3}$ *Multiply and use fundamental property.*

49. $\dfrac{y^2 + y - 2}{y^2 + 3y - 4} \div \dfrac{y+2}{y+3}$

$= \dfrac{y^2 + y - 2}{y^2 + 3y - 4} \cdot \dfrac{y+3}{y+2}$

$= \dfrac{(y+2)(y-1)}{(y+4)(y-1)} \cdot \dfrac{y+3}{y+2}$

$= \dfrac{y+3}{y+4}$ *Multiply and use fundamental property.*

51. $\dfrac{2m^2 + 7m + 3}{m^2 - 9} \cdot \dfrac{m^2 - 3m}{2m^2 + 11m + 5}$

$= \dfrac{(2m+1)(m+3)}{(m-3)(m+3)} \cdot \dfrac{m(m-3)}{(2m+1)(m+5)}$

$= \dfrac{(2m+1)(m+3)m(m-3)}{(m-3)(m+3)(2m+1)(m+5)}$

$= \dfrac{m}{m+5}$

53. $\dfrac{r^2 + rs - 12s^2}{r^2 - rs - 20s^2} \div \dfrac{r^2 - 2rs - 3s^2}{r^2 + rs - 30s^2}$

$= \dfrac{r^2 + rs - 12s^2}{r^2 - rs - 20s^2} \cdot \dfrac{r^2 + rs - 30s^2}{r^2 - 2rs - 3s^2}$

$= \dfrac{(r-3s)(r+4s)(r+6s)(r-5s)}{(r-5s)(r+4s)(r-3s)(r+s)}$

$= \dfrac{r+6s}{r+s}$

55. $\dfrac{(q-3)^4(q+2)}{q^2+3q+2} \div \dfrac{q^2-6q+9}{q^2+4q+4}$

$= \dfrac{(q-3)^4(q+2)}{q^2+3q+2} \cdot \dfrac{q^2+4q+4}{q^2-6q+9}$

$= \dfrac{(q-3)^4(q+2)(q+2)^2}{(q+2)(q+1)(q-3)^2}$

$= \dfrac{(q-3)^2(q+2)^2}{q+1}$

57. $\dfrac{x+5}{x+10} \div \left(\dfrac{x^2+10x+25}{x^2+10x} \cdot \dfrac{10x}{x^2+15x+50} \right)$

$= \dfrac{x+5}{x+10} \div \left[\dfrac{(x+5)^2 \cdot 10x}{x(x+10)(x+5)(x+10)} \right]$

$= \dfrac{x+5}{x+10} \div \left[\dfrac{10(x+5)}{(x+10)^2} \right]$

$= \dfrac{x+5}{x+10} \cdot \dfrac{(x+10)^2}{10(x+5)}$

$= \dfrac{x+10}{10}$

59. $\dfrac{3a-3b-a^2+b^2}{4a^2-4ab+b^2} \cdot \dfrac{4a^2-b^2}{2a^2-ab-b^2}$

Factor $3a-3b-a^2+b^2$ by grouping.

$3a-3b-a^2+b^2$

$= 3(a-b)-(a^2-b^2)$

$= 3(a-b)-(a-b)(a+b)$

$= (a-b)[3-(a+b)]$

$= (a-b)(3-a-b)$

Thus,

$\dfrac{3a-3b-a^2+b^2}{4a^2-4ab+b^2} \cdot \dfrac{4a^2-b^2}{2a^2-ab-b^2}$

$= \dfrac{(a-b)(3-a-b)}{(2a-b)(2a-b)} \cdot \dfrac{(2a-b)(2a+b)}{(2a+b)(a-b)}$

$= \dfrac{(a-b)(3-a-b)(2a-b)(2a+b)}{(2a-b)(2a-b)(2a+b)(a-b)}$

$= \dfrac{3-a-b}{2a-b}.$

61. $\dfrac{-x^3-y^3}{x^2-2xy+y^2} \div \dfrac{3y^2-3xy}{x^2-y^2}$

$= \dfrac{-1(x^3+y^3)}{x^2-2xy+y^2} \cdot \dfrac{x^2-y^2}{3y^2-3xy}$

$= \dfrac{-1(x+y)(x^2-xy+y^2)}{(x-y)(x-y)}$

$\cdot \dfrac{(x-y)(x+y)}{3y(y-x)}$

$= \dfrac{-1(x+y)(x^2-xy+y^2)(x-y)(x+y)}{-1(x-y)(x-y)(3y)(x-y)}$

$= \dfrac{(x+y)^2(x^2-xy+y^2)}{3y(x-y)^2}$

If we had not changed $y-x$ to $-1(x-y)$ in the denominator, we would have obtained an alternate form of the answer,

$$-\dfrac{(x+y)^2(x^2-xy+y^2)}{3y(y-x)(x-y)}.$$

63. Use the formula for the area of a rectangle with $A = \dfrac{5x^2y^3}{2pq}$ and $L = \dfrac{2xy}{p}$ to solve for W.

$$A = L \cdot W$$

$$\dfrac{5x^2y^3}{2pq} = \dfrac{2xy}{p} \cdot W$$

$$W = \dfrac{5x^2y^3}{2pq} \div \dfrac{2xy}{p}$$

$$= \dfrac{5x^2y^3}{2pq} \cdot \dfrac{p}{2xy}$$

$$= \dfrac{5x^2y^3p}{4pqxy}$$

$$= \dfrac{5xy^2}{4q}$$

Thus, the rational expression $\dfrac{5xy^2}{4q}$ represents the width of the rectangle.

65. $18 = 2 \cdot 9$

$\qquad = 2 \cdot 3 \cdot 3$

The prime factored form of 18 is $2 \cdot 3^2$.

67. $108 = 2 \cdot 54$

$\qquad = 2 \cdot 2 \cdot 27$

$\qquad = 2 \cdot 2 \cdot 3 \cdot 9$

$\qquad = 2 \cdot 2 \cdot 3 \cdot 3 \cdot 3$

The prime factored form of 108 is $2^2 \cdot 3^3$.

69. $$24m = 2^3 \cdot 3 \cdot m$$
$$18m^2 = 2 \cdot 3^2 \cdot m^2$$
$$6 = 2 \cdot 3$$

The GCF is $2 \cdot 3 = 6$.

71. $$84q^3 = 2^2 \cdot 3 \cdot 7 \cdot q^3$$
$$90q^6 = 2 \cdot 3^2 \cdot 5 \cdot q^6$$

The GCF is $2 \cdot 3 \cdot q^3 = 6q^3$.

6.3   Least Common Denominators

1. The factor a appears at most one time in any denominator as does the factor b. Thus, the LCD is the product of the two factors, ab. The correct response is **C**.

3. Since $20 = 2^2 \cdot 5$, the LCD of $\frac{11}{20}$ and $\frac{1}{2}$ must have 5 as a factor and 2^2 as a factor. Because 2 appears twice in $2^2 \cdot 5$, we don't have to include another 2 in the LCD for the number $\frac{1}{2}$. Thus, the LCD is just $2^2 \cdot 5 = 20$. Note that this is a specific case of Exercise 2 since 2 is a factor of 20. The correct response is **C**.

5. $\dfrac{7}{15}, \dfrac{21}{20}$

Factor each denominator.

$$15 = 3 \cdot 5$$
$$20 = 2 \cdot 2 \cdot 5 = 2^2 \cdot 5$$

Take each factor the greatest number of times it appears as a factor in any one of the denominators.

$$\text{LCD} = 2^2 \cdot 3 \cdot 5 = 60$$

7. $\dfrac{17}{100}, \dfrac{23}{120}, \dfrac{43}{180}$

Factor each denominator.

$$100 = 2^2 \cdot 5^2$$
$$120 = 2^3 \cdot 3 \cdot 5$$
$$180 = 2^2 \cdot 3^2 \cdot 5$$

Take each factor the greatest number of times it appears as a factor in any one of the denominators.

$$\text{LCD} = 2^3 \cdot 3^2 \cdot 5^2 = 1800$$

9. $\dfrac{9}{x^2}, \dfrac{8}{x^5}$

The greatest number of times x appears as a factor in any denominator is the greatest exponent on x, which is 5.

$$\text{LCD} = x^5$$

11. $\dfrac{-2}{5p}, \dfrac{13}{6p}$

Factor each denominator.

$$5p = 5 \cdot p$$
$$6p = 2 \cdot 3 \cdot p$$

Take each factor the greatest number of times it appears; then multiply.

$$\text{LCD} = 2 \cdot 3 \cdot 5 \cdot p = 30p$$

13. $\dfrac{17}{15y^2}, \dfrac{55}{36y^4}$

Factor each denominator.

$$15y^2 = 3 \cdot 5 \cdot y^2$$
$$36y^4 = 2^2 \cdot 3^2 \cdot y^4$$

Take each factor the greatest number of times it appears; then multiply.

$$\text{LCD} = 2^2 \cdot 3^2 \cdot 5 \cdot y^4 = 180y^4$$

15. $\dfrac{5}{21r^3}, \dfrac{7}{12r^5}$

Factor each denominator.

$$21r^3 = 3 \cdot 7 \cdot r^3$$
$$12r^5 = 2^2 \cdot 3 \cdot r^5$$

Take each factor the greatest number of times it appears; then multiply.

$$\text{LCD} = 2^2 \cdot 3 \cdot 7 \cdot r^5 = 84r^5$$

17. $\dfrac{13}{5a^2b^3}, \dfrac{29}{15a^5b}$

Factor each denominator.

$$5a^2b^3 = 5 \cdot a^2 \cdot b^3$$
$$15a^5b = 3 \cdot 5 \cdot a^5 \cdot b$$

Take each factor the greatest number of times it appears; then multiply.

$$\text{LCD} = 3 \cdot 5 \cdot a^5 \cdot b^3 = 15a^5b^3$$

19. $\dfrac{7}{6p}, \dfrac{15}{4p - 8}$

Factor each denominator.

$$6p = 2 \cdot 3 \cdot p$$
$$4p - 8 = 4(p - 2) = 2^2(p - 2)$$

Take each factor the greatest number of times it appears; then multiply.

$$\text{LCD} = 2^2 \cdot 3 \cdot p(p - 2) = 12p(p - 2)$$

21. $\dfrac{9}{28m^2}, \dfrac{3}{12m - 20}$

Factor each denominator.

$$28m^2 = 2^2 \cdot 7 \cdot m^2$$
$$12m - 20 = 4(3m - 5) = 2^2(3m - 5)$$

Take each factor the greatest number of times it appears; then multiply.

$$\text{LCD} = 2^2 \cdot 7m^2(3m - 5) = 28m^2(3m - 5)$$

23. $\dfrac{7}{5b-10}, \dfrac{11}{6b-12}$

Factor each denominator.

$$5b - 10 = 5(b - 2)$$
$$6b - 12 = 6(b - 2) = 2 \cdot 3(b - 2)$$

Take each factor the greatest number of times it appears; then multiply.

$$\text{LCD} = 2 \cdot 3 \cdot 5(b - 2) = 30(b - 2)$$

25. $\dfrac{37}{6r-12}, \dfrac{25}{9r-18}$

Factor each denominator.

$$6r - 12 = 6(r - 2) = 2 \cdot 3(r - 2)$$
$$9r - 18 = 9(r - 2) = 3^2(r - 2)$$

Take each factor the greatest number of times it appears; then multiply.

$$\text{LCD} = 2 \cdot 3^2(r - 2) = 18(r - 2)$$

27. $\dfrac{5}{12p+60}, \dfrac{17}{p^2+5p}, \dfrac{16}{p^2+10p+25}$

Factor each denominator.

$$12p + 60 = 12(p + 5) = 2^2 \cdot 3(p + 5)$$
$$p^2 + 5p = p(p + 5)$$
$$p^2 + 10p + 25 = (p + 5)(p + 5)$$

$$= 2^2 \cdot 3 \cdot p(p+5)^2 = 12p(p+5)^2$$

$$\dfrac{22}{y^2+3y+2}$$

...tor each denominator.

$$8y + 16 = 8(y + 2) = 2^3(y + 2)$$
$$y^2 + 3y + 2 = (y + 2)(y + 1)$$

$$\text{LCD} = 8(y + 2)(y + 1)$$

31. $\dfrac{5}{c-d}, \dfrac{8}{d-c}$

The denominators, $c - d$ and $d - c$, are opposites of each other since

$$-(c - d) = -c + d = d - c.$$

Therefore, either $c - d$ or $d - c$ can be used as the LCD.

33. $\dfrac{12}{m-3}, \dfrac{-4}{3-m}$

The expression $3 - m$ can be written as $-1(m - 3)$, since

$$-1(m - 3) = -m + 3 = 3 - m.$$

Because of this, either $m - 3$ or $3 - m$ can be used as the LCD.

35. $\dfrac{29}{p-q}, \dfrac{18}{q-p}$

The expression $q - p$ can be written as $-1(p - q)$, since

$$-1(p - q) = -p + q = q - p.$$

Because of this, either $p - q$ or $q - p$ can be used as the LCD.

37. $\dfrac{3}{k^2+5k}, \dfrac{2}{k^2+3k-10}$

Factor each denominator.

$$k^2 + 5k = k(k + 5)$$
$$k^2 + 3k - 10 = (k + 5)(k - 2)$$

$$\text{LCD} = k(k + 5)(k - 2)$$

39. $\dfrac{6}{a^2+6a}, \dfrac{-5}{a^2+3a-18}$

Factor each denominator.

$$a^2 + 6a = a(a + 6)$$
$$a^2 + 3a - 18 = (a + 6)(a - 3)$$

$$\text{LCD} = a(a + 6)(a - 3)$$

41. $\dfrac{5}{p^2+8p+15}, \dfrac{3}{p^2-3p-18}, \dfrac{2}{p^2-p-30}$

Factor each denominator.

$$p^2 + 8p + 15 = (p + 5)(p + 3)$$
$$p^2 - 3p - 18 = (p - 6)(p + 3)$$
$$p^2 - p - 30 = (p - 6)(p + 5)$$

$$\text{LCD} = (p + 3)(p + 5)(p - 6)$$

43. $\dfrac{-5}{k^2+2k-35}, \dfrac{-8}{k^2+3k-40}, \dfrac{9}{k^2-2k-15}$

Factor each denominator.

$$k^2 + 2k - 35 = (k + 7)(k - 5)$$
$$k^2 + 3k - 40 = (k + 8)(k - 5)$$
$$k^2 - 2k - 15 = (k - 5)(k + 3)$$

$$\text{LCD} = (k + 7)(k - 5)(k + 8)(k + 3)$$

45. $\dfrac{3}{4} = \dfrac{?}{28}$

To change 4 into 28, multiply by 7. If you multiply the denominator by 7, you must multiply the numerator by 7.

46. $\dfrac{3}{4} = \dfrac{3}{4} \cdot \dfrac{7}{7} = \dfrac{21}{28}$

Note that numerator and denominator are being multiplied by 7, so $\frac{3}{4}$ is being multiplied by the fraction $\frac{7}{7}$, which is equal to 1.

47. Since $\frac{7}{7}$ has a value of 1, the multiplier is 1. The *identity property of multiplication* is being used when we write a common fraction as an equivalent one with a larger denominator.

48. $\dfrac{2x+5}{x-4} = \dfrac{?}{7x-28} = \dfrac{?}{7(x-4)}$

The expression $7x - 28$ is factored as $7(x-4)$, so the multiplier is 7.

49. $\dfrac{2x+5}{x-4} = \dfrac{?}{7x-28} = \dfrac{?}{7(x-4)}$

To form the new denominator, 7 must be used as the multiplier for the denominator. To form an equivalent fraction, the same multiplier must be used for numerator and denominator. Thus, the multiplier is $\frac{7}{7}$, which is equal to 1.

50. The *identity property of multiplication* is being used when we write an algebraic fraction as an equivalent one with a larger denominator.

51. $\dfrac{4}{11} = \dfrac{?}{55}$

First factor the denominator on the right. Then compare the denominator on the left with the one on the right to decide what factors are missing.

$$\frac{4}{11} = \frac{?}{11 \cdot 5}$$

A factor of 5 is missing, so multiply $\frac{4}{11}$ by $\frac{5}{5}$, which is equal to 1.

$$\frac{4}{11} \cdot \frac{5}{5} = \frac{20}{55}$$

53. $\dfrac{-5}{k} = \dfrac{?}{9k}$

A factor of 9 is missing in the first fraction, so multiply numerator and denominator by 9.

$$\frac{-5}{k} \cdot \frac{9}{9} = \frac{-45}{9k}$$

55. $\dfrac{15m^2}{8k} = \dfrac{?}{32k^4}$

$32k^4 = (8k)(4k^3)$, so we must multiply the numerator and the denominator by $4k^3$.

$$\frac{15m^2}{8k} = \frac{15m^2}{8k} \cdot \frac{4k^3}{4k^3} \quad \text{\textit{Multiplicative identity property}}$$

$$= \frac{60m^2k^3}{32k^4}$$

57. $\dfrac{19z}{2z-6} = \dfrac{?}{6z-18}$

Begin by factoring each denominator.

$$2z - 6 = 2(z-3)$$
$$6z - 18 = 6(z-3)$$

The fractions may now be written as follows.

$$\frac{19z}{2(z-3)} = \frac{?}{6(z-3)}$$

Comparing the two factored forms, we see that the denominator of the fraction on the left side must be multiplied by 3; the numerator must also be multiplied by 3.

$$\frac{19z}{2z-6}$$

$$= \frac{19z}{2(z-3)} \cdot \frac{3}{3} \quad \text{\textit{Multiplicative identity property}}$$

$$= \frac{19z(3)}{2(z-3)(3)} \quad \text{\textit{Multiplication of rational expressions}}$$

$$= \frac{57z}{6z-18} \quad \text{\textit{Multiply the factors}}$$

59. $\dfrac{-2a}{9a-18} = \dfrac{?}{18a-36}$

$$\frac{-2a}{9(a-2)} = \frac{?}{18(a-2)} \quad \text{\textit{Factor each denominator}}$$

$$\frac{-2a}{9a-18} = \frac{-2a}{9(a-2)} \cdot \frac{2}{2} \quad \text{\textit{Multiplicative identity property}}$$

$$= \frac{-4a}{18a-36} \quad \text{\textit{Multiply}}$$

61. $\dfrac{6}{k^2-4k} = \dfrac{?}{k(k-4)(k+1)}$

$$\frac{6}{k(k-4)} = \frac{?}{k(k-4)(k+1)} \quad \text{\textit{Factor first denominator}}$$

$$\frac{6}{k^2-4k} = \frac{6}{k(k-4)} \cdot \frac{(k+1)}{(k+1)} \quad \text{\textit{Multiplicative identity property}}$$

$$= \frac{6(k+1)}{k(k-4)(k+1)} \quad \text{\textit{Multiply}}$$

63. $\dfrac{36r}{r^2-r-6} = \dfrac{?}{(r-3)(r+2)(r+1)}$

$$\frac{36r}{(r-3)(r+2)} = \frac{?}{(r-3)(r+2)(r+1)}$$

$$\text{\textit{Factor first denominator}}$$

$$\frac{36r}{r^2-r-6} = \frac{36r}{(r-3)(r+2)} \cdot \frac{(r+1)}{(r+1)}$$

$$\text{\textit{Multiplicative identity property}}$$

$$= \frac{36r(r+1)}{(r-3)(r+2)(r+1)}$$

65. $\dfrac{a+2b}{2a^2+ab-b^2} = \dfrac{?}{2a^3b+a^2b^2-ab^3}$

$\dfrac{a+2b}{2a^2+ab-b^2} = \dfrac{?}{ab(2a^2+ab-b^2)}$

Factor second denominator

$\dfrac{a+2b}{2a^2+ab-b^2} = \dfrac{(a+2b)}{(2a^2+ab-b^2)} \cdot \dfrac{ab}{ab}$

Multiplicative identity property

$= \dfrac{ab(a+2b)}{2a^3b+a^2b^2-ab^3}$

67. $\dfrac{4r-t}{r^2+rt+t^2} = \dfrac{?}{t^3-r^3}$

Factor the second denominator as the difference of cubes.

$t^3-r^3 = (t-r)(t^2+rt+r^2)$

$\dfrac{4r-t}{r^2+rt+t^2} = \dfrac{(4r-t)}{(r^2+rt+t^2)} \cdot \dfrac{(t-r)}{(t-r)}$

Multiplicative identity property

$= \dfrac{(4r-t)(t-r)}{t^3-r^3}$

Multiply the factors

69. $\dfrac{2(z-y)}{y^2+yz+z^2} = \dfrac{?}{y^4-z^3y}$

Factor the second denominator.

$y^4-z^3y = y(y^3-z^3) \quad GCF = y$

$= y(y-z)(y^2+yz+z^2)$

Difference of cubes

$\dfrac{2(z-y)}{y^2+yz+z^2} = \dfrac{?}{y(y-z)(y^2+yz+z^2)}$

$\dfrac{2(z-y)}{y^2+yz+z^2} = \dfrac{2(z-y)}{(y^2+yz+z^2)} \cdot \dfrac{y(y-z)}{y(y-z)}$

$= \dfrac{2y(z-y)(y-z)}{y(y-z)(y^2+yz+z^2)}$

Multiplicative identity property

$= \dfrac{2y(z-y)(y-z)}{y(y^3-z^3)}$

$= \dfrac{2y(z-y)(y-z)}{y^4-z^3y}$

or $\dfrac{-2y(y-z)^2}{y^4-z^3y}$, since $z-y = -1(y-z)$

71. $\dfrac{3}{4}+\dfrac{7}{4} = \dfrac{3+7}{4} = \dfrac{10}{4} = \dfrac{5 \cdot 2}{2 \cdot 2} = \dfrac{5}{2}$

73. $\dfrac{1}{2}+\dfrac{7}{8} = \dfrac{4}{8}+\dfrac{7}{8} = \dfrac{4+7}{8} = \dfrac{11}{8}$

75. $\dfrac{7}{5}-\dfrac{3}{4} = \dfrac{28}{20}-\dfrac{15}{20} = \dfrac{28-15}{20} = \dfrac{13}{20}$

77. $\dfrac{4}{3}-\dfrac{1}{4} = \dfrac{16}{12}-\dfrac{3}{12} = \dfrac{16-3}{12} = \dfrac{13}{12}$

6.4 Adding and Subtracting Rational Expressions

1. $\dfrac{x}{x+6}+\dfrac{6}{x+6}$

The denominators are the same, so the sum is found by adding the two numerators and keeping the same (common) denominator.

$\dfrac{x}{x+6}+\dfrac{6}{x+6} = \dfrac{x+6}{x+6} = 1$

Choice **E** is correct.

3. $\dfrac{6}{x-6}-\dfrac{x}{x-6}$

The denominators are the same, so the difference is found by subtracting the two numerators and keeping the same (common) denominator.

$\dfrac{6}{x-6}-\dfrac{x}{x-6} = \dfrac{6-x}{x-6}$

$= \dfrac{-1(x-6)}{x-6} = -1$

Choice **C** is correct.

5. $\dfrac{x}{x+6}-\dfrac{6}{x+6}$

The denominators are the same, so the difference is found by subtracting the two numerators and keeping the same (common) denominator.

$\dfrac{x}{x+6}-\dfrac{6}{x+6} = \dfrac{x-6}{x+6}$

Choice **B** is correct.

7. $\dfrac{1}{6}-\dfrac{1}{x}$

The LCD is $6x$. Now rewrite each rational expression as a fraction with the LCD as its denominator.

$\dfrac{1}{6} \cdot \dfrac{x}{x} = \dfrac{x}{6x}$

$\dfrac{1}{x} \cdot \dfrac{6}{6} = \dfrac{6}{6x}$

Since the fractions now have a common denominator, subtract the numerators and use the LCD as the denominator of the sum.

$\dfrac{1}{6}-\dfrac{1}{x} = \dfrac{x}{6x}-\dfrac{6}{6x} = \dfrac{x-6}{6x}$

Choice **G** is correct.

9. $\dfrac{4}{m} + \dfrac{7}{m}$

The denominators are the same, so the sum is found by adding the two numerators and keeping the same (common) denominator.

$$\frac{4}{m} + \frac{7}{m} = \frac{4+7}{m} = \frac{11}{m}$$

11. $\dfrac{5}{y+4} - \dfrac{1}{y+4}$

The denominators are the same, so the difference is found by subtracting the two numerators and keeping the same (common) denominator.

$$\frac{5}{y+4} - \frac{1}{y+4} = \frac{5-1}{y+4} = \frac{4}{y+4}$$

13. $\dfrac{x}{x+y} + \dfrac{y}{x+y}$

The denominators are the same, so the sum is found by adding the two numerators and keeping the same (common) denominator.

$$\frac{x}{x+y} + \frac{y}{x+y} = \frac{x+y}{x+y} = 1$$

15. $\dfrac{5m}{m+1} - \dfrac{1+4m}{m+1}$

The denominators are the same, so the difference is found by subtracting the two numerators and keeping the same (common) denominator. Don't forget the parentheses on the second numerator.

$$\frac{5m}{m+1} - \frac{1+4m}{m+1} = \frac{5m - (1+4m)}{m+1}$$
$$= \frac{5m - 1 - 4m}{m+1}$$
$$= \frac{m-1}{m+1}$$

17. $\dfrac{a+b}{2} - \dfrac{a-b}{2}$

The denominators are the same, so the difference is found by subtracting the two numerators and keeping the same (common) denominator. Don't forget the parentheses on the second numerator.

$$\frac{a+b}{2} - \frac{a-b}{2} = \frac{(a+b) - (a-b)}{2}$$
$$= \frac{a+b-a+b}{2}$$
$$= \frac{2b}{2} = b$$

19. $\dfrac{x^2}{x+5} + \dfrac{5x}{x+5} = \dfrac{x^2+5x}{x+5}$ *Add numerators.*

$$= \frac{x(x+5)}{x+5}$$ *Factor numerator.*

$$= x$$ *Lowest terms*

21. $\dfrac{y^2-3y}{y+3} + \dfrac{-18}{y+3} = \dfrac{y^2-3y-18}{y+3}$

$$= \frac{(y-6)(y+3)}{y+3}$$
$$= y-6$$

23. To add or subtract rational expressions with the same denominators, combine the numerators and keep the same denominator. For example, $\dfrac{3x+2}{x-6} + \dfrac{-2x-8}{x-6} = \dfrac{x-6}{x-6}$. Then write in lowest terms. In this example, the sum simplifies to 1.

25. $\dfrac{z}{5} + \dfrac{1}{3}$

The LCD is 15. Now rewrite each rational expression as a fraction with the LCD as its denominator.

$$\frac{z}{5} \cdot \frac{3}{3} = \frac{3z}{15}$$
$$\frac{1}{3} \cdot \frac{5}{5} = \frac{5}{15}$$

Since the fractions now have a common denominator, add the numerators and use the LCD as the denominator of the sum.

$$\frac{z}{5} + \frac{1}{3} = \frac{3z}{15} + \frac{5}{15} = \frac{3z+5}{15}$$

27. $\dfrac{5}{7} - \dfrac{r}{2} = \dfrac{5}{7} \cdot \dfrac{2}{2} - \dfrac{r}{2} \cdot \dfrac{7}{7}$ *LCD = 14*

$$= \frac{10}{14} - \frac{7r}{14}$$
$$= \frac{10 - 7r}{14}$$

29. $-\dfrac{3}{4} - \dfrac{1}{2x} = -\dfrac{3 \cdot x}{4 \cdot x} - \dfrac{1 \cdot 2}{2x \cdot 2}$ *LCD = 4x*

$$= \frac{-3x - 2}{4x}$$

31. $\dfrac{6}{5x} + \dfrac{9}{2x} = \dfrac{6}{5x} \cdot \dfrac{2}{2} + \dfrac{9}{2x} \cdot \dfrac{5}{5}$ *LCD = 10x*

$$= \frac{12}{10x} + \frac{45}{10x}$$
$$= \frac{12 + 45}{10x} = \frac{57}{10x}$$

33. $\dfrac{x+1}{6} + \dfrac{3x+3}{9}$

First reduce the second fraction.

$$\dfrac{3x+3}{9} = \dfrac{3(x+1)}{9} = \dfrac{x+1}{3}$$

Now the LCD of $\dfrac{x+1}{6}$ and $\dfrac{x+1}{3}$ is 6. Thus,

$$\dfrac{x+1}{6} + \dfrac{x+1}{3} = \dfrac{x+1}{6} + \dfrac{x+1}{3}\cdot\dfrac{2}{2}$$
$$= \dfrac{x+1+2x+2}{6}$$
$$= \dfrac{3x+3}{6}$$
$$= \dfrac{3(x+1)}{6} = \dfrac{x+1}{2}.$$

35. $\dfrac{x+3}{3x} + \dfrac{2x+2}{4x} = \dfrac{x+3}{3x} + \dfrac{2(x+1)}{4x}$
$$= \dfrac{x+3}{3x} + \dfrac{x+1}{2x} \quad Reduce.$$
$$= \dfrac{x+3}{3x}\cdot\dfrac{2}{2} + \dfrac{x+1}{2x}\cdot\dfrac{3}{3}$$
$$\qquad\qquad LCD = 6x$$
$$= \dfrac{2x+6+3x+3}{6x}$$
$$= \dfrac{5x+9}{6x}$$

37. $\dfrac{7}{3p^2} - \dfrac{2}{p} = \dfrac{7}{3p^2} - \dfrac{2}{p}\cdot\dfrac{3p}{3p} \quad LCD = 3p^2$
$$= \dfrac{7-6p}{3p^2}$$

39. $\dfrac{1}{k+4} - \dfrac{2}{k} = \dfrac{1}{k+4}\cdot\dfrac{k}{k} - \dfrac{2}{k}\cdot\dfrac{k+4}{k+4} \quad LCD = k(k+4)$
$$= \dfrac{k}{k(k+4)} - \dfrac{2(k+4)}{k(k+4)}$$
$$= \dfrac{k-2k-8}{k(k+4)}$$
$$= \dfrac{-k-8}{k(k+4)}$$

41. $\dfrac{x}{x-2} + \dfrac{-8}{x^2-4}$
$$= \dfrac{x}{x-2} + \dfrac{-8}{(x+2)(x-2)}$$
$$= \dfrac{x}{x-2}\cdot\dfrac{x+2}{x+2} + \dfrac{-8}{(x+2)(x-2)}$$
$$\qquad\qquad LCD = (x+2)(x-2)$$
$$= \dfrac{x(x+2)-8}{(x+2)(x-2)}$$
$$= \dfrac{x^2+2x-8}{(x+2)(x-2)}$$
$$= \dfrac{(x+4)(x-2)}{(x+2)(x-2)} = \dfrac{x+4}{x+2}$$

43. $\dfrac{4m}{m^2+3m+2} + \dfrac{2m-1}{m^2+6m+5}$
$$= \dfrac{4m}{(m+2)(m+1)} + \dfrac{2m-1}{(m+1)(m+5)}$$
$$= \dfrac{4m(m+5)}{(m+2)(m+1)(m+5)} + \dfrac{(2m-1)(m+2)}{(m+1)(m+5)(m+2)}$$
$$\qquad LCD = (m+2)(m+1)(m+5)$$
$$= \dfrac{(4m^2+20m)+(2m^2+3m-2)}{(m+2)(m+1)(m+5)}$$
$$= \dfrac{6m^2+23m-2}{(m+2)(m+1)(m+5)}$$

45. $\dfrac{4y}{y^2-1} - \dfrac{5}{y^2+2y+1}$
$$= \dfrac{4y}{(y+1)(y-1)} - \dfrac{5}{(y+1)(y+1)}$$
$$= \dfrac{4y(y+1)}{(y+1)^2(y-1)} - \dfrac{5(y-1)}{(y+1)^2(y-1)}$$
$$\qquad\qquad LCD = (y+1)^2(y-1)$$
$$= \dfrac{(4y^2+4y)-(5y-5)}{(y+1)^2(y-1)}$$
$$= \dfrac{4y^2-y+5}{(y+1)^2(y-1)}$$

47. $\dfrac{t}{t+2} + \dfrac{5-t}{t} - \dfrac{4}{t^2+2t}$
$$= \dfrac{t}{t+2} + \dfrac{5-t}{t} - \dfrac{4}{t(t+2)}$$
$$= \dfrac{t}{t+2}\cdot\dfrac{t}{t} + \dfrac{5-t}{t}\cdot\dfrac{t+2}{t+2}$$
$$\quad - \dfrac{4}{t(t+2)} \quad LCD = t(t+2)$$
$$= \dfrac{t\cdot t + (5-t)(t+2) - 4}{t(t+2)}$$
$$= \dfrac{t^2+5t+10-t^2-2t-4}{t(t+2)}$$
$$= \dfrac{3t+6}{t(t+2)}$$
$$= \dfrac{3(t+2)}{t(t+2)} = \dfrac{3}{t}$$

49. $\dfrac{10}{m-2} + \dfrac{5}{2-m}$

Since

$$2-m = -1(m-2),$$

either $m-2$ or $2-m$ could be used as the LCD.

51. $\dfrac{4}{x-5} + \dfrac{6}{5-x}$

The two denominators, $x - 5$ and $5 - x$, are opposites of each other, so either one may be used as the common denominator. We will work the exercise both ways and compare the answers.

$$\dfrac{4}{x-5} + \dfrac{6}{5-x} = \dfrac{4}{x-5} + \dfrac{6(-1)}{(5-x)(-1)}$$

$$LCD = x - 5$$

$$= \dfrac{4}{x-5} + \dfrac{-6}{x-5}$$

$$= \dfrac{-2}{x-5}$$

$$\dfrac{4}{x-5} + \dfrac{6}{5-x} = \dfrac{4(-1)}{(x-5)(-1)} + \dfrac{6}{5-x}$$

$$LCD = 5 - x$$

$$= \dfrac{-4}{5-x} + \dfrac{6}{5-x}$$

$$= \dfrac{2}{5-x}$$

The two answers are equivalent, since

$$\dfrac{-2}{x-5} \cdot \dfrac{-1}{-1} = \dfrac{2}{5-x}.$$

53. $\dfrac{-1}{1-y} - \dfrac{4y-3}{y-1}$

The LCD is either $1 - y$ or $y - 1$. We'll use $y - 1$.

$$\dfrac{-1}{1-y} - \dfrac{4y-3}{y-1} = \dfrac{-1 \cdot -1}{-1 \cdot (1-y)} - \dfrac{4y-3}{y-1}$$

$$= \dfrac{1 - (4y-3)}{y-1}$$

$$= \dfrac{1 - 4y + 3}{y-1}$$

$$= \dfrac{4 - 4y}{y-1}$$

$$= \dfrac{4(1-y)}{y-1} = -4$$

55. $\dfrac{2}{x-y^2} + \dfrac{7}{y^2-x}$

$LCD = x - y^2$ or $y^2 - x$

We will use $x - y^2$.

$$\dfrac{2}{x-y^2} + \dfrac{7}{y^2-x}$$

$$= \dfrac{2}{x-y^2} + \dfrac{-1(7)}{-1(y^2-x)}$$

$$= \dfrac{2}{x-y^2} + \dfrac{-7}{-y^2+x}$$

$$= \dfrac{2}{x-y^2} + \dfrac{-7}{x-y^2}$$

$$= \dfrac{2 + (-7)}{x-y^2} = \dfrac{-5}{x-y^2}$$

If $y^2 - x$ is used as the LCD, we will obtain the equivalent answer

$$\dfrac{5}{y^2-x}.$$

57. $\dfrac{x}{5x-3y} - \dfrac{y}{3y-5x}$

$LCD = 5x - 3y$ or $3y - 5x$

We will use $5x - 3y$.

$$\dfrac{x}{5x-3y} - \dfrac{y}{3y-5x}$$

$$= \dfrac{x}{5x-3y} - \dfrac{-1(y)}{-1(3y-5x)}$$

$$= \dfrac{x}{5x-3y} - \dfrac{-y}{-3y+5x}$$

$$= \dfrac{x}{5x-3y} - \dfrac{-y}{5x-3y}$$

$$= \dfrac{x - (-y)}{5x-3y} = \dfrac{x+y}{5x-3y}$$

If $3y - 5x$ is used as the LCD, we will obtain the equivalent answer

$$\dfrac{-x-y}{3y-5x}.$$

59. $\dfrac{3}{4p-5} + \dfrac{9}{5-4p}$

$LCD = 4p - 5$ or $5 - 4p$

We will use $4p - 5$.

$$\dfrac{3}{4p-5} + \dfrac{9}{5-4p}$$

$$= \dfrac{3}{4p-5} + \dfrac{-1(9)}{-1(5-4p)}$$

$$= \dfrac{3}{4p-5} + \dfrac{-9}{-5+4p}$$

$$= \dfrac{3}{4p-5} + \dfrac{-9}{4p-5}$$

$$= \dfrac{3 + (-9)}{4p-5} = \dfrac{-6}{4p-5}$$

If $5 - 4p$ is used as the LCD, we will obtain the equivalent answer

$$\dfrac{6}{5-4p}.$$

61. $\dfrac{2m}{m-n} - \dfrac{5m+n}{2m-2n}$

$= \dfrac{2m}{m-n} - \dfrac{5m+n}{2(m-n)}$ *Factor second denominator.*

$= \dfrac{2m}{m-n} \cdot \dfrac{2}{2} - \dfrac{5m+n}{2(m-n)}$ *LCD = 2(m – n)*

$= \dfrac{4m - (5m+n)}{2(m-n)}$

$= \dfrac{4m - 5m - n}{2(m-n)}$

$= \dfrac{-m - n}{2(m-n)}$

$= \dfrac{-(m+n)}{2(m-n)}$

63. $\dfrac{5}{x^2-9} - \dfrac{x+2}{x^2+4x+3}$

To find the LCD, factor the denominators.

$$x^2 - 9 = (x+3)(x-3)$$
$$x^2 + 4x + 3 = (x+3)(x+1)$$

The LCD is $(x+3)(x-3)(x+1)$.

$\dfrac{5}{x^2-9} - \dfrac{x+2}{x^2+4x+3}$

$= \dfrac{5 \cdot (x+1)}{(x+3)(x-3) \cdot (x+1)}$

$\quad - \dfrac{(x+2) \cdot (x-3)}{(x+3)(x+1) \cdot (x-3)}$

$= \dfrac{5x + 5}{(x+3)(x-3)(x+1)}$

$\quad - \dfrac{x^2 - x - 6}{(x+3)(x+1)(x-3)}$

$= \dfrac{(5x+5) - (x^2 - x - 6)}{(x+3)(x-3)(x+1)}$

$= \dfrac{5x + 5 - x^2 + x + 6}{(x+3)(x-3)(x+1)}$

$= \dfrac{-x^2 + 6x + 11}{(x+3)(x-3)(x+1)}$

65. $\dfrac{2q+1}{3q^2+10q-8} - \dfrac{3q+5}{2q^2+5q-12}$

$= \dfrac{2q+1}{(3q-2)(q+4)} - \dfrac{3q+5}{(2q-3)(q+4)}$

$= \dfrac{(2q+1) \cdot (2q-3)}{(3q-2)(q+4) \cdot (2q-3)}$

$\quad - \dfrac{(3q+5) \cdot (3q-2)}{(2q-3)(q+4) \cdot (3q-2)}$

     LCD = (3q – 2)(q + 4)(2q – 3)

$= \dfrac{(4q^2 - 4q - 3) - (9q^2 + 9q - 10)}{(3q-2)(q+4)(2q-3)}$

$= \dfrac{4q^2 - 4q - 3 - 9q^2 - 9q + 10}{(3q-2)(q+4)(2q-3)}$

$= \dfrac{-5q^2 - 13q + 7}{(3q-2)(q+4)(2q-3)}$

67. $\dfrac{4}{r^2-r} + \dfrac{6}{r^2+2r} - \dfrac{1}{r^2+r-2}$

$= \dfrac{4}{r(r-1)} + \dfrac{6}{r(r+2)} - \dfrac{1}{(r+2)(r-1)}$

$= \dfrac{4 \cdot (r+2)}{r(r-1) \cdot (r+2)} + \dfrac{6 \cdot (r-1)}{r(r+2) \cdot (r-1)}$

$\quad - \dfrac{1 \cdot r}{r \cdot (r+2)(r-1)}$

     LCD = r(r + 2)(r – 1)

$= \dfrac{4r + 8 + 6r - 6 - r}{r(r+2)(r-1)}$

$= \dfrac{9r + 2}{r(r+2)(r-1)}$

69. $\dfrac{x+3y}{x^2+2xy+y^2} + \dfrac{x-y}{x^2+4xy+3y^2}$

$= \dfrac{x+3y}{(x+y)(x+y)} + \dfrac{x-y}{(x+3y)(x+y)}$

$= \dfrac{(x+3y) \cdot (x+3y)}{(x+y)(x+y) \cdot (x+3y)}$

$\quad + \dfrac{(x-y) \cdot (x+y)}{(x+3y)(x+y) \cdot (x+y)}$

     LCD = (x + y)(x + y)(x + 3y)

$= \dfrac{(x^2 + 6xy + 9y^2) + (x^2 - y^2)}{(x+y)(x+y)(x+3y)}$

$= \dfrac{2x^2 + 6xy + 8y^2}{(x+y)(x+y)(x+3y)}$

$= \dfrac{2(x^2 + 3xy + 4y^2)}{(x+y)(x+y)(x+3y)}$

or $\dfrac{2(x^2 + 3xy + 4y^2)}{(x+y)^2(x+3y)}$

71. $\dfrac{r+y}{18r^2 + 9ry - 2y^2} + \dfrac{3r - y}{36r^2 - y^2}$

Factor the first denominator by grouping. Find integers whose product is $18(-2) = -36$ and whose sum is 9. The integers are 12 and -3.

$$18r^2 + 9ry - 2y^2$$
$$= 18r^2 + 12ry - 3ry - 2y^2$$
$$= 6r(3r + 2y) - y(3r + 2y)$$
$$= (3r + 2y)(6r - y)$$

Factor the second denominator.

$$\dfrac{r+y}{(3r+2y)(6r-y)} + \dfrac{3r-y}{(6r-y)(6r+y)}$$

Rewrite fractions with the LCD, $(3r + 2y)(6r - y)(6r + y)$.

$$= \dfrac{(r+y)\cdot(6r+y)}{(3r+2y)(6r-y)\cdot(6r+y)}$$

$$+ \dfrac{(3r-y)\cdot(3r+2y)}{(6r-y)(6r+y)\cdot(3r+2y)}$$

$$= \dfrac{6r^2 + 7ry + y^2}{(3r+2y)(6r-y)(6r+y)}$$

$$+ \dfrac{9r^2 + 3ry - 2y^2}{(3r+2y)(6r-y)(6r+y)}$$

$$= \dfrac{6r^2 + 7ry + y^2 + 9r^2 + 3ry - 2y^2}{(3r+2y)(6r-y)(6r+y)}$$

$$= \dfrac{15r^2 + 10ry - y^2}{(3r+2y)(6r-y)(6r+y)}$$

73. (a) $P = 2L + 2W$

$$= 2\left(\dfrac{3k+1}{10}\right) + 2\left(\dfrac{5}{6k+2}\right)$$

$$= 2\left(\dfrac{3k+1}{2\cdot 5}\right) + 2\left(\dfrac{5}{2(3k+1)}\right)$$

$$= \dfrac{3k+1}{5} + \dfrac{5}{3k+1}$$

To add the two fractions on the right, use $5(3k + 1)$ as the LCD.

$$P = \dfrac{(3k+1)(3k+1)}{5(3k+1)} + \dfrac{(5)(5)}{5(3k+1)}$$

$$= \dfrac{(3k+1)(3k+1) + (5)(5)}{5(3k+1)}$$

$$= \dfrac{9k^2 + 6k + 1 + 25}{5(3k+1)}$$

$$= \dfrac{9k^2 + 6k + 26}{5(3k+1)}$$

(b) $A = L \cdot W$

$$A = \dfrac{3k+1}{10} \cdot \dfrac{5}{6k+2}$$

$$= \dfrac{3k+1}{5\cdot 2} \cdot \dfrac{5}{2(3k+1)}$$

$$= \dfrac{1}{2\cdot 2} = \dfrac{1}{4}$$

75. $\dfrac{1010}{49(101-x)} - \dfrac{10}{49}$

$$= \dfrac{1010}{49(101-x)} - \dfrac{10(101-x)}{49(101-x)}$$

$$= \dfrac{1010 - 1010 + 10x}{49(101-x)}$$

$$= \dfrac{10x}{49(101-x)}$$

77. $\dfrac{\frac{5}{6}+\frac{7}{6}}{\frac{2}{3}-\frac{1}{3}} = \dfrac{\frac{12}{6}}{\frac{1}{3}} = 2 \div \dfrac{1}{3} = 2\cdot 3 = 6$

79. $\dfrac{\frac{3}{2}-\frac{5}{4}}{\frac{7}{4}+\frac{1}{3}} = \dfrac{\frac{6}{4}-\frac{5}{4}}{\frac{21}{12}+\frac{4}{12}} = \dfrac{\frac{1}{4}}{\frac{25}{12}} = \dfrac{1}{4} \div \dfrac{25}{12}$

$$= \dfrac{1}{\underset{1}{\cancel{4}}} \cdot \dfrac{\overset{3}{\cancel{12}}}{25} = \dfrac{3}{25}$$

6.5   Complex Fractions

1. (a) The LCD of $\frac{1}{2}$ and $\frac{1}{3}$ is $2\cdot 3 = 6$. The simplified form of the numerator is

$$\dfrac{1}{2} - \dfrac{1}{3} = \dfrac{3}{6} - \dfrac{2}{6} = \dfrac{1}{6}.$$

(b) The LCD of $\frac{5}{6}$ and $\frac{1}{12}$ is 12 since 12 is a multiple of 6. The simplified form of the denominator is

$$\dfrac{5}{6} - \dfrac{1}{12} = \dfrac{10}{12} - \dfrac{1}{12} = \dfrac{9}{12} = \dfrac{3}{4}.$$

(c) $\dfrac{\frac{1}{6}}{\frac{3}{4}} = \dfrac{1}{6} \div \dfrac{3}{4}$

(d) $\dfrac{1}{6} \div \dfrac{3}{4} = \dfrac{1}{6} \cdot \dfrac{4}{3}$

$$= \dfrac{2\cdot 2}{2\cdot 3\cdot 3} = \dfrac{2}{9}$$

3. $\dfrac{3-\frac{1}{2}}{2-\frac{1}{4}} = \dfrac{-3+\frac{1}{2}}{-2+\frac{1}{4}}$

Choice **D** is equivalent to the given fraction. Each term of the numerator and denominator has been multiplied by -1. Since $\frac{-1}{-1} = 1$, the fraction has been multiplied by the identity element, so its value is unchanged.

5. Method 1 indicates to write the complex fraction as a division problem, and then perform the division. For example, to simplify $\dfrac{\frac{1}{2}}{\frac{2}{3}}$, write $\frac{1}{2} \div \frac{2}{3}$.

Then simplify as $\frac{1}{2} \cdot \frac{3}{2} = \frac{3}{4}$.

In Exercises 7–36, either Method 1 or Method 2 can be used to simplify each complex fraction. Only one method will be shown for each exercise.

7. To use Method 1, divide the numerator of the complex fraction by the denominator.

$$\frac{-\frac{4}{3}}{\frac{2}{9}} = -\frac{4}{3} \div \frac{2}{9}$$

$$= -\frac{4}{3} \cdot \frac{9}{2} = -\frac{36}{6} = -6$$

9. To use Method 2, multiply the numerator and denominator of the complex fraction by the LCD, y^2.

$$\frac{\frac{x}{y^2}}{\frac{x^2}{y}} = \frac{y^2\left(\dfrac{x}{y^2}\right)}{y^2\left(\dfrac{x^2}{y}\right)}$$

$$= \frac{x}{yx^2} = \frac{1}{xy}$$

11. $\dfrac{\frac{4a^4b^3}{3a}}{\frac{2ab^4}{b^2}} = \dfrac{4a^4b^3}{3a} \div \dfrac{2ab^4}{b^2}$ *Method 1*

$$= \frac{4a^4b^3}{3a} \cdot \frac{b^2}{2ab^4}$$

$$= \frac{4a^4b^3 \cdot b^2}{3a \cdot 2ab^4}$$

$$= \frac{4a^4b^5}{6a^2b^4}$$

$$= \frac{2a^2b}{3}$$

13. To use Method 2, multiply the numerator and denominator of the complex fraction by the LCD, $3m$.

$$\frac{\frac{m+2}{3}}{\frac{m-4}{m}} = \frac{3m\left(\dfrac{m+2}{3}\right)}{3m\left(\dfrac{m-4}{m}\right)}$$

$$= \frac{m(m+2)}{3(m-4)}$$

15. $\dfrac{\frac{2}{x}-3}{\frac{2-3x}{2}} = \dfrac{2x\left(\dfrac{2}{x}-3\right)}{2x\left(\dfrac{2-3x}{2}\right)}$ *Method 2; LCD = 2x*

$$= \frac{2x\left(\dfrac{2}{x}\right) - 2x(3)}{x(2-3x)}$$

$$= \frac{4-6x}{x(2-3x)}$$

$$= \frac{2(2-3x)}{x(2-3x)}$$ *Factor.*

$$= \frac{2}{x}$$ *Lowest terms*

17. $\dfrac{\frac{1}{x}+x}{\frac{x^2+1}{8}} = \dfrac{8x\left(\dfrac{1}{x}+x\right)}{8x\left(\dfrac{x^2+1}{8}\right)}$ *Method 2; LCD = 8x*

$$= \frac{8+8x^2}{x(x^2+1)}$$ *Distributive property*

$$= \frac{8(1+x^2)}{x(x^2+1)}$$ *Factor.*

$$= \frac{8}{x}$$ *Lowest terms*

19. $\dfrac{a-\frac{5}{a}}{a+\frac{1}{a}} = \dfrac{a\left(a-\dfrac{5}{a}\right)}{a\left(a+\dfrac{1}{a}\right)}$ *Method 2; LCD = a*

$$= \frac{a^2-5}{a^2+1}$$

21. $\dfrac{\frac{5}{8}+\frac{2}{3}}{\frac{7}{3}-\frac{1}{4}} = \dfrac{24\left(\dfrac{5}{8}+\dfrac{2}{3}\right)}{24\left(\dfrac{7}{3}-\dfrac{1}{4}\right)}$ *Method 2; LCD = 24*

$$= \frac{24\left(\dfrac{5}{8}\right)+24\left(\dfrac{2}{3}\right)}{24\left(\dfrac{7}{3}\right)-24\left(\dfrac{1}{4}\right)}$$

$$= \frac{15+16}{56-6} = \frac{31}{50}$$

23. $\dfrac{\dfrac{1}{x^2} + \dfrac{1}{y^2}}{\dfrac{1}{x} - \dfrac{1}{y}}$

$= \dfrac{x^2y^2\left(\dfrac{1}{x^2} + \dfrac{1}{y^2}\right)}{x^2y^2\left(\dfrac{1}{x} - \dfrac{1}{y}\right)}$ *Method 2;*
LCD = x^2y^2

$= \dfrac{x^2y^2\left(\dfrac{1}{x^2}\right) + x^2y^2\left(\dfrac{1}{y^2}\right)}{x^2y^2\left(\dfrac{1}{x}\right) - x^2y^2\left(\dfrac{1}{y}\right)}$

$= \dfrac{y^2 + x^2}{xy^2 - x^2y} = \dfrac{y^2 + x^2}{xy(y - x)}$

25. $\dfrac{\dfrac{2}{p^2} - \dfrac{3}{5p}}{\dfrac{4}{p} + \dfrac{1}{4p}} = \dfrac{20p^2\left(\dfrac{2}{p^2} - \dfrac{3}{5p}\right)}{20p^2\left(\dfrac{4}{p} + \dfrac{1}{4p}\right)}$ *Method 2;*
LCD = $20p^2$

$= \dfrac{20p^2\left(\dfrac{2}{p^2}\right) - 20p^2\left(\dfrac{3}{5p}\right)}{20p^2\left(\dfrac{4}{p}\right) + 20p^2\left(\dfrac{1}{4p}\right)}$

$= \dfrac{40 - 12p}{80p + 5p}$

$= \dfrac{40 - 12p}{85p}$

27. $\dfrac{\dfrac{5}{x^2y} - \dfrac{2}{xy^2}}{\dfrac{3}{x^2y^2} + \dfrac{4}{xy}}$

$= \dfrac{x^2y^2\left(\dfrac{5}{x^2y} - \dfrac{2}{xy^2}\right)}{x^2y^2\left(\dfrac{3}{x^2y^2} + \dfrac{4}{xy}\right)}$ *Method 2;*
LCD = x^2y^2

$= \dfrac{x^2y^2\left(\dfrac{5}{x^2y}\right) - x^2y^2\left(\dfrac{2}{xy^2}\right)}{x^2y^2\left(\dfrac{3}{x^2y^2}\right) + x^2y^2\left(\dfrac{4}{xy}\right)}$

$= \dfrac{5y - 2x}{3 + 4xy}$

29. $\dfrac{\dfrac{1}{4} - \dfrac{1}{a^2}}{\dfrac{1}{2} + \dfrac{1}{a}}$

$= \dfrac{4a^2\left(\dfrac{1}{4} - \dfrac{1}{a^2}\right)}{4a^2\left(\dfrac{1}{2} + \dfrac{1}{a}\right)}$ *Method 2;*
LCD = $4a^2$

$= \dfrac{a^2 - 4}{2a^2 + 4a}$ *Distributive*
property

$= \dfrac{(a - 2)(a + 2)}{2a(a + 2)}$ *Factor numerator*
and denominator.

$= \dfrac{a - 2}{2a}$ *Fundamental*
property

31. $\dfrac{\dfrac{1}{z + 5}}{\dfrac{4}{z^2 - 25}}$

$= \dfrac{1}{z + 5} \div \dfrac{4}{z^2 - 25}$ *Method 1*

$= \dfrac{1}{z + 5} \cdot \dfrac{z^2 - 25}{4}$ *Multiply by reciprocal.*

$= \dfrac{1 \cdot (z^2 - 25)}{(z + 5) \cdot 4}$ *Multiply.*

$= \dfrac{(z + 5)(z - 5)}{(z + 5) \cdot 4}$ *Factor numerator.*

$= \dfrac{z - 5}{4}$ *Fundamental*
property

33. $\dfrac{\dfrac{1}{m + 1} - 1}{\dfrac{1}{m + 1} + 1}$

$= \dfrac{(m + 1)\left(\dfrac{1}{m + 1} - 1\right)}{(m + 1)\left(\dfrac{1}{m + 1} + 1\right)}$ *Method 2;*
LCD = $m + 1$

$= \dfrac{1 - 1(m + 1)}{1 + 1(m + 1)}$ *Distributive*
property

$= \dfrac{1 - m - 1}{1 + m + 1}$ *Distributive*
property

$= \dfrac{-m}{m + 2}$

35. $\dfrac{\dfrac{1}{m-1}+\dfrac{2}{m+2}}{\dfrac{2}{m+2}-\dfrac{1}{m-3}}$

$=\dfrac{(m-1)(m+2)(m-3)\left(\dfrac{1}{m-1}+\dfrac{2}{m+2}\right)}{(m-1)(m+2)(m-3)\left(\dfrac{2}{m+2}-\dfrac{1}{m-3}\right)}$

Method 2;
LCD = (m – 1)(m + 2)(m – 3)

$=\dfrac{(m+2)(m-3)+2(m-1)(m-3)}{2(m-1)(m-3)-(m-1)(m+2)}$

Distributive property

$=\dfrac{(m-3)[(m+2)+2(m-1)]}{(m-1)[2(m-3)-(m+2)]}$

Factor out m – 3 in numerator
and m – 1 in denominator.

$=\dfrac{(m-3)[m+2+2m-2]}{(m-1)[2m-6-m-2]}$

Distributive property

$=\dfrac{3m(m-3)}{(m-1)(m-8)}$ *Combine like terms.*

37. In a fraction, the fraction bar represents division. For example, $\frac{3}{5}$ can be read "3 divided by 5."

39. "The sum of $\frac{3}{8}$ and $\frac{5}{6}$, divided by 2" is written

$$\dfrac{\frac{3}{8}+\frac{5}{6}}{2}.$$

40. $\dfrac{\frac{3}{8}+\frac{5}{6}}{2}=\dfrac{\frac{9}{24}+\frac{20}{24}}{2}$ *Method 1*

$=\dfrac{\frac{29}{24}}{\frac{2}{1}}=\dfrac{29}{24}\cdot\dfrac{1}{2}=\dfrac{29}{48}$

41. $\dfrac{\frac{3}{8}+\frac{5}{6}}{2}=\dfrac{24\left(\frac{3}{8}+\frac{5}{6}\right)}{24(2)}$ *Method 2;*
LCD = 24

$=\dfrac{24\left(\frac{3}{8}\right)+24\left(\frac{5}{6}\right)}{24(2)}$

$=\dfrac{9+20}{48}=\dfrac{29}{48}$

42. Method 2 is usually shorter for more complex problems because the problem can be worked without adding and subtracting rational expressions, which can be complicated and time-consuming.

43. $1+\dfrac{1}{1+\dfrac{1}{1+1}}=1+\dfrac{1}{1+\dfrac{1}{2}}$

$=1+\dfrac{1}{\frac{2}{2}+\frac{1}{2}}$

$=1+\dfrac{1}{\frac{3}{2}}$

$=1+1\cdot\frac{2}{3}$

$=1+\frac{2}{3}$

$=\frac{3}{3}+\frac{2}{3}=\frac{5}{3}$

45. $7-\dfrac{3}{5+\dfrac{2}{4-2}}=7-\dfrac{3}{5+\dfrac{2}{2}}$

$=7-\dfrac{3}{5+1}$

$=7-\dfrac{3}{6}$

$=7-\dfrac{1}{2}$

$=\dfrac{14}{2}-\dfrac{1}{2}=\dfrac{13}{2}$

47. $r+\dfrac{r}{4-\dfrac{2}{6+2}}=r+\dfrac{r}{4-\dfrac{2}{8}}$

$=r+\dfrac{r}{4-\frac{1}{4}}$

$=r+\dfrac{r}{\frac{16}{4}-\frac{1}{4}}$

$=r+\dfrac{r}{\frac{15}{4}}$

$=r+r\cdot\dfrac{4}{15}$

$=r+\dfrac{4r}{15}$

$=\dfrac{15r}{15}+\dfrac{4r}{15}$

$=\dfrac{19r}{15}$

49. $9\left(\dfrac{4x}{3}+\dfrac{2}{9}\right)=9\left(\dfrac{4x}{3}\right)+9\left(\dfrac{2}{9}\right)$

$=3(4x)+2$

$=12x+2$

51. $-12\left(\dfrac{11p^2}{3}-\dfrac{9p}{4}\right)$

$=-12\left(\dfrac{11p^2}{3}\right)-12\left(-\dfrac{9p}{4}\right)$

$=-4\left(11p^2\right)+3(9p)$

$=-44p^2+27p$

53. $3x + 5 = 7x + 3$

$\qquad 5 = 4x + 3$ *Subtract 3x.*

$\qquad 2 = 4x$ *Subtract 3.*

$\qquad x = \frac{2}{4} = \frac{1}{2}$ *Divide by 4.*

The solution set is $\left\{ \frac{1}{2} \right\}$.

55. $6(z - 3) + 5 = 8z - 3$

$\qquad 6z - 18 + 5 = 8z - 3$

$\qquad 6z - 13 = 8z - 3$

$\qquad -13 = 2z - 3$

$\qquad -10 = 2z$

$\qquad z = \frac{-10}{2} = -5$

The solution set is $\{-5\}$.

6.6   Solving Equations with Rational Expressions

1. $\frac{7}{8}x + \frac{1}{5}x$ is the sum of two terms, so it is an *expression* to be simplified. Simplify by finding the LCD, writing each coefficient with this LCD, and combining like terms.

$$\frac{7}{8}x + \frac{1}{5}x = \frac{35}{40}x + \frac{8}{40}x \quad LCD = 40$$

$$\qquad\qquad = \frac{43}{40}x \qquad\qquad \begin{array}{l}\textit{Combine}\\\textit{like terms.}\end{array}$$

3. $\frac{7}{8}x + \frac{1}{5}x = 1$ has an equals sign, so this is an *equation* to be solved. Use the multiplication property of equality to clear fractions. The LCD is 40.

$$\frac{7}{8}x + \frac{1}{5}x = 1$$

$$40\left(\frac{7}{8}x + \frac{1}{5}x\right) = 40 \cdot 1 \quad \textit{Multiply by 40.}$$

$$40\left(\frac{7}{8}x\right) + 40\left(\frac{1}{5}x\right) = 40 \cdot 1 \quad \begin{array}{l}\textit{Distributive}\\\textit{property}\end{array}$$

$$35x + 8x = 40 \qquad \textit{Multiply.}$$

$$43x = 40 \qquad \begin{array}{l}\textit{Combine}\\\textit{like terms.}\end{array}$$

$$x = \frac{40}{43} \qquad \textit{Divide by 43.}$$

The solution set is $\left\{ \frac{40}{43} \right\}$.

5. $\frac{3}{5}x - \frac{7}{10}x$ is the difference of two terms, so it is an *expression* to be simplified.

$$\frac{3}{5}x - \frac{7}{10}x = \frac{6}{10}x - \frac{7}{10}x \quad LCD = 10$$

$$\qquad\qquad = -\frac{1}{10}x \qquad \begin{array}{l}\textit{Combine}\\\textit{like terms.}\end{array}$$

7. $\frac{3}{5}x - \frac{7}{10}x = 1$ has an equals sign, so it is an *equation* to be solved.

$$\frac{3}{5}x - \frac{7}{10}x = 1$$

$$10\left(\frac{3}{5}x - \frac{7}{10}x\right) = 10 \cdot 1 \quad LCD = 10$$

$$10\left(\frac{3}{5}x\right) - 10\left(\frac{7}{10}x\right) = 10 \cdot 1 \quad \begin{array}{l}\textit{Distributive}\\\textit{property}\end{array}$$

$$6x - 7x = 10 \qquad \textit{Multiply.}$$

$$-x = 10 \qquad \begin{array}{l}\textit{Combine}\\\textit{like terms.}\end{array}$$

$$x = -10 \quad \textit{Divide by –1.}$$

The solution set is $\{-10\}$.

9. $\frac{3}{x + 2} - \frac{5}{x} = 1$

The denominators, $x + 2$ and x, are equal to 0 for the values -2 and 0, respectively. Thus, $x \neq -2, 0$.

11. $\frac{-1}{(x + 3)(x - 4)} = \frac{1}{2x + 1}$

The denominators, $(x + 3)(x - 4)$ and $2x + 1$, are equal to 0 for the values -3, 4, and $-\frac{1}{2}$, respectively. Thus, $x \neq -3, 4, -\frac{1}{2}$.

13. $\frac{4}{x^2 + 8x - 9} + \frac{1}{x^2 - 4} = 0$

The denominators, $x^2 + 8x - 9 = (x + 9)(x - 1)$ and $x^2 - 4 = (x + 2)(x - 2)$, are equal to 0 for the values -9, 1, -2, and 2, respectively. Thus, $x \neq -9, 1, -2, 2$.

15.   We cannot solve

$$\frac{2}{3x} + \frac{1}{5x}$$

because it is an expression, not an equation. An expression cannot be solved. Only equations and inequalities are "solved."

Note: In Exercises 17–70, all proposed solutions should be checked by substituting in the original equation. It is essential to determine whether a proposed solution will make any denominator in the original equation equal to zero. Full checks will be shown here for only a few of the exercises.

17. $\frac{5}{m} - \frac{3}{m} = 8$

Multiply each side by the LCD, m.

$$m\left(\frac{5}{m} - \frac{3}{m}\right) = m \cdot 8$$

Use the distributive property to remove parentheses; then solve.

$$m\left(\frac{5}{m}\right) - m\left(\frac{3}{m}\right) = 8m$$
$$5 - 3 = 8m$$
$$2 = 8m$$
$$m = \frac{2}{8} = \frac{1}{4}$$

Check this proposed solution by replacing m with $\frac{1}{4}$ in the original equation.

$$\frac{5}{\frac{1}{4}} - \frac{3}{\frac{1}{4}} = 8 \text{ ?} \quad \textit{Let } m = \frac{1}{4}.$$
$$5 \cdot 4 - 3 \cdot 4 = 8 \text{ ?} \quad \begin{array}{l}\textit{Multiply by}\\ \textit{reciprocals.}\end{array}$$
$$20 - 12 = 8 \text{ ?}$$
$$8 = 8 \quad \textit{True}$$

Thus, the solution set is $\left\{\frac{1}{4}\right\}$.

19.
$$\frac{5}{y} + 4 = \frac{2}{y}$$
$$y\left(\frac{5}{y} + 4\right) = y\left(\frac{2}{y}\right) \quad \begin{array}{l}\textit{Multiply by}\\ \textit{LCD, y.}\end{array}$$
$$y\left(\frac{5}{y}\right) + y(4) = y\left(\frac{2}{y}\right) \quad \begin{array}{l}\textit{Distributive}\\ \textit{property}\end{array}$$
$$5 + 4y = 2$$
$$4y = -3$$
$$y = -\frac{3}{4}$$

Check $y = -\frac{3}{4}: -\frac{8}{3} = -\frac{8}{3}$ *True*

Thus, the solution set is $\left\{-\frac{3}{4}\right\}$.

21.
$$\frac{3x}{5} - 6 = x$$
$$5\left(\frac{3x}{5} - 6\right) = 5(x) \quad \begin{array}{l}\textit{Multiply by}\\ \textit{LCD, 5.}\end{array}$$
$$5\left(\frac{3x}{5}\right) - 5(6) = 5x \quad \begin{array}{l}\textit{Distributive}\\ \textit{property}\end{array}$$
$$3x - 30 = 5x$$
$$-30 = 2x$$
$$-15 = x$$

Check $x = -15: -15 = -15$ *True*

Thus, the solution set is $\{-15\}$.

23.
$$\frac{4m}{7} + m = 11$$
$$7\left(\frac{4m}{7} + m\right) = 7(11) \quad \begin{array}{l}\textit{Multiply by}\\ \textit{LCD, 7.}\end{array}$$
$$7\left(\frac{4m}{7}\right) + 7(m) = 77 \quad \begin{array}{l}\textit{Distributive}\\ \textit{property}\end{array}$$
$$4m + 7m = 77$$
$$11m = 77$$
$$m = 7$$

Check $m = 7: 11 = 11$ *True*

Thus, the solution set is $\{7\}$.

25.
$$\frac{z - 1}{4} = \frac{z + 3}{3}$$
$$12\left(\frac{z - 1}{4}\right) = 12\left(\frac{z + 3}{3}\right) \quad \begin{array}{l}\textit{Multiply by}\\ \textit{LCD, 12.}\end{array}$$
$$3(z - 1) = 4(z + 3)$$
$$3z - 3 = 4z + 12 \quad \textit{Dist. prop.}$$
$$-15 = z$$

Check $z = -15: -4 = -4$ *True*

Thus, the solution set is $\{-15\}$.

27.
$$\frac{3p + 6}{8} = \frac{3p - 3}{16}$$
$$16\left(\frac{3p + 6}{8}\right) = 16\left(\frac{3p - 3}{16}\right) \quad \begin{array}{l}\textit{Multiply by}\\ \textit{LCD, 16.}\end{array}$$
$$2(3p + 6) = 3p - 3$$
$$6p + 12 = 3p - 3 \quad \textit{Dist. prop.}$$
$$3p = -15$$
$$p = -5$$

Check $p = -5: -\frac{9}{8} = -\frac{9}{8}$ *True*

Thus, the solution set is $\{-5\}$.

29.
$$\frac{2x + 3}{x} = \frac{3}{2}$$
$$2x\left(\frac{2x + 3}{x}\right) = 2x\left(\frac{3}{2}\right) \quad \begin{array}{l}\textit{Multiply by}\\ \textit{LCD, 2x.}\end{array}$$
$$2(2x + 3) = 3x$$
$$4x + 6 = 3x \quad \textit{Dist. prop.}$$
$$x = -6$$

Check $x = -6: \frac{3}{2} = \frac{3}{2}$ *True*

Thus, the solution set is $\{-6\}$.

31.
$$\frac{k}{k - 4} - 5 = \frac{4}{k - 4}$$
$$(k - 4)\left(\frac{k}{k - 4} - 5\right) = (k - 4)\left(\frac{4}{k - 4}\right)$$
$$\textit{Multiply by LCD, k - 4.}$$
$$(k - 4)\left(\frac{k}{k - 4}\right) - 5(k - 4) = 4 \quad \begin{array}{l}\textit{Distributive}\\ \textit{property}\end{array}$$
$$k - 5k + 20 = 4$$
$$-4k = -16$$
$$k = 4$$

The proposed solution is 4. However, 4 cannot be a solution because it makes the denominator $k - 4$ equal 0. Therefore, the solution set is $\emptyset$.

33.
$$\frac{q+2}{3} + \frac{q-5}{5} = \frac{7}{3}$$

$$15\left(\frac{q+2}{3} + \frac{q-5}{5}\right) = 15\left(\frac{7}{3}\right) \quad \text{\textit{Multiply} \textit{by LCD, 15.}}$$

$$15\left(\frac{q+2}{3}\right) + 15\left(\frac{q-5}{5}\right) = 5 \cdot 7$$

$$5(q+2) + 3(q-5) = 35$$

$$5q + 10 + 3q - 15 = 35$$

$$8q - 5 = 35$$

$$8q = 40$$

$$q = 5$$

Check $q = 5$: $\frac{7}{3} = \frac{7}{3}$ *True*

Thus, the solution set is $\{5\}$.

35.
$$\frac{x}{2} = \frac{5}{4} + \frac{x-1}{4}$$

$$4\left(\frac{x}{2}\right) = 4\left(\frac{5}{4} + \frac{x-1}{4}\right) \quad \text{\textit{Multiply by} \textit{LCD, 4.}}$$

$$2(x) = 4\left(\frac{5}{4}\right) + 4\left(\frac{x-1}{4}\right)$$

$$2x = 5 + x - 1$$

$$x = 4$$

Check $x = 4$: $2 = 2$ *True*

Thus, the solution set is $\{4\}$.

37.
$$\frac{a+7}{8} - \frac{a-2}{3} = \frac{4}{3}$$

$$24\left(\frac{a+7}{8} - \frac{a-2}{3}\right) = 24\left(\frac{4}{3}\right)$$

$$\text{\textit{Multiply by LCD, 24.}}$$

$$24\left(\frac{a+7}{8}\right) - 24\left(\frac{a-2}{3}\right) = 8(4)$$

$$3(a+7) - 8(a-2) = 32$$

$$3a + 21 - 8a + 16 = 32$$

$$-5a + 37 = 32$$

$$-5a = -5$$

$$a = 1$$

Check $a = 1$: $\frac{4}{3} = \frac{4}{3}$ *True*

Thus, the solution set is $\{1\}$.

39.
$$\frac{p}{2} - \frac{p-1}{4} = \frac{5}{4}$$

$$4\left(\frac{p}{2} - \frac{p-1}{4}\right) = 4\left(\frac{5}{4}\right) \quad \text{\textit{Multiply by} \textit{LCD, 4.}}$$

$$4\left(\frac{p}{2}\right) - 4\left(\frac{p-1}{4}\right) = 5$$

$$2p - 1(p-1) = 5$$

$$2p - p + 1 = 5$$

$$p = 4$$

Check $p = 4$: $\frac{5}{4} = \frac{5}{4}$ *True*

Thus, the solution set is $\{4\}$.

41.
$$\frac{3x}{5} - \frac{x-5}{7} = 3$$

$$35\left(\frac{3x}{5} - \frac{x-5}{7}\right) = 35(3) \quad \text{\textit{Multiply by} \textit{LCD, 35.}}$$

$$35\left(\frac{3x}{5}\right) - 35\left(\frac{x-5}{7}\right) = 105$$

$$7(3x) - 5(x-5) = 105$$

$$21x - 5x + 25 = 105$$

$$16x = 80$$

$$x = 5$$

Check $x = 5$: $3 = 3$ *True*

Thus, the solution set is $\{5\}$.

43.
$$\frac{4}{x^2 - 3x} = \frac{1}{x^2 - 9}$$

$$\frac{4}{x(x-3)} = \frac{1}{(x+3)(x-3)} \quad \text{\textit{Factor} \textit{denominators.}}$$

$$x(x+3)(x-3) \cdot \frac{4}{x(x-3)}$$

$$= x(x+3)(x-3) \cdot \frac{1}{(x+3)(x-3)}$$

$$\text{\textit{Multiply by LCD, x(x + 3)(x - 3).}}$$

$$4(x+3) = x \cdot 1$$

$$4x + 12 = x$$

$$3x = -12$$

$$x = -4$$

Check $x = -4$: $\frac{1}{7} = \frac{1}{7}$ *True*

Thus, the solution set is $\{-4\}$.

45.
$$\frac{2}{m} = \frac{m}{5m + 12}$$

$$m(5m+12)\left(\frac{2}{m}\right) = m(5m+12)\left(\frac{m}{5m+12}\right)$$

$$\text{\textit{Multiply by LCD, m(5m + 12).}}$$

$$(5m+12)(2) = m(m)$$

$$10m + 24 = m^2$$

$$-m^2 + 10m + 24 = 0$$

$$m^2 - 10m - 24 = 0 \quad \text{\textit{Multiply by -1.}}$$

$$(m-12)(m+2) = 0$$

$$m - 12 = 0 \quad \text{or} \quad m + 2 = 0$$

$$m = 12 \quad \text{or} \quad m = -2$$

Check $m = 12$: $\frac{1}{6} = \frac{1}{6}$ *True*

Check $m = -2$: $-1 = -1$ *True*

Thus, the solution set is $\{-2, 12\}$.

47. $\dfrac{-2}{z+5} + \dfrac{3}{z-5} = \dfrac{20}{z^2-25}$

$\dfrac{-2}{z+5} + \dfrac{3}{z-5} = \dfrac{20}{(z+5)(z-5)}$

$(z+5)(z-5)\left(\dfrac{-2}{z+5} + \dfrac{3}{z-5}\right)$

$= (z+5)(z-5)\left(\dfrac{20}{(z+5)(z-5)}\right)$

Multiply by LCD, (z + 5)(z − 5).

$(z+5)(z-5)\left(\dfrac{-2}{z+5}\right)$

$+ (z+5)(z-5)\left(\dfrac{3}{z-5}\right) = 20$

$-2(z-5) + 3(z+5) = 20$

$-2z + 10 + 3z + 15 = 20$

$z + 25 = 20$

$z = -5$

The proposed solution, −5, cannot be a solution because it would make the denominators $z + 5$ and $z^2 - 25$ equal 0 and the corresponding fractions undefined. Since −5 cannot be a solution, the solution set is ∅.

49. $\dfrac{3}{x-1} + \dfrac{2}{4x-4} = \dfrac{7}{4}$

$\dfrac{3}{x-1} + \dfrac{2}{4(x-1)} = \dfrac{7}{4}$

$4(x-1)\left(\dfrac{3}{x-1} + \dfrac{2}{4(x-1)}\right) = 4(x-1)\left(\dfrac{7}{4}\right)$

Multiply by LCD, 4(x − 1).

$4(3) + 2 = (x-1)(7)$

$14 = 7x - 7$

$21 = 7x$

$3 = x$

Check $x = 3$: $\dfrac{7}{4} = \dfrac{7}{4}$ *True*

Thus, the solution set is $\{3\}$.

51. $\dfrac{x}{3x+3} = \dfrac{2x-3}{x+1} - \dfrac{2x}{3x+3}$

$\dfrac{x}{3(x+1)} = \dfrac{2x-3}{x+1} - \dfrac{2x}{3(x+1)}$

$3(x+1)\left(\dfrac{x}{3(x+1)}\right) =$

$3(x+1)\left[\dfrac{2x-3}{x+1} - \dfrac{2x}{3(x+1)}\right]$

Multiply by LCD, 3(x + 1).

$x = 3(x+1)\left(\dfrac{2x-3}{x+1}\right)$

$- 3(x+1)\left(\dfrac{2x}{3(x+1)}\right)$

$x = 3(2x-3) - 2x$

$x = 6x - 9 - 2x$

$x = 4x - 9$

$-3x = -9$

$x = 3$

Check:

$\dfrac{x}{3x+3} = \dfrac{2x-3}{x+1} - \dfrac{2x}{3x+3}$

$\dfrac{3}{3(3)+3} = \dfrac{2(3)-3}{3+1} - \dfrac{2(3)}{3(3)+3}$? *Let x = 3.*

$\dfrac{3}{9+3} = \dfrac{6-3}{4} - \dfrac{6}{9+3}$?

$\dfrac{3}{12} = \dfrac{3}{4} - \dfrac{6}{12}$?

$\dfrac{1}{4} = \dfrac{3}{4} - \dfrac{2}{4}$?

$\dfrac{1}{4} = \dfrac{1}{4}$ *True*

Thus, the solution set is $\{3\}$.

53. $\dfrac{2p}{p^2-1} = \dfrac{2}{p+1} - \dfrac{1}{p-1}$

$\dfrac{2p}{(p+1)(p-1)} = \dfrac{2}{p+1} - \dfrac{1}{p-1}$

$(p+1)(p-1)\left[\dfrac{2p}{(p+1)(p-1)}\right]$

$= (p+1)(p-1)\left(\dfrac{2}{p+1}\right) - (p+1)(p-1)\left(\dfrac{1}{p-1}\right)$

Multiply by LCD, (p + 1)(p − 1).

$2p = 2(p-1) - 1(p+1)$

$2p = 2p - 2 - p - 1$

$p = -3$

Check $p = -3$: $-\dfrac{3}{4} = -1 + \dfrac{1}{4}$ *True*

Thus, the solution set is $\{-3\}$.

55. $\dfrac{5x}{14x+3} = \dfrac{1}{x}$

$x(14x+3)\left(\dfrac{5x}{14x+3}\right) = x(14x+3)\left(\dfrac{1}{x}\right)$

Multiply by LCD, x(14x + 3).

$x(5x) = (14x+3)(1)$

$5x^2 = 14x + 3$

$5x^2 - 14x - 3 = 0$

$(5x+1)(x-3) = 0$

Note to reader: We may skip writing out the zero-factor property since this step can be easily performed mentally.

$x = -\dfrac{1}{5}$ or $x = 3$

Check $x = -\dfrac{1}{5}$: $-5 = -5$ *True*

Check $x = 3$: $\dfrac{1}{3} = \dfrac{1}{3}$ *True*

Thus, the solution set is $\left\{-\dfrac{1}{5}, 3\right\}$.

57. $\dfrac{2}{x-1} - \dfrac{2}{3} = \dfrac{-1}{x+1}$

$$3(x-1)(x+1)\left(\dfrac{2}{x-1} - \dfrac{2}{3}\right)$$

$$= 3(x-1)(x+1)\left(\dfrac{-1}{x+1}\right)$$

Multiply by LCD, 3(x − 1)(x + 1).

$$3(x-1)(x+1)\left(\dfrac{2}{x-1}\right)$$

$$- 3(x-1)(x+1)\left(\dfrac{2}{3}\right)$$

$$= 3(x-1)(x+1)\left(\dfrac{-1}{x+1}\right)$$

$$3(x+1)(2) - (x-1)(x+1)(2)$$
$$= 3(x-1)(-1)$$

$$6(x+1) - 2(x^2 - 1) = -3(x-1)$$
$$6x + 6 - 2x^2 + 2 = -3x + 3$$
$$-2x^2 + 9x + 5 = 0$$
$$2x^2 - 9x - 5 = 0$$
$$(2x+1)(x-5) = 0$$
$$x = -\tfrac{1}{2} \quad \text{or} \quad x = 5$$

Check $x = -\tfrac{1}{2}$: $-2 = -2$ *True*
Check $x = 5$: $-\tfrac{1}{6} = -\tfrac{1}{6}$ *True*

Thus, the solution set is $\left\{-\tfrac{1}{2}, 5\right\}$.

59. $\dfrac{x}{2x+2} = \dfrac{-2x}{4x+4} + \dfrac{2x-3}{x+1}$

$$\dfrac{x}{2(x+1)} = \dfrac{-2x}{4(x+1)} + \dfrac{2x-3}{x+1}$$

$$4(x+1)\left(\dfrac{x}{2(x+1)}\right) = 4(x+1)\left(\dfrac{-2x}{4(x+1)}\right)$$

$$+ 4(x+1)\left(\dfrac{2x-3}{x+1}\right)$$

Multiply by LCD, 4(x + 1).

$$2(x) = -2x + 4(2x-3)$$
$$2x = -2x + 8x - 12$$
$$-4x = -12$$
$$x = 3$$

Check $x = 3$: $\tfrac{3}{8} = \tfrac{3}{8}$ *True*

Thus, the solution set is $\{3\}$.

61. $\dfrac{8x+3}{x} = 3x$

$$x\left(\dfrac{8x+3}{x}\right) = x(3x) \quad \begin{array}{l}\textit{Multiply by}\\ \textit{LCD, x.}\end{array}$$

$$8x + 3 = 3x^2$$
$$0 = 3x^2 - 8x - 3$$
$$0 = (3x+1)(x-3)$$
$$x = -\tfrac{1}{3} \quad \text{or} \quad x = 3$$

Check $x = -\tfrac{1}{3}$: $-1 = -1$ *True*
Check $x = 3$: $9 = 9$ *True*

Thus, the solution set is $\left\{-\tfrac{1}{3}, 3\right\}$.

63. $\dfrac{1}{x+4} + \dfrac{x}{x-4} = \dfrac{-8}{x^2-16}$

$$(x+4)(x-4)\left(\dfrac{1}{x+4}\right) + (x+4)(x-4)\left(\dfrac{x}{x-4}\right)$$

$$= (x+4)(x-4)\left(\dfrac{-8}{x^2-16}\right)$$

Multiply by LCD, (x + 4)(x − 4).

$$1(x-4) + x(x+4) = -8$$
$$x - 4 + x^2 + 4x = -8$$
$$x^2 + 5x + 4 = 0$$
$$(x+4)(x+1) = 0$$
$$x = -4 \quad \text{or} \quad x = -1$$

$x = -4$ cannot be a solution because it would make the denominators $x + 4$ and $x^2 - 16$ equal 0 and the corresponding fractions undefined.

Check $x = -1$: $\tfrac{1}{3} + \tfrac{1}{5} = \tfrac{8}{15}$ *True*

Thus, the solution set is $\{-1\}$.

65. $\dfrac{4}{3x+6} - \dfrac{3}{x+3} = \dfrac{8}{x^2+5x+6}$

$$\dfrac{4}{3(x+2)} - \dfrac{3}{x+3} = \dfrac{8}{(x+2)(x+3)}$$

$$3(x+2)(x+3) \cdot \dfrac{4}{3(x+2)} - 3(x+2)(x+3) \cdot \dfrac{3}{x+3}$$

$$= 3(x+2)(x+3) \cdot \dfrac{8}{(x+2)(x+3)}$$

Multiply by LCD, 3(x + 2)(x + 3).

$$4(x+3) - 3(x+2)(3) = 3(8)$$
$$4x + 12 - 9x - 18 = 24$$
$$-5x = 30$$
$$x = -6$$

Check $x = -6$: $-\tfrac{1}{3} - (-1) = \tfrac{2}{3}$ *True*

Thus, the solution set is $\{-6\}$.

67. $\dfrac{3x}{x^2+5x+6}$

$=\dfrac{5x}{x^2+2x-3}-\dfrac{2}{x^2+x-2}$

$\dfrac{3x}{(x+2)(x+3)}$

$=\dfrac{5x}{(x+3)(x-1)}-\dfrac{2}{(x-1)(x+2)}$

$(x+2)(x+3)(x-1)\cdot\left[\dfrac{3x}{(x+2)(x+3)}\right]$

$=(x+2)(x+3)(x-1)\cdot\left[\dfrac{5x}{(x+3)(x-1)}\right]$

$-(x+2)(x+3)(x-1)\cdot\left[\dfrac{2}{(x-1)(x+2)}\right]$

Multiply by LCD, (x + 2)(x + 3)(x − 1).

$3x(x-1)=5x(x+2)-2(x+3)$
$3x^2-3x=5x^2+10x-2x-6$
$0=2x^2+11x-6$
$0=(2x-1)(x+6)$

$x=\frac{1}{2}$ or $x=-6$

Check $x=\frac{1}{2}$: $\quad\frac{6}{35}=-\frac{10}{7}-\left(-\frac{8}{5}\right)\quad$ *True*
Check $x=-6$: $\quad-\frac{3}{2}=-\frac{10}{7}-\frac{1}{14}\quad$ *True*

Thus, the solution set is $\left\{-6,\frac{1}{2}\right\}$.

69. $\dfrac{x+4}{x^2-3x+2}-\dfrac{5}{x^2-4x+3}$

$=\dfrac{x-4}{x^2-5x+6}$

$\dfrac{x+4}{(x-2)(x-1)}-\dfrac{5}{(x-3)(x-1)}$

$=\dfrac{x-4}{(x-3)(x-2)}$

$(x-2)(x-1)(x-3)$

$\cdot\left[\dfrac{x+4}{(x-2)(x-1)}-\dfrac{5}{(x-3)(x-1)}\right]$

$=(x-2)(x-1)(x-3)\left[\dfrac{x-4}{(x-3)(x-2)}\right]$

Multiply by LCD, (x − 2)(x − 1)(x − 3).

$(x+4)(x-3)-5(x-2)=(x-1)(x-4)$
$x^2+x-12-5x+10=x^2-5x+4$
$-4x-2=-5x+4$
$x=6$

Check $x=6$: $\frac{1}{2}-\frac{1}{3}=\frac{1}{6}$ *True*

Thus, the solution set is $\{6\}$.

71. $kr-mr=km$

If you are solving for k, put both terms with k on one side and the remaining term on the other side.

$$kr-km=mr$$

73. $m=\dfrac{kF}{a}$ for F

We need to isolate F on one side of the equation.

$m\cdot a=\left(\dfrac{kF}{a}\right)(a)\quad$ *Multiply by a.*

$ma=kF$

$\dfrac{ma}{k}=\dfrac{kF}{k}\qquad$ *Divide by k.*

$\dfrac{ma}{k}=F$

75. $m=\dfrac{kF}{a}$ for a

$m\cdot a=\left(\dfrac{kF}{a}\right)(a)\quad$ *Multiply by a.*

$ma=kF$

$\dfrac{ma}{m}=\dfrac{kF}{m}\qquad$ *Divide by m.*

$a=\dfrac{kF}{m}$

77. $I=\dfrac{E}{R+r}$ for R

We need to isolate R on one side of the equation.

$I(R+r)=\left(\dfrac{E}{R+r}\right)(R+r)\qquad$ *Multiply by R + r.*

$IR+Ir=E\qquad$ *Distributive property*

$IR=E-Ir\qquad$ *Subtract Ir.*

$R=\dfrac{E-Ir}{I},\quad$ or $\quad R=\dfrac{E}{I}-r\quad$ *Divide by I.*

79. $h=\dfrac{2A}{B+b}$ for A

$(B+b)h=(B+b)\cdot\dfrac{2A}{B+b}$

Multiply by B + b.

$h(B+b)=2A$

$\dfrac{h(B+b)}{2}=A\qquad$ *Divide by 2.*

81. $d = \dfrac{2S}{n(a + L)}$ for a

We need to isolate a on one side of the equation.

$$d \cdot n(a + L) = \dfrac{2S}{n(a + L)} \cdot n(a + L)$$

$$\qquad\qquad\qquad \textit{Multiply by } n(a + L).$$

$$nd(a + L) = 2S$$
$$and + ndL = 2S$$
$$\qquad and = 2S - ndL \quad \textit{Subtract ndL.}$$
$$\qquad a = \dfrac{2S - ndL}{nd} \quad \textit{Divide by nd.}$$
$$\text{or} \quad a = \dfrac{2S}{nd} - L$$

83. $\dfrac{1}{x} = \dfrac{1}{y} - \dfrac{1}{z}$ for y

The LCD of all the fractions in the equation is xyz, so multiply both sides by xyz.

$$xyz\left(\dfrac{1}{x}\right) = xyz\left(\dfrac{1}{y} - \dfrac{1}{z}\right)$$

$$xyz\left(\dfrac{1}{x}\right) = xyz\left(\dfrac{1}{y}\right) - xyz\left(\dfrac{1}{z}\right)$$

$$\qquad\qquad\qquad\qquad \textit{Distributive}$$
$$\qquad\qquad\qquad\qquad \textit{property}$$

$$yz = xz - xy$$

Since we are solving for y, get all terms with y on one side of the equation.

$$xy + yz = xz \quad \textit{Add xy.}$$

Factor out the common factor y on the left.

$$y(x + z) = xz$$

Finally, divide both sides by the coefficient of y, which is $x + z$.

$$y = \dfrac{xz}{x + z}$$

85. $\dfrac{2}{r} + \dfrac{3}{s} + \dfrac{1}{t} = 1$ for t

The LCD of all the fractions in the equation is rst, so multiply both sides by rst.

$$rst\left(\dfrac{2}{r} + \dfrac{3}{s} + \dfrac{1}{t}\right) = rst(1)$$

$$rst\left(\dfrac{2}{r}\right) + rst\left(\dfrac{3}{s}\right) + rst\left(\dfrac{1}{t}\right) = rst$$

$$2st + 3rt + rs = rst$$

Since we are solving for t, get all terms with t on one side of the equation.

$$2st + 3rt - rst = -rs$$

Factor out the common factor t on the left.

$$t(2s + 3r - rs) = -rs$$

Finally, divide both sides by the coefficient of t, which is $2s + 3r - rs$.

$$t = \dfrac{-rs}{2s + 3r - rs}, \quad \text{or} \quad t = \dfrac{rs}{-2s - 3r + rs}$$

87. $9x + \dfrac{3}{z} = \dfrac{5}{y}$ for z

$$yz\left(9x + \dfrac{3}{z}\right) = yz\left(\dfrac{5}{y}\right) \quad \begin{matrix}\textit{Multiply by}\\ \textit{LCD, yz.}\end{matrix}$$

$$yz(9x) + yz\left(\dfrac{3}{z}\right) = yz\left(\dfrac{5}{y}\right) \quad \begin{matrix}\textit{Distributive}\\ \textit{property}\end{matrix}$$

$$9xyz + 3y = 5z$$

$$9xyz - 5z = -3y \quad \begin{matrix}\textit{Get the z terms}\\ \textit{on one side.}\end{matrix}$$

$$z(9xy - 5) = -3y \quad \textit{Factor out z.}$$

$$z = \dfrac{-3y}{9xy - 5}, \quad \textit{Divide by 9xy} - 5.$$

$$\text{or} \quad z = \dfrac{3y}{5 - 9xy}$$

89. Using $d = rt$, we get

$$r = \dfrac{d}{t} = \dfrac{288}{t}.$$

His rate is $\dfrac{288}{t}$ mph.

91. Using $d = rt$, we get

$$t = \dfrac{d}{r} = \dfrac{289}{z}.$$

His time is $\dfrac{289}{z}$ hr.

Summary Exercises on Rational Expressions and Equations

1. No equals sign appears so this is an *expression*.

$$\dfrac{4}{p} + \dfrac{6}{p} = \dfrac{4 + 6}{p} = \dfrac{10}{p}$$

3. No equals sign appears so this is an *expression*.

$$\dfrac{1}{x^2 + x - 2} \div \dfrac{4x^2}{2x - 2}$$

$$= \dfrac{1}{x^2 + x - 2} \cdot \dfrac{2x - 2}{4x^2}$$

$$= \dfrac{1}{(x + 2)(x - 1)} \cdot \dfrac{2(x - 1)}{2 \cdot 2x^2}$$

$$= \dfrac{1}{2x^2(x + 2)}$$

5. No equals sign appears so this is an *expression*.

$$\frac{2y^2 + y - 6}{2y^2 - 9y + 9} \cdot \frac{y^2 - 2y - 3}{y^2 - 1}$$

$$= \frac{(2y-3)(y+2)(y-3)(y+1)}{(2y-3)(y-3)(y+1)(y-1)}$$

$$= \frac{y+2}{y-1}$$

7. $\dfrac{x-4}{5} = \dfrac{x+3}{6}$

There is an equals sign, so this is an *equation*.

$$30\left(\frac{x-4}{5}\right) = 30\left(\frac{x+3}{6}\right) \quad \text{\textit{Multiply by LCD, 30.}}$$
$$6(x-4) = 5(x+3)$$
$$6x - 24 = 5x + 15$$
$$x = 39$$

Check $x = 39$: $7 = 7$ *True*

The solution is set is $\{39\}$.

9. No equals sign appears so this is an *expression*.

$$\frac{4}{p+2} + \frac{1}{3p+6} = \frac{4}{p+2} + \frac{1}{3(p+2)}$$
$$= \frac{3 \cdot 4}{3(p+2)} + \frac{1}{3(p+2)}$$
$$\text{\textit{LCD} = 3(p+2)}$$
$$= \frac{12+1}{3(p+2)}$$
$$= \frac{13}{3(p+2)}$$

11. $\dfrac{3}{t-1} + \dfrac{1}{t} = \dfrac{7}{2}$

There is an equals sign, so this is an *equation*.

$$2t(t-1)\left(\frac{3}{t-1} + \frac{1}{t}\right) = 2t(t-1)\left(\frac{7}{2}\right)$$
$$\text{\textit{Multiply by LCD, 2t(t-1).}}$$
$$2t(t-1)\left(\frac{3}{t-1}\right) + 2t(t-1)\left(\frac{1}{t}\right) = 7t(t-1)$$
$$2t(3) + 2(t-1) = 7t(t-1)$$
$$6t + 2t - 2 = 7t^2 - 7t$$
$$0 = 7t^2 - 15t + 2$$
$$0 = (7t-1)(t-2)$$
$$t = \tfrac{1}{7} \quad \text{or} \quad t = 2$$

Check $t = \tfrac{1}{7}$: $-\tfrac{7}{2} + 7 = \tfrac{7}{2}$ *True*

Check $t = 2$: $3 + \tfrac{1}{2} = \tfrac{7}{2}$ *True*

The solution set is $\left\{\tfrac{1}{7}, 2\right\}$.

13. No equals sign appears so this is an *expression*.

$$\frac{5}{4z} - \frac{2}{3z} = \frac{3 \cdot 5}{3 \cdot 4z} - \frac{4 \cdot 2}{4 \cdot 3z} \quad \text{\textit{LCD} = 12z}$$
$$= \frac{15}{12z} - \frac{8}{12z}$$
$$= \frac{15-8}{12z} = \frac{7}{12z}$$

15. No equals sign appears so this is an *expression*.

$$\frac{1}{m^2 + 5m + 6} + \frac{2}{m^2 + 4m + 3}$$
$$= \frac{1}{(m+2)(m+3)} + \frac{2}{(m+1)(m+3)}$$
$$= \frac{1(m+1)}{(m+2)(m+3)(m+1)}$$
$$+ \frac{2(m+2)}{(m+1)(m+3)(m+2)}$$
$$\text{\textit{LCD} = (m+1)(m+2)(m+3)}$$
$$= \frac{(m+1) + (2m+4)}{(m+1)(m+2)(m+3)}$$
$$= \frac{3m+5}{(m+1)(m+2)(m+3)}$$

17. $\dfrac{2}{x+1} + \dfrac{5}{x-1} = \dfrac{10}{x^2-1}$

There is an equals sign, so this is an *equation*.

$$\frac{2}{x+1} + \frac{5}{x-1} = \frac{10}{(x+1)(x-1)}$$
$$(x+1)(x-1)\left(\frac{2}{x+1} + \frac{5}{x-1}\right)$$
$$= (x+1)(x-1)\left[\frac{10}{(x+1)(x-1)}\right]$$
$$\text{\textit{Multiply by LCD, (x+1)(x-1).}}$$
$$(x+1)(x-1)\left(\frac{2}{x+1}\right)$$
$$+ (x+1)(x-1)\left(\frac{5}{x-1}\right) = 10$$
$$\text{\textit{Distributive property}}$$
$$2(x-1) + 5(x+1) = 10$$
$$2x - 2 + 5x + 5 = 10$$
$$3 + 7x = 10$$
$$7x = 7$$
$$x = 1$$

Replacing x by 1 in the original equation makes the denominators $x-1$ and x^2-1 equal to 0, so the solution set is $\emptyset$.

19. No equals sign appears so this is an *expression*.

$$\frac{4t^2 - t}{6t^2 + 10t} \div \frac{8t^2 + 2t - 1}{3t^2 + 11t + 10}$$

$$= \frac{4t^2 - t}{6t^2 + 10t} \cdot \frac{3t^2 + 11t + 10}{8t^2 + 2t - 1}$$

Multiply by reciprocal.

$$= \frac{t(4t - 1)}{2t(3t + 5)} \cdot \frac{(3t + 5)(t + 2)}{(4t - 1)(2t + 1)}$$

Factor numerators and denominators.

$$= \frac{t + 2}{2(2t + 1)}$$

6.7 Applications of Rational Expressions

1. **(a)** Let $x = $ _the amount_ .

(b) An expression for "the numerator of the fraction $\frac{5}{6}$ is increased by an amount" is $\underline{5 + x}$. We could also use $\dfrac{5 + x}{6}$.

(c) An equation that can be used to solve the problem is

$$\frac{5 + x}{6} = \frac{13}{3}.$$

3. *Step 2*
Let $x = $ the numerator of the original fraction. Then $x + 6 = $ the denominator of the original fraction.

Step 3
If 3 is added to both the numerator and denominator, the resulting fraction is equivalent to $\frac{5}{7}$ translates to

$$\frac{x + 3}{(x + 6) + 3} = \frac{5}{7}.$$

Step 4
Since we have a fraction equal to another fraction, we can use cross multiplication.

$$7(x + 3) = 5[(x + 6) + 3]$$
$$7x + 21 = 5x + 45$$
$$2x = 24$$
$$x = 12$$

Step 5
The original fraction is

$$\frac{x}{x + 6} = \frac{12}{12 + 6} = \frac{12}{18}.$$

Step 6
Adding 3 to both the numerator and the denominator gives us

$$\frac{12 + 3}{18 + 3} = \frac{15}{21},$$

which is equivalent to $\frac{5}{7}$.

5. *Step 2*
Let $x = $ the denominator of the original fraction. Then $4x = $ the numerator of the original fraction.

Step 3
If 6 is added to both the numerator and the denominator, the resulting fraction is equivalent to 2 translates to

$$\frac{4x + 6}{x + 6} = 2.$$

Step 4
$$4x + 6 = 2(x + 6)$$
$$4x + 6 = 2x + 12$$
$$2x = 6$$
$$x = 3$$

Step 5
The original fraction is

$$\frac{4x}{x} = \frac{4(3)}{3} = \frac{12}{3}.$$

Step 6
Adding 6 to both the numerator and the denominator gives us

$$\frac{12 + 6}{3 + 6} = \frac{18}{9} = 2.$$

7. *Step 2*
Let $x = $ the number.

Step 3
One-third of a number is 2 more than one-sixth of the same number translates to

$$\tfrac{1}{3}x = 2 + \tfrac{1}{6}x.$$

Step 4
Multiply both sides by the LCD, 6.

$$6\left(\tfrac{1}{3}x\right) = 6\left(2 + \tfrac{1}{6}x\right)$$
$$2x = 12 + x$$
$$x = 12$$

Step 5 The number is 12.

Step 6
One-third of 12 is 4 and one-sixth of 12 is 2. So $x = 12$ checks since 4 is 2 more than 2.

9. *Step 2*
Let $x = $ the quantity.
Then $\frac{2}{3}$ of it, $\frac{1}{2}$ of it, and $\frac{1}{7}$ of it are

$$\tfrac{2}{3}x, \tfrac{1}{2}x, \text{ and } \tfrac{1}{7}x.$$

Step 3
Added together equals 33 translates to

$$x + \tfrac{2}{3}x + \tfrac{1}{2}x + \tfrac{1}{7}x = 33.$$

Step 4

Multiply both sides by the LCD of 3, 2, and 7, which is 42.

$$42\left(x + \tfrac{2}{3}x + \tfrac{1}{2}x + \tfrac{1}{7}x\right) = 42(33)$$

$$42x + 42\left(\tfrac{2}{3}x\right) + 42\left(\tfrac{1}{2}x\right) + 42\left(\tfrac{1}{7}x\right) = 42(33)$$

$$42x + 28x + 21x + 6x = 1386$$

$$97x = 1386$$

$$x = \tfrac{1386}{97}$$

Step 5

The quantity is $\tfrac{1386}{97}$. (Note that this fraction is already in lowest terms since 97 is a prime number and is not a factor of 1386.)

Step 6

Check $\tfrac{1386}{97}$ in the original problem.

$$x = \tfrac{1386}{97}, \ \tfrac{2}{3}x = \tfrac{924}{97}, \ \tfrac{1}{2}x = \tfrac{693}{97}, \ \tfrac{1}{7}x = \tfrac{198}{97}$$

Adding gives us

$$\frac{1386 + 924 + 693 + 198}{97} = \frac{3201}{97} = 33,$$

as desired.

11. We are asked to find the *time*, so we'll use the distance, rate, and time relationship

$$t = \frac{d}{r}.$$

$$t = \frac{500 \text{ meters}}{6.530 \text{ meters per second}}$$

$$\approx 76.57 \text{ seconds}$$

13. We are asked to find the average *rate*, so we'll use the distance, rate, and time relationship

$$r = \frac{d}{t}.$$

$$r = \frac{5000 \text{ meters}}{14.761 \text{ minutes}}$$

$$\approx 338.730 \text{ meters per minute}$$

15. We are asked to find the *time*, so we'll use the distance, rate, and time relationship

$$t = \frac{d}{r}.$$

$$t = \frac{500 \text{ miles}}{135.173 \text{ miles per hour}}$$

$$\approx 3.699 \text{ hours}$$

17. Use $\text{time} = \dfrac{\text{distance}}{\text{rate}}$. Since we know that the times for Stephanie and Wally are the same,

$$\text{time}_{\text{Stephanie}} = \text{time}_{\text{Wally}}$$

$$\text{or} \qquad \frac{D}{R} = \frac{d}{r}.$$

19. Let x = speed of the plane in still air. Then the speed against the wind is $x - 10$ and the speed with the wind is $x + 10$. The time flying against the wind is

$$t = \frac{d}{r} = \frac{500}{x - 10},$$

and the time flying with the wind is

$$t = \frac{d}{r} = \frac{600}{x + 10}.$$

Now complete the chart.

	d	r	t
Against the Wind	500	$x - 10$	$\dfrac{500}{x - 10}$
With the Wind	600	$x + 10$	$\dfrac{600}{x + 10}$

Since the problem states that the two times are equal, we have

$$\frac{500}{x - 10} = \frac{600}{x + 10}.$$

We would use this equation to solve the problem.

21. Let x represent the speed of the boat in still water. Then $x - 4$ is the rate against the current and $x + 4$ is the rate with the current. We fill in the chart as follows, realizing that the time column is filled in by using the formula $t = d/r$.

	d	r	t
Against the Current	20	$x - 4$	$\dfrac{20}{x - 4}$
With the Current	60	$x + 4$	$\dfrac{60}{x + 4}$

Since the times are equal, we get the following equation.

$$\frac{20}{x - 4} = \frac{60}{x + 4}$$

$$(x + 4)(x - 4)\frac{20}{x - 4} = (x + 4)(x - 4)\frac{60}{x + 4}$$

Multiply by LCD, $(x + 4)(x - 4)$

$$20(x + 4) = 60(x - 4)$$

$$20x + 80 = 60x - 240$$

$$320 = 40x$$

$$8 = x$$

The speed of the boat in still water is 8 miles per hour.

23. Let $x =$ speed of the bird in still air. Then the speed against the wind is $x - 8$ and the speed with the wind is $x + 8$. Use $t = d/r$ to complete the chart.

	d	r	t
Against the Wind	18	$x - 8$	$\dfrac{18}{x - 8}$
With the Wind	30	$x + 8$	$\dfrac{30}{x + 8}$

Since the problem states that the two times are equal, we get the following equation.

$$\frac{18}{x - 8} = \frac{30}{x + 8}$$
$$(x + 8)(x - 8)\frac{18}{x - 8} = (x + 8)(x - 8)\frac{30}{x + 8}$$
$$18(x + 8) = 30(x - 8)$$
$$18x + 144 = 30x - 240$$
$$384 = 12x$$
$$32 = x$$

The speed of the bird in still air is 32 miles per hour.

25. Let $x =$ speed of the plane in still air. Then the speed against the wind is $x - 15$ and the speed with the wind is $x + 15$. Use $t = d/r$ to complete the chart.

	d	r	t
Against the Wind	375	$x - 15$	$\dfrac{375}{x - 15}$
With the Wind	450	$x + 15$	$\dfrac{450}{x + 15}$

Since the problem states that the two times are equal, we get the following equation.

$$\frac{375}{x - 15} = \frac{450}{x + 15}$$
$$(x + 15)(x - 15)\frac{375}{x - 15} = (x + 15)(x - 15)\frac{450}{x + 15}$$
$$375(x + 15) = 450(x - 15)$$
$$375x + 5625 = 450x - 6750$$
$$12{,}375 = 75x$$
$$165 = x$$

The speed of the plane in still air is 165 miles per hour.

27. Let x represent the rate of the current of the river. Then $12 - x$ is the rate upstream (against the current) and $12 + x$ is the rate downstream (with the current). Use $t = d/r$ to complete the table.

	d	r	t
Upstream	6	$12 - x$	$\dfrac{6}{12 - x}$
Downstream	10	$12 + x$	$\dfrac{10}{12 + x}$

Since the times are equal, we get the following equation.

$$\frac{6}{12 - x} = \frac{10}{12 + x}$$
$$(12 + x)(12 - x)\frac{6}{12 - x} = (12 + x)(12 - x)\frac{10}{12 + x}$$
$$6(12 + x) = 10(12 - x)$$
$$72 + 6x = 120 - 10x$$
$$16x = 48$$
$$x = 3$$

The rate of the current of the river is 3 miles per hour.

29. Let $x =$ the average speed of the ferry.

Use the formula $t = \dfrac{d}{r}$ to make a chart.

	d	r	t
Seattle-Victoria	148	x	$\dfrac{148}{x}$
Victoria-Vancouver	74	x	$\dfrac{74}{x}$

Since the time for the Victoria-Vancouver trip is 4 hours less than the time for the Seattle-Victoria trip, solve the equation

$$\frac{74}{x} = \frac{148}{x} - 4.$$
$$x\left(\frac{74}{x}\right) = x\left(\frac{148}{x} - 4\right) \quad \textit{Multiply by LCD, x.}$$
$$74 = 148 - 4x$$
$$4x = 74$$
$$x = \frac{74}{4} = \frac{37}{2}, \text{ or } 18\tfrac{1}{2}$$

The average speed of the ferry is $18\tfrac{1}{2}$ miles per hour.

31. If it takes Elayn 10 hours to do a job, her rate is $\frac{1}{10}$ job per hour.

33. Let $t =$ the number of hours it will take Jorge and Caterina to paint the room working together.

	Rate	Time Working Together	Fractional Part of the Job Done when Working Together
Jorge	$\frac{1}{8}$	t	$\frac{1}{8}t$
Caterina	$\frac{1}{6}$	t	$\frac{1}{6}t$

part done by Jorge	+	part done by Caterina	=	1 whole job
↓	↓	↓	↓	↓
$\frac{1}{8}t$	+	$\frac{1}{6}t$	=	1

An equation that can be used to solve this problem is

$$\frac{1}{8}t + \frac{1}{6}t = 1.$$

Alternatively, we can compare the hourly rates of completion. In one hour, Jorge will complete $\frac{1}{8}$ of the job, Caterina will complete $\frac{1}{6}$ of the job, and together they will complete $\frac{1}{t}$ of the job. So another equation that can be used to solve this problem is

$$\frac{1}{8} + \frac{1}{6} = \frac{1}{t}.$$

35. Let x represent the number of hours it will take for Ms. Tseng and Jonah to grade the tests working together. Since Ms. Tseng can grade the test in 4 hours, her rate alone is $\frac{1}{4}$ job per hour. Also, since Jonah can do the job alone in 6 hours, his rate is $\frac{1}{6}$ job per hour.

	Rate	Time Working Together	Fractional Part of the Job Done when Working Together
Ms. Tseng	$\frac{1}{4}$	x	$\frac{1}{4}x$
Jonah	$\frac{1}{6}$	x	$\frac{1}{6}x$

Since together Ms. Tseng and Jonah complete 1 whole job, we must add their individual fractional parts and set the sum equal to 1.

$$\frac{1}{4}x + \frac{1}{6}x = 1$$

$$12\left(\frac{1}{4}x + \frac{1}{6}x\right) = 12(1) \qquad \textit{Multiply by the LCD, 12.}$$

$$12\left(\frac{1}{4}x\right) + 12\left(\frac{1}{6}x\right) = 12$$

$$3x + 2x = 12$$

$$5x = 12$$

$$x = \frac{12}{5}, \text{ or } 2\frac{2}{5}$$

It will take Ms. Tseng and Jonah $2\frac{2}{5}$ hours to grade the tests if they work together.

37. Let $x =$ the number of hours to pump the water using both pumps.

	Rate	Time Working Together	Fractional Part of the Job Done when Working Together
Pump 1	$\frac{1}{10}$	x	$\frac{1}{10}x$
Pump 2	$\frac{1}{12}$	x	$\frac{1}{12}x$

Since together the two pumps complete 1 whole job, we must add their individual fractional parts and set the sum equal to 1.

$$\frac{1}{10}x + \frac{1}{12}x = 1$$

$$60\left(\frac{1}{10}x + \frac{1}{12}x\right) = 60(1) \qquad \textit{Multiply by the LCD, 60.}$$

$$60\left(\frac{1}{10}x\right) + 60\left(\frac{1}{12}x\right) = 60$$

$$6x + 5x = 60$$

$$11x = 60$$

$$x = \frac{60}{11}, \text{ or } 5\frac{5}{11}$$

It would take $5\frac{5}{11}$ hours to pump out the basement if both pumps were used.

39. Let x represent the number of hours it will take the experienced employee to enter the data. Then $2x$ represents the number of hours it will take the new employee (the experienced employee takes less time). The experienced employee's rate is $\frac{1}{x}$ job per hour and the new employee's rate is $\frac{1}{2x}$ job per hour.

	Rate	Time Working Together	Fractional Part of the Job Done when Working Together
Experienced employee	$\frac{1}{x}$	2	$\frac{1}{x} \cdot 2 = \frac{2}{x}$
New employee	$\frac{1}{2x}$	2	$\frac{1}{2x} \cdot 2 = \frac{1}{x}$

Since together the two employees complete the whole job, we must add their individual fractional parts and set the sum equal to 1.

$$\frac{2}{x} + \frac{1}{x} = 1$$

$$x\left(\frac{2}{x} + \frac{1}{x}\right) = x(1) \qquad \textit{Multiply by the LCD, x.}$$

$$x\left(\frac{2}{x}\right) + x\left(\frac{1}{x}\right) = x$$

$$2 + 1 = x$$

$$3 = x$$

Working alone, it will take the experienced employee 3 hours to enter the data.

41. Let $x =$ the number of hours to fill the pool $\frac{3}{4}$ full with both pipes working together.

	Rate	Time Working Together	Fractional Part of the Job Done when Working Together
First pipe	$\frac{1}{6}$	x	$\frac{1}{6}x$
Second pipe	$\frac{1}{9}$	x	$\frac{1}{9}x$

continued

$$\begin{array}{ccccc} \text{Part done} & + & \text{Part done by} & = & \dfrac{3}{4} \text{ full} \\ \text{by first pipe} & & \text{second pipe} & & \\ \downarrow & \downarrow & \downarrow & \downarrow & \downarrow \\ \dfrac{1}{6}x & + & \dfrac{1}{9}x & = & \dfrac{3}{4} \end{array}$$

$$36\left(\tfrac{1}{6}x + \tfrac{1}{9}x\right) = 36\left(\tfrac{3}{4}\right) \quad \textit{Multiply by the}$$
$$\textit{LCD, 36.}$$
$$36\left(\tfrac{1}{6}x\right) + 36\left(\tfrac{1}{9}x\right) = 36\left(\tfrac{3}{4}\right)$$
$$6x + 4x = 27$$
$$10x = 27$$
$$x = \tfrac{27}{10}, \text{ or } 2\tfrac{7}{10}$$

It takes $2\tfrac{7}{10}$ hours to fill the pool $\tfrac{3}{4}$ full using both pipes.

Alternatively, we could solve $\tfrac{1}{6}x + \tfrac{1}{9}x = 1$ (filling the whole pool) and then multiply that answer by $\tfrac{3}{4}$.

43. Let $x =$ the number of minutes it takes to fill the sink.

In 1 minute, the cold water faucet (alone) can fill $\tfrac{1}{12}$ of the sink. In the same time, the hot water faucet (alone) can fill $\tfrac{1}{15}$ of the sink. In 1 minute, the drain (alone) empties $\tfrac{1}{25}$ of the sink. Together, they fill $\tfrac{1}{x}$ of the sink in one minute, so solve the equation

$$\tfrac{1}{12} + \tfrac{1}{15} - \tfrac{1}{25} = \tfrac{1}{x}.$$
$$300x\left(\tfrac{1}{12} + \tfrac{1}{15} - \tfrac{1}{25}\right) = 300x\left(\tfrac{1}{x}\right)$$
$$\textit{Multiply by LCD, 300x.}$$
$$25x + 20x - 12x = 300$$
$$33x = 300$$
$$x = \tfrac{300}{33} = \tfrac{100}{11}, \text{ or } 9\tfrac{1}{11}$$

It will take $9\tfrac{1}{11}$ minutes to fill the sink.

45. $\dfrac{6-2}{5-3} = \dfrac{4}{2} = 2$

47. $\dfrac{4-(-1)}{-3-(-5)} = \dfrac{4+1}{-3+5} = \dfrac{5}{2}$

49. $\dfrac{-5-(-5)}{3-2} = \dfrac{-5+5}{1} = \dfrac{0}{1} = 0$

51. $3x + 2y = 8$
$$2y = -3x + 8$$
$$y = -\tfrac{3}{2}x + 4$$

Chapter 6 Review Exercises

1. **(a)** $\dfrac{4x-3}{5x+2} = \dfrac{4(-2)-3}{5(-2)+2}$ *Let x = -2.*

$$= \dfrac{-8-3}{-10+2} = \dfrac{-11}{-8} = \dfrac{11}{8}$$

(b) $\dfrac{4x-3}{5x+2} = \dfrac{4(4)-3}{5(4)+2}$ *Let x = 4.*

$$= \dfrac{16-3}{20+2} = \dfrac{13}{22}$$

2. **(a)** $\dfrac{3x}{x^2-4} = \dfrac{3(-2)}{(-2)^2-4}$ *Let x = -2.*

$$= \dfrac{-6}{4-4} = \dfrac{-6}{0}$$

Substituting -2 for x makes the denominator zero, so the given expression is undefined when $x = -2$.

(b) $\dfrac{3x}{x^2-4} = \dfrac{3(4)}{(4)^2-4}$ *Let x = 4.*

$$= \dfrac{12}{16-4} = \dfrac{12}{12} = 1$$

3. $\dfrac{4}{x-3}$

To find the values for which this expression is undefined, set the denominator equal to zero and solve for x.

$$x - 3 = 0$$
$$x = 3$$

Because $x = 3$ will make the denominator zero, the given expression is undefined for 3. Thus, $x \neq 3$.

4. $\dfrac{y+3}{2y}$

Set the denominator equal to zero and solve for y.

$$2y = 0$$
$$y = 0$$

The given expression is undefined for 0. Thus, $y \neq 0$.

5. $\dfrac{2k+1}{3k^2+17k+10}$

Set the denominator equal to zero and solve for k.

$$3k^2 + 17k + 10 = 0$$
$$(3k+2)(k+5) = 0$$
$$k = -\tfrac{2}{3} \text{ or } k = -5$$

The given expression is undefined for -5 and $-\tfrac{2}{3}$. Thus, $k \neq -5, -\tfrac{2}{3}$.

6. Set the denominator equal to 0 and solve the equation. Any solutions are values for which the rational expression is undefined.

7. $\dfrac{5a^3b^3}{15a^4b^2} = \dfrac{b \cdot 5a^3b^2}{3a \cdot 5a^3b^2} = \dfrac{b}{3a}$

8. $\dfrac{m-4}{4-m} = \dfrac{-1(4-m)}{4-m} = -1$

9. $\dfrac{4x^2-9}{6-4x} = \dfrac{(2x+3)(2x-3)}{-2(2x-3)}$

$\qquad = \dfrac{2x+3}{-2} = \dfrac{-1(2x+3)}{2}$

$\qquad = \dfrac{-(2x+3)}{2}$

10. $\dfrac{4p^2+8pq-5q^2}{10p^2-3pq-q^2} = \dfrac{(2p-q)(2p+5q)}{(5p+q)(2p-q)}$

$\qquad = \dfrac{2p+5q}{5p+q}$

11. $-\dfrac{4x-9}{2x+3}$

Apply the negative sign to the numerator:

$$\dfrac{-(4x-9)}{2x+3}$$

Now distribute the negative sign:

$$\dfrac{-4x+9}{2x+3}$$

Apply the negative sign to the denominator:

$$\dfrac{4x-9}{-(2x+3)}$$

Again, distribute the negative sign:

$$\dfrac{4x-9}{-2x-3}$$

12. $-\dfrac{8-3x}{3-6x}$

Four equivalent forms are:

$$\dfrac{-(8-3x)}{3-6x}, \qquad \dfrac{-8+3x}{3-6x},$$

$$\dfrac{8-3x}{-(3-6x)}, \qquad \dfrac{8-3x}{-3+6x}$$

13. $\dfrac{18p^3}{6} \cdot \dfrac{24}{p^4} = \dfrac{6 \cdot 3 \cdot 24p^3}{6p^4} = \dfrac{72}{p}$

14. $\dfrac{8x^2}{12x^5} \cdot \dfrac{6x^4}{2x} = \dfrac{2 \cdot 4}{3 \cdot 4x^3} \cdot \dfrac{3x^3}{1} = 2$

15. $\dfrac{x-3}{4} \cdot \dfrac{5}{2x-6} = \dfrac{x-3}{4} \cdot \dfrac{5}{2(x-3)} = \dfrac{5}{8}$

16. $\dfrac{2r+3}{r-4} \cdot \dfrac{r^2-16}{6r+9}$

$\qquad = \dfrac{2r+3}{r-4} \cdot \dfrac{(r+4)(r-4)}{3(2r+3)}$

$\qquad = \dfrac{r+4}{3}$

17. $\dfrac{6a^2+7a-3}{2a^2-a-6} \div \dfrac{a+5}{a-2}$

$\qquad = \dfrac{6a^2+7a-3}{2a^2-a-6} \cdot \dfrac{a-2}{a+5}$

$\qquad = \dfrac{(3a-1)(2a+3)}{(2a+3)(a-2)} \cdot \dfrac{a-2}{a+5}$

$\qquad = \dfrac{3a-1}{a+5}$

18. $\dfrac{y^2-6y+8}{y^2+3y-18} \div \dfrac{y-4}{y+6}$

$\qquad = \dfrac{y^2-6y+8}{y^2+3y-18} \cdot \dfrac{y+6}{y-4}$

$\qquad = \dfrac{(y-4)(y-2)}{(y+6)(y-3)} \cdot \dfrac{y+6}{y-4}$

$\qquad = \dfrac{y-2}{y-3}$

19. $\dfrac{2p^2+13p+20}{p^2+p-12} \cdot \dfrac{p^2+2p-15}{2p^2+7p+5}$

$\qquad = \dfrac{(2p+5)(p+4)}{(p+4)(p-3)} \cdot \dfrac{(p+5)(p-3)}{(2p+5)(p+1)}$

$\qquad = \dfrac{p+5}{p+1}$

20. $\dfrac{3z^2+5z-2}{9z^2-1} \cdot \dfrac{9z^2+6z+1}{z^2+5z+6}$

$\qquad = \dfrac{(3z-1)(z+2)}{(3z-1)(3z+1)} \cdot \dfrac{(3z+1)^2}{(z+3)(z+2)}$

$\qquad = \dfrac{3z+1}{z+3}$

21. $\dfrac{4}{9y}, \dfrac{7}{12y^2}, \dfrac{5}{27y^4}$

Factor each denominator.

$$9y = 3^2y$$
$$12y^2 = 2^2 \cdot 3 \cdot y^2$$
$$27y^4 = 3^3 \cdot y^4$$

LCD $= 2^2 \cdot 3^3 \cdot y^4 = 108y^4$

22. $\dfrac{3}{x^2+4x+3}, \dfrac{5}{x^2+5x+4}$

Factor each denominator.

$$x^2+4x+3 = (x+3)(x+1)$$
$$x^2+5x+4 = (x+1)(x+4)$$

LCD $= (x+3)(x+1)(x+4)$

23. $\dfrac{3}{2a^3} = \dfrac{?}{10a^4}$

$\dfrac{3}{2a^3} = \dfrac{3}{2a^3} \cdot \dfrac{5a}{5a} = \dfrac{15a}{10a^4}$

24. $\dfrac{9}{x-3} = \dfrac{?}{18-6x} = \dfrac{?}{-6(x-3)}$

$\dfrac{9}{x-3} = \dfrac{9}{x-3} \cdot \dfrac{-6}{-6}$

$\quad = \dfrac{-54}{-6x+18}$

$\quad = \dfrac{-54}{18-6x}$

25. $\dfrac{-3y}{2y-10} = \dfrac{?}{50-10y} = \dfrac{?}{-5(2y-10)}$

$\dfrac{-3y}{2y-10} = \dfrac{-3y}{2y-10} \cdot \dfrac{-5}{-5}$

$\quad = \dfrac{15y}{-10y+50}$

$\quad = \dfrac{15y}{50-10y}$

26. $\dfrac{4b}{b^2+2b-3} = \dfrac{?}{(b+3)(b-1)(b+2)}$

$\dfrac{4b}{b^2+2b-3} = \dfrac{4b}{(b+3)(b-1)}$

$\quad = \dfrac{4b}{(b+3)(b-1)} \cdot \dfrac{b+2}{b+2}$

$\quad = \dfrac{4b(b+2)}{(b+3)(b-1)(b+2)}$

27. $\dfrac{10}{x} + \dfrac{5}{x} = \dfrac{10+5}{x} = \dfrac{15}{x}$

28. $\dfrac{6}{3p} - \dfrac{12}{3p} = \dfrac{6-12}{3p} = \dfrac{-6}{3p} = -\dfrac{2}{p}$

29. $\dfrac{9}{k} - \dfrac{5}{k-5} = \dfrac{9(k-5)}{k(k-5)} - \dfrac{5 \cdot k}{(k-5)k}$

$\qquad\qquad LCD = k(k-5)$

$\quad = \dfrac{9(k-5)-5k}{k(k-5)}$

$\quad = \dfrac{9k-45-5k}{k(k-5)}$

$\quad = \dfrac{4k-45}{k(k-5)}$

30. $\dfrac{4}{y} + \dfrac{7}{7+y} = \dfrac{4(7+y)}{y(7+y)} + \dfrac{7 \cdot y}{(7+y)y}$

$\qquad\qquad LCD = y(7+y)$

$\quad = \dfrac{28+4y+7y}{y(7+y)}$

$\quad = \dfrac{28+11y}{y(7+y)}$

31. $\dfrac{m}{3} - \dfrac{2+5m}{6} = \dfrac{m \cdot 2}{3 \cdot 2} - \dfrac{2+5m}{6}$ $LCD = 6$

$\quad = \dfrac{2m-(2+5m)}{6}$

$\quad = \dfrac{2m-2-5m}{6}$

$\quad = \dfrac{-2-3m}{6}$

32. $\dfrac{12}{x^2} - \dfrac{3}{4x} = \dfrac{12 \cdot 4}{x^2 \cdot 4} - \dfrac{3 \cdot x}{4x \cdot x}$ $LCD = 4x^2$

$\quad = \dfrac{48-3x}{4x^2}$

$\quad = \dfrac{3(16-x)}{4x^2}$

33. $\dfrac{5}{a-2b} + \dfrac{2}{a+2b}$

$\quad = \dfrac{5(a+2b)}{(a-2b)(a+2b)} + \dfrac{2(a-2b)}{(a+2b)(a-2b)}$

$\qquad\qquad LCD = (a-2b)(a+2b)$

$\quad = \dfrac{5(a+2b)+2(a-2b)}{(a-2b)(a+2b)}$

$\quad = \dfrac{5a+10b+2a-4b}{(a-2b)(a+2b)}$

$\quad = \dfrac{7a+6b}{(a-2b)(a+2b)}$

34. $\dfrac{4}{k^2-9} - \dfrac{k+3}{3k-9}$

$\quad = \dfrac{4}{(k+3)(k-3)} - \dfrac{k+3}{3(k-3)}$

$\qquad\qquad LCD = 3(k+3)(k-3)$

$\quad = \dfrac{4 \cdot 3}{(k+3)(k-3) \cdot 3} - \dfrac{(k+3)(k+3)}{3(k-3)(k+3)}$

$\quad = \dfrac{12-(k+3)(k+3)}{3(k+3)(k-3)}$

$\quad = \dfrac{12-(k^2+6k+9)}{3(k+3)(k-3)}$

$\quad = \dfrac{12-k^2-6k-9}{3(k+3)(k-3)}$

$\quad = \dfrac{-k^2-6k+3}{3(k+3)(k-3)}$

35. $\dfrac{8}{z^2 + 6z} - \dfrac{3}{z^2 + 4z - 12}$

$= \dfrac{8}{z(z + 6)} - \dfrac{3}{(z + 6)(z - 2)}$

              LCD = z(z + 6)(z − 2)

$= \dfrac{8(z - 2)}{z(z + 6)(z - 2)} - \dfrac{3 \cdot z}{(z + 6)(z - 2) \cdot z}$

$= \dfrac{8(z - 2) - 3z}{z(z + 6)(z - 2)}$

$= \dfrac{8z - 16 - 3z}{z(z + 6)(z - 2)}$

$= \dfrac{5z - 16}{z(z + 6)(z - 2)}$

36. $\dfrac{11}{2p - p^2} - \dfrac{2}{p^2 - 5p + 6}$

$= \dfrac{11}{p(2 - p)} - \dfrac{2}{(p - 3)(p - 2)}$

          LCD = p(p − 3)(p − 2)

$= \dfrac{11(-1)(p - 3)}{p(2 - p)(-1)(p - 3)}$

$\quad\quad - \dfrac{2 \cdot p}{(p - 3)(p - 2)p}$

$= \dfrac{-11(p - 3) - 2p}{p(p - 2)(p - 3)}$

$= \dfrac{-11p + 33 - 2p}{p(p - 2)(p - 3)}$

$= \dfrac{-13p + 33}{p(p - 2)(p - 3)}$

37. (a) $\dfrac{\dfrac{a^4}{b^2}}{\dfrac{a^3}{b}} = \dfrac{a^4}{b^2} \div \dfrac{a^3}{b}$

$= \dfrac{a^4}{b^2} \cdot \dfrac{b}{a^3}$

$= \dfrac{a^4 b}{a^3 b^2} = \dfrac{a}{b}$

(b) $\dfrac{\dfrac{a^4}{b^2}}{\dfrac{a^3}{b}} = \dfrac{b^2\left(\dfrac{a^4}{b^2}\right)}{b^2\left(\dfrac{a^3}{b}\right)}$ *Multiply by LCD, b^2*

$= \dfrac{a^4}{ba^3} = \dfrac{a}{b}$

(c) For this problem, the difference in the methods is negligible. In general, Method 2 is preferable because it leads to quicker simplifications.

38. $\dfrac{\frac{2}{3} - \frac{1}{6}}{\frac{1}{4} + \frac{2}{5}} = \dfrac{60\left(\frac{2}{3} - \frac{1}{6}\right)}{60\left(\frac{1}{4} + \frac{2}{5}\right)}$ *Multiply by LCD, 60*

$= \dfrac{60 \cdot \frac{2}{3} - 60 \cdot \frac{1}{6}}{60 \cdot \frac{1}{4} + 60 \cdot \frac{2}{5}}$

$= \dfrac{40 - 10}{15 + 24} = \dfrac{30}{39} = \dfrac{10}{13}$

39. $\dfrac{\dfrac{y - 3}{y}}{\dfrac{y + 3}{4y}} = \dfrac{y - 3}{y} \cdot \dfrac{4y}{y + 3} = \dfrac{4(y - 3)}{y + 3}$

40. $\dfrac{\dfrac{1}{p} - \dfrac{1}{q}}{\dfrac{1}{q - p}} = \dfrac{\left(\dfrac{1}{p} - \dfrac{1}{q}\right)pq(q - p)}{\left(\dfrac{1}{q - p}\right)pq(q - p)}$

      Multiply by LCD, $pq(q - p)$.

$= \dfrac{\dfrac{1}{p}[pq(q - p)] - \dfrac{1}{q}[pq(q - p)]}{pq}$

$= \dfrac{q(q - p) - p(q - p)}{pq}$

$= \dfrac{q^2 - pq - pq + p^2}{pq}$

$= \dfrac{q^2 - 2pq + p^2}{pq}$

$= \dfrac{(q - p)^2}{pq}$

41. $\dfrac{x + \dfrac{1}{w}}{x - \dfrac{1}{w}}$

$= \dfrac{\left(x + \dfrac{1}{w}\right) \cdot w}{\left(x - \dfrac{1}{w}\right) \cdot w}$ *Multiply by LCD, w.*

$= \dfrac{xw + \left(\dfrac{1}{w}\right)w}{xw - \left(\dfrac{1}{w}\right)w}$

$= \dfrac{xw + 1}{xw - 1}$

42. $\dfrac{\dfrac{1}{r+t} - 1}{\dfrac{1}{r+t} + 1}$

$= \dfrac{\left(\dfrac{1}{r+t} - 1\right)(r+t)}{\left(\dfrac{1}{r+t} + 1\right)(r+t)}$ *Multiply by LCD, r + t.*

$= \dfrac{\dfrac{1}{r+t}(r+t) - 1(r+t)}{\dfrac{1}{r+t}(r+t) + 1(r+t)}$

$= \dfrac{1 - r - t}{1 + r + t}$

43. When 2 is substituted for m throughout the equation, the value of the denominator in the first and third expressions is zero.

44. $\dfrac{4-z}{z} + \dfrac{3}{2} = \dfrac{-4}{z}$

Multiply each side by the LCD, $2z$.

$$2z\left(\dfrac{4-z}{z} + \dfrac{3}{2}\right) = 2z\left(-\dfrac{4}{z}\right)$$

$$2z\left(\dfrac{4-z}{z}\right) + 2z\left(\dfrac{3}{2}\right) = -8$$

$$2(4-z) + 3z = -8$$

$$8 - 2z + 3z = -8$$

$$8 + z = -8$$

$$z = -16$$

Check $z = -16$: $\quad -\frac{5}{4} + \frac{3}{2} = \frac{1}{4}$ *True*

Thus, the solution set is $\{-16\}$.

45. $\dfrac{3x-1}{x-2} = \dfrac{5}{x-2} + 1$

$(x-2)\left(\dfrac{3x-1}{x-2}\right) = (x-2)\left(\dfrac{5}{x-2} + 1\right)$
   Multiply by LCD, x − 2.

$(x-2)\left(\dfrac{3x-1}{x-2}\right) = (x-2)\left(\dfrac{5}{x-2}\right)$
$\qquad\qquad\qquad + (x-2)(1)$
   Distributive property

$$3x - 1 = 5 + x - 2$$

$$3x - 1 = 3 + x$$

$$2x = 4$$

$$x = 2$$

The solution set is $\emptyset$ because $x = 2$ makes the original denominators equal to zero.

46. $\dfrac{3}{m-2} + \dfrac{1}{m-1} = \dfrac{7}{m^2 - 3m + 2}$

$\dfrac{3}{m-2} + \dfrac{1}{m-1} = \dfrac{7}{(m-2)(m-1)}$

$(m-2)(m-1)\left(\dfrac{3}{m-2} + \dfrac{1}{m-1}\right)$

$\qquad = (m-2)(m-1) \cdot \dfrac{7}{(m-2)(m-1)}$
   Multiply by LCD, (m − 2)(m − 1).

$$3(m-1) + 1(m-2) = 7$$

$$3m - 3 + m - 2 = 7$$

$$4m - 5 = 7$$

$$4m = 12$$

$$m = 3$$

Check $m = 3$: $\quad 3 + \frac{1}{2} = \frac{7}{2}$ *True*

Thus, the solution set is $\{3\}$.

47. $m = \dfrac{Ry}{t}$ for t

$t \cdot m = t\left(\dfrac{Ry}{t}\right)$ *Multiply by t.*

$tm = Ry$

$t = \dfrac{Ry}{m}$ *Divide by m.*

48. $x = \dfrac{3y-5}{4}$ for y

$4x = 4\left(\dfrac{3y-5}{4}\right)$

$4x = 3y - 5$

$4x + 5 = 3y$

$\dfrac{4x+5}{3} = y$

49. $p^2 = \dfrac{4}{3m - q}$ for m

$(3m - q)p^2 = (3m - q)\left(\dfrac{4}{3m - q}\right)$

$3mp^2 - p^2 q = 4$

$3mp^2 = 4 + p^2 q$

$m = \dfrac{4 + p^2 q}{3p^2}$

50. Let $x =$ the numerator. Then $x - 5 =$ the denominator. Adding 5 to both the numerator and the denominator gives us a fraction that is equivalent to $\frac{5}{4}$.

$$\frac{x+5}{x-5+5} = \frac{5}{4}$$
$$\frac{x+5}{x} = \frac{5}{4}$$
$$4x\left(\frac{x+5}{x}\right) = 4x\left(\frac{5}{4}\right)$$
$$4(x+5) = x(5)$$
$$4x + 20 = 5x$$
$$20 = x$$

The numerator is 20 and the denominator is $20 - 5 = 15$, so the original fraction is $\frac{20}{15}$.

51. Let $x =$ the numerator. Then $6x =$ the denominator. Adding 3 to the numerator and subtracting 3 from the denominator gives us a fraction equivalent to $\frac{2}{5}$.

$$\frac{x+3}{6x-3} = \frac{2}{5}$$
$$5(6x-3)\left(\frac{x+3}{6x-3}\right) = 5(6x-3)\left(\frac{2}{5}\right)$$
$$5(x+3) = 2(6x-3)$$
$$5x + 15 = 12x - 6$$
$$21 = 7x$$
$$3 = x$$

The numerator is 3 and the denominator is $6 \cdot 3 = 18$, so the original fraction is $\frac{3}{18}$.

52. Let $x =$ the speed of the wind. Then the speed against the wind is $165 - x$ and the speed with the wind is $165 + x$. Complete the chart using $t = d/r$.

	d	r	t
Against the Wind	310	$165 - x$	$\dfrac{310}{165-x}$
With the Wind	350	$165 + x$	$\dfrac{350}{165+x}$

Since the times are equal, we get the following equation.

$$\frac{310}{165-x} = \frac{350}{165+x}$$
$$(165+x)(165-x)\frac{310}{165-x} = (165+x)(165-x)\frac{350}{165+x}$$
$$310(165+x) = 350(165-x)$$
$$51{,}150 + 310x = 57{,}750 - 350x$$
$$660x = 6600$$
$$x = 10$$

The speed of the wind is 10 miles per hour.

53. *Step 2*
Let $x =$ the number of hours it takes them to do the job working together.

	Rate	Time Working Together	Fractional Part of the Job Done when Working Together
Dennis	$\frac{1}{5}$	x	$\frac{1}{5}x$
Friend	$\frac{1}{8}$	x	$\frac{1}{8}x$

Step 3
Working together, they do 1 whole job, so

$$\tfrac{1}{5}x + \tfrac{1}{8}x = 1.$$

Step 4
Solve this equation by multiplying both sides by the LCD, 40.

$$40\left(\tfrac{1}{5}x + \tfrac{1}{8}x\right) = 40(1)$$
$$8x + 5x = 40$$
$$13x = 40$$
$$x = \tfrac{40}{13}, \text{ or } 3\tfrac{1}{13}$$

Step 5
Working together, it takes them $3\frac{1}{13}$ hours.

Step 6
Dennis does $\frac{1}{5}$ of the job per hour for $\frac{40}{13}$ hours:

$$\tfrac{1}{5} \cdot \tfrac{40}{13} = \tfrac{8}{13} \text{ of the job}$$

His friend does $\frac{1}{8}$ of the job per hour for $\frac{40}{13}$ hours:

$$\tfrac{1}{8} \cdot \tfrac{40}{13} = \tfrac{5}{13} \text{ of the job}$$

Together, they have done

$$\tfrac{8}{13} + \tfrac{5}{13} = \tfrac{13}{13} = 1 \text{ total job.}$$

54. *Step 2*
Let $x =$ the time needed by the head gardener to mow the lawns.
Then $2x =$ the time needed by the assistant to mow the lawns.

	Rate	Time Working Together	Fractional Part of the Job Done when Working Together
Head gardener	$\dfrac{1}{x}$	$1\frac{1}{3} = \frac{4}{3}$	$\dfrac{1}{x} \cdot \dfrac{4}{3} = \dfrac{4}{3x}$
Assistant	$\dfrac{1}{2x}$	$1\frac{1}{3} = \frac{4}{3}$	$\dfrac{1}{2x} \cdot \dfrac{4}{3} = \dfrac{2}{3x}$

Step 3

$$\frac{4}{3x} + \frac{2}{3x} = 1$$

Step 4

$$\frac{4+2}{3x} = 1$$

$$\frac{6}{3x} = 1$$

$$3x\left(\frac{6}{3x}\right) = 3x(1)$$

$$6 = 3x$$

$$2 = x$$

Step 5
It takes the head gardener 2 hours to mow the lawns.

Step 6
The head gardener does $\frac{1}{2}$ of the job per hour for $\frac{4}{3}$ hours:

$$\frac{1}{2} \cdot \frac{4}{3} = \frac{2}{3} \text{ of the job}$$

The assistant does $\frac{1}{4}$ of the job per hour for $\frac{4}{3}$ hours:

$$\frac{1}{4} \cdot \frac{4}{3} = \frac{1}{3} \text{ of the job}$$

Together, they have done

$$\frac{2}{3} + \frac{1}{3} = \frac{3}{3} = 1 \text{ total job.}$$

55. [6.4] $\dfrac{4}{m-1} - \dfrac{3}{m+1}$

To perform the indicated subtraction, use $(m-1)(m+1)$ as the LCD.

$$\frac{4}{m-1} - \frac{3}{m+1}$$

$$= \frac{4(m+1)}{(m-1)(m+1)} - \frac{3(m-1)}{(m+1)(m-1)}$$

$$= \frac{4(m+1) - 3(m-1)}{(m-1)(m+1)}$$

$$= \frac{4m+4-3m+3}{(m-1)(m+1)}$$

$$= \frac{m+7}{(m-1)(m+1)}$$

56. [6.2] $\dfrac{8p^5}{5} \div \dfrac{2p^3}{10}$

To perform the indicated division, multiply the first rational expression by the reciprocal of the second.

$$\frac{8p^5}{5} \div \frac{2p^3}{10} = \frac{8p^5}{5} \cdot \frac{10}{2p^3}$$

$$= \frac{80p^5}{10p^3}$$

$$= 8p^2$$

57. [6.2] $\dfrac{r-3}{8} \div \dfrac{3r-9}{4} = \dfrac{r-3}{8} \cdot \dfrac{4}{3r-9}$

$$= \frac{r-3}{8} \cdot \frac{4}{3(r-3)}$$

$$= \frac{4}{24} = \frac{1}{6}$$

58. [6.5] $\dfrac{\dfrac{5}{x} - 1}{\dfrac{5-x}{3x}} = \dfrac{\left(\dfrac{5}{x} - 1\right)3x}{\left(\dfrac{5-x}{3x}\right)3x}$ *Multiply by LCD, 3x.*

$$= \frac{\dfrac{5}{x}(3x) - 1(3x)}{5-x}$$

$$= \frac{15 - 3x}{5-x}$$

$$= \frac{3(5-x)}{5-x} = 3$$

59. [6.4] $\dfrac{4}{z^2 - 2z + 1} - \dfrac{3}{z^2 - 1}$

$$= \frac{4}{(z-1)^2} - \frac{3}{(z+1)(z-1)}$$

$$LCD = (z+1)(z-1)^2$$

$$= \frac{4(z+1)}{(z-1)^2(z+1)}$$

$$- \frac{3(z-1)}{(z+1)(z-1)(z-1)}$$

$$= \frac{4(z+1) - 3(z-1)}{(z+1)(z-1)^2}$$

$$= \frac{4z+4-3z+3}{(z+1)(z-1)^2}$$

$$= \frac{z+7}{(z+1)(z-1)^2}$$

60. [6.6] Solve $a = \dfrac{v-w}{t}$ for v.

$$t \cdot a = v - w \quad \textit{Multiply by t.}$$
$$at + w = v \quad \textit{Add w.}$$

61. [6.6] $\dfrac{2}{z} - \dfrac{z}{z+3} = \dfrac{1}{z+3}$

Multiply each side of the equation by the LCD, $z(z+3)$.

$$z(z+3)\left(\frac{2}{z} - \frac{z}{z+3}\right) = z(z+3)\left(\frac{1}{z+3}\right)$$

$$z(z+3)\left(\frac{2}{z}\right) - z(z+3)\left(\frac{z}{z+3}\right) = z(1)$$

$$2(z+3) - z^2 = z$$
$$2z + 6 - z^2 = z$$
$$0 = z^2 - z - 6$$
$$0 = (z-3)(z+2)$$

$$z - 3 = 0 \quad \text{or} \quad z + 2 = 0$$
$$z = 3 \quad \text{or} \qquad z = -2$$

Check $z = -2$: $-1 - (-2) = 1$ *True*

Check $z = 3$: $\quad \frac{2}{3} - \frac{1}{2} = \frac{1}{6}$ *True*

Thus, the solution set is $\{-2, 3\}$.

62. **[6.7]** Let $x =$ the number of hours it takes them to do the job working together.

	Rate	Time Working Together	Fractional Part of the Job Done when Working Together
Lizette	$\frac{1}{8}$	x	$\frac{1}{8}x$
Laura	$\frac{1}{14}$	x	$\frac{1}{14}x$

Working together, they do 1 whole job, so

$$\frac{1}{8}x + \frac{1}{14}x = 1.$$

To clear fractions, multiply both sides by the LCD, 56.

$$56\left(\frac{1}{8}x + \frac{1}{14}x\right) = 56(1)$$
$$7x + 4x = 56$$
$$11x = 56$$
$$x = \frac{56}{11}, \quad \text{or} \quad 5\frac{1}{11}$$

Working together, they can paint the house in $5\frac{1}{11}$ hours.

63. **[6.7]** Let $x =$ the speed of the plane in still air. Then the speed of the plane with the wind is $x + 50$, and the speed of the plane against the wind is $x - 50$. Use $t = \dfrac{d}{r}$ to complete the chart.

	d	r	t
With the Wind	400	$x + 50$	$\dfrac{400}{x + 50}$
Against the Wind	200	$x - 50$	$\dfrac{200}{x - 50}$

The times are the same, so

$$\frac{400}{x + 50} = \frac{200}{x - 50}.$$

To solve this equation, multiply both sides by the LCD, $(x + 50)(x - 50)$.

$$(x + 50)(x - 50) \cdot \frac{400}{x + 50}$$
$$= (x + 50)(x - 50) \cdot \frac{200}{x - 50}$$
$$400(x - 50) = 200(x + 50)$$
$$400x - 20{,}000 = 200x + 10{,}000$$
$$200x = 30{,}000$$
$$x = 150$$

The speed of the plane is 150 kilometers per hour.

64. **[6.7]** There was a different exercise in the first printing of the text. The exercise is as follows.

If the same number is added to both the numerator and denominator of the fraction $\frac{3}{7}$, the result is equivalent to $\frac{3}{4}$. Find the number.

Let $x =$ the number to be added to both the numerator and denominator of $\frac{3}{7}$.

This fraction is equivalent to $\frac{3}{4}$, so

$$\frac{3 + x}{7 + x} = \frac{3}{4}.$$

Multiply both sides by the LCD, $4(7 + x)$.

$$4(7 + x) \cdot \frac{3 + x}{7 + x} = 4(7 + x) \cdot \frac{3}{4}$$
$$4 \cdot (3 + x) = (7 + x) \cdot 3$$
$$12 + 4x = 21 + 3x$$
$$x = 9$$

The number is 9.

65. **(a)** $x + 3 = 0$
$$x = -3$$

This value of x makes the value of the denominator zero, so P will be undefined when $x = -3$.

(b) $x + 1 = 0$
$$x = -1$$

This value of x makes the value of the denominator zero, so Q will be undefined when $x = -1$.

(c) $\qquad x^2 + 4x + 3 = 0$
$$(x + 3)(x + 1) = 0$$
$$x = -3 \quad \text{or} \quad x = -1$$

These values of x make the value of the denominator zero, so the values for which R is undefined are -3 and -1.

66. $(P \cdot Q) \div R$

$$= \left(\frac{6}{x+3} \cdot \frac{5}{x+1} \right) \div \frac{4x}{x^2+4x+3}$$

$$= \frac{30}{(x+3)(x+1)} \cdot \frac{x^2+4x+3}{4x}$$

$$= \frac{30}{(x+3)(x+1)} \cdot \frac{(x+3)(x+1)}{4x}$$

$$= \frac{30}{4x} = \frac{15}{2x}$$

67. If $x = 0$, the divisor R is equal to 0, and division by 0 is undefined.

68. List the three denominators and factor if possible.

$x+3, \; x+1, \; x^2+4x+3 = (x+3)(x+1)$

The LCD for P, Q, and R is $(x+3)(x+1)$.

69. $P + Q - R$

$$= \frac{6}{x+3} + \frac{5}{x+1} - \frac{4x}{x^2+4x+3}$$

$$= \frac{6}{x+3} + \frac{5}{x+1} - \frac{4x}{(x+3)(x+1)}$$

$$LCD = (x+3)(x+1)$$

$$= \frac{6(x+1)}{(x+3)(x+1)} + \frac{5(x+3)}{(x+3)(x+1)}$$

$$- \frac{4x}{(x+3)(x+1)}$$

$$= \frac{6(x+1) + 5(x+3) - 4x}{(x+3)(x+1)}$$

$$= \frac{6x+6+5x+15-4x}{(x+3)(x+1)}$$

$$= \frac{7x+21}{(x+3)(x+1)}$$

$$= \frac{7(x+3)}{(x+3)(x+1)} = \frac{7}{x+1}$$

70. $\dfrac{P+Q}{R} = \dfrac{\dfrac{6}{x+3}+\dfrac{5}{x+1}}{\dfrac{4x}{x^2+4x+3}} = \dfrac{\dfrac{6}{x+3}+\dfrac{5}{x+1}}{\dfrac{4x}{(x+3)(x+1)}}$

To simplify this complex fraction, use Method 2. Multiply numerator and denominator by the LCD for all the fractions, $(x+3)(x+1)$.

$$\frac{(x+3)(x+1)\left(\frac{6}{x+3} + \frac{5}{x+1} \right)}{(x+3)(x+1)\left(\frac{4x}{(x+3)(x+1)} \right)}$$

$$= \frac{(x+3)(x+1)\left(\frac{6}{x+3} \right) + (x+3)(x+1)\left(\frac{5}{x+1} \right)}{4x}$$

$$= \frac{6(x+1) + 5(x+3)}{4x}$$

$$= \frac{6x+6+5x+15}{4x}$$

$$= \frac{11x+21}{4x}$$

71.
$$P + Q = R$$

$$\frac{6}{x+3} + \frac{5}{x+1} = \frac{4x}{x^2+4x+3}$$

$$\frac{6}{x+3} + \frac{5}{x+1} = \frac{4x}{(x+3)(x+1)}$$

Multiply by LCD, $(x+3)(x+1)$.

$$(x+3)(x+1)\left(\frac{6}{x+3} \right) + (x+3)(x+1)\left(\frac{5}{x+1} \right)$$

$$= (x+3)(x+1)\left(\frac{4x}{(x+3)(x+1)} \right)$$

$$6(x+1) + 5(x+3) = 4x$$

$$6x+6+5x+15 = 4x$$

$$7x = -21$$

$$x = -3$$

To check, substitute -3 for x. The denominators of the first and third fractions become zero. Reject -3 as a solution. There is no solution to the equation, so the solution set is $\emptyset$.

72. We know that -3 is not allowed because P and R are undefined for $x = -3$.

73. Solving $d = rt$ for r gives us $r = d/t$.
If $d = 6$ miles and $t = (x+3)$ minutes, then
$$r = \frac{d}{t} = \frac{6}{x+3} \text{ miles per minute. Thus,}$$

$$P = \frac{6}{x+3}$$

represents the rate of the car (in miles per minute).

74.
$$R = \frac{4x}{x^2 + 4x + 3}$$

$$\frac{40}{77} = \frac{4x}{x^2 + 4x + 3}$$

$$40(x^2 + 4x + 3) = (4x)(77) \quad \text{\textit{Cross products are equal.}}$$

$$40x^2 + 160x + 120 = 308x$$

$$40x^2 - 148x + 120 = 0$$

$$10x^2 - 37x + 30 = 0$$

$$(5x - 6)(2x - 5) = 0$$

$$x = \tfrac{6}{5} \quad \text{or} \quad x = \tfrac{5}{2}$$

Chapter 6 Test

1. **(a)** $\dfrac{6r + 1}{2r^2 - 3r - 20}$

$$= \frac{6(-2) + 1}{2(-2)^2 - 3(-2) - 20} \quad \text{\textit{Let r = -2.}}$$

$$= \frac{-12 + 1}{2 \cdot 4 + 6 - 20}$$

$$= \frac{-11}{8 + 6 - 20}$$

$$= \frac{-11}{-6} = \frac{11}{6}$$

(b) $\dfrac{6r + 1}{2r^2 - 3r - 20}$

$$= \frac{6(4) + 1}{2(4)^2 - 3(4) - 20} \quad \text{\textit{Let r = 4.}}$$

$$= \frac{24 + 1}{2 \cdot 16 - 12 - 20}$$

$$= \frac{25}{32 - 12 - 20}$$

$$= \frac{25}{20 - 20} = \frac{25}{0}$$

The expression is undefined when $r = 4$ because the denominator is 0.

2. $\dfrac{3x - 1}{x^2 - 2x - 8}$

Set the denominator equal to zero and solve for x.

$$x^2 - 2x - 8 = 0$$

$$(x + 2)(x - 4) = 0$$

$$x + 2 = 0 \quad \text{or} \quad x - 4 = 0$$

$$x = -2 \quad \text{or} \quad x = 4$$

The expression is undefined for -2 and 4, so $x \neq -2, 4$.

3. $-\dfrac{6x - 5}{2x + 3}$

Apply the negative sign to the numerator:

$$\frac{-(6x - 5)}{2x + 3}$$

Now distribute the negative sign:

$$\frac{-6x + 5}{2x + 3}$$

Apply the negative sign to the denominator:

$$\frac{6x - 5}{-(2x + 3)}$$

Again, distribute the negative sign:

$$\frac{6x - 5}{-2x - 3}$$

4. $\dfrac{-15x^6y^4}{5x^4y} = \dfrac{(5x^4y)(-3x^2y^3)}{(5x^4y)(1)}$

$$= \frac{5x^4y}{5x^4y} \cdot \frac{-3x^2y^3}{1} = -3x^2y^3$$

5. $\dfrac{6a^2 + a - 2}{2a^2 - 3a + 1} = \dfrac{(3a + 2)(2a - 1)}{(2a - 1)(a - 1)}$

$$= \frac{3a + 2}{a - 1}$$

6. $\dfrac{5(d - 2)}{9} \div \dfrac{3(d - 2)}{5}$

$$= \frac{5(d - 2)}{9} \cdot \frac{5}{3(d - 2)}$$

$$= \frac{5 \cdot 5}{9 \cdot 3} = \frac{25}{27}$$

7. $\dfrac{6k^2 - k - 2}{8k^2 + 10k + 3} \cdot \dfrac{4k^2 + 7k + 3}{3k^2 + 5k + 2}$

$$= \frac{(3k - 2)(2k + 1)}{(4k + 3)(2k + 1)} \cdot \frac{(4k + 3)(k + 1)}{(3k + 2)(k + 1)}$$

$$= \frac{3k - 2}{3k + 2}$$

8. $\dfrac{4a^2 + 9a + 2}{3a^2 + 11a + 10} \div \dfrac{4a^2 + 17a + 4}{3a^2 + 2a - 5}$

$$= \frac{4a^2 + 9a + 2}{3a^2 + 11a + 10} \cdot \frac{3a^2 + 2a - 5}{4a^2 + 17a + 4}$$

$$= \frac{(4a + 1)(a + 2)}{(3a + 5)(a + 2)} \cdot \frac{(3a + 5)(a - 1)}{(4a + 1)(a + 4)}$$

$$= \frac{a - 1}{a + 4}$$

9. $\dfrac{-3}{10p^2}, \dfrac{21}{25p^3}, \dfrac{-7}{30p^5}$

Factor each denominator.

$$10p^2 = 2 \cdot 5 \cdot p^2$$
$$25p^3 = 5^2 \cdot p^3$$
$$30p^5 = 2 \cdot 3 \cdot 5 \cdot p^5$$

$$LCD = 2 \cdot 3 \cdot 5^2 \cdot p^5 = 150p^5$$

10. $\dfrac{r+1}{2r^2 + 7r + 6}, \dfrac{-2r+1}{2r^2 - 7r - 15}$

Factor each denominator.

$$2r^2 + 7r + 6 = (2r+3)(r+2)$$
$$2r^2 - 7r - 15 = (2r+3)(r-5)$$

$$LCD = (2r+3)(r+2)(r-5)$$

11. $\dfrac{15}{4p} = \dfrac{}{64p^3} = \dfrac{}{4p \cdot 16p^2}$

$$\dfrac{15}{4p} = \dfrac{15 \cdot 16p^2}{4p \cdot 16p^2} = \dfrac{240p^2}{64p^3}$$

12. $\dfrac{3}{6m-12} = \dfrac{}{42m-84} = \dfrac{}{7(6m-12)}$

$$\dfrac{3}{6m-12} = \dfrac{3 \cdot 7}{(6m-12)7} = \dfrac{21}{42m-84}$$

13. $\dfrac{4x+2}{x+5} + \dfrac{-2x+8}{x+5}$

$$= \dfrac{(4x+2) + (-2x+8)}{x+5}$$

$$= \dfrac{2x+10}{x+5}$$

$$= \dfrac{2(x+5)}{x+5} = 2$$

14. $\dfrac{-4}{y+2} + \dfrac{6}{5y+10}$

$$= \dfrac{-4}{y+2} + \dfrac{6}{5(y+2)} \quad LCD = 5(y+2)$$

$$= \dfrac{-4 \cdot 5}{(y+2) \cdot 5} + \dfrac{6}{5(y+2)}$$

$$= \dfrac{-20+6}{5(y+2)} = \dfrac{-14}{5(y+2)}$$

15. Using LCD $= 3 - x$,

$$\dfrac{x+1}{3-x} + \dfrac{x^2}{x-3} = \dfrac{x+1}{3-x} + \dfrac{-1(x^2)}{-1(x-3)}$$

$$= \dfrac{x+1}{3-x} + \dfrac{-x^2}{-x+3}$$

$$= \dfrac{x+1}{3-x} + \dfrac{-x^2}{3-x}$$

$$= \dfrac{(x+1) + (-x^2)}{3-x}$$

$$= \dfrac{-x^2 + x + 1}{3-x}$$

If we use $x - 3$ for the LCD, we obtain the equivalent answer

$$\dfrac{x^2 - x - 1}{x - 3}.$$

16. $\dfrac{3}{2m^2 - 9m - 5} - \dfrac{m+1}{2m^2 - m - 1}$

$$= \dfrac{3}{(2m+1)(m-5)} - \dfrac{m+1}{(2m+1)(m-1)}$$

$$LCD = (2m+1)(m-5)(m-1)$$

$$= \dfrac{3(m-1)}{(2m+1)(m-5)(m-1)}$$

$$- \dfrac{(m+1)(m-5)}{(2m+1)(m-1)(m-5)}$$

$$= \dfrac{3(m-1) - (m+1)(m-5)}{(2m+1)(m-5)(m-1)}$$

$$= \dfrac{(3m-3) - (m^2 - 4m - 5)}{(2m+1)(m-5)(m-1)}$$

$$= \dfrac{3m - 3 - m^2 + 4m + 5}{(2m+1)(m-5)(m-1)}$$

$$= \dfrac{-m^2 + 7m + 2}{(2m+1)(m-5)(m-1)}$$

17. $\dfrac{\dfrac{2p}{k^2}}{\dfrac{3p^2}{k^3}} = \dfrac{2p}{k^2} \div \dfrac{3p^2}{k^3}$

$$= \dfrac{2p}{k^2} \cdot \dfrac{k^3}{3p^2}$$

$$= \dfrac{2k^3 p}{3k^2 p^2} = \dfrac{2k}{3p}$$

18. $\dfrac{\dfrac{1}{x+3} - 1}{1 + \dfrac{1}{x+3}}$

Start by multiplying the numerator and the denominator by the LCD, $x + 3$.

$$= \dfrac{(x+3)\left(\dfrac{1}{x+3} - 1\right)}{(x+3)\left(1 + \dfrac{1}{x+3}\right)}$$

$$= \dfrac{(x+3)\left(\dfrac{1}{x+3}\right) - (x+3)(1)}{(x+3)(1) + (x+3)\left(\dfrac{1}{x+3}\right)}$$

$$= \frac{1 - (x + 3)}{(x + 3) + 1}$$

$$= \frac{1 - x - 3}{x + 4}$$

$$= \frac{-2 - x}{x + 4}$$

19. $\dfrac{2x}{x - 3} + \dfrac{1}{x + 3} = \dfrac{-6}{x^2 - 9}$

$$\frac{2x}{x - 3} + \frac{1}{x + 3} = \frac{-6}{(x + 3)(x - 3)}$$

$$(x + 3)(x - 3)\left(\frac{2x}{x - 3} + \frac{1}{x + 3}\right)$$

$$= (x + 3)(x - 3)\left(\frac{-6}{(x + 3)(x - 3)}\right)$$

Multiply by LCD, (x + 3)(x − 3).

$$2x(x + 3) + 1(x - 3) = -6$$

$$2x^2 + 6x + x - 3 = -6$$

$$2x^2 + 7x + 3 = 0$$

$$(2x + 1)(x + 3) = 0$$

$$x = -\tfrac{1}{2} \quad \text{or} \quad x = -3$$

x cannot equal -3 because the denominator $x + 3$ would equal 0.

Check $x = -\tfrac{1}{2}$: $\tfrac{2}{7} + \tfrac{2}{5} = \tfrac{24}{35}$ *True*

Thus, the solution set is $\left\{-\tfrac{1}{2}\right\}$.

20. Solve $F = \dfrac{k}{d - D}$ for D.

$$(d - D)(F) = (d - D)\left(\frac{k}{d - D}\right) \quad \begin{array}{l}\textit{Multiply by}\\ \textit{LCD, } d - D.\end{array}$$

$$(d - D)(F) = k$$

$$dF - DF = k$$

$$-DF = k - dF$$

$$D = \frac{k - dF}{-F}, \quad \text{or} \quad D = \frac{dF - k}{F}$$

21. Let x = the speed of the current.

	d	**r**	**t**
Upstream	20	$7 - x$	$\dfrac{20}{7 - x}$
Downstream	50	$7 + x$	$\dfrac{50}{7 + x}$

The times are equal, so

$$\frac{20}{7 - x} = \frac{50}{7 + x}.$$

$$(7 - x)(7 + x)\left(\frac{20}{7 - x}\right) = (7 - x)(7 + x)\left(\frac{50}{7 + x}\right)$$

Multiply by LCD, (7 − x)(7 + x).

$$20(7 + x) = 50(7 - x)$$

$$140 + 20x = 350 - 50x$$

$$70x = 210$$

$$x = 3$$

The speed of the current is 3 miles per hour.

22. Let x = the time required for the couple to paint the room working together.

	Rate	**Time Working Together**	**Fractional Part of the Job Done when Working Together**
Abdalla	$\tfrac{1}{5}$	x	$\tfrac{1}{5}x$
Neighbor	$\tfrac{1}{4}$	x	$\tfrac{1}{4}x$

Working together, they do 1 whole job, so

$$\tfrac{1}{5}x + \tfrac{1}{4}x = 1.$$

$$20\left(\tfrac{1}{5}x + \tfrac{1}{4}x\right) = 20(1) \quad \begin{array}{l}\textit{Multiply by}\\ \textit{LCD, 20.}\end{array}$$

$$20\left(\tfrac{1}{5}x\right) + 20\left(\tfrac{1}{4}x\right) = 20$$

$$4x + 5x = 20$$

$$9x = 20$$

$$x = \tfrac{20}{9}, \quad \text{or} \quad 2\tfrac{2}{9}$$

They can paint the room in $2\tfrac{2}{9}$ hours.

Cumulative Review Exercises (Chapters 1–6)

1. $3 + 4\left(\tfrac{1}{2} - \tfrac{3}{4}\right)$

$$= 3 + 4\left(\tfrac{2}{4} - \tfrac{3}{4}\right)$$

$$= 3 + 4\left(-\tfrac{1}{4}\right) \qquad \textit{Parentheses}$$

$$= 3 + (-1) \qquad \textit{Multiplication}$$

$$= 2 \qquad \textit{Addition}$$

2. $3(2y - 5) = 2 + 5y$

$$6y - 15 = 2 + 5y$$

$$y = 17$$

The solution set is $\{17\}$.

3. Solve $A = \tfrac{1}{2}bh$ for b.

$$2 \cdot A = 2 \cdot \tfrac{1}{2}bh$$

$$2A = bh$$

$$\frac{2A}{h} = b$$

4. $\dfrac{2 + m}{2 - m} = \dfrac{3}{4}$

$$4(2 + m) = 3(2 - m) \quad \textit{Cross multiply.}$$

$$8 + 4m = 6 - 3m$$

$$7m = -2$$

$$m = -\tfrac{2}{7}$$

The solution set is $\left\{-\tfrac{2}{7}\right\}$.

5. $5y \leq 6y + 8$

$-y \leq 8$

$y \geq -8$ *Reverse the inequality symbol.*

The solution set is $[-8, \infty)$.

6. $5m - 9 > 2m + 3$

$3m > 12$

$m > 4$

The solution set is $(4, \infty)$.

7. $4x + 3y = -12$

(a) Let $y = 0$ to find the x-intercept.

$$4x + 3(0) = -12$$
$$4x = -12$$
$$x = -3$$

The x-intercept is $(-3, 0)$.

(b) Let $x = 0$ to find the y-intercept.

$$4(0) + 3y = -12$$
$$3y = -12$$
$$y = -4$$

The y-intercept is $(0, -4)$.

8. $y = -3x + 2$

This is an equation of a line.

If $x = 0$, $y = 2$; so the y-intercept is $(0, 2)$.
If $x = 1$, $y = -1$; and if $x = 2$, $y = -4$.

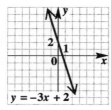

9. $y = -x^2 + 1$

This is the equation of a parabola opening downward with a y-intercept $(0, 1)$. If $x = +2$ or -2, $y = -3$.

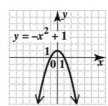

10. Through $(-5, 8)$ and $(-1, 2)$

$$m = \frac{\text{change in } y}{\text{change in } x} = \frac{2 - 8}{-1 - (-5)} = \frac{-6}{4} = -\frac{3}{2}$$

11. Perpendicular to $4x - 3y = 12$
Solve for y.

$$4x - 3y = 12$$
$$-3y = -4x + 12$$
$$y = \tfrac{4}{3}x - 4$$

The given line has slope $\frac{4}{3}$, so the slope of the line perpendicular to it is the negative reciprocal of $\frac{4}{3}$, that is, $-\frac{3}{4}$.

12. $\dfrac{(2x^3)^{-1} \cdot x}{2^3 x^5} = \dfrac{2^{-1}(x^3)^{-1} \cdot x}{2^3 x^5} = \dfrac{2^{-1} x^{-3} x}{2^3 x^5}$

$= \dfrac{2^{-1} x^{-2}}{2^3 x^5} = \dfrac{1}{2^1 \cdot 2^3 \cdot x^2 \cdot x^5}$

$= \dfrac{1}{2^4 x^7}$

13. $\dfrac{(m^{-2})^3 m}{m^5 m^{-4}} = \dfrac{m^{-6} m}{m^5 m^{-4}} = \dfrac{m \cdot m^4}{m^5 \cdot m^6}$

$= \dfrac{m^5}{m^{11}} = \dfrac{1}{m^6}$

14. $\dfrac{2p^3 q^4}{8p^5 q^3} = \dfrac{q \cdot 2p^3 q^3}{4p^2 \cdot 2p^3 q^3} = \dfrac{q}{4p^2}$

15. $(2k^2 + 3k) - (k^2 + k - 1)$

$= 2k^2 + 3k - k^2 - k + 1$

$= k^2 + 2k + 1$

16. $8x^2 y^2 (9x^4 y^5) = 72x^6 y^7$

17. $(2a - b)^2 = (2a)^2 - 2(2a)(b) + (b)^2$

$= 4a^2 - 4ab + b^2$

18. $(y^2 + 3y + 5)(3y - 1)$

Multiply vertically.

$$
\begin{array}{r}
y^2 + 3y + 5 \\
3y - 1 \\
\hline
-\ y^2 - 3y - 5 \\
3y^3 + 9y^2 + 15y \\
\hline
3y^3 + 8y^2 + 12y - 5
\end{array}
$$

19. $\dfrac{12p^3 + 2p^2 - 12p + 4}{2p - 2}$

$$
\begin{array}{r}
6p^2 + 7p + 1 \\
2p - 2 \overline{\smash{\big)}\ 12p^3 + 2p^2 - 12p + 4} \\
\underline{12p^3 - 12p^2} \\
14p^2 - 12p \\
\underline{14p^2 - 14p} \\
2p + 4 \\
\underline{2p - 2} \\
6
\end{array}
$$

The result is

$$6p^2 + 7p + 1 + \frac{6}{2p - 2}$$

$$= 6p^2 + 7p + 1 + \frac{2 \cdot 3}{2(p - 1)}$$

$$= 6p^2 + 7p + 1 + \frac{3}{p - 1}.$$

20. If one operation can be done in 1.4×10^{-7} seconds, then one trillion operations will take

$$\left(1 \times 10^{12}\right)\left(1.4 \times 10^{-7}\right) = 1.4 \times 10^{12+(-7)}$$
$$= 1.4 \times 10^5$$

or 140,000 seconds.

21. $8t^2 + 10tv + 3v^2$

$$= 8t^2 + 6tv + 4tv + 3v^2 \quad \begin{array}{l} 6 \cdot 4 = 24; \\ 6 + 4 = 10 \end{array}$$

$$= \left(8t^2 + 6tv\right) + \left(4tv + 3v^2\right)$$

$$= 2t(4t + 3v) + v(4t + 3v)$$

$$= (4t + 3v)(2t + v)$$

22. $8r^2 - 9rs + 12s^2$

To factor this polynomial by the grouping method, we must find two integers whose product is $(8)(12) = 96$ and whose sum is -9. There is no pair of integers that satisfies both of these conditions, so the polynomial is *prime*.

23. $16x^4 - 1$

$$= \left(4x^2\right)^2 - (1)^2$$

$$= \left(4x^2 + 1\right)\left(4x^2 - 1\right)$$

$$= \left(4x^2 + 1\right)\left[(2x)^2 - (1)^2\right]$$

$$= \left(4x^2 + 1\right)(2x + 1)(2x - 1)$$

24. $\qquad\qquad r^2 = 2r + 15$

$$r^2 - 2r - 15 = 0$$

$$(r + 3)(r - 5) = 0$$

$$r + 3 = 0 \qquad \text{or} \qquad r - 5 = 0$$
$$r = -3 \qquad \text{or} \qquad r = 5$$

The solution set is $\{-3, 5\}$.

25. $(r - 5)(2r + 1)(3r - 2) = 0$

$$r - 5 = 0 \quad \text{or} \quad 2r + 1 = 0 \quad \text{or} \quad 3r - 2 = 0$$
$$2r = -1 \quad \text{or} \quad 3r = 2$$
$$r = 5 \quad \text{or} \qquad r = -\tfrac{1}{2} \quad \text{or} \qquad r = \tfrac{2}{3}$$

The solution set is $\left\{5, -\tfrac{1}{2}, \tfrac{2}{3}\right\}$.

26. Let $x =$ the smaller number
Then $x + 4 =$ the larger number.

The product of the numbers is 2 less than the smaller number translates to

$$x(x + 4) = x - 2.$$
$$x^2 + 4x = x - 2$$
$$x^2 + 3x + 2 = 0$$
$$(x + 2)(x + 1) = 0$$
$$x = -2 \quad \text{or} \quad x = -1$$

The smaller number can be either -2 or -1.

27. Let $w =$ the width of the rectangle.
Then $2w - 2 =$ the length of the rectangle.

Use the formula $A = LW$ with the area $= 60$.

$$60 = (2w - 2)w$$
$$60 = 2w^2 - 2w$$
$$0 = 2w^2 - 2w - 60$$
$$0 = 2\left(w^2 - w - 30\right)$$
$$0 = 2(w - 6)(w + 5)$$
$$w = 6 \quad \text{or} \quad w = -5$$

Discard -5 because the width cannot be negative. The width of the rectangle is 6 meters.

28. All of the given expressions are equal to 1 for all real numbers for which they are defined. However, expressions **B**, **C**, and **D** all have one or more values for which the expression is undefined and therefore cannot be equal to 1 at these values. Since $k^2 + 2$ is *always* positive, the denominator in expression **A** is never equal to zero. This expression is defined and equal to 1 for all real numbers, so the correct choice is **A**.

29. The appropriate choice is **D** since

$$\frac{-(3x + 4)}{7} = \frac{-3x - 4}{7}$$

$$\neq \frac{4 - 3x}{7}.$$

30. $\dfrac{5}{q} - \dfrac{1}{q} = \dfrac{5 - 1}{q} = \dfrac{4}{q}$

31. $\dfrac{3}{7} + \dfrac{4}{r} = \dfrac{3 \cdot r}{7 \cdot r} + \dfrac{4 \cdot 7}{r \cdot 7} \quad LCD = 7r$

$$= \frac{3r + 28}{7r}$$

32. $\dfrac{4}{5q - 20} - \dfrac{1}{3q - 12}$

$$= \frac{4}{5(q - 4)} - \frac{1}{3(q - 4)}$$

$$= \frac{4 \cdot 3}{5(q - 4) \cdot 3} - \frac{1 \cdot 5}{3(q - 4) \cdot 5}$$

$$LCD = 5 \cdot 3 \cdot (q - 4) = 15(q - 4)$$

$$= \frac{12 - 5}{15(q - 4)} = \frac{7}{15(q - 4)}$$

33. $\dfrac{2}{k^2 + k} - \dfrac{3}{k^2 - k}$

$= \dfrac{2}{k(k+1)} - \dfrac{3}{k(k-1)}$

$= \dfrac{2(k-1)}{k(k+1)(k-1)} - \dfrac{3(k+1)}{k(k-1)(k+1)}$

$\qquad\qquad\qquad LCD = k(k+1)(k-1)$

$= \dfrac{2(k-1) - 3(k+1)}{k(k+1)(k-1)}$

$= \dfrac{2k - 2 - 3k - 3}{k(k+1)(k-1)}$

$= \dfrac{-k - 5}{k(k+1)(k-1)}$

34. $\dfrac{7z^2 + 49z + 70}{16z^2 + 72z - 40} \div \dfrac{3z + 6}{4z^2 - 1}$

$= \dfrac{7z^2 + 49z + 70}{16z^2 + 72z - 40} \cdot \dfrac{4z^2 - 1}{3z + 6}$

$= \dfrac{7(z^2 + 7z + 10)}{8(2z^2 + 9z - 5)} \cdot \dfrac{(2z+1)(2z-1)}{3(z+2)}$

$= \dfrac{7(z+5)(z+2)}{8(2z-1)(z+5)} \cdot \dfrac{(2z+1)(2z-1)}{3(z+2)}$

$= \dfrac{7(2z+1)}{8 \cdot 3} = \dfrac{7(2z+1)}{24}$

35. $\dfrac{\dfrac{4}{a} + \dfrac{5}{2a}}{\dfrac{7}{6a} - \dfrac{1}{5a}}$

$= \dfrac{\left(\dfrac{4}{a} + \dfrac{5}{2a}\right) \cdot 30a}{\left(\dfrac{7}{6a} - \dfrac{1}{5a}\right) \cdot 30a}$ *Multiply by*   *LCD, 30a.*

$= \dfrac{\dfrac{4}{a}(30a) + \dfrac{5}{2a}(30a)}{\dfrac{7}{6a}(30a) - \dfrac{1}{5a}(30a)}$

$= \dfrac{4 \cdot 30 + 5 \cdot 15}{7 \cdot 5 - 1 \cdot 6}$

$= \dfrac{120 + 75}{35 - 6} = \dfrac{195}{29}$

36. $\dfrac{1}{x - 4} = \dfrac{3}{2x}$

To avoid zero denominators, x cannot equal 4 or 0. Thus, $x \ne 4, 0$.

37. $\dfrac{r+2}{5} = \dfrac{r-3}{3}$

$15\left(\dfrac{r+2}{5}\right) = 15\left(\dfrac{r-3}{3}\right)$ *Multiply by*   *LCD, 15.*

$3(r+2) = 5(r-3)$

$3r + 6 = 5r - 15$

$21 = 2r$

$\dfrac{21}{2} = r$

Check $r = \dfrac{21}{2}$: $\dfrac{5}{2} = \dfrac{5}{2}$ *True*

The solution set is $\left\{\dfrac{21}{2}\right\}$.

38. $\dfrac{1}{x} = \dfrac{1}{x+1} + \dfrac{1}{2}$

$2x(x+1)\left(\dfrac{1}{x}\right) = 2x(x+1)\left(\dfrac{1}{x+1} + \dfrac{1}{2}\right)$

$\qquad\qquad$ *Multiply by LCD, $2x(x+1)$.*

$2(x+1) = 2x(x+1)\left(\dfrac{1}{x+1}\right)$

$\qquad\qquad + 2x(x+1)\left(\dfrac{1}{2}\right)$

$2(x+1) = 2x + x(x+1)$

$2x + 2 = 2x + x^2 + x$

$0 = x^2 + x - 2$

$0 = (x+2)(x-1)$

$x = -2$ or $x = 1$

Check $x = -2$: $-\dfrac{1}{2} = -1 + \dfrac{1}{2}$ *True*

Check $x = 1$: $1 = \dfrac{1}{2} + \dfrac{1}{2}$ *True*

Thus, the solution set is $\{-2, 1\}$.

39. Let $x =$ the number of hours it will take Jody and Francis to weed the yard working together.

	Rate	Time Working Together	Fractional Part of the Job Done when Working Together
Jody	$\dfrac{1}{3}$	x	$\dfrac{1}{3}x$
Francis	$\dfrac{1}{2}$	x	$\dfrac{1}{2}x$

Working together, they can do 1 whole job, so

$\dfrac{1}{3}x + \dfrac{1}{2}x = 1.$

$6\left(\dfrac{1}{3}x + \dfrac{1}{2}x\right) = 6 \cdot 1$ *LCD = 6*

$6\left(\dfrac{1}{3}x\right) + 6\left(\dfrac{1}{2}x\right) = 6$

$2x + 3x = 6$

$5x = 6$

$x = \dfrac{6}{5},$ or $1\dfrac{1}{5}$

If Jody and Francis worked together, it would take them $1\dfrac{1}{5}$ hours to weed the yard.

40. Let x represent the speed of the boat in still water. Then $x - 2$ is the rate against the current and $x + 2$ is the rate with the current. We fill in the chart as follows, realizing that the time column is filled in by using the formula $t = d/r$.

	d	r	t
Upstream	8	$x - 2$	$\dfrac{8}{x - 2}$
Downstream	11	$x + 2$	$\dfrac{11}{x + 2}$

Since the times are equal, we get the following equation.

$$\frac{8}{x - 2} = \frac{11}{x + 2}$$

$$(x + 2)(x - 2)\frac{8}{x - 2} = (x + 2)(x - 2)\frac{11}{x + 2}$$

Multiply by LCD, $(x + 2)(x - 2)$

$$8(x + 2) = 11(x - 2)$$
$$8x + 16 = 11x - 22$$
$$38 = 3x$$
$$x = \tfrac{38}{3} \quad \text{or} \quad 12\tfrac{2}{3}$$

The speed of the boat in still water is $12\tfrac{2}{3}$ miles per hour.

CHAPTER 7 EQUATIONS OF LINES; FUNCTIONS

7.1 Review of Graphs and Slopes of Lines

1. **(a)** x represents the year; y represents the revenue in billions of dollars.

(b) The dot above the year 2002 appears to be at about 1850, so the revenue in 2002 was $1850 billion.

(c) The ordered pair is $(2002, 1850)$.

(d) In 2000, federal tax revenues were about $2030 billion.

3. **(a)** The point $(1, 6)$ is located in quadrant I, since the x- and y-coordinates are both positive.

(b) The point $(-4, -2)$ is located in quadrant III, since the x- and y-coordinates are both negative.

(c) The point $(-3, 6)$ is located in quadrant II, since the x-coordinate is negative and the y-coordinate is positive.

(d) The point $(7, -5)$ is located in quadrant IV, since the x-coordinate is positive and the y-coordinate is negative.

(e) The point $(-3, 0)$ is located on the x-axis, so it does not belong to any quadrant.

(f) The point $(0, -0.5)$ is located on the y-axis, so it does not belong to any quadrant.

5. **(a)** To plot $(2, 3)$, go two units from zero to the right along the x-axis, and then go three units up parallel to the y-axis.

(b) To plot $(-3, -2)$, go three units from zero to the left along the x-axis, and then go two units down parallel to the y-axis.

(c) To plot $(0, 5)$, do not move along the x-axis at all since the x-coordinate is 0. Move five units up along the y-axis.

(d) To plot $(-2, 4)$, go two units from zero to the left along the x-axis, and then go four units up parallel to the y-axis.

(e) To plot $(-2, 0)$, go two units to the left along the x-axis. Do not move up or down since the y-coordinate is 0.

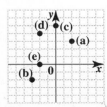

7. $x - y = 3$

(a) To complete the table, substitute the given values for x and y in the equation.

For $x = 0$: $x - y = 3$
$$0 - y = 3$$
$$y = -3 \quad (0, -3)$$

For $y = 0$: $x - y = 3$
$$x - 0 = 3$$
$$x = 3 \quad (3, 0)$$

For $x = 5$: $x - y = 3$
$$5 - y = 3$$
$$-y = -2$$
$$y = 2 \quad (5, 2)$$

For $x = 2$: $x - y = 3$
$$2 - y = 3$$
$$-y = 1$$
$$y = -1 \quad (2, -1)$$

(b) Plot the ordered pairs and draw the line through them.

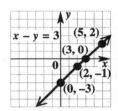

9. $y = -2x + 3$

(a)

x	$-2x$	$y = -2x + 3$
0	0	3
1	-2	1
2	-4	-1
3	-6	-3

(b) Notice that as the value of x increases by 1, the value of y decreases by 2.

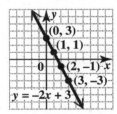

11. $2x + 3y = 12$
To find the x-intercept, let $y = 0$.
$$2x + 3y = 12$$
$$2x + 3(0) = 12$$
$$2x = 12$$
$$x = 6$$

The x-intercept is $(6, 0)$.

continued

To find the y-intercept, let $x = 0$.

$$2x + 3y = 12$$
$$2(0) + 3y = 12$$
$$3y = 12$$
$$y = 4$$

The y-intercept is $(0, 4)$.
Plot the intercepts and draw the line through them.

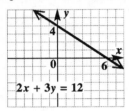

13. $x - 3y = 6$
To find the x-intercept, let $y = 0$.

$$x - 3y = 6$$
$$x - 3(0) = 6$$
$$x - 0 = 6$$
$$x = 6$$

The x-intercept is $(6, 0)$.
To find the y-intercept, let $x = 0$.

$$x - 3y = 6$$
$$0 - 3y = 6$$
$$-3y = 6$$
$$y = -2$$

The y-intercept is $(0, -2)$.
Plot the intercepts and draw the line through them.

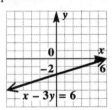

15. $\frac{2}{3}x - 3y = 7$
To find the x-intercept, let $y = 0$.

$$\frac{2}{3}x - 3(0) = 7$$
$$\frac{2}{3}x = 7$$
$$x = \frac{3}{2} \cdot 7 = \frac{21}{2}$$

The x-intercept is $\left(\frac{21}{2}, 0\right)$.
To find the y-intercept, let $x = 0$.

$$\frac{2}{3}(0) - 3y = 7$$
$$-3y = 7$$
$$y = -\frac{7}{3}$$

The y-intercept is $\left(0, -\frac{7}{3}\right)$.

Plot the intercepts and draw the line through them.

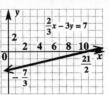

17. $y = 5$
This is a horizontal line. Every point has
y-coordinate 5, so no point has y-coordinate 0.
There is no x-intercept.
Since every point of the line has y-coordinate 5,
the y-intercept is $(0, 5)$. Draw the horizontal line
through $(0, 5)$.

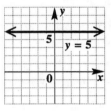

19. $x = 2$
This is a vertical line. Every point has
x-coordinate 2, so the x-intercept is $(2, 0)$.
Since every point of the line has x-coordinate 2,
no point has x-coordinate 0. There is no
y-intercept. Draw the vertical line through $(2, 0)$.

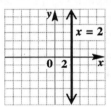

21. $x + 4 = 0$ $(x = -4)$
This is a vertical line. Every point has
x-coordinate -4, so the x-intercept is $(-4, 0)$.
Since every point of the line has x-coordinate -4,
no point has x-coordinate 0. There is no
y-intercept. Draw the vertical line through $(-4, 0)$.

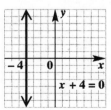

23. $x + 5y = 0$

To find the x-intercept, let $y = 0$.

$$x + 5y = 0$$
$$x + 5(0) = 0$$
$$x = 0$$

The x-intercept is $(0, 0)$, and since $x = 0$, this is also the y-intercept. Since the intercepts are the same, another point is needed to graph the line. Choose any number for y, say $y = -1$, and solve the equation for x.

$$x + 5y = 0$$
$$x + 5(-1) = 0$$
$$x = 5$$

This gives the ordered pair $(5, -1)$. Plot $(5, -1)$ and $(0, 0)$, and draw the line through them.

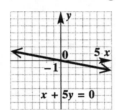

25. $2x = 3y$

If $x = 0$, then $y = 0$, so the x- and y-intercepts are $(0, 0)$. To get another point, let $x = 3$.

$$2(3) = 3y$$
$$2 = y$$

Plot $(3, 2)$ and $(0, 0)$, and draw the line through them.

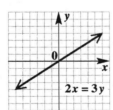

27. $-\frac{2}{3}y = x$

If $x = 0$, then $y = 0$, so the x- and y-intercepts are $(0, 0)$. To get another point, let $y = 3$.

$$-\frac{2}{3}(3) = x$$
$$-2 = x$$

Plot $(-2, 3)$ and $(0, 0)$, and draw the line through them.

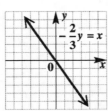

29. By the Midpoint Formula, the midpoint of the segment with endpoints $(-8, 4)$ and $(-2, -6)$ is

$$\left(\frac{-8 + (-2)}{2}, \frac{4 + (-6)}{2} \right) = \left(\frac{-10}{2}, \frac{-2}{2} \right) = (-5, -1).$$

31. By the Midpoint Formula, the midpoint of the segment with endpoints $(3, -6)$ and $(6, 3)$ is

$$\left(\frac{3 + 6}{2}, \frac{-6 + 3}{2} \right) = \left(\frac{9}{2}, \frac{-3}{2} \right) = \left(\frac{9}{2}, -\frac{3}{2} \right).$$

33. By the Midpoint Formula, the midpoint of the segment with endpoints $(-9, 3)$ and $(9, 8)$ is

$$\left(\frac{-9 + 9}{2}, \frac{3 + 8}{2} \right) = \left(\frac{0}{2}, \frac{11}{2} \right) = \left(0, \frac{11}{2} \right).$$

35. By the Midpoint Formula, the midpoint of the segment with endpoints $(2.5, 3.1)$ and $(1.7, -1.3)$ is

$$\left(\frac{2.5 + 1.7}{2}, \frac{3.1 + (-1.3)}{2} \right) = \left(\frac{4.2}{2}, \frac{1.8}{2} \right) = (2.1, 0.9).$$

37. By the Midpoint Formula, the midpoint of the segment with endpoints $\left(\frac{1}{2}, \frac{1}{3} \right)$ and $\left(\frac{3}{2}, \frac{5}{3} \right)$ is

$$\left(\frac{\frac{1}{2} + \frac{3}{2}}{2}, \frac{\frac{1}{3} + \frac{5}{3}}{2} \right) = \left(\frac{\frac{4}{2}}{2}, \frac{\frac{6}{3}}{2} \right) = \left(\frac{2}{2}, \frac{2}{2} \right) = (1, 1).$$

39. By the Midpoint Formula, the midpoint of the segment with endpoints $\left(-\frac{1}{3}, \frac{2}{7} \right)$ and $\left(-\frac{1}{2}, \frac{1}{14} \right)$ is

$$\left(\frac{-\frac{1}{3} + \left(-\frac{1}{2}\right)}{2}, \frac{\frac{2}{7} + \frac{1}{14}}{2} \right) = \left(\frac{-\frac{5}{6}}{2}, \frac{\frac{5}{14}}{2} \right) = \left(-\frac{5}{12}, \frac{5}{28} \right).$$

41. midpoint of $P(5, 8)$ and $Q(x, y) = M(8, 2)$

$$\left(\frac{5 + x}{2}, \frac{8 + y}{2} \right) = (8, 2)$$

The x- and y-coordinates must be equal.

$$\frac{5 + x}{2} = 8 \qquad \frac{8 + y}{2} = 2$$
$$5 + x = 16 \qquad 8 + y = 4$$
$$x = 11 \qquad y = -4$$

Thus, the endpoint Q is $(11, -4)$.

43. midpoint of $P\left(\frac{1}{3}, \frac{1}{5} \right)$ and $Q(x, y) = M\left(\frac{3}{2}, 1 \right)$

$$\left(\frac{\frac{1}{3} + x}{2}, \frac{\frac{1}{5} + y}{2} \right) = \left(\frac{3}{2}, 1 \right)$$

The x- and y-coordinates must be equal.

$$\frac{\frac{1}{3} + x}{2} = \frac{3}{2} \qquad \frac{\frac{1}{5} + y}{2} = 1$$
$$\frac{1}{3} + x = 3 \qquad \frac{1}{5} + y = 2$$
$$\frac{1}{3} + x = \frac{9}{3} \qquad \frac{1}{5} + y = \frac{10}{5}$$
$$x = \frac{8}{3} \qquad y = \frac{9}{5}$$

Thus, the endpoint Q is $\left(\frac{8}{3}, \frac{9}{5} \right)$.

45. The graph goes through the point $(-2, 0)$ which satisfies only equations **B** and **C**. The graph also goes through the point $(0, 3)$ which satisfies only equations **A** and **B**. Therefore, the correct equation is **B**.

47. The screen on the right is more useful because it shows the intercepts.

49. $\text{slope} = \dfrac{\text{change in vertical position}}{\text{change in horizontal position}}$

$= \dfrac{30 \text{ feet}}{100 \text{ feet}}$

Choices **A**, 0.3, **B**, $\frac{3}{10}$, and **D**, $\frac{30}{100}$ are all correct.

51. **(a)** Graph **C** indicates that sales leveled off during the second quarter.

(b) Graph **A** indicates that sales leveled off during the fourth quarter.

(c) Graph **D** indicates that sales rose sharply during the first quarter, and then fell to the original level during the second quarter.

(d) Graph **B** is the only graph that indicates that sales fell during the first two quarters.

53. $\text{slope of } BD = \dfrac{\text{rise}}{\text{run}} = \dfrac{7}{4}$

55. Let $(x_1, y_1) = (-2, -3)$ and $(x_2, y_2) = (-1, 5)$. Then

$$m = \frac{y_2 - y_1}{x_2 - x_1} = \frac{5 - (-3)}{-1 - (-2)} = \frac{8}{1} = 8.$$

The slope is 8.

57. Let $(x_1, y_1) = (-4, 1)$ and $(x_2, y_2) = (2, 6)$. Then

$$m = \frac{y_2 - y_1}{x_2 - x_1} = \frac{6 - 1}{2 - (-4)} = \frac{5}{6}.$$

The slope is $\frac{5}{6}$.

59. Let $(x_1, y_1) = (2, 4)$ and $(x_2, y_2) = (-4, 4)$. Then

$$m = \frac{y_2 - y_1}{x_2 - x_1} = \frac{4 - 4}{-4 - 2} = \frac{0}{-6} = 0.$$

The slope is 0.

61. Let $(x_1, y_1) = (1.5, 2.6)$ and $(x_2, y_2) = (0.5, 3.6)$. Then

$$m = \frac{y_2 - y_1}{x_2 - x_1} = \frac{3.6 - 2.6}{0.5 - 1.5} = \frac{1}{-1} = -1.$$

The slope is -1.

63. Let $(x_1, y_1) = \left(\frac{1}{6}, \frac{1}{2}\right)$ and $(x_2, y_2) = \left(\frac{5}{6}, \frac{9}{2}\right)$. Then

$$m = \frac{y_2 - y_1}{x_2 - x_1} = \frac{\frac{9}{2} - \frac{1}{2}}{\frac{5}{6} - \frac{1}{6}} = \frac{\frac{8}{2}}{\frac{4}{6}} = 4 \cdot \frac{3}{2} = 6.$$

The slope is 6.

65. The points shown on the line are $(-3, 3)$ and $(-1, -2)$. The slope is

$$m = \frac{-2 - 3}{-1 - (-3)} = \frac{-5}{2} = -\frac{5}{2}.$$

67. The points shown on the line are $(3, 3)$ and $(3, -3)$. The slope is

$$m = \frac{-3 - 3}{3 - 3} = \frac{-6}{0}, \text{ which is } \textit{undefined.}$$

69. "The line has positive slope" means that the line goes up from left to right. This is line B.

71. "The line has slope 0" means that there is no vertical change; that is, the line is horizontal. This is line A.

73. To find the slope of

$$x + 2y = 4,$$

first find the intercepts. Replace y with 0 to find that the x-intercept is $(4, 0)$; replace x with 0 to find that the y-intercept is $(0, 2)$. The slope is then

$$m = \frac{2 - 0}{0 - 4} = -\frac{2}{4} = -\frac{1}{2}.$$

To sketch the graph, plot the intercepts and draw the line through them.

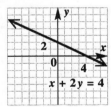

75. To find the slope of

$$5x - 2y = 10,$$

first find the intercepts. Replace y with 0 to find that the x-intercept is $(2, 0)$; replace x with 0 to find that the y-intercept is $(0, -5)$. The slope is then

$$m = \frac{-5 - 0}{0 - 2} = \frac{-5}{-2} = \frac{5}{2}.$$

To sketch the graph, plot the intercepts and draw the line through them.

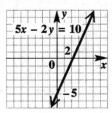

77. In the equation

$$y = 4x,$$

replace x with 0 and then x with 1 to get the ordered pairs $(0, 0)$ and $(1, 4)$, respectively. (There are other possibilities for ordered pairs.) The slope is then

$$m = \frac{4 - 0}{1 - 0} = \frac{4}{1} = 4.$$

To sketch the graph, plot the two points and draw the line through them.

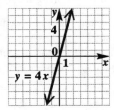

79. $x - 3 = 0 \quad (x = 3)$
The graph of $x = 3$ is the vertical line with x-intercept $(3, 0)$. The slope of a vertical line is undefined.

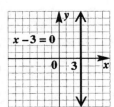

81. $y = -5$
The graph of $y = -5$ is the horizontal line with y-intercept $(0, -5)$. The slope of a horizontal line is 0.

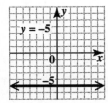

83. $2y = 3 \quad \left(y = \frac{3}{2}\right)$
The graph of $y = \frac{3}{2}$ is the horizontal line with y-intercept $\left(0, \frac{3}{2}\right)$. The slope of a horizontal line is 0.

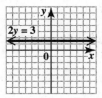

85. To graph the line through $(-4, 2)$ with slope $m = \frac{1}{2}$, locate $(-4, 2)$ on the graph. To find a second point, use the definition of slope.

$$m = \frac{\text{change in } y}{\text{change in } x} = \frac{1}{2}$$

From $(-4, 2)$, go up 1 unit. Then go 2 units to the right to get to $(-2, 3)$. Draw the line through $(-4, 2)$ and $(-2, 3)$.

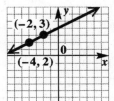

87. To graph the line through $(0, -2)$ with slope $m = -\frac{2}{3}$, locate the point $(0, -2)$ on the graph. To find a second point on the line, use the definition of slope, writing $-\frac{2}{3}$ as $\frac{-2}{3}$.

$$m = \frac{\text{change in } y}{\text{change in } x} = \frac{-2}{3}$$

From $(0, -2)$, move 2 units down and then 3 units to the right. Draw a line through this second point and $(0, -2)$. (Note that the slope could also be written as $\frac{2}{-3}$. In this case, move 2 units up and 3 units to the left to get another point on the same line.)

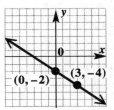

89. Locate $(-1, -2)$. Then use $m = 3 = \frac{3}{1}$ to go 3 units up and 1 unit right to $(0, 1)$.

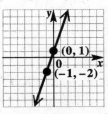

91. Locate $(2, -5)$. A slope of 0 means that the line is horizontal, so $y = -5$ at every point. Draw the horizontal line through $(2, -5)$.

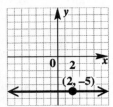

93. Locate $(-3, 1)$. Since the slope is undefined, the line is vertical. The x-value of every point is -3. Draw the vertical line through $(-3, 1)$.

95. If a line has slope $-\frac{4}{9}$, then any line parallel to it has slope $-\frac{4}{9}$ (the slope must be the same), and any line perpendicular to it has slope $\frac{9}{4}$ (the slope must be the negative reciprocal).

97. The slope of the line through $(15, 9)$ and $(12, -7)$ is

$$m = \frac{-7 - 9}{12 - 15} = \frac{-16}{-3} = \frac{16}{3}.$$

The slope of the line through $(8, -4)$ and $(5, -20)$ is

$$m = \frac{-20 - (-4)}{5 - 8} = \frac{-16}{-3} = \frac{16}{3}.$$

Since the slopes are equal, the two lines are *parallel*.

99. $x + 4y = 7$ and $4x - y = 3$
Solve the equations for y.
$$4y = -x + 7 \qquad\qquad -y = -4x + 3$$
$$y = -\tfrac{1}{4}x + \tfrac{7}{4} \qquad\qquad y = 4x - 3$$

The slopes, $-\frac{1}{4}$ and 4, are negative reciprocals of one another, so the lines are *perpendicular*.

101. $4x - 3y = 6$ and $3x - 4y = 2$
Solve the equations for y.
$$-3y = -4x + 6 \qquad\qquad -4y = -3x + 2$$
$$y = \tfrac{4}{3}x - 2 \qquad\qquad y = \tfrac{3}{4}x - \tfrac{1}{2}$$
The slopes are $\frac{4}{3}$ and $\frac{3}{4}$. The lines are *neither* parallel nor perpendicular.

103. $x = 6$ and $6 - x = 8$
The second equation can be simplified as $x = -2$. Both lines are vertical lines, so they are *parallel*.

105. $4x + y = 0$ and $5x - 8 = 2y$
Solve the equations for y.
$$y = -4x \qquad\qquad \tfrac{5}{2}x - 4 = y$$
The slopes are -4 and $\frac{5}{2}$. The lines are *neither* parallel nor perpendicular.

107. $2x = y + 3$ and $2y + x = 3$
Solve the equations for y.
$$2x - 3 = y \qquad\qquad 2y = -x + 3$$
$$\qquad\qquad\qquad\qquad y = -\tfrac{1}{2}x + \tfrac{3}{2}$$
The slopes, 2 and $-\frac{1}{2}$, are negative reciprocals of one another, so the lines are *perpendicular*.

109. Use the points $(0, 20)$ and $(4, 4)$.

average rate of change

$$= \frac{\text{change in } y}{\text{change in } x} = \frac{4 - 20}{4 - 0} = \frac{-16}{4} = -4$$

The average rate of change is $-\$4000$ per year, that is, the value of the machine is decreasing $\$4000$ each year during these years.

111. We can see that there is no change in the percent of pay raise. Thus, the average rate of change is 0% per year, that is, the percent of pay raise is not changing—it is 3% each year during these years.

113. (a) Use $(1998, 229.3)$ and $(2004, 338.8)$.

$$m = \frac{338.8 - 229.3}{2004 - 1998} = \frac{109.5}{6} = 18.25$$

The average rate of change is $\$18.25$ billion per year.

(b) The positive slope means that personal spending on recreation in the United States *increased* by an average of $\$18.25$ billion each year.

115. Use $(1997, 500)$ and $(2002, 155)$.

$$m = \frac{155 - 500}{2002 - 1997} = \frac{-345}{5} = -69$$

The average rate of change in price is $-\$69$ per year, that is, the price decreased an average of $\$69$ each year from 1997 to 2002.

117. $y - (-1) = \frac{5}{3}[x - (-4)]$
$3(y + 1) = 5(x + 4)$
$3y + 3 = 5x + 20$
$-17 = 5x - 3y$

119. $y - (-1) = -\frac{1}{2}[x - (-2)]$
$y + 1 = -\frac{1}{2}(x + 2)$
$2(y + 1) = -1(x + 2)$
$2y + 2 = -x - 2$
$x + 2y = -4$

7.2   Review of Equations of Lines; Linear Models

1. $y = 2x + 3$
This line is in slope-intercept form with slope $m = 2$ and y-intercept $(0, b) = (0, 3)$. The only graph with positive slope and with a positive y-coordinate of its y-intercept is **A**.

3. $y = -2x - 3$
This line is in slope-intercept form with slope $m = -2$ and y-intercept $(0, b) = (0, -3)$. The only graph with negative slope and with a negative y-coordinate of its y-intercept is **C**.

5. $y = 2x$
This line has slope $m = 2$ and y-intercept $(0, b) = (0, 0)$. The only graph with positive slope and with y-intercept $(0, 0)$ is **H**.

7. $y = 3$
This line is a horizontal line with y-intercept $(0, 3)$. Its y-coordinate is positive. The only graph that has these characteristics is **B**.

9.   Choice **A**, $3x - 2y = 5$, is in the form $Ax + By = C$ with $A \geq 0$ and integers A, B, and C having no common factor (except 1).

11.   Choice **A**, $y = 6x + 2$, is in the form $y = mx + b$.

13. $y + 2 = -3(x - 4)$
$y + 2 = -3x + 12$
$3x + y = 10$ *Standard form*

15. $m = 5; b = 15$
Substitute these values in the slope-intercept form.
$$y = mx + b$$
$$y = 5x + 15$$

17. $m = -\frac{2}{3}; b = \frac{4}{5}$
Substitute these values in the slope-intercept form.
$$y = mx + b$$
$$y = -\frac{2}{3}x + \frac{4}{5}$$

19.   Slope $\frac{2}{5}$; y-intercept $(0, 5)$
Here, $m = \frac{2}{5}$ and $b = 5$. Substitute these values in the slope-intercept form.
$$y = mx + b$$
$$y = \frac{2}{5}x + 5$$

21.   To get to the point $(3, 3)$ from the y-intercept $(0, 1)$, we must go up 2 units and to the right 3 units, so the slope is $\frac{2}{3}$. The slope-intercept form is
$$y = \frac{2}{3}x + 1.$$

23. $-x + y = 4$
(a) Solve for y to get the equation in slope-intercept form.
$$-x + y = 4$$
$$y = x + 4$$
(b) The slope is the coefficient of x, 1.
(c) The y-intercept is the point $(0, b)$, or $(0, 4)$.

(d)
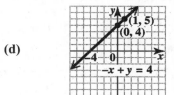

25. $6x + 5y = 30$
(a) Solve for y to get the equation in slope-intercept form.
$$6x + 5y = 30$$
$$5y = -6x + 30$$
$$y = -\frac{6}{5}x + 6$$
(b) The slope is the coefficient of x, $-\frac{6}{5}$.
(c) The y-intercept is the point $(0, b)$, or $(0, 6)$.

(d)
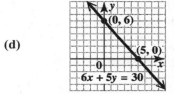

27. $4x - 5y = 20$
(a) Solve for y to get the equation in slope-intercept form.
$$4x - 5y = 20$$
$$-5y = -4x + 20$$
$$y = \frac{4}{5}x - 4$$
(b) The slope is the coefficient of x, $\frac{4}{5}$.
(c) The y-intercept is the point $(0, b)$, or $(0, -4)$.

(d)
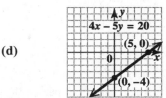

29. $x + 2y = -4$
(a) Solve for y to get the equation in slope-intercept form.
$$x + 2y = -4$$
$$2y = -x - 4$$
$$y = -\frac{1}{2}x - 2$$

(b) The slope is the coefficient of x, $-\frac{1}{2}$.

(c) The y-intercept is the point $(0, b)$, or $(0, -2)$.

(d)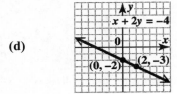

31. $-4x = -3y + 12$

(a) Solve for y to get the equation in slope-intercept form.

$$-4x = -3y + 12$$
$$3y = 4x + 12$$
$$y = \frac{4}{3}x + 4$$

(b) The slope is the coefficient of x, $\frac{4}{3}$.

(c) The y-intercept is the point $(0, b)$, or $(0, 4)$.

(d)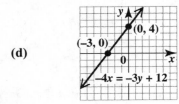

33. **(a)** Through $(5, 8)$; slope -2

Use the point-slope form with $(x_1, y_1) = (5, 8)$ and $m = -2$.

$$y - y_1 = m(x - x_1)$$
$$y - 8 = -2(x - 5)$$
$$y - 8 = -2x + 10$$
$$2x + y = 18$$

(b) Solve the last equation from part (a) for y.

$$2x + y = 18$$
$$y = -2x + 18$$

35. **(a)** Through $(-2, 4)$; slope $-\frac{3}{4}$

Use the point-slope form with $(x_1, y_1) = (-2, 4)$ and $m = -\frac{3}{4}$.

$$y - y_1 = m(x - x_1)$$
$$y - 4 = -\frac{3}{4}[x - (-2)]$$
$$4(y - 4) = -3(x + 2)$$
$$4y - 16 = -3x - 6$$
$$3x + 4y = 10$$

(b) Solve the last equation from part (a) for y.

$$3x + 4y = 10$$
$$4y = -3x + 10$$
$$y = -\frac{3}{4}x + \frac{10}{4}$$
$$y = -\frac{3}{4}x + \frac{5}{2}$$

37. **(a)** Through $(-5, 4)$; slope $\frac{1}{2}$

Use the point-slope form with $(x_1, y_1) = (-5, 4)$ and $m = \frac{1}{2}$.

$$y - y_1 = m(x - x_1)$$
$$y - 4 = \frac{1}{2}[x - (-5)]$$
$$2(y - 4) = 1(x + 5)$$
$$2y - 8 = x + 5$$
$$-x + 2y = 13$$
$$x - 2y = -13$$

(b) Solve the last equation from part (a) for y.

$$x - 2y = -13$$
$$-2y = -x - 13$$
$$y = \frac{1}{2}x + \frac{13}{2}$$

39. **(a)** Through $(3, 0)$; slope 4

Use the point-slope form with $(x_1, y_1) = (3, 0)$ and $m = 4$.

$$y - y_1 = m(x - x_1)$$
$$y - 0 = 4(x - 3)$$
$$y = 4x - 12$$
$$-4x + y = -12$$
$$4x - y = 12$$

(b) Solve the last equation from part (a) for y.

$$4x - y = 12$$
$$-y = -4x + 12$$
$$y = 4x - 12$$

41. **(a)** Through $(2, 6.8)$; slope 1.4

Use the point-slope form with $(x_1, y_1) = (2, 6.8)$ and $m = 1.4$.

$$y - y_1 = m(x - x_1)$$
$$y - 6.8 = 1.4(x - 2)$$
$$y - 6.8 = \frac{7}{5}(x - 2)$$
$$5(y - 6.8) = 7(x - 2)$$
$$5y - 34 = 7x - 14$$
$$-7x + 5y = 20$$
$$7x - 5y = -20$$

(b) Solve the last equation from part (a) for y.

$$7x - 5y = -20$$
$$-5y = -7x - 20$$
$$y = \frac{7}{5}x + 4 \quad \text{or} \quad y = 1.4x + 4$$

43. Through $(9, 5)$; slope 0

A line with slope 0 is a horizontal line. A horizontal line through the point (x, k) has equation $y = k$. Here $k = 5$, so an equation is $y = 5$.

45. Through $(9, 10)$; undefined slope

A vertical line has undefined slope and equation $x = c$. Since the x-value in $(9, 10)$ is 9, the equation is $x = 9$.

47. Through $(0.5, 0.2)$; vertical

A vertical line through the point (k, y) has equation $x = k$. Here $k = 0.5$, so the equation is $x = 0.5$.

49. Through $(-7, 8)$; horizontal

A horizontal line through the point (x, k) has equation $y = k$, so the equation is $y = 8$.

51. (a) $(3, 4)$ and $(5, 8)$

Find the slope.

$$m = \frac{8 - 4}{5 - 3} = \frac{4}{2} = 2$$

Use the point-slope form with $(x_1, y_1) = (3, 4)$ and $m = 2$.

$$
\begin{aligned}
y - y_1 &= m(x - x_1) \\
y - 4 &= 2(x - 3) \\
y - 4 &= 2x - 6 \\
-2x + y &= -2 \\
2x - y &= 2
\end{aligned}
$$

(b) Solve the last equation from part (a) for y.

$$
\begin{aligned}
2x - y &= 2 \\
-y &= -2x + 2 \\
y &= 2x - 2
\end{aligned}
$$

Note: You could use *any* of the equations in part (a) to solve for y. In this exercise, choosing

$$y - 4 = 2x - 6 \quad \text{or} \quad -2x + y = -2,$$

easily leads to the equation $y = 2x - 2$.

53. (a) $(6, 1)$ and $(-2, 5)$

Find the slope.

$$m = \frac{5 - 1}{-2 - 6} = \frac{4}{-8} = -\frac{1}{2}$$

Use the point-slope form with $(x_1, y_1) = (6, 1)$ and $m = -\frac{1}{2}$.

$$
\begin{aligned}
y - y_1 &= m(x - x_1) \\
y - 1 &= -\tfrac{1}{2}(x - 6) \\
2(y - 1) &= -1(x - 6) \\
2y - 2 &= -x + 6 \\
x + 2y &= 8
\end{aligned}
$$

(b) Solve the last equation from part (a) for y.

$$
\begin{aligned}
x + 2y &= 8 \\
2y &= -x + 8 \\
y &= -\tfrac{1}{2}x + 4
\end{aligned}
$$

55. (a) $\left(-\frac{2}{5}, \frac{2}{5}\right)$ and $\left(\frac{4}{3}, \frac{2}{3}\right)$

Find the slope.

$$
\begin{aligned}
m &= \frac{\frac{2}{3} - \frac{2}{5}}{\frac{4}{3} - \left(-\frac{2}{5}\right)} = \frac{\frac{10-6}{15}}{\frac{20+6}{15}} \\
&= \frac{\frac{4}{15}}{\frac{26}{15}} = \frac{4}{26} = \frac{2}{13}
\end{aligned}
$$

Use the point-slope form with $(x_1, y_1) = \left(-\frac{2}{5}, \frac{2}{5}\right)$ and $m = \frac{2}{13}$.

$$
\begin{aligned}
y - \tfrac{2}{5} &= \tfrac{2}{13}\left[x - \left(-\tfrac{2}{5}\right)\right] \\
13\left(y - \tfrac{2}{5}\right) &= 2\left(x + \tfrac{2}{5}\right) \\
13y - \tfrac{26}{5} &= 2x + \tfrac{4}{5} \\
-2x + 13y &= \tfrac{30}{5} \\
2x - 13y &= -6
\end{aligned}
$$

(b) Solve the last equation from part (a) for y.

$$
\begin{aligned}
2x - 13y &= -6 \\
-13y &= -2x - 6 \\
y &= \tfrac{2}{13}x + \tfrac{6}{13}
\end{aligned}
$$

57. (a) $(2, 5)$ and $(1, 5)$

Find the slope.

$$m = \frac{5 - 5}{1 - 2} = \frac{0}{-1} = 0$$

A line with slope 0 is horizontal. A horizontal line through the point (x, k) has equation $y = k$, so the equation is $y = 5$.

(b) $y = 5$ is already in the slope-intercept form.

59. (a) $(7, 6)$ and $(7, -8)$

Find the slope.

$$m = \frac{-8 - 6}{7 - 7} = \frac{-14}{0} \quad \text{Undefined}$$

A line with undefined slope is a vertical line. The equation of a vertical line is $x = k$, where k is the common x-value. So the equation is $x = 7$.

(b) It is not possible to write $x = 7$ in slope-intercept form.

61. (a) $\left(\frac{1}{2}, -3\right)$ and $\left(-\frac{2}{3}, -3\right)$

Find the slope.

$$m = \frac{-3 - (-3)}{-\frac{2}{3} - \frac{1}{2}} = \frac{0}{-\frac{7}{6}} = 0$$

A line with slope 0 is horizontal. A horizontal line through the point (x, k) has equation $y = k$, so the equation is $y = -3$.

(b) $y = -3$ is already in the slope-intercept form.

63. **(a)** Through $(7, 2)$; parallel to the graph of the line having equation $3x - y = 8$
Find the slope of $3x - y = 8$.

$$-y = -3x + 8$$
$$y = 3x - 8$$

The slope is 3, so a line parallel to it also has slope 3. Use $m = 3$ and $(x_1, y_1) = (7, 2)$ in the point-slope form.

$$y - y_1 = m(x - x_1)$$
$$y - 2 = 3(x - 7)$$
$$y - 2 = 3x - 21$$
$$y = 3x - 19$$

(b)
$$y = 3x - 19$$
$$-3x + y = -19$$
$$3x - y = 19$$

65. **(a)** Through $(-2, -2)$; parallel to $-x + 2y = 10$
Find the slope of $-x + 2y = 10$.

$$2y = x + 10$$
$$y = \tfrac{1}{2}x + 5$$

The slope is $\tfrac{1}{2}$, so a line parallel to it also has slope $\tfrac{1}{2}$. Use $m = \tfrac{1}{2}$ and $(x_1, y_1) = (-2, -2)$ in the point-slope form.

$$y - y_1 = m(x - x_1)$$
$$y - (-2) = \tfrac{1}{2}[x - (-2)]$$
$$y + 2 = \tfrac{1}{2}(x + 2)$$
$$y + 2 = \tfrac{1}{2}x + 1$$
$$y = \tfrac{1}{2}x - 1$$

(b)
$$y = \tfrac{1}{2}x - 1$$
$$2y = x - 2 \quad \textit{Multiply by 2.}$$
$$-x + 2y = -2$$
$$x - 2y = 2$$

67. **(a)** Through $(8, 5)$; perpendicular to $2x - y = 7$
Find the slope of $2x - y = 7$.

$$-y = -2x + 7$$
$$y = 2x - 7$$

The slope of the line is 2. Therefore, the slope of the line perpendicular to it is $-\tfrac{1}{2}$ since $2\left(-\tfrac{1}{2}\right) = -1$. Use $m = -\tfrac{1}{2}$ and $(x_1, y_1) = (8, 5)$ in the point-slope form.

$$y - y_1 = m(x - x_1)$$
$$y - 5 = -\tfrac{1}{2}(x - 8)$$
$$y - 5 = -\tfrac{1}{2}x + 4$$
$$y = -\tfrac{1}{2}x + 9$$

(b)
$$y = -\tfrac{1}{2}x + 9$$
$$2y = -x + 18 \quad \textit{Multiply by 2.}$$
$$x + 2y = 18$$

69. **(a)** Through $(-2, 7)$; perpendicular to $x = 9$
$x = 9$ is a vertical line so a line perpendicular to it will be a horizontal line. It goes through $(-2, 7)$ so its equation is

$$y = 7.$$

(b) $y = 7$ is already in standard form.

71. Distance = (rate)(time), so

$$y = 45x.$$

x	$y = 45x$	Ordered Pair
0	$45(0) = 0$	$(0, 0)$
5	$45(5) = 225$	$(5, 225)$
10	$45(10) = 450$	$(10, 450)$

73. Total cost = (cost/gal)(number of gallons), so

$$y = 3.01x.$$

x	$y = 3.01x$	Ordered Pair
0	$3.01(0) = 0$	$(0, 0)$
5	$3.01(5) = 15.05$	$(5, 15.05)$
10	$3.01(10) = 30.10$	$(10, 30.10)$

75. **(a)** The fixed cost is $99, so that is the value of b. The variable cost is $39, so

$$y = mx + b = 39x + 99.$$

(b) If $x = 5$, $y = 39(5) + 99 = 294$. The ordered pair is $(5, 294)$. The cost of a 5-month membership is $294.

(c) If $x = 12$, $y = 39(12) + 99 = 567$. The cost of the first year's membership is $567.

77. **(a)** The fixed cost is $19.95 + \$25 = \$44.95$, so that is the value of b. The variable cost is $35, so

$$y = mx + b = 35x + 44.95.$$

(b) If $x = 5$, $y = 35(5) + 44.95 = 219.95$. The ordered pair is $(5, 219.95)$. The cost of the plan for 5 months is $219.95.

(c) For a 1-year contract, $x = 12$, so $y = 35(12) + 44.95 = 464.95$. The cost of the plan for 1 year is $464.95.

79. **(a)** The fixed cost is $30, so that is the value of b. The variable cost is $6, so

$$y = mx + b = 6x + 30.$$

(b) If $x = 5$, $y = 6(5) + 30 = 60$. The ordered pair is $(5, 60)$. It costs $60 to rent the saw for 5 days.

(c)
$$138 = 6x + 30 \quad \textit{Let } y = 138.$$
$$108 = 6x$$
$$x = \frac{108}{6} = 18$$

The saw is rented for 18 days.

81. **(a)** Use $(0, 91)$ and $(5, 63)$.

$$m = \frac{63 - 91}{5 - 0} = \frac{-28}{5} = -5.6$$

The equation is $y = -5.6x + 91$. The slope tells us that the percent of households accessing the Internet by dial-up is decreasing 5.6% per year.

(b) The year 2006 corresponds to $x = 6$, so $y = -5.6(6) + 91 = 57.4 \approx 57\%$.

83. **(a)** Use $(5, 22{,}393)$ and $(13, 29{,}645)$.

$$m = \frac{29{,}645 - 22{,}393}{13 - 5} = \frac{7252}{8} = 906.5$$

Now use the point-slope form.

$$y - 22{,}393 = 906.5(x - 5)$$
$$y - 22{,}393 = 906.5x - 4532.5$$
$$y = 906.5x + 17{,}860.5$$

(b) The year 1999 corresponds to $x = 9$, so $y = 906.5(9) + 17{,}860.5 = 26{,}019$. This value is slightly lower than the actual value of \$27,910.

85. When $C = 0°$, $F = \underline{\ 32°\ }$, and when $C = 100°$, $F = \underline{\ 212°\ }$. These are the freezing and boiling temperatures for water.

86. The two points of the form (C, F) would be $(0, 32)$ and $(100, 212)$.

87. $m = \dfrac{212 - 32}{100 - 0} = \dfrac{180}{100} = \dfrac{9}{5}$

88. Let $m = \frac{9}{5}$ and $(x_1, y_1) = (0, 32)$.

$$y - y_1 = m(x - x_1)$$
$$F - 32 = \tfrac{9}{5}(C - 0)$$
$$F - 32 = \tfrac{9}{5}C$$
$$F = \tfrac{9}{5}C + 32$$

89.
$$F = \tfrac{9}{5}C + 32$$
$$F - 32 = \tfrac{9}{5}C$$
$$\tfrac{5}{9}(F - 32) = C$$

90. A temperature of $50°C$ corresponds to a temperature of $122°F$.

91. $y = -7x + 12$
$y = -7(3) + 12$ *Let x = 3.*
$y = -9$

93. $y = 3x - 8$
$y = 3(3) - 8$ *Let x = 3.*
$y = 1$

Summary Exercises on Slopes and Equations of Lines

1. For $3x + 5y = 9$, slope $= -\dfrac{A}{B} = -\dfrac{3}{5}$.

3. For $2x - y = 5$, slope $= -\dfrac{A}{B} = -\dfrac{2}{-1} = 2$.

5. For $0.2x + 0.8y = 0$,

$$\text{slope} = -\dfrac{A}{B} = -\dfrac{0.2}{0.8} = -0.25.$$

7. Through $(-2, 6)$ and $(4, 1)$

(a) The slope is

$$m = \frac{1 - 6}{4 - (-2)} = \frac{-5}{6} = -\frac{5}{6}.$$

Use the point-slope form.

$$y - y_1 = m(x - x_1)$$
$$y - 6 = -\tfrac{5}{6}[x - (-2)]$$
$$y - 6 = -\tfrac{5}{6}x - \tfrac{5}{3}$$
$$y = -\tfrac{5}{6}x - \tfrac{5}{3} + \tfrac{18}{3}$$
$$y = -\tfrac{5}{6}x + \tfrac{13}{3}$$

(b) $\qquad y = -\tfrac{5}{6}x + \tfrac{13}{3}$
$\qquad 6y = -5x + 26$ *Multiply by 6.*
$5x + 6y = 26$

9. Through $(0, 0)$; perpendicular to $2x - 5y = 6$

(a) Find the slope of $2x - 5y = 6$.

$$-5y = -2x + 6$$
$$y = \tfrac{2}{5}x - \tfrac{6}{5}$$

The slope of the line is $\frac{2}{5}$. Therefore, the slope of the line perpendicular to it is $-\frac{5}{2}$ since $\frac{2}{5}\left(-\frac{5}{2}\right) = -1$. Use $m = -\frac{5}{2}$ and $(x_1, y_1) = (0, 0)$ in the point-slope form.

$$y - y_1 = m(x - x_1)$$
$$y - 0 = -\tfrac{5}{2}(x - 0)$$
$$y = -\tfrac{5}{2}x$$

(b) $\qquad y = -\tfrac{5}{2}x$
$\qquad 2y = -5x$
$5x + 2y = 0$

11. Through $\left(\frac{3}{4}, -\frac{7}{9}\right)$; perpendicular to $x = \frac{2}{3}$
(a) $x = \frac{2}{3}$ is a vertical line so a line perpendicular to it will be a horizontal line. It goes through $\left(\frac{3}{4}, -\frac{7}{9}\right)$ so its equation is

$$y = -\frac{7}{9}.$$

(b) $y = -\frac{7}{9}$
$9y = -7$ *Multiply by 9.*

13. Through $(-4, 2)$; parallel to the line through $(3, 9)$ and $(6, 11)$

(a) The slope of the line through $(3, 9)$ and $(6, 11)$ is

$$m = \frac{11 - 9}{6 - 3} = \frac{2}{3}.$$

Use the point-slope form with $(x_1, y_1) = (-4, 2)$ and $m = \frac{2}{3}$ (since the slope of the desired line must equal the slope of the given line).

$$y - y_1 = m(x - x_1)$$
$$y - 2 = \tfrac{2}{3}[x - (-4)]$$
$$y - 2 = \tfrac{2}{3}(x + 4)$$
$$y - 2 = \tfrac{2}{3}x + \tfrac{8}{3}$$
$$y = \tfrac{2}{3}x + \tfrac{8}{3} + \tfrac{6}{3}$$
$$y = \tfrac{2}{3}x + \tfrac{14}{3}$$

(b)
$$y = \tfrac{2}{3}x + \tfrac{14}{3}$$
$$3y = 2x + 14$$
$$-2x + 3y = 14$$
$$2x - 3y = -14$$

15. Through $(-4, 12)$ and the midpoint of the segment with endpoints $(5, 8)$ and $(-3, 2)$

(a) The midpoint of the segment with endpoints $(5, 8)$ and $(-3, 2)$ is

$$\left(\frac{5 + (-3)}{2}, \frac{8 + 2}{2}\right) = \left(\frac{2}{2}, \frac{10}{2}\right) = (1, 5).$$

The slope of the line through $(-4, 12)$ and $(1, 5)$ is

$$m = \frac{5 - 12}{1 - (-4)} = \frac{-7}{5} = -\frac{7}{5}.$$

Use the point-slope form with $(x_1, y_1) = (-4, 12)$ and $m = -\frac{7}{5}$.

$$y - y_1 = m(x - x_1)$$
$$y - 12 = -\tfrac{7}{5}[x - (-4)]$$
$$y - 12 = -\tfrac{7}{5}(x + 4)$$
$$y - 12 = -\tfrac{7}{5}x - \tfrac{28}{5}$$
$$y = -\tfrac{7}{5}x - \tfrac{28}{5} + \tfrac{60}{5}$$
$$y = -\tfrac{7}{5}x + \tfrac{32}{5}$$

(b)
$$y = -\tfrac{7}{5}x + \tfrac{32}{5}$$
$$5y = -7x + 32$$
$$7x + 5y = 32$$

17. Through $\left(\frac{3}{7}, \frac{1}{6}\right)$; parallel to $y = \frac{1}{5}x + \frac{7}{4}$

(a) The slope of the desired line is the same as the slope of the given line, so use the point-slope form with $(x_1, y_1) = \left(\frac{3}{7}, \frac{1}{6}\right)$ and $m = \frac{1}{5}$.

$$y - y_1 = m(x - x_1)$$
$$y - \tfrac{1}{6} = \tfrac{1}{5}\left(x - \tfrac{3}{7}\right)$$
$$y - \tfrac{1}{6} = \tfrac{1}{5}x - \tfrac{3}{35}$$
$$y = \tfrac{1}{5}x - \tfrac{18}{210} + \tfrac{35}{210}$$
$$y = \tfrac{1}{5}x + \tfrac{17}{210}$$

(b)
$$y = \tfrac{1}{5}x + \tfrac{17}{210}$$
$$210y = 42x + 17 \quad \textit{Multiply by 210.}$$
$$-42x + 210y = 17$$
$$42x - 210y = -17$$

19. Through $(0.3, 1.5)$ and $(0.4, 1.7)$

(a) $m = \dfrac{1.7 - 1.5}{0.4 - 0.3} = \dfrac{0.2}{0.1} = 2$

Use the point-slope form with $(x_1, y_1) = (0.3, 1.5)$ and $m = 2$.

$$y - y_1 = m(x - x_1)$$
$$y - 1.5 = 2(x - 0.3)$$
$$y - 1.5 = 2x - 0.6$$
$$y = 2x + 0.9$$

(b)
$$y = 2x + 0.9$$
$$10y = 20x + 9 \quad \textit{Multiply by 10.}$$
$$-20x + 10y = 9$$
$$20x - 10y = -9$$

21. Slope -0.5, $b = -2$

The slope-intercept form of a line, $y = mx + b$, becomes $y = -0.5x - 2$, or $y = -\frac{1}{2}x - 2$, which is choice **B**.

23. Passes through $(4, -2)$ and $(0, 0)$

$$m = \frac{0 - (-2)}{0 - 4} = \frac{2}{-4} = -\frac{1}{2}$$

Using $m = -\frac{1}{2}$ and a y-intercept of $(0, 0)$, we get $y = -\frac{1}{2}x + 0$, which is choice **A**.

25. $m = \frac{1}{2}$, passes through the origin

Use the point-slope form with $(x_1, y_1) = (0, 0)$ and $m = \frac{1}{2}$.

$$y - y_1 = m(x - x_1)$$
$$y - 0 = \tfrac{1}{2}(x - 0)$$
$$y = \tfrac{1}{2}x \quad \text{or} \quad 2y = x$$

This is choice **E**.

7.3 Functions

1. We give one of many possible answers here. A function is a set of ordered pairs in which each first component corresponds to exactly one second component. For example, $\{(0,1),(1,2),(2,3),(3,4)\dots\}$ is a function.

3. In an ordered pair of a relation, the first element is the independent variable.

5. $\{(5,1),(3,2),(4,9),(7,6)\}$
The relation is a function since for each x-value, there is only one y-value.

7. $\{(2,4),(0,2),(2,5)\}$
The relation is not a function since the x-value 2 has two different y-values associated with it, 4 and 5.

9. $\{(-3,1),(4,1),(-2,7)\}$
The relation is a function since for each x-value, there is only one y-value.

11. $\{(1,1),(1,-1),(0,0),(2,4),(2,-4)\}$
The relation is not a function since the x-value 1 has two different y-values associated with it, 1 and -1. (A similar statement can be made for $x=2$.)

The domain is the set of x-values: $\{0,1,2\}$.
The range is the set of y-values: $\{-4,-1,0,1,4\}$.

13. The relation can be described by the set of ordered pairs

$$\{(2,1),(5,1),(11,7),(17,20),(3,20)\}.$$

The relation is a function since for each x-value, there is only one y-value.

The domain is the set of x-values: $\{2,3,5,11,17\}$.
The range is the set of y-values: $\{1,7,20\}$.

15. The relation can be described by the set of ordered pairs

$$\{(1,5),(1,2),(1,-1),(1,-4)\}.$$

The relation is not a function since the x-value 1 has four different y-values associated with it, 5, 2, -1, and -4.

The domain is the set of x-values: $\{1\}$.
The range is the set of y-values: $\{5,2,-1,-4\}$.

17. Using the vertical line test, we find any vertical line will intersect the graph at most once. This indicates that the graph represents a function. This graph extends indefinitely to the left ($-\infty$) and indefinitely to the right (∞). Therefore, the domain is $(-\infty,\infty)$. This graph extends indefinitely downward ($-\infty$), and indefinitely upward (∞). Thus, the range is $(-\infty,\infty)$.

19. Using the vertical line test, we find any vertical line will intersect the graph at most once. This indicates that the graph represents a function. This graph extends indefinitely to the left ($-\infty$) and indefinitely to the right (∞). Therefore, the domain is $(-\infty,\infty)$. This graph extends indefinitely downward ($-\infty$), and reaches a high point at $y=4$. Therefore, the range is $(-\infty,4]$.

21. Since a vertical line can intersect the graph of the relation in more than one point, the relation is not a function. The domain, the x-values of the points on the graph, is $[-4,4]$. The range, the y-values of the points on the graph, is $[-3,3]$.

23. $y=x^2$
Each value of x corresponds to one y-value. For example, if $x=3$, then $y=3^2=9$. Therefore, $y=x^2$ defines y as a function of x.
Since any x-value, positive, negative, or zero, can be squared, the domain is $(-\infty,\infty)$.

25. $x=y^6$
The ordered pairs $(64,2)$ and $(64,-2)$ both satisfy the equation. Since one value of x, 64, corresponds to two values of y, 2 and -2, the relation does not define a function. Because x is equal to the sixth power of y, the values of x must always be nonnegative. The domain is $[0,\infty)$.

27. $y=2x-6$

For any value of x, there is exactly one value of y, so this equation defines a function. The domain is the set of all real numbers, $(-\infty,\infty)$.

29. $x+y=4$

Solving for y gives us $y=-x+4$.

For any value of x, there is exactly one value of y, so this equation defines a function. The domain is the set of all real numbers, $(-\infty,\infty)$.

31. $y=\dfrac{2}{x-4}$
Given any value of x, y is found by subtracting 4, then dividing the result into 2. This process produces exactly one value of y for each x-value, so the relation represents a function. The domain includes all real numbers except those that make the denominator 0, namely 4. The domain is $(-\infty,4)\cup(4,\infty)$.

33. $y=\dfrac{1}{4x+2}$ defines y as a function of x. The domain includes all real numbers except those that make the denominator 0, namely $-\frac{1}{2}$. The domain is $\left(-\infty,-\frac{1}{2}\right)\cup\left(-\frac{1}{2},\infty\right)$.

35. $xy = 1$

Rewrite $xy = 1$ as $y = \dfrac{1}{x}$. Note that x can never equal 0, otherwise the denominator would equal 0. The domain is $(-\infty, 0) \cup (0, \infty)$.
Each nonzero x-value gives exactly one y-value. Therefore, $xy = 1$ defines y as a function of x.

37. $y = |x|$

For any value of x, there is exactly one value of y, so this equation defines a function. We can take the absolute value of any number, so the domain is the set of all real numbers, $(-\infty, \infty)$.

39. $f(3)$ is the value of the dependent variable when the independent variable is 3—choice **B**.

41. $f(x) = -3x + 4$
$f(0) = -3(0) + 4$
$ = 0 + 4$
$ = 4$

43. $g(x) = -x^2 + 4x + 1$
$g(-2) = -(-2)^2 + 4(-2) + 1$
$ = -(4) - 8 + 1$
$ = -11$

45. $f(x) = -3x + 4$
$f\left(\tfrac{1}{3}\right) = -3\left(\tfrac{1}{3}\right) + 4$
$ = -1 + 4$
$ = 3$

47. $g(x) = -x^2 + 4x + 1$
$g(0.5) = -(0.5)^2 + 4(0.5) + 1$
$ = -0.25 + 2 + 1$
$ = 2.75$

49. $f(x) = -3x + 4$
$f(p) = -3(p) + 4$
$ = -3p + 4$

51. $f(x) = -3x + 4$
$f(-x) = -3(-x) + 4$
$ = 3x + 4$

53. $f(x) = -3x + 4$
$f(x + 2) = -3(x + 2) + 4$
$ = -3x - 6 + 4$
$ = -3x - 2$

55. $g(x) = -x^2 + 4x + 1$
$g(\pi) = -\pi^2 + 4\pi + 1$

57. $f(x) = -3x + 4$
$f(x + h) = -3(x + h) + 4$
$ = -3x - 3h + 4$

59. $f(4) - g(4)$
$= [-3(4) + 4] - [-(4)^2 + 4(4) + 1]$
$= [-8] - [1]$
$= -9$

61. **(a)** When $x = 2$, $y = 2$, so $f(2) = 2$.

(b) When $x = -1$, $y = 3$, so $f(-1) = 3$.

63. **(a)** When $x = 2$, $y = 15$, so $f(2) = 15$.

(b) When $x = -1$, $y = 10$, so $f(-1) = 10$.

65. **(a)** The point $(2, 3)$ is on the graph of f, so $f(2) = 3$.

(b) The point $(-1, -3)$ is on the graph of f, so $f(-1) = -3$.

67. **(a)** Solve the equation for y.

$$x + 3y = 12$$
$$3y = 12 - x$$
$$y = \frac{12 - x}{3}$$

Since $y = f(x)$,

$$f(x) = \frac{12 - x}{3} = -\frac{1}{3}x + 4.$$

(b) $f(3) = \dfrac{12 - 3}{3} = \dfrac{9}{3} = 3$

69. **(a)** Solve the equation for y.

$$y + 2x^2 = 3$$
$$y = 3 - 2x^2$$

Since $y = f(x)$,

$$f(x) = 3 - 2x^2.$$

(b) $f(3) = 3 - 2(3)^2$
$ = 3 - 2(9)$
$ = -15$

71. **(a)** Solve the equation for y.

$$4x - 3y = 8$$
$$-3y = 8 - 4x$$
$$y = \frac{8 - 4x}{-3}$$

Since $y = f(x)$,

$$f(x) = \frac{8 - 4x}{-3} = \frac{4}{3}x - \frac{8}{3}.$$

(b) $f(3) = \dfrac{8 - 4(3)}{-3} = \dfrac{8 - 12}{-3}$
$ = \dfrac{-4}{-3} = \dfrac{4}{3}$

73. The equation $2x + y = 4$ has a straight <u>line</u> as its graph. To find y in $(3, \underline{y})$, let $x = 3$ in the equation.

$$2x + y = 4$$
$$2(3) + y = 4$$
$$6 + y = 4$$
$$y = -2$$

To use functional notation for $2x + y = 4$, solve for y to get

$$y = -2x + 4.$$

Replace y with $f(x)$ to get

$$f(x) = \underline{-2x + 4}.$$
$$f(3) = -2(3) + 4 = \underline{-2}$$

Because $y = -2$ when $x = 3$, the point $\underline{(3, -2)}$ lies on the graph of the function.

75. $f(x) = -2x + 5$
The graph will be a line. The intercepts are $(0, 5)$ and $\left(\frac{5}{2}, 0\right)$.
The domain is $(-\infty, \infty)$. The range is $(-\infty, \infty)$.

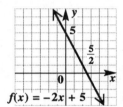

77. $h(x) = \frac{1}{2}x + 2$
The graph will be a line. The intercepts are $(0, 2)$ and $(-4, 0)$.
The domain is $(-\infty, \infty)$. The range is $(-\infty, \infty)$.

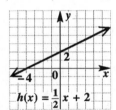

79. $G(x) = 2x$
This line includes the points $(0, 0), (1, 2)$, and $(2, 4)$. The domain is $(-\infty, \infty)$. The range is $(-\infty, \infty)$.

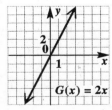

81. $g(x) = -4$
Using a y-intercept of $(0, -4)$ and a slope of $m = 0$, we graph the horizontal line. From the graph we see that the domain is $(-\infty, \infty)$. The range is $\{-4\}$.

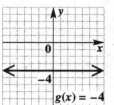

83. $f(x) = 0$
Draw the horizontal line through the point $(0, 0)$. On the horizontal line the value of x can be any real number, so the domain is $(-\infty, \infty)$. The range is $\{0\}$.

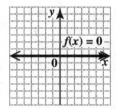

85. (a) $f(x) = 2.75x$
$$f(3) = 2.75(3)$$
$$= 8.25 \text{ (dollars)}$$

(b) 3 is the value of the independent variable, which represents a package weight of 3 pounds; $f(3)$ is the value of the dependent variable representing the cost to mail a 3-pound package.

(c) $2.75(5) = \$13.75$, the cost to mail a 5-lb package. Using function notation, we have $f(5) = 13.75$.

87. (a) Since the length of a man's femur is given, use the formula $h(r) = 69.09 + 2.24r$.

$$h(56) = 69.09 + 2.24(56) \quad \textit{Let } r = 56.$$
$$= 194.53$$

The man is 194.53 cm tall.

(b) Use the formula $h(t) = 81.69 + 2.39t$.

$$h(40) = 81.69 + 2.39(40) \quad \textit{Let } t = 40.$$
$$= 177.29$$

The man is 177.29 cm tall.

(c) Since the length of a woman's femur is given, use the formula $h(r) = 61.41 + 2.32r$.

$$h(50) = 61.41 + 2.32(50) \quad \textit{Let } r = 50.$$
$$= 177.41$$

The woman is 177.41 cm tall.

(d) Use the formula $h(t) = 72.57 + 2.53t$.

$$h(36) = 72.57 + 2.53(36) \quad \textit{Let } t = 36.$$
$$= 163.65$$

The woman is 163.65 cm tall.

89. (a) The independent variable is t, the number of hours, and the possible values are in the set $[0, 100]$. The dependent variable is g, the number of gallons, and the possible values are in the set $[0, 3000]$.

(b) The graph rises for the first 25 hours, so the water level increases for 25 hours. The graph falls for $t = 50$ to $t = 75$, so the water level decreases for 25 hours.

(c) There are 2000 gallons in the pool when $t = 90$.

(d) $f(0)$ is the number of gallons in the pool at time $t = 0$. Here, $f(0) = 0$, which means the pool is empty at time 0.

(e) $f(25) = 3000$; After 25 hours, there are 3000 gallons of water in the pool.

91. The graph shows $x = 3$ and $y = 7$. In function notation, this is

$$f(3) = 7.$$

93. $\left(15x^2 - 2x\right) + (x - 4) = 15x^2 - 2x + x - 4$
$$= 15x^2 - x - 4$$

95. $\left(2a^2 + 3a - 1\right) - \left(4a^2 + 5a - 6\right)$
$$= 2a^2 + 3a - 1 - 4a^2 - 5a + 6$$
$$= -2a^2 - 2a + 5$$

97. $(4x - 5)(3x + 1)$
$$= 4x(3x) + 4x(1) - 5(3x) - 5(1)$$
$$= 12x^2 + 4x - 15x - 5$$
$$= 12x^2 - 11x - 5$$

99. $\dfrac{27x^3 - 18x^2}{9x} = \dfrac{27x^3}{9x} - \dfrac{18x^2}{9x}$
$$= 3x^2 - 2x$$

7.4   Operations on Functions and Composition

1. $f(x) = 6x - 4$

(a) $f(-1) = 6(-1) - 4$
$$= -6 - 4 = -10$$

(b) $f(2) = 6(2) - 4$
$$= 12 - 4 = 8$$

3. $f(x) = 5x^4 - 3x^2 + 6$

(a) $f(-1) = 5(-1)^4 - 3(-1)^2 + 6$
$$= 5 \cdot 1 - 3 \cdot 1 + 6$$
$$= 5 - 3 + 6 = 8$$

(b) $f(2) = 5(2)^4 - 3(2)^2 + 6$
$$= 5 \cdot 16 - 3 \cdot 4 + 6$$
$$= 80 - 12 + 6 = 74$$

5. $f(x) = -x^2 + 2x^3 - 8$

(a) $f(-1) = -(-1)^2 + 2(-1)^3 - 8$
$$= -(1) + 2(-1) - 8$$
$$= -1 - 2 - 8 = -11$$

(b) $f(2) = -(2)^2 + 2(2)^3 - 8$
$$= -(4) + 2 \cdot 8 - 8$$
$$= -4 + 16 - 8 = 4$$

7. (a) $(f + g)(x) = f(x) + g(x)$
$$= (5x - 10) + (3x + 7)$$
$$= 8x - 3$$

(b) $(f - g)(x) = f(x) - g(x)$
$$= (5x - 10) - (3x + 7)$$
$$= (5x - 10) + (-3x - 7)$$
$$= 2x - 17$$

9. (a) $(f + g)(x)$
$$= f(x) + g(x)$$
$$= (4x^2 + 8x - 3) + (-5x^2 + 4x - 9)$$
$$= -x^2 + 12x - 12$$

(b) $(f - g)(x)$
$$= f(x) - g(x)$$
$$= (4x^2 + 8x - 3) - (-5x^2 + 4x - 9)$$
$$= (4x^2 + 8x - 3) + (5x^2 - 4x + 9)$$
$$= 9x^2 + 4x + 6$$

For Exercises 11–25, let $f(x) = x^2 - 9$, $g(x) = 2x$, and $h(x) = x - 3$.

11. $(f + g)(x) = f(x) + g(x)$
$$= \left(x^2 - 9\right) + (2x)$$
$$= x^2 + 2x - 9$$

13. $(f + g)(3) = f(3) + g(3)$
$$= \left(3^2 - 9\right) + 2(3)$$
$$= 0 + 6 = 6$$

Alternatively, we could evaluate the polynomial in Exercise 11, $x^2 + 2x - 9$, using $x = 3$.

15. $(f - h)(x) = f(x) - h(x)$
$$= \left(x^2 - 9\right) - (x - 3)$$
$$= x^2 - 9 - x + 3$$
$$= x^2 - x - 6$$

17. $(f - h)(-3) = f(-3) - h(-3)$
$$= \left[(-3)^2 - 9\right] - \left[(-3) - 3\right]$$
$$= (9 - 9) - (-6)$$
$$= 0 + 6 = 6$$

19. $(g + h)(-10) = g(-10) + h(-10)$
$$= 2(-10) + \left[(-10) - 3\right]$$
$$= -20 + (-13)$$
$$= -33$$

21. $(g - h)(-3) = g(-3) - h(-3)$
$$= 2(-3) - \left[(-3) - 3\right]$$
$$= -6 - (-6)$$
$$= -6 + 6 = 0$$

23. $(g + h)\left(\frac{1}{4}\right) = g\left(\frac{1}{4}\right) + h\left(\frac{1}{4}\right)$
$$= 2\left(\frac{1}{4}\right) + \left(\frac{1}{4} - 3\right)$$
$$= \frac{1}{2} + \left(-\frac{11}{4}\right)$$
$$= -\frac{9}{4}$$

25. $(g + h)\left(-\frac{1}{2}\right) = g\left(-\frac{1}{2}\right) + h\left(-\frac{1}{2}\right)$
$$= 2\left(-\frac{1}{2}\right) + \left(-\frac{1}{2} - 3\right)$$
$$= -1 + \left(-\frac{7}{2}\right)$$
$$= -\frac{9}{2}$$

27. Answers will vary. Let $f(x) = x^3$ and $g(x) = x^4$.
$$(f - g)(x) = f(x) - g(x)$$
$$= x^3 - x^4$$
$$(g - f)(x) = g(x) - f(x)$$
$$= x^4 - x^3$$

Because the two differences are not equal, subtraction of polynomial functions is not commutative.

29. $f(x) = 2x, \ g(x) = 5x - 1$
$$(fg)(x) = f(x) \cdot g(x)$$
$$= 2x(5x - 1)$$
$$= 10x^2 - 2x$$

31. $f(x) = x + 1, \ g(x) = 2x - 3$
$$(fg)(x) = f(x) \cdot g(x)$$
$$= (x + 1)(2x - 3)$$
$$= 2x^2 - 3x + 2x - 3$$
$$= 2x^2 - x - 3$$

33. $f(x) = 2x - 3, \ g(x) = 4x^2 + 6x + 9$
$$(fg)(x) = f(x) \cdot g(x)$$
$$= (2x - 3)(4x^2 + 6x + 9)$$

Multiply vertically.

$$
\begin{array}{r}
4x^2 + 6x + 9 \\
2x - 3 \\
\hline
-12x^2 - 18x - 27 \\
8x^3 + 12x^2 + 18x \qquad\quad \\
\hline
8x^3 \qquad\qquad\qquad - 27
\end{array}
$$

For Exercises 35–46, let $f(x) = x^2 - 9$, $g(x) = 2x$, and $h(x) = x - 3$.

35. $(fg)(x) = f(x) \cdot g(x)$
$$= (x^2 - 9)(2x)$$
$$= 2x^3 - 18x$$

37. $(fg)(2) = f(2) \cdot g(2)$
$$= (2^2 - 9)[2(2)]$$
$$= -5 \cdot 4 = -20$$

39. $(gh)(x) = g(x) \cdot h(x)$
$$= (2x)(x - 3)$$
$$= 2x^2 - 6x$$

41. $(gh)(-3) = g(-3) \cdot h(-3)$
$$= [2(-3)] \cdot [(-3) - 3]$$
$$= -6(-6) = 36$$

43. $(fg)\left(-\frac{1}{2}\right) = f\left(-\frac{1}{2}\right) \cdot g\left(-\frac{1}{2}\right)$
$$= \left[\left(-\frac{1}{2}\right)^2 - 9\right] \cdot \left[2\left(-\frac{1}{2}\right)\right]$$
$$= \left(\frac{1}{4} - \frac{36}{4}\right)(-1)$$
$$= -\frac{35}{4}(-1) = \frac{35}{4}$$

45. $(fh)\left(-\frac{1}{4}\right) = f\left(-\frac{1}{4}\right) \cdot h\left(-\frac{1}{4}\right)$
$$= \left[\left(-\frac{1}{4}\right)^2 - 9\right] \cdot \left[-\frac{1}{4} - 3\right]$$
$$= \left(\frac{1}{16} - \frac{144}{16}\right)\left(-\frac{1}{4} - \frac{12}{4}\right)$$
$$= -\frac{143}{16}\left(-\frac{13}{4}\right) = \frac{1859}{64}$$

47. $\left(\dfrac{f}{g}\right)(x) = \dfrac{f(x)}{g(x)} = \dfrac{10x^2 - 2x}{2x}$
$$= \frac{10x^2}{2x} - \frac{2x}{2x}$$
$$= 5x - 1$$

The x-values that are not in the domain of the quotient function are found by solving $g(x) = 0$.
$$2x = 0$$
$$x = 0$$

49.

$$
\begin{array}{r}
2x - 3 \\
x + 1 \overline{\smash{\big)}\, 2x^2 - x - 3} \\
\underline{2x^2 + 2x} \\
-3x - 3 \\
\underline{-3x - 3} \\
0
\end{array}
$$

Quotient: $2x - 3$

$$g(x) = 0$$
$$x + 1 = 0$$
$$x = -1$$

51.

$$
\begin{array}{r}
4x^2 + 6x + 9 \\
2x - 3 \overline{\smash{\big)}\, 8x^3 + 0x^2 + 0x - 27} \\
\underline{8x^3 - 12x^2} \\
12x^2 + 0x \\
\underline{12x^2 - 18x} \\
18x - 27 \\
\underline{18x - 27} \\
0
\end{array}
$$

Quotient: $4x^2 + 6x + 9$

$$g(x) = 0$$
$$2x - 3 = 0$$
$$2x = 3$$
$$x = \tfrac{3}{2}$$

For Exercises 53–64, let $f(x) = x^2 - 9$, $g(x) = 2x$, and $h(x) = x - 3$.

53. $\left(\dfrac{f}{g}\right)(x) = \dfrac{f(x)}{g(x)} = \dfrac{x^2 - 9}{2x}$

We must exclude any values of x that make the denominator equal to zero, that is, $x \neq 0$.

55. $\left(\dfrac{f}{g}\right)(2) = \dfrac{f(2)}{g(2)} = \dfrac{2^2 - 9}{2(2)} = \dfrac{-5}{4} = -\dfrac{5}{4}$

57. $\left(\dfrac{h}{g}\right)(x) = \dfrac{h(x)}{g(x)} = \dfrac{x - 3}{2x},\ x \neq 0$

59. $\left(\dfrac{h}{g}\right)(3) = \dfrac{h(3)}{g(3)} = \dfrac{(3) - 3}{2(3)} = \dfrac{0}{6} = 0$

61. $\left(\dfrac{f}{g}\right)\left(\dfrac{1}{2}\right) = \dfrac{f\left(\frac{1}{2}\right)}{g\left(\frac{1}{2}\right)} = \dfrac{\left(\frac{1}{2}\right)^2 - 9}{2\left(\frac{1}{2}\right)} = \dfrac{\frac{1}{4} - \frac{36}{4}}{1}$

$ = -\dfrac{35}{4}$

63. $\left(\dfrac{h}{g}\right)\left(-\dfrac{1}{2}\right) = \dfrac{h\left(-\frac{1}{2}\right)}{g\left(-\frac{1}{2}\right)} = \dfrac{-\frac{1}{2} - 3}{2\left(-\frac{1}{2}\right)} = \dfrac{-\frac{7}{2}}{-1} = \dfrac{7}{2}$

For Exercises 65–80,
$f(x) = x^2 + 4$, $g(x) = 2x + 3$, and $h(x) = x + 5$.

65. $(h \circ g)(4) = h(g(4))$
$ = h(2 \cdot 4 + 3)$
$ = h(11)$
$ = 11 + 5$
$ = 16$

67. $(g \circ f)(6) = g(f(6))$
$ = g(6^2 + 4)$
$ = g(40)$
$ = 2 \cdot 40 + 3$
$ = 83$

69. $(f \circ h)(-2) = f(h(-2))$
$ = f(-2 + 5)$
$ = f(3)$
$ = 3^2 + 4$
$ = 13$

71. $(f \circ g)(x) = f(g(x))$
$ = f(2x + 3)$
$ = (2x + 3)^2 + 4$
$ = 4x^2 + 12x + 9 + 4$
$ = 4x^2 + 12x + 13$

73. $(f \circ h)(x) = f(h(x))$
$ = f(x + 5)$
$ = (x + 5)^2 + 4$
$ = x^2 + 10x + 25 + 4$
$ = x^2 + 10x + 29$

75. $(h \circ g)(x) = h(g(x))$
$ = h(2x + 3)$
$ = 2x + 3 + 5$
$ = 2x + 8$

77. $(f \circ h)\left(\dfrac{1}{2}\right) = f\left(h\left(\dfrac{1}{2}\right)\right)$
$ = f\left(\dfrac{1}{2} + 5\right)$
$ = f\left(\dfrac{11}{2}\right)$
$ = \left(\dfrac{11}{2}\right)^2 + 4$
$ = \dfrac{121}{4} + \dfrac{16}{4} = \dfrac{137}{4}$

79. $(f \circ g)\left(-\dfrac{1}{2}\right) = f\left(g\left(-\dfrac{1}{2}\right)\right)$
$ = f\left(2\left(-\dfrac{1}{2}\right) + 3\right)$
$ = f(2)$
$ = 2^2 + 4$
$ = 4 + 4 = 8$

81. $f(x) = 12x$, $g(x) = 5280x$

$(f \circ g)(x) = f[g(x)]$
$ = f(5280x)$
$ = 12(5280x)$
$ = 63{,}360x$

$(f \circ g)(x)$ computes the number of inches in x mi.

83. $r(t) = 2t$, $A(r) = \pi r^2$

$$(A \circ r)(t) = A[r(t)]$$
$$= A(2t)$$
$$= \pi(2t)^2$$
$$= 4\pi t^2$$

This is the area of the circular layer as a function of time.

85. $y = kx$

$1 = k(3)$ *Let y = 1, x = 3.*

$\frac{1}{3} = k$

87. $y = \dfrac{k}{x}$

$1 = \dfrac{k}{3}$ *Let y = 1, x = 3.*

$1(3) = k$

$3 = k$

7.5   Variation

1. This suggests *direct* variation.
As the number of tickets you buy increases, so does the probability that you will win.

3. This suggests *direct* variation.
As the pressure put on the accelerator increases, so does the speed of the car.

5. This suggests *inverse* variation.
As your age gets larger, the probability that you believe in Santa Claus gets smaller.

7. This suggests *inverse* variation.
As the number of days gets smaller, the number of home runs increases.

9. The equation $y = \dfrac{3}{x}$ represents *inverse* variation.
y varies inversely as x because x is in the denominator.

11. The equation $y = 10x^2$ represents *direct* variation. The number 10 is the constant of variation, and y varies directly as the square of x.

13. The equation $y = 3xz^4$ represents *joint* variation.
y varies directly as x and z^4.

15. The equation $y = \dfrac{4x}{wz}$ represents *combined* variation. In the numerator, 4 is the constant of variation, and y varies directly as x. In the denominator, y varies inversely as w and z.

17. For $k > 0$, if y varies directly as x (then $y = kx$), when x increases, y *increases*, and when x decreases, y *decreases*.

19. "x varies directly as y" means

$$x = ky$$

for some constant k.
Substitute $x = 9$ and $y = 3$ in the equation and solve for k.

$$x = ky$$
$$9 = k(3)$$
$$k = \tfrac{9}{3} = 3$$

So $x = 3y$.

To find x when $y = 12$, substitute 12 for y in the equation.

$$x = 3y$$
$$x = 3(12)$$
$$x = 36$$

21. "a varies directly as the square of b" means

$$a = kb^2$$

for some constant k.
Substitute $a = 4$ and $b = 3$ in the equation and solve for k.

$$a = kb^2$$
$$4 = k(3)^2$$
$$k = \tfrac{4}{9}$$

So $a = \tfrac{4}{9}b^2$.

To find a when $b = 2$, substitute 2 for b in the equation.

$$a = \tfrac{4}{9}b^2$$
$$a = \tfrac{4}{9}(2)^2$$
$$a = \frac{4 \cdot 4}{9} = \frac{16}{9}$$

23. "z varies inversely as w" means

$$z = \frac{k}{w}$$

for some constant k. Since $z = 10$ when $w = 0.5$, substitute these values in the equation and solve for k.

$$z = \frac{k}{w}$$
$$10 = \frac{k}{0.5}$$
$$k = 10(0.5) = 5$$

So $z = \dfrac{5}{w}$.

To find z when $w = 8$, substitute 8 for w in the equation.

$$z = \frac{5}{w}$$
$$z = \tfrac{5}{8} \text{ or } 0.625$$

25. "m varies inversely as p^2" means

$$m = \frac{k}{p^2}$$

for some constant k. Since $m = 20$ when $p = 2$, substitute these values in the equation and solve for k.

$$m = \frac{k}{p^2}$$
$$20 = \frac{k}{2^2}$$
$$k = 20(4) = 80$$

So $m = \dfrac{80}{p^2}$. Now let $p = 5$.

$$m = \frac{80}{p^2}$$
$$m = \frac{80}{5^2} = \frac{16}{5}$$

27. "p varies jointly as q and r^2" means

$$p = kqr^2$$

for some constant k. Given that $p = 200$ when $q = 2$ and $r = 3$, solve for k.

$$p = kqr^2$$
$$200 = k(2)(3)^2$$
$$200 = 18k$$
$$k = \frac{200}{18} = \frac{100}{9}$$

So $p = \dfrac{100}{9}qr^2$. Now let $q = 5$ and $r = 2$.

$$p = \frac{100}{9}qr^2$$
$$p = \frac{100}{9}(5)(2)^2$$
$$= \frac{100}{9}(20)$$
$$= \frac{2000}{9} \quad \text{or} \quad 222\frac{2}{9}$$

29. Let $x = $ the number of gallons he bought and let $C = $ the cost.
C varies directly as x, so

$$C = kx.$$

Since $C = 43.79$ when $x = 15$,

$$43.79 = k(15)$$
$$k = \frac{43.79}{15} \approx 2.919.$$

The price per gallon is $\$2.91\frac{9}{10}$.

31. Let $y = $ the weight of an object on earth and $x = $ the weight of the object on the moon.
y varies directly as x, so

$$y = kx$$

for some constant k. Since $y = 200$ when $x = 32$, substitute these values in the equation and solve for k.

$$y = kx$$
$$200 = k(32)$$
$$k = \frac{200}{32} = 6.25$$

So $y = 6.25x$.
To find x when $y = 50$, substitute 50 for y in the equation

$$y = 6.25x.$$
$$50 = 6.25x$$
$$x = \frac{50}{6.25} = 8$$

The dog would weigh 8 lb on the moon.

33. Let $V = $ the volume of the can and let $h = $ the height of the can.
V varies directly as h, so

$$V = kh.$$

Since $V = 300$ when $h = 10.62$,

$$300 = k(10.62)$$
$$k = \frac{300}{10.62} \approx 28.25.$$

So $V = 28.25h$. Now let $h = 15.92$.

$$V = 28.25h$$
$$V = 28.25(15.92) = 449.74$$

The volume is about 450 cubic inches.

35. Let $d = $ the distance and $t = $ the time.
d varies directly as the square of t, so $d = kt^2$.
Let $d = -576$ and $t = 6$. (You could also use $d = 576$, but the negative sign indicates the direction of the body.)

$$-576 = k(6)^2$$
$$-576 = 36k$$
$$-16 = k$$

So $d = -16t^2$. Now let $t = 4$.

$$d = -16(4)^2 = -256$$

The object fell 256 feet in the first 4 seconds.

37. Let $s = $ the speed and $t = $ the time.
The speed varies inversely with time, so there is a constant k such that $s = k/t$. Find the value of k by replacing s with 160 and t with $\frac{1}{2}$.

$$s = \frac{k}{t}$$
$$160 = \frac{k}{\frac{1}{2}}$$
$$k = 160\left(\frac{1}{2}\right) = 80$$

So $s = \dfrac{80}{t}$. Now let $t = \frac{3}{4}$.

$$s = \frac{80}{\frac{3}{4}}$$

$$s = \frac{80}{1} \cdot \frac{4}{3} = \frac{320}{3} \quad \text{or} \quad 106\frac{2}{3}$$

A speed $106\frac{2}{3}$ miles per hour is needed to go the same distance in $\frac{3}{4}$ minute.

39. Let f = the frequency of a string in cycles per second and s = the length in feet.
f varies inversely as s, so

$$f = \frac{k}{s}$$

for some constant k. Since $f = 250$ when $s = 2$, substitute these values in the equation and solve for k.

$$f = \frac{k}{s}$$
$$250 = \frac{k}{2}$$
$$k = 250(2) = 500$$

So $f = \frac{500}{s}$. Now let $s = 5$.

$$f = \frac{500}{5} = 100.$$

The string would have a frequency of 100 cycles per second.

41. Let I = the illumination produced by a light source and d = the distance from the source.
I varies inversely as d^2, so

$$I = \frac{k}{d^2}$$

for some constant k. Since $I = 768$ when $d = 1$, substitute these values in the equation and solve for k.

$$I = \frac{k}{d^2}$$
$$768 = \frac{k}{1^2}$$
$$768 = k$$

So $I = \frac{768}{d^2}$. Now let $d = 6$.

$$I = \frac{768}{d^2}$$
$$I = \frac{768}{6^2} = \frac{768}{36} = \frac{64}{3} \quad \text{or} \quad 21\frac{1}{3}$$

The illumination produced by the light source is $21\frac{1}{3}$ foot-candles.

43. Let I = the simple interest, P the principal, and t the time.
Since I varies jointly as the principal and time, there is a constant k such that $I = kPt$. Find k by replacing I with 280, P with 2000, and t with 4.

$$I = kPt$$
$$280 = k(2000)(4)$$
$$k = \frac{280}{8000} = 0.035$$

So $I = 0.035Pt$. Now let $t = 6$.

$$I = 0.035(2000)(6)$$
$$= 420$$

The interest would be $420.

45. The weight W of a bass varies jointly as its girth G and the square of its length L, so

$$W = kGL^2$$

for some constant k. Substitute 22.7 for W, 21 for G, and 36 for L.

$$22.7 = k(21)(36)^2$$
$$k = \frac{22.7}{27,216} \approx 0.000834$$

So $W = 0.000834GL^2$. Now let $G = 18$ and $L = 28$.

$$W = 0.000834GL^2$$
$$= 0.000834(18)(28)^2$$
$$\approx 11.8$$

The bass would weigh about 11.8 pounds.

47. Let F = the force, w = the weight of the car, s = the speed, and r = the radius.
The force varies inversely as the radius and jointly as the weight and the square of the speed, so

$$F = \frac{kws^2}{r}.$$

Let $F = 242$, $w = 2000$, $r = 500$, and $s = 30$.

$$242 = \frac{k(2000)(30)^2}{500}$$
$$k = \frac{242(500)}{2000(900)} = \frac{121}{1800}$$

So $F = \frac{121ws^2}{1800r}.$
Let $r = 750$, $s = 50$, and $w = 2000$.

$$F = \frac{121(2000)(50)^2}{1800(750)} \approx 448.1$$

Approximately 448.1 pounds of force would be needed.

49. Let N = the number of long distance calls,
p_1 = the population of City 1,
p_2 = the population of City 2,
and d = the distance between them.

$$N = \frac{kp_1p_2}{d}$$

Let $N = 80{,}000$, $p_1 = 70{,}000$, $p_2 = 100{,}000$, and $d = 400$.

$$80{,}000 = \frac{k(70{,}000)(100{,}000)}{400}$$
$$80{,}000 = 17{,}500{,}000k$$
$$k = \frac{80{,}000}{17{,}500{,}000} = \frac{4}{875}$$
$$N = \frac{4}{875}\left(\frac{p_1p_2}{d}\right)$$

Let $p_1 = 50{,}000$, $p_2 = 75{,}000$, and $d = 250$.

$$N = \frac{4}{875}\left(\frac{50{,}000 \cdot 75{,}000}{250}\right)$$
$$= \frac{480{,}000}{7} = 68{,}571\tfrac{3}{7}$$

Rounded to the nearest hundred, there are approximately 68,600 calls.

51. Use the BMI from Example 7, with $k = 694$.

$$B = \frac{694w}{h^2}$$

Substitute your weight in pounds for w and your height in inches for h to determine your BMI. Answers will vary. A BMI from 19 to 25 is considered desirable.

53. If y varies inversely as x, then x is the denominator; however, if y varies directly as x, then x is in the numerator. If $k > 0$, then, with inverse variation, as x increases, y decreases. With direct variation, y increases as x increases.

55. $2x + 3y = 12$

If $x = 0$, $y = 4$. If $y = 0$, $x = 6$. Graph a line through the intercepts, $(0, 4)$ and $(6, 0)$.

$4x - 2y = 8$

If $x = 0$, $y = -4$. If $y = 0$, $x = 2$. Graph a line through the intercepts, $(0, -4)$ and $(2, 0)$.

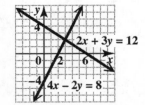

57. $x + y = 4$

If $x = 0$, $y = 4$. If $y = 0$, $x = 4$. Graph a line through the intercepts, $(0, 4)$ and $(4, 0)$.

$2x = 8 - 2y$

If $x = 0$, $y = 4$. If $y = 0$, $x = 4$. The intercepts are the same as the intercepts for the graph of $x + y = 4$, so the graph is the same as the first graph.

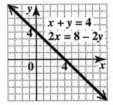

Chapter 7 Review Exercises

1. $3x + 2y = 10$
For $x = 0$:

$$3(0) + 2y = 10$$
$$2y = 10$$
$$y = 5 \quad \mathbf{(0, 5)}$$

For $y = 0$:

$$3x + 2(0) = 10$$
$$3x = 10$$
$$x = \tfrac{10}{3} \quad \left(\tfrac{10}{3}, 0\right)$$

For $x = 2$:

$$3(2) + 2y = 10$$
$$6 + 2y = 10$$
$$2y = 4$$
$$y = 2 \quad \mathbf{(2, 2)}$$

For $y = -2$:

$$3x + 2(-2) = 10$$
$$3x - 4 = 10$$
$$3x = 14$$
$$x = \tfrac{14}{3} \quad \left(\tfrac{14}{3}, -2\right)$$

Plot the ordered pairs, and draw the line through them.

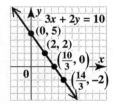

2. $x - y = 8$
For $x = 2$:

$$2 - y = 8$$
$$-y = 6$$
$$y = -6 \quad \textbf{(2, −6)}$$

For $y = -3$:

$$x - (-3) = 8$$
$$x + 3 = 8$$
$$x = 5 \quad \textbf{(5, −3)}$$

For $x = 3$:

$$3 - y = 8$$
$$-y = 5$$
$$y = -5 \quad \textbf{(3, −5)}$$

For $y = -2$:

$$x - (-2) = 8$$
$$x + 2 = 8$$
$$x = 6 \quad \textbf{(6, −2)}$$

Plot the ordered pairs, and draw the line through them.

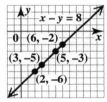

3. $4x - 3y = 12$
To find the x-intercept, let $y = 0$.

$$4x - 3y = 12$$
$$4x - 3(0) = 12$$
$$4x = 12$$
$$x = 3$$

The x-intercept is $(3, 0)$.
To find the y-intercept, let $x = 0$.

$$4x - 3y = 12$$
$$4(0) - 3y = 12$$
$$-3y = 12$$
$$y = -4$$

The y-intercept is $(0, -4)$.
Plot the intercepts and draw the line through them.

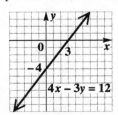

4. $5x + 7y = 28$
To find the x-intercept, let $y = 0$.

$$5x + 7y = 28$$
$$5x + 7(0) = 28$$
$$5x = 28$$
$$x = \frac{28}{5}$$

The x-intercept is $\left(\frac{28}{5}, 0\right)$.
To find the y-intercept, let $x = 0$.

$$5x + 7y = 28$$
$$5(0) + 7y = 28$$
$$7y = 28$$
$$y = 4$$

The y-intercept is $(0, 4)$.
Plot the intercepts and draw the line through them.

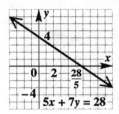

5. $2x + 5y = 20$
To find the x-intercept, let $y = 0$.

$$2x + 5y = 20$$
$$2x + 5(0) = 20$$
$$2x = 20$$
$$x = 10$$

The x-intercept is $(10, 0)$.
To find the y-intercept, let $x = 0$.

$$2x + 5y = 20$$
$$2(0) + 5y = 20$$
$$5y = 20$$
$$y = 4$$

The y-intercept is $(0, 4)$.
Plot the intercepts and draw the line through them.

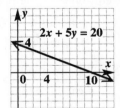

6. $x - 4y = 8$

To find the x-intercept, let $y = 0$.

$$x - 4y = 8$$
$$x - 4(0) = 8$$
$$x = 8$$

The x-intercept is $(8, 0)$.

To find the y-intercept, let $x = 0$.

$$0 - 4y = 8$$
$$-4y = 8$$
$$y = -2$$

The y-intercept is $(0, -2)$.

Plot the intercepts and draw the line through them.

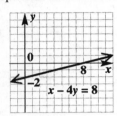

7. If both coordinates are positive, the point lies in quadrant I. If the first coordinate is negative and the second is positive, the point lies in quadrant II. To lie in quadrant III, the point must have both coordinates negative. To lie in quadrant IV, the first coordinate must be positive and the second must be negative.

8. Through $(-1, 2)$ and $(4, -5)$

$$m = \frac{\text{change in } y}{\text{change in } x} = \frac{-5 - 2}{4 - (-1)} = \frac{-7}{5} = -\frac{7}{5}$$

9. Through $(0, 3)$ and $(-2, 4)$

Let $(x_1, y_1) = (0, 3)$ and $(x_2, y_2) = (-2, 4)$.

$$m = \frac{y_2 - y_1}{x_2 - x_1} = \frac{4 - 3}{-2 - 0} = \frac{1}{-2} = -\frac{1}{2}$$

10. The slope of $y = 2x + 3$ is 2, the coefficient of x.

11. $3x - 4y = 5$

Write the equation in slope-intercept form.

$$-4y = -3x + 5$$
$$y = \frac{3}{4}x - \frac{5}{4}$$

The slope is $\frac{3}{4}$.

12. $x = 5$ is a vertical line and has *undefined* slope.

13. Parallel to $3y = 2x + 5$

Write the equation in slope-intercept form.

$$3y = 2x + 5$$
$$y = \frac{2}{3}x + \frac{5}{3}$$

The slope of $3y = 2x + 5$ is $\frac{2}{3}$; all lines parallel to it will also have a slope of $\frac{2}{3}$.

14. Perpendicular to $3x - y = 4$

Solve for y.

$$y = 3x - 4$$

The slope is 3; the slope of a line perpendicular to it is $-\frac{1}{3}$ since

$$3\left(-\frac{1}{3}\right) = -1.$$

15. Through $(-1, 5)$ and $(-1, -4)$

$$m = \frac{\Delta y}{\Delta x} = \frac{-4 - 5}{-1 - (-1)} = \frac{-9}{0} \quad \textit{Undefined}$$

This is a vertical line; it has undefined slope.

16. Through $(3, -1)$ and $(-3, 1)$

$$m = \frac{\Delta y}{\Delta x} = \frac{1 - (-1)}{-3 - 3} = \frac{2}{-6} = -\frac{1}{3}.$$

17. The x-intercept is $(2, 0)$ and the y-intercept is $(0, 2)$. The slope is

$$m = \frac{\text{change in } y}{\text{change in } x} = \frac{2 - 0}{0 - 2} = \frac{2}{-2} = -1.$$

18. The line goes up from left to right, so it has positive slope.

19. The line goes down from left to right, so it has negative slope.

20. The line is vertical, so it has *undefined* slope.

21. The line is horizontal, so it has 0 slope.

22. The slope is $\frac{2}{10}$ which can be written as 0.2, 20%, $\frac{20}{100}$, or $\frac{1}{5}$.

The correct responses are **A, B, C, D,** and **F**.

23. To rise 1 foot, we must move 4 feet in the horizontal direction. To rise 3 feet, we must move $3(4) = 12$ feet in the horizontal direction.

24. Let $(x_1, y_1) = (1980, 21{,}000)$ and $(x_2, y_2) = (2003, 52{,}700)$. Then

$$m = \frac{\Delta y}{\Delta x} = \frac{52{,}700 - 21{,}000}{2003 - 1980} = \frac{31{,}700}{23}$$
$$\approx 1378.$$

The average rate of change is $1378 per year.

25. **(a)** Slope $-\frac{1}{3}$, y-intercept $(0, -1)$

Use the slope-intercept form with $m = -\frac{1}{3}$ and $b = -1$.

$$y = mx + b$$
$$y = -\frac{1}{3}x - 1$$

(b)
$$y = -\frac{1}{3}x - 1$$
$$3y = -x - 3$$
$$x + 3y = -3$$

26. **(a)** Slope 0, y-intercept $(0, -2)$

Use the slope-intercept form with $m = 0$ and $b = -2$.

$$y = mx + b$$
$$y = (0)x - 2$$
$$y = -2$$

(b) $y = -2$ is already in standard form.

27. **(a)** Slope $-\frac{4}{3}$, through $(2, 7)$

Use the point-slope form with $m = -\frac{4}{3}$ and $(x_1, y_1) = (2, 7)$.

$$y - y_1 = m(x - x_1)$$
$$y - 7 = -\frac{4}{3}(x - 2)$$
$$y - 7 = -\frac{4}{3}x + \frac{8}{3}$$
$$y = -\frac{4}{3}x + \frac{29}{3}$$

(b)
$$y = -\frac{4}{3}x + \frac{29}{3}$$
$$3y = -4x + 29$$
$$4x + 3y = 29$$

28. **(a)** Slope 3, through $(-1, 4)$

Use the point-slope form with $m = 3$ and $(x_1, y_1) = (-1, 4)$.

$$y - y_1 = m(x - x_1)$$
$$y - 4 = 3[x - (-1)]$$
$$y - 4 = 3(x + 1)$$
$$y - 4 = 3x + 3$$
$$y = 3x + 7$$

(b)
$$y = 3x + 7$$
$$-3x + y = 7$$
$$3x - y = -7$$

29. **(a)** Vertical, through $(2, 5)$

The equation of any vertical line is in the form $x = k$. Since the line goes through $(2, 5)$, the equation is $x = 2$. (Slope-intercept form is not possible.)

(b) $x = 2$ is already in standard form.

30. **(a)** Through $(2, -5)$ and $(1, 4)$

Find the slope.

$$m = \frac{\Delta y}{\Delta x} = \frac{4 - (-5)}{1 - 2} = \frac{9}{-1} = -9$$

Use the point-slope form with $m = -9$ and $(x_1, y_1) = (2, -5)$.

$$y - y_1 = m(x - x_1)$$
$$y - (-5) = -9(x - 2)$$
$$y + 5 = -9x + 18$$
$$y = -9x + 13$$

(b)
$$y = -9x + 13$$
$$9x + y = 13$$

31. **(a)** Through $(-3, -1)$ and $(2, 6)$

Find the slope.

$$m = \frac{\Delta y}{\Delta x} = \frac{6 - (-1)}{2 - (-3)} = \frac{7}{5}$$

Use the point-slope form with $m = \frac{7}{5}$ and $(x_1, y_1) = (2, 6)$.

$$y - y_1 = m(x - x_1)$$
$$y - 6 = \frac{7}{5}(x - 2)$$
$$y - 6 = \frac{7}{5}x - \frac{14}{5}$$
$$y = \frac{7}{5}x + \frac{16}{5}$$

(b)
$$y = \frac{7}{5}x + \frac{16}{5}$$
$$5y = 7x + 16$$
$$-7x + 5y = 16$$
$$7x - 5y = -16$$

32. **(a)** From Exercise 17, we have $m = -1$ and a y-intercept of $(0, 2)$. The slope-intercept form is

$$y = -1x + 2 \quad \text{or} \quad y = -x + 2.$$

(b)
$$y = -x + 2$$
$$x + y = 2$$

33. **(a)** Parallel to $4x - y = 3$ and through $(7, -1)$

Writing $4x - y = 3$ in slope-intercept form gives us $y = 4x - 3$, which has slope 4. Lines parallel to it will also have slope 4. The line with slope 4 through $(7, -1)$ is :

$$y - y_1 = m(x - x_1)$$
$$y - (-1) = 4(x - 7)$$
$$y + 1 = 4x - 28$$
$$y = 4x - 29$$

(b)
$$y = 4x - 29$$
$$-4x + y = -29$$
$$4x - y = 29$$

34. **(a)** Perpendicular to $2x - 5y = 7$ and through $(4, 3)$

Write the equation in slope-intercept form.

$$2x - 5y = 7$$
$$-5y = -2x + 7$$
$$y = \frac{2}{5}x - \frac{7}{5}$$

$y = \frac{2}{5}x - \frac{7}{5}$ has slope $\frac{2}{5}$ and is perpendicular to lines with slope $-\frac{5}{2}$.

The line with slope $-\frac{5}{2}$ through $(4, 3)$ is

$$y - y_1 = m(x - x_1)$$
$$y - 3 = -\frac{5}{2}(x - 4)$$
$$y - 3 = -\frac{5}{2}x + 10$$
$$y = -\frac{5}{2}x + 13$$

(b)
$$y = -\frac{5}{2}x + 13$$
$$2y = -5x + 26$$
$$5x + 2y = 26$$

35. (a) The fixed cost is $159, so that is the value of b. The variable cost is $57, so

$$y = mx + b = 57x + 159.$$

The cost of a 1-year membership can be found by substituting 12 for x.

$$y = 57(12) + 159$$
$$= 684 + 159 = 843$$

The cost is $843.

(b) As in part (a),

$$y = 47x + 159.$$
$$y = 47(12) + 159$$
$$= 564 + 159 = 723$$

The cost is $723.

36. (a) Use $(6, 12.6)$ and $(12, 35.0)$.

$$m = \frac{\Delta y}{\Delta x} = \frac{35.0 - 12.6}{12 - 6} = \frac{22.4}{6} \approx 3.73$$

Use the point-slope form of a line.

$$y - y_1 = m(x - x_1)$$
$$y - 12.6 = 3.73(x - 6)$$
$$y - 12.6 = 3.73x - 22.38$$
$$y = 3.73x - 9.78$$
$$\text{or} \quad y = 3.73x - 9.8$$

The slope, 3.73, indicates that the number of e-filing taxpayers increased by 3.73% each year from 1996 to 2002.

(b) The year 2005 corresponds to $x = 15$.

$$y = 3.73(15) - 9.8$$
$$= 55.95 - 9.8 = 46.15$$

According to the equation from part (a), 46.15% of tax returns will be filed electronically in 2005.

37. $\{(-4, 2), (-4, -2), (1, 5), (1, -5)\}$
The domain, the set of x-values, is $\{-4, 1\}$.
The range, the set of y-values, is $\{2, -2, 5, -5\}$.
Since each x-value has more than one y-value, the relation is not a function.

38. The relation can be described by the set of ordered pairs

$$\{(9, 32), (11, 47), (4, 47), (17, 69), (25, 14)\}.$$

The relation is a function since for each x-value, there is only one y-value.

The domain is the set of x-values: $\{9, 11, 4, 17, 25\}$.
The range is the set of y-values: $\{32, 47, 69, 14\}$.

39. The domain, the x-values of the points on the graph, is $[-4, 4]$. The range, the y-values of the points on the graph, is $[0, 2]$. Since a vertical line intersects the graph of the relation in at most one point, the relation is a function.

40. The x-values are negative or zero, so the domain is $(-\infty, 0]$. The y-values can be any real number, so the range is $(-\infty, \infty)$. A vertical line, such as $x = -3$, will intersect the graph twice, so by the vertical line test, the relation is not a function.

41. $y = 3x - 3$
For any value of x, there is exactly one value of y, so the equation defines a function, actually a linear function. The domain is the set of all real numbers, $(-\infty, \infty)$.

42. $x = y^2$
The ordered pairs $(4, 2)$ and $(4, -2)$ both satisfy the equation. Since one value of x, 4, corresponds to two values of y, 2 and -2, the equation does not define a function. Because x is equal to the square of y, the values of x must always be nonnegative. The domain is $[0, \infty)$.

43. $y = \dfrac{7}{x - 6}$
Given any value of x, y is found by subtracting 6, then dividing the result into 7. This process produces exactly one value of y for each x-value, so the equation defines a function. The domain includes all real numbers except those that make the denominator 0, namely 6. The domain is $(-\infty, 6) \cup (6, \infty)$.

44. If no vertical line intersects the graph in more than one point, then it is the graph of a function.

In Exercises 45–48, use

$$f(x) = -2x^2 + 3x - 6.$$

45. $f(0) = -2(0)^2 + 3(0) - 6 = -6$

46. $f(2.1) = -2(2.1)^2 + 3(2.1) - 6$
$$= -8.82 + 6.3 - 6 = -8.52$$

47. $f\left(-\frac{1}{2}\right) = -2\left(-\frac{1}{2}\right)^2 + 3\left(-\frac{1}{2}\right) - 6$
$$= -\frac{1}{2} - \frac{3}{2} - 6 = -8$$

48. $f(k) = -2k^2 + 3k - 6$

49. (a) For each year, there is exactly one life expectancy associated with the year, so the table defines a function.

(b) The domain is the set of years, that is, $\{1943, 1953, 1963, 1973, 1983, 1993, 2003\}$. The range is the set of life expectancies, that is, $\{63.3, 68.8, 69.9, 71.4, 74.6, 75.5, 77.6\}$.

(c) Answers will vary. Two possible answers are $(1943, 63.3)$ and $(1953, 68.8)$.

(d) $f(1973) = 71.4$. In 1973, life expectancy at birth was 71.4 yr.

(e) Since $f(1993) = 75.5$, $x = 1993$.

50. $2x^2 - y = 0$

$$-y = -2x^2$$
$$y = 2x^2$$

Since $y = f(x)$,

$$f(x) = 2x^2,$$

and $f(3) = 2(3)^2 = 2(9) = 18$.

51.   Solve for y in terms of x.

$$2x - 5y = 7$$
$$2x - 7 = 5y$$
$$\tfrac{2}{5}x - \tfrac{7}{5} = y$$

Thus, choice **C** is correct.

52.   No, because the equation of a line with an undefined slope is $x = a$. The ordered pairs have the form (a, y), where a is a constant and y is a variable. Thus, the number a corresponds to an infinite number of values of y.

53. $f(x) = -2x^2 + 5x + 7$

(a) $f(-2) = -2(-2)^2 + 5(-2) + 7$
$$= -2(4) - 10 + 7$$
$$= -8 - 10 + 7$$
$$= -18 + 7 = -11$$

(b) $f(3) = -2(3)^2 + 5(3) + 7$
$$= -2(9) + 15 + 7$$
$$= -18 + 15 + 7$$
$$= -3 + 7 = 4$$

54. $f(x) = 2x + 3$, $g(x) = 5x^2 - 3x + 2$

(a) $(f + g)(x) = f(x) + g(x)$
$$= (2x + 3) + (5x^2 - 3x + 2)$$
$$= 5x^2 + 2x - 3x + 3 + 2$$
$$= 5x^2 - x + 5$$

(b) $(f - g)(x) = f(x) - g(x)$
$$= (2x + 3) - (5x^2 - 3x + 2)$$
$$= 2x + 3 - 5x^2 + 3x - 2$$
$$= -5x^2 + 5x + 1$$

(c) $(f + g)(-1)$
$$= f(-1) + g(-1)$$
$$= [2(-1) + 3] + [5(-1)^2 - 3(-1) + 2]$$
$$= [1] + [10]$$
$$= 11$$

(d) $(f - g)(-1) = f(-1) - g(-1)$
$$= 1 - 10 \quad \textit{from part (c)}$$
$$= -9$$

55.   **(a)** $(fg)(x) = f(x) \cdot g(x)$
$$= (12x^2 - 3x)(3x)$$
$$= 12x^2(3x) - 3x(3x)$$
$$= 36x^3 - 9x^2$$

(b) $\left(\dfrac{f}{g}\right)(x) = \dfrac{f(x)}{g(x)} = \dfrac{12x^2 - 3x}{3x}$
$$= \frac{12x^2}{3x} - \frac{3x}{3x}$$
$$= 4x - 1, \; x \neq 0$$

(c) $(fg)(-1)$
$$= 36(-1)^3 - 9(-1)^2 \quad \textit{from part (a)}$$
$$= -36 - 9$$
$$= -45$$

(d) $\left(\dfrac{f}{g}\right)(2) = 4(2) - 1 \quad \textit{from part (b)}$
$$= 8 - 1 = 7$$

56. $f(x) = 3x^2 + 2x - 1$, $g(x) = 5x + 7$

(a) $(g \circ f)(3) = g(f(3))$
$$= g(3 \cdot 3^2 + 2 \cdot 3 - 1)$$
$$= g(32)$$
$$= 5 \cdot 32 + 7$$
$$= 167$$

(b) $(f \circ g)(3) = f(g(3))$
$$= f(5 \cdot 3 + 7)$$
$$= f(22)$$
$$= 3 \cdot 22^2 + 2 \cdot 22 - 1$$
$$= 1495$$

(c) $(f \circ g)(-2) = f(g(-2))$
$$= f[5(-2) + 7]$$
$$= f(-3)$$
$$= 3(-3)^2 + 2(-3) - 1$$
$$= 20$$

(d) $(g \circ f)(-2) = g(f(-2))$
$$= g[3(-2)^2 + 2(-2) - 1]$$
$$= g(7)$$
$$= 5 \cdot 7 + 7$$
$$= 42$$

(e) $(f \circ g)(x)$
$$= f(g(x))$$
$$= f(5x + 7)$$
$$= 3(5x + 7)^2 + 2(5x + 7) - 1$$
$$= 3(25x^2 + 70x + 49) + 10x + 14 - 1$$
$$= 75x^2 + 210x + 147 + 10x + 13$$
$$= 75x^2 + 220x + 160$$

(f) $(g \circ f)(x) = g(f(x))$
$$= g(3x^2 + 2x - 1)$$
$$= 5(3x^2 + 2x - 1) + 7$$
$$= 15x^2 + 10x - 5 + 7$$
$$= 15x^2 + 10x + 2$$

57. If y varies inversely as x, then $y = \dfrac{k}{x}$, for some constant k. This form fits choice **C**.

58. Let $v =$ the viewing distance and $e =$ the amount of enlargement.
v varies directly as e, so
$$v = ke$$
for some constant k. Since $v = 250$ when $e = 5$, substitute these values in the equation and solve for k.
$$v = ke$$
$$250 = k(5)$$
$$50 = k$$
So $v = 50e$. Now let $e = 8.6$.
$$v = 50(8.6) = 430$$
It should be viewed from 430 mm.

59. The frequency f of a vibrating guitar string varies inversely as its length L, so
$$f = \frac{k}{L}$$
for some constant k. Substitute 0.65 for L and 4.3 for f.
$$4.3 = \frac{k}{6.5}$$
$$k = 4.3(0.65) = 2.795$$
So $f = \dfrac{2.795}{L}$. Now let $L = 0.5$.
$$f = \frac{2.795}{0.5} = 5.59$$
The frequency would be 5.59 vibrations per second.

60. The volume V of a rectangular box of a given height is proportional to its width W and length L, so
$$V = kWL$$
for some constant k. Substitute 2 for W, 4 for L, and 12 for V.
$$12 = k(2)(4)$$
$$k = \frac{12}{8} = \frac{3}{2}$$

So $V = \frac{3}{2}WL$. Now let $W = 3$ and $L = 5$.
$$V = \frac{3}{2}(3)(5) = \frac{45}{2}$$
The volume is 22.5 cubic feet.

Chapter 7 Test

1. $2x - 3y = 12$
For $x = 1$:
$$2(1) - 3y = 12$$
$$2 - 3y = 12$$
$$-3y = 10$$
$$y = -\frac{10}{3} \quad \left(1, -\frac{10}{3}\right)$$

For $x = 3$:
$$2(3) - 3y = 12$$
$$6 - 3y = 12$$
$$-3y = 6$$
$$y = -2 \quad (3, -2)$$

For $y = -4$:
$$2x - 3(-4) = 12$$
$$2x + 12 = 12$$
$$2x = 0$$
$$x = 0 \quad (0, -4)$$

2. $3x - 2y = 20$
To find the x-intercept, let $y = 0$.
$$3x - 2(0) = 20$$
$$3x = 20$$
$$x = \frac{20}{3}$$
The x-intercept is $\left(\frac{20}{3}, 0\right)$.
To find the y-intercept, let $x = 0$.
$$3(0) - 2y = 20$$
$$-2y = 20$$
$$y = -10$$
The y-intercept is $(0, -10)$.
Draw the line through these two points.

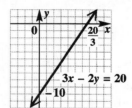

3. The graph of $y = 5$ is the horizontal line with slope 0 and y-intercept $(0, 5)$. There is no x-intercept.

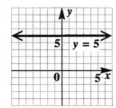

4. The graph of $x = 2$ is the vertical line with x-intercept at $(2, 0)$. There is no y-intercept.

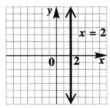

5. Through $(6, 4)$ and $(-4, -1)$

$$m = \frac{\Delta y}{\Delta x} = \frac{-1 - 4}{-4 - 6} = \frac{-5}{-10} = \frac{1}{2}$$

The slope of the line is $\frac{1}{2}$.

6. The graph of a line with undefined slope is the graph of a vertical line.

7. Find the slope of each line.

$$5x - y = 8$$
$$-y = -5x + 8$$
$$y = 5x - 8$$

The slope is 5.

$$5y = -x + 3$$
$$y = -\tfrac{1}{5}x + \tfrac{3}{5}$$

The slope is $-\frac{1}{5}$.
Since $5\left(-\frac{1}{5}\right) = -1$, the two slopes are negative reciprocals and the lines are perpendicular.

8. Find the slope of each line.

$$2y = 3x + 12$$
$$y = \tfrac{3}{2}x + 6$$

The slope is $\frac{3}{2}$.

$$3y = 2x - 5$$
$$y = \tfrac{2}{3}x - \tfrac{5}{3}$$

The slope is $\frac{2}{3}$.
The lines are neither parallel nor perpendicular.

9. Use the points $(1980, 119{,}000)$ and $(2005, 89{,}000)$.

average rate of change

$$= \frac{\text{change in } y}{\text{change in } x} = \frac{89{,}000 - 119{,}000}{2005 - 1980}$$
$$= \frac{-30{,}000}{25} = -1200$$

The average rate of change is about -1200 farms per year, that is, the number of farms decreased by about 1200 each year from 1980 to 2005.

10. Through $(4, -1)$; $m = -5$

(a) Let $m = -5$ and $(x_1, y_1) = (4, -1)$ in the point-slope form.

$$y - y_1 = m(x - x_1)$$
$$y - (-1) = -5(x - 4)$$
$$y + 1 = -5x + 20$$
$$y = -5x + 19$$

(b) $y = -5x + 19$ *From part (a)*
      $5x + y = 19$ *Standard form*

11. Through $(-3, 14)$; horizontal

(a) A horizontal line has equation $y = k$. Here $k = 14$, so the line has equation $y = 14$.

(b) $y = 14$ is already in standard form.

12. Through $(-7, 2)$ and parallel to $3x + 5y = 6$

(a) To find the slope of $3x + 5y = 6$, write the equation in slope-intercept form by solving for y.

$$3x + 5y = 6$$
$$5y = -3x + 6$$
$$y = -\tfrac{3}{5}x + \tfrac{6}{5}$$

The slope is $-\frac{3}{5}$, so a line parallel to it also has slope $-\frac{3}{5}$. Let $m = -\frac{3}{5}$ and $(x_1, y_1) = (-7, 2)$ in the point-slope form.

$$y - y_1 = m(x - x_1)$$
$$y - 2 = -\tfrac{3}{5}[x - (-7)]$$
$$y - 2 = -\tfrac{3}{5}(x + 7)$$
$$y - 2 = -\tfrac{3}{5}x - \tfrac{21}{5}$$
$$y = -\frac{3}{5}x - \frac{11}{5}$$

(b) $y = -\frac{3}{5}x - \frac{11}{5}$ *From part (a)*

   $\frac{3}{5}x + y = -\frac{11}{5}$ *Variable terms on one side*

  $5\left(\frac{3}{5}x + y\right) = 5\left(-\frac{11}{5}\right)$ *Multiply by 5.*

   $3x + 5y = -11$ *Standard form*

13. Through $(-7, 2)$ and perpendicular to $y = 2x$

(a) Since $y = 2x$ is in slope-intercept form ($b = 0$), the slope, m, of $y = 2x$ is 2. A line perpendicular to it has a slope that is the negative reciprocal of 2, that is, $-\frac{1}{2}$. Let $m = -\frac{1}{2}$ and $(x_1, y_1) = (-7, 2)$ in the point-slope form.

$$y - y_1 = m(x - x_1)$$
$$y - 2 = -\tfrac{1}{2}(x + 7)$$
$$y - 2 = -\tfrac{1}{2}x - \tfrac{7}{2}$$
$$y = -\tfrac{1}{2}x - \tfrac{3}{2}$$

(b)
$$y = -\tfrac{1}{2}x - \tfrac{3}{2} \quad \textit{From part (a)}$$
$$\tfrac{1}{2}x + y = -\tfrac{3}{2} \quad \begin{array}{l}\textit{Variable terms}\\ \textit{on one side}\end{array}$$
$$2\left(\tfrac{1}{2}x + y\right) = 2\left(-\tfrac{3}{2}\right) \quad \textit{Multiply by 2.}$$
$$x + 2y = -3 \quad \textit{Standard form}$$

14. Through $(-2, 3)$ and $(6, -1)$

(a) First find the slope.

$$m = \frac{\Delta y}{\Delta x} = \frac{-1 - 3}{6 - (-2)} = \frac{-4}{8} = -\frac{1}{2}$$

Use $m = -\frac{1}{2}$ and $(x_1, y_1) = (-2, 3)$ in the point-slope form.

$$y - y_1 = m(x - x_1)$$
$$y - 3 = -\tfrac{1}{2}[x - (-2)]$$
$$y - 3 = -\tfrac{1}{2}(x + 2)$$
$$y - 3 = -\tfrac{1}{2}x - 1$$
$$y = -\tfrac{1}{2}x + 2$$

(b)
$$y = -\tfrac{1}{2}x + 2 \quad \textit{From part (a)}$$
$$\tfrac{1}{2}x + y = 2 \quad \begin{array}{l}\textit{Variable terms}\\ \textit{on one side}\end{array}$$
$$2\left(\tfrac{1}{2}x + y\right) = 2(2) \quad \textit{Multiply by 2.}$$
$$x + 2y = 4 \quad \textit{Standard form}$$

15. Through $(5, -6)$; vertical

(a) The equation of any vertical line is in the form $x = k$. Since the line goes through $(5, -6)$, the equation is $x = 5$. Writing $x = 5$ in slope-intercept form is *not possible* since there is no y-term.

(b) From part (a), the standard form is $x = 5$.

16. Positive slope means that the line goes up from left to right. The only line that has positive slope and a negative y-coordinate for its y-intercept is choice **B**.

17. (a) Use the points $(5, 22{,}860)$ and $(11, 33{,}565)$.

$$m = \frac{\Delta y}{\Delta x} = \frac{33{,}565 - 22{,}860}{11 - 5} = \frac{10{,}705}{6}$$
$$\approx 1784.17$$

Use the point-slope form with $m = \frac{10{,}705}{6}$ and $(x_1, y_1) = (5, 22{,}860)$.

$$y - 22{,}860 = \tfrac{10{,}705}{6}(x - 5)$$
$$y - 22{,}860 = \tfrac{10{,}705}{6}x - \tfrac{53{,}525}{6}$$
$$y \approx 1784.17x + 13{,}939.17$$

(b) $y = 1784.17(9) + 13{,}939.17$ *Let $x = 9$.*
$$= 29{,}996.7 \approx \$29{,}997,$$
which is slightly less than the actual value.

18. Choice **D** is the only graph that passes the vertical line test.

19. Choice **D** does not define a function, since its domain (input) element 0 is paired with two different range (output) elements, 1 and 2.

20. (a) The x-values are greater than or equal to zero, so the domain is $[0, \infty)$. Since y can be any value, the range is $(-\infty, \infty)$.

(b) The domain is the set of x-values: $\{0, -2, 4\}$. The range is the set of y-values: $\{1, 3, 8\}$.

21. $f(x) = -x^2 + 2x - 1$

(a) $f(1) = -(1)^2 + 2(1) - 1$
$$= -1 + 2 - 1$$
$$= 0$$

(b) $f(a) = -a^2 + 2a - 1$

22. $f(x) = \frac{2}{3}x - 1$
This function represents a line with y-intercept $(0, -1)$ and x-intercept $\left(\frac{3}{2}, 0\right)$.
Draw the line through these two points.
The domain is $(-\infty, \infty)$, and the range is $(-\infty, \infty)$.

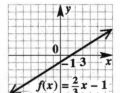

23. $f(x) = -2x^2 + 5x - 6$, $g(x) = 7x - 3$

(a) $f(x) = -2x^2 + 5x - 6$
$$f(4) = -2(4)^2 + 5(4) - 6$$
$$= -2 \cdot 16 + 20 - 6$$
$$= -32 + 20 - 6$$
$$= -12 - 6 = -18$$

(b) $(f+g)(x) = f(x) + g(x)$
$$= (-2x^2 + 5x - 6) + (7x - 3)$$
$$= -2x^2 + 12x - 9$$

(c) $(f-g)(x) = f(x) - g(x)$
$$= (-2x^2 + 5x - 6) - (7x - 3)$$
$$= -2x^2 + 5x - 6 - 7x + 3$$
$$= -2x^2 - 2x - 3$$

(d) Using the answer in part (c), we have
$$(f-g)(-2) = -2(-2)^2 - 2(-2) - 3$$
$$= -8 + 4 - 3 = -7.$$

24. $f(x) = 3x + 5, g(x) = x^2 + 2$

(a) $(f \circ g)(-2) = f[g(-2)]$
$$= f\left[(-2)^2 + 2\right]$$
$$= f(6)$$
$$= 3 \cdot 6 + 5$$
$$= 23$$

(b) $(f \circ g)(x) = f[g(x)]$
$$= f\left(x^2 + 2\right)$$
$$= 3\left(x^2 + 2\right) + 5$$
$$= 3x^2 + 6 + 5$$
$$= 3x^2 + 11$$

(c) $(g \circ f)(x) = g[f(x)]$
$$= g(3x + 5)$$
$$= (3x + 5)^2 + 2$$
$$= 9x^2 + 30x + 25 + 2$$
$$= 9x^2 + 30x + 27$$

25. $f(x) = x^2 + 3x + 2, g(x) = x + 1$

(a) $(fg)(x) = f(x) \cdot g(x)$
$$= \left(x^2 + 3x + 2\right)(x + 1)$$
$$= \left(x^2 + 3x + 2\right)(x)$$
$$+ \left(x^2 + 3x + 2\right)(1)$$
$$= x^3 + 3x^2 + 2x + x^2 + 3x + 2$$
$$= x^3 + 4x^2 + 5x + 2$$

(b) $(fg)(-2) = f(-2) \cdot g(-2)$
$$= \left[(-2)^2 + 3(-2) + 2\right] \cdot [(-2) + 1]$$
$$= [4 - 6 + 2] \cdot [-1]$$
$$= 0(-1) = 0$$

Alternatively, we could have substituted -2 for x into our answer from part (a).

26. (a) $\left(\dfrac{f}{g}\right)(x) = \dfrac{f(x)}{g(x)} = \dfrac{x^2 + 3x + 2}{x + 1}$

$$\begin{array}{r} x + 2 \\ x + 1 \overline{\smash{)}\, x^2 + 3x + 2} \\ \underline{x^2 + x } \\ 2x + 2 \\ \underline{2x + 2} \\ 0 \end{array}$$

Thus, $\left(\dfrac{f}{g}\right)(x) = x + 2$ if $x + 1 \neq 0$, that is, $x \neq -1$.

(b) Using our answer from part (a),
$$\left(\dfrac{f}{g}\right)(-2) = (-2) + 2 = 0.$$

27. The current I is inversely proportional to the resistance R, so
$$I = \dfrac{k}{R}$$
for some constant k. Let $I = 80$ and $R = 30$. Find k.
$$80 = \dfrac{k}{30}$$
$$k = 80(30) = 2400$$
So $I = \dfrac{2400}{R}$. Now let $R = 12$.
$$I = \tfrac{2400}{12} = 200$$
The current is 200 amperes.

28. The force F of the wind blowing on a vertical surface varies jointly as the area A of the surface and the square of the velocity V, so
$$F = kAV^2$$
for some constant k. Let $F = 50$, $A = 500$, and $V = 40$. Find k.
$$50 = k(500)(40)^2$$
$$k = \dfrac{50}{500(1600)} = \dfrac{1}{16{,}000}$$
So $F = \tfrac{1}{16{,}000} AV^2$. Now let $A = 2$ and $V = 80$.
$$F = \dfrac{1}{16{,}000}(2)(80)^2 = 0.8$$
The force of the wind is 0.8 pounds.

Cumulative Review Exercises (Chapters 1–7)

1. The absolute value of a negative number is a positive number and the additive inverse of the same negative number is the same positive number. For example, suppose the negative number is -5:

$$|-5| = -(-5) = 5$$
$$\text{and} \quad -(-5) = 5$$

The statement is *always true*.

2. The statement is *always true*; in fact, it is the definition of a rational number.

3. The sum of two negative numbers is another negative number, so the statement is *never true*.

4. The statement is *sometimes true*. For example,

$$3 + (-3) = 0,$$
$$\text{but} \quad 3 + (-1) = 2 \neq 0.$$

5. $-|-2| - 4 + |-3| + 7 = -2 - 4 + 3 + 7$
$$= -6 + 3 + 7$$
$$= -3 + 7$$
$$= 4$$

6. $\dfrac{(4^2 - 4) - (-1)7}{4 + (-6)} = \dfrac{(16 - 4) - (-7)}{-2}$

$$= \dfrac{12 + 7}{-2} = -\dfrac{19}{2}$$

7. Let $p = -4$ and $q = \frac{1}{2}$.

$-3(2q - 3p) = -3\left[2\left(\tfrac{1}{2}\right) - 3(-4)\right]$
$$= -3(1 + 12)$$
$$= -3(13)$$
$$= -39$$

8. Let $p = -4$ and $r = 16$.

$\dfrac{r}{8p + 2r} = \dfrac{16}{8(-4) + 2(16)}$

$$= \dfrac{16}{-32 + 32}$$

$$= \dfrac{16}{0}, \text{ which is } undefined.$$

9. $2z - 5 + 3z = 2 - z$
$$5z - 5 = 2 - z$$
$$6z = 7$$
$$z = \tfrac{7}{6}$$

The solution set is $\left\{\tfrac{7}{6}\right\}$.

10. $\dfrac{3a - 1}{5} + \dfrac{a + 2}{2} = -\dfrac{3}{10}$

Multiply both sides by the LCD, 10.

$10\left(\dfrac{3a - 1}{5} + \dfrac{a + 2}{2}\right) = 10\left(-\dfrac{3}{10}\right)$

$$2(3a - 1) + 5(a + 2) = -3$$
$$6a - 2 + 5a + 10 = -3$$
$$11a + 8 = -3$$
$$11a = -11$$
$$a = -1$$

The solution set is $\{-1\}$.

11. Solve $V = \dfrac{1}{3}\pi r^2 h$ for h.

$3V = \pi r^2 h$ *Multiply by 3.*

$\dfrac{3V}{\pi r^2} = h$ *Divide by πr^2.*

12. Let x denote the side of the original square and $4x$ the perimeter. Now $x + 4$ is the side of the new square and $4(x + 4)$ is its perimeter.
"The perimeter would be 8 inches less than twice the perimeter of the original square " translates as

$$4(x + 4) = 2(4x) - 8.$$
$$4x + 16 = 8x - 8$$
$$24 = 4x$$
$$6 = x$$

The length of a side of the original square is 6 inches.

13. Let x = the time it takes for the planes to be 2100 miles apart.
Make a table. Use the formula $d = rt$.

	r	t	d
Eastbound Plane	550	x	$550x$
Westbound Plane	500	x	$500x$

The total distance is 2100 miles.

$$550x + 500x = 2100$$
$$1050x = 2100$$
$$x = 2$$

It will take 2 hr for the planes to be 2100 mi apart.

14. $-4 < 3 - 2k < 9$

$-7 < -2k < 6$

Divide by -2; reverse the inequality symbols.

$\frac{7}{2} > k > -3$ or $-3 < k < \frac{7}{2}$

The solution set is $\left(-3, \frac{7}{2}\right)$.

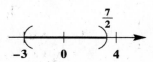

15. $-0.3x + 2.1(x - 4) \le -6.6$

$-3x + 21(x - 4) \le -66$

Multiply by 10.

$-3x + 21x - 84 \le -66$

$18x - 84 \le -66$

$18x \le 18$

$x \le 1$

The solution set is $(-\infty, 1]$.

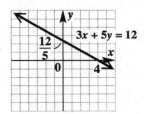

16. $3x + 5y = 12$

To find the x-intercept, let $y = 0$.

$3x + 5(0) = 12$

$3x = 12$

$x = 4$

The x-intercept is $(4, 0)$.
To find the y-intercept, let $x = 0$.

$3(0) + 5y = 12$

$5y = 12$

$y = \frac{12}{5}$

The y-intercept is $\left(0, \frac{12}{5}\right)$.
Plot the intercepts and draw the line through them.

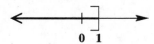

17. $A(-2, 1)$ and $B(3, -5)$

(a) The slope of line AB is

$$m = \frac{\Delta y}{\Delta x} = \frac{-5 - 1}{3 - (-2)} = \frac{-6}{5} = -\frac{6}{5}.$$

(b) The slope of a line perpendicular to line AB is the negative reciprocal of $-\frac{6}{5}$, which is $\frac{5}{6}$.

18.   **(a)** Slope $-\frac{3}{4}$; y-intercept $(0, -1)$
To write an equation of this line, let $m = -\frac{3}{4}$ and $b = -1$ in the slope-intercept form.

$$y = mx + b$$
$$y = -\frac{3}{4}x - 1$$

(b) $y = -\frac{3}{4}x - 1$

$4y = -3x - 4$

$3x + 4y = -4$

19.   **(a)** Horizontal; through $(2, -2)$
A horizontal line through the point (c, d) has equation $y = d$. Here $d = -2$, so the equation of the line is $y = -2$.

(b) $y = -2$ is already in standard form.

20.   **(a)** Through $(4, -3)$ and $(1, 1)$
First find the slope of the line.

$$m = \frac{\Delta y}{\Delta x} = \frac{1 - (-3)}{1 - 4} = \frac{4}{-3} = -\frac{4}{3}$$

Now substitute $(x_1, y_1) = (4, -3)$ and $m = -\frac{4}{3}$ in the point-slope form. Then solve for y.

$$y - y_1 = m(x - x_1)$$
$$y - (-3) = -\frac{4}{3}(x - 4)$$
$$y + 3 = -\frac{4}{3}x + \frac{16}{3}$$
$$y = -\frac{4}{3}x + \frac{7}{3}$$

(b) $y = -\frac{4}{3}x + \frac{7}{3}$

$3y = -4x + 7$

$4x + 3y = 7$

21. $\left(3x^2 y^{-1}\right)^{-2} \left(2x^{-3}y\right)^{-1}$

$= 3^{-2} x^{-4} y^2 \cdot 2^{-1} x^3 y^{-1}$

$= 3^{-2} 2^{-1} x^{-1} y$

$= \dfrac{y}{3^2 \cdot 2x} = \dfrac{y}{18x}$

22. $\dfrac{5m^{-2}y^3}{3m^{-3}y^{-1}} = \dfrac{5}{3} \cdot \dfrac{m^{-2}}{m^{-3}} \cdot \dfrac{y^3}{y^{-1}}$

$= \dfrac{5}{3} m^{-2-(-3)} y^{3-(-1)}$

$= \dfrac{5}{3} m^1 y^4 \quad \text{or} \quad \dfrac{5my^4}{3}$

23. $\left(3x^3 + 4x^2 - 7\right) - \left(2x^3 - 8x^2 + 3x\right)$

$= 3x^3 + 4x^2 - 7 - 2x^3 + 8x^2 - 3x$

$= x^3 + 12x^2 - 3x - 7$

24. $(7x + 3y)^2 = (7x)^2 + 2(7x)(3y) + (3y)^2$

$= 49x^2 + 42xy + 9y^2$

25. $(2p + 3)\left(5p^2 - 4p - 8\right)$

$= 10p^3 - 8p^2 - 16p + 15p^2 - 12p - 24$

$= 10p^3 + 7p^2 - 28p - 24$

26. $\dfrac{m^3 - 3m^2 + 5m - 3}{m - 1}$

$$
\begin{array}{r}
m^2 - 2m + 3 \\
m - 1 \overline{\smash{\big)}\, m^3 - 3m^2 + 5m - 3} \\
\underline{m^3 - m^2 } \\
-2m^2 + 5m \\
\underline{-2m^2 + 2m } \\
3m - 3 \\
\underline{3m - 3} \\
0
\end{array}
$$

The quotient is $m^2 - 2m + 3$.

27. $16w^2 + 50wz - 21z^2$

Two integer factors whose product is $(16)(-21) = -336$ and whose sum is 50 are 56 and -6.

$$
\begin{aligned}
&= 16w^2 + 56wz - 6wz - 21z^2 \\
&= 8w(2w + 7z) - 3z(2w + 7z) \\
&= (2w + 7z)(8w - 3z)
\end{aligned}
$$

28. $4x^2 - 4x + 1 - y^2$

Group the first three terms.

$$
\begin{aligned}
&= \left(4x^2 - 4x + 1\right) - y^2 \\
&= (2x - 1)^2 - y^2 \\
&= [(2x - 1) + y][(2x - 1) - y] \\
&= (2x - 1 + y)(2x - 1 - y)
\end{aligned}
$$

29. $4y^2 - 36y + 81$

$$
\begin{aligned}
&= (2y)^2 - 2(2y)(9) + 9^2 \\
&= (2y - 9)^2
\end{aligned}
$$

30. $100x^4 - 81 = \left(10x^2\right)^2 - 9^2$

$$
= \left(10x^2 + 9\right)\left(10x^2 - 9\right)
$$

31. $8p^3 + 27$

$$
\begin{aligned}
&= (2p)^3 + 3^3 \\
&= (2p + 3)\left[(2p)^2 - (2p)(3) + 3^2\right] \\
&= (2p + 3)\left(4p^2 - 6p + 9\right)
\end{aligned}
$$

32. $(p - 1)(2p + 3)(p + 4) = 0$

$$
\begin{array}{lll}
p - 1 = 0 \quad \text{or} & 2p + 3 = 0 \quad \text{or} & p + 4 = 0 \\
p = 1 & 2p = -3 & p = -4 \\
& p = -\frac{3}{2} &
\end{array}
$$

The solution set is $\left\{-4, -\frac{3}{2}, 1\right\}$.

33.
$$
\begin{aligned}
9q^2 &= 6q - 1 \\
9q^2 - 6q + 1 &= 0 \\
(3q - 1)^2 &= 0 \\
3q - 1 &= 0 \\
3q &= 1 \\
q &= \tfrac{1}{3}
\end{aligned}
$$

The solution set is $\left\{\tfrac{1}{3}\right\}$.

34. Let x be the length of the base. Then $x + 3$ will be the height. The area is 14 square feet. Use the formula $A = \frac{1}{2}bh$, and substitute 14 for A, x for b, and $x + 3$ for h.

$$
\begin{aligned}
\tfrac{1}{2}bh &= A \\
\tfrac{1}{2}(x)(x + 3) &= 14 \\
x(x + 3) &= 28 \quad \textit{Multiply by 2.} \\
x^2 + 3x &= 28 \\
x^2 + 3x - 28 &= 0 \\
(x + 7)(x - 4) &= 0
\end{aligned}
$$

$$
\begin{array}{lll}
x + 7 = 0 & \text{or} & x - 4 = 0 \\
x = -7 & \text{or} & x = 4
\end{array}
$$

The length cannot be negative, so reject -7 as a solution. The only possible solution is 4. The base is 4 feet long.

35. Let x be the distance between the longer sides. (This is actually the width.) Then $x + 2$ will be the length of the longer side. The area of the rectangle is 288 in^2. Use the formula $LW = A$. Substitute 288 for A, $x + 2$ for L, and x for W.

$$
\begin{aligned}
(x + 2)x &= 288 \\
x^2 + 2x &= 288 \\
x^2 + 2x - 288 &= 0 \\
(x + 18)(x - 16) &= 0
\end{aligned}
$$

$$
\begin{array}{lll}
x + 18 = 0 & \text{or} & x - 16 = 0 \\
x = -18 & \text{or} & x = 16
\end{array}
$$

The distance cannot be negative, so reject -18 as a solution. The only possible solution is 16. The distance between the longer sides is 16 inches, and the length of the longer sides is $16 + 2 = 18$ inches.

36. $\dfrac{8}{x + 1} - \dfrac{2}{x + 3}$

$$
\begin{aligned}
&= \frac{8(x + 3)}{(x + 1)(x + 3)} - \frac{2(x + 1)}{(x + 3)(x + 1)} \\
&= \frac{8x + 24 - 2x - 2}{(x + 1)(x + 3)} \\
&= \frac{6x + 22}{(x + 1)(x + 3)}
\end{aligned}
$$

37. $\dfrac{x^2 - 25}{3x + 6} \cdot \dfrac{4x + 8}{x^2 + 10x + 25}$

$$
\begin{aligned}
&= \frac{(x + 5)(x - 5)}{3(x + 2)} \cdot \frac{4(x + 2)}{(x + 5)(x + 5)} \\
&= \frac{4(x - 5)}{3(x + 5)}
\end{aligned}
$$

38.
$$\frac{x^2 + 5x + 6}{3x} \div \frac{x^2 - 4}{x^2 + x - 6}$$
$$= \frac{(x+2)(x+3)}{3x} \cdot \frac{(x+3)(x-2)}{(x+2)(x-2)}$$
$$= \frac{(x+3)^2}{3x}$$

39.
$$\frac{\dfrac{12}{x+6}}{\dfrac{4}{2x+12}} = \frac{12}{x+6} \div \frac{4}{2x+12}$$
$$= \frac{12}{x+6} \cdot \frac{2(x+6)}{4}$$
$$= \frac{12 \cdot 2}{4} = 6$$

40.
$$\frac{2}{x-1} = \frac{5}{x-1} - \frac{3}{4}$$
Multiply each side by the LCD, $4(x-1)$.
$$4(x-1)\left(\frac{2}{x-1}\right) = 4(x-1)\left(\frac{5}{x-1} - \frac{3}{4}\right)$$
$$4 \cdot 2 = 4 \cdot 5 - 3(x-1)$$
$$8 = 20 - 3x + 3$$
$$3x = 15$$
$$x = 5$$

Check $x = 5$: $\frac{2}{4} = \frac{5}{4} - \frac{3}{4}$ *True*

Solution set: $\{5\}$

41. The domain of the relation consists of the elements in the leftmost figure; that is, $\{14, 91, 75, 23\}$.

The range of the relation consists of the elements in the rightmost figure; that is, $\{9, 70, 56, 5\}$.

Since the element 75 in the domain is paired with two different values, 70 and 56, in the range, the relation is not a function.

42. $f(x) = -4x + 10$

(a) The variable x can be any real number, so the domain is $(-\infty, \infty)$. The function is a non-constant linear function, so its range is $(-\infty, \infty)$.

(b) $f(-3) = -4(-3) + 10 = 12 + 10 = 22$

43. Use $(1997, 12{,}000)$ and $(2004, 86{,}000)$.

$$m = \frac{\Delta y}{\Delta x} = \frac{86{,}000 - 12{,}000}{2004 - 1997} = \frac{74{,}000}{7}$$
$$\approx 10{,}571$$

So the average rate of change is 10,571 per year; that is, the number of motor scooters sold in the United States increased by an average of 10,571 per year from 1997 to 2004.

44. Let $C =$ the cost of a pizza and $r =$ the radius of the pizza. C varies directly as r^2, so
$$C = kr^2$$
for some constant k. Since $C = 6$ when $r = 7$, substitute these values in the equation and solve for k.
$$C = kr^2$$
$$6 = k(7)^2$$
$$6 = 49k$$
$$\frac{6}{49} = k$$

So $C = \frac{6}{49}r^2$. Now let $r = 9$.
$$C = \frac{6}{49}(9)^2 = \frac{6}{49}(81) = \frac{486}{49} \approx 9.92.$$

A pizza with a 9-inch radius should cost $9.92.

CHAPTER 8   SYSTEMS OF LINEAR EQUATIONS

8.1   Solving Systems of Linear Equations by Graphing

1. $(2, -3)$

$$x + y = -1$$
$$2x + 5y = 19$$

To decide whether $(2, -3)$ is a solution of the system, substitute 2 for x and -3 for y in each equation.

$$x + y = -1$$
$$2 + (-3) = -1 \text{ ?}$$
$$-1 = -1 \quad \textit{True}$$

$$2x + 5y = 19$$
$$2(2) + 5(-3) = 19 \text{ ?}$$
$$4 + (-15) = 19 \text{ ?}$$
$$-11 = 19 \quad \textit{False}$$

The ordered pair $(2, -3)$ satisfies the first equation but not the second. Because it does not satisfy *both* equations, it is not a solution of the system.

3. $(-1, -3)$

$$3x + 5y = -18$$
$$4x + 2y = -10$$

Substitute -1 for x and -3 for y in each equation.

$$3x + 5y = -18$$
$$3(-1) + 5(-3) = -18 \text{ ?}$$
$$-3 - 15 = -18 \text{ ?}$$
$$-18 = -18 \quad \textit{True}$$

$$4x + 2y = -10$$
$$4(-1) + 2(-3) = -10 \text{ ?}$$
$$-4 - 6 = -10 \text{ ?}$$
$$-10 = -10 \quad \textit{True}$$

Since $(-1, -3)$ satisfies both equations, it is a solution of the system.

5. $(7, -2)$

$$4x = 26 - y$$
$$3x = 29 + 4y$$

Substitute 7 for x and -2 for y in each equation.

$$4x = 26 - y$$
$$4(7) = 26 - (-2) \text{ ?}$$
$$28 = 26 + 2 \quad \text{ ?}$$
$$28 = 28 \quad \quad \textit{True}$$

$$3x = 29 + 4y$$
$$3(7) = 29 + 4(-2) \text{ ?}$$
$$21 = 29 - 8 \quad \text{ ?}$$
$$21 = 21 \quad \quad \textit{True}$$

Since $(7, -2)$ satisfies both equations, it is a solution of the system.

7. $(6, -8)$

$$-2y = x + 10$$
$$3y = 2x + 30$$

Substitute 6 for x and -8 for y in each equation.

$$-2y = x + 10$$
$$-2(-8) = 6 + 10 \quad \text{ ?}$$
$$16 = 16 \quad \quad \textit{True}$$

$$3y = 2x + 30$$
$$3(-8) = 2(6) + 30 \quad \text{ ?}$$
$$-24 = 12 + 30 \quad \quad \text{ ?}$$
$$-24 = 42 \quad \quad \textit{False}$$

The ordered pair $(6, -8)$ satisfies the first equation but not the second. Because it does not satisfy *both* equations, it is not a solution of the system.

9. $(0, 0)$

$$4x + 2y = 0$$
$$x + y = 0$$

Substitute 0 for x and 0 for y in each equation.

$$4x + 2y = 0$$
$$4(0) + 2(0) = 0 \quad \text{ ?}$$
$$0 = 0 \quad \quad \textit{True}$$

$$x + y = 0$$
$$0 + 0 = 0 \quad \text{ ?}$$
$$0 = 0 \quad \quad \textit{True}$$

Since $(0, 0)$ satisfies both equations, it is a solution of the system.

11. **(a)** $(3, 4)$ is in quadrant I—choice **B** is correct.

(b) $(-2, 3)$ is in quadrant II—choice **C** is correct since choice **D** is the only other choice with intersection point in quadrant II, but it's x-coordinate of the intersection point is -3, not -2.

(c) $(-3, 2)$ is in quadrant II—choice **D** is correct.

(d) $(5, -2)$ is in quadrant IV—choice **A** is correct.

13. The two lines intersect at a point on the y-axis below the origin. Therefore, an ordered pair can satisfy the system only if its x-coordinate is 0 and its y-coordinate is negative. The only such pair among the given choices is **D**, $(0, -2)$.

15. $x - y = 2$
$x + y = 6$

To graph the equations, find the intercepts.

$x - y = 2$:   Let $y = 0$; then $x = 2$.
Let $x = 0$; then $y = -2$.

Plot the intercepts, $(2, 0)$ and $(0, -2)$, and draw the line through them.

$x + y = 6$:   Let $y = 0$; then $x = 6$.
Let $x = 0$; then $y = 6$.

Plot the intercepts, $(6, 0)$ and $(0, 6)$, and draw the line through them.

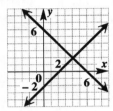

It appears that the lines intersect at the point $(4, 2)$. Check this by substituting 4 for x and 2 for y in both equations. Since $(4, 2)$ satisfies both equations, the solution set of this system is $\{(4, 2)\}$.

17. $x + y = 4$
$y - x = 4$

To graph the equations, find the intercepts.

$x + y = 4$:   Let $y = 0$; then $x = 4$.
Let $x = 0$; then $y = 4$.

Plot the intercepts, $(4, 0)$ and $(0, 4)$, and draw the line through them.

$y - x = 4$:   Let $y = 0$; then $x = -4$.
Let $x = 0$; then $y = 4$.

Plot the intercepts, $(-4, 0)$ and $(0, 4)$, and draw the line through them.

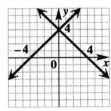

The lines intersect at their common y-intercept, $(0, 4)$, so $\{(0, 4)\}$ is the solution set of the system.

19. $x - 2y = 6$
$x + 2y = 2$

To graph the equations, find the intercepts.

$x - 2y = 6$:   Let $y = 0$; then $x = 6$.
Let $x = 0$; then $y = -3$.

Plot the intercepts, $(6, 0)$ and $(0, -3)$, and draw the line through them.

$x + 2y = 2$:   Let $y = 0$; then $x = 2$.
Let $x = 0$; then $y = 1$.

Plot the intercepts, $(2, 0)$ and $(0, 1)$, and draw the line through them.

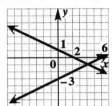

It appears that the lines intersect at the point $(4, -1)$. Since $(4, -1)$ satisfies both equations, the solution set of this system is $\{(4, -1)\}$.

21. $3x - 2y = -3$
$-3x - y = -6$

To graph the equations, find the intercepts.

$3x - 2y = -3$:   Let $y = 0$; then $x = -1$.
Let $x = 0$; then $y = \frac{3}{2}$.

Plot the intercepts, $(-1, 0)$ and $\left(0, \frac{3}{2}\right)$, and draw the line through them.

$-3x - y = -6$:   Let $y = 0$; then $x = 2$.
Let $x = 0$; then $y = 6$.

Plot the intercepts, $(2, 0)$ and $(0, 6)$, and draw the line through them.

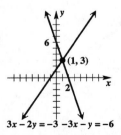

It appears that the lines intersect at the point $(1, 3)$. Since $(1, 3)$ satisfies both equations, the solution set of this system is $\{(1, 3)\}$.

23. $2x - 3y = -6$
$y = -3x + 2$

To graph the first line, find the intercepts.

$2x - 3y = -6$: Let $y = 0$; then $x = -3$.
        Let $x = 0$; then $y = 2$.

Plot the intercepts, $(-3, 0)$ and $(0, 2)$, and draw the line through them.

To graph the second line, start by plotting the y-intercept, $(0, 2)$. From this point, go 3 units down and 1 unit to the right (because the slope is -3) to reach the point $(1, -1)$. Draw the line through $(0, 2)$ and $(1, -1)$.

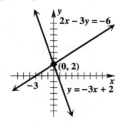

The lines intersect at their common y-intercept, $(0, 2)$, so $\{(0, 2)\}$ is the solution set of the system.

25. $2x - y = 6$
$4x - 2y = 8$

Graph the line $2x - y = 6$ through its intercepts, $(3, 0)$ and $(0, -6)$.

Graph the line $4x - 2y = 8$ through its intercepts, $(2, 0)$ and $(0, -4)$.

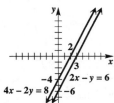

The lines each have slope 2, and hence, are parallel. Since they do not intersect, there is no solution. This is an inconsistent system and the solution set is $\emptyset$.

27. $3x + y = 5$
$6x + 2y = 10$

$3x + y = 5$: Let $y = 0$; then $x = \frac{5}{3}$.
        Let $x = 0$; then $y = 5$.

Plot these intercepts, $\left(\frac{5}{3}, 0\right)$ and $(0, 5)$, and draw the line through them.

$6x + 2y = 10$: Let $y = 0$; then $x = \frac{5}{3}$.
        Let $x = 0$; then $y = 5$.

Plot these intercepts, $\left(\frac{5}{3}, 0\right)$ and $(0, 5)$, and draw the line through them.

Since both equations have the same intercepts, they are equations of the same line.

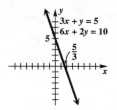

There is an infinite number of solutions. The equations are dependent equations and the solution set contains an infinite number of ordered pairs. The solution set is $\{(x, y) \mid 3x + y = 5\}$.

29. $3x - 4y = 24$
$y = -\frac{3}{2}x + 3$

Graph the line $3x - 4y = 24$ through its intercepts, $(8, 0)$ and $(0, -6)$.

To graph the line $y = -\frac{3}{2}x + 3$, plot the y-intercept $(0, 3)$ and then go 3 units down and 2 units to the right $\left(\text{because the slope is } -\frac{3}{2}\right)$ to reach the point $(2, 0)$. Draw the line through $(0, 3)$ and $(2, 0)$.

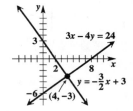

It appears that the lines intersect at the point $(4, -3)$. Since $(4, -3)$ satisfies both equations, the solution set of this system is $\{(4, -3)\}$.

31. If the coordinates of the point of intersection are not integers, the solution will be difficult to determine from a graph.

33. $y - x = -5$
$x + y = 1$

Write the equations in slope-intercept form.

$y - x = -5$	$x + y = 1$
$y = x - 5$	$y = -x + 1$
$m = 1$	$m = -1$

The lines have different slopes.

(a) The system is consistent because it has a solution. The equations are independent because they have different graphs. Therefore, the answer is "neither."

(b) The graph is a pair of intersecting lines.

(c) The system has one solution.

35. $x + 2y = 0$

$4y = -2x$

Write the equations in slope-intercept form.

$x + 2y = 0$	$4y = -2x$
$2y = -x$	$y = -\frac{1}{2}x$
$y = -\frac{1}{2}x$	

For both lines, $m = -\frac{1}{2}$ and $b = 0$.

(a) Since the equations have the same slope and y-intercept, they are dependent.

(b) The graph is one line.

(c) The system has an infinite number of solutions.

37. $5x + 4y = 7$

$10x + 8y = 4$

Write the equations in slope-intercept form.

$5x + 4y = 7$	$10x + 8y = 4$
$4y = -5x + 7$	$8y = -10x + 4$
$y = -\frac{5}{4}x + \frac{7}{4}$	$y = -\frac{10}{8}x + \frac{4}{8}$
$m = -\frac{5}{4}, b = \frac{7}{4}$	$y = -\frac{5}{4}x + \frac{1}{2}$
	$m = -\frac{5}{4}, b = \frac{1}{2}$

The lines have the same slope but different y-intercepts.

(a) The system is inconsistent because it has no solution.

(b) The graph is a pair of parallel lines.

(c) The system has no solution.

39. $x - 3y = 5$

$2x + y = 8$

Write the equations in slope-intercept form.

$x - 3y = 5$	$2x + y = 8$
$-3y = -x + 5$	$y = -2x + 8$
$y = \frac{1}{3}x - \frac{5}{3}$	$m = -2$
$m = \frac{1}{3}$	

The lines have different slopes.

(a) The system is consistent because it has a solution. The equations are independent because they have different graphs. Therefore, the answer is "neither."

(b) The graph is a pair of intersecting lines.

(c) The system has one solution.

41. $\frac{1}{5}x + \frac{1}{5}y = -1$

$-\frac{1}{3}x + \frac{1}{3}y = 1$

Write the equations in slope-intercept form.

$\frac{1}{5}x + \frac{1}{5}y = -1$	$-\frac{1}{3}x + \frac{1}{3}y = 1$
$\frac{1}{5}y = -\frac{1}{5}x - 1$	$\frac{1}{3}y = \frac{1}{3}x + 1$
$y = -x - 5$	$y = x + 3$
$m = -1$	$m = 1$

The lines have different slopes.

(a) The system is consistent because it has a solution. The equations are independent because they have different graphs. Therefore, the answer is "neither."

(b) The graph is a pair of intersecting lines.

(c) The system has one solution.

43. The graphs intersect when the year is about 2001. The intersection point is just below 800, so there were about 800 newspapers of each type in 2001.

45. Supply equals demand at the point where the two lines intersect, or when $p = 30$.

47. (a) The graph representing ATM use is above the graphs representing credit card *and* debit card use for the years 1997–2002.

(b) The graph representing debit card use rises above the graph for credit card use in about 2001.

(c) The graph representing debit card use rises above the graph for ATM use in about 2002.

(d) The y-value is approximately 30, so the ordered pair is $(1998, 30)$.

(e) During the period 1997–2004, debit card use went from least popular to most popular of the three methods depicted.

49. For $x + y = 4$, the x-intercept is $(4, 0)$ and the y-intercept is $(0, 4)$. The only screen with a graph having those intercepts is choice **B**. The point of intersection, $(3, 1)$ is listed at the bottom of the screen. Check $(3, 1)$ in the system of equations. Since it satisfies the equations, $\{(3, 1)\}$ is the solution set.

51. The graph of $x - y = 0$ goes through the origin. The only screen with a graph that goes through the origin is choice **A**.

53. $3x + y = 2$

$2x - y = -7$

First, solve the equations for y.

$3x + y = 2$	$2x - y = -7$
$y = -3x + 2$	$2x + 7 = y$

Now enter the equations.

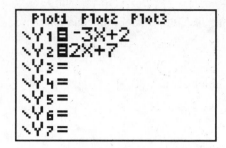

Next, graph the equations using a standard window. On the TI-82/3/4, just press $\boxed{\text{ZOOM}}$ $\boxed{6}$.

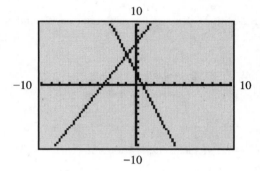

Now we'll let the calculator find the coordinates of the point of intersection of the graphs. Press $\boxed{\text{2nd}}$ $\boxed{\text{CALC}}$ $\boxed{5}$ $\boxed{\text{ENTER}}$ $\boxed{\text{ENTER}}$ to indicate the graphs for which we're trying to find the point of intersection. Now press the left cursor key, $\boxed{\triangleleft}$, four times to get close to the point of intersection. Lastly, press $\boxed{\text{ENTER}}$ to produce the following graph.

(On the TI-86, use the key sequence $\boxed{\text{MORE}}$ $\boxed{\text{F1}}$ $\boxed{\text{MORE}}$ $\boxed{\text{F3}}$ $\boxed{\text{ENTER}}$ $\boxed{\text{ENTER}}$ to indicate the graphs.)

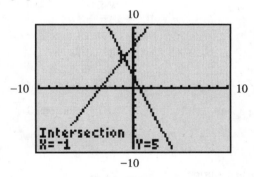

The display at the bottom of the last figure indicates that the solution set is $\{(-1, 5)\}$.

55. $8x + 4y = 0$
$4x - 2y = 2$

First, solve the equations for y.

$$8x + 4y = 0 \qquad 4x - 2y = 2$$
$$4y = -8x \qquad -2y = -4x + 2$$
$$y = -2x \qquad y = 2x - 1$$

Now enter, graph, and solve the system as in Exercise 53.

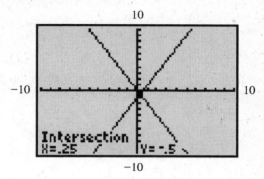

The solution set is $\{(0.25, -0.5)\}$.

57. $3x + y = 4$
$\quad y = -3x + 4$

59. $9x - 2y = 4$
$\quad -2y = -9x + 4$
$\quad y = \frac{9}{2}x - 2$

61. $-2(x - 2) + 5x = 10$
$\quad -2x + 4 + 5x = 10$
$\quad \quad 3x + 4 = 10$
$\quad \quad \quad 3x = 6$
$\quad \quad \quad x = 2$

Check $x = 2$: $0 + 10 = 10$

The solution set is $\{2\}$.

63. $p + 4(6 - 2p) = 3$
$\quad p + 24 - 8p = 3$
$\quad \quad -7p + 24 = 3$
$\quad \quad \quad -7p = -21$
$\quad \quad \quad p = 3$

Check $p = 3$: $3 + 0 = 3$

The solution set is $\{3\}$.

65. $4x - 2(1 - 3x) = 6$
$\quad 4x - 2 + 6x = 6$
$\quad \quad 10x - 2 = 6$
$\quad \quad \quad 10x = 8$
$\quad \quad \quad x = \frac{8}{10} = \frac{4}{5}$

Check $x = \frac{4}{5}$: $\frac{16}{5} - 2\left(-\frac{7}{5}\right) = 6$
$\quad \quad \quad \frac{16}{5} + \frac{14}{5} = 6$

The solution set is $\left\{\frac{4}{5}\right\}$.

8.2 Solving Systems of Linear Equations by Substitution

1. No, it is not correct, because the solution set is $\{(3, 0)\}$. The y-value in the ordered pair must also be determined.

In this section, all solutions should be checked by substituting in *both* equations of the original system. Checks will not be shown here.

3. $x + y = 12$ (1)
  $y = 3x$ (2)

Equation (2) is already solved for y. Substitute $3x$ for y in equation (1) and solve the resulting equation for x.

$$x + y = 12$$
$$x + 3x = 12 \quad \textit{Let } y = 3x.$$
$$4x = 12$$
$$x = 3$$

To find the y-value of the solution, substitute 3 for x in equation (2).

$$y = 3x$$
$$y = 3(3) \quad \textit{Let } x = 3.$$
$$= 9$$

The solution set is $\{(3, 9)\}$.

To check this solution, substitute 3 for x and 9 for y in both equations of the given system.

5. $3x + 2y = 27$ (1)
  $x = y + 4$ (2)

Equation (2) is already solved for x. Substitute $y + 4$ for x in equation (1).

$$3x + 2y = 27$$
$$3(y + 4) + 2y = 27$$
$$3y + 12 + 2y = 27$$
$$5y = 15$$
$$y = 3$$

To find x, substitute 3 for y in equation (2).

$$x = y + 4$$
$$x = 3 + 4 = 7$$

The solution set is $\{(7, 3)\}$.

7. $3x + 5y = 25$ (1)
  $x - 2y = -10$ (2)

Solve equation (2) for x since its coefficient is 1.

$$x - 2y = -10$$
$$x = 2y - 10 \quad (3)$$

Substitute $2y - 10$ for x in equation (1) and solve for y.

$$3x + 5y = 25$$
$$3(2y - 10) + 5y = 25$$
$$6y - 30 + 5y = 25$$
$$11y = 55$$
$$y = 5$$

To find x, substitute 5 for y in equation (3).

$$x = 2y - 10$$
$$x = 2(5) - 10 = 0$$

The solution set is $\{(0, 5)\}$.

9. $3x + 4 = -y$ (1)
  $2x + y = 0$ (2)

Solve equation (1) for y.

$$3x + 4 = -y$$
$$y = -3x - 4 \quad (3)$$

Substitute $-3x - 4$ for y in equation (2) and solve for x.

$$2x + y = 0$$
$$2x + (-3x - 4) = 0$$
$$-x - 4 = 0$$
$$-x = 4$$
$$x = -4$$

To find y, substitute -4 for x in equation (3).

$$y = -3x - 4$$
$$= -3(-4) - 4 = 8$$

The solution set is $\{(-4, 8)\}$.

11. $7x + 4y = 13$ (1)
  $x + y = 1$ (2)

Solve equation (2) for y.

$$x + y = 1$$
$$y = 1 - x \quad (3)$$

Substitute $1 - x$ for y in equation (1).

$$7x + 4y = 13$$
$$7x + 4(1 - x) = 13$$
$$7x + 4 - 4x = 13$$
$$3x + 4 = 13$$
$$3x = 9$$
$$x = 3$$

To find y, substitute 3 for x in equation (3).

$$y = 1 - x$$
$$y = 1 - 3 = -2$$

The solution set is $\{(3, -2)\}$.

13. $3x - y = 5$ (1)
  $y = 3x - 5$ (2)

Equation (2) is already solved for y, so we substitute $3x - 5$ for y in equation (1).

$$3x - (3x - 5) = 5$$
$$3x - 3x + 5 = 5$$
$$5 = 5 \quad \textit{True}$$

This true result means that every solution of one equation is also a solution of the other, so the system has an infinite number of solutions. The solution set is $\{(x, y) \mid 3x - y = 5\}$.

15. $2x + y = 0$ $\quad$ (1)
$\quad$ $4x - 2y = 2$ $\quad$ (2)

Solve equation (1) for y.

$$2x + y = 0$$
$$y = -2x \quad (3)$$

Substitute $-2x$ for y in equation (2) and solve for x.

$$4x - 2(-2x) = 2$$
$$4x + 4x = 2$$
$$8x = 2$$
$$x = \tfrac{1}{4}$$

To find y, let $x = \tfrac{1}{4}$ in equation (3).

$$y = -2x$$
$$y = -2\left(\tfrac{1}{4}\right)$$
$$= -\tfrac{1}{2}$$

The solution set is $\left\{\left(\tfrac{1}{4}, -\tfrac{1}{2}\right)\right\}$.

17. $2x + 8y = 3$ $\quad$ (1)
$\quad$ $x = 8 - 4y$ $\quad$ (2)

Equation (2) is already solved for x, so substitute $8 - 4y$ for x in equation (1).

$$2(8 - 4y) + 8y = 3$$
$$16 - 8y + 8y = 3$$
$$16 = 3 \quad \textit{False}$$

This false result means that the system is inconsistent and its solution set is $\emptyset$.

19. $\quad$ $2y = 4x + 24$ $\quad\quad$ (1)
$\quad$ $2x - y = -12$ $\quad\quad$ (2)

Solve equation (2) for y.

$$2x - y = -12$$
$$2x = y - 12$$
$$2x + 12 = y$$

Substitute $2x + 12$ for y in (1) and solve for x.

$$2(2x + 12) = 4x + 24$$
$$4x + 24 = 4x + 24 \quad \textit{True}$$

This true result means that every solution of one equation is also a solution of the other, so the system has an infinite number of solutions. The solution set is $\{(x, y) \mid 2x - y = -12\}$.

21. **(a)** If you arrive at a false statement such as $0 = 5$ when using the substitution method, then there is no solution.

(b) If you arrive at a true statement such as $0 = 0$ when using the substitution method, then the number of solutions is infinite.

23. $\tfrac{1}{2}x + \tfrac{1}{3}y = 3$ $\quad$ (1)
$\quad$ $y = 3x$ $\quad\quad\quad$ (2)

First, clear all fractions in equation (1).

$$6\left(\tfrac{1}{2}x + \tfrac{1}{3}y\right) = 6(3) \quad \textit{Multiply by the LCD, 6.}$$
$$6\left(\tfrac{1}{2}x\right) + 6\left(\tfrac{1}{3}y\right) = 18 \quad \textit{Distributive property}$$
$$3x + 2y = 18 \quad (3)$$

From equation (2), substitute $3x$ for y in equation (3).

$$3x + 2(3x) = 18$$
$$3x + 6x = 18$$
$$9x = 18$$
$$x = 2$$

To find y, let $x = 2$ in equation (2).

$$y = 3(2) = 6$$

The solution set is $\{(2, 6)\}$.

25. $\tfrac{1}{2}x + \tfrac{1}{3}y = -\tfrac{1}{3}$ $\quad$ (1)
$\quad$ $\tfrac{1}{2}x + 2y = -7$ $\quad$ (2)

First, clear all fractions.

Equation (1):

$$6\left(\tfrac{1}{2}x + \tfrac{1}{3}y\right) = 6\left(-\tfrac{1}{3}\right) \quad \textit{Multiply by the LCD, 6.}$$
$$6\left(\tfrac{1}{2}x\right) + 6\left(\tfrac{1}{3}y\right) = -2 \quad \textit{Distributive property}$$
$$3x + 2y = -2 \quad (3)$$

Equation (2):

$$2\left(\tfrac{1}{2}x + 2y\right) = 2(-7) \quad \textit{Multiply by 2.}$$
$$x + 4y = -14 \quad (4)$$

The system has been simplified to

$$3x + 2y = -2 \quad (3)$$
$$x + 4y = -14 \quad (4)$$

Solve this system by the substitution method.

$$x = -4y - 14 \quad (5) \quad \textit{Solve (4) for x.}$$
$$3(-4y - 14) + 2y = -2 \quad \textit{Substitute for x in (3).}$$
$$-12y - 42 + 2y = -2$$
$$-10y - 42 = -2$$
$$-10y = 40$$
$$y = -4$$

To find x, let $y = -4$ in equation (5).

$$x = -4(-4) - 14 = 16 - 14 = 2$$

The solution set is $\{(2, -4)\}$.

27. $\frac{x}{5} + 2y = \frac{8}{5}$ (1)

$\frac{3x}{5} + \frac{y}{2} = -\frac{7}{10}$ (2)

First, clear all fractions.

Equation (1):

$$5\left(\frac{x}{5} + 2y\right) = 5\left(\frac{8}{5}\right) \quad \textit{Multiply by 5.}$$

$$x + 10y = 8 \quad (3)$$

Equation (2):

$$10\left(\frac{3x}{5} + \frac{y}{2}\right) = 10\left(-\frac{7}{10}\right) \quad \textit{Multiply by 10.}$$

$$6x + 5y = -7 \quad (4)$$

The system has been simplified to

$$x + 10y = 8 \quad (3)$$
$$6x + 5y = -7. \quad (4)$$

Solve this system by the substitution method. Solve equation (3) for x.

$$x = 8 - 10y \quad (5)$$

Substitute $8 - 10y$ for x in equation (4).

$$6(8 - 10y) + 5y = -7$$
$$48 - 60y + 5y = -7$$
$$-55y = -55$$
$$y = 1$$

To find x, let $y = 1$ in equation (5).

$$x = 8 - 10(1) = -2$$

The solution set is $\{(-2, 1)\}$.

29. $\frac{x}{5} + y = \frac{6}{5}$ (1)

$\frac{x}{10} + \frac{y}{3} = \frac{5}{6}$ (2)

First, clear all fractions.

Equation (1):

$$5\left(\frac{x}{5} + y\right) = 5\left(\frac{6}{5}\right) \quad \textit{Multiply by 5.}$$

$$x + 5y = 6 \quad (3)$$

Equation (2):

$$30\left(\frac{x}{10} + \frac{y}{3}\right) = 30\left(\frac{5}{6}\right) \quad \textit{Multiply by the LCD, 30.}$$

$$3x + 10y = 25 \quad (4)$$

The system has been simplified to

$$x + 5y = 6 \quad (3)$$
$$3x + 10y = 25 \quad (4)$$

Solve this system by the substitution method.

$$x = -5y + 6 \quad (5) \textit{ Solve (3) for } x.$$

$3(-5y + 6) + 10y = 25$ *Substitute for x in (4).*

$$-15y + 18 + 10y = 25$$
$$-5y + 18 = 25$$
$$-5y = 7$$
$$y = -\frac{7}{5}$$

To find x, let $y = -\frac{7}{5}$ in equation (5).

$$x = -5\left(-\frac{7}{5}\right) + 6 = 7 + 6 = 13$$

The solution set is $\left\{\left(13, -\frac{7}{5}\right)\right\}$.

31. $\frac{1}{6}x + \frac{1}{3}y = 8$ (1)

$\frac{1}{4}x + \frac{1}{2}y = 12$ (2)

Multiply equation (1) by 6.

$$6\left(\frac{1}{6}x + \frac{1}{3}y\right) = 6(8)$$
$$x + 2y = 48 \quad (3)$$

Multiply equation (2) by 4.

$$4\left(\frac{1}{4}x + \frac{1}{2}y\right) = 4(12)$$
$$x + 2y = 48 \quad (4)$$

Equations (3) and (4) are identical.

This means that every solution of one equation is also a solution of the other, so the system has an infinite number of solutions. The solution set is $\{(x, y) \mid x + 2y = 48\}$.

33. To find the total cost, multiply the number of bicycles (x) by the cost per bicycle (400), and add the fixed cost (5000). Thus, $y_1 = 400x + 5000$ gives the total cost (in dollars).

34. Since each bicycle sells for $600, the total revenue for selling x bicycles is $600x$ (in dollars). Thus, $y_2 = 600x$ gives the total revenue.

35. $y_1 = 400x + 5000$ (1)
$y_2 = 600x$ (2)

To solve this system by the substitution method, substitute $600x$ for y_1 in equation (1).

$$600x = 400x + 5000$$
$$200x = 5000$$
$$x = 25$$

If $x = 25$, $y_2 = 600(25) = 15,000$.

The solution set is $\{(25, 15,000)\}$.

36. The value of x from Exercise 35 is the number of bikes it takes to break even. When __25__ bikes are sold, the break-even point is reached. At that point, you have spent __15,000__ dollars and taken in __15,000__ dollars.

37. $y = 6 - x$ (1)

     $y = 2x$ (2)

Substitute $2x$ for y in equation (1).

$$2x = 6 - x$$
$$3x = 6$$
$$x = 2$$

Substituting 2 for x in either of the original equations gives $y = 4$.

The solution set is $\{(2, 4)\}$.

Input the equations $Y_1 = 6 - X$ and $Y_2 = 2X$ and then use the intersection feature to obtain the following graph. See the solution to Exercise 53 in Section 8.1 for more specifics.

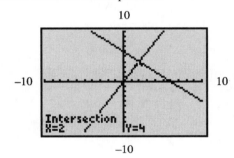

39. $y = -\frac{4}{3}x + \frac{19}{3}$ (1)

     $y = \frac{15}{2}x - \frac{5}{2}$ (2)

Substitute the expression from equation (2) into equation (1).

$$\tfrac{15}{2}x - \tfrac{5}{2} = -\tfrac{4}{3}x + \tfrac{19}{3}$$

Multiply by 6 to clear fractions.

$$6\left(\tfrac{15}{2}x - \tfrac{5}{2}\right) = 6\left(-\tfrac{4}{3}x + \tfrac{19}{3}\right)$$
$$45x - 15 = -8x + 38$$
$$53x = 53$$
$$x = 1$$

To find y, let $x = 1$ in equation (1).

$$y = -\tfrac{4}{3}(1) + \tfrac{19}{3}$$
$$= -\tfrac{4}{3} + \tfrac{19}{3} = \tfrac{15}{3} = 5$$

The solution set is $\{(1, 5)\}$.

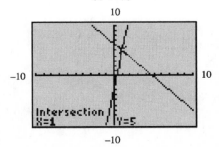

41. $4x + 5y = 5$ (1)

     $2x + 3y = 1$ (2)

Solve equation (2) for y.

$$2x + 3y = 1$$
$$3y = 1 - 2x$$
$$y = \frac{1 - 2x}{3} \quad (3)$$

Substitute for y in equation (1).

$$4x + 5\left(\frac{1 - 2x}{3}\right) = 5$$

$$4x + \frac{5(1 - 2x)}{3} = 5$$

Multiply by 3 to clear fractions.

$$3\left[4x + \frac{5(1 - 2x)}{3}\right] = 3(5)$$
$$12x + 5(1 - 2x) = 15$$
$$12x + 5 - 10x = 15$$
$$2x = 10$$
$$x = 5$$

To find y, let $x = 5$ in equation (3).

$$y = \frac{1 - 2(5)}{3} = \frac{-9}{3} = -3$$

The solution set is $\{(5, -3)\}$.

To graph the original system on a graphing calculator, each equation must be solved for y.

Equation (1):

$$4x + 5y = 5$$
$$5y = 5 - 4x$$
$$y = \frac{5 - 4x}{5}$$

Equation (2) was solved for y — see (3).

Thus, the equations to input are

$$Y_1 = \frac{5 - 4X}{5} = (5 - 4X)/5$$

and

$$Y_2 = \frac{1 - 2X}{3} = (1 - 2X)/3.$$

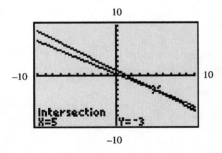

43. If the point of intersection does not appear on your screen, you will need to adjust the viewing window. Change the x- and y-minimum values or the x- and y-maximum values as necessary.

45. $(14x - 3y) + (2x + 3y)$
$= 14x - 3y + 2x + 3y$
$= (14 + 2)x + (-3 + 3)y$
$= 16x + 0y = 16x$

47. $(-x + 7y) + (3y + x)$
$= -x + 7y + 3y + x$
$= (-1 + 1)x + (7 + 3)y$
$= 0x + 10y = 10y$

49. To get a sum of 0, we must add the additive inverse of $-4x$, that is, $4x$.

51. The desired number times $4y$ must equal $-8y$, which is the additive inverse of $8y$. By trial and error, -2 times $4y$ equals $-8y$, so $4y$ must be multiplied by -2.

8.3 Solving Systems of Linear Equations by Elimination

1. The statement is *false*; we should multiply the bottom equation by -3, not 3. Then we will have $12y + (-12y) = 0$, so the y terms will be eliminated.

3. It is impossible to have two numbers whose sum is both 1 and 2, so the given statement is *true*.

In Exercises 5–41, check your answers by substituting into *both* of the original equations. The check will be shown only for Exercise 5.

5. To eliminate y, add equations (1) and (2).

$$\begin{array}{rcll} x & - & y & = & -2 & (1) \\ x & + & y & = & 10 & (2) \\ \hline 2x & & & = & 8 & \text{Add (1) and (2).} \\ & & x & = & 4 \end{array}$$

This result gives the x-value of the solution. To find the y-value of the solution, substitute 4 for x in either equation. We will use equation (2).

$$\begin{array}{rl} x + y = 10 & \\ 4 + y = 10 & \textit{Let x = 4.} \\ y = 6 & \end{array}$$

Check by substituting 4 for x and 6 for y in both equations of the original system.

Check Equation (1)

$$\begin{array}{rl} x - y = -2 & \\ 4 - 6 = -2 \, ? & \textit{Let x = 4, y = 6.} \\ -2 = -2 & \textit{True} \end{array}$$

Check Equation (2)

$$\begin{array}{rl} x + y = 10 & \\ 4 + 6 = 10 \, ? & \textit{Let x = 4, y = 6.} \\ 10 = 10 & \textit{True} \end{array}$$

The solution set is $\{(4, 6)\}$.

7.
$$\begin{array}{rcll} 2x & + & y & = & -5 & (1) \\ x & - & y & = & 2 & (2) \\ \hline 3x & & & = & -3 & \text{Add (1) and (2).} \\ & & x & = & -1 \end{array}$$

Substitute -1 for x in equation (1) to find the y-value of the solution.

$$\begin{array}{rl} 2x + y = -5 & \\ 2(-1) + y = -5 & \textit{Let x = -1.} \\ -2 + y = -5 & \\ y = -3 & \end{array}$$

The solution set is $\{(-1, -3)\}$.

9. First, rewrite both equations in the standard form, $Ax + By = C$. We can write the first equation, $2y = -3x$, as $3x + 2y = 0$ by adding $3x$ to each side.

$$\begin{array}{rcll} 3x & + & 2y & = & 0 & (1) \\ -3x & - & y & = & 3 & (2) \\ \hline & & y & = & 3 & \text{Add (1) and (2).} \end{array}$$

Substitute 3 for y in equation (1).

$$\begin{array}{rl} 3x + 2y = 0 & \\ 3x + 2(3) = 0 & \\ 3x + 6 = 0 & \\ 3x = -6 & \\ x = -2 & \end{array}$$

The solution set is $\{(-2, 3)\}$.

11.
$$\begin{array}{l} 4x - 3y = 1 \\ 8x = 3 + 6y \end{array}$$

Rewrite in standard form.

$$\begin{array}{rl} 4x - 3y = 1 & (1) \\ 8x - 6y = 3 & (2) \end{array}$$

To eliminate x, multiply equation (1) by -2 and add the result to equation (2).

$$\begin{array}{rcll} -8x & + & 6y & = & -2 & (3) \\ 8x & - & 6y & = & 3 & (2) \\ \hline & & 0 & = & 1 & \text{Add (3) and (2).} \end{array}$$

The false statement, $0 = 1$, indicates that the given system has solution set $\emptyset$.

13.
$$x + 3y = 6$$
$$-2x + 12 = 6y$$

Rewrite in standard form.

$$x + 3y = 6 \quad (1)$$
$$-2x - 6y = -12 \quad (2)$$

To eliminate x, multiply equation (1) by 2 and add the result to equation (2).

$$
\begin{array}{rrrrll}
2x & + & 6y & = & 12 & (3) \\
-2x & - & 6y & = & -12 & (2) \\
\hline
 & & 0 & = & 0 & \text{Add (3) and (2).}
\end{array}
$$

The true statement, $0 = 0$, indicates that there are an infinite number of solutions and the solution set is $\{(x, y) \mid x + 3y = 6\}$.

15.
$$6x - y = -1$$
$$5y = 17 + 6x$$

Rewrite in standard form.

$$
\begin{array}{rrrrll}
6x & - & y & = & -1 & (1) \\
-6x & + & 5y & = & 17 & (2) \\
\hline
 & & 4y & = & 16 & \text{Add (1) and (2).} \\
 & & y & = & 4 & \text{Solve for } y.
\end{array}
$$

Substitute 4 for y in equation (1).

$$6x - y = -1$$
$$6x - 4 = -1$$
$$6x = 3$$
$$x = \tfrac{3}{6} = \tfrac{1}{2}$$

The solution set is $\left\{ \left(\tfrac{1}{2}, 4 \right) \right\}$.

17.
$$2x - y = 12 \quad (1)$$
$$3x + 2y = -3 \quad (2)$$

If we simply add the equations, we will not eliminate either variable. To eliminate y, multiply equation (1) by 2 and add the result to equation (2).

$$
\begin{array}{rrrrll}
4x & - & 2y & = & 24 & (3) \\
3x & + & 2y & = & -3 & (2) \\
\hline
7x & & & = & 21 & \text{Add (3) and (2).} \\
 & & x & = & 3 &
\end{array}
$$

Substitute 3 for x in equation (1).

$$2x - y = 12$$
$$2(3) - y = 12$$
$$-y = 6$$
$$y = -6$$

The solution set is $\{(3, -6)\}$.

19.
$$x + 4y = 16 \quad (1)$$
$$3x + 5y = 20 \quad (2)$$

To eliminate x, multiply equation (1) by -3 and add the result to equation (2).

$$
\begin{array}{rrrrll}
-3x & - & 12y & = & -48 & (3) \\
3x & + & 5y & = & 20 & (2) \\
\hline
 & & -7y & = & -28 & \text{Add (3) and (2).} \\
 & & y & = & 4 &
\end{array}
$$

Substitute 4 for y in equation (1).

$$x + 4y = 16$$
$$x + 4(4) = 16$$
$$x + 16 = 16$$
$$x = 0$$

The solution set is $\{(0, 4)\}$.

21.
$$2x - 8y = 0 \quad (1)$$
$$4x + 5y = 0 \quad (2)$$

To eliminate x, multiply equation (1) by -2 and add the result to equation (2).

$$
\begin{array}{rrrrll}
-4x & + & 16y & = & 0 & (3) \\
4x & + & 5y & = & 0 & (2) \\
\hline
 & & 21y & = & 0 & \text{Add (3) and (2).} \\
 & & y & = & 0 &
\end{array}
$$

Substitute 0 for y in equation (1).

$$2x - 8y = 0$$
$$2x - 8(0) = 0$$
$$2x = 0$$
$$x = 0$$

The solution set is $\{(0, 0)\}$.

23.
$$3x + 3y = 33 \quad (1)$$
$$5x - 2y = 27 \quad (2)$$

To eliminate y, multiply equation (1) by 2 and equation (2) by 3.

$$
\begin{array}{rrrrll}
6x & + & 6y & = & 66 & (3) \\
15x & - & 6y & = & 81 & (4) \\
\hline
21x & & & = & 147 & \text{Add (3) and (4).} \\
 & & x & = & 7 &
\end{array}
$$

Substitute 7 for x in equation (2).

$$5x - 2y = 27$$
$$5(7) - 2y = 27$$
$$35 - 2y = 27$$
$$-2y = -8$$
$$y = 4$$

The solution set is $\{(7, 4)\}$.

Note that we could have reduced equation (1) by dividing by 3 in the beginning.

25. $5x + 4y = 12 \quad (1)$
$\quad\ \ 3x + 5y = 15 \quad (2)$

To eliminate x, we could multiply equation (1) by $-\frac{3}{5}$, but that would introduce fractions and make the solution more complicated. Instead, we'll work with the least common multiple of the coefficients of x, which is 15, and choose suitable multipliers of these coefficients so that the new coefficients are opposites.

In this case, we could pick -3 times equation (1) and 5 times equation (2) *or* 3 times equation (1) and -5 times equation (2). If we wanted to eliminate y, we could multiply equation (1) by -5 and equation (2) by 4 *or* equation (1) by 5 and equation (2) by -4.

$$
\begin{array}{rcll}
-15x \ - \ 12y &=& -36 & (3)\ -3 \times \text{Eq.}(1) \\
15x \ + \ 25y &=& 75 & (4)\ \ 5 \times \text{Eq.}(2) \\
\hline
13y &=& 39 & \text{Add } (3) \text{ and } (4). \\
y &=& 3 &
\end{array}
$$

Substitute 3 for y in equation (1).

$$
\begin{aligned}
5x + 4y &= 12 \\
5x + 4(3) &= 12 \\
5x + 12 &= 12 \\
5x &= 0 \\
x &= 0
\end{aligned}
$$

The solution set is $\{(0, 3)\}$.

27. $5x - 4y = 15 \qquad (1)$
$\quad\ \ -3x + 6y = -9 \quad (2)$

$$
\begin{array}{rcll}
15x \ - \ 12y &=& 45 & (3)\ \ 3 \times \text{Eq.}(1) \\
-15x \ + \ 30y &=& -45 & (4)\ \ 5 \times \text{Eq.}(2) \\
\hline
18y &=& 0 & \text{Add } (3) \text{ and } (4). \\
y &=& 0 &
\end{array}
$$

Substitute 0 for y in equation (1).

$$
\begin{aligned}
5x - 4y &= 15 \\
5x - 4(0) &= 15 \\
5x &= 15 \\
x &= 3
\end{aligned}
$$

The solution set is $\{(3, 0)\}$.

29. $3x = 3 + 2y \qquad (1)$
$\quad\ \ -\frac{4}{3}x + y = \frac{1}{3} \quad (2)$

Rewrite equation (1) in standard form and multiply equation (2) by 3 to clear fractions.

$$
\begin{array}{rll}
3x - 2y = 3 & (3) \\
-4x + 3y = 1 & (4)
\end{array}
$$

$$
\begin{array}{rcll}
12x \ - \ 8y &=& 12 & (5)\ \ 4 \times \text{Eq.}(3) \\
-12x \ + \ 9y &=& 3 & (6)\ \ 3 \times \text{Eq.}(4) \\
\hline
y &=& 15 & \text{Add } (5) \text{ and } (6).
\end{array}
$$

$$
\begin{array}{rcll}
9x \ - \ 6y &=& 9 & (7) \quad 3 \times \text{Eq.}(3) \\
-8x \ + \ 6y &=& 2 & (8) \quad 2 \times \text{Eq.}(4) \\
\hline
x &=& 11 & \text{Add } (7) \text{ and } (8).
\end{array}
$$

The solution set is $\{(11, 15)\}$.

31. $-x + 3y = 4 \qquad (1)$
$\quad\ \ -2x + 6y = 8 \qquad (2)$

$$
\begin{array}{rcll}
2x \ - \ 6y &=& -8 & (3) \quad -2 \times \text{Eq.}(1) \\
-2x \ + \ 6y &=& 8 & (2) \\
\hline
0 &=& 0 & \text{Add } (3) \text{ and } (2).
\end{array}
$$

Since $0 = 0$ is a *true* statement, the equations are equivalent. This result indicates that every solution of one equation is also a solution of the other; there are an *infinite number of solutions*. The solution set is $\{(x, y) \mid x - 3y = -4\}$.

33. $5x - 2y = 3 \qquad (1)$
$\quad\ \ 10x - 4y = 5 \qquad (2)$

$$
\begin{array}{rcll}
-10x \ + \ 4y &=& -6 & (3) \quad -2 \times \text{Eq.}(1) \\
10x \ - \ 4y &=& 5 & (2) \\
\hline
0 &=& -1 & \text{Add } (3) \text{ and } (2).
\end{array}
$$

Since $0 = -1$ is a *false* statement, there are no solutions of the system and the solution set is $\emptyset$.

35. $6x - 2y = -22 \quad (1)$
$\quad\ \ -3x + 4y = 17 \qquad (2)$

$$
\begin{array}{rcll}
6x \ - \ 2y &=& -22 & (1) \\
-6x \ + \ 8y &=& 34 & (3) \quad 2 \times \text{Eq.}(2) \\
\hline
6y &=& 12 & \text{Add } (1) \text{ and } (3). \\
y &=& 2 &
\end{array}
$$

Substitute 2 for y in equation (1).

$$
\begin{aligned}
6x - 2y &= -22 \\
6x - 2(2) &= -22 \\
6x - 4 &= -22 \\
6x &= -18 \\
x &= -3
\end{aligned}
$$

The solution set is $\{(-3, 2)\}$.

37. $4x = 3y - 2$
$\quad\ \ 5x + 3 = 2y$

Rewrite in standard form.

$$
\begin{array}{rll}
-4x + 3y = 2 & (1) \\
5x - 2y = -3 & (2)
\end{array}
$$

$$
\begin{array}{rcll}
-8x \ + \ 6y &=& 4 & (3)\ \ 2 \times \text{Eq.}(1) \\
15x \ - \ 6y &=& -9 & (4)\ \ 3 \times \text{Eq.}(2) \\
\hline
7x &=& -5 & \text{Add } (3) \text{ and } (4). \\
x &=& -\frac{5}{7} &
\end{array}
$$

Rather than substitute $-\frac{5}{7}$ for x in (1) or (2), we will eliminate x by multiplying equation (1) by 5, equation (2) by 4, and adding the results.

$$\begin{array}{rrrll} -20x & + & 15y & = & 10 \quad (5) \quad 5 \times \text{Eq.}(1) \\ 20x & - & 8y & = & -12 \quad (6) \quad 4 \times \text{Eq.}(2) \\ \hline & & 7y & = & -2 \quad \text{Add (5) and (6).} \\ & & y & = & -\frac{2}{7} \end{array}$$

Solving the system in this fashion reduces the chance of making an arithmetic error.

The solution set is $\left\{\left(-\frac{5}{7}, -\frac{2}{7}\right)\right\}$.

When you get a solution that has non-integer components, it is sometimes more difficult to check the problem than it was to solve it. A graphing calculator can be very helpful in this case. Just store the values for x and y in their respective memory locations, and then type the expressions as shown in the following screen. The results 2 and -3 (the right sides of the *equations* (1) and (2)) indicate that we have found the correct solution.

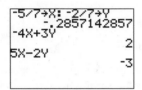

39. $\quad 24x + 12y = -7 \qquad (1)$
$\quad 16x - 18y = 17 \qquad (2)$

$$\begin{array}{rrrll} 48x & + & 24y & = & -14 \qquad (3) \quad 2 \times \text{Eq.}(1) \\ -48x & + & 54y & = & -51 \qquad (4) \quad -3 \times \text{Eq.}(2) \\ \hline & & 78y & = & -65 \qquad \text{Add (3) and (4).} \\ & & y & = & \frac{-65}{78} = -\frac{5}{6} \\ 72x & + & 36y & = & -21 \qquad (5) \quad 3 \times \text{Eq.}(1) \\ 32x & - & 36y & = & 34 \qquad (6) \quad 2 \times \text{Eq.}(2) \\ \hline 104x & & & = & 13 \qquad \text{Add (5) and (6).} \\ & & x & = & \frac{13}{104} = \frac{1}{8} \end{array}$$

The solution set is $\left\{\left(\frac{1}{8}, -\frac{5}{6}\right)\right\}$.

41. $\quad 7x - 4y = 0$
$\quad 3x = 2y$

Rewrite in standard form.

$$\begin{array}{rl} 7x - 4y = 0 & (1) \\ 3x - 2y = 0 & (2) \end{array}$$

$$\begin{array}{rrrll} 7x & - & 4y & = & 0 \qquad (1) \\ -6x & + & 4y & = & 0 \qquad (3) \quad -2 \times \text{Eq.}(2) \\ \hline x & & & = & 0 \qquad \text{Add (1) and (3).} \end{array}$$

Substitute 0 for x in equation (1).

$$\begin{array}{rl} 7x - 4y & = 0 \\ 7(0) - 4y & = 0 \\ -4y & = 0 \\ y & = 0 \end{array}$$

The solution set is $\{(0, 0)\}$.

43. $\quad y = ax + b$
$\quad 1339 = a(1996) + b \quad \textit{Let } x = 1996, y = 1339.$
$\quad 1339 = 1996a + b$

44. As in Exercise 43,

$$1536 = 2004a + b.$$

45. $\quad 1996a + b = 1339 \qquad (1)$
$\quad 2004a + b = 1536 \qquad (2)$

Multiply equation (1) by -1 and add the result to equation (2),

$$\begin{array}{rrrll} -1996a & - & b & = & -1339 \\ 2004a & + & b & = & 1536 \\ \hline 8a & & & = & 197 \\ & & a & = & 24.625 \end{array}$$

Substitute 24.625 for a in equation (1).

$$\begin{array}{rl} 1996(24.625) + b & = 1339 \\ 49{,}151.5 + b & = 1339 \\ b & = -47{,}812.5 \end{array}$$

The solution set is $\{(24.625, -47{,}812.5)\}$.

46. An equation of the segment PQ is

$$y = 24.625x - 47{,}812.5$$

for $1996 \le x \le 2004$.

47. $\quad y = 24.625x - 47{,}812.5$
$\quad y = 24.625(2002) - 47{,}812.5 \qquad \textit{Let } x = 2002.$
$\quad\quad = 49{,}299.25 - 47{,}812.5$
$\quad\quad = 1486.75 \approx 1486.8 \text{ (million)}$

This is quite a bit less than the actual figure.

48. Since the data do not lie in a perfectly straight line, the quantity obtained from an equation determined in this way will probably be "off" a bit. We cannot put too much faith in models such as this one, because not all sets of data points are linear in nature.

49. $\quad ax + by = c \qquad (1)$
$\quad ax - 2by = c \qquad (2)$

To eliminate y, multiply (1) by 2 and add the result to equation (2).

$$\begin{array}{rrrll} 2ax & + & 2by & = & 2c \qquad (3) \quad 2 \times (1) \\ ax & - & 2by & = & c \qquad (2) \\ \hline 3ax & & & = & 3c \\ & & x & = & \dfrac{3c}{3a} = \dfrac{c}{a} \end{array}$$

Substitute $\dfrac{c}{a}$ for x in (1).

continued

$$a\left(\frac{c}{a}\right) + by = c$$
$$c + by = c$$
$$by = 0$$
$$y = 0$$

The solution $\left(\frac{c}{a}, 0\right)$ checks.

The solution set is $\left\{\left(\frac{c}{a}, 0\right)\right\}$.

51. $2ax - y = 3$ (1)
    $y = 5ax$ (2)

Substitute $5ax$ for y in (1).

$$2ax - 5ax = 3$$
$$-3ax = 3$$
$$x = \frac{3}{-3a} = -\frac{1}{a}$$

Substitute $-\frac{1}{a}$ for x in (2).

$$y = 5a\left(-\frac{1}{a}\right) = -5$$

The solution $\left(-\frac{1}{a}, -5\right)$ checks.

The solution set is $\left\{\left(-\frac{1}{a}, -5\right)\right\}$.

53. $4(2x - 3y + z) = 4(5)$
    $8x - 12y + 4z = 20$

55. $x + 2y + 3z = 9$
    $1 + 2(-2) + 3z = 9$ *Let x = 1, y = −2.*
    $1 - 4 + 3z = 9$
    $3z = 12$
    $z = 4$

57. Multiplying the first equation by -3 will give us $-3x$, which when added to $3x$ in the second equation, gives us 0.

Summary Exercises on Solving Systems of Linear Equations

1. **(a)** $3x + 5y = 69$
        $y = 4x$

Use substitution since the second equation is solved for y.

(b) $3x + y = -7$
        $x - y = -5$

Use elimination since the coefficients of the y-terms are opposites.

(c) $3x - 2y = 0$
        $9x + 8y = 7$

Use elimination since the equations are in standard form with no coefficients of 1 or -1. Solving by substitution would involve fractions.

3. $4x - 3y = -8$ (1)
    $x + 3y = 13$ (2)

(a) Solve the system by the elimination method.

$$
\begin{array}{rcrl}
4x & - & 3y & = -8 \quad (1) \\
x & + & 3y & = 13 \quad (2) \\
\hline
5x & & & = 5 \quad \text{Add (1) and (2).} \\
& & x & = 1
\end{array}
$$

To find y, let $x = 1$ in equation (2).

$$x + 3y = 13$$
$$1 + 3y = 13$$
$$3y = 12$$
$$y = 4$$

The solution set is $\{(1, 4)\}$.

(b) To solve this system by the substitution method, begin by solving equation (2) for x.

$$x + 3y = 13$$
$$x = -3y + 13$$

Substitute $-3y + 13$ for x in equation (1).

$$4(-3y + 13) - 3y = -8$$
$$-12y + 52 - 3y = -8$$
$$-15y = -60$$
$$y = 4$$

To find x, let $y = 4$ in equation (2).

$$x + 3y = 13$$
$$x + 3(4) = 13$$
$$x + 12 = 13$$
$$x = 1$$

The solution set is $\{(1, 4)\}$.

(c) For this particular system, the elimination method is preferable because both equations are already written in the form $Ax + By = C$, and the equations can be added without multiplying either by a constant. Comparing solutions by the two methods, we see that the elimination method requires fewer steps than the substitution method for this system.

5. $3x + 5y = 69$ (1)
$y = 4x$ (2)

Equation (2) is already solved for y, so we'll use the substitution method. Substitute $4x$ for y in equation (1) and solve the resulting equation for x.

$$3x + 5y = 69$$
$$3x + 5(4x) = 69 \qquad \textit{Let y = 4x.}$$
$$3x + 20x = 69$$
$$23x = 69$$
$$x = \tfrac{69}{23} = 3$$

To find the y-value of the solution, substitute 3 for x in equation (2).

$$y = 4x$$
$$y = 4(3) \quad \textit{Let x = 3.}$$
$$= 12$$

The solution set is $\{(3, 12)\}$.

7. $3x - 2y = 0$ (1)
$9x + 8y = 7$ (2)

$$
\begin{array}{rcll}
12x & - & 8y & = & 0 & (3) \;\; 4 \times \text{Eq. (1)} \\
9x & + & 8y & = & 7 & (2) \\
\hline
21x & & & = & 7 & \textit{Add (3) and (2).} \\
& & x & = & \tfrac{7}{21} = \tfrac{1}{3} &
\end{array}
$$

$$
\begin{array}{rcll}
-9x & + & 6y & = & 0 & (4) \;\; -3 \times \text{Eq.(1)} \\
9x & + & 8y & = & 7 & (2) \\
\hline
& & 14y & = & 7 & \textit{Add (4) and (2).} \\
& & y & = & \tfrac{7}{14} = \tfrac{1}{2} &
\end{array}
$$

The solution set is $\left\{ \left(\tfrac{1}{3}, \tfrac{1}{2} \right) \right\}$.

9. $6x + 7y = 4$ (1)
$5x + 8y = -1$ (2)

$$
\begin{array}{rcll}
-30x & - & 35y & = & -20 & (3) \;\; -5 \times \text{Eq.(1)} \\
30x & + & 48y & = & -6 & (4) \;\;\; 6 \times \text{Eq.(2)} \\
\hline
& & 13y & = & -26 & \textit{Add (3) and (4).} \\
& & y & = & -2 &
\end{array}
$$

Substitute -2 for y in equation (1).

$$6x + 7y = 4$$
$$6x + 7(-2) = 4 \quad \textit{Let y = -2.}$$
$$6x - 14 = 4$$
$$6x = 18$$
$$x = 3$$

The solution set is $\{(3, -2)\}$.

11. $4x - 6y = 10$ (1)
$-10x + 15y = -25$ (2)

Divide equation (1) by 2 and equation (2) by 5.

$$
\begin{array}{rcll}
2x & - & 3y & = & 5 & (3) \\
-2x & + & 3y & = & -5 & (4) \\
\hline
& & 0 & = & 0 & \textit{Add (3) and (4).}
\end{array}
$$

Since $0 = 0$ is a *true* statement, there are an *infinite number of solutions*. The solution set is $\{(x, y) \mid 2x - 3y = 5\}$.

13. $5x = 7 + 2y$
$5y = 5 - 3x$

Rewrite in standard form.

$5x - 2y = 7$ (1)
$3x + 5y = 5$ (2)

$$
\begin{array}{rcll}
25x & - & 10y & = & 35 & (3) \;\; 5 \times \text{Eq.(1)} \\
6x & + & 10y & = & 10 & (4) \;\; 2 \times \text{Eq.(2)} \\
\hline
31x & & & = & 45 & \textit{Add (3) and (4).} \\
& & x & = & \tfrac{45}{31} &
\end{array}
$$

$$
\begin{array}{rcll}
-15x & + & 6y & = & -21 & (5) \;\; -3 \times \text{Eq.(1)} \\
15x & + & 25y & = & 25 & (6) \;\;\; 5 \times \text{Eq.(2)} \\
\hline
& & 31y & = & 4 & \textit{Add (5) and (6).} \\
& & y & = & \tfrac{4}{31} &
\end{array}
$$

The solution set is $\left\{ \left(\tfrac{45}{31}, \tfrac{4}{31} \right) \right\}$.

15. $2x - 3y = 7$ (1)
$-4x + 6y = 14$ (2)

$$
\begin{array}{rcll}
4x & - & 6y & = & 14 & (3) \;\; 2 \times \text{Eq.(1)} \\
-4x & + & 6y & = & 14 & (2) \\
\hline
& & 0 & = & 28 & \textit{Add (3) and (2).}
\end{array}
$$

Since $0 = 28$ is *false*, the solution set is $\emptyset$.

17. $2x + 5y = 4$ (1)
$x + y = -1$ (2)

Solve equation (2) for x.

$$x = -y - 1 \quad (3)$$

Substitute $-y - 1$ for x in equation (1).

$$2(-y - 1) + 5y = 4 \quad \textit{Let x = -y - 1.}$$
$$-2y - 2 + 5y = 4$$
$$3y - 2 = 4$$
$$3y = 6$$
$$y = 2$$

Substitute 2 for y in equation (3).

$$x = -(2) - 1 \quad \textit{Let y = 2.}$$
$$= -3$$

The solution set is $\{(-3, 2)\}$.

19. $\frac{1}{3}x + \frac{1}{2}y = 4$ (1)

$-\frac{1}{2}x + \frac{1}{4}y = -2$ (2)

We could clear all fractions first, but we'll simply use elimination for this system.

To eliminate y, multiply equation (2) by -2 and add the resulting equation to equation (1).

$$\begin{array}{rcll} \frac{1}{3}x & + & \frac{1}{2}y & = & 4 & (1) \\ x & - & \frac{1}{2}y & = & 4 & (3) \;\; -2 \times \text{Eq.}(2) \\ \hline \frac{4}{3}x & & & = & 8 & \textit{Add} (1) \textit{ and } (3). \\ & & x & = & 6 & \textit{Multiply by } \frac{3}{4}. \end{array}$$

Substitute 6 for x in equation (1).

$$\frac{1}{3}x + \frac{1}{2}y = 4$$
$$\frac{1}{3}(6) + \frac{1}{2}y = 4$$
$$2 + \frac{1}{2}y = 4$$
$$\frac{1}{2}y = 2$$
$$y = 4$$

The solution set is $\{(6, 4)\}$.

21. $\frac{1}{5}x + \frac{2}{3}y = -\frac{8}{5}$ (1)

$3x - y = 9$ (2)

Multiply each side of equation (1) by 15 to clear fractions.

$$15\left(\frac{1}{5}x + \frac{2}{3}y\right) = 15\left(-\frac{8}{5}\right)$$
$$15\left(\frac{1}{5}x\right) + 15\left(\frac{2}{3}y\right) = -24$$
$$3x + 10y = -24 \quad (3)$$

Now use the elimination method with equations (2) and (3).

$$\begin{array}{rcll} 3x & + & 10y & = & -24 & (3) \\ -3x & + & y & = & -9 & (4) \;\; -1 \times \text{Eq.}(2) \\ \hline & & 11y & = & -33 & \textit{Add} (3) \textit{ and } (4). \\ & & y & = & -3 \end{array}$$

Substitute -3 for y in equation (2).

$$3x - (-3) = 9 \;\; \textit{Let } y = -3.$$
$$3x + 3 = 9$$
$$3x = 6$$
$$x = 2$$

The solution set is $\{(2, -3)\}$.

23. $\frac{x}{2} - \frac{y}{3} = 9$ (1)

$\frac{x}{5} - \frac{y}{4} = 5$ (2)

To clear fractions, multiply each side of equation (1) by the LCD, 6.

$$6\left(\frac{x}{2} - \frac{y}{3}\right) = 6(9)$$
$$3x - 2y = 54 \quad (3)$$

To clear fractions, multiply each side of equation (2) by the LCD, 20.

$$20\left(\frac{x}{5} - \frac{y}{4}\right) = 20(5)$$
$$4x - 5y = 100 \quad (4)$$

We now have the simplified system

$$\begin{array}{rl} 3x - 2y = 54 & (3) \\ 4x - 5y = 100. & (4) \end{array}$$

To solve this system by the elimination method, multiply equation (3) by 5 and equation (4) by -2; then add the results.

$$\begin{array}{rcccr} 15x & - & 10y & = & 270 \\ -8x & + & 10y & = & -200 \\ \hline 7x & & & = & 70 \\ & & x & = & 10 \end{array}$$

To find y, let $x = 10$ in equation (3).

$$3x - 2y = 54$$
$$3(10) - 2y = 54$$
$$30 - 2y = 54$$
$$-2y = 24$$
$$y = -12$$

The solution set is $\{(10, -12)\}$.

25. $\frac{x}{5} + 2y = \frac{16}{5}$ (1)

$\frac{3x}{5} + \frac{y}{2} = -\frac{7}{5}$ (2)

Multiply each side of equation (1) by 5.

$$5\left(\frac{x}{5} + 2y\right) = 5\left(\frac{16}{5}\right)$$
$$x + 10y = 16 \quad (3)$$

Multiply each side of equation (2) by 10.

$$10\left(\frac{3x}{5} + \frac{y}{2}\right) = 10\left(-\frac{7}{5}\right)$$
$$6x + 5y = -14 \quad (4)$$

We now have the simplified system

$$\begin{array}{rl} x + 10y = 16 & (3) \\ 6x + 5y = -14 & (4) \end{array}$$

To solve this system by the substitution method, solve equation (3) for x.

$$x = 16 - 10y \quad (5)$$

Substitute $16 - 10y$ for x in equation (4).

$$6x + 5y = -14$$
$$6(16 - 10y) + 5y = -14$$
$$96 - 60y + 5y = -14$$
$$-55y = -110$$
$$y = 2$$

To find x, let $y = 2$ in equation (5).

$$x = 16 - 10y$$
$$= 16 - 10(2) = -4$$

The solution set is $\{(-4, 2)\}$.

8.4 Systems of Linear Equations in Three Variables

1. Substitute 1 for x, 2 for y, and 3 for z in $3x + 2y - z$, which is the left side of each equation.

$$3(1) + 2(2) - (3) = 3 + 4 - 3$$
$$= 7 - 3$$
$$= 4$$

Choice **B** is correct since its right side is 4.

3.
$$
\begin{aligned}
2x - 5y + 3z &= -1 \quad (1) \\
x + 4y - 2z &= 9 \quad (2) \\
x - 2y - 4z &= -5 \quad (3)
\end{aligned}
$$

Eliminate x by adding equation (1) to -2 times equation (2).

$$
\begin{array}{llll}
2x - & 5y + 3z = & -1 & (1) \\
-2x - & 8y + 4z = & -18 & -2 \times (2) \\
\hline
& -13y + 7z = & -19 & (4)
\end{array}
$$

Now eliminate x by adding equation (2) to -1 times equation (3).

$$
\begin{array}{lll}
x + 4y - 2z = & 9 & (2) \\
-x + 2y + 4z = & 5 & -1 \times (3) \\
\hline
6y + 2z = & 14 & (5)
\end{array}
$$

Use equations (4) and (5) to eliminate z. Multiply equation (4) by -2 and add the result to 7 times equation (5).

$$
\begin{array}{lll}
26y - 14z = & 38 & -2 \times (4) \\
42y + 14z = & 98 & 7 \times (5) \\
\hline
68y = & 136 & \\
y = & 2 &
\end{array}
$$

Substitute 2 for y in equation (5) to find z.

$$
\begin{aligned}
6y + 2z &= 14 \quad (5) \\
6(2) + 2z &= 14 \\
12 + 2z &= 14 \\
2z &= 2 \\
z &= 1
\end{aligned}
$$

Substitute 2 for y and 1 for z in equation (3) to find x.

$$
\begin{aligned}
x - 2y - 4z &= -5 \quad (3) \\
x - 2(2) - 4(1) &= -5 \\
x - 4 - 4 &= -5 \\
x - 8 &= -5 \\
x &= 3
\end{aligned}
$$

The solution $(3, 2, 1)$ checks in all three of the original equations.
The solution set is $\{(3, 2, 1)\}$.

5.
$$
\begin{aligned}
3x + 2y + z &= 8 \quad (1) \\
2x - 3y + 2z &= -16 \quad (2) \\
x + 4y - z &= 20 \quad (3)
\end{aligned}
$$

Eliminate z by adding equations (1) and (3).

$$
\begin{array}{lll}
3x + 2y + z = & 8 & (1) \\
x + 4y - z = & 20 & (3) \\
\hline
4x + 6y = & 28 & (4)
\end{array}
$$

To get another equation without z, multiply equation (3) by 2 and add the result to equation (2).

$$
\begin{array}{lll}
2x + 8y - 2z = & 40 & 2 \times (3) \\
2x - 3y + 2z = & -16 & (2) \\
\hline
4x + 5y = & 24 & (5)
\end{array}
$$

Use equations (4) and (5) to eliminate x. Multiply equation (4) by -1 and add the result to equation (5).

$$
\begin{array}{lll}
-4x - 6y = & -28 & -1 \times (4) \\
4x + 5y = & 24 & (5) \\
\hline
-y = & -4 & \\
y = & 4 &
\end{array}
$$

Substitute 4 for y in equation (5) to find x.

$$
\begin{aligned}
4x + 5y &= 24 \quad (5) \\
4x + 5(4) &= 24 \\
4x + 20 &= 24 \\
4x &= 4 \\
x &= 1
\end{aligned}
$$

Substitute 1 for x and 4 for y in equation (3) to find z.

$$
\begin{aligned}
x + 4y - z &= 20 \quad (3) \\
1 + 4(4) - z &= 20 \\
1 + 16 - z &= 20 \\
17 - z &= 20 \\
-z &= 3 \\
z &= -3
\end{aligned}
$$

The solution $(1, 4, -3)$ checks in all three of the original equations.
The solution set is $\{(1, 4, -3)\}$.

7.
$$2x + 5y + 2z = 0 \quad (1)$$
$$4x - 7y - 3z = 1 \quad (2)$$
$$3x - 8y - 2z = -6 \quad (3)$$

Add equations (1) and (3) to eliminate z.

$$\begin{array}{r} 2x + 5y + 2z = 0 \quad (1) \\ 3x - 8y - 2z = -6 \quad (3) \\ \hline 5x - 3y = -6 \quad (4) \end{array}$$

Multiply equation (1) by 3 and equation (2) by 2. Then add the results to eliminate z again.

$$\begin{array}{r} 6x + 15y + 6z = 0 \quad 3 \times (1) \\ 8x - 14y - 6z = 2 \quad 2 \times (2) \\ \hline 14x + y = 2 \quad (5) \end{array}$$

Solve the system

$$5x - 3y = -6 \quad (4)$$
$$14x + y = 2. \quad (5)$$

Multiply equation (5) by 3 then add this result to (4).

$$\begin{array}{r} 5x - 3y = -6 \quad (4) \\ 42x + 3y = 6 \quad 3 \times (5) \\ \hline 47x = 0 \\ x = 0 \end{array}$$

To find y, substitute $x = 0$ into equation (4).

$$5x - 3y = -6 \quad (4)$$
$$5(0) - 3y = -6$$
$$y = 2$$

To find z, substitute $x = 0$ and $y = 2$ in equation (1).

$$2x + 5y + 2z = 0 \quad (1)$$
$$2(0) + 5(2) + 2z = 0$$
$$10 + 2z = 0$$
$$2z = -10$$
$$z = -5$$

The solution set is $\{(0, 2, -5)\}$.

9.
$$x + 2y + z = 4 \quad (1)$$
$$2x + y - z = -1 \quad (2)$$
$$x - y - z = -2 \quad (3)$$

Add (1) and (2) to eliminate z.

$$\begin{array}{r} x + 2y + z = 4 \quad (1) \\ 2x + y - z = -1 \quad (2) \\ \hline 3x + 3y = 3 \quad (4) \end{array}$$

Add (1) and (3) to eliminate z.

$$\begin{array}{r} x + 2y + z = 4 \quad (1) \\ x - y - z = -2 \quad (3) \\ \hline 2x + y = 2 \quad (5) \end{array}$$

Multiply equation (5) by -3.

$$\begin{array}{r} -6x - 3y = -6 \quad (6) \;\; -3 \times (5) \\ 3x + 3y = 3 \quad (4) \\ \hline -3x = -3 \\ x = 1 \end{array}$$

Substitute 1 for x in equation (5) and solve for y.

$$2(1) + y = 2$$
$$y = 0$$

Substitute 1 for x and 0 for y in (1).

$$(1) + 2(0) + z = 4$$
$$z = 3$$

The solution set is $\{(1, 0, 3)\}$.

11.
$$\tfrac{1}{3}x + \tfrac{1}{6}y - \tfrac{2}{3}z = -1 \quad (1)$$
$$-\tfrac{3}{4}x - \tfrac{1}{3}y - \tfrac{1}{4}z = 3 \quad (2)$$
$$\tfrac{1}{2}x + \tfrac{3}{2}y + \tfrac{3}{4}z = 21 \quad (3)$$

Eliminate all fractions.

$$2x + y - 4z = -6 \quad (4) \quad 6 \times (1)$$
$$-9x - 4y - 3z = 36 \quad (5) \quad 12 \times (2)$$
$$2x + 6y + 3z = 84 \quad (6) \quad 4 \times (3)$$

Multiply (4) by 4.

$$\begin{array}{r} 8x + 4y - 16z = -24 \\ -9x - 4y - 3z = 36 \quad (5) \\ \hline -x - 19z = 12 \quad (7) \end{array}$$

Multiply (4) by -6.

$$\begin{array}{r} -12x - 6y + 24z = 36 \\ 2x + 6y + 3z = 84 \quad (6) \\ \hline -10x + 27z = 120 \quad (8) \end{array}$$

Multiply (7) by -10.

$$\begin{array}{r} 10x + 190z = -120 \\ -10x + 27z = 120 \quad (8) \\ \hline 217z = 0 \\ z = 0 \end{array}$$

Substituting 0 for z in (7) gives us $x = -12$. Substitute -12 for x and 0 for z in (4).

$$2(-12) + y - 4(0) = -6$$
$$-24 + y = -6$$
$$y = 18$$

The solution set is $\{(-12, 18, 0)\}$.

13.
$$-x + 2y + 6z = 2 \quad (1)$$
$$3x + 2y + 6z = 6 \quad (2)$$
$$x + 4y - 3z = 1 \quad (3)$$

Eliminate y and z by adding equation (1) to -1 times equation (2).

$$-x + 2y + 6z = 2 \quad (1)$$
$$\underline{-3x - 2y - 6z = -6} \qquad -1 \times (2)$$
$$-4x \qquad\qquad\quad = -4$$
$$x = 1$$

Eliminate z by adding equation (2) to 2 times equation (3).

$$3x + 2y + 6z = 6 \quad (2)$$
$$\underline{2x + 8y - 6z = 2} \qquad 2 \times (3)$$
$$5x + 10y \qquad\;\; = 8 \quad (4)$$

Substitute 1 for x in equation (4).

$$5x + 10y = 8 \qquad (4)$$
$$5(1) + 10y = 8$$
$$10y = 3$$
$$y = \tfrac{3}{10}$$

Substitute 1 for x and $\tfrac{3}{10}$ for y in equation (1).

$$-x + 2y + 6z = 2 \qquad (1)$$
$$-1 + 2\left(\tfrac{3}{10}\right) + 6z = 2$$
$$\tfrac{3}{5} + 6z = 3$$
$$6z = \tfrac{12}{5}$$
$$z = \tfrac{2}{5}$$

The solution set is $\left\{\left(1, \tfrac{3}{10}, \tfrac{2}{5}\right)\right\}$.

15.
$$x + y - z = -2 \qquad (1)$$
$$2x - y + z = -5 \qquad (2)$$
$$-x + 2y - 3z = -4 \qquad (3)$$

Eliminate y and z by adding equations (1) and (2).

$$x + y - z = -2 \quad (1)$$
$$\underline{2x - y + z = -5} \quad (2)$$
$$3x \qquad\qquad = -7$$
$$x = -\tfrac{7}{3}$$

To get another equation without y, multiply equation (2) by 2 and add the result to equation (3).

$$4x - 2y + 2z = -10 \qquad 2 \times (2)$$
$$\underline{-x + 2y - 3z = -4} \quad (3)$$
$$3x \qquad\; - z = -14 \quad (4)$$

Substitute $-\tfrac{7}{3}$ for x in equation (4) to find z.

$$3x - z = -14 \quad (4)$$
$$3\left(-\tfrac{7}{3}\right) - z = -14$$
$$-7 - z = -14$$
$$-z = -7$$
$$z = 7$$

Substitute $-\tfrac{7}{3}$ for x and 7 for z in equation (1) to find y.

$$x + y - z = -2 \quad (1)$$
$$-\tfrac{7}{3} + y - 7 = -2$$
$$-7 + 3y - 21 = -6 \quad \textit{Multiply by 3.}$$
$$3y - 28 = -6$$
$$3y = 22$$
$$y = \tfrac{22}{3}$$

The solution set is $\left\{\left(-\tfrac{7}{3}, \tfrac{22}{3}, 7\right)\right\}$.

A calculator check reduces the probability of making any arithmetic errors and is highly recommended. The following screen shows the substitution of the solution for x, y, and z along with the left sides of the three original equations. The evaluation of the three expressions, -2, -5, and -4 (the right sides of the three equations), indicates that we have found the correct solution.

```
-7/3→X:22/3→Y:7→
Z:X+Y-Z
                    -2
2X-Y+Z
                    -5
-X+2Y-3Z
                    -4
```

17.
$$2x - 3y + 2z = -1 \qquad (1)$$
$$x + 2y + z = 17 \qquad (2)$$
$$2y - z = 7 \qquad (3)$$

Multiply equation (2) by -2, and add the result to equation (1).

$$2x - 3y + 2z = -1 \quad (1)$$
$$\underline{-2x - 4y - 2z = -34} \qquad -2 \times (2)$$
$$-7y \qquad\qquad = -35$$
$$y = 5$$

To find z, substitute 5 for y in equation (3).

$$2y - z = 7 \qquad (3)$$
$$2(5) - z = 7$$
$$10 - z = 7$$
$$-z = -3$$
$$z = 3$$

To find x, substitute $y = 5$ and $z = 3$ into equation (1).

$$2x - 3y + 2z = -1 \qquad (1)$$
$$2x - 3(5) + 2(3) = -1$$
$$2x - 9 = -1$$
$$2x = 8$$
$$x = 4$$

The solution set is $\{(4, 5, 3)\}$.

19.
$$4x + 2y - 3z = 6 \quad (1)$$
$$x - 4y + z = -4 \quad (2)$$
$$-x \quad + 2z = 2 \quad (3)$$

Equation (3) is missing y. Eliminate y again by multiplying equation (1) by 2 and adding the result to equation (2).

$$8x + 4y - 6z = 12 \quad 2 \times (1)$$
$$\underline{x - 4y + z = -4} \quad (2)$$
$$9x \quad - 5z = 8 \quad (4)$$

Use equations (3) and (4) to eliminate x. Multiply equation (3) by 9 and add the result to equation (4).

$$-9x + 18z = 18 \quad 9 \times (3)$$
$$\underline{9x - 5z = 8} \quad (4)$$
$$13z = 26$$
$$z = 2$$

Substitute 2 for z in equation (3) to find x.

$$-x + 2z = 2 \quad (3)$$
$$-x + 2(2) = 2$$
$$-x + 4 = 2$$
$$-x = -2$$
$$x = 2$$

Substitute 2 for x and 2 for z in equation (2) to find y.

$$x - 4y + z = -4 \quad (2)$$
$$2 - 4y + 2 = -4$$
$$-4y + 4 = -4$$
$$-4y = -8$$
$$y = 2$$

The solution set is $\{(2, 2, 2)\}$.

21.
$$2x + y \quad = 6 \quad (1)$$
$$3y - 2z = -4 \quad (2)$$
$$3x \quad - 5z = -7 \quad (3)$$

To eliminate y, multiply equation (1) by -3 and add the result to equation (2).

$$-6x - 3y \quad = -18 \quad -3 \times (1)$$
$$\underline{3y - 2z = -4} \quad (2)$$
$$-6x \quad - 2z = -22 \quad (4)$$

Since equation (3) does not have a y-term, we can multiply equation (3) by 2 and add the result to equation (4) to eliminate x and solve for z.

$$6x - 10z = -14 \quad 2 \times (3)$$
$$\underline{-6x - 2z = -22} \quad (4)$$
$$-12z = -36$$
$$z = 3$$

To find x, substitute 3 for z into equation (3).

$$3x - 5z = -7 \quad (3)$$
$$3x - 5(3) = -7$$
$$3x = 8$$
$$x = \frac{8}{3}$$

To find y, substitute 3 for z into equation (2).

$$3y - 2z = -4 \quad (2)$$
$$3y - 2(3) = -4$$
$$3y = 2$$
$$y = \frac{2}{3}$$

The solution set is $\left\{\left(\frac{8}{3}, \frac{2}{3}, 3\right)\right\}$.

23.
$$-5x + 2y + z = 5 \quad (1)$$
$$-3x - 2y - z = 3 \quad (2)$$
$$-x + 6y \quad = 1 \quad (3)$$

Add (1) and (2) to eliminate y and z.

$$-8x = 8$$
$$x = -1$$

Substitute -1 for x in equation (3).

$$-x + 6y = 1 \quad (3)$$
$$-(-1) + 6y = 1$$
$$6y = 0$$
$$y = 0$$

Substitute -1 for x and 0 for y in (1).

$$-5x + 2y + z = 5 \quad (1)$$
$$-5(-1) + 2(0) + z = 5$$
$$5 + z = 5$$
$$z = 0$$

The solution set is $\{(-1, 0, 0)\}$.

25.
$$4x \quad - z = -6 \quad (1)$$
$$\frac{3}{5}y + \frac{1}{2}z = 0 \quad (2)$$
$$\frac{1}{3}x \quad + \frac{2}{3}z = -5 \quad (3)$$

Eliminate fractions first.

$$6y + 5z = 0 \quad (4) \quad 10 \times (2)$$
$$x + 2z = -15 \quad (5) \quad 3 \times (3)$$

Eliminate z by adding (5) to 2 times (1).

$$8x - 2z = -12 \quad 2 \times (1)$$
$$\underline{x + 2z = -15} \quad (5)$$
$$9x \quad = -27$$
$$x = -3$$

Substitute -3 for x in (5).

$$x + 2z = -15 \quad (5)$$
$$-3 + 2z = -15$$
$$2z = -12$$
$$z = -6$$

Substitute -6 for z in (4).

$$6y + 5z = 0 \quad (4)$$
$$6y + 5(-6) = 0$$
$$6y = 30$$
$$y = 5$$

The solution set is $\{(-3, 5, -6)\}$.

27.
$$2x + 2y - 6z = 5 \quad (1)$$
$$-3x + y - z = -2 \quad (2)$$
$$-x - y + 3z = 4 \quad (3)$$

Multiply equation (3) by 2 and add the result to equation (1).

$$2x + 2y - 6z = 5 \quad (1)$$
$$\underline{-2x - 2y + 6z = 8} \qquad 2 \times (3)$$
$$0 = 13 \quad \textit{False}$$

The solution set is $\emptyset$; inconsistent system.

29.
$$-5x + 5y - 20z = -40 \quad (1)$$
$$x - y + 4z = 8 \quad (2)$$
$$3x - 3y + 12z = 24 \quad (3)$$

Dividing equation (1) by -5 gives equation (2). Dividing equation (3) by 3 also gives equation (2). The resulting equations are the same, so the three equations are dependent.
The solution set is $\{(x, y, z) \mid x - y + 4z = 8\}$.

31.
$$x + 5y - 2z = -1 \quad (1)$$
$$-2x + 8y + z = -4 \quad (2)$$
$$3x - y + 5z = 19 \quad (3)$$

Eliminate x by adding (2) to 2 times (1).

$$2x + 10y - 4z = -2 \qquad 2 \times (1)$$
$$\underline{-2x + 8y + z = -4} \quad (2)$$
$$18y - 3z = -6 \quad (4)$$

Eliminate x by adding (3) to -3 times (1).

$$-3x - 15y + 6z = 3 \qquad -3 \times (1)$$
$$\underline{3x - y + 5z = 19} \quad (3)$$
$$-16y + 11z = 22 \quad (5)$$

To eliminate z from equations (4) and (5), we could first divide equation (4) by 3 and then multiply the resulting equation by 11, or we could simply multiply equation (4) by $\frac{11}{3}$ and add it to equation (5).

$$66y - 11z = -22 \qquad \frac{11}{3} \times (4)$$
$$\underline{-16y + 11z = 22} \quad (5)$$
$$50y = 0$$
$$y = 0$$

Substitute 0 for y in equation (5).

$$-16y + 11z = 22 \quad (5)$$
$$-16(0) + 11z = 22$$
$$11z = 22$$
$$z = 2$$

Substitute 0 for y and 2 for z in equation (1).

$$x + 5y - 2z = -1 \quad (1)$$
$$x + 5(0) - 2(2) = -1$$
$$x - 4 = -1$$
$$x = 3$$

The solution set is $\{(3, 0, 2)\}$.

33.
$$2x + y - z = 6 \quad (1)$$
$$4x + 2y - 2z = 12 \quad (2)$$
$$-x - \tfrac{1}{2}y + \tfrac{1}{2}z = -3 \quad (3)$$

Multiplying equation (1) by 2 gives equation (2). Multiplying equation (3) by -4 also gives equation (2). The resulting equations are the same, so the three equations are dependent.
The solution set is $\{(x, y, z) \mid 2x + y - z = 6\}$.

35.
$$x + y - 2z = 0 \quad (1)$$
$$3x - y + z = 0 \quad (2)$$
$$4x + 2y - z = 0 \quad (3)$$

Eliminate z by adding equations (2) and (3).

$$3x - y + z = 0 \quad (2)$$
$$\underline{4x + 2y - z = 0} \quad (3)$$
$$7x + y = 0 \quad (4)$$

To get another equation without z, multiply equation (2) by 2 and add the result to equation (1).

$$6x - 2y + 2z = 0 \qquad 2 \times (2)$$
$$\underline{x + y - 2z = 0} \quad (1)$$
$$7x - y = 0 \quad (5)$$

Add equations (4) and (5) to find x.

$$7x + y = 0 \quad (4)$$
$$\underline{7x - y = 0} \quad (5)$$
$$14x = 0$$
$$x = 0$$

Substitute 0 for x in equation (4) to find y.

$$7x + y = 0 \quad (4)$$
$$7(0) + y = 0$$
$$0 + y = 0$$
$$y = 0$$

Substitute 0 for x and 0 for y in equation (1) to find z.

$$x + y - 2z = 0 \quad (1)$$
$$0 + 0 - 2z = 0$$
$$-2z = 0$$
$$z = 0$$

The solution set is $\{(0, 0, 0)\}$.

37.
$$
\begin{aligned}
x - 2y + \tfrac{1}{3}z &= 4 \quad (1) \\
3x - 6y + z &= 12 \quad (2) \\
-6x + 12y - 2z &= -3 \quad (3)
\end{aligned}
$$

The coefficients of z are $\frac{1}{3}$, 1, and -2. We'll multiply the equations by values that will eliminate fractions and make the coefficients of z easy to compare.

$$
\begin{aligned}
-6x + 12y - 2z &= -24 \quad (4) \; -6 \times (1) \\
-6x + 12y - 2z &= -24 \quad (5) \; -2 \times (2) \\
-6x + 12y - 2z &= -3 \quad (3)
\end{aligned}
$$

We can now easily see that (4) and (5) are dependent equations (they have the same graph—in fact, their graph is the same plane). Equation (3) has the same coefficients, but a different constant term, so its graph is a plane parallel to the other plane—that is, there are no points in common. Thus, the system is inconsistent and the solution set is $\varnothing$.

39.
$$
\begin{aligned}
x + y + z - w &= 5 \quad (1) \\
2x + y - z + w &= 3 \quad (2) \\
x - 2y + 3z + w &= 18 \quad (3) \\
-x - y + z + 2w &= 8 \quad (4)
\end{aligned}
$$

Eliminate w. Add equations (1) and (2).

$$
\begin{array}{rl}
x + y + z - w = 5 & (1) \\
\underline{2x + y - z + w = 3} & (2) \\
3x + 2y \qquad\qquad = 8 & (5)
\end{array}
$$

Eliminate w again. Add equations (1) and (3).

$$
\begin{array}{rl}
x + y + z - w = 5 & (1) \\
\underline{x - 2y + 3z + w = 18} & (3) \\
2x - y + 4z \qquad = 23 & (6)
\end{array}
$$

Eliminate w again. Multiply equation (2) by -2. Add the result to equation (4).

$$
\begin{array}{rl}
-4x - 2y + 2z - 2w = -6 & -2 \times (2) \\
\underline{-x - y + z + 2w = 8} & (4) \\
-5x - 3y + 3z \qquad = 2 & (7)
\end{array}
$$

Equations (5), (6), and (7) do not contain a w-term. Since (5) does not have a z-term, we will find another equation without a z-term.

Eliminate z. Multiply equation (6) by 3 and equation (7) by -4. Then add the results.

$$
\begin{array}{rl}
6x - 3y + 12z = 69 & 3 \times (6) \\
\underline{20x + 12y - 12z = -8} & -4 \times (7) \\
26x + 9y \qquad = 61 & (8)
\end{array}
$$

Eliminate z again. Multiply equation (5) by 9 and equation (8) by -2. Then add the results.

$$
\begin{array}{rl}
27x + 18y = 72 & 9 \times (5) \\
\underline{-52x - 18y = -122} & -2 \times (8) \\
-25x \qquad = -50 & \\
x = 2 &
\end{array}
$$

To find y, substitute $x = 2$ into equation (5).

$$
\begin{aligned}
3x + 2y &= 8 \quad (5) \\
3(2) + 2y &= 8 \\
2y &= 2 \\
y &= 1
\end{aligned}
$$

To find z, substitute $x = 2$ and $y = 1$ into equation (6).

$$
\begin{aligned}
2x - y + 4z &= 23 \quad (6) \\
2(2) - 1 + 4z &= 23 \\
4z &= 20 \\
z &= 5
\end{aligned}
$$

To find w, substitute $x = 2$, $y = 1$, and $z = 5$ into equation (1).

$$
\begin{aligned}
x + y + z - w &= 5 \quad (1) \\
2 + 1 + 5 - w &= 5 \\
-w &= -3 \\
w &= 3
\end{aligned}
$$

The solution set is $\{(2, 1, 5, 3)\}$.

41.
$$
\begin{aligned}
3x + y - z + w &= -3 \quad (1) \\
2x + 4y + z - w &= -7 \quad (2) \\
-2x + 3y - 5z + w &= 3 \quad (3) \\
5x + 4y - 5z + 2w &= -7 \quad (4)
\end{aligned}
$$

Eliminate w. Add equations (1) and (2).

$$
\begin{array}{rl}
3x + y - z + w = -3 & (1) \\
\underline{2x + 4y + z - w = -7} & (2) \\
5x + 5y \qquad = -10 & (5)
\end{array}
$$

Eliminate w again. Add equations (2) and (3).

$$
\begin{array}{rl}
2x + 4y + z - w = -7 & (2) \\
\underline{-2x + 3y - 5z + w = 3} & (3) \\
7y - 4z \qquad = -4 & (6)
\end{array}
$$

Eliminate w again. Multiply equation (2) by 2. Add the result to equation (4).

$$
\begin{array}{rl}
4x + 8y + 2z - 2w = -14 & 2 \times (2) \\
\underline{5x + 4y - 5z + 2w = -7} & (4) \\
9x + 12y - 3z \qquad = -21 & (7)
\end{array}
$$

Equations (5), (6), and (7) do not contain a w-term. Since (5) does not have a z-term, we will find another equation without a z-term.

Eliminate z. Multiply equation (6) by 3 and equation (7) by -4. Then add the results.

$$21y - 12z = -12 \qquad 3 \times (6)$$
$$\underline{-36x - 48y + 12z = 84} \qquad -4 \times (7)$$
$$-36x - 27y = 72 \quad (8)$$

Eliminate y using (5) and (8).

$$3x + 3y = -6 \qquad \tfrac{3}{5} \times (5)$$
$$\underline{-4x - 3y = 8} \qquad (8) \div 9$$
$$-x = 2$$
$$x = -2$$

Substitute -2 for x in (5).

$$5(-2) + 5y = -10$$
$$5y = 0$$
$$y = 0$$

Substitute 0 for y in (6).

$$7(0) - 4z = -4$$
$$-4z = -4$$
$$z = 1$$

Substitute -2 for x, 0 for y, and 1 for z in (1).

$$3(-2) + (0) - (1) + w = -3$$
$$-6 - 1 + w = -3$$
$$w = 4$$

The solution set is $\{(-2, 0, 1, 4)\}$.

43. Let $x =$ the length of the longest side.
Then $\frac{5}{6}x =$ the length of the shortest side,
and $x - 17 =$ the length of the medium side.

The perimeter is 323, so

$$x + \tfrac{5}{6}x + (x - 17) = 323.$$
$$6x + 5x + 6(x - 17) = 6(323) \quad \textit{Multiply by 6.}$$
$$6x + 5x + 6x - 102 = 1938$$
$$17x = 2040$$
$$x = 120$$

Since $x = 120$, $\frac{5}{6}x = 100$, and $x - 17 = 103$. The lengths of the sides are 100 inches, 103 inches, and 120 inches.

45. Let $x =$ the least number.
Then $-3x =$ the greatest number,
and $-3x - 4 =$ the middle number.

The sum is 16, so

$$x + (-3x) + (-3x - 4) = 16.$$
$$-5x - 4 = 16$$
$$-5x = 20$$
$$x = -4$$

Since $x = -4$, $-3x = 12$, and $-3x - 4 = 8$. The three numbers are $-4, 8,$ and 12.

8.5 Applications of Systems of Linear Equations

1. *Step 2*
Let $x =$ the number of games that the White Sox won and let $y =$ the number of games that they lost.

Step 3
They played 162 games, so

$$x + y = 162. \quad (1)$$

They won 36 more games than they lost, so

$$x = 36 + y. \quad (2)$$

Step 4
Substitute $36 + y$ for x in (1).

$$(36 + y) + y = 162$$
$$36 + 2y = 162$$
$$2y = 126$$
$$y = 63$$

Substitute 63 for y in (2).

$$x = 36 + 63 = 99$$

Step 5
The White Sox win-loss record was 99 wins and 63 losses.

Step 6
99 is 36 more than 63 and the sum of 99 and 63 is 162.

3. Let $W =$ the width of the tennis court
and $L =$ the length of the court.

Since the length is 42 ft more than the width,

$$L = W + 42. \quad (1)$$

The perimeter of a rectangle is given by

$$2W + 2L = P.$$

With perimeter $P = 228$ ft,

$$2W + 2L = 228. \quad (2)$$

Substitute $W + 42$ for L in equation (2).

$$2W + 2(W + 42) = 228$$
$$2W + 2W + 84 = 228$$
$$4W = 144$$
$$W = 36$$

Substitute $W = 36$ into equation (1).

$$L = W + 42 \quad (1)$$
$$L = 36 + 42$$
$$= 78$$

The length is 78 ft and the width is 36 ft.

5. Let x = the revenue for Wal-Mart and y = the revenue for ExxonMobil (both in billions of dollars).

The total revenue was $656 billion.

$$x + y = 656 \quad (1)$$

ExxonMobil's revenue was $24 billion more than that of Wal-Mart.

$$y = x + 24 \quad (2)$$

Substitute $x + 24$ for y in equation (1).

$$x + (x + 24) = 656$$
$$2x + 24 = 656$$
$$2x = 632$$
$$x = 316$$

Substitute 316 for x in equation (2).

$$y = x + 24 = 316 + 24 = 340$$

Wal-Mart's revenue was $316 billion and ExxonMobil's revenue was $340 billion.

7. From the figure in the text, the angles marked y and $3x + 10$ are supplementary, so

$$(3x + 10) + y = 180. \quad (1)$$

Also, the angles x and y are complementary, so

$$x + y = 90. \quad (2)$$

Solve equation (2) for y to get

$$y = 90 - x. \quad (3)$$

Substitute $90 - x$ for y in equation (1).

$$(3x + 10) + (90 - x) = 180$$
$$2x + 100 = 180$$
$$2x = 80$$
$$x = 40$$

Substitute $x = 40$ into equation (3) to get

$$y = 90 - x = 90 - 40 = 50.$$

The angles measure 40° and 50°.

9. Let x = the hockey FCI and y = the basketball FCI.

The sum is $514.69, so

$$x + y = 514.69. \quad (1)$$

The hockey FCI was $20.05 less than the basketball FCI, so

$$x = y - 20.05. \quad (2)$$

From (2), substitute $y - 20.05$ for x in (1).

$$(y - 20.05) + y = 514.69$$
$$2y - 20.05 = 514.69$$
$$2y = 534.74$$
$$y = 267.37$$

From (2),

$$x = y - 20.05 = 267.37 - 20.05 = 247.32.$$

The hockey FCI was $247.32 and the basketball FCI was $267.37.

11. Let x = the cost of a single Regular Roast Beef sandwich, and y = the cost of a single Large Roast Beef sandwich.

15 Regular Roast Beef sandwiches and 10 Large Roast Beef sandwiches cost $77.75.

$$15x + 10y = 77.75 \quad (1)$$

30 Regular Roast Beef sandwiches and 5 Large Roast Beef sandwiches cost $92.65.

$$30x + 5y = 92.65 \quad (2)$$

Multiply equation (1) by -2 and add to equation (2).

$$
\begin{array}{rrll}
-30x & - \ 20y & = \ -155.50 & -2 \times (1) \\
30x & + \ \ 5y & = \ \ \ \ 92.65 & (2) \\
\hline
& -15y & = \ -62.85 & \\
& y & = \ 4.19 &
\end{array}
$$

Substitute 4.19 for y in equation (1).

$$15x + 10y = 77.75$$
$$15x + 10(4.19) = 77.75$$
$$15x + 41.90 = 77.75$$
$$15x = 35.85$$
$$x = 2.39$$

A single Regular Roast Beef sandwich costs $2.39 and a single Large Roast Beef sandwich costs $4.19.

13. Use the formula (rate of percent) • (base amount) = amount (percentage) of pure acid to compute parts (a) – (d).

(a) $0.10(60) = 6$ oz

(b) $0.25(60) = 15$ oz

(c) $0.40(60) = 24$ oz

(d) $0.50(60) = 30$ oz

15. The cost is the price per pound, $1.29, times the number of pounds, x, or $1.29x$.

17. Let x = the amount of 25% alcohol solution, and y = the amount of 35% alcohol solution.

Make a table. The percent times the amount of solution gives the amount of pure alcohol in the third column.

Gallons of Solution	Percent (as a decimal)	Gallons of Pure Alcohol
x	$25\% = 0.25$	$0.25x$
y	$35\% = 0.35$	$0.35y$
20	$32\% = 0.32$	$0.32(20) = 6.4$

The third row gives the total amounts of solution and pure alcohol. From the columns in the table, write a system of equations.

$$x + \quad y = 20 \qquad (1)$$
$$0.25x + 0.35y = 6.4 \qquad (2)$$

Solve the system. Multiply equation (1) by -25 and equation (2) by 100. Then add the results.

$$
\begin{array}{rl}
-25x - 25y = -500 & -25 \times (1) \\
25x + 35y = 640 & 100 \times (2) \\
\hline
10y = 140 & \\
y = 14 &
\end{array}
$$

Substitute $y = 14$ into equation (1).

$$x + y = 20 \quad (1)$$
$$x + 14 = 20$$
$$x = 6$$

Mix 6 gal of 25% solution and 14 gal of 35% solution.

19. Let $x =$ the amount of pure acid
and $y =$ the amount of 10% acid.

Make a table.

Liters of Solution	Percent (as a decimal)	Liters of Pure Acid
x	$100\% = 1$	$1.00x = x$
y	$10\% = 0.10$	$0.10y$
54	$20\% = 0.20$	$0.20(54) = 10.8$

Solve the following system.

$$x + \quad y = 54 \qquad (1)$$
$$x + 0.10y = 10.8 \qquad (2)$$

Multiply equation (2) by 10 to clear the decimals.

$$10x + y = 108 \qquad (3)$$

To eliminate y, multiply equation (1) by -1 and add the result to equation (3).

$$
\begin{array}{rl}
-x - y = -54 & -1 \times (1) \\
10x + y = 108 & (3) \\
\hline
9x = 54 & \\
x = 6 &
\end{array}
$$

Since $x = 6$,

$$x + y = 54 \quad (1)$$
$$6 + y = 54$$
$$y = 48.$$

Use 6 L of pure acid and 48 L of 10% acid.

21. Complete the table.

	Number of Kilograms	Price per Kilogram	Value
Nuts	x	2.50	$2.50x$
Cereal	y	1.00	$1.00y$
Mixture	30	1.70	$1.70(30) = 51$

From the "Number of Kilograms" column,

$$x + y = 30. \quad (1)$$

From the "Value" column,

$$2.50x + 1.00y = 51. \quad (2)$$

Solve the system.

$$
\begin{array}{rl}
-10x - 10y = -300 & -10 \times (1) \\
25x + 10y = 510 & 10 \times (2) \\
\hline
15x = 210 & \\
x = 14 &
\end{array}
$$

From (1), $14 + y = 30$, so $y = 16$.
The party mix should be made from 14 kg of nuts and 16 kg of cereal.

23. From the "Principal" column in the text,

$$x + y = 3000. \quad (1)$$

From the "Interest" column in the text,

$$0.02x + 0.04y = 100. \quad (2)$$

Multiply equation (2) by 100 to clear the decimals.

$$2x + 4y = 10{,}000 \quad (3)$$

To eliminate x, multiply equation (1) by -2 and add the result to equation (3).

$$
\begin{array}{rl}
-2x - 2y = -6000 & -2 \times (1) \\
2x + 4y = 10{,}000 & (3) \\
\hline
2y = 4000 & \\
y = 2000 &
\end{array}
$$

From (1), $x + 2000 = 3000$, so $x = 1000$.
$1000 is invested at 2%, and $2000 is invested at 4%.

25. **(a)** The speed of the boat going upstream is *decreased* by the speed of the current, so it is $(10 - x)$ mph.

(b) The speed of the boat going downstream is *increased* by the speed of the current, so it is $(10 + x)$ mph.

27. Let x = the speed of the train
and y = the speed of the plane.

	r	t	d
Train	x	$\dfrac{150}{x}$	150
Plane	y	$\dfrac{400}{y}$	400

The times are equal, so

$$\frac{150}{x} = \frac{400}{y}. \quad (1)$$

The speed of the plane is 20 km per hour less than 3 times the speed of the train, so

$$3x - 20 = y. \quad (2)$$

Multiply (1) by xy (or use cross products).

$$150y = 400x \quad (3)$$

From (2), substitute $3x - 20$ for y in (3).

$$150(3x - 20) = 400x$$
$$450x - 3000 = 400x$$
$$50x = 3000$$
$$x = 60$$

From (2), $y = 3(60) - 20 = 160$.
The speed of the train is 60 km/hr, and the speed of the plane is 160 km/hr.

29. Let x = the speed of the boat in still water
and y = the speed of the current.

Furthermore,

$$\text{rate upstream} = x - y$$
and $$\text{rate downstream} = x + y.$$

Use these rates and the information in the problem to make a table.

	r	t	d
Upstream	$x - y$	2	36
Downstream	$x + y$	1.5	36

From the table, use the formula $d = rt$ to write a system of equations.

$$36 = 2(x - y)$$
$$36 = 1.5(x + y)$$

Remove the parentheses and move the variables to the left side.

$$2x - 2y = 36 \quad (1)$$
$$1.5x + 1.5y = 36 \quad (2)$$

Solve the system. Multiply equation (1) by -3 and equation (2) by 4. Then add the results.

$$
\begin{array}{rll}
-6x + 6y = & -108 & -3 \times (1) \\
\underline{6x + 6y = } & \underline{144} & 4 \times (2) \\
12y = & 36 & \\
y = & 3 &
\end{array}
$$

Substitute $y = 3$ into equation (1).

$$
\begin{array}{rl}
2x - 2y = 36 & (1) \\
2x - 2(3) = 36 & \\
2x = 42 & \\
x = 21 &
\end{array}
$$

The speed of the boat is 21 mph, and the speed of the current is 3 mph.

31. Let x = the number of pounds of the $0.75-per-lb candy and y = the number of pounds of the $1.25-per-lb candy.

Make a table.

	Price per Pound	Number of Pounds	Value
Less Expensive Candy	$0.75	x	$0.75x
More Expensive Candy	$1.25	y	$1.25y
Mixture	$0.96	9	$0.96(9) = $8.64

From the "Number of Pounds" column,

$$x + y = 9. \quad (1)$$

From the "Value" column,

$$0.75x + 1.25y = 8.64. \quad (2)$$

Solve the system.

$$
\begin{array}{rll}
-75x - 75y = & -675 & -75 \times (1) \\
\underline{75x + 125y = } & \underline{864} & 100 \times (2) \\
50y = & 189 & \\
y = & 3.78 &
\end{array}
$$

From (1), $x + 3.78 = 9$, so $x = 5.22$.
Mix 5.22 pounds of the $0.75-per-lb candy with 3.78 pounds of the $1.25-per-lb candy to obtain 9 pounds of a mixture that sells for $0.96 per pound.

33. Let x = the number of general admission tickets and y = the number of student tickets.

Make a table.

Ticket	Number	Value of Tickets
General	x	$5 \cdot x = 5x$
Student	y	$4 \cdot y = 4y$
Totals	184	812

Solve the system.

$$x + y = 184 \quad (1)$$
$$5x + 4y = 812 \quad (2)$$

To eliminate y, multiply equation (1) by -4 and add the result to equation (2).

$$
\begin{array}{rcrl}
-4x & - & 4y = & -736 \quad -4 \times (1) \\
5x & + & 4y = & 812 \quad (2) \\
\hline
x & & = & 76
\end{array}
$$

From (1), $76 + y = 184$, so $y = 108$.

76 general admission tickets and 108 student tickets were sold.

35. Let $x =$ the price for a citron and let $y =$ the price for a wood apple.

"9 citrons and 7 fragrant wood apples is 107" gives us

$$9x + 7y = 107. \quad (1)$$

"7 citrons and 9 fragrant wood apples is 101" gives us

$$7x + 9y = 101. \quad (2)$$

Multiply equation (1) by -7 and equation (2) by 9. Then add.

$$
\begin{array}{rcrl}
-63x & - & 49y = & -749 \quad -7 \times (1) \\
63x & + & 81y = & 909 \quad 9 \times (2) \\
\hline
& & 32y = & 160 \\
& & y = & 5
\end{array}
$$

Substitute 5 for y in equation (1).

$$9x + 7(5) = 107$$
$$9x + 35 = 107$$
$$9x = 72$$
$$x = 8$$

The prices are 8 for a citron and 5 for a wood apple.

37. Let $x =$ the measure of one angle, $y =$ the measure of another angle, and $z =$ the measure of the last angle.

Two equations are given, so

$$z = x + 10$$
$$\text{or} \quad -x + z = 10 \quad (1)$$
$$\text{and} \quad x + y = 100. \quad (2)$$

Since the sum of the measures of the angles of a triangle is $180°$, the third equation of the system is

$$x + y + z = 180. \quad (3)$$

Equation (1) is missing y. To eliminate y again, multiply equation (2) by -1 and add the result to equation (3).

$$
\begin{array}{rcrl}
-x & - & y & = -100 \quad -1 \times (2) \\
x & + & y + z = & 180 \quad (3) \\
\hline
& & z = & 80
\end{array}
$$

Since $z = 80$,

$$-x + z = 10 \quad (1)$$
$$-x + 80 = 10$$
$$-x = -70$$
$$x = 70.$$

From (2), $70 + y = 100$, so $y = 30$.
The measures of the angles are $70°$, $30°$, and $80°$.

39. Let $x =$ the measure of the first angle, $y =$ the measure of the second angle, and $z =$ the measure of the third angle.

The sum of the angles in a triangle equals $180°$, so

$$x + y + z = 180. \quad (1)$$

The measure of the second angle is $10°$ more than 3 times that of the first angle, so

$$y = 3x + 10. \quad (2)$$

The third angle is equal to the sum of the other two, so

$$z = x + y. \quad (3)$$

Solve the system. Substitute z for $x + y$ in equation (1).

$$(x + y) + z = 180 \quad (1)$$
$$z + z = 180$$
$$2z = 180$$
$$z = 90$$

Substitute 90 for z and $3x + 10$ for y in equation (3).

$$z = x + y \quad (3)$$
$$90 = x + (3x + 10)$$
$$80 = 4x$$
$$20 = x$$

Substitute $x = 20$ and $z = 90$ into equation (3).

$$z = x + y \quad (3)$$
$$90 = 20 + y$$
$$70 = y$$

The three angles have measures of $20°$, $70°$, and $90°$.

41. Let $x =$ the length of the longest side,
$y =$ the length of the middle side,
and $z =$ the length of the shortest side.

Perimeter is the sum of the measures of the sides, so

$$x + y + z = 70. \quad (1)$$

The longest side is 4 cm less than the sum of the other sides, so

$$x = y + z - 4$$
$$\text{or} \quad x - y - z = -4. \qquad (2)$$

Twice the shortest side is 9 cm less than the longest side, so

$$2z = x - 9$$
$$\text{or} \quad -x + 2z = -9 \qquad (3)$$

Add equations (1) and (2) to eliminate y and z .

$$\begin{array}{rcrcrcr} x & + & y & + & z & = & 70 \quad (1) \\ x & - & y & - & z & = & -4 \quad (2) \\ \hline 2x & & & & & = & 66 \\ & & & & x & = & 33 \end{array}$$

Substitute 33 for x in (3).

$$-x + 2z = -9 \quad (3)$$
$$-33 + 2z = -9$$
$$2z = 24$$
$$z = 12$$

Substitute 33 for x and 12 for z in (1).

$$x + y + z = 70 \quad (1)$$
$$33 + y + 12 = 70$$
$$y + 45 = 70$$
$$y = 25$$

The shortest side is 12 cm long, the middle side is 25 cm long, and the longest side is 33 cm long.

43. Let $x =$ the number of gold medals, $y =$ the number of silver medals, and $z =$ the number of bronze medals.

The U.S. earned 6 more gold medals than bronze, so

$$x = 6 + z. \quad (1)$$

The number of silver medals earned was 19 less than twice the number of bronze medals, so

$$y = 2z - 19. \quad (2)$$

The total number of medals earned was 103, so

$$x + y + z = 103. \quad (3)$$

Substitute $6 + z$ for x and $2z - 19$ for y in (3).

$$(6 + z) + (2z - 17) + z = 103$$
$$4z - 13 = 103$$
$$4z = 116$$
$$z = 29$$

From (1), $x = 6 + 29 = 35$.
From (2), $y = 2(29) - 19 = 58 - 19 = 39$.

The United States earned 35 gold, 39 silver, and 29 bronze medals.

45. Let $x =$ the number of \$14 tickets, $y =$ the number of \$20 tickets, and $z =$ the number of VIP \$50 tickets.

Five times as many \$14 tickets have been sold as VIP tickets, so

$$x = 5z. \quad (1)$$

The number of \$14 tickets is 15 more than the sum of the number of \$20 tickets and the number of VIP tickets, so

$$x = 15 + y + z. \quad (2)$$

Since x is in terms of z in (1), we'll substitute $5z$ for x in (2) and then get y in terms of z .

$$5z = 15 + y + z$$
$$4z - 15 = y \qquad (3)$$

Sales of these tickets totaled \$11,700.

$$14x + 20y + 50z = 11,700$$
$$14(5z) + 20(4z - 15) + 50z = 11,700$$
$$70z + 80z - 300 + 50z = 11,700$$
$$200z = 12,000$$
$$z = 60$$

From (1), $x = 5(60) = 300$.
From (3), $y = 4(60) - 15 = 225$.

There were 300 \$14 tickets, 225 \$20 tickets, and 60 \$50 tickets sold.

47. Let $x =$ the number of T-shirts shipped to bookstore A,
$y =$ the number of T-shirts shipped to bookstore B,
and $z =$ the number of T-shirts shipped to bookstore C.

Twice as many T-shirts were shipped to bookstore B as to bookstore A, so

$$y = 2x. \quad (1)$$

The number shipped to bookstore C was 40 less than the sum of the numbers shipped to the other two bookstores, so

$$z = x + y - 40. \quad (2)$$

Substitute $2x$ for y [from (1)] into equation (2) to get z in terms of x .

$$z = x + y - 40 \qquad (2)$$
$$z = x + (2x) - 40$$
$$z = 3x - 40 \qquad (3)$$

The total number of T-shirts shipped was 800, so

$$x + y + z = 800. \qquad (4)$$

Substitute $2x$ for y and $3x - 40$ for z in (4).

$$x + (2x) + (3x - 40) = 800$$
$$6x - 40 = 800$$
$$6x = 840$$
$$x = 140$$

From (1), $y = 2(140) = 280$.
From (3), $z = 3(140) - 40 = 380$.
The number of T-shirts shipped to bookstores A, B, and C was 140, 280, and 380, respectively.

49. Let x, y, and z denote the number of kilograms of the first, second, and third chemicals, respectively. The mix must include 60% of the first and second chemicals, so

$$x + y = 0.60(750) = 450. \qquad (1)$$

The second and third chemicals must be in a ratio of 4 to 3 by weight, so

$$y = \tfrac{4}{3}z. \qquad (2)$$

From (1),

$$x = 450 - y. \qquad (3)$$

From (2),

$$z = \tfrac{3}{4}y. \qquad (4)$$

The total is 750, so

$$x + y + z = 750.$$
$$(450 - y) + y + \tfrac{3}{4}y = 750$$
$$\tfrac{3}{4}y = 300$$
$$y = 400$$

From (3), $x = 450 - 400 = 50$.
From (4), $z = \tfrac{3}{4}(400) = 300$.

Use 50 kg of the first chemical, 400 kg of the second chemical, and 300 kg of the third chemical to make the plant food.

51. *Step 2*
Let x = the number of wins,
y = the number of losses,
and z = the number of ties.

Step 3
They played 82 games, so

$$x + y + z = 82. \qquad (1)$$

Their wins and losses totaled 71, so

$$x + y = 71. \qquad (2)$$

They tied 14 fewer games than they lost, so

$$z = y - 14. \qquad (3)$$

Step 4
Multiply (2) by -1 and add to (1).

$$
\begin{array}{rcll}
x + y + z & = & 82 & (1) \\
-x - y \phantom{{}+z} & = & -71 & -1 \times (2) \\
\hline
z & = & 11 &
\end{array}
$$

Substitute 11 for z in (3).

$$11 = y - 14$$
$$25 = y$$

Substitute 25 for y in (2).

$$x + 25 = 71$$
$$x = 46$$

Step 5
The Flames won 46 games, lost 25 games, and tied 11 games.

Step 6
Adding 46, 25, and 11 gives 82 total games. The wins and losses add up to 71, and there were 14 fewer ties than losses. The solution is correct.

53. (a) The additive inverse of -6 is $-(-6) = 6$.

(b) The multiplicative inverse (reciprocal) of -6 is $\frac{1}{-6} = -\frac{1}{6}$.

55. (a) The additive inverse of $\frac{7}{8}$ is $-\frac{7}{8}$.

(b) The multiplicative inverse (reciprocal) of $\frac{7}{8}$ is $\frac{1}{7/8} = \frac{8}{7}$.

8.6 Solving Systems of Linear Equations by Matrix Methods

1.
$$\begin{bmatrix} -2 & 3 & 1 \\ 0 & 5 & -3 \\ 1 & 4 & 8 \end{bmatrix}$$

(a) The elements of the second row are 0, 5, and -3.

(b) The elements of the third column are 1, -3, and 8.

(c) The matrix is square since the number of rows (three) is the same as the number of columns.

(d) The matrix obtained by interchanging the first and third rows is

$$\begin{bmatrix} 1 & 4 & 8 \\ 0 & 5 & -3 \\ -2 & 3 & 1 \end{bmatrix}.$$

(e) The matrix obtained by multiplying the first row by $-\frac{1}{2}$ is

$$\begin{bmatrix} -2\left(-\frac{1}{2}\right) & 3\left(-\frac{1}{2}\right) & 1\left(-\frac{1}{2}\right) \\ 0 & 5 & -3 \\ 1 & 4 & 8 \end{bmatrix} = \begin{bmatrix} 1 & -\frac{3}{2} & -\frac{1}{2} \\ 0 & 5 & -3 \\ 1 & 4 & 8 \end{bmatrix}.$$

(f) The matrix obtained by multiplying the third row by 3 and adding to the first row is

$$\begin{bmatrix} -2+3(1) & 3+3(4) & 1+3(8) \\ 0 & 5 & -3 \\ 1 & 4 & 8 \end{bmatrix} = \begin{bmatrix} 1 & 15 & 25 \\ 0 & 5 & -3 \\ 1 & 4 & 8 \end{bmatrix}.$$

3. $\quad 4x + 8y = 44$
$\qquad 2x - \ y = -3$

$$\begin{bmatrix} 4 & 8 & | & 44 \\ 2 & -1 & | & -3 \end{bmatrix}$$

$$\begin{bmatrix} 1 & 2 & | & 11 \\ 2 & -1 & | & -3 \end{bmatrix} \qquad \frac{1}{4}R_1$$

$$\begin{bmatrix} 1 & 2 & | & 11 \\ 0 & -5 & | & -25 \end{bmatrix} \qquad -2R_1 + R_2$$

Note: $\begin{cases} -2(2) + (-1) = -5 \\ -2(11) + (-3) = -25 \end{cases}$

$$\begin{bmatrix} 1 & 2 & | & 11 \\ 0 & 1 & | & 5 \end{bmatrix} \qquad -\frac{1}{5}R_2$$

This represents the system

$$x + 2y = 11$$
$$y = 5.$$

Substitute $y = 5$ in the first equation.

$$x + 2y = 11$$
$$x + 2(5) = 11$$
$$x + 10 = 11$$
$$x = 1$$

The solution set is $\{(1, 5)\}$.

5. $\quad x + y = 5$
$\qquad x - y = 3$

Write the augmented matrix for this system.

$$\begin{bmatrix} 1 & 1 & | & 5 \\ 1 & -1 & | & 3 \end{bmatrix}$$

$$\begin{bmatrix} 1 & 1 & | & 5 \\ 0 & -2 & | & -2 \end{bmatrix} \qquad -1R_1 + R_2$$

$$\begin{bmatrix} 1 & 1 & | & 5 \\ 0 & 1 & | & 1 \end{bmatrix} \qquad -\frac{1}{2}R_2$$

This matrix gives the system

$$x + y = 5$$
$$y = 1.$$

Substitute $y = 1$ in the first equation.

$$x + y = 5$$
$$x + 1 = 5$$
$$x = 4$$

The solution set is $\{(4, 1)\}$.

7. $\quad 2x + 4y = 6$
$\qquad 3x - \ y = 2$

Write the augmented matrix.

$$\begin{bmatrix} 2 & 4 & | & 6 \\ 3 & -1 & | & 2 \end{bmatrix}$$

The easiest way to get a 1 in the first row, first column position is to multiply the elements in the first row by $\frac{1}{2}$.

$$\begin{bmatrix} 1 & 2 & | & 3 \\ 3 & -1 & | & 2 \end{bmatrix} \qquad \frac{1}{2}R_1$$

To get a 0 in row two, column 1, we need to subtract 3 from the 3 that is in that position. To do this we will multiply row 1 by -3 and add the result to row 2.

$$\begin{bmatrix} 1 & 2 & | & 3 \\ 0 & -7 & | & -7 \end{bmatrix} \qquad -3R_1 + R_2$$

$$\begin{bmatrix} 1 & 2 & | & 3 \\ 0 & 1 & | & 1 \end{bmatrix} \qquad -\frac{1}{7}R_2$$

This matrix gives the system

$$x + 2y = 3$$
$$y = 1.$$

Substitute $y = 1$ in the first equation.

$$x + 2y = 3$$
$$x + 2(1) = 3$$
$$x = 1$$

The solution set is $\{(1, 1)\}$.

9. $\quad 3x + 4y = \ \ 13$
$\qquad 2x - 3y = -14$

Write the augmented matrix.

$$\begin{bmatrix} 3 & 4 & | & 13 \\ 2 & -3 & | & -14 \end{bmatrix}$$

$$\begin{bmatrix} 1 & 7 & | & 27 \\ 2 & -3 & | & -14 \end{bmatrix} \qquad -1R_2 + R_1$$

$$\begin{bmatrix} 1 & 7 & | & 27 \\ 0 & -17 & | & -68 \end{bmatrix} \qquad -2R_1 + R_2$$

$$\begin{bmatrix} 1 & 7 & | & 27 \\ 0 & 1 & | & 4 \end{bmatrix} \qquad -\frac{1}{17}R_2$$

This matrix gives the system

$$x + 7y = 27$$
$$y = 4.$$

Substitute $y = 4$ in the first equation.

$$x + 7y = 27$$
$$x + 7(4) = 27$$
$$x + 28 = 27$$
$$x = -1$$

The solution set is $\{(-1, 4)\}$.

11. $\quad -4x + 12y = 36$
$\qquad\quad x - 3y = 9$

Write the augmented matrix.

$$\begin{bmatrix} -4 & 12 & | & 36 \\ 1 & -3 & | & 9 \end{bmatrix}$$

$$\begin{bmatrix} -1 & 3 & | & 9 \\ 1 & -3 & | & 9 \end{bmatrix} \quad \tfrac{1}{4}R_1$$

$$\begin{bmatrix} -1 & 3 & | & 9 \\ 0 & 0 & | & 18 \end{bmatrix} \quad R_1 + R_2$$

The corresponding system is

$$-x + 3y = 9$$
$$0 = 18 \quad \text{False}$$

which is inconsistent and has no solution.

The solution set is $\emptyset$.

13. $\quad 2x + y = 4$
$\qquad 4x + 2y = 8$

Write the augmented matrix.

$$\begin{bmatrix} 2 & 1 & | & 4 \\ 4 & 2 & | & 8 \end{bmatrix}$$

$$\begin{bmatrix} 1 & \tfrac{1}{2} & | & 2 \\ 4 & 2 & | & 8 \end{bmatrix} \quad \tfrac{1}{2}R_1$$

$$\begin{bmatrix} 1 & \tfrac{1}{2} & | & 2 \\ 0 & 0 & | & 0 \end{bmatrix} \quad -4R_1 + R_2$$

Row 2, $0 = 0$, indicates that the system has dependent equations.

The solution set is $\{(x, y) \mid 2x + y = 4\}$.

15. $\quad x + y - z = -3$
$\qquad 2x + y + z = 4$
$\qquad 5x - y + 2z = 23$

Write the augmented matrix.

$$\begin{bmatrix} 1 & 1 & -1 & | & -3 \\ 2 & 1 & 1 & | & 4 \\ 5 & -1 & 2 & | & 23 \end{bmatrix}$$

$$\begin{bmatrix} 1 & 1 & -1 & | & -3 \\ 0 & -1 & 3 & | & 10 \\ 0 & -6 & 7 & | & 38 \end{bmatrix} \quad \begin{array}{l} -2R_1 + R_2 \\ -5R_1 + R_3 \end{array}$$

$$\begin{bmatrix} 1 & 1 & -1 & | & -3 \\ 0 & 1 & -3 & | & -10 \\ 0 & -6 & 7 & | & 38 \end{bmatrix} \quad -1R_2$$

$$\begin{bmatrix} 1 & 1 & -1 & | & -3 \\ 0 & 1 & -3 & | & -10 \\ 0 & 0 & -11 & | & -22 \end{bmatrix} \quad 6R_2 + R_3$$

$$\begin{bmatrix} 1 & 1 & -1 & | & -3 \\ 0 & 1 & -3 & | & -10 \\ 0 & 0 & 1 & | & 2 \end{bmatrix} \quad -\tfrac{1}{11}R_3$$

This matrix gives the system

$$x + y - z = -3$$
$$y - 3z = -10$$
$$z = 2.$$

Substitute $z = 2$ in the second equation.

$$y - 3z = -10$$
$$y - 3(2) = -10$$
$$y - 6 = -10$$
$$y = -4$$

Substitute $y = -4$ and $z = 2$ in the first equation.

$$x + y - z = -3$$
$$x - 4 - 2 = -3$$
$$x - 6 = -3$$
$$x = 3$$

The solution set is $\{(3, -4, 2)\}$.

17. $\quad x + y - 3z = 1$
$\qquad 2x - y + z = 9$
$\qquad 3x + y - 4z = 8$

Write the augmented matrix.

$$\begin{bmatrix} 1 & 1 & -3 & | & 1 \\ 2 & -1 & 1 & | & 9 \\ 3 & 1 & -4 & | & 8 \end{bmatrix}$$

$$\begin{bmatrix} 1 & 1 & -3 & | & 1 \\ 0 & -3 & 7 & | & 7 \\ 0 & -2 & 5 & | & 5 \end{bmatrix} \quad \begin{array}{l} -2R_1 + R_2 \\ -3R_1 + R_3 \end{array}$$

$$\begin{bmatrix} 1 & 1 & -3 & | & 1 \\ 0 & 1 & -\tfrac{7}{3} & | & -\tfrac{7}{3} \\ 0 & -2 & 5 & | & 5 \end{bmatrix} \quad -\tfrac{1}{3}R_2$$

$$\begin{bmatrix} 1 & 1 & -3 & | & 1 \\ 0 & 1 & -\tfrac{7}{3} & | & -\tfrac{7}{3} \\ 0 & 0 & \tfrac{1}{3} & | & \tfrac{1}{3} \end{bmatrix} \quad 2R_2 + R_3$$

$$\begin{bmatrix} 1 & 1 & -3 & | & 1 \\ 0 & 1 & -\tfrac{7}{3} & | & -\tfrac{7}{3} \\ 0 & 0 & 1 & | & 1 \end{bmatrix} \quad 3R_3$$

continued

This matrix gives the system

$$x + y - 3z = 1$$
$$y - \tfrac{7}{3}z = -\tfrac{7}{3}$$
$$z = 1.$$

Substitute $z = 1$ in the second equation.

$$y - \tfrac{7}{3}z = -\tfrac{7}{3}$$
$$y - \tfrac{7}{3}(1) = -\tfrac{7}{3}$$
$$y = 0$$

Substitute $y = 0$ and $z = 1$ in the first equation.

$$x + y - 3z = 1$$
$$x + 0 - 3(1) = 1$$
$$x - 3 = 1$$
$$x = 4$$

The solution set is $\{(4, 0, 1)\}$.

19.
$$x + y - z = 6$$
$$2x - y + z = -9$$
$$x - 2y + 3z = 1$$

Write the augmented matrix.

$$\begin{bmatrix} 1 & 1 & -1 & 6 \\ 2 & -1 & 1 & -9 \\ 1 & -2 & 3 & 1 \end{bmatrix}$$

$$\begin{bmatrix} 1 & 1 & -1 & 6 \\ 0 & -3 & 3 & -21 \\ 0 & -3 & 4 & -5 \end{bmatrix} \begin{matrix} \\ -2R_1 + R_2 \\ -1R_1 + R_3 \end{matrix}$$

$$\begin{bmatrix} 1 & 1 & -1 & 6 \\ 0 & 1 & -1 & 7 \\ 0 & -3 & 4 & -5 \end{bmatrix} \begin{matrix} \\ -\tfrac{1}{3}R_2 \\ \\ \end{matrix}$$

$$\begin{bmatrix} 1 & 1 & -1 & 6 \\ 0 & 1 & -1 & 7 \\ 0 & 0 & 1 & 16 \end{bmatrix} \begin{matrix} \\ \\ 3R_2 + R_3 \end{matrix}$$

This matrix gives the system

$$x + y - z = 6$$
$$y - z = 7$$
$$z = 16.$$

Substitute $z = 16$ in the second equation.

$$y - 16 = 7$$
$$y = 23$$

Substitute $y = 23$ and $z = 16$ in the first equation.

$$x + 23 - 16 = 6$$
$$x + 7 = 6$$
$$x = -1$$

The solution set is $\{(-1, 23, 16)\}$.

21.
$$x - y = 1$$
$$y - z = 6$$
$$x + z = -1$$

Write the augmented matrix.

$$\begin{bmatrix} 1 & -1 & 0 & 1 \\ 0 & 1 & -1 & 6 \\ 1 & 0 & 1 & -1 \end{bmatrix}$$

$$\begin{bmatrix} 1 & -1 & 0 & 1 \\ 0 & 1 & -1 & 6 \\ 0 & 1 & 1 & -2 \end{bmatrix} \begin{matrix} \\ \\ -1R_1 + R_3 \end{matrix}$$

$$\begin{bmatrix} 1 & -1 & 0 & 1 \\ 0 & 1 & -1 & 6 \\ 0 & 0 & 2 & -8 \end{bmatrix} \begin{matrix} \\ \\ -1R_2 + R_3 \end{matrix}$$

$$\begin{bmatrix} 1 & -1 & 0 & 1 \\ 0 & 1 & -1 & 6 \\ 0 & 0 & 1 & -4 \end{bmatrix} \begin{matrix} \\ \\ \tfrac{1}{2}R_3 \end{matrix}$$

This matrix gives the system

$$x - y = 1$$
$$y - z = 6$$
$$z = -4.$$

Substitute $z = -4$ in the second equation.

$$y - z = 6$$
$$y - (-4) = 6$$
$$y = 2$$

Substitute $y = 2$ in the first equation.

$$x - y = 1$$
$$x - 2 = 1$$
$$x = 3$$

The solution set is $\{(3, 2, -4)\}$.

23.
$$x - 2y + z = 4$$
$$3x - 6y + 3z = 12$$
$$-2x + 4y - 2z = -8$$

Write the augmented matrix.

$$\begin{bmatrix} 1 & -2 & 1 & 4 \\ 3 & -6 & 3 & 12 \\ -2 & 4 & -2 & -8 \end{bmatrix}$$

$$\begin{bmatrix} 1 & -2 & 1 & 4 \\ 1 & -2 & 1 & 4 \\ -1 & 2 & -1 & -4 \end{bmatrix} \begin{matrix} \\ \tfrac{1}{3}R_2 \\ \tfrac{1}{2}R_3 \end{matrix}$$

$$\begin{bmatrix} 1 & -2 & 1 & 4 \\ 0 & 0 & 0 & 0 \\ 0 & 0 & 0 & 0 \end{bmatrix} \begin{matrix} \\ -1R_1 + R_2 \\ R_1 + R_3 \end{matrix}$$

This augmented matrix represents a system of dependent equations.
The solution set is $\{(x, y, z) \mid x - 2y + z = 4\}$.

25. $\begin{aligned} x + 2y + 3z &= -2 \\ 2x + 4y + 6z &= -5 \\ x - y + 2z &= 6 \end{aligned}$

Write the augmented matrix.

$$\begin{bmatrix} 1 & 2 & 3 & | & -2 \\ 2 & 4 & 6 & | & -5 \\ 1 & -1 & 2 & | & 6 \end{bmatrix}$$

$$\begin{bmatrix} 1 & 2 & 3 & | & -2 \\ 0 & 0 & 0 & | & -1 \\ 0 & -3 & -1 & | & 8 \end{bmatrix} \quad \begin{aligned} -2R_1 + R_2 \\ -R_1 + R_3 \end{aligned}$$

From the second row, $0 = -1$, we see that the system is inconsistent.
The solution set is $\emptyset$.

27. $\begin{aligned} 4x + y &= 5 \\ 2x + y &= 3 \end{aligned}$

Enter the augmented matrix as $[A]$.

$$\begin{bmatrix} 4 & 1 & 5 \\ 2 & 1 & 3 \end{bmatrix}$$

The TI-83 screen for A follows. (Use MATRX EDIT.)

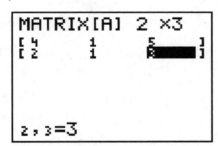

Now use the reduced row echelon form (rref) command to simplify the system. (Use MATRX MATH ALPHA B for rref and MATRX 1 for $[A]$.)

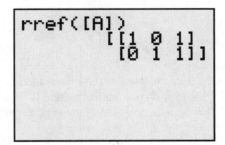

This matrix gives the system

$$\begin{aligned} 1x + 0y &= 1 \\ 0x + 1y &= 1, \end{aligned}$$

or simply, $x = 1$ and $y = 1$.
The solution set is $\{(1, 1)\}$.

29. $\begin{aligned} 5x + y - 3z &= -6 \\ 2x + 3y + z &= 5 \\ -3x - 2y + 4z &= 3 \end{aligned}$

Enter the augmented matrix as $[C]$.

$$\begin{bmatrix} 5 & 1 & -3 & -6 \\ 2 & 3 & 1 & 5 \\ -3 & -2 & 4 & 3 \end{bmatrix}$$

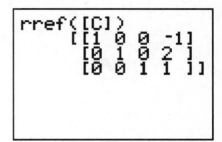

The solution set is $\{(-1, 2, 1)\}$.

31. $\begin{aligned} x + z &= -3 \\ y + z &= 3 \\ x + y &= 8 \end{aligned}$

Enter the augmented matrix as $[E]$.

$$\begin{bmatrix} 1 & 0 & 1 & -3 \\ 0 & 1 & 1 & 3 \\ 1 & 1 & 0 & 8 \end{bmatrix}$$

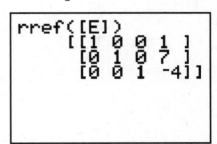

The solution set is $\{(1, 7, -4)\}$.

33. $\begin{aligned} x - 4 &\geq 12 \\ x &\geq 16 \quad \textit{Add 4.} \end{aligned}$

Check that the solution set is the interval $[16, \infty)$.

35. $\begin{aligned} -5z + 18 &> -2 \\ -5z &> -20 \end{aligned}$

Divide by -5; reverse the inequality symbol.
$$z < 4$$

Check that the solution set is the interval $(-\infty, 4)$.

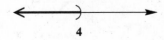

4

Chapter 8 Review Exercises

1. $(3, 4)$

$$4x - 2y = 4$$
$$5x + y = 19$$

To decide whether $(3, 4)$ is a solution of the system, substitute 3 for x and 4 for y in each equation.

$$4x - 2y = 4$$
$$4(3) - 2(4) = 4 \text{ ?}$$
$$12 - 8 = 4 \text{ ?}$$
$$4 = 4 \quad \textit{True}$$

$$5x + y = 19$$
$$5(3) + 4 = 19 \text{ ?}$$
$$15 + 4 = 19 \text{ ?}$$
$$19 = 19 \quad \textit{True}$$

Since $(3, 4)$ satisfies both equations, it is a solution of the system.

2. $(-5, 2)$

$$x - 4y = -13 \quad (1)$$
$$2x + 3y = 4 \quad (2)$$

Substitute -5 for x and 2 for y in equation (2).

$$2x + 3y = 4$$
$$2(-5) + 3(2) = 4 \text{ ?}$$
$$-10 + 6 = 4 \text{ ?}$$
$$-4 = 4 \quad \textit{False}$$

Since $(-5, 2)$ is not a solution of the second equation, it cannot be a solution of the system.

3. $x + y = 4$
   $2x - y = 5$

To graph the equations, find the intercepts.

$x + y = 4$: Let $y = 0$;   then $x = 4$.
　　　　　　Let $x = 0$;   then $y = 4$.

Plot the intercepts, $(4, 0)$ and $(0, 4)$, and draw the line through them.

$2x - y = 5$: Let $y = 0$;   then $x = \frac{5}{2}$.
　　　　　　Let $x = 0$;   then $y = -5$.

Plot the intercepts, $\left(\frac{5}{2}, 0\right)$ and $(0, -5)$, and draw the line through them.

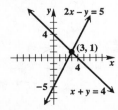

It appears that the lines intersect at the point $(3, 1)$. Check this by substituting 3 for x and 1 for y in both equations. Since $(3, 1)$ satisfies both equations, the solution set of this system is $\{(3, 1)\}$.

4. $x - 2y = 4$
   $2x + y = -2$

To graph the equations, find the intercepts.

$x - 2y = 4$: Let $y = 0$;   then $x = 4$.
　　　　　　Let $x = 0$;   then $y = -2$.

Plot the intercepts, $(4, 0)$ and $(0, -2)$, and draw the line through them.

$2x + y = -2$: Let $y = 0$;   then $x = -1$.
　　　　　　Let $x = 0$;   then $y = -2$.

Plot the intercepts, $(-1, 0)$ and $(0, -2)$, and draw the line through them.

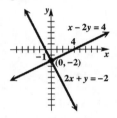

The lines intersect at their common y-intercept, $(0, -2)$, so $\{(0, -2)\}$ is the solution set of the system.

5. $2x + 4 = 2y$
   $y - x = -3$

Graph the line $2x + 4 = 2y$ through its intercepts, $(-2, 0)$ and $(0, 2)$. Graph the line $y - x = -3$ through its intercepts, $(3, 0)$ and $(0, -3)$.

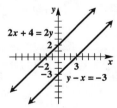

The lines are parallel (both lines have slope 1). Since they do not intersect, there is no solution. This is an inconsistent system and the solution set is $\emptyset$.

6. No, this is not correct. A false statement indicates that the solution set is $\emptyset$.

7. No, two lines cannot intersect in exactly three points.

8. $3x + y = 7 \quad (1)$
   $x = 2y \quad (2)$

Substitute $2y$ for x in equation (1) and solve the resulting equation for y.

$$3x + y = 7$$
$$3(2y) + y = 7$$
$$6y + y = 7$$
$$7y = 7$$
$$y = 1$$

To find x, let $y = 1$ in equation (2).

$$x = 2y$$
$$x = 2(1) = 2$$

The solution set is $\{(2, 1)\}$.

9. $2x - 5y = -19$ (1)
    $y = x + 2$ (2)

Substitute $x + 2$ for y in equation (1).

$$2x - 5y = -19$$
$$2x - 5(x + 2) = -19$$
$$2x - 5x - 10 = -19$$
$$-3x - 10 = -19$$
$$-3x = -9$$
$$x = 3$$

To find y, let $x = 3$ in equation (2).

$$y = x + 2$$
$$y = 3 + 2 = 5$$

The solution set is $\{(3, 5)\}$.

10. $5x + 15y = 30$ (1)
    $x + 3y = 6$ (2)

Solve equation (2) for x.

$$x + 3y = 6$$
$$x = 6 - 3y \quad (3)$$

Substitute $6 - 3y$ for x in equation (1).

$$5x + 15y = 30$$
$$5(6 - 3y) + 15y = 30$$
$$30 - 15y + 15y = 30$$
$$30 = 30 \quad True$$

This true result means that every solution of one equation is also a solution of the other, so the system has an infinite number of solutions. The solution set is $\{(x, y) \mid x + 3y = 6\}$.

11. His answer was incorrect since the system has infinitely many solutions (as indicated by the true statement $0 = 0$).

12. It would be easiest to solve for x in the second equation because its coefficient is -1. No fractions would be involved.

13. If we simply add the given equations without first multiplying one or both equations by a constant, choice **C** is the only system in which a variable will be eliminated. If we add the equations in **C** we get $3x = 17$. (The variable y was eliminated.)

14. $2x + 12y = 7$
    $3x + 4y = 1$

(a) If we multiply the first equation by -3, the first term will become $-6x$. To eliminate x, we need to change the first term on the left side of the second equation from $3x$ to $6x$. In order to do this, we must multiply the second equation by 2.

(b) If we multiply the first equation by -3, the second term will become $-36y$. To eliminate y, we need to change the second term on the left side of the second equation from $4y$ to $36y$. In order to do this, we must multiply the second equation by 9.

15.
$$
\begin{array}{rcll}
2x - y &=& 13 & (1) \\
x + y &=& 8 & (2) \\
\hline
3x &=& 21 & Add\ (1)\ and\ (2). \\
x &=& 7 &
\end{array}
$$

From (2), $y = 1$.

The solution set is $\{(7, 1)\}$.

16. $-4x + 3y = 25$ (1)
    $6x - 5y = -39$ (2)

Multiply equation (1) by 3 and equation (2) by 2; then add the results.

$$
\begin{array}{rcl}
-12x + 9y &=& 75 \\
12x - 10y &=& -78 \\
\hline
-y &=& -3 \\
y &=& 3
\end{array}
$$

To find x, let $y = 3$ in equation (1).

$$-4x + 3y = 25$$
$$-4x + 3(3) = 25$$
$$-4x + 9 = 25$$
$$-4x = 16$$
$$x = -4$$

The solution set is $\{(-4, 3)\}$.

17. $3x - 4y = 9$ (1)
    $6x - 8y = 18$ (2)

Multiply equation (1) by -2 and add the result to equation (2).

$$
\begin{array}{rcl}
-6x + 8y &=& -18 \\
6x - 8y &=& 18 \\
\hline
0 &=& 0 \quad True
\end{array}
$$

This result indicates that all solutions of equation (1) are also solutions of equation (2). The given system has an infinite number of solutions. The solution set is $\{(x, y) \mid 3x - 4y = 9\}$.

18. $2x + y = 3$ $\qquad$ (1)
$\qquad$ $-4x - 2y = 6$ $\qquad$ (2)

Multiply equation (1) by 2 and add the result to equation (2).

$$
\begin{array}{rrrcr}
4x & + & 2y & = & 6 \\
-4x & - & 2y & = & 6 \\
\hline
& & 0 & = & 12 \quad \textit{False}
\end{array}
$$

This result indicates that the given system has solution set $\emptyset$.

19. $2x + 3y = -5$ $\qquad$ (1)
$\qquad$ $3x + 4y = -8$ $\qquad$ (2)

Multiply equation (1) by -3 and equation (2) by 2; then add the results.

$$
\begin{array}{rrrcr}
-6x & - & 9y & = & 15 \\
6x & + & 8y & = & -16 \\
\hline
& & -y & = & -1 \\
& & y & = & 1
\end{array}
$$

To find x, let $y = 1$ in equation (1).

$$
\begin{aligned}
2x + 3y &= -5 \\
2x + 3(1) &= -5 \\
2x + 3 &= -5 \\
2x &= -8 \\
x &= -4
\end{aligned}
$$

The solution set is $\{(-4, 1)\}$.

20. $6x - 9y = 0$ $\qquad$ (1)
$\qquad$ $2x - 3y = 0$ $\qquad$ (2)

Multiply equation (2) by -3 and add the result to equation (1).

$$
\begin{array}{rrrcr}
6x & - & 9y & = & 0 \\
-6x & + & 9y & = & 0 \\
\hline
& & 0 & = & 0 \quad \textit{True}
\end{array}
$$

This result indicates that the system has an infinite number of solutions. The solution set is $\{(x, y) \mid 2x - 3y = 0\}$.

21. $x - 2y = 5$ $\qquad$ (1)
$\qquad$ $y = x - 7$ $\qquad$ (2)

From (2), substitute $x - 7$ for y in equation (1).

$$
\begin{aligned}
x - 2y &= 5 \\
x - 2(x - 7) &= 5 \\
x - 2x + 14 &= 5 \\
-x &= -9 \\
x &= 9
\end{aligned}
$$

Let $x = 9$ in equation (2) to find y.

$$
y = 9 - 7 = 2
$$

The solution set is $\{(9, 2)\}$.

22. $\dfrac{x}{2} + \dfrac{y}{3} = 7$ $\qquad$ (1)

$\dfrac{x}{4} + \dfrac{2y}{3} = 8$ $\qquad$ (2)

Multiply equation (1) by 6 to clear fractions.

$$
\begin{aligned}
6\left(\frac{x}{2} + \frac{y}{3}\right) &= 6(7) \\
3x + 2y &= 42 \qquad (3)
\end{aligned}
$$

Multiply equation (2) by 12 to clear fractions.

$$
\begin{aligned}
12\left(\frac{x}{4} + \frac{2y}{3}\right) &= 12(8) \\
3x + 8y &= 96 \qquad (4)
\end{aligned}
$$

To solve this system by the elimination method, multiply equation (3) by -1 and add the result to equation (4).

$$
\begin{array}{rrrcr}
-3x & - & 2y & = & -42 \\
3x & + & 8y & = & 96 \\
\hline
& & 6y & = & 54 \\
& & y & = & 9
\end{array}
$$

To find x, let $y = 9$ in equation (3).

$$
\begin{aligned}
3x + 2y &= 42 \\
3x + 2(9) &= 42 \\
3x + 18 &= 42 \\
3x &= 24 \\
x &= 8
\end{aligned}
$$

The solution set is $\{(8, 9)\}$.

23. The three methods of solving a system of equations are graphing, substitution, and elimination.

Answers to the second part of this exercise may vary. The following is one possible answer.

Consider the method of graphing. One advantage of this method of solution is that it is fast and that we can easily see if the system has one solution, no solution, or infinitely many solutions. One drawback is that we cannot always read the exact coordinates of the point of intersection.

24. System **B** is easier to solve by the substitution method than system **A** because the bottom equation in system **B** is already solved for y.

Solving system **A** would require our solving one of the equations for one of the variables before substituting, and the expression to be substituted would involve fractions.

25.
$$
\begin{aligned}
2x + 3y - z &= -16 \quad (1) \\
x + 2y + 2z &= -3 \quad (2) \\
-3x + y + z &= -5 \quad (3)
\end{aligned}
$$

To eliminate z, add equations (1) and (3).

$$
\begin{array}{r}
2x + 3y - z = -16 \quad (1) \\
-3x + y + z = -5 \quad (3) \\
\hline
-x + 4y = -21 \quad (4)
\end{array}
$$

To eliminate z again, multiply equation (1) by 2 and add the result to equation (2).

$$
\begin{array}{lr}
4x + 6y - 2z = -32 & 2 \times (1) \\
x + 2y + 2z = -3 & (2) \\
\hline
5x + 8y = -35 & (5)
\end{array}
$$

Use equations (4) and (5) to eliminate x. Multiply equation (4) by 5 and add the result to equation (5).

$$
\begin{array}{lr}
-5x + 20y = -105 & 5 \times (4) \\
5x + 8y = -35 & (5) \\
\hline
28y = -140 & \\
y = -5 &
\end{array}
$$

Substitute -5 for y in equation (4) to find x.

$$
\begin{aligned}
-x + 4y &= -21 \quad (4) \\
-x + 4(-5) &= -21 \\
-x - 20 &= -21 \\
-x &= -1 \\
x &= 1
\end{aligned}
$$

Substitute 1 for x and -5 for y in equation (2) to find z.

$$
\begin{aligned}
x + 2y + 2z &= -3 \quad (2) \\
1 + 2(-5) + 2z &= -3 \\
1 - 10 + 2z &= -3 \\
2z &= 6 \\
z &= 3
\end{aligned}
$$

The solution set is $\{(1, -5, 3)\}$.

26.
$$
\begin{aligned}
4x - y &= 2 \quad (1) \\
3y + z &= 9 \quad (2) \\
x + 2z &= 7 \quad (3)
\end{aligned}
$$

To eliminate y, multiply equation (1) by 3 and add the result to equation (2).

$$
\begin{array}{lr}
12x - 3y = 6 & 3 \times (1) \\
3y + z = 9 & (2) \\
\hline
12x + z = 15 & (4)
\end{array}
$$

To eliminate z, multiply equation (4) by -2 and add the result to equation (3).

$$
\begin{array}{lr}
-24x - 2z = -30 & -2 \times (4) \\
x + 2z = 7 & (3) \\
\hline
-23x = -23 & \\
x = 1 &
\end{array}
$$

Substitute 1 for x in equation (3) to find z.

$$
\begin{aligned}
x + 2z &= 7 \quad (3) \\
1 + 2z &= 7 \\
2z &= 6 \\
z &= 3
\end{aligned}
$$

Substitute 1 for x in equation (1) to find y.

$$
\begin{aligned}
4x - y &= 2 \quad (1) \\
4(1) - y &= 2 \\
4 - y &= 2 \\
-y &= -2 \\
y &= 2
\end{aligned}
$$

The solution set is $\{(1, 2, 3)\}$.

27.
$$
\begin{aligned}
3x - y - z &= -8 \quad (1) \\
4x + 2y + 3z &= 15 \quad (2) \\
-6x + 2y + 2z &= 10 \quad (3)
\end{aligned}
$$

To eliminate y, multiply equation (1) by 2 and add the result to equation (3).

$$
\begin{array}{lr}
6x - 2y - 2z = -16 & 2 \times (1) \\
-6x + 2y + 2z = 10 & (3) \\
\hline
0 = -6 & \textit{False}
\end{array}
$$

Since a false statement results, equations (1) and (3) have no common solution. The system is *inconsistent*. The solution set is $\emptyset$.

28. Let $x =$ the width of the rink, and $y =$ the length of the rink.

The length is 30 ft longer than twice the width.

$$
y = 2x + 30 \quad (1)
$$

The perimeter is 570 ft.

$$
2x + 2y = 570 \quad (2)
$$

Substitute $2x + 30$ for y in equation (2).

$$
\begin{aligned}
2x + 2y &= 570 \quad (2) \\
2x + 2(2x + 30) &= 570 \\
2x + 4x + 60 &= 570 \\
6x &= 510 \\
x &= 85
\end{aligned}
$$

From (1), $y = 2(85) + 30 = 200$.

The width is 85 ft and the length is 200 ft.

29. Let $x =$ the average price for a Red Sox ticket and $y =$ the average price for a Cubs ticket.

From the given information, we get the following system of equations:

$$
\begin{aligned}
4x + 4y &= 276.88 \quad (1) \\
2x + 6y &= 252.24 \quad (2)
\end{aligned}
$$

Divide (1) by 4 and (2) by -2.

continued

$$\begin{array}{rl} x + y = & 69.22 \quad (3) \ (1) \div 4 \\ -x - 3y = & -126.12 \quad (4) \ (2) \div (-2) \\ \hline -2y = & -56.90 \quad Add \ (3) \ and \ (4). \\ y = & 28.45 \end{array}$$

From (3), $x + 28.45 = 69.22$, so $x = 40.77$. The average price for a Red Sox ticket was $40.77 and the average price for a Cubs ticket was $28.45.

30. Let $x =$ the speed of the plane and $y =$ the speed of the wind.

Make a chart.

	r	t	d
With Wind	$x+y$	1.75	$1.75(x+y)$
Against Wind	$x-y$	2	$2(x-y)$

The distance each way is 560 miles. From the chart,

$$1.75(x+y) = 560.$$

Divide by 1.75.

$$x + y = 320 \quad (1)$$

From the chart,

$$2(x-y) = 560$$
$$x - y = 280. \quad (2)$$

Solve the system by adding equations (1) and (2) to eliminate y.

$$\begin{array}{rl} x + y = & 320 \quad (1) \\ x - y = & 280 \quad (2) \\ \hline 2x = & 600 \\ x = & 300 \end{array}$$

From (1), $300 + y = 320$, so $y = 20$.

The speed of the plane was 300 mph, and the speed of the wind was 20 mph.

31. Let $x =$ the amount of $2-per-pound nuts and $y =$ the amount of $1-per-pound chocolate candy.

Make a chart.

	Number of Pounds	Price per Pound	Value
Nuts	x	2	$2x$
Chocolate	y	1	$1y = y$
Mixture	100	1.30	$1.30(100) = 130$

Solve the system formed from the first and third columns.

$$x + y = 100 \quad (1)$$
$$2x + y = 130 \quad (2)$$

Solve equation (1) for y.

$$y = 100 - x \quad (3)$$

Substitute $100 - x$ for y in equation (2).

$$2x + (100 - x) = 130$$
$$x = 30$$

From (3), $y = 100 - 30 = 70$.

She should use 30 lb of $2-per-pound nuts and 70 lb of $1-per-pound chocolate candy.

32. Let $x =$ the measure of the largest angle,
$y =$ the measure of the middle-sized angle,
and $z =$ the measure of the smallest angle.

Since the sum of the measures of the angles of a triangle is $180°$,

$$x + y + z = 180. \quad (1)$$

Since the largest angle measures $10°$ less than the sum of the other two,

$$x = y + z - 10$$
$$\text{or} \quad x - y - z = -10. \quad (2)$$

Since the measure of the middle-sized angle is the average of the other two,

$$y = \frac{x+z}{2}$$
$$2y = x + z$$
$$-x + 2y - z = 0. \quad (3)$$

Solve the system.

$$\begin{array}{rrrrl} x + & y + & z = & 180 & (1) \\ x - & y - & z = & -10 & (2) \\ -x + & 2y - & z = & 0 & (3) \end{array}$$

Add equations (1) and (2) to find x.

$$\begin{array}{rrrrl} x + & y + & z = & 180 & (1) \\ x - & y - & z = & -10 & (2) \\ \hline 2x & & = & 170 \\ & & x = & 85 \end{array}$$

Add equations (1) and (3), to find y.

$$\begin{array}{rrrrl} x + & y + & z = & 180 & (1) \\ -x + & 2y - & z = & 0 & (3) \\ \hline & 3y & = & 180 \\ & y & = & 60 \end{array}$$

Substitute 85 for x and 60 for y in equation (1) to find z.

$$x + y + z = 180 \quad (1)$$
$$85 + 60 + z = 180$$
$$145 + z = 180$$
$$z = 35$$

The three angle measures are $85°$, $60°$, and $35°$.

33. Let $x =$ the value of sales at 10%,
$y =$ the value of sales at 6%,
and $z =$ the value of sales at 5%.

Since her total sales were $280,000,

$$x + y + z = 280{,}000 \quad (1)$$

Since her commissions on the sales totaled $17,000,

$$0.10x + 0.06y + 0.05z = 17{,}000.$$

Multiply by 100 to clear the decimals, so

$$10x + 6y + 5z = 1{,}700{,}000. \quad (2)$$

Since the 5% sale amounted to the sum of the other two sales,

$$z = x + y. \quad (3)$$

Solve the system.

$$
\begin{array}{rcrcrcr}
x & + & y & + & z & = & 280{,}000 \quad (1) \\
10x & + & 6y & + & 5z & = & 1{,}700{,}000 \quad (2) \\
& & & & z & = & x + y \quad (3)
\end{array}
$$

Since equation (3) is given in terms of z, substitute $x + y$ for z in equations (1) and (2).

$$
\begin{aligned}
x + y + z &= 280{,}000 \quad (1) \\
x + y + (x + y) &= 280{,}000 \\
2x + 2y &= 280{,}000 \\
x + y &= 140{,}000 \quad (4)
\end{aligned}
$$

$$
\begin{aligned}
10x + 6y + 5z &= 1{,}700{,}000 \quad (2) \\
10x + 6y + 5(x + y) &= 1{,}700{,}000 \\
10x + 6y + 5x + 5y &= 1{,}700{,}000 \\
15x + 11y &= 1{,}700{,}000 \quad (5)
\end{aligned}
$$

To eliminate x, multiply equation (4) by -11 and add the result to equation (5).

$$
\begin{array}{rcrl}
-11x & - & 11y = -1{,}540{,}000 & \quad -11 \times (4) \\
15x & + & 11y = 1{,}700{,}000 & \quad (5) \\
\hline
4x & & = 160{,}000 & \\
x & & = 40{,}000 &
\end{array}
$$

From (4), $y = 100{,}000$.
From (3), $z = 40{,}000 + 100{,}000 = 140{,}000$.
He sold $40,000 at 10%, $100,000 at 6%, and $140,000 at 5%.

34. Let $x =$ the number of liters of 8% solution,
$y =$ the number of liters of 10% solution,
and $z =$ the number of liters of 20% solution.

Since the amount of the mixture will be 8 L,

$$x + y + z = 8. \quad (1)$$

Since the final solution will be 12.5% hydrogen peroxide,

$$0.08x + 0.10y + 0.20z = 0.125(8).$$

Multiply by 100 to clear the decimals.

$$8x + 10y + 20z = 100 \quad (2)$$

Since the amount of 8% solution used must be 2 L more than the amount of 20% solution,

$$x = z + 2. \quad (3)$$

Solve the system.

$$
\begin{array}{rcrcrcr}
x & + & y & + & z & = & 8 \quad (1) \\
8x & + & 10y & + & 20z & = & 100 \quad (2) \\
x & & & = & z + 2 & & (3)
\end{array}
$$

Since equation (3) is given in terms of x, substitute $z + 2$ for x in equations (1) and (2).

$$
\begin{aligned}
x + y + z &= 8 \quad (1) \\
(z + 2) + y + z &= 8 \\
y + 2z &= 6 \quad (4)
\end{aligned}
$$

$$
\begin{aligned}
8x + 10y + 20z &= 100 \quad (2) \\
8(z + 2) + 10y + 20z &= 100 \\
8z + 16 + 10y + 20z &= 100 \\
10y + 28z &= 84 \quad (5)
\end{aligned}
$$

To eliminate y, multiply equation (4) by -10 and add the result to equation (5).

$$
\begin{array}{rcrl}
-10y & - & 20z = -60 & \quad -10 \times (4) \\
10y & + & 28z = 84 & \quad (5) \\
\hline
& & 8z = 24 & \\
& & z = 3 &
\end{array}
$$

From (3), $x = z + 2 = 3 + 2 = 5$.
From (4), $y = 6 - 2z = 6 - 2(3) = 0$.
Mix 5 L of 8% solution, none of 10% solution, and 3 L of 20% solution.

35. Let $x =$ the number of home runs hit by Mantle,
$y =$ the number of home runs hit by Maris,
and $z =$ the number of home runs hit by Blanchard.

They combined for 136 home runs, so

$$x + y + z = 136. \quad (1)$$

Mantle hit 7 fewer than Maris, so

$$x = y - 7. \quad (2)$$

Maris hit 40 more than Blanchard, so

$$y = z + 40 \quad \text{or} \quad z = y - 40. \quad (3)$$

Substitute $y - 7$ for x and $y - 40$ for z in (1).

$$
\begin{aligned}
x + y + z &= 136 \quad (1) \\
(y - 7) + y + (y - 40) &= 136 \\
3y - 47 &= 136 \\
3y &= 183 \\
y &= 61
\end{aligned}
$$

From (2), $x = y - 7 = 61 - 7 = 54$.
From (3), $z = y - 40 = 61 - 40 = 21$.
Mantle hit 54 home runs, Maris hit 61 home runs, and Blanchard hit 21 home runs.

36. $2x + 5y = -4$
$4x - y = 14$

Write the augmented matrix.

$$\begin{bmatrix} 2 & 5 & | & -4 \\ 4 & -1 & | & 14 \end{bmatrix}$$

$$\begin{bmatrix} 2 & 5 & | & -4 \\ 0 & -11 & | & 22 \end{bmatrix} \quad -2R_1 + R_2$$

$$\begin{bmatrix} 2 & 5 & | & -4 \\ 0 & 1 & | & -2 \end{bmatrix} \quad -\frac{1}{11}R_2$$

This matrix gives the system

$$2x + 5y = -4$$
$$y = -2.$$

Substitute $y = -2$ in the first equation.

$$2x + 5y = -4$$
$$2x + 5(-2) = -4$$
$$2x - 10 = -4$$
$$2x = 6$$
$$x = 3$$

The solution set is $\{(3, -2)\}$.

37. $6x + 3y = 9$
$-7x + 2y = 17$

Write the augmented matrix.

$$\begin{bmatrix} 6 & 3 & | & 9 \\ -7 & 2 & | & 17 \end{bmatrix}$$

$$\begin{bmatrix} 1 & -5 & | & -26 \\ -7 & 2 & | & 17 \end{bmatrix} \quad -R_1 - R_2$$

$$\begin{bmatrix} 1 & -5 & | & -26 \\ 0 & -33 & | & -165 \end{bmatrix} \quad 7R_1 + R_2$$

$$\begin{bmatrix} 1 & -5 & | & -26 \\ 0 & 1 & | & 5 \end{bmatrix} \quad -\frac{1}{33}R_2$$

This matrix gives the system

$$x - 5y = -26$$
$$y = 5.$$

Substitute $y = 5$ in the first equation.

$$x - 5y = -26$$
$$x - 5(5) = -26$$
$$x - 25 = -26$$
$$x = -1$$

The solution set is $\{(-1, 5)\}$.

38. $x + 2y - z = 1$
$3x + 4y + 2z = -2$
$-2x - y + z = -1$

$$\begin{bmatrix} 1 & 2 & -1 & | & 1 \\ 3 & 4 & 2 & | & -2 \\ -2 & -1 & 1 & | & -1 \end{bmatrix}$$

$$\begin{bmatrix} 1 & 2 & -1 & | & 1 \\ 0 & -2 & 5 & | & -5 \\ 0 & 3 & -1 & | & 1 \end{bmatrix} \quad \begin{matrix} -3R_1 + R_2 \\ 2R_1 + R_3 \end{matrix}$$

$$\begin{bmatrix} 1 & 2 & -1 & | & 1 \\ 0 & 1 & 4 & | & -4 \\ 0 & 3 & -1 & | & 1 \end{bmatrix} \quad R_3 + R_2$$

$$\begin{bmatrix} 1 & 2 & -1 & | & 1 \\ 0 & 1 & 4 & | & -4 \\ 0 & 0 & -13 & | & 13 \end{bmatrix} \quad -3R_2 + R_3$$

$$\begin{bmatrix} 1 & 2 & -1 & | & 1 \\ 0 & 1 & 4 & | & -4 \\ 0 & 0 & 1 & | & -1 \end{bmatrix} \quad -\frac{1}{13}R_3$$

This matrix gives the system

$$x + 2y - z = 1$$
$$y + 4z = -4$$
$$z = -1.$$

Substitute $z = -1$ in the second equation.

$$y + 4z = -4$$
$$y + 4(-1) = -4$$
$$y = 0$$

Substitute $y = 0$ and $z = -1$ in the first equation.

$$x + 2y - z = 1$$
$$x + 2(0) - (-1) = 1$$
$$x + 1 = 1$$
$$x = 0$$

The solution set is $\{(0, 0, -1)\}$.

39. $x + 3y = 7$
$3x + z = 2$
$y - 2z = 4$

$$\begin{bmatrix} 1 & 3 & 0 & | & 7 \\ 3 & 0 & 1 & | & 2 \\ 0 & 1 & -2 & | & 4 \end{bmatrix}$$

$$\begin{bmatrix} 1 & 3 & 0 & | & 7 \\ 0 & -9 & 1 & | & -19 \\ 0 & 1 & -2 & | & 4 \end{bmatrix} \quad -3R_1 + R_2$$

$$\begin{bmatrix} 1 & 3 & 0 & | & 7 \\ 0 & 1 & -2 & | & 4 \\ 0 & -9 & 1 & | & -19 \end{bmatrix} \quad R_2 \leftrightarrow R_3$$

We use $\leftrightarrow$ to represent the interchanging of 2 rows.

$$\begin{bmatrix} 1 & 3 & 0 & | & 7 \\ 0 & 1 & -2 & | & 4 \\ 0 & 0 & -17 & | & 17 \end{bmatrix} \quad 9R_2 + R_3$$

$$\begin{bmatrix} 1 & 3 & 0 & | & 7 \\ 0 & 1 & -2 & | & 4 \\ 0 & 0 & 1 & | & -1 \end{bmatrix} \quad -\frac{1}{17}R_3$$

This matrix gives the system

$$x + 3y = 7$$
$$y - 2z = 4$$
$$z = -1.$$

Substitute $z = -1$ in the second equation.

$$y - 2z = 4$$
$$y - 2(-1) = 4$$
$$y + 2 = 4$$
$$y = 2$$

Substitute $y = 2$ in the first equation.

$$x + 3y = 7$$
$$x + 3(2) = 7$$
$$x + 6 = 7$$
$$x = 1$$

The solution set is $\{(1, 2, -1)\}$.

40. **[8.1]** **(a)** Rising graphs indicate population growth, so Houston, Phoenix, and Dallas will experience population growth.

(b) Philadelphia's graph indicates that it will experience population decline.

(c) In the year 2000, the city populations from least to greatest are Dallas, Phoenix, Philadelphia, and Houston.

(d) The graphs for Dallas and Philadelphia intersect in the year 2010. The population for each city will be about 1.45 million.

(e) The graphs for Houston and Phoenix appear to intersect in the year 2025. The population for each city will be about 2.8 million. This can be represented by the ordered pair $(2025, 2.8)$.

41. **[8.3]** $\frac{2}{3}x + \frac{1}{6}y = \frac{19}{2}$ (1)
         $\frac{1}{3}x - \frac{2}{9}y = 2$ (2)

Multiply equation (1) by 6 and equation (2) by 9 to clear the fractions.

$$4x + y = 57 \quad (3) \; 6 \times (1)$$
$$3x - 2y = 18 \quad (4) \; 9 \times (2)$$

To eliminate y, multiply equation (3) by 2 and add the result to equation (4).

$$8x + 2y = 114 \qquad 2 \times (3)$$
$$\underline{3x - 2y = 18 \quad (4)}$$
$$11x \quad\quad = 132$$
$$x = 12$$

Substitute 12 for x in equation (3) to find y.

$$4x + y = 57 \quad (3)$$
$$4(12) + y = 57$$
$$48 + y = 57$$
$$y = 9$$

The solution set is $\{(12, 9)\}$.

42. **[8.4]** $2x + 5y - z = 12$ (1)
         $-x + y - 4z = -10$ (2)
         $-8x - 20y + 4z = 31$ (3)

Multiply equation (1) by 4 and add the result to equation (3).

$$8x + 20y - 4z = 48 \qquad 4 \times (1)$$
$$\underline{-8x - 20y + 4z = 31 \quad (3)}$$
$$0 = 79 \; \textit{False}$$

Since a false statement results, the system is *inconsistent*. The solution set is $\emptyset$.

43. **[8.2]** $x = 7y + 10$ (1)
         $2x + 3y = 3$ (2)

Since equation (1) is given in terms of x, substitute $7y + 10$ for x in equation (2) and solve for y.

$$2(7y + 10) + 3y = 3$$
$$14y + 20 + 3y = 3$$
$$17y = -17$$
$$y = -1$$

From (1), $x = 7(-1) + 10 = 3$.

The solution set is $\{(3, -1)\}$.

44. **[8.3]** $x + 4y = 17$ (1)
         $-3x + 2y = -9$ (2)

To eliminate x, multiply equation (1) by 3 and add the result to equation (2).

$$3x + 12y = 51 \qquad 3 \times (1)$$
$$\underline{-3x + 2y = -9 \quad (2)}$$
$$14y = 42$$
$$y = 3$$

Substitute 3 for y in equation (1) to find x.

$$x + 4y = 17 \quad (1)$$
$$x + 4(3) = 17$$
$$x + 12 = 17$$
$$x = 5$$

The solution set is $\{(5, 3)\}$.

45. **[8.3]** $-7x + 3y = 12$ (1)
         $5x + 2y = 8$ (2)

To eliminate y, multiply equation (1) by 2 and equation (2) by -3. Then add the results.

$$-14x + 6y = 24 \quad 2 \times (1)$$
$$\underline{-15x - 6y = -24 \; -3 \times (2)}$$
$$-29x \quad\quad = 0$$
$$x = 0$$

continued

Substitute 0 for x in equation (1) to find y.

$$-7x + 3y = 12 \quad (1)$$
$$-7(0) + 3y = 12$$
$$3y = 12$$
$$y = 4$$

The solution set is $\{(0, 4)\}$.

46. **[8.3]** $2x - 5y = 8 \quad (1)$
$\qquad\quad\; 3x + 4y = 10 \quad (2)$

To eliminate y, multiply equation (1) by 4 and equation (2) by 5 and add the results.

$$8x - 20y = 32 \quad 4 \times (1)$$
$$\underline{15x + 20y = 50} \quad 5 \times (2)$$
$$23x \qquad\quad = 82$$
$$x = \tfrac{82}{23}$$

Instead of substituting to find y, we'll choose different multipliers and eliminate x from the original system.

$$6x - 15y = 24 \quad 3 \times (1)$$
$$\underline{-6x - 8y = -20} \quad -2 \times (2)$$
$$-23y = 4$$
$$y = -\tfrac{4}{23}$$

The solution set is $\left\{ \left(\tfrac{82}{23}, -\tfrac{4}{23} \right) \right\}$.

47. **[8.5]** Let $x =$ the number of liters of 5% solution and $y =$ the number of liters of 10% solution.

Liters of Solution	Percent (as a decimal)	Liters of Pure Acid
x	0.05	$0.05x$
10	0.20	$0.20(10) = 2$
y	0.10	$0.10y$

Solve the system formed from the first and third columns.

$$x + 10 = y \qquad (1)$$
$$0.05x + 2 = 0.10y \qquad (2)$$

Multiply equation (2) by 100 to clear the decimals.

$$5x + 200 = 10y \quad (3)$$

Substitute $x + 10$ for y in equation (3) and solve for x.

$$5x + 200 = 10y \qquad (3)$$
$$5x + 200 = 10(x + 10)$$
$$5x + 200 = 10x + 100$$
$$100 = 5x$$
$$20 = x$$

He should use 20 L of 5% solution.

48. **[8.5]** Let x, y, and z denote the number of medals won by Germany, the United States, and Canada, respectively.

The total number of medals won was 78, so

$$x + y + z = 78. \quad (1)$$

Germany won four more medals than the United States, so

$$x = y + 4. \quad (2)$$

Canada won one fewer medal than the United States, so

$$z = y - 1. \quad (3)$$

Substitute $y + 4$ for x and $y - 1$ for z in (1).

$$(y + 4) + y + (y - 1) = 78$$
$$3y + 3 = 78$$
$$3y = 75$$
$$y = 25$$

From (2), $x = 25 + 4 = 29$.
From (3), $z = 25 - 1 = 24$.

Germany won 29 medals, the United States won 25 medals, and Canada won 24 medals.

Chapter 8 Test

1. The lines intersect at $(8, 3000)$, so cost equals revenue at $x = 8$ (which is 800 parts).

2. The revenue is $3000.

3. $2x + y = -3 \quad (1)$
$\quad\; x - y = -9 \quad (2)$

(a) $(1, -5)$

Substitute 1 for x and -5 for y in (1) and (2).

$$(1) \quad 2(1) + (-5) = -3 \,?$$
$$-3 = -3 \quad True$$

$$(2) \quad (1) - (-5) = -9 \,?$$
$$6 = -9 \quad False$$

Since $(1, -5)$ does not satisfy both equations, it *is not* a solution of the system.

(b) $(1, 10)$

$$(1) \quad 2(1) + (10) = -3 \,?$$
$$12 = -3 \quad False$$

The false result indicates that $(1, 10)$ *is not* a solution of the system.

(c) $(-4, 5)$

(1) $\quad 2(-4) + (5) = -3$?
$\qquad -3 = -3 \quad$ *True*

(2) $\quad (-4) - (5) = -9$?
$\qquad -9 = -9 \quad$ *True*

Since $(-4, 5)$ satisfies both equations, it *is* a solution of the system.

4. When each equation of the system

$$x + y = 7$$
$$x - y = 5$$

is graphed, the point of intersection appears to be $(6, 1)$. To check, substitute 6 for x and 1 for y in each of the equations. Since $(6, 1)$ makes both equations true, the solution set of the system is $\{(6, 1)\}$.

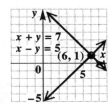

5. $\quad 2x - 3y = 24 \qquad (1)$
$\qquad\qquad y = -\frac{2}{3}x \qquad (2)$

Since equation (2) is solved for y, substitute $-\frac{2}{3}x$ for y in equation (1) and solve for x.

$$2x - 3y = 24 \quad (1)$$
$$2x - 3\left(-\frac{2}{3}x\right) = 24$$
$$2x + 2x = 24$$
$$4x = 24$$
$$x = 6$$

From (2), $y = -\frac{2}{3}(6) = -4$.

The solution set is $\{(6, -4)\}$.

6. $\quad 3x - y = -8 \qquad (1)$
$\qquad 2x + 6y = 3 \qquad (2)$

To eliminate x, multiply equation (1) by 6. Then add that equation and equation (2).

$$18x - 6y = -48 \qquad 6 \times (1)$$
$$\underline{2x + 6y = 3 \qquad (2)}$$
$$20x \qquad = -45$$
$$x = -\frac{45}{20} = -\frac{9}{4}$$

To find y, substitute $-\frac{9}{4}$ for x in equation (2). (We could also use elimination by adding $2 \times (1)$ and $-3 \times (2)$.)

$$2x + 6y = 3 \qquad (2)$$
$$2\left(-\frac{9}{4}\right) + 6y = 3$$
$$-\frac{9}{2} + 6y = 3$$
$$6y = \frac{15}{2}$$
$$y = \frac{15}{12} = \frac{5}{4}$$

The solution set is $\left\{-\frac{9}{4}, \frac{5}{4}\right\}$.

7. $\quad 12x - 5y = 8 \qquad (1)$
$\qquad\quad 3x = \frac{5}{4}y + 2$
$\text{or} \qquad x = \frac{5}{12}y + \frac{2}{3} \qquad (2)$

Substitute $\frac{5}{12}y + \frac{2}{3}$ for x in equation (1) and solve for y.

$$12x - 5y = 8 \quad (1)$$
$$12\left(\frac{5}{12}y + \frac{2}{3}\right) - 5y = 8$$
$$5y + 8 - 5y = 8$$
$$8 = 8 \quad \text{True}$$

Equations (1) and (2) are dependent.
The solution set is $\{(x, y) \mid 12x - 5y = 8\}$.

8. $\quad 3x + y = 12 \qquad (1)$
$\qquad 2x - y = 3 \qquad (2)$

To eliminate y, add equations (1) and (2).

$$3x + y = 12 \quad (1)$$
$$\underline{2x - y = 3 \quad (2)}$$
$$5x \qquad = 15$$
$$x = 3$$

Substitute 3 for x in equation (1) to find y.

$$3x + y = 12 \quad (1)$$
$$3(3) + y = 12$$
$$9 + y = 12$$
$$y = 3$$

The solution set is $\{(3, 3)\}$.

9. $\quad -5x + 2y = -4 \qquad (1)$
$\qquad\quad 6x + 3y = -6 \qquad (2)$

To eliminate x, multiply equation (1) by 6 and equation (2) by 5. Then add the results.

$$-30x + 12y = -24 \quad 6 \times (1)$$
$$\underline{30x + 15y = -30 \quad 5 \times (2)}$$
$$27y = -54$$
$$y = -2$$

Substitute -2 for y in equation (1) to find x.

$$-5x + 2y = -4 \quad (1)$$
$$-5x + 2(-2) = -4$$
$$-5x - 4 = -4$$
$$-5x = 0$$
$$x = 0$$

The solution set is $\{(0, -2)\}$.

10.
$$3x + 4y = 8 \quad (1)$$
$$8y = 7 - 6x$$
or $\quad 6x + 8y = 7 \quad (2)$

Multiply equation (1) by -2 and add the result to equation (2).

$$\begin{array}{rl} -6x - 8y = -16 & -2 \times (1) \\ 6x + 8y = 7 & (2) \\ \hline 0 = -9 & \textit{False} \end{array}$$

Since a false statement results, the system is *inconsistent*. The solution set is $\emptyset$.

11. $\frac{6}{5}x - \frac{1}{3}y = -20 \quad (1)$
$-\frac{2}{3}x + \frac{1}{6}y = 11 \quad (2)$

To clear fractions, multiply by the LCD for each equation. Multiply equation (1) by 15.

$$15\left(\frac{6}{5}x - \frac{1}{3}y\right) = 15(-20)$$
$$18x - 5y = -300 \quad (3)$$

Multiply equation (2) by 6.

$$6\left(-\frac{2}{3}x + \frac{1}{6}y\right) = 6(11)$$
$$-4x + y = 66 \quad (4)$$

To solve this system by the elimination method, multiply equation (4) by 5 and add the result to equation (3).

$$\begin{array}{rl} 18x - 5y = -300 \\ -20x + 5y = 330 \\ \hline -2x = 30 \\ x = -15 \end{array}$$

To find y, let $x = -15$ in equation (4).

$$-4x + y = 66$$
$$-4(-15) + y = 66$$
$$60 + y = 66$$
$$y = 6$$

The solution set is $\{(-15, 6)\}$.

12. $3x + 5y + 3z = 2 \quad (1)$
$6x + 5y + z = 0 \quad (2)$
$3x + 10y - 2z = 6 \quad (3)$

To eliminate x, multiply equation (1) by -1 and add the result to equation (3).

$$\begin{array}{rl} -3x - 5y - 3z = -2 & -1 \times (1) \\ 3x + 10y - 2z = 6 & (3) \\ \hline 5y - 5z = 4 & (4) \end{array}$$

To eliminate x again, multiply equation (1) by -2 and add the result to equation (2).

$$\begin{array}{rl} -6x - 10y - 6z = -4 & -2 \times (1) \\ 6x + 5y + z = 0 & (2) \\ \hline -5y - 5z = -4 & (5) \end{array}$$

To eliminate y, add equations (4) and (5).

$$\begin{array}{rl} 5y - 5z = 4 & (4) \\ -5y - 5z = -4 & (5) \\ \hline -10z = 0 \\ z = 0 \end{array}$$

Substitute 0 for z in equation (4) to find y.

$$5y - 5z = 4 \quad (4)$$
$$5y - 5(0) = 4$$
$$5y - 0 = 4$$
$$5y = 4$$
$$y = \frac{4}{5}$$

Substitute $\frac{4}{5}$ for y and 0 for z in equation (1) to find x.

$$3x + 5y + 3z = 2 \quad (1)$$
$$3x + 5\left(\frac{4}{5}\right) + 3(0) = 2$$
$$3x + 4 + 0 = 2$$
$$3x = -2$$
$$x = -\frac{2}{3}$$

The solution set is $\left\{\left(-\frac{2}{3}, \frac{4}{5}, 0\right)\right\}$.

13. $4x + y + z = 11 \quad (1)$
$x - y - z = 4 \quad (2)$
$y + 2z = 0 \quad (3)$

To eliminate x, multiply equation (2) by -4 and add the result to equation (1).

$$\begin{array}{rl} 4x + y + z = 11 & (1) \\ -4x + 4y + 4z = -16 & -4 \times (2) \\ \hline 5y + 5z = -5 & (4) \end{array}$$

To eliminate y, divide equation (4) by -5 and add the result to equation (3).

$$\begin{array}{rl} -y - z = 1 & (4) \div (-5) \\ y + 2z = 0 & (3) \\ \hline z = 1 \end{array}$$

From (3), $y + 2(1) = 0$, so $y = -2$.
From (2), $x - (-2) - 1 = 4$, so $x = 3$.

The solution set is $\{(3, -2, 1)\}$.

14. Let $x =$ the gross (in millions of dollars) for *Ocean's Eleven*, and $y =$ the gross (in millions of dollars) for *Runaway Bride*.

Together the movies grossed $335.5 million, so

$$x + y = 335.5. \quad (1)$$

Runaway Bride grossed $31.3 million less than *Ocean's Eleven*, so

$$y = x - 31.3. \quad (2)$$

Substitute $x - 31.3$ for y in equation (1).

$$x + y = 335.5 \quad (1)$$
$$x + (x - 31.3) = 335.5$$
$$2x = 366.8$$
$$x = 183.4$$

From (2), $y = 183.4 - 31.3 = 152.1$.

Ocean's Eleven grossed $183.4 million and *Runaway Bride* grossed $152.1 million.

15. Let $x =$ the speed of the faster car and $y =$ the speed of the slower car.

    Make a table.

	r	t	d
Faster Car	x	3.5	$3.5x$
Slower Car	y	3.5	$3.5y$

    Since the slow car travels 30 mph slower than the fast car,

    $$x - y = 30. \quad (1)$$

    Since the cars travel a total of 420 miles,

    $$3.5x + 3.5y = 420.$$

    Multiply by 10 to clear the decimals.

    $$35x + 35y = 4200 \quad (2)$$

    To eliminate y, multiply equation (1) by 35 and add the result to equation (2).

    $$\begin{array}{rl} 35x - 35y = 1050 & 35 \times (1) \\ 35x + 35y = 4200 & (2) \\ \hline 70x \qquad = 5250 & \\ x = 75 & \end{array}$$

    Substitute 75 for x in equation (1) to find y.

    $$x - y = 30 \quad (1)$$
    $$75 - y = 30$$
    $$-y = -45$$
    $$y = 45$$

    The faster car is traveling at 75 mph, and the slower car is traveling at 45 mph.

16. Let $x =$ the number of liters of 20% solution and $y =$ the number of liters of 50% solution.

    Make a table.

Liters of Solution	Percent (as a decimal)	Liters of Pure Alcohol
x	0.20	$0.20x$
y	0.50	$0.50y$
12	0.40	$0.40(12) = 4.8$

    Since 12 L of the mixture are needed,

    $$x + y = 12. \quad (1)$$

Since the amount of pure alcohol in the 20% solution plus the amount of pure alcohol in the 50% solution must equal the amount of alcohol in the mixture,

$$0.2x + 0.5y = 4.8.$$

Multiply by 10 to clear the decimals.

$$2x + 5y = 48 \quad (2)$$

Multiply equation (1) by -2 and add the result to equation (2).

$$\begin{array}{rl} -2x - 2y = -24 & -2 \times (1) \\ 2x + 5y = 48 & (2) \\ \hline 3y = 24 & \\ y = 8 & \end{array}$$

From (1), $x + 8 = 12$, so $x = 4$.
4 L of 20% solution and 8 L of 50% solution are needed.

17. Let $x =$ the price of an AC adaptor and $y =$ the price of a rechargeable flashlight.

    Since 7 AC adaptors and 2 rechargeable flashlights cost $86,

    $$7x + 2y = 86. \quad (1)$$

    Since 3 AC adaptors and 4 rechargeable flashlights cost $84,

    $$3x + 4y = 84. \quad (2)$$

    Solve the system.

    $$\begin{array}{rl} 7x + 2y = 86 & (1) \\ 3x + 4y = 84 & (2) \end{array}$$

    To eliminate y, multiply equation (1) by -2 and add the result to equation (2).

    $$\begin{array}{rl} -14x - 4y = -172 & -2 \times (1) \\ 3x + 4y = 84 & (2) \\ \hline -11x = -88 & \\ x = 8 & \end{array}$$

    Substitute 8 for x in equation (1) to find y.

    $$7x + 2y = 86 \quad (1)$$
    $$7(8) + 2y = 86$$
    $$56 + 2y = 86$$
    $$2y = 30$$
    $$y = 15$$

    An AC adaptor costs $8, and a rechargeable flashlight costs $15.

18. Let x = the amount of Orange Pekoe, y = the amount of Irish Breakfast, and z = the amount of Earl Grey.

The owner wants 100 oz of tea, so

$$x + y + z = 100. \quad (1)$$

An equation which relates the prices of the tea is

$$0.80x + 0.85y + 0.95z = 0.83(100).$$

Multiply by 100 to clear the decimals.

$$80x + 85y + 95z = 8300 \quad (2)$$

The mixture must contain twice as much Orange Pekoe as Irish Breakfast, so

$$x = 2y. \quad (3)$$

To eliminate z, multiply equation (1) by -95 and add the result to equation (2).

$$
\begin{array}{rrrrl}
-95x & - & 95y & - & 95z & = & -9500 & \quad -95 \times (1) \\
80x & + & 85y & + & 95z & = & 8300 & \quad (2) \\
\hline
-15x & - & 10y & & & = & -1200 & \quad (4)
\end{array}
$$

Divide equation (4) by -5.

$$3x + 2y = 240 \quad (5)$$

Substitute $2y$ for x in equation (5) to find y.

$$
\begin{aligned}
3x + 2y &= 240 \quad (5) \\
3(2y) + 2y &= 240 \\
8y &= 240 \\
y &= 30
\end{aligned}
$$

From (3), $x = 2(30) = 60$.
Substitute 60 for x and 30 for y in equation (1) to find z.

$$
\begin{aligned}
x + y + z &= 100 \quad (1) \\
60 + 30 + z &= 100 \\
z &= 10
\end{aligned}
$$

He should use 60 oz of Orange Pekoe, 30 oz of Irish Breakfast, and 10 oz of Earl Grey.

19.
$$
\begin{aligned}
3x + 2y &= 4 \\
5x + 5y &= 9
\end{aligned}
$$

Write the augmented matrix.

$$
\left[\begin{array}{cc|c}
3 & 2 & 4 \\
5 & 5 & 9
\end{array}\right]
$$

We could divide row 1 by 3, but to avoid working with fractions, we'll multiply row 1 by 2 and then multiply row 2 by -1 and add to row 1 to obtain a "1" in the first row.

$$
\left[\begin{array}{cc|c}
6 & 4 & 8 \\
5 & 5 & 9
\end{array}\right] \quad 2R_1
$$

$$
\left[\begin{array}{cc|c}
1 & -1 & -1 \\
5 & 5 & 9
\end{array}\right] \quad -R_2 + R_1
$$

$$
\left[\begin{array}{cc|c}
1 & -1 & -1 \\
0 & 10 & 14
\end{array}\right] \quad -5R_1 + R_2
$$

$$
\left[\begin{array}{cc|c}
1 & -1 & -1 \\
0 & 1 & \frac{7}{5}
\end{array}\right] \quad \frac{1}{10}R_2
$$

This matrix gives the system

$$
\begin{aligned}
x - y &= -1 \\
y &= \tfrac{7}{5}.
\end{aligned}
$$

Substitute $y = \frac{7}{5}$ in the first equation.

$$
\begin{aligned}
x - \tfrac{7}{5} &= -1 \\
x &= -1 + \tfrac{7}{5} \\
&= -\tfrac{5}{5} + \tfrac{7}{5} = \tfrac{2}{5}
\end{aligned}
$$

The solution set is $\left\{ \left(\frac{2}{5}, \frac{7}{5} \right) \right\}$.

20.
$$
\begin{array}{rrrrl}
x & + & 3y & + & 2z & = & 11 \\
3x & + & 7y & + & 4z & = & 23 \\
5x & + & 3y & - & 5z & = & -14
\end{array}
$$

Write the augmented matrix.

$$
\left[\begin{array}{ccc|c}
1 & 3 & 2 & 11 \\
3 & 7 & 4 & 23 \\
5 & 3 & -5 & -14
\end{array}\right]
$$

$$
\left[\begin{array}{ccc|c}
1 & 3 & 2 & 11 \\
0 & -2 & -2 & -10 \\
0 & -12 & -15 & -69
\end{array}\right] \quad \begin{array}{l} -3R_1 + R_2 \\ -5R_1 + R_3 \end{array}
$$

$$
\left[\begin{array}{ccc|c}
1 & 3 & 2 & 11 \\
0 & 1 & 1 & 5 \\
0 & -12 & -15 & -69
\end{array}\right] \quad -\tfrac{1}{2}R_2
$$

$$
\left[\begin{array}{ccc|c}
1 & 3 & 2 & 11 \\
0 & 1 & 1 & 5 \\
0 & 0 & -3 & -9
\end{array}\right] \quad 12R_2 + R_3
$$

$$
\left[\begin{array}{ccc|c}
1 & 3 & 2 & 11 \\
0 & 1 & 1 & 5 \\
0 & 0 & 1 & 3
\end{array}\right] \quad -\tfrac{1}{3}R_3
$$

This matrix gives the system

$$
\begin{aligned}
x + 3y + 2z &= 11 \\
y + z &= 5 \\
z &= 3.
\end{aligned}
$$

Substitute $z = 3$ in the second equation.

$$
\begin{aligned}
y + z &= 5 \\
y + 3 &= 5 \\
y &= 2
\end{aligned}
$$

Substitute $y = 2$ and $z = 3$ in the first equation.

$$
\begin{aligned}
x + 3y + 2z &= 11 \\
x + 3(2) + 2(3) &= 11 \\
x + 6 + 6 &= 11 \\
x &= -1
\end{aligned}
$$

The solution set is $\{(-1, 2, 3)\}$.

Cumulative Review Exercises
(Chapters 1–8)

1. The integer factors of 40 are $-1, 1, -2, 2, -4, 4,$
$-5, 5, -8, 8, -10, 10, -20, 20, -40,$ and $40.$

2. $\dfrac{3x^2 + 2y^2}{10y + 3} = \dfrac{3 \cdot 1^2 + 2 \cdot 5^2}{10(5) + 3}$ *Let x = 1, y = 5.*

$\qquad = \dfrac{3 \cdot 1 + 2 \cdot 25}{50 + 3}$

$\qquad = \dfrac{3 + 50}{50 + 3} = \dfrac{53}{53} = 1$

3. $5 + (-4) = (-4) + 5$

The order of the numbers has been changed, so
this is an example of the commutative property of
addition.

4. $r(s - k) = rs - rk$

This is an example of the distributive property.

5. $-\frac{2}{3} + \frac{2}{3} = 0$

The numbers $-\frac{2}{3}$ and $\frac{2}{3}$ are additive inverses (or
opposites) of each other. This is an example of the
inverse property for addition.

6. $-2 + 6[3 - (4 - 9)] = -2 + 6[3 - (-5)]$
$\qquad\qquad\qquad\qquad = -2 + 6(8)$
$\qquad\qquad\qquad\qquad = -2 + 48$
$\qquad\qquad\qquad\qquad = 46$

7. $2 - 3(6x + 2) = 4(x + 1) + 18$
$\quad 2 - 18x - 6 = 4x + 4 + 18$
$\quad\; -18x - 4 = 4x + 22$
$\qquad\quad -22x = 26$
$\qquad\qquad x = \frac{26}{-22} = -\frac{13}{11}$

The solution set is $\left\{ -\frac{13}{11} \right\}$.

8. $\frac{3}{2}\left(\frac{1}{3}x + 4 \right) = 6\left(\frac{1}{4} + x \right)$

Multiply each side by 2.

$2\left[\frac{3}{2}\left(\frac{1}{3}x + 4 \right) \right] = 2\left[6\left(\frac{1}{4} + x \right) \right]$
$\qquad 3\left(\frac{1}{3}x + 4 \right) = 12\left(\frac{1}{4} + x \right)$

Use the distributive property to remove
parentheses.

$\qquad x + 12 = 3 + 12x$
$\qquad\; -11x = -9$
$\qquad\qquad x = \frac{9}{11}$

The solution set is $\left\{ \frac{9}{11} \right\}$.

9. $\qquad P = \dfrac{kT}{V}$ for T

$\quad PV = kT$ *Multiply by V.*

$\quad \dfrac{PV}{k} = \dfrac{kT}{k}$ *Divide by k.*

$\quad \dfrac{PV}{k} = T$ or $T = \dfrac{PV}{k}$

10. $-\frac{5}{6}x < 15$

Multiply each side by the reciprocal of $-\frac{5}{6}$, which
is $-\frac{6}{5}$, and reverse the direction of the inequality
symbol.

$-\frac{6}{5}\left(-\frac{5}{6}x \right) > -\frac{6}{5}(15)$
$\qquad\qquad x > -18$

The solution set is $(-18, \infty)$.

11. $\quad -8 < 2x + 3$
$\quad -11 < 2x$
$\quad -\frac{11}{2} < x$ or $x > -\frac{11}{2}$

The solution set is $\left(-\frac{11}{2}, \infty \right)$.

12. 80.4% of 2500 is $0.804(2500) = 2010$
72.5% of 2500 is $0.725(2500) = 1812.5 \approx 1813$

$\dfrac{1570}{2500} = 0.628$ or 62.8%

$\dfrac{1430}{2500} = 0.572$ or 57.2%

Product or Company	Percent	Actual Number
Charmin	80.4%	2010
Wheaties	72.5%	1813
Budweiser	62.8%	1570
State Farm	57.2%	1430

13. Let x = number of "against" votes;
$\quad y$ = number of "in favor" votes.

There were 100 total votes and there were 56 more
"in favor" votes than "against" votes, so we have:

$\qquad x + y = 100 \qquad (1)$
$\qquad\quad\; y = x + 56 \quad (2)$

Substitute $x + 56$ for y in equation (1).

$\qquad x + (x + 56) = 100$
$\qquad\quad\; 2x + 56 = 100$
$\qquad\qquad\quad 2x = 44$
$\qquad\qquad\qquad x = 22$

From (2), $y = 22 + 56 = 78$.
There were 78 "in favor" votes and 22 "against"
votes.

14. Let x = the measure of the equal angles.
Then $2x - 4$ = the measure of the third angle.

The sum of the measures of the angles in a triangle is $180°$.

$$x + x + (2x - 4) = 180$$
$$4x - 4 = 180$$
$$4x = 184$$
$$x = 46$$

The third angle has measure

$$2x - 4 = 2(46) - 4 = 92 - 4 = 88.$$

The measures of the angles are $46°$, $46°$, and $88°$.

15. Let L = the length of the book.
Then $L - 2.58$ = the width of the book.

$$P = 2L + 2W$$
$$37.8 = 2L + 2(L - 2.58)$$
$$37.8 = 2L + 2L - 5.16$$
$$42.96 = 4L$$
$$10.74 = L$$

The width is $10.74 - 2.58 = 8.16$ inches and the length is 10.74 inches.

16. $x - y = 4$

To graph this line, find the intercepts.

If $y = 0$, $x = 4$, so the x-intercept is $(4, 0)$.
If $x = 0$, $y = -4$, so the y-intercept is $(0, -4)$.

Graph the line through these intercepts. A third point, such as $(5, 1)$, may be used as a check.

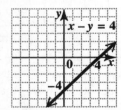

17. $3x + y = 6$

If $y = 0$, $x = 2$, so the x-intercept is $(2, 0)$.
If $x = 0$, $y = 6$, so the y-intercept is $(0, 6)$.

Graph the line through these intercepts. A third point, such as $(1, 3)$, may be used as a check.

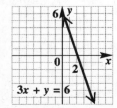

18. The slope m of the line passing through the points $(-5, 6)$ and $(1, -2)$ is

$$m = \frac{y_2 - y_1}{x_2 - x_1} = \frac{-2 - 6}{1 - (-5)} = \frac{-8}{6} = -\frac{4}{3}.$$

19. The slope of the line $y = 4x - 3$ is 4. The slope of the line whose graph is perpendicular to that of $y = 4x - 3$ is the negative reciprocal of 4, namely $-\frac{1}{4}$.

20. Through $(-4, 1)$ with slope $\frac{1}{2}$
Use the point-slope form of a line.

$$y - y_1 = m(x - x_1)$$
$$y - 1 = \frac{1}{2}[x - (-4)]$$
$$y - 1 = \frac{1}{2}(x + 4)$$
$$y - 1 = \frac{1}{2}x + 2$$
$$y = \frac{1}{2}x + 3 \qquad \textit{Slope-intercept form}$$

21. Through the points $(1, 3)$ and $(-2, -3)$
Find the slope.

$$m = \frac{y_2 - y_1}{x_2 - x_1} = \frac{-3 - 3}{-2 - 1} = \frac{-6}{-3} = 2$$

Use the point-slope form of a line.

$$y - y_1 = m(x - x_1)$$
$$y - 3 = 2(x - 1)$$
$$y - 3 = 2x - 2$$
$$y = 2x + 1 \qquad \textit{Slope-intercept form}$$

22. **(a)** On the vertical line through $(9, -2)$, the x-coordinate of every point is 9. Therefore, an equation of this line is $x = 9$.

(b) On the horizontal line through $(4, -1)$, the y-coordinate of every point is -1. Therefore, an equation of this line is $y = -1$.

23. $$\left(\frac{a^{-3}b^4}{a^2b^{-1}}\right)^{-2} = \left(\frac{b^4b^1}{a^2a^3}\right)^{-2} = \left(\frac{b^5}{a^5}\right)^{-2}$$
$$= \left(\frac{a^5}{b^5}\right)^2 = \frac{a^{10}}{b^{10}}$$

24. $$\left(\frac{m^{-4}n^2}{m^2n^{-3}}\right) \cdot \left(\frac{m^5n^{-1}}{m^{-2}n^5}\right)$$
$$= \left(\frac{n^2n^3}{m^2m^4}\right) \cdot \left(\frac{m^5m^2}{n^1n^5}\right)$$
$$= \left(\frac{n^5}{m^6}\right) \cdot \left(\frac{m^7}{n^6}\right)$$
$$= \frac{n^5m^7}{n^6m^6} = \frac{m}{n}$$

25. $$\left(3y^2 - 2y + 6\right) - \left(-y^2 + 5y + 12\right)$$
$$= 3y^2 - 2y + 6 + y^2 - 5y - 12$$
$$= 4y^2 - 7y - 6$$

26. $$(4f + 3)(3f - 1) = 12f^2 - 4f + 9f - 3$$
$$= 12f^2 + 5f - 3$$

27. $\left(\frac{1}{4}x + 5\right)^2$

Use the formula for the square of a binomial, $(a + b)^2 = a^2 + 2ab + b^2$.

$$= \left(\frac{1}{4}x\right)^2 + 2\left(\frac{1}{4}x\right)(5) + 5^2$$

$$= \frac{1}{16}x^2 + \frac{5}{2}x + 25$$

28. $(3x^3 + 13x^2 - 17x - 7) \div (3x + 1)$

$$
\require{enclose}
\begin{array}{r}
x^2 + 4x - 7 \\
3x + 1 \enclose{longdiv}{3x^3 + 13x^2 - 17x - 7} \\
\underline{3x^3 + x^2} \\
12x^2 - 17x \\
\underline{12x^2 + 4x} \\
-21x - 7 \\
\underline{-21x - 7} \\
0
\end{array}
$$

Answer: $x^2 + 4x - 7$

29. $2x^2 - 13x - 45 = (2x + 5)(x - 9)$

30. $100t^4 - 25 = 25\left(4t^4 - 1\right)$

$$= 25\left[\left(2t^2\right)^2 - 1^2\right]$$

$$= 25\left(2t^2 + 1\right)\left(2t^2 - 1\right)$$

31. Use the sum of cubes formula,

$$x^3 + y^3 = (x + y)\left(x^2 - xy + y^2\right).$$

$$8p^3 + 125 = (2p)^3 + 5^3$$

$$= (2p + 5)\left[(2p)^2 - (2p)(5) + 5^2\right]$$

$$= (2p + 5)\left(4p^2 - 10p + 25\right)$$

32.
$$3x^2 + 4x = 7$$
$$3x^2 + 4x - 7 = 0$$
$$(3x + 7)(x - 1) = 0$$

$$3x + 7 = 0 \quad \text{or} \quad x - 1 = 0$$
$$3x = -7$$
$$x = -\frac{7}{3} \quad \text{or} \quad x = 1$$

The solution set is $\left\{-\frac{7}{3}, 1\right\}$.

33. $\dfrac{y^2 - 16}{y^2 - 8y + 16} = \dfrac{(y + 4)(y - 4)}{(y - 4)(y - 4)}$

$$= \dfrac{y + 4}{y - 4}$$

34. $\dfrac{2a^2}{a + b} \cdot \dfrac{a - b}{4a} = \dfrac{2a^2(a - b)}{4a(a + b)}$

$$= \dfrac{a(a - b)}{2(a + b)}$$

35. $\dfrac{2x}{2x - 1} + \dfrac{4}{2x + 1} + \dfrac{8}{4x^2 - 1}$

$$= \dfrac{2x}{2x - 1} + \dfrac{4}{2x + 1} + \dfrac{8}{(2x + 1)(2x - 1)}$$

The LCD is $(2x + 1)(2x - 1)$.

$$= \dfrac{2x(2x + 1)}{(2x - 1)(2x + 1)} + \dfrac{4(2x - 1)}{(2x + 1)(2x - 1)}$$

$$+ \dfrac{8}{(2x + 1)(2x - 1)}$$

$$= \dfrac{2x(2x + 1) + 4(2x - 1) + 8}{(2x + 1)(2x - 1)}$$

$$= \dfrac{4x^2 + 2x + 8x - 4 + 8}{(2x + 1)(2x - 1)}$$

$$= \dfrac{4x^2 + 10x + 4}{(2x + 1)(2x - 1)}$$

$$= \dfrac{2(2x^2 + 5x + 2)}{(2x + 1)(2x - 1)}$$

$$= \dfrac{2(2x + 1)(x + 2)}{(2x + 1)(2x - 1)}$$

$$= \dfrac{2(x + 2)}{2x - 1}$$

36.
$$\dfrac{-3x}{x + 1} + \dfrac{4x + 1}{x} = \dfrac{-3}{x^2 + x}$$

$$\dfrac{-3x}{x + 1} + \dfrac{4x + 1}{x} = \dfrac{-3}{x(x + 1)}$$

Multiply by the LCD, $x(x + 1)$. $(x \neq -1, 0)$

$$x(x + 1)\left(\dfrac{-3x}{x + 1} + \dfrac{4x + 1}{x}\right)$$

$$= x(x + 1)\left(\dfrac{-3}{x(x + 1)}\right)$$

$$x(-3x) + (x + 1)(4x + 1) = -3$$
$$-3x^2 + 4x^2 + x + 4x + 1 = -3$$
$$x^2 + 5x + 4 = 0$$
$$(x + 4)(x + 1) = 0$$

$$x + 4 = 0 \quad \text{or} \quad x + 1 = 0$$
$$x = -4 \quad \text{or} \quad x = -1$$

The number -1 is not allowed as a solution because substituting it in the original equation results in division by 0.

Check $x = -4$: $-4 + \frac{15}{4} = -\frac{1}{4}$ *True*

The solution set is $\{-4\}$.

37. (a) Solve the equation for y.

$$5x - 3y = 8$$
$$5x - 8 = 3y$$
$$\dfrac{5x - 8}{3} = y$$

So $f(x) = \dfrac{5x - 8}{3}$ or $f(x) = \dfrac{5}{3}x - \dfrac{8}{3}$.

(b) $f(1) = \dfrac{5(1) - 8}{3} = \dfrac{-3}{3} = -1$

38. Since no year is repeated, the relation defines a function.
The domain is
$\{1990, 1992, 1994, 1996, 1998, 2000, 2002\}$.
The range is
$\{1.25, 1.61, 1.80, 1.21, 1.94, 2.26, 2.60\}$.

39. The relation is not a function since a vertical line may intersect the graph in more than one point. The domain is the set of x-values, $[-2, \infty)$. The range is the set of y-values, $(-\infty, \infty)$.

40. **(a)** $(f + g)(x) = f(x) + g(x)$
$$= (x^2 + 2x - 3)$$
$$+ (2x^3 - 3x^2 + 4x - 1)$$
$$= 2x^3 - 2x^2 + 6x - 4$$

(b) $(g - f)(x) = g(x) - f(x)$
$$= (2x^3 - 3x^2 + 4x - 1)$$
$$- (x^2 + 2x - 3)$$
$$= 2x^3 - 3x^2 + 4x - 1$$
$$- x^2 - 2x + 3$$
$$= 2x^3 - 4x^2 + 2x + 2$$

(c) Using part (a),
$(f + g)(-1) = 2(-1)^3 - 2(-1)^2 + 6(-1) - 4$
$$= -2 - 2 - 6 - 4$$
$$= -14$$

(d) $(f \circ h)(x) = f(h(x))$
$$= f(x^2)$$
$$= (x^2)^2 + 2(x^2) - 3$$
$$= x^4 + 2x^2 - 3$$

41. $2x - y = -8$ (1)
$x + 2y = 11$ (2)

To solve this system by the elimination method, multiply equation (1) by 2 and add the result to equation (2) to eliminate y.

$$
\begin{array}{rrrr}
4x & - & 2y & = & -16 \\
x & + & 2y & = & 11 \\
\hline
5x & & & = & -5 \\
& & x & = & -1
\end{array}
$$

To find y, let $x = -1$ in equation (2).

$$-1 + 2y = 11$$
$$2y = 12$$
$$y = 6$$

The solution set is $\{(-1, 6)\}$.

42. $4x + 5y = -8$ (1)
$3x + 4y = -7$ (2)

Multiply equation (1) by -3 and equation (2) by 4. Add the resulting equations to eliminate x.

$$
\begin{array}{rrrr}
-12x & - & 15y & = & 24 \\
12x & + & 16y & = & -28 \\
\hline
& & y & = & -4
\end{array}
$$

To find x, let $y = -4$ in equation (1).

$$4x + 5y = -8$$
$$4x + 5(-4) = -8$$
$$4x - 20 = -8$$
$$4x = 12$$
$$x = 3$$

The solution set is $\{(3, -4)\}$.

43. $3x + 4y = 2$ (1)
$6x + 8y = 1$ (2)

Multiply equation (1) by -2 and add the result to equation (2).

$$
\begin{array}{rrrrl}
-6x & - & 8y & = & -4 \\
6x & + & 8y & = & 1 \\
\hline
& & 0 & = & -3 \;\; \textit{False}
\end{array}
$$

Since $0 = -3$ is a false statement, the solution set is $\emptyset$.

44. *Step 2*
Let $x =$ the length of each side of the equal sides;
$y =$ the length of the shorter third side.

Step 3
The third side measures 4 inches less than each of the equal sides, so

$$y = x - 4. \quad (1)$$

The perimeter of the triangle is 53 inches, so

$$x + x + y = 53$$
$$\text{or} \quad 2x + y = 53. \quad (2)$$

Step 4
Substitute $x - 4$ for y in equation (2).

$$2x + (x - 4) = 53$$
$$3x - 4 = 53$$
$$3x = 57$$
$$x = 19$$

From (1), $y = 19 - 4 = 15$.

Step 5
The lengths of the sides are 19 inches, 19 inches, and 15 inches.

Step 6
15 is 4 less than 19 and the perimeter is $19 + 19 + 15 = 53$ inches, as required.

45. Let $x =$ the number of liters of 20%
   alcohol solution;
   $y =$ the number of liters of 50%
   alcohol solution.

Liters of Solution	Percent (as a decimal)	Liters of Pure Alcohol
x	0.20	$0.20x$
y	0.50	$0.50y$
12	0.40	$0.40(12) = 4.8$

Since 12 L of the mixture are needed,

$$x + y = 12. \quad (1)$$

Since the amount of pure alcohol in the 20% solution plus the amount of pure alcohol in the 50% solution must equal the amount of pure alcohol in the mixture,

$$0.20x + 0.50y = 4.8.$$

Multiply by 10 to clear the decimals.

$$2x + 5y = 48 \quad (2)$$

Multiply equation (1) by -2 and add the result to equation (2).

$$
\begin{aligned}
-2x - 2y &= -24 \\
\underline{2x + 5y} &= \underline{48} \\
3y &= 24 \\
y &= 8
\end{aligned}
$$

From (1), $x = 12 - 8 = 4$.

4 L of 20% alcohol solution and 8 L of 50% alcohol solution will be needed.

46. Let $x =$ the average cost of the original Tickle
   Me Elmo
and $y =$ the recommended cost of T.M.X.

The original's cost was $12.37 less than T.M.X.'s, so

$$x = y - 12.37. \quad (1)$$

Together they cost $67.63, so

$$x + y = 67.63. \quad (2)$$

From (1), substitute $y - 12.37$ for x in equation (2).

$$
\begin{aligned}
(y - 12.37) + y &= 67.63 \\
2y - 12.37 &= 67.63 \\
2y &= 80 \\
y &= 40
\end{aligned}
$$

From (1), $x = y - 12.37 = 40 - 12.37 = 27.63$.
The average cost of the original Elmo was $27.63 and the recommended cost of a T.M.X. is $40.

CHAPTER 9 INEQUALITIES AND ABSOLUTE VALUE

9.1 Set Operations and Compound Inequalities

1. This statement is *true*. The solution set of $x + 1 = 5$ is $\{4\}$. The solution set of $x + 1 > 5$ is $(4, \infty)$. The solution set of $x + 1 < 5$ is $(-\infty, 4)$. Taken together we have the set of real numbers. (See Section 2.8, Exercises 73–76, for a discussion of this concept.)

3. This statement is *false*. The union is $(-\infty, 8) \cup (8, \infty)$. The only real number that is *not* in the union is 8.

5. This statement is *false* since 0 is a rational number but not an irrational number. The sets of rational numbers and irrational numbers have no common elements so their intersection is $\emptyset$.

In Exercises 7–14, let
$$A = \{1, 2, 3, 4, 5, 6\}, B = \{1, 3, 5\}, C = \{1, 6\},$$
and $D = \{4\}$.

7. The intersection of sets B and A contains only those elements in both sets B and A.
$$B \cap A = \{1, 3, 5\} \text{ or set } B$$

9. The intersection of sets A and D is the set of all elements in both set A and D. Therefore,
$$A \cap D = \{4\} \text{ or set } D.$$

11. The intersection of set B and the set of no elements (empty set), $B \cap \emptyset$, is the set of no elements or $\emptyset$.

13. The union of sets A and B is the set of all elements that are in either set A or set B or both sets A and B. Since all numbers in set B are also in set A, the set $A \cup B$ will be the same as set A.
$$A \cup B = \{1, 2, 3, 4, 5, 6\} \text{ or set } A$$

15. The first graph represents the set $(-\infty, 2)$. The second graph represents the set $(-3, \infty)$. The intersection includes the elements common to both sets, that is, $(-3, 2)$.

17. The first graph represents the set $(-\infty, 5]$. The second graph represents the set $(-\infty, 2]$. The intersection includes the elements common to both sets, that is, $(-\infty, 2]$.

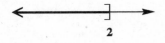

19. $x < 2$ and $x > -3$

The graph of the solution set will be all numbers that are both less than 2 and greater than -3. The solution set is $(-3, 2)$.

21. $x \le 2$ and $x \le 5$

The graph of the solution set will be all numbers that are both less than or equal to 2 and less than or equal to 5. The overlap is the numbers less than or equal to 2. The solution set is $(-\infty, 2]$.

23. $x \le 3$ and $x \ge 6$

The graph of the solution set will be all numbers that are both less than or equal to 3 and greater than or equal to 6. There are no such numbers. The solution set is $\emptyset$.

25. $x - 3 \le 6 \quad$ and $\quad x + 2 \ge 7$
$\quad\quad x \le 9 \quad$ and $\quad\quad x \ge 5$

The graph of the solution set is all numbers that are both less than or equal to 9 and greater than or equal to 5. This is the intersection. The elements common to both sets are the numbers between 5 and 9, including the endpoints. The solution set is $[5, 9]$.

27. $-3x > 3 \quad$ and $\quad x + 3 > 0$
$\quad\quad x < -1 \quad$ and $\quad\quad x > -3$

The graph of the solution set is all numbers that are both less than -1 and greater than -3. This is the intersection. The elements common to both sets are the numbers between -3 and -1, not including the endpoints. The solution set is $(-3, -1)$.

29. $3x - 4 \le 8 \quad$ and $\quad -4x + 1 \ge -15$
$\quad 3x \le 12 \quad$ and $\quad\quad -4x \ge -16$
$\quad\quad x \le 4 \quad$ and $\quad\quad\quad x \le 4$

Since both inequalities are identical, the graph of the solution set is the same as the graph of one of the inequalities. The solution set is $(-\infty, 4]$.

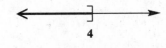

31. The first graph represents the set $(-\infty, 2]$. The second graph represents the set $[4, \infty)$. The union includes all elements in either set, or in both, that is, $(-\infty, 2] \cup [4, \infty)$.

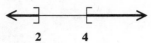

33. The first graph represents the set $[1, \infty)$. The second graph represents the set $(-\infty, 8]$. The union includes all elements in either set, or in both, that is, $(-\infty, \infty)$.

35. $x \leq 1$ or $x \leq 8$

The word "or" means to take the union of both sets. The graph of the solution set is all numbers that are either less than or equal to 1 *or* less than or equal to 8, or both. This is all numbers less than or equal to 8. The solution set is $(-\infty, 8]$.

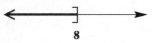

37. $x \geq -2$ or $x \geq 5$

The graph of the solution set will be all numbers that are either greater than or equal to -2 or greater than or equal to 5.
The solution set is $[-2, \infty)$.

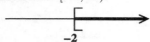

39. $x \geq -2$ or $x \leq 4$

The graph of the solution set will be all numbers that are either greater than or equal to -2 or less than or equal to 4. This is the set of all real numbers. The solution set is $(-\infty, \infty)$.

41. $x + 2 > 7$ or $1 - x > 6$
$$-x > 5$$
$x > 5$ or $x < -5$
The graph of the solution set is all numbers either greater than 5 or less than -5. This is the union. The solution set is $(-\infty, -5) \cup (5, \infty)$.

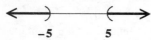

43. $x + 1 > 3$ or $-4x + 1 > 5$
$$-4x > 4$$
$x > 2$ or $x < -1$
The graph of the solution set is all numbers either less than -1 or greater than 2. This is the union. The solution set is $(-\infty, -1) \cup (2, \infty)$.

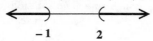

45. $4x + 1 \geq -7$ or $-2x + 3 \geq 5$
$4x \geq -8$ or $-2x \geq 2$
$x \geq -2$ or $x \leq -1$
The graph of the solution set is all numbers either greater than or equal to -2 or less than or equal to -1. This is the set of all real numbers. The solution set is $(-\infty, \infty)$.

47. $(-\infty, -1] \cap [-4, \infty)$
The intersection is the set of numbers less than or equal to -1 and greater than or equal to -4. The numbers common to both original sets are between, and including, -4 and -1. The simplest interval form is $[-4, -1]$.

49. $(-\infty, -6] \cap [-9, \infty)$
The intersection is the set of numbers less than or equal to -6 and greater than or equal to -9. The numbers common to both original sets are between, and including, -9 and -6. The simplest interval form is $[-9, -6]$.

51. $(-\infty, 3) \cup (-\infty, -2)$
The union is the set of numbers that are either less than 3 or less than -2, or both. This is all numbers less than 3. The simplest interval form is $(-\infty, 3)$.

53. $[3, 6] \cup (4, 9)$
The union is the set of numbers between, and including, 3 and 6, or between, but not including, 4 and 9. This is the set of numbers greater than or equal to 3 and less than 9. The simplest interval form is $[3, 9)$.

55. $x < -1$ and $x > -5$
The word "and" means to take the intersection of both sets. $x < -1$ and $x > -5$ is true only when
$$-5 < x < -1.$$
The graph of the solution set is all numbers greater than -5 *and* less than -1. This is all numbers between -5 and -1, not including -5 or -1. The solution set is $(-5, -1)$.

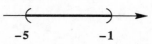

57. $x < 4$ or $x < -2$

The word "or" means to take the union of both sets. The graph of the solution set is all numbers that are either less than 4 *or* less than -2, or both. This is all numbers less than 4. The solution set is $(-\infty, 4)$.

59. $-3x \le -6$ or $-3x \ge 0$

$x \ge 2$ or $x \le 0$

The word "or" means to take the union of both sets. The graph of the solution set is all numbers that are either greater than or equal to 2 *or* less than or equal to 0. The solution set is $(-\infty, 0] \cup [2, \infty)$.

61. $x + 1 \ge 5$ and $x - 2 \le 10$

$x \ge 4$ and $x \le 12$

The word "and" means to take the intersection of both sets. The graph of the solution set is all numbers that are both greater than or equal to 4 *and* less than or equal to 12. This is all numbers between, and including, 4 and 12. The solution set is $[4, 12]$.

63. The set of expenses that are less than $3000 for public schools *and* are greater than $5000 for private schools is {Tuition and fees}.

65. The set of expenses that are greater than $2900 for public schools *or* are greater than $5000 for private schools is {Tuition and fees, Dormitory charges}.

67. Find "the yard can be fenced *and* the yard can be sodded."

A yard that can be fenced has $P \le 150$. Maria and Joe qualify.

A yard that can be sodded has $A \le 1400$. Again, Maria and Joe qualify.

Find the intersection. Maria's and Joe's yards are common to both sets, so Maria and Joe can have their yards both fenced and sodded.

68. Find "the yard can be fenced *and* the yard cannot be sodded."

A yard that can be fenced has $P \le 150$. Maria and Joe qualify.

A yard that cannot be sodded has $A > 1400$. Luigi and Than qualify.

Find the intersection. There are no yards common to both sets, so none of them qualify.

69. Find "the yard cannot be fenced *and* the yard can be sodded."

A yard that cannot be fenced has $P > 150$. Luigi and Than qualify.

A yard that can be sodded has $A \le 1400$. Maria and Joe qualify.

Find the intersection. There are no yards common to both sets, so none of the qualify.

70. Find "the yard cannot be fenced *and* the yard cannot be sodded."

A yard that cannot be fenced has $P > 150$. Luigi and Than qualify.

A yard that cannot be sodded has $A > 1400$. Again, Luigi and Than qualify.

Find the intersection. Luigi's and Than's yards are common to both sets, so Luigi and Than qualify.

71. Find "the yard can be fenced *or* the yard can be sodded." From Exercise 67, Maria's and Joe's yards qualify for both conditions, so the union is Maria and Joe.

72. Find "the yard cannot be fenced *or* the yard can be sodded." From Exercise 69, Luigi's and Than's yards cannot be fenced, and Maria's and Joe's yards can be sodded. The union includes all of them.

73.
$$2y - 4 \le 3y + 2$$
$$-y - 4 \le 2$$
$$-y \le 6$$
$$y \ge -6$$

The solution set is $[-6, \infty)$.

75.
$$-5 < 2r + 1 < 5$$
$$-6 < 2r < 4$$
$$-3 < r < 2$$

The solution set is $(-3, 2)$.

77.
$$-|6| - |-11| + (-4) = -6 - (11) + (-4)$$
$$= -17 + (-4)$$
$$= -21$$

9.2   Absolute Value Equations and Inequalities

1. $|x| = 5$ has two solutions, $x = 5$ or $x = -5$. The graph is Choice **E**.

$|x| < 5$ is written $-5 < x < 5$. Notice that -5 and 5 are not included. The graph is Choice **C**, which uses parentheses.

$|x| > 5$ is written $x < -5$ or $x > 5$. The graph is Choice **D**, which uses parentheses.

$|x| \leq 5$ is written $-5 \leq x \leq 5$. This time -5 and 5 are included. The graph is Choice **B**, which uses brackets.

$|x| \geq 5$ is written $x \leq -5$ or $x \geq 5$. The graph is Choice **A**, which uses brackets.

3.   **(a)** $|ax + b| = k, k = 0$
This means the distance from $ax + b$ to 0 is 0, so $ax + b = 0$, which has one solution.

(b) $|ax + b| = k, k > 0$
This means the distance from $ax + b$ to 0 is a positive number, so $ax + b = k$ or $ax + b = -k$. There are two solutions.

(c) $|ax + b| = k, k < 0$
This means the distance from $ax + b$ to 0 is a negative number, which is impossible because distance is always positive. There are no solutions.

5. $|x| = 12$
$x = 12$ or $x = -12$
The solution set is $\{-12, 12\}$.

7. $|4x| = 20$
$4x = 20$ or $4x = -20$
$x = 5$ or $x = -5$
The solution set is $\{-5, 5\}$.

9. $|y - 3| = 9$
$y - 3 = 9$ or $y - 3 = -9$
$y = 12$ or $y = -6$
The solution set is $\{-6, 12\}$.

11. $|2x - 1| = 11$
$2x - 1 = 11$ or $2x - 1 = -11$
$2x = 12$ $2x = -10$
$x = 6$ or $x = -5$
The solution set is $\{-5, 6\}$.

13. $|4r - 5| = 17$
$4r - 5 = 17$ or $4r - 5 = -17$
$4r = 22$ $4r = -12$
$r = \frac{22}{4} = \frac{11}{2}$ or $r = -3$
The solution set is $\{-3, \frac{11}{2}\}$.

15. $|2y + 5| = 14$
$2y + 5 = 14$ or $2y + 5 = -14$
$2y = 9$ $2y = -19$
$y = \frac{9}{2}$ or $y = -\frac{19}{2}$
The solution set is $\{-\frac{19}{2}, \frac{9}{2}\}$.

17. $|\frac{1}{2}x + 3| = 2$
$\frac{1}{2}x + 3 = 2$ or $\frac{1}{2}x + 3 = -2$
$\frac{1}{2}x = -1$ $\frac{1}{2}x = -5$
$x = -2$ or $x = -10$
The solution set is $\{-10, -2\}$.

19. $|1 + \frac{3}{4}k| = 7$
$1 + \frac{3}{4}k = 7$ or $1 + \frac{3}{4}k = -7$
Multiply each side by 4.
$4 + 3k = 28$ or $4 + 3k = -28$
$3k = 24$ $3k = -32$
$k = 8$ or $k = \frac{-32}{3}$
The solution set is $\{-\frac{32}{3}, 8\}$.

21. $|x| > 3$
$x > 3$ or $x < -3$
The solution set is $(-\infty, -3) \cup (3, \infty)$.

23. $|k| \geq 4$
$k \geq 4$ or $k \leq -4$
The solution set is $(-\infty, -4] \cup [4, \infty)$.

25. $|r + 5| \geq 20$
$r + 5 \leq -20$ or $r + 5 \geq 20$
$r \leq -25$ or $r \geq 15$
The solution set is $(-\infty, -25] \cup [15, \infty)$.

27. $|t + 2| > 10$
$t + 2 > 10$ or $t + 2 < -10$
$t > 8$ or $t < -12$
The solution set is $(-\infty, -12) \cup (8, \infty)$.

29. $|3 - x| > 5$

$3 - x > 5$ or $3 - x < -5$

$-x > 2$ or $-x < -8$

Multiply by -1,

and reverse the inequality symbols.

$x < -2$ or $x > 8$

The solution set is $(-\infty, -2) \cup (8, \infty)$.

31. $|-5x + 3| \geq 12$

$-5x + 3 \geq 12$ or $-5x + 3 \leq -12$

$-5x \geq 9$ $-5x \leq -15$

$x \leq -\frac{9}{5}$ or $x \geq 3$

The solution set is $\left(-\infty, -\frac{9}{5}\right] \cup [3, \infty)$.

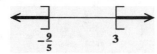

33. (a) $|2x + 1| < 9$

The graph of the solution set will be all numbers between -5 and 4, since the absolute value is less than 9.

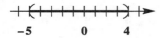

(b) $|2x + 1| > 9$

The graph of the solution set will be all numbers less than -5 or greater than 4, since the absolute value is greater than 9.

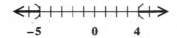

35. $|x| \leq 3$

$-3 \leq x \leq 3$

The solution set is $[-3, 3]$.

37. $|k| < 4$

$-4 < k < 4$

The solution set is $(-4, 4)$.

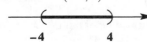

39. $|r + 5| \leq 20$

$-20 \leq r + 5 \leq 20$

$-25 \leq r \leq 15$ *Subtract 5.*

The solution set is $[-25, 15]$.

41. $|t + 2| \leq 10$

$-10 \leq t + 2 \leq 10$

$-12 \leq t \leq 8$

The solution set is $[-12, 8]$.

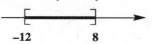

43. $|3 - x| \leq 5$

$-5 \leq 3 - x \leq 5$

$-8 \leq -x \leq 2$

Multiply by -1, and reverse the inequality symbols.

$8 \geq x \geq -2$ or $-2 \leq x \leq 8$

The solution set is $[-2, 8]$.

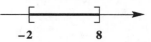

45. $|-5x + 3| \leq 12$

$-12 \leq -5x + 3 \leq 12$

$-15 \leq \quad -5x \quad \leq 9$

Divide by -5, and reverse the inequality symbols.

$3 \geq x \geq -\frac{9}{5}$ or $-\frac{9}{5} \leq x \leq 3$

The solution set is $\left[-\frac{9}{5}, 3\right]$.

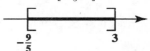

47. $|-4 + k| > 9$

$-4 + k > 9$ or $-4 + k < -9$

$k > 13$ or $k < -5$

The solution set is $(-\infty, -5) \cup (13, \infty)$.

49. $|r + 5| > 20$

$r + 5 > 20$ or $r + 5 < -20$

$r > 15$ or $r < -25$

The solution set is $(-\infty, -25) \cup (15, \infty)$.

51. $|7 + 2z| = 5$

$7 + 2z = 5$ or $7 + 2z = -5$

$2z = -2$ $2z = -12$

$z = -1$ or $z = -6$

The solution set is $\{-6, -1\}$.

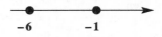

53. $|3r - 1| \leq 11$

$-11 \leq 3r - 1 \leq 11$

$-10 > 3r \leq 12$

$-\frac{10}{3} \leq r \leq 4$

The solution set is $\left[-\frac{10}{3}, 4\right]$.

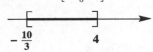

55. $|-6x - 6| \leq 1$

$-1 \leq -6x - 6 \leq 1$

$5 \leq -6x \leq 7$

Divide by -6, and reverse the inequality symbols.

$-\frac{5}{6} \geq x \geq -\frac{7}{6}$ or $-\frac{7}{6} \leq x \leq -\frac{5}{6}$

The solution set is $\left[-\frac{7}{6}, -\frac{5}{6}\right]$.

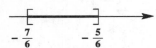

57. $|2x - 1| \geq 7$

$2x - 1 \geq 7$ or $2x - 1 \leq -7$

$2x \geq 8$ or $2x \leq -6$

$x \geq 4$ or $x \leq -3$

The solution set is $(-\infty, -3] \cup [4, \infty)$.

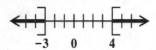

59. $|x + 2| = 3$

$x + 2 = 3$ or $x + 2 = -3$

$x = 1$ or $x = -5$

The solution set is $\{-5, 1\}$.

61. $|x - 6| = 3$

$x - 6 = 3$ or $x - 6 = -3$

$x = 9$ or $x = 3$

The solution set is $\{3, 9\}$.

63. $|x| - 1 = 4$

$|x| = 5$

$x = 5$ or $x = -5$

The solution set is $\{-5, 5\}$.

65. $|x + 4| + 1 = 2$

$|x + 4| = 1$

$x + 4 = 1$ or $x + 4 = -1$

$x = -3$ or $x = -5$

The solution set is $\{-5, -3\}$.

67. $|2x + 1| + 3 > 8$

$|2x + 1| > 5$

$2x + 1 > 5$ or $2x + 1 < -5$

$2x > 4$ $2x < -6$

$x > 2$ or $x < -3$

The solution set is $(-\infty, -3) \cup (2, \infty)$.

69. $|x + 5| - 6 \leq -1$

$|x + 5| \leq 5$

$-5 \leq x + 5 \leq 5$

$-10 \leq x \leq 0$

The solution set is $[-10, 0]$.

71. $|3x + 1| = |2x + 4|$

$3x + 1 = 2x + 4$ or $3x + 1 = -(2x + 4)$

$3x + 1 = -2x - 4$

$5x = -5$

$x = 3$ or $x = -1$

The solution set is $\{-1, 3\}$.

73. $\left|m - \frac{1}{2}\right| = \left|\frac{1}{2}m - 2\right|$

$m - \frac{1}{2} = \frac{1}{2}m - 2$ or $m - \frac{1}{2} = -\left(\frac{1}{2}m - 2\right)$

Multiply by 2. $m - \frac{1}{2} = -\frac{1}{2}m + 2$

$2m - 1 = m - 4$ $2m - 1 = -m + 4$

$3m = 5$

$m = -3$ or $m = \frac{5}{3}$

The solution set is $\left\{-3, \frac{5}{3}\right\}$.

75. $|6x| = |9x + 1|$

$6x = 9x + 1$ or $6x = -(9x + 1)$

$-3x = 1$ $6x = -9x - 1$

$15x = -1$

$x = -\frac{1}{3}$ or $x = -\frac{1}{15}$

The solution set is $\left\{-\frac{1}{3}, -\frac{1}{15}\right\}$.

77. $|2p - 6| = |2p + 11|$

$2p - 6 = 2p + 11$ or $2p - 6 = -(2p + 11)$

$-6 = 11$ *False* $2p - 6 = -2p - 11$

$4p = -5$

No solution or $p = -\frac{5}{4}$

The solution set is $\left\{-\frac{5}{4}\right\}$.

79. $|x| \geq -10$

The absolute value of a number is always greater than or equal to 0. Therefore, the inequality is true for all real numbers.

The solution set is $(-\infty, \infty)$.

81. $|12t - 3| = -8$

Since the absolute value of an expression can never be negative, there are no solutions for this equation.

The solution set is $\emptyset$.

83. $|4x + 1| = 0$

The expression $4x + 1$ will equal 0 *only* for the solution of the equation

$$4x + 1 = 0.$$
$$4x = -1$$
$$x = \tfrac{-1}{4} \text{ or } -\tfrac{1}{4}$$

The solution set is $\left\{-\tfrac{1}{4}\right\}$.

85. $|2q - 1| = -6$

Since the absolute value of an expression can never be negative, there are no solutions for this equation.

The solution set is $\emptyset$.

87. $|x + 5| > -9$

Since the absolute value of an expression is always nonnegative (positive or zero), the inequality is true for any real number x.

The solution set is $(-\infty, \infty)$.

89. $|7x + 3| \leq 0$

The absolute value of an expression is always nonnegative (positive or zero), so this inequality is true only when

$$7x + 3 = 0$$
$$7x = -3$$
$$x = -\tfrac{3}{7}.$$

The solution set is $\left\{-\tfrac{3}{7}\right\}$.

91. $|5x - 2| = 0$

The expression $5x - 2$ will equal 0 *only* for the solution of the equation

$$5x - 2 = 0.$$
$$5x = 2$$
$$x = \tfrac{2}{5}$$

The solution set is $\left\{\tfrac{2}{5}\right\}$.

93. $|x - 2| + 3 \geq 2$

$$|x - 2| \geq -1$$

Since the absolute value of an expression is always nonnegative (positive or zero), the inequality is true for any real number x.

The solution set is $(-\infty, \infty)$.

95. $|10z + 7| + 3 < 1$

$$|10z + 7| < -2$$

There is no number whose absolute value is less than -2, so this inequality has no solution.

The solution set is $\emptyset$.

97. Let x represent the calcium intake for a specific female. For x to be within 100 mg of 1000 mg, we must have

$$|x - 1000| \leq 100.$$

$$-100 \leq x - 1000 \leq 100$$
$$900 \leq \quad x \quad \leq 1100$$

99. Add the given heights with a calculator to get 4756. There are 10 numbers, so divide the sum by 10.

$$\frac{4756}{10} = 475.6$$

The average height is 475.6 ft.

100. $|x - k| < 50$

Substitute 475.6 for k and solve the inequality.

$$|x - 475.6| < 50$$

$$-50 < x - 475.6 < 50$$
$$425.6 < \quad x \quad < 525.6$$

The buildings with heights between 425.6 ft and 525.6 ft are the 1201 Walnut, City Hall, Fidelity Bank and Trust Building, Kansas City Power and Light, and the Hyatt Regency Crown Center.

101. $|x - k| < 75$

Substitute 475.6 for k and solve the inequality.

$$|x - 475.6| < 75$$

$$-75 < x - 475.6 < 75$$
$$400.6 < \quad x \quad < 550.6$$

The buildings with heights between 400.6 ft and 550.6 ft are City Center Square, Commerce Tower, Federal Office Building, 1201 Walnut, City Hall, Fidelity Bank and Trust Building, Kansas City Power and Light, and the Hyatt Regency Crown Center.

102. (a) This would be the opposite of the inequality in Exercise 101, that is,

$$|x - 475.6| \geq 75.$$

(b) $|x - 475.6| \geq 75$

$x - 475.6 \geq 75$ or $x - 475.6 \leq -75$
$\qquad x \geq 550.6$ or $\qquad\qquad x \leq 400.6$

(c) The buildings that are not within 75 ft of the average have height less than or equal to 400.6 or greater than or equal to 550.6. They are Town Pavillion and One Kansas City Place.

(d) The answer makes sense because it includes all the buildings *not* listed earlier which had heights within 75 ft of the average.

103. $2x + 5 < 9$

$\qquad 2x < 4$

$\qquad\quad x < 2$

The solution set is $(-\infty, 2)$.

105. $5 - 3x \geq 9$

$\qquad -3x \geq 4$

$\qquad\quad x \leq -\frac{4}{3}$

The solution set is $\left(-\infty, -\frac{4}{3}\right]$.

Summary Exercises on Solving Linear and Absolute Value Equations and Inequalities

1. $4z + 1 = 49$

$\qquad 4z = 48$

$\qquad\ z = 12$

The solution set is $\{12\}$.

3. $6q - 9 = 12 + 3q$

$\qquad 3q = 21$

$\qquad\ q = 7$

The solution set is $\{7\}$.

5. $|a + 3| = -4$

Since the absolute value of an expression is always nonnegative, there is no number that makes this statement true. Therefore, the solution set is $\emptyset$.

7. $8r + 2 \geq 5r$

$\qquad 3r \geq -2$

$\qquad\ r \geq -\frac{2}{3}$

The solution set is $\left[-\frac{2}{3}, \infty\right)$.

9. $2q - 1 = -7$

$\qquad 2q = -6$

$\qquad\ q = -3$

The solution set is $\{-3\}$.

11. $6z - 5 \leq 3z + 10$

$\qquad 3z \leq 15$

$\qquad\ z \leq 5$

The solution set is $(-\infty, 5]$.

13. $9x - 3(x + 1) = 8x - 7$

$\qquad 9x - 3x - 3 = 8x - 7$

$\qquad\quad 6x - 3 = 8x - 7$

$\qquad\qquad\quad 4 = 2x$

$\qquad\qquad\quad 2 = x$

The solution set is $\{2\}$.

15. $9x - 5 \geq 9x + 3$

$\qquad\quad -5 \geq 3$ *False*

This is a false statement, so the inequality is a contradiction.

The solution set is $\emptyset$.

17. $|q| < 5.5$

$\quad -5.5 < q < 5.5$

The solution set is $(-5.5, 5.5)$.

19. $\frac{2}{3}x + 8 = \frac{1}{4}x$

$\quad 8x + 96 = 3x$ *Multiply by 12.*

$\qquad\ 5x = -96$

$\qquad\quad x = -\frac{96}{5}$

The solution set is $\left\{-\frac{96}{5}\right\}$.

21. $\frac{1}{4}p < -6$

$\quad 4\left(\frac{1}{4}p\right) < 4(-6)$

$\qquad\quad p < -24$

The solution set is $(-\infty, -24)$.

23. $\frac{3}{5}q - \frac{1}{10} = 2$

$\quad 6q - 1 = 20$ *Multiply by 10.*

$\qquad 6q = 21$

$\qquad\ q = \frac{21}{6} = \frac{7}{2}$

The solution set is $\left\{\frac{7}{2}\right\}$.

25. $r + 9 + 7r = 4(3 + 2r) - 3$

$\qquad 8r + 9 = 12 + 8r - 3$

$\qquad 8r + 9 = 8r + 9$

$\qquad\qquad 0 = 0$ *True*

The last statement is true for any real number r.

The solution set is $(-\infty, \infty)$.

27. $|2p - 3| > 11$

$\ 2p - 3 > 11$ or $2p - 3 < -11$

$\quad 2p > 14$ $\qquad\qquad 2p < -8$

$\qquad p > 7$ or $\qquad\ p < -4$

The solution set is $(-\infty, -4) \cup (7, \infty)$.

29. $|5a + 1| \leq 0$

The expression $|5a + 1|$ is never less than 0 since an absolute value expression must be nonnegative. However, $|5a + 1| = 0$ if

$$5a + 1 = 0$$

$$5a = -1$$

$$a = \frac{-1}{5} = -\frac{1}{5}$$

The solution set is $\left\{-\frac{1}{5}\right\}$.

31. $-2 \leq 3x - 1 \leq 8$

$\quad -1 \leq 3x \leq 9$

$\quad -\frac{1}{3} \leq x \leq 3$

The solution set is $\left[-\frac{1}{3}, 3\right]$.

33. $|7z - 1| = |5z + 3|$

$7z - 1 = 5z + 3$ or $7z - 1 = -(5z + 3)$

$2z = 4$ $\qquad$ $7z - 1 = -5z - 3$

$\qquad\qquad\qquad$ $12z = -2$

$z = 2$ or $\qquad$ $z = \frac{-2}{12} = -\frac{1}{6}$

The solution set is $\left\{-\frac{1}{6}, 2\right\}$.

35. $|1 - 3x| \geq 4$

$1 - 3x \geq 4$ or $\quad 1 - 3x \leq -4$

$-3x \geq 3$ $\qquad\qquad$ $-3x \leq -5$

$x \leq -1$ or $\qquad\quad$ $x \geq \frac{5}{3}$

The solution set is $(-\infty, -1] \cup \left[\frac{5}{3}, \infty\right)$.

37. $-(m + 4) + 2 = 3m + 8$

$-m - 4 + 2 = 3m + 8$

$-m - 2 = 3m + 8$

$-10 = 4m$

$m = \frac{-10}{4} = -\frac{5}{2}$

The solution set is $\left\{-\frac{5}{2}\right\}$.

39. $-6 \leq \frac{3}{2} - x \leq 6$

$-12 \leq 3 - 2x \leq 12$

$-15 \leq -2x \leq 9$

$\frac{15}{2} \geq x \geq -\frac{9}{2}$ or $-\frac{9}{2} \leq x \leq \frac{15}{2}$

The solution set is $\left[-\frac{9}{2}, \frac{15}{2}\right]$.

41. $|x - 1| \geq -6$

The absolute value of an expression is always nonnegative, so the inequality is true for any real number x.

The solution set is $(-\infty, \infty)$.

43. $8q - (1 - q) = 3(1 + 3q) - 4$

$8q - 1 + q = 3 + 9q - 4$

$9q - 1 = 9q - 1$ *True*

This is an identity.

The solution set is $(-\infty, \infty)$.

45. $|r - 5| = |r + 9|$

$r - 5 = r + 9$ $\qquad$ or $\qquad$ $r - 5 = -(r + 9)$

$-5 = 9$ *False* $\qquad\qquad$ $r - 5 = -r - 9$

$\qquad\qquad\qquad\qquad\qquad$ $2r = -4$

$\quad$ *No solution* $\qquad$ or $\qquad\qquad$ $r = -2$

The solution set is $\{-2\}$.

47. $2x + 1 > 5$ or $\quad 3x + 4 < 1$

$2x > 4$ $\qquad\qquad$ $3x < -3$

$x > 2$ or $\qquad\quad$ $x < -1$

The solution set is $(-\infty, -1) \cup (2, \infty)$.

9.3 Linear Inequalities in Two Variables

1. The boundary of the graph of $y \leq -x + 2$ will be a _solid_ line (since the inequality involves $\leq$), and the shading will be _below_ the line (since the inequality sign is $\leq$ or $<$).

3. The boundary of the graph of $y > -x + 2$ will be a _dashed_ line (since the inequality involves $>$), and the shading will be _above_ the line (since the inequality sign is $\geq$ or $>$).

5. The graph of $Ax + By = C$ divides the plane into two regions. In one of these regions, the ordered pairs satisfy $Ax + By < C$; in the other, they satisfy $Ax + By > C$.

7. $x + y \leq 2$

Graph the line $x + y = 2$ by drawing a solid line (since the inequality involves $\leq$) through the intercepts $(2, 0)$ and $(0, 2)$.

Test a point not on this line, such as $(0, 0)$.

$x + y \leq 2$

$0 + 0 \leq 2$?

$0 \leq 2$ $\quad$ *True*

Shade that side of the line containing the test point $(0, 0)$.

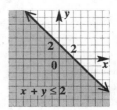

9. $4x - y < 4$

Graph the line $4x - y = 4$ by drawing a dashed line (since the inequality involves $<$) through the intercepts $(1, 0)$ and $(0, -4)$. Instead of using a test point, we will solve the inequality for y.

$-y < -4x + 4$

$y > 4x - 4$

Since we have "$y > $" in the last inequality, shade the region *above* the boundary line.

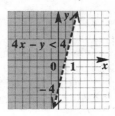

11. $x + 3y \geq -2$

Graph the solid line $x + 3y = -2$ (since the inequality involves $\geq$) through the intercepts $(-2, 0)$ and $\left(0, -\frac{2}{3}\right)$.

Test a point not on this line such as $(0, 0)$.

$0 + 3(0) \geq -2$?

$0 \geq -2$ $\quad$ *True*

Shade that side of the line containing the test point $(0, 0)$.

continued

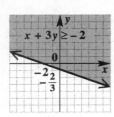

13. $x + y > 0$

Graph the line $x + y = 0$, which includes the points $(0, 0)$ and $(2, -2)$, as a dashed line (since the inequality involves $>$). Solving the inequality for y gives us

$$y > -x,$$

So shade the region above the boundary line.

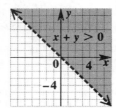

15. $x - 3y \leq 0$

Graph the solid line $x - 3y = 0$ through the points $(0, 0)$ and $(3, 1)$.
Solve the inequality for y.

$$-3y \leq -x$$
$$y \geq \tfrac{1}{3}x$$

Shade the region above the boundary line.

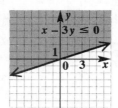

17. $y \geq 4x$

The boundary has the equation $y = 4x$. This line goes through the points $(0, 0)$ and $(1, 4)$. Make the line solid because of the $\geq$ sign. Because the boundary passes through the origin, we cannot use $(0, 0)$ as a test point.

Using $(2, 0)$ as a test point will result in the inequality $0 \geq 8$, which is false. Shade the region *not* containing $(2, 0)$. The solid line shows that the boundary is part of the graph.

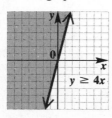

19. $x + y \leq 1$ and $x \geq 1$

Graph the solid line $x + y = 1$ through $(0, 1)$ and $(1, 0)$. The inequality $x + y \leq 1$ can be written as $y \leq -x + 1$, so shade the region below the boundary line.
Graph the solid vertical line $x = 1$ through $(1, 0)$ and shade the region to the right. The required graph is the common shaded area as well as the portions of the lines that bound it.

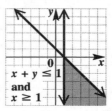

21. $2x - y \geq 2$ and $y < 4$

Graph the solid line $2x - y = 2$ through the intercepts $(1, 0)$ and $(0, -2)$. Test $(0, 0)$ to get $0 \geq 2$, a false statement. Shade that side of the graph not containing $(0, 0)$. To graph $y < 4$ on the same axes, graph the dashed horizontal line through $(0, 4)$. Test $(0, 0)$ to get $0 < 4$, a true statement. Shade that side of the dashed line containing $(0, 0)$.
The word "and" indicates the intersection of the two graphs. The final solution set consists of the region where the two shaded regions overlap.

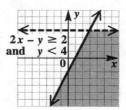

23. $x + y > -5$ and $y < -2$

Graph $x + y = -5$, which has intercepts $(-5, 0)$ and $(0, -5)$, as a dashed line. Test $(0, 0)$, which yields $0 > -5$, a true statement. Shade the region that includes $(0, 0)$.
Graph $y = -2$ as a dashed horizontal line. Shade the region below $y = -2$. The required graph of the intersection is the region common to both graphs.

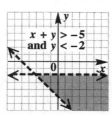

25. $|x| < 3$ can be rewritten as $-3 < x < 3$. The boundaries are the dashed vertical lines $x = -3$ and $x = 3$. Since x is between -3 and 3, the graph includes all points between the lines.

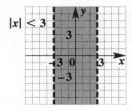

27. $|x + 1| < 2$ can be rewritten as

$$-2 < x + 1 < 2$$
$$-3 < x < 1.$$

The boundaries are the dashed vertical lines $x = -3$ and $x = 1$. Since x is between -3 and 1, the graph includes all points between the lines.

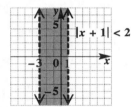

29. $x - y \geq 1$ or $y \geq 2$
Graph the solid line $x - y = 1$, which crosses the y-axis at -1 and the x-axis at 1. Use $(0, 0)$ as a test point, which yields $0 \geq 1$, a false statement. Shade the region that does not include $(0, 0)$.
Now graph the solid line $y = 2$. Since the inequality is $y \geq 2$, shade above this line.
The required graph of the union includes all the shaded regions, that is, all the points that satisfy either inequality.

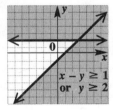

31. $x - 2 > y$ or $x < 1$
Graph $x - 2 = y$, which has intercepts $(2, 0)$ and $(0, -2)$, as a dashed line. Test $(0, 0)$, which yields $-2 > 0$, a false statement. Shade the region that does not include $(0, 0)$.
Graph $x = 1$ as a dashed vertical line. Shade the region to the left of $x = 1$.
The required graph of the union includes all the shaded regions, that is, all the points that satisfy either inequality.

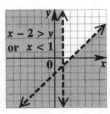

33. $3x + 2y < 6$ or $x - 2y > 2$
Graph $3x + 2y = 6$, which has intercepts $(2, 0)$ and $(0, 3)$, as a dashed line. Test $(0, 0)$, which yields $0 < 6$, a true statement. Shade the region that includes $(0, 0)$.
Graph $x - 2y = 2$, which has intercepts $(2, 0)$ and $(0, -1)$, as a dashed line. Test $(0, 0)$, which yields $0 > 2$, a false statement. Shade the region that does not include $(0, 0)$.
The required graph of the union includes all the shaded regions, that is, all the points that satisfy either inequality.

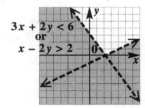

35. $y \leq 3x - 6$
The boundary line, $y = 3x - 6$, has slope 3 and y-intercept -6. This would be graph **B** or graph **C**. Since we want the region less than or equal to $3x - 6$, we want the region on or below the boundary line. The answer is graph **C**.

37. $y \leq -3x - 6$
The slope of the boundary line $y = -3x - 6$ is -3, and the y-intercept is -6. This would be graph **A** or graph **D**. The inequality sign is $\leq$, so we want the region on or below the boundary line. The answer is graph **A**.

39. **(a)** The x-intercept is $(-4, 0)$, so the solution set for $y = 0$ is $\{-4\}$.

(b) The solution set for $y < 0$ is $(-\infty, -4)$, since the graph is below the x-axis for these values of x.

(c) The solution set for $y > 0$ is $(-4, \infty)$, since the graph is above the x-axis for these values of x.

41. **(a)** The x-intercept is $(3.5, 0)$, so the solution set for $y = 0$ is $\{3.5\}$.

(b) The solution set for $y < 0$ is $(3.5, \infty)$, since the graph is below the x-axis for these values of x.

(c) The solution set for $y > 0$ is $(-\infty, 3.5)$, since the graph is above the x-axis for these values of x.

43. **(a)** $5x + 3 = 0$
$$5x = -3$$
$$x = -\tfrac{3}{5} = -0.6$$
The solution set is $\{-0.6\}$.

(b) $5x + 3 > 0$
$$5x > -3$$
$$x > -\tfrac{3}{5} \text{ or } -0.6$$
The solution set is $(-0.6, \infty)$.

(c) $5x + 3 < 0$

$$5x < -3$$

$$x < -\frac{3}{5} \text{ or } -0.6$$

The solution set is $(-\infty, -0.6)$.

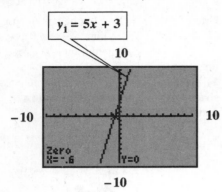

The x-intercept is $(-0.6, 0)$, as in part (a). The graph is above the x-axis for $x > -0.6$, as in part (b), and below the x-axis for $x < -0.6$, as in part (c).

45. (a) $-8x - (2x + 12) = 0$

$$-8x - 2x - 12 = 0$$

$$-10x - 12 = 0$$

$$-10x = 12$$

$$x = -1.2$$

The solution set is $\{-1.2\}$.

(b) $-8x - (2x + 12) \geq 0$

$$-8x - 2x - 12 \geq 0$$

$$-10x - 12 \geq 0$$

$$-10x \geq 12$$

$$x \leq -1.2$$

The solution set is $(-\infty, -1.2]$.

(c) $-8x - (2x + 12) \leq 0$

$$-8x - 2x - 12 \leq 0$$

$$-10x - 12 \leq 0$$

$$-10x \leq 12$$

$$x \geq -1.2$$

The solution set is $[-1.2, \infty)$.

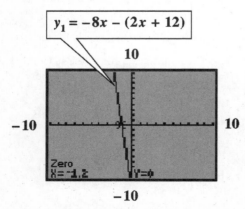

The x-intercept is $(-1.2, 0)$, as in part (a). The graph is on or above the x-axis for $(-\infty, -1.2]$, as in part (b), and on or below the x-axis for $[-1.2, \infty)$, as in part (c).

47. $x + y \geq 500$

(a) Graph the inequality.

Step 1
Graph the line $x + y = 500$.

If $x = 0$, then $y = 500$, so the y-intercept is $(0, 500)$.
If $y = 0$, then $x = 500$, so the x-intercept is $(500, 0)$.

Graph the line with these intercepts.

The line is solid because of the $\geq$ sign.

Step 2
Use $(0, 0)$ as a test point.

$$x + y \geq 500 \qquad \textit{Original inequality}$$
$$0 + 0 \geq 500 ? \qquad \textit{Let x = 0, y = 0.}$$
$$0 \geq 500 \qquad \textit{False}$$

Since $0 \geq 500$ is false, shade the side of the boundary not containing $(0, 0)$. Because of the restrictions $x \geq 0$ and $y \geq 0$ in this applied problem, only the portion of the graph that lies in quadrant I is included.

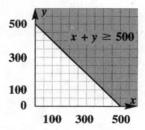

(b) Any point in the shaded region satisfies the inequality. Some ordered pairs are $(500, 0)$, $(200, 400)$, and $(400, 200)$. There are many other ordered pairs that will also satisfy the inequality.

49. "A factory can have *no more than* 200 workers on a shift, but must have *at least* 100" can be translated as $x \leq 200$ and $x \geq 100$. "Must manufacture *at least* 3000 units" can be translated as $y \geq 3000$.

50.

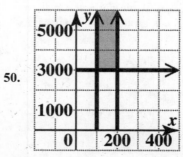

51. The total daily cost C consists of $50 per worker and $100 to manufacture 1 unit, so
$C = 50x + 100y$.

52. Some examples of points in the shaded region are $(150, 4000)$, $(120, 3500)$, and $(180, 6000)$. Some examples of points on the boundary are $(100, 5000)$, $(150, 3000)$, and $(200, 4000)$. The corner points are $(100, 3000)$ and $(200, 3000)$.

53.

(x, y)	$50x + 100y = C$
$(150, 4000)$	$50(150) + 100(4000) = 407{,}500$
$(120, 3500)$	$50(120) + 100(3500) = 356{,}000$
$(180, 6000)$	$50(180) + 100(6000) = 609{,}000$
$(100, 5000)$	$50(100) + 100(5000) = 505{,}000$
$(150, 3000)$	$50(150) + 100(3000) = 307{,}500$
$(200, 4000)$	$50(200) + 100(4000) = 410{,}000$
$(100, 3000)$	$50(100) + 100(3000) = 305{,}000$
	(least value)
$(200, 3000)$	$50(200) + 100(3000) = 310{,}000$

54. The company should use 100 workers and manufacture 3000 units to achieve the least possible cost.

55. $8^2 = 8 \cdot 8 = 64$

57. $-12^2 = -1 \cdot 12 \cdot 12 = -144$

59. $a^2 + b^2 = 5^2 + 12^2$
$= 25 + 144$
$= 169$

Chapter 9 Review Exercises

For Exercises 1–4, let $A = \{a, b, c, d\}$, $B = \{a, c, e, f\}$, and $C = \{a, e, f, g\}$.

1. $A \cap B = \{a, b, c, d\} \cap \{a, c, e, f\}$
$= \{a, c\}$

2. $A \cap C = \{a, b, c, d\} \cap \{a, e, f, g\}$
$= \{a\}$

3. $B \cup C = \{a, c, e, f\} \cup \{a, e, f, g\}$
$= \{a, c, e, f, g\}$

4. $A \cup C = \{a, b, c, d\} \cup \{a, e, f, g\}$
$= \{a, b, c, d, e, f, g\}$

5. $x > 6$ and $x < 9$

The graph of the solution set will be all numbers which are both greater than 6 and less than 9. The overlap is the numbers between 6 and 9, not including the endpoints.

The solution set is $(6, 9)$.

6. $x + 4 > 12$ and $x - 2 < 12$
$x > 8$ and $x < 14$

The graph of the solution set will be all numbers between 8 and 14, not including the endpoints.

The solution set is $(8, 14)$.

7. $x > 5$ or $x \leq -3$

The graph of the solution set will be all numbers that are either greater than 5 or less than or equal to -3.

The solution set is $(-\infty, -3] \cup (5, \infty)$.

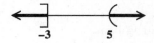

8. $x \geq -2$ or $x < 2$

The graph of the solution set will be all numbers that are either greater than or equal to -2 or less than 2. All real numbers satisfy these criteria.

The solution set is $(-\infty, \infty)$.

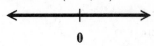

9. $x - 4 > 6$ and $x + 3 \leq 10$
$x > 10$ and $x \leq 7$

The graph of the solution set will be all numbers that are both greater than 10 and less than or equal to 7. There are no real numbers satisfying these criteria.

The solution set is $\emptyset$.

10. $-5x + 1 \geq 11$ or $3x + 5 \geq 26$
$-5x \geq 10$ $3x \geq 21$
$x \leq -2$ or $x \geq 7$

The graph of the solution set will be all numbers that are either less than or equal to -2 or greater than or equal to 7.

The solution set is $(-\infty, -2] \cup [7, \infty)$.

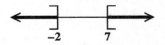

11. $(-3, \infty) \cap (-\infty, 4)$
$(-3, \infty)$ includes all real numbers greater than -3.
$(-\infty, 4)$ includes all real numbers less than 4. Find the intersection. The numbers common to both sets are greater than -3 and less than 4.

$$-3 < x < 4$$

The solution set is $(-3, 4)$.

12. $(-\infty, 6) \cap (-\infty, 2)$

$(-\infty, 6)$ includes all real numbers less than 6.
$(-\infty, 2)$ includes all real numbers less than 2.
Find the intersection. The numbers common to both sets are less than 2.

The solution set is $(-\infty, 2)$.

13. $(4, \infty) \cup (9, \infty)$

$(4, \infty)$ includes all real numbers greater than 4.
$(9, \infty)$ includes all real numbers greater than 9.
Find the union. The numbers in the first set, the second set, or in both sets are all the real numbers that are greater than 4.

The solution set is $(4, \infty)$.

14. $(1, 2) \cup (1, \infty)$

$(1, 2)$ includes the real numbers between 1 and 2, not including 1 and 2.
$(1, \infty)$ includes all real numbers greater than 1.
Find the union. The numbers in the first set, the second set, or in both sets are all real numbers greater than 1.

The solution set is $(1, \infty)$.

15. $|x| = 7$

$x = 7$ or $x = -7$

The solution set is $\{-7, 7\}$.

16. $|x + 2| = 9$

$x + 2 = 9$ or $x + 2 = -9$
$x = 7$ or $x = -11$

The solution set is $\{-11, 7\}$.

17. $|3k - 7| = 8$

$3k - 7 = 8$ or $3k - 7 = -8$
$3k = 15$ $\qquad$ $3k = -1$
$k = 5$ or $\qquad$ $k = -\frac{1}{3}$

The solution set is $\{-\frac{1}{3}, 5\}$.

18. $|z - 4| = -12$

Since the absolute value of an expression can never be negative, there are no solutions for this equation.

The solution set is $\emptyset$.

19. $|2k - 7| + 4 = 11$

$|2k - 7| = 7$
$2k - 7 = 7$ or $2k - 7 = -7$
$2k = 14$ $\qquad$ $2k = 0$
$k = 7$ or $\qquad$ $k = 0$

The solution set is $\{0, 7\}$.

20. $|4a + 2| - 7 = -3$

$|4a + 2| = 4$
$4a + 2 = 4$ or $4a + 2 = -4$
$4a = 2$ $\qquad$ $4a = -6$
$a = \frac{2}{4}$ $\qquad$ $a = -\frac{6}{4}$
$a = \frac{1}{2}$ or $\qquad$ $a = -\frac{3}{2}$

The solution set is $\{-\frac{3}{2}, \frac{1}{2}\}$.

21. $|3p + 1| = |p + 2|$

$3p + 1 = p + 2$ or $3p + 1 = -(p + 2)$
$2p = 1$ $\qquad$ $3p + 1 = -p - 2$
$\qquad$ $\qquad$ $4p = -3$
$p = \frac{1}{2}$ or $\qquad$ $p = -\frac{3}{4}$

The solution set is $\{-\frac{3}{4}, \frac{1}{2}\}$.

22. $|2m - 1| = |2m + 3|$

$2m - 1 = 2m + 3$ or $2m - 1 = -(2m + 3)$
$-1 = 3$ *False* $\qquad$ $2m - 1 = -2m - 3$
$\qquad$ $\qquad$ $4m = -2$
No solution or $\qquad$ $m = -\frac{2}{4} = -\frac{1}{2}$

The solution set is $\{-\frac{1}{2}\}$.

23. $|p| < 14$

$-14 < p < 14$

The solution set is $(-14, 14)$.

24. $|-t + 6| \le 7$

$-7 \le -t + 6 \le 7$
$-13 \le -t \le 1$

Multiply by -1; reverse the inequality symbols.
$13 \ge t \ge -1$ or $-1 \le t \le 13$

The solution set is $[-1, 13]$.

25. $|2p + 5| \le 1$

$-1 \le 2p + 5 \le 1$
$-6 \le 2p \le -4$
$-3 \le p \le -2$

The solution set is $[-3, -2]$.

26. $|x + 1| \ge -3$

Since the absolute value of an expression is always nonnegative (positive or zero), the inequality is *true* for any real number x.

The solution set is $(-\infty, \infty)$.

27. $|5r - 1| > 9$

$5r - 1 > 9$ or $5r - 1 < -9$
$5r > 10$ $\qquad$ $5r < -8$
$r > 2$ or $\qquad$ $r < -\frac{8}{5}$

The solution set is $\left(-\infty, -\frac{8}{5}\right) \cup (2, \infty)$.

28. $|11x - 3| \leq -2$

There is no number whose absolute value is less than or equal to -2, so this inequality has no solution.

The solution set is $\emptyset$.

29. $|11x - 3| \geq -2$

Since the absolute value of an expression is always nonnegative (positive or zero), the inequality is true for any real number x.

The solution set is $(-\infty, \infty)$.

30. $|11x - 3| \leq 0$

The absolute value of an expression is always nonnegative (positive or zero), so this inequality is true only when

$$11x - 3 = 0$$
$$11x = 3$$
$$x = \tfrac{3}{11}.$$

The solution set is $\left\{\tfrac{3}{11}\right\}$.

31. $3x - 2y \leq 12$
Graph $3x - 2y = 12$ as a solid line through $(0, -6)$ and $(4, 0)$. Use $(0, 0)$ as a test point. Since $(0, 0)$ satisfies the inequality, shade the region on the side of the line containing $(0, 0)$.

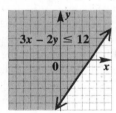

32. $5x - y > 6$
Graph $5x - y = 6$ as a dashed line through $(0, -6)$ and $\left(\tfrac{6}{5}, 0\right)$. Use $(0, 0)$ as a test point. Since $(0, 0)$ does not satisfy the inequality, shade the region on the side of the line that does not contain $(0, 0)$.

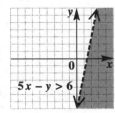

33. $2x + y \leq 1$ and $x \geq 2y$
Graph $2x + y = 1$ as a solid line through $\left(\tfrac{1}{2}, 0\right)$ and $(0, 1)$, and shade the region on the side containing $(0, 0)$ since it satisfies the inequality. Next, graph $x = 2y$ as a solid line through $(0, 0)$ and $(2, 1)$, and shade the region on the side containing $(2, 0)$ since $2 > 2(0)$ or $2 > 0$ is true.

The intersection is the region where the graphs overlap.

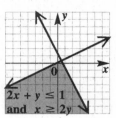

34. $x \geq 2$ or $y \geq 2$

Graph $x = 2$ as a solid vertical line through $(2, 0)$. Shade the region to the right of $x = 2$.

Graph $y = 2$ as a solid horizontal line through $(0, 2)$. Shade the region above $y = 2$. The graph of

$$x \geq 2 \quad \text{or} \quad y \geq 2$$

includes all the shaded regions.

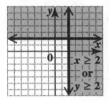

35. In $y < 4x + 3$, the " $<$ " symbol indicates that the graph has a dashed boundary line and that the shading is below the line, so the correct choice is **D**.

36. **[9.1]** $x < 3$ and $x \geq -2$
The real numbers that are common to both sets are the numbers greater than or equal to -2 and less than 3.

$$-2 \leq x < 3$$

The solution set is $[-2, 3)$.

37. **[9.2]** $|3k + 6| \geq 0$
The absolute value of an expression is always nonnegative, so the inequality is true for any real number k.
The solution set is $(-\infty, \infty)$.

38. **[9.2]** $|3x + 2| + 4 = 9$
$$|3x + 2| = 5$$

$$3x + 2 = 5 \quad \text{or} \quad 3x + 2 = -5$$
$$3x = 3 \qquad\qquad 3x = -7$$
$$x = 1 \quad \text{or} \qquad x = -\tfrac{7}{3}$$

The solution set is $\left\{-\tfrac{7}{3}, 1\right\}$.

39. **[9.2]** $|m + 3| \leq 13$
$$-13 \leq m + 3 \leq 13$$
$$-16 \leq \quad m \quad \leq 10$$

The solution set is $[-16, 10]$.

40. **[9.2]** $|5r - 1| > 14$

$5r - 1 > 14$ or $5r - 1 < -14$

$5r > 15$ $5r < -13$

$r > 3$ or $r < -\frac{13}{5}$

The solution set is $\left(-\infty, -\frac{13}{5}\right) \cup (3, \infty)$.

41. **[9.1]** $x \geq -2$ or $x < 4$

The solution set includes all numbers either greater than or equal to -2 or all numbers less than 4. This is the union and is the set of all real numbers. The solution set is $(-\infty, \infty)$.

42. **[9.2]** $|m - 1| = |2m + 3|$

$m - 1 = 2m + 3$ or $m - 1 = -(2m + 3)$

$m - 1 = -2m - 3$

$3m = -2$

$-4 = m$ or $m = -\frac{2}{3}$

The solution set is $\left\{-4, -\frac{2}{3}\right\}$.

43. **[9.2]** $|m + 3| \leq 1$

$-1 \leq m + 3 \leq 1$

$-4 \leq m \leq -2$

The solution set is $[-4, -2]$.

44. **[9.2]** $|3k - 7| = 4$

$3k - 7 = 4$ or $3k - 7 = -4$

$3k = 11$ $3k = 3$

$k = \frac{11}{3}$ or $k = 1$

The solution set is $\left\{1, \frac{11}{3}\right\}$.

45. **[9.1]** $-5x + 1 \geq 11$ or $3x + 5 \geq 26$

$-5x \geq 10$ $3x \geq 21$

$x \leq -2$ or $x \geq 7$

The graph of the solution set is all numbers either less than or equal to -2 *or* greater than or equal to 7. This is the union. The solution set is $(-\infty, -2] \cup [7, \infty)$.

46. **[9.1]** $x > 6$ and $x < 8$

The graph of the solution set is all numbers both greater than 6 *and* less than 8. This is the intersection. The elements common to both sets are the numbers between 6 and 8, not including the endpoints. The solution set is $(6, 8)$.

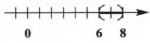

47. **[9.3]** $2x - 3y > -6$

Use intercepts to graph the boundary, $2x - 3y = -6$.

If $y = 0$, $x = -3$, so the x-intercept is $(-3, 0)$.

If $x = 0$, $y = 2$, so the y-intercept is $(0, 2)$.

Draw a dashed line through $(-3, 0)$ and $(0, 2)$.

Using $(0, 0)$ as a test point will result in the inequality $0 > -6$, which is true. Shade the region containing the origin. This is the region below the line. The dashed line shows that the boundary is not part of the graph.

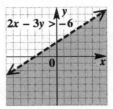

48. **[9.3]** $3x + 5y > 9$

To graph the boundary, which is the line $3x + 5y = 9$, find its intercepts.

$3x + 5y = 9$ $3x + 5y = 9$

$3x + 5(0) = 9$ $3(0) + 5y = 9$

Let y = 0. *Let x = 0.*

$3x = 9$ $5y = 9$

$x = 3$ $y = \frac{9}{5}$

The x-intercept is $(3, 0)$ and the y-intercept is $\left(0, \frac{9}{5}\right)$. (A third point may be used as a check.) Draw a dashed line through these points. In order to determine which side of the line should be shaded, use $(0, 0)$ as a test point. Substituting 0 for x and 0 for y will result in the inequality $0 > 9$, which is false. Shade the region *not* containing the origin. This is the region above the line. The dashed line shows that the boundary is not part of the graph.

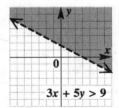

49. **[9.1] (a)** The set of states with less than 1 million female workers is {Maine, Oregon, Utah}. The set of states with more than 1 million male workers is {Illinois, North Carolina, Oregon, Wisconsin}. Oregon is the only state in both sets, so the set of states with less than 1 million female workers *and* more than 1 million male workers is {Oregon}.

(b) The set of states with less than 1 million female workers *or* more than 2 million male workers is {Illinois, Maine, North Carolina, Oregon, Utah}.

(c) It is easy to see that the sum of the female and male workers for each state doesn't exceed 7 million, so the set of states with a total of more than 7 million civilian workers is $\{\}$, or $\emptyset$.

50. **[9.2]** **(a)** The endpoints on the number line are $-\frac{11}{3}$ and 1. The solution set of $|3x + 4| \geq 7$ would be

$$\left(-\infty, -\tfrac{11}{3}\right] \cup [1, \infty).$$

(b) The solution set of $|3x + 4| \leq 7$ would be

$$\left[-\tfrac{11}{3}, 1\right].$$

Chapter 9 Test

1. There are *no* occupations with median earnings for men less than \$900 and for women greater than \$500. This set is denoted by $\emptyset$.

2. *All* of the listed occupations have median earnings for men greater than \$600 or for women less than \$400. They are: Managerial/Professional, Technical/Sales/Administrative Support, Service, Operators/Fabricators/Laborers.

3. $A \cap B = \{1, 2, 5, 7\} \cap \{1, 5, 9, 12\}$
$ = \{1, 5\}$

4. $A \cup B = \{1, 2, 5, 7\} \cup \{1, 5, 9, 12\}$
$ = \{1, 2, 5, 7, 9, 12\}$

5. $3k \geq 6$ and $k - 4 < 5$
$k \geq 2$ and $k < 9$
The solution set is all numbers both greater than or equal to 2 *and* less than 9. This is the intersection. The numbers common to both sets are between 2 and 9, including 2 but not 9. The solution set is $[2, 9)$.

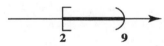

6. $-4x \leq -24$ or $4x - 2 < 10$
$\phantom{-4x \leq -24 \text{ or } }4x < 12$
$x \geq 6$ or $x < 3$
The solution set is all numbers less than 3 or greater than or equal to 6. This is the union. The solution set is $(-\infty, 3) \cup [6, \infty)$.

7. $|4x - 3| = 7$
$4x - 3 = 7$ or $4x - 3 = -7$
$4x = 10$ $4x = -4$
$x = \frac{10}{4} = \frac{5}{2}$ or $x = -1$

The solution set is $\left\{-1, \frac{5}{2}\right\}$.

8. $|5 - 6x| > 12$

$5 - 6x > 12$ or $5 - 6x < -12$
$-6x > 7$ $-6x < -17$
$x < -\frac{7}{6}$ or $x > \frac{17}{6}$

The solution set is $\left(-\infty, -\frac{7}{6}\right) \cup \left(\frac{17}{6}, \infty\right)$.

9. $|7 - x| \leq -1$

Since the absolute value of an expression is always nonnegative (positive or zero), the inequality is *false* for any real number x.

The solution set is $\emptyset$.

10. $|3 - 5x| = |2x + 8|$
$3 - 5x = 2x + 8$ or $3 - 5x = -(2x + 8)$
${-7x} = 5$ $3 - 5x = -2x - 8$
${-3x} = -11$
$x = -\frac{5}{7}$ or $x = \frac{11}{3}$

The solution set is $\left\{-\frac{5}{7}, \frac{11}{3}\right\}$.

11. $|-3x + 4| - 4 < -1$
$|-3x + 4| < 3$
$-3 < -3x + 4 < 3$
$-7 < -3x < -1$
$\frac{7}{3} > x > \frac{1}{3}$ or $\frac{1}{3} < x < \frac{7}{3}$

The solution set is $\left(\frac{1}{3}, \frac{7}{3}\right)$.

12. $|12t + 7| \geq 0$
Since the absolute value of an expression is always nonnegative (positive or zero), the inequality is true for any real number t.

The solution set is $(-\infty, \infty)$.

13. **(a)** $|5x + 3| < k$
If $k < 0$, then $|5x + 3|$ would be less than a negative number. Since the absolute value of an expression is always nonnegative (positive or zero), the solution set is $\emptyset$.

(b) $|5x + 3| > k$
If $k < 0$, then $|5x + 3|$ would be greater than a negative number. Since the absolute value of an expression is always nonnegative (positive or zero), the solution set is the set of all real numbers, $(-\infty, \infty)$.

(c) $|5x + 3| = k$
If $k < 0$, then $|5x + 3|$ would be equal to a negative number. Since the absolute value of an expression is always nonnegative (positive or zero), the solution set is $\emptyset$.

14. $3x - 2y > 6$

Graph the line $3x - 2y = 6$, which has intercepts $(2, 0)$ and $(0, -3)$, as a dashed line since the inequality involves $>$. Test $(0, 0)$, which yields $0 > 6$, a false statement. Shade the region that does not include $(0, 0)$.

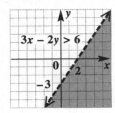

15. $3x - y > 0$

The boundary, $3x - y = 0$, goes through the origin, so both intercepts are $(0, 0)$. Two other points on this line are $(1, 3)$ and $(-1, -3)$. Draw the boundary as a dashed line.

Choose a test point which is not on the boundary. Using $(3, 0)$ results in the true statement $9 > 0$, so shade the region containing $(3, 0)$. This is the region below the line. The dashed line shows that the boundary is not part of the graph.

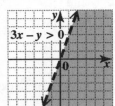

16. $y < 2x - 1$ and $x - y < 3$

First graph $y = 2x - 1$ as a dashed line through $(2, 3)$ and $(0, -1)$. Test $(0, 0)$, which yields $0 < -1$, a false statement. Shade the side of the line not containing $(0, 0)$.

Next, graph $x - y = 3$ as a dashed line through $(3, 0)$ and $(0, -3)$. Test $(0, 0)$, which yields $0 < 3$, a true statement. Shade the side of the line containing $(0, 0)$. The intersection is the region where the graphs overlap.

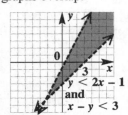

17. $x - 2 \geq y$ or $y \geq 3$

Graph $x - 2 = y$, which has intercepts $(2, 0)$ and $(0, -2)$, as a solid line. Test $(0, 0)$, which yields $-2 \geq 0$, a false statement. Shade the region that does not include $(0, 0)$.

Graph $y = 3$ as a solid horizontal line through $(0, 3)$. Shade the region above the line.

The required graph of the union includes all the shaded regions, that is, all the points that satisfy either inequality.

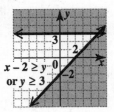

Cumulative Review Exercises (Chapters 1–9)

1. **(a)** 34 is a natural number, so it is also a whole number, an integer, a rational number, and a real number. **A, B, C, D, F**

(b) 0 is a whole number, so it is also an integer, a rational number, and a real number. **B, C, D, F**

(c) 2.16 is a rational number, so it is also a real number. **D, F**

(d) $-\pi$ is an irrational number, so it is also a real number. **E, F**

(e) $\sqrt{3}$ is an irrational number, so it is also a real number. **E, F**

(f) $-\frac{4}{5}$ is a rational number, so it is also a real number. **D, F**

2. $9 \cdot 4 - 16 \div 4 = (9 \cdot 4) - (16 \div 4) = 36 - 4 = 32$

3. $-|8 - 13| - |-4| + |-9| = -|-5| - 4 + 9$
$$= -5 - 4 + 9$$
$$= -9 + 9 = 0$$

4. $-5(8 - 2z) + 4(7 - z) = 7(8 + z) - 3$
$-40 + 10z + 28 - 4z = 56 + 7z - 3$ *Distributive property*
$6z - 12 = 7z + 53$ *Combine like terms*
$-65 = z$ *Subtract 6z; 53*

Thus, the solution set is $\{-65\}$.

5. $3(x + 2) - 5(x + 2) = -2x - 4$
$3x + 6 - 5x - 10 = -2x - 4$
$-2x - 4 = -2x - 4$

The last statement is true for all real numbers, so the solution set is $(-\infty, \infty)$.

6. Solve $A = p + prt$ for t.

$$A = p + prt$$
$A - p = prt$ *Subtract p*
$\dfrac{A - p}{pr} = t$ *Divide by pr*

7.
$$2(m+5) - 3m + 1 > 5$$
$$2m + 10 - 3m + 1 > 5$$
$$-m + 11 > 5$$
$$-m > -6$$
$$m < 6$$

The solution set is $(-\infty, 6)$.

8. Personal computer: $\dfrac{480}{1500} = 32\%$

Pacemaker:

$$26\% \text{ of } 1500 = .26(1500) = 390$$

Wireless communication:

$$18\% \text{ of } 1500 = .18(1500) = 270$$

Television: $\dfrac{150}{1500} = 10\%$

9. The sum of the measures of the angles of any triangle is $180°$, so

$$(x + 15) + (6x + 10) + (x - 5) = 180.$$

Solve this equation.

$$8x + 20 = 180$$
$$8x = 160$$
$$x = 20$$

Substitute 20 for x to find the measures of the angles.

$$x - 5 = 20 - 5 = 15$$
$$x + 15 = 20 + 15 = 35$$
$$6x + 10 = 6(20) + 10 = 130$$

The measures of the angles of the triangle are $15°$, $35°$, and $130°$.

10. Through $(-4, 5)$ and $(2, -3)$

Use the definition of slope with $x_1 = -4$, $y_1 = 5$, $x_2 = 2$, and $y_2 = -3$.

$$m = \frac{y_2 - y_1}{x_2 - x_1} = \frac{-3 - 5}{2 - (-4)} = \frac{-8}{6} = -\frac{4}{3}$$

11. Horizontal, through $(4, 5)$

The slope of every horizontal line is 0.

12. Through $(4, -1)$, $m = -4$

Use the point-slope form with $x_1 = 4$, $y_1 = -1$, and $m = -4$.

$$y - y_1 = m(x - x_1)$$
$$y - (-1) = -4(x - 4)$$
$$y + 1 = -4x + 16$$
$$y = -4x + 15$$

13. Through $(0, 0)$ and $(1, 4)$

Find the slope.

$$m = \frac{4 - 0}{1 - 0} = 4$$

Because the slope is 4 and the y-intercept is 0, the equation of the line in slope-intercept form is

$$y = 4x.$$

14. $-3x + 4y = 12$

If $y = 0$, $x = -4$, so the x-intercept is $(-4, 0)$.

If $x = 0$, $y = 3$, so the y-intercept is $(0, 3)$.

Draw a line through these intercepts. A third point may be used as a check.

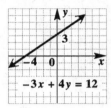

15.
$$\left(\frac{2m^3 n}{p^2}\right)^3 = \frac{2^3 (m^3)^3 n^3}{(p^2)^3}$$
$$= \frac{8m^9 n^3}{p^6}$$

16.
$$\frac{x^{-6} y^3 z^{-1}}{x^7 y^{-4} z} = \frac{y^4 y^3}{x^6 x^7 z^1 z} = \frac{y^7}{x^{13} z^2}$$

17.
$$2(3x^2 - 8x + 1) - 4(x^2 - 3x - 9)$$
$$= 6x^2 - 16x + 2 - 4x^2 + 12x + 36$$

Distributive property

$$= (6x^2 - 4x^2) + (-16x + 12x) + (2 + 36)$$

Combine like terms

$$= 2x^2 - 4x + 38$$

18. $(3x + 2y)(5x - y)$

$$\qquad \mathbf{F} \qquad \mathbf{O} \qquad \mathbf{I} \qquad \mathbf{L}$$
$$= 3x(5x) + 3x(-y) + 2y(5x) + 2y(-y)$$
$$= 15x^2 - 3xy + 10xy - 2y^2$$
$$= 15x^2 + 7xy - 2y^2$$

19. $(8m + 5n)(8m - 5n) = (8m)^2 - (5n)^2$
$$= 64m^2 - 25n^2$$

20. $\dfrac{m^3 - 3m^2 + 5m - 3}{m - 1}$

$$m - 1 \overline{\smash{\big)}\, \begin{array}{r} m^2 \quad -2m \quad +3 \\ m^3 - 3m^2 + 5m - 3 \\ \underline{m^3 - m^2} \\ -2m^2 + 5m \\ \underline{-2m^2 + 2m} \\ 3m - 3 \\ \underline{3m - 3} \\ 0 \end{array}}$$

The remainder is 0. The answer is the quotient,

$$m^2 - 2m + 3.$$

21. $m^2 + 12m + 32 = (m + 8)(m + 4)$

22. $25t^4 - 36 = \left(5t^2\right)^2 - (6)^2$
$$= \left(5t^2 + 6\right)\left(5t^2 - 6\right)$$

23. $81z^2 + 72z + 16$
$$= (9z)^2 + 2(9z)(4) + 4^2$$
$$= (9z + 4)^2$$

24. $(x + 4)(x - 1) = -6$
$$x^2 + 3x - 4 = -6$$
$$x^2 + 3x + 2 = 0$$
$$(x + 2)(x + 1) = 0$$
$$x = -2 \quad \text{or} \quad x = -1$$

The solution set is $\{-2, -1\}$.

25. $\dfrac{3}{x^2 + 5x - 14} = \dfrac{3}{(x + 7)(x - 2)}$

The expression is undefined when x is -7 or 2, because those values make the denominator equal zero.

26. $\dfrac{x^2 - 3x - 4}{x^2 + 3x} \cdot \dfrac{x^2 + 2x - 3}{x^2 - 5x + 4}$

$$= \dfrac{(x - 4)(x + 1)}{x(x + 3)} \cdot \dfrac{(x - 1)(x + 3)}{(x - 4)(x - 1)} \quad \textit{Factor}$$

$$= \dfrac{x + 1}{x} \qquad \textit{Lowest terms}$$

27. $\dfrac{t^2 + 4t - 5}{t + 5} \div \dfrac{t - 1}{t^2 + 8t + 15}$

$$= \dfrac{t^2 + 4t - 5}{t + 5} \cdot \dfrac{t^2 + 8t + 15}{t - 1}$$

$$\textit{Multiply by the reciprocal}$$

$$= \dfrac{(t + 5)(t - 1)}{t + 5} \cdot \dfrac{(t + 5)(t + 3)}{t - 1} \quad \textit{Factor}$$

$$= (t + 5)(t + 3), \qquad \textit{Lowest terms}$$

or $t^2 + 8t + 15$

28. $\dfrac{2}{x + 3} - \dfrac{4}{x - 1}$

$$= \dfrac{2(x - 1)}{(x + 3)(x - 1)} - \dfrac{4(x + 3)}{(x - 1)(x + 3)}$$

$$= \dfrac{2(x - 1) - 4(x + 3)}{(x + 3)(x - 1)}$$

$$= \dfrac{2x - 2 - 4x - 12}{(x + 3)(x - 1)}$$

$$= \dfrac{-2x - 14}{(x + 3)(x - 1)}$$

29. $\dfrac{\dfrac{2}{3} + \dfrac{1}{2}}{\dfrac{1}{9} - \dfrac{1}{6}} = \dfrac{\dfrac{4}{6} + \dfrac{3}{6}}{\dfrac{2}{18} - \dfrac{3}{18}}$

$$= \dfrac{\dfrac{7}{6}}{\dfrac{-1}{18}} = \dfrac{7}{6} \div \dfrac{-1}{18}$$

$$= \dfrac{7}{6} \cdot \dfrac{18}{-1} = 7 \cdot (-3) = -21$$

30. $\dfrac{x}{x + 8} - \dfrac{3}{x - 8} = \dfrac{128}{x^2 - 64}$
$$\textit{Multiply by the LCD, } (x + 8)(x - 8)$$
$$x(x - 8) - 3(x + 8) = 128$$
$$x^2 - 8x - 3x - 24 = 128$$
$$x^2 - 11x - 152 = 0$$
$$(x + 8)(x - 19) = 0$$
$$x = -8 \quad \text{or} \quad x = 19$$

Check $x = 19$: $\frac{19}{27} - \frac{3}{11} = \frac{128}{297}$ *True*

We cannot have -8 for a solution because that would result in division by zero. Thus, the solution set is $\{19\}$.

31. **(a)** $m = \dfrac{128{,}665 - 96{,}445}{10 - 0} = \dfrac{32{,}220}{10} = 3222$

The average rate of change is 3222 twin births per year, that is, the number of twin births increased an average of 3222 per year.

(b) $y = mx + b$
$$y = 3222x + 96{,}445$$

(c) If $x = 2006 - 1993 = 13$, then

$$y = 3222(13) + 96{,}445$$
$$= 138{,}331$$

The model predicts about 138,331 twin births in 2006.

32. $\{(-4, -2), (-1, 0), (2, 0), (5, 2)\}$

The domain is the set of first components, that is, $\{-4, -1, 2, 5\}$.

The range is the set of second components, that is, $\{-2, 0, 2\}$.

The relation is a function since each first component is paired with a unique second component.

33. $g(x) = -x^2 - 2x + 6$

$$g(3) = -3^2 - 2(3) + 6$$
$$= -9 - 6 + 6$$
$$= -9$$

34. $f(x) = x^2 + 6x + 1, \quad g(x) = 2x$

$$(f \circ g)(-2) = f(g(-2))$$
$$= f(2(-2))$$
$$= f(-4)$$
$$= (-4)^2 + 6(-4) + 1$$
$$= 16 - 24 + 1$$
$$= -7$$

35. $\quad 3x - 4y = 1 \qquad (1)$
$\quad 2x + 3y = 12 \qquad (2)$

To eliminate y, multiply equation (1) by 3 and equation (2) by 4. Then add the results.

$$
\begin{array}{rcll}
9x - 12y &=& 3 & 3 \times (1) \\
8x + 12y &=& 48 & 4 \times (2) \\
\hline
17x \quad\;\; &=& 51 & \\
x &=& 3 &
\end{array}
$$

Since $x = 3$,

$$
\begin{array}{rcl}
3x - 4y &=& 1 \qquad (1) \\
3(3) - 4y &=& 1 \\
9 - 4y &=& 1 \\
-4y &=& -8 \\
y &=& 2
\end{array}
$$

The solution set is $\{(3, 2)\}$.

36. $\quad 3x - 2y = 4 \qquad (1)$
$\quad -6x + 4y = 7 \qquad (2)$

Multiply equation (1) by 2 and add the result to equation (2).

$$
\begin{array}{rcll}
6x - 4y &=& 8 & 2 \times (1) \\
-6x + 4y &=& 7 & (2) \\
\hline
0 &=& 15 & \textit{False}
\end{array}
$$

Since a false statement results, the system is *inconsistent*. The solution set is $\emptyset$.

37. $\quad x + 3y - 6z = \;\; 7 \quad (1)$
$\quad 2x - y + z = \;\; 1 \quad (2)$
$\quad x + 2y + 2z = -1 \quad (3)$

To eliminate x, multiply equation (1) by -2 and add the result to equation (2).

$$
\begin{array}{rcll}
-2x - 6y + 12z &=& -14 & -2 \times (1) \\
2x - y + z &=& 1 & (2) \\
\hline
-7y + 13z &=& -13 & (4)
\end{array}
$$

To eliminate x again, multiply equation (3) by -2 and add the result to equation (2).

$$
\begin{array}{rcll}
2x - y + z &=& 1 & (2) \\
-2x - 4y - 4z &=& 2 & -2 \times (3) \\
\hline
-5y - 3z &=& 3 & (5)
\end{array}
$$

Use equations (4) and (5) to eliminate z. Multiply equation (4) by 3 and add the result to 13 times equation (5).

$$
\begin{array}{rcll}
-21y + 39z &=& -39 & 3 \times (4) \\
-65y - 39z &=& 39 & 13 \times (5) \\
\hline
-86y &=& 0 & \\
y &=& 0 &
\end{array}
$$

From (5), $-3z = 3$, so $z = -1$.
From (3), $x - 2 = -1$, so $x = 1$.

The solution set is $\{(1, 0, -1)\}$.

38. The length L of the rectangular flag measured 12 feet more than its width W, so

$$L = W + 12. \quad (1)$$

The perimeter is 144 feet.

$$P = 2L + 2W \quad (2)$$

Substitute $W + 12$ for L into equation (2).

$$
\begin{array}{rcl}
144 &=& 2(W + 12) + 2W \\
144 &=& 2W + 24 + 2W \\
120 &=& 4W \\
30 &=& W
\end{array}
$$

From (1), $L = 30 + 12 = 42$.

The length is 42 feet and the width is 30 feet.

39. Make a chart.

Number of Liters	Percent (as a decimal)	Pure Liters of Alcohol
x	0.15	$0.15x$
y	0.30	$0.30y$
9	0.20	$0.20(9) = 1.8$

From the first and third columns, we have the following system:

$$x + \quad y = \quad 9 \quad (1)$$
$$0.15x + 0.30y = 1.8 \quad (2)$$

To eliminate x, multiply equation (1) by -15 and add the result to 100 times equation (2).

$$
\begin{aligned}
-15x \; - \; 15y &= -135 \\
15x \; + \; 30y &= \quad 180 \\
\hline
15y &= 45 \\
y &= 3
\end{aligned}
$$

From (1), $x + 3 = 9$, so $x = 6$.

She should use 6 L of 15% solution and 3 L of 30% solution.

40. $x > -4$ and $x < 4$

The numbers that are greater than -4 *and* less than 4 are in the interval $(-4, 4)$.

The solution set is $(-4, 4)$.

41.
$$
\begin{aligned}
2x + 1 > 5 \quad &\text{or} \quad 2 - x \geq 2 \\
2x > 4 \quad & \qquad -x \geq 0 \\
x > 2 \quad &\text{or} \quad \; x \leq 0
\end{aligned}
$$

The solution set is $(-\infty, 0] \cup (2, \infty)$.

42. $|3x - 1| = 2$

$$
\begin{aligned}
3x - 1 = 2 \quad &\text{or} \quad 3x - 1 = -2 \\
3x = 3 \quad & \qquad 3x = -1 \\
x = 1 \quad &\text{or} \quad \; x = -\tfrac{1}{3}
\end{aligned}
$$

The solution set is $\left\{-\tfrac{1}{3}, 1\right\}$.

43. $|3z + 1| \geq 7$

$$
\begin{aligned}
3z + 1 \geq 7 \quad &\text{or} \quad 3z + 1 \leq -7 \\
3z \geq 6 \quad & \qquad 3z \leq -8 \\
z \geq 2 \quad &\text{or} \quad \; z \leq -\tfrac{8}{3}
\end{aligned}
$$

The solution set is $\left(-\infty, -\tfrac{8}{3}\right] \cup [2, \infty)$.

44. $y \leq 2x - 6$

Graph the boundary, $y = 2x - 6$, as a solid line through the intercepts $(3, 0)$ and $(0, -6)$. A third point such as $(1, -4)$ can be used as a check. Using $(0, 0)$ as a test point results in the false inequality $0 \leq -6$, so shade the region *not* containing the origin. This is the region below the line. The solid line shows that the boundary is part of the graph.

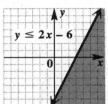

45. $3x + 2y < 0$

Graph the boundary, $3x + 2y = 0$, as a dashed line through $(0, 0)$, $(-2, 3)$, and $(2, -3)$. Choose a test point not on the line. Using $(1, 1)$ results in the false statement $5 < 0$, so shade the region *not* containing $(1, 1)$. This is the region below the line. The dashed line shows that the boundary is not part of the graph.

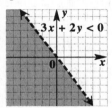

46. $x - y \geq 3$ and $3x + 4y \leq 12$

Graph the solid line $x - y = 3$ through $(3, 0)$ and $(0, -3)$. The inequality $x - y \geq 3$ can be written as $y \leq x - 3$, so shade the region below the boundary line.

Graph the solid line $3x + 4y = 12$ through $(4, 0)$ and $(0, 3)$. The inequality $3x + 4y \leq 12$ can be written as $y \leq -\tfrac{3}{4}x + 3$, so shade the region below the boundary line.

The required graph is the common shaded area as well as the portions of the lines that bound it.

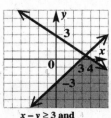

$x - y \geq 3$ and
$3x + 4y \leq 12$

CHAPTER 10 ROOTS, RADICALS, AND ROOT FUNCTIONS

10.1 Radical Expressions and Graphs

1. Every positive number has two real square roots. This statement is *true*. One of the real square roots is a positive number and the other is its opposite.

3. Every nonnegative number has two real square roots. This statement is *false* since zero is a nonnegative number that has only one square root, namely 0.

5. The cube root of every real number has the same sign as the number itself. This statement is *true*. The cube root of a positive real number is positive and the cube root of a negative real number is negative. The cube root of 0 is 0.

7. The square roots of 9 are -3 and 3 because $(-3)(-3) = 9$ and $3 \cdot 3 = 9$.

9. The square roots of 64 are -8 and 8 because $(-8)(-8) = 64$ and $8 \cdot 8 = 64$.

11. The square roots of 144 are -12 and 12 because $(-12)(-12) = 144$ and $12 \cdot 12 = 144$.

13. The square roots of $\frac{25}{196}$ are $-\frac{5}{14}$ and $\frac{5}{14}$ because
$$\left(-\tfrac{5}{14}\right)\left(-\tfrac{5}{14}\right) = \tfrac{25}{196} \quad \text{and} \quad \tfrac{5}{14} \cdot \tfrac{5}{14} = \tfrac{25}{196}.$$

15. The square roots of 900 are -30 and 30 because $(-30)(-30) = 900$ and $30 \cdot 30 = 900$.

17. $\sqrt{1}$ represents the positive square root of 1. Since $1 \cdot 1 = 1$,
$$\sqrt{1} = 1.$$

19. $\sqrt{49}$ represents the positive square root of 49. Since $7 \cdot 7 = 49$,
$$\sqrt{49} = 7.$$

21. $-\sqrt{121}$ represents the negative square root of 121. Since $11 \cdot 11 = 121$,
$$-\sqrt{121} = -11.$$

23. $-\sqrt{\frac{144}{121}}$ represents the negative square root of $\frac{144}{121}$. Since $\frac{12}{11} \cdot \frac{12}{11} = \frac{144}{121}$,
$$-\sqrt{\tfrac{144}{121}} = -\tfrac{12}{11}.$$

25. $\sqrt{-121}$ is not a real number because there is no real number whose square is -121.

27. The square of $\sqrt{19}$ is
$$\left(\sqrt{19}\right)^2 = 19,$$
by the definition of square root.

29. The square of $-\sqrt{19}$ is
$$\left(-\sqrt{19}\right)^2 = 19,$$
since the square of a negative number is positive.

31. The square of $\sqrt{\frac{2}{3}}$ is
$$\left(\sqrt{\tfrac{2}{3}}\right)^2 = \tfrac{2}{3},$$
by the definition of square root.

33. The square of $\sqrt{3x^2 + 4}$ is
$$\left(\sqrt{3x^2 + 4}\right)^2 = 3x^2 + 4.$$

35. For the statement "$\sqrt{a}$ represents a positive number" to be true, a must be positive because the square root of a negative number is not a real number and $\sqrt{0} = 0$.

37. For the statement "$\sqrt{a}$ is not a real number" to be true, a must be negative.

39. $\sqrt{25}$

The number 25 is a perfect square, 5^2, so $\sqrt{25}$ is a *rational* number.
$$\sqrt{25} = 5$$

41. $\sqrt{29}$

Because 29 is not a perfect square, $\sqrt{29}$ is *irrational*. Using a calculator, we obtain
$$\sqrt{29} \approx 5.385.$$

43. $-\sqrt{64}$

The number 64 is a perfect square, 8^2, so $-\sqrt{64}$ is *rational*.
$$-\sqrt{64} = -8$$

45. $-\sqrt{300}$

The number 300 is not a perfect square, so $-\sqrt{300}$ is *irrational*. Using a calculator, we obtain
$$-\sqrt{300} \approx -17.321.$$

47. $\sqrt{-29}$

There is no real number whose square is -29. Therefore, $\sqrt{-29}$ is *not a real number*.

49. $\sqrt{1200}$

Because 1200 is not a perfect square, $\sqrt{1200}$ is *irrational*. Using a calculator, we obtain

$$\sqrt{1200} \approx 34.641.$$

51. Since $81 < 94 < 100$, $\sqrt{81} = 9$, and $\sqrt{100} = 10$, we conclude that $\sqrt{94}$ is between 9 and 10.

53. Since $49 < 51 < 64$, $\sqrt{49} = 7$, and $\sqrt{64} = 8$, we conclude that $\sqrt{51}$ is between 7 and 8.

55. Since $36 < 40 < 49$, $\sqrt{36} = 6$, and $\sqrt{49} = 7$, we conclude that $-\sqrt{40}$ is between -7 and -6.

57. Since $16 < 23.2 < 25$, $\sqrt{16} = 4$, and $\sqrt{25} = 5$, we conclude that $\sqrt{23.2}$ is between 4 and 5.

59. (a) $-\sqrt{16} = -(4) = -4$ **(E)**

(b) $\sqrt{-16}$ is not a real number (the index, 2, is even and the radicand, -16, is negative). **(F)**

(c) $\sqrt[3]{-27} = -3$, because $(-3)^3 = -27$. **(D)**

(d) $\sqrt[5]{-32} = -2$, because $(-2)^5 = -32$. **(B)**

(e) $\sqrt[4]{81} = 3$, because $3^4 = 81$. **(A)**

(f) $\sqrt[3]{8} = 2$, because $2^3 = 8$. **(C)**

61. $\sqrt{67.8} \approx \sqrt{64} = 8$, because $8^2 = 64$. **(B)**

63. The length $\sqrt{98}$ is closer to $\sqrt{100} = 10$ than to $\sqrt{81} = 9$. The width $\sqrt{26}$ is closer to $\sqrt{25} = 5$ than to $\sqrt{36} = 6$. Use the estimates $L = 10$ and $W = 5$ in $P = 2L + 2W$ to find an estimate of the perimeter.

$$P \approx 2(10) + 2(5) = 30$$

Choice **D** is the best estimate.

65. $\sqrt{81} = 9$, because $9^2 = 81$. Notice that -9 is not an answer because the symbol $\sqrt{81}$ means the *nonnegative* square root of 81. However, the negative in front of the radical does lead to a negative answer since

$$-\sqrt{81} = -(9) = -9.$$

67. $\sqrt[3]{216} = 6$, because $6^3 = 216$.

69. $\sqrt[3]{-64} = -4$, because $(-4)^3 = -64$.

71. $\sqrt[3]{512} = 8$, because $8^3 = 512$, so $-\sqrt[3]{512} = -8$.

73. $\sqrt[4]{1296} = 6$, because $6^4 = 1296$.

75. $\sqrt[4]{256} = 4$, because $4^4 = 256$, so $-\sqrt[4]{256} = -4$.

77. $\sqrt[4]{-256}$ is not a real number since no real number to the fourth power equals -256.

79. $\sqrt[6]{64} = 2$, because $2^6 = 64$.

81. $\sqrt[6]{-32}$ is not a real number, because the index, 6, is even and the radicand, -32, is negative.

83. $\sqrt{\frac{64}{81}} = \frac{8}{9}$, because $\left(\frac{8}{9}\right)^2 = \frac{64}{81}$.

85. $\sqrt[3]{\frac{64}{27}} = \frac{4}{3}$, because $\left(\frac{4}{3}\right)^3 = \frac{64}{27}$.

87. $\sqrt[6]{\frac{1}{64}} = \frac{1}{2}$, because $\left(\frac{1}{2}\right)^6 = \frac{1}{64}$, so $-\sqrt[6]{\frac{1}{64}} = -\frac{1}{2}$.

89. $\sqrt{0.49} = 0.7$, because $(0.7)^2 = 0.49$.

91. $\sqrt[3]{0.001} = 0.1$, because $(0.1)^3 = 0.001$.

93. $f(x) = \sqrt{x + 3}$

For the radicand to be nonnegative, we must have

$$x + 3 \geq 0 \quad \text{or} \quad x \geq -3.$$

Thus, the domain is $[-3, \infty)$.
The function values are positive or zero (the result of the radical), so the range is $[0, \infty)$.

x	$f(x) = \sqrt{x + 3}$
-3	$\sqrt{-3 + 3} = 0$
-2	$\sqrt{-2 + 3} = 1$
1	$\sqrt{1 + 3} = 2$

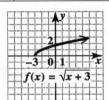

95. $f(x) = \sqrt{x} - 2$

For the radicand to be nonnegative, we must have

$$x \geq 0.$$

Note that the "-2" does not affect the domain, which is $[0, \infty)$.
The result of the radical is positive or zero, but the function values are 2 less than those values, so the range is $[-2, \infty)$.

x	$f(x) = \sqrt{x} - 2$
0	$\sqrt{0} - 2 = -2$
1	$\sqrt{1} - 2 = -1$
4	$\sqrt{4} - 2 = 0$

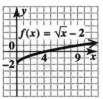

97. $f(x) = \sqrt[3]{x} - 3$

Since we can take the cube root of any real number, the domain is $(-\infty, \infty)$.

The result of a cube root can be any real number, so the range is $(-\infty, \infty)$. (The "-3" does not affect the range.)

x	$f(x) = \sqrt[3]{x} - 3$
-8	$\sqrt[3]{-8} - 3 = -5$
-1	$\sqrt[3]{-1} - 3 = -4$
0	$\sqrt[3]{0} - 3 = -3$
1	$\sqrt[3]{1} - 3 = -2$
8	$\sqrt[3]{8} - 3 = -1$

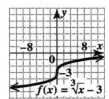

99. $f(x) = \sqrt[3]{x} - 3$

Both the domain and range are $(-\infty, \infty)$.

x	$f(x) = \sqrt[3]{x} - 3$
-5	$\sqrt[3]{-5} - 3 = -2$
2	$\sqrt[3]{2} - 3 = -1$
3	$\sqrt[3]{3} - 3 = 0$
4	$\sqrt[3]{4} - 3 = 1$
11	$\sqrt[3]{11} - 3 = 2$

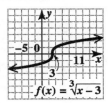

101. $\sqrt{12^2} = |12| = 12$

103. $\sqrt{(-10)^2} = |-10| = -(-10) = 10$

105. Since 6 is an even positive integer, $\sqrt[6]{a^6} = |a|$, so $\sqrt[6]{(-2)^6} = |-2| = 2$.

107. Since 5 is odd, $\sqrt[5]{a^5} = a$, so $\sqrt[5]{(-9)^5} = -9$.

109. $\sqrt[6]{(-5)^6} = |-5| = 5$, so $-\sqrt[6]{(-5)^6} = -5$.

111. Since the index is even, $\sqrt{x^2} = |x|$.

113. $\sqrt{(-z)^2} = |-z| = |z|$

115. Since the index is odd, $\sqrt[3]{x^3} = x$.

117. $\sqrt[3]{x^{15}} = \sqrt[3]{(x^5)^3} = x^5$ (3 is odd)

119. $\sqrt[6]{x^{30}} = \sqrt[6]{(x^5)^6} = |x^5|$,   or $|x|^5$ (6 is even)

In Exercises 121–128, use a calculator and round to three decimal places

121. $\sqrt{9483} \approx 97.381$

123. $-\sqrt{82} \approx -9.055$

125. $\sqrt[3]{423} \approx 7.507$

127. $\sqrt[5]{23.8} \approx 1.885$

129. $f = \dfrac{1}{2\pi\sqrt{LC}}$

$= \dfrac{1}{2\pi\sqrt{(7.237 \times 10^{-5})(2.5 \times 10^{-10})}}$

$\approx 1{,}183{,}235$

or about 1,183,000 cycles per second.

131. Since $H = 44 + 6 = 50$ ft, substitute 50 for H in the formula.

$$D = \sqrt{2H} = \sqrt{2 \cdot 50} = \sqrt{100} = 10$$

She will be able to see about 10 miles.

133. Let $a = 850$, $b = 925$, and $c = 1300$. First find s.

$$s = \tfrac{1}{2}(a + b + c)$$
$$= \tfrac{1}{2}(850 + 925 + 1300)$$
$$= \tfrac{3075}{2} = 1537.5$$

Now find the area using Heron's formula.

$$A = \sqrt{s(s-a)(s-b)(s-c)}$$
$$= \sqrt{1537.5(687.5)(612.5)(237.5)}$$
$$\approx 392{,}128.8$$

The area of the Bermuda Triangle is about 392,000 square miles.

135. $I = \sqrt{\dfrac{2P}{L}}$

$= \sqrt{\dfrac{2(120)}{80}} = \sqrt{3} \approx 1.732$ amps

137. $x^5 \cdot x^{-1} \cdot x^{-3} = x^{5+(-1)+(-3)} = x^1$,   or x

139. $(13x^0 y^5)(13x^4 y^3) = 13^{1+1} x^{0+4} y^{5+3}$
$$= 13^2 x^4 y^8, \text{ or } 169x^4 y^8$$

141. $\dfrac{5}{5^{-1}} = 5^{1-(-1)} = 5^2$,   or   25

10.2 Rational Exponents

1. $2^{1/2} = \sqrt[2]{2^1} = \sqrt{2}$ **(C)**

3. $-16^{1/2} = -\sqrt{16} = -(4) = -4$ **(A)**

5. $(-32)^{1/5} = \left[(-2)^5\right]^{1/5} = (-2)^1 = -2$ **(H)**

7. $4^{3/2} = \left[(2^2)^{1/2}\right]^3 = 2^3 = 8$ **(B)**

9. $-6^{2/4} = -\left(6^{1/2}\right) = -\sqrt{6}$ **(D)**

11. $169^{1/2} = \left(13^2\right)^{1/2} = 13^1 = 13$

We could use radical notation as follows:
$$169^{1/2} = \sqrt{169} = 13$$

13. $729^{1/3} = \left(9^3\right)^{1/3} = 9$

15. $16^{1/4} = \left(2^4\right)^{1/4} = 2$

17. $\left(\frac{64}{81}\right)^{1/2} = \left[\left(\frac{8}{9}\right)^2\right]^{1/2} = \frac{8}{9}$

19. $(-27)^{1/3} = \left[(-3)^3\right]^{1/3} = -3$

21. $(-144)^{1/2}$ is not a real number, because no real number squared equals -144.

23. $100^{3/2} = \left(100^{1/2}\right)^3 = 10^3 = 1000$

25. $81^{3/4} = \left(81^{1/4}\right)^3 = \left(\sqrt[4]{81}\right)^3 = 3^3 = 27$

27. $-16^{5/2} = -\left(16^{1/2}\right)^5 = -(4)^5 = -1024$

29. $(-8)^{4/3} = \left[(-8)^{1/3}\right]^4 = \left(\sqrt[3]{-8}\right)^4$
$$= (-2)^4 = 16$$

31. $32^{-3/5} = \frac{1}{32^{3/5}} = \frac{1}{(32^{1/5})^3} = \frac{1}{\left(\sqrt[5]{32}\right)^3}$
$$= \frac{1}{2^3} = \frac{1}{8}$$

33. $64^{-3/2} = \frac{1}{64^{3/2}} = \frac{1}{(64^{1/2})^3} = \frac{1}{8^3} = \frac{1}{512}$

35. $\left(\frac{125}{27}\right)^{-2/3} = \left(\frac{27}{125}\right)^{2/3}$
$$= \left(\left[\left(\frac{3}{5}\right)^3\right]^{1/3}\right)^2$$
$$= \left(\frac{3}{5}\right)^2 = \frac{9}{25}$$

37. $10^{1/2} = \left(\sqrt[2]{10}\right)^1 = \sqrt{10}$

39. $8^{3/4} = \left(8^{1/4}\right)^3 = \left(\sqrt[4]{8}\right)^3$

41. $(9q)^{5/8} - (2x)^{2/3}$
$$= \left[(9q)^{1/8}\right]^5 - \left[(2x)^{1/3}\right]^2$$
$$= \left(\sqrt[8]{9q}\right)^5 - \left(\sqrt[3]{2x}\right)^2$$

43. $(2m)^{-3/2} = \left[(2m)^{1/2}\right]^{-3}$
$$= \left(\sqrt{2m}\right)^{-3}$$
$$= \frac{1}{\left(\sqrt{2m}\right)^3}$$

45. $(2y + x)^{2/3} = \left[(2y + x)^{1/3}\right]^2$
$$= \left(\sqrt[3]{2y + x}\right)^2$$

47. $\left(3m^4 + 2k^2\right)^{-2/3}$
$$= \frac{1}{\left(3m^4 + 2k^2\right)^{2/3}}$$
$$= \frac{1}{\left[\left(3m^4 + 2k^2\right)^{1/3}\right]^2}$$
$$= \frac{1}{\left(\sqrt[3]{3m^4 + 2k^2}\right)^2}$$

49. We are to show that, in general,
$$\sqrt{a^2 + b^2} \neq a + b.$$
When $a = 3$ and $b = 4$,
$$\sqrt{a^2 + b^2} = \sqrt{3^2 + 4^2}$$
$$= \sqrt{9 + 16}$$
$$= \sqrt{25} = 5,$$
but
$$a + b = 3 + 4 = 7.$$
Since $5 \neq 7$, $\sqrt{a^2 + b^2} \neq a + b$.

51. $\sqrt{2^{12}} = (2^{12})^{1/2} = 2^{12/2} = 2^6 = 64$

53. $\sqrt[3]{4^9} = 4^{9/3} = 4^3 = 64$

55. $\sqrt{x^{20}} = x^{20/2} = x^{10}$

57. $\sqrt[3]{x} \cdot \sqrt{x} = x^{1/3} \cdot x^{1/2} = x^{1/3+1/2} = x^{2/6+3/6}$
$$= x^{5/6} = \sqrt[6]{x^5}$$

59. $\frac{\sqrt[3]{t^4}}{\sqrt[5]{t^4}} = \frac{t^{4/3}}{t^{4/5}} = \frac{t^{20/15}}{t^{12/15}} = t^{20/15 - 12/15}$
$$= t^{8/15} = \sqrt[15]{t^8}$$

61. $3^{1/2} \cdot 3^{3/2} = 3^{1/2 + 3/2} = 3^{4/2} = 3^2 = 9$

63. $\frac{64^{5/3}}{64^{4/3}} = 64^{5/3 - 4/3} = 64^{1/3} = \sqrt[3]{64} = 4$

65. $y^{7/3} \cdot y^{-4/3} = y^{7/3 + (-4/3)} = y^{3/3} = y$

67. $x^{2/3} \cdot x^{-1/4} = x^{2/3 + (-1/4)} = x^{8/12 - 3/12} = x^{5/12}$

69. $\frac{k^{1/3}}{k^{2/3} \cdot k^{-1}} = \frac{k^{1/3}}{k^{-1/3}} = k^{1/3 - (-1/3)} = k^{2/3}$

71. $\frac{\left(x^{1/4} y^{2/5}\right)^{20}}{x^2} = \frac{x^{(1/4) \cdot 20} y^{(2/5) \cdot 20}}{x^2}$
$$= \frac{x^5 y^8}{x^2}$$
$$= x^3 y^8$$

73. $\frac{\left(x^{2/3}\right)^2}{\left(x^2\right)^{7/3}} = \frac{x^{4/3}}{x^{14/3}} = x^{4/3 - 14/3}$
$$= x^{-10/3} = \frac{1}{x^{10/3}}$$

75. $\dfrac{m^{3/4}n^{-1/4}}{(m^2n)^{1/2}} = \dfrac{m^{3/4}n^{-1/4}}{m^1n^{1/2}}$

$= m^{3/4-1}n^{-1/4-1/2}$

$= m^{3/4-4/4}n^{-1/4-2/4}$

$= m^{-1/4}n^{-3/4}$

$= \dfrac{1}{m^{1/4}n^{3/4}}$

77. $\dfrac{p^{1/5}p^{7/10}p^{1/2}}{(p^3)^{-1/5}} = \dfrac{p^{2/10+7/10+5/10}}{p^{-3/5}}$

$= \dfrac{p^{14/10}}{p^{-6/10}}$

$= p^{14/10-(-6/10)}$

$= p^{20/10} = p^2$

79. $\left(\dfrac{b^{-3/2}}{c^{-5/3}}\right)^2 (b^{-1/4}c^{-1/3})^{-1}$

$= \left(\dfrac{c^{5/3}}{b^{3/2}}\right)^2 (b^{1/4}c^{1/3})$

$= \dfrac{c^{10/3}}{b^3}(b^{1/4}c^{1/3})$

$= \dfrac{c^{10/3}b^{1/4}c^{1/3}}{b^3}$

$= c^{10/3+1/3}b^{1/4-3}$

$= c^{11/3}b^{-11/4}$

$= \dfrac{c^{11/3}}{b^{11/4}}$

81. $\left(\dfrac{p^{-1/4}q^{-3/2}}{3^{-1}p^{-2}q^{-2/3}}\right)^{-2}$

$= \dfrac{p^{1/2}q^3}{3^2p^4q^{4/3}}$

$= \dfrac{q^{3-4/3}}{9p^{4-1/2}}$

$= \dfrac{q^{5/3}}{9p^{7/2}}$

83. $p^{2/3}(p^{1/3} + 2p^{4/3})$

$= p^{2/3}p^{1/3} + p^{2/3}(2p^{4/3})$

$= p^{2/3+1/3} + 2p^{2/3+4/3}$

$= p^{3/3} + 2p^{6/3}$

$= p^1 + 2p^2$

$= p + 2p^2$

85. $k^{1/4}(k^{3/2} - k^{1/2})$

$= k^{1/4+3/2} - k^{1/4+1/2}$

$= k^{1/4+6/4} - k^{1/4+2/4}$

$= k^{7/4} - k^{3/4}$

87. $6a^{7/4}(a^{-7/4} + 3a^{-3/4})$

$= 6a^{7/4+(-7/4)} + 18a^{7/4+(-3/4)}$

$= 6a^0 + 18a^{4/4}$

$= 6(1) + 18a^1 = 6 + 18a$

89. $\sqrt[5]{x^3} \cdot \sqrt[4]{x} = x^{3/5} \cdot x^{1/4}$

$= x^{3/5+1/4}$

$= x^{12/20+5/20}$

$= x^{17/20}$

91. $\dfrac{\sqrt{x^5}}{\sqrt{x^8}} = \dfrac{(x^5)^{1/2}}{(x^8)^{1/2}}$

$= x^{5/2-(8/2)}$

$= x^{-3/2} = \dfrac{1}{x^{3/2}}$

93. $\sqrt{y} \cdot \sqrt[3]{yz} = y^{1/2} \cdot (yz)^{1/3}$

$= y^{1/2}y^{1/3}z^{1/3}$

$= y^{1/2+1/3}z^{1/3}$

$= y^{3/6+2/6}z^{1/3}$

$= y^{5/6}z^{1/3}$

95. $\sqrt[4]{\sqrt[3]{m}} = \sqrt[4]{m^{1/3}} = (m^{1/3})^{1/4} = m^{1/12}$

97. $\sqrt{\sqrt[3]{\sqrt[4]{x}}} = \sqrt{\sqrt[3]{x^{1/4}}}$

$= \sqrt{(x^{1/4})^{1/3}}$

$= (x^{1/12})^{1/2}$

$= x^{1/24}$

99. Use $T(D) = 0.07D^{3/2}$ with $D = 16$.

$T(16) = 0.07(16)^{3/2} = 0.07(16^{1/2})^3$

$= 0.07(4)^3 = 0.07(64) = 4.48$

To the nearest tenth of an hour, time T is 4.5 hours.

101. Use a calculator and the formula

$W = 35.74 + 0.6215T - 35.75V^{4/25} + 0.4275TV^{4/25}$.

For $T = 30°F$ and $V = 15$ mph, $W \approx 19.0°F$. The table gives 19.0°F.

103. $\sqrt{25} \cdot \sqrt{36} = 5 \cdot 6 = 30$

$\sqrt{25 \cdot 36} = \sqrt{900} = 30$

The results are the same.

105. $\dfrac{\sqrt[3]{27}}{\sqrt[3]{729}} = \dfrac{3}{9} = \dfrac{1}{3}$

$\sqrt[3]{\dfrac{27}{729}} = \sqrt[3]{\dfrac{1}{27}} = \dfrac{1}{3}$

The results are the same.

10.3 Simplifying Radical Expressions

1. Does $2\sqrt{12} = \sqrt{48}$?

$2\sqrt{12} = 2\sqrt{4 \cdot 3} = 2\sqrt{4} \cdot \sqrt{3} = 2 \cdot 2 \cdot \sqrt{3}$

$\qquad = 4\sqrt{3}$

$\sqrt{48} = \sqrt{16 \cdot 3} = \sqrt{16} \cdot \sqrt{3} = 4\sqrt{3}$

The calculator approximation for each expression is 6.92820323. The statement is true.

3. Does $3\sqrt{8} = 2\sqrt{18}$?

$3\sqrt{8} = 3\sqrt{4 \cdot 2} = 3\sqrt{4}\sqrt{2} = 3 \cdot 2\sqrt{2} = 6\sqrt{2}$

$2\sqrt{18} = 2\sqrt{9 \cdot 2} = 2\sqrt{9}\sqrt{2} = 2 \cdot 3\sqrt{2} = 6\sqrt{2}$

The calculator approximation for each expression is 8.485281374. The statement is true.

5. **A.** $0.5 = \frac{1}{2}$, so $\sqrt{0.5} = \sqrt{\frac{1}{2}}$

B. $\frac{2}{4} = \frac{1}{2}$, so $\sqrt{\frac{2}{4}} = \sqrt{\frac{1}{2}}$

C. $\frac{3}{6} = \frac{1}{2}$, so $\sqrt{\frac{3}{6}} = \sqrt{\frac{1}{2}}$

D. $\frac{\sqrt{4}}{\sqrt{16}} = \sqrt{\frac{4}{16}} = \sqrt{\frac{1}{4}} \neq \sqrt{\frac{1}{2}}$

Choice **D** is not equal to $\sqrt{\frac{1}{2}}$.

7. $\sqrt{5} \cdot \sqrt{6} = \sqrt{5 \cdot 6} = \sqrt{30}$

9. $\sqrt{14} \cdot \sqrt{x} = \sqrt{14 \cdot x} = \sqrt{14x}$

11. $\sqrt{14} \cdot \sqrt{3pqr} = \sqrt{14 \cdot 3pqr} = \sqrt{42pqr}$

13. $\sqrt[3]{7x} \cdot \sqrt[3]{2y} = \sqrt[3]{7x \cdot 2y} = \sqrt[3]{14xy}$

15. $\sqrt[4]{11} \cdot \sqrt[4]{3} = \sqrt[4]{11 \cdot 3} = \sqrt[4]{33}$

17. $\sqrt[4]{2x} \cdot \sqrt[4]{3x^2} = \sqrt[4]{2x \cdot 3x^2} = \sqrt[4]{6x^3}$

19. $\sqrt[3]{7} \cdot \sqrt[4]{3}$ cannot be multiplied using the product rule, because the indexes (3 and 4) are different.

21. To multiply two radical expressions with the same index, multiply the radicands and keep the same index. For example, $\sqrt[3]{3} \cdot \sqrt[3]{5} = \sqrt[3]{15}$.

23. $\sqrt{\frac{64}{121}} = \frac{\sqrt{64}}{\sqrt{121}} = \frac{8}{11}$

25. $\sqrt{\frac{3}{25}} = \frac{\sqrt{3}}{\sqrt{25}} = \frac{\sqrt{3}}{5}$

27. $\sqrt{\frac{x}{25}} = \frac{\sqrt{x}}{\sqrt{25}} = \frac{\sqrt{x}}{5}$

29. $\sqrt{\frac{p^6}{81}} = \frac{\sqrt{p^6}}{\sqrt{81}} = \frac{\sqrt{(p^3)^2}}{9} = \frac{p^3}{9}$

31. $\sqrt[3]{-\frac{27}{64}} = \sqrt[3]{\frac{-27}{64}} = \frac{\sqrt[3]{-27}}{\sqrt[3]{64}} = \frac{-3}{4} = -\frac{3}{4}$

33. $\sqrt[3]{\frac{r^2}{8}} = \frac{\sqrt[3]{r^2}}{\sqrt[3]{8}} = \frac{\sqrt[3]{r^2}}{2}$

35. $-\sqrt[4]{\frac{81}{x^4}} = -\frac{\sqrt[4]{3^4}}{\sqrt[4]{x^4}} = -\frac{3}{x}$

37. $\sqrt[5]{\frac{1}{x^{15}}} = \frac{\sqrt[5]{1}}{\sqrt[5]{(x^3)^5}} = \frac{1}{x^3}$

39. $\sqrt{12} = \sqrt{4 \cdot 3} = \sqrt{4} \cdot \sqrt{3} = 2\sqrt{3}$

41. $\sqrt{288} = \sqrt{144 \cdot 2} = \sqrt{144} \cdot \sqrt{2} = 12\sqrt{2}$

43. $-\sqrt{32} = -\sqrt{16 \cdot 2} = -\sqrt{16} \cdot \sqrt{2} = -4\sqrt{2}$

45. $-\sqrt{28} = -\sqrt{4 \cdot 7} = -\sqrt{4} \cdot \sqrt{7} = -2\sqrt{7}$

47. $\sqrt{30}$ cannot be simplified further.

49. $\sqrt[3]{128} = \sqrt[3]{64 \cdot 2} = \sqrt[3]{64} \cdot \sqrt[3]{2} = 4\sqrt[3]{2}$

51. $\sqrt[3]{-16} = \sqrt[3]{-8 \cdot 2} = \sqrt[3]{-8} \cdot \sqrt[3]{2} = -2\sqrt[3]{2}$

53. $\sqrt[3]{40} = \sqrt[3]{8 \cdot 5} = \sqrt[3]{8} \cdot \sqrt[3]{5} = 2\sqrt[3]{5}$

55. $-\sqrt[4]{512} = -\sqrt[4]{256 \cdot 2} = -\sqrt[4]{4^4} \cdot \sqrt[4]{2} = -4\sqrt[4]{2}$

57. $\sqrt[5]{64} = \sqrt[5]{32 \cdot 2} = \sqrt[5]{2^5} \cdot \sqrt[5]{2} = 2\sqrt[5]{2}$

59. His reasoning was incorrect. The radicand 14 must be written as a product of two factors (not a sum of two terms) where one of the two factors is a perfect cube.

61. $\sqrt{72k^2} = \sqrt{36k^2 \cdot 2} = \sqrt{36k^2} \cdot \sqrt{2} = 6k\sqrt{2}$

63. $\sqrt{144x^3y^9} = \sqrt{144x^2y^8 \cdot xy}$

$\qquad = \sqrt{(12xy^4)^2} \cdot \sqrt{xy}$

$\qquad = 12xy^4\sqrt{xy}$

65. $\sqrt{121x^6} = \sqrt{(11x^3)^2} = 11x^3$

67. $-\sqrt[3]{27t^{12}} = -\sqrt[3]{(3t^4)^3} = -3t^4$

69. $-\sqrt{100m^8z^4} = -\sqrt{(10m^4z^2)^2} = -10m^4z^2$

71. $-\sqrt[3]{-125a^6b^9c^{12}} = -\sqrt[3]{(-5a^2b^3c^4)^3}$

$\qquad = -(-5a^2b^3c^4)$

$\qquad = 5a^2b^3c^4$

73. $\sqrt[4]{\frac{1}{16}r^8t^{20}} = \sqrt[4]{\left(\frac{1}{2}r^2t^5\right)^4} = \frac{1}{2}r^2t^5$

75. $\sqrt{50x^3} = \sqrt{25x^2 \cdot 2x} = \sqrt{(5x)^2} \cdot \sqrt{2x}$

$\qquad = 5x\sqrt{2x}$

77. $-\sqrt{500r^{11}} = -\sqrt{100r^{10} \cdot 5r}$

$\qquad = -\sqrt{(10r^5)^2} \cdot \sqrt{5r} = -10r^5\sqrt{5r}$

79. $\sqrt{13x^7y^8} = \sqrt{(x^6y^8)(13x)}$
$= \sqrt{x^6y^8} \cdot \sqrt{13x}$
$= x^3y^4\sqrt{13x}$

81. $\sqrt[3]{8z^6w^9} = \sqrt[3]{(2z^2w^3)^3} = 2z^2w^3$

83. $\sqrt[3]{-16z^5t^7} = \sqrt[3]{(-2^3z^3t^6)(2z^2t)}$
$= -2zt^2\sqrt[3]{2z^2t}$

85. $\sqrt[4]{81x^{12}y^{16}} = \sqrt[4]{(3x^3y^4)^4} = 3x^3y^4$

87. $-\sqrt[4]{162r^{15}s^{10}} = -\sqrt[4]{81r^{12}s^8(2r^3s^2)}$
$= -\sqrt[4]{81r^{12}s^8} \cdot \sqrt[4]{2r^3s^2}$
$= -3r^3s^2\sqrt[4]{2r^3s^2}$

89. $\sqrt{\dfrac{y^{11}}{36}} = \dfrac{\sqrt{y^{11}}}{\sqrt{36}} = \dfrac{\sqrt{y^{10} \cdot y}}{6} = \dfrac{y^5\sqrt{y}}{6}$

91. $\sqrt[3]{\dfrac{x^{16}}{27}} = \dfrac{\sqrt[3]{x^{15} \cdot x}}{\sqrt[3]{27}} = \dfrac{x^5\sqrt[3]{x}}{3}$

93. $\sqrt[4]{48^2} = 48^{2/4} = 48^{1/2} = \sqrt{48} = \sqrt{16 \cdot 3}$
$= \sqrt{16} \cdot \sqrt{3} = 4\sqrt{3}$

95. $\sqrt[4]{25} = 25^{1/4} = (5^2)^{1/4} = 5^{2/4} = 5^{1/2} = \sqrt{5}$

97. $\sqrt[10]{x^{25}} = x^{25/10} = x^{5/2} = \sqrt{x^5}$
$= \sqrt{x^4 \cdot x} = x^2\sqrt{x}$

99. $\sqrt[3]{4} \cdot \sqrt{3}$
The least common index of 3 and 2 is 6. Write each radical as a sixth root.
$$\sqrt[3]{4} = 4^{1/3} = 4^{2/6} = \sqrt[6]{4^2} = \sqrt[6]{16}$$
$$\sqrt{3} = 3^{1/2} = 3^{3/6} = \sqrt[6]{3^3} = \sqrt[6]{27}$$
Therefore,
$$\sqrt[3]{4} \cdot \sqrt{3} = \sqrt[6]{16} \cdot \sqrt[6]{27}$$
$$= \sqrt[6]{16 \cdot 27} = \sqrt[6]{432}.$$

101. $\sqrt[4]{3} \cdot \sqrt[3]{4}$
The least common index of 4 and 3 is 12. Write each radical as a twelfth root.
$$\sqrt[4]{3} = 3^{1/4} = 3^{3/12} = \sqrt[12]{3^3} = \sqrt[12]{27}$$
$$\sqrt[3]{4} = 4^{1/3} = 4^{4/12} = \sqrt[12]{4^4} = \sqrt[12]{256}$$
Therefore,
$$\sqrt[4]{3} \cdot \sqrt[3]{4} = \sqrt[12]{27} \cdot \sqrt[12]{256}$$
$$= \sqrt[12]{27 \cdot 256} = \sqrt[12]{6912}.$$

103. $\sqrt{x} = x^{1/2} = x^{3/6} = \sqrt[6]{x^3}$
$\sqrt[3]{x} = x^{1/3} = x^{2/6} = \sqrt[6]{x^2}$
So $\sqrt{x} \cdot \sqrt[3]{x} = \sqrt[6]{x^3} \cdot \sqrt[6]{x^2}$
$= \sqrt[6]{x^3 \cdot x^2} = \sqrt[6]{x^5}.$

105. Substitute 3 for a and 4 for b in the Pythagorean formula to find c.
$$c^2 = a^2 + b^2$$
$$c = \sqrt{a^2 + b^2} = \sqrt{3^2 + 4^2}$$
$$= \sqrt{9 + 16} = \sqrt{25} = 5$$
The length of the hypotenuse is 5.

107. Substitute 12 for c and 4 for a in the Pythagorean formula to find b.
$$a^2 + b^2 = c^2$$
$$b = \sqrt{c^2 - a^2} = \sqrt{12^2 - 4^2}$$
$$= \sqrt{144 - 16} = \sqrt{128}$$
$$= \sqrt{64}\sqrt{2} = 8\sqrt{2}$$
The length of the unknown leg is $8\sqrt{2}$.

In Exercises 109–120, use the distance formula
$$d = \sqrt{(x_2 - x_1)^2 + (y_2 - y_1)^2}.$$

109. $(6, 13)$ and $(1, 1)$
$$d = \sqrt{(6 - 1)^2 + (13 - 1)^2}$$
$$= \sqrt{5^2 + 12^2} = \sqrt{25 + 144}$$
$$= \sqrt{169} = 13$$

111. $(-6, 5)$ and $(3, -4)$
$$d = \sqrt{(-6 - 3)^2 + [5 - (-4)]^2}$$
$$= \sqrt{(-9)^2 + (9)^2} = \sqrt{81 + 81}$$
$$= \sqrt{162} = \sqrt{81} \cdot \sqrt{2} = 9\sqrt{2}$$

113. $(-8, 2)$ and $(-4, 1)$
$$d = \sqrt{[-8 - (-4)]^2 + (2 - 1)^2}$$
$$= \sqrt{(-4)^2 + 1^2}$$
$$= \sqrt{16 + 1} = \sqrt{17}$$

115. $(4.7, 2.3)$ and $(1.7, -1.7)$
$$d = \sqrt{(4.7 - 1.7)^2 + [2.3 - (-1.7)]^2}$$
$$= \sqrt{3^2 + 4^2} = \sqrt{9 + 16}$$
$$= \sqrt{25} = 5$$

117. $\left(\sqrt{2}, \sqrt{6}\right)$ and $\left(-2\sqrt{2}, 4\sqrt{6}\right)$
$$d = \sqrt{\left[\sqrt{2} - \left(-2\sqrt{2}\right)\right]^2 + \left(\sqrt{6} - 4\sqrt{6}\right)^2}$$
$$= \sqrt{\left(3\sqrt{2}\right)^2 + \left(-3\sqrt{6}\right)^2}$$
$$= \sqrt{9 \cdot 2 + 9 \cdot 6} = \sqrt{18 + 54}$$
$$= \sqrt{72} = \sqrt{36} \cdot \sqrt{2} = 6\sqrt{2}$$

119. $(x + y, y)$ and $(x - y, x)$

$$d = \sqrt{[(x + y) - (x - y)]^2 + (y - x)^2}$$
$$= \sqrt{(2y)^2 + (y - x)^2}$$
$$= \sqrt{4y^2 + y^2 - 2xy + x^2}$$
$$= \sqrt{5y^2 - 2xy + x^2}$$

121. Since $\sqrt{a} = a^{1/2}$, the distance formula

$$d = \sqrt{(x_2 - x_1)^2 + (y_2 - y_1)^2}$$

may be expressed as

$$d = \left[(x_2 - x_1)^2 + (y_2 - y_1)^2\right]^{1/2}.$$

123. To find the lengths of the three sides of the triangle, use the distance formula to find the distance between each pair of points. Then add the distances to find the perimeter.

$$P = \sqrt{(-3 - 2)^2 + (-3 - 6)^2}$$
$$+ \sqrt{(2 - 6)^2 + (6 - 2)^2}$$
$$+ \sqrt{[6 - (-3)]^2 + [2 - (-3)]^2}$$
$$= \sqrt{(-5)^2 + (-9)^2} + \sqrt{(-4)^2 + 4^2}$$
$$+ \sqrt{9^2 + 5^2}$$
$$= \sqrt{25 + 81} + \sqrt{16 + 16} + \sqrt{81 + 25}$$
$$= \sqrt{106} + \sqrt{32} + \sqrt{106}$$
$$= 2\sqrt{106} + \sqrt{16} \cdot \sqrt{2}$$
$$= 2\sqrt{106} + 4\sqrt{2}$$

125. $d = 1.224\sqrt{h}$
$$= 1.224\sqrt{156} \approx 15.3 \text{ miles}$$

127. Substitute 21.7 for a and 16 for b in the Pythagorean formula to find c.

$$c^2 = a^2 + b^2 = (21.7)^2 + 16^2$$
$$c = \sqrt{726.89} \approx 26.96$$

To the nearest tenth of an inch, the length of the diagonal of the screen is 27.0 inches.

129. Use $f_1 = f_2\sqrt{\dfrac{F_1}{F_2}}$ with $F_1 = 300$, $F_2 = 60$, and $f_2 = 260$ to find f_1.

$$f_1 = 260\sqrt{\tfrac{300}{60}} = 260\sqrt{5} \approx 581$$

131. $13x^4 - 12x^3 + 9x^4 + 2x^3$
$$= 13x^4 + 9x^4 - 12x^3 + 2x^3$$
$$= 22x^4 - 10x^3$$

133. $9q^2 + 2q - 5q - q^2$
$$= 9q^2 - q^2 + 2q - 5q$$
$$= 8q^2 - 3q$$

10.4 Adding and Subtracting Radical Expressions

1. Only choice **B** has like radical terms, so it can be simplified without first simplifying the individual radical expressions.

$$3\sqrt{6} + 9\sqrt{6} = 12\sqrt{6}$$

3. $\sqrt{64} + \sqrt[3]{125} + \sqrt[4]{16} = \sqrt{8^2} + \sqrt[3]{5^3} + \sqrt[4]{2^4}$
$$= 8 + 5 + 2 = 15$$

This sum can be found easily since each radicand has a whole number power corresponding to the index of the radical; that is, each radical expression simplifies to a whole number.

5. Simplify each radical and subtract.

$$\sqrt{36} - \sqrt{100} = 6 - 10 = -4$$

7. $-2\sqrt{48} + 3\sqrt{75}$
$$= -2\sqrt{16 \cdot 3} + 3\sqrt{25 \cdot 3}$$
$$= -2 \cdot 4\sqrt{3} + 3 \cdot 5\sqrt{3}$$
$$= -8\sqrt{3} + 15\sqrt{3} = 7\sqrt{3}$$

9. $\sqrt[3]{16} + 4\sqrt[3]{54}$
$$= \sqrt[3]{8 \cdot 2} + 4\sqrt[3]{27 \cdot 2}$$
$$= \sqrt[3]{8}\sqrt[3]{2} + 4\sqrt[3]{27}\sqrt[3]{2}$$
$$= 2\sqrt[3]{2} + 4 \cdot 3\sqrt[3]{2}$$
$$= 2\sqrt[3]{2} + 12\sqrt[3]{2} = 14\sqrt[3]{2}$$

11. $\sqrt[4]{32} + 3\sqrt[4]{2}$
$$= \sqrt[4]{16 \cdot 2} + 3\sqrt[4]{2}$$
$$= \sqrt[4]{16}\sqrt[4]{2} + 3\sqrt[4]{2}$$
$$= 2\sqrt[4]{2} + 3\sqrt[4]{2} = 5\sqrt[4]{2}$$

13. $6\sqrt{18} - \sqrt{32} + 2\sqrt{50}$
$$= 6\sqrt{9 \cdot 2} - \sqrt{16 \cdot 2} + 2\sqrt{25 \cdot 2}$$
$$= 6 \cdot 3\sqrt{2} - 4\sqrt{2} + 2 \cdot 5\sqrt{2}$$
$$= 18\sqrt{2} - 4\sqrt{2} + 10\sqrt{2}$$
$$= 24\sqrt{2}$$

15. $5\sqrt{6} + 2\sqrt{10}$

The radicals differ and are already simplified, so the expression cannot be simplified further.

17. $2\sqrt{5} + 3\sqrt{20} + 4\sqrt{45}$
$$= 2\sqrt{5} + 3\sqrt{4 \cdot 5} + 4\sqrt{9 \cdot 5}$$
$$= 2\sqrt{5} + 3 \cdot 2\sqrt{5} + 4 \cdot 3\sqrt{5}$$
$$= 2\sqrt{5} + 6\sqrt{5} + 12\sqrt{5}$$
$$= 20\sqrt{5}$$

19. $\sqrt{72x} - \sqrt{8x}$

$= \sqrt{36 \cdot 2x} - \sqrt{4 \cdot 2x}$

$= 6\sqrt{2x} - 2\sqrt{2x}$

$= 4\sqrt{2x}$

21. $3\sqrt{72m^2} - 5\sqrt{32m^2} - 3\sqrt{18m^2}$

$= 3\sqrt{36m^2 \cdot 2} - 5\sqrt{16m^2 \cdot 2} - 3\sqrt{9m^2 \cdot 2}$

$= 3 \cdot 6m\sqrt{2} - 5 \cdot 4m\sqrt{2} - 3 \cdot 3m\sqrt{2}$

$= 18m\sqrt{2} - 20m\sqrt{2} - 9m\sqrt{2}$

$= (18m - 20m - 9m)\sqrt{2} = -11m\sqrt{2}$

23. $2\sqrt[3]{16} - \sqrt[3]{54} = 2\sqrt[3]{8 \cdot 2} - \sqrt[3]{27 \cdot 2}$

$= 2 \cdot 2\sqrt[3]{2} - 3\sqrt[3]{2}$

$= 4\sqrt[3]{2} - 3\sqrt[3]{2}$

$= \sqrt[3]{2}$

25. $2\sqrt[3]{27x} - 2\sqrt[3]{8x}$

$= 2\sqrt[3]{27 \cdot x} - 2\sqrt[3]{8 \cdot x}$

$= 2 \cdot 3\sqrt[3]{x} - 2 \cdot 2\sqrt[3]{x}$

$= 6\sqrt[3]{x} - 4\sqrt[3]{x}$

$= 2\sqrt[3]{x}$

27. $\sqrt[3]{x^2y} - \sqrt[3]{8x^2y} = \sqrt[3]{x^2y} - \sqrt[3]{8}\sqrt[3]{x^2y}$

$= 1\sqrt[3]{x^2y} - 2\sqrt[3]{x^2y}$

$= (1 - 2)\sqrt[3]{x^2y}$

$= -\sqrt[3]{x^2y}$

29. $3x\sqrt[3]{xy^2} - 2\sqrt[3]{8x^4y^2}$

$= 3x\sqrt[3]{xy^2} - 2\sqrt[3]{8x^3} \cdot \sqrt[3]{xy^2}$

$= 3x\sqrt[3]{xy^2} - 2 \cdot 2x \cdot \sqrt[3]{xy^2}$

$= (3x - 4x)\sqrt[3]{xy^2} = -x\sqrt[3]{xy^2}$

31. $5\sqrt[4]{32} + 3\sqrt[4]{162} = 5\sqrt[4]{16 \cdot 2} + 3\sqrt[4]{81 \cdot 2}$

$= 5 \cdot 2\sqrt[4]{2} + 3 \cdot 3\sqrt[4]{2}$

$= 10\sqrt[4]{2} + 9\sqrt[4]{2}$

$= 19\sqrt[4]{2}$

33. $3\sqrt[4]{x^5y} - 2x\sqrt[4]{xy}$

$= 3\sqrt[4]{x^4 \cdot xy} - 2x\sqrt[4]{xy}$

$= 3x\sqrt[4]{xy} - 2x\sqrt[4]{xy}$

$= (3x - 2x)\sqrt[4]{xy} = x\sqrt[4]{xy}$

35. $2\sqrt[4]{32a^3} + 5\sqrt[4]{2a^3}$

$= 2\sqrt[4]{16} \cdot \sqrt[4]{2a^3} + 5\sqrt[4]{2a^3}$

$= 2 \cdot 2 \cdot \sqrt[4]{2a^3} + 5\sqrt[4]{2a^3}$

$= (4 + 5)\sqrt[4]{2a^3} = 9\sqrt[4]{2a^3}$

37. $\sqrt[3]{64xy^2} + \sqrt[3]{27x^4y^5}$

$= \sqrt[3]{64 \cdot xy^2} + \sqrt[3]{27x^3y^3 \cdot xy^2}$

$= 4\sqrt[3]{xy^2} + 3xy\sqrt[3]{xy^2}$

$= (4 + 3xy)\sqrt[3]{xy^2}$

39. $4\sqrt[3]{x} - 6\sqrt{x}$ cannot be simplified further.

41. $2\sqrt[3]{8x^4} + 3\sqrt[4]{16x^5}$

$= 2\sqrt[3]{8x^3 \cdot x} + 3\sqrt[4]{16x^4 \cdot x}$

$= 2 \cdot 2x\sqrt[3]{x} + 3 \cdot 2x\sqrt[4]{x}$

$= 4x\sqrt[3]{x} + 6x\sqrt[4]{x}$

43. $\sqrt{8} - \dfrac{\sqrt{64}}{\sqrt{16}} = \sqrt{4 \cdot 2} - \dfrac{8}{4}$

$= \sqrt{4} \cdot \sqrt{2} - 2$

$= 2\sqrt{2} - 2$

45. $\dfrac{2\sqrt{5}}{3} + \dfrac{\sqrt{5}}{6} = \dfrac{4\sqrt{5}}{6} + \dfrac{1\sqrt{5}}{6}$

$= \dfrac{4\sqrt{5} + 1\sqrt{5}}{6}$

$= \dfrac{5\sqrt{5}}{6}$

47. $\sqrt{\dfrac{8}{9}} + \sqrt{\dfrac{18}{36}} = \dfrac{\sqrt{8}}{\sqrt{9}} + \dfrac{\sqrt{18}}{\sqrt{36}}$

$= \dfrac{\sqrt{4}\sqrt{2}}{3} + \dfrac{\sqrt{9}\sqrt{2}}{6}$

$= \dfrac{2\sqrt{2}}{3} + \dfrac{3\sqrt{2}}{6}$

$= \dfrac{4\sqrt{2}}{6} + \dfrac{3\sqrt{2}}{6}$

$= \dfrac{4\sqrt{2} + 3\sqrt{2}}{6} = \dfrac{7\sqrt{2}}{6}$

49. $\dfrac{\sqrt{32}}{3} + \dfrac{2\sqrt{2}}{3} - \dfrac{\sqrt{2}}{\sqrt{9}}$

$= \dfrac{\sqrt{16}\sqrt{2}}{3} + \dfrac{2\sqrt{2}}{3} - \dfrac{\sqrt{2}}{3}$

$= \dfrac{4\sqrt{2} + 2\sqrt{2} - \sqrt{2}}{3} = \dfrac{5\sqrt{2}}{3}$

51. $3\sqrt{\dfrac{50}{9}} + 8\dfrac{\sqrt{2}}{\sqrt{8}} = 3\dfrac{\sqrt{50}}{\sqrt{9}} + 8\dfrac{\sqrt{2}}{2\sqrt{2}}$

$= 3 \cdot \dfrac{5\sqrt{2}}{3} + 8 \cdot \dfrac{1}{2}$

$= 5\sqrt{2} + 4$

53. $\sqrt{\dfrac{25}{x^8}} - \sqrt{\dfrac{9}{x^6}} = \dfrac{\sqrt{25}}{\sqrt{x^8}} - \dfrac{\sqrt{9}}{\sqrt{x^6}}$

$\qquad\qquad = \dfrac{5}{x^4} - \dfrac{3}{x^3}$

$\qquad\qquad = \dfrac{5}{x^4} - \dfrac{3 \cdot x}{x^3 \cdot x} \qquad LCD = x^4$

$\qquad\qquad = \dfrac{5 - 3x}{x^4}$

55. $3\sqrt[3]{\dfrac{m^5}{27}} - 2m\sqrt[3]{\dfrac{m^2}{64}}$

$\qquad = \dfrac{3\sqrt[3]{m^5}}{\sqrt[3]{27}} - \dfrac{2m\sqrt[3]{m^2}}{\sqrt[3]{64}}$

$\qquad = \dfrac{3\sqrt[3]{m^3}\sqrt[3]{m^2}}{3} - \dfrac{2m\sqrt[3]{m^2}}{4}$

$\qquad = \dfrac{m\sqrt[3]{m^2}}{1} - \dfrac{m\sqrt[3]{m^2}}{2}$

$\qquad = \dfrac{2m\sqrt[3]{m^2} - m\sqrt[3]{m^2}}{2} = \dfrac{m\sqrt[3]{m^2}}{2}$

57. $3\sqrt[3]{\dfrac{2}{x^6}} - 4\sqrt[3]{\dfrac{5}{x^9}} = 3\dfrac{\sqrt[3]{2}}{\sqrt[3]{x^6}} - 4\dfrac{\sqrt[3]{5}}{\sqrt[3]{x^9}}$

$\qquad\qquad = 3\dfrac{\sqrt[3]{2}}{x^2} - 4\dfrac{\sqrt[3]{5}}{x^3}$

$\qquad\qquad = \dfrac{3 \cdot x \cdot \sqrt[3]{2}}{x^2 \cdot x} - \dfrac{4\sqrt[3]{5}}{x^3}$

$\qquad\qquad\qquad\qquad LCD = x^3$

$\qquad\qquad = \dfrac{3x\sqrt[3]{2} - 4\sqrt[3]{5}}{x^3}$

59. $3\sqrt{32} - 2\sqrt{8} \approx 11.3137085$

$\qquad 8\sqrt{2} \approx 11.3137085$

Both calculator approximations are the same, supporting (but not proving) the truth of the statement.

61. Let $L = \sqrt{192} \approx \sqrt{196} = 14$ and $W = \sqrt{48} \approx \sqrt{49} = 7$. An estimate of the perimeter is $2L + 2W = 2(14) + 2(7) = 42$ meters. **(A)**

63. The perimeter, P, of a triangle is the sum of the measures of the sides.

$P = 3\sqrt{20} + 2\sqrt{45} + \sqrt{75}$

$\quad = 3\sqrt{4 \cdot 5} + 2\sqrt{9 \cdot 5} + \sqrt{25 \cdot 3}$

$\quad = 3 \cdot 2\sqrt{5} + 2 \cdot 3\sqrt{5} + 5\sqrt{3}$

$\quad = 6\sqrt{5} + 6\sqrt{5} + 5\sqrt{3}$

$\quad = 12\sqrt{5} + 5\sqrt{3}$

The perimeter is $\left(12\sqrt{5} + 5\sqrt{3}\right)$ inches.

65. To find the perimeter, add the lengths of the sides.

$4\sqrt{18} + \sqrt{108} + 2\sqrt{72} + 3\sqrt{12}$

$\quad = 4\sqrt{9}\sqrt{2} + \sqrt{36}\sqrt{3} + 2\sqrt{36}\sqrt{2} + 3\sqrt{4}\sqrt{3}$

$\quad = 4 \cdot 3\sqrt{2} + 6\sqrt{3} + 2 \cdot 6\sqrt{2} + 3 \cdot 2\sqrt{3}$

$\quad = 12\sqrt{2} + 6\sqrt{3} + 12\sqrt{2} + 6\sqrt{3}$

$\quad = 24\sqrt{2} + 12\sqrt{3}$

The perimeter is $\left(24\sqrt{2} + 12\sqrt{3}\right)$ inches.

67. $5xy\left(2x^2y^3 - 4x\right) = 5xy\left(2x^2y^3\right) - 5xy(4x)$

$\qquad\qquad\qquad\qquad = 10x^3y^4 - 20x^2y$

69. $\left(a^2 + b\right)\left(a^2 - b\right) = \left(a^2\right)^2 - b^2$

$\qquad\qquad\qquad\qquad = a^4 - b^2$

71. $\left(4x^3 + 3\right)^2 = \left(4x^3\right)^2 + 2 \cdot 4x^3 \cdot 3 + 3^2$

$\qquad\qquad\qquad = 16x^6 + 24x^3 + 9$

Now multiply by $4x^3 + 3$.

$$
\begin{array}{r}
16x^6 + 24x^3 + 9 \\
4x^3 + 3 \\
\hline
48x^6 + 72x^3 + 27 \\
64x^9 + 96x^6 + 36x^3 \\
\hline
64x^9 + 144x^6 + 108x^3 + 27
\end{array}
$$

Thus, $\left(4x^3 + 3\right)^3 = 64x^9 + 144x^6 + 108x^3 + 27$.

73. $\dfrac{8x^2 - 10x}{6x^2} = \dfrac{2x(4x - 5)}{2x \cdot 3x}$

$\qquad\qquad = \dfrac{4x - 5}{3x}$

10.5 Multiplying and Dividing Radical Expressions

1. $\left(x + \sqrt{y}\right)\left(x - \sqrt{y}\right)$

$\quad = x^2 - x\sqrt{y} + x\sqrt{y} - \left(\sqrt{y}\right)^2$

$\quad = x^2 - y$ **(E)**

3. $\left(\sqrt{x} + \sqrt{y}\right)\left(\sqrt{x} - \sqrt{y}\right)$

$\quad = \left(\sqrt{x}\right)^2 - \left(\sqrt{y}\right)^2$

$\quad = x - y$ **(A)**

5. $\left(\sqrt{x} - \sqrt{y}\right)^2$

$\quad = \left(\sqrt{x}\right)^2 - 2\sqrt{x}\sqrt{y} + \left(\sqrt{y}\right)^2$

$\quad = x - 2\sqrt{xy} + y$ **(D)**

7. $\sqrt{6}\left(3 + \sqrt{2}\right) = 3\sqrt{6} + \sqrt{6} \cdot \sqrt{2}$

$\qquad\qquad\quad = 3\sqrt{6} + \sqrt{12}$

$\qquad\qquad\quad = 3\sqrt{6} + \sqrt{4} \cdot \sqrt{3}$

$\qquad\qquad\quad = 3\sqrt{6} + 2\sqrt{3}$

9. $5\left(\sqrt{72} - \sqrt{8}\right) = 5\sqrt{72} - 5\sqrt{8}$
$$= 5\sqrt{36}\cdot\sqrt{2} - 5\sqrt{4}\cdot\sqrt{2}$$
$$= 5\cdot6\cdot\sqrt{2} - 5\cdot2\cdot\sqrt{2}$$
$$= 30\sqrt{2} - 10\sqrt{2}$$
$$= 20\sqrt{2}$$

11. $\left(\sqrt{7}+3\right)\left(\sqrt{7}-3\right) = \left(\sqrt{7}\right)^2 - 3^2$
$$= 7 - 9 = -2$$

13. $\left(\sqrt{2}-\sqrt{3}\right)\left(\sqrt{2}+\sqrt{3}\right) = \left(\sqrt{2}\right)^2 - \left(\sqrt{3}\right)^2$
$$= 2 - 3 = -1$$

15. $\left(\sqrt{8}-\sqrt{2}\right)\left(\sqrt{8}+\sqrt{2}\right) = \left(\sqrt{8}\right)^2 - \left(\sqrt{2}\right)^2$
$$= 8 - 2 = 6$$

17. $\left(\sqrt{2}+1\right)\left(\sqrt{3}-1\right)$
$$\quad\mathbf{F}\qquad\mathbf{O}\qquad\mathbf{I}\qquad\mathbf{L}$$
$$= \sqrt{2}\cdot\sqrt{3} - 1\sqrt{2} + 1\sqrt{3} - 1\cdot1$$
$$= \sqrt{6} - \sqrt{2} + \sqrt{3} - 1$$

19. $\left(\sqrt{11}-\sqrt{7}\right)\left(\sqrt{2}+\sqrt{5}\right)$
$$\quad\mathbf{F}\qquad\quad\mathbf{O}\qquad\quad\mathbf{I}\qquad\quad\mathbf{L}$$
$$= \sqrt{11}\cdot\sqrt{2} + \sqrt{11}\cdot\sqrt{5} - \sqrt{7}\cdot\sqrt{2} - \sqrt{7}\cdot\sqrt{5}$$
$$= \sqrt{22} + \sqrt{55} - \sqrt{14} - \sqrt{35}$$

21. $\left(2\sqrt{3}+\sqrt{5}\right)\left(3\sqrt{3}-2\sqrt{5}\right)$
$$= \left(2\sqrt{3}\right)\left(3\sqrt{3}\right) + \left(2\sqrt{3}\right)\left(-2\sqrt{5}\right)$$
$$\; + \left(\sqrt{5}\right)\left(3\sqrt{3}\right) + \left(\sqrt{5}\right)\left(-2\sqrt{5}\right)$$
$$= 2\cdot3\cdot3 - 2\cdot2\sqrt{3\cdot5} + 3\sqrt{5\cdot3} - 2\cdot5$$
$$= 18 - 4\sqrt{15} + 3\sqrt{15} - 10$$
$$= 8 - \sqrt{15}$$

23. $\left(\sqrt{5}+2\right)^2 = \left(\sqrt{5}\right)^2 + 2\cdot\sqrt{5}\cdot2 + 2^2$
$$= 5 + 4\sqrt{5} + 4$$
$$= 9 + 4\sqrt{5}$$

25. $\left(\sqrt{21}-\sqrt{5}\right)^2$
$$= \left(\sqrt{21}\right)^2 - 2\cdot\sqrt{21}\cdot\sqrt{5} + \left(\sqrt{5}\right)^2$$
$$= 21 - 2\sqrt{105} + 5$$
$$= 26 - 2\sqrt{105}$$

27. $\left(2+\sqrt[3]{6}\right)\left(2-\sqrt[3]{6}\right) = 2^2 - \left(\sqrt[3]{6}\right)^2$
$$= 4 - \sqrt[3]{6}\cdot\sqrt[3]{6}$$
$$= 4 - \sqrt[3]{36}$$

29. $\left(2+\sqrt[3]{2}\right)\left(4-2\sqrt[3]{2}+\sqrt[3]{4}\right)$
$$= 2\cdot4 - 2\cdot2\sqrt[3]{2} + 2\sqrt[3]{4}$$
$$\; + 4\sqrt[3]{2} - 2\sqrt[3]{2}\cdot\sqrt[3]{2} + \sqrt[3]{2}\cdot\sqrt[3]{4}$$
$$= 8 - 4\sqrt[3]{2} + 2\sqrt[3]{4} + 4\sqrt[3]{2} - 2\sqrt[3]{4} + \sqrt[3]{8}$$
$$= 8 + 2$$
$$= 10$$

31. $\left(3\sqrt{x}-\sqrt{5}\right)\left(2\sqrt{x}+1\right)$
$$= \left(3\sqrt{x}\right)\left(2\sqrt{x}\right) + 3\sqrt{x} - 2\sqrt{5}\sqrt{x} - \sqrt{5}$$
$$= 6x + 3\sqrt{x} - 2\sqrt{5x} - \sqrt{5}$$

33. $\left(3\sqrt{r}-\sqrt{s}\right)\left(3\sqrt{r}+\sqrt{s}\right)$
$$= 3\sqrt{r}\cdot3\sqrt{r} + 3\sqrt{r}\cdot\sqrt{s}$$
$$\; - 3\sqrt{r}\cdot\sqrt{s} - \sqrt{s}\cdot\sqrt{s}$$
$$= 9r - s$$

35. $\left(\sqrt[3]{2y}-5\right)\left(4\sqrt[3]{2y}+1\right)$
$$= \sqrt[3]{2y}\cdot4\sqrt[3]{2y} + \sqrt[3]{2y} - 5\cdot4\sqrt[3]{2y} - 5$$
$$= 4\sqrt[3]{4y^2} + \sqrt[3]{2y} - 20\sqrt[3]{2y} - 5$$
$$= 4\sqrt[3]{4y^2} - 19\sqrt[3]{2y} - 5$$

37. $\left(\sqrt{3x}+2\right)\left(\sqrt{3x}-2\right) = \left(\sqrt{3x}\right)^2 - 2^2$
$$= 3x - 4$$

39. $\left(2\sqrt{x}+\sqrt{y}\right)\left(2\sqrt{x}-\sqrt{y}\right)$
$$= \left(2\sqrt{x}\right)^2 - \left(\sqrt{y}\right)^2$$
$$= 2^2\left(\sqrt{x}\right)^2 - y$$
$$= 4x - y$$

41. $\left[\left(\sqrt{2}+\sqrt{3}\right) - \sqrt{6}\right]\left[\left(\sqrt{2}+\sqrt{3}\right) + \sqrt{6}\right]$
$$= \left(\sqrt{2}+\sqrt{3}\right)^2 - \left(\sqrt{6}\right)^2$$
$$= \left[\left(\sqrt{2}\right)^2 + 2\sqrt{2}\sqrt{3} + \left(\sqrt{3}\right)^2\right] - 6$$
$$= \left(2 + 2\sqrt{6} + 3\right) - 6$$
$$= 2\sqrt{6} - 1$$

43. $\dfrac{7}{\sqrt{7}} = \dfrac{7\cdot\sqrt{7}}{\sqrt{7}\cdot\sqrt{7}} = \dfrac{7\sqrt{7}}{7} = \sqrt{7}$

45. $\dfrac{15}{\sqrt{3}} = \dfrac{15\cdot\sqrt{3}}{\sqrt{3}\cdot\sqrt{3}} = \dfrac{15\sqrt{3}}{3} = 5\sqrt{3}$

47. $\dfrac{\sqrt{3}}{\sqrt{2}} = \dfrac{\sqrt{3}\cdot\sqrt{2}}{\sqrt{2}\cdot\sqrt{2}} = \dfrac{\sqrt{6}}{2}$

49. $\dfrac{9\sqrt{3}}{\sqrt{5}} = \dfrac{9\sqrt{3}\cdot\sqrt{5}}{\sqrt{5}\cdot\sqrt{5}} = \dfrac{9\sqrt{15}}{5}$

51. $\dfrac{-6}{\sqrt{18}} = \dfrac{-6}{\sqrt{9\cdot2}} = \dfrac{-6}{3\sqrt{2}} = \dfrac{-2}{\sqrt{2}} = \dfrac{-2\cdot\sqrt{2}}{\sqrt{2}\cdot\sqrt{2}}$
$$= \dfrac{-2\sqrt{2}}{2} = -\sqrt{2}$$

53. $\sqrt{\dfrac{7}{2}} = \dfrac{\sqrt{7}}{\sqrt{2}} = \dfrac{\sqrt{7}\cdot\sqrt{2}}{\sqrt{2}\cdot\sqrt{2}} = \dfrac{\sqrt{14}}{2}$

55. $-\sqrt{\dfrac{7}{50}} = -\dfrac{\sqrt{7}}{\sqrt{25\cdot2}} = -\dfrac{\sqrt{7}}{5\sqrt{2}}$
$$= -\dfrac{\sqrt{7}\cdot\sqrt{2}}{5\sqrt{2}\cdot\sqrt{2}} = -\dfrac{\sqrt{14}}{5\cdot2} = -\dfrac{\sqrt{14}}{10}$$

57. $\sqrt{\dfrac{24}{x}} = \dfrac{\sqrt{24}}{\sqrt{x}} = \dfrac{\sqrt{4 \cdot 6}}{\sqrt{x}} = \dfrac{2\sqrt{6}}{\sqrt{x}}$

$\qquad = \dfrac{2\sqrt{6} \cdot \sqrt{x}}{\sqrt{x} \cdot \sqrt{x}} = \dfrac{2\sqrt{6x}}{x}$

59. $\dfrac{-8\sqrt{3}}{\sqrt{k}} = \dfrac{-8\sqrt{3} \cdot \sqrt{k}}{\sqrt{k} \cdot \sqrt{k}} = \dfrac{-8\sqrt{3k}}{k}$

61. $-\sqrt{\dfrac{150m^5}{n^3}} = \dfrac{-\sqrt{150m^5}}{\sqrt{n^3}}$

$\qquad = \dfrac{-\sqrt{25m^4 \cdot 6m}}{\sqrt{n^2 \cdot n}} = \dfrac{-5m^2\sqrt{6m}}{n\sqrt{n}}$

$\qquad = \dfrac{-5m^2\sqrt{6m} \cdot \sqrt{n}}{n\sqrt{n} \cdot \sqrt{n}}$

$\qquad = \dfrac{-5m^2\sqrt{6mn}}{n \cdot n}$

$\qquad = \dfrac{-5m^2\sqrt{6mn}}{n^2}$

63. $\sqrt{\dfrac{288x^7}{y^9}} = \dfrac{\sqrt{288x^7}}{\sqrt{y^9}} = \dfrac{\sqrt{144x^6 \cdot 2x}}{\sqrt{y^8 \cdot y}}$

$\qquad = \dfrac{12x^3\sqrt{2x}}{y^4\sqrt{y}} = \dfrac{12x^3\sqrt{2x} \cdot \sqrt{y}}{y^4\sqrt{y} \cdot \sqrt{y}}$

$\qquad = \dfrac{12x^3\sqrt{2xy}}{y^4 \cdot y} = \dfrac{12x^3\sqrt{2xy}}{y^5}$

65. $\dfrac{5\sqrt{2m}}{\sqrt{y^3}} = \dfrac{5\sqrt{2m}}{\sqrt{y^3}} \cdot \dfrac{\sqrt{y}}{\sqrt{y}}$

$\qquad = \dfrac{5\sqrt{2my}}{\sqrt{y^4}}$

$\qquad = \dfrac{5\sqrt{2my}}{y^2}$

67. $-\sqrt{\dfrac{48k^2}{z}} = -\dfrac{\sqrt{16k^2 \cdot 3}}{\sqrt{z}} \cdot \dfrac{\sqrt{z}}{\sqrt{z}}$

$\qquad = -\dfrac{4k\sqrt{3z}}{z}$

69. $\sqrt[3]{\dfrac{2}{3}} = \dfrac{\sqrt[3]{2} \cdot \sqrt[3]{9}}{\sqrt[3]{3} \cdot \sqrt[3]{9}} = \dfrac{\sqrt[3]{18}}{\sqrt[3]{27}} = \dfrac{\sqrt[3]{18}}{3}$

71. $\sqrt[3]{\dfrac{4}{9}} = \dfrac{\sqrt[3]{4}}{\sqrt[3]{9}} = \dfrac{\sqrt[3]{4}}{\sqrt[3]{3^2}} = \dfrac{\sqrt[3]{4} \cdot \sqrt[3]{3}}{\sqrt[3]{3^2} \cdot \sqrt[3]{3}}$

$\qquad = \dfrac{\sqrt[3]{12}}{\sqrt[3]{3^3}} = \dfrac{\sqrt[3]{12}}{3}$

73. $\sqrt[3]{\dfrac{9}{32}} = \dfrac{\sqrt[3]{9}}{\sqrt[3]{32}} = \dfrac{\sqrt[3]{9}}{\sqrt[3]{8} \cdot \sqrt[3]{4}} \cdot \dfrac{\sqrt[3]{2}}{\sqrt[3]{2}}$

$\qquad = \dfrac{\sqrt[3]{9 \cdot 2}}{2 \cdot \sqrt[3]{8}}$

$\qquad = \dfrac{\sqrt[3]{18}}{4}$

75. $-\sqrt[3]{\dfrac{2p}{r^2}} = -\dfrac{\sqrt[3]{2p}}{\sqrt[3]{r^2}} = -\dfrac{\sqrt[3]{2p} \cdot \sqrt[3]{r}}{\sqrt[3]{r^2} \cdot \sqrt[3]{r}}$

$\qquad = -\dfrac{\sqrt[3]{2pr}}{\sqrt[3]{r^3}} = -\dfrac{\sqrt[3]{2pr}}{r}$

77. $\sqrt[3]{\dfrac{x^6}{y}} = \dfrac{\sqrt[3]{x^6}}{\sqrt[3]{y}} = \dfrac{x^2}{\sqrt[3]{y}} \cdot \dfrac{\sqrt[3]{y^2}}{\sqrt[3]{y^2}}$

$\qquad = \dfrac{x^2\sqrt[3]{y^2}}{\sqrt[3]{y^3}}$

$\qquad = \dfrac{x^2\sqrt[3]{y^2}}{y}$

79. $\sqrt[4]{\dfrac{16}{x}} = \dfrac{\sqrt[4]{16}}{\sqrt[4]{x}} = \dfrac{2}{\sqrt[4]{x}} = \dfrac{2 \cdot \sqrt[4]{x^3}}{\sqrt[4]{x} \cdot \sqrt[4]{x^3}}$

$\qquad = \dfrac{2\sqrt[4]{x^3}}{\sqrt[4]{x^4}} = \dfrac{2\sqrt[4]{x^3}}{x}$

81. $\sqrt[4]{\dfrac{2y}{z}} = \dfrac{\sqrt[4]{2y}}{\sqrt[4]{z}} \cdot \dfrac{\sqrt[4]{z^3}}{\sqrt[4]{z^3}}$

$\qquad = \dfrac{\sqrt[4]{2yz^3}}{z}$

83. $\dfrac{3}{4 + \sqrt{5}}$

Multiply both the numerator and denominator by the conjugate of the denominator, $4 - \sqrt{5}$.

$\qquad = \dfrac{3(4 - \sqrt{5})}{(4 + \sqrt{5})(4 - \sqrt{5})}$

$\qquad = \dfrac{3(4 - \sqrt{5})}{16 - 5} = \dfrac{3(4 - \sqrt{5})}{11}$

85. $\dfrac{\sqrt{8}}{3 - \sqrt{2}}$

Multiply both the numerator and denominator by the conjugate of the denominator, $3 + \sqrt{2}$.

$\qquad = \dfrac{\sqrt{4}\sqrt{2}(3 + \sqrt{2})}{(3 - \sqrt{2})(3 + \sqrt{2})}$

$\qquad = \dfrac{2\sqrt{2}(3 + \sqrt{2})}{3^2 - 2}$

$\qquad = \dfrac{2 \cdot 3\sqrt{2} + 2 \cdot 2}{7}$

$\qquad = \dfrac{6\sqrt{2} + 4}{7}$

87. $\dfrac{2}{3\sqrt{5}+2\sqrt{3}}$

Multiply both the numerator and denominator by the conjugate of the denominator, $3\sqrt{5}-2\sqrt{3}$.

$$= \frac{2\left(3\sqrt{5}-2\sqrt{3}\right)}{\left(3\sqrt{5}+2\sqrt{3}\right)\left(3\sqrt{5}-2\sqrt{3}\right)}$$

$$= \frac{2\left(3\sqrt{5}-2\sqrt{3}\right)}{3^2\cdot 5 - 2^2\cdot 3}$$

$$= \frac{2\left(3\sqrt{5}-2\sqrt{3}\right)}{45-12} = \frac{2\left(3\sqrt{5}-2\sqrt{3}\right)}{33}$$

89. $\dfrac{\sqrt{2}-\sqrt{3}}{\sqrt{6}-\sqrt{5}}$

Multiply both the numerator and denominator by the conjugate of the denominator, $\sqrt{6}+\sqrt{5}$.

$$= \frac{\left(\sqrt{2}-\sqrt{3}\right)\left(\sqrt{6}+\sqrt{5}\right)}{\left(\sqrt{6}-\sqrt{5}\right)\left(\sqrt{6}+\sqrt{5}\right)}$$

$$= \frac{\sqrt{12}+\sqrt{10}-\sqrt{18}-\sqrt{15}}{\left(\sqrt{6}\right)^2-\left(\sqrt{5}\right)^2}$$

$$= \frac{\sqrt{4}\cdot\sqrt{3}+\sqrt{10}-\sqrt{9}\cdot\sqrt{2}-\sqrt{15}}{6-5}$$

$$= 2\sqrt{3}+\sqrt{10}-3\sqrt{2}-\sqrt{15}$$

91. $\dfrac{m-4}{\sqrt{m}+2}$

Multiply both the numerator and denominator by the conjugate of the denominator, $\sqrt{m}-2$.

$$= \frac{(m-4)\left(\sqrt{m}-2\right)}{\left(\sqrt{m}+2\right)\left(\sqrt{m}-2\right)}$$

$$= \frac{(m-4)\left(\sqrt{m}-2\right)}{m-4}$$

$$= \sqrt{m}-2$$

93. $\dfrac{4}{\sqrt{x}-2\sqrt{y}}$

Multiply both the numerator and denominator by the conjugate of the denominator, $\sqrt{x}+2\sqrt{y}$.

$$= \frac{4\left(\sqrt{x}+2\sqrt{y}\right)}{\left(\sqrt{x}-2\sqrt{y}\right)\left(\sqrt{x}+2\sqrt{y}\right)}$$

$$= \frac{4\left(\sqrt{x}+2\sqrt{y}\right)}{x-4y}$$

95. $\dfrac{\sqrt{x}-\sqrt{y}}{\sqrt{x}+\sqrt{y}}$

Multiply both the numerator and denominator by the conjugate of the denominator, $\sqrt{x}-\sqrt{y}$.

$$= \frac{\left(\sqrt{x}-\sqrt{y}\right)\left(\sqrt{x}-\sqrt{y}\right)}{\left(\sqrt{x}+\sqrt{y}\right)\left(\sqrt{x}-\sqrt{y}\right)}$$

$$= \frac{\left(\sqrt{x}\right)^2 - 2\sqrt{x}\sqrt{y}+\left(\sqrt{y}\right)^2}{\left(\sqrt{x}\right)^2-\left(\sqrt{y}\right)^2}$$

$$= \frac{x-2\sqrt{xy}+y}{x-y}$$

97. $\dfrac{5\sqrt{k}}{2\sqrt{k}+\sqrt{q}}$

Multiply both the numerator and denominator by the conjugate of the denominator, $2\sqrt{k}-\sqrt{q}$.

$$= \frac{5\sqrt{k}\left(2\sqrt{k}-\sqrt{q}\right)}{\left(2\sqrt{k}+\sqrt{q}\right)\left(2\sqrt{k}-\sqrt{q}\right)}$$

$$= \frac{5\sqrt{k}\left(2\sqrt{k}-\sqrt{q}\right)}{4k-q}$$

99. $\dfrac{30-20\sqrt{6}}{10} = \dfrac{10\left(3-2\sqrt{6}\right)}{10} = 3-2\sqrt{6}$

101. $\dfrac{3-3\sqrt{5}}{3} = \dfrac{3\left(1-\sqrt{5}\right)}{3} = 1-\sqrt{5}$

103. $\dfrac{16-4\sqrt{8}}{12} = \dfrac{16-4\left(2\sqrt{2}\right)}{12} = \dfrac{16-8\sqrt{2}}{12}$

$$= \frac{4\left(4-2\sqrt{2}\right)}{4\cdot 3} = \frac{4-2\sqrt{2}}{3}$$

105. $\dfrac{6p+\sqrt{24p^3}}{3p}$

$$= \frac{6p+\sqrt{4p^2\cdot 6p}}{3p} = \frac{6p+2p\sqrt{6p}}{3p}$$

$$= \frac{p\left(6+2\sqrt{6p}\right)}{3p} = \frac{6+2\sqrt{6p}}{3}$$

107. $\dfrac{1}{\sqrt{x+y}} = \dfrac{1}{\sqrt{x+y}}\cdot\dfrac{\sqrt{x+y}}{\sqrt{x+y}}$

$$= \frac{1\cdot\sqrt{x+y}}{\left(\sqrt{x+y}\right)^2}$$

$$= \frac{\sqrt{x+y}}{x+y}$$

109. $\dfrac{p}{\sqrt{p+2}} = \dfrac{p}{\sqrt{p+2}}\cdot\dfrac{\sqrt{p+2}}{\sqrt{p+2}}$

$$= \frac{p\sqrt{p+2}}{p+2}$$

111. $\dfrac{1}{\sqrt{2}} \cdot \dfrac{\sqrt{3}}{2} - \dfrac{1}{\sqrt{2}} \cdot \dfrac{1}{2} = \dfrac{\sqrt{3}}{2\sqrt{2}} - \dfrac{1}{2\sqrt{2}}$

$\qquad = \dfrac{\sqrt{3}-1}{2\sqrt{2}} = \dfrac{(\sqrt{3}-1)\sqrt{2}}{(2\sqrt{2})\sqrt{2}}$

$\qquad = \dfrac{\sqrt{6}-\sqrt{2}}{2\cdot 2} = \dfrac{\sqrt{6}-\sqrt{2}}{4}$

Using a calculator,

$\dfrac{1}{\sqrt{2}} \cdot \dfrac{\sqrt{3}}{2} - \dfrac{1}{\sqrt{2}} \cdot \dfrac{1}{2} \approx 0.2588190451$ and

$\dfrac{\sqrt{6}-\sqrt{2}}{4} \approx 0.2588190451.$

113. $\dfrac{6-\sqrt{2}}{4} = \dfrac{6-\sqrt{2}}{4} \cdot \dfrac{6+\sqrt{2}}{6+\sqrt{2}}$

$\qquad = \dfrac{6^2 - (\sqrt{2})^2}{4(6+\sqrt{2})}$

$\qquad = \dfrac{36-2}{4(6+\sqrt{2})}$

$\qquad = \dfrac{34}{4(6+\sqrt{2})}$

$\qquad = \dfrac{17\cdot 2}{2\cdot 2(6+\sqrt{2})}$

$\qquad = \dfrac{17}{2(6+\sqrt{2})}$

115. $\dfrac{3\sqrt{a}+\sqrt{b}}{b} = \dfrac{3\sqrt{a}+\sqrt{b}}{b} \cdot \dfrac{3\sqrt{a}-\sqrt{b}}{3\sqrt{a}-\sqrt{b}}$

$\qquad = \dfrac{(3\sqrt{a})^2 - (\sqrt{b})^2}{b(3\sqrt{a}-\sqrt{b})}$

$\qquad = \dfrac{9a-b}{b(3\sqrt{a}-\sqrt{b})}$

117. $-8t + 7 = 4$

$\qquad -8t = -3$

$\qquad t = \dfrac{3}{8}$

The solution set is $\left\{\dfrac{3}{8}\right\}$.

119. $\qquad 6x^2 - 7x = 3$

$\qquad 6x^2 - 7x - 3 = 0$

$\qquad (3x+1)(2x-3) = 0$

$3x + 1 = 0 \qquad$ or $\qquad 2x - 3 = 0$

$\qquad 3x = -1 \qquad$ or $\qquad 2x = 3$

$\qquad x = -\dfrac{1}{3} \qquad$ or $\qquad x = \dfrac{3}{2}$

The solution set is $\left\{-\dfrac{1}{3}, \dfrac{3}{2}\right\}$.

121. $(2x+5)^2 = (2x)^2 + 2\cdot 2x\cdot 5 + 5^2$

$\qquad = 4x^2 + 20x + 25$

123. $\left(\sqrt{x^4 + 2x^2 + 5}\right)^2 = x^4 + 2x^2 + 5$

125. $\sqrt{4-x} = x+2$

$\quad \sqrt{4-0} = 0+2 \quad$? $\quad$ *Let x = 0.*

$\qquad \sqrt{4} = 2 \qquad$?

$\qquad 2 = 2 \qquad\qquad$ *True*

127. $\sqrt{x^2 - 4x + 9} = x - 1$

$\sqrt{4^2 - 4(4) + 9} = 4 - 1 \quad$? $\quad$ *Let x = 4.*

$\qquad \sqrt{16 - 16 + 9} = 3 \qquad$?

$\qquad\qquad \sqrt{9} = 3 \qquad$?

$\qquad\qquad 3 = 3 \qquad\qquad$ *True*

Summary Exercises on Operations with Radicals and Rational Exponents

1. $6\sqrt{10} - 12\sqrt{10} = (6-12)\sqrt{10}$

$\qquad\qquad = -6\sqrt{10}$

3. $(1 - \sqrt{3})(2 + \sqrt{6})$

$\qquad$ **F** $\quad$ **O** $\quad$ **I** $\quad$ **L**

$\qquad = 2 + \sqrt{6} - 2\sqrt{3} - \sqrt{18}$

$\qquad = 2 + \sqrt{6} - 2\sqrt{3} - \sqrt{9}\cdot\sqrt{2}$

$\qquad = 2 + \sqrt{6} - 2\sqrt{3} - 3\sqrt{2}$

5. $(3\sqrt{5} + 2\sqrt{7})^2$

$\qquad = (3\sqrt{5})^2 + 2(3\sqrt{5})(2\sqrt{7}) + (2\sqrt{7})^2$

$\qquad = 9\cdot 5 + 12\sqrt{35} + 4\cdot 7$

$\qquad = 45 + 12\sqrt{35} + 28$

$\qquad = 73 + 12\sqrt{35}$

7. $\dfrac{8}{\sqrt{7}+\sqrt{5}} = \dfrac{8}{\sqrt{7}+\sqrt{5}} \cdot \dfrac{\sqrt{7}-\sqrt{5}}{\sqrt{7}-\sqrt{5}}$

$\qquad = \dfrac{8(\sqrt{7}-\sqrt{5})}{7-5}$

$\qquad = \dfrac{8(\sqrt{7}-\sqrt{5})}{2}$

$\qquad = 4(\sqrt{7}-\sqrt{5})$

9. $(\sqrt{5}+7)(\sqrt{5}-7) = (\sqrt{5})^2 - 7^2$

$\qquad\qquad = 5 - 49$

$\qquad\qquad = -44$

11. $\sqrt[3]{8a^3b^5c^9} = \sqrt[3]{8a^3b^3c^9} \cdot \sqrt[3]{b^2}$

$\qquad = 2abc^3\sqrt[3]{b^2}$

13. $\dfrac{3}{\sqrt{5}+2} = \dfrac{3}{\sqrt{5}+2} \cdot \dfrac{\sqrt{5}-2}{\sqrt{5}-2}$

$\qquad = \dfrac{3(\sqrt{5}-2)}{5-4}$

$\qquad = 3(\sqrt{5}-2)$

15. $\dfrac{16\sqrt{3}}{5\sqrt{12}} = \dfrac{16\sqrt{3}}{5\cdot\sqrt{4}\cdot\sqrt{3}}$

$\qquad = \dfrac{16}{5\cdot 2}$

$\qquad = \dfrac{8}{5}$

17. $\dfrac{-10}{\sqrt[3]{10}} = \dfrac{-10}{\sqrt[3]{10}}\cdot\dfrac{\sqrt[3]{100}}{\sqrt[3]{100}}$

$\qquad = \dfrac{-10\sqrt[3]{100}}{\sqrt[3]{1000}}$

$\qquad = \dfrac{-10\sqrt[3]{100}}{10} = -\sqrt[3]{100}$

19. $\sqrt{12x} - \sqrt{75x} = \sqrt{4}\cdot\sqrt{3x} - \sqrt{25}\cdot\sqrt{3x}$

$\qquad = 2\sqrt{3x} - 5\sqrt{3x}$

$\qquad = -3\sqrt{3x}$

21. $\sqrt[3]{\dfrac{13}{81}} = \dfrac{\sqrt[3]{13}}{\sqrt[3]{81}}$

$\qquad = \dfrac{\sqrt[3]{13}}{\sqrt[3]{27}\cdot\sqrt[3]{3}}\cdot\dfrac{\sqrt[3]{9}}{\sqrt[3]{9}}$

$\qquad = \dfrac{\sqrt[3]{13\cdot 9}}{3\cdot\sqrt[3]{27}}$

$\qquad = \dfrac{\sqrt[3]{117}}{3\cdot 3} = \dfrac{\sqrt[3]{117}}{9}$

23. $\dfrac{6}{\sqrt[4]{3}} = \dfrac{6}{\sqrt[4]{3}}\cdot\dfrac{\sqrt[4]{3^3}}{\sqrt[4]{3^3}}$

$\qquad = \dfrac{6\sqrt[4]{27}}{3} = 2\sqrt[4]{27}$

25. $\sqrt[3]{\dfrac{x^2 y}{x^{-3}y^4}} = \sqrt[3]{x^{2-(-3)}y^{1-4}}$

$\qquad = \sqrt[3]{x^5 y^{-3}}$

$\qquad = \dfrac{\sqrt[3]{x^5}}{\sqrt[3]{y^3}}$

$\qquad = \dfrac{\sqrt[3]{x^3\cdot x^2}}{y} = \dfrac{x\sqrt[3]{x^2}}{y}$

27. $\dfrac{x^{-2/3}y^{4/5}}{x^{-5/3}y^{-2/5}} = x^{-2/3-(-5/3)}y^{4/5-(-2/5)}$

$\qquad = x^{3/3}y^{6/5}$

$\qquad = xy^{6/5}$

29. $(125x^3)^{-2/3} = \dfrac{1}{(125x^3)^{2/3}}$

$\qquad = \dfrac{1}{\left(\sqrt[3]{125x^3}\right)^2}$

$\qquad = \dfrac{1}{(5x)^2}$

$\qquad = \dfrac{1}{25x^2}$

31. $\sqrt[3]{16x^2} - \sqrt[3]{54x^2} + \sqrt[3]{128x^2}$

$\quad = \sqrt[3]{8}\cdot\sqrt[3]{2x^2} - \sqrt[3]{27}\cdot\sqrt[3]{2x^2} + \sqrt[3]{64}\cdot\sqrt[3]{2x^2}$

$\quad = 2\sqrt[3]{2x^2} - 3\sqrt[3]{2x^2} + 4\sqrt[3]{2x^2}$

$\quad = (2-3+4)\sqrt[3]{2x^2}$

$\quad = 3\sqrt[3]{2x^2}$

33. $\left(\sqrt{74}-\sqrt{73}\right)\left(\sqrt{74}+\sqrt{73}\right) = 74 - 73 = 1$

35. $\left(3x^{-2/3}y^{1/2}\right)\left(-2x^{5/8}y^{-1/3}\right)$

$\quad = 3(-2)x^{-2/3+5/8}y^{1/2+(-1/3)}$

$\quad = -6x^{-16/24+15/24}y^{3/6-2/6}$

$\quad = -6x^{-1/24}y^{1/6}$

$\quad = -\dfrac{6y^{1/6}}{x^{1/24}}$

37. **(a)** $\sqrt{64} = 8$

     (b) $x^2 = 64$

$\qquad x = -\sqrt{64} \quad$ or $\quad x = \sqrt{64}$

$\qquad x = -8 \quad\;\;$ or $\quad x = 8$

     The solution set is $\{-8, 8\}$.

39. **(a)** $x^2 = 16$

$\qquad x = -\sqrt{16} \quad$ or $\quad x = \sqrt{16}$

$\qquad x = -4 \quad\;\;$ or $\quad x = 4$

     The solution set is $\{-4, 4\}$.

     (b) $-\sqrt{16} = -\left(\sqrt{16}\right) = -4$

41. **(a)** Since $\left(\frac{9}{11}\right)^2 = \frac{81}{121}$, $-\sqrt{\frac{81}{121}} = -\frac{9}{11}$.

     (b) $x^2 = \frac{81}{121}$

$\qquad x = -\sqrt{\frac{81}{121}} \quad$ or $\quad x = \sqrt{\frac{81}{121}}$

$\qquad x = -\frac{9}{11} \quad\;\;$ or $\quad x = \frac{9}{11}$

     The solution set is $\left\{-\frac{9}{11}, \frac{9}{11}\right\}$.

43. **(a)** $x^2 = 0.04$

$\qquad x = -\sqrt{0.04} \quad$ or $\quad x = \sqrt{0.04}$

$\qquad x = -0.2 \quad\;\;$ or $\quad x = 0.2$

     The solution set is $\{-0.2, 0.2\}$.

     (b) Since $(0.2)^2 = 0.04$, $\sqrt{0.04} = 0.2$.

45. $\qquad\qquad x^2 = 36$

$\qquad\qquad x^2 - 36 = 0$

$\qquad (x+6)(x-6) = 0$

$\qquad x+6 = 0 \quad$ or $\quad x-6 = 0$

$\qquad\quad x = -6 \quad$ or $\qquad\;\; x = 6$

     The solution set is $\{-6, 6\}$.

10.6 Solving Equations with Radicals

1. $\sqrt{3x + 18} = x$

(a) Check $x = 6$:

$$\sqrt{3(6) + 18} \overset{?}{=} 6$$
$$\sqrt{18 + 18} \overset{?}{=} 6$$
$$\sqrt{36} \overset{?}{=} 6$$
$$6 = 6 \quad \textit{True}$$

The number 6 is a solution.

(b) Check $x = -3$:

$\sqrt{3(-3) + 18} = -3$ is a false statement since the principal square root of a number is nonnegative. The number -3 is not a solution.

3. $\sqrt{x + 2} = \sqrt{9x - 2} - 2\sqrt{x - 1}$

(a) Check $x = 2$:

$$\sqrt{2 + 2} \overset{?}{=} \sqrt{9(2) - 2} - 2\sqrt{2 - 1}$$
$$\sqrt{4} \overset{?}{=} \sqrt{16} - 2\sqrt{1}$$
$$2 \overset{?}{=} 4 - 2$$
$$2 = 2 \quad \textit{True}$$

The number 2 is a solution.

(b) Check $x = 7$:

$$\sqrt{7 + 2} \overset{?}{=} \sqrt{9(7) - 2} - 2\sqrt{7 - 1}$$
$$\sqrt{9} \overset{?}{=} \sqrt{61} - 2\sqrt{6}$$
$$3 = \sqrt{61} - 2\sqrt{6} \quad \textit{False}$$

The number 7 is not a solution.

5. $\sqrt{9} = 3$, not -3. There is no solution of $\sqrt{x} = -3$ since the value of a principal square root cannot equal a negative number.

In Exercises 7–34, check each solution in the original equation.

7. $\sqrt{r - 2} = 3$

Square both sides.

$$\left(\sqrt{r - 2}\right)^2 = 3^2$$
$$r - 2 = 9$$
$$r = 11$$

Check the proposed solution, 11.

Check $r = 11$: $\sqrt{9} = 3$ $\quad \textit{True}$
The solution set is $\{11\}$.

9. $\sqrt{6k - 1} = 1$

Square both sides.

$$\left(\sqrt{6k - 1}\right)^2 = 1^2$$
$$6k - 1 = 1$$
$$6k = 2$$
$$k = \tfrac{2}{6} = \tfrac{1}{3}$$

Check the proposed solution, $\tfrac{1}{3}$.

Check $k = \tfrac{1}{3}$: $\sqrt{1} = 1$ $\quad \textit{True}$
The solution set is $\left\{\tfrac{1}{3}\right\}$.

11. $\sqrt{4r + 3} + 1 = 0$

Isolate the radical.

$$\sqrt{4r + 3} = -1$$

This equation has no solution, because $\sqrt{4r + 3}$ cannot be negative.

The solution set is $\emptyset$.

13. $\sqrt{3k + 1} - 4 = 0$

Isolate the radical.

$$\sqrt{3k + 1} = 4$$

Square both sides.

$$\left(\sqrt{3k + 1}\right)^2 = 4^2$$
$$3k + 1 = 16$$
$$3k = 15$$
$$k = 5$$

Check $k = 5$: $\sqrt{16} - 4 = 0$ $\quad \textit{True}$
The solution set is $\{5\}$.

15. $4 - \sqrt{x - 2} = 0$

$$4 = \sqrt{x - 2} \qquad \textit{Isolate.}$$
$$4^2 = \left(\sqrt{x - 2}\right)^2 \qquad \textit{Square.}$$
$$16 = x - 2$$
$$18 = x$$

Check $x = 18$: $4 - \sqrt{16} = 0$ $\quad \textit{True}$
The solution set is $\{18\}$.

17. $\sqrt{9a - 4} = \sqrt{8a + 1}$

$$\left(\sqrt{9a - 4}\right)^2 = \left(\sqrt{8a + 1}\right)^2 \qquad \textit{Square.}$$
$$9a - 4 = 8a + 1$$
$$a = 5$$

Check $a = 5$: $\sqrt{41} = \sqrt{41}$ $\quad \textit{True}$
The solution set is $\{5\}$.

19.
$$2\sqrt{x} = \sqrt{3x+4}$$
$$\left(2\sqrt{x}\right)^2 = \left(\sqrt{3x+4}\right)^2 \quad \textit{Square.}$$
$$4x = 3x+4$$
$$x = 4$$

Check $x = 4$: $\ 4 = \sqrt{16}$ *True*
The solution set is $\{4\}$.

21.
$$3\sqrt{z-1} = 2\sqrt{2z+2}$$
$$\left(3\sqrt{z-1}\right)^2 = \left(2\sqrt{2z+2}\right)^2 \quad \textit{Square.}$$
$$9(z-1) = 4(2z+2)$$
$$9z-9 = 8z+8$$
$$z = 17$$

Check $z = 17$: $\ 3(4) = 2(6) \quad$ *True*
The solution set is $\{17\}$.

23.
$$k = \sqrt{k^2+4k-20}$$
$$k^2 = \left(\sqrt{k^2+4k-20}\right)^2 \quad \textit{Square.}$$
$$k^2 = k^2+4k-20$$
$$20 = 4k$$
$$5 = k$$

Check $k = 5$: $\ 5 = \sqrt{25} \quad$ *True*
The solution set is $\{5\}$.

25.
$$a = \sqrt{a^2+3a+9}$$
$$a^2 = \left(\sqrt{a^2+3a+9}\right)^2 \quad \textit{Square.}$$
$$a^2 = a^2+3a+9$$
$$-3a = 9$$
$$a = -3$$

Substituting -3 for a makes the left side of the original equation negative, but the right side is nonnegative, so the solution set is $\emptyset$.

27.
$$\sqrt{9-x} = x+3$$
$$\left(\sqrt{9-x}\right)^2 = (x+3)^2 \quad \textit{Square.}$$
$$9-x = x^2+6x+9$$
$$0 = x^2+7x$$
$$0 = x(x+7)$$
$$x = 0 \quad \text{or} \quad x+7 = 0$$
$$x = -7$$

Check $x = -7$: $\ \sqrt{16} = -4$ *False*
Check $x = 0$: $\ \ \sqrt{9} = 3 \quad$ *True*

The solution set is $\{0\}$.

29.
$$\sqrt{k^2+2k+9} = k+3$$
$$\left(\sqrt{k^2+2k+9}\right)^2 = (k+3)^2 \quad \textit{Square.}$$
$$k^2+2k+9 = k^2+6k+9$$
$$0 = 4k$$
$$0 = k$$

Check $k = 0$: $\ \sqrt{9} = 3$ *True*
The solution set is $\{0\}$.

31.
$$\sqrt{r^2+9r+3} = -r$$
$$\left(\sqrt{r^2+9r+3}\right)^2 = (-r)^2 \quad \textit{Square.}$$
$$r^2+9r+3 = r^2$$
$$9r = -3$$
$$r = -\tfrac{1}{3}$$

Check $r = -\tfrac{1}{3}$: $\ \sqrt{\tfrac{1}{9}} = \tfrac{1}{3}$ *True*

The solution set is $\left\{-\tfrac{1}{3}\right\}$.

33. $\sqrt{z^2+12z-4}+4-z = 0$
$$\sqrt{z^2+12z-4} = z-4 \qquad \textit{Isolate.}$$
$$\left(\sqrt{z^2+12z-4}\right)^2 = (z-4)^2 \qquad \textit{Square.}$$
$$z^2+12z-4 = z^2-8z+16$$
$$20z = 20$$
$$z = 1$$

Substituting 1 for z makes the left side of the original equation positive, but the right side is zero, so the solution set is $\emptyset$.

35. $\sqrt{3x+4} = 8-x$

$(8-x)^2$ equals $64-16x+x^2$, not $64+x^2$. The first step should be
$$3x+4 = 64-16x+x^2.$$
Then we have
$$0 = x^2-19x+60.$$
$$0 = (x-4)(x-15)$$
$$x-4 = 0 \quad \text{or} \quad x-15 = 0$$
$$x = 4 \quad \text{or} \qquad x = 15$$

Check $x = 4$: $\ \sqrt{16} = 8-4 \quad$ *True*
Check $x = 15$: $\ \sqrt{49} = 8-15$ *False*
The solution set is $\{4\}$.

37. $\sqrt[3]{2x+5} = \sqrt[3]{6x+1}$
Cube both sides.
$$\left(\sqrt[3]{2x+5}\right)^3 = \left(\sqrt[3]{6x+1}\right)^3$$
$$2x+5 = 6x+1$$
$$4 = 4x$$
$$1 = x$$

Check $x = 1$: $\ \sqrt[3]{7} = \sqrt[3]{7} \quad$ *True*
The solution set is $\{1\}$.

39. $\sqrt[3]{a^2 + 5a + 1} = \sqrt[3]{a^2 + 4a}$

$\left(\sqrt[3]{a^2 + 5a + 1}\right)^3 = \left(\sqrt[3]{a^2 + 4a}\right)^3$ *Cube.*

$a^2 + 5a + 1 = a^2 + 4a$

$a = -1$

Check $a = -1$: $\sqrt[3]{-3} = \sqrt[3]{-3}$ *True*

The solution set is $\{-1\}$.

41. $\sqrt[3]{2m - 1} = \sqrt[3]{m + 13}$

$\left(\sqrt[3]{2m - 1}\right)^3 = \left(\sqrt[3]{m + 13}\right)^3$ *Cube.*

$2m - 1 = m + 13$

$m = 14$

Check $m = 14$: $\sqrt[3]{27} = \sqrt[3]{27}$ *True*

The solution set is $\{14\}$.

43. $\sqrt[4]{a + 8} = \sqrt[4]{2a}$

Raise each side to the fourth power.

$\left(\sqrt[4]{a + 8}\right)^4 = \left(\sqrt[4]{2a}\right)^4$

$a + 8 = 2a$

$8 = a$

Check $a = 8$: $\sqrt[4]{16} = \sqrt[4]{16}$ *True*

The solution set is $\{8\}$.

45. $\sqrt[3]{x - 8} + 2 = 0$

$\sqrt[3]{x - 8} = -2$ *Isolate.*

$\left(\sqrt[3]{x - 8}\right)^3 = (-2)^3$ *Cube.*

$x - 8 = -8$

$x = 0$

Check $x = 0$: $\sqrt[3]{-8} + 2 = 0$ *True*

The solution set is $\{0\}$.

47. $\sqrt[4]{2k - 5} + 4 = 0$

$\sqrt[4]{2k - 5} = -4$ *Isolate.*

This equation has no solution, because $\sqrt[4]{2k - 5}$ cannot be negative.

The solution set is $\emptyset$.

49. $\sqrt{k + 2} - \sqrt{k - 3} = 1$

Get one radical on each side of the equals sign.

$\sqrt{k + 2} = 1 + \sqrt{k - 3}$

$\left(\sqrt{k + 2}\right)^2 = \left(1 + \sqrt{k - 3}\right)^2$ *Square.*

$k + 2 = 1 + 2\sqrt{k - 3} + k - 3$

$4 = 2\sqrt{k - 3}$ *Isolate.*

$2 = \sqrt{k - 3}$ *Divide by 2.*

$2^2 = \left(\sqrt{k - 3}\right)^2$ *Square again.*

$4 = k - 3$

$7 = k$

Check $k = 7$: $\sqrt{9} - \sqrt{4} = 1$ *True*

The solution set is $\{7\}$.

51. $\sqrt{2r + 11} - \sqrt{5r + 1} = -1$

Get one radical on each side of the equals sign.

$\sqrt{2r + 11} = -1 + \sqrt{5r + 1}$

Square both sides.

$\left(\sqrt{2r + 11}\right)^2 = \left(-1 + \sqrt{5r + 1}\right)^2$

$2r + 11 = 1 - 2\sqrt{5r + 1} + 5r + 1$

Isolate the remaining radical.

$2\sqrt{5r + 1} = 3r - 9$

Square both sides again.

$\left(2\sqrt{5r + 1}\right)^2 = (3r - 9)^2$

$4(5r + 1) = 9r^2 - 54r + 81$

$20r + 4 = 9r^2 - 54r + 81$

$0 = 9r^2 - 74r + 77$

$0 = (9r - 11)(r - 7)$

$9r - 11 = 0$ or $r - 7 = 0$

$r = \frac{11}{9}$ $r = 7$

Check $r = \frac{11}{9}$: $\frac{11}{3} - \frac{8}{3} = -1$ *False*

Check $r = 7$: $5 - 6 = -1$ *True*

The solution set is $\{7\}$.

53. $\sqrt{3p + 4} - \sqrt{2p - 4} = 2$

Get one radical on each side of the equals sign.

$\sqrt{3p + 4} = 2 + \sqrt{2p - 4}$

Square both sides.

$\left(\sqrt{3p + 4}\right)^2 = \left(2 + \sqrt{2p - 4}\right)^2$

$3p + 4 = 4 + 4\sqrt{2p - 4} + 2p - 4$

Isolate the remaining radical.

$p + 4 = 4\sqrt{2p - 4}$

Square both sides again.

$(p + 4)^2 = \left(4\sqrt{2p - 4}\right)^2$

$p^2 + 8p + 16 = 16(2p - 4)$

$p^2 + 8p + 16 = 32p - 64$

$p^2 - 24p + 80 = 0$

$(p - 4)(p - 20) = 0$

$p - 4 = 0$ or $p - 20 = 0$

$p = 4$ $p = 20$

Check $p = 4$: $\sqrt{16} - \sqrt{4} = 2$ *True*

Check $p = 20$: $\sqrt{64} - \sqrt{36} = 2$ *True*

The solution set is $\{4, 20\}$.

55.
$$\sqrt{3-3p}-3=\sqrt{3p+2}$$
Square both sides.
$$\left(\sqrt{3-3p}-3\right)^2=\left(\sqrt{3p+2}\right)^2$$
$$3-3p-6\sqrt{3-3p}+9=3p+2$$
Isolate the remaining radical.
$$-6\sqrt{3-3p}=6p-10$$
$$-3\sqrt{3-3p}=3p-5$$
Square both sides again.
$$\left(-3\sqrt{3-3p}\right)^2=(3p-5)^2$$
$$9(3-3p)=9p^2-30p+25$$
$$27-27p=9p^2-30p+25$$
$$0=9p^2-3p-2$$
$$0=(3p+1)(3p-2)$$

$$3p+1=0 \quad\text{or}\quad 3p-2=0$$
$$p=-\tfrac{1}{3} \qquad\qquad p=\tfrac{2}{3}$$

Check $p=-\tfrac{1}{3}$: $\quad\sqrt{4}-3=\sqrt{1}$ *False*
Check $p=\tfrac{2}{3}$: $\quad\sqrt{1}-3=\sqrt{4}$ *False*
The solution set is $\emptyset$.

57.
$$\sqrt{2\sqrt{x+11}}=\sqrt{4x+2}$$
$$2\sqrt{x+11}=4x+2 \qquad\qquad \text{Square.}$$
$$\left(2\sqrt{x+11}\right)^2=(4x+2)^2 \qquad \text{Square again.}$$
$$4(x+11)=16x^2+16x+4$$
$$4x+44=16x^2+16x+4$$
$$0=16x^2+12x-40$$
$$0=4x^2+3x-10$$
$$0=(x+2)(4x-5)$$

$$x+2=0 \quad\text{or}\quad 4x-5=0$$
$$x=-2 \qquad\qquad x=\tfrac{5}{4}$$

Check $x=-2$: $\quad\sqrt{6}=\sqrt{-6}$ *False*
Check $x=\tfrac{5}{4}$: $\quad\sqrt{7}=\sqrt{7}$ *True*
The solution set is $\left\{\tfrac{5}{4}\right\}$.

59. Graph the functions
$$Y_1=\sqrt{3-3x} \quad\text{and}\quad Y_2=3+\sqrt{3x+2}$$

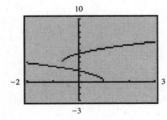

The graphs do not intersect, so the solution set is $\emptyset$.
To find the domain of
$$y=\sqrt{3-3x}-3-\sqrt{3x+2}$$

we must have
$$3-3x\ge 0 \quad\text{and}\quad 3x+2\ge 0$$
$$-3x\ge -3 \qquad\qquad 3x\ge -2$$
$$x\le 1 \quad\text{and}\quad x\ge -\tfrac{2}{3}.$$

The domain is $\left[-\tfrac{2}{3},1\right]$. This is the region where the graphs of Y_1 and Y_2 vertically overlap.

61.
$$(2x-9)^{1/2}=2+(x-8)^{1/2}$$
$$\sqrt{2x-9}=2+\sqrt{x-8}$$
$$\left(\sqrt{2x-9}\right)^2=\left(2+\sqrt{x-8}\right)^2$$
$$2x-9=4+4\sqrt{x-8}+x-8$$
$$x-5=4\sqrt{x-8}$$
$$(x-5)^2=\left(4\sqrt{x-8}\right)^2$$
$$x^2-10x+25=16(x-8)$$
$$x^2-10x+25=16x-128$$
$$x^2-26x+153=0$$
$$(x-9)(x-17)=0$$
$$x=9 \quad\text{or}\quad x=17$$
Check $x=9$: $9^{1/2}\overset{?}{=}2+1^{1/2}$
$$3=2+1 \quad \text{True}$$
Check $x=17$: $25^{1/2}\overset{?}{=}2+9^{1/2}$
$$5=2+3 \quad \text{True}$$
The solution set is $\{9,17\}$.

63.
$$(2w-1)^{2/3}-w^{1/3}=0$$
$$\sqrt[3]{(2w-1)^2}=\sqrt[3]{w}$$
$$\left[\sqrt[3]{(2w-1)^2}\right]^3=\left(\sqrt[3]{w}\right)^3$$
$$(2w-1)^2=w$$
$$4w^2-4w+1=w$$
$$4w^2-5w+1=0$$
$$(4w-1)(w-1)=0$$

$$4w-1=0 \quad\text{or}\quad w-1=0$$
$$w=\tfrac{1}{4} \quad\text{or}\qquad w=1$$

Check $w=\tfrac{1}{4}$: $\left(-\tfrac{1}{2}\right)^{2/3}-\left(\tfrac{1}{4}\right)^{1/3}\overset{?}{=}0$
$$\left(\tfrac{1}{4}\right)^{1/3}-\left(\tfrac{1}{4}\right)^{1/3}=0 \quad \text{True}$$
Check $w=1$: $\qquad 1^{2/3}-1^{1/3}=0$ *True*
The solution set is $\left\{\tfrac{1}{4},1\right\}$.

65. Solve $V=\sqrt{\dfrac{2K}{m}}$ for K.
$$(V)^2=\left(\sqrt{\dfrac{2K}{m}}\right)^2 \qquad \text{Square.}$$
$$V^2=\dfrac{2K}{m} \qquad\qquad \text{Multiply by } \dfrac{m}{2}.$$
$$\dfrac{V^2m}{2}=K$$

67. Solve $f = \dfrac{1}{2\pi\sqrt{LC}}$ for L.

$$2\pi f\sqrt{LC} = 1$$
$$\left(2\pi f\sqrt{LC}\right)^2 = 1^2$$
$$4\pi^2 f^2 LC = 1$$
$$L = \dfrac{1}{4\pi^2 f^2 C}$$

69. Solve $N = \dfrac{1}{2\pi}\sqrt{\dfrac{a}{r}}$ for r.

$$2\pi N = \sqrt{\dfrac{a}{r}}$$
$$(2\pi N)^2 = \left(\sqrt{\dfrac{a}{r}}\right)^2$$
$$4\pi^2 N^2 = \dfrac{a}{r}$$
$$4\pi^2 N^2 r = a$$
$$r = \dfrac{a}{4\pi^2 N^2}$$

71. $(5 + 9x) + (-4 - 8x)$
$$= 5 - 4 + 9x - 8x$$
$$= 1 + x$$

73. $(x + 3)(2x - 5) = 2x^2 - 5x + 6x - 15$
$$= 2x^2 + x - 15$$

75. $\dfrac{-7}{5 - \sqrt{2}} = \dfrac{-7}{5 - \sqrt{2}} \cdot \dfrac{5 + \sqrt{2}}{5 + \sqrt{2}}$

$$= \dfrac{-7\left(5 + \sqrt{2}\right)}{5^2 - \left(\sqrt{2}\right)^2}$$

$$= \dfrac{-7\left(5 + \sqrt{2}\right)}{25 - 2} = \dfrac{-7\left(5 + \sqrt{2}\right)}{23}$$

10.7 Complex Numbers

1. $\sqrt{-1} = i$

3. $i^2 = -1$

5. $\dfrac{1}{i} = \dfrac{1}{i} \cdot \dfrac{-i}{-i}$ $-i$ *is the conjugate of* i

$$= \dfrac{-i}{-i^2} = \dfrac{i}{i^2} = \dfrac{i}{-1} = -i$$

7. $\sqrt{-169} = i\sqrt{169} = 13i$

9. $-\sqrt{-144} = -i\sqrt{144} = -12i$

11. $\sqrt{-5} = i\sqrt{5}$

13. $\sqrt{-48} = i\sqrt{48} = i\sqrt{16 \cdot 3} = 4i\sqrt{3}$

15. $\sqrt{-7} \cdot \sqrt{-15} = i\sqrt{7} \cdot i\sqrt{15} = i^2\sqrt{7 \cdot 15}$
$$= -1\sqrt{105} = -\sqrt{105}$$

17. $\sqrt{-4} \cdot \sqrt{-25} = i\sqrt{4} \cdot i\sqrt{25} = 2i \cdot 5i = 10i^2$
$$= 10(-1) = -10$$

19. $\sqrt{-3} \cdot \sqrt{11} = i\sqrt{3} \cdot \sqrt{11}$
$$= i\sqrt{33}$$

21. $\dfrac{\sqrt{-300}}{\sqrt{-100}} = \dfrac{i\sqrt{300}}{i\sqrt{100}} = \sqrt{\dfrac{300}{100}} = \sqrt{3}$

23. $\dfrac{\sqrt{-75}}{\sqrt{3}} = \dfrac{i\sqrt{75}}{\sqrt{3}} = i\sqrt{\dfrac{75}{3}} = i\sqrt{25} = 5i$

25. **(a)** Since any real number can be written as $a + bi$, where $b = 0$, every real number is also a complex number.

(b) Not every complex number is a real number. For example, any number $a + bi$, $b \neq 0$, such as $3 + 7i$, is a complex number that is not a real number.

27. $(3 + 2i) + (-4 + 5i)$
$$= [(3 + (-4)] + (2 + 5)i$$
$$= -1 + 7i$$

29. $(5 - i) + (-5 + i)$
$$= (5 - 5) + (-1 + 1)i$$
$$= 0$$

31. $(4 + i) - (-3 - 2i)$
$$= [(4 - (-3)] + [(1 - (-2)]i$$
$$= 7 + 3i$$

33. $(-3 - 4i) - (-1 - 4i)$
$$= [-3 - (-1)] + [-4 - (-4)]i$$
$$= -2$$

35. $(-4 + 11i) + (-2 - 4i) + (7 + 6i)$
$$= (-4 - 2 + 7) + (11 - 4 + 6)i$$
$$= 1 + 13i$$

37. $[(7 + 3i) - (4 - 2i)] + (3 + i)$
Work inside the brackets first.
$$= [(7 - 4) + (3 + 2)i] + (3 + i)$$
$$= (3 + 5i) + (3 + i)$$
$$= (3 + 3) + (5 + 1)i$$
$$= 6 + 6i$$

39. If $a - c = b$, then $b + c = a$.
So, $(4 + 2i) - (3 + i) = 1 + i$ implies that
$$(1 + i) + (3 + i) = \underline{\,4 + 2i\,}.$$

41. $(3i)(27i) = 81i^2 = 81(-1) = -81$

43. $(-8i)(-2i) = 16i^2 = 16(-1) = -16$

45. $5i(-6 + 2i) = (5i)(-6) + (5i)(2i)$
$$= -30i + 10i^2$$
$$= -30i + 10(-1)$$
$$= -10 - 30i$$

47. $(4 + 3i)(1 - 2i)$

$$\qquad \mathbf{F} \qquad \mathbf{O} \qquad \mathbf{I} \qquad \mathbf{L}$$
$$= (4)(1) + 4(-2i) + (3i)(1) + (3i)(-2i)$$
$$= 4 - 8i + 3i - 6i^2$$
$$= 4 - 5i - 6(-1)$$
$$= 4 - 5i + 6 = 10 - 5i$$

49. $(4 + 5i)^2 = 4^2 + 2(4)(5i) + (5i)^2$
$$= 16 + 40i + 25i^2$$
$$= 16 + 40i + 25(-1)$$
$$= 16 + 40i - 25$$
$$= -9 + 40i$$

51. $2i(-4 - i)^2$
$$= 2i\left[(-4)^2 - 2(-4)(i) + i^2\right]$$
$$= 2i(16 + 8i - 1)$$
$$= 2i(15 + 8i) = 30i + 16i^2$$
$$= 30i + 16(-1) = -16 + 30i$$

53. $(12 + 3i)(12 - 3i)$
$$= 12^2 - (3i)^2 = 144 - 9i^2$$
$$= 144 - 9(-1) = 144 + 9 = 153$$

55. $(4 + 9i)(4 - 9i)$
$$= 4^2 - (9i)^2 = 16 - 81i^2$$
$$= 16 - 81(-1) = 16 + 81 = 97$$

57. The conjugate of $a + bi$ is $a - bi$.

59. $\dfrac{2}{1 - i}$

Multiply the numerator and the denominator by the conjugate of the denominator, $1 + i$.

$$= \frac{2(1 + i)}{(1 - i)(1 + i)} = \frac{2(1 + i)}{1^2 - i^2}$$
$$= \frac{2(1 + i)}{1 - (-1)} = \frac{2(1 + i)}{2} = 1 + i$$

61. $\dfrac{-7 + 4i}{3 + 2i}$

Multiply the numerator and the denominator by the conjugate of the denominator, $3 - 2i$.

$$= \frac{(-7 + 4i)(3 - 2i)}{(3 + 2i)(3 - 2i)}$$

In the denominator, we make use of the fact that $(a + bi)(a - bi) = a^2 + b^2$.

$$= \frac{-21 + 14i + 12i + 8}{3^2 + 2^2}$$
$$= \frac{-13 + 26i}{13} = \frac{13(-1 + 2i)}{13} = -1 + 2i$$

63. $\dfrac{8i}{2 + 2i}$

Write in lowest terms.

$$= \frac{2 \cdot 4i}{2(1 + i)} = \frac{4i}{1 + i}$$

Multiply the numerator and the denominator by the conjugate of the denominator, $1 - i$.

$$= \frac{4i(1 - i)}{(1 + i)(1 - i)} = \frac{4(i - i^2)}{1^2 + 1^2}$$
$$= \frac{4(i + 1)}{2} = 2(i + 1) = 2 + 2i$$

65. $\dfrac{2 - 3i}{2 + 3i}$

Multiply the numerator and the denominator by the conjugate of the denominator, $2 - 3i$.

$$= \frac{(2 - 3i)(2 - 3i)}{(2 + 3i)(2 - 3i)} = \frac{2^2 - 2(2)(3i) + (3i)^2}{2^2 + 3^2}$$
$$= \frac{4 - 12i + 9i^2}{4 + 9} = \frac{4 - 12i - 9}{13}$$
$$= \frac{-5 - 12i}{13} = -\frac{5}{13} - \frac{12}{13}i$$

67. $\dfrac{3 + i}{i}$

Multiply the numerator and the denominator by the conjugate of the denominator, $-i$.

$$= \frac{(3 + i)(-i)}{i(-i)} = \frac{-3i - i^2}{-i^2}$$
$$= \frac{-3i - (-1)}{-(-1)} = \frac{-3i + 1}{1}$$
$$= 1 - 3i$$

69. $i^{18} = i^{16} \cdot i^2 = \left(i^4\right)^4 \cdot i^2$
$$= 1^4 \cdot (-1) = 1 \cdot (-1) = -1$$

71. $i^{89} = i^{88} \cdot i = \left(i^4\right)^{22} \cdot i = 1^{22} \cdot i$
$$= 1 \cdot i = i$$

73. $i^{38} = i^{36} \cdot i^2$
$$= \left(i^4\right)^9 \cdot i^2 = 1^9(-1) = -1$$

75. $i^{43} = i^{40} \cdot i^2 \cdot i$
$$= \left(i^4\right)^{10} \cdot (-1) \cdot i$$
$$= 1^{10} \cdot (-i) = -i$$

77. $i^{-5} = \dfrac{1}{i^5} = \dfrac{1}{i^4 \cdot i} = \dfrac{1}{1 \cdot i} = \dfrac{1}{i}$

From Exercise 5, $\dfrac{1}{i} = -i$.

79. Since $i^{20} = \left(i^4\right)^5 = 1^5 = 1$, the student multiplied by 1, which is justified by the identity property for multiplication.

81. $I = \dfrac{E}{R + (X_L - X_c)i}$

Substitute $2 + 3i$ for E, 5 for R, 4 for X_L, and 3 for X_c.

$$I = \dfrac{2 + 3i}{5 + (4 - 3)i} = \dfrac{2 + 3i}{5 + i}$$

$$= \dfrac{(2 + 3i)(5 - i)}{(5 + i)(5 - i)} = \dfrac{10 - 2i + 15i - 3i^2}{5^2 + 1^2}$$

$$= \dfrac{10 + 3 + 13i}{25 + 1} = \dfrac{13 + 13i}{26}$$

$$= \dfrac{13(1 + i)}{13 \cdot 2} = \dfrac{1 + i}{2} = \dfrac{1}{2} + \dfrac{1}{2}i$$

83. To check that $1 + 5i$ is a solution of the equation, substitute $1 + 5i$ for x.

$$x^2 - 2x + 26 = 0$$

$$(1 + 5i)^2 - 2(1 + 5i) + 26 \overset{?}{=} 0$$

$$(1 + 10i + 25i^2) - 2 - 10i + 26 \overset{?}{=} 0$$

$$1 + 10i - 25 - 2 - 10i + 26 \overset{?}{=} 0$$

$$(1 - 25 - 2 + 26) + (10 - 10)i \overset{?}{=} 0$$

$$0 = 0$$

$$\textit{True}$$

Thus, $1 + 5i$ is a solution of

$$x^2 - 2x + 26 = 0.$$

Now substitute $1 - 5i$ for x.

$$(1 - 5i)^2 - 2(1 - 5i) + 26 \overset{?}{=} 0$$

$$(1 - 10i + 25i^2) - 2 + 10i + 26 \overset{?}{=} 0$$

$$1 - 10i - 25 - 2 + 10i + 26 \overset{?}{=} 0$$

$$(1 - 25 - 2 + 26) + (-10 + 10)i \overset{?}{=} 0$$

$$0 = 0$$

$$\textit{True}$$

Thus, $1 - 5i$ is also a solution of the given equation.

85. $\dfrac{3}{2 - i} + \dfrac{5}{1 + i}$

$$= \dfrac{3(2 + i)}{(2 - i)(2 + i)} + \dfrac{5(1 - i)}{(1 + i)(1 - i)}$$

$$= \dfrac{6 + 3i}{4 + 1} + \dfrac{5 - 5i}{1 + 1}$$

$$= \dfrac{2(6 + 3i)}{2 \cdot 5} + \dfrac{5(5 - 5i)}{5 \cdot 2}$$

$$= \dfrac{12 + 6i}{10} + \dfrac{25 - 25i}{10}$$

$$= \dfrac{37 - 19i}{10} = \dfrac{37}{10} - \dfrac{19}{10}i$$

87. $\dfrac{2 + i}{2 - i} + \dfrac{i}{1 + i}$

$$= \dfrac{(2 + i)(2 + i)}{(2 - i)(2 + i)} + \dfrac{i(1 - i)}{(1 + i)(1 - i)}$$

$$= \dfrac{4 + 4i + i^2}{4 + 1} + \dfrac{i - i^2}{1 + 1}$$

$$= \dfrac{3 + 4i}{5} + \dfrac{1 + i}{2}$$

$$= \dfrac{2(3 + 4i)}{2 \cdot 5} + \dfrac{5(1 + i)}{5 \cdot 2}$$

$$= \dfrac{6 + 8i}{10} + \dfrac{5 + 5i}{10}$$

$$= \dfrac{11 + 13i}{10} = \dfrac{11}{10} + \dfrac{13}{10}i$$

Thus, $\left(\dfrac{2 + i}{2 - i} + \dfrac{i}{1 + i} \right)i$

$$= \left(\dfrac{11}{10} + \dfrac{13}{10}i \right)i$$

$$= \dfrac{11}{10}i + \dfrac{13}{10}i^2$$

$$= -\dfrac{13}{10} + \dfrac{11}{10}i.$$

89. $6x + 13 = 0$

$$6x = -13$$

$$x = -\dfrac{13}{6}$$

The solution set is $\left\{ -\dfrac{13}{6} \right\}$.

91. $x(x + 3) = 40$

$$x^2 + 3x = 40$$

$$x^2 + 3x - 40 = 0$$

$$(x + 8)(x - 5) = 0$$

$$x + 8 = 0 \quad \text{or} \quad x - 5 = 0$$

$$x = -8 \quad \text{or} \quad x = 5$$

Check $x = -8$: $\quad -8(-5) = 40 \quad \textit{True}$

Check $x = 5$: $\quad\quad 5(8) = 40 \quad \textit{True}$

The solution set is $\{-8, 5\}$.

93. $5x^2 - 3x = 2$

$$5x^2 - 3x - 2 = 0$$

$$(5x + 2)(x - 1) = 0$$

$$5x + 2 = 0 \quad \text{or} \quad x - 1 = 0$$

$$x = -\dfrac{2}{5} \quad \text{or} \quad x = 1$$

Check $x = -\dfrac{2}{5}$: $\quad \dfrac{4}{5} + \dfrac{6}{5} = 2 \quad \textit{True}$

Check $x = 1$: $\quad\quad 5 - 3 = 2 \quad \textit{True}$

The solution set is $\left\{ -\dfrac{2}{5}, 1 \right\}$.

Chapter 10 Review Exercises

1. $\sqrt{1764} = 42$, because $42^2 = 1764$.

2. $-\sqrt{289} = -(17) = -17$, since $17^2 = 289$.

3. $\sqrt[3]{216} = 6$, because $6^3 = 216$.

4. $\sqrt[3]{-125} = -5$, because $(-5)^3 = -125$.

5. $-\sqrt[3]{27} = -(3) = -3$, since $3^3 = 27$.

6. $\sqrt[5]{-32} = -2$, because $(-2)^5 = -32$.

7. $\sqrt[n]{a}$ is not a real number if n is even and a is negative.

8. **(a)** $\sqrt{x^2} = |x|$

   (b) $-\sqrt{x^2} = -|x|$

   (c) $\sqrt[3]{x^3} = x$

9. $-\sqrt{47} \approx -6.856$

10. $\sqrt[3]{-129} \approx -5.053$

11. $\sqrt[4]{605} \approx 4.960$

12. $500^{-3/4} \approx 0.009$

13. $-500^{4/3} \approx -3968.503$

14. $-28^{-1/2} \approx -0.189$

15. $f(x) = \sqrt{x-1}$
For the radicand to be nonnegative, we must have

$$x - 1 \geq 0 \quad \text{or} \quad x \geq 1.$$

Thus, the domain is $[1, \infty)$.
The function values are positive or zero (the result of the radical), so the range is $[0, \infty)$.

x	$f(x) = \sqrt{x-1}$
1	$\sqrt{1-1} = 0$
2	$\sqrt{2-1} = 1$
5	$\sqrt{5-1} = 2$

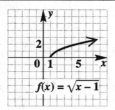

$f(x) = \sqrt{x-1}$

16. $f(x) = \sqrt[3]{x} + 4$
Since we can take the cube root of any real number, the domain is $(-\infty, \infty)$.
The result of a cube root can be any real number, so the range is $(-\infty, \infty)$. (The "+4" does not affect that range.)

x	$f(x) = \sqrt[3]{x} + 4$
-8	$\sqrt[3]{-8} + 4 = 2$
-1	$\sqrt[3]{-1} + 4 = 3$
0	$\sqrt[3]{0} + 4 = 4$
1	$\sqrt[3]{1} + 4 = 5$
8	$\sqrt[3]{8} + 4 = 6$

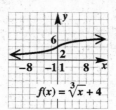

$f(x) = \sqrt[3]{x} + 4$

17. The base $\sqrt{38}$ is closest to $\sqrt{36} = 6$. The height $\sqrt{99}$ is closest to $\sqrt{100} = 10$. Use the estimates $b = 6$ and $h = 10$ in $A = \frac{1}{2}bh$ to find an estimate of the area.

$$A \approx \tfrac{1}{2}(6)(10) = 30$$

Choice **B** is the best estimate.

18. One way to evaluate $8^{2/3}$ is to first find the _cube (or third)_ root of _8_, which is _2_. Then raise that result to the _second_ power, to get an answer of _4_. Therefore, $8^{2/3} = $ _4_.

19. **A.** $(-27)^{2/3} = \left[(-27)^{1/3}\right]^2$
This number is a square, so it is a positive number.
B. $(-64)^{5/3} = \left[(-64)^{1/3}\right]^5$
This number is the odd power of an odd root of a negative number, so it is a negative number.

C. $(-100)^{1/2}$
This number is the square root of a negative number, so it is not a real number.

D. $(-32)^{1/5}$
This number is an odd root of a negative number, so it is a negative number.
The only positive number is choice **A**.

20. $a^{m/n} = \sqrt[n]{a^m}$
Since n is odd, $\sqrt[n]{a^m}$ is positive if a^m is positive and negative if a^m is negative. Since a is negative, a^m is positive if m is even and negative if m is odd.

(a) If a is negative and n is odd, then $a^{m/n}$ is positive if m is even.

(b) If a is negative and n is odd, then $a^{m/n}$ is negative if m is odd.

21. If a is negative and n is even, then $a^{1/n}$ is not a real number. An example is $(-4)^{1/2}$, which is not a real number.

22. $49^{1/2} = (7^2)^{1/2} = 7$

23. $-121^{1/2} = -(11^2)^{1/2} = -11$

24. $16^{5/4} = \left[(2^4)^{1/4}\right]^5 = 2^5 = 32$

25. $-8^{2/3} = -\left[(2^3)^{1/3}\right]^2 = -2^2 = -4$

26. $-\left(\frac{36}{25}\right)^{3/2} = -\left(\left[\left(\frac{6}{5}\right)^2\right]^{1/2}\right)^3$

$= -\left(\frac{6}{5}\right)^3 = -\frac{216}{125}$

27. $\left(-\frac{1}{8}\right)^{-5/3} = (-8)^{5/3} = \left(\left[(-2)^3\right]^{1/3}\right)^5$

$= (-2)^5 = -32$

28. $\left(\frac{81}{10,000}\right)^{-3/4} = \left(\frac{10,000}{81}\right)^{3/4}$

$= \left(\left[\left(\frac{10}{3}\right)^4\right]^{1/4}\right)^3$

$= \left(\frac{10}{3}\right)^3 = \frac{1000}{27}$

29. The base, -16, is negative and the index, 4, is even, so $(-16)^{3/4}$ is not a real number.

30. Solve $a^2 + b^2 = c^2$ for b. $(b > 0)$

$b^2 = c^2 - a^2$

$b = \sqrt{c^2 - a^2}$

31. The expression with fractional exponents, $a^{m/n}$, is equivalent to the radical expression, $\sqrt[n]{a^m}$. The denominator of the exponent is the index of the radical. For example, $\sqrt[3]{8^2} = \sqrt[3]{64} = 4$, and $8^{2/3} = (8^{1/3})^2 = 2^2 = 4$.

32. $(m + 3n)^{1/2} = \sqrt{m + 3n}$

33. $(3a + b)^{-5/3} = \dfrac{1}{(3a + b)^{5/3}}$

$= \dfrac{1}{((3a + b)^{1/3})^5}$

$= \dfrac{1}{\left(\sqrt[3]{3a + b}\right)^5}$

or $\dfrac{1}{\sqrt[3]{(3a + b)^5}}$

34. $\sqrt{7^9} = (7^9)^{1/2} = 7^{9/2}$

35. $\sqrt[5]{p^4} = (p^4)^{1/5} = p^{4/5}$

36. $5^{1/4} \cdot 5^{7/4} = 5^{1/4 + 7/4} = 5^{8/4}$

$= 5^2$, or 25

37. $\dfrac{96^{2/3}}{96^{-1/3}} = 96^{2/3 - (-1/3)} = 96^1 = 96$

38. $\dfrac{\left(a^{1/3}\right)^4}{a^{2/3}} = \dfrac{a^{4/3}}{a^{2/3}} = a^{4/3 - 2/3} = a^{2/3}$

39. $\dfrac{y^{-1/3} \cdot y^{5/6}}{y} = \dfrac{y^{-2/6} y^{5/6}}{y^{6/6}}$

$= y^{-2/6 + 5/6 - 6/6}$

$= y^{-3/6} = y^{-1/2} = \dfrac{1}{y^{1/2}}$

40. $\left(\dfrac{z^{-1} x^{-3/5}}{2^{-2} z^{-1/2} x}\right)^{-1} = \dfrac{z^1 x^{3/5}}{2^2 z^{1/2} x^{-1}}$

$= \frac{1}{4} z^{1 - 1/2} x^{3/5 - (-1)}$

$= \dfrac{z^{1/2} x^{8/5}}{4}$

41. $r^{-1/2}\left(r + r^{3/2}\right)$

$= r^{-1/2}(r) + r^{-1/2}\left(r^{3/2}\right)$

$= r^{-1/2 + 1} + r^{-1/2 + 3/2}$

$= r^{1/2} + r^{2/2}$

$= r^{1/2} + r$

42. $\sqrt[8]{s^4} = \left(s^4\right)^{1/8} = s^{4/8} = s^{1/2}$

43. $\sqrt[6]{r^9} = \left(r^9\right)^{1/6} = r^{9/6} = r^{3/2}$

44. $\dfrac{\sqrt{p^5}}{p^2} = \dfrac{p^{5/2}}{p^2} = p^{5/2 - 2}$

$= p^{5/2 - 4/2} = p^{1/2}$

45. $\sqrt[4]{k^3} \cdot \sqrt{k^3} = \left(k^3\right)^{1/4} \left(k^3\right)^{1/2}$

$= k^{3/4} k^{3/2} = k^{3/4 + 3/2}$

$= k^{3/4 + 6/4} = k^{9/4}$

46. $\sqrt[3]{m^5} \cdot \sqrt[3]{m^8} = \left(m^5\right)^{1/3} \left(m^8\right)^{1/3}$

$= m^{5/3} m^{8/3} = m^{5/3 + 8/3} = m^{13/3}$

47. $\sqrt[4]{\sqrt[3]{z}} = \sqrt[4]{z^{1/3}} = \left(z^{1/3}\right)^{1/4} = z^{1/12}$

48. $\sqrt{\sqrt{\sqrt{x}}} = \sqrt{\sqrt{x^{1/2}}} = \sqrt{\left(x^{1/2}\right)^{1/2}}$

$= \sqrt{x^{1/4}} = \left(x^{1/4}\right)^{1/2} = x^{1/8}$

49. $\sqrt[3]{\sqrt[5]{x}} = \sqrt[3]{x^{1/5}} = \left(x^{1/5}\right)^{1/3} = x^{1/15}$

50. $\sqrt{\sqrt[6]{\sqrt[3]{x}}} = \sqrt{\sqrt[6]{x^{1/3}}} = \sqrt{\left(x^{1/3}\right)^{1/6}}$

$= \sqrt{x^{1/18}} = \left(x^{1/18}\right)^{1/2} = x^{1/36}$

51. The product rule for exponents applies only if the bases are the same.

52. $\sqrt{6} \cdot \sqrt{11} = \sqrt{6 \cdot 11} = \sqrt{66}$

53. $\sqrt{5} \cdot \sqrt{r} = \sqrt{5 \cdot r} = \sqrt{5r}$

54. $\sqrt[3]{6} \cdot \sqrt[3]{5} = \sqrt[3]{6 \cdot 5} = \sqrt[3]{30}$

55. $\sqrt[4]{7} \cdot \sqrt[4]{3} = \sqrt[4]{7 \cdot 3} = \sqrt[4]{21}$

56. $\sqrt{20} = \sqrt{4 \cdot 5} = \sqrt{4}\sqrt{5} = 2\sqrt{5}$

57. $\sqrt{75} = \sqrt{25 \cdot 3} = \sqrt{25}\sqrt{3} = 5\sqrt{3}$

58. $-\sqrt{125} = -\sqrt{25 \cdot 5} = -5\sqrt{5}$

59. $\sqrt[3]{-108} = \sqrt[3]{-27 \cdot 4} = -3\sqrt[3]{4}$

60. $\sqrt{100y^7} = \sqrt{100y^6 \cdot y} = 10y^3\sqrt{y}$

61. $\sqrt[3]{64p^4q^6} = \sqrt[3]{64p^3q^6 \cdot p} = 4pq^2\sqrt[3]{p}$

62. $\sqrt[3]{108a^8b^5} = \sqrt[3]{27a^6b^3 \cdot 4a^2b^2}$
$$= 3a^2b\sqrt[3]{4a^2b^2}$$

63. $\sqrt[3]{632r^8t^4} = \sqrt[3]{8r^6t^3 \cdot 79r^2t}$
$$= 2r^2t\sqrt[3]{79r^2t}$$

64. $\sqrt{\dfrac{y^3}{144}} = \dfrac{\sqrt{y^3}}{\sqrt{144}} = \dfrac{\sqrt{y^2 \cdot y}}{12} = \dfrac{y\sqrt{y}}{12}$

65. $\sqrt[3]{\dfrac{m^{15}}{27}} = \dfrac{\sqrt[3]{m^{15}}}{\sqrt[3]{27}} = \dfrac{\sqrt[3]{(m^5)^3}}{\sqrt[3]{3^3}} = \dfrac{m^5}{3}$

66. $\sqrt[3]{\dfrac{r^2}{8}} = \dfrac{\sqrt[3]{r^2}}{\sqrt[3]{8}} = \dfrac{\sqrt[3]{r^2}}{2}$

67. $\sqrt[4]{\dfrac{a^9}{81}} = \dfrac{\sqrt[4]{a^9}}{\sqrt[4]{81}} = \dfrac{\sqrt[4]{a^8 \cdot a}}{3} = \dfrac{a^2\sqrt[4]{a}}{3}$

68. $\sqrt[6]{15^3} = 15^{3/6} = 15^{1/2} = \sqrt{15}$

69. $\sqrt[4]{p^6} = \left(p^6\right)^{1/4} = p^{6/4} = p^{3/2}$
$$= p^{2/2}p^{1/2} = p\sqrt{p}$$

70. $\sqrt[3]{2} \cdot \sqrt[4]{5} = 2^{1/3} \cdot 5^{1/4}$
$$= 2^{4/12} \cdot 5^{3/12}$$
$$= \left(2^4 \cdot 5^3\right)^{1/12}$$
$$= \sqrt[12]{16 \cdot 125} = \sqrt[12]{2000}$$

71. $\sqrt{x} \cdot \sqrt[5]{x} = x^{1/2} \cdot x^{1/5}$
$$= x^{5/10} \cdot x^{2/10}$$
$$= x^{7/10} = \sqrt[10]{x^7}$$

72. Substitute 8 for a and 6 for b in the Pythagorean formula to find the hypotenuse, c.
$$c^2 = a^2 + b^2$$
$$c = \sqrt{a^2 + b^2} = \sqrt{8^2 + 6^2}$$
$$= \sqrt{64 + 36} = \sqrt{100} = 10$$

The length of the hypotenuse is 10.

73. $(-4, 7)$ and $(10, 6)$
$$d = \sqrt{(x_2 - x_1)^2 + (y_2 - y_1)^2}$$
$$= \sqrt{[10 - (-4)]^2 + (6 - 7)^2}$$
$$= \sqrt{14^2 + (-1)^2}$$
$$= \sqrt{196 + 1} = \sqrt{197}$$

74. $2\sqrt{8} - 3\sqrt{50} = 2\sqrt{4 \cdot 2} - 3\sqrt{25 \cdot 2}$
$$= 2 \cdot 2\sqrt{2} - 3 \cdot 5\sqrt{2}$$
$$= 4\sqrt{2} - 15\sqrt{2} = -11\sqrt{2}$$

75. $8\sqrt{80} - 3\sqrt{45} = 8\sqrt{16 \cdot 5} - 3\sqrt{9 \cdot 5}$
$$= 8 \cdot 4\sqrt{5} - 3 \cdot 3\sqrt{5}$$
$$= 32\sqrt{5} - 9\sqrt{5} = 23\sqrt{5}$$

76. $-\sqrt{27y} + 2\sqrt{75y} = -\sqrt{9 \cdot 3y} + 2\sqrt{25 \cdot 3y}$
$$= -3\sqrt{3y} + 2 \cdot 5\sqrt{3y}$$
$$= -3\sqrt{3y} + 10\sqrt{3y}$$
$$= 7\sqrt{3y}$$

77. $2\sqrt{54m^3} + 5\sqrt{96m^3}$
$$= 2\sqrt{9m^2 \cdot 6m} + 5\sqrt{16m^2 \cdot 6m}$$
$$= 2 \cdot 3m\sqrt{6m} + 5 \cdot 4m\sqrt{6m}$$
$$= 6m\sqrt{6m} + 20m\sqrt{6m} = 26m\sqrt{6m}$$

78. $3\sqrt[3]{54} + 5\sqrt[3]{16} = 3\sqrt[3]{27 \cdot 2} + 5\sqrt[3]{8 \cdot 2}$
$$= 3 \cdot 3\sqrt[3]{2} + 5 \cdot 2\sqrt[3]{2}$$
$$= 9\sqrt[3]{2} + 10\sqrt[3]{2} = 19\sqrt[3]{2}$$

79. $-6\sqrt[4]{32} + \sqrt[4]{512} = -6\sqrt[4]{16 \cdot 2} + \sqrt[4]{256 \cdot 2}$
$$= -6 \cdot 2\sqrt[4]{2} + 4\sqrt[4]{2}$$
$$= -12\sqrt[4]{2} + 4\sqrt[4]{2} = -8\sqrt[4]{2}$$

80. $\dfrac{3}{\sqrt{16}} - \dfrac{\sqrt{5}}{2} = \dfrac{3}{4} - \dfrac{2\sqrt{5}}{2 \cdot 2} = \dfrac{3 - 2\sqrt{5}}{4}$

81. $\dfrac{4}{\sqrt{25}} + \dfrac{\sqrt{5}}{4} = \dfrac{4}{5} + \dfrac{\sqrt{5}}{4}$
$$= \dfrac{16}{20} + \dfrac{5\sqrt{5}}{20} = \dfrac{16 + 5\sqrt{5}}{20}$$

82. Add the measures of the sides.
$$P = a + b + c + d$$
$$P = 4\sqrt{8} + 6\sqrt{12} + 8\sqrt{2} + 3\sqrt{48}$$
$$= 4\sqrt{4 \cdot 2} + 6\sqrt{4 \cdot 3} + 8\sqrt{2} + 3\sqrt{16 \cdot 3}$$
$$= 4 \cdot 2\sqrt{2} + 6 \cdot 2\sqrt{3} + 8\sqrt{2} + 3 \cdot 4\sqrt{3}$$
$$= 8\sqrt{2} + 12\sqrt{3} + 8\sqrt{2} + 12\sqrt{3}$$
$$= 16\sqrt{2} + 24\sqrt{3}$$

The perimeter is $\left(16\sqrt{2} + 24\sqrt{3}\right)$ feet.

83. Add the measures of the sides.
$$P = a + b + c$$
$$P = 2\sqrt{27} + \sqrt{108} + \sqrt{50}$$
$$= 2\sqrt{9 \cdot 3} + \sqrt{36 \cdot 3} + \sqrt{25 \cdot 2}$$
$$= 2 \cdot 3\sqrt{3} + 6\sqrt{3} + 5\sqrt{2}$$
$$= 6\sqrt{3} + 6\sqrt{3} + 5\sqrt{2}$$
$$= 12\sqrt{3} + 5\sqrt{2}$$

The perimeter is $\left(12\sqrt{3} + 5\sqrt{2}\right)$ feet.

84. $\left(\sqrt{3} + 1\right)\left(\sqrt{3} - 2\right) = 3 - 2\sqrt{3} + \sqrt{3} - 2$
$$= 1 - \sqrt{3}$$

85. $\left(\sqrt{7}+\sqrt{5}\right)\left(\sqrt{7}-\sqrt{5}\right)=\left(\sqrt{7}\right)^2-\left(\sqrt{5}\right)^2$
$$= 7 - 5 = 2$$

86. $\left(3\sqrt{2}+1\right)\left(2\sqrt{2}-3\right)$
$$= 6\cdot 2 - 9\sqrt{2} + 2\sqrt{2} - 3$$
$$= 12 - 7\sqrt{2} - 3 = 9 - 7\sqrt{2}$$

87. $\left(\sqrt{13}-\sqrt{2}\right)^2$
$$= \left(\sqrt{13}\right)^2 - 2\cdot\sqrt{13}\cdot\sqrt{2} + \left(\sqrt{2}\right)^2$$
$$= 13 - 2\sqrt{26} + 2 = 15 - 2\sqrt{26}$$

88. $\left(\sqrt[3]{2}+3\right)\left(\sqrt[3]{4}-3\sqrt[3]{2}+9\right)$
$$= \sqrt[3]{2}\cdot\sqrt[3]{4} - \sqrt[3]{2}\cdot 3\sqrt[3]{2} + 9\sqrt[3]{2}$$
$$\quad + 3\sqrt[3]{4} - 3\cdot 3\sqrt[3]{2} + 27$$
$$= \sqrt[3]{8} - 3\sqrt[3]{4} + 9\sqrt[3]{2} + 3\sqrt[3]{4} - 9\sqrt[3]{2} + 27$$
$$= 2 + 27 = 29$$

89. $\left(\sqrt[3]{4y}-1\right)\left(\sqrt[3]{4y}+3\right)$
$$= \sqrt[3]{16y^2} + 3\sqrt[3]{4y} - \sqrt[3]{4y} - 3$$
$$= \sqrt[3]{8\cdot 2y^2} + 2\sqrt[3]{4y} - 3$$
$$= 2\sqrt[3]{2y^2} + 2\sqrt[3]{4y} - 3$$

90. Show that $15 - 2\sqrt{26} \neq 13\sqrt{26}$.

Find a calculator approximation of each term.
$$4.801960973 \neq 66.28725368$$
Therefore, $15 - 2\sqrt{26} \neq 13\sqrt{26}$.

91. Multiplying by $\sqrt[3]{6}$ still results in an expression with a radical in the denominator.
$$\frac{5\sqrt[3]{6}}{\sqrt[3]{6}\cdot\sqrt[3]{6}} = \frac{5\sqrt[3]{6}}{\sqrt[3]{36}}$$

To rationalize the denominator, multiply by $\sqrt[3]{6^2}$ or $\sqrt[3]{36}$.
$$\frac{5\sqrt[3]{6^2}}{\sqrt[3]{6}\cdot\sqrt[3]{6^2}} = \frac{5\sqrt[3]{36}}{\sqrt[3]{6^3}} = \frac{5\sqrt[3]{36}}{6}$$

92. $\dfrac{\sqrt{6}}{\sqrt{5}} = \dfrac{\sqrt{6}\cdot\sqrt{5}}{\sqrt{5}\cdot\sqrt{5}} = \dfrac{\sqrt{30}}{5}$

93. $\dfrac{-6\sqrt{3}}{\sqrt{2}} = \dfrac{-6\sqrt{3}\cdot\sqrt{2}}{\sqrt{2}\cdot\sqrt{2}} = \dfrac{-6\sqrt{6}}{2} = -3\sqrt{6}$

94. $\dfrac{3\sqrt{7p}}{\sqrt{y}} = \dfrac{3\sqrt{7p}\cdot\sqrt{y}}{\sqrt{y}\cdot\sqrt{y}} = \dfrac{3\sqrt{7py}}{y}$

95. $\sqrt{\dfrac{11}{8}} = \dfrac{\sqrt{11}}{\sqrt{8}} = \dfrac{\sqrt{11}}{\sqrt{4\cdot 2}} = \dfrac{\sqrt{11}}{2\sqrt{2}} = \dfrac{\sqrt{11}\cdot\sqrt{2}}{2\sqrt{2}\cdot\sqrt{2}}$
$$= \dfrac{\sqrt{22}}{2\cdot 2} = \dfrac{\sqrt{22}}{4}$$

96. $-\sqrt[3]{\dfrac{9}{25}} = -\dfrac{\sqrt[3]{9}}{\sqrt[3]{5^2}} = -\dfrac{\sqrt[3]{9}\cdot\sqrt[3]{5}}{\sqrt[3]{5^2}\cdot\sqrt[3]{5}}$
$$= -\dfrac{\sqrt[3]{45}}{\sqrt[3]{5^3}} = -\dfrac{\sqrt[3]{45}}{5}$$

97. $\sqrt[3]{\dfrac{108m^3}{n^5}} = \dfrac{\sqrt[3]{108m^3}}{\sqrt[3]{n^5}} = \dfrac{\sqrt[3]{27m^3\cdot 4}}{\sqrt[3]{n^3\cdot n^2}}$
$$= \dfrac{3m\sqrt[3]{4}}{n\sqrt[3]{n^2}} = \dfrac{3m\sqrt[3]{4}\cdot\sqrt[3]{n}}{n\sqrt[3]{n^2}\cdot\sqrt[3]{n}}$$
$$= \dfrac{3m\sqrt[3]{4n}}{n\cdot n} = \dfrac{3m\sqrt[3]{4n}}{n^2}$$

98. $\dfrac{1}{\sqrt{2}+\sqrt{7}}$

Multiply the numerator and denominator by the conjugate of the denominator, $\sqrt{2}-\sqrt{7}$.
$$= \dfrac{1\left(\sqrt{2}-\sqrt{7}\right)}{\left(\sqrt{2}+\sqrt{7}\right)\left(\sqrt{2}-\sqrt{7}\right)}$$
$$= \dfrac{\sqrt{2}-\sqrt{7}}{2-7} = \dfrac{\sqrt{2}-\sqrt{7}}{-5}$$

99. $\dfrac{-5}{\sqrt{6}-3}$

Multiply the numerator and denominator by the conjugate of the denominator, $\sqrt{6}+3$.
$$= \dfrac{-5\left(\sqrt{6}+3\right)}{\left(\sqrt{6}-3\right)\left(\sqrt{6}+3\right)} = \dfrac{-5\left(\sqrt{6}+3\right)}{6-9}$$
$$= \dfrac{-5\left(\sqrt{6}+3\right)}{-3} = \dfrac{5\left(\sqrt{6}+3\right)}{3}$$

100. $\dfrac{2-2\sqrt{5}}{8} = \dfrac{2\left(1-\sqrt{5}\right)}{2\cdot 4} = \dfrac{1-\sqrt{5}}{4}$

101. $\dfrac{4-8\sqrt{8}}{12} = \dfrac{4\left(1-2\sqrt{8}\right)}{3\cdot 4} = \dfrac{1-2\sqrt{8}}{3}$
$$= \dfrac{1-2\sqrt{4\cdot 2}}{3} = \dfrac{1-4\sqrt{2}}{3}$$

102. $\dfrac{-18+\sqrt{27}}{6} = \dfrac{-18+\sqrt{9}\cdot\sqrt{3}}{6} = \dfrac{-18+3\sqrt{3}}{6}$
$$= \dfrac{3\left(-6+\sqrt{3}\right)}{3\cdot 2} = \dfrac{-6+\sqrt{3}}{2}$$

103. $\sqrt{8x+9} = 5$
$$\left(\sqrt{8x+9}\right)^2 = 5^2 \quad \textit{Square.}$$
$$8x + 9 = 25$$
$$8x = 16$$
$$x = 2$$

Check $x = 2$: $\sqrt{25} = 5$ *True*
The solution set is $\{2\}$.

104. $\sqrt{2z-3} - 3 = 0$

$\qquad \sqrt{2z-3} = 3 \qquad$ *Isolate.*

$\qquad \left(\sqrt{2z-3}\right)^2 = 3^2 \qquad$ *Square.*

$\qquad 2z - 3 = 9$

$\qquad 2z = 12$

$\qquad z = 6$

Check $z = 6$: $\sqrt{9} - 3 = 0$ *True*

The solution set is $\{6\}$.

105. $\sqrt{3m+1} - 2 = -3$

$\qquad \sqrt{3m+1} = -1 \qquad$ *Isolate.*

This equation has no solution, because $\sqrt{3m+1}$ cannot be negative.

The solution set is $\emptyset$.

106. $\sqrt{7z+1} = z + 1$

$\qquad \left(\sqrt{7z+1}\right)^2 = (z+1)^2 \qquad$ *Square.*

$\qquad 7z + 1 = z^2 + 2z + 1$

$\qquad 0 = z^2 - 5z$

$\qquad 0 = z(z-5)$

$\qquad z = 0 \quad$ or $\quad z = 5$

Check $z = 0$: $\sqrt{1} = 1$ *True*

Check $z = 5$: $\sqrt{36} = 6$ *True*

The solution set is $\{0, 5\}$.

107. $3\sqrt{m} = \sqrt{10m - 9}$

$\qquad \left(3\sqrt{m}\right)^2 = \left(\sqrt{10m-9}\right)^2 \qquad$ *Square.*

$\qquad 9m = 10m - 9$

$\qquad 9 = m$

Check $m = 9$: $3\sqrt{9} = \sqrt{81}$ *True*

The solution set is $\{9\}$.

108. $\sqrt{p^2 + 3p + 7} = p + 2$

$\qquad \left(\sqrt{p^2+3p+7}\right)^2 = (p+2)^2 \qquad$ *Square.*

$\qquad p^2 + 3p + 7 = p^2 + 4p + 4$

$\qquad 3 = p$

Check $p = 3$: $\sqrt{25} = 5$ *True*

The solution set is $\{3\}$.

109. $\sqrt{a+2} - \sqrt{a-3} = 1$

Get one radical on each side of the equals sign.

$\qquad \sqrt{a+2} = 1 + \sqrt{a-3}$

Square both sides.

$\qquad \left(\sqrt{a+2}\right)^2 = \left(1 + \sqrt{a-3}\right)^2$

$\qquad a + 2 = 1 + 2\sqrt{a-3} + a - 3$

$\qquad 4 = 2\sqrt{a-3}$

$\qquad 2 = \sqrt{a-3}$

Square both sides again.

$\qquad 2^2 = \left(\sqrt{a-3}\right)^2$

$\qquad 4 = a - 3$

$\qquad 7 = a$

Check $a = 7$: $\sqrt{9} - \sqrt{4} = 1$ *True*

The solution set is $\{7\}$.

110. $\sqrt[3]{5m-1} = \sqrt[3]{3m-2}$

Cube both sides.

$\qquad \left(\sqrt[3]{5m-1}\right)^3 = \left(\sqrt[3]{3m-2}\right)^3$

$\qquad 5m - 1 = 3m - 2$

$\qquad 2m = -1$

$\qquad m = -\frac{1}{2}$

Check $m = -\frac{1}{2}$: $\sqrt[3]{-\frac{7}{2}} = \sqrt[3]{-\frac{7}{2}}$ *True*

The solution set is $\left\{-\frac{1}{2}\right\}$.

111. $\sqrt[3]{2x^2 + 3x - 7} = \sqrt[3]{2x^2 + 4x + 6}$

Cube both sides.

$\qquad \left(\sqrt[3]{2x^2+3x-7}\right)^3 = \left(\sqrt[3]{2x^2+4x+6}\right)^3$

$\qquad 2x^2 + 3x - 7 = 2x^2 + 4x + 6$

$\qquad -13 = x$

Check $x = -13$: $\sqrt[3]{292} = \sqrt[3]{292}$ *True*

The solution set is $\{-13\}$.

112. $\sqrt[3]{3y^2 - 4y + 6} = \sqrt[3]{3y^2 - 2y + 8}$

$\qquad \left(\sqrt[3]{3y^2-4y+6}\right)^3 = \left(\sqrt[3]{3y^2-2y+8}\right)^3$

$\qquad 3y^2 - 4y + 6 = 3y^2 - 2y + 8$

$\qquad -2 = 2y$

$\qquad -1 = y$

Check $y = -1$: $\sqrt[3]{13} = \sqrt[3]{13}$ *True*

The solution set is $\{-1\}$.

113. $\sqrt[3]{1 - 2k} - \sqrt[3]{-k - 13} = 0$

$\qquad \sqrt[3]{1-2k} = \sqrt[3]{-k-13}$

Cube both sides.

$\qquad \left(\sqrt[3]{1-2k}\right)^3 = \left(\sqrt[3]{-k-13}\right)^3$

$\qquad 1 - 2k = -k - 13$

$\qquad 14 = k$

Check $k = 14$: $\sqrt[3]{-27} - \sqrt[3]{-27} = 0$ *True*

The solution set is $\{14\}$.

114. $\sqrt[3]{11 - 2t} - \sqrt[3]{-1 - 5t} = 0$

$\qquad \sqrt[3]{11-2t} = \sqrt[3]{-1-5t}$

$\qquad \left(\sqrt[3]{11-2t}\right)^3 = \left(\sqrt[3]{-1-5t}\right)^3$

$\qquad 11 - 2t = -1 - 5t$

$\qquad 3t = -12$

$\qquad t = -4$

Check $t = -4$: $\sqrt[3]{19} - \sqrt[3]{19} = 0$ *True*

The solution set is $\{-4\}$.

115. $\sqrt[4]{x - 1} + 2 = 0$

$\qquad \sqrt[4]{x-1} = -2$

This equation has no solution, because $\sqrt[4]{x-1}$ cannot be negative.

The solution set is $\emptyset$.

116. $\sqrt[4]{2k + 3} + 1 = 0$

$\qquad \sqrt[4]{2k + 3} = -1$

This equation has no solution, because $\sqrt[4]{2k + 3}$ cannot be negative.

The solution set is $\emptyset$.

117. $\qquad \sqrt[4]{x + 7} = \sqrt[4]{2x}$

Raise each side to the fourth power.

$\left(\sqrt[4]{x + 7} \right)^4 = \left(\sqrt[4]{2x} \right)^4$

$\qquad x + 7 = 2x$

$\qquad 7 = x$

Check $x = 7$: $\sqrt[4]{14} = \sqrt[4]{14}$ *True*

The solution set is $\{7\}$.

118. $\qquad \sqrt[4]{x + 8} = \sqrt[4]{3x}$

Raise each side to the fourth power.

$\left(\sqrt[4]{x + 8} \right)^4 = \left(\sqrt[4]{3x} \right)^4$

$\qquad x + 8 = 3x$

$\qquad 8 = 2x$

$\qquad 4 = x$

Check $x = 4$: $\sqrt[4]{12} = \sqrt[4]{12}$ *True*

The solution set is $\{4\}$.

119. (a) Solve $L = \sqrt{H^2 + W^2}$ for H.

$\qquad L^2 = H^2 + W^2$

$\qquad L^2 - W^2 = H^2$

$\qquad \sqrt{L^2 - W^2} = H$

(b) Substitute 12 for L and 9 for W in $H = \sqrt{L^2 - W^2}$.

$\qquad H = \sqrt{12^2 - 9^2}$

$\qquad = \sqrt{144 - 81} = \sqrt{63}$

To the nearest tenth of a foot, the height is approximately 7.9 feet.

120. $\sqrt{-25} = i\sqrt{25} = 5i$

121. $\sqrt{-200} = i\sqrt{100 \cdot 2} = 10i\sqrt{2}$

122. If a is a positive real number, then $-a$ is negative. So, $\sqrt{-a}$ is not a real number. Therefore, $-\sqrt{-a}$ is not a real number either.

123. $(-2 + 5i) + (-8 - 7i)$

$\qquad = [-2 + (-8)] + [5 + (-7)]i$

$\qquad = -10 - 2i$

124. $(5 + 4i) - (-9 - 3i)$

$\qquad = [(5 - (-9)] + [(4 - (-3)]i$

$\qquad = 14 + 7i$

125. $\sqrt{-5} \cdot \sqrt{-7} = i\sqrt{5} \cdot i\sqrt{7} = i^2\sqrt{35}$

$\qquad = -1\left(\sqrt{35} \right) = -\sqrt{35}$

126. $\sqrt{-25} \cdot \sqrt{-81} = 5i \cdot 9i = 45i^2$

$\qquad = 45(-1) = -45$

127. $\dfrac{\sqrt{-72}}{\sqrt{-8}} = \dfrac{i\sqrt{72}}{i\sqrt{8}} = \sqrt{\dfrac{72}{8}} = \sqrt{9} = 3$

128. $(2 + 3i)(1 - i) = 2 - 2i + 3i - 3i^2$

$\qquad = 2 + i - 3(-1)$

$\qquad = 2 + i + 3$

$\qquad = 5 + i$

129. $(6 - 2i)^2 = 6^2 - 2 \cdot 6 \cdot 2i + (2i)^2$

$\qquad = 36 - 24i + 4i^2$

$\qquad = 36 - 24i + 4(-1)$

$\qquad = 36 - 24i - 4$

$\qquad = 32 - 24i$

130. $\dfrac{3 - i}{2 + i}$

Multiply by the conjugate of the denominator, $2 - i$.

$\qquad = \dfrac{(3 - i)(2 - i)}{(2 + i)(2 - i)}$

$\qquad = \dfrac{6 - 3i - 2i + i^2}{4 - i^2}$

$\qquad = \dfrac{6 - 5i - 1}{4 - (-1)} = \dfrac{5 - 5i}{5}$

$\qquad = \dfrac{5(1 - i)}{5} = 1 - i$

131. $\dfrac{5 + 14i}{2 + 3i}$

Multiply by the conjugate of the denominator, $2 - 3i$.

$\qquad = \dfrac{(5 + 14i)(2 - 3i)}{(2 + 3i)(2 - 3i)}$

$\qquad = \dfrac{10 - 15i + 28i - 42i^2}{4 - 9i^2}$

$\qquad = \dfrac{10 + 13i - 42(-1)}{4 - 9(-1)} = \dfrac{52 + 13i}{13}$

$\qquad = \dfrac{13(4 + i)}{13} = 4 + i$

132. $i^{11} = i^8 \cdot i^3 = \left(i^4 \right)^2 \cdot i^2 \cdot i$

$\qquad = 1^2 \cdot (-1) \cdot i = -i$

133. $i^{36} = \left(i^4 \right)^9 = 1^9 = 1$

134. $i^{-10} = \dfrac{1}{i^{10}} = \dfrac{1}{i^8 \cdot i^2} = \dfrac{1}{\left(i^4 \right)^2 \cdot (-1)}$

$\qquad = \dfrac{1}{1^2 \cdot (-1)} = \dfrac{1}{-1} = -1$

Another method:

$i^{-10} = i^{-10} \cdot i^{12} = i^2 = -1$

Note that $i^{12} = \left(i^4 \right)^3 = 1^3 = 1$.

135. $i^{-8} = \dfrac{1}{i^8} = \dfrac{1}{\left(i^4 \right)^2} = \dfrac{1}{1^2} = \dfrac{1}{1} = 1$

136. [10.1] $-\sqrt[4]{256} = -\left(\sqrt[4]{4^4}\right) = -(4) = -4$

137. [10.2] $1000^{-2/3} = \dfrac{1}{1000^{2/3}} = \dfrac{1}{\left[(10^3)^{1/3}\right]^2}$

$$= \dfrac{1}{10^2} = \dfrac{1}{100}$$

138. [10.2] $\dfrac{z^{-1/5} \cdot z^{3/10}}{z^{7/10}} = z^{-2/10 + 3/10 - 7/10}$

$$= z^{-6/10} = \dfrac{1}{z^{6/10}} = \dfrac{1}{z^{3/5}}$$

139. [10.2] $\sqrt[4]{k^{24}} = k^{24/4} = k^6$

140. [10.3] $\sqrt[3]{54z^9t^8} = \sqrt[3]{27z^9t^6 \cdot 2t^2} = 3z^3t^2\sqrt[3]{2t^2}$

141. [10.4]
$$-5\sqrt{18} + 12\sqrt{72} = -5\sqrt{9 \cdot 2} + 12\sqrt{36 \cdot 2}$$
$$= -5 \cdot 3\sqrt{2} + 12 \cdot 6\sqrt{2}$$
$$= -15\sqrt{2} + 72\sqrt{2} = 57\sqrt{2}$$

142. [10.5] $\dfrac{-1}{\sqrt{12}} = \dfrac{-1}{\sqrt{4 \cdot 3}} = \dfrac{-1}{2\sqrt{3}} = \dfrac{-1 \cdot \sqrt{3}}{2\sqrt{3} \cdot \sqrt{3}}$

$$= \dfrac{-\sqrt{3}}{2 \cdot 3} = \dfrac{-\sqrt{3}}{6}$$

143. [10.5] $\sqrt[3]{\dfrac{12}{25}} = \dfrac{\sqrt[3]{12}}{\sqrt[3]{25}} = \dfrac{\sqrt[3]{12}}{\sqrt[3]{5^2}}$

$$= \dfrac{\sqrt[3]{12} \cdot \sqrt[3]{5}}{\sqrt[3]{5^2} \cdot \sqrt[3]{5}} = \dfrac{\sqrt[3]{60}}{\sqrt[3]{5^3}} = \dfrac{\sqrt[3]{60}}{5}$$

144. [10.7] $i^{-1000} = \dfrac{1}{i^{1000}} = \dfrac{1}{(i^4)^{250}}$

$$= \dfrac{1}{1^{250}} = \dfrac{1}{1} = 1$$

145. [10.7] $\sqrt{-49} = i\sqrt{49} = 7i$

146. [10.7] $(4 - 9i) + (-1 + 2i)$
$$= (4 - 1) + (-9 + 2)i$$
$$= 3 - 7i$$

147. [10.7] $\dfrac{\sqrt{50}}{\sqrt{-2}} = \dfrac{\sqrt{25 \cdot 2}}{i\sqrt{2}} = \dfrac{5\sqrt{2}}{i\sqrt{2}} = \dfrac{5}{i}$

The conjugate of i is $-i$.
$$\dfrac{5(-i)}{i(-i)} = \dfrac{-5i}{-i^2} = \dfrac{-5i}{-(-1)} = -5i$$

148. [10.5] $\dfrac{3 + \sqrt{54}}{6} = \dfrac{3 + \sqrt{9 \cdot 6}}{6} = \dfrac{3 + 3\sqrt{6}}{6}$

$$= \dfrac{3(1 + \sqrt{6})}{2 \cdot 3} = \dfrac{1 + \sqrt{6}}{2}$$

149. [10.7] $(3 + 2i)^2 = 3^2 + 2 \cdot 3 \cdot 2i + (2i)^2$
$$= 9 + 12i + 4i^2$$
$$= 9 + 12i + 4(-1)$$
$$= 5 + 12i$$

150. [10.4] $8\sqrt[3]{x^3y^2} - 2x\sqrt[3]{y^2} = 8x\sqrt[3]{y^2} - 2x\sqrt[3]{y^2}$
$$= 6x\sqrt[3]{y^2}$$

151. [10.4] $9\sqrt{5} - 4\sqrt{15}$ cannot be simplified further.

152. [10.5] $\left(\sqrt{5} - \sqrt{3}\right)\left(\sqrt{7} + \sqrt{3}\right)$
$$= \sqrt{35} + \sqrt{15} - \sqrt{21} - 3$$

153. [10.6] $\sqrt{x + 4} = x - 2$
$$\left(\sqrt{x + 4}\right)^2 = (x - 2)^2 \qquad \textit{Square.}$$
$$x + 4 = x^2 - 4x + 4$$
$$0 = x^2 - 5x$$
$$0 = x(x - 5)$$

$$x = 0 \quad \text{or} \quad x - 5 = 0$$
$$x = 5$$

Check $x = 0$: $\quad \sqrt{4} = -2 \quad \textit{False}$
Check $x = 5$: $\quad \sqrt{9} = 3 \quad \textit{True}$
The solution set is $\{5\}$.

154. [10.6] $\sqrt[3]{2x - 9} = \sqrt[3]{5x + 3}$
Cube both sides.
$$\left(\sqrt[3]{2x - 9}\right)^3 = \left(\sqrt[3]{5x + 3}\right)^3$$
$$2x - 9 = 5x + 3$$
$$-3x = 12$$
$$x = -4$$
Check $x = -4$: $\sqrt[3]{-17} = \sqrt[3]{-17} \quad \textit{True}$
The solution set is $\{-4\}$.

155. [10.6] $\sqrt{6 + 2x} - 1 = \sqrt{7 - 2x}$
Square both sides.
$$\left(\sqrt{6 + 2x} - 1\right)^2 = \left(\sqrt{7 - 2x}\right)^2$$
$$6 + 2x - 2\sqrt{6 + 2x} + 1 = 7 - 2x$$
$$4x = 2\sqrt{6 + 2x}$$
$$2x = \sqrt{6 + 2x}$$
Square both sides again.
$$(2x)^2 = \left(\sqrt{6 + 2x}\right)^2$$
$$4x^2 = 6 + 2x$$
$$4x^2 - 2x - 6 = 0$$
$$2x^2 - x - 3 = 0$$
$$(2x - 3)(x + 1) = 0$$

$$2x - 3 = 0 \quad \text{or} \quad x + 1 = 0$$
$$x = \tfrac{3}{2} \quad \text{or} \quad x = -1$$

Check $x = \tfrac{3}{2}$: $\quad \sqrt{9} - 1 = \sqrt{4} \quad \textit{True}$
Check $x = -1$: $\quad \sqrt{4} - 1 = \sqrt{9} \quad \textit{False}$
The solution set is $\left\{\tfrac{3}{2}\right\}$.

156. [10.6] $\sqrt{7x + 11} - 5 = 0$

$$\sqrt{7x + 11} = 5$$
$$\left(\sqrt{7x + 11}\right)^2 = 5^2$$
$$7x + 11 = 25$$
$$7x = 14$$
$$x = 2$$

Check $x = 2$: $\sqrt{25} - 5 = 0$ *True*

The solution set is $\{2\}$.

157. [10.6] $\sqrt{6x + 2} - \sqrt{5x + 3} = 0$

Get one radical on each side of the equals sign.
$$\sqrt{6x + 2} = \sqrt{5x + 3}$$

Square both sides.
$$\left(\sqrt{6x + 2}\right)^2 = \left(\sqrt{5x + 3}\right)^2$$
$$6x + 2 = 5x + 3$$
$$x = 1$$

Check $x = 1$: $\sqrt{8} - \sqrt{8} = 0$ *True*

The solution set is $\{1\}$.

158. [10.6] $\sqrt{3 + 5x} - \sqrt{x + 11} = 0$

$$\sqrt{3 + 5x} = \sqrt{x + 11}$$
$$\left(\sqrt{3 + 5x}\right)^2 = \left(\sqrt{x + 11}\right)^2$$
$$3 + 5x = x + 11$$
$$4x = 8$$
$$x = 2$$

Check $x = 2$: $\sqrt{13} - \sqrt{13} = 0$ *True*

The solution set is $\{2\}$.

159. [10.6] $3\sqrt{x} = \sqrt{8x + 9}$

$$\left(3\sqrt{x}\right)^2 = \left(\sqrt{8x + 9}\right)^2 \quad \text{Square.}$$
$$9x = 8x + 9$$
$$x = 9$$

Check $x = 9$: $3\sqrt{9} = \sqrt{81}$ *True*

The solution set is $\{9\}$.

160. [10.6] $6\sqrt{p} = \sqrt{30p + 24}$

$$\left(6\sqrt{p}\right)^2 = \left(\sqrt{30p + 24}\right)^2$$
$$36p = 30p + 24$$
$$6p = 24$$
$$p = 4$$

Check $p = 4$: $6\sqrt{4} = \sqrt{144}$ *True*

The solution set is $\{4\}$.

161. [10.6] $\sqrt{11 + 2x} + 1 = \sqrt{5x + 1}$

Square both sides.
$$\left(\sqrt{11 + 2x} + 1\right)^2 = \left(\sqrt{5x + 1}\right)^2$$
$$\left(\sqrt{11 + 2x}\right)^2 + 2 \cdot \sqrt{11 + 2x} \cdot 1 + 1$$
$$= 5x + 1$$
$$11 + 2x + 2\sqrt{11 + 2x} + 1 = 5x + 1$$

Isolate the remaining radical.
$$2\sqrt{11 + 2x} = 3x - 11$$

Square both sides again.
$$\left(2\sqrt{11 + 2x}\right)^2 = (3x - 11)^2$$
$$4(11 + 2x) = 9x^2 - 66x + 121$$
$$44 + 8x = 9x^2 - 66x + 121$$
$$0 = 9x^2 - 74x + 77$$
$$0 = (9x - 11)(x - 7)$$

$$9x - 11 = 0 \quad \text{or} \quad x - 7 = 0$$
$$x = \tfrac{11}{9} \quad \text{or} \quad x = 7$$

Check $x = \tfrac{11}{9}$: $\sqrt{\tfrac{121}{9}} + 1 = \sqrt{\tfrac{64}{9}}$ *False*

Check $x = 7$: $\sqrt{25} + 1 = \sqrt{36}$ *True*

The solution set is $\{7\}$.

162. [10.6]

$$\sqrt{5x + 6} - \sqrt{x + 3} = 3$$

Get one radical on each side.
$$\sqrt{5x + 6} = 3 + \sqrt{x + 3}$$
$$\left(\sqrt{5x + 6}\right)^2 = \left(3 + \sqrt{x + 3}\right)^2$$
$$5x + 6 = 9 + 6\sqrt{x + 3} + x + 3$$

Isolate the remaining radical.
$$4x - 6 = 6\sqrt{x + 3}$$
$$2x - 3 = 3\sqrt{x + 3}$$
$$(2x - 3)^2 = \left(3\sqrt{x + 3}\right)^2$$
$$4x^2 - 12x + 9 = 9(x + 3)$$
$$4x^2 - 12x + 9 = 9x + 27$$
$$4x^2 - 21x - 18 = 0$$
$$(4x + 3)(x - 6) = 0$$

$$4x + 3 = 0 \quad \text{or} \quad x - 6 = 0$$
$$x = -\tfrac{3}{4} \qquad\qquad x = 6$$

Check $x = -\tfrac{3}{4}$: $\sqrt{\tfrac{9}{4}} - \sqrt{\tfrac{9}{4}} = 3$ *False*

Check $x = 6$: $\sqrt{36} - \sqrt{9} = 3$ *True*

The solution set is $\{6\}$.

Chapter 10 Test

1. $\sqrt{841} = 29$, because $29^2 = 841$, so $-\sqrt{841} = -29$.

2. $\sqrt[3]{-512} = -8$, because $(-8)^3 = -512$.

3. $125^{1/3} = \sqrt[3]{125} = 5$, because $5^3 = 125$.

4. $\sqrt{146.25} \approx \sqrt{144} = 12$, because $12^2 = 144$, so choice **C** is the best estimate.

5. $\sqrt{478} \approx 21.863$

6. $\sqrt[3]{-832} \approx -9.405$

7. $f(x) = \sqrt{x + 6}$

For the radicand to be nonnegative, we must have

$$x + 6 \geq 0 \quad \text{or} \quad x \geq -6.$$

Thus, the domain is $[-6, \infty)$.
The function values are positive or zero (the result of the radical), so the range is $[0, \infty)$.

x	$f(x) = \sqrt{x + 6}$
-6	$\sqrt{-6 + 6} = 0$
-5	$\sqrt{-5 + 6} = 1$
-2	$\sqrt{-2 + 6} = 2$

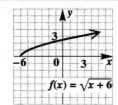

8. $\left(\dfrac{16}{25}\right)^{-3/2} = \left(\dfrac{25}{16}\right)^{3/2} = \dfrac{25^{3/2}}{16^{3/2}}$

$$= \dfrac{\left[(5^2)^{1/2}\right]^3}{\left[(4^2)^{1/2}\right]^3} = \dfrac{5^3}{4^3} = \dfrac{125}{64}$$

9. $(-64)^{-4/3} = \dfrac{1}{(-64)^{4/3}} = \dfrac{1}{\left(\left[(-4)^3\right]^{1/3}\right)^4}$

$$= \dfrac{1}{(-4)^4} = \dfrac{1}{256}$$

10. $\dfrac{3^{2/5}x^{-1/4}y^{2/5}}{3^{-8/5}x^{7/4}y^{1/10}}$

$$= 3^{2/5 - (-8/5)}x^{-1/4 - 7/4}y^{2/5 - 1/10}$$

$$= 3^{10/5}x^{-8/4}y^{4/10 - 1/10}$$

$$= 3^2 x^{-2} y^{3/10} = \dfrac{9y^{3/10}}{x^2}$$

11. $\left(\dfrac{x^{-4}y^{-6}}{x^{-2}y^3}\right)^{-2/3} = \left(\dfrac{x^2}{x^4y^6y^3}\right)^{-2/3}$

$$= \left(\dfrac{1}{x^2y^9}\right)^{-2/3} = \left(x^2y^9\right)^{2/3}$$

$$= \left(x^2\right)^{2/3}\left(y^9\right)^{2/3} = x^{4/3}y^6$$

12. $7^{3/4} \cdot 7^{-1/4} = 7^{3/4 + (-1/4)}$

$$= 7^{1/2}, \quad \text{or} \quad \sqrt{7}$$

13. $\sqrt[3]{a^4} \cdot \sqrt[3]{a^7} = a^{4/3} \cdot a^{7/3} = a^{4/3 + 7/3} = a^{11/3}$

$$= \sqrt[3]{a^{11}} = \sqrt[3]{(a^3)^3 \cdot a^2} = a^3 \sqrt[3]{a^2}$$

14. $a^2 + b^2 = c^2$

$12^2 + b^2 = 17^2 \qquad \textit{Let } a = 12,\ c = 17.$

$144 + b^2 = 289$

$b^2 = 145$

$b = \sqrt{145}$

15. $(-4, 2)$ and $(2, 10)$

$$d = \sqrt{(x_2 - x_1)^2 + (y_2 - y_1)^2}$$

$$= \sqrt{[2 - (-4)]^2 + (10 - 2)^2}$$

$$= \sqrt{6^2 + 8^2} = \sqrt{36 + 64} = \sqrt{100} = 10$$

16. $\sqrt{54x^5y^6} = \sqrt{9x^4y^6 \cdot 6x} = 3x^2y^3\sqrt{6x}$

17. $\sqrt[4]{32a^7b^{13}} = \sqrt[4]{16a^4b^{12} \cdot 2a^3b}$

$$= 2ab^3\sqrt[4]{2a^3b}$$

18. $\sqrt{2} \cdot \sqrt[3]{5} = 2^{1/2} \cdot 5^{1/3} = 2^{3/6} \cdot 5^{2/6}$

$$= \left(2^3 \cdot 5^2\right)^{1/6} = \sqrt[6]{2^3 \cdot 5^2}$$

$$= \sqrt[6]{8 \cdot 25} = \sqrt[6]{200}$$

19. $3\sqrt{20} - 5\sqrt{80} + 4\sqrt{500}$

$$= 3\sqrt{4 \cdot 5} - 5\sqrt{16 \cdot 5} + 4\sqrt{100 \cdot 5}$$

$$= 3 \cdot 2\sqrt{5} - 5 \cdot 4\sqrt{5} + 4 \cdot 10\sqrt{5}$$

$$= 6\sqrt{5} - 20\sqrt{5} + 40\sqrt{5} = 26\sqrt{5}$$

20. $\sqrt[3]{16t^3s^5} - \sqrt[3]{54t^6s^2}$

$$= \sqrt[3]{8t^3s^3} \cdot \sqrt[3]{2s^2} - \sqrt[3]{27t^6} \cdot \sqrt[3]{2s^2}$$

$$= 2ts\sqrt[3]{2s^2} - 3t^2\sqrt[3]{2s^2}$$

$$= \left(2ts - 3t^2\right)\sqrt[3]{2s^2}, \quad \text{or} \quad t(2s - 3t)\sqrt[3]{2s^2}$$

21. $\left(7\sqrt{5} + 4\right)\left(2\sqrt{5} - 1\right)$

$$= 14 \cdot 5 - 7\sqrt{5} + 8\sqrt{5} - 4$$

$$= 70 + \sqrt{5} - 4 = 66 + \sqrt{5}$$

22. $\left(\sqrt{3} - 2\sqrt{5}\right)^2$

$$= \left(\sqrt{3}\right)^2 - 2 \cdot \sqrt{3} \cdot 2\sqrt{5} + \left(2\sqrt{5}\right)^2$$

$$= 3 - 4\sqrt{15} + 4 \cdot 5$$

$$= 3 - 4\sqrt{15} + 20 = 23 - 4\sqrt{15}$$

23. $\dfrac{-5}{\sqrt{40}} = \dfrac{-5}{\sqrt{4 \cdot 10}} = \dfrac{-5}{2\sqrt{10}} = \dfrac{-5 \cdot \sqrt{10}}{2\sqrt{10} \cdot \sqrt{10}}$

$$= \dfrac{-5\sqrt{10}}{2 \cdot 10} = \dfrac{-5\sqrt{10}}{20} = -\dfrac{\sqrt{10}}{4}$$

24. $\dfrac{2}{\sqrt[3]{5}} = \dfrac{2 \cdot \sqrt[3]{5^2}}{\sqrt[3]{5}\sqrt[3]{5^2}} = \dfrac{2\sqrt[3]{25}}{5}$

25. $\dfrac{-4}{\sqrt{7} + \sqrt{5}}$

Multiply the numerator and denominator by the conjugate of the denominator, $\sqrt{7} - \sqrt{5}$.

$$= \dfrac{-4\left(\sqrt{7} - \sqrt{5}\right)}{\left(\sqrt{7} + \sqrt{5}\right)\left(\sqrt{7} - \sqrt{5}\right)}$$

$$= \dfrac{-4\left(\sqrt{7} - \sqrt{5}\right)}{7 - 5}$$

$$= \dfrac{-4\left(\sqrt{7} - \sqrt{5}\right)}{2} = -2\left(\sqrt{7} - \sqrt{5}\right)$$

26. $\dfrac{6+\sqrt{24}}{2} = \dfrac{6+\sqrt{4\cdot 6}}{2} = \dfrac{6+2\sqrt{6}}{2}$

$\qquad = \dfrac{2(3+\sqrt{6})}{2} = 3+\sqrt{6}$

27. **(a)** Substitute 50 for V_0, 0.01 for k, and 30 for T in the formula.

$$V = \dfrac{V_0}{\sqrt{1-kT}}$$

$$V = \dfrac{50}{\sqrt{1-(0.01)(30)}}$$

$$= \dfrac{50}{\sqrt{1-0.3}} = \dfrac{50}{\sqrt{0.7}} \approx 59.8$$

The velocity is about 59.8.

(b) $\qquad V = \dfrac{V_0}{\sqrt{1-kT}}$

$\qquad\qquad V^2 = \dfrac{V_0^2}{1-kT}$ *Square.*

$\qquad V^2(1-kT) = V_0^2$

$\qquad V^2 - V^2kT = V_0^2$

$\qquad V^2 - V_0^2 = V^2kT$

$\qquad\qquad T = \dfrac{V^2 - V_0^2}{V^2 k}$

or $\qquad\qquad T = \dfrac{V_0^2 - V^2}{-V^2 k}$

28. $\sqrt[3]{5x} = \sqrt[3]{2x-3}$

$\left(\sqrt[3]{5x}\right)^3 = \left(\sqrt[3]{2x-3}\right)^3$

$\qquad 5x = 2x-3$

$\qquad 3x = -3$

$\qquad x = -1$

Check $x = -1$: $\sqrt[3]{-5} = \sqrt[3]{-5}$ *True*

The solution set is $\{-1\}$.

29. $x + \sqrt{x+6} = 9 - x$

Isolate the radical.

$\qquad \sqrt{x+6} = 9 - 2x$

$\qquad \left(\sqrt{x+6}\right)^2 = (9-2x)^2$ *Square.*

$\qquad x + 6 = 81 - 36x + 4x^2$

$\qquad 0 = 4x^2 - 37x + 75$

$\qquad 0 = (x-3)(4x-25)$

$x - 3 = 0$ or $4x - 25 = 0$

$\qquad x = 3 \qquad\qquad x = \frac{25}{4}$

Check $x = 3$: $3 + \sqrt{9} = 6$ *True*

Check $x = \frac{25}{4}$: $\frac{25}{4} + \sqrt{\frac{49}{4}} = \frac{11}{4}$ *False*

The solution set is $\{3\}$.

30. $\sqrt{x+4} - \sqrt{1-x} = -1$

$\qquad \sqrt{x+4} = -1 + \sqrt{1-x}$

$\qquad \left(\sqrt{x+4}\right)^2 = \left(-1+\sqrt{1-x}\right)^2$

$\qquad x + 4 = 1 - 2\sqrt{1-x} + 1 - x$

$\qquad 2x + 2 = -2\sqrt{1-x}$

$\qquad x + 1 = -\sqrt{1-x}$

$\qquad (x+1)^2 = \left(-\sqrt{1-x}\right)^2$

$\qquad x^2 + 2x + 1 = 1 - x$

$\qquad x^2 + 3x = 0$

$\qquad x(x+3) = 0$

$x = 0$ or $x + 3 = 0$

$\qquad\qquad\qquad\quad x = -3$

Check $x = -3$: $1 - 2 = -1$ *True*

Check $x = 0$: $2 - 1 = -1$ *False*

The solution set is $\{-3\}$.

31. $(-2+5i) - (3+6i) - 7i$

$= (-2-3) + (5-6-7)i$

$= -5 - 8i$

32. $(1+5i)(3+i)$

$= 3 + i + 15i + 5i^2$

$= 3 + 16i + 5(-1)$

$= -2 + 16i$

33. $\dfrac{7+i}{1-i}$

Multiply the numerator and denominator by the conjugate of the denominator, $1 + i$.

$= \dfrac{(7+i)(1+i)}{(1-i)(1+i)}$

$= \dfrac{7 + 7i + i + i^2}{1 - i^2}$

$= \dfrac{7 + 8i - 1}{1 - (-1)}$

$= \dfrac{6 + 8i}{2} = \dfrac{2(3+4i)}{2} = 3 + 4i$

34. $i^{37} = i^{36} \cdot i = \left(i^4\right)^9 \cdot i = 1^9 \cdot i = 1 \cdot i = i$

35. **(a)** $i^2 = -1$ is *true.*

(b) $i = \sqrt{-1}$ is *true.*

(c) $i = -1$ is *false;* $i = \sqrt{-1}$.

(d) $\sqrt{-3} = i\sqrt{3}$ is *true.*

Cumulative Review Exercises
(Chapters 1–10)

In Exercises 1 and 2, $a = -3$, $b = 5$, and $c = -4$.

1. $\left|2a^2 - 3b + c\right|$

$= \left|2(-3)^2 - 3(5) + (-4)\right|$

$= \left|2(9) - 15 - 4\right|$

$= \left|18 - 15 - 4\right| = \left|-1\right| = 1$

2. $\dfrac{(a+b)(a+c)}{3b-6}$

$= \dfrac{(-3+5)[(-3+(-4)]}{3(5)-6}$

$= \dfrac{(2)(-7)}{15-6} = -\dfrac{14}{9}$

3. $3(x+2) - 4(2x+3) = -3x + 2$

$3x + 6 - 8x - 12 = -3x + 2$

$-5x - 6 = -3x + 2$

$-2x = 8$

$x = -4$

Check $x = -4$: $-6 + 20 = 14$ *True*

The solution set is $\{-4\}$.

4. $\frac{1}{3}x + \frac{1}{4}(x+8) = x + 7$

Multiply by the LCD, 12.

$12\left[\frac{1}{3}x + \frac{1}{4}(x+8)\right] = 12(x+7)$

$4x + 3(x+8) = 12x + 84$

$4x + 3x + 24 = 12x + 84$

$7x + 24 = 12x + 84$

$-5x = 60$

$x = -12$

Check $x = -12$: $-4 - 1 = -5$ *True*

The solution set is $\{-12\}$.

5. $0.04x + 0.06(100 - x) = 5.88$

Multiply both sides by 100 to clear decimals.

$4x + 6(100 - x) = 588$

$4x + 600 - 6x = 588$

$-2x = -12$

$x = 6$

Check $x = 6$: $0.24 + 5.64 = 5.88$ *True*

The solution set is $\{6\}$.

6. $-5 - 3(m-2) < 11 - 2(m+2)$

$-5 - 3m + 6 < 11 - 2m - 4$

$1 - 3m < 7 - 2m$

$-m < 6$

Multiply by -1; reverse the inequality.

$m > -6$

The solution set is $(-6, \infty)$.

7. $2k + 4 < 10$ and $3k - 1 > 5$

$\qquad 2k < 6 \qquad\qquad\qquad 3k > 6$

$\qquad k < 3 \qquad$ and $\qquad k > 2$

The solution set is $(2, 3)$.

8. $2k + 4 > 10$ or $3k - 1 < 5$

$\qquad 2k > 6 \qquad\qquad\qquad 3k < 6$

$\qquad k > 3 \qquad$ or $\qquad k < 2$

The solution set is $(-\infty, 2) \cup (3, \infty)$.

9. $|6x + 7| = 13$

$6x + 7 = 13$ or $6x + 7 = -13$

$\quad 6x = 6 \qquad\qquad\qquad 6x = -20$

$\quad\; x = 1$ or $x = -\frac{20}{6} = -\frac{10}{3}$

Check $x = 1$: $|13| = 13$ *True*

Check $x = -\frac{10}{3}$: $|-13| = 13$ *True*

The solution set is $\left\{-\frac{10}{3}, 1\right\}$.

10. $|-2x + 4| = |-2x - 3|$

$-2x + 4 = -2x - 3$ or $-2x + 4 = -(-2x - 3)$

$\qquad 4 = -3 \qquad\qquad\qquad -2x + 4 = 2x + 3$

$\qquad$ *False* $\qquad\qquad\qquad\qquad -4x = -1$

$\qquad\qquad\qquad\qquad\qquad\qquad\quad x = \frac{1}{4}$

Check $x = \frac{1}{4}$: $\left|\frac{7}{2}\right| = \left|-\frac{7}{2}\right|$ *True*

The solution set is $\left\{\frac{1}{4}\right\}$.

11. $|2p - 5| \geq 9$

$2p - 5 \geq 9$ or $2p - 5 \leq -9$

$\quad 2p \geq 14 \qquad\qquad\qquad 2p \leq -4$

$\quad\; p \geq 7 \qquad$ or $\qquad\;\; p \leq -2$

The solution set is $(-\infty, -2] \cup [7, \infty)$.

12.   The two angles have the same measure, so

$$10x - 70 = 7x - 25$$
$$3x = 45$$
$$x = 15.$$

Then,

$$10x - 70 = 10(15) - 70$$
$$= 150 - 70 = 80.$$

Each angle measures $80°$.

13.   Let $x =$ the number of nickels and then $50 - x =$ the number of quarters.

Value of nickels	+	value of quarters	=	total value.
$0.05x$	+	$0.25(50 - x)$	=	8.90

Multiply by 100 to clear the decimals.

$$5x + 25(50 - x) = 890$$
$$5x + 1250 - 25x = 890$$
$$-20x = -360$$
$$x = 18$$

Since $x = 18$, $50 - x = 50 - 18 = 32$.

There are 18 nickels and 32 quarters.

14. Let $x =$ the amount of pure alcohol.
Make a table.

Number of Liters	Percent (as a decimal)	Liters of Pure Alcohol
x	$100\% = 1$	$1 \cdot x = x$
40	$18\% = 0.18$	$0.18(40) = 7.2$
$x + 40$	$22\% = 0.22$	$0.22(x + 40)$

From the last column:

$$x + 7.2 = 0.22x + 8.8$$
$$0.78x = 1.6$$
$$x = \tfrac{1.6}{0.78} = \tfrac{160}{78} = \tfrac{80}{39} \text{ or } 2\tfrac{2}{39}$$

The required amount is $\frac{80}{39}$ or $2\frac{2}{39}$ liters of pure alcohol.

15. $4x - 3y = 12$
Let $x = 0$ to find the y-intercept, $(0, -4)$.
Let $y = 0$ to find the x-intercept, $(3, 0)$.
Draw a line through the intercepts.

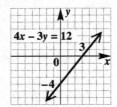

16. $(-4, 6)$ and $(2, -3)$
$$m = \frac{-3 - 6}{2 - (-4)} = \frac{-9}{6} = -\frac{3}{2}$$
Use the point-slope form:
$$y - 6 = -\tfrac{3}{2}[x - (-4)]$$
$$y - 6 = -\tfrac{3}{2}(x + 4)$$
$$y - 6 = -\tfrac{3}{2}x - 6$$
$$y = -\tfrac{3}{2}x$$

17. $\left(3k^3 - 5k^2 + 8k - 2\right) - \left(4k^3 + 11k + 7\right)$
$+ \left(2k^2 - 5k\right)$
$= 3k^3 - 4k^3 - 5k^2 + 2k^2$
$\quad + 8k - 11k - 5k - 2 - 7$
$= -k^3 - 3k^2 - 8k - 9$

18. $(8x - 7)(x + 3)$
$= 8x^2 + 24x - 7x - 21$
$= 8x^2 + 17x - 21$

19. $\dfrac{8z^3 - 16z^2 + 24z}{8z^2} = \dfrac{8z^3}{8z^2} - \dfrac{16z^2}{8z^2} + \dfrac{24z}{8z^2}$
$\qquad\qquad = z - 2 + \dfrac{3}{z}$

20.

$$
\begin{array}{r}
3y^3 - 3y^2 + 4y + 1 \\
2y + 1 \overline{\smash{\big)}\ 6y^4 - 3y^3 + 5y^2 + 6y - 9} \\
\underline{6y^4 + 3y^3} \\
-6y^3 + 5y^2 \\
\underline{-6y^3 - 3y^2} \\
8y^2 + 6y \\
\underline{8y^2 + 4y} \\
2y - 9 \\
\underline{2y + 1} \\
-10
\end{array}
$$

The answer is

$$3y^3 - 3y^2 + 4y + 1 + \frac{-10}{2y + 1}.$$

21. $2p^2 - 5pq + 3q^2 = (2p - 3q)(p - q)$

22. $3k^4 + k^2 - 4 = \left(3k^2 + 4\right)\left(k^2 - 1\right)$
$\qquad\qquad\quad = \left(3k^2 + 4\right)(k + 1)(k - 1)$

23. $x^3 + 512 = x^3 + 8^3$
$\qquad\qquad = (x + 8)\left(x^2 - 8x + 64\right)$

24. $2x^2 + 11x + 15 = 0$
$(x + 3)(2x + 5) = 0$
$\quad x + 3 = 0 \quad$ or $\quad 2x + 5 = 0$
$\qquad x = -3 \quad$ or $\qquad x = -\tfrac{5}{2}$
Check $x = -3$: $\ 18 - 33 + 15 = 0 \quad$ *True*
Check $x = -\tfrac{5}{2}$: $\ \tfrac{25}{2} - \tfrac{55}{2} + \tfrac{30}{2} = 0 \quad$ *True*
The solution set is $\left\{-3, -\tfrac{5}{2}\right\}$.

25. $\qquad 5t(t - 1) = 2(1 - t)$
$\qquad\quad 5t^2 - 5t = 2 - 2t$
$\quad 5t^2 - 3t - 2 = 0$
$\ (5t + 2)(t - 1) = 0$
$\quad 5t + 2 = 0 \quad$ or $\quad t - 1 = 0$
$\qquad t = -\tfrac{2}{5} \quad$ or $\qquad t = 1$
Check $x = -\tfrac{2}{5}$: $\ -2\left(-\tfrac{7}{5}\right) = 2\left(\tfrac{7}{5}\right) \quad$ *True*
Check $x = 1$: $\qquad 5(0) = 2(0) \quad$ *True*
The solution set is $\left\{-\tfrac{2}{5}, 1\right\}$.

26. $\dfrac{2}{x^2 - 9} = \dfrac{2}{(x + 3)(x - 3)}$
The numbers -3 and 3 make the denominator 0 so they are the values of the variable that make the rational expression undefined.

27. $\dfrac{y^2 + y - 12}{y^3 + 9y^2 + 20y} \div \dfrac{y^2 - 9}{y^3 + 3y^2}$
$= \dfrac{y^2 + y - 12}{y(y^2 + 9y + 20)} \cdot \dfrac{y^3 + 3y^2}{y^2 - 9}$
$= \dfrac{(y + 4)(y - 3)}{y(y + 4)(y + 5)} \cdot \dfrac{y^2(y + 3)}{(y + 3)(y - 3)}$
$= \dfrac{y}{y + 5}$

28. $\dfrac{1}{x+y} + \dfrac{3}{x-y}$ *The LCD is*
$$(x+y)(x-y).$$

$$= \dfrac{1(x-y)}{(x+y)(x-y)} + \dfrac{3(x+y)}{(x-y)(x+y)}$$

$$= \dfrac{(x-y) + 3(x+y)}{(x+y)(x-y)}$$

$$= \dfrac{x-y+3x+3y}{(x+y)(x-y)}$$

$$= \dfrac{4x+2y}{(x+y)(x-y)}$$

29. $\dfrac{\dfrac{-6}{x-2}}{\dfrac{8}{3x-6}} = \dfrac{-6}{x-2} \div \dfrac{8}{3x-6}$

$$= \dfrac{-6}{x-2} \cdot \dfrac{3x-6}{8}$$

$$= \dfrac{-6}{x-2} \cdot \dfrac{3(x-2)}{8}$$

$$= \dfrac{-2 \cdot 3 \cdot 3}{2 \cdot 4} = -\dfrac{9}{4}$$

30. $\dfrac{\dfrac{1}{a} - \dfrac{1}{b}}{\dfrac{a}{b} - \dfrac{b}{a}}$ *The LCD of both the numerator and the denominator is ab.*

$$= \dfrac{\dfrac{b-a}{ab}}{\dfrac{a^2 - b^2}{ab}}$$

$$= \dfrac{b-a}{ab} \div \dfrac{a^2 - b^2}{ab}$$

$$= \dfrac{b-a}{ab} \cdot \dfrac{ab}{a^2 - b^2}$$

$$= \dfrac{b-a}{a^2 - b^2}$$

$$= \dfrac{-(a-b)}{(a-b)(a+b)} = \dfrac{-1}{a+b}$$

31. Let x = Mike's speed.
Then $x + 4$ = Cecily's speed.

Use $d = rt$, or $t = \dfrac{d}{r}$, to make a table.

	Distance	Rate	Time
Mike	24	x	$\dfrac{24}{x}$
Cecily	48	$x+4$	$\dfrac{48}{x+4}$

Since the times are the same,

$$\dfrac{24}{x} = \dfrac{48}{x+4}.$$

Multiply by the LCD, $x(x+4)$.

$$x(x+4)\left(\dfrac{24}{x}\right) = x(x+4)\left(\dfrac{48}{x+4}\right)$$
$$24(x+4) = 48x$$
$$24x + 96 = 48x$$
$$-24x = -96$$
$$x = 4$$

Since $x = 4$, $x + 4 = 4 + 4 = 8$.
Mike's speed is 4 mph; Cecily's speed is 8 mph.

32. $$\dfrac{p+1}{p-3} = \dfrac{4}{p-3} + 6$$

Multiply by the LCD, $p - 3$.

$$(p-3)\left(\dfrac{p+1}{p-3}\right) = (p-3)\left(\dfrac{4}{p-3} + 6\right)$$
$$p + 1 = 4 + 6(p-3)$$
$$p + 1 = 4 + 6p - 18$$
$$p + 1 = -14 + 6p$$
$$-5p = -15$$
$$p = 3$$

Substituting 3 in the original equation results in division by zero, so 3 cannot be a solution. The solution set is $\emptyset$.

33. $$f(x) = 3x - 7$$
$$f(-10) = 3(-10) - 7 = -30 - 7 = -37$$

34. The speed s of a pulley varies inversely as its diameter d, so

$$s = \dfrac{k}{d}.$$

$$450 = \dfrac{k}{9} \quad \text{Let } s = 450, \, d = 9.$$
$$k = 450 \cdot 9 = 4050$$

So $s = \dfrac{4050}{d}$ and when $d = 10$,

$$s = \dfrac{4050}{10} = 405.$$

The speed of a pulley with diameter 10 inches is 405 revolutions per minute.

35. $3x - y = 23$ (1)
$2x + 3y = 8$ (2)
To eliminate y, multiply equation (1) by 3 and add the result to equation (2).

$$\begin{array}{rl} 9x - 3y = 69 & 3 \times (1) \\ \underline{2x + 3y = 8} & (2) \\ 11x \quad\quad = 77 & \\ x = 7 & \end{array}$$

Substitute 7 for x in (2),

$$2(7) + 3y = 8$$
$$14 + 3y = 8$$
$$3y = -6$$
$$y = -2$$

The solution set is $\{(7, -2)\}$.

36.
$$x + y + z = 1$$
$$x - y - z = -3$$
$$x + y - z = -1$$

Write the augmented matrix.

$$\begin{bmatrix} 1 & 1 & 1 & | & 1 \\ 1 & -1 & -1 & | & -3 \\ 1 & 1 & -1 & | & -1 \end{bmatrix}$$

$$\begin{bmatrix} 1 & 1 & 1 & | & 1 \\ 0 & -2 & -2 & | & -4 \\ 0 & 0 & -2 & | & -2 \end{bmatrix} \begin{matrix} \\ -R_1 + R_2 \\ -R_1 + R_3 \end{matrix}$$

$$\begin{bmatrix} 1 & 1 & 1 & | & 1 \\ 0 & 1 & 1 & | & 2 \\ 0 & 0 & 1 & | & 1 \end{bmatrix} \begin{matrix} \\ -\frac{1}{2}R_2 \\ -\frac{1}{2}R_3 \end{matrix}$$

This matrix gives the system

$$\begin{aligned} x + y + z &= 1 \quad (1) \\ y + z &= 2 \quad (2) \\ z &= 1. \end{aligned}$$

Substitute 1 for z in (2).

$$y + 1 = 2$$
$$y = 1$$

Substitute 1 for y and 1 for z in (1).

$$x + 1 + 1 = 1$$
$$x = -1$$

The solution set is $\{(-1, 1, 1)\}$.

37. Let $x =$ the number of 2-ounce letters and
$y =$ the number of 3-ounce letters.

$$\begin{aligned} 5x + 3y &= 5.76 \quad (1) \\ 3x + 5y &= 6.24 \quad (2) \end{aligned}$$

To eliminate x, multiply (1) by -3 and (2) by 5 and add the results.

$$\begin{array}{rcl} -15x - 9y &=& -17.28 \quad -3 \times (1) \\ \underline{15x + 25y} &=& \underline{31.20} \quad\;\; 5 \times (2) \\ 16y &=& 13.92 \\ y &=& 0.87 \end{array}$$

Substitute $y = 0.87$ in (1).

$$\begin{aligned} 5x + 3y &= 5.76 \quad (1) \\ 5x + 3(0.87) &= 5.76 \\ 5x + 2.61 &= 5.76 \\ 5x &= 3.15 \\ x &= 0.63 \end{aligned}$$

The 2006 postage rate for a 2-ounce letter was $0.63 and for a 3-ounce letter, $0.87.

38. $27^{-2/3} = \dfrac{1}{27^{2/3}} = \dfrac{1}{\left[(3^3)^{1/3}\right]^2} = \dfrac{1}{3^2} = \dfrac{1}{9}$

39. $\sqrt{200x^4} = \sqrt{100x^4 \cdot 2}$
$$= \sqrt{10^2(x^2)^2 \cdot 2} = 10x^2\sqrt{2}$$

40. $\sqrt[3]{16x^2y} \cdot \sqrt[3]{3x^3y}$
$$= \sqrt[3]{8} \cdot \sqrt[3]{2x^2y} \cdot \sqrt[3]{x^3} \cdot \sqrt[3]{3y}$$
$$= 2\sqrt[3]{2x^2y} \cdot x\sqrt[3]{3y}$$
$$= 2x\sqrt[3]{6x^2y^2}$$

41. $\sqrt{50} + \sqrt{8} = \sqrt{25 \cdot 2} + \sqrt{4 \cdot 2}$
$$= 5\sqrt{2} + 2\sqrt{2} = 7\sqrt{2}$$

42. $\dfrac{1}{\sqrt{10} - \sqrt{8}}$

Multiply the numerator and denominator by the conjugate of the denominator, $\sqrt{10} + \sqrt{8}$.

$$= \frac{1(\sqrt{10} + \sqrt{8})}{(\sqrt{10} - \sqrt{8})(\sqrt{10} + \sqrt{8})}$$
$$= \frac{\sqrt{10} + \sqrt{8}}{10 - 8} = \frac{\sqrt{10} + \sqrt{4 \cdot 2}}{2}$$
$$= \frac{\sqrt{10} + 2\sqrt{2}}{2}$$

43. $(2\sqrt{x} + \sqrt{y})(-3\sqrt{x} - 4\sqrt{y})$
$$= -6x - 8\sqrt{xy} - 3\sqrt{xy} - 4y$$
$$= -6x - 11\sqrt{xy} - 4y$$

44. $(-4, 4)$ and $(-2, 9)$
$$d = \sqrt{[(-2) - (-4)]^2 + (9 - 4)^2}$$
$$= \sqrt{2^2 + 5^2}$$
$$= \sqrt{4 + 25} = \sqrt{29}$$

45. $\sqrt{3r - 8} = r - 2$

Square both sides.

$$\left(\sqrt{3r - 8}\right)^2 = (r - 2)^2$$
$$3r - 8 = r^2 - 4r + 4$$
$$0 = r^2 - 7r + 12$$
$$0 = (r - 3)(r - 4)$$

$$\begin{array}{ccc} r - 3 = 0 & \text{or} & r - 4 = 0 \\ r = 3 & \text{or} & r = 4 \end{array}$$

Check $r = 3$: $\sqrt{1} = 1$ *True*

Check $r = 4$: $\sqrt{4} = 2$ *True*

The solution set is $\{3, 4\}$.

46. Substitute 32 for D and 5 for h in the given formula.

$$S = \frac{2.74D}{\sqrt{h}} = \frac{2.74(32)}{\sqrt{5}} = \frac{87.68}{\sqrt{5}} \approx 39.2$$

The fall speed is about 39.2 mph.

47. $(5 + 7i) - (3 - 2i)$
$$= (5 - 3) + [7 - (-2)]i$$
$$= 2 + 9i$$

48. $\dfrac{6 - 2i}{1 - i}$

Multiply the numerator and denominator by the conjugate of the denominator, $1 + i$.

$$= \frac{(6 - 2i)(1 + i)}{(1 - i)(1 + i)}$$

$$= \frac{6 + 6i - 2i - 2i^2}{1^2 + 1^2}$$

$$= \frac{6 + 4i + 2}{1 + 1}$$

$$= \frac{8 + 4i}{2} = \frac{2(4 + 2i)}{2} = 4 + 2i$$

CHAPTER 11 QUADRATIC EQUATIONS, INEQUALITIES, AND FUNCTIONS

11.1 Solving Quadratic Equations by the Square Root Property

1. By the square root property, if $x^2 = 16$, then

$$x = \pm\sqrt{16}$$

Thus, the equation is also true for $x = -4$.

Solution set: $\{\pm 4\}$

3. **(a)** A quadratic equation in standard form has a second-degree polynomial in decreasing powers equal to 0.

(b) The zero-factor property states that if a product equals 0, then at least one of the factors equals 0.

(c) The square root property states that if the square of a quantity equals a number, then the quantity equals the positive or negative square root of the number.

5. $x^2 = 81$

$$x = 9 \quad \text{or} \quad x = -9$$

Solution set: $\{\pm 9\}$

7. $x^2 = 17$

$$x = \sqrt{17} \quad \text{or} \quad x = -\sqrt{17}$$

Solution set: $\left\{\pm\sqrt{17}\right\}$

9. $x^2 = 32$

$$x = \sqrt{32} \quad \text{or} \quad x = -\sqrt{32}$$
$$x = 4\sqrt{2} \quad \text{or} \quad x = -4\sqrt{2}$$

Solution set: $\left\{\pm 4\sqrt{2}\right\}$

11. $x^2 = \frac{25}{4}$

$$x = \sqrt{\frac{25}{4}} \quad \text{or} \quad x = -\sqrt{\frac{25}{4}}$$
$$x = \frac{5}{2} \quad \text{or} \quad x = -\frac{5}{2}$$

Solution set: $\left\{\pm\frac{5}{2}\right\}$

13. $z^2 = 2.25$

$$z = \sqrt{2.25} \quad \text{or} \quad z = -\sqrt{2.25}$$
$$z = 1.5 \quad \text{or} \quad z = -1.5$$

Solution set: $\{\pm 1.5\}$

15. $r^2 - 3 = 0$
$$r^2 = 3$$

$$r = \sqrt{3} \quad \text{or} \quad r = -\sqrt{3}$$

Solution set: $\left\{\pm\sqrt{3}\right\}$

17. $x^2 - 20 = 0$
$$x^2 = 20$$

$$x = \sqrt{20} \quad \text{or} \quad x = -\sqrt{20}$$
$$x = 2\sqrt{5} \quad \text{or} \quad x = -2\sqrt{5}$$

Solution set: $\left\{\pm 2\sqrt{5}\right\}$

19. $3n^2 - 72 = 0$
$$3n^2 = 72$$
$$n = 24$$

$$n = \sqrt{24} \quad \text{or} \quad n = -\sqrt{24}$$
$$n = 2\sqrt{6} \quad \text{or} \quad n = -2\sqrt{6}$$

Solution set: $\left\{\pm 2\sqrt{6}\right\}$

21. $5a^2 + 4 = 8$
$$5a^2 = 4 \quad \textit{Subtract 4.}$$
$$a^2 = \tfrac{4}{5} \quad \textit{Divide by 5.}$$

$$a = \sqrt{\tfrac{4}{5}} \qquad \text{or} \qquad a = -\sqrt{\tfrac{4}{5}}$$
$$= \frac{\sqrt{4}}{\sqrt{5}} \cdot \frac{\sqrt{5}}{\sqrt{5}} \qquad \qquad = -\frac{\sqrt{4}}{\sqrt{5}} \cdot \frac{\sqrt{5}}{\sqrt{5}}$$
$$= \frac{2\sqrt{5}}{5} \qquad \qquad = -\frac{2\sqrt{5}}{5}$$

Solution set: $\left\{\pm\frac{2\sqrt{5}}{5}\right\}$

23. $2t^2 + 7 = 61$
$$2t^2 = 54$$
$$t^2 = 27$$
$$t = \pm\sqrt{27} = \pm\sqrt{9 \cdot 3} = \pm 3\sqrt{3}$$

Solution set: $\left\{\pm 3\sqrt{3}\right\}$

25. $(x - 3)^2 = 25$

Use the square root property.

$$x - 3 = \sqrt{25} \quad \text{or} \quad x - 3 = -\sqrt{25}$$
$$x - 3 = 5 \quad \text{or} \quad x - 3 = -5$$
$$x = 8 \quad \text{or} \quad x = -2$$

Solution set: $\{-2, 8\}$

27. $(x - 8)^2 = 27$

Begin by using the square root property.

$$x - 8 = \sqrt{27} \quad \text{or} \quad x - 8 = -\sqrt{27}$$

Now simplify the radical.

$$\sqrt{27} = \sqrt{9} \cdot \sqrt{3} = 3\sqrt{3}$$

$$x - 8 = 3\sqrt{3} \quad \text{or} \quad x - 8 = -3\sqrt{3}$$
$$x = 8 + 3\sqrt{3} \quad \text{or} \quad x = 8 - 3\sqrt{3}$$

Solution set: $\left\{ 8 \pm 3\sqrt{3} \right\}$

29. $(3k + 2)^2 = 49$

$$3k + 2 = \sqrt{49} \quad \text{or} \quad 3k + 2 = -\sqrt{49}$$
$$3k + 2 = 7 \quad \text{or} \quad 3k + 2 = -7$$
$$3k = 5 \quad \text{or} \quad 3k = -9$$
$$k = \tfrac{5}{3} \quad \text{or} \quad k = -3$$

Solution set: $\left\{ -3, \tfrac{5}{3} \right\}$

31. $(4x - 3)^2 = 9$

$$4x - 3 = \sqrt{9} \quad \text{or} \quad 4x - 3 = -\sqrt{9}$$
$$4x - 3 = 3 \quad \text{or} \quad 4x - 3 = -3$$
$$4x = 6 \quad \text{or} \quad 4x = 0$$
$$x = \tfrac{6}{4} = \tfrac{3}{2} \quad \text{or} \quad x = 0$$

Solution set: $\left\{ 0, \tfrac{3}{2} \right\}$

33. $(x - 4)^2 = 3$

$$x - 4 = \sqrt{3} \quad \text{or} \quad x - 4 = -\sqrt{3}$$
$$x = 4 + \sqrt{3} \quad \text{or} \quad x = 4 - \sqrt{3}$$

Solution set: $\left\{ 4 \pm \sqrt{3} \right\}$

35. $(t + 5)^2 = 48$

$$t + 5 = \sqrt{48} \quad \text{or} \quad t + 5 = -\sqrt{48}$$
$$t + 5 = 4\sqrt{3} \quad \quad t + 5 = -4\sqrt{3}$$
$$t = -5 + 4\sqrt{3} \quad \text{or} \quad t = -5 - 4\sqrt{3}$$

Solution set: $\left\{ -5 \pm 4\sqrt{3} \right\}$

37. $(3x - 1)^2 = 7$

$$3x - 1 = \sqrt{7} \quad \text{or} \quad 3x - 1 = -\sqrt{7}$$
$$3x = 1 + \sqrt{7} \quad \quad 3x = 1 - \sqrt{7}$$
$$x = \frac{1 + \sqrt{7}}{3} \quad \text{or} \quad x = \frac{1 - \sqrt{7}}{3}$$

Solution set: $\left\{ \frac{1 \pm \sqrt{7}}{3} \right\}$

39. $(4p + 1)^2 = 24$

$$4p + 1 = \sqrt{24} \quad \text{or} \quad 4p + 1 = -\sqrt{24}$$
$$4p + 1 = 2\sqrt{6} \quad \quad 4p + 1 = -2\sqrt{6}$$
$$4p = -1 + 2\sqrt{6} \quad \quad 4p = -1 - 2\sqrt{6}$$
$$p = \frac{-1 + 2\sqrt{6}}{4} \quad \text{or} \quad p = \frac{-1 - 2\sqrt{6}}{4}$$

Solution set: $\left\{ \frac{-1 \pm 2\sqrt{6}}{4} \right\}$

41. $(5 - 2x)^2 = 30$

$$5 - 2x = \sqrt{30} \quad \text{or} \quad 5 - 2x = -\sqrt{30}$$
$$-2x = -5 + \sqrt{30} \quad \text{or} \quad -2x = -5 - \sqrt{30}$$
$$x = \frac{-5 + \sqrt{30}}{-2} \quad \text{or} \quad x = \frac{-5 - \sqrt{30}}{-2}$$
$$x = \frac{-5 + \sqrt{30}}{-2} \cdot \frac{-1}{-1} \quad \text{or} \quad x = \frac{-5 - \sqrt{30}}{-2} \cdot \frac{-1}{-1}$$
$$x = \frac{5 - \sqrt{30}}{2} \quad \text{or} \quad x = \frac{5 + \sqrt{30}}{2}$$

Solution set: $\left\{ \frac{5 \pm \sqrt{30}}{2} \right\}$

43. $(3k + 1)^2 = 18$

$$3k + 1 = \sqrt{18} \quad \text{or} \quad 3k + 1 = -\sqrt{18}$$
$$3k = -1 + 3\sqrt{2} \quad \text{or} \quad 3k = -1 - 3\sqrt{2}$$

Note that $\sqrt{18} = \sqrt{9 \cdot 2} = 3\sqrt{2}.$

$$k = \frac{-1 + 3\sqrt{2}}{3} \quad \text{or} \quad k = \frac{-1 - 3\sqrt{2}}{3}$$

Solution set: $\left\{ \frac{-1 \pm 3\sqrt{2}}{3} \right\}$

45. $\left(\tfrac{1}{2}x + 5 \right)^2 = 12$

$$\tfrac{1}{2}x + 5 = \sqrt{12} \quad \text{or} \quad \tfrac{1}{2}x + 5 = -\sqrt{12}$$
$$\tfrac{1}{2}x = -5 + 2\sqrt{3} \quad \text{or} \quad \tfrac{1}{2}x = -5 - 2\sqrt{3}$$

Note that $\sqrt{12} = \sqrt{4 \cdot 3} = 2\sqrt{3}.$

$$x = 2\left(-5 + 2\sqrt{3} \right) \quad \text{or} \quad x = 2\left(-5 - 2\sqrt{3} \right)$$
$$x = -10 + 4\sqrt{3} \quad \text{or} \quad x = -10 - 4\sqrt{3}$$

Solution set: $\left\{ -10 \pm 4\sqrt{3} \right\}$

47. $(4k - 1)^2 - 48 = 0$

$$(4k - 1)^2 = 48$$

$$4k - 1 = \sqrt{48} \quad \text{or} \quad 4k - 1 = -\sqrt{48}$$
$$4k - 1 = 4\sqrt{3} \quad \text{or} \quad 4k - 1 = -4\sqrt{3}$$
$$4k = 1 + 4\sqrt{3} \quad \text{or} \quad 4k = 1 - 4\sqrt{3}$$
$$k = \frac{1 + 4\sqrt{3}}{4} \quad \text{or} \quad k = \frac{1 - 4\sqrt{3}}{4}$$

Solution set: $\left\{ \frac{1 \pm 4\sqrt{3}}{4} \right\}$

49. Johnny's first solution, $\dfrac{5 + \sqrt{30}}{2}$, is equivalent to

Linda's second solution, $\dfrac{-5 - \sqrt{30}}{-2}$. This can be

verified by multiplying $\dfrac{5 + \sqrt{30}}{2}$ by 1 in the form

$\dfrac{-1}{-1}$. Similarly, Johnny's second solution is
equivalent to Linda's first one.

51. $(k + 2.14)^2 = 5.46$

$k + 2.14 = \sqrt{5.46}$ or $k + 2.14 = -\sqrt{5.46}$
$k = -2.14 + \sqrt{5.46}$ or $k = -2.14 - \sqrt{5.46}$
$\quad k \approx 0.20$ or $\quad\quad k \approx -4.48$

To the nearest hundredth, the solution set is
$\{-4.48, 0.20\}$.

53. $(2.11p + 3.42)^2 = 9.58$
$\quad 2.11p + 3.42 = \pm\sqrt{9.58}$
Remember that this represents two equations.
$\quad\quad 2.11p = -3.42 \pm \sqrt{9.58}$
$\quad\quad\quad p = \dfrac{-3.42 \pm \sqrt{9.58}}{2.11}$
$\quad\quad\quad\quad \approx -3.09 \text{ or } -0.15$

To the nearest hundredth, the solution set is
$\{-3.09, -0.15\}$.

55. $x^2 = -12$

$\quad x = \sqrt{-12}$ or $x = -\sqrt{-12}$
$\quad x = i\sqrt{12}$ $\quad\quad x = -i\sqrt{12}$
$\quad x = 2i\sqrt{3}$ or $\quad x = -2i\sqrt{3}$

Solution set: $\left\{ \pm 2i\sqrt{3} \right\}$

57. $(r - 5)^2 = -4$

$\quad r - 5 = \sqrt{-4}$ or $r - 5 = -\sqrt{-4}$
$\quad\quad r = 5 + 2i$ or $\quad\quad r = 5 - 2i$

Solution set: $\{5 \pm 2i\}$

59. $(6k - 1)^2 = -8$

$6k - 1 = \sqrt{-8}$ or $6k - 1 = -\sqrt{-8}$
$6k - 1 = i\sqrt{8}$ $\quad\quad 6k - 1 = -i\sqrt{8}$
$6k - 1 = 2i\sqrt{2}$ $\quad\quad 6k - 1 = -2i\sqrt{2}$
$\quad 6k = 1 + 2i\sqrt{2}$ $\quad\quad 6k = 1 - 2i\sqrt{2}$
$\quad k = \dfrac{1 + 2i\sqrt{2}}{6}$ or $\quad k = \dfrac{1 - 2i\sqrt{2}}{6}$

Solution set: $\left\{ \frac{1}{6} \pm \frac{\sqrt{2}}{3} i \right\}$

61. $\quad d = 16t^2$
$\quad 500 = 16t^2$
$\quad\quad t^2 = \frac{500}{16} = 31.25$

$\quad\quad t = \sqrt{31.25}$ or $\quad t = -\sqrt{31.25}$
$\quad\quad t \approx 5.6$ or $\quad t \approx -5.6$

$t \approx 5.6$ seconds (time must be positive)

63. $\quad A = \pi r^2$
$\quad 81\pi = \pi r^2$ *Let $A = 81\pi$.*
$\quad\quad 81 = r^2$ *Divide by π.*
$r = 9$ or $r = -9$

Discard -9 since the radius cannot be negative.
The radius is 9 inches.

65. Let $A = 110.25$ and $P = 100$.
$$A = P(1 + r)^2$$
$$110.25 = 100(1 + r)^2$$
$$(1 + r)^2 = \frac{110.25}{100} = 1.1025$$
$$1 + r = \pm\sqrt{1.1025}$$
$$1 + r = \pm 1.05$$
$$r = -1 \pm 1.05$$

So $r = -1 + 1.05 = 0.05$ or
$r = -1 - 1.05 = -2.05$. Reject the solution
-2.05. The rate is $r = 0.05$ or 5%.

67. $\dfrac{4}{5} + \sqrt{\dfrac{48}{25}} = \dfrac{4}{5} + \dfrac{\sqrt{48}}{\sqrt{25}} = \dfrac{4}{5} + \dfrac{\sqrt{16} \cdot \sqrt{3}}{5}$

$\quad\quad\quad\quad\quad = \dfrac{4 + 4\sqrt{3}}{5}$

69. $\dfrac{6 + \sqrt{24}}{8} = \dfrac{6 + \sqrt{4} \cdot \sqrt{6}}{8} = \dfrac{6 + 2\sqrt{6}}{8}$

$\quad\quad = \dfrac{2\left(3 + \sqrt{6}\right)}{2 \cdot 4} = \dfrac{3 + \sqrt{6}}{4}$

71. $x^2 - 10x + 25 = x^2 - 2 \cdot 5 \cdot x + 5^2$
$\quad\quad\quad\quad\quad = (x - 5)^2$

73. $x^2 - 7x + \frac{49}{4} = x^2 - 2 \cdot \frac{7}{2} \cdot x + \left(\frac{7}{2}\right)^2$
$\quad\quad\quad\quad\quad = \left(x - \frac{7}{2}\right)^2$

11.2 Solving Quadratic Equations by Completing the Square

1. $(2x + 1)^2 = 5$ is more suitable for solving by the
square root property. $x^2 + 4x = 12$ is more
suitable for solving by completing the square.

3. $x^2 + 6x +$ ___

We need to add the square of half the coefficient of x to get a perfect square trinomial.

$$\tfrac{1}{2}(6) = 3 \quad \text{and} \quad 3^2 = 9$$

Add 9 to $x^2 + 6x$ to get a perfect square trinomial.

$$x^2 + 6x + 9 = (x + 3)^2$$

5. $p^2 - 12p +$ ___

$$\tfrac{1}{2}(-12) = -6 \quad \text{and} \quad (-6)^2 = 36$$
$$p^2 - 12p + 36 = (p - 6)^2$$

7. $q^2 + 9q +$ ___

$$\tfrac{1}{2}(9) = \tfrac{9}{2} \quad \text{and} \quad \left(\tfrac{9}{2}\right)^2 = \tfrac{81}{4}$$
$$q^2 + 9q + \tfrac{81}{4} = \left(q + \tfrac{9}{2}\right)^2$$

9. $x^2 + \tfrac{1}{4}x +$ ___

$$\tfrac{1}{2}\left(\tfrac{1}{4}\right) = \tfrac{1}{8} \quad \text{and} \quad \left(\tfrac{1}{8}\right)^2 = \tfrac{1}{64}$$
$$x^2 + \tfrac{1}{4}x + \tfrac{1}{64} = \left(x + \tfrac{1}{8}\right)^2$$

11. $x^2 - 0.8x +$ ___

$$\tfrac{1}{2}(-0.8) = -0.4 \quad \text{and} \quad (-0.4)^2 = 0.16$$
$$x^2 - 0.8x + 0.16 = (x - 0.4)^2$$

13. $x^2 + 4x - 2 = 0$
$$x^2 + 4x = 2$$
$$\left[\tfrac{1}{2}(4)\right]^2 = 2^2 = 4$$

15. $x^2 + 10x + 18 = 0$
$$x^2 + 10x = -18$$
$$\left[\tfrac{1}{2}(10)\right]^2 = 5^2 = 25$$

17. $3w^2 - w - 24 = 0$
$$w^2 - \tfrac{1}{3}w - 8 = 0 \quad \textit{Divide by 3.}$$
$$w^2 - \tfrac{1}{3}w = 8$$
$$\left[\tfrac{1}{2}\left(-\tfrac{1}{3}\right)\right]^2 = \left(-\tfrac{1}{6}\right)^2 = \tfrac{1}{36}$$

19. $x^2 - 4x = -3$

Take half of the coefficient of x and square it. Half of -4 is -2, and $(-2)^2 = 4$. Add 4 to each side of the equation, and write the left side as a perfect square.

$$x^2 - 4x + 4 = -3 + 4$$
$$(x - 2)^2 = 1$$

Use the square root property.

$$x - 2 = \sqrt{1} \quad \text{or} \quad x - 2 = -\sqrt{1}$$
$$x - 2 = 1 \quad \text{or} \quad x - 2 = -1$$
$$x = 3 \quad \text{or} \quad x = 1$$

A check verifies that the solution set is $\{1, 3\}$.

21. $x^2 + 2x - 5 = 0$

Add 5 to each side.

$$x^2 + 2x = 5$$

Take half the coefficient of x and square it.

$$\tfrac{1}{2}(2) = 1, \quad \text{and} \quad 1^2 = 1.$$

Add 1 to each side of the equation, and write the left side as a perfect square.

$$x^2 + 2x + 1 = 5 + 1$$
$$(x + 1)^2 = 6$$

Use the square root property.

$$x + 1 = \sqrt{6} \qquad \text{or} \qquad x + 1 = -\sqrt{6}$$
$$x = -1 + \sqrt{6} \quad \text{or} \qquad x = -1 - \sqrt{6}$$

A check verifies that the solution set is $\left\{-1 \pm \sqrt{6}\right\}$. Using a calculator for your check is highly recommended.

23. $x^2 - 8x = -4$
$$x^2 - 8x + 16 = -4 + 16 \quad \left[\tfrac{1}{2}(-8)\right]^2 = 16$$
$$(x - 4)^2 = 12 \qquad \textit{Factor.}$$
$$x - 4 = \sqrt{12} \quad \text{or} \quad x - 4 = -\sqrt{12}$$
$$x - 4 = \sqrt{4} \cdot \sqrt{3} \quad \text{or} \quad x - 4 = -\sqrt{4} \cdot \sqrt{3}$$
$$x = 4 + 2\sqrt{3} \quad \text{or} \qquad x = 4 - 2\sqrt{3}$$

Solution set: $\left\{4 \pm 2\sqrt{3}\right\}$

25. $x^2 + 7x - 1 = 0$
$$x^2 + 7x = 1$$
$$x^2 + 7x + \tfrac{49}{4} = 1 + \tfrac{49}{4} \quad \left[\tfrac{1}{2}(7)^2\right] = \tfrac{49}{4}$$
$$\left(x + \tfrac{7}{2}\right)^2 = \tfrac{53}{4}$$
$$x + \tfrac{7}{2} = \sqrt{\tfrac{53}{4}} \qquad \text{or} \quad x + \tfrac{7}{2} = -\sqrt{\tfrac{53}{4}}$$
$$x = -\tfrac{7}{2} + \tfrac{\sqrt{53}}{2} \qquad\qquad x = -\tfrac{7}{2} - \tfrac{\sqrt{53}}{2}$$
$$x = \tfrac{-7 + \sqrt{53}}{2} \quad \text{or} \qquad x = \tfrac{-7 - \sqrt{53}}{2}$$

Solution set: $\left\{\tfrac{-7 \pm \sqrt{53}}{2}\right\}$

27. $4x^2 + 4x = 3$

Divide each side by 4 so that the coefficient of x^2 is 1.

$$x^2 + x = \tfrac{3}{4}$$

The coefficient of x is 1. Take half of 1, square the result, and add this square to each side.

$$\tfrac{1}{2}(1) = \tfrac{1}{2} \quad \text{and} \quad \left(\tfrac{1}{2}\right)^2 = \tfrac{1}{4}$$
$$x^2 + x + \tfrac{1}{4} = \tfrac{3}{4} + \tfrac{1}{4}$$

The left-hand side can then be written as a perfect square.

$$\left(x + \tfrac{1}{2}\right)^2 = 1$$

Use the square root property.

$$x + \tfrac{1}{2} = 1 \qquad \text{or} \qquad x + \tfrac{1}{2} = -1$$
$$x = -\tfrac{1}{2} + 1 \qquad \text{or} \qquad x = -\tfrac{1}{2} - 1$$
$$x = \tfrac{1}{2} \qquad \text{or} \qquad x = -\tfrac{3}{2}$$

A check verifies that the solution set is $\left\{-\tfrac{3}{2}, \tfrac{1}{2}\right\}$.

29.
$$3w^2 - w = 24$$
$$w^2 - \tfrac{1}{3}w = 8 \qquad \textit{Divide by 3.}$$

Complete the square by taking half of $\tfrac{1}{3}$, the coefficient of w, and squaring the result.

$$\left[\tfrac{1}{2}\left(-\tfrac{1}{3}\right)\right]^2 = \left(-\tfrac{1}{6}\right)^2 = \tfrac{1}{36}$$

Add $\tfrac{1}{36}$ to each side.

$$w^2 - \tfrac{1}{3}w + \tfrac{1}{36} = 8 + \tfrac{1}{36}$$
$$\left(w - \tfrac{1}{6}\right)^2 = \tfrac{288}{36} + \tfrac{1}{36}$$
$$\left(w - \tfrac{1}{6}\right)^2 = \tfrac{289}{36}$$

$$w - \tfrac{1}{6} = \sqrt{\tfrac{289}{36}} \quad \text{or} \quad w - \tfrac{1}{6} = -\sqrt{\tfrac{289}{36}}$$
$$w = \tfrac{1}{6} + \tfrac{\sqrt{289}}{\sqrt{36}} \qquad\qquad w = \tfrac{1}{6} - \tfrac{\sqrt{289}}{\sqrt{36}}$$
$$w = \tfrac{1}{6} + \tfrac{17}{6} \qquad\qquad w = \tfrac{1}{6} - \tfrac{17}{6}$$
$$w = \tfrac{18}{6} \qquad\qquad\qquad w = -\tfrac{16}{6}$$
$$w = 3 \qquad \text{or} \qquad w = -\tfrac{8}{3}$$

Solution set: $\left\{-\tfrac{8}{3}, 3\right\}$

31.
$$2k^2 + 5k - 2 = 0$$
$$2k^2 + 5k = 2$$
$$k^2 + \tfrac{5}{2}k = 1 \qquad \textit{Divide by 2.}$$

Complete the square.

$$\left(\tfrac{1}{2} \cdot \tfrac{5}{2}\right)^2 = \left(\tfrac{5}{4}\right)^2 = \tfrac{25}{16}$$

Add $\tfrac{25}{16}$ to each side.

$$k^2 + \tfrac{5}{2}k + \tfrac{25}{16} = 1 + \tfrac{25}{16}$$
$$\left(k + \tfrac{5}{4}\right)^2 = \tfrac{41}{16}$$

$$k + \tfrac{5}{4} = \sqrt{\tfrac{41}{16}} \quad \text{or} \quad k + \tfrac{5}{4} = -\sqrt{\tfrac{41}{16}}$$
$$k = -\tfrac{5}{4} + \tfrac{\sqrt{41}}{4} \qquad\qquad k = -\tfrac{5}{4} - \tfrac{\sqrt{41}}{4}$$
$$k = \tfrac{-5+\sqrt{41}}{4} \qquad \text{or} \qquad k = \tfrac{-5-\sqrt{41}}{4}$$

Solution set: $\left\{\tfrac{-5\pm\sqrt{41}}{4}\right\}$

33.
$$5x^2 - 10x + 2 = 0$$
$$5x^2 - 10x = -2$$
$$x^2 - 2x = -\tfrac{2}{5} \quad \textit{Divide by 5.}$$

Complete the square.

$$\left[\tfrac{1}{2}(-2)\right]^2 = (-1)^2 = 1$$

Add 1 to each side.

$$x^2 - 2x + 1 = -\tfrac{2}{5} + 1$$
$$(x - 1)^2 = \tfrac{3}{5}$$

$$x - 1 = \sqrt{\tfrac{3}{5}} \qquad \text{or} \quad x - 1 = -\sqrt{\tfrac{3}{5}}$$
$$x - 1 = \tfrac{\sqrt{3}}{\sqrt{5}} \cdot \tfrac{\sqrt{5}}{\sqrt{5}} \qquad x - 1 = -\tfrac{\sqrt{3}}{\sqrt{5}} \cdot \tfrac{\sqrt{5}}{\sqrt{5}}$$
$$x = \tfrac{5}{5} + \tfrac{\sqrt{15}}{5} \qquad\qquad x = \tfrac{5}{5} - \tfrac{\sqrt{15}}{5}$$
$$x = \tfrac{5+\sqrt{15}}{5} \qquad \text{or} \qquad x = \tfrac{5-\sqrt{15}}{5}$$

Solution set: $\left\{\tfrac{5\pm\sqrt{15}}{5}\right\}$

35.
$$9x^2 - 24x = -13$$
$$x^2 - \tfrac{24}{9}x = \tfrac{-13}{9} \qquad \textit{Divide by 9.}$$
$$x^2 - \tfrac{8}{3}x = \tfrac{-13}{9}$$

Complete the square.

$$\left[\tfrac{1}{2}\left(-\tfrac{8}{3}\right)\right]^2 = \left(-\tfrac{4}{3}\right)^2 = \tfrac{16}{9}$$

Add $\tfrac{16}{9}$ to each side.

$$x^2 - \tfrac{8}{3}x + \tfrac{16}{9} = \tfrac{-13}{9} + \tfrac{16}{9}$$
$$\left(x - \tfrac{4}{3}\right)^2 = \tfrac{3}{9}$$

$$x - \tfrac{4}{3} = \sqrt{\tfrac{3}{9}} \qquad \text{or} \quad x - \tfrac{4}{3} = -\sqrt{\tfrac{3}{9}}$$
$$x = \tfrac{4}{3} + \tfrac{\sqrt{3}}{3} \qquad\qquad x = \tfrac{4}{3} - \tfrac{\sqrt{3}}{3}$$
$$x = \tfrac{4+\sqrt{3}}{3} \qquad \text{or} \qquad x = \tfrac{4-\sqrt{3}}{3}$$

Solution set: $\left\{\tfrac{4\pm\sqrt{3}}{3}\right\}$

37.
$$(x + 3)(x - 1) = 5$$
$$x^2 + 2x - 3 = 5$$
$$x^2 + 2x = 8$$
$$x^2 + 2x + 1 = 8 + 1$$
$$(x + 1)^2 = 9$$

$$x + 1 = 3 \qquad \text{or} \qquad x + 1 = -3$$
$$x = 2 \qquad \text{or} \qquad x = -4$$

A check verifies that the solution set is $\{-4, 2\}$.

39.
$$(r - 3)(r - 5) = 2$$
$$r^2 - 8r + 15 = 2$$
$$r^2 - 8r = -13$$
$$r^2 - 8r + 16 = -13 + 16$$
$$(r - 4)^2 = 3$$

$$r - 4 = \sqrt{3} \qquad \text{or} \qquad r - 4 = -\sqrt{3}$$
$$r = 4 + \sqrt{3} \qquad \text{or} \qquad r = 4 - \sqrt{3}$$

A check verifies that the solution set is $\left\{4 \pm \sqrt{3}\right\}$.

41. $z^2 - \frac{4}{3}z = -\frac{1}{9}$

Complete the square.

$$\left[\frac{1}{2}\left(-\frac{4}{3}\right)\right]^2 = \left(-\frac{2}{3}\right)^2 = \frac{4}{9}$$

Add $\frac{4}{9}$ to each side.

$$z^2 - \frac{4}{3}z + \frac{4}{9} = -\frac{1}{9} + \frac{4}{9}$$

$$\left(z - \frac{2}{3}\right)^2 = \frac{3}{9}$$

$$z - \frac{2}{3} = \sqrt{\frac{3}{9}} \quad \text{or} \quad z - \frac{2}{3} = -\sqrt{\frac{3}{9}}$$

$$z = \frac{2}{3} + \frac{\sqrt{3}}{3} \qquad\qquad z = \frac{2}{3} - \frac{\sqrt{3}}{3}$$

$$z = \frac{2+\sqrt{3}}{3} \quad \text{or} \quad z = \frac{2-\sqrt{3}}{3}$$

Solution set: $\left\{\frac{2\pm\sqrt{3}}{3}\right\}$

43. $0.1x^2 - 0.2x - 0.1 = 0$

Multiply each side by 10 to clear the decimals.

$$x^2 - 2x - 1 = 0$$

$$x^2 - 2x = 1$$

Complete the square.

$$\left[\frac{1}{2}(-2)\right]^2 = (-1)^2 = 1$$

Add 1 to each side.

$$x^2 - 2x + 1 = 1 + 1$$

$$(x-1)^2 = 2$$

$$x - 1 = \sqrt{2} \quad \text{or} \quad x - 1 = -\sqrt{2}$$

$$x = 1 + \sqrt{2} \quad \text{or} \quad x = 1 - \sqrt{2}$$

Solution set: $\left\{1 \pm \sqrt{2}\right\}$

45. $3r^2 - 2 = 6r + 3$

$$3r^2 - 6r = 5$$

$$r^2 - 2r = \frac{5}{3}$$

$$r^2 - 2r + 1 = \frac{5}{3} + 1$$

$$(r-1)^2 = \frac{8}{3}$$

$$r - 1 = \pm\sqrt{\frac{8}{3}}$$

Simplify the radical.

$$\sqrt{\frac{8}{3}} = \frac{\sqrt{8}}{\sqrt{3}} = \frac{2\sqrt{2}}{\sqrt{3}} \cdot \frac{\sqrt{3}}{\sqrt{3}} = \frac{2\sqrt{6}}{3}$$

$$r = 1 \pm \frac{2\sqrt{6}}{3}$$

$$r = \frac{3}{3} \pm \frac{2\sqrt{6}}{3} = \frac{3\pm2\sqrt{6}}{3}$$

(a) The solution set with *exact* values is $\left\{\frac{3\pm2\sqrt{6}}{3}\right\}$.

(b) $\frac{3+2\sqrt{6}}{3} \approx 2.633$

$\frac{3-2\sqrt{6}}{3} \approx -0.633$

The solution set with *approximate* values is $\{-0.633, 2.633\}$.

47. $(x+1)(x+3) = 2$

$$x^2 + 3x + x + 3 = 2$$

$$x^2 + 4x = -1$$

$$x^2 + 4x + 4 = -1 + 4$$

$$(x+2)^2 = 3$$

$$x + 2 = \pm\sqrt{3}$$

$$x = -2 \pm \sqrt{3}$$

(a) The solution set with *exact* values is $\left\{-2 \pm \sqrt{3}\right\}$.

(b) $-2 + \sqrt{3} \approx -0.268$

$-2 - \sqrt{3} \approx -3.732$

The solution set with *approximate* values is $\{-3.732, -0.268\}$.

49. $m^2 + 4m + 13 = 0$

$$m^2 + 4m = -13$$

Complete the square.

$$\left(\frac{1}{2}\cdot 4\right)^2 = 2^2 = 4$$

Add 4 to each side.

$$m^2 + 4m + 4 = -13 + 4$$

$$(m+2)^2 = -9$$

$$m + 2 = \sqrt{-9} \quad \text{or} \quad m + 2 = -\sqrt{-9}$$

$$m = -2 + 3i \quad \text{or} \quad m = -2 - 3i$$

Solution set: $\{-2 \pm 3i\}$

51. $3r^2 + 4r + 4 = 0$

$$3r^2 + 4r = -4$$

$$r^2 + \frac{4}{3}r = \frac{-4}{3} \qquad \textit{Divide by 3.}$$

Complete the square.

$$\left(\frac{1}{2}\cdot\frac{4}{3}\right)^2 = \left(\frac{2}{3}\right)^2 = \frac{4}{9}$$

Add $\frac{4}{9}$ to each side.

$$r^2 + \frac{4}{3}r + \frac{4}{9} = \frac{-4}{3} + \frac{4}{9}$$

$$\left(r + \frac{2}{3}\right)^2 = \frac{-8}{9}$$

$$r + \frac{2}{3} = \frac{\sqrt{-8}}{\sqrt{9}} \quad \text{or} \quad r + \frac{2}{3} = -\frac{\sqrt{-8}}{\sqrt{9}}$$

$$r = -\frac{2}{3} + \frac{2i\sqrt{2}}{3} \qquad r = -\frac{2}{3} - \frac{2i\sqrt{2}}{3}$$

Solution set: $\left\{-\frac{2}{3} \pm \frac{2\sqrt{2}}{3}i\right\}$

53. $-m^2 - 6m - 12 = 0$

Multiply each side by -1.

$$m^2 + 6m + 12 = 0$$

$$m^2 + 6m = -12$$

Complete the square.

$$\left(\frac{1}{2}\cdot 6\right)^2 = 3^2 = 9$$

Add 9 to each side.

$$m^2 + 6m + 9 = -12 + 9$$
$$(m + 3)^2 = -3$$
$$m + 3 = \sqrt{-3} \quad \text{or} \quad m + 3 = -\sqrt{-3}$$
$$m = -3 + i\sqrt{3} \quad \text{or} \quad m = -3 - i\sqrt{3}$$

Solution set: $\left\{ -3 \pm i\sqrt{3} \right\}$

55. The area of the original square is $x \cdot x$, or x^2.

56. Each rectangular strip has length x and width 1, so each strip has an area of $x \cdot 1$, or x.

57. From Exercise 56, the area of a rectangular strip is x. The area of 6 rectangular strips is $6x$.

58. These are 1 by 1 squares, so each has an area of $1 \cdot 1$, or 1.

59. There are 9 small squares, each with area 1 (from Exercise 58), so the total area is $9 \cdot 1$, or 9.

60. The area of the larger square is $(x + 3)^2$. Using the results from Exercises 55–59,

$$(x + 3)^2, \quad \text{or} \quad x^2 + 6x + 9.$$

61. $x^2 - b = 0$
$$x^2 = b$$
$$x = \sqrt{b} \quad \text{or} \quad x = -\sqrt{b}$$

Solution set: $\left\{ \pm\sqrt{b} \right\}$

63. $4x^2 = b^2 + 16$
$$x^2 = \frac{b^2 + 16}{4}$$
$$x = \sqrt{\frac{b^2 + 16}{4}} \quad \text{or} \quad x = -\sqrt{\frac{b^2 + 16}{4}}$$
$$x = \frac{\sqrt{b^2 + 16}}{2} \quad \text{or} \quad x = -\frac{\sqrt{b^2 + 16}}{2}$$

Solution set: $\left\{ \pm \dfrac{\sqrt{b^2 + 16}}{2} \right\}$

65. $(5x - 2b)^2 = 3a$
$$5x - 2b = \sqrt{3a} \quad \text{or } 5x - 2b = -\sqrt{3a}$$
$$5x = 2b + \sqrt{3a} \qquad 5x = 2b - \sqrt{3a}$$
$$x = \frac{2b + \sqrt{3a}}{5} \quad \text{or} \qquad x = \frac{2b - \sqrt{3a}}{5}$$

Solution set: $\left\{ \dfrac{2b \pm \sqrt{3a}}{5} \right\}$

67. $a = 3, b = 1, c = -1$
$$\sqrt{b^2 - 4ac} = \sqrt{1^2 - 4(3)(-1)}$$
$$= \sqrt{1 + 12}$$
$$= \sqrt{13}$$

69. $a = 6, b = 7, c = 2$
$$\sqrt{b^2 - 4ac} = \sqrt{7^2 - 4(6)(2)}$$
$$= \sqrt{49 - 48}$$
$$= \sqrt{1} = 1$$

71. $a = 3, b = 1, c = -1$ (Exercise 67)
$$\frac{-b + \sqrt{b^2 - 4ac}}{2a} = \frac{-1 + \sqrt{13}}{2(3)}$$
$$= \frac{-1 + \sqrt{13}}{6}$$

11.3 Solving Quadratic Equations by the Quadratic Formula

1. The patron forgot the $\pm$ sign in the numerator. The correct formula is
$$x = \frac{-b \pm \sqrt{b^2 - 4ac}}{2a}.$$

3. No, the quadratic formula can be used to solve *any* quadratic equation. Here, the quadratic formula can be used with $a = 2$, $b = 0$, and $c = -5$.

5. $x^2 - 8x + 15 = 0$
Here $a = 1$, $b = -8$, and $c = 15$.
$$x = \frac{-b \pm \sqrt{b^2 - 4ac}}{2a}$$
$$x = \frac{-(-8) \pm \sqrt{(-8)^2 - 4(1)(15)}}{2(1)}$$
$$= \frac{8 \pm \sqrt{64 - 60}}{2}$$
$$= \frac{8 \pm \sqrt{4}}{2} = \frac{8 \pm 2}{2}$$
$$x = \frac{8 + 2}{2} = \frac{10}{2} = 5 \text{ or}$$
$$x = \frac{8 - 2}{2} = \frac{6}{2} = 3$$

Solution set: $\{3, 5\}$

7. $2x^2 + 4x + 1 = 0$
Here $a = 2$, $b = 4$, and $c = 1$.
$$x = \frac{-b \pm \sqrt{b^2 - 4ac}}{2a}$$
$$x = \frac{-4 \pm \sqrt{4^2 - 4(2)(1)}}{2(2)}$$
$$= \frac{-4 \pm \sqrt{16 - 8}}{4}$$
$$= \frac{-4 \pm \sqrt{8}}{4} = \frac{-4 \pm 2\sqrt{2}}{4}$$
$$= \frac{2\left(-2 \pm \sqrt{2}\right)}{2 \cdot 2} = \frac{-2 \pm \sqrt{2}}{2}$$

Solution set: $\left\{ \dfrac{-2 + \sqrt{2}}{2}, \dfrac{-2 - \sqrt{2}}{2} \right\}$

9.
$$2x^2 - 2x = 1$$
$$2x^2 - 2x - 1 = 0$$
Here $a = 2$, $b = -2$, and $c = -1$.
$$x = \frac{-b \pm \sqrt{b^2 - 4ac}}{2a}$$
$$x = \frac{-(-2) \pm \sqrt{(-2)^2 - 4(2)(-1)}}{2(2)}$$
$$= \frac{2 \pm \sqrt{4 + 8}}{4} = \frac{2 \pm \sqrt{12}}{4} = \frac{2 \pm 2\sqrt{3}}{4}$$
$$= \frac{2\left(1 \pm \sqrt{3}\right)}{2 \cdot 2} = \frac{1 \pm \sqrt{3}}{2}$$
Solution set: $\left\{ \dfrac{1 + \sqrt{3}}{2}, \dfrac{1 - \sqrt{3}}{2} \right\}$

11.
$$x^2 + 18 = 10x$$
$$x^2 - 10x + 18 = 0$$
Here $a = 1$, $b = -10$, and $c = 18$.
$$x = \frac{-b \pm \sqrt{b^2 - 4ac}}{2a}$$
$$x = \frac{-(-10) \pm \sqrt{(-10)^2 - 4(1)(18)}}{2(1)}$$
$$= \frac{10 \pm \sqrt{100 - 72}}{2} = \frac{10 \pm \sqrt{28}}{2}$$
$$= \frac{10 \pm 2\sqrt{7}}{2} = \frac{2\left(5 \pm \sqrt{7}\right)}{2} = 5 \pm \sqrt{7}$$
Solution set: $\left\{ 5 + \sqrt{7}, 5 - \sqrt{7} \right\}$

13. $4k^2 + 4k - 1 = 0$
Here $a = 4$, $b = 4$, and $c = -1$.
$$k = \frac{-b \pm \sqrt{b^2 - 4ac}}{2a}$$
$$k = \frac{-4 \pm \sqrt{4^2 - 4(4)(-1)}}{2(4)}$$
$$= \frac{-4 \pm \sqrt{16 + 16}}{8} = \frac{-4 \pm \sqrt{32}}{8}$$
$$= \frac{-4 \pm 4\sqrt{2}}{8} = \frac{4\left(-1 \pm \sqrt{2}\right)}{2 \cdot 4} = \frac{-1 \pm \sqrt{2}}{2}$$
Solution set: $\left\{ \dfrac{-1 + \sqrt{2}}{2}, \dfrac{-1 - \sqrt{2}}{2} \right\}$

15. $2 - 2x = 3x^2$
$$0 = 3x^2 + 2x - 2$$
Here $a = 3$, $b = 2$, and $c = -2$.
$$x = \frac{-b \pm \sqrt{b^2 - 4ac}}{2a}$$
$$x = \frac{-2 \pm \sqrt{2^2 - 4(3)(-2)}}{2(3)}$$
$$= \frac{-2 \pm \sqrt{4 + 24}}{6} = \frac{-2 \pm \sqrt{28}}{6}$$

$$= \frac{-2 \pm 2\sqrt{7}}{6} = \frac{2\left(-1 \pm \sqrt{7}\right)}{2 \cdot 3} = \frac{-1 \pm \sqrt{7}}{3}$$
Solution set: $\left\{ \dfrac{-1 + \sqrt{7}}{3}, \dfrac{-1 - \sqrt{7}}{3} \right\}$

17.
$$\frac{x^2}{4} - \frac{x}{2} = 1$$
$$\frac{x^2}{4} - \frac{x}{2} - 1 = 0$$
$$x^2 - 2x - 4 = 0 \quad \textit{Multiply by 4.}$$
Here $a = 1$, $b = -2$, and $c = -4$.
$$x = \frac{-b \pm \sqrt{b^2 - 4ac}}{2a}$$
$$x = \frac{-(-2) \pm \sqrt{(-2)^2 - 4(1)(-4)}}{2(1)}$$
$$= \frac{2 \pm \sqrt{4 + 16}}{2} = \frac{2 \pm \sqrt{20}}{2}$$
$$= \frac{2 \pm 2\sqrt{5}}{2} = 1 \pm \sqrt{5}$$
Solution set: $\left\{ 1 + \sqrt{5}, 1 - \sqrt{5} \right\}$

19.
$$-2t(t + 2) = -3$$
$$-2t^2 - 4t = -3$$
$$-2t^2 - 4t + 3 = 0$$
Here $a = -2$, $b = -4$, and $c = 3$.
$$t = \frac{-b \pm \sqrt{b^2 - 4ac}}{2a}$$
$$t = \frac{-(-4) \pm \sqrt{(-4)^2 - 4(-2)(3)}}{2(-2)}$$
$$= \frac{4 \pm \sqrt{16 + 24}}{-4} = \frac{4 \pm \sqrt{40}}{-4}$$
$$= \frac{4 \pm 2\sqrt{10}}{-4} = \frac{2\left(2 \pm \sqrt{10}\right)}{-2 \cdot 2}$$
$$= \frac{2 \pm \sqrt{10}}{-2} \cdot \frac{-1}{-1} = \frac{-2 \mp \sqrt{10}}{2}$$
$$= \frac{-2 \pm \sqrt{10}}{2}$$
Solution set: $\left\{ \dfrac{-2 + \sqrt{10}}{2}, \dfrac{-2 - \sqrt{10}}{2} \right\}$

21. $(r - 3)(r + 5) = 2$
$$r^2 + 2r - 15 = 2$$
$$r^2 + 2r - 17 = 0$$
Here $a = 1$, $b = 2$, and $c = -17$.
$$r = \frac{-b \pm \sqrt{b^2 - 4ac}}{2a}$$
$$r = \frac{-2 \pm \sqrt{2^2 - 4(1)(-17)}}{2(1)}$$

$$= \frac{-2 \pm \sqrt{4 + 68}}{2} = \frac{-2 \pm \sqrt{72}}{2}$$

$$= \frac{-2 \pm 6\sqrt{2}}{2} = \frac{2\left(-1 \pm 3\sqrt{2}\right)}{2}$$

$$= -1 \pm 3\sqrt{2}$$

Solution set: $\left\{-1 + 3\sqrt{2}, -1 - 3\sqrt{2}\right\}$

23. $(x + 2)(x - 3) = 1$

$$x^2 - x - 6 = 1$$
$$x^2 - x - 7 = 0$$

Here $a = 1$, $b = -1$, and $c = -7$.

$$x = \frac{-b \pm \sqrt{b^2 - 4ac}}{2a}$$

$$x = \frac{-(-1) \pm \sqrt{(-1)^2 - 4(1)(-7)}}{2(1)}$$

$$= \frac{1 \pm \sqrt{1 + 28}}{2} = \frac{1 \pm \sqrt{29}}{2}$$

Solution set: $\left\{\dfrac{1 + \sqrt{29}}{2}, \dfrac{1 - \sqrt{29}}{2}\right\}$

25.
$$p = \frac{5(5 - p)}{3(p + 1)}$$
$$3p(p + 1) = 5(5 - p)$$
$$3p^2 + 3p = 25 - 5p$$
$$3p^2 + 8p - 25 = 0$$

Here $a = 3$, $b = 8$, and $c = -25$.

$$p = \frac{-b \pm \sqrt{b^2 - 4ac}}{2a}$$

$$p = \frac{-8 \pm \sqrt{8^2 - 4(3)(-25)}}{2(3)}$$

$$= \frac{-8 \pm \sqrt{64 + 300}}{6} = \frac{-8 \pm \sqrt{364}}{6}$$

$$= \frac{-8 \pm 2\sqrt{91}}{6} = \frac{2\left(-4 \pm \sqrt{91}\right)}{2 \cdot 3} = \frac{-4 \pm \sqrt{91}}{3}$$

Solution set: $\left\{\dfrac{-4 + \sqrt{91}}{3}, \dfrac{-4 - \sqrt{91}}{3}\right\}$

27.
$$(2x + 1)^2 = x + 4$$
$$4x^2 + 4x + 1 = x + 4$$
$$4x^2 + 3x - 3 = 0$$

Here $a = 4$, $b = 3$, and $c = -3$.

$$x = \frac{-b \pm \sqrt{b^2 - 4ac}}{2a}$$

$$x = \frac{-3 \pm \sqrt{3^2 - 4(4)(-3)}}{2(4)}$$

$$= \frac{-3 \pm \sqrt{9 + 48}}{8}$$

$$= \frac{-3 \pm \sqrt{57}}{8}$$

Solution set: $\left\{\dfrac{-3 + \sqrt{57}}{8}, \dfrac{-3 - \sqrt{57}}{8}\right\}$

29. $x^2 - 3x + 6 = 0$

Here $a = 1$, $b = -3$, and $c = 6$.

$$x = \frac{-b \pm \sqrt{b^2 - 4ac}}{2a}$$

$$x = \frac{-(-3) \pm \sqrt{(-3)^2 - 4(1)(6)}}{2(1)}$$

$$= \frac{3 \pm \sqrt{9 - 24}}{2}$$

$$= \frac{3 \pm \sqrt{-15}}{2}$$

$$= \frac{3 \pm i\sqrt{15}}{2} = \frac{3}{2} \pm \frac{\sqrt{15}}{2}i$$

Solution set: $\left\{\dfrac{3}{2} + \dfrac{\sqrt{15}}{2}i, \dfrac{3}{2} - \dfrac{\sqrt{15}}{2}i\right\}$

31. $r^2 - 6r + 14 = 0$

Here $a = 1$, $b = -6$, and $c = 14$.

$$r = \frac{-b \pm \sqrt{b^2 - 4ac}}{2a}$$

$$r = \frac{-(-6) \pm \sqrt{(-6)^2 - 4(1)(14)}}{2(1)}$$

$$= \frac{6 \pm \sqrt{36 - 56}}{2}$$

$$= \frac{6 \pm \sqrt{-20}}{2} = \frac{6 \pm 2i\sqrt{5}}{2}$$

$$= \frac{2\left(3 \pm i\sqrt{5}\right)}{2} = 3 \pm i\sqrt{5}$$

Solution set: $\left\{3 + i\sqrt{5}, 3 - i\sqrt{5}\right\}$

33.
$$4x^2 - 4x = -7$$
$$4x^2 - 4x + 7 = 0$$

Here $a = 4$, $b = -4$, and $c = 7$.

$$x = \frac{-b \pm \sqrt{b^2 - 4ac}}{2a}$$

$$x = \frac{-(-4) \pm \sqrt{(-4)^2 - 4(4)(7)}}{2(4)}$$

$$= \frac{4 \pm \sqrt{16 - 112}}{8} = \frac{4 \pm \sqrt{-96}}{8}$$

$$= \frac{4 \pm 4i\sqrt{6}}{8} = \frac{4\left(1 \pm i\sqrt{6}\right)}{2 \cdot 4}$$

$$= \frac{1 \pm i\sqrt{6}}{2} = \frac{1}{2} \pm \frac{\sqrt{6}}{2}i$$

Solution set: $\left\{\dfrac{1}{2} + \dfrac{\sqrt{6}}{2}i, \dfrac{1}{2} - \dfrac{\sqrt{6}}{2}i\right\}$

35.
$$x(3x + 4) = -2$$
$$3x^2 + 4x = -2$$
$$3x^2 + 4x + 2 = 0$$
Here $a = 3$, $b = 4$, and $c = 2$.

$$x = \frac{-b \pm \sqrt{b^2 - 4ac}}{2a}$$

$$x = \frac{-4 \pm \sqrt{4^2 - 4(3)(2)}}{2(3)}$$

$$= \frac{-4 \pm \sqrt{16 - 24}}{6} = \frac{-4 \pm \sqrt{-8}}{6}$$

$$= \frac{-4 \pm 2i\sqrt{2}}{6} = \frac{2\left(-2 \pm i\sqrt{2}\right)}{2 \cdot 3}$$

$$= \frac{-2 \pm i\sqrt{2}}{3} = -\frac{2}{3} \pm \frac{\sqrt{2}}{3}i$$

Solution set: $\left\{ -\frac{2}{3} + \frac{\sqrt{2}}{3}i, -\frac{2}{3} - \frac{\sqrt{2}}{3}i \right\}$

37.
$$(2x - 1)(8x - 4) = -1$$
$$16x^2 - 16x + 4 = -1$$
$$16x^2 - 16x + 5 = 0$$
Here $a = 16$, $b = -16$, and $c = 5$.

$$x = \frac{-b \pm \sqrt{b^2 - 4ac}}{2a}$$

$$x = \frac{-(-16) \pm \sqrt{(-16)^2 - 4(16)(5)}}{2(16)}$$

$$= \frac{16 \pm \sqrt{256 - 320}}{32}$$

$$= \frac{16 \pm \sqrt{-64}}{32} = \frac{16 \pm 8i}{32}$$

$$= \frac{16}{32} \pm \frac{8}{32}i = \frac{1}{2} \pm \frac{1}{4}i$$

Solution set: $\left\{ \frac{1}{2} + \frac{1}{4}i, \frac{1}{2} - \frac{1}{4}i \right\}$

Note: We could also solve this equation without the quadratic formula as follows:
$$(2x - 1)(8x - 4) = -1$$
$$(2x - 1)4(2x - 1) = -1$$
$$(2x - 1)^2 = -\frac{1}{4}$$
$$2x - 1 = \pm \sqrt{-\frac{1}{4}}$$
$$2x = 1 \pm \frac{1}{2}i$$
$$x = \frac{1}{2} \pm \frac{1}{4}i$$

39. $25x^2 + 70x + 49 = 0$

Here $a = 25$, $b = 70$, and $c = 49$, so the discriminant is

$$b^2 - 4ac = 70^2 - 4(25)(49)$$
$$= 4900 - 4900$$
$$= 0.$$

Since the discriminant is 0, the quantity under the radical in the quadratic formula is 0, and there is only one rational solution. The answer is **B**.

41. $x^2 + 4x + 2 = 0$

Here $a = 1$, $b = 4$, and $c = 2$, so the discriminant is

$$b^2 - 4ac = 4^2 - 4(1)(2)$$
$$= 16 - 8$$
$$= 8.$$

Since the discriminant is positive, but not a perfect square, there are two distinct irrational number solutions. The answer is **C**.

43.
$$3x^2 = 5x + 2$$
$$3x^2 - 5x - 2 = 0$$
Here $a = 3$, $b = -5$, and $c = -2$, so the discriminant is

$$b^2 - 4ac = (-5)^2 - 4(3)(-2)$$
$$= 25 + 24$$
$$= 49.$$

Since the discriminant is a perfect square, there are two distinct rational solutions. The answer is **A**.

45. $3m^2 - 10m + 15 = 0$

Here $a = 3$, $b = -10$, and $c = 15$, so the discriminant is

$$b^2 - 4ac = (-10)^2 - 4(3)(15)$$
$$= 100 - 180$$
$$= -80.$$

Since the discriminant is negative, there are two distinct nonreal complex number solutions. The answer is **D**.

47. $0.5x^2 + 10x + 50 = 0$

Here $a = 0.5$, $b = 10$, and $c = 50$, so the discriminant is

$$b^2 - 4ac = 10^2 - 4(0.5)(50)$$
$$= 100 - 100$$
$$= 0.$$

Since the discriminant is 0, the quantity under the radical in the quadratic formula is 0, and there is only one rational solution. The answer is **B**.

49.
$$25x^2 + 70x + 49 = 0$$
$$(5x + 7)^2 = 0$$
$$5x + 7 = 0$$
$$x = -\frac{7}{5}$$

Solution set: $\left\{ -\frac{7}{5} \right\}$

51.
$$3x^2 = 5x + 2$$
$$3x^2 - 5x - 2 = 0$$
$$(3x + 1)(x - 2) = 0$$

$$3x + 1 = 0 \quad \text{or} \quad x - 2 = 0$$
$$x = -\tfrac{1}{3} \quad \text{or} \quad x = 2$$

Solution set: $\left\{-\tfrac{1}{3}, 2\right\}$

53. (a)
$$3k^2 + 13k = -12$$
$$3k^2 + 13k + 12 = 0$$

Here $a = 3$, $b = 13$, and $c = 12$, so the discriminant is

$$b^2 - 4ac = 13^2 - 4(3)(12)$$
$$= 169 - 144$$
$$= 25.$$

The discriminant is a perfect square, so the equation can be solved by factoring.

$$3k^2 + 13k + 12 = 0$$
$$(3k + 4)(k + 3) = 0$$

$$3k + 4 = 0 \quad \text{or} \quad k + 3 = 0$$
$$k = -\tfrac{4}{3} \qquad\qquad k = -3$$

Solution set: $\left\{-\tfrac{4}{3}, -3\right\}$

(b)
$$2x^2 + 19 = 14x$$
$$2x^2 - 14x + 19 = 0$$

Here $a = 2$, $b = -14$, and $c = 19$, so the discriminant is

$$b^2 - 4ac = (-14)^2 - 4(2)(19)$$
$$= 196 - 152$$
$$= 44.$$

The discriminant is not a perfect square, so use the quadratic formula.

$$x = \frac{-b \pm \sqrt{b^2 - 4ac}}{2a}$$

$$x = \frac{-(-14) \pm \sqrt{44}}{2 \cdot 2}$$

$$= \frac{14 \pm 2\sqrt{11}}{2 \cdot 2} = \frac{7 \pm \sqrt{11}}{2}$$

Solution set: $\left\{\dfrac{7 + \sqrt{11}}{2}, \dfrac{7 - \sqrt{11}}{2}\right\}$

55. $p^2 + bp + 25 = 0$
For there to be only one rational solution, $b^2 - 4ac$ must equal zero.
Since $a = 1$ and $c = 25$,

$$b^2 - 4(1)(25) = 0$$
$$b^2 - 100 = 0$$
$$b^2 = 100$$
$$b = \pm\sqrt{100}$$
$$b = 10 \quad \text{or} \quad b = -10.$$

57. $am^2 + 8m + 1 = 0$
For there to be only one rational solution, $b^2 - 4ac$ must equal zero.
Since $b = 8$ and $c = 1$,

$$8^2 - 4(a)(1) = 0$$
$$64 - 4a = 0$$
$$-4a = -64$$
$$a = 16.$$

59. $9x^2 - 30x + c = 0$
For there to be only one rational solution, $b^2 - 4ac$ must equal zero.
Since $a = 9$ and $b = -30$,

$$(-30)^2 - 4(9)(c) = 0$$
$$900 - 36c = 0$$
$$-36c = -900$$
$$c = 25.$$

61. Substitute $-\tfrac{5}{2}$ for x and solve for b.

$$4x^2 + bx - 3 = 0$$
$$4\left(-\tfrac{5}{2}\right)^2 + b\left(-\tfrac{5}{2}\right) - 3 = 0$$
$$25 - \tfrac{5}{2}b - 3 = 0$$
$$22 = \tfrac{5}{2}b$$
$$\tfrac{44}{5} = b$$

So the equation is

$$4x^2 + \tfrac{44}{5}x - 3 = 0.$$

Now multiply by 5.

$$20x^2 + 44x - 15 = 0$$

Since $-\tfrac{5}{2}$ is a solution, $2x + 5$ must be one factor.

$$(2x + 5)(10x - 3) = 0$$

So the other solution is $\tfrac{3}{10}$.

63. $(7z + 3)^2 + 4(7z + 3) - 5$
$$= u^2 + 4u - 5 \qquad \textit{Let } u = 7z + 3.$$
$$= (u + 5)(u - 1)$$

65. $\tfrac{3}{4}x + \tfrac{1}{2}x = -10$
$$3x + 2x = -40 \quad \textit{Multiply by 4.}$$
$$5x = -40$$
$$x = -8$$

Solution set: $\{-8\}$.

67.
$$\sqrt{2x+6} = x-1$$
$$\left(\sqrt{2x+6}\right)^2 = (x-1)^2$$
$$2x+6 = x^2 - 2x + 1$$
$$0 = x^2 - 4x - 5$$
$$0 = (x-5)(x+1)$$

$$x-5=0 \quad \text{or} \quad x+1=0$$
$$x=5 \quad \text{or} \quad x=-1$$

Check $x=5$: $\sqrt{16} = 4$ *True*
Check $x=-1$: $\sqrt{4} = -2$ *False*

Solution set: $\{5\}$

11.4 Equations Quadratic in Form

1. $(2x+3)^2 = 4$

Since the equation has the form $(ax+b)^2 = c$, use the *square root property*.

3. $x^2 + 5x - 8 = 0$

The discriminant is
$$b^2 - 4ac = 5^2 - 4(1)(-8)$$
$$= 25 + 32 = 57.$$

Since the discriminant is not a perfect square, use the *quadratic formula*.

5.
$$3x^2 = 2 - 5x$$
$$3x^2 + 5x - 2 = 0$$

The discriminant is
$$b^2 - 4ac = 5^2 - 4(3)(-2)$$
$$= 25 + 24 = 49.$$

Since the discriminant is a perfect square, use *factoring*.

7. $\dfrac{14}{x} = x - 5$

This is a rational equation, so multiply both sides by the LCD, x.

9. $\left(x^2 + x\right)^2 - 8\left(x^2 + x\right) + 12 = 0$

This is quadratic in form, so substitute a variable for $x^2 + x$.

11. The proposed solution -1 does not check.
Solution set: $\{4\}$

13.
$$\frac{14}{x} = x - 5$$

To clear the fraction, multiply each term by the LCD, x.
$$x\left(\frac{14}{x}\right) = x(x) - 5x$$
$$14 = x^2 - 5x$$
$$-x^2 + 5x + 14 = 0$$
$$x^2 - 5x - 14 = 0$$
$$(x+2)(x-7) = 0$$

$$x+2=0 \quad \text{or} \quad x-7=0$$
$$x=-2 \quad \text{or} \quad x=7$$

Check $x=-2$: $-7 = -7$ *True*
Check $x=7$: $2 = 2$ *True*

Solution set: $\{-2, 7\}$

15.
$$1 - \frac{3}{x} - \frac{28}{x^2} = 0$$

Multiply by the LCD, x^2.
$$x^2(1) - x^2\left(\frac{3}{x}\right) - x^2\left(\frac{28}{x^2}\right) = x^2 \cdot 0$$
$$x^2 - 3x - 28 = 0$$
$$(x+4)(x-7) = 0$$

$$x+4=0 \quad \text{or} \quad x-7=0$$
$$x=-4 \quad \text{or} \quad x=7$$

Check $x=-4$: $1 + \frac{3}{4} - \frac{7}{4} = 0$ *True*
Check $x=7$: $1 - \frac{3}{7} - \frac{4}{7} = 0$ *True*

Solution set: $\{-4, 7\}$

17. $3 - \dfrac{1}{t} = \dfrac{2}{t^2}$

Multiply each term by the LCD, t^2.
$$t^2(3) - t^2\left(\frac{1}{t}\right) = t^2\left(\frac{2}{t^2}\right)$$
$$3t^2 - t = 2$$
$$3t^2 - t - 2 = 0$$
$$(3t+2)(t-1) = 0$$

$$3t+2=0 \quad \text{or} \quad t-1=0$$
$$t = -\frac{2}{3} \quad \text{or} \quad t=1$$

Check $t=-\frac{2}{3}$: $3 + \frac{3}{2} = \frac{9}{2}$ *True*
Check $t=1$: $3 - 1 = 2$ *True*
Solution set: $\left\{-\frac{2}{3}, 1\right\}$

19.
$$\frac{1}{x} + \frac{2}{x+2} = \frac{17}{35}$$

Multiply by the LCD, $35x(x+2)$.
$$35x(x+2)\left(\frac{1}{x}\right) + 35x(x+2)\left(\frac{2}{x+2}\right)$$
$$= 35x(x+2)\left(\tfrac{17}{35}\right)$$
$$35(x+2) + 35x(2) = 17x(x+2)$$
$$35x + 70 + 70x = 17x^2 + 34x$$
$$70 + 105x = 17x^2 + 34x$$
$$0 = 17x^2 - 71x - 70$$
$$0 = (17x+14)(x-5)$$

$$17x+14=0 \quad \text{or} \quad x-5=0$$
$$x = -\frac{14}{17} \quad \text{or} \quad x=5$$

Check $x = -\frac{14}{17}$: $-\frac{17}{14} + \frac{17}{10} = \frac{17}{35}$ *True*
Check $x=5$: $\frac{1}{5} + \frac{2}{7} = \frac{17}{35}$ *True*
Solution set: $\left\{-\frac{14}{17}, 5\right\}$

21. $\dfrac{2}{x+1} + \dfrac{3}{x+2} = \dfrac{7}{2}$

Multiply by the LCD, $2(x+1)(x+2)$.

$$2(x+1)(x+2)\left(\dfrac{2}{x+1} + \dfrac{3}{x+2}\right)$$
$$= 2(x+1)(x+2)\left(\tfrac{7}{2}\right)$$

$$2(x+2)(2) + 2(x+1)(3)$$
$$= (x+1)(x+2)(7)$$
$$4x + 8 + 6x + 6 = \left(x^2 + 3x + 2\right)(7)$$
$$10x + 14 = 7x^2 + 21x + 14$$
$$0 = 7x^2 + 11x$$
$$0 = x(7x + 11)$$

$$x = 0 \quad \text{or} \quad 7x + 11 = 0$$
$$x = -\tfrac{11}{7}$$

Check $x = -\tfrac{11}{7}$: $\quad -\tfrac{7}{2} + 7 = \tfrac{7}{2} \quad$ *True*

Check $x = 0$: $\quad\quad 2 + \tfrac{3}{2} = \tfrac{7}{2} \quad$ *True*

Solution set: $\left\{-\tfrac{11}{7}, 0\right\}$

23. $\dfrac{3}{2x} - \dfrac{1}{2(x+2)} = 1$

Multiply by the LCD, $2x(x+2)$.

$$2x(x+2)\left(\dfrac{3}{2x} - \dfrac{1}{2(x+2)}\right)$$
$$= 2x(x+2) \cdot 1$$
$$3(x+2) - x(1) = 2x(x+2)$$
$$3x + 6 - x = 2x^2 + 4x$$
$$0 = 2x^2 + 2x - 6$$
$$0 = x^2 + x - 3$$

Use $a = 1$, $b = 1$, $c = -3$ in the quadratic formula.

$$x = \dfrac{-b \pm \sqrt{b^2 - 4ac}}{2a}$$

$$x = \dfrac{-1 \pm \sqrt{1^2 - 4(1)(-3)}}{2(1)}$$

$$= \dfrac{-1 \pm \sqrt{1 + 12}}{2} = \dfrac{-1 \pm \sqrt{13}}{2}$$

Use a calculator to check both proposed solutions. Both solutions check.

Solution set: $\left\{\dfrac{-1 \pm \sqrt{13}}{2}\right\}$

25. $3 = \dfrac{1}{t+2} + \dfrac{2}{(t+2)^2}$

Multiply by the LCD, $(t+2)^2$.

$$3(t+2)^2 = 1(t+2) + 2$$
$$3\left(t^2 + 4t + 4\right) = t + 2 + 2$$
$$3t^2 + 12t + 12 = t + 4$$
$$3t^2 + 11t + 8 = 0$$
$$(3t + 8)(t + 1) = 0$$

$$3t + 8 = 0 \quad \text{or} \quad t + 1 = 0$$
$$t = -\tfrac{8}{3} \quad\quad\quad t = -1$$

Check $t = -\tfrac{8}{3}$: $\quad 3 = -\tfrac{3}{2} + \tfrac{9}{2} \quad$ *True*

Check $t = -1$: $\quad 3 = 1 + 2 \quad$ *True*

Solution set: $\left\{-\tfrac{8}{3}, -1\right\}$

27. $\dfrac{6}{p} = 2 + \dfrac{p}{p+1}$

Multiply by the LCD, $p(p+1)$.

$$6(p+1) = 2p(p+1) + p \cdot p$$
$$6p + 6 = 2p^2 + 2p + p^2$$
$$0 = 3p^2 - 4p - 6$$

Use $a = 3$, $b = -4$, and $c = -6$ in the quadratic formula.

$$p = \dfrac{-b \pm \sqrt{b^2 - 4ac}}{2a}$$

$$p = \dfrac{-(-4) \pm \sqrt{(-4)^2 - 4(3)(-6)}}{2(3)}$$

$$= \dfrac{4 \pm \sqrt{16 + 72}}{2(3)} = \dfrac{4 \pm \sqrt{88}}{2(3)}$$

$$= \dfrac{4 \pm 2\sqrt{22}}{2(3)} = \dfrac{2 \pm \sqrt{22}}{3}$$

Use a calculator to check both proposed solutions. Both solutions check.

Solution set: $\left\{\dfrac{2 \pm \sqrt{22}}{3}\right\}$

29. $1 - \dfrac{1}{2x+1} - \dfrac{1}{(2x+1)^2} = 0$

Multiply by the LCD, $(2x+1)^2$.

$$(2x+1)^2 - 1(2x+1) - 1 = 0$$
$$4x^2 + 4x + 1 - 2x - 1 - 1 = 0$$
$$4x^2 + 2x - 1 = 0$$

Use $a = 4$, $b = 2$, $c = -1$ in the quadratic formula.

$$x = \dfrac{-b \pm \sqrt{b^2 - 4ac}}{2a}$$

$$x = \dfrac{-2 \pm \sqrt{2^2 - 4(4)(-1)}}{2(4)}$$

$$= \dfrac{-2 \pm \sqrt{4 + 16}}{2(4)} = \dfrac{-2 \pm \sqrt{20}}{2(4)}$$

$$= \dfrac{-2 \pm 2\sqrt{5}}{2(4)} = \dfrac{-1 \pm \sqrt{5}}{4}$$

Use a calculator to check both proposed solutions. Both solutions check.

Solution set: $\left\{\dfrac{-1 \pm \sqrt{5}}{4}\right\}$

31. Rate in still water: 20 mph
Rate of current: t mph

(a) When the boat travels upstream, the current works against the rate of the boat in still water, so the rate is $(20 - t)$ mph.

(b) When the boat travels downstream, the current works with the rate of the boat in still water, so the rate is $(20 + t)$ mph.

33. Let x = rate of the boat in still water.
With the speed of the current at 15 mph, then
$x - 15$ = rate going upstream and
$x + 15$ = rate going downstream.
Complete a table using the information in the problem, the rates given above, and the formula
$d = rt$ or $t = \dfrac{d}{r}$.

	d	r	t
Upstream	4	$x - 15$	$\dfrac{4}{x - 15}$
Downstream	16	$x + 15$	$\dfrac{16}{x + 15}$

The time, 48 min, is written as $\frac{48}{60} = \frac{4}{5}$ hr. The time upstream plus the time downstream equals $\frac{4}{5}$.
So, from the table, the equation is written as
$\dfrac{4}{x - 15} + \dfrac{16}{x + 15} = \dfrac{4}{5}$.
Multiply by the LCD, $5(x - 15)(x + 15)$.

$5(x - 15)(x + 15)\left(\dfrac{4}{x - 15} + \dfrac{16}{x + 15}\right)$
$\qquad = 5(x - 15)(x + 15) \cdot \dfrac{4}{5}$

$20(x + 15) + 80(x - 15)$
$\qquad = 4(x - 15)(x + 15)$

$20x + 300 + 80x - 1200$
$\qquad = 4(x^2 - 225)$

$100x - 900 = 4x^2 - 900$

$0 = 4x^2 - 100x$

$0 = 4x(x - 25)$

$4x = 0 \quad \text{or} \quad x - 25 = 0$
$x = 0 \quad \text{or} \qquad x = 25$

Reject $x = 0$ mph as a possible boat speed.
Eduardo's boat had a top speed of 25 mph.

35. Let x = Rico's speed from Jackson to Lodi.
Then $x + 10$ = his speed from Lodi to Manteca.

Make a table. Use $t = \frac{d}{r}$.

	d	r	t
Jackson to Lodi	40	x	$\dfrac{40}{x}$
Lodi to Manteca	40	$x + 10$	$\dfrac{40}{x + 10}$

Driving time for the entire trip was 88 minutes, or $\frac{88}{60} = \frac{22}{15}$ hours.

$\dfrac{40}{x} + \dfrac{40}{x + 10} = \dfrac{22}{15}$

Multiply by the LCD, $15x(x + 10)$.

$15x(x + 10)\left(\dfrac{40}{x} + \dfrac{40}{x + 10}\right)$
$\qquad = 15x(x + 10)\left(\tfrac{22}{15}\right)$

$600(x + 10) + 600x = 22x(x + 10)$

$600x + 6000 + 600x = 22x^2 + 220x$

$0 = 22x^2 - 980x - 6000$

$0 = 11x^2 - 490x - 3000$

$0 = (11x + 60)(x - 50)$

$11x + 60 = 0 \qquad \text{or} \quad x - 50 = 0$
$x = -\tfrac{60}{11} \quad \text{or} \qquad x = 50$

Reject $-\frac{60}{11}$ since speed cannot be negative. Rico's speed from Jackson to Lodi is 50 mph.

37. Let x be the time in hours required for the faster person to cut the lawn. Then the slower person requires $x + 1$ hours.
Complete the chart.

	Rate	Time Working Together	Fractional Part of the Job Done
Faster Worker	$\dfrac{1}{x}$	2	$\dfrac{2}{x}$
Slower Worker	$\dfrac{1}{x + 1}$	2	$\dfrac{2}{x + 1}$

$$\begin{array}{ccccc} \text{Part done by} \\ \text{faster person} \end{array} + \begin{array}{c} \text{Part done by} \\ \text{slower person} \end{array} = \begin{array}{c} \text{one whole} \\ \text{job.} \end{array}$$

$\qquad \dfrac{2}{x} \qquad + \qquad \dfrac{2}{x + 1} \qquad = \qquad 1$

Multiply each side by the LCD, $x(x + 1)$.

$x(x + 1)\left(\dfrac{2}{x} + \dfrac{2}{x + 1}\right) = x(x + 1) \cdot 1$

$2(x + 1) + 2x = x^2 + x$

$2x + 2 + 2x = x^2 + x$

$0 = x^2 - 3x - 2$

Solve for x using the quadratic formula with $a = 1$, $b = -3$, and $c = -2$.

$x = \dfrac{-(-3) \pm \sqrt{(-3)^2 - 4(1)(-2)}}{2(1)}$

$\quad = \dfrac{3 \pm \sqrt{9 + 8}}{2} = \dfrac{3 \pm \sqrt{17}}{2}$

$x = \dfrac{3 + \sqrt{17}}{2} \quad \text{or} \quad x = \dfrac{3 - \sqrt{17}}{2}$

$x \approx 3.6 \qquad \text{or} \quad x \approx -0.6$

Discard -0.6 as a solution since time cannot be negative.

It would take the faster person approximately 3.6 hours.

39. Let x represent the time in hours it takes Nancy to plant the flowers. Then $x + 2$ is the time it takes Rusty.

Organize the information in a chart.

Worker	Rate	Time Working Together	Fractional Part of the Job Done
Nancy	$\dfrac{1}{x}$	12	$\dfrac{12}{x}$
Rusty	$\dfrac{1}{x+2}$	12	$\dfrac{12}{x+2}$

$$\begin{array}{ccccc} \text{Part done} & & \text{part done} & & \text{one whole} \\ \text{by Nancy} & + & \text{by Rusty} & = & \text{job.} \\ \dfrac{12}{x} & + & \dfrac{12}{x+2} & = & 1 \end{array}$$

Multiply each side by the LCD, $x(x + 2)$.

$$x(x+2)\left(\frac{12}{x} + \frac{12}{x+2}\right) = x(x+2) \cdot 1$$
$$12(x+2) + 12x = x^2 + 2x$$
$$12x + 24 + 12x = x^2 + 2x$$
$$0 = x^2 - 22x - 24$$

Solve for x using the quadratic formula with $a = 1$, $b = -22$, and $c = -24$.

$$x = \frac{-(-22) \pm \sqrt{(-22)^2 - 4(1)(-24)}}{2(1)}$$
$$= \frac{22 \pm \sqrt{580}}{2}$$
$$x = \frac{22 + \sqrt{580}}{2} \approx 23.0 \ \text{ or}$$
$$x = \frac{22 - \sqrt{580}}{2} \approx -1.0$$

Since x represents time, discard the negative solution.

Nancy takes about 23.0 hours planting flowers alone while Rusty takes about 25.0 hours planting alone.

41. Let $x =$ the number of hours it takes for the faster pipe alone to fill the tank.

$x + 3 =$ the number of hours it takes for the slower pipe alone to fill the tank.

Working together, both pipes can fill the tank in 2 hours. Make a chart.

Pipe	Rate	Time	Fractional Part of Tank Filled
Faster	$\dfrac{1}{x}$	2	$\dfrac{2}{x}$
Slower	$\dfrac{1}{x+3}$	2	$\dfrac{2}{x+3}$

Since together the faster and slower pipes fill one tank, the sum of their fractional parts is 1; that is,

$$\frac{2}{x} + \frac{2}{x+3} = 1.$$

Multiply by the LCD, $x(x + 3)$.

$$2(x+3) + 2(x) = x(x+3)$$
$$2x + 6 + 2x = x^2 + 3x$$
$$0 = x^2 - x - 6$$
$$0 = (x-3)(x+2)$$

$$x - 3 = 0 \quad \text{or} \quad x + 2 = 0$$
$$x = 3 \quad \text{or} \quad x = -2$$

Reject -2. The faster pipe takes 3 hours to fill the tank alone and the slower pipe takes 6 hours to fill the tank alone.

43.
$$x = \sqrt{7x - 10}$$
$$(x)^2 = \left(\sqrt{7x-10}\right)^2$$
$$x^2 = 7x - 10$$
$$x^2 - 7x + 10 = 0$$
$$(x-2)(x-5) = 0$$

$$x - 2 = 0 \quad \text{or} \quad x - 5 = 0$$
$$x = 2 \quad \text{or} \quad x = 5$$

Check $x = 2$: $2 = \sqrt{4}$ *True*
Check $x = 5$: $5 = \sqrt{25}$ *True*
Solution set: $\{2, 5\}$

45.
$$2x = \sqrt{11x + 3}$$
$$(2x)^2 = \left(\sqrt{11x+3}\right)^2$$
$$4x^2 = 11x + 3$$
$$4x^2 - 11x - 3 = 0$$
$$(4x+1)(x-3) = 0$$

$$4x + 1 = 0 \quad \text{or} \quad x - 3 = 0$$
$$x = -\tfrac{1}{4} \quad \text{or} \quad x = 3$$

Check $x = -\tfrac{1}{4}$: $-\tfrac{1}{2} = \sqrt{\tfrac{1}{4}}$ *False*
Check $x = 3$: $6 = \sqrt{36}$ *True*

Solution set: $\{3\}$

47.
$$3x = \sqrt{16 - 10x}$$
$$(3x)^2 = \left(\sqrt{16 - 10x}\right)^2$$
$$9x^2 = 16 - 10x$$
$$9x^2 + 10x - 16 = 0$$
$$(9x - 8)(x + 2) = 0$$
$$9x - 8 = 0 \quad \text{or} \quad x + 2 = 0$$
$$x = \tfrac{8}{9} \quad \text{or} \quad x = -2$$

Check $x = \tfrac{8}{9}$: $\tfrac{8}{3} = \sqrt{\tfrac{64}{9}}$ *True*

Check $x = -2$: $-6 = \sqrt{36}$ *False*

Solution set: $\left\{\tfrac{8}{9}\right\}$

49. $k + \sqrt{k} = 12$
$$\sqrt{k} = 12 - k$$
$$\left(\sqrt{k}\right)^2 = (12 - k)^2$$
$$k = 144 - 24k + k^2$$
$$0 = k^2 - 25k + 144$$
$$0 = (k - 9)(k - 16)$$
$$k - 9 = 0 \quad \text{or} \quad k - 16 = 0$$
$$k = 9 \qquad\qquad k = 16$$

Check $k = 9$: $9 + 3 = 12$ *True*

Check $k = 16$: $16 + 4 = 12$ *False*

Solution set: $\{9\}$

51.
$$m = \sqrt{\frac{6 - 13m}{5}}$$
$$m^2 = \frac{6 - 13m}{5}$$
$$5m^2 = 6 - 13m$$
$$5m^2 + 13m - 6 = 0$$
$$(5m - 2)(m + 3) = 0$$
$$5m - 2 = 0 \quad \text{or} \quad m + 3 = 0$$
$$m = \tfrac{2}{5} \quad \text{or} \quad m = -3$$

Check $m = \tfrac{2}{5}$: $\tfrac{2}{5} = \sqrt{\tfrac{4}{25}}$ *True*

Check $m = -3$: $-3 = \sqrt{9}$ *False*

Solution set: $\left\{\tfrac{2}{5}\right\}$

53.
$$-x = \sqrt{\frac{8 - 2x}{3}}$$
$$(-x)^2 = \left(\sqrt{\frac{8 - 2x}{3}}\right)^2$$
$$x^2 = \frac{8 - 2x}{3}$$
$$3x^2 = 8 - 2x$$
$$3x^2 + 2x - 8 = 0$$
$$(3x - 4)(x + 2) = 0$$
$$3x - 4 = 0 \quad \text{or} \quad x + 2 = 0$$
$$x = \tfrac{4}{3} \quad \text{or} \quad x = -2$$

Check $x = \tfrac{4}{3}$: $-\tfrac{4}{3} = \sqrt{\tfrac{16}{9}}$ *False*

Check $x = -2$: $2 = \sqrt{4}$ *True*

Solution set: $\{-2\}$

55. $x^4 - 29x^2 + 100 = 0$
  Let $u = x^2$ and $u^2 = x^4$ to get
$$u^2 - 29u + 100 = 0$$
$$(u - 4)(u - 25) = 0$$
$$u - 4 = 0 \quad \text{or} \quad u - 25 = 0$$
$$u = 4 \qquad \text{or} \qquad u = 25$$
To find x, substitute x^2 for u.
$$x^2 = 4 \qquad \text{or} \qquad x^2 = 25$$
$$x = \pm 2 \quad \text{or} \qquad x = \pm 5$$

Check $x = \pm 2$: $16 - 116 + 100 = 0$ *True*

Check $x = \pm 5$: $625 - 725 + 100 = 0$ *True*

Solution set: $\{\pm 5, \pm 2\}$

57. $4k^4 - 13k^2 + 9 = 0$
  Let $u = k^2$ and $u^2 = k^4$ to get
$$4u^2 - 13u + 9 = 0$$
$$(4u - 9)(u - 1) = 0$$
$$4u - 9 = 0 \quad \text{or} \quad u - 1 = 0$$
$$u = \tfrac{9}{4} \quad \text{or} \qquad u = 1.$$
To find k, substitute k^2 for u.
$$k^2 = \tfrac{9}{4} \qquad \text{or} \quad k^2 = 1$$
$$k = \pm \tfrac{3}{2} \quad \text{or} \quad k = \pm 1$$

Check $k = \pm \tfrac{3}{2}$: $\tfrac{81}{4} - \tfrac{117}{4} + 9 = 0$ *True*

Check $k = \pm 1$: $4 - 13 + 9 = 0$ *True*

Solution set: $\left\{\pm \tfrac{3}{2}, \pm 1\right\}$

59.
$$x^4 + 48 = 16x^2$$
$$x^4 - 16x^2 + 48 = 0$$
Let $u = x^2$, so $u^2 = x^4$. The equation becomes
$$u^2 - 16u + 48 = 0.$$
$$(u - 4)(u - 12) = 0$$
$$u - 4 = 0 \quad \text{or} \quad u - 12 = 0$$
$$u = 4 \quad \text{or} \qquad u = 12$$
To find x substitute x^2 for u.
$$x^2 = 4 \qquad\qquad x^2 = 12$$
$$x = \pm \sqrt{4} \qquad\quad x = \pm \sqrt{12}$$
$$x = \pm 2 \quad \text{or} \quad x = \pm 2\sqrt{3}$$

Check $x = \pm 2$: $16 + 48 = 64$ *True*

Check $x = \pm 2\sqrt{3}$: $144 + 48 = 192$ *True*

Solution set: $\left\{\pm 2, \pm 2\sqrt{3}\right\}$

61. $(x+3)^2 + 5(x+3) + 6 = 0$

Let $u = x + 3$, so $u^2 = (x+3)^2$.

$$u^2 + 5u + 6 = 0$$
$$(u+3)(u+2) = 0$$

$$u + 3 = 0 \quad \text{or} \quad u + 2 = 0$$
$$u = -3 \quad \text{or} \quad u = -2$$

To find x, substitute $x + 3$ for u.

$$x + 3 = -3 \quad \text{or} \quad x + 3 = -2$$
$$x = -6 \quad \text{or} \quad x = -5$$

Check $x = -6$: $9 - 15 + 6 = 0$ *True*
Check $x = -5$: $4 - 10 + 6 = 0$ *True*

Solution set: $\{-6, -5\}$

63. $3(m+4)^2 - 8 = 2(m+4)$

Let $x = m + 4$, so $x^2 = (m+4)^2$.

$$3x^2 - 8 = 2x$$
$$3x^2 - 2x - 8 = 0$$
$$(3x+4)(x-2) = 0$$

$$3x + 4 = 0 \quad \text{or} \quad x - 2 = 0$$
$$x = -\tfrac{4}{3} \qquad\qquad x = 2$$

$$m + 4 = -\tfrac{4}{3} \quad \text{or} \quad m + 4 = 2$$
$$m = -\tfrac{16}{3} \qquad\qquad m = -2$$

Check $m = -\tfrac{16}{3}$: $\tfrac{16}{3} - 8 = -\tfrac{8}{3}$ *True*
Check $m = -2$: $12 - 8 = 4$ *True*

Solution set: $\left\{-\tfrac{16}{3}, -2\right\}$

65. $2 + \dfrac{5}{3k-1} = \dfrac{-2}{(3k-1)^2}$

Let $u = 3k - 1$, so $u^2 = (3k-1)^2$.

$$2 + \frac{5}{u} = -\frac{2}{u^2}$$

Multiply by the LCD, u^2.

$$u^2\left(2 + \frac{5}{u}\right) = u^2\left(-\frac{2}{u^2}\right)$$
$$2u^2 + 5u = -2$$
$$2u^2 + 5u + 2 = 0$$
$$(2u+1)(u+2) = 0$$

$$2u + 1 = 0 \quad \text{or} \quad u + 2 = 0$$
$$u = -\tfrac{1}{2} \qquad\qquad u = -2$$

To find k, substitute $3k - 1$ for u.

$$3k - 1 = -\tfrac{1}{2} \quad \text{or} \quad 3k - 1 = -2$$
$$3k = \tfrac{1}{2} \qquad\qquad 3k = -1$$
$$k = \tfrac{1}{6} \quad \text{or} \quad k = -\tfrac{1}{3}$$

Check $k = \tfrac{1}{6}$: $2 - 10 = -8$ *True*
Check $k = -\tfrac{1}{3}$: $2 - \tfrac{5}{2} = -\tfrac{1}{2}$ *True*

Solution set: $\left\{-\tfrac{1}{3}, \tfrac{1}{6}\right\}$

67. $2 - 6(m-1)^{-2} = (m-1)^{-1}$

Let $u = m - 1$ to get

$$2 - 6u^{-2} = u^{-1}$$

or $\quad 2 - \dfrac{6}{u^2} = \dfrac{1}{u}.$

Multiply by the LCD, u^2.

$$2u^2 - 6 = u$$
$$2u^2 - u - 6 = 0$$
$$(2u+3)(u-2) = 0$$

$$2u + 3 = 0 \quad \text{or} \quad u - 2 = 0$$
$$u = -\tfrac{3}{2} \quad \text{or} \quad u = 2$$

To find m, substitute $m - 1$ for u.

$$m - 1 = -\tfrac{3}{2} \quad \text{or} \quad m - 1 = 2$$
$$m = -\tfrac{1}{2} \quad \text{or} \quad m = 3$$

Check $m = -\tfrac{1}{2}$: $2 - \tfrac{8}{3} = -\tfrac{2}{3}$ *True*
Check $m = 3$: $2 - \tfrac{3}{2} = \tfrac{1}{2}$ *True*

Solution set: $\left\{-\tfrac{1}{2}, 3\right\}$

69. $x^{2/3} + x^{1/3} - 2 = 0$

Let $u = x^{1/3}$, so $u^2 = x^{2/3}$.

$$u^2 + u - 2 = 0$$
$$(u+2)(u-1) = 0$$

$$u + 2 = 0 \quad \text{or} \quad u - 1 = 0$$
$$u = -2 \quad \text{or} \quad u = 1$$

To find x, substitute $x^{1/3}$ for u.

$$x^{1/3} = -2 \quad \text{or} \quad x^{1/3} = 1$$

Cube both sides of each equation.

$$\left(x^{1/3}\right)^3 = (-2)^3 \qquad \left(x^{1/3}\right)^3 = 1^3$$
$$x = -8 \qquad \text{or} \qquad x = 1$$

Check $x = -8$: $4 - 2 - 2 = 0$ *True*
Check $x = 1$: $1 + 1 - 2 = 0$ *True*

Solution set: $\{-8, 1\}$

71. $r^{2/3} + r^{1/3} - 12 = 0$

Let $u = r^{1/3}$, so $u^2 = r^{2/3}$. The equation becomes

$$u^2 + u - 12 = 0.$$
$$(u+4)(u-3) = 0$$

$$u + 4 = 0 \quad \text{or} \quad u - 3 = 0$$
$$u = -4 \quad \text{or} \quad u = 3$$

To find r, substitute $r^{1/3}$ for u.

$$r^{1/3} = -4 \qquad \text{or} \qquad r^{1/3} = 3$$
$$\left(r^{1/3}\right)^3 = (-4)^3 \qquad \left(r^{1/3}\right)^3 = 3^3$$
$$r = -64 \qquad \text{or} \qquad r = 27$$

continued

Check $r = -64$: $16 - 4 - 12 = 0$ *True*
Check $r = 27$: $9 + 3 - 12 = 0$ *True*
Solution set: $\{-64, 27\}$

73. $4k^{4/3} - 13k^{2/3} + 9 = 0$
Let $x = k^{2/3}$, so $x^2 = k^{4/3}$.
$$4x^2 - 13x + 9 = 0$$
$$(4x - 9)(x - 1) = 0$$

$4x - 9 = 0$ $\qquad$ or $\qquad$ $x - 1 = 0$

$x = \frac{9}{4}$ $\qquad$ or $\qquad$ $x = 1$

$k^{2/3} = \frac{9}{4}$ $\qquad$ or $\qquad$ $k^{2/3} = 1$

$(k^{2/3})^{1/2} = (\frac{9}{4})^{1/2}$ or $(k^{2/3})^{1/2} = 1^{1/2}$

$k^{1/3} = \pm \frac{3}{2}$ $\qquad$ or $\qquad$ $k^{1/3} = \pm 1$

$(k^{1/3})^3 = (\pm \frac{3}{2})^3$ or $(k^{1/3})^3 = (\pm 1)^3$

$k = \pm \frac{27}{8}$ $\qquad$ or $\qquad$ $k = \pm 1$

Check $k = \pm \frac{27}{8}$:
$$4(\tfrac{81}{16}) - 13(\tfrac{9}{4}) + 9 \overset{?}{=} 0$$
$$\tfrac{81}{4} - \tfrac{117}{4} + \tfrac{36}{4} = 0 \quad \textit{True}$$
Check $k = \pm 1$:
$$4 - 13 + 9 = 0 \quad \textit{True}$$

Solution set: $\left\{ \pm \frac{27}{8}, \pm 1 \right\}$

75. $2(1 + \sqrt{r})^2 = 13(1 + \sqrt{r}) - 6$
Let $u = 1 + \sqrt{r}$.
$$2u^2 = 13u - 6$$
$$2u^2 - 13u + 6 = 0$$
$$(2u - 1)(u - 6) = 0$$

$2u - 1 = 0$ $\quad$ or $\quad$ $u - 6 = 0$

$u = \frac{1}{2}$ $\quad$ or $\quad$ $u = 6$

Replace u with $1 + \sqrt{r}$.

$1 + \sqrt{r} = \frac{1}{2}$ $\quad$ or $\quad$ $1 + \sqrt{r} = 6$

$\sqrt{r} = -\frac{1}{2}$ $\qquad\qquad$ $\sqrt{r} = 5$

Not possible, $\qquad\qquad$ $r = 25$

since $\sqrt{r} \geq 0$.

Check $r = 25$: $72 = 78 - 6$ *True*
Solution set: $\{25\}$

77. $2x^4 + x^2 - 3 = 0$
Let $m = x^2$, so $m^2 = x^4$.
$$2m^2 + m - 3 = 0$$
$$(2m + 3)(m - 1) = 0$$

$2m + 3 = 0$ $\quad$ or $\quad$ $m - 1 = 0$

$m = -\frac{3}{2}$ $\quad$ or $\quad$ $m = 1$

To find x, substitute x^2 for m.

$x^2 = -\frac{3}{2}$ $\quad$ or $\quad$ $x^2 = 1$

$x^2 = -\frac{3}{2}$ $\qquad$ or $\quad$ $x^2 = 1$

$x = \pm \sqrt{-\frac{3}{2}}$ $\qquad$ $x = \pm \sqrt{1}$

$x = \pm \frac{\sqrt{3}}{\sqrt{2}} \cdot \frac{\sqrt{2}}{\sqrt{2}} i$ $\qquad$ $x = \pm 1$

$x = \pm \frac{\sqrt{6}}{2} i$

Check $x = \pm \frac{\sqrt{6}}{2} i$: $\frac{9}{2} - \frac{3}{2} - 3 = 0$ *True*
Check $x = \pm 1$: $2 + 1 - 3 = 0$ *True*

Solution set: $\left\{ \pm 1, \pm \frac{\sqrt{6}}{2} i \right\}$

79. $12x^4 - 11x^2 + 2 = 0$
Let $u = x^2$, so $u^2 = x^4$.
$$12u^2 - 11u + 2 = 0$$
$$(4u - 1)(3u - 2) = 0$$

$4u - 1 = 0$ $\quad$ or $\quad$ $3u - 2 = 0$

$u = \frac{1}{4}$ $\quad$ or $\quad$ $u = \frac{2}{3}$

$x^2 = \frac{1}{4}$ $\qquad\qquad$ $x^2 = \frac{2}{3}$

$x = \pm \frac{1}{2}$ $\qquad\qquad$ $x = \pm \sqrt{\frac{2}{3}}$

Note: $x = \pm \sqrt{\frac{2}{3}} = \pm \frac{\sqrt{2}}{\sqrt{3}} \cdot \frac{\sqrt{3}}{\sqrt{3}} = \pm \frac{\sqrt{6}}{3}$

Check $x = \pm \frac{1}{2}$: $\frac{3}{4} - \frac{11}{4} + 2 = 0$ *True*
Check $x = \pm \frac{\sqrt{6}}{3}$: $\frac{16}{3} - \frac{22}{3} + 2 = 0$ *True*

Solution set: $\left\{ \pm \frac{1}{2}, \pm \frac{\sqrt{6}}{3} \right\}$

81. $\sqrt{2x + 3} = 2 + \sqrt{x - 2}$
Square both sides.
$$\left(\sqrt{2x + 3}\right)^2 = \left(2 + \sqrt{x - 2}\right)^2$$
$$2x + 3 = 4 + 4\sqrt{x - 2} + (x - 2)$$
$$2x + 3 = x + 2 + 4\sqrt{x - 2}$$
Isolate the radical term on one side.
$$x + 1 = 4\sqrt{x - 2}$$
Square both sides again.
$$(x + 1)^2 = \left(4\sqrt{x - 2}\right)^2$$
$$x^2 + 2x + 1 = 16(x - 2)$$
$$x^2 + 2x + 1 = 16x - 32$$
$$x^2 - 14x + 33 = 0$$
$$(x - 11)(x - 3) = 0$$

$x - 11 = 0$ $\quad$ or $\quad$ $x - 3 = 0$

$x = 11$ $\quad$ or $\qquad$ $x = 3$

Check $x = 11$: $\sqrt{25} = 2 + \sqrt{9}$ *True*
Check $x = 3$: $\sqrt{9} = 2 + \sqrt{1}$ *True*

Solution set: $\{3, 11\}$

83. $2m^6 + 11m^3 + 5 = 0$

Let $y = m^3$, so $y^2 = m^6$.

$2y^2 + 11y + 5 = 0$

$(2y + 1)(y + 5) = 0$

$2y + 1 = 0 \quad$ or $\quad y + 5 = 0$

$y = -\frac{1}{2} \quad$ or $\quad y = -5$

To find m, substitute m^3 for y.

$$m^3 = -\frac{1}{2} \quad \text{or} \quad m^3 = -5$$

Take the cube root of both sides of each equation.

$m = \sqrt[3]{-\frac{1}{2}} \qquad$ or $\quad m = \sqrt[3]{-5}$

$m = -\sqrt[3]{\frac{1}{2}} \qquad\qquad m = -\sqrt[3]{5}$

$\quad = -\dfrac{\sqrt[3]{1}}{\sqrt[3]{2}} \cdot \dfrac{\sqrt[3]{2^2}}{\sqrt[3]{2^2}}$

$\quad = -\dfrac{\sqrt[3]{4}}{2}$

Check $m = -\dfrac{\sqrt[3]{4}}{2}$: $\dfrac{1}{2} - \dfrac{11}{2} + 5 = 0$ *True*

Check $m = -\sqrt[3]{5}$: $50 - 55 + 5 = 0$ *True*

Solution set: $\left\{ -\sqrt[3]{5}, -\dfrac{\sqrt[3]{4}}{2} \right\}$

85. $6 = 7(2w - 3)^{-1} + 3(2w - 3)^{-2}$

Let $x = (2w - 3)^{-1}$, so $x^2 = (2w - 3)^{-2}$.

$6 = 7x + 3x^2$

$0 = 3x^2 + 7x - 6$

$0 = (3x - 2)(x + 3)$

$3x - 2 = 0 \qquad$ or $\qquad x + 3 = 0$

$x = \frac{2}{3} \qquad$ or $\qquad x = -3$

$(2w - 3)^{-1} = \frac{2}{3} \qquad (2w - 3)^{-1} = -3$

$\dfrac{1}{2w - 3} = \dfrac{2}{3} \qquad \dfrac{1}{2w - 3} = -3$

$3 = 2(2w - 3) \qquad 1 = -3(2w - 3)$

$3 = 4w - 6 \qquad\quad 1 = -6w + 9$

$9 = 4w \qquad\qquad -8 = -6w$

$\dfrac{9}{4} = w \qquad\qquad \dfrac{4}{3} = w$

Check $w = \frac{4}{3}$: $6 = -21 + 27$ *True*

Check $w = \frac{9}{4}$: $6 = \frac{14}{3} + \frac{4}{3}$ *True*

Solution set: $\left\{ \frac{4}{3}, \frac{9}{4} \right\}$

87. $2x^4 - 9x^2 = -2$

$2x^4 - 9x^2 + 2 = 0$

Let $u = x^2$, so $u^2 = x^4$.

$2u^2 - 9u + 2 = 0$

Use $a = 2$, $b = -9$, and $c = 2$ in the quadratic formula.

$u = \dfrac{-b \pm \sqrt{b^2 - 4ac}}{2a}$

$u = \dfrac{-(-9) \pm \sqrt{(-9)^2 - 4(2)(2)}}{2(2)}$

$\quad = \dfrac{9 \pm \sqrt{81 - 16}}{4}$

$\quad = \dfrac{9 \pm \sqrt{65}}{4}$

To find x, substitute x^2 for u.

$x^2 = \dfrac{9 \pm \sqrt{65}}{4}$

$x = \pm \sqrt{\dfrac{9 \pm \sqrt{65}}{4}}$

$\quad = \pm \dfrac{\sqrt{9 \pm \sqrt{65}}}{2}$

Note: the last expression represents four numbers. All four proposed solutions check.

Solution set: $\left\{ \pm \dfrac{\sqrt{9 + \sqrt{65}}}{2}, \pm \dfrac{\sqrt{9 - \sqrt{65}}}{2} \right\}$

89. $\qquad P = 2L + 2W$ for W

$P - 2L = 2W$

$\dfrac{P - 2L}{2} = W$, or $W = \dfrac{P}{2} - L$

91. $\qquad F = \frac{9}{5}C + 32$ for C

$F - 32 = \frac{9}{5}C$

$\frac{5}{9}(F - 32) = C$

Summary Exercises on Solving Quadratic Equations

1. $p^2 = 7$

$p = \sqrt{7} \quad$ or $\quad p = -\sqrt{7}$

Solution set: $\left\{ \pm \sqrt{7} \right\}$

3. $n^2 + 6n + 4 = 0$

$n^2 + 6n = -4$

$n^2 + 6n + 9 = -4 + 9 \quad \left[\frac{1}{2}(6) \right]^2 = 9$

$(n + 3)^2 = 5$

$n + 3 = \sqrt{5} \qquad$ or $\quad n + 3 = -\sqrt{5}$

$n = -3 + \sqrt{5} \quad$ or $\qquad n = -3 - \sqrt{5}$

Solution set: $\left\{ -3 \pm \sqrt{5} \right\}$

5. $\dfrac{5}{m} + \dfrac{12}{m^2} = 2$

Multiply by the LCD, m^2.

$5m + 12 = 2m^2$

$0 = 2m^2 - 5m - 12$

$0 = (2m + 3)(m - 4)$

$2m + 3 = 0$ or $m - 4 = 0$

$m = -\dfrac{3}{2}$ or $m = 4$

Solution set: $\left\{-\dfrac{3}{2}, 4\right\}$

7. $2r^2 - 4r + 1 = 0$

Use $a = 2$, $b = -4$, and $c = 1$ in the quadratic formula.

$r = \dfrac{-b \pm \sqrt{b^2 - 4ac}}{2a}$

$r = \dfrac{-(-4) \pm \sqrt{(-4)^2 - 4(2)(1)}}{2(2)}$

$= \dfrac{4 \pm \sqrt{16 - 8}}{2(2)}$

$= \dfrac{4 \pm \sqrt{8}}{2(2)} = \dfrac{4 \pm 2\sqrt{2}}{2(2)}$

$= \dfrac{2 \pm \sqrt{2}}{2}$

Solution set: $\left\{\dfrac{2 \pm \sqrt{2}}{2}\right\}$

9. $x\sqrt{2} = \sqrt{5x - 2}$

$\left(x\sqrt{2}\right)^2 = \left(\sqrt{5x - 2}\right)^2$

$x^2 \cdot 2 = 5x - 2$

$2x^2 - 5x + 2 = 0$

$(2x - 1)(x - 2) = 0$

$2x - 1 = 0$ or $x - 2 = 0$

$x = \dfrac{1}{2}$ $x = 2$

Check $x = \dfrac{1}{2}$: $\dfrac{1}{2}\sqrt{2} = \sqrt{\dfrac{1}{2}}$ *True*

Check $x = 2$: $2\sqrt{2} = \sqrt{8}$ *True*

Solution set: $\left\{\dfrac{1}{2}, 2\right\}$

11. $(2k + 3)^2 = 8$

$2k + 3 = \sqrt{8}$ or $2k + 3 = -\sqrt{8}$

$2k = -3 + 2\sqrt{2}$ $2k = -3 - 2\sqrt{2}$

$k = \dfrac{-3 + 2\sqrt{2}}{2}$ or $k = \dfrac{-3 - 2\sqrt{2}}{2}$

Solution set: $\left\{\dfrac{-3 \pm 2\sqrt{2}}{2}\right\}$

13. $t^4 + 14 = 9t^2$

$t^4 - 9t^2 + 14 = 0$

$(t^2 - 2)(t^2 - 7) = 0$

$t^2 - 2 = 0$ or $t^2 - 7 = 0$

$t^2 = 2$ $t^2 = 7$

$t = \pm\sqrt{2}$ or $t = \pm\sqrt{7}$

Solution set: $\left\{\pm\sqrt{2}, \pm\sqrt{7}\right\}$

15. $z^2 + z + 1 = 0$

Use $a = 1$, $b = 1$, and $c = 1$ in the quadratic formula.

$z = \dfrac{-b \pm \sqrt{b^2 - 4ac}}{2a}$

$z = \dfrac{-1 \pm \sqrt{1^2 - 4(1)(1)}}{2(1)}$

$= \dfrac{-1 \pm \sqrt{1 - 4}}{2} = \dfrac{-1 \pm \sqrt{-3}}{2}$

$= \dfrac{-1 \pm i\sqrt{3}}{2} = -\dfrac{1}{2} \pm \dfrac{\sqrt{3}}{2}i$

Solution set: $\left\{-\dfrac{1}{2} \pm \dfrac{\sqrt{3}}{2}i\right\}$

17. $4t^2 - 12t + 9 = 0$

$(2t - 3)(2t - 3) = 0$

$(2t - 3)^2 = 0$

$2t - 3 = 0$

$t = \dfrac{3}{2}$

Solution set: $\left\{\dfrac{3}{2}\right\}$

19. $r^2 - 72 = 0$

$r^2 = 72$

$r = \pm\sqrt{72} = \pm 6\sqrt{2}$

Solution set: $\left\{\pm 6\sqrt{2}\right\}$

21. $x^2 - 5x - 36 = 0$

$(x + 4)(x - 9) = 0$

$x + 4 = 0$ or $x - 9 = 0$

$x = -4$ or $x = 9$

Solution set: $\{-4, 9\}$

23. $3p^2 = 6p - 4$

$3p^2 - 6p + 4 = 0$

Use $a = 3$, $b = -6$, and $c = 4$ in the quadratic formula.

$p = \dfrac{-b \pm \sqrt{b^2 - 4ac}}{2a}$

$p = \dfrac{-(-6) \pm \sqrt{(-6)^2 - 4(3)(4)}}{2(3)}$

$$= \frac{6 \pm \sqrt{36 - 48}}{2(3)} = \frac{6 \pm \sqrt{-12}}{2(3)}$$

$$= \frac{6 \pm 2i\sqrt{3}}{2(3)} = \frac{3 \pm i\sqrt{3}}{3}$$

$$= 1 \pm \frac{\sqrt{3}}{3} i$$

Solution set: $\left\{ 1 \pm \frac{\sqrt{3}}{3} i \right\}$

25. $\quad \frac{4}{r^2} + 3 = \frac{1}{r}$

Multiply by the LCD, r^2.

$$4 + 3r^2 = r$$

$$3r^2 - r + 4 = 0$$

Use $a = 3$, $b = -1$, and $c = 4$ in the quadratic formula.

$$r = \frac{-b \pm \sqrt{b^2 - 4ac}}{2a}$$

$$r = \frac{-(-1) \pm \sqrt{(-1)^2 - 4(3)(4)}}{2(3)}$$

$$= \frac{1 \pm \sqrt{1 - 48}}{6} = \frac{1 \pm \sqrt{-47}}{6}$$

$$= \frac{1 \pm i\sqrt{47}}{6} = \frac{1}{6} \pm \frac{\sqrt{47}}{6} i$$

Solution set: $\left\{ \frac{1}{6} \pm \frac{\sqrt{47}}{6} i \right\}$

11.5 Formulas and Further Applications

1. The first step in solving a formula that has the specified variable in the denominator is to multiply both sides by the LCD to clear the equation of fractions.

3. We must recognize that a formula like

$$gw^2 = kw + 24$$

is quadratic in w. So the first step is to write the formula in standard form (with 0 on one side, in decreasing powers of w). This allows us to apply the quadratic formula to solve for w.

5. Since the triangle is a right triangle, use the Pythagorean formula with legs m and n and hypotenuse p.

$$m^2 + n^2 = p^2$$
$$m^2 = p^2 - n^2$$
$$m = \sqrt{p^2 - n^2}$$

Only the positive square root is given since m represents the side of a triangle.

7. Solve $d = kt^2$ for t.

$$kt^2 = d$$

$$t^2 = \frac{d}{k} \qquad \text{Divide by } k.$$

$$t = \pm\sqrt{\frac{d}{k}} \qquad \begin{array}{l}\textit{Use square root}\\\textit{property.}\end{array}$$

$$= \frac{\pm\sqrt{d}}{\sqrt{k}} \cdot \frac{\sqrt{k}}{\sqrt{k}} \qquad \begin{array}{l}\textit{Rationalize}\\\textit{denominator.}\end{array}$$

$$t = \frac{\pm\sqrt{dk}}{k} \qquad \textit{Simplify.}$$

9. Solve $I = \frac{ks}{d^2}$ for d.

$$Id^2 = ks \qquad \textit{Multiply by } d^2.$$

$$d^2 = \frac{ks}{I} \qquad \textit{Divide by } I.$$

$$d = \pm\sqrt{\frac{ks}{I}} \qquad \textit{Use square root property.}$$

$$= \pm\frac{\sqrt{ks}}{\sqrt{I}} \cdot \frac{\sqrt{I}}{\sqrt{I}} \qquad \textit{Rationalize denominator.}$$

$$d = \frac{\pm\sqrt{ksI}}{I} \qquad \textit{Simplify.}$$

11. Solve $F = \frac{kA}{v^2}$ for v.

$$v^2 F = kA \qquad \textit{Multiply by } v^2.$$

$$v^2 = \frac{kA}{F} \qquad \textit{Divide by } F.$$

$$v = \pm\sqrt{\frac{kA}{F}} \qquad \begin{array}{l}\textit{Use square root}\\\textit{property.}\end{array}$$

$$= \frac{\pm\sqrt{kA}}{\sqrt{F}} \cdot \frac{\sqrt{F}}{\sqrt{F}} \qquad \begin{array}{l}\textit{Rationalize}\\\textit{denominator.}\end{array}$$

$$v = \frac{\pm\sqrt{kAF}}{F} \qquad \textit{Simplify.}$$

13. Solve $V = \frac{1}{3}\pi r^2 h$ for r.

$$3V = \pi r^2 h \qquad \textit{Multiply by } 3.$$

$$\frac{3V}{\pi h} = r^2 \qquad \textit{Divide by } \pi h.$$

$$r = \pm\sqrt{\frac{3V}{\pi h}} \qquad \begin{array}{l}\textit{Use square root}\\\textit{property.}\end{array}$$

$$= \frac{\pm\sqrt{3V} \cdot \sqrt{\pi h}}{\sqrt{\pi h} \cdot \sqrt{\pi h}} \qquad \begin{array}{l}\textit{Rationalize}\\\textit{denominator.}\end{array}$$

$$r = \frac{\pm\sqrt{3\pi V h}}{\pi h} \qquad \textit{Simplify.}$$

15. Solve $At^2 + Bt = -C$ for t.

$$At^2 + Bt + C = 0$$

Use the quadratic formula.

$$t = \frac{-B \pm \sqrt{B^2 - 4AC}}{2A}$$

17. Solve $D = \sqrt{kh}$ for h.

$$D^2 = kh \qquad \textit{Square both sides.}$$
$$\frac{D^2}{k} = h \qquad \textit{Divide by k.}$$

19. Solve $p = \sqrt{\dfrac{k\ell}{g}}$ for ℓ.

$$p^2 = \frac{k\ell}{g} \qquad \textit{Square both sides.}$$
$$p^2 g = k\ell \qquad \textit{Multiply by g.}$$
$$\frac{p^2 g}{k} = \ell \qquad \textit{Divide by k.}$$

21. Solve $S = 4\pi r^2$ for r.

$$\frac{S}{4\pi} = r^2 \qquad \textit{Divide by } 4\pi.$$
$$r = \pm\sqrt{\frac{S}{4\pi}} \qquad \begin{array}{l}\textit{Use square root}\\ \textit{property.}\end{array}$$
$$= \frac{\pm\sqrt{S}\cdot\sqrt{\pi}}{\sqrt{4\pi}\cdot\sqrt{\pi}} \qquad \begin{array}{l}\textit{Rationalize}\\ \textit{denominator.}\end{array}$$
$$r = \frac{\pm\sqrt{S\pi}}{2\pi} \qquad \textit{Simplify.}$$

23. Solve $p = \dfrac{E^2 R}{(r+R)^2}$ $(E > 0)$ for R.

$$p(r+R)^2 = E^2 R$$
$$p\left(r^2 + 2rR + R^2\right) = E^2 R$$
$$pr^2 + 2prR + pR^2 = E^2 R$$
$$pR^2 + 2prR - E^2 R + pr^2 = 0$$
$$pR^2 + \left(2pr - E^2\right)R + pr^2 = 0$$

Here $a = p$, $b = 2pr - E^2$, and $c = pr^2$.

$$R = \frac{-(2pr - E^2) \pm \sqrt{(2pr - E^2)^2 - 4p\bullet pr^2}}{2p}$$
$$= \frac{E^2 - 2pr \pm \sqrt{4p^2 r^2 - 4pr E^2 + E^4 - 4p^2 r^2}}{2p}$$
$$= \frac{E^2 - 2pr \pm \sqrt{E^4 - 4pr E^2}}{2p}$$
$$= \frac{E^2 - 2pr \pm \sqrt{E^2(E^2 - 4pr)}}{2p}$$
$$R = \frac{E^2 - 2pr \pm E\sqrt{E^2 - 4pr}}{2p}$$

25. Solve $10p^2 c^2 + 7pcr = 12r^2$ for r.

$$0 = 12r^2 - 7pcr - 10p^2 c^2$$

Here $a = 12$, $b = -7pc$, and $c = -10p^2 c^2$.

$$r = \frac{-(-7pc) \pm \sqrt{(-7pc)^2 - 4(12)(-10p^2 c^2)}}{2(12)}$$
$$= \frac{7pc \pm \sqrt{49p^2 c^2 + 480p^2 c^2}}{24}$$

$$= \frac{7pc \pm \sqrt{529p^2 c^2}}{24} = \frac{7pc \pm 23pc}{24}$$
$$r = \frac{7pc + 23pc}{24} = \frac{30pc}{24} = \frac{5pc}{4} \text{ or}$$
$$r = \frac{7pc - 23pc}{24} = \frac{-16pc}{24} = -\frac{2pc}{3}$$

27. Solve $LI^2 + RI + \dfrac{1}{c} = 0$ for I.

$$cLI^2 + cRI + 1 = 0 \quad \textit{Multiply by c.}$$

Here $a = cL$, $b = cR$, and $c = 1$.

$$I = \frac{-cR \pm \sqrt{(cR)^2 - 4(cL)(1)}}{2(cL)}$$
$$= \frac{-cR \pm \sqrt{c^2 R^2 - 4cL}}{2cL}$$

29. Apply the Pythagorean formula.

$$(x+4)^2 = x^2 + (x+1)^2$$
$$x^2 + 8x + 16 = x^2 + x^2 + 2x + 1$$
$$0 = x^2 - 6x - 15$$

Here $a = 1$, $b = -6$, and $c = -15$.

$$x = \frac{-(-6) \pm \sqrt{(-6)^2 - 4(1)(-15)}}{2(1)}$$
$$= \frac{6 \pm \sqrt{36 + 60}}{2} = \frac{6 \pm \sqrt{96}}{2}$$
$$x = \frac{6 + \sqrt{96}}{2} \approx 7.9 \text{ or}$$
$$x = \frac{6 - \sqrt{96}}{2} \approx -1.9$$

Reject the negative solution.
If $x = 7.9$, then

$$x + 4 = 11.9 \quad \text{and} \quad x + 1 = 8.9.$$

The lengths of the sides of the triangle are approximately 7.9, 8.9, and 11.9.

31. Let $x =$ the distance traveled by the eastbound ship. Then $x + 70 =$ the distance traveled by the southbound ship.

Since the ships are traveling at right angles to one another, the distance d between them can be found using the Pythagorean formula.

$$c^2 = a^2 + b^2$$
$$d^2 = x^2 + (x+70)^2$$

Let $d = 170$, and solve for x.

$$170^2 = x^2 + (x+70)^2$$
$$28{,}900 = x^2 + x^2 + 140x + 4900$$
$$0 = 2x^2 + 140x - 24{,}000$$
$$0 = x^2 + 70x - 12{,}000$$
$$0 = (x+150)(x-80)$$

$$x + 150 = 0 \quad \text{or} \quad x - 80 = 0$$
$$x = -150 \quad \text{or} \quad x = 80$$

Distance cannot be negative, so reject -150. If $x = 80$, then $x + 70 = 150$. The eastbound ship traveled 80 miles, and the southbound ship traveled 150 miles.

33. Let $\quad x = $ length of the shorter leg;
$$2x - 1 = \text{length of the longer leg};$$
$$2x - 1 + 2 = \text{length of the hypotenuse}.$$

Use the Pythagorean formula.
$$x^2 + (2x - 1)^2 = (2x + 1)^2$$
$$x^2 + 4x^2 - 4x + 1 = 4x^2 + 4x + 1$$
$$5x^2 - 4x + 1 = 4x^2 + 4x + 1$$
$$x^2 - 8x = 0$$
$$x(x - 8) = 0$$
$$x = 0 \quad \text{or} \quad x - 8 = 0$$
$$x = 8$$

Since x represents length, discard 0 as a solution. If $x = 8$, then
$$2x - 1 = 2(8) - 1 = 15 \text{ and}$$
$$2x - 1 + 2 = 2(8) + 1 = 17.$$

The lengths are 8 inches, 15 inches, and 17 inches.

35. Let $x = $ the width of the rug.
Then $2x + 4 = $ the length of the rug.

Use the Pythagorean formula.
$$x^2 + (2x + 4)^2 = 26^2$$
$$x^2 + 4x^2 + 16x + 16 = 676$$
$$5x^2 + 16x - 660 = 0$$
$$(5x + 66)(x - 10) = 0$$
$$5x + 66 = 0 \quad \text{or} \quad x - 10 = 0$$
$$x = -\tfrac{66}{5} \quad \text{or} \quad x = 10$$

Discard the negative solution.
If $x = 10$, then $2x + 4 = 24$.
The width of the rug is 10 feet, and the length is 24 feet.

37. Let $x = $ the width of the border. Then the width of the pool and the border is $30 + 2x$ ft and the length of the pool and border is $40 + 2x$ ft. Since the area of the pool is $30 \cdot 40 = 1200$ ft^2, we can write an equation using the total area of the pool and the border.

The area of the pool and border	is	the area of the pool	plus	the area of the border.
$(30 + 2x)(40 + 2x)$	$=$	1200	$+$	296

$$1200 + 140x + 4x^2 = 1496$$
$$4x^2 + 140x - 296 = 0$$
$$x^2 + 35x - 74 = 0$$
$$(x - 2)(x + 37) = 0$$
$$x - 2 = 0 \quad \text{or} \quad x + 37 = 0$$
$$x = 2 \quad \text{or} \quad x = -37$$

Discard the negative solution. The strip can be 2 feet wide.

39. Let $x = $ original width of the rectangle. Then $2x - 2$ represents the original length of the rectangle.

Now $x + 5$ is the new width that makes the rectangle a square. Thus, the new width must equal the original length since the sides of a square are equal.

$$x + 5 = 2x - 2$$
$$7 = x$$

The dimensions of the original rectangle are 7 meters and $2(7) - 2 = 12$ meters. Note that the area of the square is $(12)(12) = 144$ m^2.

41. Let x be the width of the sheet metal. Then the length is $2x - 4$.

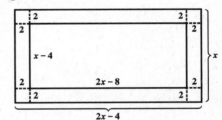

By cutting out 2-inch squares from each corner we get a rectangle with width $x - 4$ and length $(2x - 4) - 4 = 2x - 8$. The uncovered box then has height 2 inches, length $2x - 8$ inches, and width $x - 4$ inches.
Use the formula $V = LWH$ or $V = HLW$.

$$256 = 2(2x - 8)(x - 4)$$
$$256 = 4(x - 4)(x - 4) \qquad \textit{Factor out 2.}$$
$$64 = (x - 4)^2 \qquad \textit{Divide by 4.}$$

Use the square root property.
$$\pm 8 = x - 4$$

$$x - 4 = 8 \quad \text{or} \quad x - 4 = -8$$
$$x = 12 \quad \text{or} \quad x = -4$$

Since x represents width, discard the negative solution.
The width is 12 inches, and the length is $2(12) - 4 = 20$ inches.

43.
$$s = 144t - 16t^2$$
$$128 = 144t - 16t^2 \qquad \textit{Let s = 128.}$$
$$0 = -16t^2 + 144t - 128$$
$$0 = t^2 - 9t + 8 \qquad \textit{Divide by } -16.$$
$$0 = (t - 8)(t - 1)$$
$$t - 8 = 0 \quad \text{or} \quad t - 1 = 0$$
$$t = 8 \quad \text{or} \qquad t = 1$$

The object will be 128 feet above the ground at two times, going up and coming down, or at 1 second and at 8 seconds.

45. Let $s = 213$ in the equation.
$$s = -16t^2 + 128t$$
$$213 = -16t^2 + 128t$$
$$0 = -16t^2 + 128t - 213$$

Here $a = -16$, $b = 128$, and $c = -213$.

$$t = \frac{-b \pm \sqrt{b^2 - 4ac}}{2a}$$
$$t = \frac{-128 \pm \sqrt{128^2 - 4(-16)(-213)}}{2(-16)}$$
$$= \frac{-128 \pm \sqrt{16,384 - 13,632}}{-32}$$
$$= \frac{-128 \pm \sqrt{2752}}{-32}$$
$$t = \frac{-128 + \sqrt{2752}}{-32} \approx 2.4 \quad \text{or}$$
$$t = \frac{-128 - \sqrt{2752}}{-32} \approx 5.6$$

The ball will be 213 feet from the ground after 2.4 seconds and again after 5.6 seconds.

47.
$$D(t) = 13t^2 - 100t$$
$$180 = 13t^2 - 100t \qquad \textit{Let D(t) = 180.}$$
$$0 = 13t^2 - 100t - 180$$

Here $a = 13$, $b = -100$, and $c = -180$.

$$t = \frac{-b \pm \sqrt{b^2 - 4ac}}{2a}$$
$$t = \frac{-(-100) \pm \sqrt{(-100)^2 - 4(13)(-180)}}{2(13)}$$
$$= \frac{100 \pm \sqrt{10,000 + 9360}}{2(13)}$$
$$= \frac{100 \pm \sqrt{19,360}}{2(13)}$$
$$= \frac{100 \pm 44\sqrt{10}}{2(13)} = \frac{50 \pm 22\sqrt{10}}{13}$$
$$t = \frac{50 + 22\sqrt{10}}{13} \approx 9.2 \quad \text{or}$$
$$t = \frac{50 - 22\sqrt{10}}{13} \approx -1.5$$

Discard the negative solution. The car will skid 180 feet in approximately 9.2 seconds.

49.
$$s(t) = -16t^2 + 160t$$
$$400 = -16t^2 + 160t \qquad \textit{Let s(t) = 400.}$$
$$0 = -16t^2 + 160t - 400$$
$$0 = t^2 - 10t + 25 \qquad \textit{Divide by } -16.$$
$$0 = (t - 5)(t - 5)$$
$$0 = (t - 5)^2$$
$$0 = t - 5$$
$$5 = t$$

The rock reaches a height of 400 feet after 5 seconds. This is its maximum height since this is the only time it reaches 400 feet.

51.
$$V = 3(x - 6)^2$$
$$432 = 3(x - 6)^2 \qquad \textit{Let V = 432.}$$
$$144 = (x - 6)^2 \qquad \textit{Divide by 3.}$$

Use the square root property.
$$\pm \sqrt{144} = x - 6$$
$$6 \pm 12 = x$$
$$x = 6 + 12 = 18$$
$$\text{or} \qquad x = 6 - 12 = -6$$

Discard the negative solution since x represents the length.
The original length is 18 inches.

53. Let F denote the Froude number. Solve
$$F = \frac{v^2}{g\ell}$$
for v.
$$v^2 = Fg\ell$$
$$v = \pm \sqrt{Fg\ell}$$
v is positive, so
$$v = \sqrt{Fg\ell}.$$
For the rhinoceros, $\ell = 1.2$ and $F = 2.57$.
$$v = \sqrt{(2.57)(9.8)(1.2)} \approx 5.5$$
or 5.5 meters per second.

55. Write a proportion.
$$\frac{x - 4}{3x - 19} = \frac{4}{x - 3}$$
Multiply by the LCD, $(3x - 19)(x - 3)$.
$$(3x - 19)(x - 3)\left(\frac{x - 4}{3x - 19}\right)$$
$$= (3x - 19)(x - 3)\left(\frac{4}{x - 3}\right)$$
$$(x - 3)(x - 4) = (3x - 19)4$$
$$x^2 - 7x + 12 = 12x - 76$$
$$x^2 - 19x + 88 = 0$$
$$(x - 8)(x - 11) = 0$$
$$x - 8 = 0 \quad \text{or} \quad x - 11 = 0$$
$$x = 8 \quad \text{or} \qquad x = 11$$

If $x = 8$, then

$$3x - 19 = 3(8) - 19 = 5.$$

If $x = 11$, then

$$3x - 19 = 3(11) - 19 = 14.$$

Thus, $AC = 5$ or $AC = 14$.

57. **(a)** From the graph, the number of miles traveled in 2000 appears to be 2750 billion (to the nearest ten billion).

(b) For 2000, $x = 2000 - 1994 = 6$.

$$f(x) = -1.705x^2 + 75.93x + 2351$$
$$f(6) = -1.705(6)^2 + 75.93(6) + 2351$$
$$= 2745.2$$

To the nearest ten billion, the model gives 2750 billion, the same as the estimate in part (a).

59. Use $f(x) = -1.705x^2 + 75.93x + 2351$ with $f(x) = 2800$.

$$2800 = -1.705x^2 + 75.93x + 2351$$
$$0 = -1.705x^2 + 75.93x - 449$$

Here $a = -1.705$, $b = 75.93$, and $c = -449$.

$$x = \frac{-b \pm \sqrt{b^2 - 4ac}}{2a}$$

$$x = \frac{-75.93 \pm \sqrt{(75.93)^2 - 4(-1.705)(-449)}}{2(-1.705)}$$

$$= \frac{-75.93 \pm \sqrt{2703.1849}}{-3.41}$$

$$\approx 7.02 \text{ or } 37.51$$

The model indicates that the number of miles traveled was 2800 billion in the year $1994 + 7 = 2001$. The other value represents a future year. The graph indicates that vehicle-miles reached 2800 billion in 2001.

61. $f(x) = x^2 + 4x - 3$

$$f(2) = 2^2 + 4(2) - 3$$
$$= 4 + 8 - 3$$
$$= 9$$

63. $f(x) = ax^2 + bx + c$

$$f\left(\frac{-b}{2a}\right) = a\left(\frac{-b}{2a}\right)^2 + b\left(\frac{-b}{2a}\right) + c$$

$$= a\left(\frac{b^2}{4a^2}\right) - \frac{b^2}{2a} + c$$

$$= \frac{b^2}{4a} - \frac{2b^2}{4a} + \frac{4ac}{4a}$$

$$= \frac{-b^2 + 4ac}{4a}$$

65. $f(x) = (x - 4)^2$

$$9 = (x - 4)^2$$
$$x - 4 = \pm\sqrt{9}$$
$$x = 4 \pm 3 = 7 \text{ or } 1$$

Solution set: $\{1, 7\}$

11.6 Graphs of Quadratic Functions

1. A parabola with equation $f(x) = a(x - h)^2 + k$ has vertex $V(h, k)$. We'll identify the vertex for each quadratic function.

(a) $f(x) = (x + 2)^2 - 1$

$V(-2, -1)$, choice **B**

(b) $f(x) = (x + 2)^2 + 1$

$V(-2, 1)$, choice **C**

(c) $f(x) = (x - 2)^2 - 1$

$V(2, -1)$, choice **A**

(d) $f(x) = (x - 2)^2 + 1$

$V(2, 1)$, choice **D**

For Exercises 3–11, we write $f(x)$ in the form $f(x) = a(x - h)^2 + k$ and then list the vertex (h, k).

3. $f(x) = -3x^2 = -3(x - 0)^2 + 0$
The vertex (h, k) is $(0, 0)$.

5. $f(x) = x^2 + 4 = 1(x - 0)^2 + 4$
The vertex (h, k) is $(0, 4)$.

7. $f(x) = (x - 1)^2 = 1(x - 1)^2 + 0$
The vertex (h, k) is $(1, 0)$.

9. $f(x) = (x + 3)^2 - 4 = 1[x - (-3)]^2 - 4$
The vertex (h, k) is $(-3, -4)$.

11. $f(x) = -(x - 5)^2 + 6 = -1(x - 5)^2 + 6$
The vertex (h, k) is $(5, 6)$.

13. $f(x) = -\frac{2}{5}x^2$

Since $a = -\frac{2}{5} < 0$, the graph opens down. Since $|a| = \left|-\frac{2}{5}\right| = \frac{2}{5} < 1$, the graph is wider than the graph of $f(x) = x^2$.

15. $f(x) = 3x^2 + 1$

Since $a = 3 > 0$, the graph opens up.
Since $|a| = |3| = 3 > 1$, the graph is narrower than the graph of $f(x) = x^2$.

17. Consider $f(x) = a(x - h)^2 + k$.

(a) If $h > 0$ and $k > 0$ in $f(x) = a(x - h)^2 + k$, the shift is to the right and upward, so the vertex is in quadrant I.

(b) If $h > 0$ and $k < 0$, the shift is to the right and downward, so the vertex is in quadrant IV.

(c) If $h < 0$ and $k > 0$, the shift is to the left and upward, so the vertex is in quadrant II.

(d) If $h < 0$ and $k < 0$, the shift is to the left and downward, so the vertex is in quadrant III.

19. (a) $f(x) = (x - 4)^2 - 2 = 1(x - 4)^2 - 2$ has vertex $(4, -2)$. Because $a = 1 > 0$, the graph opens up. The correct answer is **D**.

(b) $f(x) = (x - 2)^2 - 4 = 1(x - 2)^2 - 4$ has vertex $(2, -4)$. Because $a = 1 > 0$, the graph opens up. The correct answer is **B**.

(c) $f(x) = -(x - 4)^2 - 2 = -1(x - 4)^2 - 2$ has vertex $(4, -2)$. Because $a = -1 < 0$, the graph opens down. The correct answer is **C**.

(d) $f(x) = -(x - 2)^2 - 4 = -1(x - 2)^2 - 4$ has vertex $(2, -4)$. Because $a = -1 < 0$, the graph opens down. The correct answer is **A**.

21. $f(x) = -2x^2$ written in the form
$f(x) = a(x - h)^2 + k$ is
$f(x) = -2(x - 0)^2 + 0$.

Here, $h = 0$ and $k = 0$, so the vertex (h, k) is $(0, 0)$. Since $a = -2 < 0$, the graph opens down. Since $|a| = |-2| = 2 > 1$, the graph is narrower than the graph of $f(x) = x^2$. By evaluating the function with $x = 2$ and $x = -2$, we see that the points $(2, -8)$ and $(-2, -8)$ are on the graph.

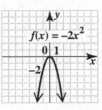

23. $f(x) = x^2 - 1$ written in the form
$f(x) = a(x - h)^2 + k$ is
$f(x) = 1(x - 0)^2 + (-1)$.

Here, $h = 0$ and $k = -1$, so the vertex is $(0, -1)$. The graph opens up and has the same shape as $f(x) = x^2$ because $a = 1$. Two other points on the graph are $(-2, 3)$ and $(2, 3)$.

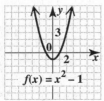

25. $f(x) = -x^2 + 2$ written in the form
$f(x) = a(x - h)^2 + k$ is
$f(x) = -1(x - 0)^2 + 2$.

Here, $h = 0$ and $k = 2$, so the vertex (h, k) is $(0, 2)$. Since $a = -1 < 0$, the graph opens down. Since $|a| = |-1| = 1$, the graph has the same shape as $f(x) = x^2$. The points $(2, -2)$ and $(-2, -2)$ are on the graph.

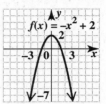

27. $f(x) = (x - 4)^2$ written in the form
$f(x) = a(x - h)^2 + k$ is
$f(x) = 1(x - 4)^2 + 0$.

Here, $h = 4$ and $k = 0$, so the vertex (h, k) is $(4, 0)$ and the axis is $x = 4$. The graph opens up since a is positive and has the same shape as $f(x) = x^2$ because $|a| = 1$. Two other points on the graph are $(2, 4)$ and $(6, 4)$. We can substitute any value for x, so the domain is $(-\infty, \infty)$. The range is $[0, \infty)$ since the smallest y-value is 0.

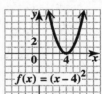

29. $f(x) = (x + 2)^2 - 1$ written in the form
$f(x) = a(x - h)^2 + k$ is
$f(x) = 1[x - (-2)]^2 + (-1)$.

Since $h = -2$ and $k = -1$, the vertex (h, k) is $(-2, -1)$ and the axis is $x = -2$. Here, $a = 1$, so the graph opens up and has the same shape as $f(x) = x^2$. The points $(-1, 0)$ and $(-3, 0)$ are on the graph. The domain is $(-\infty, \infty)$. The range is $[-1, \infty)$ since the smallest y-value is -1.

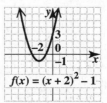

31. $f(x) = 2(x - 2)^2 - 4$ written in the form
$f(x) = a(x - h)^2 + k$ is
$f(x) = 2(x - 2)^2 + (-4)$.

Here, $h = 2$ and $k = -4$, so the vertex (h, k) is
$(2, -4)$ and the axis is $x = 2$. The graph opens up
and is narrower than $f(x) = x^2$ because
$|a| = |2| > 1$. Two other points on the graph are
$(0, 4)$ and $(4, 4)$.
We can substitute any value for x, so the domain
is $(-\infty, \infty)$. The value of y is greater than or
equal to -4, so the range is $[-4, \infty)$.

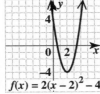

33. $f(x) = -\frac{1}{2}(x + 1)^2 + 2$ written in the form
$f(x) = a(x - h)^2 + k$ is
$f(x) = -\frac{1}{2}[x - (-1)]^2 + 2$.

Since $h = -1$ and $k = 2$, the vertex (h, k) is
$(-1, 2)$ and the axis is $x = -1$. Here,
$a = -0.5 < 0$, so the graph opens down. Also,
$|a| = |-0.5| = 0.5 < 1$, so the graph is wider than
the graph of $f(x) = x^2$. The points $(1, 0)$ and
$(-3, 0)$ are on the graph.
We can substitute any value for x, so the domain
is $(-\infty, \infty)$. The value of y is less than or equal to
2, so the range is $(-\infty, 2]$.

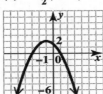

35. $f(x) = 2(x - 2)^2 - 3$ written in the form
$f(x) = a(x - h)^2 + k$ is
$f(x) = 2(x - 2)^2 + (-3)$.

Here, $h = 2$ and $k = -3$, so the vertex (h, k) is
$(2, -3)$ and the axis is $x = 2$. The graph opens up
and is narrower than $f(x) = x^2$ because
$|a| = |2| > 1$. Two other points on the graph are
$(3, -1)$ and $(1, -1)$.
We can substitute any value for x, so the domain
is $(-\infty, \infty)$. The value of y is greater than or
equal to -3, so the range is $[-3, \infty)$.

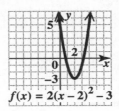

37. The graph of $F(x) = x^2 + 6$ would be shifted 6
units upward from the graph of $f(x) = x^2$.

38. To graph $G(x) = x + 6$, plot the intercepts
$(-6, 0)$ and $(0, 6)$, and draw the line through them.

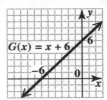

39. When considering the graph of $G(x) = x + 6$, the
y-intercept is 6. The graph of $g(x) = x$ has
y-intercept 0. Therefore, the graph of
$G(x) = x + 6$ is shifted 6 units upward compared
to the graph of $g(x) = x$.

40. The graph of $F(x) = (x - 6)^2$ is shifted 6 units to
the right compared to the graph of $f(x) = x^2$.

41. To graph $G(x) = x - 6$, plot the intercepts $(6, 0)$
and $(0, -6)$, and draw the line through them.

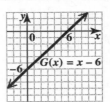

42. When considering the graph of $G(x) = x - 6$, its
x-intercept is 6 as compared to the graph of
$g(x) = x$ with x-intercept 0. The graph of
$G(x) = x - 6$ is shifted 6 units to the right
compared to the graph of $g(x) = x$.

43. The points appear to lie on a line, so a *linear*
function would be a more appropriate model. The
line would rise, so it would have a *positive* slope.

45. The points appear to lie on a parabola, so a
quadratic function would be a more appropriate
model. The parabola would open up, so a would
be *positive*.

47. Since the arrangement of the data points is
approximately parabolic, a quadratic function
would be the more appropriate model for the data
set. The coefficient of x^2 should be negative, since
the roughly parabolic shape of the graphed data set
opens downward.

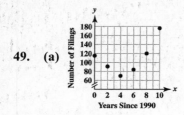

49. (a)

(b) Since the arrangement of the data points is approximately parabolic, a quadratic function would be the more appropriate model for the data set. The coefficient of x^2 should be positive, since the roughly parabolic shape of the graphed data set opens upward.

(c) Use $ax^2 + bx + c = y$ with $(0, 115)$, $(4, 70)$, and $(8, 120)$.

$$0a + 0b + c = 115 \quad (1)$$
$$16a + 4b + c = 70 \quad (2)$$
$$64a + 8b + c = 120 \quad (3)$$

From (1), $c = 115$, so the system becomes

$$16a + 4b = -45 \quad (4)$$
$$64a + 8b = 5 \quad (5)$$

Now eliminate b.

$$\begin{array}{rll} -32a - 8b &= 90 & -2 \times (4) \\ 64a + 8b &= 5 & \\ \hline 32a &= 95 & \\ a &= \frac{95}{32} = 2.96875 \approx 2.969 \end{array}$$

From (5) with $a = \frac{95}{32}$,

$$64\left(\tfrac{95}{32}\right) + 8b = 5$$
$$190 + 8b = 5$$
$$8b = -185$$
$$b = -\frac{185}{8} = -23.125$$

The quadratic function is approximately

$$y = f(x) = 2.969x^2 - 23.125x + 115.$$

(d) $x = 2002 - 1990 = 12$ and $f(12) \approx 265$.

(e) No. About 16 companies filed for bankruptcy each month, so at this rate, filings for 2002 would be about 192. The approximation from the model seems high.

51. (a) $y = f(x) = 0.2455x^2 - 1.856x + 30.7105$
$x = 2002 - 1970 = 32$ and
$f(32) = 222.7105 \approx 222.7$ (per 100,000)

(b) The approximation using the model is high.

53. $x^2 - x - 20 = 0$
From the screens, we see that the x-values of the x-intercepts are -4 and 5, so the solution set is $\{-4, 5\}$.

55.
$$x^2 + 6x - 3 = 0$$
$$x^2 + 6x = 3$$
$$x^2 + 6x + 9 = 3 + 9$$
$$(x + 3)^2 = 12$$
$$x + 3 = \pm\sqrt{12}$$
$$x = -3 \pm 2\sqrt{3}$$

Solution set: $\left\{-3 \pm 2\sqrt{3}\right\}$

57.
$$2x^2 - 12x = 5$$
$$x^2 - 6x = \tfrac{5}{2}$$
$$x^2 - 6x + 9 = \tfrac{5}{2} + 9$$
$$(x - 3)^2 = \tfrac{23}{2}$$
$$x - 3 = \pm\sqrt{\tfrac{46}{4}}$$
$$x = 3 \pm \frac{\sqrt{46}}{2} = \frac{6 \pm \sqrt{46}}{2}$$

Solution set: $\left\{\frac{6 \pm \sqrt{46}}{2}\right\}$

11.7 More About Parabolas and Their Applications

1. If there is an x^2-term in the equation, the axis is vertical. If there is a y^2-term, the axis is horizontal.

3. Use the discriminant, $b^2 - 4ac$, of the function. If it is positive, there are two x-intercepts. If it is zero, there is one x-intercept (at the vertex), and if it is negative, there is no x-intercept.

5. As in Example 1, we'll complete the square to find the vertex.

$$f(x) = x^2 + 8x + 10$$
$$= x^2 + 8x + \underline{16} + 10 - \underline{16} \quad \left[\tfrac{1}{2}(8)\right]^2 = 16$$
$$= (x + 4)^2 - 6$$

The vertex is $(-4, -6)$.

7. As in Example 2, we'll complete the square to find the vertex.

$$f(x) = -2x^2 + 4x - 5$$
$$= -2(x^2 - 2x) - 5$$
$$= -2(x^2 - 2x + 1 - 1) - 5$$
$$= -2(x^2 - 2x + 1) + (-2)(-1) - 5$$
$$= -2(x - 1)^2 - 3$$

The vertex is $(1, -3)$.

9. As in Example 3, we'll use the vertex formula to find the vertex.

$$f(x) = x^2 + x - 7$$

The x-coordinate of the vertex is

$$\frac{-b}{2a} = \frac{-1}{2(1)} = -\frac{1}{2}.$$

The y-coordinate of the vertex is

$$f\left(-\tfrac{1}{2}\right) = \tfrac{1}{4} - \tfrac{1}{2} - 7 = -\tfrac{29}{4}.$$

The vertex is $\left(-\tfrac{1}{2}, -\tfrac{29}{4}\right)$.

11. As in Example 2, we'll complete the square to find the vertex.

$$\begin{aligned}
f(x) &= 2x^2 + 4x + 5 \\
&= 2\left(x^2 + 2x\right) + 5 \\
&= 2\left(x^2 + 2x + 1 - 1\right) + 5 \\
&= 2\left(x^2 + 2x + 1\right) + 2(-1) + 5 \\
&= 2(x+1)^2 - 2 + 5 \\
f(x) &= 2(x+1)^2 + 3
\end{aligned}$$

The vertex is $(-1, 3)$.
Because $a = 2 > 1$, the graph opens up and is narrower than the graph of $y = x^2$.

For $y = f(x) = 2x^2 + 4x + 5$, $a = 2$, $b = 4$, and $c = 5$. The discriminant is

$$\begin{aligned}
b^2 - 4ac &= 4^2 - 4(2)(5) \\
&= 16 - 40 = -24.
\end{aligned}$$

The discriminant is negative, so the parabola has no x-intercepts.

13. $f(x) = -x^2 + 5x + 3$

Use the vertex formula with $a = -1$ and $b = 5$.

The x-coordinate of the vertex is

$$\frac{-b}{2a} = \frac{-5}{2(-1)} = \frac{5}{2}.$$

The y-coordinate of the vertex is

$$\begin{aligned}
f\left(\frac{-b}{2a}\right) = f\left(\frac{5}{2}\right) \\
= -\left(\tfrac{5}{2}\right)^2 + 5\left(\tfrac{5}{2}\right) + 3 \\
= -\tfrac{25}{4} + \tfrac{25}{2} + 3 \\
= \frac{-25 + 50 + 12}{4} = \frac{37}{4}.
\end{aligned}$$

The vertex is

$$\left(\frac{-b}{2a}, f\left(\frac{-b}{2a}\right)\right) = \left(\frac{5}{2}, \frac{37}{4}\right).$$

Because $a = -1$, the parabola opens down and has the same shape as the graph of $y = x^2$.

$$\begin{aligned}
b^2 - 4ac &= 5^2 - 4(-1)(3) \\
&= 25 + 12 = 37
\end{aligned}$$

The discriminant is positive, so the parabola has two x-intercepts.

15. Complete the square on the y-terms to find the vertex.

$$\begin{aligned}
x &= \tfrac{1}{3}y^2 + 6y + 24 \\
&= \tfrac{1}{3}\left(y^2 + 18y\right) + 24 \\
&= \tfrac{1}{3}\left(y^2 + 18y + 81 - 81\right) + 24 \\
&= \tfrac{1}{3}\left(y^2 + 18y + 81\right) + \tfrac{1}{3}(-81) + 24 \\
&= \tfrac{1}{3}(y+9)^2 - 27 + 24 \\
x &= \tfrac{1}{3}(y+9)^2 - 3
\end{aligned}$$

The vertex is $(-3, -9)$.
The graph is a horizontal parabola. The graph opens to the right since $a = 0.\overline{3} > 0$ and is wider than the graph of $y = x^2$ since $|a| = \left|0.\overline{3}\right| < 1$.

17. The graph of $y = 2x^2 + 4x - 3$ is a vertical parabola opening up, so choice F is correct. **(F)**

19. The graph of $y = -\tfrac{1}{2}x^2 - x + 1$ is a vertical parabola opening down, so choices A and C are possibilities. The graph in C is wider than the graph in A, so it must correspond to $a = -\tfrac{1}{2}$ while the graph in A must correspond to $a = -1$. **(C)**

21. The graph of $x = -y^2 - 2y + 4$ is a horizontal parabola opening to the left, so choice D is correct. **(D)**

23. $y = f(x) = x^2 + 8x + 10$

Step 1
Since $a = 1 > 0$, the graph opens up and is the same shape as the graph of $y = x^2$.

Step 2
From Exercise 5, the vertex is $(-4, -6)$. Since the graph opens up, the axis goes through the x-coordinate of the vertex—its equation is $x = -4$.

Step 3
To find the y-intercept, let $x = 0$.
$f(0) = 10$, so the y-intercept is $(0, 10)$.
To find the x-intercepts, let $y = 0$.

$$0 = x^2 + 8x + 10$$

$$\begin{aligned}
x &= \frac{-8 \pm \sqrt{64 - 40}}{2} = \frac{-8 \pm \sqrt{24}}{2} \\
&= \frac{-8 \pm 2\sqrt{6}}{2} = -4 \pm \sqrt{6}
\end{aligned}$$

The x-intercepts are approximately $(-6.45, 0)$ and $(-1.55, 0)$.

Step 4
For an additional point on the graph, let $x = -2$ (two units to the right of the axis) to get $f(-2) = -2$. So the point $(-2, -2)$ is on the graph. By symmetry, the point $(-6, -2)$ (two units to the left of the axis) is on the graph.

continued

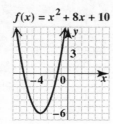

$$f(x) = x^2 + 8x + 10$$

From the graph, we see that the domain is $(-\infty, \infty)$ and the range is $[-6, \infty)$.

25. $y = f(x) = -2x^2 + 4x - 5$

Step 1
Since $a = -2$, the graph opens down and is narrower than the graph of $y = x^2$.

Step 2
From Exercise 7, the vertex is $(1, -3)$. Since the graph opens down, the axis goes through the x-coordinate of the vertex—its equation is $x = 1$.

Step 3
If $x = 0$, $y = -5$, so the y-intercept is $(0, -5)$.
To find the x-intercepts, let $y = 0$.

$$0 = -2x^2 + 4x - 5$$
$$x = \frac{-4 \pm \sqrt{16 - 40}}{2(-2)}$$

The discriminant is negative, so there are no x-intercepts.

Step 4
By symmetry, $(2, -5)$ is also on the graph.

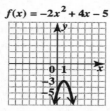

$$f(x) = -2x^2 + 4x - 5$$

From the graph, we see that the domain is $(-\infty, \infty)$ and the range is $(-\infty, -3]$.

27. $x = (y + 2)^2 + 1 = y^2 + 4y + 5$

The roles of x and y are reversed, so this is a horizontal parabola.

Step 1
The coefficient of y^2 is $1 > 0$, so the graph opens to the right.

Step 2
The vertex can be identified from the given form of the equation. When $y = -2$, $x = 1$, so the vertex is $(1, -2)$. Since the graph opens right, the axis goes through the y-coordinate of the vertex—its equation is $y = -2$.

Step 3
To find the x-intercept, let $y = 0$.
$x = 0^2 + 4(0) + 5 = 5$, so the x-intercept is $(5, 0)$.

To find the y-intercepts, let $x = 0$.

$$0 = (y + 2)^2 + 1$$
$$-1 = (y + 2)^2$$

Since $(y + 2)^2$ cannot be negative, there are no y-intercepts.

Step 4
For an additional point on the graph, let $y = -4$ (two units below the axis) to get
$x = (-4 + 2)^2 + 1 = 5.$

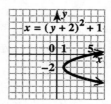

$$x = (y + 2)^2 + 1$$

From the graph, we see the domain is $[1, \infty)$ and the range is $(-\infty, \infty)$.

29. $x = -\frac{1}{5}y^2 + 2y - 4$

The roles of x and y are reversed, so this is a horizontal parabola.

Step 1
Since $a = -\frac{1}{5} < 0$, the graph opens to the left and is wider than the graph of $y = x^2$.

Step 2
The y-coordinate of the vertex is

$$\frac{-b}{2a} = \frac{-2}{2\left(-\frac{1}{5}\right)} = \frac{-2}{-\frac{2}{5}} = 5.$$

The x-coordinate of the vertex is

$$-\frac{1}{5}(5)^2 + 2(5) - 4 = -5 + 10 - 4 = 1.$$

Thus, the vertex is $(1, 5)$. Since the graph opens left, the axis goes through the y-coordinate of the vertex—its equation is $y = 5$.

Step 3
To find the x-intercept, let $y = 0$.
If $y = 0$, $x = -4$, so the x-intercept is $(-4, 0)$.
To find the y-intercepts, let $x = 0$.

$$0 = -\frac{1}{5}y^2 + 2y - 4$$
$$0 = y^2 - 10y + 20 \quad \textit{Multiply by } -5.$$
$$y = \frac{10 \pm \sqrt{100 - 80}}{2} = \frac{10 \pm \sqrt{20}}{2}$$
$$= \frac{10 \pm 2\sqrt{5}}{2} = 5 \pm \sqrt{5}$$

The y-intercepts are approximately $(0, 7.2)$ and $(0, 2.8)$.

Step 4
For an additional point on the graph, let $y = 7$ (two units above the axis) to get $x = \frac{1}{5}$. So the point $\left(\frac{1}{5}, 7\right)$ is on the graph. By symmetry, the point $\left(\frac{1}{5}, 3\right)$ (two units below the axis) is on the graph.

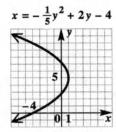

$$x = -\tfrac{1}{5}y^2 + 2y - 4$$

From the graph, we see that the domain is $(-\infty, 1]$ and the range is $(-\infty, \infty)$.

31. $x = 3y^2 + 12y + 5$

The roles of x and y are reversed, so this is a horizontal parabola.

Step 1
Since $a = 3 > 0$, the graph opens to the right and is narrower than the graph of $y = x^2$.

Step 2
Use the formula to find the y-value of the vertex.

$$\frac{-b}{2a} = \frac{-12}{2(3)} = -2$$

If $y = -2$, $x = -7$, so the vertex is $(-7, -2)$. Since the graph opens right, the axis goes through the y-coordinate of the vertex—its equation is $y = -2$.

Step 3
If $y = 0$, $x = 5$, so the x-intercept is $(5, 0)$. To find the y-intercepts, let $x = 0$.

$$0 = 3y^2 + 12y + 5$$
$$y = \frac{-12 \pm \sqrt{144 - 60}}{6} = \frac{-12 \pm \sqrt{84}}{6}$$
$$= \frac{-12 \pm 2\sqrt{21}}{6} = \frac{-6 \pm \sqrt{21}}{3}$$

The y-intercepts are approximately $(0, -0.5)$ and $(0, -3.5)$.

Step 4
By symmetry, $(5, -4)$ is also on the graph.

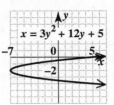

$$x = 3y^2 + 12y + 5$$

From the graph, we see that the domain is $[-7, \infty)$ and the range is $(-\infty, \infty)$.

33. Let $x =$ one number, $40 - x =$ the other number, and $P =$ the product.

$$P = x(40 - x)$$
$$= 40x - x^2 \text{ or } -x^2 + 40x$$

This parabola opens down so the maximum occurs at the vertex.

Here $a = -1$, $b = 40$, and $c = 0$.

$$\frac{-b}{2a} = \frac{-40}{2(-1)} = 20$$

$x = 20$ when the product is a maximum.

Since $x = 20$, $40 - x = 20$, and the two numbers are 20 and 20.

35. Let x represent the length of the two equal sides, and let $280 - 2x$ represent the length of the remaining side (the side parallel to the highway). Then substitute x for L and $280 - 2x$ for W in the formula for the area of a rectangle, $A = LW$.

$$A = x(280 - 2x)$$
$$= 280x - 2x^2 \text{ or } -2x^2 + 280x$$

The maximum area will occur at the vertex.

$$x = \frac{-b}{2a} = \frac{-280}{2(-2)} = 70$$

When $x = 70$, $280 - 2x = 140$, and $A = 9800$. Thus, the dimensions of the lot with maximum area are 140 feet by 70 feet and the maximum area is 9800 square feet.

37. $s(t) = -16t^2 + 32t$

Here, $a = -16 < 0$, so the parabola opens down. The time it takes to reach the maximum height and the maximum height are given by the vertex of the parabola. Use the vertex formula to find that

$$t = \frac{-b}{2a} = \frac{-32}{2(-16)} = \frac{-32}{-32} = 1,$$

and $$s(t) = -16(1)^2 + 32(1)$$
$$= -16 + 32 = 16.$$

The vertex is $(1, 16)$, so the maximum height is 16 feet which occurs when the time is 1 second. The object hits the ground when $s = 0$.

continued

$$0 = -16t^2 + 32t$$
$$0 = -16t(t - 2)$$

$$-16t = 0 \quad \text{or} \quad t - 2 = 0$$
$$t = 0 \quad \text{or} \quad t = 2$$

It takes 2 seconds for the object to hit the ground.

39. The graph of the height of the projectile,

$$s(t) = -16t^2 + 64t + 3,$$

is a parabola that opens down since $a = -16 < 0$. The time at which the cork reaches its maximum height and the maximum height are the t and s coordinates of the vertex.

$$t = \frac{-b}{2a} = \frac{-64}{2(-16)} = 2$$

$$s(2) = -16(2)^2 + 64(2) + 3$$
$$= -64 + 128 + 3 = 67$$

The cork reaches a maximum height of 67 feet after 2 seconds.

41. $f(x) = -0.0334x^2 + 0.2351x + 12.79$

(a) Since the graph opens down, the vertex is a maximum.

(b) The x-value of the vertex is given by

$$x = \frac{-b}{2a} = \frac{-0.2351}{2(-0.0334)} \approx 3.519 \approx 3.5$$

The year was $1990 + 3 = 1993$.

$f(3.5) \approx 13.2\%$, which is the maximum percent of births in the U.S. to teenage mothers.

43. $f(x) = -20.57x^2 + 758.9x - 3140$

(a) The coefficient of x^2 is negative because a parabola that models the data must open down.

(b) Use the vertex formula.

$$x = \frac{-b}{2a} = \frac{-758.9}{2(-20.57)} \approx 18.45$$
$$f(18.45) \approx 3860$$

The vertex is approximately $(18.45, 3860)$.

(c) 18 corresponds to 2018, so in 2018 social security assets will reach their maximum value of $3860 billion.

45. The number of people on the plane is $100 - x$ since x is the number of unsold seats. The price per seat is $200 + 4x$.

(a) The total revenue received for the flight is found by multiplying the number of seats by the price per seat. Thus, the revenue is

$$R(x) = (100 - x)(200 + 4x)$$
$$= 20{,}000 + 200x - 4x^2.$$

(b) Use the formula for the vertex.

$$x = \frac{-b}{2a} = \frac{-200}{2(-4)} = 25$$

$R(25) = 22{,}500$, so the vertex is $(25, 22{,}500)$. $R(0) = 20{,}000$, so the R-intercept is $(0, 20{,}000)$. From the factored form for R, we see that the positive x-intercept is $(100, 0)$. (The factor $200 + 4x$ leads to a negative x-intercept, meaningless in this problem.)

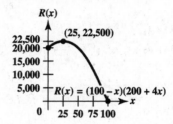

(c) The number of unsold seats x that produce the maximum revenue is 25, the x-value of the vertex.

(d) The maximum revenue is $22{,}500, the y-value of the vertex.

47. $f(x) = x^2 - 8x + 18$

$$\frac{-b}{2a} = \frac{-(-8)}{2(1)} = 4$$

$f(4) = 2$, so the vertex is $(4, 2)$, which matches choice **B**.

49. $f(x) = x^2 - 8x + 14$

$$\frac{-b}{2a} = \frac{-(-8)}{2(1)} = 4$$

$f(4) = -2$, so the vertex is $(4, -2)$, which matches choice **A**.

51. (a) $|x - (-p)| = |x + p|$

(b) The focus should have coordinates $(p, 0)$ because the distance from the focus to the origin should equal the distance from the directrix to the origin.

(c) The distance from (x, y) to $(p, 0)$ is

$$\sqrt{(x - p)^2 + (y - 0)^2} = \sqrt{(x - p)^2 + y^2}.$$

(d) Using the results from parts (a) and (c), these distances should be equal.

$$\sqrt{(x - p)^2 + y^2} = |x + p|$$
Square both sides.
$$(x - p)^2 + y^2 = (x + p)^2$$
$$x^2 - 2px + p^2 + y^2 = x^2 + 2px + p^2$$
$$y^2 = 4px$$

53. $[1, 5] \Leftrightarrow 1 \le x \le 5$

55. $(-\infty, 1] \cup [5, \infty) \Leftrightarrow x \le 1$ or $x \ge 5$

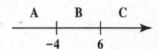

57. $-2x + 1 < 4$
   $-2x < 3$
   $x > -\frac{3}{2}$

Solution set: $\left(-\frac{3}{2}, \infty\right)$

11.8   Quadratic and Rational Inequalities

1.   (a) The x-intercepts determine the solutions of the equation $x^2 - 4x + 3 = 0$. From the graph, the solution set is $\{1, 3\}$.

(b) The x-values of the points on the graph that are *above* the x-axis form the solution set of the inequality $x^2 - 4x + 3 > 0$. From the graph, the solution set is $(-\infty, 1) \cup (3, \infty)$.

(c) The x-values of the points on the graph that are *below* the x-axis form the solution set of the inequality $x^2 - 4x + 3 < 0$. From the graph, the solution set is $(1, 3)$.

3.   (a) The x-intercepts determine the solutions of the equation $-2x^2 - x + 15 = 0$. From the graph, the solution set is $\left\{-3, \frac{5}{2}\right\}$.

(b) The x-values of the points on the graph that are *above* the x-axis form the solution set of the inequality $-2x^2 - x + 15 > 0$. From the graph, the solution set for $-2x^2 - x + 15 \ge 0$ is $\left[-3, \frac{5}{2}\right]$.

(c) The x-values of the points on the graph that are *below* the x-axis form the solution set of the inequality $-2x^2 - x + 15 < 0$. From the graph, the solution set for $-2x^2 - x + 15 \le 0$ is $(-\infty, -3] \cup \left[\frac{5}{2}, \infty\right)$.

5. Include the endpoints if the symbol is $\le$ or $\ge$. Exclude the endpoints if the symbol is $<$ or $>$.

7. $(x + 1)(x - 5) > 0$
Solve the equation
$(x + 1)(x - 5) = 0$.

$x + 1 = 0$ or $x - 5 = 0$
$x = -1$ or $x = 5$

The numbers -1 and 5 divide a number line into three intervals: A, B, and C.

```
        A      B      C
    ──────┼──────┼──────▶
         -1      5
```

Test a number from each interval in the original inequality.

Interval A: Let $x = -2$.
   $(x + 1)(x - 5) > 0$
   $(-2 + 1)(-2 - 5) > 0$?
   $-1(-7) > 0$?
   $7 > 0$ *True*

Interval B: Let $x = 0$.
   $(0 + 1)(0 - 5) > 0$?
   $-5 > 0$ *False*

Interval C: Let $x = 6$.
   $(6 + 1)(6 - 5) > 0$?
   $7 > 0$ *True*

The solution set includes the numbers in Intervals A and C, excluding -1 and 5 because of $>$.

Solution set: $(-\infty, -1) \cup (5, \infty)$

```
    ◀──────)──────(──────▶
          -1      5
```

9. $(x + 4)(x - 6) < 0$
Solve the equation
$(x + 4)(x - 6) = 0$.

$x + 4 = 0$ or $x - 6 = 0$
$x = -4$ or $x = 6$

These numbers divide a number line into three intervals: A, B, and C.

```
        A      B      C
    ──────┼──────┼──────▶
         -4      6
```

Test a number from each interval in the original inequality.

Interval A: Let $x = -5$.
   $(x + 4)(x - 6) < 0$
   $(-5 + 4)(-5 - 6) < 0$?
   $-1(-11) < 0$?
   $11 < 0$ *False*

Interval B: Let $x = 0$.
   $4(-6) < 0$?
   $-24 < 0$ *True*

Interval C: Let $x = 7$.
   $(7 + 4)(7 - 6) < 0$?
   $11(1) < 0$?
   $11 < 0$ *False*

The solution set includes Interval B, where the expression is negative.

Solution set: $(-4, 6)$

```
    ◀──(──────────)──▶
       -4          6
```

11. $x^2 - 4x + 3 \geq 0$

Solve the equation

$x^2 - 4x + 3 = 0.$

$(x - 1)(x - 3) = 0$

$x - 1 = 0 \quad \text{or} \quad x - 3 = 0$

$x = 1 \quad \text{or} \quad x = 3$

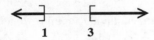

Interval A: Let $x = 0$.

$3 \geq 0 \qquad$ *True*

Interval B: Let $x = 2$.

$2^2 - 4(2) + 3 \geq 0 \quad ?$

$-1 \geq 0 \qquad$ *False*

Interval C: Let $x = 4$.

$4^2 - 4(4) + 3 \geq 0 \quad ?$

$3 \geq 0 \qquad$ *True*

The solution set includes the numbers in Intervals A and C, including 1 and 3 because of $\geq$.

Solution set: $(-\infty, 1] \cup [3, \infty)$

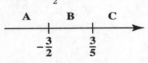

13. $10x^2 + 9x \geq 9$

$10x^2 + 9x - 9 \geq 0$

Solve the equation

$10x^2 + 9x - 9 = 0.$

$(2x + 3)(5x - 3) = 0$

$2x + 3 = 0 \quad \text{or} \quad 5x - 3 = 0$

$x = -\frac{3}{2} \quad \text{or} \quad x = \frac{3}{5}$

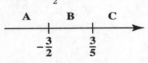

Test a number from each interval in the original inequality.

Interval A: Let $x = -2$.

$10(-2)^2 + 9(-2) \geq 9 \quad ?$

$40 - 18 \geq 9 \quad ?$

$22 \geq 9 \qquad$ *True*

Interval B: Let $x = 0$.

$0 \geq 9 \qquad$ *False*

Interval C: Let $x = 1$.

$10(1)^2 + 9(1) \geq 9 \quad ?$

$10 + 9 \geq 9 \quad ?$

$19 \geq 9 \qquad$ *True*

The solution set includes the numbers in Intervals A and C, including $-\frac{3}{2}$ and $\frac{3}{5}$ because of $\geq$.

Solution set: $\left(-\infty, -\frac{3}{2}\right] \cup \left[\frac{3}{5}, \infty\right)$

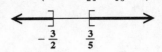

15. $4x^2 - 9 \leq 0$

Solve the equation

$4x^2 - 9 = 0.$

$(2x + 3)(2x - 3) = 0$

$2x + 3 = 0 \quad \text{or} \quad 2x - 3 = 0$

$x = -\frac{3}{2} \quad \text{or} \quad x = \frac{3}{2}$

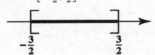

Test a number from each interval in the inequality

$4x^2 - 9 \leq 0.$

Interval A: Let $x = -2$.

$4(-2)^2 - 9 \leq 0 \quad ?$

$7 \leq 0 \qquad$ *False*

Interval B: Let $x = 0$.

$-9 \leq 0 \qquad$ *True*

Interval C: Let $x = 2$.

$4(2)^2 - 9 \leq 0 \quad ?$

$7 \leq 0 \qquad$ *False*

The solution set includes Interval B, including the endpoints.

Solution set: $\left[-\frac{3}{2}, \frac{3}{2}\right]$

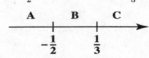

17. $6x^2 + x \geq 1$

$6x^2 + x - 1 \geq 0$

Solve the equation

$6x^2 + x - 1 = 0.$

$(2x + 1)(3x - 1) = 0$

$2x + 1 = 0 \quad \text{or} \quad 3x - 1 = 0$

$x = -\frac{1}{2} \quad \text{or} \quad x = \frac{1}{3}$

Test a number from each interval in the inequality.

$6x^2 + x \geq 1.$

Interval A: Let $x = -1$.

$$6(-1)^2 + (-1) \geq 1 \qquad ?$$
$$5 \geq 1 \qquad \textit{True}$$

Interval B: Let $x = 0$.

$$0 \geq 1 \qquad \textit{False}$$

Interval C: Let $x = 1$.

$$6(1)^2 + 1 \geq 1 \qquad ?$$
$$7 \geq 1 \qquad \textit{True}$$

The solution set includes the numbers in Intervals A and C, including $-\frac{1}{2}$ and $\frac{1}{3}$ because of $\geq$.

Solution set: $\left(-\infty, -\frac{1}{2}\right] \cup \left[\frac{1}{3}, \infty\right)$

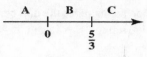

19. $z^2 - 4z \geq 0$

Solve the equation

$$z^2 - 4z = 0.$$
$$z(z - 4) = 0$$

$$z = 0 \quad \text{or} \quad z - 4 = 0$$
$$z = 4$$

Interval A: Let $z = -1$.

$$(-1)^2 - 4(-1) \geq 0 \qquad ?$$
$$5 \geq 0 \qquad \textit{True}$$

Interval B: Let $z = 2$.

$$2^2 - 4(2) \geq 0 \qquad ?$$
$$-4 \geq 0 \qquad \textit{False}$$

Interval C: Let $z = 5$.

$$5^2 - 4(5) \geq 0 \qquad ?$$
$$5 \geq 0 \qquad \textit{True}$$

The solution set includes the numbers in Intervals A and C, including 0 and 4 because of $\geq$.

Solution set: $(-\infty, 0] \cup [4, \infty)$

21. $3k^2 - 5k \leq 0$

Solve the equation

$$3k^2 - 5k = 0.$$
$$k(3k - 5) = 0$$

$$k = 0 \quad \text{or} \quad 3k - 5 = 0$$
$$k = \frac{5}{3}$$

Interval A: Let $k = -1$.

$$3(-1)^2 - 5(-1) \leq 0 \qquad ?$$
$$8 \leq 0 \qquad \textit{False}$$

Interval B: Let $k = 1$.

$$3(1)^2 - 5(1) \leq 0 \qquad ?$$
$$-2 \leq 0 \qquad \textit{True}$$

Interval C: Let $k = 2$.

$$3(2)^2 - 5(2) \leq 0 \qquad ?$$
$$2 \leq 0 \qquad \textit{False}$$

The solution set includes the numbers in Interval B, including the endpoints.

Solution set: $\left[0, \frac{5}{3}\right]$

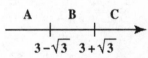

23. $x^2 - 6x + 6 \geq 0$

Solve the equation

$$x^2 - 6x + 6 = 0.$$

Since $x^2 - 6x + 6$ does not factor, let $a = 1$, $b = -6$, and $c = 6$ in the quadratic formula.

$$x = \frac{-(-6) \pm \sqrt{(-6)^2 - 4(1)(6)}}{2(1)}$$

$$= \frac{6 \pm \sqrt{12}}{2} = \frac{6 \pm 2\sqrt{3}}{2}$$

$$= \frac{2\left(3 \pm \sqrt{3}\right)}{2} = 3 \pm \sqrt{3}$$

$$x = 3 + \sqrt{3} \approx 4.7 \text{ or}$$
$$x = 3 - \sqrt{3} \approx 1.3$$

Test a number from each interval in the inequality

$$x^2 - 6x + 6 \geq 0.$$

Interval A: Let $x = 0$.

$$6 \geq 0 \qquad \textit{True}$$

Interval B: Let $x = 3$.

$$3^2 - 6(3) + 6 \geq 0 \qquad ?$$
$$-3 \geq 0 \qquad \textit{False}$$

Interval C: Let $x = 5$.

$$5^2 - 6(5) + 6 \geq 0 \qquad ?$$
$$1 \geq 0 \qquad \textit{True}$$

continued

The solution set includes the numbers in Intervals A and C, including $3 - \sqrt{3}$ and $3 + \sqrt{3}$ because of $\geq$.

Solution set: $\left(-\infty, 3 - \sqrt{3}\right] \cup \left[3 + \sqrt{3}, \infty\right)$

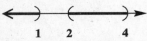

25. $(4 - 3x)^2 \geq -2$

Since $(4 - 3x)^2$ is either 0 or positive, $(4 - 3x)^2$ will always be greater than -2. Therefore, the solution set is $(-\infty, \infty)$.

27. $(3x + 5)^2 \leq -4$

Since $(3x + 5)^2$ is never negative, $(3x + 5)^2$ will never be less than or equal to a negative number. Therefore, the solution set is $\emptyset$.

29. $(x - 1)(x - 2)(x - 4) < 0$

The numbers 1, 2, and 4 are solutions of the cubic equation

$$(x - 1)(x - 2)(x - 4) = 0.$$

These numbers divide a number line into four intervals.

```
A   B    C      D
----+----+------+---->
    1    2      4
```

Test a number from each interval in the inequality

$$(x - 1)(x - 2)(x - 4) < 0.$$

Interval A: Let $x = 0$.
$$-1(-2)(-4) < 0 \quad ?$$
$$-8 < 0 \qquad \textit{True}$$

Interval B: Let $x = 1.5$.
$$(1.5 - 1)(1.5 - 2)(1.5 - 4) < 0 \quad ?$$
$$0.5(-0.5)(-2.5) < 0 \quad ?$$
$$0.625 < 0 \qquad \textit{False}$$

Interval C: Let $x = 3$.
$$(3 - 1)(3 - 2)(3 - 4) < 0 \quad ?$$
$$2(1)(-1) < 0 \quad ?$$
$$-2 < 0 \qquad \textit{True}$$

Interval D: Let $x = 5$.
$$(5 - 1)(5 - 2)(5 - 4) < 0 \quad ?$$
$$4(3)(1) < 0 \quad ?$$
$$12 < 0 \qquad \textit{False}$$

The numbers in Intervals A and C, not including 1, 2, or 4, are solutions.

Solution set: $(-\infty, 1) \cup (2, 4)$

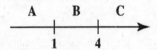

31. $(x - 4)(2x + 3)(3x - 1) \geq 0$

The numbers 4, $-\frac{3}{2}$, and $\frac{1}{3}$ are solutions of the cubic equation

$$(x - 4)(2x + 3)(3x - 1) = 0.$$

These numbers divide a number line into 4 intervals.

```
  A    B   C        D
----+----+--------+---->
  -3/2  1/3        4
```

Interval A: Let $x = -2$.
$$-6(-1)(-7) \geq 0 \quad ?$$
$$-42 \geq 0 \qquad \textit{False}$$

Interval B: Let $x = 0$.
$$-4(3)(-1) \geq 0 \quad ?$$
$$12 \geq 0 \qquad \textit{True}$$

Interval C: Let $x = 1$.
$$-3(5)(2) \geq 0 \quad ?$$
$$-30 \geq 0 \qquad \textit{False}$$

Interval D: Let $x = 5$.
$$1(13)(14) \geq 0 \quad ?$$
$$182 \geq 0 \qquad \textit{True}$$

The solution set includes numbers in Intervals B and D, including the endpoints.

Solution set: $\left[-\frac{3}{2}, \frac{1}{3}\right] \cup [4, \infty)$

```
  [----]        [------->
-3/2  1/3        4
```

33. $\dfrac{x - 1}{x - 4} > 0$

The number 1 makes the numerator 0, and 4 makes the denominator 0. These two numbers determine three intervals.

```
   A     B      C
----+------+------->
    1      4
```

Test a number from each interval in the inequality

$$\frac{x - 1}{x - 4} > 0.$$

Interval A: Let $x = 0$.
$$\frac{0 - 1}{0 - 4} > 0 \quad ?$$
$$\frac{1}{4} > 0 \qquad \textit{True}$$

Interval B: Let $x = 2$.
$$\frac{2 - 1}{2 - 4} > 0 \quad ?$$
$$\frac{1}{-2} > 0 \qquad \textit{False}$$

Interval C: Let $x = 5$.
$$\frac{5 - 1}{5 - 4} > 0 \quad ?$$
$$4 > 0 \qquad \textit{True}$$

The solution set includes numbers in Intervals A and C, excluding endpoints.

Solution set: $(-\infty, 1) \cup (4, \infty)$

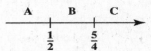

35. $\dfrac{2x + 3}{x - 5} \le 0$

The number $-\frac{3}{2}$ makes the numerator 0, and 5 makes the denominator 0. These two numbers determine three intervals.

Test a number from each interval in the inequality
$$\frac{2x + 3}{x - 5} \le 0.$$

Interval A: Let $x = -2$.
$$\frac{2(-2) + 3}{(-2) - 5} \le 0 \qquad ?$$
$$\tfrac{1}{7} \le 0 \qquad\qquad \textit{False}$$

Interval B: Let $x = 0$.
$$\frac{2(0) + 3}{0 - 5} \le 0 \qquad ?$$
$$-\tfrac{3}{5} \le 0 \qquad\qquad \textit{True}$$

Interval C: Let $x = 6$.
$$\frac{2(6) + 3}{6 - 5} \le 0 \qquad ?$$
$$15 \le 0 \qquad\qquad \textit{False}$$

The solution set includes the points in Interval B. The endpoint 5 is not included since it makes the left side undefined. The endpoint $-\frac{3}{2}$ is included because it makes the left side equal to 0.
Solution set: $\left[-\frac{3}{2}, 5\right)$

37. $\dfrac{8}{x - 2} \ge 2$

Write the inequality so that 0 is on one side.
$$\frac{8}{x - 2} - 2 \ge 0$$
$$\frac{8}{x - 2} - \frac{2(x - 2)}{x - 2} \ge 0$$

$$\frac{8 - 2x + 4}{x - 2} \ge 0$$
$$\frac{-2x + 12}{x - 2} \ge 0$$

The number 6 makes the numerator 0, and 2 makes the denominator 0. These two numbers determine three intervals.

Test a number from each interval in the inequality
$$\frac{8}{x - 2} \ge 2.$$

Interval A: Let $x = 0$.
$$\frac{8}{0 - 2} \ge 2 \qquad ?$$
$$-4 \ge 2 \qquad\qquad \textit{False}$$

Interval B: Let $x = 3$.
$$\frac{8}{3 - 2} \ge 2 \qquad ?$$
$$8 \ge 2 \qquad\qquad \textit{True}$$

Interval C: Let $x = 7$.
$$\frac{8}{7 - 2} \ge 2 \qquad ?$$
$$\tfrac{8}{5} \ge 2 \qquad\qquad \textit{False}$$

The solution set includes numbers in Interval B, including 6 but excluding 2, which makes the fraction undefined.
Solution set: $(2, 6]$

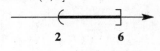

39. $\dfrac{3}{2x - 1} < 2$

Write the inequality so that 0 is on one side.
$$\frac{3}{2x - 1} - 2 < 0$$
$$\frac{3}{2x - 1} - \frac{2(2x - 1)}{2x - 1} < 0$$
$$\frac{3 - 4x + 2}{2x - 1} < 0$$
$$\frac{-4x + 5}{2x - 1} < 0$$

The number $\frac{5}{4}$ makes the numerator 0, and $\frac{1}{2}$ makes the denominator 0. These two numbers determine three intervals.

continued

Test a number from each interval in the inequality

$$\frac{3}{2x-1} < 2.$$

Interval A: Let $x = 0$.

$$\frac{3}{2(0)-1} < 2 \quad ?$$

$$-3 < 2 \qquad \textit{True}$$

Interval B: Let $x = 1$.

$$\frac{3}{2(1)-1} < 2 \quad ?$$

$$3 < 2 \qquad \textit{False}$$

Interval C: Let $x = 2$.

$$\frac{3}{2(2)-1} < 2 \quad ?$$

$$1 < 2 \qquad \textit{True}$$

The solution set includes numbers in Intervals A and C, excluding endpoints.

Solution set: $\left(-\infty, \frac{1}{2}\right) \cup \left(\frac{5}{4}, \infty\right)$

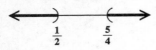

41. $\qquad \dfrac{x-3}{x+2} \geq 2$

Write the inequality so that 0 is on one side.

$$\frac{x-3}{x+2} - 2 \geq 0$$

$$\frac{x-3}{x+2} - \frac{2(x+2)}{x+2} \geq 0$$

$$\frac{x-3-2x-4}{x+2} \geq 0$$

$$\frac{-x-7}{x+2} \geq 0$$

The number -7 makes the numerator 0, and -2 makes the denominator 0. These two numbers determine three intervals.

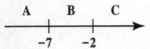

Test a number from each interval in the inequality

$$\frac{x-3}{x+2} \geq 2.$$

Interval A: Let $x = -8$.

$$\frac{-11}{-6} \geq 2 \quad ?$$

$$\frac{11}{6} \geq 2 \qquad \textit{False}$$

Interval B: Let $x = -4$.

$$\frac{-7}{-2} \geq 2 \quad ?$$

$$\frac{7}{2} \geq 2 \qquad \textit{True}$$

Interval C: Let $x = 0$.

$$\frac{-3}{2} \geq 2 \qquad \textit{False}$$

The solution set includes numbers in Interval B, including -7 but excluding -2, which makes the fraction undefined.

Solution set: $[-7, -2)$

43. $\qquad \dfrac{x-8}{x-4} < 3$

Write the inequality so that 0 is on one side.

$$\frac{x-8}{x-4} - 3 < 0$$

$$\frac{x-8}{x-4} - \frac{3(x-4)}{x-4} < 0$$

$$\frac{x-8-3x+12}{x-4} < 0$$

$$\frac{-2x+4}{x-4} < 0$$

The number 2 makes the numerator 0, and 4 makes the denominator 0. These two numbers determine three intervals.

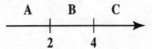

Test a number from each interval in the inequality

$$\frac{x-8}{x-4} < 3.$$

Interval A: Let $x = 0$.

$$\frac{-8}{-4} < 3 \quad ?$$

$$2 < 3 \qquad \textit{True}$$

Interval B: Let $x = 3$.

$$\frac{-5}{-1} < 3 \quad ?$$

$$5 < 3 \qquad \textit{False}$$

Interval C: Let $x = 5$.

$$\frac{-3}{1} < 3 \qquad \textit{True}$$

The solution set includes numbers in Intervals A and C, excluding endpoints.

Solution set: $(-\infty, 2) \cup (4, \infty)$

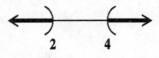

45. $\qquad \dfrac{4k}{2k-1} < k$

Write the inequality so that 0 is on one side.

$$\frac{4k}{2k-1} - k < 0$$

$$\frac{4k}{2k-1} - \frac{k(2k-1)}{2k-1} < 0$$

$$\frac{4k - 2k^2 + k}{2k - 1} < 0$$

$$\frac{-2k^2 + 5k}{2k - 1} < 0$$

$$\frac{k(-2k + 5)}{2k - 1} < 0$$

The numbers 0 and $\frac{5}{2}$ make the numerator 0, and $\frac{1}{2}$ makes the denominator 0. These three numbers determine four intervals.

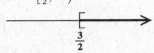

Test a number from each interval in the inequality

$$\frac{4k}{2k - 1} < k.$$

Interval A: Let $k = -1$.

$$\frac{4(-1)}{2(-1) - 1} < -1 \quad ?$$

$$\frac{4}{3} < -1 \qquad \textit{False}$$

Interval B: Let $k = \frac{1}{4}$.

$$\frac{4\left(\frac{1}{4}\right)}{2\left(\frac{1}{4}\right) - 1} < \frac{1}{4} \quad ?$$

$$-2 < \frac{1}{4} \qquad \textit{True}$$

Interval C: Let $k = 1$.

$$\frac{4(1)}{2(1) - 1} < 1 \quad ?$$

$$4 < 1 \qquad \textit{False}$$

Interval D: Let $k = 3$.

$$\frac{4(3)}{2(3) - 1} < 3 \quad ?$$

$$\frac{12}{5} < 3 \qquad \textit{True}$$

The solution set includes numbers in Intervals B and D. None of the endpoints are included.

Solution set: $\left(0, \frac{1}{2}\right) \cup \left(\frac{5}{2}, \infty\right)$

47. $\dfrac{2x - 3}{x^2 + 1} \geq 0$

The denominator is positive for all real numbers x, so it has no effect on the solution set for the inequality.

$$2x - 3 \geq 0$$
$$2x \geq 3$$
$$x \geq \frac{3}{2}$$

Solution set: $\left[\frac{3}{2}, \infty\right)$

49. $\dfrac{(3x - 5)^2}{x + 2} > 0$

The numerator is positive for all real numbers x except $x = \frac{5}{3}$, which makes it equal to 0. If we solve the inequality $x + 2 > 0$, then we only have to be sure to exclude $\frac{5}{3}$ from that solution set to determine the solution set of the original inequality.

$$x + 2 > 0$$
$$x > -2$$

Solution set: $\left(-2, \frac{5}{3}\right) \cup \left(\frac{5}{3}, \infty\right)$

51. $\left\{\left(-3, \frac{1}{8}\right), \left(-2, \frac{1}{4}\right), \left(-1, \frac{1}{2}\right), (0, 1), (1, 2), (2, 4), (3, 8)\right\}$

The domain is the set of x-values:
$\{-3, -2, -1, 0, 1, 2, 3\}$.
The range is the set of y-values:
$\left\{\frac{1}{8}, \frac{1}{4}, \frac{1}{2}, 1, 2, 4, 8\right\}$.

53.   Using the vertical line test, we find any vertical line will intersect the graph at most once. This indicates that the graph represents a function.

Chapter 11 Review Exercises

1. $t^2 = 121$

$$t = 11 \quad \text{or} \quad t = -11$$

Solution set: $\{\pm 11\}$

2. $p^2 = 3$

$$p = \sqrt{3} \quad \text{or} \quad p = -\sqrt{3}$$

Solution set: $\left\{\pm \sqrt{3}\right\}$

3. $(2x + 5)^2 = 100$

$$2x + 5 = 10 \quad \text{or} \quad 2x + 5 = -10$$
$$2x = 5 \qquad\qquad 2x = -15$$
$$x = \frac{5}{2} \quad \text{or} \qquad x = -\frac{15}{2}$$

Solution set: $\left\{-\frac{15}{2}, \frac{5}{2}\right\}$

4. $(3k - 2)^2 = -25$

$$3k - 2 = \sqrt{-25} \quad \text{or} \quad 3k - 2 = -\sqrt{-25}$$
$$3k - 2 = 5i \qquad\qquad 3k - 2 = -5i$$
$$3k = 2 + 5i \qquad\qquad 3k = 2 - 5i$$
$$k = \frac{2 + 5i}{3} \qquad\qquad k = \frac{2 - 5i}{3}$$
$$k = \frac{2}{3} + \frac{5}{3}i \quad \text{or} \qquad k = \frac{2}{3} - \frac{5}{3}i$$

Solution set: $\left\{\frac{2}{3} \pm \frac{5}{3}i\right\}$

5. By the square root property, the first step should be

$$x = \sqrt{12} \quad \text{or} \quad x = -\sqrt{12}.$$

Solution set: $\left\{ \pm 2\sqrt{3} \right\}$

6. $16t^2 = d$

$16t^2 = 150$ *Let $d = 150$.*

$t^2 = \frac{150}{16}$

$t = \sqrt{\frac{150}{16}}$ $t \geq 0$

$t \approx 3.1$ seconds

It would take about 3.1 seconds for the wallet to fall 150 feet.

7. $x^2 + 4x = 15$

Complete the square.

$\left(\frac{1}{2} \cdot 4 \right)^2 = 2^2 = 4$

Add 4 to each side.

$x^2 + 4x + 4 = 15 + 4$

$(x + 2)^2 = 19$

$x + 2 = \sqrt{19}$ or $x + 2 = -\sqrt{19}$

$x = -2 + \sqrt{19}$ or $x = -2 - \sqrt{19}$

Solution set: $\left\{ -2 \pm \sqrt{19} \right\}$

8. $2m^2 - 3m = -1$

$m^2 - \frac{3}{2}m = -\frac{1}{2}$ *Divide by 2.*

Complete the square.

$\left[\frac{1}{2} \left(-\frac{3}{2} \right) \right]^2 = \left(-\frac{3}{4} \right)^2 = \frac{9}{16}$

Add $\frac{9}{16}$ to each side.

$m^2 - \frac{3}{2}m + \frac{9}{16} = -\frac{1}{2} + \frac{9}{16}$

$\left(m - \frac{3}{4} \right)^2 = -\frac{8}{16} + \frac{9}{16}$

$\left(m - \frac{3}{4} \right)^2 = \frac{1}{16}$

$m - \frac{3}{4} = \sqrt{\frac{1}{16}}$ or $m - \frac{3}{4} = -\sqrt{\frac{1}{16}}$

$m - \frac{3}{4} = \frac{1}{4}$ $m - \frac{3}{4} = -\frac{1}{4}$

$m = \frac{3}{4} + \frac{1}{4}$ $m = \frac{3}{4} - \frac{1}{4}$

$m = 1$ or $m = \frac{1}{2}$

Solution set: $\left\{ \frac{1}{2}, 1 \right\}$

9. $2z^2 + 8z - 3 = 0$

$z^2 + 4z - \frac{3}{2} = 0$ *Divide by 2.*

Rewrite the equation with the variable terms on one side and the constant on the other side.

$$z^2 + 4z = \frac{3}{2}$$

Take half the coefficient of z and square it.

$$\frac{1}{2}(4) = 2, \quad \text{and} \quad 2^2 = 4.$$

Add 4 to both sides of the equation.

$z^2 + 4z + 4 = \frac{3}{2} + 4$

$(z + 2)^2 = \frac{11}{2}$

$z + 2 = \pm \sqrt{\frac{11}{2}}$

$z + 2 = \pm \frac{\sqrt{11}}{\sqrt{2}} \cdot \frac{\sqrt{2}}{\sqrt{2}}$

$z + 2 = \pm \frac{\sqrt{22}}{2}$

$z = -2 \pm \frac{\sqrt{22}}{2}$

$z = \frac{-4}{2} \pm \frac{\sqrt{22}}{2}$

$z = \frac{-4 \pm \sqrt{22}}{2}$

Solution set: $\left\{ \dfrac{-4 \pm \sqrt{22}}{2} \right\}$

10. $(4a + 1)(a - 1) = -7$

Multiply on the left side and then simplify. Get all variable terms on one side and the constant on the other side.

$4a^2 - 4a + a - 1 = -7$

$4a^2 - 3a = -6$

Divide both sides by 4 so that the coefficient of a^2 will be 1.

$$a^2 - \frac{3}{4}a = -\frac{6}{4} = -\frac{3}{2}$$

Square half the coefficient of a and add it to both sides.

$a^2 - \frac{3}{4}a + \frac{9}{64} = -\frac{3}{2} + \frac{9}{64}$

$\left(a - \frac{3}{8} \right)^2 = -\frac{96}{64} + \frac{9}{64}$

$\left(a - \frac{3}{8} \right)^2 = -\frac{87}{64}$

$a - \frac{3}{8} = \pm \sqrt{-\frac{87}{64}}$

$a - \frac{3}{8} = \pm \frac{i\sqrt{87}}{8}$

$a = \frac{3}{8} \pm \frac{\sqrt{87}}{8}i$

Solution set: $\left\{ \dfrac{3}{8} \pm \dfrac{\sqrt{87}}{8}i \right\}$

11. (a) $x^2 + 5x + 2 = 0$

Here $a = 1$, $b = 5$, and $c = 2$.

$b^2 - 4ac = 5^2 - 4(1)(2)$

$= 25 - 8 = 17$

Since the discriminant is positive, but not a perfect square, there are two distinct irrational number solutions. The answer is **C**.

(b) $$4t^2 = 3 - 4t$$
$$4t^2 + 4t - 3 = 0$$
Here $a = 4$, $b = 4$, and $c = -3$.
$$b^2 - 4ac = 4^2 - 4(4)(-3)$$
$$= 16 + 48$$
$$= 64 \text{ or } 8^2$$

Since the discriminant is positive, and a perfect square, there are two distinct rational number solutions. The answer is **A**.

12. (a) $$4x^2 = 6x - 8$$
$$4x^2 - 6x + 8 = 0$$
Here $a = 4$, $b = -6$, and $c = 8$.
$$b^2 - 4ac = (-6)^2 - 4(4)(8)$$
$$= 36 - 128 = -92$$

Since the discriminant is negative, there are two distinct nonreal complex number solutions. The answer is **D**.

(b) $9z^2 + 30z + 25 = 0$
Here $a = 9$, $b = 30$, and $c = 25$.
$$b^2 - 4ac = 30^2 - 4(9)(25)$$
$$= 900 - 900 = 0$$

Since the discriminant is zero, there is exactly one rational number solution. The answer is **B**.

13. $2x^2 + x - 21 = 0$
Here $a = 2$, $b = 1$, and $c = -21$.
$$x = \frac{-b \pm \sqrt{b^2 - 4ac}}{2a}$$
$$x = \frac{-1 \pm \sqrt{1^2 - 4(2)(-21)}}{2(2)}$$
$$= \frac{-1 \pm \sqrt{1 + 168}}{4}$$
$$= \frac{-1 \pm \sqrt{169}}{4} = \frac{-1 \pm 13}{4}$$
$$x = \frac{-1 + 13}{4} = \frac{12}{4} = 3 \text{ or}$$
$$x = \frac{-1 - 13}{4} = -\frac{14}{4} = -\frac{7}{2}$$
Solution set: $\left\{ -\frac{7}{2}, 3 \right\}$

14. $$k^2 + 5k = 7$$
$$k^2 + 5k - 7 = 0$$
Here $a = 1$, $b = 5$, and $c = -7$.
$$k = \frac{-b \pm \sqrt{b^2 - 4ac}}{2a}$$
$$k = \frac{-5 \pm \sqrt{5^2 - 4(1)(-7)}}{2(1)}$$
$$= \frac{-5 \pm \sqrt{25 + 28}}{2}$$
$$= \frac{-5 \pm \sqrt{53}}{2}$$

Solution set: $\left\{ \dfrac{-5 \pm \sqrt{53}}{2} \right\}$

15. $$(t + 3)(t - 4) = -2$$
$$t^2 - t - 12 = -2$$
$$t^2 - t - 10 = 0$$

Here $a = 1$, $b = -1$, and $c = -10$.
$$t = \frac{-b \pm \sqrt{b^2 - 4ac}}{2a}$$
$$t = \frac{-(-1) \pm \sqrt{(-1)^2 - 4(1)(-10)}}{2(1)}$$
$$= \frac{1 \pm \sqrt{1 + 40}}{2} = \frac{1 \pm \sqrt{41}}{2}$$
Solution set: $\left\{ \dfrac{1 \pm \sqrt{41}}{2} \right\}$

16. $2x^2 + 3x + 4 = 0$
Here $a = 2$, $b = 3$, and $c = 4$.
$$x = \frac{-b \pm \sqrt{b^2 - 4ac}}{2a}$$
$$x = \frac{-3 \pm \sqrt{3^2 - 4(2)(4)}}{2(2)}$$
$$= \frac{-3 \pm \sqrt{9 - 32}}{4}$$
$$= \frac{-3 \pm \sqrt{-23}}{4}$$
$$= \frac{-3 \pm i\sqrt{23}}{4} = -\frac{3}{4} \pm \frac{\sqrt{23}}{4}i$$
Solution set: $\left\{ -\frac{3}{4} \pm \frac{\sqrt{23}}{4}i \right\}$

17. $$3p^2 = 2(2p - 1)$$
$$3p^2 = 4p - 2$$
$$3p^2 - 4p + 2 = 0$$
Here $a = 3$, $b = -4$, and $c = 2$.
$$p = \frac{-b \pm \sqrt{b^2 - 4ac}}{2a}$$
$$p = \frac{-(-4) \pm \sqrt{(-4)^2 - 4(3)(2)}}{2(3)}$$
$$= \frac{4 \pm \sqrt{16 - 24}}{6} = \frac{4 \pm \sqrt{-8}}{6}$$
$$= \frac{4 \pm 2i\sqrt{2}}{6} = \frac{2\left(2 \pm i\sqrt{2}\right)}{6}$$
$$= \frac{2 \pm i\sqrt{2}}{3} = \frac{2}{3} \pm \frac{\sqrt{2}}{3}i$$
Solution set: $\left\{ \frac{2}{3} \pm \frac{\sqrt{2}}{3}i \right\}$

18. $m(2m-7) = 3m^2 + 3$

$2m^2 - 7m = 3m^2 + 3$

$0 = m^2 + 7m + 3$

Here $a = 1$, $b = 7$, and $c = 3$.

$$m = \frac{-b \pm \sqrt{b^2 - 4ac}}{2a}$$

$$m = \frac{-7 \pm \sqrt{7^2 - 4(1)(3)}}{2(1)}$$

$$= \frac{-7 \pm \sqrt{49 - 12}}{2}$$

$$= \frac{-7 \pm \sqrt{37}}{2}$$

Solution set: $\left\{ \frac{-7 \pm \sqrt{37}}{2} \right\}$

19. $\frac{15}{x} = 2x - 1$

$x\left(\frac{15}{x} \right) = x(2x - 1)$ *Multiply by the LCD, x.*

$15 = 2x^2 - x$

$0 = 2x^2 - x - 15$

$0 = (2x + 5)(x - 3)$

$2x + 5 = 0$ or $x - 3 = 0$

$x = -\frac{5}{2}$ or $x = 3$

Check $x = -\frac{5}{2}$: $-6 = -5 - 1$ *True*

Check $x = 3$: $5 = 6 - 1$ *True*

Solution set: $\left\{ -\frac{5}{2}, 3 \right\}$

20. $\frac{1}{n} + \frac{2}{n+1} = 2$

$n(n+1)\left(\frac{1}{n} + \frac{2}{n+1} \right)$ *Multiply by the LCD, n(n+1).*

$= n(n+1) \cdot 2$

$(n+1) + 2n = 2n^2 + 2n$

$0 = 2n^2 - n - 1$

$0 = (2n + 1)(n - 1)$

$2n + 1 = 0$ or $n - 1 = 0$

$n = -\frac{1}{2}$ or $n = 1$

Check $n = -\frac{1}{2}$: $-2 + 4 = 2$ *True*

Check $n = 1$: $1 + 1 = 2$ *True*

Solution set: $\left\{ -\frac{1}{2}, 1 \right\}$

21. $-2r = \sqrt{\frac{48 - 20r}{2}}$

Square both sides.

$$(-2r)^2 = \left(\sqrt{\frac{48 - 20r}{2}} \right)^2$$

$$4r^2 = \frac{48 - 20r}{2}$$

$$4r^2 = 24 - 10r$$

$4r^2 + 10r - 24 = 0$

$2r^2 + 5r - 12 = 0$

$(r + 4)(2r - 3) = 0$

$r + 4 = 0$ or $2r - 3 = 0$

$r = -4$ or $r = \frac{3}{2}$

Check $r = -4$: $8 = \sqrt{64}$ *True*

Check $r = \frac{3}{2}$: $-3 = \sqrt{9}$ *False*

Solution set: $\{-4\}$

22. $8(3x + 5)^2 + 2(3x + 5) - 1 = 0$

Let $u = 3x + 5$. The equation becomes

$8u^2 + 2u - 1 = 0.$

$(2u + 1)(4u - 1) = 0$

$2u + 1 = 0$ or $4u - 1 = 0$

$u = -\frac{1}{2}$ or $u = \frac{1}{4}$

To find x, substitute $3x + 5$ for u.

$3x + 5 = -\frac{1}{2}$ or $3x + 5 = \frac{1}{4}$

$3x = -\frac{11}{2}$ $3x = -\frac{19}{4}$

$x = -\frac{11}{6}$ or $x = -\frac{19}{12}$

Check $x = -\frac{11}{6}$: $2 - 1 - 1 = 0$ *True*

Check $x = -\frac{19}{12}$: $0.5 + 0.5 - 1 = 0$ *True*

Solution set: $\left\{ -\frac{11}{6}, -\frac{19}{12} \right\}$

23. $2x^{2/3} - x^{1/3} - 28 = 0$

Let $u = x^{1/3}$, so $u^2 = \left(x^{1/3} \right)^2 = x^{2/3}$.

The equation becomes

$2u^2 - u - 28 = 0.$

$(2u + 7)(u - 4) = 0$

$2u + 7 = 0$ or $u - 4 = 0$

$u = -\frac{7}{2}$ or $u = 4$

To find x, substitute $x^{1/3}$ for u.

$x^{1/3} = -\frac{7}{2}$ or $x^{1/3} = 4$

$\left(x^{1/3} \right)^3 = \left(-\frac{7}{2} \right)^3$ $\left(x^{1/3} \right)^3 = 4^3$

$x = -\frac{343}{8}$ or $x = 64$

Check $x = -\frac{343}{8}$: $24.5 + 3.5 - 28 = 0$ *True*

Check $x = 64$: $32 - 4 - 28 = 0$ *True*

Solution set: $\left\{ -\frac{343}{8}, 64 \right\}$

24. $p^4 - 10p^2 + 9 = 0$

Let $x = p^2$, so $x^2 = p^4$.

$$x^2 - 10x + 9 = 0$$
$$(x - 1)(x - 9) = 0$$

$$x - 1 = 0 \quad \text{or} \quad x - 9 = 0$$
$$x = 1 \quad \text{or} \quad x = 9$$

To find p, substitute p^2 for x.

$$p^2 = 1 \quad \text{or} \quad p^2 = 9$$
$$p = \pm\sqrt{1} \quad\quad p = \pm\sqrt{9}$$
$$p = \pm 1 \quad \text{or} \quad p = \pm 3$$

Check $p = \pm 1$: $1 - 10 + 9 = 0$ *True*

Check $p = \pm 3$: $81 - 90 + 9 = 0$ *True*

Solution set: $\{\pm 3, \pm 1\}$

25. Let $x =$ Phong's speed.

Make a table. Use $d = rt$, or $t = \dfrac{d}{r}$.

	Distance	Rate	Time
Upstream	20	$x - 3$	$\dfrac{20}{x - 3}$
Downstream	20	$x + 3$	$\dfrac{20}{x + 3}$

$$\begin{array}{ccccc}
\text{Time} & & \text{time} & & \\
\text{upstream} & \text{plus} & \text{downstream} & \text{equals} & \text{7 hr.} \\
\downarrow & \downarrow & \downarrow & \downarrow & \downarrow \\
\dfrac{20}{x - 3} & + & \dfrac{20}{x + 3} & = & 7
\end{array}$$

Multiply each side by the LCD, $(x + 3)(x - 3)$.

$$(x + 3)(x - 3)\left(\frac{20}{x - 3} + \frac{20}{x + 3}\right) = (x + 3)(x - 3)(7)$$
$$20(x + 3) + 20(x - 3) = 7(x^2 - 9)$$
$$20x + 60 + 20x - 60 = 7x^2 - 63$$
$$0 = 7x^2 - 40x - 63$$
$$0 = (x - 7)(7x + 9)$$

$$x - 7 = 0 \quad \text{or} \quad 7x + 9 = 0$$
$$x = 7 \quad \text{or} \quad x = -\frac{9}{7}$$

Reject $-\frac{9}{7}$ since speed can't be negative. Phong's speed was 7 mph.

26. Let $x =$ Maureen's speed on the trip to pick up Laurie.
Make a chart. Use $d = rt$, or $t = \frac{d}{r}$.

	Distance	Rate	Time
To Laurie	8	x	$\dfrac{8}{x}$
To the Mall	11	$x + 15$	$\dfrac{11}{x + 15}$

$$\begin{array}{ccccc}
\text{Time to pick} & + & \text{time to} & = & \text{24 min} \\
\text{up Laurie} & & \text{mall} & & \text{(or 0.4 hr).} \\
\dfrac{8}{x} & + & \dfrac{11}{x + 15} & = & 0.4
\end{array}$$

Multiply each side by the LCD, $x(x + 15)$.

$$x(x + 15)\left(\frac{8}{x} + \frac{11}{x + 15}\right) = x(x + 15)(0.4)$$
$$8(x + 15) + 11x = 0.4x(x + 15)$$
$$8x + 120 + 11x = 0.4x^2 + 6x$$
$$0 = 0.4x^2 - 13x - 120$$

Multiply by 5 to clear the decimal.
$$0 = 2x^2 - 65x - 600$$
$$0 = (x - 40)(2x + 15)$$

$$x - 40 = 0 \quad \text{or} \quad 2x + 15 = 0$$
$$x = 40 \quad \text{or} \quad x = -\frac{15}{2}$$

Speed cannot be negative, so $-\frac{15}{2}$ is not a solution. Maureen's speed on the trip to pick up Laurie was 40 mph.

27. Let $x =$ the amount of time for the old machine alone and

$x - 1 =$ the amount of time for the new machine alone.

Make a chart.

Machine	Rate	Time Together	Fractional Part of the Job Done
Old	$\dfrac{1}{x}$	2	$\dfrac{2}{x}$
New	$\dfrac{1}{x - 1}$	2	$\dfrac{2}{x - 1}$

$$\begin{array}{ccccc}
\text{Part done by} & + & \text{Part done by} & = & \text{1 whole} \\
\text{old machine} & & \text{new machine} & & \text{job.} \\
\dfrac{2}{x} & + & \dfrac{2}{x - 1} & = & 1
\end{array}$$

Multiply by the LCD, $x(x - 1)$.

$$x(x - 1)\left(\frac{2}{x} + \frac{2}{x - 1}\right) = x(x - 1) \cdot 1$$
$$2(x - 1) + 2x = x^2 - x$$
$$2x - 2 + 2x = x^2 - x$$
$$0 = x^2 - 5x + 2$$

Use the quadratic formula.

$$x = \frac{-b \pm \sqrt{b^2 - 4ac}}{2a}$$
$$x = \frac{-(-5) \pm \sqrt{(-5)^2 - 4(1)(2)}}{2(1)}$$

continued

$$= \frac{5 \pm \sqrt{25 - 8}}{2} = \frac{5 \pm \sqrt{17}}{2}$$

$$x = \frac{5 + \sqrt{17}}{2} \approx 4.6 \text{ or}$$

$$x = \frac{5 - \sqrt{17}}{2} \approx 0.4$$

Reject 0.4 as the time for the old machine, because that would yield a negative time for the new machine. Thus, the old machine takes about 4.6 hours.

28. Let $x =$ the time for Carter alone and $x - 1 =$ the time for Greg alone.

Worker	Rate	Time Together	Fractional Part of the Job Done
Carter	$\frac{1}{x}$	1.5	$\frac{1.5}{x}$
Greg	$\frac{1}{x-1}$	1.5	$\frac{1.5}{x-1}$

$$\begin{array}{ccccc} \text{Part by} & & \text{part} & & \text{1 whole} \\ \text{Carter} & + & \text{by Greg} & = & \text{job.} \\ \frac{1.5}{x} & + & \frac{1.5}{x-1} & = & 1 \end{array}$$

Multiply by the LCD, $x(x - 1)$.

$$x(x-1)\left(\frac{1.5}{x} + \frac{1.5}{x-1}\right) = x(x-1) \cdot 1$$

$$1.5(x-1) + 1.5x = x^2 - x$$

$$1.5x - 1.5 + 1.5x = x^2 - x$$

$$0 = x^2 - 4x + 1.5$$

Use the quadratic formula.

$$x = \frac{-b \pm \sqrt{b^2 - 4ac}}{2a}$$

$$x = \frac{-(-4) \pm \sqrt{(-4)^2 - 4(1)(1.5)}}{2(1)}$$

$$= \frac{4 \pm \sqrt{16 - 6}}{2} = \frac{4 \pm \sqrt{10}}{2}$$

$$= \frac{4 \pm \sqrt{10}}{2}$$

$$x = \frac{4 + \sqrt{10}}{2} \approx 3.6 \text{ or}$$

$$x = \frac{4 - \sqrt{10}}{2} \approx 0.4$$

Reject 0.4 as Carter's time, because that would make Greg's time negative. Thus, Carter's time alone is about 3.6 hours and Greg's time alone is about $x - 1 = 2.6$ hours.

29. Solve $k = \dfrac{rF}{wv^2}$ for v.

Multiply both sides by v^2, then divide by k.

$$v^2 = \frac{rF}{kw}$$

$$v = \pm \sqrt{\frac{rF}{kw}} = \frac{\pm \sqrt{rF}}{\sqrt{kw}}$$

$$= \frac{\pm \sqrt{rF}}{\sqrt{kw}} \cdot \frac{\sqrt{kw}}{\sqrt{kw}}$$

$$v = \frac{\pm \sqrt{rFkw}}{kw}$$

30. Solve $p = \sqrt{\dfrac{yz}{6}}$ for y.

Square both sides.

$$p^2 = \left(\sqrt{\frac{yz}{6}}\right)^2$$

$$p^2 = \frac{yz}{6}$$

$$\frac{6p^2}{z} = y \text{ or } y = \frac{6p^2}{z}$$

31. Solve $mt^2 = 3mt + 6$ for t.

$$mt^2 - 3mt - 6 = 0$$

Use the quadratic formula with $a = m$, $b = -3m$, and $c = -6$.

$$t = \frac{-b \pm \sqrt{b^2 - 4ac}}{2a}$$

$$t = \frac{3m \pm \sqrt{(-3m)^2 - 4(m)(-6)}}{2m}$$

$$= \frac{3m \pm \sqrt{9m^2 + 24m}}{2m}$$

32. Let $x =$ the length of the longer leg;
$\frac{3}{4}x =$ the length of the shorter leg;
$2x - 9 =$ the length of the hypotenuse.

Use the Pythagorean formula.

$$c^2 = a^2 + b^2$$

$$(2x - 9)^2 = x^2 + \left(\tfrac{3}{4}x\right)^2$$

$$4x^2 - 36x + 81 = x^2 + \tfrac{9}{16}x^2$$

$$16\left(4x^2 - 36x + 81\right) = 16\left(x^2 + \tfrac{9}{16}x^2\right)$$

$$64x^2 - 576x + 1296 = 16x^2 + 9x^2$$

$$39x^2 - 576x + 1296 = 0$$

$$13x^2 - 192x + 432 = 0 \qquad \textit{Divide by 3.}$$

$$(13x - 36)(x - 12) = 0$$

$$13x - 36 = 0 \quad \text{or} \quad x - 12 = 0$$

$$x = \tfrac{36}{13} \quad \text{or} \quad x = 12$$

Reject $x = \frac{36}{13}$ since $2\left(\frac{36}{13}\right) - 9$ is negative.

If $x = 12$, then
$$\tfrac{3}{4}x = \tfrac{3}{4}(12) = 9$$
and
$$2x - 9 = 2(12) - 9 = 15.$$
The lengths of the three sides are 9 feet, 12 feet, and 15 feet.

33. Let x = the amount removed from one dimension.

The area of the square is 256 cm², so the length of one side is $\sqrt{256}$ or 16 cm. The dimensions of the new rectangle are $16 + x$ and $16 - x$ cm. The area of the new rectangle is 16 cm² less than the area of the square.

$$(16 + x)(16 - x) = 256 - 16$$
$$256 - x^2 = 240$$
$$-x^2 = -16$$
$$x^2 = 16$$
$$x = \pm\sqrt{16} = \pm 4$$

Length cannot be negative, so reject -4. If $x = 4$, then $16 + x = 20$, and $16 - x = 12$. The dimensions are 20 cm by 12 cm.

34. Let x = the width of the border.

Area of mat = length $\cdot$ width
$$352 = (2x + 20)(2x + 14)$$
$$352 = 4x^2 + 68x + 280$$
$$0 = 4x^2 + 68x - 72$$
$$0 = x^2 + 17x - 18$$
$$0 = (x + 18)(x - 1)$$

$$x + 18 = 0 \quad \text{or} \quad x - 1 = 0$$
$$x = -18 \quad \text{or} \quad x = 1$$

Reject the negative answer for length. The mat is 1 inch wide.

35. $f(t) = 100t^2 - 300t$ is the distance of the light from the starting point at t minutes. When the light returns to the starting point, the value of $f(t)$ will be 0.

$$0 = 100t^2 - 300t$$
$$0 = t^2 - 3t \qquad \textit{Divide by 100.}$$
$$0 = t(t - 3)$$

$$t = 0 \quad \text{or} \quad t - 3 = 0$$
$$t = 3$$

Since $t = 0$ represents the starting time, the light will return to the starting point in 3 minutes.

36. $f(t) = -16t^2 + 45t + 400$
$$200 = -16t^2 + 45t + 400 \quad \textit{Let f(t) = 200.}$$
$$0 = -16t^2 + 45t + 200$$
$$0 = 16t^2 - 45t - 200$$

Here $a = 16$, $b = -45$, and $c = -200$.

$$t = \frac{-b \pm \sqrt{b^2 - 4ac}}{2a}$$
$$t = \frac{-(-45) \pm \sqrt{(-45)^2 - 4(16)(-200)}}{2(16)}$$
$$= \frac{45 \pm \sqrt{2025 + 12{,}800}}{32}$$
$$= \frac{45 \pm \sqrt{14{,}825}}{32}$$
$$t = \frac{45 + \sqrt{14{,}825}}{32} \approx 5.2 \quad \text{or}$$
$$t = \frac{45 - \sqrt{14{,}825}}{32} \approx -2.4$$

Reject the negative solution since time cannot be negative. The ball will reach a height of 200 ft above the ground after about 5.2 seconds.

37. $s(t) = -16t^2 + 75t + 407$
$$450 = -16t^2 + 75t + 407 \quad \textit{Let s(t) = 450.}$$
$$0 = -16t^2 + 75t - 43$$

Here $a = -16$, $b = 75$, and $c = -43$.

$$t = \frac{-b \pm \sqrt{b^2 - 4ac}}{2a}$$
$$t = \frac{-75 \pm \sqrt{75^2 - 4(-16)(-43)}}{2(-16)}$$
$$= \frac{-75 \pm \sqrt{5625 - 2752}}{-32}$$
$$= \frac{-75 \pm \sqrt{2873}}{-32}$$
$$t = \frac{-75 + \sqrt{2873}}{-32} \approx 0.7 \quad \text{or}$$
$$t = \frac{-75 - \sqrt{2873}}{-32} \approx 4.0$$

The ball will be 450 ft above the ground after 0.7 second (going up) and again after 4.0 seconds (coming down).

38. Let p = the price.

Demand = Supply
$$\frac{25}{p} = 70p + 15$$
$$p\left(\frac{25}{p}\right) = p(70p + 15) \qquad \textit{Multiply by p.}$$
$$25 = 70p^2 + 15p$$
$$0 = 70p^2 + 15p - 25$$
$$0 = 14p^2 + 3p - 5 \qquad \textit{Divide by 5.}$$
$$0 = (2p - 1)(7p + 5)$$

$$2p - 1 = 0 \quad \text{or} \quad 7p + 5 = 0$$
$$p = \tfrac{1}{2} \quad \text{or} \qquad p = -\tfrac{5}{7}$$

Reject the negative answer for price. Supply and demand are equal when p is $\tfrac{1}{2}$ dollar or \$0.50.

39. Let $A = 10{,}920.25$, $P = 10{,}000$, and solve for r.

$$A = P(1 + r)^2$$
$$10{,}920.25 = 10{,}000(1 + r)^2$$
$$1.092025 = (1 + r)^2$$

$1 + r = \sqrt{1.092025}$ or $1 + r = -\sqrt{1.092025}$
$1 + r = 1.045$ or $1 + r = -1.045$
$\quad\; r = 0.045$ or $r = -2.045$

Reject the negative rate. The required interest rate is $0.045 = 4.5\%$.

40. **(a)** Use $f(x) = 3.29x^2 - 10.4x + 21.6$ with $x = 11$.

$$f(11) = 3.29(11)^2 - 10.4(11) + 21.6$$
$$f(11) \approx 305$$

The value of 305 is close to the number shown on the graph.

(b) $200 = 3.29x^2 - 10.4x + 21.6$ *Let f(x) = 200.*
$\qquad 0 = 3.29x^2 - 10.4x - 178.4$

Here $a = 3.29$, $b = -10.4$, and $c = -178.4$.

$$x = \frac{-b \pm \sqrt{b^2 - 4ac}}{2a}$$

$$x = \frac{-(-10.4) \pm \sqrt{(-10.4)^2 - 4(3.29)(-178.4)}}{2(3.29)}$$

$$= \frac{10.4 \pm \sqrt{2455.904}}{6.58}$$

$$x = \frac{10.4 + \sqrt{2455.904}}{6.58} \approx 9.1 \;\; \text{or}$$

$$x = \frac{10.4 - \sqrt{2455.904}}{6.58} \approx -6.0$$

Since time cannot be negative, the correct answer is $x \approx 9$, which represents 1999. Based on the graph, the number of e-mail boxes reached 200 million in 2000.

41. $y = 6 - 2x^2$
Write in $y = a(x - h)^2 + k$ form as
$y = -2(x - 0)^2 + 6$. The vertex (h, k) is $(0, 6)$.

42. $f(x) = -(x - 1)^2$
Write in $y = a(x - h)^2 + k$ form as
$y = -1(x - 1)^2 + 0$. The vertex (h, k) is $(1, 0)$.

43. $y = (x - 3)^2 + 7$
The equation is in the form $y = a(x - h)^2 + k$, so the vertex (h, k) is $(3, 7)$.

44. $y = f(x) = -3x^2 + 4x - 2$

Use the vertex formula with $a = -3$ and $b = 4$.

The x-coordinate of the vertex is

$$\frac{-b}{2a} = \frac{-4}{2(-3)} = \frac{2}{3}.$$

The y-coordinate of the vertex is

$$f\left(\frac{-b}{2a}\right) = f\left(\frac{2}{3}\right)$$
$$= -3\left(\tfrac{2}{3}\right)^2 + 4\left(\tfrac{2}{3}\right) - 2$$
$$= -\tfrac{4}{3} + \tfrac{8}{3} - 2$$
$$= -\tfrac{2}{3}.$$

The vertex is $\left(\tfrac{2}{3}, -\tfrac{2}{3}\right)$.

45. $x = (y - 3)^2 - 4$

When $y = 3$, $x = -4$, so the vertex is $(-4, 3)$.

46. If the discriminant is negative, there are no x-intercepts.

47. $y = 2(x - 2)^2 - 3$

The graph opens up since $a = 2 > 0$.

The vertex is $(2, -3)$ and the axis is $x = 2$. The domain is $(-\infty, \infty)$. The smallest y-value is -3, so the range is $[-3, \infty)$.

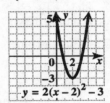

48. $f(x) = -2x^2 + 8x - 5$
Complete the square to find the vertex.

$$f(x) = -2\left(x^2 - 4x\right) - 5$$
$$= -2\left(x^2 - 4x + 4 - 4\right) - 5$$
$$= -2\left(x^2 - 4x + 4\right) - 2(-4) - 5$$
$$= -2(x - 2)^2 + 8 - 5$$
$$= -2(x - 2)^2 + 3$$

The equation is in the form $y = a(x - h)^2 + k$, so the vertex (h, k) is $(2, 3)$ and the axis is $x = 2$. Here, $a = -2 < 0$, so the parabola opens down.

Also, $|a| = |-2| = 2 > 1$, so the graph is narrower than the graph of $y = x^2$. The points $(0, -5)$, $(1, 1)$, and $(3, 1)$ are on the graph.

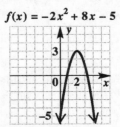

The domain is $(-\infty, \infty)$. The largest y-value is 3, so the range is $(-\infty, 3]$.

49. $x = 2(y + 3)^2 - 4$

Since the roles of x and y are reversed, this is a horizontal parabola.

$x = 2[y - (-3)]^2 + (-4)$

The equation is in the form

$$x = a(y - k)^2 + h,$$

so the vertex (h, k) is $(-4, -3)$ and the axis is $y = -3$. Here, $a = 2 > 0$, so the parabola opens to the right and is narrower than the graph of $y = x^2$.

Two other points on the graph are $(4, -1)$ and $(4, -5)$.

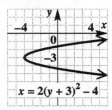

$x = 2(y + 3)^2 - 4$

The smallest x-value is -4, so the domain is $[-4, \infty)$. The range is $(-\infty, \infty)$.

50. $x = -\frac{1}{2}y^2 + 6y - 14$

Since the roles of x and y are reversed, this is a horizontal parabola. Complete the square to find the vertex.

$$\begin{aligned}
x &= -\tfrac{1}{2}y^2 + 6y - 14 \\
&= -\tfrac{1}{2}(y^2 - 12y) - 14 \\
&= -\tfrac{1}{2}(y^2 - 12y + 36 - 36) - 14 \\
&= -\tfrac{1}{2}(y^2 - 12y + 36) - \tfrac{1}{2}(-36) - 14 \\
&= -\tfrac{1}{2}(y - 6)^2 + 18 - 14 \\
x &= -\tfrac{1}{2}(y - 6)^2 + 4
\end{aligned}$$

The equation is in the form $x = a(y - k)^2 + h$, so the vertex (h, k) is $(4, 6)$ and the axis is $y = 6$. Here, $a = -\frac{1}{2} < 0$, so the parabola opens to the left.
Also, $|a| = \left|-\frac{1}{2}\right| = \frac{1}{2} < 1$, so the graph is wider than the graph of $y = x^2$. The points $(-14, 0)$, $(2, 4)$, and $(2, 8)$ are on the graph.

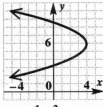

$x = -\dfrac{1}{2}\, y^2 + 6y - 14$

The largest x-value is 4, so the domain is $(-\infty, 4]$. The range is $(-\infty, \infty)$.

51.   **(a)** Use $ax^2 + bx + c = y$ with $(1, 11.47)$, $(4, 24.45)$, and $(8, 29.78)$.

$$\begin{aligned}
a +\ b + c &= 11.47 \quad (1) \\
16a + 4b + c &= 24.45 \quad (2) \\
64a + 8b + c &= 29.78 \quad (3)
\end{aligned}$$

(b) Add $-1 \times (1)$ to (2).

$$15a + 3b = 12.98 \quad (4)$$

Add $-1 \times (1)$ to (3).

$$63a + 7b = 18.31 \quad (5)$$

Now eliminate b.

$$\begin{array}{rcll}
-105a\ -\ 21b &=& -90.86 & -7 \times (4) \\
189a\ +\ 21b &=& 54.93 & 3 \times (5) \\
\hline
84a\ &=& -35.93 & \\
a &=& -\frac{35.93}{84} & \approx -0.4277
\end{array}$$

Use (4) to find b.

$$\begin{aligned}
15a + 3b &= 12.98 \quad (4) \\
15\left(-\tfrac{35.93}{84}\right) + 3b &= 12.98 \\
3b &\approx 19.39607 \\
b &\approx 6.4654
\end{aligned}$$

Use (1) to find c.

$$\begin{aligned}
c &= 11.47 - a - b \\
&\approx 11.47 + 0.4277 - 6.4654 \\
&= 5.4323
\end{aligned}$$

Rounding slightly, we get the quadratic function

$$f(x) = -0.428x^2 + 6.47x + 5.43.$$

(c) For 2000, $x = 2000 - 1995 = 5$.

$$\begin{aligned}
f(5) &= -0.428(5)^2 + 6.47(5) + 5.43 \\
&= 27.08
\end{aligned}$$

From the model, approximate consumer spending for home videos was \$27.08, which is slightly higher than the actual value of \$25.89.

52. $s(t) = -16t^2 + 160t$
The equation represents a parabola. Since $a = -16 < 0$, the parabola opens down. The time and maximum height occur at the vertex (h, k) of the parabola, given by

$$\left(\frac{-b}{2a},\, s\!\left(\frac{-b}{2a}\right)\right).$$

Using the standard form of the equation, $a = -16$ and $b = 160$, so

$$h = \frac{-b}{2a} = \frac{-160}{2(-16)} = 5,$$

and

$$\begin{aligned}
k = s(h) &= -16(5)^2 + 160(5) \\
&= -400 + 800 = 400.
\end{aligned}$$

The vertex is $(5, 400)$. The time at which the maximum height is reached is 5 seconds. The maximum height is 400 feet.

53. Let L = the length of the rectangle and W = the width.

The perimeter of the rectangle is 200 m, so

$$2L + 2W = 200$$
$$2W = 200 - 2L$$
$$W = 100 - L.$$

Since the area is length times width, substitute $100 - L$ for W.

$$A = LW$$
$$= L(100 - L)$$
$$= 100L - L^2 \quad \text{or} \quad -L^2 + 100L$$

Use the vertex formula.

$$L = \frac{-b}{2a} = \frac{-100}{2(-1)} = 50$$

So $L = 50$ meters and
$W = 100 - L = 100 - 50 = 50$ meters.
The maximum area is

$$50 \cdot 50 = 2500 \text{ m}^2.$$

54. $(x - 4)(2x + 3) > 0$
Solve the equation
$(x - 4)(2x + 3) = 0.$

$$x - 4 = 0 \quad \text{or} \quad 2x + 3 = 0$$
$$x = 4 \quad \text{or} \quad x = -\frac{3}{2}$$

The numbers $-\frac{3}{2}$ and 4 divide a number line into three intervals.

Test a number from each interval in the inequality

$$(x - 4)(2x + 3) > 0.$$

Interval A: Let $x = -2$.
$$-6(-1) > 0 \quad ?$$
$$6 > 0 \quad\quad True$$

Interval B: Let $x = 0$.
$$-4(3) > 0 \quad ?$$
$$-12 > 0 \quad\quad False$$

Interval C: Let $x = 5$.
$$1(13) > 0 \quad ?$$
$$13 > 0 \quad\quad True$$

The solution set includes numbers in Intervals A and C, excluding endpoints.

Solution set: $\left(-\infty, -\frac{3}{2}\right) \cup (4, \infty)$

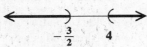

55. $$x^2 + x \le 12$$
Solve the equation
$$x^2 + x = 12.$$
$$x^2 + x - 12 = 0$$
$$(x + 4)(x - 3) = 0$$

$$x + 4 = 0 \quad \text{or} \quad x - 3 = 0$$
$$x = -4 \quad \text{or} \quad x = 3$$

The numbers -4 and 3 divide a number line into three intervals.

Test a number from each interval in the inequality

$$x^2 + x \le 12.$$

Interval A: Let $x = -5$.
$$25 - 5 \le 12 \quad ?$$
$$20 \le 12 \quad\quad False$$

Interval B: Let $x = 0$.
$$0 \le 12 \quad\quad True$$

Interval C: Let $x = 4$.
$$16 + 4 \le 12 \quad ?$$
$$20 \le 12 \quad\quad False$$

The numbers in Interval B, including -4 and 3, are solutions.

Solution set: $[-4, 3]$

56. $(x + 2)(x - 3)(x + 5) \le 0$
The numbers -2, 3, and -5 are solutions of the cubic equation

$$(x + 2)(x - 3)(x + 5) = 0.$$

These numbers divide a number line into four intervals.

Test a number from each interval in the inequality

$$(x + 2)(x - 3)(x + 5) \le 0.$$

Interval A: Let $x = -6$.
$$(-4)(-9)(-1) \le 0 \quad True$$

Interval B: Let $x = -3$.
$$(-5)(-6)(2) \le 0 \quad False$$

Interval C: Let $x = 0$.
$$(2)(-3)(5) \le 0 \quad True$$

Interval D: Let $x = 4$.
$$(6)(1)(9) \le 0 \quad False$$

The numbers in Intervals A and C, including -5, -2, and 3, are solutions.

Solution set: $(-\infty, -5] \cup [-2, 3]$

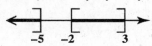

57. $(4m + 3)^2 \leq -4$

The square of a real number is never negative. So, the solution set of this inequality is $\emptyset$.

58.
$$\frac{6}{2z - 1} < 2$$

Write the inequality so that 0 is on one side.

$$\frac{6}{2z - 1} - 2 < 0$$

$$\frac{6}{2z - 1} - \frac{2(2z - 1)}{2z - 1} < 0$$

$$\frac{6 - 4z + 2}{2z - 1} < 0$$

$$\frac{-4z + 8}{2z - 1} < 0$$

The number 2 makes the numerator 0, and $\frac{1}{2}$ makes the denominator 0. These two numbers determine three intervals.

Test a number from each interval in the inequality

$$\frac{6}{2z - 1} < 2.$$

Interval A:	Let $z = 0$.		
	$-6 < 2$		*True*
Interval B:	Let $z = 1$.		
	$6 < 2$		*False*
Interval C:	Let $z = 3$.		
	$\frac{6}{5} < 2$		*True*

The solution set includes numbers in Intervals A and C, excluding endpoints.

Solution set: $\left(-\infty, \frac{1}{2}\right) \cup (2, \infty)$

59.
$$\frac{3t + 4}{t - 2} \leq 1$$

Write the inequality so that 0 is on one side.

$$\frac{3t + 4}{t - 2} - 1 \leq 0$$

$$\frac{3t + 4}{t - 2} - \frac{1(t - 2)}{t - 2} \leq 0$$

$$\frac{3t + 4 - t + 2}{t - 2} \leq 0$$

$$\frac{2t + 6}{t - 2} \leq 0$$

The number -3 makes the numerator 0, and 2 makes the denominator 0. These two numbers determine three intervals.

Test a number from each interval in the inequality

$$\frac{3t + 4}{t - 2} \leq 1.$$

Interval A:	Let $t = -4$.		
	$\frac{-8}{-6} \leq 1$	?	
	$\frac{4}{3} \leq 1$		*False*
Interval B:	Let $t = 0$.		
	$\frac{4}{-2} \leq 1$	?	
	$-2 \leq 1$		*True*
Interval C:	Let $t = 3$.		
	$\frac{13}{1} \leq 1$	?	
	$13 \leq 1$		*False*

The numbers in Interval B, including -3 but not 2, are solutions.

Solution set: $[-3, 2)$

60. **[11.5]** Solve $V = r^2 + R^2 h$ for R.

$$V - r^2 = R^2 h$$

$$R^2 h = V - r^2$$

$$R^2 = \frac{V - r^2}{h}$$

$$R = \pm\sqrt{\frac{V - r^2}{h}} = \frac{\pm\sqrt{V - r^2}}{\sqrt{h}}$$

$$= \frac{\pm\sqrt{V - r^2}}{\sqrt{h}} \cdot \frac{\sqrt{h}}{\sqrt{h}}$$

$$= \frac{\pm\sqrt{Vh - r^2 h}}{h}$$

61. **[11.3]** $3t^2 - 6t = -4$

$$3t^2 - 6t + 4 = 0$$

Use the quadratic formula.

$$t = \frac{-b \pm \sqrt{b^2 - 4ac}}{2a}$$

$$t = \frac{-(-6) \pm \sqrt{(-6)^2 - 4(3)(4)}}{2(3)}$$

continued

$$= \frac{6 \pm \sqrt{-12}}{6} = \frac{6 \pm 2i\sqrt{3}}{6}$$

$$= \frac{2\left(3 \pm i\sqrt{3}\right)}{6} = \frac{3 \pm i\sqrt{3}}{3} = 1 \pm \frac{\sqrt{3}}{3}i$$

Solution set: $\left\{ 1 \pm \frac{\sqrt{3}}{3}i \right\}$

62. [11.4]
$$x^4 - 1 = 0$$
$$\left(x^2 + 1\right)\left(x^2 - 1\right) = 0$$

$$x^2 + 1 = 0 \qquad \text{or } x^2 - 1 = 0$$
$$x^2 = -1 \qquad\qquad x^2 = 1$$
$$x = \pm\sqrt{-1} \qquad\quad x = \pm\sqrt{1}$$
$$x = \pm i \quad \text{ or } \quad x = \pm 1$$

Solution set: $\{ \pm i, \, \pm 1 \}$

63. [11.4] $\left(x^2 - 2x\right)^2 = 11\left(x^2 - 2x\right) - 24$
Let $u = x^2 - 2x$. The equation becomes

$$u^2 = 11u - 24.$$

$$u^2 - 11u + 24 = 0$$
$$(u - 8)(u - 3) = 0$$

$$u - 8 = 0 \quad \text{or} \quad u - 3 = 0$$
$$u = 8 \quad \text{or} \qquad u = 3$$

To find x, substitute $x^2 - 2x$ for u.

$$x^2 - 2x = 8 \qquad \text{or} \qquad x^2 - 2x = 3$$
$$x^2 - 2x - 8 = 0 \qquad\quad x^2 - 2x - 3 = 0$$
$$(x - 4)(x + 2) = 0 \qquad (x - 3)(x + 1) = 0$$
$$x - 4 = 0 \text{ or } x + 2 = 0 \qquad x - 3 = 0 \text{ or } x + 1 = 0$$
$$x = 4 \ \text{ or } \ x = -2 \ \text{ or } \ x = 3 \ \text{ or } \ x = -1$$

The proposed solutions all check.

Solution set: $\{-2, -1, 3, 4\}$

64. [11.8] $(r - 1)(2r + 3)(r + 6) < 0$
Solve the equation
$$(r - 1)(2r + 3)(r + 6) = 0.$$

$$r - 1 = 0 \quad \text{or} \quad 2r + 3 = 0 \quad \text{or} \quad r + 6 = 0$$
$$r = 1 \quad \text{or} \qquad r = -\tfrac{3}{2} \quad \text{or} \qquad r = -6$$

The numbers -6, $-\frac{3}{2}$, and 1 divide a number line into four intervals.

```
      A    B      C       D
  ────┼────┼──────┼───────┼────▶
      -6        -3/2    1
```

Test a number from each interval in the inequality
$$(r - 1)(2r + 3)(r + 6) < 0.$$

Interval A: Let $r = -7$.
$$-8(-11)(-1) < 0 \qquad ?$$
$$-88 < 0 \qquad\qquad \textit{True}$$
Interval B: Let $r = -2$.
$$-3(-1)(4) < 0 \qquad ?$$
$$12 < 0 \qquad\qquad \textit{False}$$
Interval C: Let $r = 0$.
$$-1(3)(6) < 0 \qquad ?$$
$$-18 < 0 \qquad\qquad \textit{True}$$
Interval D: Let $r = 2$.
$$1(7)(8) < 0 \qquad ?$$
$$56 < 0 \qquad\qquad \textit{False}$$

The numbers in Intervals A and C, not including -6, $-\frac{3}{2}$, or 1, are solutions.

Solution set: $(-\infty, -6) \cup \left(-\frac{3}{2}, 1\right)$

65. [11.4]
$$2x - \sqrt{x} = 6$$
$$2x - \sqrt{x} - 6 = 0$$
Let $u = \sqrt{x}$, so $u^2 = x$.
$$2u^2 - u - 6 = 0$$
$$(2u + 3)(u - 2) = 0$$

$$2u + 3 = 0 \qquad \text{or} \quad u - 2 = 0$$
$$u = -\tfrac{3}{2} \quad \text{or} \qquad u = 2$$
$$\sqrt{x} = -\tfrac{3}{2} \quad \text{or} \quad \sqrt{x} = 2$$

Since $\sqrt{x} \geq 0$, we must have $\sqrt{x} = 2$, or $x = 4$.

Check $x = 4$: $8 - 2 = 6$ *True*
Solution set: $\{4\}$

66. [11.1] $(3k + 11)^2 = 7$

$$3k + 11 = \sqrt{7} \qquad \text{or } 3k + 11 = -\sqrt{7}$$
$$3k = -11 + \sqrt{7} \qquad\quad 3k = -11 - \sqrt{7}$$
$$k = \frac{-11 + \sqrt{7}}{3} \ \text{or} \qquad k = \frac{-11 - \sqrt{7}}{3}$$

Solution set: $\left\{ \dfrac{-11 \pm \sqrt{7}}{3} \right\}$

67. [11.5] Solve $S = \dfrac{Id^2}{k}$ for d.
Multiply both sides by k, then divide by I.
$$\frac{Sk}{I} = d^2$$
$$d = \pm\sqrt{\frac{Sk}{I}} = \frac{\pm\sqrt{Sk}}{\sqrt{I}}$$
$$= \frac{\pm\sqrt{Sk}}{\sqrt{I}} \cdot \frac{\sqrt{I}}{\sqrt{I}}$$
$$d = \frac{\pm\sqrt{SkI}}{I}$$

68. **[11.8]** $(8k - 7)^2 \geq -1$

The square of any real number is always greater than or equal to 0, so any real number satisfies this inequality. Solution set: $(-\infty, \infty)$

69. **[11.4]** $\qquad 6 + \dfrac{15}{s^2} = -\dfrac{19}{s}$

Multiply by the LCD, s^2.

$$s^2\left(6 + \frac{15}{s^2}\right) = s^2\left(-\frac{19}{s}\right)$$

$$6s^2 + 15 = -19s$$

$$6s^2 + 19s + 15 = 0$$

$$(3s + 5)(2s + 3) = 0$$

$$3s + 5 = 0 \quad \text{or} \quad 2s + 3 = 0$$

$$s = -\tfrac{5}{3} \quad \text{or} \qquad s = -\tfrac{3}{2}$$

Check $s = -\tfrac{5}{3}$: $6 + \tfrac{27}{5} = \tfrac{57}{5}$ *True*

Check $s = -\tfrac{3}{2}$: $6 + \tfrac{20}{3} = \tfrac{38}{3}$ *True*

Solution set: $\left\{-\tfrac{5}{3}, -\tfrac{3}{2}\right\}$

70. **[11.4]** $\qquad x^4 - 8x^2 = -1$

$$x^4 - 8x^2 + 1 = 0$$

Use the quadratic formula with $a = 1$, $b = -8$, and $c = 1$ to solve for x^2, not x.

$$x^2 = \frac{-b \pm \sqrt{b^2 - 4ac}}{2a}$$

$$x^2 = \frac{-(-8) \pm \sqrt{(-8)^2 - 4(1)(1)}}{2(1)}$$

$$= \frac{8 \pm \sqrt{64 - 4}}{2} = \frac{8 \pm \sqrt{60}}{2}$$

$$= \frac{8 \pm 2\sqrt{15}}{2} = 4 \pm \sqrt{15}$$

Both $4 + \sqrt{15}$ and $4 - \sqrt{15}$ are positive and can be set equal to x^2.

$$x^2 = 4 + \sqrt{15} \quad \text{or} \quad x^2 = 4 - \sqrt{15}$$

$$x = \pm\sqrt{4 + \sqrt{15}} \quad \text{or} \quad x = \pm\sqrt{4 - \sqrt{15}}$$

The proposed solutions all check.

Solution set: $\left\{\pm\sqrt{4 + \sqrt{15}}, \pm\sqrt{4 - \sqrt{15}}\right\}$

71. **[11.8]**

$$\frac{-2}{x + 5} \leq -5$$

Write the inequality so that 0 is on one side.

$$\frac{-2}{x + 5} + 5 \leq 0$$

$$\frac{-2}{x + 5} + \frac{5(x + 5)}{x + 5} \leq 0$$

$$\frac{-2 + 5x + 25}{x + 5} \leq 0$$

$$\frac{5x + 23}{x + 5} \leq 0$$

The number $-\tfrac{23}{5}$ makes the numerator 0, and -5 makes the denominator 0. These two numbers determine three intervals.

Test a number from each interval in the inequality

$$\frac{-2}{x + 5} \leq -5.$$

Interval A: Let $x = -6$.

$$\frac{-2}{-1} \leq -5 \quad ?$$

$$2 \leq -5 \qquad \textit{False}$$

Interval B: Let $x = -\tfrac{24}{5}$.

$$\frac{-2}{\frac{1}{5}} \leq -5 \quad ?$$

$$-10 \leq -5 \qquad \textit{True}$$

Interval C: Let $x = 0$.

$$\tfrac{-2}{5} \leq -5 \qquad \textit{False}$$

The numbers in Interval B, including $-\tfrac{23}{5}$ but not -5, are solutions.

Solution set: $\left(-5, -\tfrac{23}{5}\right]$

72. **[11.6]** $g(x) = x^2 - 5$ written in the form

$$g(x) = a(x - h)^2 + k \text{ is}$$

$$g(x) = 1(x - 0)^2 + (-5).$$

Here, $h = 0$ and $k = -5$, so the vertex (h, k) is $(0, -5)$. The graph is shifted 5 units down from the graph of $f(x) = x^2$. Since $a = 1 > 0$, the graph opens up. The correct figure is **F**.

73. **[11.6]** $h(x) = -x^2 + 4$

Because the coefficient of x^2 is negative, the parabola opens down. Because $k = 4$, the shift is upward 4 units. The figure that most closely resembles the graph of $h(x)$ is Choice **B**.

74. **[11.6]** $F(x) = (x - 1)^2$ written in the form

$$F(x) = a(x - h)^2 + k \text{ is}$$

$$F(x) = 1(x - 1)^2 + 0.$$

Here, $h = 1$ and $k = 0$, so the vertex (h, k) is $(1, 0)$. The graph is shifted 1 unit to the right of the graph of $f(x) = x^2$. Since $a = 1 > 0$, the graph opens up. The correct figure is **C**.

75. **[11.6]** $G(x) = (x + 1)^2$
Because the coefficient of the x^2-term is positive, the graph opens up. Because $h = -1$, the shift is to the left 1 unit. The figure that most closely resembles the graph of $G(x)$ is Choice **A**.

76. **[11.6]** $H(x) = (x - 1)^2 + 1$ is written in the form
$$H(x) = a(x - h)^2 + k.$$

Here, $h = 1$ and $k = 1$, so the vertex (h, k) is $(1, 1)$. The graph is shifted 1 unit to the right and 1 unit up from the graph of $f(x) = x^2$. Since $a = 1 > 0$, the graph opens up. The correct figure is **E**.

77. **[11.6]** $K(x) = (x + 1)^2 + 1$
Because $h = -1$ and $k = 1$, the shift is to the left 1 unit and up 1 unit. The figure that most closely resembles the graph of $K(x)$ is choice **D**.

78. **[11.7]** $y = f(x) = 4x^2 + 4x - 2$
Complete the square to find the vertex.
$$y = 4(x^2 + x) - 2$$
$$= 4\left(x^2 + x + \tfrac{1}{4} - \tfrac{1}{4}\right) - 2$$
$$= 4\left(x^2 + x + \tfrac{1}{4}\right) + 4\left(-\tfrac{1}{4}\right) - 2$$
$$= 4\left(x + \tfrac{1}{2}\right)^2 - 1 - 2$$
$$= 4\left(x + \tfrac{1}{2}\right)^2 - 3$$
$$y = 4\left(x + \tfrac{1}{2}\right)^2 - 3$$
The equation is now in the form
$y = a(x - h)^2 + k$, so the vertex (h, k) is $\left(-\tfrac{1}{2}, -3\right)$ and the axis is $x = -\tfrac{1}{2}$. Since $a = 4 > 0$, the parabola opens up. Also, $|a| = |4| = 4 > 1$, so the graph is narrower than the graph of $y = x^2$. The points $(-2, 6)$, $(0, -2)$, and $(1, 6)$ are on the graph.

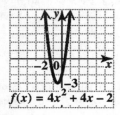

$$f(x) = 4x^2 + 4x - 2$$

The domain is $(-\infty, \infty)$.
The smallest y-value is -3, so the range is $[-3, \infty)$.

79. **[11.7]** $-0.001x^2 + 0.295x + 0.227$

(a) For 2002, $x = 2002 - 1970 = 32$.
$f(32) = -0.001(32)^2 + 0.295(32) + 0.227$
$$= 8.643$$

The nuclear power consumption in the U.S. in 2002 was approximately 8.64 quadrillion Btu.

(b) The result using the model is high.

(c)
$$f(x) = 10$$
$$-0.001x^2 + 0.295x + 0.227 = 10$$
$$-0.001x^2 + 0.295x - 9.773 = 0$$
$$x = \frac{-0.295 \pm \sqrt{(0.295)^2 - 4(-0.001)(-9.773)}}{2(-0.001)}$$
$$= \frac{-0.295 \pm \sqrt{0.047933}}{-0.002}$$
$$\approx 38.03 \quad \text{or} \quad 256.97$$
Rounding down, the year is $1970 + 38 = 2008$. (The other value is outside the practical domain of this model.)

80. **[11.5]** Let $x =$ the speed of the boat in still water. Use $t = d/r$ to make a table.

	Distance	Rate	Time
Upstream	15	$x - 5$	$\dfrac{15}{x - 5}$
Downstream	15	$x + 5$	$\dfrac{15}{x + 5}$

The total time is 4 hours.
$$\frac{15}{x - 5} + \frac{15}{x + 5} = 4$$
Multiply by the LCD, $(x + 5)(x - 5)$.
$$15(x + 5) + 15(x - 5) = 4(x + 5)(x - 5)$$
$$15x + 75 + 15x - 75 = 4x^2 - 100$$
$$0 = 4x^2 - 30x - 100$$
$$0 = 2x^2 - 15x - 50$$
$$0 = (2x + 5)(x - 10)$$

$$2x + 5 = 0 \quad \text{or} \quad x - 10 = 0$$
$$x = -\tfrac{5}{2} \quad \text{or} \quad x = 10$$

The speed must be positive, so check $x = 10$. The 15-mile trip upstream would take 3 hours at 5 mph. The trip downstream would take 1 hour at 15 mph. Thus, the total time would be 4 hours and we have the speed of the boat in still water equal to 10 mph.

81. **[11.5]** Let $x =$ the length of the wire.
Use the Pythagorean formula.
$$x^2 = 100^2 + 400^2 \qquad \textit{Height=400.}$$
$$= 10,000 + 160,000$$
$$= 170,000$$
$$x = \pm\sqrt{170,000} \approx \pm 412.3$$

Reject the negative solution. The length of the ire is about 412.3 feet.

82. **[11.5]** Let x = the width of the rectangle.
Then $2x - 1$ = the length of the rectangle.

Use the Pythagorean formula.

$$x^2 + (2x-1)^2 = (2.5)^2$$
$$x^2 + 4x^2 - 4x + 1 = 6.25$$
$$5x^2 - 4x - 5.25 = 0$$

Multiply by 4.

$$20x^2 - 16x - 21 = 0$$
$$(2x - 3)(10x + 7) = 0$$

$$2x - 3 = 0 \quad \text{or} \quad 10x + 7 = 0$$
$$x = \tfrac{3}{2} \quad \text{or} \quad x = -\tfrac{7}{10}$$

Discard the negative solution.
If $x = 1.5$, then

$$2x - 1 = 2(1.5) - 1 = 2.$$

The width of the rectangle is 1.5 cm, and the length is 2 cm.

Chapter 11 Test

1. $t^2 = 54$

$$t = \sqrt{54} \quad \text{or} \quad t = -\sqrt{54}$$
$$t = 3\sqrt{6} \quad \text{or} \quad t = -3\sqrt{6}$$

Solution set: $\left\{ \pm 3\sqrt{6} \right\}$

2. $(7x + 3)^2 = 25$

$$7x + 3 = 5 \quad \text{or} \quad 7x + 3 = -5$$
$$7x = 2 \qquad\qquad 7x = -8$$
$$x = \tfrac{2}{7} \quad \text{or} \quad x = -\tfrac{8}{7}$$

Solution set: $\left\{ -\tfrac{8}{7}, \tfrac{2}{7} \right\}$

3. $2x^2 + 4x = 8$

$$x^2 + 2x = 4 \qquad \textit{Divide by 2.}$$
$$x^2 + 2x + 1 = 4 + 1$$
$$(x + 1)^2 = 5$$

$$x + 1 = \sqrt{5} \qquad \text{or} \quad x + 1 = -\sqrt{5}$$
$$x = -1 + \sqrt{5} \quad \text{or} \qquad x = -1 - \sqrt{5}$$

Solution set: $\left\{ -1 \pm \sqrt{5} \right\}$

4. $2x^2 - 3x - 1 = 0$

Here $a = 2$, $b = -3$, and $c = -1$.

$$x = \frac{-b \pm \sqrt{b^2 - 4ac}}{2a}$$
$$x = \frac{-(-3) \pm \sqrt{(-3)^2 - 4(2)(-1)}}{2(2)}$$
$$= \frac{3 \pm \sqrt{17}}{4}$$

Solution set: $\left\{ \dfrac{3 \pm \sqrt{17}}{4} \right\}$

5.
$$3t^2 - 4t = -5$$
$$3t^2 - 4t + 5 = 0$$

Here $a = 3$, $b = -4$, and $c = 5$.

$$t = \frac{-b \pm \sqrt{b^2 - 4ac}}{2a}$$
$$t = \frac{-(-4) \pm \sqrt{(-4)^2 - 4(3)(5)}}{2(3)}$$
$$= \frac{4 \pm \sqrt{-44}}{6} = \frac{4 \pm 2i\sqrt{11}}{6}$$
$$= \frac{2\left(2 \pm i\sqrt{11}\right)}{6} = \frac{2 \pm i\sqrt{11}}{3}$$
$$= \frac{2}{3} \pm \frac{\sqrt{11}}{3} i$$

Solution set: $\left\{ \dfrac{2}{3} \pm \dfrac{\sqrt{11}}{3} i \right\}$

6. If k is a negative number, then $4k$ is also negative, so the equation $x^2 = 4k$ will have two nonreal complex solutions. The answer is **A**.

7. $2x^2 - 8x - 3 = 0$

$$b^2 - 4ac = (-8)^2 - 4(2)(-3)$$
$$= 64 + 24 = 88$$

The discriminant, 88, is positive but not a perfect square, so there will be two distinct irrational number solutions.

8.
$$3x = \sqrt{\frac{9x + 2}{2}}$$

Square both sides.

$$9x^2 = \frac{9x + 2}{2}$$
$$18x^2 = 9x + 2$$
$$18x^2 - 9x - 2 = 0$$

Use the quadratic formula with $a = 18$, $b = -9$, and $c = -2$.

$$x = \frac{-b \pm \sqrt{b^2 - 4ac}}{2a}$$
$$x = \frac{-(-9) \pm \sqrt{(-9)^2 - 4(18)(-2)}}{2(18)}$$
$$= \frac{9 \pm \sqrt{225}}{36} = \frac{9 \pm 15}{36}$$
$$x = \frac{9 + 15}{36} = \frac{24}{36} = \frac{2}{3} \text{ or}$$
$$x = \frac{9 - 15}{36} = \frac{-6}{36} = -\frac{1}{6}$$

Check $x = \tfrac{2}{3}$: $\quad 2 = \sqrt{4} \quad$ *True*

Check $x = -\tfrac{1}{6}$: $\;-\tfrac{1}{2} = \sqrt{\tfrac{1}{4}} \quad$ *False*

Solution set: $\left\{ \tfrac{2}{3} \right\}$

9. $3 - \dfrac{16}{x} - \dfrac{12}{x^2} = 0$

Multiply by the LCD, x^2.

$$x^2 \left(3 - \dfrac{16}{x} - \dfrac{12}{x^2} \right) = x^2 \cdot 0$$
$$3x^2 - 16x - 12 = 0$$
$$(3x + 2)(x - 6) = 0$$
$$3x + 2 = 0 \quad \text{or} \quad x - 6 = 0$$
$$x = -\tfrac{2}{3} \quad \text{or} \quad x = 6$$

Check $x = -\tfrac{2}{3}$: $3 + 24 - 27 = 0$ *True*

Check $x = 6$: $3 - \tfrac{8}{3} - \tfrac{1}{3} = 0$ *True*

Solution set: $\left\{ -\tfrac{2}{3}, 6 \right\}$

10. $4x^2 + 7x - 3 = 0$

Use the quadratic formula with $a = 4$, $b = 7$, and $c = -3$.

$$x = \dfrac{-b \pm \sqrt{b^2 - 4ac}}{2a}$$
$$x = \dfrac{-7 \pm \sqrt{7^2 - 4(4)(-3)}}{2(4)}$$
$$= \dfrac{-7 \pm \sqrt{97}}{8}$$

Solution set: $\left\{ \dfrac{-7 \pm \sqrt{97}}{8} \right\}$

11. $9x^4 + 4 = 37x^2$

$9x^4 - 37x^2 + 4 = 0$

Let $u = x^2$, so $u^2 = (x^2)^2 = x^4$.
The equation becomes

$$9u^2 - 37u + 4 = 0.$$
$$(9u - 1)(u - 4) = 0$$
$$9u - 1 = 0 \quad \text{or} \quad u - 4 = 0$$
$$u = \tfrac{1}{9} \quad \text{or} \quad u = 4$$

To find x, substitute x^2 for u.

$$x^2 = \tfrac{1}{9} \quad \text{or} \quad x^2 = 4$$
$$x = \pm\sqrt{\tfrac{1}{9}} \qquad x = \pm\sqrt{4}$$
$$x = \pm\tfrac{1}{3} \quad \text{or} \quad x = \pm 2$$

Check $x = \pm\tfrac{1}{3}$: $\tfrac{1}{9} + 4 = \tfrac{37}{9}$ *True*

Check $x = \pm 2$: $144 + 4 = 37(4)$ *True*

Solution set: $\left\{ \pm 2, \pm\tfrac{1}{3} \right\}$

12. $12 = (2n + 1)^2 + (2n + 1)$

Let $u = 2n + 1$. The equation becomes

$$12 = u^2 + u.$$
$$0 = u^2 + u - 12$$
$$0 = (u + 4)(u - 3)$$
$$u + 4 = 0 \quad \text{or} \quad u - 3 = 0$$
$$u = -4 \quad \text{or} \quad u = 3$$

To find n, substitute $2n + 1$ for u.

$$2n + 1 = -4 \quad \text{or} \quad 2n + 1 = 3$$
$$2n = -5 \qquad\qquad 2n = 2$$
$$n = -\tfrac{5}{2} \quad \text{or} \quad n = 1$$

Check $n = -\tfrac{5}{2}$: $12 = 16 - 4$ *True*

Check $n = 1$: $12 = 9 + 3$ *True*

Solution set: $\left\{ -\tfrac{5}{2}, 1 \right\}$

13. Solve $S = 4\pi r^2$ for r.

$$\dfrac{S}{4\pi} = r^2$$
$$r = \pm\sqrt{\dfrac{S}{4\pi}} = \dfrac{\pm\sqrt{S}}{2\sqrt{\pi}}$$
$$= \dfrac{\pm\sqrt{S}}{2\sqrt{\pi}} \cdot \dfrac{\sqrt{\pi}}{\sqrt{\pi}}$$
$$r = \dfrac{\pm\sqrt{\pi S}}{2\pi}$$

14. Let $x =$ Andrew's time alone.
Then $x - 2 =$ Kent's time alone.

Make a table.

	Rate	Time Together	Fractional Part of the Job Done
Andrew	$\dfrac{1}{x}$	5	$\dfrac{5}{x}$
Kent	$\dfrac{1}{x-2}$	5	$\dfrac{5}{x-2}$

Part done   part done   1 whole
by Andrew plus by Kent equals   job.
  ↓   ↓   ↓   ↓   ↓
$\dfrac{5}{x}$ $+$ $\dfrac{5}{x-2}$ $=$ 1

Multiply both sides by the LCD, $x(x - 2)$.

$$x(x - 2)\left(\dfrac{5}{x} + \dfrac{5}{x - 2} \right) = x(x - 2) \cdot 1$$
$$5x - 10 + 5x = x^2 - 2x$$
$$0 = x^2 - 12x + 10$$

Use the quadratic formula with $a = 1$, $b = -12$, and $c = 10$.

$$x = \frac{-b \pm \sqrt{b^2 - 4ac}}{2a}$$

$$x = \frac{-(-12) \pm \sqrt{(-12)^2 - 4(1)(10)}}{2(1)}$$

$$= \frac{12 \pm \sqrt{104}}{2} = \frac{12 \pm 2\sqrt{26}}{2}$$

$$= \frac{2\left(6 \pm \sqrt{26}\right)}{2} = 6 \pm \sqrt{26}$$

$$x = 6 + \sqrt{26} \approx 11.1 \text{ or}$$
$$x = 6 - \sqrt{26} \approx 0.9$$

Reject 0.9 for Andrew's time, because that would yield a negative time for Kent. Thus, Andrew's time is about 11.1 hours and Kent's time is $x - 2 \approx 9.1$ hours.

15. Let x = Abby's rate.

Make a table. Use $d = rt$, or $t = \dfrac{d}{r}$.

	Distance	Rate	Time
Upstream	10	$x - 3$	$\dfrac{10}{x - 3}$
Downstream	10	$x + 3$	$\dfrac{10}{x + 3}$

Time   Time
upstream plus downstream equals 3.5 hr.

$$\begin{array}{ccccc} \downarrow & \downarrow & \downarrow & \downarrow & \downarrow \\ \dfrac{10}{x-3} & + & \dfrac{10}{x+3} & = & \dfrac{7}{2} \end{array}$$

Multiply both sides by the LCD, $2(x + 3)(x - 3)$.

$$2(x + 3)(x - 3)\left(\frac{10}{x-3} + \frac{10}{x+3}\right) = 2(x+3)(x-3)\left(\frac{7}{2}\right)$$

$$20(x + 3) + 20(x - 3) = 7\left(x^2 - 9\right)$$
$$20x + 60 + 20x - 60 = 7x^2 - 63$$
$$0 = 7x^2 - 40x - 63$$
$$0 = (x - 7)(7x + 9)$$
$$x - 7 = 0 \quad \text{or} \quad 7x + 9 = 0$$
$$x = 7 \quad \text{or} \quad x = -\frac{9}{7}$$

Reject $-\frac{9}{7}$ since the rate can't be negative. Abby's rate was 7 mph.

16. Let x = the width of the walk.
The area of the walk is equal to the area of the outer figure minus the area of the pool.

$$152 = (10 + 2x)(24 + 2x) - (24)(10)$$
$$152 = 240 + 68x + 4x^2 - 240$$
$$0 = 4x^2 + 68x - 152$$
$$0 = x^2 + 17x - 38$$
$$0 = (x + 19)(x - 2)$$

$$x + 19 = 0 \quad \text{or} \quad x - 2 = 0$$
$$x = -19 \quad \text{or} \quad x = 2$$

Reject -19 since width can't be negative. The walk is 2 feet wide.

17. Let x = the height of the tower. Then $2x + 2$ = the distance from the point to the top.

The distance from the base to the point is 30 m. These three segments form a right triangle, so the Pythagorean formula applies.

$$a^2 + b^2 = c^2$$
$$x^2 + 30^2 = (2x + 2)^2$$
$$x^2 + 900 = 4x^2 + 8x + 4$$
$$0 = 3x^2 + 8x - 896$$
$$0 = (x - 16)(3x + 56)$$

$$x - 16 = 0 \quad \text{or} \quad 3x + 56 = 0$$
$$x = 16 \quad \text{or} \quad x = -\frac{56}{3}$$

Reject $-\frac{56}{3}$ since height can't be negative. The tower is 16 m high.

18. $f(x) = a(x - h)^2 + k$
Since $a < 0$, the parabola opens down. Since $h > 0$ and $k < 0$, the x-coordinate is positive and the y-coordinate is negative. Therefore, the vertex is in quadrant IV. The correct graph is **A**.

19. $f(x) = \frac{1}{2}x^2 - 2$
$f(x) = \frac{1}{2}(x - 0)^2 - 2$
The graph is a parabola in $f(x) = a(x - h)^2 + k$ form with vertex (h, k) at $(0, -2)$. The axis is $x = 0$. Since $a = \frac{1}{2} > 0$, the parabola opens up. Also, $|a| = \left|\frac{1}{2}\right| = \frac{1}{2} < 1$, so the graph of the parabola is wider than the graph of $f(x) = x^2$. The points $(2, 0)$ and $(-2, 0)$ are on the graph.

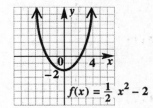

From the graph, we see that the x-values can be any real number, so the domain is $(-\infty, \infty)$. The y-values are greater than or equal to -2, so the range is $[-2, \infty)$.

20. $f(x) = -x^2 + 4x - 1$

The x-coordinate of the vertex is

$$x = \frac{-b}{2a} = \frac{-4}{2(-1)} = 2.$$

The y-coordinate of the vertex is

$$f(2) = -4 + 8 - 1 = 3.$$

The graph is a parabola with vertex (h, k) at $(2, 3)$ and axis $x = 2$. Since $a = -1 < 0$, the parabola opens down.
Also, $|a| = |-1| = 1$, so the graph has the same shape as the graph of $f(x) = x^2$. The points $(0, -1)$ and $(4, -1)$ are on the graph.

$f(x) = -x^2 + 4x - 1$

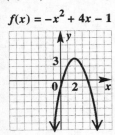

From the graph, we see that the x-values can be any real number, so the domain is $(-\infty, \infty)$.

The y-values are less than or equal to 3, so the range is $(-\infty, 3]$.

21. $x = -(y - 2)^2 + 2$

The equation is in $x = a(y - k)^2 + h$ form. The graph is a horizontal parabola with vertex (h, k) at $(2, 2)$ and axis $y = 2$. Since $a = -1 < 0$, the graph opens to the left. Also, $|a| = |-1| = 1$, so the graph has the same shape as the graph of $y = x^2$. The points $(-2, 0)$ and $(-2, 4)$ are on the same graph.

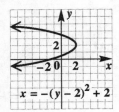

$x = -(y - 2)^2 + 2$

The largest value of x is 2, so the domain is $(-\infty, 2]$.

The y-values can be any real number, so the range is $(-\infty, \infty)$.

22. (a) Use $f(x) = 0.156x^2 - 2.05x + 10.2$ with $x = 11$.

$$f(11) = 0.156(11)^2 - 2.05(11) + 10.2$$
$$\approx 6.5 \text{ (to the nearest tenth)}$$

In 2001, there was a 6.5% increase in tuition.

(b) Find the vertex.

$$\frac{-b}{2a} = \frac{-(-2.05)}{2(0.156)} \approx 6 \text{ (rounding down)}$$

$$f(6) = 0.156(6)^2 - 2.05(6) + 10.2$$
$$\approx 3.5 \text{ (to the nearest tenth)}$$

The minimum tuition increase was 3.5%, which occurred in $1990 + 6 = 1996$.

23. Let $x =$ the width of the lot.
Then $640 - 2x =$ the length of the lot.

Area A is length times width.

$$A = x(640 - 2x)$$
$$A(x) = 640x - 2x^2 = -2x^2 + 640x$$

Use the vertex formula.

$$x = \frac{-b}{2a} = \frac{-640}{2(-2)} = 160$$

$$A(160) = -2(160)^2 + 640(160)$$
$$= -51{,}200 + 102{,}400$$
$$= 51{,}200$$

The graph is a parabola that opens down, so the maximum occurs at the vertex $(160, 51{,}200)$. The maximum area is $51{,}200$ ft^2 if the width x is 160 feet and the length is
$640 - 2x = 640 - 2(160) = 320$ feet.

24. $$2x^2 + 7x > 15$$
$$2x^2 + 7x - 15 > 0$$
Solve the equation

$$2x^2 + 7x - 15 = 0.$$
$$(2x - 3)(x + 5) = 0$$

$$2x - 3 = 0 \quad \text{or} \quad x + 5 = 0$$
$$x = \tfrac{3}{2} \quad \text{or} \qquad x = -5$$

The numbers -5 and $\frac{3}{2}$ divide a number line into three intervals.

```
      A       B     C
  ────┼───────┼──────→
     -5       3
             ─
             2
```

Test a number from each interval in the inequality

$$2x^2 + 7x > 15.$$

Interval A: Let $x = -6$.
$$72 - 42 > 15 \quad ?$$
$$30 > 15 \qquad \text{True}$$

Interval B: Let $x = 0$.
$$0 > 15 \qquad \text{False}$$

Interval C: Let $x = 2$.
$$8 + 14 > 15 \quad ?$$
$$22 > 15 \qquad \text{True}$$

The numbers in Intervals A and C, not including -5 and $\frac{3}{2}$, are solutions.

Solution set: $(-\infty, -5) \cup \left(\frac{3}{2}, \infty\right)$

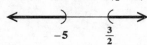

25. $\dfrac{5}{t-4} \le 1$

Write the inequality so that 0 is on one side.

$$\frac{5}{t-4} - 1 \le 0$$

$$\frac{5}{t-4} - \frac{1(t-4)}{t-4} \le 0$$

$$\frac{5 - t + 4}{t-4} \le 0$$

$$\frac{-t+9}{t-4} \le 0$$

The number 9 makes the numerator 0, and 4 makes the denominator 0. These two numbers determine three intervals.

$$
\begin{array}{ccc}
\textbf{A} & \textbf{B} & \textbf{C}
\end{array}
$$

Test a number from each interval in the inequality

$$\frac{5}{t-4} \le 1.$$

Interval A: Let $t = 0$.

      $\frac{5}{-4} \le 1$ *True*

Interval B: Let $t = 7$.

      $\frac{5}{3} \le 1$ *False*

Interval C: Let $t = 10$.

      $\frac{5}{6} \le 1$ *True*

The numbers in Intervals A and C, including 9 but not 4, are solutions.

Solution set: $(-\infty, 4) \cup [9, \infty)$

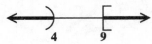

Cumulative Review Exercises (Chapters 1–11)

1. $S = \left\{ -\frac{7}{3}, -2, -\sqrt{3}, 0, 0.7, \sqrt{12}, \sqrt{-8}, 7, \frac{32}{3} \right\}$

   (a) The elements of S that are integers are $-2, 0$, and 7.

   (b) The elements of S that are rational numbers are $-\frac{7}{3}, -2, 0, 0.7, 7$, and $\frac{32}{3}$.

   (c) All the elements of S except $\sqrt{-8}$ are real numbers.

   (d) All the elements of S are complex numbers.

2. $|-3| + 8 - |-9| - (-7 + 3) = 3 + 8 - 9 - (-4)$

                             $= 3 + 8 - 9 + 4$

                             $= 6$

3. $2(-3)^2 + (-8)(-5) + (-17)$

    $= 2(9) + 40 - 17$

    $= 18 + 40 - 17 = 41$

4. $7 - (4 + 3t) + 2t = -6(t-2) - 5$

     $7 - 4 - 3t + 2t = -6t + 12 - 5$

          $3 - t = -6t + 7$

             $5t = 4$

              $t = \dfrac{4}{5}$

Check $t = \frac{4}{5}$: $\frac{11}{5} = \frac{11}{5}$ *True*

Solution set: $\left\{ \frac{4}{5} \right\}$

5. $|6x - 9| = |-4x + 2|$

   $6x - 9 = -4x + 2$ or $6x - 9 = -(-4x + 2)$

     $10x = 11$ $6x - 9 = 4x - 2$

                           $2x = 7$

    $x = \frac{11}{10}$ or $x = \frac{7}{2}$

Check $x = \frac{11}{10}$: $\left| -\frac{24}{10} \right| = \left| -\frac{24}{10} \right|$ *True*

Check $x = \frac{7}{2}$: $|12| = |-12|$ *True*

Solution set: $\left\{ \frac{11}{10}, \frac{7}{2} \right\}$

6. $2x = \sqrt{\dfrac{5x+2}{3}}$

Square both sides.

$$(2x)^2 = \left(\sqrt{\frac{5x+2}{3}} \right)^2$$

$$4x^2 = \frac{5x+2}{3}$$

$$12x^2 = 5x + 2$$

$$12x^2 - 5x - 2 = 0$$

$$(3x - 2)(4x + 1) = 0$$

   $3x - 2 = 0$ or $4x + 1 = 0$

      $x = \frac{2}{3}$ or $x = -\frac{1}{4}$

Check $x = \frac{2}{3}$: $\frac{4}{3} = \sqrt{\frac{16}{9}}$ *True*

Check $x = -\frac{1}{4}$: $-\frac{1}{2} = \sqrt{\frac{1}{4}}$ *False*

Solution set: $\left\{ \frac{2}{3} \right\}$

7.
$$\frac{3}{x-3} - \frac{2}{x-2} = \frac{3}{x^2 - 5x + 6}$$
$$\frac{3}{x-3} - \frac{2}{x-2} = \frac{3}{(x-3)(x-2)}$$

Multiply by the LCD, $(x-3)(x-2)$.

$$(x-3)(x-2)\left(\frac{3}{x-3} - \frac{2}{x-2}\right)$$
$$= (x-3)(x-2)\left[\frac{3}{(x-3)(x-2)}\right]$$
$$3(x-2) - 2(x-3) = 3$$
$$3x - 6 - 2x + 6 = 3$$
$$x = 3$$

The number 3 is not allowed as a solution since it makes the denominator 0. Solution set: $\emptyset$

8.
$$(r-5)(2r+3) = 1$$
$$2r^2 - 7r - 15 = 1$$
$$2r^2 - 7r - 16 = 0$$
Use the quadratic formula.
$$r = \frac{-b \pm \sqrt{b^2 - 4ac}}{2a}$$
$$r = \frac{-(-7) \pm \sqrt{(-7)^2 - 4(2)(-16)}}{2(2)}$$
$$= \frac{7 \pm \sqrt{49 + 128}}{4} = \frac{7 \pm \sqrt{177}}{4}$$

Solution set: $\left\{\dfrac{7 \pm \sqrt{177}}{4}\right\}$

9.
$$x^4 - 5x^2 + 4 = 0$$
Let $u = x^2$, so $u^2 = \left(x^2\right)^2 = x^4$.
The equation becomes
$$u^2 - 5u + 4 = 0.$$
$$(u-4)(u-1) = 0$$
$$u - 4 = 0 \quad \text{or} \quad u - 1 = 0$$
$$u = 4 \quad \text{or} \quad u = 1$$

To find x, substitute x^2 for u.
$$x^2 = 4 \qquad \text{or} \qquad x^2 = 1$$
$$x = 2 \text{ or } x = -2 \quad \text{or} \quad x = 1 \text{ or } x = -1$$

The proposed solutions check.
Solution set: $\{\pm 2, \pm 1\}$

10.
$$-2x + 4 \leq -x + 3$$
$$-x \leq -1$$
Multiply by -1, and reverse the direction of the inequality.
$$x \geq 1$$
Solution set: $[1, \infty)$

11.
$$|3x - 7| \leq 1$$
$$-1 \leq 3x - 7 \leq 1$$
$$6 \leq 3x \leq 8$$
$$2 \leq x \leq \frac{8}{3}$$
Solution set: $\left[2, \frac{8}{3}\right]$

12.
$$x^2 - 4x + 3 < 0$$
Solve the equation
$$x^2 - 4x + 3 = 0.$$
$$(x-3)(x-1) = 0$$

$$x - 3 = 0 \quad \text{or} \quad x - 1 = 0$$
$$x = 3 \quad \text{or} \quad x = 1$$

The numbers 1 and 3 divide a number line into three intervals.

$$
\begin{array}{ccc}
\text{A} & \text{B} & \text{C} \\
\hline
\quad\quad | \quad\quad | \quad\quad \longrightarrow \\
1 \quad\quad 3
\end{array}
$$

Test a number from each interval in the inequality
$$x^2 - 4x + 3 < 0.$$

Interval A: Let $x = 0$.
$$3 < 0 \qquad \textit{False}$$
Interval B: Let $x = 2$.
$$4 - 8 + 3 < 0 \quad ?$$
$$-1 < 0 \qquad \textit{True}$$
Interval C: Let $x = 4$.
$$16 - 16 + 3 < 0 \quad ?$$
$$3 < 0 \qquad \textit{False}$$

The numbers from Interval B, not including 1 or 3, are solutions.

Solution set: $(1, 3)$

13.
$$\frac{3}{p+2} > 1$$
Write the inequality so that 0 is on one side.
$$\frac{3}{p+2} - 1 > 0$$
$$\frac{3}{p+2} - \frac{1(p+2)}{p+2} > 0$$
$$\frac{3 - p - 2}{p+2} > 0$$
$$\frac{-p + 1}{p+2} > 0$$

The number 1 makes the numerator 0, and -2 makes the denominator 0. These two numbers determine three intervals.

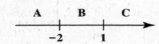

Test a number from each interval in the inequality

$$\frac{3}{p+2} > 1.$$

Interval A: Let $p = -4$.

$\frac{3}{-2} > 1$ *False*

Interval B: Let $p = 0$.

$\frac{3}{2} > 1$ *True*

Interval C: Let $p = 2$.

$\frac{3}{4} > 1$ *False*

The numbers from Interval B, not including -2 or 1, are solutions.

Solution set: $(-2, 1)$

14. $4x - 5y = 15$

Draw the line through its intercepts, $\left(\frac{15}{4}, 0\right)$ and $(0, -3)$. The graph passes the vertical line test, so the relation is a function. As with any line that is not horizontal or vertical, the domain and range are both $(-\infty, \infty)$.

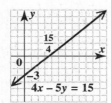

Solve the equation for y.

$$4x - 5y = 15$$
$$-5y = 15 - 4x$$
$$y = \frac{15 - 4x}{-5}, \quad \text{or} \quad \frac{4x - 15}{5}$$

Thus, $f(x) = \frac{4}{5}x - 3$.

15. $4x - 5y < 15$

Draw a dashed line through the points $\left(\frac{15}{4}, 0\right)$ and $(0, -3)$. Check the origin:

$$4(0) - 5(0) < 15 \quad ?$$
$$0 < 15 \qquad \textit{True}$$

Shade the region that contains the origin.

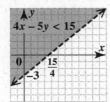

The relation is not a function since for any value of x, there is more than one value of y.

16. $y = -2(x - 1)^2 + 3$

The equation is in $f(x) = a(x - h)^2 + k$ form, so the graph is a parabola with vertex (h, k) at $(1, 3)$. Since $a = -2 < 0$, the parabola opens down. Also $|a| = |-2| = 2 > 1$, so the graph is narrower than the graph of $f(x) = x^2$. The points $(0, 1)$ and $(2, 1)$ are on the graph.

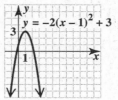

The relation is a function since it passes the vertical line test. The domain is $(-\infty, \infty)$. The largest value of y is 3, so the range is $(-\infty, 3]$. The equation is already solved for y, so using function notation we have
$f(x) = -2(x - 1)^2 + 3$.

17. $-2x + 7y = 16$

Solve the equation for y.

$$7y = 2x + 16$$
$$y = \frac{2}{7}x + \frac{16}{7}$$

So the slope is $\frac{2}{7}$ and the y-intercept is $\left(0, \frac{16}{7}\right)$. Let $y = 0$ in $-2x + 7y = 16$ to find the x-intercept.

$$-2x + 7(0) = 16$$
$$-2x = 16$$
$$x = -8$$

The x-intercept is $(-8, 0)$.

18. **(a)** Solve $5x + 2y = 6$ for y.
$$2y = -5x + 6$$
$$y = -\frac{5}{2}x + 3$$
So the slope of the given line and the desired line is $-\frac{5}{2}$. The required form is

$$y = -\frac{5}{2}x + b.$$
$$-3 = -\frac{5}{2}(2) + b \quad \textit{Let x = 2, y = -3.}$$
$$-3 = -5 + b$$
$$2 = b$$

The equation is $y = -\frac{5}{2}x + 2$.

(b) The negative reciprocal of the slope in part (a) is

$$-\frac{1}{-\frac{5}{2}} = \frac{2}{5},$$

which is the slope of the line perpendicular to the given line.

continued

So $y = \frac{2}{5}x + b$.

$1 = \frac{2}{5}(-4) + b$ *Let x = −4, y = 1.*

$1 = -\frac{8}{5} + b$

$\frac{13}{5} = b$

The equation is $y = \frac{2}{5}x + \frac{13}{5}$.

19. $\left(\dfrac{x^{-3}y^2}{x^5y^{-2}}\right)^{-1} = \left(x^{-3-5}y^{2-(-2)}\right)^{-1}$

$\qquad = \left(x^{-8}y^4\right)^{-1}$

$\qquad = x^8y^{-4}$

$\qquad = \dfrac{x^8}{y^4}$

20. $\dfrac{(4x^{-2})^2(2y^3)}{8x^{-3}y^5} = \dfrac{16x^{-4}(2y^3)}{8x^{-3}y^5}$

$\qquad = \dfrac{4x^{-4}y^3}{x^{-3}y^5}$

$\qquad = 4x^{-4-(-3)}y^{3-5}$

$\qquad = 4x^{-1}y^{-2}$

$\qquad = \dfrac{4}{xy^2}$

21. $(7x + 4)(2x - 3)$

$\qquad$ **F**   **O**   **I**   **L**

$\qquad = 14x^2 - 21x + 8x - 12$

$\qquad = 14x^2 - 13x - 12$

22. $\left(\frac{2}{3}t + 9\right)^2 = \left(\frac{2}{3}t\right)^2 + 2\left(\frac{2}{3}t\right)(9) + 9^2$

$\qquad = \frac{4}{9}t^2 + 12t + 81$

23. $\left(3t^3 + 5t^2 - 8t + 7\right) - \left(6t^3 + 4t - 8\right)$

$\qquad = 3t^3 + 5t^2 - 8t + 7 - 6t^3 - 4t + 8$

$\qquad = -3t^3 + 5t^2 - 12t + 15$

24. Divide $4x^3 + 2x^2 - x + 26$ by $x + 2$.

$$
\begin{array}{r}
4x^2 \quad - \quad 6x \quad + \quad 11 \\
x + 2 \overline{)\, 4x^3 + \quad 2x^2 - \quad x + 26} \\
\underline{4x^3 + \quad 8x^2} \qquad\qquad\qquad \\
-6x^2 - \quad x \qquad\quad \\
\underline{-6x^2 - \quad 12x} \qquad\quad \\
11x + 26 \\
\underline{11x + 22} \\
4
\end{array}
$$

The answer is

$$4x^2 - 6x + 11 + \frac{4}{x + 2}.$$

25. $16x - x^3 = x\left(16 - x^2\right)$

$\qquad\qquad\quad = x(4 + x)(4 - x)$

26. $24m^2 + 2m - 15$

The two integers whose product is $24(-15) = -360$ and whose sum is 2 are 20 and -18.

$\qquad 24m^2 + 2m - 15$

$\qquad = 24m^2 + 20m - 18m - 15$

$\qquad = 4m(6m + 5) - 3(6m + 5)$

$\qquad = (6m + 5)(4m - 3)$

27. $8x^3 + 27y^3$

Use the sum of cubes formula,

$$a^3 + b^3 = (a + b)\left(a^2 - ab + b^2\right),$$

with $a = 2x$ and $b = 3y$.

$\qquad 8x^3 + 27y^3$

$\qquad = (2x + 3y)\left[(2x)^2 - (2x)(3y) + (3y)^2\right]$

$\qquad = (2x + 3y)\left(4x^2 - 6xy + 9y^2\right)$

28. $9x^2 - 30xy + 25y^2$

Use the perfect square formula,

$$a^2 - 2ab + b^2 = (a - b)^2,$$

with $a = 3x$ and $b = 5y$.

$\qquad 9x^2 - 30xy + 25y^2$

$\qquad = \left[(3x)^2 - 2(3x)(5y) + (5y)^2\right]$

$\qquad = (3x - 5y)^2$

29. $\dfrac{x^2 - 3x - 10}{x^2 + 3x + 2} \cdot \dfrac{x^2 - 2x - 3}{x^2 + 2x - 15}$

$\qquad = \dfrac{(x - 5)(x + 2)}{(x + 2)(x + 1)} \cdot \dfrac{(x - 3)(x + 1)}{(x + 5)(x - 3)}$

$\qquad = \dfrac{x - 5}{x + 5}$

30. $\dfrac{3}{2 - k} - \dfrac{5}{k} + \dfrac{6}{k^2 - 2k}$

$\qquad = \dfrac{3}{2 - k} - \dfrac{5}{k} + \dfrac{6}{k(k - 2)}$

$\qquad = \dfrac{-3}{k - 2} - \dfrac{5}{k} + \dfrac{6}{k(k - 2)}$

The LCD is $k(k - 2)$.

$\qquad = \dfrac{-3k}{(k - 2)k} - \dfrac{5(k - 2)}{k(k - 2)} + \dfrac{6}{k(k - 2)}$

$\qquad = \dfrac{-3k - 5(k - 2) + 6}{k(k - 2)}$

$\qquad = \dfrac{-3k - 5k + 10 + 6}{k(k - 2)}$

$\qquad = \dfrac{-8k + 16}{k(k - 2)}$

$\qquad = \dfrac{-8(k - 2)}{k(k - 2)} = -\dfrac{8}{k}$

31. $\dfrac{\dfrac{r}{s} - \dfrac{s}{r}}{\dfrac{r}{s} + 1}$

Multiply the numerator and denominator by the LCD of all the fractions, rs.

$$= \frac{\left(\dfrac{r}{s} - \dfrac{s}{r}\right)rs}{\left(\dfrac{r}{s} + 1\right)rs} = \frac{r^2 - s^2}{r^2 + rs}$$

$$= \frac{(r-s)(r+s)}{r(r+s)} = \frac{r-s}{r}$$

32. (a) We are given the y-intercept, $(0, 600)$, so the slope-intercept form of the line is

$$y = mx + 600.$$

Substitute 1340 for y and 6 for x to solve for the slope m.

$$1340 = m(6) + 600$$
$$740 = 6m$$
$$m = \tfrac{740}{6} = 123.\overline{3}$$

Rounding to the nearest whole number gives us the linear model

$$y = 123x + 600.$$

(b) For 2002, use $x = 7$.

$$y = 123(7) + 600$$
$$= 861 + 600$$
$$= 1461$$

From the model, the approximate sales were $1461 million, which is a little high compared to the table value of $1400 million.

33. No. The graph is a vertical line, which is not the graph of a function by the vertical line test. Also, the only domain value, 5, can have infinitely many range values paired with it.

34. $f(x) = 2(x-1)^2 - 5$

(a) $f(-2) = 2(-2-1)^2 - 5$
$\quad = 2(-3)^2 - 5$
$\quad = 2(9) - 5 = 13$

(b) Any value can be substituted for x, so the domain is $(-\infty, \infty)$. The graph of f is a parabola that opens up with vertex $(1, -5)$. The vertex is a minimum point so the y-values are all greater than or equal to -5. Thus, the range is $[-5, \infty)$.

35. $2x - 4y = 10$ (1)
$9x + 3y = 3$ (2)
Simplify the equations.

$x - 2y = 5$ (3) $\frac{1}{2} \times (1)$
$3x + y = 1$ (4) $\frac{1}{3} \times (2)$

To eliminate y, multiply (4) by 2 and add the result to (3).

$x - 2y = 5$ (3)
$6x + 2y = 2$ $2 \times (4)$
$\overline{7x \quad\quad = 7}$
$x = 1$

Substitute $x = 1$ into (4).

$3x + y = 1$ (4)
$3(1) + y = 1$
$y = -2$

Solution set: $\{(1, -2)\}$

36. $x + y + 2z = 3$ (1)
$-x + y + z = -5$ (2)
$2x + 3y - z = -8$ (3)
Eliminate z by adding (2) and (3).

$-x + y + z = -5$ (2)
$2x + 3y - z = -8$ (3)
$\overline{x + 4y \quad = -13}$ (4)

To get another equation without z, multiply equation (3) by 2 and add the result to equation (1).

$x + y + 2z = 3$ (1)
$4x + 6y - 2z = -16$ $2 \times (3)$
$\overline{5x + 7y \quad = -13}$ (5)

To eliminate x, multiply (4) by -5 and add the result to (5).

$-5x - 20y = 65$ $-5 \times (4)$
$5x + 7y = -13$ (5)
$\overline{-13y = 52}$
$y = -4$

Use (4) to find x.

$x + 4y = -13$ (4)
$x + 4(-4) = -13$
$x - 16 = -13$
$x = 3$

Use (2) to find z.

$-x + y + z = -5$ (2)
$-3 - 4 + z = -5$
$-7 + z = -5$
$z = 2$

Solution set: $\{(3, -4, 2)\}$

37. **(a)** Let $x =$ the amount of sales (in billions) for AOL and $y =$ the amount of sales for Time Warner.

The combined sales were \$34.2 billion, so

$$x + y = 34.2. \quad (1)$$

Sales for AOL were \$.3 billion less than 4 times the sales of Time Warner, so

$$x = 4y - 0.3. \quad (2)$$

(b) Substitute $4y - 0.3$ for x in (1), then solve for y.

$$(4y - 0.3) + y = 34.2$$
$$5y = 34.5$$
$$y = 6.9$$

From (2), $x = 4(6.9) - 0.3 = 27.3$. Thus, AOL had \$27.3 billion in sales and Time Warner had \$6.9 billion in sales.

38. $\sqrt[3]{\dfrac{27}{16}} = \dfrac{\sqrt[3]{27}}{\sqrt[3]{16}} = \dfrac{\sqrt[3]{3^3}}{\sqrt[3]{8 \cdot 2}} = \dfrac{3}{2\sqrt[3]{2}}$

$\qquad = \dfrac{3 \cdot \sqrt[3]{4}}{2\sqrt[3]{2} \cdot \sqrt[3]{4}} = \dfrac{3\sqrt[3]{4}}{2\sqrt[3]{8}}$

$\qquad = \dfrac{3\sqrt[3]{4}}{2 \cdot 2} = \dfrac{3\sqrt[3]{4}}{4}$

39. $\dfrac{2}{\sqrt{7} - \sqrt{5}} = \dfrac{2(\sqrt{7} + \sqrt{5})}{(\sqrt{7} - \sqrt{5})(\sqrt{7} + \sqrt{5})}$

$\qquad = \dfrac{2(\sqrt{7} + \sqrt{5})}{7 - 5}$

$\qquad = \dfrac{2(\sqrt{7} + \sqrt{5})}{2} = \sqrt{7} + \sqrt{5}$

40. Let $x =$ Tri's rate on the bicycle.
Then $x - 10 =$ Tri's rate while walking.

Make a chart. Use $d = rt$, or $t = \dfrac{d}{r}$.

	Distance	Rate	Time
Bicycle	12	x	$\dfrac{12}{x}$
Walking	8	$x - 10$	$\dfrac{8}{x - 10}$

$$\begin{matrix} \text{Tri's time on} \\ \text{the bicycle} \end{matrix} + \begin{matrix} \text{Tri's time} \\ \text{walking} \end{matrix} = \begin{matrix} 5 \\ \text{hours.} \end{matrix}$$

$$\dfrac{12}{x} + \dfrac{8}{x - 10} = 5$$

Multiply by the LCD, $x(x - 10)$.

$$x(x - 10)\left(\dfrac{12}{x} + \dfrac{8}{x - 10}\right) = x(x - 10) \cdot 5$$
$$12(x - 10) + 8x = 5x(x - 10)$$
$$12x - 120 + 8x = 5x^2 - 50x$$
$$0 = 5x^2 - 70x + 120$$
$$0 = x^2 - 14x + 24$$
$$0 = (x - 12)(x - 2)$$

$$x - 12 = 0 \quad \text{or} \quad x - 2 = 0$$
$$x = 12 \quad \text{or} \quad x = 2$$

Reject 2 for Tri's bicycle speed, since it would yield a negative walking speed. Thus, his bicycle speed was 12 mph, and his walking speed was $x - 10 = 2$ mph.

41. Let $\quad x =$ the distance traveled by the southbound car and
$2x - 38 =$ the distance traveled by the eastbound car.

Since the cars are traveling at right angles with one another, the Pythagorean formula can be applied.

$$a^2 + b^2 = c^2$$
$$x^2 + (2x - 38)^2 = 95^2$$
$$x^2 + 4x^2 - 152x + 1444 = 9025$$
$$5x^2 - 152x - 7581 = 0$$

$$x = \dfrac{-(-152) \pm \sqrt{(-152)^2 - 4(5)(-7581)}}{2(5)}$$

$$= \dfrac{152 \pm \sqrt{174{,}724}}{10} = \dfrac{152 \pm 418}{10}$$

Thus, $x = \dfrac{152 + 418}{10} = 57$ (the other value is negative). The southbound car traveled 57 miles, and the eastbound car traveled $2x - 38 = 2(57) - 38 = 76$ miles.

42. Since 31% of the people in the United States curse the ATM, the number of people in the United States who curse the ATM in a sample of 4000 is

$$0.31(4000) = 1240.$$

43. $0.24(4000) = 960$

44. $0.33(4000) = 1320$

45. German cursers: $0.53(4000) = 2120$
United States thankers: $0.22(4000) = 880$

The difference is

$$2120 - 880 = 1240.$$

CHAPTER 12 INVERSE, EXPONENTIAL, AND LOGARITHMIC FUNCTIONS

12.1 Inverse Functions

1. This function is not one-to-one because both France and the United States are paired with the same trans fat percentage, 11. Also both Hungary and Poland are paired with the same trans fat percentage, 8.

3. The function in the table that pairs a city with a distance is a one-to-one function because for each city there is one distance and each distance has only one city to which it is paired.

If the distance from Indianapolis to Denver had 1 mile added to it, it would be $1058 + 1 = 1059$ mi, the same as the distance from Los Angeles to Denver. In this case, one distance would have two cities to which it is paired, and the function would not be one-to-one.

5. If a function is made up of ordered pairs in such a way that the same y-value appears in a correspondence with two different x-values, then the function is not one-to-one. Choice **B**

7. All of the graphs pass the vertical line test, so they all represent functions. The graph in choice **A** is the only one that passes the horizontal line test, so it is the one-to-one function.

9. $\{(3, 6), (2, 10), (5, 12)\}$ is a one-to-one function, since each x-value corresponds to only one y-value and each y-value corresponds to only one x-value. To find the inverse, interchange x and y in each ordered pair. The inverse is

$$\{(6, 3), (10, 2), (12, 5)\}.$$

11. $\{(-1, 3), (2, 7), (4, 3), (5, 8)\}$ is not a one-to-one function. The ordered pairs $(-1, 3)$ and $(4, 3)$ have the same y-value for two different x-values.

13. The graph of $f(x) = 2x + 4$ is a nonvertical, nonhorizontal line. By the horizontal line test, $f(x)$ is a one-to-one function. To find the inverse, replace $f(x)$ with y.

$$y = 2x + 4$$
Interchange x and y.
$$x = 2y + 4$$
Solve for y.
$$2y = x - 4$$
$$y = \frac{x - 4}{2}$$

Replace y with $f^{-1}(x)$.
$$f^{-1}(x) = \frac{x - 4}{2}, \quad \text{or} \quad f^{-1}(x) = \frac{1}{2}x - 2$$

15. Write $g(x) = \sqrt{x - 3}$ as $y = \sqrt{x - 3}$.

Since $x \geq 3$, $y \geq 0$. The graph of g is half of a horizontal parabola that opens to the right. The graph passes the horizontal line test, so g is one-to-one. To find the inverse, interchange x and y to get

$$x = \sqrt{y - 3}.$$

Note that now $y \geq 3$, so $x \geq 0$.
Solve for y by squaring both sides.

$$x^2 = y - 3$$
$$x^2 + 3 = y$$

Replace y with $g^{-1}(x)$.
$$g^{-1}(x) = x^2 + 3, \, x \geq 0$$

17. $f(x) = 3x^2 + 2$ is not a one-to-one function because two x-values, such as 1 and -1, both have the same y-value, in this case 5. The graph of this function is a vertical parabola which does not pass the horizontal line test.

19. The graph of $f(x) = x^3 - 4$ is the graph of $g(x) = x^3$ shifted down 4 units. (The graph of $g(x) = x^3$ is the elongated S-shaped curve.) The graph of f passes the horizontal line test, so f is one-to-one.

Replace $f(x)$ with y.
$$y = x^3 - 4$$
Interchange x and y.
$$x = y^3 - 4$$
Solve for y.
$$x + 4 = y^3$$
Take the cube root of each side.
$$\sqrt[3]{x + 4} = y$$
Replace y with $f^{-1}(x)$.
$$f^{-1}(x) = \sqrt[3]{x + 4}$$

In Exercises 21–24, $f(x) = 2^x$ is a one-to-one function.

21. (a) To find $f(3)$, substitute 3 for x.

$f(x) = 2^x$, so $f(3) = 2^3 = 8$.

(b) Since f is one-to-one and $f(3) = 8$, it follows that $f^{-1}(8) = 3$.

23. (a) To find $f(0)$, substitute 0 for x.

$f(x) = 2^x$, so $f(0) = 2^0 = 1$.

(b) Since f is one-to-one and $f(0) = 1$, it follows that $f^{-1}(1) = 0$.

25. (a) The function is one-to-one since any horizontal line intersects the graph at most once.

(b) In the graph, the two points marked on the line are $(-1, 5)$ and $(2, -1)$. Interchange x and y in each ordered pair to get $(5, -1)$ and $(-1, 2)$. Plot these points, then draw a dashed line through them to obtain the graph of the inverse function.

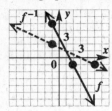

27. **(a)** The function is not one-to-one since there are horizontal lines that intersect the graph more than once. For example, the line $y = 1$ intersects the graph twice.

29. **(a)** The function is one-to-one since any horizontal line intersects the graph at most once.

(b) In the graph, the four points marked on the curve are $(-4, 2)$, $(-1, 1)$, $(1, -1)$, and $(4, -2)$. Interchange x and y in each ordered pair to get $(2, -4)$, $(1, -1)$, $(-1, 1)$, and $(-2, 4)$. Plot these points, then draw a dashed curve (symmetric to the original graph about the line $y = x$) through them to obtain the graph of the inverse.

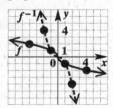

31. $f(x) = 2x - 1$ or $y = 2x - 1$
The graph is a line through $(-2, -5)$, $(0, -1)$, and $(3, 5)$. Plot these points and draw the solid line through them. Then the inverse will be a line through $(-5, -2)$, $(-1, 0)$, and $(5, 3)$. Plot these points and draw the dashed line through them.

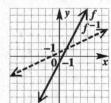

33. $g(x) = -4x$ or $y = -4x$
The graph is a line through $(0, 0)$ and $(1, -4)$. For the inverse, interchange x and y in each ordered pair to get the points $(0, 0)$ and $(-4, 1)$. Draw a dashed line through these points to obtain the graph of the inverse function.

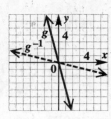

35. $f(x) = \sqrt{x}, \ x \geq 0$
Complete the table of values.

x	$f(x)$
0	0
1	1
4	2

Plot these points and connect them with a solid smooth curve.
Since $f(x)$ is one-to-one, make a table of values for $f^{-1}(x)$ by interchanging x and y.

x	$f^{-1}(x)$
0	0
1	1
2	4

Plot these points and connect them with a dashed smooth curve.

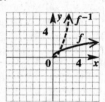

37. $f(x) = x^3 - 2$
Complete the table of values.

x	$f(x)$
-1	-3
0	-2
1	-1
2	6

Plot these points and connect them with a solid smooth curve.
Make a table of values for f^{-1}.

x	$f^{-1}(x)$
-3	-1
-2	0
-1	1
6	2

Plot these points and connect them with a dashed smooth curve.

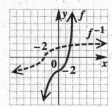

39.
$$f(x) = 4x - 5$$
Replace $f(x)$ with y.
$$y = 4x - 5$$
Interchange x and y.
$$x = 4y - 5$$
Solve for y.
$$x + 5 = 4y$$
$$\frac{x + 5}{4} = y$$
Replace y with $f^{-1}(x)$.
$$\frac{x + 5}{4} = f^{-1}(x),$$
$$\text{or} \quad f^{-1}(x) = \frac{1}{4}x + \frac{5}{4}$$

40. Insert each number in the inverse function found in Exercise 39,
$$f^{-1}(x) = \frac{x + 5}{4}.$$
$$f^{-1}(47) = \frac{47 + 5}{4} = \frac{52}{4} = 13 = \text{M},$$
$$f^{-1}(95) = \frac{95 + 5}{4} = \frac{100}{4} = 25 = \text{Y},$$
and so on.

The decoded message is as follows:
My graphing calculator is the greatest thing since sliced bread.

41. A one-to-one code is essential to this process because if the code is not one-to-one, an encoded number would refer to two different letters.

42. Answers will vary according to the student's name. For example, Jane Doe is encoded as follows:

1004 5 2748 129 68 3379 129.

43. $Y_1 = f(x) = 2x - 7$
Replace $f(x)$ with y.
$$y = 2x - 7$$
Interchange x and y.
$$x = 2y - 7$$
Solve for y.
$$x + 7 = 2y$$
$$\frac{x + 7}{2} = y$$
Replace y with $f^{-1}(x)$.
$$\frac{x + 7}{2} = f^{-1}(x) = Y_2$$

Now graph Y_1 and Y_2.

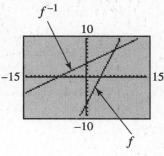

45. $Y_1 = f(x) = x^3 + 5$
Replace $f(x)$ with y.
$$y = x^3 + 5$$
Interchange x and y.
$$x = y^3 + 5$$
Solve for y.
$$x - 5 = y^3$$
Take the cube root of each side.
$$\sqrt[3]{x - 5} = y$$
Replace y with $f^{-1}(x)$.
$$\sqrt[3]{x - 5} = f^{-1}(x) = Y_2$$

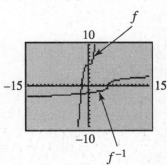

47. $Y_1 = X^2 + 3X + 4$
Graph Y_1 and its inverse in the same square window on a graphing calculator. On a TI-83, graph Y_1 and then enter

$$\text{DrawInv } Y_1$$

on the home screen. DrawInv is choice 8 under the DRAW menu. Y_1 is choice 1 under VARS, Y-VARS, Function.

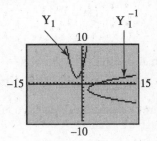

49. $f(x) = 4^x$, so $f(3) = 4^3 = 64$.

51. $f(x) = 4^x$, so $f\left(\frac{1}{2}\right) = 4^{1/2} = \sqrt{4} = 2$.

53. $f(x) = 4^x$, so $f(2.73) = 4^{2.73} \approx 44.02$.

12.2   Exponential Functions

1.   Since the graph of $F(x) = a^x$ always contains the point $(0, 1)$, the correct response is **C**.

3.   Since the graph of $F(x) = a^x$ always approaches the x-axis, the correct response is **A**.

5. $f(x) = 3^x$
Make a table of values.
$$f(-2) = 3^{-2} = \frac{1}{3^2} = \frac{1}{9},$$
$$f(-1) = 3^{-1} = \frac{1}{3^1} = \frac{1}{3}, \text{ and so on.}$$

x	-2	-1	0	1	2
$f(x)$	$\frac{1}{9}$	$\frac{1}{3}$	1	3	9

Plot the points from the table and draw a smooth curve through them.

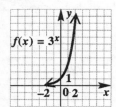

7. $g(x) = \left(\frac{1}{3}\right)^x$
Make a table of values.
$$g(-2) = \left(\frac{1}{3}\right)^{-2} = \left(\frac{3}{1}\right)^2 = 9,$$
$$g(-1) = \left(\frac{1}{3}\right)^{-1} = \left(\frac{3}{1}\right)^1 = 3, \text{ and so on.}$$

x	-2	-1	0	1	2
$g(x)$	9	3	1	$\frac{1}{3}$	$\frac{1}{9}$

Plot the points from the table and draw a smooth curve through them.

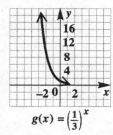
$$g(x) = \left(\frac{1}{3}\right)^x$$

9. $y = 4^{-x}$
This equation can be rewritten as
$$y = (4^{-1})^x = \left(\frac{1}{4}\right)^x,$$
which shows that it is *falling* from left to right. Make a table of values.

x	-2	-1	0	1	2
y	16	4	1	$\frac{1}{4}$	$\frac{1}{16}$

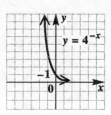

11. $y = 2^{2x-2}$
Make a table of values. It will help to find values for $2x - 2$ before you find y.

x	-2	-1	0	1	2	3
$2x - 2$	-6	-4	-2	0	2	4
y	$\frac{1}{64}$	$\frac{1}{16}$	$\frac{1}{4}$	1	4	16

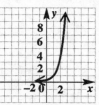
$$y = 2^{2x-2}$$

13.   **(a)** For an exponential function defined by $f(x) = a^x$, if $a > 1$, the graph <u>rises</u> from left to right. (See Example 1, $f(x) = 2^x$, in your text.) If $0 < a < 1$, the graph <u>falls</u> from left to right. (See Example 2, $g(x) = \left(\frac{1}{2}\right)^x = 2^{-x}$, in your text.)

(b) An exponential function defined by $f(x) = a^x$ is one-to-one and has an inverse, since each value of $f(x)$ corresponds to one and only one value of x.

15. $$6^x = 36$$
Write each side as a power of 6.
$$6^x = 6^2$$
For $a > 0$ and $a \neq 1$, if $a^x = a^y$, then $x = y$. Set the exponents equal to each other.
$$x = 2$$
Check $x = 2$: $6^2 = 36$ *True*
The solution set is $\{2\}$.

17. $$100^x = 1000$$
Write each side as a power of 10.
$$\left(10^2\right)^x = 10^3$$
$$10^{2x} = 10^3$$
For $a > 0$ and $a \neq 1$, if $a^x = a^y$, then $x = y$. Set the exponents equal to each other.
$$2x = 3$$
$$x = \frac{3}{2}$$
Check $x = \frac{3}{2}$: $100^{3/2} = 1000$ *True*
The solution set is $\left\{\frac{3}{2}\right\}$.

19. $16^{2x+1} = 64^{x+3}$

Write each side as a power of 4.

$\left(4^2\right)^{2x+1} = \left(4^3\right)^{x+3}$

$4^{4x+2} = 4^{3x+9}$

Set the exponents equal.

$4x + 2 = 3x + 9$

$x = 7$

Check $x = 7$: $16^{15} = 64^{10}$ *True*

The solution set is $\{7\}$.

21. $5^x = \frac{1}{125}$

$5^x = \left(\frac{1}{5}\right)^3$

Write each side as a power of 5.

$5^x = 5^{-3}$

Set the exponents equal.

$x = -3$

Check $x = -3$: $5^{-3} = \frac{1}{125}$ *True*

The solution set is $\{-3\}$.

23. $5^x = 0.2$

$5^x = \frac{2}{10} = \frac{1}{5}$

Write each side as a power of 5.

$5^x = 5^{-1}$

Set the exponents equal.

$x = -1$

Check $x = -1$: $5^{-1} = 0.2$ *True*

The solution set is $\{-1\}$.

25. $\left(\frac{3}{2}\right)^x = \frac{8}{27}$

$\left(\frac{3}{2}\right)^x = \left(\frac{2}{3}\right)^3$

Write each side as a power of $\frac{3}{2}$.

$\left(\frac{3}{2}\right)^x = \left(\frac{3}{2}\right)^{-3}$

Set the exponents equal.

$x = -3$

Check $x = -3$: $\left(\frac{3}{2}\right)^{-3} = \frac{8}{27}$ *True*

The solution set is $\{-3\}$.

27. $12^{2.6} \approx 639.545$

29. $0.5^{3.921} \approx 0.066$

31. $2.718^{2.5} \approx 12.179$

33. **(a)** The increase for the exponential-type curve in the year 2000 is about $0.5°C$.

(b) The increase for the linear graph in the year 2000 is about $0.35°C$.

35. **(a)** The increase for the exponential-type curve in the year 2020 is about $1.6°C$.

(b) The increase for the linear graph in the year 2020 is about $0.5°C$.

37. $f(x) = 220{,}717(1.0217)^{-x}$

(a) 1970 corresponds to $x = 0$.

$f(0) = 220{,}717(1.0217)^0$

$= 220{,}717(1) = 220{,}717$

The answer has units in thousands of tons.

(b) 1995 corresponds to $x = 25$.

$f(25) = 220{,}717(1.0217)^{-25} \approx 129{,}048$

(c) 2002 corresponds to $x = 32$.

$f(32) = 220{,}717(1.0217)^{-32} \approx 111{,}042$

The actual amount, 112,049 thousand tons, is greater than the 111,042 thousand tons that the model provides.

39. $V(t) = 5000(2)^{-0.15t}$

(a) The original value is found when $t = 0$.

$V(0) = 5000(2)^{-0.15(0)}$

$= 5000(2)^0$

$= 5000(1) = 5000$

The original value is 5000.

(b) The value after 5 years is found when $t = 5$.

$V(5) = 5000(2)^{-0.15(5)}$

$= 5000(2)^{-0.75} \approx 2973.02$

The value after 5 years is about 2973.

(c) The value after 10 years is found when $t = 10$.

$V(10) = 5000(2)^{-0.15(10)}$

$= 5000(2)^{-1.5} \approx 1767.77$

The value after 10 years is about 1768.

(d) Use the results of parts (a) – (c) to make a table of values.

t	0	5	10
$V(t)$	5000	2973	1768

Plot the points from the table and draw a smooth curve through them.

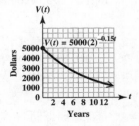

41. $V(t) = 5000(2)^{-0.15t}$

$2500 = 5000(2)^{-0.15t}$ *Let V(t) = 2500.*

$\frac{1}{2} = (2)^{-0.15t}$ *Divide by 5000.*

$2^{-1} = 2^{-0.15t}$

$-1 = -0.15t$ *Equate exponents.*

$t = \dfrac{-1}{-0.15} \approx 6.67$

The value of the machine will be $2500 in approximately 6.67 years after it was purchased.

43. $16 = 2 \cdot 2 \cdot 2 \cdot 2 = 2^4$, so $\square = 4$.

45. $2^0 = 1$, so $\square = 0$.

12.3 Logarithmic Functions

1. **(a)** $\log_{1/3} 3 = -1$ is equivalent to $\left(\frac{1}{3}\right)^{-1} = 3$. **(B)**

   (b) $\log_5 1 = 0$ is equivalent to $5^0 = 1$. **(E)**

   (c) $\log_2 \sqrt{2} = \frac{1}{2}$ is equivalent to $2^{1/2} = \sqrt{2}$. **(D)**

   (d) $\log_{10} 1000 = 3$ is equivalent to $10^3 = 1000$. **(F)**

   (e) $\log_8 \sqrt[3]{8} = \frac{1}{3}$ is equivalent to $8^{1/3} = \sqrt[3]{8}$ **(A)**

   (f) $\log_4 4 = 1$ is equivalent to $4^1 = 4$. **(C)**

3. The base is 4, the exponent (logarithm) is 5, and the number is 1024, so $4^5 = 1024$ becomes $\log_4 1024 = 5$ in logarithmic form.

5. $\frac{1}{2}$ is the base and -3 is the exponent, so $\left(\frac{1}{2}\right)^{-3} = 8$ becomes $\log_{1/2} 8 = -3$ in logarithmic form.

7. The base is 10, the exponent (logarithm) is -3, and the number is 0.001, so $10^{-3} = 0.001$ becomes $\log_{10} 0.001 = -3$ in logarithmic form.

9. $\sqrt[4]{625} = 625^{1/4} = 5$

The base is 625, the exponent (logarithm) is $\frac{1}{4}$, and the number is 5, so $\sqrt[4]{625} = 5$ becomes $\log_{625} 5 = \frac{1}{4}$ in logarithmic form.

11. In $\log_4 64 = 3$, 4 is the base and 3 is the logarithm (exponent), so $\log_4 64 = 3$ becomes $4^3 = 64$ in exponential form.

13. In $\log_{10} \frac{1}{10,000} = -4$, the base is 10, the logarithm (exponent) is -4, and the number is $\frac{1}{10,000}$, so $\log_{10} \frac{1}{10,000} = -4$ becomes $10^{-4} = \frac{1}{10,000}$ in exponential form.

15. In $\log_6 1 = 0$, 6 is the base and 0 is the logarithm (exponent), so $\log_6 1 = 0$ becomes $6^0 = 1$ in exponential form.

17. In $\log_9 3 = \frac{1}{2}$, the base is 9, the logarithm (exponent) is $\frac{1}{2}$, and the number is 3, so $\log_9 3 = \frac{1}{2}$ becomes $9^{1/2} = 3$ in exponential form.

19. Use the properties of logarithms,

$$\log_b b = 1 \quad \text{and} \quad \log_b 1 = 0,$$

for $b > 0$, $b \neq 1$.

   (a) $\log_8 8 = 1$ **(C)**

   (b) $\log_{16} 1 = 0$ **(B)**

   (c) $\log_{0.3} 1 = 0$ **(B)**

   (d) $\log_{\sqrt{7}} \sqrt{7} = 1$ **(C)**

21. $x = \log_{27} 3$

Write in exponential form.

$$27^x = 3$$

Write each side as a power of 3.

$$(3^3)^x = 3$$
$$3^{3x} = 3^1$$

Set the exponents equal.

$$3x = 1$$
$$x = \frac{1}{3}$$

Check $x = \frac{1}{3}$: $\frac{1}{3} = \log_{27} 3$ since $27^{1/3} = 3$.

The solution set is $\left\{\frac{1}{3}\right\}$.

23. $\log_x 9 = \frac{1}{2}$

Change to exponential form.

$$x^{1/2} = 9$$
$$(x^{1/2})^2 = 9^2 \quad \textit{Square.}$$
$$x^1 = 81$$
$$x = 81$$

$x = 81$ is an acceptable base since it is a positive number (not equal to 1).

Check $x = 81$: $\quad 81^{1/2} = 9$ *True*

The solution set is $\{81\}$.

25. $\log_x 125 = -3$

Write in exponential form.

$$x^{-3} = 125$$
$$\frac{1}{x^3} = 125$$
$$1 = 125(x^3)$$
$$\frac{1}{125} = x^3$$

Take the cube root of each side.

$$\sqrt[3]{\frac{1}{125}} = \sqrt[3]{x^3}$$

$$x = \sqrt[3]{\frac{1}{5^3}} = \frac{1}{5}$$

$x = \frac{1}{5}$ is an acceptable base since it is a positive number (not equal to 1).

The solution set is $\left\{\frac{1}{5}\right\}$.

27. $\log_{12} x = 0$

Write in exponential form.

$$12^0 = x$$
$$1 = x$$

The argument (the input of the logarithm) must be a positive number, so $x = 1$ is acceptable.

The solution set is $\{1\}$.

29. $\log_x x = 1$

Write in exponential form.

$$x^1 = x$$

This equation is true for all the numbers x that are allowed as the base of a logarithm; that is, all positive numbers x, $x \neq 1$.

The solution set is $\{x \mid x > 0, \; x \neq 1\}$.

31. $\log_x \dfrac{1}{25} = -2$

Write in exponential form.

$$x^{-2} = \frac{1}{25}$$
$$\frac{1}{x^2} = \frac{1}{25}$$
$$x^2 = 25 \qquad \textit{Denominators must be equal.}$$
$$x = \pm 5$$

Reject $x = -5$ since the base of a logarithm must be positive and not equal to 1.

The solution set is $\{5\}$.

33. $\log_8 32 = x$

$$8^x = 32 \quad \textit{Exponential form}$$

Write each side as a power of 2.

$$\left(2^3\right)^x = 2^5$$
$$2^{3x} = 2^5$$
$$3x = 5 \quad \textit{Equate exponents.}$$
$$x = \frac{5}{3}$$

Check $x = \frac{5}{3}$: $\log_8 32 = \frac{5}{3}$ since $8^{5/3} = 2^5 = 32$.

The solution set is $\left\{\frac{5}{3}\right\}$.

35. $\log_\pi \pi^4 = x$

$$\pi^x = \pi^4 \quad \textit{Exponential form}$$
$$x = 4 \quad \textit{Equate exponents.}$$

Check $x = 4$: $\log_\pi \pi^4 = 4$ since $\pi^4 = \pi^4$.

The solution set is $\{4\}$.

37. $\log_6 \sqrt{216} = x$

$$\log_6 216^{1/2} = x \qquad \textit{Equivalent form}$$
$$6^x = 216^{1/2} \qquad \textit{Exponential form}$$
$$6^x = \left(6^3\right)^{1/2} \qquad \textit{Same base}$$
$$6^x = 6^{3/2}$$
$$x = \frac{3}{2} \qquad \textit{Equate exponents.}$$

Check $x = \frac{3}{2}$:

$$\log_6 \sqrt{216} = \frac{3}{2} \text{ since } 6^{3/2} = \sqrt{6^3} = \sqrt{216}.$$

The solution set is $\left\{\frac{3}{2}\right\}$.

39. $\log_4 (2x + 4) = 3$

$$2x + 4 = 4^3 \qquad \textit{Exponential form}$$
$$2x = 64 - 4$$
$$2x = 60$$
$$x = 30$$

Check $x = 30$:

$$\log_4 (2 \cdot 30 + 4) = \log_4 64 = 3.$$

The solution set is $\{30\}$.

41. $$y = \log_3 x$$

Change to exponential form.

$$3^y = x$$

Refer to Section 12.2, Exercise 5, for the graph of $f(x) = 3^x$. Since $y = \log_3 x$ (or $3^y = x$) is the inverse of $f(x) = y = 3^x$, its graph is symmetric about the line $y = x$ to the graph of $f(x) = 3^x$. The graph can be plotted by reversing the ordered pairs in the table of values belonging to $f(x) = 3^x$.

x	$\frac{1}{9}$	$\frac{1}{3}$	1	3	9
y	-2	-1	0	1	2

Plot the points, and draw a smooth curve through them.

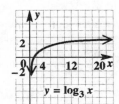

43. $y = \log_{1/3} x$

Change to exponential form.

$$\left(\frac{1}{3}\right)^y = x$$

Refer to Section 12.2, Exercise 7, for the graph of $g(x) = \left(\frac{1}{3}\right)^x$. Since $y = \log_{1/3} x$ (or $\left(\frac{1}{3}\right)^y = x$) is the inverse of $y = \left(\frac{1}{3}\right)^x$, its graph is symmetric about the line $y = x$ to the graph of $y = \left(\frac{1}{3}\right)^x$. The graph can be plotted by reversing the ordered pairs in the table of values belonging to $g(x) = \left(\frac{1}{3}\right)^x$.

x	9	3	1	$\frac{1}{3}$	$\frac{1}{9}$
y	-2	-1	0	1	2

Plot the points, and draw a smooth curve through them.

continued

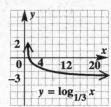

$y = \log_{1/3} x$

45. The number 1 is not used as a base for a logarithmic function since the function would look like $x = 1^y$ in exponential form. Then, for any real value of y, the statement $1 = 1$ would always be the result since every power of 1 is equal to 1.

47. The range of $F(x) = a^x$ is the domain of $G(x) = \log_a x$, that is, $\underline{(0, \infty)}$.
The domain of $F(x) = a^x$ is the range of $G(x) = \log_a x$, that is, $\underline{(-\infty, \infty)}$.

49. The values of t are on the horizontal axis, and the values of $f(t)$ are on the vertical axis. Read the value of $f(t)$ from the graph for the given value of t. At $t = 0$, $f(0) = 8$.

51. To find $f(60)$, find 60 on the t-axis, then go up to the graph and across to the $f(t)$ axis to read the value of $f(60)$. At $t = 60$, $f(60) = 24$.

53. $f(x) = 3800 + 585 \log_2 x$

(a) $x = 1982 - 1980 = 2$

$$f(2) = 3800 + 585 \log_2 2$$
$$= 3800 + 585(1)$$
$$= 4385$$

The model gives an approximate withdrawal of 4385 billion ft³ of natural gas from crude oil wells in the United States for 1982.

(b) $x = 1988 - 1980 = 8$

$$f(8) = 3800 + 585 \log_2 8$$
$$= 3800 + 585(3)$$
$$= 5555$$

The model gives an approximate withdrawal of 5555 billion ft³ of natural gas from crude oil wells in the United States for 1988.

(c) $x = 1996 - 1980 = 16$

$$f(16) = 3800 + 585 \log_2 16$$
$$= 3800 + 585(4)$$
$$= 6140$$

The model gives an approximate withdrawal of 6140 billion ft³ of natural gas from crude oil wells in the United States for 1996.

55. $S(t) = 100 + 30 \log_3 (2t + 1)$

(a) $S(1) = 100 + 30 \log_3 (2 \cdot 1 + 1)$
$$= 100 + 30 \log_3 (3)$$
$$= 100 + 30(1) = 130$$

After 1 year, the sales were 130 thousand units.

(b) $S(13) = 100 + 30 \log_3 (2 \cdot 13 + 1)$
$$= 100 + 30 \log_3 (27)$$
$$= 100 + 30(3) = 190$$

After 13 years, the sales were 190 thousand units.

(c) Make a table of values, plot the points they represent, and draw a smooth curve through them.

To make the table, find values of t such that $2t + 1 = 3^k$, where $k = 0, 1, 2, 3, 4$.

k	0	1	2	3	4
3^k	1	3	9	27	81
$2t + 1$	1	3	9	27	81
$2t$	0	2	8	26	80
t	0	1	4	13	40
$S(t)$	100	130	160	190	220

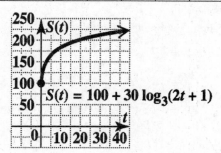

$$S(t) = 100 + 30 \log_3 (2t + 1)$$

57. $R = \log_{10} \dfrac{x}{x_0}$

Change to exponential form.

$$10^R = \frac{x}{x_0}, \text{ so } x = x_0 \, 10^R.$$

Let $R = 6.7$ for the Northridge earthquake, with intensity x_1.

$$x_1 = x_0 10^{6.7}$$

Let $R = 7.3$ for the Landers earthquake, with intensity x_2.

$$x_2 = x_0 10^{7.3}$$

The ratio of x_2 to x_1 is

$$\frac{x_2}{x_1} = \frac{x_0 10^{7.3}}{x_0 10^{6.7}} = 10^{0.6} \approx 3.98.$$

The Landers earthquake was about 4 times more powerful than the Northridge earthquake.

59. $g(x) = \log_3 x$
On a TI-83, assign $3^\wedge x$ to Y_1. Then enter

$$\text{DrawInv } Y_1$$

on the home screen to obtain the figure that follows. (See Exercise 47 in Section 12.1 for TI-83 specifics.)

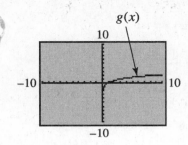

61. $g(x) = \log_{1/3} x$

Assign (1/3)^x to Y_1 and enter DrawInv Y_1.

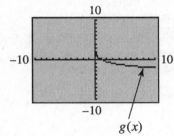

$g(x)$

63. $4^7 \cdot 4^2 = 4^{7+2} = 4^9$

65. $\dfrac{5^{-3}}{5^8} = 5^{-3-8} = 5^{-11} = \dfrac{1}{5^{11}}$

67. $\left(9^3\right)^{-2} = 9^{3(-2)} = 9^{-6} = \dfrac{1}{9^6}$

12.4 Properties of Logarithms

1. By the product rule,

$$\log_{10}(3 \cdot 4) = \log_{10} 3 + \log_{10} 4.$$

3. By a special property (see page 856 in the text),

$$3^{\log_3 4} = 4.$$

5. By a special property (see page 856 in the text),

$$\log_3 3^4 = 4.$$

7. Use the product rule for logarithms.

$$\log_7(4 \cdot 5) = \log_7 4 + \log_7 5$$

9. Use the quotient rule for logarithms.

$$\log_5 \tfrac{8}{3} = \log_5 8 - \log_5 3$$

11. Use the power rule for logarithms.

$$\log_4 6^2 = 2 \log_4 6$$

13. $\log_3 \dfrac{\sqrt[3]{4}}{x^2 y} = \log_3 \dfrac{4^{1/3}}{x^2 y}$

Use the quotient rule for logarithms.

$$= \log_3 4^{1/3} - \log_3 \left(x^2 y\right)$$

Use the product rule for logarithms.

$$= \log_3 4^{1/3} - \left(\log_3 x^2 + \log_3 y\right)$$

$$= \log_3 4^{1/3} - \log_3 x^2 - \log_3 y$$

Use the power rule for logarithms.

$$= \tfrac{1}{3} \log_3 4 - 2 \log_3 x - \log_3 y$$

15. $\log_3 \sqrt{\dfrac{xy}{5}}$

$$= \log_3 \left(\dfrac{xy}{5}\right)^{1/2}$$

$$= \tfrac{1}{2} \log_3 \left(\dfrac{xy}{5}\right) \qquad \textit{Power rule}$$

$$= \tfrac{1}{2} \left[\log_3(xy) - \log_3 5\right] \qquad \textit{Quotient rule}$$

$$= \tfrac{1}{2} \left(\log_3 x + \log_3 y - \log_3 5\right) \qquad \textit{Product rule}$$

$$= \tfrac{1}{2} \log_3 x + \tfrac{1}{2} \log_3 y - \tfrac{1}{2} \log_3 5$$

17. $\log_2 \dfrac{\sqrt[3]{x} \cdot \sqrt[5]{y}}{r^2}$

$$= \log_2 \dfrac{x^{1/3} y^{1/5}}{r^2}$$

$$= \log_2 \left(x^{1/3} y^{1/5}\right) - \log_2 r^2 \qquad \textit{Quotient rule}$$

$$= \log_2 x^{1/3} + \log_2 y^{1/5} - \log_2 r^2 \quad \textit{Product rule}$$

$$= \tfrac{1}{3} \log_2 x + \tfrac{1}{5} \log_2 y - 2 \log_2 r \quad \textit{Power rule}$$

19. The distributive property tells us that the *product* $a(x + y)$ equals the sum $ax + ay$. In the notation $\log_a(x + y)$, the parentheses do not indicate multiplication. They indicate that $x + y$ is the result of raising a to some power.

21. By the product rule for logarithms,

$$\log_b x + \log_b y = \log_b xy.$$

23. By the quotient rule for logarithms,

$$\log_a m - \log_a n = \log_a \dfrac{m}{n}.$$

25. $\left(\log_a r - \log_a s\right) + 3 \log_a t$

Use the quotient and power rules for logarithms.

$$= \log_a \dfrac{r}{s} + \log_a t^3$$

$$= \log_a \dfrac{rt^3}{s} \qquad \textit{Product rule}$$

27. $3 \log_a 5 - 4 \log_a 3$

$$= \log_a 5^3 - \log_a 3^4 \qquad \textit{Power rule}$$

$$= \log_a \dfrac{5^3}{3^4} \qquad \textit{Quotient rule}$$

$$= \log_a \dfrac{125}{81}$$

29. $\log_{10}(x + 3) + \log_{10}(x - 3)$

$$= \log_{10}(x + 3)(x - 3) \qquad \textit{Product rule}$$

$$= \log_{10}\left(x^2 - 9\right)$$

31. By the power rule for logarithms,

$$3\log_p x + \tfrac{1}{2}\log_p y - \tfrac{3}{2}\log_p z - 3\log_p a$$
$$= \log_p x^3 + \log_p y^{1/2} - \log_p z^{3/2} - \log_p a^3$$

Group the terms into sums.

$$= (\log_p x^3 + \log_p y^{1/2}) - (\log_p z^{3/2} + \log_p a^3)$$
$$= \log_p x^3 y^{1/2} - \log_p z^{3/2} a^3 \quad \textit{Product rule}$$
$$= \log_p \frac{x^3 y^{1/2}}{z^{3/2} a^3} \qquad \textit{Quotient rule}$$

In Exercises 33–44, $\log_{10} 2 \approx 0.3010$ and $\log_{10} 9 \approx 0.9542$.

33. By the product rule for logarithms,

$$\log_{10} 18 = \log_{10}(2 \cdot 9)$$
$$= \log_{10} 2 + \log_{10} 9$$
$$\approx 0.3010 + 0.9542$$
$$= 1.2552.$$

35. By the quotient rule for logarithms,

$$\log_{10} \tfrac{2}{9} = \log_{10} 2 - \log_{10} 9$$
$$\approx 0.3010 - 0.9542$$
$$= -0.6532.$$

37. By the product and power rules for logarithms,

$$\log_{10} 36 = \log_{10} 2^2 \cdot 9$$
$$= 2\log_{10} 2 + \log_{10} 9$$
$$\approx 2(0.3010) + 0.9542$$
$$= 1.5562.$$

39.
$$\log_{10} 3 = \log_{10} 9^{1/2} \quad \textit{Rename 3}$$
$$= \tfrac{1}{2}\log_{10} 9 \quad \textit{Power rule}$$
$$\approx \tfrac{1}{2}(0.9542)$$
$$= 0.4771$$

41.
$$\log_{10} \sqrt[4]{9} = \log_{10} 9^{1/4}$$
$$= \tfrac{1}{4}\log_{10} 9 \qquad \textit{Power rule}$$
$$\approx \tfrac{1}{4}(0.9542)$$
$$= 0.23855 \approx 0.2386$$

43.
$$\log_{10} 9^5 = 5\log_{10} 9 \quad \textit{Power rule}$$
$$\approx 5(0.9542)$$
$$= 4.7710$$

45. $\text{LS} = \log_2(8 + 32) = \log_2 40$
$\text{RS} = \log_2 8 + \log_2 32 = \log_2(8 \cdot 32)$
$$= \log_2 256$$

$\text{LS} \neq \text{RS}$, so the statement is *false*.

47. $\log_3 7 + \log_3 7^{-1} = \log_3 7 + (-1)\log_3 7$
$$= 0$$

The statement is *true*.

49. $\log_6 60 - \log_6 10 = \log_6 \frac{60}{10}$
$$= \log_6 6 = 1$$
The statement is *true*.

51.
$$\frac{\log_{10} 7}{\log_{10} 14} \stackrel{?}{=} \frac{1}{2}$$
$$2\log_{10} 7 \stackrel{?}{=} 1\log_{10}(7 \cdot 2)$$
$$\qquad\qquad \textit{Cross products are equal}$$
$$2\log_{10} 7 \stackrel{?}{=} \log_{10} 7 + \log_{10} 2$$
$$\log_{10} 7 \stackrel{?}{=} \log_{10} 2 \quad \textit{Subtract } \log_{10} 7.$$
The statement is *false*.

53. The exponent of a quotient is the difference between the exponent of the numerator and the exponent of the denominator.

55. $\log_2 8 - \log_2 4 = \log_2 \tfrac{8}{4}$
$$= \log_2 2 = 1$$

57. $10^4 = 10,000$ becomes $\log_{10} 10,000 = 4$.

59. $10^{-2} = 0.01$ becomes $\log_{10} 0.01 = -2$.

61. $\log_{10} 1 = 0$ becomes $10^0 = 1$.

12.5 Common and Natural Logarithms

1. Since $\log x = \log_{10} x$, the base is 10. The correct response is **C**.

3. $10^0 = 1$ and $10^1 = 10$, so $\log 1 = 0$ and $\log 10 = 1$. Thus, the value of $\log 5.6$ must lie between 0 and 1. The correct response is **C**.

5. $\log 10^{19.2} = \log_{10} 10^{19.2} = 19.2$ by the special property, $\log_b b^x = x$.

7. To four decimal places,
$$\log 43 \approx 1.6335.$$

9. $\log 328.4 \approx 2.5164$

11. $\log 0.0326 \approx -1.4868$

13. $\log(4.76 \times 10^9) \approx 9.6776$
On a TI-83, enter

$$\boxed{\text{LOG}}\ 4.76\ \boxed{\text{2nd}}\ \boxed{\text{EE}}\ 9\ ).$$

15. $\ln 7.84 \approx 2.0592$

17. $\ln 0.0556 \approx -2.8896$

19. $\ln 388.1 \approx 5.9613$

21. $\ln(8.59 \times e^2) \approx 4.1506$
On a TI-83, enter

$$\boxed{\text{LN}}\ 8.59\ \boxed{\text{X}}\ \boxed{\text{2nd}}\ \boxed{e^x}\ 2\ )\ ).$$

23. $\ln 10 \approx 2.3026$

25. (a) $\log 356.8 \approx 2.552\,424\,846$

(b) $\log 35.68 \approx 1.552\,424\,846$

(c) $\log 3.568 \approx 0.552\,424\,846$

(d) The whole number part of the answers (2, 1, or 0) varies, whereas the decimal part (0.552 424 846) remains the same, indicating that the whole number part corresponds to the placement of the decimal point and the decimal part corresponds to the digits 3, 5, 6, and 8.

27. When you try to find $\log(-1)$ on a calculator, an error message is displayed. This is because the domain of the logarithmic function is $(0, \infty)$; -1 is not in the domain.

29. $\text{pH} = -\log[\text{H}_3\text{O}^+]$
$= -\log(2.5 \times 10^{-2}) \approx 1.6$
Since the pH is less than 3.0, the wetland is classified as a *bog*.

31. Ammonia has a hydronium ion concentration of 2.5×10^{-12}.
$\text{pH} = -\log[\text{H}_3\text{O}^+]$
$\text{pH} = -\log(2.5 \times 10^{-12}) \approx 11.6$

33. Grapes have a hydronium ion concentration of 5.0×10^{-5}.
$\text{pH} = -\log[\text{H}_3\text{O}^+]$
$\text{pH} = -\log(5.0 \times 10^{-5}) \approx 4.3$

35. Human blood plasma has a pH of 7.4.
$$\text{pH} = -\log[\text{H}_3\text{O}^+]$$
$$7.4 = -\log[\text{H}_3\text{O}^+]$$
$$\log_{10}[\text{H}_3\text{O}^+] = -7.4$$
$$[\text{H}_3\text{O}^+] = 10^{-7.4} \approx 4.0 \times 10^{-8}$$

37. Spinach has a pH value of 5.4.
$$\text{pH} = -\log[\text{H}_3\text{O}^+]$$
$$5.4 = -\log[\text{H}_3\text{O}^+]$$
$$\log_{10}[\text{H}_3\text{O}^+] = -5.4$$
$$[\text{H}_3\text{O}^+] = 10^{-5.4} \approx 4.0 \times 10^{-6}$$

39. $D = 10\log\left(\dfrac{I}{I_0}\right)$

(a) $D = 10\log\left(\dfrac{5.012 \times 10^{10} I_0}{I_0}\right)$
$= 10\log(5.012 \times 10^{10}) \approx 107$

The average decibel level for *Spider-Man 2* is about 107 dB.

(b) $D = 10\log\left(\dfrac{10^{10} I_0}{I_0}\right)$
$= 10\log 10^{10} = 10 \cdot 10 = 100$

The average decibel level for *Finding Nemo* is 100 dB.

(c) $D = 10\log\left(\dfrac{6{,}310{,}000{,}000\, I_0}{I_0}\right)$
$= 10\log(6{,}310{,}000{,}000) \approx 98$

The average decibel level for *Saving Private Ryan* is about 98 dB.

41. $N(r) = -5000\ln r$

(a) 85% (or 0.85)

$N(0.85) = -5000\ln 0.85 \approx 813 \approx 800$

The number of years since the split for 85% is about 800 years.

(b) 35% (or 0.35)

$N(0.35) = -5000\ln 0.35 \approx 5249 \approx 5200$

The number of years since the split for 35% is about 5200 years.

(c) 10% (or 0.10)

$N(0.10) = -5000\ln 0.10 \approx 11{,}513 \approx 11{,}500$

The number of years since the split for 10% is about 11,500 years.

43. $f(x) = -1317 + 304\ln x$

(a) $x = 1998 - 1900 = 98$

$f(98) = -1317 + 304\ln(98) \approx 77$

The prediction for 1998 is 77%.

(b) $f(x) = 50$
$50 = -1317 + 304\ln x$
$1367 = 304\ln x$
$\ln x = \frac{1367}{304}$
$e^{\ln x} = e^{1367/304}$
$x \approx 89.72$

The outpatient surgeries reached 50% in 1989.

45. $T = -0.642 - 189\ln(1-p)$

(a) $T = -0.642 - 189\ln(1 - 0.25)$
$= -0.642 - 189\ln(0.75) \approx 53.73$

About $54 per ton will reduce emissions by 25%.

(b) If $p = 0$, then $\ln(1-p) = \ln 1 = 0$, so T would be negative. If $p = 1$, then $\ln(1-p) = \ln 0$, but the domain of $\ln x$ is $(0, \infty)$.

47. The change-of-base rule is

$$\log_a x = \frac{\log_b x}{\log_b a}.$$

Use common logarithms ($b = 10$).

$$\log_3 12 = \frac{\log_{10} 12}{\log_{10} 3} = \frac{\log 12}{\log 3} \approx 2.2619$$

49. Use natural logarithms ($b = e$).

$$\log_5 3 = \frac{\log_e 3}{\log_e 5} = \frac{\ln 3}{\ln 5} \approx 0.6826$$

51. $\log_3 \sqrt{2} = \dfrac{\ln \sqrt{2}}{\ln 3} \approx 0.3155$

53. $\log_\pi e = \dfrac{\ln e}{\ln \pi} = \dfrac{1}{\ln \pi} \approx 0.8736$

55. $\log_e 12 = \dfrac{\ln 12}{\ln e} = \dfrac{\ln 12}{1} \approx 2.4849$

57. Let $m =$ the number of letters in your first name and $n =$ the number of letters in your last name.

Answers will vary, but suppose the name is Paul Bunyan, with $m = 4$ and $n = 6$.

(a) $\log_m n = \log_4 6$ is the exponent to which 4 must be raised in order to obtain 6.

(b) Use the change-of-base rule.

$$\log_4 6 = \frac{\log 6}{\log 4}$$
$$\approx 1.292\,481\,25$$

(c) Here, $m = 4$. Use the power key (y^x, x^y, $\wedge$) on your calculator.

$$4^{1.292\,481\,25} \approx 6$$

The result is 6, the value of n.

59. $f(x) = 3800 + 585 \log_2 x$

$x = 2003 - 1980 = 23$

$$f(23) = 3800 + 585 \log_2 23$$
$$= 3800 + 585\left(\frac{\log 23}{\log 2}\right)$$
$$\approx 6446$$

The model gives an approximate withdrawal of 6446 billion ft^3 of natural gas from crude oil wells in the United States for 2003.

61. To graph $g(x) = \log_3 x$, assign either

$\dfrac{\log X}{\log 3}$ or $\dfrac{\ln X}{\ln 3}$ to Y_1.

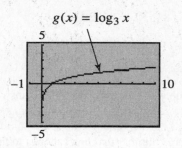

$g(x) = \log_3 x$

63. To graph $g(x) = \log_{1/3} x$, assign either

$\dfrac{\log X}{\log \frac{1}{3}}$ or $\dfrac{\ln X}{\ln \frac{1}{3}}$ to Y_1.

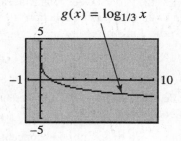

$g(x) = \log_{1/3} x$

65.
$$4^{2x} = 8^{3x+1}$$
$$\left(2^2\right)^{2x} = \left(2^3\right)^{3x+1}$$
$$2^{4x} = 2^{9x+3}$$
$$4x = 9x + 3$$
$$-3 = 5x$$
$$x = -\tfrac{3}{5}$$

The solution set is $\left\{-\tfrac{3}{5}\right\}$.

67. $\log_3 (x + 4) = 2$
$$x + 4 = 3^2$$
$$x + 4 = 9$$
$$x = 5$$

The solution set is $\{5\}$.

69. $\log_{1/2} 8 = x$
$$\left(\tfrac{1}{2}\right)^x = 8$$
$$\left(2^{-1}\right)^x = 2^3$$
$$2^{-x} = 2^3$$
$$-x = 3$$
$$x = -3$$

The solution set is $\{-3\}$.

71. $\log (x + 2) + \log (x - 3)$
$\quad = \log (x + 2)(x - 3), \quad \text{or} \quad \log (x^2 - x - 6)$

12.6 Exponential and Logarithmic Equations; Further Applications

1.
$$5^x = 125$$
$$\log 5^x = \log 125$$

3.
$$\frac{x \log 5}{\log 5} = \frac{\log 125}{\log 5}$$
$$x = \frac{\log 125}{\log 5}$$

5.
$$7^x = 5$$
Take the logarithm of each side.
$$\log 7^x = \log 5$$
Use the power rule for logarithms.
$$x \log 7 = \log 5$$
$$x = \frac{\log 5}{\log 7} \approx 0.827$$
The solution set is $\{0.827\}$.

7.
$$9^{-x+2} = 13$$
$$\log 9^{-x+2} = \log 13$$
$$(-x + 2) \log 9 = \log 13 \; (*)$$
$$-x \log 9 + 2 \log 9 = \log 13$$
$$-x \log 9 = \log 13 - 2 \log 9$$
$$x \log 9 = 2 \log 9 - \log 13$$
$$x = \frac{2 \log 9 - \log 13}{\log 9}$$
$$\approx 0.833$$

$(*)$ Alternative solution steps:
$$(-x + 2) \log 9 = \log 13$$
$$-x + 2 = \frac{\log 13}{\log 9}$$
$$2 - \frac{\log 13}{\log 9} = x$$

The solution set is $\{0.833\}$.

9.
$$3^{2x} = 14$$
$$\log 3^{2x} = \log 14$$
$$2x \log 3 = \log 14$$
$$x = \frac{\log 14}{2 \log 3} \approx 1.201$$
The solution set is $\{1.201\}$.

11.
$$2^{x+3} = 5^x$$
$$\log 2^{x+3} = \log 5^x$$
$$(x + 3) \log 2 = x \log 5$$
$$x \log 2 + 3 \log 2 = x \log 5$$
Get x-terms on one side.
$$x \log 2 - x \log 5 = -3 \log 2$$
$$x (\log 2 - \log 5) = -3 \log 2 \quad \textit{Factor out x.}$$

$$x = \frac{-3 \log 2}{\log 2 - \log 5}$$
$$\approx 2.269$$
The solution set is $\{2.269\}$.

13.
$$2^{x+3} = 3^{x-4}$$
$$\log 2^{x+3} = \log 3^{x-4}$$
$$(x + 3) \log 2 = (x - 4) \log 3$$
$$x \log 2 + 3 \log 2 = x \log 3 - 4 \log 3$$
$$\textit{Distributive property}$$
Get x-terms on one side.
$$x \log 2 - x \log 3 = -3 \log 2 - 4 \log 3$$
Factor out x.
$$x (\log 2 - \log 3) = -3 \log 2 - 4 \log 3$$
$$x = \frac{-3 \log 2 - 4 \log 3}{\log 2 - \log 3}$$
$$\approx 15.967$$

The solution set is $\{15.967\}$.

15.
$$e^{0.012x} = 23$$
$$\ln e^{0.012x} = \ln 23$$
$$0.012x (\ln e) = \ln 23$$
$$0.012x = \ln 23 \quad \textit{ln e = 1}$$
$$x = \frac{\ln 23}{0.012} \approx 261.291$$
The solution set is $\{261.291\}$.

17.
$$e^{-0.205x} = 9$$
$$\ln e^{-0.205x} = \ln 9$$
$$-0.205x (\ln e) = \ln 9$$
$$-0.205x = \ln 9 \quad \textit{ln e = 1}$$
$$x = \frac{\ln 9}{-0.205} \approx -10.718$$
The solution set is $\{-10.718\}$.

19.
$$\ln e^{3x} = 9$$
$$3x = 9$$
$$x = 3$$

The solution set is $\{3\}$.

21.
$$\ln e^{0.45x} = \sqrt{7}$$
$$0.45x = \sqrt{7}$$
$$x = \frac{\sqrt{7}}{0.45} \approx 5.879$$

The solution set is $\{5.879\}$.

23. Let's try Exercise 14.
$$e^{0.006x} = 30$$
$$\log e^{0.006x} = \log 30$$
$$0.006x (\log e) = \log 30$$
$$x = \frac{\log 30}{0.006 \log e} \approx 566.866$$

The natural logarithm is a better choice because $\ln e = 1$, whereas $\log e$ needs to be calculated.

25. $\log_3 (6x + 5) = 2$

$\qquad 6x + 5 = 3^2 \qquad$ *Exponential form*

$\qquad 6x + 5 = 9$

$\qquad 6x = 4$

$\qquad x = \frac{4}{6} = \frac{2}{3}$

Check $x = \frac{2}{3}$: $\log_3 9 = \log_3 3^2 = 2$

The solution set is $\left\{\frac{2}{3}\right\}$.

27. $\log_2 (2x - 1) = 5$

$\qquad 2x - 1 = 2^5 \quad$ *Exponential form*

$\qquad 2x - 1 = 32$

$\qquad 2x = 33$

$\qquad x = \frac{33}{2}$

Check $x = \frac{33}{2}$: $\log_2 32 = \log_2 2^5 = 5$

The solution set is $\left\{\frac{33}{2}\right\}$.

29. $\log_7 (x + 1)^3 = 2$

$\qquad (x + 1)^3 = 7^2 \qquad$ *Exponential form*

$\qquad x + 1 = \sqrt[3]{49} \qquad$ *Cube root*

$\qquad x = -1 + \sqrt[3]{49}$

Check $x = -1 + \sqrt[3]{49}$: $\log_7 49 = \log_7 7^2 = 2$

The solution set is $\left\{-1 + \sqrt[3]{49}\right\}$.

31. 2 cannot be a solution because
$\log (2 - 3) = \log (-1)$, and -1 is not in the domain of $\log x$.

33. $\log (6x + 1) = \log 3$

$\qquad 6x + 1 = 3 \qquad$ *Property 4*

$\qquad 6x = 2$

$\qquad x = \frac{2}{6} = \frac{1}{3}$

Check $x = \frac{1}{3}$: $\log (2 + 1) = \log 3 \quad$ *True*

The solution set is $\left\{\frac{1}{3}\right\}$.

35. $\log_5 (3t + 2) - \log_5 t = \log_5 4$

$\qquad \log_5 \dfrac{3t + 2}{t} = \log_5 4$

$\qquad \dfrac{3t + 2}{t} = 4$

$\qquad 3t + 2 = 4t$

$\qquad 2 = t$

Check $t = 2$: $\log_5 8 - \log_5 2 = \log_5 \frac{8}{2} = \log_5 4$

The solution set is $\{2\}$.

37. $\log 4x - \log (x - 3) = \log 2$

$\qquad \log \dfrac{4x}{x - 3} = \log 2$

$\qquad \dfrac{4x}{x - 3} = 2$

$\qquad 4x = 2(x - 3)$

$\qquad 4x = 2x - 6$

$\qquad 2x = -6$

$\qquad x = -3$

Reject $x = -3$, because $4x = -12$, which yields an equation in which the logarithm of a negative number must be found.

The solution set is $\emptyset$.

39. $\log_2 x + \log_2 (x - 7) = 3$

$\qquad \log_2 [x(x - 7)] = 3$

$\qquad x(x - 7) = 2^3 \quad$ *Exponential form*

$\qquad x^2 - 7x = 8$

$\qquad x^2 - 7x - 8 = 0$

$\qquad (x - 8)(x + 1) = 0$

$\qquad x - 8 = 0 \quad$ or $\quad x + 1 = 0$

$\qquad x = 8 \quad$ or $\qquad x = -1$

Reject $x = -1$, because it yields an equation in which the logarithm of a negative number must be found.

Check $x = 8$: $\log_2 8 + \log_2 1 = \log_2 2^3 + 0 = 3$

The solution set is $\{8\}$.

41. $\log 5x - \log (2x - 1) = \log 4$

$\qquad \log \dfrac{5x}{2x - 1} = \log 4$

$\qquad \dfrac{5x}{2x - 1} = 4$

$\qquad 5x = 8x - 4$

$\qquad 4 = 3x$

$\qquad \frac{4}{3} = x$

Check $x = \frac{4}{3}$: $\log \frac{20}{3} - \log \frac{5}{3} = \log \frac{20/3}{5/3} = \log 4$

The solution set is $\left\{\frac{4}{3}\right\}$.

43. $\log_2 x + \log_2 (x - 6) = 4$

$\qquad \log_2 [x(x - 6)] = 4$

$\qquad x(x - 6) = 2^4 \quad$ *Exponential form*

$\qquad x^2 - 6x = 16$

$\qquad x^2 - 6x - 16 = 0$

$\qquad (x - 8)(x + 2) = 0$

$\qquad x - 8 = 0 \quad$ or $\quad x + 2 = 0$

$\qquad x = 8 \quad$ or $\qquad x = -2$

Reject $x = -2$, because it yields an equation in which the logarithm of a negative number must be found.

Check $x = 8$:

$\qquad \log_2 8 + \log_2 2 = \log_2 16 = \log_2 2^4 = 4$

The solution set is $\{8\}$.

45. (a) Use the formula $A = P\left(1 + \dfrac{r}{n}\right)^{nt}$ with $P = 2000$, $r = 0.04$, $n = 4$, and $t = 6$.

$$A = 2000\left(1 + \tfrac{0.04}{4}\right)^{4 \cdot 6}$$
$$= 2000(1.01)^{24} \approx 2539.47$$

The account will contain $2539.47.

(b)
$$3000 = 2000\left(1 + \tfrac{0.04}{4}\right)^{4t}$$
$$\frac{3000}{2000} = (1.01)^{4t}$$
$$\log\left(\tfrac{3}{2}\right) = \log(1.01)^{4t}$$
$$\log\left(\tfrac{3}{2}\right) = 4t\log(1.01)$$
$$t = \frac{\log\left(\tfrac{3}{2}\right)}{4\log(1.01)} \approx 10.2$$

It will take about 10.2 years for the account to grow to $3000.

47. (a) Use the formula $A = Pe^{rt}$ with $P = 4000$, $r = 0.035$, and $t = 6$.

$$A = 4000e^{(0.035)(6)}$$
$$= 4000e^{0.21} \approx 4934.71$$

There will be $4934.71 in the account.

(b) If the initial amount doubles, then $A = 2P$, or $8000.

$$8000 = 4000e^{0.035t}$$
$$2 = e^{0.035t} \qquad \textit{Divide by 4000.}$$
$$\ln 2 = \ln e^{0.035t}$$
$$\ln 2 = 0.035t$$
$$t = \frac{\ln 2}{0.035} \approx 19.8$$

The initial amount will double in about 19.8 years.

49. Use $A = P\left(1 + \dfrac{r}{n}\right)^{nt}$ with $P = 5000$, $r = 0.07$, and $t = 12$.

(a) If the interest is compounded annually, $n = 1$.

$$A = 5000\left(1 + \tfrac{0.07}{1}\right)^{1 \cdot 12}$$
$$= 5000(1.07)^{12} \approx 11{,}260.96$$

There will be $11,260.96 in the account.

(b) If the interest is compounded semiannually, $n = 2$.

$$A = 5000\left(1 + \tfrac{0.07}{2}\right)^{2 \cdot 12}$$
$$= 5000(1.035)^{24} \approx 11{,}416.64$$

There will be $11,416.64 in the account.

(c) If the interest is compounded quarterly, $n = 4$.

$$A = 5000\left(1 + \tfrac{0.07}{4}\right)^{4 \cdot 12}$$
$$= 5000(1.0175)^{48} \approx 11{,}497.99$$

There will be $11,497.99 in the account.

(d) If the interest is compounded daily, $n = 365$.

$$A = 5000\left(1 + \tfrac{0.07}{365}\right)^{365 \cdot 12}$$
$$\approx 11{,}580.90$$

There will be $11,580.90 in the account.

(e) Use the continuous compound interest formula.

$$A = Pe^{rt}$$
$$A = 5000e^{0.07(12)}$$
$$= 5000e^{0.84} \approx 11{,}581.83$$

There will be $11,581.83 in the account.

51. In the continuous compound interest formula, let $A = 1850$, $r = 0.065$, and $t = 40$.

$$A = Pe^{rt}$$
$$1850 = Pe^{0.065(40)}$$
$$1850 = Pe^{2.6}$$
$$P = \frac{1850}{e^{2.6}} \approx 137.41$$

Deposit $137.41 today.

53. $f(x) = 15.80e^{0.0708x}$

(a) 1980 corresponds to $x = 0$.

$$f(0) = 15.80e^{0.0708(0)}$$
$$= 15.80(1) = 15.8$$

The approximate volume of materials recovered from municipal solid waste collections in the United States in 1980 was 15.8 million tons.

(b) 1985 corresponds to $x = 5$.

$$f(5) \approx 22.5 \text{ million tons}$$

(c) 1995 corresponds to $x = 15$.

$$f(15) \approx 45.7 \text{ million tons}$$

(d) 2003 corresponds to $x = 23$.

$$f(23) \approx 80.5 \text{ million tons}$$

55. $B(x) = 27{,}190e^{0.0448x}$
2004 corresponds to $x = 9$.
$B(9) = 27{,}190e^{0.0448(9)}$
$$\approx 40{,}693$$
The approximate value of consumer expenditures for 2004 was 40,693 million dollars.

57. $A(t) = 2.00e^{-0.053t}$

(a) Let $t = 4$.

$$A(4) = 2.00e^{-0.053(4)}$$
$$= 2.00e^{-0.212}$$
$$\approx 1.62$$

About 1.62 grams would be present.

(b) $A(10) = 2.00e^{-0.053(10)}$

≈ 1.18

About 1.18 grams would be present.

(c) $A(20) = 2.00e^{-0.053(20)}$

≈ 0.69

About 0.69 grams would be present.

(d) The initial amount is the amount $A(t)$ present at time $t = 0$.

$A(0) = 2.00e^{-0.053(0)}$

$= 2.00e^0 = 2.00(1) = 2.00$

Initially, 2.00 grams were present.

59. (a) Find $A(t) = 400e^{-0.032t}$ when $t = 25$.

$A(25) = 400e^{-0.032(25)}$

$= 400e^{-0.8} \approx 179.73$

About 179.73 grams of lead will be left.

(b) Use $A(t) = 400e^{-0.032t}$, with $A(t) = \frac{1}{2}(400) = 200$.

$200 = 400e^{-0.032t}$

$\frac{200}{400} = e^{-0.032t}$

$0.5 = e^{-0.032t}$

$\ln 0.5 = \ln e^{-0.032t}$

$\ln 0.5 = -0.032t(\ln e)$

$t = \frac{\ln 0.5}{-0.032} \approx 21.66$

It would take about 21.66 years for the sample to decay to half its original amount.

61.
$f(x) = 15.80e^{0.0708x}$

$100 = 15.8e^{0.0708x}$

$e^{0.0708x} = \frac{100}{15.8}$

$\ln e^{0.0708x} = \ln \frac{100}{15.8}$

$0.0708x \, (\ln e) = \ln \frac{100}{15.8}$

$0.0708x = \ln \frac{100}{15.8}$

$x = \frac{\ln \frac{100}{15.8}}{0.0708} \approx 26$

Since $x = 0$ corresponds to 1980, $x = 26$ corresponds to 2006.

63. $f(t) = 300e^{0.4t}$

The original number of ants, at $t = 0$, is 300. Replace $f(t)$ with 600, (double the number of ants), and solve for t.

$600 = 300e^{0.4t}$

$2 = e^{0.4t}$

$\ln 2 = \ln e^{0.4t}$

$\ln 2 = 0.4t \, (\ln e)$

$\ln 2 = 0.4t$

$t = \frac{\ln 2}{0.4} \approx 1.733$

It will take about 1.733 days for the number of ants to double.

65. (a) The expression $\frac{1}{x}$ in the base cannot be evaluated since division by 0 is not defined.

(b) When $X = 100,000$, $Y_1 \approx 2.7183$. The decimal approximation appears to be close to the decimal approximation for e, $2.71828\ldots$.

(c) $\left(1 + \frac{1}{1,000,000}\right)^{1,000,000} \approx 2.718280469$

$e = e^1 \approx 2.718281828$

The two values differ in the sixth decimal place.

(d) As the values of x approach infinity, the value of $\left(1 + \frac{1}{x}\right)^x$ approaches $\underline{e}$.

67. $f(x) = 2x^2$

The graph of f is narrower than the graph of $y = x^2$.

x	0	± 1	± 2
y	0	2	8

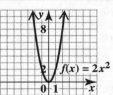

69. $f(x) = (x + 1)^2$

The graph of f is the graph of $y = x^2$ shifted *left* 1 unit.

x	-3	-2	-1	0	1
y	4	1	0	1	4

Chapter 12 Review Exercises

1. Since a horizontal line intersects the graph in two points, the function is not one-to-one.

2. Since every horizontal line intersects the graph in no more than one point, the function is one-to-one.

3. This function is not one-to-one because two sodas in the list have 41 mg of caffeine.

4. The function $f(x) = -3x + 7$ is a linear function. By the horizontal line test, it is a one-to-one function. To find the inverse, replace $f(x)$ with y.

$$y = -3x + 7$$

Interchange x and y.

$$x = -3y + 7$$

Solve for y.

$$x - 7 = -3y$$

$$\frac{x - 7}{-3} = y \quad \text{or} \quad \frac{7 - x}{3} = y$$

Replace y with $f^{-1}(x)$.

$$f^{-1}(x) = \frac{x - 7}{-3}, \quad \text{or} \quad f^{-1}(x) = -\frac{1}{3}x + \frac{7}{3}$$

5. $f(x) = \sqrt[3]{6x - 4}$

The function is one-to-one since each $f(x)$-value corresponds to exactly one x-value.

To find the inverse, replace $f(x)$ with y.

$$y = \sqrt[3]{6x - 4}$$

Cube both sides.

$$y^3 = 6x - 4$$

Interchange x and y.

$$x^3 = 6y - 4$$

Solve for y.

$$x^3 + 4 = 6y$$

$$\frac{x^3 + 4}{6} = y$$

Replace y with $f^{-1}(x)$

$$\frac{x^3 + 4}{6} = f^{-1}(x)$$

6. $f(x) = -x^2 + 3$

This is an equation of a vertical parabola which opens down.

Since a horizontal line will intersect the graph in two points, the function is not one-to-one.

7. The graph is a linear function through $(0, 1)$ and $(3, 0)$. The graph of $f^{-1}(x)$ will include the points $(1, 0)$ and $(0, 3)$, found by interchanging x and y. Plot these points, and draw a straight line through them.

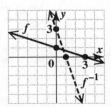

8. The graph is a curve through $(1, 2)$, $(0, 1)$, and $(-1, \frac{1}{2})$. Interchange x and y to get $(2, 1)$, $(1, 0)$, and $(\frac{1}{2}, -1)$, which are on the graph of $f^{-1}(x)$. Plot these points, and draw a smooth curve through them.

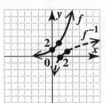

9. $f(x) = 3^x$

Make a table of values.

x	-2	-1	0	1	2
$f(x)$	$\frac{1}{9}$	$\frac{1}{3}$	1	3	9

Plot the points from the table and draw a smooth curve through them.

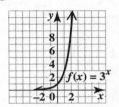

10. $f(x) = \left(\frac{1}{3}\right)^x$

Make a table of values.

x	-2	-1	0	1	2
$f(x)$	9	3	1	$\frac{1}{3}$	$\frac{1}{9}$

Plot the points from the table and draw a smooth curve through them.

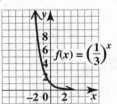

11. $y = 3^{x+1}$

Make a table of values.

x	-3	-2	-1	0	1
y	$\frac{1}{9}$	$\frac{1}{3}$	1	3	9

Plot the points from the table and draw a smooth curve through them.

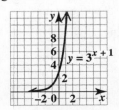

12. $y = 2^{2x+3}$

Make a table of values.

x	-2	$-\frac{3}{2}$	-1	0	$\frac{1}{2}$
y	$\frac{1}{2}$	1	2	8	16

Plot the points from the table and draw a smooth curve through them.

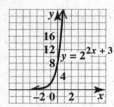

13. $4^{3x} = 8^{x+4}$

Write each side as a power of 2.

$\left(2^2\right)^{3x} = \left(2^3\right)^{x+4}$

$2^{6x} = 2^{3x+12}$

$6x = 3x + 12$ *Equate exponents.*

$3x = 12$

$x = 4$

Check $x = 4$: $4^{12} = 8^8$ *True*

The solution set is $\{4\}$.

14. $\left(\frac{1}{27}\right)^{x-1} = 9^{2x}$

$\left[\left(\frac{1}{3}\right)^3\right]^{x-1} = \left(3^2\right)^{2x}$

Write each side as a power of 3.

$\left(3^{-3}\right)^{x-1} = \left(3^2\right)^{2x}$

$3^{-3x+3} = 3^{4x}$

$-3x + 3 = 4x$ *Equate exponents.*

$3 = 7x$

$\frac{3}{7} = x$

Check $x = \frac{3}{7}$: $\left(\frac{1}{27}\right)^{-4/7} = 9^{6/7}$ *True*

The solution set is $\left\{\frac{3}{7}\right\}$.

15. $W(x) = 7.77(1.059)^x$

(a) $x = 1985 - 1980 = 5$

$W(5) = 7.77(1.059)^5$

≈ 10.3

The approximate amount of plastic waste in 1985 was 10.3 million tons.

(b) $x = 1995 - 1980 = 15$

$W(15) \approx 18.4$ million tons

(c) $x = 2000 - 1980 = 20$

$W(20) \approx 24.5$ million tons

16. $g(x) = \log_3 x$

Replace $g(x)$ with y, and write in exponential form.

$y = \log_3 x$

$3^y = x$

Make a table of values. Since $x = 3^y$ is the inverse of $f(x) = y = 3^x$ in Exercise 9, simply reverse the ordered pairs in the table of values belonging to $f(x) = 3^x$.

x	$\frac{1}{9}$	$\frac{1}{3}$	1	3	9
y	-2	-1	0	1	2

Plot the points from the table and draw a smooth curve through them.

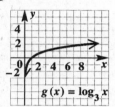

17. $g(x) = \log_{1/3} x$

Replace $g(x)$ with y, and write in exponential form.

$y = \log_{1/3} x$

$\left(\frac{1}{3}\right)^y = x$

Make a table of values. Since $x = \left(\frac{1}{3}\right)^y$ is the inverse of $f(x) = y = \left(\frac{1}{3}\right)^x$ in Exercise 10, simply reverse the ordered pairs in the table of values belonging to $f(x) = \left(\frac{1}{3}\right)^x$.

x	9	3	1	$\frac{1}{3}$	$\frac{1}{9}$
y	-2	-1	0	1	2

Plot the points from the table and draw a smooth curve through them.

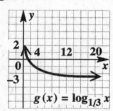

18. $\log_8 64 = x$

$8^x = 64$ *Exponential form*

Write each side as a power of 8.

$8^x = 8^2$

$x = 2$ *Equate exponents.*

The solution set is $\{2\}$.

19. $\log_2 \sqrt{8} = x$

$2^x = \sqrt{8}$ *Exponential form*

$2^x = 8^{1/2}$

Write each side as a power of 2.

$2^x = \left(2^3\right)^{1/2}$

$2^x = 2^{3/2}$

$x = \frac{3}{2}$ *Equate exponents.*

The solution set is $\left\{\frac{3}{2}\right\}$.

20. $\log_x \left(\dfrac{1}{49}\right) = -2$

$x^{-2} = \dfrac{1}{49}$ *Exponential form*

$\dfrac{1}{x^2} = \dfrac{1}{49}$

$x^2 = 49$

$x = \pm 7$

Since x is the base, we cannot have a negative number.
The solution set is $\{7\}$.

21. $\log_4 x = \frac{3}{2}$

$x = 4^{3/2}$ *Exponential form*

$x = \left(\sqrt{4}\right)^3 = 2^3 = 8$

The solution set is $\{8\}$.

22. $\log_k 4 = 1$

$k^1 = 4$ *Exponential form*

$k = 4$

The solution set is $\{4\}$.

23. $\log_b b^2 = 2$

$b^2 = b^2$ *Exponential form*

This is an identity. Thus, b can be any real number, $b > 0$ and $b \neq 1$.
The solution set is $\{b \mid b > 0,\ b \neq 1\}$.

24. $\log_b a$ is the exponent to which b must be raised to obtain a.

25.   From Exercise 24,

$$b^{\log_b a} = a.$$

26. $S(x) = 100 \log_2 (x + 2)$

(a) When $x = 6$,

$$S(6) = 100 \log_2 (6 + 2)$$
$$= 100(3) = 300.$$

After 6 weeks the sales were 300 thousand dollars or $300,000.

(b) To graph the function, make a table of values that includes the ordered pair from above.

x	0	2	6
$S(x)$	100	200	300

Plot the ordered pairs and draw the graph through them.

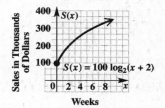

27. $\log_2 3xy^2$

$= \log_2 3 + \log_2 x + \log_2 y^2$ *Product rule*

$= \log_2 3 + \log_2 x + 2\log_2 y$ *Power rule*

28. $\log_4 \dfrac{\sqrt{x} \cdot w^2}{z}$

$= \log_4 \left(\sqrt{x} \cdot w^2\right) - \log_4 z$ *Quotient rule*

$= \log_4 x^{1/2} + \log_4 w^2 - \log_4 z$ *Product rule*

$= \frac{1}{2} \log_4 x + 2\log_4 w - \log_4 z$ *Power rule*

29. $\log_b 3 + \log_b x - 2\log_b y$

Use the product and power rules for logarithms.

$= \log_b (3 \cdot x) - \log_b y^2$

$= \log_b \dfrac{3x}{y^2}$ *Quotient rule*

30. $\log_3 (x + 7) - \log_3 (4x + 6)$

$= \log_3 \left(\dfrac{x + 7}{4x + 6}\right)$ *Quotient rule*

31. $\log 28.9 \approx 1.4609$

32. $\log 0.257 \approx -0.5901$

33. $\ln 28.9 \approx 3.3638$

34. $\ln 0.257 \approx -1.3587$

35. $\log_{16} 13 = \dfrac{\log 13}{\log 16} \approx 0.9251$

36. $\log_4 12 = \dfrac{\log 12}{\log 4} \approx 1.7925$

37.   Milk has a hydronium ion concentration of 4.0×10^{-7}.

$pH = -\log [H_3O^+]$

$pH = -\log \left(4.0 \times 10^{-7}\right) \approx 6.4$

38. Crackers have a hydronium ion concentration of 3.8×10^{-9}.

$$\text{pH} = -\log\left[\text{H}_3\text{O}^+\right]$$
$$\text{pH} = -\log\left(3.8 \times 10^{-9}\right) \approx 8.4$$

39. Orange juice has a pH of 4.6.

$$\text{pH} = -\log\left[\text{H}_3\text{O}^+\right]$$
$$4.6 = -\log\left[\text{H}_3\text{O}^+\right]$$
$$\log_{10}\left[\text{H}_3\text{O}^+\right] = -4.6$$
$$\left[\text{H}_3\text{O}^+\right] = 10^{-4.6} \approx 2.5 \times 10^{-5}$$

40. $Q(t) = 500e^{-0.05t}$

(a) Let $t = 0$.

$$Q(0) = 500e^{-0.05(0)}$$
$$= 500e^0 = 500(1) = 500$$

There are 500 grams.

(b) Let $t = 4$.

$$Q(4) = 500e^{-0.05(4)}$$
$$= 500e^{-0.2} \approx 409.4$$

There will be about 409 grams in 4 days.

41. $t(r) = \dfrac{\ln 2}{\ln(1 + r)}$

(a) $4\% = 0.04$; $t(0.04) = \dfrac{\ln 2}{\ln(1 + 0.04)} \approx 18$

At 4%, it would take about 18 years.

(b) $6\% = 0.06$; $t(0.06) = \dfrac{\ln 2}{\ln(1 + 0.06)} \approx 12$

At 6%, it would take about 12 years.

(c) $10\% = 0.10$; $t(0.10) = \dfrac{\ln 2}{\ln(1 + 0.10)} \approx 7$

At 10%, it would take about 7 years.

(d) $12\% = 0.12$; $t(0.12) = \dfrac{\ln 2}{\ln(1 + 0.12)} \approx 6$

At 12%, it would take about 6 years.

(e) Each comparison shows approximately the same number. For example, in part (a) the doubling time is 18 yr (rounded) and $\frac{72}{4} = 18$.

Thus, the formula $t = \dfrac{72}{100r}$ (called the *rule of 72*) is an excellent approximation of the doubling time formula. (It is used by bankers for that purpose.)

42.
$$3^x = 9.42$$
$$\log 3^x = \log 9.42$$
$$x \log 3 = \log 9.42$$
$$x = \frac{\log 9.42}{\log 3} \approx 2.042$$

Check $x = 2.042$: $3^{2.042} \approx 9.425$

The solution set is $\{2.042\}$.

43.
$$2^{x-1} = 15$$
$$\log 2^{x-1} = \log 15$$
$$(x - 1)\log 2 = \log 15$$
$$x - 1 = \frac{\log 15}{\log 2}$$
$$x = \frac{\log 15}{\log 2} + 1 \approx 4.907$$

Check $x = 4.907$: $2^{4.907} \approx 15.0$

The solution set is $\{4.907\}$.

44.
$$e^{0.06x} = 3$$

Take base e logarithms on both sides.

$$\ln e^{0.06x} = \ln 3$$
$$0.06x \ln e = \ln 3$$
$$0.06x = \ln 3 \qquad \qquad \textit{ln e = 1}$$
$$x = \frac{\ln 3}{0.06} \approx 18.310$$

Check $x = 18.310$: $e^{1.0986} \approx 3.0$

The solution set is $\{18.310\}$.

45.
$$\log_3(9x + 8) = 2$$
$$9x + 8 = 3^2 \quad \textit{Exponential form}$$
$$9x + 8 = 9$$
$$9x = 1$$
$$x = \tfrac{1}{9}$$

Check $x = \tfrac{1}{9}$: $\log_3 9 = \log_3 3^2 = 2$

The solution set is $\left\{\tfrac{1}{9}\right\}$.

46. $\log_5(y + 6)^3 = 2$

Change to exponential form.

$$(y + 6)^3 = 5^2$$
$$(y + 6)^3 = 25$$

Take the cube root of each side.

$$y + 6 = \sqrt[3]{25}$$
$$y = \sqrt[3]{25} - 6$$

Check $y = -6 + \sqrt[3]{25}$: $\log_5 25 = \log_5 5^2 = 2$

The solution set is $\left\{-6 + \sqrt[3]{25}\right\}$.

47. $\log_3(p + 2) - \log_3 p = \log_3 2$

$$\log_3 \frac{p + 2}{p} = \log_3 2 \quad \textit{Quotient rule}$$
$$\frac{p + 2}{p} = 2 \qquad \qquad \textit{Property 4}$$
$$p + 2 = 2p$$
$$2 = p$$

Check $p = 2$: $\log_3 4 - \log_3 2 = \log_3 \tfrac{4}{2} = \log_3 2$

The solution set is $\{2\}$.

48.
$$\log (2x + 3) = 1 + \log x$$
$$\log (2x + 3) - \log x = 1$$
$$\log_{10} \frac{2x + 3}{x} = 1 \qquad \textit{Quotient rule}$$
$$10^1 = \frac{2x + 3}{x} \quad \begin{array}{l} \textit{Exponential} \\ \textit{form} \end{array}$$
$$10x = 2x + 3$$
$$8x = 3$$
$$x = \tfrac{3}{8}$$

Check $x = \frac{3}{8}$:

LS $= \log \left(\frac{3}{4} + 3\right) = \log \frac{15}{4}$

RS $= \log 10 + \log \frac{3}{8} = \log \frac{30}{8} = \log \frac{15}{4}$

The solution set is $\left\{\frac{3}{8}\right\}$.

49.
$$\log_4 x + \log_4 (8 - x) = 2$$
$$\log_4 [x(8 - x)] = 2 \qquad \textit{Product rule}$$
$$x(8 - x) = 4^2 \qquad \textit{Exponential form}$$
$$8x - x^2 = 16$$
$$x^2 - 8x + 16 = 0$$
$$(x - 4)(x - 4) = 0$$
$$x - 4 = 0$$
$$x = 4$$

Check $x = 4$: $\log_4 4 + \log_4 4 = 1 + 1 = 2$

The solution set is $\{4\}$.

50.
$$\log_2 x + \log_2 (x + 15) = \log_2 16$$
$$\log_2 [x(x + 15)] = \log_2 16 \quad \begin{array}{l} \textit{Product} \\ \textit{rule} \end{array}$$
$$x^2 + 15x = 16 \qquad \textit{Property 4}$$
$$x^2 + 15x - 16 = 0$$
$$(x + 16)(x - 1) = 0$$
$$x + 16 = 0 \qquad \text{or} \qquad x - 1 = 0$$
$$x = -16 \quad \text{or} \qquad x = 1$$

Reject $x = -16$, because it yields an equation in which the logarithm of a negative number must be found.

Check $x = 1$:

$\log_2 1 + \log_2 16 = 0 + \log_2 16 = \log_2 16$

The solution set is $\{1\}$.

51. When the power rule was applied in the second step, the domain was changed from $\{x \mid x \neq 0\}$ to $\{x \mid x > 0\}$. Instead of using the power rule for logarithms, we can change the original equation to the exponential form $x^2 = 10^2$ and get $x = \pm 10$. As you can see in the erroneous solution, the valid solution -10 was "lost." The solution set is $\{\pm 10\}$.

52. $A = P\left(1 + \dfrac{r}{n}\right)^{nt}$

Let $P = 20{,}000$, $r = 0.07$, and $t = 5$. For $n = 4$ (quarterly compounding),

$$A = 20{,}000\left(1 + \frac{0.07}{4}\right)^{4 \cdot 5} \approx 28{,}295.56.$$

There will be \$28,295.56 in the account after 5 years.

53. In the continuous compounding formula, let $P = 10{,}000$, $r = 0.06$, and $t = 3$.

$$A = Pe^{rt}$$
$$A = 10{,}000e^{0.06(3)} \approx 11{,}972.17$$

There will be \$11,972.17 in the account after 3 years.

54. Use $A = P\left(1 + \dfrac{r}{n}\right)^{nt}$.

Plan A:

Let $P = 1000$, $r = 0.04$, $n = 4$, and $t = 3$.

$$A = 1000\left(1 + \frac{0.04}{4}\right)^{4 \cdot 3} \approx 1126.83$$

Plan B:

Let $P = 1000$, $r = 0.039$, $n = 12$, and $t = 3$.

$$A = 1000\left(1 + \frac{0.039}{12}\right)^{12 \cdot 3} \approx 1123.91$$

Plan A is the better plan by \$2.92.

55. Let $Q(t) = \frac{1}{2}(500)$ to find the half-life of the radioactive substance.

$$Q(t) = 500e^{-0.05t}$$
$$\tfrac{1}{2}(500) = 500e^{-0.05t}$$
$$0.5 = e^{-0.05t}$$
$$\ln 0.5 = \ln e^{-0.05t}$$
$$\ln 0.5 = -0.05t \, (\ln e)$$
$$-0.05t = \ln 0.5$$
$$t = \frac{\ln 0.5}{-0.05} \approx 13.9$$

The half-life is about 13.9 days.

56. $S = C(1 - r)^n$

(a) Let $C = 30{,}000$, $r = 0.15$, and $n = 12$.

$$S = 30{,}000(1 - 0.15)^{12}$$
$$= 30{,}000(0.85)^{12} \approx 4267$$

The scrap value is about \$4267.

(b) Let $S = \frac{1}{2}C$ and $n = 6$.

$$S = C(1-r)^n$$
$$\tfrac{1}{2}C = C(1-r)^6$$
$$0.5 = (1-r)^6 \;(*)$$
$$\ln 0.5 = \ln (1-r)^6$$
$$\ln 0.5 = 6 \ln (1-r)$$
$$\ln (1-r) = \frac{\ln 0.5}{6}$$
$$\ln (1-r) \approx -0.1155$$
$$1 - r = e^{-0.1155}$$
$$1 - r \approx 0.89$$
$$r = 0.11$$

The rate is approximately 11%.

$(*)$ Alternative solution steps without logarithms:

$$0.5 = (1-r)^6$$
$$\sqrt[6]{0.5} = 1 - r$$
$$r = 1 - \sqrt[6]{0.5} \approx 0.11$$

Note that $1 - r$ must be positive, so $\pm \sqrt[6]{0.5}$ is not needed.

57. $N(r) = -5000 \ln r$
Replace $N(r)$ with 2000, and solve for r.

$$2000 = -5000 \ln r$$
$$\ln r = -\tfrac{2000}{5000}$$
$$\log_e r = -0.4$$

Change to exponential form and approximate.

$$r = e^{-0.4} \approx 0.67$$

About 67% of the words are common to both of the evolving languages.

58. **A.** Solve $7^x = 23$ by using the power rule with common logarithms.

$$7^x = 23$$
$$\log 7^x = \log 23$$
$$x \log 7 = \log 23$$
$$x = \frac{\log 23}{\log 7}$$

B. Solve $7^x = 23$ by using the power rule with natural logarithms.

$$7^x = 23$$
$$\ln 7^x = \ln 23$$
$$x \ln 7 = \ln 23$$
$$x = \frac{\ln 23}{\ln 7}$$

C. Use the change-of-base rule with the solution from **A.**

$$x = \frac{\log 23}{\log 7} = \log_7 23$$

D. $x = \dfrac{\log 23}{\log 7} \neq \log_{23} 7$

The answer is **D.**

59. **[12.4]** $\log_2 128 = \log_2 2^7 = 7$, by a special property of logarithms.

60. **[12.4]** By a special property of logarithms,

$$5^{\log_5 36} = 36.$$

61. **[12.5]** $e^{\ln 4} = 4$ since $e^{\ln x} = x$.

62. **[12.5]** $10^{\log e} = e$ since $10^{\log x} = x$.

63. **[12.4]** $\log_3 3^{-5} = -5$ since $\log_a a^x = x$.

64. **[12.5]** $\ln e^{5.4} = 5.4$ since $\ln e^x = x$.

65. **[12.6]** $\log_3 (x + 9) = 4$

$$x + 9 = 3^4 \quad \textit{Exponential form}$$
$$x + 9 = 81$$
$$x = 72$$

Check $x = 72$: $\log_3 81 = \log_3 3^4 = 4$

The solution set is $\{72\}$.

66. **[12.6]** $\ln e^x = 3$

$$x = 3$$

The solution set is $\{3\}$.

67. **[12.3]** $\log_x \frac{1}{81} = 2$

$$x^2 = \tfrac{1}{81} \quad \textit{Exponential form}$$
$$x^2 = \left(\tfrac{1}{9}\right)^2$$
$$x = \pm \tfrac{1}{9} \quad \textit{Square root property}$$

The base x cannot be negative.

Check $x = \frac{1}{9}$: $\log_{1/9} \frac{1}{81} = 2$ since $\left(\frac{1}{9}\right)^2 = \frac{1}{81}$

The solution set is $\left\{\frac{1}{9}\right\}$.

68. **[12.2]** $27^x = 81$

Write each side as a power of 3.

$$(3^3)^x = 3^4$$
$$3^{3x} = 3^4$$
$$3x = 4 \quad \textit{Equate exponents.}$$
$$x = \tfrac{4}{3}$$

Check $x = \frac{4}{3}$: $27^{4/3} = 3^4 = 81$

The solution set is $\left\{\frac{4}{3}\right\}$.

69. **[12.2]** $2^{2x-3} = 8$

Write each side as a power of 2.

$$2^{2x-3} = 2^3$$
$$2x - 3 = 3 \quad \textit{Equate exponents.}$$
$$2x = 6$$
$$x = 3$$

Check $x = 3$: $2^3 = 8$ \quad \textit{True}

The solution set is $\{3\}$.

70. **[12.2]** $5^{x+2} = 25^{2x+1}$

$5^{x+2} = (5^2)^{2x+1}$ *Equal bases*

$5^{x+2} = 5^{4x+2}$

$x + 2 = 4x + 2$ *Equate exponents.*

$0 = 3x$

$0 = x$

Check $x = 0$: $5^2 = 25^1$ *True*

The solution set is $\{0\}$.

71. **[12.6]**

$\log_3 (x + 1) - \log_3 x = 2$

$\log_3 \dfrac{x + 1}{x} = 2$ *Quotient rule*

$\dfrac{x + 1}{x} = 3^2$ *Exponential form*

$9x = x + 1$

$8x = 1$

$x = \tfrac{1}{8}$

Check $x = \tfrac{1}{8}$:

$\log_3 \tfrac{9}{8} - \log_3 \tfrac{1}{8} = \log_3 9 = \log_3 3^2 = 2$

The solution set is $\left\{\tfrac{1}{8}\right\}$.

72. **[12.6]** $\log (3x - 1) = \log 10$

$3x - 1 = 10$

$3x = 11$

$x = \tfrac{11}{3}$

Check $x = \tfrac{11}{3}$: $\log (11 - 1) = \log 10$ *True*

The solution set is $\left\{\tfrac{11}{3}\right\}$.

73. **[12.6]** $\ln (x^2 + 3x + 4) = \ln 2$

$x^2 + 3x + 4 = 2$

$x^2 + 3x + 2 = 0$

$(x + 2)(x + 1) = 0$

$x + 2 = 0$ or $x + 1 = 0$

$x = -2$ or $x = -1$

Check $x = -2$: $\ln (4 - 6 + 4) = \ln 2$ *True*

Check $x = -1$: $\ln (1 - 3 + 4) = \ln 2$ *True*

The solution set is $\{-2, -1\}$.

74. **[12.6] (a)** $\log (2x + 3) = \log x + 1$

$\log (2x + 3) - \log x = 1$

$\log_{10} \dfrac{2x + 3}{x} = 1$ *Quotient rule*

$\dfrac{2x + 3}{x} = 10^1$ *Exponential form*

$2x + 3 = 10x$

$3 = 8x$

$\tfrac{3}{8} = x$

Check $x = \tfrac{3}{8}$:

LS $= \log \left(\tfrac{3}{4} + 3\right) = \log \tfrac{15}{4}$

RS $= \log \tfrac{3}{8} + \log 10 = \log \tfrac{30}{8} = \log \tfrac{15}{4}$

The solution set is $\left\{\tfrac{3}{8}\right\}$.

(b) From the graph, the x-value of the x-intercept is 0.375, the decimal equivalent of $\tfrac{3}{8}$. Note that the solutions of $Y_1 - Y_2 = 0$ are the same as the solutions of $Y_1 = Y_2$.

75. **[12.6]** $x = 2000 - 1980 = 20$

$R(x) = 10.001e^{0.0521x}$

$R(20) = 10.001e^{0.0521(20)}$

≈ 28.35

In 2000, about 28.35% of municipal solid waste was recovered.

76. **[12.5] (a)** There are 90 of one species and 10 of another, so

$p_1 = \tfrac{90}{100} = 0.9$ and $p_2 = \tfrac{10}{100} = 0.1$.

Thus, the index of diversity is

$-(p_1 \ln p_1 + p_2 \ln p_2)$

$= -(0.9 \ln 0.9 + 0.1 \ln 0.1)$

≈ 0.325.

(b) There are 60 of one species and 40 of another, so

$p_1 = \tfrac{60}{100} = 0.6$ and $p_2 = \tfrac{40}{100} = 0.4$.

Thus, the index of diversity is

$-(p_1 \ln p_1 + p_2 \ln p_2)$

$= -(0.6 \ln 0.6 + 0.4 \ln 0.4)$

≈ 0.673.

Chapter 12 Test

1. **(a)** $f(x) = x^2 + 9$

This function is not one-to-one. The graph of $f(x)$ is a vertical parabola. A horizontal line will intersect the graph more than once.

(b) This function is one-to-one. A horizontal line will not intersect the graph in more than one point.

2.
$$f(x) = \sqrt[3]{x+7}$$

Replace $f(x)$ with y.
$$y = \sqrt[3]{x+7}$$

Interchange x and y.
$$x = \sqrt[3]{y+7}$$

Solve for y.
$$x^3 = y + 7$$
$$x^3 - 7 = y$$

Replace y with $f^{-1}(x)$.
$$f^{-1}(x) = x^3 - 7$$

3. By the horizontal line test, $f(x)$ is a one-to-one function and has an inverse. Choose some points on the graph of $f(x)$, such as $(4, 0)$, $(3, -1)$, and $(0, -2)$. To graph the inverse, interchange the x- and y-values to get $(0, 4)$, $(-1, 3)$, and $(-2, 0)$. Plot these points and draw a smooth curve through them.

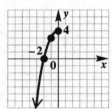

4. $f(x) = 6^x$

Make a table of values.

x	-2	-1	0	1
$f(x)$	$\frac{1}{36}$	$\frac{1}{6}$	1	6

Plot these points and draw a smooth exponential curve through them.

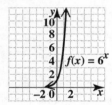

5. $g(x) = \log_6 x$

Make a table of values.

Powers of 6	6^{-2}	6^{-1}	6^0	6^1
x	$\frac{1}{36}$	$\frac{1}{6}$	1	6
$g(x)$	-2	-1	0	1

Plot these points and draw a smooth logarithmic curve through them.

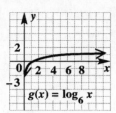

6. $y = 6^x$ and $y = \log_6 x$ are inverse functions. To use the graph from Exercise 4 to obtain the graph of the function in Exercise 5, interchange the x- and y-coordinates of the ordered pairs $\left(-2, \frac{1}{36}\right)$, $\left(-1, \frac{1}{6}\right)$, $(0, 1)$, and $(1, 6)$ to get $\left(\frac{1}{36}, -2\right)$, $\left(\frac{1}{6}, -1\right)$, $(1, 0)$, and $(6, 1)$. Plot these points and draw a smooth logarithmic curve through them.

7.
$$5^x = \frac{1}{625}$$
$$5^x = \left(\frac{1}{5}\right)^4$$

Write each side as a power of 5.
$$5^x = 5^{-4}$$
$$x = -4 \quad \textit{Equate exponents.}$$

The solution set is $\{-4\}$.

8.
$$2^{3x-7} = 8^{2x+2}$$

Write each side as a power of 2.
$$2^{3x-7} = \left(2^3\right)^{2x+2}$$
$$3x - 7 = 3(2x + 2) \quad \textit{Equate exponents.}$$
$$3x - 7 = 6x + 6$$
$$-13 = 3x$$
$$-\frac{13}{3} = x$$

Check $x = -\frac{13}{3}$:
$$\text{LS} = 2^{-13-7} = 2^{-20}$$
$$\text{RS} = 8^{-20/3} = \left(2^3\right)^{-20/3} = 2^{-20}$$

The solution set is $\left\{-\frac{13}{3}\right\}$.

9. $f(x) = 26.7e^{0.023x}$

(a) 2010: $x = 2010 - 1995 = 15$
$$f(15) = 26.7e^{0.023(15)} \approx 37.7$$

The U.S. Hispanic population estimate for 2010 is 37.7 million.

(b) 2015: $x = 2015 - 1995 = 20$
$$f(20) = 26.7e^{0.023(20)} \approx 42.3$$

The U.S. Hispanic population estimate for 2015 is 42.3 million.

10. The base is 4, the exponent (logarithm) is -2, and the number is 0.0625, so $4^{-2} = 0.0625$ becomes $\log_4 0.0625 = -2$ in logarithmic form.

11. The base is 7, the logarithm (exponent) is 2, and the number is 49, so $\log_7 49 = 2$ becomes $7^2 = 49$ in exponential form.

12. $\log_{1/2} x = -5$

$\qquad x = \left(\frac{1}{2}\right)^{-5}$ *Exponential form*

$\qquad x = \left(\frac{2}{1}\right)^5 = 32$

The argument (the input of the logarithm) must be a positive number, so $x = 32$ is acceptable.

Check $x = 32$:

$\quad \log_{1/2} 32 = -5$ since $\left(\frac{1}{2}\right)^{-5} = 2^5 = 32$

The solution set is $\{32\}$.

13. $\qquad\qquad x = \log_9 3$

$\qquad\qquad 9^x = 3$ *Exponential form*

Write each side as a power of 3.

$\qquad\qquad \left(3^2\right)^x = 3$

$\qquad\qquad 3^{2x} = 3^1$

$\qquad\qquad 2x = 1$ *Equate exponents.*

$\qquad\qquad x = \frac{1}{2}$

Check $x = \frac{1}{2}$:

$\quad \frac{1}{2} = \log_9 3$ since $9^{1/2} = \sqrt{9} = 3$

The solution set is $\left\{\frac{1}{2}\right\}$.

14. $\log_x 16 = 4$

$\qquad x^4 = 16$ *Exponential form*

$\qquad x^2 = \pm 4$ *Square root property*

Reject -4 since $x^2 \geq 0$.

$\qquad x^2 = 4$

$\qquad x = \pm 2$ *Square root property*

Reject -2 since the base cannot be negative.

Check $x = 2$: $\log_2 16 = 4$ since $2^4 = 16$

The solution set is $\{2\}$.

15. The value of $\log_2 32$ is 5 . This means that if we raise 2 to the 5th power, the result is 32 .

16. $\log_3 x^2 y$

$\quad = \log_3 x^2 + \log_3 y$ *Product rule*

$\quad = 2 \log_3 x + \log_3 y$ *Power rule*

17. $\log_5 \left(\dfrac{\sqrt{x}}{yz}\right)$

$\quad = \log_5 \sqrt{x} - \log_5 yz$ *Quotient rule*

$\quad = \log_5 x^{1/2} - \left(\log_5 y + \log_5 z\right)$ *Product rule*

$\quad = \frac{1}{2} \log_5 x - \log_5 y - \log_5 z$ *Power rule*

18. $3 \log_b s - \log_b t$

$\quad = \log_b s^3 - \log_b t$ *Power rule*

$\quad = \log_b \dfrac{s^3}{t}$ *Quotient rule*

19. $\frac{1}{4} \log_b r + 2 \log_b s - \frac{2}{3} \log_b t$

Use the power rule for logarithms.

$\quad = \log_b r^{1/4} + \log_b s^2 - \log_b t^{2/3}$

Use the product and quotient rules for logarithms.

$\quad = \log_b \dfrac{r^{1/4} s^2}{t^{2/3}}$

20. **(a)** $\log 23.1 \approx 1.3636$

(b) $\ln 0.82 \approx -0.1985$

21. **(a)** $\log_3 19 = \dfrac{\log_{10} 19}{\log_{10} 3} = \dfrac{\log 19}{\log 3}$

(b) $\log_3 19 = \dfrac{\log_e 19}{\log_e 3} = \dfrac{\ln 19}{\ln 3}$

(c) The four-decimal-place approximation of either fraction is 2.6801.

22. $\qquad 3^x = 78$

$\quad \ln 3^x = \ln 78$

$\quad x \ln 3 = \ln 78$ *Power rule*

$\qquad x = \dfrac{\ln 78}{\ln 3} \approx 3.9656$

Check $x = 3.9656$: $3^{3.9656} \approx 78.0$

The solution set is $\{3.9656\}$.

23. $\log_8 (x + 5) + \log_8 (x - 2) = 1$

Use the product rule for logarithms.

$\quad \log_8 [(x + 5)(x - 2)] = 1$

$\qquad (x + 5)(x - 2) = 8^1$ *Exp. form*

$\qquad x^2 + 3x - 10 = 8$

$\qquad x^2 + 3x - 18 = 0$

$\qquad (x + 6)(x - 3) = 0$

$\quad x + 6 = 0 \quad$ or $\quad x - 3 = 0$

$\qquad x = -6 \quad$ or $\qquad x = 3$

Reject $x = -6$, because $x + 5 = -1$, which yields an equation in which the logarithm of a negative number must be found.

Check $x = 3$: $\log_8 8 + \log_8 1 = 1 + 0 = 1$

The solution set is $\{3\}$.

24. $A = P \left(1 + \dfrac{r}{n}\right)^{nt}$

$A = 10{,}000 \left(1 + \dfrac{0.045}{4}\right)^{4 \cdot 5} \approx 12{,}507.51$

\$10,000 invested at 4.5% annual interest, compounded quarterly, will increase to \$12,507.51 in 5 years.

25. $A = P e^{rt}$

(a) $A = 15{,}000 e^{0.05(5)} \approx 19{,}260.38$

There will be \$19,260.38 in the account.

(b) Let $A = 2(15,000)$ and solve for t.

$$2(15,000) = 15,000e^{0.05t}$$
$$2 = e^{0.05t}$$
$$\ln 2 = \ln e^{0.05t}$$
$$\ln 2 = 0.05t\,(\ln e)$$
$$0.05t = \ln 2$$
$$t = \frac{\ln 2}{0.05} \approx 13.9$$

The principal will double in about 13.9 years.

Cumulative Review Exercises (Chapters 1–12)

For Exercises 1–4,

$$S = \left\{ -\tfrac{9}{4}, -2, -\sqrt{2}, 0, 0.6, \sqrt{11}, \sqrt{-8}, 6, \tfrac{30}{3} \right\}.$$

1. The integers are -2, 0, 6, and $\frac{30}{3}$ (or 10).

2. The rational numbers are $-\frac{9}{4}$, -2, 0, 0.6, 6, and $\frac{30}{3}$ (or 10). Each can be expressed as a quotient of two integers.

3. The irrational numbers are $-\sqrt{2}$ and $\sqrt{11}$.

4. All are real numbers except $\sqrt{-8}$.

5. $|-8| + 6 - |-2| - (-6 + 2)$
$= 8 + 6 - 2 - (-4)$
$= 14 - 2 + 4 = 16$

6. $-12 - |-3| - 7 - |-5|$
$= -12 - 3 - 7 - 5 = -27$

7. $2(-5) + (-8)(4) - (-3)$
$= -10 - 32 + 3 = -39$

8. $7 - (3 + 4a) + 2a = -5(a - 1) - 3$
$7 - 3 - 4a + 2a = -5a + 5 - 3$
$4 - 2a = -5a + 2$
$3a = -2$
$a = -\frac{2}{3}$

The solution set is $\left\{ -\frac{2}{3} \right\}$.

9. $2m + 2 \le 5m - 1$
$-3m \le -3$
Divide by -3; reverse the inequality.
$m \ge 1$

The solution set is $[1, \infty)$.

10. $(2p + 3)(3p - 1) = 6p^2 - 2p + 9p - 3$
$= 6p^2 + 7p - 3$

11. $(4k - 3)^2 = (4k)^2 - 2(4k)(3) + 3^2$
$= 16k^2 - 24k + 9$

12. $(3m^3 + 2m^2 - 5m) - (8m^3 + 2m - 4)$
$= 3m^3 + 2m^2 - 5m - 8m^3 - 2m + 4$
$= 3m^3 - 8m^3 + 2m^2 - 5m - 2m + 4$
$= -5m^3 + 2m^2 - 7m + 4$

13.

$$
\begin{array}{r}
2t^3 + 5t^2 - 3t + 4 \\
3t + 1 \overline{) 6t^4 + 17t^3 - 4t^2 + 9t + 4} \\
\underline{6t^4 + 2t^3} \\
15t^3 - 4t^2 \\
\underline{15t^3 + 5t^2} \\
-9t^2 + 9t \\
\underline{-9t^2 - 3t} \\
12t + 4 \\
\underline{12t + 4} \\
0
\end{array}
$$

The quotient is $2t^3 + 5t^2 - 3t + 4$.

14. $8x + x^3$
Factor out the GCF, x.

$$8x + x^3 = x(8 + x^2)$$

15. $24y^2 - 7y - 6$
Factor by trial and error.

$$24y^2 - 7y - 6 = (8y + 3)(3y - 2)$$

16. $5z^3 - 19z^2 - 4z$
Factor out the GCF, z, and then factor by trial and error.

$$5z^3 - 19z^2 - 4z = z(5z^2 - 19z - 4)$$
$$= z(5z + 1)(z - 4)$$

17. $16a^2 - 25b^4$

Use the difference of squares formula,

$$x^2 - y^2 = (x + y)(x - y),$$

where $x = 4a$ and $y = 5b^2$.

$$16a^2 - 25b^4 = (4a + 5b^2)(4a - 5b^2)$$

18. $8c^3 + d^3$

Use the sum of cubes formula,

$$x^3 + y^3 = (x + y)(x^2 - xy + y^2),$$

where $x = 2c$ and $y = d$.

$$8c^3 + d^3 = (2c + d)(4c^2 - 2cd + d^2)$$

19. $16r^2 + 56rq + 49q^2$
$= (4r)^2 + 2(4r)(7q) + (7q)^2$

Use the perfect square formula,

$$x^2 + 2xy + y^2 = (x + y)^2,$$

where $x = 4r$ and $y = 7q$.

$$16r^2 + 56rq + 49q^2 = (4r + 7q)^2$$

20. $\dfrac{(5p^3)^4\,(-3p^7)}{2p^2(4p^4)} = \dfrac{(5^4p^{12})(-3p^7)}{8p^6}$

$\quad\quad = \dfrac{(625)(-3)p^{19}}{8p^6}$

$\quad\quad = -\dfrac{1875p^{13}}{8}$

21. $\dfrac{x^2 - 9}{x^2 + 7x + 12} \div \dfrac{x - 3}{x + 5}$

Multiply by the reciprocal.

$\quad = \dfrac{x^2 - 9}{x^2 + 7x + 12} \cdot \dfrac{x + 5}{x - 3}$

$\quad = \dfrac{(x+3)(x-3)}{(x+3)(x+4)} \cdot \dfrac{(x+5)}{(x-3)}$ *Factor*

$\quad = \dfrac{x + 5}{x + 4}$

22. $\dfrac{2}{k + 3} - \dfrac{5}{k - 2}$

The LCD is $(k+3)(k-2)$.

$\quad = \dfrac{2(k-2)}{(k+3)(k-2)} - \dfrac{5(k+3)}{(k-2)(k+3)}$

$\quad = \dfrac{2k - 4 - 5k - 15}{(k+3)(k-2)}$

$\quad = \dfrac{-3k - 19}{(k+3)(k-2)}$

23. $\dfrac{3}{p^2 - 4p} - \dfrac{4}{p^2 + 2p}$

$\quad = \dfrac{3}{p(p-4)} - \dfrac{4}{p(p+2)}$

The LCD is $p(p-4)(p+2)$.

$\quad = \dfrac{3(p+2)}{p(p-4)(p+2)} - \dfrac{4(p-4)}{p(p+2)(p-4)}$

$\quad = \dfrac{3p + 6 - 4p + 16}{p(p-4)(p+2)}$

$\quad = \dfrac{22 - p}{p(p-4)(p+2)}$

24. $5x + 2y = 10$

Find the x- and y-intercepts. To find the x-intercept, let $y = 0$.

$\quad\quad 5x + 2(0) = 10$

$\quad\quad\quad\quad 5x = 10$

$\quad\quad\quad\quad\ x = 2$

The x-intercept is $(2, 0)$.
To find the y-intercept, let $x = 0$.

$\quad\quad 5(0) + 2y = 10$

$\quad\quad\quad\quad 2y = 10$

$\quad\quad\quad\quad\ y = 5$

The y-intercept is $(0, 5)$.
Plot the intercepts and draw the line through them.

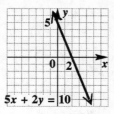

25. $-4x + y \le 5$

Graph the line $-4x + y = 5$, which has intercepts $(0, 5)$ and $\left(-\tfrac{5}{4}, 0\right)$, as a solid line because the inequality involves $\le$. Test $(0, 0)$, which yields $0 \le 5$, a true statement. Shade the region that includes $(0, 0)$.

$-4x + y \le 5$

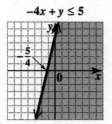

26. **(a)** Yes, this graph is the graph of a function because it passes the vertical line test.

(b) $(x_1, y_1) = (2000,\ 61{,}327)$ and $(x_2, y_2) = (2003,\ 56{,}250)$.

$\quad m = \dfrac{y_2 - y_1}{x_2 - x_1} = \dfrac{56{,}250 - 61{,}327}{2003 - 2000}$

$\quad\quad = \dfrac{-5077}{3} = -1692.\overline{3}$

The slope of the line in the graph is about -1692 and can be interpreted as follows: The number of U.S. travelers to international countries decreased by approximately 1692 thousand per year during 2000-2003.

27. Through $(5, -1)$; parallel to $3x - 4y = 12$
Find the slope of

$\quad\quad 3x - 4y = 12$

$\quad\quad\quad -4y = -3x + 12$

$\quad\quad\quad\quad\ y = \tfrac{3}{4}x - 3.$

The slope is $\tfrac{3}{4}$, so a line parallel to it also has slope $\tfrac{3}{4}$. Let $m = \tfrac{3}{4}$ and $(x_1, y_1) = (5, -1)$ in the point-slope form.

$\quad\quad y - y_1 = m(x - x_1)$

$\quad y - (-1) = \tfrac{3}{4}(x - 5)$

$\quad\quad y + 1 = \tfrac{3}{4}x - \tfrac{15}{4}$

$\quad\quad\quad\ y = \tfrac{3}{4}x - \tfrac{19}{4}$

28.
$5x - 3y = 14$ (1)
$2x + 5y = 18$ (2)

Multiply equation (1) by 5 and equation (2) by 3. Then add the results.

$$
\begin{array}{llll}
25x & - & 15y & = & 70 & \quad 5 \times (1) \\
6x & + & 15y & = & 54 & \quad 3 \times (2) \\
\hline
31x & & & = & 124 & \quad \textit{Add} \\
& & x & = & 4
\end{array}
$$

Substitute 4 for x in equation (1) to find y.

$$
\begin{aligned}
5x - 3y &= 14 \quad (1) \\
5(4) - 3y &= 14 \\
20 - 3y &= 14 \\
-3y &= -6 \\
y &= 2
\end{aligned}
$$

The solution set is $\{(4, 2)\}$.

29.
$2x - 7y = 8$ (1)
$4x - 14y = 3$ (2)

Multiply equation (1) by -2 and add the result to equation (2).

$$
\begin{array}{llll}
-4x & + & 14y & = & -16 & \quad -2 \times (1) \\
4x & - & 14y & = & 3 & \quad (2) \\
\hline
& & 0 & = & -13 & \quad \textit{Add}
\end{array}
$$

The statement $0 = -13$ is false, so the solution set is $\emptyset$.

30.
$x + 2y + 3z = 11$ (1)
$3x - y + z = 8$ (2)
$2x + 2y - 3z = -12$ (3)

To eliminate z, add equations (1) and (3).

$$
\begin{array}{llll}
x & + & 2y & + & 3z & = & 11 & \quad (1) \\
2x & + & 2y & - & 3z & = & -12 & \quad (3) \\
\hline
3x & + & 4y & & & = & -1 & \quad (4)
\end{array}
$$

To eliminate z again, multiply equation (2) by 3 and add the result to equation (3).

$$
\begin{array}{llll}
9x & - & 3y & + & 3z & = & 24 & \quad 3 \times (2) \\
2x & + & 2y & - & 3z & = & -12 & \quad (3) \\
\hline
11x & - & y & & & = & 12 & \quad (5)
\end{array}
$$

Multiply equation (5) by 4 and add the result to equation (4).

$$
\begin{array}{llll}
44x & - & 4y & = & 48 & \quad 4 \times (5) \\
3x & + & 4y & = & -1 & \quad (4) \\
\hline
47x & & & = & 47 \\
& & x & = & 1
\end{array}
$$

Substitute 1 for x in equation (5) to find y.

$$
\begin{aligned}
11x - y &= 12 \quad (5) \\
11(1) - y &= 12 \\
11 - y &= 12 \\
-y &= 1 \\
y &= -1
\end{aligned}
$$

Substitute 1 for x and -1 for y in equation (2) to find z.

$$
\begin{aligned}
3x - y + z &= 8 \quad (2) \\
3(1) - (-1) + z &= 8 \\
3 + 1 + z &= 8 \\
4 + z &= 8 \\
z &= 4
\end{aligned}
$$

The solution set is $\{(1, -1, 4)\}$.

31. Let $x =$ the amount of candy at \$1.00 per pound.

	Number of Pounds	Price per Pound	Value
First Candy	x	\$1.00	$1.00x$
Second Candy	10	\$1.96	$1.96(10)$
Mixture	$x + 10$	\$1.60	$1.60(x + 10)$

The sum of the values of each candy must equal the value of the mixture.

$$1.00x + 1.96(10) = 1.60(x + 10)$$

Multiply by 10 to clear the decimals.

$$
\begin{aligned}
10x + 196 &= 16(x + 10) \\
10x + 196 &= 16x + 160 \\
36 &= 6x \\
6 &= x
\end{aligned}
$$

Use 6 pounds of the \$1.00 candy.

32.
$|2x - 5| = 9$

$$
\begin{array}{lll}
2x - 5 = 9 & \text{or} & 2x - 5 = -9 \\
2x = 14 & & 2x = -4 \\
x = 7 & \text{or} & x = -2
\end{array}
$$

The solution set is $\{-2, 7\}$.

33.
$|3p| - 4 = 12$
$|3p| = 16$

$$
\begin{array}{lll}
3p = 16 & \text{or} & 3p = -16 \\
p = \frac{16}{3} & \text{or} & p = -\frac{16}{3}
\end{array}
$$

The solution set is $\left\{ \pm \frac{16}{3} \right\}$.

34.
$|3k - 8| \le 1$

$$
\begin{aligned}
-1 &\le 3k - 8 \le 1 \\
7 &\le 3k \le 9 \\
\tfrac{7}{3} &\le k \le 3
\end{aligned}
$$

The solution set is $\left[\frac{7}{3}, 3 \right]$.

35. $|4m + 2| > 10$

$4m + 2 > 10 \quad \text{or} \quad 4m + 2 < -10$

$\quad 4m > 8 \qquad\qquad 4m < -12$

$\quad\; m > 2 \quad \text{or} \qquad m < -3$

The solution set is $(-\infty, -3) \cup (2, \infty)$.

36. $\sqrt{288} = \sqrt{144 \cdot 2} = \sqrt{144}\sqrt{2} = 12\sqrt{2}$

37. $2\sqrt{32} - 5\sqrt{98} = 2\sqrt{16 \cdot 2} - 5\sqrt{49 \cdot 2}$

$\qquad\qquad\qquad = 2 \cdot 4\sqrt{2} - 5 \cdot 7\sqrt{2}$

$\qquad\qquad\qquad = 8\sqrt{2} - 35\sqrt{2}$

$\qquad\qquad\qquad = -27\sqrt{2}$

38. $\sqrt{2x + 1} - \sqrt{x} = 1$

$\qquad \sqrt{2x + 1} = 1 + \sqrt{x}$

$\qquad \left(\sqrt{2x + 1}\right)^2 = \left(1 + \sqrt{x}\right)^2$

$\qquad 2x + 1 = 1 + 2\sqrt{x} + x$

$\qquad\qquad x = 2\sqrt{x}$

$\qquad (x)^2 = \left(2\sqrt{x}\right)^2$

$\qquad\qquad x^2 = 4x$

$\qquad x^2 - 4x = 0$

$\qquad x(x - 4) = 0$

$\qquad x = 0 \quad \text{or} \quad x = 4$

Check $x = 0$: $\quad \sqrt{1} - \sqrt{0} = 1 \quad$ *True*

Check $x = 4$: $\quad \sqrt{9} - \sqrt{4} = 1 \quad$ *True*

The solution set is $\{0, 4\}$.

39. $(5 + 4i)(5 - 4i) = 5^2 - (4i)^2$

$\qquad\qquad\qquad\quad = 25 - 16i^2$

$\qquad\qquad\qquad\quad = 25 - 16(-1)$

$\qquad\qquad\qquad\quad = 25 + 16 = 41$

40. $3x^2 - x - 1 = 0$

Here $a = 3$, $b = -1$, and $c = -1$.

Use the quadratic formula.

$$x = \frac{-b \pm \sqrt{b^2 - 4ac}}{2a}$$

$$x = \frac{-(-1) \pm \sqrt{(-1)^2 - 4(3)(-1)}}{2(3)}$$

$$= \frac{1 \pm \sqrt{1 + 12}}{6} = \frac{1 \pm \sqrt{13}}{6}$$

The solution set is $\left\{ \dfrac{1 + \sqrt{13}}{6}, \dfrac{1 - \sqrt{13}}{6} \right\}$.

41. $k^2 + 2k - 8 > 0$

Solve the equation

$$k^2 + 2k - 8 = 0.$$
$$(k + 4)(k - 2) = 0$$

$k + 4 = 0 \quad \text{or} \quad k - 2 = 0$

$\quad k = -4 \quad \text{or} \qquad k = 2$

The numbers -4 and 2 divide a number line into three intervals.

$$
\begin{array}{ccc}
\mathbf{A} & \mathbf{B} & \mathbf{C}
\end{array}
$$

(number line with points at -4 and 2)

Test a number from each interval in the inequality

$$k^2 + 2k - 8 > 0.$$

Interval A: Let $k = -5$.

$\qquad 25 - 10 - 8 > 0 \qquad ?$

$\qquad\qquad\qquad 7 > 0 \qquad$ *True*

Interval B: Let $k = 0$.

$\qquad\qquad\qquad -8 > 0 \qquad$ *False*

Interval C: Let $k = 3$.

$\qquad 9 + 6 - 8 > 0 \qquad ?$

$\qquad\qquad\qquad 7 > 0 \qquad$ *True*

The numbers in Intervals A and C, not including -4 or 2 because of $>$, are solutions.
The solution set is $(-\infty, -4) \cup (2, \infty)$.

42. $x^4 - 5x^2 + 4 = 0$

Let $u = x^2$, so $u^2 = \left(x^2\right)^2 = x^4$.

$\qquad u^2 - 5u + 4 = 0$

$\qquad (u - 1)(u - 4) = 0$

$\qquad u - 1 = 0 \quad \text{or} \quad u - 4 = 0$

$\qquad\qquad u = 1 \quad \text{or} \qquad u = 4$

To find x, substitute x^2 for u.

$\qquad x^2 = 1 \qquad \text{or} \quad x^2 = 4$

$\qquad x = \pm 1 \quad \text{or} \qquad x = \pm 2$

The solution set is $\{\pm 1, \pm 2\}$.

43. Let $x =$ one of the numbers.
Then $300 - x =$ the other number.

The product of the two numbers is given by

$$P = x(300 - x).$$

Writing this equation in standard form gives us

$$P = -x^2 + 300x.$$

Finding the maximum of the product is the same as finding the vertex of the graph of P.
The x-value of the vertex is

$$x = -\frac{b}{2a} = -\frac{300}{2(-1)} = 150.$$

If x is 150, then $300 - x$ must also be 150. The two numbers are 150 and 150 and the product is $150 \cdot 150 = 22{,}500$.

44. $f(x) = \frac{1}{3}(x-1)^2 + 2$ is in
$f(x) = a(x-h)^2 + k$ form. The graph is a
vertical parabola with vertex (h, k) at $(1, 2)$. Since
$a = \frac{1}{3} > 0$, the graph opens up.
Also, $|a| = \left|\frac{1}{3}\right| = \frac{1}{3} < 1$, so the graph is wider
than the graph of $f(x) = x^2$. The points $\left(0, 2\frac{1}{3}\right)$,
$(-2, 5)$, and $(4, 5)$ are also on the graph.

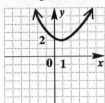

$$f(x) = \frac{1}{3}(x-1)^2 + 2$$

45. $f(x) = 2^x$
Make a table of values.

x	-2	-1	0	1	2
$f(x)$	$\frac{1}{4}$	$\frac{1}{2}$	1	2	4

Plot the ordered pairs from the table, and draw a
smooth exponential curve through the points.

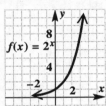

$$f(x) = 2^x$$

46.
$$5^{x+3} = \left(\frac{1}{25}\right)^{3x+2}$$
$$5^{x+3} = \left[\left(\frac{1}{5}\right)^2\right]^{3x+2}$$

Write each side to the power of 5.
$$5^{x+3} = \left(5^{-2}\right)^{(3x+2)}$$
$$5^{x+3} = 5^{-2(3x+2)}$$
$$x + 3 = -2(3x+2)$$
$$x + 3 = -6x - 4$$
$$7x = -7$$
$$x = -1$$

Check $x = -1$: $5^2 = \left(\frac{1}{25}\right)^{-1}$ *True*
The solution set is $\{-1\}$.

47. $f(x) = \log_3 x$
Make a table of values.

Powers of 3	3^{-2}	3^{-1}	3^0	3^1	3^2
x	$\frac{1}{9}$	$\frac{1}{3}$	1	3	9
y	-2	-1	0	1	2

Plot the ordered pairs and draw a smooth
logarithmic curve through the points.

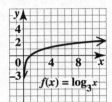

$$f(x) = \log_3 x$$

48.
$$\log_2 81 = \log_2 9^2 = 2\log_2 9$$
$$\approx 2(3.1699) = 6.3398$$

49. $\log \dfrac{x^3 \sqrt{y}}{z}$

$$= \log \frac{x^3 y^{1/2}}{z}$$
$$= \log\left(x^3 y^{1/2}\right) - \log z \qquad \textit{Quotient rule}$$
$$= \log x^3 + \log y^{1/2} - \log z \qquad \textit{Product rule}$$
$$= 3\log x + \frac{1}{2}\log y - \log z \qquad \textit{Power rule}$$

50. $B(t) = 25{,}000e^{0.2t}$

(a) At noon, $t = 0$.

$$B(0) = 25{,}000e^{0.2(0)}$$
$$= 25{,}000e^0 = 25{,}000(1) = 25{,}000$$

25,000 bacteria are present at noon.

(b) At 1 P.M., $t = 1$.

$$B(1) = 25{,}000e^{0.2(1)} \approx 30{,}535$$

About 30,500 bacteria are present at 1 P.M.

(c) At 2 P.M., $t = 2$.

$$B(2) = 25{,}000e^{0.2(2)} \approx 37{,}296$$

About 37,300 bacteria are present at 2 P.M.

(d) The population doubles when $B(t) = 50{,}000$.

$$50{,}000 = 25{,}000e^{0.2t}$$
$$2 = e^{0.2t}$$
$$\ln 2 = \ln e^{0.2t}$$
$$\ln 2 = 0.2t \ln e$$
$$t = \frac{\ln 2}{0.2} \approx 3.5$$

The population will double in about 3.5 hours, or
at about 3:30 P.M.

CHAPTER 13 NONLINEAR FUNCTIONS, CONIC SECTIONS, AND NONLINEAR SYSTEMS

13.1   Additional Graphs of Functions

1. For the reciprocal function defined by $f(x) = 1/x$, $\underline{0}$ is the only real number not in the domain since division by 0 is undefined.

3. The lowest point on the graph of $f(x) = |x|$ has coordinates ($\underline{0}$, $\underline{0}$).

5. $f(x) = |x - 2| + 2$

The graph of this function has its "vertex" at $(2, 2)$, so the correct graph is **B**.

7. $f(x) = |x - 2| - 2$

The graph of this function has its "vertex" at $(2, -2)$, so the correct graph is **A**.

9. $f(x) = |x + 1|$

Since x can be any real number, the domain is $(-\infty, \infty)$.

The value of y is always greater than or equal to 0, so the range is $[0, \infty)$.

The graph of $y = |x + 1|$ looks like the graph of the absolute value function $y = |x|$, but the graph is translated 1 unit to the left. The x-value of its "vertex" is obtained by setting $x + 1 = 0$ and solving for x:

$$x + 1 = 0$$
$$x = -1.$$

Since the corresponding y-value is 0, the "vertex" is $(-1, 0)$. Some additional points are $(-3, 2)$, $(-2, 1)$, $(0, 1)$, and $(1, 2)$.

11. $f(x) = \dfrac{1}{x} + 1$

The graph of this function is similar to the graph of $g(x) = \dfrac{1}{x}$, except that each point is translated 1 unit upward. Just as with $g(x) = \dfrac{1}{x}$, $x = 0$ is the vertical asymptote, but this graph has $y = 1$ as its horizontal asymptote.

The domain is all real numbers except 0, that is, $(-\infty, 0) \cup (0, \infty)$.

The range is all real numbers except 1, that is, $(-\infty, 1) \cup (1, \infty)$.

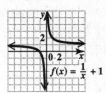

13. $f(x) = \sqrt{x - 2}$

The graph is found by shifting the graph of $y = \sqrt{x}$ two units to the right. The following table of ordered pairs gives some specific points the graph passes through.

x	2	3	6
y	0	1	2

The domain of the function is $[2, \infty)$ and its range is $[0, \infty)$.

15. $f(x) = \dfrac{1}{x - 2}$

This is the graph of the reciprocal function, $g(x) = \dfrac{1}{x}$, shifted 2 units to the right. Since $x \neq 2$ (or a denominator of 0 results), the line $x = 2$ is a vertical asymptote. Since $\dfrac{1}{x - 2} \neq 0$, the line $y = 0$ is a horizontal asymptote.

x	-1	0	1	$\frac{3}{2}$
y	$-\frac{1}{3}$	$-\frac{1}{2}$	-1	-2

x	$\frac{5}{2}$	3	4	5
y	2	1	$\frac{1}{2}$	$\frac{1}{3}$

The domain of the function is $(-\infty, 2) \cup (2, \infty)$ and its range is $(-\infty, 0) \cup (0, \infty)$.

17. $f(x) = \sqrt{x+3} - 3$

The graph is found by shifting the graph of $y = \sqrt{x}$ three units to the left and three units down. The following table of ordered pairs gives some specific points the graph passes through.

x	-3	-2	1
y	-3	-2	-1

$f(x) = \sqrt{x+3} - 3$

The domain of the function is $[-3, \infty)$ and its range is $[-3, \infty)$.

In Exercises 18–25, the answer is the greatest integer that is less than or equal to the number in the greatest integer symbol.

19. $[\![3]\!] = 3 \quad (3 \le 3)$

21. $[\![\frac{1}{2}]\!] = 0 \quad (0 \le \frac{1}{2})$

23. $[\![-14]\!] = -14 \quad (-14 \le -14)$

25. $[\![-10.1]\!] = -11 \quad (-11 \le -10.1)$

27. $f(x) = [\![x - 3]\!]$

This is the graph of the greatest integer function, $g(x) = [\![x]\!]$, shifted 3 units to the right.

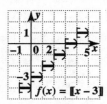

$f(x) = [\![x - 3]\!]$

29. For any portion of the first ounce, the cost will be one 39¢ stamp. If the weight exceeds one ounce (up to two ounces), an additional 24¢ stamp is required. The following table summarizes the weight of a letter, x, and the number of stamps required, $p(x)$, on the interval $(0, 5]$.

x	$(0, 1]$	$(1, 2]$	$(2, 3]$	$(3, 4]$	$(4, 5]$
$p(x)$	1	2	3	4	5

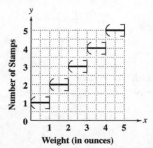

In Exercises 31–34, use the distance formula

$$d = \sqrt{(x_2 - x_1)^2 + (y_2 - y_1)^2}.$$

31. $(2, -1)$ and $(4, 3)$

$$\begin{aligned} d &= \sqrt{(4-2)^2 + [3-(-1)]^2} \\ &= \sqrt{2^2 + 4^2} = \sqrt{4 + 16} \\ &= \sqrt{20} = \sqrt{4} \cdot \sqrt{5} = 2\sqrt{5} \end{aligned}$$

33. (x, y) and $(-2, 5)$

$$\begin{aligned} d &= \sqrt{[x-(-2)]^2 + (y-5)^2} \\ &= \sqrt{(x+2)^2 + (y-5)^2} \end{aligned}$$

13.2 The Circle and the Ellipse

1. **(a)** $x^2 + y^2 = 25$ can be written in the center-radius form as

$$(x-0)^2 + (y-0)^2 = 5^2.$$

The center is the point $(0, 0)$.

(b) The radius is 5.

(c) The x-intercepts are $(5, 0)$ and $(-5, 0)$. The y-intercepts are $(0, 5)$ and $(0, -5)$.

$x^2 + y^2 = 25$

3. $(x-3)^2 + (y-2)^2 = 25$ is an equation of a circle with center $(3, 2)$ and radius 5, choice **B**.

5. $(x+3)^2 + (y-2)^2 = 25$ is an equation of a circle with center $(-3, 2)$ and radius 5, choice **D**.

7. Center: $(-4, 3)$; radius: 2

Substitute $h = -4$, $k = 3$, and $r = 2$ in the center-radius form of the equation of a circle.

$$\begin{aligned} (x-h)^2 + (y-k)^2 &= r^2 \\ [x-(-4)]^2 + (y-3)^2 &= 2^2 \\ (x+4)^2 + (y-3)^2 &= 4 \end{aligned}$$

9. Center: $(-8, -5)$; radius: $\sqrt{5}$

Substitute $h = -8$, $k = -5$, and $r = \sqrt{5}$ in the center-radius form of the equation of a circle.

$$\begin{aligned} (x-h)^2 + (y-k)^2 &= r^2 \\ [x-(-8)]^2 + [y-(-5)]^2 &= \left(\sqrt{5}\right)^2 \\ (x+8)^2 + (y+5)^2 &= 5 \end{aligned}$$

11. $x^2 + y^2 + 4x + 6y + 9 = 0$

Rewrite the equation keeping only the variable terms on the left and grouping the x-terms and y-terms.

$$x^2 + 4x + y^2 + 6y = -9$$

Complete both squares on the left, and add the same constants to the right.

$$\left(x^2 + 4x + \underline{4}\right) + \left(y^2 + 6y + \underline{9}\right) = -9 + \underline{4} + \underline{9}$$
$$(x + 2)^2 + (y + 3)^2 = 4$$

From the form $(x - h)^2 + (y - k)^2 = r^2$, we have $h = -2$, $k = -3$, and $r = 2$. The center is $(-2, -3)$, and the radius r is 2.

13. $x^2 + y^2 + 10x - 14y - 7 = 0$

$$\left(x^2 + 10x \quad\right) + \left(y^2 - 14y \quad\right) = 7$$
$$\left(x^2 + 10x + \underline{25}\right) + \left(y^2 - 14y + \underline{49}\right)$$
$$= 7 + \underline{25} + \underline{49}$$
$$(x + 5)^2 + (y - 7)^2 = 81$$

The center is $(-5, 7)$, and the radius is $\sqrt{81} = 9$.

15. $3x^2 + 3y^2 - 12x - 24y + 12 = 0$

$$3\left(x^2 - 4x \quad\right) + 3\left(y^2 - 8y \quad\right) = -12$$

Divide by 3.

$$\left(x^2 - 4x \quad\right) + \left(y^2 - 8y \quad\right) = -4$$
$$\left(x^2 - 4x + \underline{4}\right) + \left(y^2 - 8y + \underline{16}\right)$$
$$= -4 + \underline{4} + \underline{16}$$
$$(x - 2)^2 + (y - 4)^2 = 16$$

The center is $(2, 4)$, and the radius is $\sqrt{16} = 4$.

17. $x^2 + y^2 = 9$
$$(x - 0)^2 + (y - 0)^2 = 3^2$$

Here, $h = 0$, $k = 0$, and $r = 3$, so the graph is a circle with center $(0, 0)$ and radius 3.

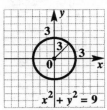

19. $2y^2 = 10 - 2x^2$
$$2x^2 + 2y^2 = 10$$
$$x^2 + y^2 = 5 \qquad \textit{Divide by 2.}$$
$$(x - 0)^2 + (y - 0)^2 = \left(\sqrt{5}\right)^2$$

Here, $h = 0$, $k = 0$, and $r = \sqrt{5} \approx 2.2$, so the graph is a circle with center $(0, 0)$ and radius $\sqrt{5}$.

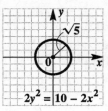

$$2y^2 = 10 - 2x^2$$

21. $(x + 3)^2 + (y - 2)^2 = 9$

Here, $h = -3$, $k = 2$, and $r = \sqrt{9} = 3$. The graph is a circle with center $(-3, 2)$ and radius 3.

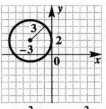

$$(x + 3)^2 + (y - 2)^2 = 9$$

23. $x^2 + y^2 - 4x - 6y + 9 = 0$
$$\left(x^2 - 4x \quad\right) + \left(y^2 - 6y \quad\right) = -9$$

Complete the square for each variable.

$$\left(x^2 - 4x + \underline{4}\right) + \left(y^2 - 6y + \underline{9}\right)$$
$$= -9 + \underline{4} + \underline{9}$$
$$(x - 2)^2 + (y - 3)^2 = 4$$

Here, $h = 2$, $k = 3$, and $r = \sqrt{4} = 2$. The graph is a circle with center $(2, 3)$ and radius 2.

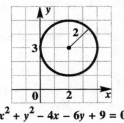

$$x^2 + y^2 - 4x - 6y + 9 = 0$$

25. $x^2 + y^2 + 6x - 6y + 9 = 0$

$$\left(x^2 + 6x \quad\right) + \left(y^2 - 6y \quad\right) = -9$$
$$\left(x^2 + 6x + \underline{9}\right) + \left(y^2 - 6y + \underline{9}\right) = -9 + \underline{9} + \underline{9}$$
$$(x + 3)^2 + (y - 3)^2 = 9$$

The center is $(-3, 3)$, and the radius is $\sqrt{9} = 3$.

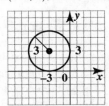

$$x^2 + y^2 + 6x - 6y + 9 = 0$$

27. This method works because the pencil is always the same distance from the fastened end. The fastened end works as the center, and the length of the string from the fastened end to the pencil is the radius.

29. The equation $\dfrac{x^2}{9} + \dfrac{y^2}{25} = 1$ is of the form

$\dfrac{x^2}{a^2} + \dfrac{y^2}{b^2} = 1$. The graph is an ellipse with

$a^2 = 9$ and $b^2 = 25$, so $a = 3$ and $b = 5$. The x-intercepts are $(3, 0)$ and $(-3, 0)$. The y-intercepts are $(0, 5)$ and $(0, -5)$. Plot the intercepts, and draw the ellipse through them.

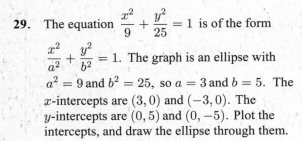

31. $\dfrac{x^2}{36} = 1 - \dfrac{y^2}{16}$

$\dfrac{x^2}{36} + \dfrac{y^2}{16} = 1$ is in the form $\dfrac{x^2}{a^2} + \dfrac{y^2}{b^2} = 1$.

The graph is an ellipse with $a^2 = 36$ and $b^2 = 16$, so $a = 6$ and $b = 4$. The x-intercepts are $(6, 0)$ and $(-6, 0)$. The y-intercepts are $(0, 4)$ and $(0, -4)$. Plot the intercepts, and draw the ellipse through them.

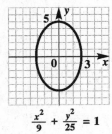

33. $\dfrac{y^2}{25} = 1 - \dfrac{x^2}{49}$

$\dfrac{x^2}{49} + \dfrac{y^2}{25} = 1$ is in the form $\dfrac{x^2}{a^2} + \dfrac{y^2}{b^2} = 1$.

The graph is an ellipse with $a^2 = 49$ and $b^2 = 25$, so $a = 7$ and $b = 5$. The x-intercepts are $(7, 0)$ and $(-7, 0)$. The y-intercepts are $(0, 5)$ and $(0, -5)$. Plot the intercepts, and draw the ellipse through them.

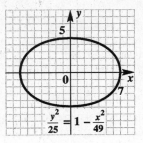

35. $\dfrac{x^2}{16} + \dfrac{y^2}{4} = 1$ is in the form

$$\dfrac{x^2}{a^2} + \dfrac{y^2}{b^2} = 1,$$

so $a = 4$ and $b = 2$. Its x-intercepts are $(4, 0)$ and $(-4, 0)$, and its y-intercepts are $(0, 2)$ and $(0, -2)$. Plot the intercepts, and draw the ellipse through them.

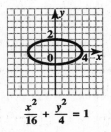

37. $\dfrac{(x + 1)^2}{64} + \dfrac{(y - 2)^2}{49} = 1$

This equation is of the form

$$\dfrac{(x - h)^2}{a^2} + \dfrac{(y - k)^2}{b^2} = 1,$$

so the center of the ellipse is at $(-1, 2)$. Since $a^2 = 64$, $a = 8$. Since $b^2 = 49$, $b = 7$. Add ± 8 to -1, and add ± 7 to 2 to find the points $(7, 2)$, $(-9, 2)$, $(-1, 9)$, and $(-1, -5)$. Plot the points, and draw the ellipse through them.

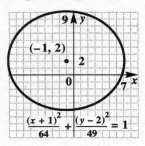

39. $\dfrac{(x - 2)^2}{16} + \dfrac{(y - 1)^2}{9} = 1$

The center of the ellipse is at $(2, 1)$. Since $a^2 = 16$, $a = 4$. Since $b^2 = 9$, $b = 3$. Add ± 4 to 2, and add ± 3 to 1 to find the points $(6, 1)$, $(-2, 1)$, $(2, 4)$, and $(2, -2)$. Plot the points, and draw the ellipse through them.

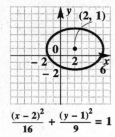

$$\frac{(x-2)^2}{16} + \frac{(y-1)^2}{9} = 1$$

41. By the vertical line test the set is not a function, because a vertical line may intersect the graph of an ellipse in two points.

43. $(x+2)^2 + (y-4)^2 = 16$
$$(y-4)^2 = 16 - (x+2)^2$$

Take the square root of each side.

$$y - 4 = \pm\sqrt{16 - (x+2)^2}$$
$$y = 4 \pm \sqrt{16 - (x+2)^2}$$

Therefore, the two functions used to obtain the graph were

$$y_1 = 4 + \sqrt{16 - (x+2)^2} \quad \text{and}$$
$$y_2 = 4 - \sqrt{16 - (x+2)^2}.$$

45. $x^2 + y^2 = 36$
$$y^2 = 36 - x^2$$

Take the square root of both sides.

$$y = \pm\sqrt{36 - x^2}$$

Therefore,

$$y_1 = \sqrt{36 - x^2} \quad \text{and} \quad y_2 = -\sqrt{36 - x^2}.$$

Use these two functions to obtain the graph.

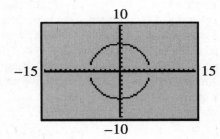

47. $\frac{x^2}{16} + \frac{y^2}{4} = 1$
$$\frac{y^2}{4} = 1 - \frac{x^2}{16}$$
$$y^2 = 4\left(1 - \frac{x^2}{16}\right)$$

Take the square root of both sides.

$$y = \pm 2\sqrt{1 - \frac{x^2}{16}}$$

Therefore,

$$y_1 = 2\sqrt{1 - \frac{x^2}{16}} \quad \text{and} \quad y_2 = -2\sqrt{1 - \frac{x^2}{16}}.$$

Use these two functions to obtain the graph.

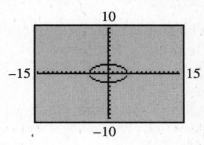

49. $\frac{x^2}{36} + \frac{y^2}{9} = 1$

$$c^2 = a^2 - b^2 = 36 - 9 = 27, \text{ so}$$
$$c = \sqrt{27} = 3\sqrt{3}.$$

The kidney stone and the source of the beam must be placed $3\sqrt{3}$ units from the center of the ellipse.

51. **(a)** $100x^2 + 324y^2 = 32,400$

$$\frac{x^2}{324} + \frac{y^2}{100} = 1 \qquad \textit{Divide by 32,400.}$$
$$\frac{x^2}{18^2} + \frac{y^2}{10^2} = 1$$

The height in the center is the y-coordinate of the positive y-intercept. The height is 10 meters.

(b) The width of the ellipse is the distance between the x-intercepts, $(-18,0)$ and $(18,0)$. The width across the bottom of the arch is $18 + 18 = 36$ meters.

53. $\frac{x^2}{141.7^2} + \frac{y^2}{141.1^2} = 1$

(a) $c^2 = a^2 - b^2$, so
$$c = \sqrt{a^2 - b^2} = \sqrt{141.7^2 - 141.1^2}$$
$$= \sqrt{169.68} \approx 13.0$$

From the figure, the *apogee* is
$a + c = 141.7 + 13.0 = 154.7$ million miles.

(b) The *perigee* is $a - c = 141.7 - 13.0 = 128.7$ million miles.

55. Plot the points $(3, 4), (-3, 4), (3, -4)$, and $(-3, -4)$.

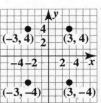

57. $4x + 3y = 12$

If $y = 0$, then $4x = 12$, so $x = 3$ and $(3, 0)$ is the x-intercept.
If $x = 0$, then $3y = 12$, so $y = 4$ and $(0, 4)$ is the y-intercept.

13.3 The Hyperbola and Functions Defined by Radicals

1. $\dfrac{x^2}{25} + \dfrac{y^2}{9} = 1$

This is the standard form for the equation of an ellipse with x-intercepts $(5, 0)$ and $(-5, 0)$ and y-intercepts $(0, 3)$ and $(0, -3)$. This is graph **C**.

3. $\dfrac{x^2}{9} - \dfrac{y^2}{25} = 1$

This is the standard form for the equation of a hyperbola that opens left and right. Its x-intercepts are $(3, 0)$ and $(-3, 0)$. This is graph **D**.

5. If the equation of a hyperbola is in standard form (that is, equal to one), the hyperbola would open to the left and right if the x^2-term was positive. It would open up and down if the y^2-term was positive.

7. The equation $\dfrac{x^2}{16} - \dfrac{y^2}{9} = 1$ is in the form $\dfrac{x^2}{a^2} - \dfrac{y^2}{b^2} = 1$. The graph is a hyperbola with $a = 4$ and $b = 3$. The x-intercepts are $(4, 0)$ and $(-4, 0)$. There are no y-intercepts. The vertices of the fundamental rectangle are $(4, 3)$, $(4, -3)$, $(-4, -3)$, and $(-4, 3)$. Extend the diagonals of the rectangle through these points to get the asymptotes. Graph a branch of the hyperbola through each intercept and approaching the asymptotes.

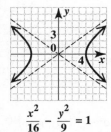

$$\dfrac{x^2}{16} - \dfrac{y^2}{9} = 1$$

9. $\dfrac{y^2}{4} - \dfrac{x^2}{25} = 1$ is in the form $\dfrac{y^2}{b^2} - \dfrac{x^2}{a^2} = 1$, where $b = 2$ and $a = 5$. The branches open upward and downward, so the hyperbola has y-intercepts $(0, 2)$ and $(0, -2)$. The fundamental rectangle has vertices $(5, 2)$, $(5, -2)$, $(-5, -2)$, and $(-5, 2)$.

Sketch the extended diagonals, which are the asymptotes of this hyperbola. Graph a branch of the hyperbola through each intercept and approaching the asymptotes.

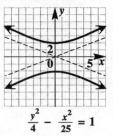

$$\dfrac{y^2}{4} - \dfrac{x^2}{25} = 1$$

11. $\dfrac{x^2}{25} - \dfrac{y^2}{36} = 1$ is a hyperbola with $a = 5$ and $b = 6$. The x-intercepts are $(5, 0)$ and $(-5, 0)$. There are no y-intercepts. To sketch the graph, draw the extended diagonals of the fundamental rectangle with vertices $(5, 6)$, $(5, -6)$, $(-5, -6)$ and $(-5, 6)$. Graph a branch of the hyperbola through each intercept and approaching the asymptotes.

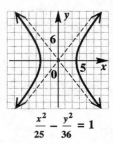

$$\dfrac{x^2}{25} - \dfrac{y^2}{36} = 1$$

13. $\dfrac{y^2}{16} - \dfrac{x^2}{16} = 1$ is in the form $\dfrac{y^2}{b^2} - \dfrac{x^2}{a^2} = 1$, where $b = 4$ and $a = 4$. The branches open upward and downward, so the hyperbola has y-intercepts $(0, 4)$ and $(0, -4)$. The fundamental rectangle has vertices $(4, 4)$, $(4, -4)$, $(-4, -4)$, and $(-4, 4)$. Sketch the extended diagonals, which are the asymptotes of this hyperbola. Graph a branch of the hyperbola through each intercept and approaching the asymptotes.

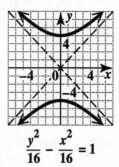

$$\dfrac{y^2}{16} - \dfrac{x^2}{16} = 1$$

15. $x^2 - y^2 = 16$

$\dfrac{x^2}{16} - \dfrac{y^2}{16} = 1$ *Divide by 16.*

This equation is in the form $\dfrac{x^2}{a^2} - \dfrac{y^2}{b^2} = 1$ with $a = 4$ and $b = 4$. The graph is a hyperbola with x-intercepts $(4, 0)$ and $(-4, 0)$ and no y-intercepts. One asymptote passes through $(4, 4)$ and $(-4, -4)$. The other asymptote passes through $(-4, 4)$ and $(4, -4)$. Sketch the graph through the intercepts and approaching the asymptotes.

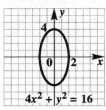

17. $4x^2 + y^2 = 16$

$\dfrac{x^2}{4} + \dfrac{y^2}{16} = 1$ *Divide by 16.*

This equation is in the form $\dfrac{x^2}{a^2} + \dfrac{y^2}{b^2} = 1$ with $a = 2$ and $b = 4$. The graph is an ellipse. The x-intercepts $(2, 0)$ and $(-2, 0)$. The y-intercepts are $(0, 4)$ and $(0, -4)$. Plot the intercepts and draw the ellipse through them.

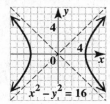

19. $x^2 - 2y = 0$

$\quad x^2 = 2y$

$\quad \tfrac{1}{2}x^2 = y$

The equation is in the form

$$f(x) = a(x - h)^2 + k,$$

with $a = \tfrac{1}{2}$, $h = 0$, and $k = 0$. The graph is a parabola that opens up and is wider than the graph of $f(x) = x^2$ because $|a| = \tfrac{1}{2} < 1$. The vertex is at $(0, 0)$.

x	1	2	3
y	$\tfrac{1}{2}$	2	4.5

Use symmetry about $x = 0$ to draw the graph.

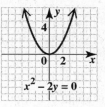

21. $\quad\quad 9x^2 = 144 + 16y^2$

$9x^2 - 16y^2 = 144$

$\dfrac{x^2}{16} - \dfrac{y^2}{9} = 1$ *Divide by 144.*

The equation is a hyperbola in the form $\dfrac{x^2}{a^2} - \dfrac{y^2}{b^2} = 1$ with $a = 4$ and $b = 3$. The x-intercepts are $(4, 0)$ and $(-4, 0)$. There are no y-intercepts. To sketch the graph, draw the extended diagonals of the fundamental rectangle with vertices $(4, 3)$, $(4, -3)$, $(-4, -3)$ and $(-4, 3)$. These are the asymptotes. Graph a branch of the hyperbola through each intercept approaching the asymptotes.

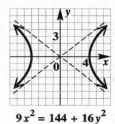

23. $\quad\quad y^2 = 4 + x^2$

$\quad y^2 - x^2 = 4$

$\dfrac{y^2}{4} - \dfrac{x^2}{4} = 1$ *Divide by 4.*

The graph is a hyperbola with y-intercepts $(0, 2)$ and $(0, -2)$. One asymptote passes through $(2, 2)$ and $(-2, -2)$. The other asymptote passes through $(-2, 2)$ and $(2, -2)$.

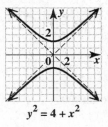

25. $f(x) = \sqrt{16 - x^2}$

Replace $f(x)$ with y and square both sides to get the equation

$$y^2 = 16 - x^2 \quad \text{or} \quad x^2 + y^2 = 16.$$

This is the graph of a circle with center $(0, 0)$ and radius 4. Since $f(x)$, or y, represents a principal square root in the original equation, $f(x)$ must be nonnegative. This restricts the graph to the upper half of the circle.

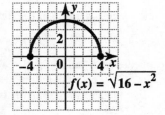

The domain is $[-4, 4]$, and the range is $[0, 4]$.

27. $$f(x) = -\sqrt{36 - x^2}$$

Replace $f(x)$ with y, and square both sides of the equation.

$$y = -\sqrt{36 - x^2}$$
$$y^2 = 36 - x^2$$
$$x^2 + y^2 = 36$$

This is a circle centered at the origin with radius $\sqrt{36} = 6$. Since $f(x)$, or y, represents a nonpositive square root in the original equation, $f(x)$ must be nonpositive. This restricts the graph to the bottom half of the circle.

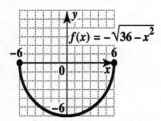

The domain is $[-6, 6]$, and the range is $[-6, 0]$.

29. $$\frac{y}{3} = \sqrt{1 + \frac{x^2}{9}}$$

Square both sides.

$$\frac{y^2}{9} = 1 + \frac{x^2}{9}$$
$$\frac{y^2}{9} - \frac{x^2}{9} = 1$$

This is a hyperbola opening up and down with y-intercepts $(0, 3)$ and $(0, -3)$. The four points $(3, 3)$, $(3, -3)$, $(-3, 3)$, and $(-3, -3)$ are the vertices of the rectangle that determine the asymptotes. Since $f(x)$, or y, represents a square root in the original equation, $f(x)$ must be

nonnegative. This restricts the graph to the upper half of the hyperbola.

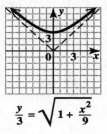

$$\frac{y}{3} = \sqrt{1 + \frac{x^2}{9}}$$

The domain is $(-\infty, \infty)$, and the range is $[3, \infty)$.

31. $$\frac{(x - 2)^2}{4} - \frac{(y + 1)^2}{9} = 1$$

is a hyperbola centered at $(2, -1)$, with $a = 2$ and $b = 3$. The x-intercepts are $(2 \pm 2, -1)$ or $(4, -1)$ and $(0, -1)$. The asymptotes are the extended diagonals of the rectangle with vertices $(2, 3)$, $(2, -3)$, $(-2, -3)$ and $(-2, 3)$ shifted 2 units right and 1 unit down, or $(4, 2)$, $(4, -4)$, $(0, -4)$ and $(0, 2)$. Draw the hyperbola.

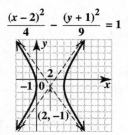

33. $$\frac{y^2}{36} - \frac{(x - 2)^2}{49} = 1$$

is a hyperbola centered at $(2, 0)$ with $a = 7$, and $b = 6$. The asymptotes are the extended diagonals of the rectangle with vertices $(7, 6)$, $(7, -6)$, $(-7, -6)$, and $(-7, 6)$ shifted right 2 units, or $(9, 6)$, $(9, -6)$, $(-5, -6)$, and $(-5, 6)$. Draw the hyperbola.

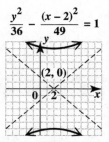

35. **(a)** $400x^2 - 625y^2 = 250{,}000$

$$\frac{x^2}{625} - \frac{y^2}{400} = 1 \quad \textit{Divide by 250,000.}$$

$$\frac{x^2}{25^2} - \frac{y^2}{20^2} = 1$$

The x-intercepts are $(25, 0)$ and $(-25, 0)$. The distance between the buildings is the distance between the x-intercepts. The buildings are $25 + 25 = 50$ meters apart at their closest point.

(b) At $x = 50$, $y = \dfrac{d}{2}$, so $d = 2y$.

$$400(50)^2 - 625y^2 = 250,000$$
$$1,000,000 - 625y^2 = 250,000$$
$$-625y^2 = -750,000$$
$$y^2 = 1200$$
$$y = \sqrt{1200}$$

The distance d is $2\sqrt{1200} \approx 69.3$ meters.

37. $\dfrac{x^2}{9} - y^2 = 1$

$$-y^2 = 1 - \dfrac{x^2}{9}$$

$$y^2 = \dfrac{x^2}{9} - 1 \qquad \textit{Multiply by } -1.$$

Take the square root of both sides.

$$y = \pm\sqrt{\dfrac{x^2}{9} - 1}$$

The two functions used to obtain the graph were

$$y_1 = \sqrt{\dfrac{x^2}{9} - 1} \quad \text{and} \quad y_2 = -\sqrt{\dfrac{x^2}{9} - 1}.$$

39. $\dfrac{x^2}{25} - \dfrac{y^2}{49} = 1$

$$-\dfrac{y^2}{49} = 1 - \dfrac{x^2}{25}$$

$$\dfrac{y^2}{49} = \dfrac{x^2}{25} - 1 \qquad \textit{Multiply by } -1.$$

$$y^2 = 49\left(\dfrac{x^2}{25} - 1\right)$$

Take the square root of both sides.

$$y = \pm 7\sqrt{\dfrac{x^2}{25} - 1}$$

To obtain the graph, use the two functions

$$y_1 = 7\sqrt{\dfrac{x^2}{25} - 1} \quad \text{and} \quad y_2 = -7\sqrt{\dfrac{x^2}{25} - 1}.$$

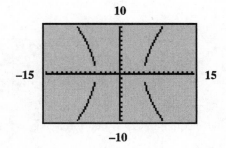

41. $y^2 - 9x^2 = 9$

$$y^2 = 9 + 9x^2$$
$$y^2 = 9(1 + x^2)$$

Take the square root of both sides.

$$y = \pm 3\sqrt{1 + x^2}$$

To obtain the graphs, use the two functions

$$y_1 = 3\sqrt{1 + x^2} \quad \text{and} \quad y_2 = -3\sqrt{1 + x^2}.$$

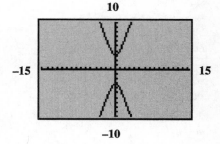

43. $\quad 2x + y = 13 \quad (1)$
$\qquad\quad y = 3x + 3 \quad (2)$

Substitute $3x + 3$ for y in (1).

$$2x + (3x + 3) = 13$$
$$5x + 3 = 13$$
$$5x = 10$$
$$x = 2$$

Substitute 2 for x in (2).

$$y = 3(2) + 3 = 9$$

The solution set is $\{(2, 9)\}$.

45. $\quad 9x + 2y = 10 \quad (1)$
$\qquad\quad x - y = -5 \quad (2)$

Solve (2) for x.

$$x = y - 5 \quad (3)$$

Substitute $y - 5$ for x in (1).

$$9(y - 5) + 2y = 10$$
$$9y - 45 + 2y = 10$$
$$11y = 55$$
$$y = 5$$

Substitute 5 for y in (3).

$$x = 5 - 5 = 0$$

The solution set is $\{(0, 5)\}$.

47. $2x^4 - 5x^2 - 3 = 0$

Let $u = x^2$, so that $u^2 = x^4$.

$2u^2 - 5u - 3 = 0$

$(2u + 1)(u - 3) = 0$

$2u + 1 = 0$ or $u - 3 = 0$

$u = -\frac{1}{2}$ or $u = 3$

$x^2 = -\frac{1}{2}$ or $x^2 = 3$

$x = \pm\sqrt{-\frac{2}{4}}$ or $x = \pm\sqrt{3}$

$x = \pm\frac{\sqrt{2}}{2}i$ or $x = \pm\sqrt{3}$

The solution set is $\left\{-\sqrt{3}, \sqrt{3}, -\frac{\sqrt{2}}{2}i, \frac{\sqrt{2}}{2}i\right\}$.

49. $x^4 - 7x^2 + 12 = 0$

Let $u = x^2$, so that $u^2 = x^4$.

$u^2 - 7u + 12 = 0$

$(u - 4)(u - 3) = 0$

$u - 4 = 0$ or $u - 3 = 0$

$u = 4$ or $u = 3$

$x^2 = 4$ or $x^2 = 3$

$x = \pm\sqrt{4}$ or $x = \pm\sqrt{3}$

$x = \pm 2$ or $x = \pm\sqrt{3}$

The solution set is $\left\{-2, -\sqrt{3}, \sqrt{3}, 2\right\}$.

13.4 Nonlinear Systems of Equations

1. Substitute $x - 1$ for y in the first equation. Then solve for x. Find the corresponding y-values by substituting back into $y = x - 1$. In the first equation, both variables are squared and in the second, both variables are to the first power, so the elimination method is not appropriate.

3. The line intersects the ellipse in exactly one point, so there is *one* point in the solution set of the system.

5. The line does not intersect the hyperbola, so there are no points in the solution set of the system.

7. A line and a circle; no points

Draw any circle, and then draw a line that does not cross the circle.

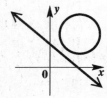

9. A line and a hyperbola; one point

The line is tangent to the hyperbola.

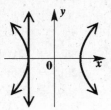

11. A circle and an ellipse; four points

Draw any ellipse, and then draw a circle with the same center whose radius is just large enough so that there are four points of intersection. (If the radius of the circle is too large or too small, there may be fewer points of intersection.)

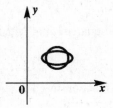

13. A parabola and an ellipse; four points

Draw any parabola, and then draw an ellipse large enough so that there are four points of intersection. (If the ellipse is too large or too small, there may be fewer points of intersection.)

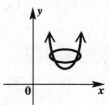

15. $y = 4x^2 - x$ (1)

$y = x$ (2)

Substitute x for y in equation (1).

$$y = 4x^2 - x \quad (1)$$
$$x = 4x^2 - x$$
$$0 = 4x^2 - 2x$$
$$0 = 2x(2x - 1)$$

$2x = 0$ or $2x - 1 = 0$

$x = 0$ or $x = \frac{1}{2}$

Use equation (2) to find y for each x-value.

If $x = 0$, then $y = 0$.

If $x = \frac{1}{2}$, then $y = \frac{1}{2}$.

The solution set is $\left\{(0, 0), \left(\frac{1}{2}, \frac{1}{2}\right)\right\}$.

17.
$$y = x^2 + 6x + 9 \quad (1)$$
$$x + y = 3 \quad\quad\quad (2)$$

Substitute $x^2 + 6x + 9$ for y in equation (2).

$$x + y = 3 \quad (2)$$
$$x + (x^2 + 6x + 9) = 3$$
$$x^2 + 7x + 9 = 3$$
$$x^2 + 7x + 6 = 0$$
$$(x + 6)(x + 1) = 0$$

$$x + 6 = 0 \quad \text{or} \quad x + 1 = 0$$
$$x = -6 \quad \text{or} \quad\quad x = -1$$

Substitute these values for x in equation (2) and solve for y.

If $x = -6$, then
$$x + y = 3 \quad (2)$$
$$-6 + y = 3$$
$$y = 9.$$

If $x = -1$, then
$$x + y = 3 \quad (2)$$
$$-1 + y = 3$$
$$y = 4.$$

The solution set is $\{(-6, 9), (-1, 4)\}$.

19. $x^2 + y^2 = 2 \quad (1)$
$2x + y = 1 \quad (2)$

Solve equation (2) for y.

$$y = 1 - 2x \quad (3)$$

Substitute $1 - 2x$ for y in equation (1).

$$x^2 + y^2 = 2 \quad (1)$$
$$x^2 + (1 - 2x)^2 = 2$$
$$x^2 + 1 - 4x + 4x^2 = 2$$
$$5x^2 - 4x - 1 = 0$$
$$(5x + 1)(x - 1) = 0$$

$$5x + 1 = 0 \quad \text{or} \quad x - 1 = 0$$
$$x = -\tfrac{1}{5} \quad \text{or} \quad\quad x = 1$$

Use equation (3) to find y for each x-value.

If $x = -\tfrac{1}{5}$, then
$$y = 1 - 2\left(-\tfrac{1}{5}\right) = 1 + \tfrac{2}{5} = \tfrac{7}{5}.$$

If $x = 1$, then
$$y = 1 - 2(1) = -1.$$

The solution set is $\left\{\left(-\tfrac{1}{5}, \tfrac{7}{5}\right), (1, -1)\right\}$.

21.
$$xy = 4 \quad\quad (1)$$
$$3x + 2y = -10 \quad (2)$$

Solve equation (1) for y to get $y = \dfrac{4}{x}$.

Substitute $\dfrac{4}{x}$ for y in equation (2) to find x.

$$3x + 2y = -10 \quad (2)$$
$$3x + 2\left(\frac{4}{x}\right) = -10$$

Multiply by the LCD, x.

$$3x^2 + 8 = -10x$$
$$3x^2 + 10x + 8 = 0$$
$$(3x + 4)(x + 2) = 0$$

$$3x + 4 = 0 \quad \text{or} \quad x + 2 = 0$$
$$x = -\tfrac{4}{3} \quad \text{or} \quad\quad x = -2$$

Since $y = \dfrac{4}{x}$, if $x = -\dfrac{4}{3}$, then $y = \dfrac{4}{-\frac{4}{3}} = -3$.

If $x = -2$, then
$$y = \tfrac{4}{-2} = -2.$$

The solution set is $\left\{(-2, -2), \left(-\tfrac{4}{3}, -3\right)\right\}$.

23.
$$xy = -3 \quad (1)$$
$$x + y = -2 \quad (2)$$

Solve equation (2) for y.

$$y = -x - 2 \quad (3)$$

Substitute $-x - 2$ for y in equation (1).

$$xy = -3 \quad (1)$$
$$x(-x - 2) = -3$$
$$-x^2 - 2x = -3$$
$$-x^2 - 2x + 3 = 0$$
$$x^2 + 2x - 3 = 0 \quad\quad \textit{Multiply by } -1.$$
$$(x + 3)(x - 1) = 0$$

$$x + 3 = 0 \quad \text{or} \quad x - 1 = 0$$
$$x = -3 \quad \text{or} \quad\quad x = 1$$

Use equation (3) to find y for each x-value.

If $x = -3$, then
$$y = -(-3) - 2 = 1.$$

If $x = 1$, then
$$y = -(1) - 2 = -3.$$

The solution set is $\{(-3, 1), (1, -3)\}$.

25. $y = 3x^2 + 6x$ (1)
$y = x^2 - x - 6$ (2)

Substitute $x^2 - x - 6$ for y in equation (1) to find x.

$$y = 3x^2 + 6x \qquad (1)$$
$$x^2 - x - 6 = 3x^2 + 6x$$
$$0 = 2x^2 + 7x + 6$$
$$0 = (2x + 3)(x + 2)$$

$$2x + 3 = 0 \quad \text{or} \quad x + 2 = 0$$
$$x = -\tfrac{3}{2} \quad \text{or} \qquad x = -2$$

Substitute $-\tfrac{3}{2}$ for x in equation (1) to find y.

$$y = 3x^2 + 6x \qquad (1)$$
$$y = 3\left(-\tfrac{3}{2}\right)^2 + 6\left(-\tfrac{3}{2}\right)$$
$$= 3\left(\tfrac{9}{4}\right) + 6\left(-\tfrac{6}{4}\right)$$
$$= \tfrac{27}{4} - \tfrac{36}{4} = -\tfrac{9}{4}$$

Substitute -2 for x in equation (1) to find y.

$$y = 3x^2 + 6x \qquad (1)$$
$$y = 3(-2)^2 + 6(-2)$$
$$= 12 - 12 = 0$$

The solution set is $\left\{\left(-\tfrac{3}{2}, -\tfrac{9}{4}\right), (-2, 0)\right\}$.

27. $2x^2 - y^2 = 6$ (1)
$y = x^2 - 3$ (2)

Substitute $x^2 - 3$ for y in equation (1).

$$2x^2 - y^2 = 6 \quad (1)$$
$$2x^2 - \left(x^2 - 3\right)^2 = 6$$
$$2x^2 - \left(x^4 - 6x^2 + 9\right) = 6$$
$$-x^4 + 8x^2 - 9 = 6$$
$$-x^4 + 8x^2 - 15 = 0$$
$$x^4 - 8x^2 + 15 = 0 \qquad \textit{Multiply by } -1.$$

Let $z = x^2$, so $z^2 = x^4$.

$$z^2 - 8z + 15 = 0$$
$$(z - 3)(z - 5) = 0$$

$$z - 3 = 0 \quad \text{or} \quad z - 5 = 0$$
$$z = 3 \quad \text{or} \qquad z = 5$$

Since $z = x^2$,

$$x^2 = 3 \quad \text{or} \quad x^2 = 5.$$

Use the square root property.

$$x = \pm\sqrt{3} \quad \text{or} \quad x = \pm\sqrt{5}$$

Use equation (2) to find y for each x-value.

If $x = \sqrt{3}$ or $-\sqrt{3}$, then

$$y = \left(\pm\sqrt{3}\right)^2 - 3 = 0.$$

(Note: We could substitute 3 for x^2 in (2) to obtain the same result.)

If $x = \sqrt{5}$ or $-\sqrt{5}$, then

$$y = \left(\pm\sqrt{5}\right)^2 - 3 = 2.$$

The solution set is

$$\left\{\left(-\sqrt{3}, 0\right), \left(\sqrt{3}, 0\right), \left(-\sqrt{5}, 2\right), \left(\sqrt{5}, 2\right)\right\}.$$

29. $x^2 - xy + y^2 = 0$ (1)
$x - 2y = 1$ (2)

Solve equation (2) for x.

$$x = 2y + 1 \qquad (3)$$

Substitute $2y + 1$ for x in equation (1).

$$(2y + 1)^2 - (2y + 1)y + y^2 = 0$$
$$4y^2 + 4y + 1 - 2y^2 - y + y^2 = 0$$
$$3y^2 + 3y + 1 = 0$$

Use the quadratic formula with $a = 3$, $b = 3$, and $c = 1$.

$$y = \frac{-b \pm \sqrt{b^2 - 4ac}}{2a}$$

$$y = \frac{-3 \pm \sqrt{9 - 12}}{6}$$

$$= \frac{-3 \pm \sqrt{-3}}{6} = \frac{-3 \pm i\sqrt{3}}{6}$$

Use equation (3) to solve for x.

$$x = 2y + 1 \qquad\qquad (3)$$

$$= 2\left(\frac{-3 \pm i\sqrt{3}}{6}\right) + 1$$

$$= \frac{-3 \pm i\sqrt{3}}{3} + \frac{3}{3}$$

$$= \pm\frac{i\sqrt{3}}{3}$$

The solution set is

$$\left\{\left(\tfrac{\sqrt{3}}{3}i, -\tfrac{1}{2} + \tfrac{\sqrt{3}}{6}i\right), \left(-\tfrac{\sqrt{3}}{3}i, -\tfrac{1}{2} - \tfrac{\sqrt{3}}{6}i\right)\right\}.$$

31. $3x^2 + 2y^2 = 12$ (1)
$x^2 + 2y^2 = 4$ (2)

Multiply equation (2) by -1 and add the result to equation (1).

$$
\begin{array}{rcll}
3x^2 + 2y^2 & = & 12 & (1) \\
-x^2 - 2y^2 & = & -4 & -1 \times (2) \\
\hline
2x^2 & = & 8 & \\
x^2 & = & 4 & \\
x & = & \pm 2 &
\end{array}
$$

Substitute ± 2 for x in equation (2) to find y.

$$x^2 + 2y^2 = 4 \quad (2)$$
$$(\pm 2)^2 + 2y^2 = 4$$
$$4 + 2y^2 = 4$$
$$2y^2 = 0$$
$$y^2 = 0$$
$$y = 0$$

The solution set is $\{(-2, 0), (2, 0)\}$.

33. $2x^2 + 3y^2 = 6 \quad (1)$
$\quad\ \ x^2 + 3y^2 = 3 \quad (2)$

Multiply equation (2) by -1 and add the result to equation (1).

$$\begin{array}{rcll} 2x^2 + 3y^2 &=& 6 & (1) \\ -x^2 - 3y^2 &=& -3 & -1 \times (2) \\ \hline x^2 &=& 3 & \\ x &=& \pm\sqrt{3} & \end{array}$$

Substitute $\pm\sqrt{3}$ for x in equation (2).

$$x^2 + 3y^2 = 3 \quad (2)$$
$$\left(\pm\sqrt{3}\right)^2 + 3y^2 = 3$$
$$3 + 3y^2 = 3$$
$$y^2 = 0$$
$$y = 0$$

The solution set is $\left\{\left(\sqrt{3}, 0\right), \left(-\sqrt{3}, 0\right)\right\}$.

35. $2x^2 + \ y^2 = 28 \quad (1)$
$\quad\ 4x^2 - 5y^2 = 28 \quad (2)$

Multiply (1) by 5 and add the result to equation (2).

$$\begin{array}{rcll} 10x^2 + 5y^2 &=& 140 & 5 \times (1) \\ 4x^2 - 5y^2 &=& 28 & (2) \\ \hline 14x^2 &=& 168 & \\ x^2 &=& 12 & \end{array}$$
$$x = \pm\sqrt{12} = \pm 2\sqrt{3}$$

Let $x = \pm 2\sqrt{3}$ in (1).

$$2\left(\pm 2\sqrt{3}\right)^2 + y^2 = 28$$
$$2(12) + y^2 = 28$$
$$y^2 = 4$$
$$y = \pm 2$$

The solution set is $\left\{\left(2\sqrt{3}, 2\right), \left(2\sqrt{3}, -2\right),\right.$
$\left.\left(-2\sqrt{3}, 2\right), \left(-2\sqrt{3}, -2\right)\right\}$.

37. $2x^2 = 8 - 2y^2 \quad (1)$
$\quad\ 3x^2 = 24 - 4y^2 \quad (2)$

Multiply equation (1) by -2 and add the result to equation (2).

$$\begin{array}{rcll} -4x^2 &=& -16 + 4y^2 & -2 \times (1) \\ 3x^2 &=& 24 - 4y^2 & (2) \\ \hline -x^2 &=& 8 & \\ x^2 &=& -8 & \end{array}$$
$$x = \pm\sqrt{-8} = \pm 2i\sqrt{2}$$

Substitute $\pm 2i\sqrt{2}$ for x in equation (1).

$$2x^2 = 8 - 2y^2 \qquad (1)$$
$$2\left(\pm 2i\sqrt{2}\right)^2 = 8 - 2y^2$$
$$2(-8) = 8 - 2y^2$$
$$-16 = 8 - 2y^2$$
$$2y^2 = 24$$
$$y^2 = 12$$
$$y = \pm\sqrt{12} = \pm 2\sqrt{3}$$

Since $2i\sqrt{2}$ can be paired with either $2\sqrt{3}$ or $-2\sqrt{3}$ and $-2i\sqrt{2}$ can be paired with either $2\sqrt{3}$ or $-2\sqrt{3}$, there are four possible solutions.

The solution set is

$$\left\{\left(-2i\sqrt{2}, -2\sqrt{3}\right), \left(-2i\sqrt{2}, 2\sqrt{3}\right),\right.$$
$$\left.\left(2i\sqrt{2}, -2\sqrt{3}\right), \left(2i\sqrt{2}, 2\sqrt{3}\right)\right\}.$$

39. $x^2 + xy + y^2 = 15 \quad (1)$
$\qquad\quad\ x^2 + y^2 = 10 \quad (2)$

Multiply equation (2) by -1 and add the result to equation (1).

$$\begin{array}{rcll} x^2 + xy + y^2 &=& 15 & (1) \\ -x^2 \quad\ - y^2 &=& -10 & -1 \times (2) \\ \hline xy &=& 5 & \end{array}$$
$$y = \frac{5}{x}$$

Substitute $\dfrac{5}{x}$ for y in equation (2).

$$x^2 + y^2 = 10 \qquad (2)$$
$$x^2 + \left(\frac{5}{x}\right)^2 = 10$$
$$x^2 + \frac{25}{x^2} = 10$$
$$x^4 + 25 = 10x^2 \quad \textit{Multiply by } x^2.$$
$$x^4 - 10x^2 + 25 = 0$$

continued

Let $z = x^2$, so $z^2 = x^4$.

$$z^2 - 10z + 25 = 0$$
$$(z - 5)^2 = 0$$
$$z - 5 = 0$$
$$z = 5$$

Since $z = x^2$,

$$x^2 = 5, \text{ and } x = \pm\sqrt{5}.$$

Using the equation $y = \dfrac{5}{x}$, we get the following.

If $x = -\sqrt{5}$, then

$$y = \frac{5}{-\sqrt{5}} = \frac{5 \cdot \sqrt{5}}{-\sqrt{5} \cdot \sqrt{5}} = \frac{5\sqrt{5}}{-5} = -\sqrt{5}.$$

Similarly, if $x = \sqrt{5}$, then $y = \sqrt{5}$.

The solution set is $\left\{ \left(-\sqrt{5}, -\sqrt{5}\right), \left(\sqrt{5}, \sqrt{5}\right) \right\}$.

41. $\quad 3x^2 + 2xy - 3y^2 = 5 \quad (1)$
$\quad\quad -x^2 - 3xy + y^2 = 3 \quad (2)$

Multiply equation (2) by 3 and add the result to equation (1).

$$
\begin{array}{rcll}
3x^2 + 2xy - 3y^2 &=& 5 & (1) \\
-3x^2 - 9xy + 3y^2 &=& 9 & 3 \times (2) \\
\hline
-7xy &=& 14 & \\
\end{array}
$$
$$x = \frac{14}{-7y} = -\frac{2}{y}$$

Substitute $-\dfrac{2}{y}$ for x in equation (2).

$$-x^2 - 3xy + y^2 = 3 \qquad (2)$$
$$-\left(-\frac{2}{y}\right)^2 - 3\left(-\frac{2}{y}\right)y + y^2 = 3$$
$$-\left(\frac{4}{y^2}\right) + 6 + y^2 = 3$$
$$y^2 + 3 - \frac{4}{y^2} = 0$$
$$y^4 + 3y^2 - 4 = 0 \quad \textit{Multiply by } y^2.$$
$$(y^2 + 4)(y^2 - 1) = 0$$

$$
\begin{array}{ccc}
y^2 + 4 = 0 & \text{or} & y^2 - 1 = 0 \\
y^2 = -4 & & y^2 = 1 \\
y = \pm 2i & \text{or} & y = \pm 1 \\
\end{array}
$$

Since $x = -\dfrac{2}{y}$, substitute these values for y to find the values of x.

If $y = 2i$, then

$$x = -\frac{2}{2i} = -\frac{1}{i} = -\frac{1}{i} \cdot \frac{i}{i} = \frac{-i}{-1} = i.$$

If $y = -2i$, then

$$x = -\frac{2}{-2i} = \frac{1}{i} = \frac{1}{i} \cdot \frac{i}{i} = \frac{i}{-1} = -i.$$

If $y = 1$, then

$$x = -\frac{2}{1} = -2.$$

If $y = -1$, then

$$x = -\frac{2}{-1} = 2.$$

The solution set is
$\{(i, 2i), (-i, -2i), (2, -1), (-2, 1)\}$.

43. $\quad xy = -6 \quad (1)$
$\quad\quad x + y = -1 \quad (2)$

Solve both equations for y.

$$y = -\frac{6}{x} \quad \text{and} \quad y = -x - 1$$

Graph

$$Y_1 = -\frac{6}{X} \quad \text{and} \quad Y_2 = -X - 1$$

on a graphing calculator to obtain the solution set $\{(2, -3), (-3, 2)\}$.

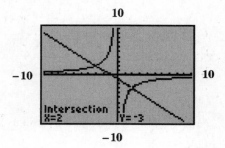

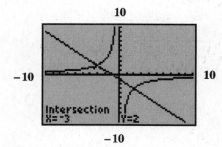

Now solve the system algebraically using substitution.

$$xy = -6 \quad (1)$$
$$x + y = -1 \quad (2)$$

Solve equation (2) for y.

$$y = -1 - x \qquad (3)$$

Substitute $-1 - x$ for y in equation (1).

$$x(-1 - x) = -6$$
$$-x - x^2 = -6$$
$$0 = x^2 + x - 6$$
$$0 = (x + 3)(x - 2)$$

$$x + 3 = 0 \quad \text{or} \quad x - 2 = 0$$
$$x = -3 \quad \text{or} \quad x = 2$$

Substitute these values for x in (3) to find y.

If $x = -3$, then $y = -1 - (-3) = 2$.

If $x = 2$, then $y = -1 - 2 = -3$.

The solution set, $\{(2, -3), (-3, 2)\}$, is the same as that obtained using a graphing calculator.

45. Let $W =$ the width, and
$L =$ the length.

The formula for the area of a rectangle is $LW = A$, so

$$LW = 84. \quad (1)$$

The perimeter of a rectangle is given by $2L + 2W = P$, so

$$2L + 2W = 38. \quad (2)$$

Solve equation (2) for L to get

$$L = 19 - W. \quad (3)$$

Substitute $19 - W$ for L in equation (1).

$$LW = 84 \quad (1)$$
$$(19 - W)W = 84$$
$$19W - W^2 = 84$$
$$-W^2 + 19W - 84 = 0$$
$$W^2 - 19W + 84 = 0 \qquad \textit{Multiply by } -1.$$
$$(W - 7)(W - 12) = 0$$

$$W - 7 = 0 \quad \text{or} \quad W - 12 = 0$$
$$W = 7 \quad \text{or} \quad W = 12$$

Using equation (3), with $W = 7$,

$$L = 19 - 7 = 12.$$

If $W = 12$, then $L = 7$, which are the same two numbers. Length must be greater than width, so the length is 12 feet and the width is 7 feet.

47.
$$px = 16 \qquad (1)$$
$$p = 10x + 12 \qquad (2)$$

Substitute $10x + 12$ for p in equation (1).

$$px = 16 \quad (1)$$
$$(10x + 12)x = 16$$
$$10x^2 + 12x = 16$$
$$10x^2 + 12x - 16 = 0$$
$$5x^2 + 6x - 8 = 0$$
$$(5x - 4)(x + 2) = 0$$

$$5x - 4 = 0 \quad \text{or} \quad x + 2 = 0$$
$$x = \tfrac{4}{5} \quad \text{or} \quad x = -2$$

Since x cannot be negative, eliminate -2 as a value of x. Substitute $\tfrac{4}{5}$ for x in equation (2) to find p.

$$p = 10x + 12 \qquad (2)$$
$$p = 10\left(\tfrac{4}{5}\right) + 12$$
$$= 8 + 12 = 20$$

Since the demand x is in thousands,

$$\text{demand} = \tfrac{4}{5}(1000) = 800.$$

The equilibrium price is $20. The supply/demand at that price is 800 calculators.

49. $2x - y \leq 4$

Draw a solid line through the intercepts $(2, 0)$ and $(0, -4)$. Checking $(0, 0)$ gives us $0 \leq 4$, a true statement, so shade the side containing $(0, 0)$.

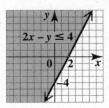

51. $-5x + 3y \leq 15$

Draw a solid line through the intercepts $(-3, 0)$ and $(0, 5)$. Checking $(0, 0)$ gives us $0 \leq 15$, a true statement, so shade the side containing $(0, 0)$.

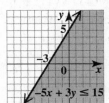

13.5 Second-Degree Inequalities and Systems of Inequalities

1. To graph the solution set of a nonlinear inequality, first graph the corresponding equality. This graph should be a dashed curve for $<$ or $>$ inequalities or a solid curve for $\leq$ or $\geq$ inequalities. Next, decide which region to shade by substituting any point not on the boundary (usually $(0, 0)$ is the easiest) into the inequality. If the statement is true, then shade that area. If the statement is false, then shade the other area.

3. $x^2 + y^2 < 25$
$\quad\quad y > -2$

The boundary, $x^2 + y^2 = 25$, is a circle (dashed because of the $<$ sign) with center $(0,0)$ and radius 5. When $(0,0)$ is tested, a true statement, $0 < 25$, results, so the inside of the circle is shaded. The graph of $y = -2$ is a horizontal line through $(0, -2)$ with shading above the dashed line, since $y > -2$. The correct answer is **C**.

5. $y \geq x^2 + 4$

This is an inequality whose boundary is a solid parabola, opening up, with vertex $(0, 4)$. The inside of the parabola is shaded since $(0,0)$ makes the inequality false. This is graph **B**.

7. $y < x^2 + 4$

This is an inequality whose boundary is a dashed parabola, opening up, with vertex $(0, 4)$. The outside of the parabola is shaded since $(0,0)$ makes the inequality true. This is graph **A**.

9. $y^2 > 4 + x^2$
$\quad\quad y^2 - x^2 > 4$
$\quad\quad \dfrac{y^2}{4} - \dfrac{x^2}{4} > 1$

The boundary, $\dfrac{y^2}{4} - \dfrac{x^2}{4} = 1$, is a hyperbola with y-intercepts $(0, 2)$ and $(0, -2)$ and asymptotes formed by the extended diagonals of the fundamental rectangle with vertices at $(2, 2)$, $(2, -2)$, $(-2, -2)$, and $(-2, 2)$. The hyperbola has dashed branches because of $>$. Test $(0, 0)$.

$$0^2 > 4 + 0^2 \quad ?$$
$$0 > 4 \quad\quad \textit{False}$$

Shade the sides of the hyperbola that do not contain $(0,0)$. These are the regions inside the branches of the hyperbola.

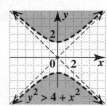

11. $y \geq x^2 - 2$
$\quad\quad y \geq (x - 0)^2 - 2$

Graph the solid vertical parabola $y = x^2 - 2$ with vertex $(0, -2)$. Two other points on the parabola are $(2, 2)$ and $(-2, 2)$. Test a point not on the parabola, say $(0, 0)$, in $y \geq x^2 - 2$ to get $0 \geq -2$, a true statement.

Shade that portion of the graph that contains the point $(0, 0)$. This is the region inside the parabola.

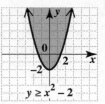

13. $2y^2 \geq 8 - x^2$
$\quad\quad x^2 + 2y^2 \geq 8$
$\quad\quad \dfrac{x^2}{8} + \dfrac{y^2}{4} \geq 1$

The boundary, $\dfrac{x^2}{8} + \dfrac{y^2}{4} = 1$, is the ellipse with intercepts $(2\sqrt{2}, 0)$, $(-2\sqrt{2}, 0)$, $(0, 2)$, and $(0, -2)$, drawn as a solid curve because of $\geq$. Test $(0, 0)$.

$$2(0)^2 \geq 8 - 0^2 \quad ?$$
$$0 \geq 8 \quad\quad \textit{False}$$

Shade the region of the ellipse that does not contain $(0, 0)$. This is the region outside the ellipse.

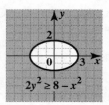

15. $y \leq x^2 + 4x + 2$

Graph the solid vertical parabola $y = x^2 + 4x + 2$. Use the vertex formula $x = \dfrac{-b}{2a}$ to obtain the vertex $(-2, -2)$. Two other points on the parabola are $(0, 2)$ and $(1, 7)$. Test a point not on the parabola, say $(0, 0)$, in $y \leq x^2 + 4x + 2$ to get $0 \leq 2$, a true statement. Shade outside the parabola, since this region contains $(0, 0)$.

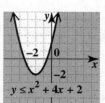

17.
$$9x^2 > 16y^2 + 144$$
$$9x^2 - 16y^2 > 144$$
$$\frac{x^2}{16} - \frac{y^2}{9} > 1$$

The boundary, $\dfrac{x^2}{16} - \dfrac{y^2}{9} = 1$, is a hyperbola with x-intercepts $(4, 0)$ and $(-4, 0)$ and asymptotes formed by the extended diagonals of the fundamental rectangle with vertices at $(4, 3)$, $(4, -3)$, $(-4, -3)$, and $(-4, 3)$. The hyperbola has dashed branches because of $>$. Test $(0, 0)$.

$$9(0)^2 > 16(0)^2 + 144 \quad ?$$
$$0 > 144 \qquad \textit{False}$$

Shade the sides of the hyperbola that do not contain $(0, 0)$. These are the regions inside the branches of the hyperbola.

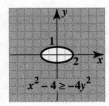

19.
$$x^2 - 4 \ge -4y^2$$
$$x^2 + 4y^2 \ge 4$$
$$\frac{x^2}{4} + \frac{y^2}{1} \ge 1$$

Graph the solid ellipse $\dfrac{x^2}{4} + \dfrac{y^2}{1} = 1$ through the x-intercepts $(2, 0)$ and $(-2, 0)$ and y-intercepts $(0, 1)$ and $(0, -1)$. Test a point not on the ellipse, say $(0, 0)$, in $x^2 - 4 \ge -4y^2$ to get $-4 \ge 0$, a false statement. Shade outside the ellipse, since this region does *not* include $(0, 0)$.

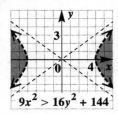

21. $x \le -y^2 + 6y - 7$

Complete the square to find the vertex.

$$x = -y^2 + 6y - 7$$
$$= -(y^2 - 6y) - 7$$
$$= -(y^2 - 6y + 9) + 9 - 7$$
$$x = -(y - 3)^2 + 2$$

The boundary, $x = -(y - 3)^2 + 2$, is a solid horizontal parabola with vertex at $(2, 3)$ that opens to the left. Test $(0, 0)$.

$$0 \le -0^2 + 6(0) - 7 \quad ?$$
$$0 \le -7 \qquad \textit{False}$$

Shade the region of the parabola that does not contain $(0, 0)$. This is the region inside the parabola.

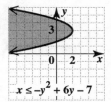

23.
$$2x + 5y < 10$$
$$x - 2y < 4$$

Graph $2x + 5y = 10$ as a dashed line through $(5, 0)$ and $(0, 2)$. Test $(0, 0)$.

$$2x + 5y < 10$$
$$2(0) + 5(0) < 10 \quad ?$$
$$0 < 10 \quad \textit{True}$$

Shade the region containing $(0, 0)$. Graph $x - 2y = 4$ as a dashed line through $(4, 0)$ and $(0, -2)$. Test $(0, 0)$.

$$x - 2y < 4$$
$$0 - 2(0) < 4 \quad ?$$
$$0 < 4 \quad \textit{True}$$

Shade the region containing $(0, 0)$. The graph of the system is the intersection of the two shaded regions.

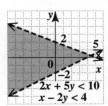

25.
$$5x - 3y \le 15$$
$$4x + y \ge 4$$

The boundary, $5x - 3y = 15$, is a solid line with intercepts $(3, 0)$ and $(0, -5)$. Test $(0, 0)$.

$$5(0) - 3(0) \le 15 \quad ?$$
$$0 \le 15 \quad \textit{True}$$

Shade the side of the line that contains $(0, 0)$. The boundary, $4x + y = 4$, is a solid line with intercepts $(1, 0)$ and $(0, 4)$. Test $(0, 0)$.

$$4(0) + 0 \ge 4 \quad ?$$
$$0 \ge 4 \quad \textit{False}$$

continued

Shade the side of the line that does not contain $(0, 0)$.
The graph of the system is the intersection of the two shaded regions.

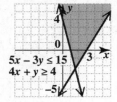

27. $x \leq 5$
  $y \leq 4$

Graph $x = 5$ as a solid vertical line through $(5, 0)$.
Since $x \leq 5$, shade the left side of the line.
Graph $y = 4$ as a solid horizontal line through $(0, 4)$. Since $y \leq 4$, shade below the line.
The graph of the system is the intersection of the two shaded regions.

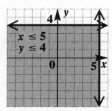

29. $y > x^2 - 4$
  $y < -x^2 + 3$

The boundary, $y = x^2 - 4$, is a dashed parabola with vertex $(0, -4)$ that opens up. Test $(0, 0)$.

$$0 > 0^2 - 4 \quad ?$$
$$0 > -4 \qquad True$$

Shade the side of the parabola that contains $(0, 0)$. This is the region inside the parabola.
The boundary, $y = -x^2 + 3$, is a dashed parabola with vertex $(0, 3)$ that opens down. Test $(0, 0)$.

$$0 < -0^2 + 3 \quad ?$$
$$0 < 3 \qquad True$$

Shade the side of the parabola that contains $(0, 0)$. This is the region inside the parabola.
The graph of the system is the intersection of the two shaded regions.

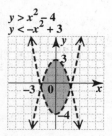

31. $x^2 + y^2 \geq 4 \quad (1)$
  $x + y \leq 5 \quad (2)$
  $x \geq 0 \quad (3)$
  $y \geq 0 \quad (4)$

The graph of (1) is the region outside the solid circle boundary $x^2 + y^2 = 4$ with center $(0, 0)$ and radius 2.
The graph of (2) is the region below the solid line $x + y = 5$ with intercepts $(5, 0)$ and $(0, 5)$.
The intersection of the graphs of (3) and (4) is quadrant I where both x and y are positive. The positive x- and y-axes are included.
The graph of the system is the intersection of the graphs of (1) and (2) in quadrant I.

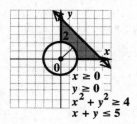

33. $y \leq -x^2$
  $y \geq x - 3$
  $y \leq -1$
  $x < 1$

The boundary, $y = -x^2$, is a solid parabola with vertex at $(0, 0)$ that opens down. Test $(0, -1)$.

$$-1 \leq -(0)^2 \quad ?$$
$$-1 \leq 0 \qquad True$$

Shade the side of the parabola that contains $(0, -1)$. This is the region inside the parabola.
The boundary, $y = x - 3$, is a solid line with intercepts $(3, 0)$ and $(0, -3)$. Test $(0, 0)$.

$$0 \geq 0 - 3 \quad ?$$
$$0 \geq -3 \qquad True$$

Shade the side of the line that contains $(0, 0)$.
For $y \leq -1$, shade below the solid horizontal line $y = -1$.
For $x < 1$, shade to the left of the dashed vertical line $x = 1$.
The intersection of the four shaded regions is the graph of the system.

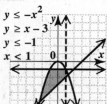

35. $x^2 + y^2 > 36,\ x \geq 0$

This is a circle of radius 6 centered at the origin. The graph is a dashed curve. Since $(0,0)$ does not satisfy the inequality, the shading will be outside the circle. By including the restriction $x \geq 0$, we consider only the shading to the right of, and including, the y-axis.

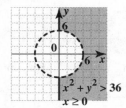

37. $x < y^2 - 3,\ x < 0$

Consider the equation.

$$x = y^2 - 3,\ \text{or}$$
$$x = (y - 0)^2 - 3.$$

This is a parabola with vertex $(-3, 0)$, opening to the right having the same shape as $x = y^2$. The graph is a dashed curve. The shading will be outside the parabola, since $(0,0)$ does not satisfy the inequality. By including the restriction $x < 0$, we consider only the shading to the left of, but not including, the y-axis. The y-axis is a dashed line.

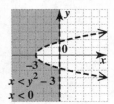

39. $4x^2 - y^2 > 16,\ x < 0$

Consider the equation

$$4x^2 - y^2 = 16,\ \text{or}$$
$$\frac{x^2}{4} - \frac{y^2}{16} = 1.$$

This is a hyperbola with x-intercepts $(2, 0)$ and $(-2, 0)$. The graph is a dashed curve. The shading will be to the left of the left branch and to the right of the right branch of the hyperbola, since $(0,0)$ does not satisfy the inequality. By including the restriction $x < 0$, we consider only the shading to the left of the y-axis.

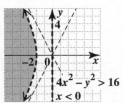

41. $x^2 + 4y^2 \geq 1,\ x \geq 0,\ y \geq 0$

Consider the equation

$$x^2 + 4y^2 = 1,\ \text{or}$$
$$\frac{x^2}{1} + \frac{y^2}{\frac{1}{4}} = 1.$$

The graph is a solid ellipse with x-intercepts $(1, 0)$ and $(-1, 0)$ and y-intercepts $\left(0, \frac{1}{2}\right)$ and $\left(0, -\frac{1}{2}\right)$. The shading will be outside the ellipse, since $(0,0)$ does not satisfy the inequality. By including the restrictions $x \geq 0$ and $y \geq 0$, we consider only the shading in quadrant I and portions of the x- and y-axis.

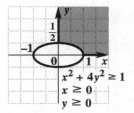

43. $y \geq x - 3$
$y \leq -x + 4$

The graphs of both inequalities include the points on the lines as part of the solution because of $\geq$ and $\leq$.

To produce the graph of the system on the TI-83, make the following Y-assignments:

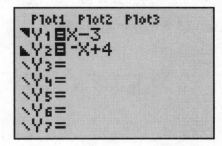

To get the upper and lower darkened triangles to the left of Y_1 and Y_2, simply place the cursor in that spot and press $\boxed{\text{ENTER}}$ to cycle through the graphing choices.

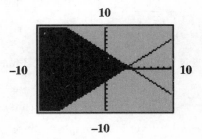

45. $y < x^2 + 4x + 4$

$y > -3$

The graphs do *not* include the points on the parabola or the line as part of the solution because of $<$ and $>$.

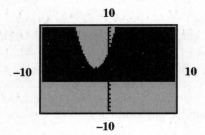

47. $\dfrac{n+5}{n}$

  (a) $[n = 1]$ $\frac{1+5}{1} = \frac{6}{1} = 6$

  (b) $[n = 2]$ $\frac{2+5}{2} = \frac{7}{2}$

  (c) $[n = 3]$ $\frac{3+5}{3} = \frac{8}{3}$

  (d) $[n = 4]$ $\frac{4+5}{4} = \frac{9}{4}$

49. $n^2 - n$

  (a) $[n = 1]$ $1^2 - 1 = 1 - 1 = 0$

  (b) $[n = 2]$ $2^2 - 2 = 4 - 2 = 2$

  (c) $[n = 3]$ $3^2 - 3 = 9 - 3 = 6$

  (d) $[n = 4]$ $4^2 - 4 = 16 - 4 = 12$

Chapter 13 Review Exercises

1. $f(x) = |x + 4|$

This is the graph of the absolute value function $g(x) = |x|$ shifted 4 units to the left (since $x + 4 = 0$ if $x = -4$).

x	y
-6	2
-5	1
-4	0
-3	1
-2	2

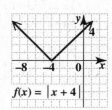

$f(x) = |x + 4|$

2. $f(x) = \dfrac{1}{x - 4}$

This is the graph of the reciprocal function, $g(x) = \dfrac{1}{x}$, shifted 4 units to the right. Since $x \ne 4$ (or a denominator of 0 results), the line $x = 4$ is a vertical asymptote. Since $\dfrac{1}{x - 4} \ne 0$, the line $y = 0$ is a horizontal asymptote.

x	y
2	$-\frac{1}{2}$
3	-1
$3\frac{1}{2}$	-2
$4\frac{1}{2}$	2
5	1
6	$\frac{1}{2}$

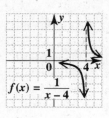

$f(x) = \dfrac{1}{x - 4}$

3. $f(x) = \sqrt{x} + 3$

This is the graph of the square root function, $g(x) = \sqrt{x}$, shifted 3 units upward.

x	y
0	3
1	4
4	5

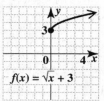

$f(x) = \sqrt{x} + 3$

4. $f(x) = [\![x]\!] - 2$

This is the graph of the greatest integer function, $g(x) = [\![x]\!]$, shifted down 2 units.

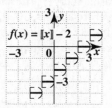

$f(x) = [\![x]\!] - 2$

5.   Center $(-2, 4)$, $r = 3$

Here $h = -2$, $k = 4$, and $r = 3$, so an equation of the circle is

$$(x - h)^2 + (y - k)^2 = r^2$$
$$[x - (-2)]^2 + (y - 4)^2 = 3^2$$
$$(x + 2)^2 + (y - 4)^2 = 9.$$

6.   Center $(-1, -3)$, $r = 5$

Here $h = -1$, $k = -3$, and $r = 5$, so an equation of the circle is

$$(x - h)^2 + (y - k)^2 = r^2$$
$$[x - (-1)]^2 + [y - (-3)]^2 = 5^2$$
$$(x + 1)^2 + (y + 3)^2 = 25.$$

7.   Center $(4, 2)$, $r = 6$

Here $h = 4$, $k = 2$, and $r = 6$, so an equation of the circle is

$$(x - h)^2 + (y - k)^2 = r^2$$
$$(x - 4)^2 + (y - 2)^2 = 6^2$$
$$(x - 4)^2 + (y - 2)^2 = 36.$$

8. $x^2 + y^2 + 6x - 4y - 3 = 0$

Write the equation in center-radius form,

$$(x - h)^2 + (y - k)^2 = r^2,$$

by completing the squares on x and y.

$$\left(x^2 + 6x \quad\right) + \left(y^2 - 4y \quad\right) = 3$$
$$\left(x^2 + 6x + \underline{9}\right) + \left(y^2 - 4y + \underline{4}\right)$$
$$= 3 + \underline{9} + \underline{4}$$
$$(x + 3)^2 + (y - 2)^2 = 16$$
$$[x - (-3)]^2 + (y - 2)^2 = 16$$

The circle has center (h, k) at $(-3, 2)$ and radius $\sqrt{16} = 4$.

9. $x^2 + y^2 - 8x - 2y + 13 = 0$

Write the equation in center-radius form,

$$(x - h)^2 + (y - k)^2 = r^2,$$

by completing the squares on x and y.

$$\left(x^2 - 8x \quad\right) + \left(y^2 - 2y \quad\right) = -13$$
$$\left(x^2 - 8x + \underline{16}\right) + \left(y^2 - 2y + \underline{1}\right)$$
$$= -13 + \underline{16} + \underline{1}$$
$$(x - 4)^2 + (y - 1)^2 = 4$$

The circle has center (h, k) at $(4, 1)$ and radius $\sqrt{4} = 2$.

10. $2x^2 + 2y^2 + 4x + 20y = -34$
$$x^2 + y^2 + 2x + 10y = -17$$
Write the equation in center-radius form,
$$(x - h)^2 + (y - k)^2 = r^2,$$
by completing the squares on x and y.

$$\left(x^2 + 2x \quad\right) + \left(y^2 + 10y \quad\right) = -17$$
$$\left(x^2 + 2x + \underline{1}\right) + \left(y^2 + 10y + \underline{25}\right)$$
$$= -17 + \underline{1} + \underline{25}$$
$$(x + 1)^2 + (y + 5)^2 = 9$$
$$[x - (-1)]^2 + [y - (-5)]^2 = 9$$

The circle has center (h, k) at $(-1, -5)$ and radius $\sqrt{9} = 3$.

11. $4x^2 + 4y^2 - 24x + 16y = 48$
$$x^2 + y^2 - 6x + 4y = 12$$
Write the equation in center-radius form,
$$(x - h)^2 + (y - k)^2 = r^2,$$
by completing the squares on x and y.

$$\left(x^2 - 6x \quad\right) + \left(y^2 + 4y \quad\right) = 12$$
$$\left(x^2 - 6x + \underline{9}\right) + \left(y^2 + 4y + \underline{4}\right)$$
$$= 12 + \underline{9} + \underline{4}$$
$$(x - 3)^2 + (y + 2)^2 = 25$$
$$(x - 3)^2 + [y - (-2)]^2 = 25$$

The circle has center (h, k) at $(3, -2)$ and radius $\sqrt{25} = 5$.

12.
$$x^2 + y^2 = 16$$
$$(x - 0)^2 + (y - 0)^2 = 4^2$$

This is a circle with center $(0, 0)$ and radius 4.

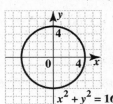

13. $\dfrac{x^2}{16} + \dfrac{y^2}{9} = 1$ is in $\dfrac{x^2}{a^2} + \dfrac{y^2}{b^2} = 1$ form with $a = 4$ and $b = 3$. The graph is an ellipse with x-intercepts $(4, 0)$ and $(-4, 0)$ and y-intercepts $(0, 3)$ and $(0, -3)$. Plot the intercepts, and draw the ellipse through them.

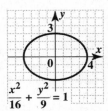

14. $\dfrac{x^2}{49} + \dfrac{y^2}{25} = 1$ is in $\dfrac{x^2}{a^2} + \dfrac{y^2}{b^2} = 1$ form with $a = 7$ and $b = 5$. The graph is an ellipse with x-intercepts $(7, 0)$ and $(-7, 0)$ and y-intercepts $(0, 5)$ and $(0, -5)$. Plot the intercepts, and draw the ellipse through them.

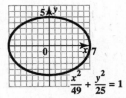

15. The total distance on the horizontal axis is $160 + 16{,}000 = 16{,}160$ km. This represents $2a$, so $a = \frac{1}{2}(16{,}160) = 8080$. The distance from Earth to the center of the ellipse is

$$8080 - 160 = 7920,$$

which is the value of c. From Exercise 50 in Section 11.2, we know that $c^2 = a^2 - b^2$, so

$$b^2 = a^2 - c^2.$$
$$b^2 = 8080^2 - 7920^2$$
$$= 2{,}560{,}000$$

Thus, $b = \sqrt{2{,}560{,}000} = 1600$ and the equation is

$$\frac{x^2}{8080^2} + \frac{y^2}{1600^2} = 1$$

or $\dfrac{x^2}{65{,}286{,}400} + \dfrac{y^2}{2{,}560{,}000} = 1.$

16. **(a)** The distance between the foci is $2c$, where c can be found using the relationship

$$c = \sqrt{a^2 - b^2}.$$
$$= \sqrt{310^2 - (513/2)^2}$$
$$\approx 174.1 \text{ feet}$$

So the distance is about 348.2 feet.

(b) The approximate circumference of the Roman Colosseum is

$$C \approx 2\pi\sqrt{\frac{a^2 + b^2}{2}}.$$
$$= 2\pi\sqrt{\frac{310^2 + (513/2)^2}{2}}$$
$$\approx 1787.6 \text{ feet}$$

17. $\dfrac{x^2}{16} - \dfrac{y^2}{25} = 1$ is in $\dfrac{x^2}{a^2} - \dfrac{y^2}{b^2} = 1$ form with $a = 4$ and $b = 5$. The graph is a hyperbola with x-intercepts $(4, 0)$ and $(-4, 0)$ and asymptotes that are the extended diagonals of the rectangle with vertices $(4, 5)$, $(4, -5)$, $(-4, -5)$, and $(-4, 5)$. Graph a branch of the hyperbola through each intercept approaching the asymptotes.

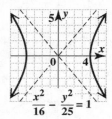

18. $\dfrac{y^2}{25} - \dfrac{x^2}{4} = 1$ is in $\dfrac{y^2}{b^2} - \dfrac{x^2}{a^2} = 1$ form with $a = 2$ and $b = 5$. The graph is a hyperbola with y-intercepts $(0, 5)$ and $(0, -5)$ and asymptotes that are the extended diagonals of the rectangle with vertices $(2, 5)$, $(2, -5)$, $(-2, -5)$, and $(-2, 5)$. Graph a branch of the hyperbola through each intercept approaching the asymptotes.

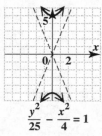

19. $$f(x) = -\sqrt{16 - x^2}$$
Replace $f(x)$ with y.
$$y = -\sqrt{16 - x^2}$$
Square both sides.
$$y^2 = 16 - x^2$$
$$x^2 + y^2 = 16$$

This equation is the graph of a circle with center $(0, 0)$ and radius 4. Since $f(x)$ represents a nonpositive square root, $f(x)$ is nonpositive and its graph is the lower half of the circle.

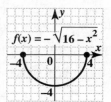

20. $$x^2 + y^2 = 64$$
$$(x - 0)^2 + (y - 0)^2 = 8^2$$

The last equation is in $(x - h)^2 + (y - k)^2 = r^2$ form. The graph is a *circle*.

21. $y = 2x^2 - 3$
$y = 2(x - 0)^2 - 3$

The last equation is in $y = a(x - h)^2 + k$ form. The graph is a *parabola*.

22. $$y^2 = 2x^2 - 8$$
$$2x^2 - y^2 = 8$$
$$\frac{x^2}{4} - \frac{y^2}{8} = 1$$

The last equation is in $\dfrac{x^2}{a^2} - \dfrac{y^2}{b^2} = 1$ form, so the graph is a *hyperbola*.

23. $$y^2 = 8 - 2x^2$$
$$2x^2 + y^2 = 8$$
$$\frac{x^2}{4} + \frac{y^2}{8} = 1$$

The last equation is in $\dfrac{x^2}{a^2} + \dfrac{y^2}{b^2} = 1$ form, so the graph is an *ellipse*.

24. $x = y^2 + 4$
$x = (y - 0)^2 + 4$

The last equation is in $x = a(y - k)^2 + h$ form, so the graph is a *parabola*.

25. $x^2 - y^2 = 64$

$\dfrac{x^2}{64} - \dfrac{y^2}{64} = 1$

The last equation is in $\dfrac{x^2}{a^2} - \dfrac{y^2}{b^2} = 1$ form, so the graph is a *hyperbola*.

26. A hyperbola is defined as the set of all points in a plane such that the absolute value of the *difference* of the distances from two fixed points (called *foci*) is constant. The hyperbola shown in the text will have an equation of the form

$$\frac{x^2}{a^2} - \frac{y^2}{b^2} = 1. \quad (1)$$

The constant difference for this hyperbola is

$$|d_1 - d_2| = |80 - 30| = 50.$$

Let Q be the point on the right branch of the hyperbola that is on $\overline{MS}$. Let $x = \overline{MQ}$ and $y = \overline{QS}$. Since $\overline{MS} = 100$ and $\overline{MQ} - \overline{QS}$ must be 50, we have the following system.

$$\begin{array}{rcll} x + y &=& 100 & (2)\\ x - y &=& 50 & \\ \hline 2x &=& 150 & \textit{Add.}\\ x &=& 75 & \end{array}$$

From (2), we see that $y = 25$. The distance from the center of the hyperbola, C, to S, is 50, so $a = \overline{CQ} = 50 - 25 = 25$.
To find b^2, we'll find a point on the hyperbola, substitute for a, x, and y, and then solve for b^2. Let P have coordinates (d, e). If we draw a perpendicular line from P to $\overline{MS}$, we see two right triangles, PQM and PQS. Using the Pythagorean formula, we get the following system of equations.

$$\begin{array}{rcl} (50 + d)^2 + e^2 &=& 80^2 \\ (50 - d)^2 + e^2 &=& 30^2 \quad (3)\\ \hline (2500 + 100d + d^2) - (2500 - 100d + d^2) & & \\ &=& 6400 - 900 \quad \textit{Subtract.}\\ 200d &=& 5500 \\ d &=& 27.5 \end{array}$$

From (3) with $d = 27.5$, we get $e^2 = 30^2 - 22.5^2$, so $e = \sqrt{393.75}$. Now substitute 25 for a, 27.5 for x, and $\sqrt{393.75}$ for y in (1) to solve for b^2.

$$\frac{27.5^2}{25^2} - \frac{393.75}{b^2} = 1$$

$$\frac{756.25}{625} - \frac{625}{625} = \frac{393.75}{b^2}$$

$$\frac{131.25}{625} = \frac{393.75}{b^2}$$

$$\frac{625}{131.25} = \frac{b^2}{393.75}$$

$$b^2 = \frac{625}{131.25}(393.75)$$

$$b^2 = 1875$$

Thus, equation (1) becomes

$$\frac{x^2}{625} - \frac{y^2}{1875} = 1.$$

27. $\quad 2y = 3x - x^2 \quad (1)$
$\quad x + 2y = -12 \qquad (2)$

Substitute $3x - x^2$ for $2y$ in equation (2).

$$\begin{array}{rcl} x + 2y &=& -12 \quad (2)\\ x + (3x - x^2) &=& -12 \\ -x^2 + 4x + 12 &=& 0 \\ x^2 - 4x - 12 &=& 0 \\ (x - 6)(x + 2) &=& 0 \end{array}$$

$$\begin{array}{rcl} x - 6 = 0 & \text{or} & x + 2 = 0 \\ x = 6 & \text{or} & x = -2 \end{array}$$

Substitute these values for x in equation (2) to find y.

If $x = 6$, then

$$\begin{array}{rcl} x + 2y &=& -12 \quad (2)\\ 6 + 2y &=& -12 \\ 2y &=& -18 \\ y &=& -9. \end{array}$$

If $x = -2$, then

$$\begin{array}{rcl} x + 2y &=& -12 \quad (2)\\ -2 + 2y &=& -12 \\ 2y &=& -10 \\ y &=& -5. \end{array}$$

The solution set is $\{(6, -9), (-2, -5)\}$.

28. $\quad y + 1 = x^2 + 2x \quad (1)$
$\quad y + 2x = 4 \qquad\quad (2)$

Solve equation (2) for y.

$$y = 4 - 2x \quad (3)$$

Substitute $4 - 2x$ for y in (1).

$$\begin{array}{rcl} (4 - 2x) + 1 &=& x^2 + 2x \\ 0 &=& x^2 + 4x - 5 \\ 0 &=& (x + 5)(x - 1) \end{array}$$

$$\begin{array}{rcl} x + 5 = 0 & \text{or} & x - 1 = 0 \\ x = -5 & \text{or} & x = 1 \end{array}$$

Substitute these values for x in equation (3) to find y.

If $x = -5$, then $y = 4 - 2(-5) = 14$.
If $x = 1$, then $y = 4 - 2(1) = 2$.

The solution set is $\{(1, 2), (-5, 14)\}$.

29. $x^2 + 3y^2 = 28$ (1)
      $y - x = -2$ (2)

Solve equation (2) for y.

$$y = x - 2$$

Substitute $x - 2$ for y in equation (1).

$$x^2 + 3y^2 = 28 \quad (1)$$
$$x^2 + 3(x - 2)^2 = 28$$
$$x^2 + 3(x^2 - 4x + 4) - 28 = 0$$
$$x^2 + 3x^2 - 12x + 12 - 28 = 0$$
$$4x^2 - 12x - 16 = 0$$
$$4(x^2 - 3x - 4) = 0$$
$$4(x - 4)(x + 1) = 0$$

$$x - 4 = 0 \quad \text{or} \quad x + 1 = 0$$
$$x = 4 \quad \text{or} \quad x = -1$$

Since $y = x - 2$, if $x = 4$, then $y = 4 - 2 = 2$.
If $x = -1$, then $y = -1 - 2 = -3$.
The solution set is $\{(4, 2), (-1, -3)\}$.

30. $xy = 8$ (1)
      $x - 2y = 6$ (2)

Solve equation (2) for x.

$$x = 2y + 6 \quad (3)$$

Substitute $2y + 6$ for x in equation (1) to find y.

$$xy = 8 \quad (1)$$
$$(2y + 6)y = 8$$
$$2y^2 + 6y - 8 = 0$$
$$2(y^2 + 3y - 4) = 0$$
$$2(y + 4)(y - 1) = 0$$

$$y + 4 = 0 \quad \text{or} \quad y - 1 = 0$$
$$y = -4 \quad \text{or} \quad y = 1$$

Substitute these values for y in equation (3) to find x.
If $y = -4$, then $x = 2(-4) + 6 = -2$.
If $y = 1$, then $x = 2(1) + 6 = 8$.

The solution set is $\{(-2, -4), (8, 1)\}$.

31. $x^2 + y^2 = 6$ (1)
      $x^2 - 2y^2 = -6$ (2)

Multiply equation (2) by -1 and add the result to equation (1).

$$\begin{array}{rcll} x^2 + y^2 &=& 6 & (1) \\ -x^2 + 2y^2 &=& 6 & -1 \times (2) \\ \hline 3y^2 &=& 12 & \\ y^2 &=& 4 & \end{array}$$

$$y = 2 \quad \text{or} \quad y = -2$$

Substitute these values for y in equation (1) to find x.
If $y = \pm 2$, then

$$x^2 + y^2 = 6 \quad (1)$$
$$x^2 + (\pm 2)^2 = 6$$
$$x^2 + 4 = 6$$
$$x^2 = 2.$$

$$x = \sqrt{2} \quad \text{or} \quad x = -\sqrt{2}$$

Since each value of x can be paired with each value of y, there are four points and the solution set is
$$\left\{ (\sqrt{2}, 2), (-\sqrt{2}, 2), (\sqrt{2}, -2), (-\sqrt{2}, -2) \right\}.$$

32. $3x^2 - 2y^2 = 12$ (1)
      $x^2 + 4y^2 = 18$ (2)

Multiply equation (1) by 2 and add the result to equation (2).

$$\begin{array}{rcll} 6x^2 - 4y^2 &=& 24 & 2 \times (1) \\ x^2 + 4y^2 &=& 18 & (2) \\ \hline 7x^2 &=& 42 & \\ x^2 &=& 6 & \end{array}$$

$$x = \sqrt{6} \quad \text{or} \quad x = -\sqrt{6}$$

Substitute these values for x in equation (2) to find y.
If $x = \pm \sqrt{6}$, then

$$x^2 + 4y^2 = 18. \quad (2)$$
$$(\pm \sqrt{6})^2 + 4y^2 = 18$$
$$6 + 4y^2 = 18$$
$$4y^2 = 12$$
$$y^2 = 3$$

$$y = \sqrt{3} \quad \text{or} \quad y = -\sqrt{3}$$

The solution set is
$$\left\{ (\sqrt{6}, \sqrt{3}), (\sqrt{6}, -\sqrt{3}), \right.$$
$$\left. (-\sqrt{6}, \sqrt{3}), (-\sqrt{6}, -\sqrt{3}) \right\}.$$

33. A circle and a line can intersect in zero, one, or two points, so zero, one, or two solutions are possible.

34. A parabola and a hyperbola can intersect in zero, one, two, three, or four points, so zero, one, two, three, or four solutions are possible.

35.
$$9x^2 \geq 16y^2 + 144$$
$$9x^2 - 16y^2 \geq 144$$
$$\frac{x^2}{16} - \frac{y^2}{9} \geq 1$$

The boundary, $\dfrac{x^2}{16} - \dfrac{y^2}{9} = 1$, is a solid hyperbola with x-intercepts $(4, 0)$ and $(-4, 0)$. The asymptotes are the extended diagonals of the rectangle with vertices $(4, 3)$, $(4, -3)$, $(-4, -3)$, and $(-4, 3)$. Test $(0, 0)$.

$$9(0)^2 \geq 16(0)^2 + 144 \quad ?$$
$$0 \geq 144 \qquad \textit{False}$$

Shade the sides of the hyperbola that do not contain $(0, 0)$. These are the regions inside the branches of the hyperbola.

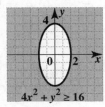

36. $4x^2 + y^2 \geq 16$
$$\dfrac{x^2}{4} + \dfrac{y^2}{16} \geq 1$$

The boundary, $\dfrac{x^2}{4} + \dfrac{y^2}{16} = 1$, is a solid ellipse with intercepts $(2, 0)$, $(-2, 0)$, $(0, 4)$, and $(0, -4)$. Test $(0, 0)$.

$$4(0)^2 + 0^2 \geq 16 \quad ?$$
$$0 \geq 16 \qquad \textit{False}$$

Shade the side of ellipse that does not contain $(0, 0)$. This is the region outside the ellipse.

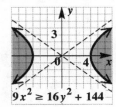

37. $y < -(x + 2)^2 + 1$

The boundary, $y = -(x + 2)^2 + 1$, is a dashed vertical parabola with vertex $(-2, 1)$. Since $a = -1 < 0$, the parabola opens down. Also, $|a| = |-1| = 1$, so the graph has the same shape as the graph of $y = x^2$. Test $(0, 0)$.

$$0 < -(0 + 2)^2 + 1 \quad ?$$
$$0 < -(4) + 1 \quad ?$$
$$0 < -3 \qquad \textit{False}$$

Shade the side of the parabola that does not contain $(0, 0)$. This is the region inside the parabola.

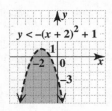

38. $2x + 5y \leq 10$
$3x - y \leq 6$

The boundary, $2x + 5y = 10$ is a solid line with intercepts $(5, 0)$ and $(0, 2)$. Test $(0, 0)$.

$$2(0) + 5(0) \leq 10 \quad ?$$
$$0 \leq 10 \qquad \textit{True}$$

Shade the side of the line that contains $(0, 0)$. The boundary, $3x - y = 6$, is a solid line with intercepts $(2, 0)$ and $(0, -6)$. Test $(0, 0)$.

$$3(0) - 0 \leq 6 \quad ?$$
$$0 \leq 6 \qquad \textit{True}$$

Shade the side of the line that contains $(0, 0)$. The graph of the system is the intersection of the two shaded regions.

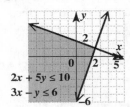

39. $|x| \leq 2$
$|y| > 1$
$4x^2 + 9y^2 \leq 36$

The equation of the boundary, $|x| = 2$, can be written as

$$x = -2 \text{ or } x = 2.$$

The graph is these two solid vertical lines. Since $|0| \leq 2$ is true, the region between the lines, containing $(0, 0)$, is shaded. The boundary, $|y| = 1$, consists of the two dashed horizontal lines $y = 1$ and $y = -1$. Since $|0| > 1$ is false, the regions above $y = 1$ and below $y = -1$, not containing $(0, 0)$, are shaded. The boundary given by

$$4x^2 + 9y^2 = 36$$
$$\text{or} \quad \dfrac{x^2}{9} + \dfrac{y^2}{4} = 1$$

is graphed as a solid ellipse with intercepts $(3, 0)$, $(-3, 0)$, $(0, 2)$, and $(0, -2)$. Test $(0, 0)$.

$$4(0)^2 + 9(0)^2 \leq 36 \quad ?$$
$$0 \leq 36 \qquad \textit{True}$$

continued

The region inside the ellipse, containing $(0,0)$, is shaded.

The graph of the system consists of the regions that include the common points of the three shaded regions.

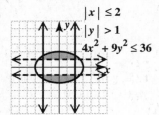

$$|x| \le 2$$
$$|y| > 1$$
$$4x^2 + 9y^2 \le 36$$

40. $9x^2 \le 4y^2 + 36$
$$x^2 + y^2 \le 16$$

The equation of the first boundary is

$$9x^2 = 4y^2 + 36$$
$$9x^2 - 4y^2 = 36$$
$$\frac{x^2}{4} - \frac{y^2}{9} = 1.$$

The graph is a solid hyperbola with x-intercepts $(2,0)$ and $(-2,0)$. The asymptotes are the extended diagonals of the rectangle with vertices $(2,3)$, $(2,-3)$, $(-2,-3)$, and $(-2,3)$. Test $(0,0)$.

$$9(0)^2 \le 4(0)^2 + 36 \quad ?$$
$$0 \le 36 \qquad \textit{True}$$

Shade the region between the branches of the hyperbola that contains $(0,0)$.

The equation of the second boundary is $x^2 + y^2 = 16$. This is a solid circle with center $(0,0)$ and radius 4. Test $(0,0)$.

$$0^2 + 0^2 \le 16 \quad ?$$
$$0 \le 16 \qquad \textit{True}$$

Shade the region inside the circle.

The graph of the system is the intersection of the shaded regions which is between the two branches of the hyperbola and inside the circle.

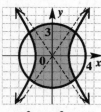

$$9x^2 \le 4y^2 + 36$$
$$x^2 + y^2 \le 16$$

41.   **[13.2]** $\frac{x^2}{64} + \frac{y^2}{25} = 1$ is in $\frac{x^2}{a^2} + \frac{y^2}{b^2} = 1$ form with $a = 8$ and $b = 5$. The graph is an ellipse with intercepts $(8,0)$, $(-8,0)$, $(0,5)$, and $(0,-5)$.

Plot the intercepts, and draw the ellipse through them.

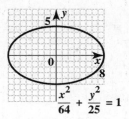

$$\frac{x^2}{64} + \frac{y^2}{25} = 1$$

42.   **[13.3]** $\frac{y^2}{4} - 1 = \frac{x^2}{9}$

$$\frac{y^2}{4} - \frac{x^2}{9} = 1$$

The equation is in $\frac{y^2}{b^2} - \frac{x^2}{a^2} = 1$ form with $a = 3$ and $b = 2$. The graph is a hyperbola with y-intercepts $(0,2)$ and $(0,-2)$ and asymptotes that are the extended diagonals of the rectangle with vertices $(3,2)$, $(3,-2)$, $(-3,-2)$, and $(-3,2)$. Draw a branch of the hyperbola through each intercept and approaching the asymptotes.

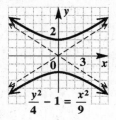

$$\frac{y^2}{4} - 1 = \frac{x^2}{9}$$

43.   **[13.2]** $x^2 + y^2 = 25$ is in $(x-h)^2 + (y-k)^2 = r^2$ form. The graph is a circle with center at $(0,0)$ and radius 5.

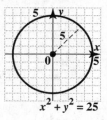

$$x^2 + y^2 = 25$$

44.   **[13.2]** $x^2 + 9y^2 = 9$ in $\frac{x^2}{a^2} + \frac{y^2}{b^2} = 1$ form is

$\frac{x^2}{9} + \frac{y^2}{1} = 1$ with $a = 3$ and $b = 1$. The graph is an ellipse with x-intercepts $(3,0)$ and $(-3,0)$ and y-intercepts $(0,1)$ and $(0,-1)$. Plot the intercepts, and draw the ellipse through them.

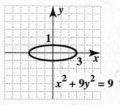

$$x^2 + 9y^2 = 9$$

45. **[13.3]** $x^2 - 9y^2 = 9$

$$\frac{x^2}{9} - \frac{y^2}{1} = 1$$

The equation is in the $\dfrac{x^2}{a^2} - \dfrac{y^2}{b^2} = 1$ form with $a = 3$ and $b = 1$. The graph is a hyperbola with x-intercepts $(3, 0)$ and $(-3, 0)$ and asymptotes that are the extended diagonals of the rectangle with vertices $(3, 1)$, $(3, -1)$, and $(-3, -1)$, and $(-3, 1)$. Graph a branch of the hyperbola through each intercept and approaching the asymptotes.

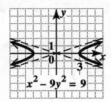

46. **[13.3]** $f(x) = \sqrt{4 - x}$
Replace $f(x)$ with y.
$$y = \sqrt{4 - x}$$
Square both sides.
$$y^2 = 4 - x$$
$$x = -y^2 + 4$$
$$x = -1(y - 0)^2 + 4$$

This equation is the graph of a horizontal parabola with vertex $(4, 0)$. Since $a = -1 < 0$, the graph opens to the left. Also, $|a| = |-1| = 1$, so the graph has the same shape as the graph of $y = x^2$. The points $(0, 2)$ and $(3, 1)$ are on the graph. Since $f(x)$ represents a square root, $f(x)$ is nonnegative and its graph is the upper half of the parabola.

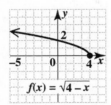

47. **[13.5]** $3x + 2y \geq 0$
$$y \leq 4$$
$$x \leq 4$$

The boundary $3x + 2y = 0$ is a solid line through $(0, 0)$ and $(2, -3)$. Test $(0, 1)$.

$$3(0) + 2(1) \geq 0 \quad ?$$
$$2 \geq 0 \quad \textit{True}$$

Shade the side of the line that contains $(0, 1)$. The boundary $y = 4$ is a solid horizontal line through $(0, 4)$. Since $y \leq 4$, shade below the line.

The boundary $x = 4$ is a solid vertical line through $(4, 0)$. Since $x \leq 4$, shade the region to the left of

the line.
The graph of the system is the intersection of the three shaded regions.

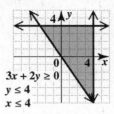

$3x + 2y \geq 0$
$y \leq 4$
$x \leq 4$

48. **[13.5]** $4y > 3x - 12$
$$x^2 < 16 - y^2$$

The boundary $4y = 3x - 12$ is a dashed line with intercepts $(0, -3)$ and $(4, 0)$. Test $(0, 0)$.

$$4(0) > 3(0) - 12 \quad ?$$
$$0 > -12 \quad \textit{True}$$

Shade the side of the line that contains $(0, 0)$. The boundary $x^2 = 16 - y^2$, or $x^2 + y^2 = 16$, is a dashed circle with center at $(0, 0)$ and radius 4. Test $(0, 0)$.

$$0^2 < 16 - 0^2 \quad ?$$
$$0 < 16 \quad \textit{True}$$

Shade the region inside the circle.
The graph of the system is the intersection of the two shaded regions.

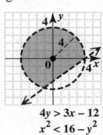

$4y > 3x - 12$
$x^2 < 16 - y^2$

Chapter 13 Test

1.   **(a)** $f(x) = \sqrt{x - 2}$ is the graph of $g(x) = \sqrt{x}$ shifted 2 units right. The graph is **C**.

(b) $f(x) = \sqrt{x + 2}$ is the graph of $g(x) = \sqrt{x}$ shifted 2 units left. The graph is **A**.

(c) $f(x) = \sqrt{x} + 2$ is the graph of $g(x) = \sqrt{x}$ shifted 2 units up. The graph is **D**.

(d) $f(x) = \sqrt{x} - 2$ is the graph of $g(x) = \sqrt{x}$ shifted 2 units down. The graph is **B**.

2. $f(x) = |x - 3| + 4$

This is the graph of the absolute value function, $g(x) = |x|$, shifted 3 units to the right (since $x - 3 = 0$ if $x = 3$) and 4 units upward (because of the $+4$). Its lowest point is at $(3, 4)$.

x	0	1	2	3	4	5	6
y	7	6	5	4	5	6	7

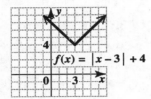

3. $(x - 2)^2 + (y + 3)^2 = 16$
$(x - 2)^2 + [y - (-3)]^2 = 4^2$

The graph is a circle with center $(2, -3)$ and radius 4.

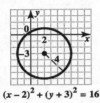

$(x - 2)^2 + (y + 3)^2 = 16$

4. $x^2 + y^2 + 8x - 2y = 8$

To find the center and radius, complete the squares on x and y.

$$\left(x^2 + 8x \quad\right) + \left(y^2 - 2y \quad\right) = 8$$
$$\left(x^2 + 8x + \underline{16}\right) + \left(y^2 - 2y + \underline{1}\right) = 8 + \underline{16} + \underline{1}$$
$$(x + 4)^2 + (y - 1)^2 = 25$$

The graph is a circle with center $(-4, 1)$ and radius $\sqrt{25} = 5$.

5. $f(x) = \sqrt{9 - x^2}$

Replace $f(x)$ with y.
$$y = \sqrt{9 - x^2}$$
Square both sides.
$$y^2 = 9 - x^2$$
$$x^2 + y^2 = 9$$

The graph of $x^2 + y^2 = 9$ is a circle of radius $\sqrt{9} = 3$ centered at the origin. Since $f(x)$ is nonnegative, only the top half of the circle is graphed.

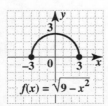

$f(x) = \sqrt{9 - x^2}$

6. $4x^2 + 9y^2 = 36$
$$\frac{x^2}{9} + \frac{y^2}{4} = 1$$

The equation is in $\dfrac{x^2}{a^2} + \dfrac{y^2}{b^2} = 1$ form with $a = 3$ and $b = 2$. The graph is an ellipse with intercepts $(3, 0)$, $(-3, 0)$, $(0, 2)$, and $(0, -2)$. Plot these intercepts, and draw the ellipse through them.

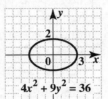

$4x^2 + 9y^2 = 36$

7. $16y^2 - 4x^2 = 64$
$$\frac{y^2}{4} - \frac{x^2}{16} = 1$$

The equation is in $\dfrac{y^2}{b^2} - \dfrac{x^2}{a^2} = 1$ form with $a = 4$ and $b = 2$. The graph is a hyperbola with y-intercepts $(0, 2)$ and $(0, -2)$ and asymptotes that are the extended diagonals of the rectangle with vertices $(4, 2)$, $(4, -2)$, $(-4, -2)$, and $(-4, 2)$.

Draw a branch of the hyperbola through each intercept and approaching the asymptotes.

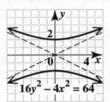

$16y^2 - 4x^2 = 64$

8. $\dfrac{y}{2} = -\sqrt{1 - \dfrac{x^2}{9}}$

Square both sides.
$$\frac{y^2}{4} = 1 - \frac{x^2}{9}$$
$$\frac{x^2}{9} + \frac{y^2}{4} = 1$$

This is an ellipse with x-intercepts $(3, 0)$ and $(-3, 0)$ and y-intercepts $(0, 2)$ and $(0, -2)$. Since y represents a negative square root in the original equation, y must be nonpositive. This restricts the graph to the lower half of the ellipse.

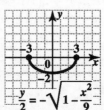

9. $6x^2 + 4y^2 = 12$

We have the *sum* of squares with different coefficients equal to a positive number, so this is an equation of an *ellipse*.

10.
$$16x^2 = 144 + 9y^2$$
$$16x^2 - 9y^2 = 144$$

We have the *difference* of squares equal to a positive number, so this is an equation of a *hyperbola*.

11. $4y^2 + 4x = 9$
$$4x = -4y^2 + 9$$

We have an x-term and a y^2-term, so this is an equation of a horizontal *parabola*.

12. $2x - y = 9$ (1)
        $xy = 5$ (2)

Solve equation (1) for y.
$$y = 2x - 9 \quad (3)$$

Substitute $2x - 9$ for y in equation (2).
$$xy = 5 \quad (2)$$
$$x(2x - 9) = 5$$
$$2x^2 - 9x = 5$$
$$2x^2 - 9x - 5 = 0$$
$$(2x + 1)(x - 5) = 0$$

$2x + 1 = 0$ or $x - 5 = 0$
    $x = -\frac{1}{2}$ or $x = 5$

Substitute these values for x in equation (3) to find y.

If $x = -\frac{1}{2}$, then $y = 2\left(-\frac{1}{2}\right) - 9 = -10$.

If $x = 5$, then $y = 2(5) - 9 = 1$.

The solution set is $\left\{\left(-\frac{1}{2}, -10\right), (5, 1)\right\}$.

13. $x - 4 = 3y$ (1)
   $x^2 + y^2 = 8$ (2)

Solve equation (1) for x.
$$x = 3y + 4$$

Substitute $3y + 4$ for x in equation (2).
$$x^2 + y^2 = 8 \quad (2)$$
$$(3y + 4)^2 + y^2 = 8$$
$$9y^2 + 24y + 16 + y^2 = 8$$
$$10y^2 + 24y + 8 = 0$$
$$2(5y^2 + 12y + 4) = 0$$
$$2(5y + 2)(y + 2) = 0$$

$5y + 2 = 0$ or $y + 2 = 0$
    $y = -\frac{2}{5}$ or $y = -2$

Since $x = 3y + 4$, substitute these values for y to find x.

If $y = -\frac{2}{5}$, then
$$x = 3\left(-\frac{2}{5}\right) + 4 = -\frac{6}{5} + 4 = \frac{14}{5}.$$

If $y = -2$, then $x = 3(-2) + 4 = -2$.

The solution set is $\left\{(-2, -2), \left(\frac{14}{5}, -\frac{2}{5}\right)\right\}$.

14. $x^2 + y^2 = 25$ (1)
     $x^2 - 2y^2 = 16$ (2)

Multiply equation (1) by 2 and add the result to equation (2).

$$
\begin{array}{rl}
2x^2 + 2y^2 = 50 & \quad 2 \times (1) \\
x^2 - 2y^2 = 16 & \quad (2) \\
\hline
3x^2 = 66 & \\
x^2 = 22 &
\end{array}
$$

$$x = \sqrt{22} \quad \text{or} \quad x = -\sqrt{22}$$

Substitute 22 for x^2 in equation (1).
$$x^2 + y^2 = 25 \quad (1)$$
$$22 + y^2 = 25$$
$$y^2 = 3$$
$$y = \sqrt{3} \quad \text{or} \quad y = -\sqrt{3}$$

The solution set is
$$\left\{\left(\sqrt{22}, \sqrt{3}\right), \left(\sqrt{22}, -\sqrt{3}\right),\right.$$
$$\left.\left(-\sqrt{22}, \sqrt{3}\right), \left(-\sqrt{22}, -\sqrt{3}\right)\right\}.$$

15. $y < x^2 - 2$

The boundary, $y = x^2 - 2$, is a dashed parabola in $y = a(x - h)^2 + k$ form with vertex (h, k) at $(0, -2)$. Since $a = 1 > 0$, the parabola opens up. It also has the same shape as $y = x^2$. The points $(2, 2)$ and $(-2, 2)$ are on the graph. Test $(0, 0)$.

$$0 \leq (0)^2 - 2 \quad ?$$
$$0 \leq -2 \quad\quad \textit{False}$$

Shade the side of the parabola that does not contain $(0, 0)$. This is the region outside the parabola.

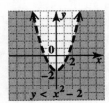

16. $x^2 + 25y^2 \le 25$
$x^2 + y^2 \le 9$

The first boundary, $\dfrac{x^2}{25} + \dfrac{y^2}{1} = 1$, is a solid ellipse with intercepts $(5,0)$, $(-5,0)$, $(0,1)$, and $(0,-1)$. Test $(0,0)$.

$$0^2 + 25 \cdot 0^2 \le 25 \quad ?$$
$$0 \le 25 \quad True$$

Shade the region inside the ellipse. The second boundary, $x^2 + y^2 = 9$, is a solid circle with center $(0,0)$ and radius 3. Test $(0,0)$.

$$0^2 + 0^2 \le 9 \quad ?$$
$$0 \le 9 \quad True$$

Shade the region inside the circle. The solution of the system is the intersection of the two shaded regions.

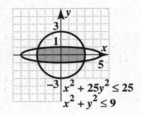

Cumulative Review Exercises (Chapters 1–13)

1. $-10 + |-5| - |3| + 4$
$= -10 + 5 - 3 + 4$
$= -5 - 3 + 4$
$= -8 + 4 = -4$

2. $4 - (2x + 3) + x = 5x - 3$
$4 - 2x - 3 + x = 5x - 3$
$-x + 1 = 5x - 3$
$-6x = -4$
$x = \frac{2}{3}$

The solution set is $\left\{ \frac{2}{3} \right\}$.

3. $-4k + 7 \ge 6k + 1$
$-10k \ge -6$

Divide by -10; reverse the direction of the inequality.

$$k \le \frac{-6}{-10}$$
$$k \le \frac{3}{5}$$

The solution set is $\left(-\infty, \frac{3}{5} \right]$.

4. Let $(x_1, y_1) = (2, 5)$ and $(x_2, y_2) = (-4, 1)$.

$$m = \frac{y_2 - y_1}{x_2 - x_1} = \frac{1 - 5}{-4 - 2} = \frac{-4}{-6} = \frac{2}{3}$$

5. Through $(-3, -2)$; perpendicular to $2x - 3y = 7$

Write $2x - 3y = 7$ in slope-intercept form.

$$-3y = -2x + 7$$
$$y = \frac{2}{3}x - \frac{7}{3}$$

The slope is $\frac{2}{3}$. Perpendicular lines have slopes that are negative reciprocals of each other, so a line perpendicular to the given line will have slope $-\frac{3}{2}$. Let $m = -\frac{3}{2}$ and $(x_1, y_1) = (-3, -2)$ in the point-slope form.

$$y - y_1 = m(x - x_1)$$
$$y - (-2) = -\frac{3}{2}[x - (-3)]$$
$$y + 2 = -\frac{3}{2}(x + 3)$$

Multiply by 2 to clear the fraction.

$$2y + 4 = -3(x + 3)$$
$$2y + 4 = -3x - 9$$
$$3x + 2y = -13$$

6. $(5y - 3)^2 = (5y)^2 - 2(5y)3 + 3^2$
$= 25y^2 - 30y + 9$

7. $(2r + 7)(6r - 1)$
$= 12r^2 - 2r + 42r - 7$
$= 12r^2 + 40r - 7$

8. $\dfrac{8x^4 - 4x^3 + 2x^2 + 13x + 8}{2x + 1}$

$$
\require{enclose}
\begin{array}{r}
4x^3 - 4x^2 + 3x + 5 \\
2x + 1 \enclose{longdiv}{8x^4 - 4x^3 + 2x^2 + 13x + 8} \\
\underline{8x^4 + 4x^3 } \\
-8x^3 + 2x^2 \\
\underline{-8x^3 - 4x^2 } \\
6x^2 + 13x \\
\underline{6x^2 + 3x } \\
10x + 8 \\
\underline{10x + 5} \\
3
\end{array}
$$

The answer is

$$4x^3 - 4x^2 + 3x + 5 + \frac{3}{2x + 1}.$$

9. $12x^2 - 7x - 10 = (4x - 5)(3x + 2)$

10. $2y^4 + 5y^2 - 3$

Let $p = y^2$, so $p^2 = y^4$.

$$2y^4 + 5y^2 - 3 = 2p^2 + 5p - 3$$
$$= (2p - 1)(p + 3)$$

Now substitute y^2 for p.

$$= (2y^2 - 1)(y^2 + 3)$$

11. $z^4 - 1 = (z^2 + 1)(z^2 - 1)$
$= (z^2 + 1)(z + 1)(z - 1)$

12. $a^3 - 27b^3 = a^3 - (3b)^3$

$$= (a - 3b)(a^2 + 3ab + 9b^2)$$

13. $\dfrac{5x - 15}{24} \cdot \dfrac{64}{3x - 9} = \dfrac{5(x - 3)}{3 \cdot 8} \cdot \dfrac{8 \cdot 8}{3(x - 3)}$

$$= \dfrac{5 \cdot 8}{3 \cdot 3} = \dfrac{40}{9}$$

14. $\dfrac{y^2 - 4}{y^2 - y - 6} \div \dfrac{y^2 - 2y}{y - 1}$

Multiply by the reciprocal.

$$= \dfrac{y^2 - 4}{y^2 - y - 6} \cdot \dfrac{y - 1}{y^2 - 2y}$$

Factor and simplify.

$$= \dfrac{(y + 2)(y - 2)}{(y - 3)(y + 2)} \cdot \dfrac{(y - 1)}{y(y - 2)}$$

$$= \dfrac{y - 1}{y(y - 3)}$$

15. $\dfrac{5}{c + 5} - \dfrac{2}{c + 3}$

The LCD is $(c + 5)(c + 3)$.

$$= \dfrac{5(c + 3)}{(c + 5)(c + 3)} - \dfrac{2(c + 5)}{(c + 3)(c + 5)}$$

$$= \dfrac{5c + 15 - 2c - 10}{(c + 5)(c + 3)}$$

$$= \dfrac{3c + 5}{(c + 5)(c + 3)}$$

16. $\dfrac{p}{p^2 + p} + \dfrac{1}{p^2 + p} = \dfrac{p + 1}{p^2 + p}$

$$= \dfrac{p + 1}{p(p + 1)} = \dfrac{1}{p}$$

17.   Let $x =$ the time to do the job working together.

Make a chart.

Worker	Rate	Time Together	Fractional Part of the Job Done
Kareem	$\dfrac{1}{3}$	x	$\dfrac{x}{3}$
Jamal	$\dfrac{1}{2}$	x	$\dfrac{x}{2}$

$$\underset{\substack{\text{Part done} \\ \text{by Kareem}}}{\dfrac{x}{3}} \;\; \underset{\text{plus}}{+} \;\; \underset{\substack{\text{part done} \\ \text{by Jamal}}}{\dfrac{x}{2}} \;\; \underset{\text{equals}}{=} \;\; \underset{\substack{\text{1 whole} \\ \text{job.}}}{1}$$

Multiply by the LCD, 6.

$$6\left(\dfrac{x}{3} + \dfrac{x}{2}\right) = 6 \cdot 1$$

$$2x + 3x = 6$$

$$5x = 6$$

$$x = \dfrac{6}{5} \;\; \text{or} \;\; 1\tfrac{1}{5}$$

It takes $\dfrac{6}{5}$ or $1\tfrac{1}{5}$ hours to do the job together.

18. $f(x) = 0.07x + 135$

$f(2000) = 0.07(2000) + 135 = 275$

The weekly fee is \$275 if \$2000 is taken in for the week.

19. $3x - y = 12$ (1)

     $2x + 3y = -3$ (2)

Multiply equation (1) by 3 and add the result to equation (2).

$$\begin{array}{r} 9x - 3y = 36 \qquad 3 \times (1) \\ \underline{2x + 3y = -3 \quad (2)} \\ 11x = 33 \\ x = 3 \end{array}$$

Substitute 3 for x in equation (1) to find y.

$$3x - y = 12 \quad (1)$$

$$3(3) - y = 12$$

$$9 - y = 12$$

$$-y = 3$$

$$y = -3$$

The solution set is $\{(3, -3)\}$.

20. $x + y - 2z = 9$ (1)

     $2x + y + z = 7$ (2)

     $3x - y - z = 13$ (3)

Add equation (2) and equation (3).

$$\begin{array}{r} 2x + y + z = 7 \quad (2) \\ \underline{3x - y - z = 13 \quad (3)} \\ 5x = 20 \\ x = 4 \end{array}$$

Multiply equation (1) by -1 and add the result to equation (2).

$$\begin{array}{r} -x - y + 2z = -9 \\ \underline{2x + y + z = 7 \quad (2)} \\ x + 3z = -2 \quad (4) \end{array}$$

Substitute 4 for x in equation (4) to find z.

$$x + 3z = -2 \quad (4)$$

$$4 + 3z = -2$$

$$3z = -6$$

$$z = -2$$

Substitute 4 for x and -2 for z in equation (2) to find y.

$$2x + y + z = 7 \quad (2)$$

$$2(4) + y - 2 = 7$$

$$y + 6 = 7$$

$$y = 1$$

The solution set is $\{(4, 1, -2)\}$.

21.
$$xy = -5 \quad (1)$$
$$2x + y = 3 \quad (2)$$

Solve equation (2) for y.

$$y = -2x + 3 \quad (3)$$

Substitute $-2x + 3$ for y in equation (1).

$$xy = -5 \quad (1)$$
$$x(-2x + 3) = -5$$
$$-2x^2 + 3x = -5$$
$$-2x^2 + 3x + 5 = 0$$
$$2x^2 - 3x - 5 = 0$$
$$(2x - 5)(x + 1) = 0$$

$$2x - 5 = 0 \quad \text{or} \quad x + 1 = 0$$
$$x = \tfrac{5}{2} \quad \text{or} \quad x = -1$$

Substitute these values for x in equation (3) to find y.

If $x = \tfrac{5}{2}$, then $y = -2\left(\tfrac{5}{2}\right) + 3 = -2$.

If $x = -1$, then $y = -2(-1) + 3 = 5$.

The solution set is $\left\{(-1, 5), \left(\tfrac{5}{2}, -2\right)\right\}$.

22. Let $s =$ Al's speed, $2s =$ Bev's speed, $t =$ Bev's time, and $t + \tfrac{1}{2} =$ Al's time.

	Distance	Rate	Time
Al	20	s	$t + \tfrac{1}{2}$
Bev	20	$2s$	t

Since $d = rt$, the system of equations is

$$s\left(t + \tfrac{1}{2}\right) = 20 \quad (1)$$
$$2st = 20. \quad (2)$$

Solve equation (2) for s.

$$2st = 20 \quad (2)$$
$$s = \frac{20}{2t} = \frac{10}{t}$$

Substitute $\dfrac{10}{t}$ for s in equation (1) to find t.

$$s\left(t + \tfrac{1}{2}\right) = 20 \quad (1)$$
$$\frac{10}{t}\left(t + \frac{1}{2}\right) = 20$$
$$10 + \frac{5}{t} = 20$$

Multiply each term by the LCD, t.

$$10t + 5 = 20t$$
$$5 = 10t$$
$$t = \tfrac{5}{10} = \tfrac{1}{2}$$

Since $s = \dfrac{10}{t}$ and $t = \dfrac{1}{2}$,

$$s = \frac{10}{\frac{1}{2}} = 20.$$

So Al's speed was 20 mph and Bev's speed was $2 \cdot 20 = 40$ mph.

Another solution:

$$st + \tfrac{1}{2}s = 20 \quad (1)$$
$$2st = 20 \quad (2)$$

Multiply equation (1) by -2 and add the result to equation (2).

$$
\begin{array}{rcl r}
-2st - s &=& -40 & -2 \times (1) \\
2st &=& 20 & (2) \\
\hline
-s &=& -20 & \\
s &=& 20 &
\end{array}
$$

Substitute 20 for s in equation (2) to find $t = \tfrac{1}{2}$.

23. $\left(\tfrac{4}{3}\right)^{-1} = \left(\tfrac{3}{4}\right)^{1} = \tfrac{3}{4}$

24.
$$\frac{(2a)^{-2}a^4}{a^{-3}} = \frac{2^{-2}a^{-2}a^4}{a^{-3}} = \frac{2^{-2}a^2}{a^{-3}}$$
$$= \frac{a^2 a^3}{2^2} = \frac{a^5}{4}$$

25.
$$4\sqrt[3]{16} - 2\sqrt[3]{54} = 4\sqrt[3]{8 \cdot 2} - 2\sqrt[3]{27 \cdot 2}$$
$$= 4 \cdot 2\sqrt[3]{2} - 2 \cdot 3\sqrt[3]{2}$$
$$= 8\sqrt[3]{2} - 6\sqrt[3]{2} = 2\sqrt[3]{2}$$

26.
$$\frac{3\sqrt{5x}}{\sqrt{2x}} = \frac{3\sqrt{5x} \cdot \sqrt{2x}}{\sqrt{2x} \cdot \sqrt{2x}} = \frac{3\sqrt{10x^2}}{2x}$$
$$= \frac{3x\sqrt{10}}{2x} = \frac{3\sqrt{10}}{2}$$

27. $\dfrac{5 + 3i}{2 - i}$

Multiply the numerator and denominator by the conjugate of the denominator.

$$= \frac{(5 + 3i)(2 + i)}{(2 - i)(2 + i)}$$
$$= \frac{10 + 5i + 6i + 3i^2}{4 - i^2}$$
$$= \frac{10 + 11i + 3(-1)}{4 - (-1)}$$
$$= \frac{7 + 11i}{5} = \frac{7}{5} + \frac{11}{5}i$$

28.
$$|5m| - 6 = 14$$
$$|5m| = 20$$

$$5m = 20 \quad \text{or} \quad 5m = -20$$
$$m = 4 \quad \text{or} \quad m = -4$$

The solution set is $\{-4, 4\}$.

29. $|2p - 5| > 15$

$2p - 5 > 15 \quad \text{or} \quad 2p - 5 < -15$

$\qquad 2p > 20 \qquad\qquad 2p < -10$

$\qquad p > 10 \quad \text{or} \qquad p < -5$

The solution set is $(-\infty, -5) \cup (10, \infty)$.

30. $2\sqrt{k} = \sqrt{5k + 3}$

$\quad 4k = 5k + 3 \qquad \textit{Square both sides.}$

$\quad -k = 3$

$\quad k = -3$

Since k must be nonnegative so that $\sqrt{k}$ is a real number, -3 cannot be a solution. The solution set is $\emptyset$.

31.
$$10q^2 + 13q = 3$$
$$10q^2 + 13q - 3 = 0$$
$$(5q - 1)(2q + 3) = 0$$

$5q - 1 = 0 \quad \text{or} \quad 2q + 3 = 0$

$\quad q = \frac{1}{5} \quad \text{or} \qquad q = -\frac{3}{2}$

The solution set is $\left\{ \frac{1}{5}, -\frac{3}{2} \right\}$.

32. $(4x - 1)^2 = 8$

$\quad 4x - 1 = \pm\sqrt{8}$

$\quad 4x = 1 \pm 2\sqrt{2}$

$\quad x = \dfrac{1 \pm 2\sqrt{2}}{4}$

The solution set is $\left\{ \frac{1+2\sqrt{2}}{4}, \frac{1-2\sqrt{2}}{4} \right\}$.

33. $3k^2 - 3k - 2 = 0$

Use the quadratic formula with $a = 3$, $b = -3$, and $c = -2$.

$$k = \frac{-b \pm \sqrt{b^2 - 4ac}}{2a}$$

$$k = \frac{-(-3) \pm \sqrt{(-3)^2 - 4(3)(-2)}}{2(3)}$$

$$= \frac{3 \pm \sqrt{9 + 24}}{6} = \frac{3 \pm \sqrt{33}}{6}$$

The solution set is $\left\{ \frac{3+\sqrt{33}}{6}, \frac{3-\sqrt{33}}{6} \right\}$.

34. $2(x^2 - 3)^2 - 5(x^2 - 3) = 12$

Let $u = (x^2 - 3)$.

$$2u^2 - 5u = 12$$
$$2u^2 - 5u - 12 = 0$$
$$(2u + 3)(u - 4) = 0$$

$2u + 3 = 0 \quad \text{or} \quad u - 4 = 0$

$\quad u = -\frac{3}{2} \quad \text{or} \qquad u = 4$

Substitute $x^2 - 3$ for u to find x.

If $u = -\frac{3}{2}$, then

$x^2 - 3 = -\frac{3}{2}$

$\quad x^2 = \frac{3}{2}$

$\quad x = \pm\sqrt{\dfrac{3}{2}} = \pm\dfrac{\sqrt{3}}{\sqrt{2}} \cdot \dfrac{\sqrt{2}}{\sqrt{2}} = \pm\dfrac{\sqrt{6}}{2}.$

If $u = 4$, then

$$x^2 - 3 = 4$$
$$x^2 = 7$$
$$x = \pm\sqrt{7}.$$

The solution set is $\left\{ -\frac{\sqrt{6}}{2}, \frac{\sqrt{6}}{2}, -\sqrt{7}, \sqrt{7} \right\}$.

35. Solve $F = \dfrac{kwv^2}{r}$ for v.

$$Fr = kwv^2$$
$$v^2 = \frac{Fr}{kw}$$

Take the square root of each side.

$$v = \pm\sqrt{\frac{Fr}{kw}} = \frac{\pm\sqrt{Fr}}{\sqrt{kw}} \cdot \frac{\sqrt{kw}}{\sqrt{kw}} = \frac{\pm\sqrt{Frkw}}{kw}$$

36. $f(x) = y = x^3 + 4$

Interchange x and y and solve for y.

$$x = y^3 + 4$$
$$y^3 = x - 4$$
$$y = \sqrt[3]{x - 4}$$
$$f^{-1}(x) = \sqrt[3]{x - 4}$$

37. $a^{\log_a x} = x$, so $3^{\log_3 4} = 4$.

38. $e^{\ln x} = x$, so $e^{\ln 7} = 7$.

39. $2\log (3x + 7) - \log 4 = \log (3x + 7)^2 - \log 4$

$$= \log \frac{(3x + 7)^2}{4}$$

40. $\log (x + 2) + \log (x - 1) = 1$

$\log_{10} [(x + 2)(x - 1)] = 1$

$\quad (x + 2)(x - 1) = 10^1$

$\qquad x^2 + x - 2 = 10$

$\qquad x^2 + x - 12 = 0$

$\qquad (x + 4)(x - 3) = 0$

$x + 4 = 0 \quad \text{or} \quad x - 3 = 0$

$\quad x = -4 \quad \text{or} \qquad x = 3$

The original equation is undefined when $x = -4$.

The solution set is $\{3\}$.

41. $P = 10,000, r = 0.05, t = 4$

(a) $A = P\left(1 + \dfrac{r}{n}\right)^{nt}$

$= 10,000\left(1 + \dfrac{0.05}{4}\right)^{4 \cdot 4}$ *Let n = 4.*

$= 10,000(1.0125)^{16}$

$\approx 12,198.90$

There will be $12,198.90 in the account.

(b) $A = Pe^{rt}$

$= 10,000e^{(0.05)(4)}$

$\approx 12,214.03$

There will be $12,214.03 in the account.

42. $y = 1.38(1.65)^x$

$y = 1.38(1.65)^5$ *Let x = 5.*

≈ 16.9 billion dollars

The sales are estimated to be $16.9 billion in 2000.

43. $2003 - 1995 = 8$

$y = 1.38(1.65)^8$ *Let x = 8.*

≈ 75.8 billion dollars

The sales are estimated to be $75.8 billion in 2003.

44. $f(x) = |x - 3|$

The function is defined for any value of x, so the domain is $(-\infty, \infty)$. The result of the absolute value is nonnegative, so the range is $[0, \infty)$.

45. $f(x) = -3x + 5$

The equation is in slope-intercept form, so the y-intercept is $(0, 5)$ and $m = -3$ or $\frac{-3}{1}$.

Plot $(0, 5)$. From $(0, 5)$, move down 3 units and right 1 unit. Draw the line through these two points.

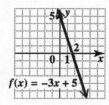

46. $f(x) = -2(x - 1)^2 + 3$

The graph is a parabola that has been shifted 1 unit to the right and 3 units upward from $(0, 0)$, so its vertex is at $(1, 3)$. Since $a = -2 < 0$, the parabola opens down. Also $|a| = |-2| = 2 > 1$, so the graph is narrower than the graph of $f(x) = x^2$. The points $(0, 1)$ and $(2, 1)$ are on the graph.

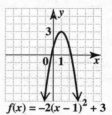

47. $\dfrac{x^2}{25} + \dfrac{y^2}{16} \leq 1$

The boundary, $\dfrac{x^2}{25} + \dfrac{y^2}{16} = 1$, is a solid ellipse in $\dfrac{x^2}{a^2} + \dfrac{y^2}{b^2} = 1$ form with intercepts $(5, 0)$, $(-5, 0)$, $(0, 4)$, and $(0, -4)$. Test $(0, 0)$.

$$\dfrac{0^2}{25} + \dfrac{0^2}{16} \leq 1 \quad ?$$
$$0 \leq 1 \quad \text{True}$$

Shade the region inside the ellipse.

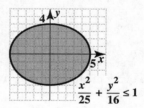

48. $f(x) = \sqrt{x - 2}$

This is the graph of $g(x) = \sqrt{x}$ shifted 2 units right.

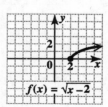

49. $\dfrac{x^2}{4} - \dfrac{y^2}{16} = 1$ is in $\dfrac{x^2}{a^2} - \dfrac{y^2}{b^2} = 1$ form.

The graph is a hyperbola with x-intercepts $(2,0)$ and $(-2,0)$ and asymptotes that are the extended diagonals of the rectangle with vertices $(2,4)$, $(2,-4)$, $(-2,-4)$, and $(-2,4)$. Draw a branch of the hyperbola through each intercept approaching the asymptotes.

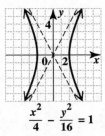

$$\dfrac{x^2}{4} - \dfrac{y^2}{16} = 1$$

50. $f(x) = 3^x$

The graph of f is an increasing exponential.

x	-1	0	1	2
$f(x)$	$\frac{1}{3}$	1	3	9

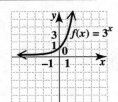

CHAPTER 14 SEQUENCES AND SERIES

14.1 Sequences and Series

1. $a_n = n + 1$
To get a_1, the first term, replace n with 1.
$$a_1 = 1 + 1 = 2$$
To get a_2, the second term, replace n with 2.
$$a_2 = 2 + 1 = 3$$
To get a_3, the third term, replace n with 3.
$$a_3 = 3 + 1 = 4$$
To get a_4, the fourth term, replace n with 4.
$$a_4 = 4 + 1 = 5$$
To get a_5, the fifth term, replace n with 5.
$$a_5 = 5 + 1 = 6$$
Answer: 2, 3, 4, 5, 6

3. $a_n = \dfrac{n + 3}{n}$
To get a_1, the first term, replace n with 1.
$$a_1 = \frac{1 + 3}{1} = \frac{4}{1} = 4$$
To get a_2, the second term, replace n with 2.
$$a_2 = \frac{2 + 3}{2} = \frac{5}{2}$$
To get a_3, the third term, replace n with 3.
$$a_3 = \frac{3 + 3}{3} = \frac{6}{3} = 2$$
To get a_4, the fourth term, replace n with 4.
$$a_4 = \frac{4 + 3}{4} = \frac{7}{4}$$
To get a_5, the fifth term, replace n with 5.
$$a_5 = \frac{5 + 3}{5} = \frac{8}{5}$$
Answer: $4, \frac{5}{2}, 2, \frac{7}{4}, \frac{8}{5}$

5. $a_n = 3^n$
$$a_1 = 3^1 = 3$$
$$a_2 = 3^2 = 9$$
$$a_3 = 3^3 = 27$$
$$a_4 = 3^4 = 81$$
$$a_5 = 3^5 = 243$$
Answer: 3, 9, 27, 81, 243

7. $a_n = \dfrac{1}{n^2}$
$$a_1 = \frac{1}{1^2} = 1$$
$$a_2 = \frac{1}{2^2} = \frac{1}{4}$$
$$a_3 = \frac{1}{3^2} = \frac{1}{9}$$
$$a_4 = \frac{1}{4^2} = \frac{1}{16}$$
$$a_5 = \frac{1}{5^2} = \frac{1}{25}$$
Answer: $1, \frac{1}{4}, \frac{1}{9}, \frac{1}{16}, \frac{1}{25}$

9. $a_n = (-1)^n$
$$a_1 = (-1)^1 = -1$$
$$a_2 = (-1)^2 = 1$$
$$a_3 = (-1)^3 = -1$$
$$a_4 = (-1)^4 = 1$$
$$a_5 = (-1)^5 = -1$$
Answer: $-1, 1, -1, 1, -1$

11. $a_n = -9n + 2$
To find a_8, replace n with 8.
$$a_8 = -9(8) + 2 = -72 + 2 = -70$$

13. $a_n = \dfrac{3n + 7}{2n - 5}$
$$a_{14} = \frac{3(14) + 7}{2(14) - 5} = \frac{42 + 7}{28 - 5} = \frac{49}{23}$$

15. $a_n = (n + 1)(2n + 3)$
$$a_8 = (8 + 1)[2(8) + 3]$$
$$= 9(19)$$
$$= 171$$

17. 4, 8, 12, 16, ... can be written as
$4 \cdot 1, 4 \cdot 2, 4 \cdot 3, 4 \cdot 4, \ldots$, so $a_n = 4n$.

19. $\frac{1}{3}, \frac{1}{9}, \frac{1}{27}, \frac{1}{81}, \ldots$ can be written as
$\dfrac{1}{3^1}, \dfrac{1}{3^2}, \dfrac{1}{3^3}, \dfrac{1}{3^4}, \ldots$, so $a_n = \dfrac{1}{3^n}$.

21. Make a table as follows:

Month	Interest	Payment	Unpaid balance
0			1000
1	$1000(0.01) = 10$	$100 + 10 = 110$	$1000 - 100 = 900$
2	$900(0.01) = 9$	$100 + 9 = 109$	$900 - 100 = 800$
3	$800(0.01) = 8$	$100 + 8 = 108$	$800 - 100 = 700$
4	$700(0.01) = 7$	$100 + 7 = 107$	$700 - 100 = 600$
5	$600(0.01) = 6$	$100 + 6 = 106$	$600 - 100 = 500$
6	$500(0.01) = 5$	$100 + 5 = 105$	$500 - 100 = 400$

The payments are $110, $109, $108, $107, $106, and $105; the unpaid balance is $400.

23. When new, the car is worth \$20,000.
Let a_n = the value of the car after the nth year.
The car retains $\frac{4}{5}$ of its value each year.

$$a_1 = \tfrac{4}{5}(20{,}000) = \$16{,}000$$
$$\text{(value after the first year)}$$
$$a_2 = \tfrac{4}{5}(16{,}000) = \$12{,}800$$
$$a_3 = \tfrac{4}{5}(12{,}800) = \$10{,}240$$
$$a_4 = \tfrac{4}{5}(10{,}240) = \$8192$$
$$a_5 = \tfrac{4}{5}(8192) = \$6553.60$$

The value of the car after 5 years is about \$6554.

25. $\displaystyle\sum_{i=1}^{5}(i+3)$

$$= (1+3)+(2+3)+(3+3)+(4+3)+(5+3)$$
$$\textit{Let } i = 1 \textit{ to 5 and add terms.}$$
$$= 4+5+6+7+8$$
$$= 30$$

27. $\displaystyle\sum_{i=1}^{3}(i^2+2)$

$$= \left(1^2+2\right)+\left(2^2+2\right)+\left(3^2+2\right)$$
$$\textit{Let } i = 1 \textit{ to 3 and add terms.}$$
$$= 3+6+11$$
$$= 20$$

29. $\displaystyle\sum_{i=1}^{6}(-1)^i$

$$= (-1)^1+(-1)^2+(-1)^3+(-1)^4$$
$$\quad +(-1)^5+(-1)^6$$
$$= -1+1-1+1-1+1$$
$$= 0$$

31. $\displaystyle\sum_{i=3}^{7}(i-3)(i+2)$

$$= (3-3)(3+2)+(4-3)(4+2)$$
$$\quad +(5-3)(5+2)+(6-3)(6+2)$$
$$\quad +(7-3)(7+2)$$
$$= 0(5)+1(6)+2(7)+3(8)+4(9)$$
$$= 0+6+14+24+36$$
$$= 80$$

33. $\displaystyle\sum_{i=1}^{5}2x\cdot i$

$$= 2x\cdot 1+2x\cdot 2+2x\cdot 3+2x\cdot 4$$
$$\quad +2x\cdot 5$$
$$= 2x+4x+6x+8x+10x$$

35. $\displaystyle\sum_{i=1}^{5}i\cdot x^i$

$$= 1\cdot x^1+2\cdot x^2+3\cdot x^3+4\cdot x^4+5\cdot x^5$$
$$= x+2x^2+3x^3+4x^4+5x^5$$

37. $3+4+5+6+7$

$$= (1+2)+(2+2)+(3+2)$$
$$\quad +(4+2)+(5+2)$$
$$= \sum_{i=1}^{5}(i+2)$$

39. $\frac{1}{2}+\frac{1}{3}+\frac{1}{4}+\frac{1}{5}+\frac{1}{6}$

$$= \frac{1}{1+1}+\frac{1}{2+1}+\frac{1}{3+1}+\frac{1}{4+1}$$
$$\quad +\frac{1}{5+1}$$
$$= \sum_{i=1}^{5}\frac{1}{i+1}$$

41. $1+4+9+16+25$

$$= 1^2+2^2+3^2+4^2+5^2$$
$$= \sum_{i=1}^{5}i^2$$

43. The similarities are that both are defined by the same linear expression and that points satisfying both lie in a straight line. The difference is that the domain of f consists of all real numbers, but the domain of the sequence is $\{1,2,3,\dots\}$. An example of a similarity is that $f(1)=6$ and $a_1=6$. An example of a difference is that $f\left(\frac{3}{2}\right)=7$, but $a_{3/2}$ is not defined.

45. A sequence is a list of terms in a specific order, while a series is the indicated sum of the terms of a sequence.

47. $\displaystyle\overline{x} = \frac{\displaystyle\sum_{i=1}^{n}x_i}{n} = \frac{\displaystyle\sum_{i=1}^{7}x_i}{7}$

$$= \frac{8+11+14+9+3+6+8}{7} = \frac{59}{7}$$

49. $\displaystyle\overline{x} = \frac{5+9+8+2+4+7+3+2}{8} = \frac{40}{8} = 5$

51. $\displaystyle\overline{x} = \frac{8155+8305+8244+8126+8044}{5}$

$$= \frac{40{,}874}{5} = 8174.8$$

The average number of funds available for this five-year period was about 8175.

53.
$$a + 3d = 12 \quad (1)$$
$$a + 8d = 22 \quad (2)$$

Multiply (1) by -1 and add the result to (2).

$$
\begin{array}{rcll}
-a - 3d &=& -12 & -1 \times (1) \\
a + 8d &=& 22 & \\
\hline
5d &=& 10 & \textit{Add.} \\
d &=& 2 &
\end{array}
$$

Substitute 2 for d in (1).

$$a + 3d = 12 \quad (1)$$
$$a + 3(2) = 12$$
$$a + 6 = 12$$
$$a = 6$$

Thus, $a = 6$ and $d = 2$.

55. Given $a = -2$, $n = 5$, and $d = 3$,

$$a + (n-1)d = -2 + (5-1)3$$
$$= -2 + (4)3$$
$$= -2 + 12$$
$$= 10.$$

14.2 Arithmetic Sequences

1. An arithmetic sequence is a sequence (list) of numbers in a specific order such that there is a common difference between any two successive terms. For example, the sequence 1, 5, 9, 13, ... is arithmetic with difference $d = 5 - 1 = 9 - 5 = 13 - 9 = 4$. As another example, 2, -1, -4, -7, ... is an arithmetic sequence with $d = -3$.

3. 1, 2, 3, 4, 5, ...

d is the difference between any two adjacent terms. Choose the terms 3 and 2.

$$d = 3 - 2 = 1$$

The terms 2 and 1 would give

$$d = 2 - 1 = 1,$$

the same result. Therefore, the common difference is $d = 1$.

Note: You should find the difference for all pairs of adjacent terms to determine if the sequence is arithmetic.

5. 2, -4, 6, -8, 10, -12, ...

The difference between the first two terms is $-4 - 2 = -6$, but the difference between the second and third terms is $6 - (-4) = 10$. The differences are not the same so the sequence is *not arithmetic*.

7. -10, -5, 0, 5, 10, ...

Choose the terms 10 and 5, and find the difference.

$$d = 10 - 5 = 5$$

The terms -5 and -10 would give

$$d = -5 - (-10) = -5 + 10 = 5,$$

the same result. Therefore, the common difference is $d = 5$.

9. $a_1 = 5$
$a_2 = a_1 + d = 5 + 4 = 9$
$a_3 = a_2 + d = 9 + 4 = 13$
$a_4 = a_3 + d = 13 + 4 = 17$
$a_5 = a_4 + d = 17 + 4 = 21$

11. $a_1 = -2$
$a_2 = a_1 + d = -2 + (-4) = -6$
$a_3 = a_2 + d = -6 + (-4) = -10$
$a_4 = a_3 + d = -10 + (-4) = -14$
$a_5 = a_4 + d = -14 + (-4) = -18$

13. $a_1 = 2$, $d = 5$
$a_n = a_1 + (n-1)d$
$ = 2 + (n-1)5$
$ = 2 + 5n - 5$
$ = 5n - 3$

15. $3, \frac{15}{4}, \frac{9}{2}, \frac{21}{4}, \ldots$

To find d, subtract any two adjacent terms.

$$d = \frac{15}{4} - 3 = \frac{15}{4} - \frac{12}{4} = \frac{3}{4}$$

The first term is $a_1 = 3$. Now find a_n.

$$a_n = a_1 + (n-1)d$$
$$= 3 + (n-1)\left(\frac{3}{4}\right)$$
$$= 3 + \frac{3}{4}n - \frac{3}{4}$$
$$= \frac{3}{4}n + \frac{9}{4}$$

17. $-3, 0, 3, \ldots$

To find d, subtract any two adjacent terms.

$$d = 0 - (-3) = 3$$

The first term is $a_1 = -3$. Now find a_n.

$$a_n = a_1 + (n-1)d$$
$$= -3 + (n-1)3$$
$$= -3 + 3n - 3$$
$$= 3n - 6$$

19. Given $a_1 = 4$ and $d = 3$, find a_{25}.

$$a_n = a_1 + (n-1)d$$
$$a_{25} = 4 + (25-1)3$$
$$= 4 + 72$$
$$= 76$$

21. Given 2, 4, 6, ..., find a_{24}.

Here, $a_1 = 2$ and $d = 4 - 2 = 2$.

$$a_n = a_1 + (n-1)d$$
$$a_{24} = 2 + (24-1)2$$
$$= 2 + 46$$
$$= 48$$

23. Given $a_{12} = -45$ and $a_{10} = -37$, find a_1.

Use $a_n = a_1 + (n-1)d$ to write a system of equations.

$$a_{12} = a_1 + (12-1)d$$
$$-45 = a_1 + 11d \qquad (1)$$
$$a_{10} = a_1 + (10-1)d$$
$$-37 = a_1 + 9d \qquad (2)$$

To eliminate d, multiply equation (1) by -9 and equation (2) by 11. Then add the results.

$$
\begin{array}{rll}
405 &= -9a_1 - 99d & -9 \times (1) \\
-407 &= 11a_1 + 99d & 11 \times (2) \\
\hline
-2 &= 2a_1 & \\
-1 &= a_1 &
\end{array}
$$

25. $3, 5, 7, \ldots, 33$

Let n represent the number of terms in the sequence. So, $a_n = 33$, $a_1 = 3$, and $d = 5 - 3 = 2$.

$$a_n = a_1 + (n-1)d$$
$$33 = 3 + (n-1)2$$
$$33 = 3 + 2n - 2$$
$$33 = 2n + 1$$
$$32 = 2n$$
$$n = 16$$

The sequence has 16 terms.

27. $\frac{3}{4}, 3, \frac{21}{4}, \ldots, 12$

Let n represent the number of terms in the sequence. So, $a_n = 12$, $a_1 = \frac{3}{4}$, and $d = 3 - \frac{3}{4} = \frac{9}{4}$.

$$a_n = a_1 + (n-1)d$$
$$12 = \frac{3}{4} + (n-1)\left(\frac{9}{4}\right)$$
$$\frac{45}{4} = (n-1)\left(\frac{9}{4}\right)$$
$$5 = n - 1 \qquad \textit{Multiply by } \tfrac{4}{9}.$$
$$6 = n$$

The sequence has 6 terms.

29. n represents the number of terms.

31. Find S_6 given $a_1 = 6$, $d = 3$, and $n = 6$.

$$S_n = \frac{n}{2}\big[2a_1 + (n-1)d\big]$$
$$S_6 = \frac{6}{2}\big[2 \cdot 6 + (6-1)3\big]$$
$$= 3(12 + 5 \cdot 3)$$
$$= 3(27)$$
$$= 81$$

33. Find S_6 given $a_1 = 7$, $d = -3$, and $n = 6$.

$$S_n = \frac{n}{2}\big[2a_1 + (n-1)d\big]$$
$$S_6 = \frac{6}{2}\big[2 \cdot 7 + (6-1)(-3)\big]$$
$$= 3\big[14 + 5(-3)\big]$$

$$= 3(14 - 15)$$
$$= 3(-1)$$
$$= -3$$

35. Find S_6 given $a_n = 4 + 3n$.

Find the first and last terms.

$$a_1 = 4 + 3(1) = 7$$
$$a_6 = 4 + 3(6) = 22$$

Now find the sum.

$$S_n = \frac{n}{2}(a_1 + a_n)$$
$$S_6 = \frac{6}{2}(7 + 22)$$
$$= 3(29)$$
$$= 87$$

37. $\displaystyle\sum_{i=1}^{10} (8i - 5)$

$$a_n = 8n - 5$$
$$a_1 = 8(1) - 5 = 3$$
$$a_{10} = 8(10) - 5 = 75$$

Use $S_n = \dfrac{n}{2}(a_1 + a_n)$ with $n = 10$, $a_1 = 3$, and $a_{10} = 75$.

$$S_{10} = \frac{10}{2}(3 + 75)$$
$$= 5(78)$$
$$= 390$$

39. $\displaystyle\sum_{i=1}^{20} (2i - 5)$

$$a_n = 2n - 5$$
$$a_1 = 2(1) - 5 = -3$$
$$a_{20} = 2(20) - 5 = 35$$

Use $S_n = \dfrac{n}{2}(a_1 + a_n)$ with $n = 20$, $a_1 = -3$, and $a_{20} = 35$.

$$S_{20} = \frac{20}{2}(-3 + 35)$$
$$= 10(32)$$
$$= 320$$

41. $\displaystyle\sum_{i=1}^{250} i$

Here, $a_n = n$, $a_1 = 1$, and $a_{250} = 250$.
Use $S_n = \dfrac{n}{2}(a_1 + a_n)$ with $n = 250$, $a_1 = 1$, and $a_{250} = 250$.

$$S_{250} = \frac{250}{2}(1 + 250)$$
$$= 125(251)$$
$$= 31{,}375$$

43. The sequence is 1, 2, 3, ..., 30.

$$S_n = \frac{n}{2}(a_1 + a_n)$$
$$S_{30} = \frac{30}{2}(1 + 30)$$
$$= 15(31)$$
$$= 465$$

The account will have $465 deposited in it over the entire month.

45. Your salaries at six-month intervals form an arithmetic sequence with $a_1 = 1600$ and $d = 50$. Since your salary is increased every 6 months, or $2(5) = 10$ times, after 5 years your salary will equal the term a_{11}.

$$a_n = a_1 + (n-1)d$$
$$a_{11} = 1600 + (11-1)50$$
$$= 1600 + 500$$
$$= 2100$$

Your salary will be $2100/month.

47. Given the sequence 20, 22, 24, ..., for 25 terms, find a_{25}. Here, $a_1 = 20$, $d = 22 - 20 = 2$, and $n = 25$.

$$a_n = a_1 + (n-1)d$$
$$a_{25} = 20 + (25-1)2$$
$$= 20 + 48$$
$$= 68$$

There are 68 seats in the last row.
Now find S_{25}.

$$S_n = \frac{n}{2}(a_1 + a_n)$$
$$S_{25} = \frac{25}{2}(20 + 68)$$
$$= \frac{25}{2}(88)$$
$$= 25(44)$$
$$= 1100$$

There are 1100 seats in the section.

49. Given the sequence 35, 31, 27, ..., can the sequence end in 1? If not, find the last positive value. If the sequence ends in 1, we can find n, a whole number.

$$d = 31 - 35 = -4$$
$$a_n = a_1 + (n-1)d$$
$$1 = 35 + (n-1)(-4)$$
$$1 = 35 - 4n + 4$$
$$-38 = -4n$$
$$9.5 = n$$

Since n is not a whole number, the sequence cannot end in 1. The largest n possible is $n = 9$.

$$a_n = a_1 + (n-1)d$$
$$a_9 = 35 + (9-1)(-4)$$
$$= 35 - 32 = 3$$

She can build 9 rows. There are 3 blocks in the last row.

51. $a = 2, r = 3, n = 2$
$$ar^n = 2(3)^2 = 2(9) = 18$$

53. $a = 4, r = \frac{1}{2}, n = 3$
$$ar^n = 4\left(\frac{1}{2}\right)^3 = 4\left(\frac{1}{8}\right) = \frac{1}{2}$$

14.3 Geometric Sequences

1. A geometric sequence is an ordered list of numbers such that each term after the first is obtained by multiplying the previous term by a constant, r, called the common ratio. For example, if the first term is 3 and $r = 4$, then the sequence is 3, 12, 48, 192, ... If the first term is 2 and $r = -1$, then the sequence is 2, -2, 2, -2, ...

3. 4, 8, 16, 32, ...

To find r, choose any two adjacent terms and divide the second one by the first one.

$$r = \frac{8}{4} = 2$$

Notice that any two other adjacent terms could have been used with the same result. The common ratio is $r = 2$.

Note: You should find the ratio for all pairs of adjacent terms to determine if the sequence is geometric.

5. $\frac{1}{3}, \frac{2}{3}, \frac{3}{3}, \frac{4}{3}, \frac{5}{3}, ...$

Choose any two adjacent terms and divide the second by the first.

$$r = \frac{\frac{2}{3}}{\frac{1}{3}} = \frac{2}{3} \cdot 3 = 2$$

Confirm this result with any two other adjacent terms.

$$r = \frac{\frac{3}{3}}{\frac{2}{3}} = 1 \cdot \frac{3}{2} = \frac{3}{2}$$

Since $2 \neq \frac{3}{2}$, the ratios are not the same. The sequence is *not geometric*.

7. 1, -3, 9, -27, 81, ...

[1st and 2nd] $r = \frac{-3}{1} = -3$
[2nd and 3rd] $r = \frac{9}{-3} = -3$

The common ratio is $r = -3$.

9. $1, -\frac{1}{2}, \frac{1}{4}, -\frac{1}{8}, \frac{1}{16}, ...$

[1st and 2nd] $r = \frac{-\frac{1}{2}}{1} = -\frac{1}{2}$

[2nd and 3rd] $r = \frac{\frac{1}{4}}{-\frac{1}{2}} = \frac{1}{4} \cdot (-2) = -\frac{1}{2}$

The common ratio is $r = -\frac{1}{2}$.

11. Find a general term for $5, 10, \ldots$.

First, find r.

$$r = \tfrac{10}{5} = 2$$

Use $a_1 = 5$ and $r = 2$ to find a_n.

$$a_n = a_1 r^{n-1}$$
$$a_n = 5(2)^{n-1}$$

13. Find a general term for $\tfrac{1}{9}, \tfrac{1}{3}, \ldots$.

Here, $a_1 = \tfrac{1}{9}$. Find r.

$$r = \frac{\tfrac{1}{3}}{\tfrac{1}{9}} = \frac{1}{3} \cdot \frac{9}{1} = 3$$

Now find a_n.

$$a_n = a_1 r^{n-1}$$
$$a_n = \frac{1}{9}(3)^{n-1}, \text{ or } a_n = \frac{3^{n-1}}{9}$$

15. Find a general term for $10, -2, \ldots$.

Here, $a_1 = 10$. Find r.

$$r = \frac{-2}{10} = -\frac{1}{5}$$

Now find a_n.

$$a_n = a_1 r^{n-1}$$
$$a_n = 10\left(-\tfrac{1}{5}\right)^{n-1}$$

17. Substitute $a_1 = 2$, $r = 5$, and $n = 10$ in the nth-term formula to find a_{10}.

$$a_n = a_1 r^{n-1}$$
$$a_{10} = a_1(r)^{10-1}$$
$$= 2(5)^9 = 3{,}906{,}250$$

19. Given $\tfrac{1}{2}, \tfrac{1}{6}, \tfrac{1}{18}, \ldots$, find a_{12}.

First find the common ratio.

$$r = \frac{\tfrac{1}{6}}{\tfrac{1}{2}} = \frac{1}{6} \cdot 2 = \frac{1}{3}$$

Substitute $a_1 = \tfrac{1}{2}$, $r = \tfrac{1}{3}$, and $n = 12$ in the nth-term formula.

$$a_n = a_1 r^{n-1}$$
$$a_{12} = a_1(r)^{12-1}$$
$$= \tfrac{1}{2}\left(\tfrac{1}{3}\right)^{11}$$

21. Given $a_3 = \tfrac{1}{2}$ and $a_7 = \tfrac{1}{32}$, find a_{25}.

Find a_1 and r using the general term $a_n = a_1 r^{n-1}$.

$$a_3 = a_1 r^{3-1}$$
$$\tfrac{1}{2} = a_1 r^2 \quad (1)$$
$$a_7 = a_1 r^{7-1}$$
$$\tfrac{1}{32} = a_1 r^6 \quad (2)$$

Solve (1) for a_1.

$$a_1 = \frac{1}{2r^2}$$

Substitute $\tfrac{1}{2r^2}$ for a_1 in (2).

$$\frac{1}{32} = \frac{1}{2r^2} r^6$$
$$\tfrac{1}{16} = r^4$$
$$r^2 = \pm \tfrac{1}{4}$$

Since r^2 is positive,

$$r^2 = \tfrac{1}{4} \quad \text{and} \quad r = \pm \tfrac{1}{2}.$$

Substitute $\tfrac{1}{4}$ for r^2 in (1).

$$\tfrac{1}{2} = a_1\left(\tfrac{1}{4}\right)$$
$$2 = a_1$$

Use $a_1 = 2$ and $r = \tfrac{1}{2}$ (or $-\tfrac{1}{2}$) to find a_{25}.

$$a_{25} = a_1(r)^{25-1}$$
$$= 2\left(\tfrac{1}{2}\right)^{24}$$
$$= \frac{1}{2^{23}}$$

23. $a_1 = 2$, $r = 3$; use $a_n = a_1 r^{n-1}$.

$$a_2 = a_1 r^1 = 2(3) = 6$$
$$a_3 = a_1 r^2 = 2(3)^2 = 18$$
$$a_4 = a_1 r^3 = 2(3)^3 = 54$$
$$a_5 = a_1 r^4 = 2(3)^4 = 162$$

Answer: $2, 6, 18, 54, 162$

25. $a_1 = 5$, $r = -\tfrac{1}{5}$; use $a_n = a_1 r^{n-1}$.

$$a_2 = a_1 r^1 = 5\left(-\tfrac{1}{5}\right) = -1$$
$$a_3 = a_1 r^2 = 5\left(-\tfrac{1}{5}\right)^2 = \tfrac{1}{5}$$
$$a_4 = a_1 r^3 = 5\left(-\tfrac{1}{5}\right)^3 = -\tfrac{1}{25}$$
$$a_5 = a_1 r^4 = 5\left(-\tfrac{1}{5}\right)^4 = \tfrac{1}{125}$$

Answer: $5, -1, \tfrac{1}{5}, -\tfrac{1}{25}, \tfrac{1}{125}$

27. $\tfrac{1}{3}, \tfrac{1}{9}, \tfrac{1}{27}, \tfrac{1}{81}, \tfrac{1}{243}$

Here, $a_1 = \tfrac{1}{3}$, $n = 5$, and

$$r = \frac{\tfrac{1}{9}}{\tfrac{1}{3}} = \frac{1}{9} \cdot 3 = \frac{1}{3}.$$

$$S_n = \frac{a_1(r^n - 1)}{r - 1}$$
$$S_5 = \frac{\tfrac{1}{3}\left[\left(\tfrac{1}{3}\right)^5 - 1\right]}{\tfrac{1}{3} - 1}$$
$$= \frac{\tfrac{1}{3}\left(\tfrac{1}{243} - 1\right)}{-\tfrac{2}{3}}$$
$$= \frac{\tfrac{1}{3}\left(-\tfrac{242}{243}\right)}{-\tfrac{2}{3}} = \frac{121}{243}$$

29. $-\frac{4}{3}, -\frac{4}{9}, -\frac{4}{27}, -\frac{4}{81}, -\frac{4}{243}, -\frac{4}{729}$

Here, $a_1 = -\frac{4}{3}$, $n = 6$, and

$$r = \frac{-\frac{4}{9}}{-\frac{4}{3}} = -\frac{4}{9} \cdot \left(-\frac{3}{4}\right) = \frac{1}{3}.$$

$$S_n = \frac{a_1(r^n - 1)}{r - 1}$$

$$S_6 = \frac{-\frac{4}{3}\left[\left(\frac{1}{3}\right)^6 - 1\right]}{\frac{1}{3} - 1}$$

$$= \frac{-\frac{4}{3}\left(\frac{1}{729} - 1\right)}{-\frac{2}{3}}$$

$$= \frac{-\frac{4}{3}\left(-\frac{728}{729}\right)}{-\frac{2}{3}} = -\frac{1456}{729} \approx -1.997$$

31. $\sum_{i=1}^{7} 4\left(\frac{2}{5}\right)^i$

Use $a_1 = 4\left(\frac{2}{5}\right) = \frac{8}{5}$, $n = 7$, and $r = \frac{2}{5}$.

$$S_n = \frac{a_1(r^n - 1)}{r - 1}$$

$$S_7 = \frac{\frac{8}{5}\left[\left(\frac{2}{5}\right)^7 - 1\right]}{\frac{2}{5} - 1}$$

$$= \frac{\frac{8}{5}\left[\left(\frac{2}{5}\right)^7 - 1\right]}{-\frac{3}{5}}$$

$$= -\frac{8}{3}\left[\left(\frac{2}{5}\right)^7 - 1\right] \approx 2.662$$

33. $\sum_{i=1}^{10} (-2)\left(\frac{3}{5}\right)^i$

Use $a_1 = (-2)\left(\frac{3}{5}\right) = -\frac{6}{5}$, $n = 10$, and $r = \frac{3}{5}$.

$$S_n = \frac{a_1(r^n - 1)}{r - 1}$$

$$S_{10} = \frac{-\frac{6}{5}\left[\left(\frac{3}{5}\right)^{10} - 1\right]}{\frac{3}{5} - 1}$$

$$= \frac{-\frac{6}{5}\left[\left(\frac{3}{5}\right)^{10} - 1\right]}{-\frac{2}{5}}$$

$$= 3\left[\left(\frac{3}{5}\right)^{10} - 1\right] \approx -2.982$$

35. There are 22 deposits, so $n = 22$.

$$S = R\left[\frac{(1+i)^n - 1}{i}\right]$$

$$= 1000\left[\frac{(1 + 0.065)^{22} - 1}{0.065}\right]$$

$$= 46,101.64$$

There will be $46,101.64 in the account.

37. Quarterly deposits for 10 years give us $n = 4 \cdot 10 = 40$. The interest rate per period is $i = \frac{0.07}{4} = 0.0175$.

$$S = R\left[\frac{(1+i)^n - 1}{i}\right]$$

$$= 1200\left[\frac{(1 + 0.0175)^{40} - 1}{0.0175}\right]$$

$$= 68,680.96$$

We now use the compound interest formula to determine the value of this money after 5 more years.

$$A = P\left(1 + \frac{r}{n}\right)^{nt}$$

$$= 68,680.96\left(1 + \frac{0.09}{12}\right)^{12(5)}$$

$$= 107,532.48$$

The woman is also saving $300 per month, so we use the annuity formula to determine that value.

$$S = R\left[\frac{(1+i)^n - 1}{i}\right]$$

$$= 300\left[\frac{\left(1 + \frac{0.09}{12}\right)^{12(5)} - 1}{\frac{0.09}{12}}\right]$$

$$= 22,627.24$$

Adding $22,627.24 to $107,532.48 gives a total of $130,159.72 in the account.

39. Find the sum if $a_1 = 6$ and $r = \frac{1}{3}$.
Since $|r| < 1$, the sum exists.

$$S = \frac{a_1}{1 - r} = \frac{6}{1 - \frac{1}{3}} = \frac{6}{\frac{2}{3}} = 6 \cdot \frac{3}{2} = 9$$

41. Find the sum if $a_1 = 1000$ and $r = -\frac{1}{10}$.
Since $|r| < 1$, the sum exists.

$$S = \frac{a_1}{1 - r} = \frac{1000}{1 - \left(-\frac{1}{10}\right)} = \frac{1000}{\frac{11}{10}}$$

$$= 1000 \cdot \frac{10}{11} = \frac{10,000}{11}$$

43. $\sum_{i=1}^{\infty} \frac{9}{8}\left(-\frac{2}{3}\right)^i$

$a_1 = \frac{9}{8}\left(-\frac{2}{3}\right)^1 = -\frac{3}{4}$ and $r = -\frac{2}{3}$.
Since $|r| < 1$, the sum exists.

$$S = \frac{a_1}{1 - r} = \frac{-\frac{3}{4}}{1 - \left(-\frac{2}{3}\right)} = \frac{-\frac{3}{4}}{\frac{5}{3}}$$

$$= -\frac{3}{4} \cdot \frac{3}{5} = -\frac{9}{20}$$

45. $\sum_{i=1}^{\infty} \frac{12}{5}\left(\frac{5}{4}\right)^i$

Since $|r| = \frac{5}{4} > 1$, the sum *does not exist*.

47. The ball is dropped from a height of 10 feet and will rebound $\frac{3}{5}$ of its original height.

Let $a_n =$ the ball's height on the nth rebound.

$$a_1 = 10 \quad \text{and} \quad r = \frac{3}{5}.$$

Since we must find the height after the fourth bounce, $n = 5$ (since a_1 is the starting point). Use $a_n = a_1 r^{n-1}$.

$$a_5 = 10\left(\frac{3}{5}\right)^{5-1} = 10\left(\frac{3}{5}\right)^4 \approx 1.3$$

The ball will rebound approximately 1.3 feet after the fourth bounce.

49. This exercise can be modeled by a geometric sequence with $a_1 = 256$ and $r = \frac{1}{2}$. First we need to find n so that $a_n = 32$.

$$32 = a_1 r^{n-1}$$
$$32 = 256\left(\frac{1}{2}\right)^{n-1}$$
$$\frac{1}{8} = \left(\frac{1}{2}\right)^{n-1}$$
$$\left(\frac{1}{2}\right)^3 = \left(\frac{1}{2}\right)^{n-1}$$
$$3 = n - 1$$
$$4 = n$$

Since n is 4, this means that 32 grams will be present on the day which corresponds to the 4th term of the sequence. That would be on day 3. To find what is left after the tenth day, we need to find a_{11} since we started with a_1.

$$a_{11} = a_1 r^{11-1} = 256\left(\frac{1}{2}\right)^{10} = \frac{256}{1024} = \frac{1}{4}$$

There will be $\frac{1}{4}$ gram of the substance after 10 days.

51. **(a)** Here, $a_1 = 1.1$ billion and $r = 106\% = 1.06$. Since we must find the consumption after 5 years, $n = 6$ (since a_1 is the starting point).

$$a_6 = a_1 r^{n-1}$$
$$a_6 = 1.1(1.06)^{6-1}$$
$$= 1.1(1.06)^5 \approx 1.5$$

The community will use about 1.5 billion units 5 years from now.

(b) If consumption doubles, then the consumption would be $2a_1$.

$$2a_1 = a_1(1.06)^{n-1}$$
$$2 = (1.06)^{n-1}$$
$$\ln 2 = \ln(1.06)^{n-1}$$
$$\ln 2 = (n-1)\ln(1.06)$$
$$n - 1 = \frac{\ln 2}{\ln 1.06}$$
$$n - 1 \approx 12$$
$$n \approx 13$$

Since n is about 13, that would represent the 13th term of the sequence, which represents about 12 years after the start.

53. Since the machine depreciates by $\frac{1}{4}$ of its value, it retains $1 - \frac{1}{4} = \frac{3}{4}$ of its value. Since the cost of the machine new is \$50,000, $a_1 = 50,000$. We want the value after 8 years so since the original cost is a_1, we need to find a_9.

$$a_9 = a_1 r^{9-1} = 50,000\left(\frac{3}{4}\right)^8 \approx 5006$$

The machine's value after 8 years is about \$5000.

55. $\frac{1}{3} = 0.33333\ldots$

56. $\frac{2}{3} = 0.66666\ldots$

57.
$$\begin{array}{r} 0.33333\ldots \\ + 0.66666\ldots \\ \hline 0.99999\ldots \end{array}$$

58. $S = \dfrac{a_1}{1-r} = \dfrac{0.9}{1 - 0.1} = \dfrac{0.9}{0.9} = 1$

Therefore, $0.99999\ldots = 1$.

59. From Exercise 58, $0.99999\ldots = 1$, so choice **B** is correct.

60. $0.49999\ldots = 0.4 + 0.09999\ldots$
$$= \frac{4}{10} + \frac{1}{10}(0.9999\ldots)$$
$$= \frac{4}{10} + \frac{1}{10}(1)$$
$$= \frac{5}{10} = \frac{1}{2}$$

61. $(3x + 2y)^2 = (3x)^2 + 2(3x)(2y) + (2y)^2$
$$= 9x^2 + 12xy + 4y^2$$

63. $(a - b)^2 = a^2 - 2ab + b^2$
$(a - b)^3$
$$= (a-b)^2(a-b)^1$$
$$= \left(a^2 - 2ab + b^2\right)(a - b)$$
$$= a^3 - 2a^2b + ab^2 - a^2b + 2ab^2 - b^3$$
$$= a^3 - 3a^2b + 3ab^2 - b^3$$

14.4   The Binomial Theorem

1. $6! = 6 \cdot 5 \cdot 4 \cdot 3 \cdot 2 \cdot 1 = 720$

3. $\dfrac{6!}{4!\,2!} = \dfrac{6 \cdot 5 \cdot 4 \cdot 3 \cdot 2 \cdot 1}{(4 \cdot 3 \cdot 2 \cdot 1)(2 \cdot 1)} = \dfrac{6 \cdot 5}{2 \cdot 1} = 15$

5. $_6C_2 = \dfrac{6!}{2!\,(6-2)!} = \dfrac{6!}{2!\,4!} = 15,$

by Exercise 3.

7. $\dfrac{4!}{0!\,4!} = \dfrac{4!}{(1)(4!)} = \dfrac{1}{1} = 1$

9. $4! \cdot 5 = (4 \cdot 3 \cdot 2 \cdot 1) \cdot 5$
$$= 24 \cdot 5 = 120$$

11. $_{13}C_{11} = \dfrac{13!}{11!\,(13-11)!} = \dfrac{13!}{11!\,2!}$

$\quad = \dfrac{13 \cdot 12}{2 \cdot 1} \qquad$ *11! cancels*

$\quad = 13 \cdot 6 = 78$

13. $(m+n)^4$

$\quad = m^4 + \dfrac{4!}{3!\,1!}m^3n^1 + \dfrac{4!}{2!\,2!}m^2n^2$

$\qquad + \dfrac{4!}{1!\,3!}m^1n^3 + n^4$

$\quad = m^4 + 4m^3n + 6m^2n^2 + 4mn^3 + n^4$

15. $(a-b)^5$

$\quad = \big[a + (-b)\big]^5$

$\quad = a^5 + \dfrac{5!}{4!\,1!}a^4(-b)^1 + \dfrac{5!}{3!\,2!}a^3(-b)^2$

$\qquad + \dfrac{5!}{2!\,3!}a^2(-b)^3 + \dfrac{5!}{1!\,4!}a^1(-b)^4 + (-b)^5$

$\quad = a^5 - 5a^4b + 10a^3b^2 - 10a^2b^3 + 5ab^4 - b^5$

17. $(2x+3)^3$

$\quad = (2x)^3 + \dfrac{3!}{2!\,1!}(2x)^2(3)^1 + \dfrac{3!}{1!\,2!}(2x)(3)^2$

$\qquad + (3)^3$

$\quad = 8x^3 + 36x^2 + 54x + 27$

19. $\left(\dfrac{x}{2} - y\right)^4$

$\quad = \left[\left(\dfrac{x}{2}\right) + (-y)\right]^4$

$\quad = \left(\dfrac{x}{2}\right)^4 + \dfrac{4!}{3!\,1!}\left(\dfrac{x}{2}\right)^3(-y)^1 + \dfrac{4!}{2!\,2!}\left(\dfrac{x}{2}\right)^2(-y)^2$

$\qquad + \dfrac{4!}{1!\,3!}\left(\dfrac{x}{2}\right)^1(-y)^3 + (-y)^4$

$\quad = \dfrac{x^4}{16} - \dfrac{x^3y}{2} + \dfrac{3x^2y^2}{2} - 2xy^3 + y^4$

21. $(mx - n^2)^3$

$\quad = \big[mx + (-n^2)\big]^3$

$\quad = (mx)^3 + \dfrac{3!}{2!\,1!}(mx)^2(-n^2)^1$

$\qquad + \dfrac{3!}{1!\,2!}(mx)^1(-n^2)^2 + (-n^2)^3$

$\quad = m^3x^3 - 3m^2n^2x^2 + 3mn^4x - n^6$

23. $(r + 2s)^{12}$

$\quad = r^{12} + \dfrac{12!}{11!\,1!}r^{11}(2s)^1 + \dfrac{12!}{10!\,2!}r^{10}(2s)^2$

$\qquad + \dfrac{12!}{9!\,3!}r^9(2s)^3 + \cdots$

The first four terms are

$\quad r^{12} + 24r^{11}s + 264r^{10}s^2 + 1760r^9s^3.$

25. $(3x - y)^{14}$

$\quad = \big[3x + (-y)\big]^{14}$

$\quad = (3x)^{14} + \dfrac{14!}{13!\,1!}(3x)^{13}(-y)^1$

$\qquad + \dfrac{14!}{12!\,2!}(3x)^{12}(-y)^2$

$\qquad + \dfrac{14!}{11!\,3!}(3x)^{11}(-y)^3 + \cdots$

The first four terms are

$\quad 3^{14}x^{14} - 14\big(3^{13}\big)x^{13}y + 91\big(3^{12}\big)x^{12}y^2$

$\qquad - 364\big(3^{11}\big)x^{11}y^3.$

27. $(t^2 + u^2)^{10}$

$\quad = \big(t^2\big)^{10} + \dfrac{10!}{9!\,1!}\big(t^2\big)^9\big(u^2\big)^1$

$\qquad + \dfrac{10!}{8!\,2!}\big(t^2\big)^8\big(u^2\big)^2$

$\qquad + \dfrac{10!}{7!\,3!}\big(t^2\big)^7\big(u^2\big)^3 + \cdots$

The first four terms are

$\quad t^{20} + 10t^{18}u^2 + 45t^{16}u^4 + 120t^{14}u^6.$

29. The rth term of the expansion of $(x+y)^n$ is

$$\dfrac{n!}{[n - (r-1)]!\,(r-1)!}(x)^{n-(r-1)}(y)^{r-1}.$$

Start with the exponent on y, which is 1 less than the term number r. In this case, we are looking for the fourth term, so $r = 4$ and $r - 1 = 3$. Thus, the fourth term of $(2m + n)^{10}$ is

$$\dfrac{10!}{(10-3)!\,3!}(2m)^{10-3}(n)^3$$

$$= \dfrac{10!}{7!\,3!}2^7m^7n^3$$

$$= 120\big(2^7\big)m^7n^3.$$

31. The seventh term of $\left(x + \dfrac{y}{2}\right)^8$ is

$$\dfrac{8!}{(8-6)!\,6!}(x)^{8-6}\left(\dfrac{y}{2}\right)^6$$

$$= \dfrac{8!}{6!\,2!}x^2\,\dfrac{y^6}{2^6}$$

$$= \dfrac{7x^2y^6}{16}.$$

33. The third term of $(k-1)^9$ is

$$\dfrac{9!}{(9-2)!\,2!}k^{9-2}(-1)^2 = \dfrac{9!}{7!\,2!}k^7 = 36k^7.$$

35. The expansion of $(x^2 - 2y)^6$ has seven terms, so the middle term is the fourth. The fourth term of $(x^2 - 2y)^6$ is

$$\frac{6!}{(6-3)!\,3!}(x^2)^{6-3}(-2y)^3$$
$$= \frac{6!}{3!\,3!}(x^2)^3(-8y^3)$$
$$= 20x^6(-8y^3) = -160x^6y^3.$$

37. The term of the expansion of $(3x^3 - 4y^2)^5$ with x^9y^4 in it is the term with $(3x^3)^3(-4y^2)^2$, since $(x^3)^3(y^2)^2 = x^9y^4$. The term is

$$\frac{5!}{3!\,2!}(3x^3)^3(-4y^2)^2$$
$$= 10(27x^9)(16y^4) = 4320x^9y^4.$$

Chapter 14 Review Exercises

1. $a_n = 2n - 3$
$a_1 = 2(1) - 3 = -1$
$a_2 = 2(2) - 3 = 1$
$a_3 = 2(3) - 3 = 3$
$a_4 = 2(4) - 3 = 5$

Answer: $-1, 1, 3, 5$

2. $a_n = \dfrac{n-1}{n}$

$a_1 = \dfrac{1-1}{1} = 0$

$a_2 = \dfrac{2-1}{2} = \dfrac{1}{2}$

$a_3 = \dfrac{3-1}{3} = \dfrac{2}{3}$

$a_4 = \dfrac{4-1}{4} = \dfrac{3}{4}$

Answer: $0, \frac{1}{2}, \frac{2}{3}, \frac{3}{4}$

3. $a_n = n^2$
$a_1 = (1)^2 = 1$
$a_2 = (2)^2 = 4$
$a_3 = (3)^2 = 9$
$a_4 = (4)^2 = 16$

Answer: $1, 4, 9, 16$

4. $a_n = \left(\frac{1}{2}\right)^n$

$a_1 = \left(\frac{1}{2}\right)^1 = \frac{1}{2}$

$a_2 = \left(\frac{1}{2}\right)^2 = \frac{1}{4}$

$a_3 = \left(\frac{1}{2}\right)^3 = \frac{1}{8}$

$a_4 = \left(\frac{1}{2}\right)^4 = \frac{1}{16}$

Answer: $\frac{1}{2}, \frac{1}{4}, \frac{1}{8}, \frac{1}{16}$

5. $a_n = (n+1)(n-1)$
$a_1 = (1+1)(1-1) = 2(0) = 0$
$a_2 = (2+1)(2-1) = 3(1) = 3$
$a_3 = (3+1)(3-1) = 4(2) = 8$
$a_4 = (4+1)(4-1) = 5(3) = 15$

Answer: $0, 3, 8, 15$

6. $\displaystyle\sum_{i=1}^{5} i^2 x$
$= 1^2x + 2^2x + 3^2x + 4^2x + 5^2x$
$= x + 4x + 9x + 16x + 25x$

7. $\displaystyle\sum_{i=1}^{6} (i+1)x^i$
$= (1+1)x^1 + (2+1)x^2 + (3+1)x^3$
$\quad + (4+1)x^4 + (5+1)x^5 + (6+1)x^6$
$= 2x + 3x^2 + 4x^3 + 5x^4 + 6x^5 + 7x^6$

8. $\displaystyle\sum_{i=1}^{4} (i+2)$
$= (1+2) + (2+2) + (3+2) + (4+2)$
$= 3 + 4 + 5 + 6$
$= 18$

9. $\displaystyle\sum_{i=1}^{6} 2^i$
$= 2^1 + 2^2 + 2^3 + 2^4 + 2^5 + 2^6$
$= 2 + 4 + 8 + 16 + 32 + 64$
$= 126$

10. $\displaystyle\sum_{i=4}^{7} \dfrac{i}{i+1}$
$= \frac{4}{4+1} + \frac{5}{5+1} + \frac{6}{6+1} + \frac{7}{7+1}$
$= \frac{4}{5} + \frac{5}{6} + \frac{6}{7} + \frac{7}{8}$ $LCD = 2^3 \cdot 3 \cdot 5 \cdot 7 = 840$
$= \frac{672}{840} + \frac{700}{840} + \frac{720}{840} + \frac{735}{840} = \frac{2827}{840}$

11. $\overline{x} = \dfrac{2535 + 2478 + 2342 + 2078 + 2662}{5}$

$\overline{x} = \dfrac{12{,}095}{5} = 2419$

The average mutual fund retirement assets were 2419 billion dollars for the five given years.

12. $2, 5, 8, 11, \ldots$ is an *arithmetic* sequence with

$$d = 5 - 2 = 3.$$

Note: You should find the difference for all pairs of adjacent terms to determine if the sequence is arithmetic.

13. $-6, -2, 2, 6, 10, \ldots$ is an *arithmetic* sequence with

$$d = -2 - (-6) = 4.$$

14. $\frac{2}{3}, -\frac{1}{3}, \frac{1}{6}, -\frac{1}{12}, \ldots$ is a *geometric* sequence with

$$r = \frac{-\frac{1}{3}}{\frac{2}{3}} = -\frac{1}{3} \cdot \frac{3}{2} = -\frac{1}{2}.$$

Note: You should find the ratio for all pairs of adjacent terms to determine if the sequence is geometric.

15. $-1, 1, -1, 1, -1, \ldots$ is a *geometric* sequence with

$$r = \frac{1}{-1} = -1.$$

16. $64, 32, 8, \frac{1}{2}, \ldots$

Find two differences:
$32 - 64 = -32$ and $8 - 32 = -24$, so the sequence is not arithmetic.

Find two ratios:
$\frac{32}{64} \neq \frac{8}{32}$, so the sequence is not geometric.

Therefore, the sequence is *neither*.

17. $64, 32, 16, 8, \ldots$ is a *geometric* sequence with

$$r = \frac{32}{64} = \frac{1}{2}.$$

18. $10, 8, 6, 4, \ldots$ is an *arithmetic* sequence with

$$d = 8 - 10 = -2.$$

19. Given $a_1 = -2$ and $d = 5$, find a_{16}.

$$a_n = a_1 + (n-1)d$$
$$a_{16} = -2 + (16-1)5$$
$$= -2 + 15(5)$$
$$= -2 + 75 = 73$$

20. Given $a_6 = 12$ and $a_8 = 18$, find a_{25}.

$$a_n = a_1 + (n-1)d$$
$$a_6 = a_1 + 5d, \quad \text{so} \quad 12 = a_1 + 5d \quad (1)$$
$$a_8 = a_1 + 7d, \quad \text{so} \quad 18 = a_1 + 7d \quad (2)$$

Multiply equation (1) by -1 and add the result to equation (2).

$$-12 = -a_1 - 5d \qquad -1 \times (1)$$
$$\underline{18 = \quad a_1 + 7d \quad (2)}$$
$$6 = \qquad 2d$$
$$3 = d$$

To find a_1, substitute $d = 3$ in equation (1).

$$12 = a_1 + 5d \quad (1)$$
$$12 = a_1 + 5(3)$$
$$12 = a_1 + 15$$
$$-3 = a_1$$

Use $a_1 = -3$ and $d = 3$ to find a_{25}.

$$a_n = a_1 + (n-1)d$$
$$a_{25} = -3 + (25-1)3$$
$$= -3 + 24(3)$$
$$= -3 + 72 = 69$$

21. $a_1 = -4, \ d = -5$

$$a_n = a_1 + (n-1)d$$
$$a_n = -4 + (n-1)(-5)$$
$$= -4 - 5n + 5$$
$$= -5n + 1$$

22. $6, 3, 0, -3, \ldots$

To get the general term, $a_n,$ first find d.
$d = 3 - 6 = -3$

$$a_n = a_1 + (n-1)d$$
$$a_n = 6 + (n-1)(-3)$$
$$= 6 - 3n + 3$$
$$= -3n + 9$$

23. $7, 10, 13, \ldots, 49$

Here, $a_1 = 7$ and $d = 10 - 7 = 3$.
Now find n, the number of terms.

$$a_n = a_1 + (n-1)d$$
$$49 = 7 + (n-1)(3)$$
$$42 = 3(n-1)$$
$$14 = n - 1 \qquad \textit{Divide by 3.}$$
$$15 = n$$

There are 15 terms in this sequence.

24. $5, 1, -3, \ldots, -79$

Here, $a_1 = 5$ and $d = 1 - 5 = -4$.
Now find n, the number of terms.

$$a_n = a_1 + (n-1)d$$
$$-79 = 5 + (n-1)(-4)$$
$$-79 = 9 - 4n$$
$$-88 = -4n$$
$$n = 22$$

There are 22 terms in this sequence.

25. Find S_8 if $a_1 = -2$ and $d = 6$. Find a_8 first.

$$a_8 = a_1 + (8-1)d$$
$$= -2 + 7(6)$$
$$= -2 + 42 = 40$$

Now find the sum.

$$S_n = \frac{n}{2}(a_1 + a_n)$$
$$S_8 = \frac{8}{2}(-2 + 40)$$
$$= 4(38) = 152$$

26. Find S_8 if $a_n = -2 + 5n$.
Find the first and last terms.

$$a_1 = -2 + 5(1) = 3$$
$$a_8 = -2 + 5(8) = 38$$

Now find the sum.

$$S_n = \frac{n}{2}(a_1 + a_n)$$
$$S_8 = \frac{8}{2}(3 + 38)$$
$$= 4(41)$$
$$= 164$$

27. Find the general term for the geometric sequence $-1, -4, \ldots$.

$$a_1 = -1 \quad \text{and} \quad r = \frac{-4}{-1} = 4.$$
$$a_n = a_1 r^{n-1}$$
$$a_n = -1(4)^{n-1}$$

28. $\frac{2}{3}, \frac{2}{15}, \ldots$

$$a_1 = \frac{2}{3} \quad \text{and} \quad r = \frac{\frac{2}{15}}{\frac{2}{3}} = \frac{2}{15} \cdot \frac{3}{2} = \frac{1}{5}.$$
$$a_n = a_1 r^{n-1}$$
$$a_n = \frac{2}{3}\left(\frac{1}{5}\right)^{n-1}$$

29. Find a_{11} for $2, -6, 18, \ldots$.

$$a_1 = 2 \quad \text{and} \quad r = \frac{-6}{2} = -3.$$
$$a_n = a_1 r^{n-1}$$
$$a_{11} = 2(-3)^{11-1}$$
$$= 2(-3)^{10} = 118{,}098$$

30. Given $a_3 = 20$ and $a_5 = 80$, find a_{10}.
$a_n = a_1 r^{n-1}$
For a_3, $\quad a_3 = a_1 r^{3-1}$
$\qquad 20 = a_1 r^2.$
For a_5, $\quad a_5 = a_1 r^{5-1}$
$\qquad 80 = a_1 r^4.$
The ratio of a_5 to a_3 is

$$\frac{80}{20} = \frac{a_1 r^4}{a_1 r^2}$$
$$4 = r^2$$
$$r = \pm 2.$$

Since $20 = a_1 r^2$, and $r^2 = 4$,
$$20 = a_1(4)$$
$$5 = a_1.$$
Now find a_{10}.
$$a_n = a_1 r^{n-1}$$
$$a_{10} = 5(\pm 2)^{10-1}$$
$$= 5(\pm 2)^9$$

Two answers are possible for a_{10}:
$$5(2)^9 = 2560 \quad \text{or} \quad 5(-2)^9 = -2560.$$

31. $\sum_{i=1}^{5} \left(\frac{1}{4}\right)^i$
$a_1 = \frac{1}{4}, r = \frac{1}{4}$, and $n = 5$.

$$S_n = \frac{a_1(1 - r^n)}{1 - r}$$
$$S_5 = \frac{\frac{1}{4}\left[1 - \left(\frac{1}{4}\right)^5\right]}{1 - \frac{1}{4}}$$
$$= \frac{\frac{1}{4}\left(1 - \frac{1}{1024}\right)}{\frac{3}{4}}$$
$$= \frac{1}{3}\left(\frac{1023}{1024}\right) = \frac{341}{1024}$$

32. $\sum_{i=1}^{8} \frac{3}{4}(-1)^i$
$a_1 = -\frac{3}{4}, r = -1$, and $n = 8$.

$$S_n = \frac{a_1(1 - r^n)}{1 - r}$$
$$S_8 = \frac{-\frac{3}{4}\left[1 - (-1)^8\right]}{1 - (-1)}$$
$$= \frac{-\frac{3}{4}(1 - 1)}{2}$$
$$= \frac{-\frac{3}{4}(0)}{2} = 0$$

33. $\sum_{i=1}^{\infty} 4\left(\frac{1}{5}\right)^i$
The terms are the terms of an infinite geometric sequence with $a_1 = \frac{4}{5}$ and $r = \frac{1}{5}$.

$$S = \frac{a_1}{1 - r} = \frac{\frac{4}{5}}{1 - \frac{1}{5}} = \frac{\frac{4}{5}}{\frac{4}{5}} = 1$$

34. $\sum_{i=1}^{\infty} 2(3)^i$
The terms are the terms of an infinite geometric sequence with $a_1 = 6$ and $r = 3$.

$$S = \frac{a_1}{1 - r} \quad \text{if} \quad |r| < 1,$$

but $r = 3$ so S does not exist, and, thus, the sum does not exist.

35. $(2p - q)^5$
$$= [2p + (-q)]^5$$
$$= (2p)^5 + \frac{5!}{4!\,1!}(2p)^4(-q)^1$$
$$+ \frac{5!}{3!\,2!}(2p)^3(-q)^2 + \frac{5!}{2!\,3!}(2p)^2(-q)^3$$
$$+ \frac{5!}{1!\,4!}(2p)^1(-q)^4 + (-q)^5$$
$$= 32p^5 + 5(16p^4)(-q) + 10(8p^3)q^2$$
$$+ 10(4p^2)(-q^3) + 5(2p)q^4 - q^5$$
$$= 32p^5 - 80p^4q + 80p^3q^2 - 40p^2q^3$$
$$+ 10pq^4 - q^5$$

36. $(x^2 + 3y)^4$

$$= (x^2)^4 + \frac{4!}{3! \, 1!} (x^2)^3 (3y)^1$$

$$+ \frac{4!}{2! \, 2!} (x^2)^2 (3y)^2$$

$$+ \frac{4!}{1! \, 3!} (x^2)^1 (3y)^3 + (3y)^4$$

$$= x^8 + 4(x^6)(3y) + 6(x^4)(9y^2)$$

$$+ 4(x^2)(27y^3) + 81y^4$$

$$= x^8 + 12x^6y + 54x^4y^2 + 108x^2y^3$$

$$+ 81y^4$$

37. $\left(\sqrt{m} + \sqrt{n}\right)^4$

$$= \left(\sqrt{m}\right)^4 + \frac{4!}{3! \, 1!} \left(\sqrt{m}\right)^3 \left(\sqrt{n}\right)^1$$

$$+ \frac{4!}{2! \, 2!} \left(\sqrt{m}\right)^2 \left(\sqrt{n}\right)^2$$

$$+ \frac{4!}{1! \, 3!} \left(\sqrt{m}\right)^1 \left(\sqrt{n}\right)^3 + \left(\sqrt{n}\right)^4$$

$$= m^2 + 4(m\sqrt{m})\left(\sqrt{n}\right) + 6(m)(n)$$

$$+ 4\left(\sqrt{m}\right)(n\sqrt{n}) + n^2$$

$$= m^2 + 4m\sqrt{mn} + 6mn + 4n\sqrt{mn} + n^2$$

38. The fourth term ($r = 4$, so $r - 1 = 3$) of $(3a + 2b)^{19}$ is

$$\frac{19!}{16! \, 3!} (3a)^{16} (2b)^3$$

$$= 969(3)^{16}(a^{16})(8)b^3$$

$$= 7752(3)^{16} a^{16} b^3.$$

39. The twenty-third term ($r = 23$, so $r - 1 = 22$) of $(-2k + 3)^{25}$ is

$$= \frac{25!}{3! \, 22!} (-2k)^3 (3)^{22}$$

$$= -18{,}400(3)^{22} k^3.$$

40. **[14.2]** The arithmetic sequence $1, 7, 13, \ldots$ has $a_1 = 1$ and $d = 7 - 1 = 6$. First find a_{40}.

$$a_n = a_1 + (n - 1)d$$

$$a_{40} = 1 + (40 - 1)6$$

$$= 1 + (39)6$$

$$= 1 + 234 = 235$$

Now find the tenth term so that we can find the sum of the first 10 terms.

$$a_{10} = 1 + (10 - 1)6$$

$$= 1 + (9)6 = 55$$

Now find the sum.

$$S_n = \frac{n}{2}(a_1 + a_n)$$

$$S_{10} = \frac{10}{2}(1 + 55)$$

$$= 5(56) = 280$$

41. **[14.3]** The geometric sequence $-3, \, 6, \, -12, \ldots$ has $a_1 = -3$ and $r = \frac{6}{-3} = -2$. First find a_{10}.

$$a_n = a_1 r^{n-1}$$

$$a_{10} = -3(-2)^9$$

$$= -3(-512) = 1536$$

Now find the sum of the first ten terms.

$$S_n = \frac{a_1(1 - r^n)}{1 - r}$$

$$S_{10} = \frac{-3[1 - (-2)^{10}]}{1 - (-2)}$$

$$= \frac{-3(1 - 1024)}{3}$$

$$= -(-1023) = 1023$$

42. **[14.3]** The geometric sequence has $a_1 = 1$ and $r = -3$. First find a_9.

$$a_n = a_1 r^{n-1}$$

$$a_9 = 1(-3)^{9-1}$$

$$= (-3)^8 = 6561$$

Now find the sum of the first ten terms.

$$S_n = \frac{a_1(r^n - 1)}{r - 1}$$

$$S_{10} = \frac{1[(-3)^{10} - 1]}{-3 - 1}$$

$$= -\tfrac{1}{4}(3^{10} - 1)$$

$$= -\tfrac{1}{4}(59{,}049 - 1)$$

$$= -14{,}762$$

43. **[14.2]** The arithmetic sequence with $a_1 = -4$ and $d = 3$ is $-4, -1, 2, 5, \ldots$. First find a_{15}.

$$a_n = a_1 + (n - 1)d$$

$$a_{15} = -4 + (15 - 1)3$$

$$= -4 + 42 = 38$$

Now find the sum of the first ten terms.

$$S_n = \frac{n}{2}[2a_1 + (n - 1)d]$$

$$S_{10} = \tfrac{10}{2}[2(-4) + (10 - 1)3]$$

$$= 5(-8 + 27)$$

$$= 5(19) = 95$$

44. **[14.2]** $2, 7, 12, \ldots$
This is an arithmetic sequence with $a_1 = 2$ and $d = 7 - 2 = 5$.

$$a_n = a_1 + (n - 1)d$$

$$a_n = 2 + (n - 1)5$$

$$= 2 + 5n - 5$$

$$= 5n - 3$$

45. [14.3] $2, 8, 32, \ldots$

This is a geometric sequence with $a_1 = 2$ and $r = \frac{8}{2} = 4$.

$$a_n = a_1 r^{n-1}$$
$$a_n = 2(4)^{n-1}$$

46. [14.3] $27, 9, 3, \ldots$

This is a geometric sequence with $a_1 = 27$ and $r = \frac{9}{27} = \frac{1}{3}$.

$$a_n = a_1 r^{n-1}$$
$$a_n = 27\left(\frac{1}{3}\right)^{n-1}$$

47. [14.2] $12, 9, 6, \ldots$

This is an arithmetic sequence with $a_1 = 12$ and $d = -3$.

$$a_n = a_1 + (n-1)d$$
$$a_n = 12 + (n-1)(-3)$$
$$= 12 - 3n + 3$$
$$= -3n + 15$$

48. [14.2] The distances traveled in successive seconds are

$$3, 7, 11, 15, 19, \ldots.$$

This is an arithmetic sequence with $a_1 = 3$ and $d = 4$. Since we know a_1 and d, we'll use the second formula for the sum of an arithmetic sequence with $S_n = 210$.

$$S_n = \frac{n}{2}\left[2a_1 + (n-1)d\right]$$
$$210 = \frac{n}{2}\left[2(3) + (n-1)4\right]$$
$$420 = n(6 + 4n - 4)$$
$$420 = 6n + 4n^2 - 4n$$
$$0 = 4n^2 + 2n - 420$$
$$0 = 2n^2 + n - 210$$
$$0 = (2n + 21)(n - 10)$$
$$2n + 21 = 0 \quad \text{or} \quad n - 10 = 0$$
$$n = -\frac{21}{2} \quad \text{or} \quad n = 10$$

Discard $-\frac{21}{2}$ since time cannot be negative. It takes her 10 seconds.

49. [14.3] Use the formula for the future value of an ordinary annuity with $R = 672$, $i = \frac{0.06}{4} = 0.015$, and $n = 7(4) = 28$.

$$S = R\left[\frac{(1+i)^n - 1}{i}\right]$$
$$S = 672\left[\frac{(1+0.015)^{28} - 1}{0.015}\right]$$
$$= 23{,}171.55$$

The future value of the annuity is $23,171.55.

50. [14.1] Since $100\% - 3\% = 97\% = 0.97$, the population after 1 year is $0.97(50{,}000)$, after 2 years is $0.97\left[0.97(50{,}000)\right]$ or $(0.97)^2(50{,}000)$, and after n years is $(0.97)^n(50{,}000)$. After 6 years, the population is

$$(0.97)^6(50{,}000) \approx 41{,}649 \approx 42{,}000.$$

51. [14.1] $\left(\frac{1}{2}\right)^n$ is left after n strokes. So $\left(\frac{1}{2}\right)^7 = \frac{1}{128} = 0.0078125$ is left after 7 strokes.

52. [14.3] **(a)** We can write the repeating decimal number $0.55555\ldots$ as an infinite geometric sequence as follows:

$$\frac{5}{10} + \frac{5}{10}\left(\frac{1}{10}\right) + \frac{5}{10}\left(\frac{1}{10}\right)^2 + \frac{5}{10}\left(\frac{1}{10}\right)^3 + \cdots$$

(b) The common ratio r is

$$\frac{\frac{5}{10}\left(\frac{1}{10}\right)}{\frac{5}{10}} = \frac{1}{10}.$$

(c) Since $|r| < 1$, the sum exists.

$$S = \frac{a_1}{1 - r} = \frac{\frac{5}{10}}{1 - \frac{1}{10}} = \frac{\frac{5}{10}}{\frac{9}{10}} = \frac{5}{9}$$

53. [14.3] No, the sum cannot be found, because $r = 2$ and this value of r does not satisfy $|r| < 1$.

54. [14.3] No, the terms must be successive, such as the first and second or the second and third.

Chapter 14 Test

1. $a_n = (-1)^n + 1$

$a_1 = (-1)^1 + 1 = 0$

$a_2 = (-1)^2 + 1 = 1 + 1 = 2$

$a_3 = (-1)^3 + 1 = -1 + 1 = 0$

$a_4 = (-1)^4 + 1 = 1 + 1 = 2$

$a_5 = (-1)^5 + 1 = -1 + 1 = 0$

Answer: $0, 2, 0, 2, 0$

2. $a_1 = 4$, $d = 2$

$a_2 = a_1 + d = 4 + 2 = 6$

$a_3 = a_2 + d = 6 + 2 = 8$

$a_4 = a_3 + d = 8 + 2 = 10$

$a_5 = a_4 + d = 10 + 2 = 12$

Answer: $4, 6, 8, 10, 12$

3. $a_4 = 6$, $r = \frac{1}{2}$

First find a_1.

$$a_n = a_1 r^{n-1}$$
$$a_4 = a_1\left(\frac{1}{2}\right)^{4-1}$$
$$6 = a_1\left(\frac{1}{8}\right)$$
$$a_1 = 48$$

Now find the remaining terms.

$$a_2 = \tfrac{1}{2}a_1 = \tfrac{1}{2}(48) = 24$$
$$a_3 = \tfrac{1}{2}a_2 = \tfrac{1}{2}(24) = 12$$
$$a_4 = \tfrac{1}{2}a_3 = \tfrac{1}{2}(12) = 6$$
$$a_5 = \tfrac{1}{2}a_4 = \tfrac{1}{2}(6) = 3$$

Answer: 48, 24, 12, 6, 3

4.   Given $a_1 = 6$ and $d = -2$, find a_4.

$$a_n = a_1 + (n-1)d$$
$$a_4 = a_1 + (4-1)d$$
$$= 6 + (3)(-2)$$
$$= 6 - 6 = 0$$

5.   Given $a_5 = 16$ and $a_7 = 9$, find a_4. This is a geometric sequence, so

$$a_6 = a_5 r \quad \text{and} \quad a_7 = a_6 r.$$

From the last equation, $a_6 = \dfrac{a_7}{r}$, so by substitution,

$$a_5 r = \frac{a_7}{r}.$$

Multiply by r and divide by a_5 to get

$$r^2 = \frac{a_7}{a_5}$$
$$r^2 = \frac{9}{16}$$
$$r = \pm\sqrt{\frac{9}{16}} = \pm\frac{3}{4}$$

Since $a_5 = a_4 r$, we know that $a_4 = \dfrac{16}{r}$. Substituting each value of r in the last equation gives us two values of a_4.

$$a_4 = \frac{16}{\frac{3}{4}} = \frac{64}{3} \quad \text{or} \quad a_4 = \frac{16}{-\frac{3}{4}} = -\frac{64}{3}$$

6.   Given the arithmetic sequence with $a_2 = 12$ and $a_3 = 15$, find S_5. First find d.

$$d = a_3 - a_2 = 15 - 12 = 3$$

Now find the first and fifth terms.

$$a_1 = a_2 - 3 = 12 - 3 = 9$$
$$a_5 = a_4 + 3 = a_3 + 6 = 15 + 6 = 21$$

Now find the sum of the first five terms.

$$S_n = \frac{n}{2}(a_1 + a_n)$$
$$S_5 = \tfrac{5}{2}(9 + 21)$$
$$= \tfrac{5}{2}(30) = 75$$

7.   Given the geometric sequence with $a_5 = 4$ and $a_7 = 1$, find S_5.

$$r^2 = \frac{a_7}{a_5} = \frac{1}{4}, \text{ so } r = \frac{1}{2} \text{ or } r = -\frac{1}{2}.$$

Use $a_7 = a_1 r^6$ to get $1 = a_1\left(\pm\tfrac{1}{2}\right)^6$, and so $a_1 = 64$.

Use $S_n = \dfrac{a_1(r^n - 1)}{r - 1}$ or $S_n = \dfrac{a_1(1 - r^n)}{1 - r}$.

$$S_5 = \frac{64\left[1 - \left(\frac{1}{2}\right)^5\right]}{1 - \frac{1}{2}} \quad \text{or} \quad S_5 = \frac{64\left[1 - \left(-\frac{1}{2}\right)^5\right]}{1 - \left(-\frac{1}{2}\right)}$$

$$= \frac{64}{\frac{1}{2}}\left(1 - \frac{1}{32}\right) \qquad = \frac{64}{\frac{3}{2}}\left(1 + \frac{1}{32}\right)$$

$$= 128\left(\tfrac{31}{32}\right) \qquad\qquad = \tfrac{128}{3}\left(\tfrac{33}{32}\right)$$

$$= 124 \qquad\qquad\qquad = 44$$

8. $\overline{x} = \dfrac{\text{total}}{5}$, where total $=$

$$71{,}911 + 72{,}458 + 73{,}527 + 74{,}638 + 76{,}579.$$

Thus, $\overline{x} = \dfrac{369{,}113}{5} = 73{,}822.6 \approx 73{,}823.$

The average number of banks for this five-year period was 73,823.

9. $S = R\left[\dfrac{(1+i)^n - 1}{i}\right]$

$$= 4000\left[\frac{\left(1 + \frac{0.06}{4}\right)^{4(7)} - 1}{\frac{0.06}{4}}\right]$$

$$= 4000\left[\frac{(1.015)^{28} - 1}{0.015}\right]$$

$$\approx 137{,}925.91$$

The account will have $137,925.91 at the end of this term.

10.   An infinite geometric series has a sum if $|r| < 1$, where r is the common ratio.

11. $\displaystyle\sum_{i=1}^{5}(2i + 8)$

$$= [2(1) + 8] + [2(2) + 8] + [2(3) + 8]$$
$$+ [2(4) + 8] + [2(5) + 8]$$
$$= 10 + 12 + 14 + 16 + 18$$
$$= 70$$

12. $\displaystyle\sum_{i=1}^{6}(3i - 5)$

Find the first and sixth terms.

$$a_1 = 3(1) - 5 = -2$$
$$a_6 = 3(6) - 5 = 13$$

Now find the sum of the first six terms.

$$S_6 = \frac{n}{2}(a_1 + a_6)$$
$$S_6 = \tfrac{6}{2}(-2 + 13)$$
$$= 3(11) = 33$$

13. $\displaystyle\sum_{i=1}^{500} i$

Use the formula $S_{500} = \dfrac{n}{2}(a_1 + a_{500})$ with $a_1 = 1$ and $a_{500} = 500$.

$$S_{500} = \tfrac{500}{2}(1 + 500)$$
$$= 250(501) = 125{,}250$$

14. $\displaystyle\sum_{i=1}^{3} \tfrac{1}{2}(4^i) = \tfrac{1}{2}\sum_{i=1}^{3}(4^i)$
$$= \tfrac{1}{2}(4^1 + 4^2 + 4^3)$$
$$= \tfrac{1}{2}(4 + 16 + 64)$$
$$= \tfrac{1}{2}(84) = 42$$

15. $\displaystyle\sum_{i=1}^{\infty} \left(\tfrac{1}{4}\right)^i$

This is an infinite geometric series with $a_1 = \tfrac{1}{4}$ and $r = \tfrac{1}{4}$.
Since $|r| = \tfrac{1}{4} < 1$, the sum exists.

$$S = \frac{a_1}{1-r} = \frac{\tfrac{1}{4}}{1 - \tfrac{1}{4}} = \frac{\tfrac{1}{4}}{\tfrac{3}{4}} = \frac{1}{4}\cdot\frac{4}{3} = \frac{1}{3}$$

16. $\displaystyle\sum_{i=1}^{\infty} 6\left(\tfrac{3}{2}\right)^i$

This is an infinite geometric series with $a_1 = 6\left(\tfrac{3}{2}\right)^1 = 9$ and $r = \tfrac{3}{2}$.
Since $|r| = \tfrac{3}{2} > 1$, the sum does not exist.

17. $8! = 8\cdot7\cdot6\cdot5\cdot4\cdot3\cdot2\cdot1$
$$= 40{,}320$$

18. By definition, $0! = 1$.

19. $\dfrac{6!}{4!\,2!} = \dfrac{6\cdot5\cdot4\cdot3\cdot2\cdot1}{(4\cdot3\cdot2\cdot1)(2\cdot1)} = \dfrac{6\cdot5}{2\cdot1} = 15$

20. $_{12}C_{10} = \dfrac{12!}{10!\,(12-10)!} = \dfrac{12\cdot11\cdot10!}{10!\,(2!)}$
$$= \dfrac{12\cdot11}{2} = 66$$

21. $(3k - 5)^4$
$$= (3k)^4 + \frac{4!}{3!\,1!}(3k)^3(-5)^1$$
$$+ \frac{4!}{2!\,2!}(3k)^2(-5)^2 + \frac{4!}{1!\,3!}(3k)^1(-5)^3$$
$$+ (-5)^4$$
$$= 81k^4 - 4(27k^3)(5) + 6(9k^2)(25)$$
$$- 4(3k)(125) + 625$$
$$= 81k^4 - 540k^3 + 1350k^2 - 1500k + 625$$

22. The fifth term ($r = 5$, so $r - 1 = 4$) of $\left(2x - \dfrac{y}{3}\right)^{12}$ is

$$\frac{12!}{(12-4)!\,4!}(2x)^{12-4}\left(-\frac{y}{3}\right)^4$$
$$= \frac{12!}{8!\,4!}(2x)^8\left(-\frac{y}{3}\right)^4$$
$$= \frac{14{,}080x^8y^4}{9}.$$

23. The amounts of unpaid balance during 15 months form an arithmetic sequence

$$300, 280, 260, \ldots, 40, 20,$$

which is the sequence with $n = 15$, $a_1 = 300$, and $a_{15} = 20$. Find the sum of these balances.

$$S_n = \frac{n}{2}(a_1 + a_n)$$
$$S_{15} = \frac{15}{2}(300 + 20)$$
$$= \frac{15}{2}(320) = 2400$$

Since 1% interest is paid on this total, the interest paid is 1% of \$2400 or \$24. The sewing machine cost \$300 (paid monthly at \$20), so the total cost is \$300 + \$24 = \$324.

24. The weekly populations form a geometric sequence with $a_1 = 20$ and $r = 3$ since the colony begins with 20 insects and triples each week. Find the general term of this geometric sequence.

$$a_n = a_1 r^{n-1}$$
$$a_n = 20(3)^{n-1}$$

We're assuming that from the beginning of July to the end of September is 12 weeks, so find a_{12}.

$$a_n = 20(3)^{n-1}$$
$$a_{12} = 20(3)^{11}$$

At the end of September, $20(3)^{11} = 3{,}542{,}940$ insects will be present in the colony.

Cumulative Review Exercises (Chapters 1–14)

1. $|-7| + 6 - |-10| - (-8 + 3)$
$$= 7 + 6 - 10 - (-5)$$
$$= 13 - 10 + 5 = 8$$

2. $-15 - |-4| - 10 - |-6|$
$$= -15 - 4 - 10 - 6 = -35$$

3. $4(-6) + (-8)(5) - (-9)$
$$= -24 - 40 + 9 = -55$$

In Exercises 4–7, let

$$P = \left\{-\tfrac{8}{3}, 10, 0, \sqrt{13}, -\sqrt{3}, \tfrac{45}{15}, \sqrt{-7}, 0.82, -3\right\}.$$

4. The integers are 10, 0, $\tfrac{45}{15}$ (or 3), and -3.

5. The rational numbers are
$$-\tfrac{8}{3}, 10, 0, \tfrac{45}{15} \text{ (or } 3\text{), } 0.82, \text{ and } -3.$$

6. The irrational numbers are $\sqrt{13}$ and $-\sqrt{3}$.

7. All are real numbers except $\sqrt{-7}$.

8.
$$9 - (5 + 3a) + 5a = -4(a - 3) - 7$$
$$9 - 5 - 3a + 5a = -4a + 12 - 7$$
$$4 + 2a = -4a + 5$$
$$6a = 1$$
$$a = \tfrac{1}{6}$$

The solution set is $\left\{\tfrac{1}{6}\right\}$.

9.
$$7m + 18 \le 9m - 2$$
$$-2m \le -20$$
Divide by -2; reverse the direction of the inequality symbol.
$$m \ge 10$$
The solution set is $[10, \infty)$.

10. $|4x - 3| = 21$

$4x - 3 = 21$	or	$4x - 3 = -21$
$4x = 24$		$4x = -18$
$x = 6$	or	$x = -\tfrac{18}{4} = -\tfrac{9}{2}$

The solution set is $\left\{-\tfrac{9}{2}, 6\right\}$.

11. $\dfrac{x + 3}{12} - \dfrac{x - 3}{6} = 0$

Multiply by the LCD, 12.
$$12\left(\frac{x + 3}{12} - \frac{x - 3}{6}\right) = 12(0)$$
$$x + 3 - 2(x - 3) = 0$$
$$x + 3 - 2x + 6 = 0$$
$$9 - x = 0$$
$$9 = x$$

Check $x = 9$: $1 - 1 = 0$ *True*
The solution set is $\{9\}$.

12.
$2x > 8$	or	$-3x > 9$
$x > 4$	or	$x < -3$

The solution set is $(-\infty, -3) \cup (4, \infty)$.

13. $|2m - 5| \ge 11$

$2m - 5 \ge 11$	or	$2m - 5 \le -11$
$2m \ge 16$		$2m \le -6$
$m \ge 8$	or	$m \le -3$

The solution set is $(-\infty, -3] \cup [8, \infty)$.

14. Let $(x_1, y_1) = (4, -5)$
and $(x_2, y_2) = (-12, -17)$. Then
$$m = \frac{y_2 - y_1}{x_2 - x_1} = \frac{-17 - (-5)}{-12 - 4} = \frac{-12}{-16} = \frac{3}{4}.$$
The slope is $\tfrac{3}{4}$.

15. $(4p + 2)(5p - 3)$
$$\overset{\textstyle\text{F}\qquad\text{O}\qquad\text{I}\qquad\text{L}}{= 20p^2 - 12p + 10p - 6}$$
$$= 20p^2 - 2p - 6$$

16. $(3k - 7)^2 = (3k)^2 - 2(3k)(7) + 7^2$
$$= 9k^2 - 42k + 49$$

17. $(2m^3 - 3m^2 + 8m) - (7m^3 + 5m - 8)$
$$= 2m^3 - 7m^3 - 3m^2 + 8m - 5m + 8$$
$$= -5m^3 - 3m^2 + 3m + 8$$

18.

$$
\begin{array}{r}
2t^3 + 3t^2 - 4t + 2 \\
3t - 2\,\overline{\smash{\big)}\,6t^4 + 5t^3 - 18t^2 + 14t - 1} \\
\underline{6t^4 - 4t^3} \\
9t^3 - 18t^2 \\
\underline{9t^3 - 6t^2} \\
-12t^2 + 14t \\
\underline{-12t^2 + 8t} \\
6t - 1 \\
\underline{6t - 4} \\
3
\end{array}
$$

Remainder

Answer: $\quad 2t^3 + 3t^2 - 4t + 2 + \dfrac{3}{3t - 2}$

19. $7x + x^3 = x(7 + x^2)$

20. $14y^2 + 13y - 12$
Look for two integers whose product is $(14)(-12) = -168$ and whose sum is 13. The required numbers are 21 and -8.
$$14y^2 + 13y - 12$$
$$= 14y^2 + 21y - 8y - 12$$
$$= 7y(2y + 3) - 4(2y + 3)$$
$$= (2y + 3)(7y - 4)$$

21. $6z^3 + 5z^2 - 4z = z\left(6z^2 + 5z - 4\right)$
$$= z(3z + 4)(2z - 1)$$

22. $49a^4 - 9b^2 = \left(7a^2\right)^2 - (3b)^2$
$$= \left(7a^2 + 3b\right)\left(7a^2 - 3b\right)$$

23. $c^3 + 27d^3 = c^3 + (3d)^3$
$$= (c + 3d)\left(c^2 - 3cd + 9d^2\right)$$

24. $64r^2 + 48rq + 9q^2$
$$= (8r)^2 + 2(8r)(3q) + (3q)^2$$
$$= (8r + 3q)^2$$

25.
$$2x^2 + x = 10$$
$$2x^2 + x - 10 = 0$$
$$(2x + 5)(x - 2) = 0$$

$$2x + 5 = 0 \quad \text{or} \quad x - 2 = 0$$
$$x = -\tfrac{5}{2} \quad \text{or} \quad x = 2$$

Check $x = -\tfrac{5}{2}$: $\tfrac{25}{2} - \tfrac{5}{2} = 10$ *True*
Check $x = 2$: $8 + 2 = 10$ *True*

The solution set is $\left\{-\tfrac{5}{2}, 2\right\}$.

26. $k^2 - k - 6 \leq 0$

Solve the equation
$$k^2 - k - 6 = 0.$$
$$(k - 3)(k + 2) = 0$$

$$k - 3 = 0 \quad \text{or} \quad k + 2 = 0$$
$$k = 3 \quad \text{or} \quad k = -2$$

The numbers -2 and 3 divide a number line into three intervals.

```
      A       B       C
  ────+───────+────────▶
     -2       3
```

Test a number from each interval in the original inequality.

$$k^2 - k - 6 \leq 0$$
Interval A: Let $k = -3$.
$$(-3)^2 - (-3) - 6 \leq 0 \quad ?$$
$$6 \leq 0 \qquad \textit{False}$$
Interval B: Let $k = 0$.
$$0^2 - 0 - 6 \leq 0 \quad ?$$
$$-6 \leq 0 \qquad \textit{True}$$
Interval C: Let $k = 4$.
$$4^2 - 4 - 6 \leq 0 \quad ?$$
$$6 \leq 0 \qquad \textit{False}$$

The numbers in Interval B, including the endpoints -2 and 3 because of $\leq$, are solutions.

The solution set is $[-2, 3]$.

27. $\left(\tfrac{2}{3}\right)^{-2} = \left(\tfrac{3}{2}\right)^{2} = \tfrac{3}{2} \cdot \tfrac{3}{2} = \tfrac{9}{4}$

28.
$$\frac{(3p^2)^3(-2p^6)}{4p^3(5p^7)} = \frac{3^3 p^6(-2)p^6}{20p^{10}}$$
$$= \frac{-54p^{12}}{20p^{10}}$$
$$= -\frac{27}{10}p^{12-10}$$
$$= -\frac{27p^2}{10}$$

29. $\dfrac{2}{x^2 - 81} = \dfrac{2}{(x + 9)(x - 9)}$

The numbers -9 and 9 make the denominator 0 so they are the values of the variable that make the rational expression undefined.

30.
$$\frac{x^2 - 16}{x^2 + 2x - 8} \div \frac{x - 4}{x + 7}$$
$$= \frac{x^2 - 16}{x^2 + 2x - 8} \cdot \frac{x + 7}{x - 4}$$
$$= \frac{(x + 4)(x - 4)(x + 7)}{(x + 4)(x - 2)(x - 4)}$$
$$= \frac{x + 7}{x - 2}$$

31.
$$\frac{5}{p^2 + 3p} - \frac{2}{p^2 - 4p}$$
$$= \frac{5}{p(p + 3)} - \frac{2}{p(p - 4)}$$

The LCD is $p(p + 3)(p - 4)$.

$$= \frac{5(p - 4)}{p(p + 3)(p - 4)} - \frac{2(p + 3)}{p(p - 4)(p + 3)}$$
$$= \frac{5p - 20 - 2p - 6}{p(p + 3)(p - 4)}$$
$$= \frac{3p - 26}{p(p + 3)(p - 4)}$$

32.
$$\frac{4}{x - 3} - \frac{6}{x + 3} = \frac{24}{x^2 - 9}$$
$$\frac{4}{x - 3} - \frac{6}{x + 3} = \frac{24}{(x + 3)(x - 3)}$$

Multiply by the LCD, $(x + 3)(x - 3)$. $(x \neq \pm 3)$

$$4(x + 3) - 6(x - 3) = 24$$
$$4x + 12 - 6x + 18 = 24$$
$$-2x + 30 = 24$$
$$-2x = -6$$
$$x = 3$$

But $x \neq 3$.
The solution set is $\emptyset$.

33.
$$6x^2 + 5x = 8$$
$$6x^2 + 5x - 8 = 0$$

Use the quadratic formula with $a = 6$, $b = 5$, and $c = -8$.

$$x = \frac{-b \pm \sqrt{b^2 - 4ac}}{2a}$$
$$x = \frac{-5 \pm \sqrt{5^2 - 4(6)(-8)}}{2(6)}$$
$$= \frac{-5 \pm \sqrt{25 + 192}}{12}$$
$$= \frac{-5 \pm \sqrt{217}}{12}$$

Solution set: $\left\{\dfrac{-5 + \sqrt{217}}{12}, \dfrac{-5 - \sqrt{217}}{12}\right\}$

34. $\sqrt{3x-2} = x$

$\qquad 3x - 2 = x^2$ *Square.*

$\qquad 0 = x^2 - 3x + 2$

$\qquad 0 = (x-1)(x-2)$

$\qquad x - 1 = 0 \quad \text{or} \quad x - 2 = 0$

$\qquad\quad x = 1 \quad \text{or} \qquad x = 2$

Check $x = 1$: $\sqrt{1} = 1$ *True*

Check $x = 2$: $\sqrt{4} = 2$ *True*

The solution set is $\{1, 2\}$.

35. To find the equation of the line through $(-2, 10)$ and parallel to $3x + y = 7$, find the slope of

$$3x + y = 7$$
$$y = -3x + 7.$$

The slope is -3, so a line parallel to it also has slope -3. Use $m = -3$ and $(x_1, y_1) = (-2, 10)$ in the point-slope form.

$$y - y_1 = m(x - x_1)$$
$$y - 10 = -3[x - (-2)]$$
$$y - 10 = -3(x + 2)$$

Write in standard form.

$$y - 10 = -3x - 6$$
$$3x + y = 4$$

Alternative solution: The line must be of the form $3x + y = k$ since it is parallel to $3x + y = 7$. Substitute -2 for x and 10 for y to find k.

$$3(-2) + 10 = k$$
$$4 = k$$

The equation is $3x + y = 4$.

36. $x - 3y = 6$

Find the x- and y-intercepts. To find the x-intercept, let $y = 0$.

$$x - 3(0) = 6$$
$$x = 6$$

The x-intercept is $(6, 0)$.

To find the y-intercept, let $x = 0$.

$$0 - 3y = 6$$
$$y = -2$$

The y-intercept is $(0, -2)$.

Plot the intercepts and draw the line through them.

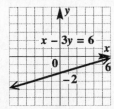

37. $4x - y < 4$

Graph the line $4x - y = 4$, which has intercepts $(0, -4)$ and $(1, 0)$, as a dashed line because the inequality involves $<$. Test $(0, 0)$, which yields $0 < 4$, a true statement. Shade the region on the side of the line that includes $(0, 0)$.

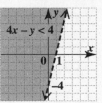

38. $\{(-3, 2), (-2, 6), (0, 4), (1, 2), (2, 6)\}$

(a) The set of ordered pairs is a function since every first coordinate is paired with a unique second coordinate.

(b) The domain is the set of first coordinates, that is,

$$\{-3, -2, 0, 1, 2\}.$$

(c) The range is the set of second coordinates, that is,

$$\{2, 6, 4\}.$$

39. $\quad 2x + 5y = -19 \quad (1)$

$\quad -3x + 2y = -19 \quad (2)$

To eliminate x, multiply equation (1) by 3 and equation (2) by 2. Then add the results.

$$\begin{array}{rl} 6x + 15y = -57 & \quad 3 \times (1) \\ \underline{-6x + 4y = -38} & \quad 2 \times (2) \\ 19y = -95 & \\ y = -5 & \end{array}$$

Substitute -5 for y in equation (1) to find x.

$$2x + 5y = -19 \qquad (1)$$
$$2x + 5(-5) = -19$$
$$2x - 25 = -19$$
$$2x = 6$$
$$x = 3$$

The solution set is $\{(3, -5)\}$.

40. $y = 5x + 3 \qquad (1)$

$\quad 2x + 3y = -8 \quad (2)$

From equation (1), substitute $5x + 3$ for y in equation (2). Then solve for x.

$$2x + 3(5x + 3) = -8$$
$$2x + 15x + 9 = -8$$
$$17x = -17$$
$$x = -1$$

From (1), $y = 5(-1) + 3 = -2$.

The solution $(-1, -2)$ checks.

The solution set is $\{(-1, -2)\}$.

41.
$$x + 2y + z = 8$$
$$2x - y + 3z = 15$$
$$-x + 3y - 3z = -11$$

Write the augmented matrix.

$$\begin{bmatrix} 1 & 2 & 1 & | & 8 \\ 2 & -1 & 3 & | & 15 \\ -1 & 3 & -3 & | & -11 \end{bmatrix}$$

$$\begin{bmatrix} 1 & 2 & 1 & | & 8 \\ 0 & -5 & 1 & | & -1 \\ 0 & 5 & -2 & | & -3 \end{bmatrix} \quad \begin{array}{l} -2R_1 + R_2 \\ R_1 + R_3 \end{array}$$

$$\begin{bmatrix} 1 & 2 & 1 & | & 8 \\ 0 & 1 & -\frac{1}{5} & | & \frac{1}{5} \\ 0 & 5 & -2 & | & -3 \end{bmatrix} \quad -\frac{1}{5}R_2$$

$$\begin{bmatrix} 1 & 2 & 1 & | & 8 \\ 0 & 1 & -\frac{1}{5} & | & \frac{1}{5} \\ 0 & 0 & -1 & | & -4 \end{bmatrix} \quad -5R_2 + R_3$$

$$\begin{bmatrix} 1 & 2 & 1 & | & 8 \\ 0 & 1 & -\frac{1}{5} & | & \frac{1}{5} \\ 0 & 0 & 1 & | & 4 \end{bmatrix} \quad -R_3$$

This matrix gives the system

$$x + 2y + z = 8$$
$$y - \tfrac{1}{5}z = \tfrac{1}{5}$$
$$z = 4.$$

Substitute $z = 4$ in the second equation.

$$y - \tfrac{1}{5}(4) = \tfrac{1}{5}$$
$$y = \tfrac{1}{5} + \tfrac{4}{5} = 1$$

Substitute $y = 1$ and $z = 4$ in the first equation.

$$x + 2(1) + 4 = 8$$
$$x + 6 = 8$$
$$x = 2$$

The solution set is $\{(2, 1, 4)\}$.

42. Let $x =$ the number of pounds of $3 per
pound nuts.

	Number of Pounds	Price per Pound	Value
$3/lb nuts	x	3	$3x$
$4.25/lb nuts	8	4.25	4.25(8)
Mixture	$x + 8$	4	$4(x + 8)$

The last column gives the equation.

$$3x + 4.25(8) = 4(x + 8)$$
$$3x + 34 = 4x + 32$$
$$2 = x$$

Use 2 pounds of the $3 per pound nuts.

43. $5\sqrt{72} - 4\sqrt{50} = 5\sqrt{36 \cdot 2} - 4\sqrt{25 \cdot 2}$
$$= 5 \cdot 6\sqrt{2} - 4 \cdot 5\sqrt{2}$$
$$= 30\sqrt{2} - 20\sqrt{2}$$
$$= 10\sqrt{2}$$

44. $(8 + 3i)(8 - 3i) = 8^2 - (3i)^2$
$$= 64 - 9i^2$$
$$= 64 - 9(-1)$$
$$= 64 + 9 = 73$$

45. The graph of $f(x) = 9x + 5$ is a line. To find the
inverse, replace $f(x)$ with y.

$$y = 9x + 5$$

Interchange x and y.

$$x = 9y + 5$$

Solve for y.

$$x - 5 = 9y$$
$$\frac{x - 5}{9} = y$$

Replace y with $f^{-1}(x)$.

$$f^{-1}(x) = \frac{x - 5}{9}, \quad \text{or} \quad f^{-1}(x) = \frac{1}{9}x - \frac{5}{9}$$

46. Graph $g(x) = \left(\frac{1}{3}\right)^x$.

Make a table of values.

x	-2	-1	0	1	2
$g(x)$	9	3	1	$\frac{1}{3}$	$\frac{1}{9}$

Plot these points, and draw a smooth decreasing
exponential curve through them.

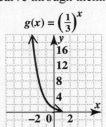

47. $3^{2x-1} = 81$
$$3^{2x-1} = 3^4$$
$$2x - 1 = 4 \qquad \textit{Equate exponents.}$$
$$2x = 5$$
$$x = \tfrac{5}{2}$$

Check $x = \frac{5}{2}$: $3^{5-1} = 3^4 = 81$

The solution set is $\left\{\frac{5}{2}\right\}$.

48. Graph $y = \log_{1/3} x$.

Change to exponential form.

$$\left(\tfrac{1}{3}\right)^y = x$$

This is the inverse of the graph of $g(x) = y = \left(\tfrac{1}{3}\right)^x$ in Exercise 46. To find points on the graph, interchange the x- and y-values in the table.

x	9	3	1	$\frac{1}{3}$	$\frac{1}{9}$
y	-2	-1	0	1	2

Plot these points, and draw a smooth decreasing logarithmic curve through them.

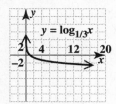

49. $\log_8 x + \log_8 (x + 2) = 1$

Use the product rule for logarithms.

$$\log_8 x(x + 2) = 1$$

Change to exponential form.

$$x(x + 2) = 8^1$$
$$x^2 + 2x - 8 = 0$$
$$(x + 4)(x - 2) = 0$$
$$x + 4 = 0 \quad \text{or} \quad x - 2 = 0$$
$$x = -4 \quad \text{or} \quad x = 2$$

$x \ne -4$ because $\log_8 (-4)$ does not exist.

Check $x = 2$: $\log_8 2 + \log_8 4 = \log_8 (2 \cdot 4) = 1$

The solution set is $\{2\}$.

50. $f(x) = 2(x - 2)^2 - 3$ is in
$f(x) = a(x - h)^2 + k$ form.

The graph is a vertical parabola with vertex (h, k) at $(2, -3)$. Since $a = 2 > 0$, the graph opens up. Also, $|a| = |2| = 2 > 1$, so the graph is narrower than the graph of $f(x) = x^2$. The points $(0, 5)$ and $(4, 5)$ are on the graph.

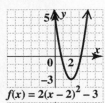

$f(x) = 2(x - 2)^2 - 3$

51. $\dfrac{x^2}{9} + \dfrac{y^2}{25} = 1$ is in $\dfrac{x^2}{a^2} + \dfrac{y^2}{b^2} = 1$ form with $a = 3$ and $b = 5$. The graph is an ellipse centered at $(0, 0)$ with x-intercepts $(3, 0)$ and $(-3, 0)$ and y-intercepts $(0, 5)$ and $(0, -5)$. Plot the intercepts and draw the ellipse through them.

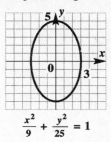

$\dfrac{x^2}{9} + \dfrac{y^2}{25} = 1$

52. $x^2 - y^2 = 9$
$$\dfrac{x^2}{9} - \dfrac{y^2}{9} = 1$$

The graph is a hyperbola centered at $(0, 0)$ with x-intercepts $(3, 0)$ and $(-3, 0)$. The asymptotes are $y = \pm x$. Draw the right and left branches through the intercepts and approaching the asymptotes.

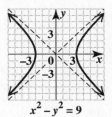

$x^2 - y^2 = 9$

53. $\begin{aligned} xy &= -5 \quad (1) \\ 2x + y &= 3 \quad (2) \end{aligned}$

Solve equation (2) for y.

$$y = -2x + 3 \quad (3)$$

Substitute $-2x + 3$ for y in equation (1).

$$xy = -5 \quad (1)$$
$$x(-2x + 3) = -5$$
$$-2x^2 + 3x = -5$$
$$-2x^2 + 3x + 5 = 0$$
$$2x^2 - 3x - 5 = 0$$
$$(2x - 5)(x + 1) = 0$$

$$2x - 5 = 0 \quad \text{or} \quad x + 1 = 0$$
$$x = \tfrac{5}{2} \quad \text{or} \quad x = -1$$

Substitute these values for x in equation (3) to find y.

If $x = \tfrac{5}{2}$, then $y = -2\left(\tfrac{5}{2}\right) + 3 = -2$.

If $x = -1$, then $y = -2(-1) + 3 = 5$.

The solution set is $\left\{(-1, 5), \left(\tfrac{5}{2}, -2\right)\right\}$.

54. Center at $(-5, 12)$; radius 9

Use the equation of a circle with $h = -5$, $k = 12$, and $r = 9$.

$$(x - h)^2 + (y - k)^2 = r^2$$
$$[x - (-5)]^2 + (y - 12)^2 = 9^2$$
$$(x + 5)^2 + (y - 12)^2 = 81$$

55. $a_n = 5n - 12$

$a_1 = 5(1) - 12 = 5 - 12 = -7$
$a_2 = 5(2) - 12 = 10 - 12 = -2$
$a_3 = 5(3) - 12 = 15 - 12 = 3$
$a_4 = 5(4) - 12 = 20 - 12 = 8$
$a_5 = 5(5) - 12 = 25 - 12 = 13$

Answer: $-7, -2, 3, 8, 13$

56. **(a)** $a_1 = 8$, $d = 2$

$$S_n = \frac{n}{2}\big[2a_1 + (n-1)d\big]$$
$$S_6 = \frac{6}{2}\big[2(8) + (6-1)2\big]$$
$$= 3(16 + 10) = 3(26) = 78$$

(b) $15 - 6 + \frac{12}{5} - \frac{24}{25} + \cdots$

This is an infinite geometric series with $a_1 = 15$ and $r = \frac{-6}{15} = -\frac{2}{5}$. The sum is

$$S = \frac{a_1}{1 - r} = \frac{15}{1 - \left(-\frac{2}{5}\right)}$$
$$= \frac{15}{\frac{7}{5}} = 15 \cdot \frac{5}{7} = \frac{75}{7}.$$

57. $\displaystyle\sum_{i=1}^{4} 3i = 3\sum_{i=1}^{4} i = 3(1 + 2 + 3 + 4)$
$$= 3(10) = 30$$

58. $9! = 9 \cdot 8 \cdot 7 \cdot 6 \cdot 5 \cdot 4 \cdot 3 \cdot 2 \cdot 1$
$$= 362{,}880$$

59. $(2a - 1)^5$

$$= (2a)^5 + \frac{5!}{4!\,1!}(2a)^4(-1)^1$$
$$+ \frac{5!}{3!\,2!}(2a)^3(-1)^2 + \frac{5!}{2!\,3!}(2a)^2(-1)^3$$
$$+ \frac{5!}{1!\,4!}(2a)^1(-1)^4 + (-1)^5$$
$$= 32a^5 + 5\big(16a^4\big)(-1) + 10\big(8a^3\big)(1)$$
$$+ 10\big(4a^2\big)(-1) + 5(2a)(1) + (-1)$$
$$= 32a^5 - 80a^4 + 80a^3 - 40a^2 + 10a - 1$$

60. The fourth term ($r = 4$, so $r - 1 = 3$) of $\left(3x^4 - \frac{1}{2}y^2\right)^5$ is

$$\frac{5!}{(5-3)!\,3!}\big(3x^4\big)^{5-3}\left(-\frac{1}{2}y^2\right)^3$$
$$= \frac{5!}{2!\,3!}\big(3x^4\big)^2\left(-\frac{1}{2}\right)^3\big(y^2\big)^3$$
$$= 10\big(9x^8\big)\big(-\tfrac{1}{8}\big)y^6$$
$$= -\frac{45x^8y^6}{4}.$$

APPENDIX B   REVIEW OF DECIMALS AND PERCENTS

1. $14.23 + 9.81 + 74.63 + 18.715$

Add in columns.

$$
\begin{array}{r}
14.230 \\
9.810 \\
74.630 \\
+\ 18.715 \\
\hline
117.385
\end{array}
$$ *Line up decimal points and attach zeros.*

3. $19.74 - 6.53$

Subtract in columns.

$$
\begin{array}{r}
19.74 \\
-\ 6.53 \\
\hline
13.21
\end{array}
$$ *Line up decimal points.*

5. $219 - 68.51$

$$
\begin{array}{r}
219.00 \\
-\ 68.51 \\
\hline
150.49
\end{array}
$$ *Line up decimal points and attach zeros.*

7.

$$
\begin{array}{r}
48.960 \\
37.421 \\
+\ 9.720 \\
\hline
96.101
\end{array}
$$ *Attach zeros.*

9.

$$
\begin{array}{r}
8.600 \\
-\ 3.751 \\
\hline
4.849
\end{array}
$$ *Attach zeros.*

11. 39.6×4.2

Multiply as if the numbers were whole numbers.

$$
\begin{array}{r}
3\,9.6 \\
\times\quad 4.2 \\
\hline
7\,9\,2 \\
158\,4\quad \\
\hline
166.3\,2
\end{array}
$$ *1 decimal place*
1 decimal place
2 decimal places

13. 42.1×3.9

$$
\begin{array}{r}
4\,2.1 \\
\times\quad 3.9 \\
\hline
37\,8\,9 \\
126\,3\quad \\
\hline
164.1\,9
\end{array}
$$ *1 decimal place*
1 decimal place
2 decimal places

15.

$$
\begin{array}{r}
0.042 \\
\times\quad 32 \\
\hline
84 \\
1\,26\quad \\
\hline
1.344
\end{array}
$$ *3 decimal places*
no decimal places
3 decimal places

17. $24.84 \div 6$

The divisor is a whole number, so place the decimal point in the quotient above the decimal point in the dividend.

$$
\begin{array}{r}
4\,.\,1\,4 \\
6\,\overline{)\,24\,.\,8\,4} \\
\underline{24} \\
8 \\
\underline{6} \\
2\,4 \\
\underline{2\,4} \\
0
\end{array}
$$

19. $7.6266 \div 3.42$

Move the decimal point in 3.42 two places to the right to get a whole number. Then move the decimal point in 7.6266 two places to the right. Move the decimal point straight up and divide as with whole numbers.

$$
\begin{array}{r}
2\,.\,2\,3 \\
3.42.\,\overline{)\,7\,.62.\,66} \\
\underline{6\,84} \\
78\,6 \\
\underline{68\,4} \\
10\,26 \\
\underline{10\,26} \\
0
\end{array}
$$

21. $2496 \div 0.52$

Move the decimal point in 0.52 and 2496 two places to the right. Bring the decimal point straight up and divide as with whole numbers.

$$
\begin{array}{r}
48\,00. \\
0.52.\,\overline{)\,2496.00} \\
\underline{208} \\
416 \\
\underline{416} \\
0
\end{array}
$$

23. $53\% = 53 \cdot 1\% = 53(0.01) = 0.53$

25. $129\% = 129 \cdot 1\% = 129(0.01) = 1.29$

27. $96\% = 96 \cdot 1\% = 96(0.01) = 0.96$

29. $0.9\% = 0.9 \cdot 1\% = 0.9(0.01) = 0.009$

31. $0.80 = 80(0.01) = 80 \cdot 1\% = 80\%$

33. $0.007 = 0.7(0.01) = 0.7 \cdot 1\% = 0.7\%$

35. $0.67 = 67(0.01) = 67 \cdot 1\% = 67\%$

37. $0.125 = 12.5(0.01) = 12.5 \cdot 1\% = 12.5\%$

39.   What is 14% of 780?

14% of $780 = (0.14)(780) = 109.2$

41. Find 22% of 1086.

$22\% \cdot 1086 = (0.22)(1086) = 238.92$

43. 4 is what percent of 80?

As in Example 5(c), we can translate this sentence to symbols word by word.

$4 = p \cdot 80$
$p = \frac{4}{80} = 0.05$ in decimal form.

Now convert 0.05 to a percent.

$0.05 = 5(0.01) = 5 \cdot 1\% = 5\%$

4 is 5% of 80.

45. What percent of 5820 is 6402?

$p \cdot 5820 = 6402$
$p = \frac{6402}{5820} = 1.1$ *In decimal form*

Now convert 1.1 to a percent.

$1.1 = 110(0.01) = 110 \cdot 1\% = 110\%$

6402 is 110% of 5820.

47. 121 is what percent of 484?

$121 = p \cdot 484$
$p = \frac{121}{484} = 0.25 = 25\%$

121 is 25% of 484.

49. Find 118% of 125.8.

118% of 125.8

$= (1.18)(125.8)$
$= 148.444$
≈ 148.44 *Round to the nearest hundredth.*

51. What is 91.72% of 8546.95?

91.72% of 8546.95
$= (0.9172)(8546.95)$
$= 7839.26254$
≈ 7839.26

53. What percent of 198.72 is 14.68?

$p \cdot 198.72 = 14.68$
$p = \frac{14.68}{198.72} \approx 0.0739$ *Round to nearest ten-thousandth.*

Now convert 0.0739 to a percent.

$0.0739 = 7.39(0.01) = 7.39 \cdot 1\% = 7.39\%$

14.68 is 7.39% of 198.72.

55. 12% of $23,000 = (0.12)(23,000)$
$= 2760$

She is earning $2760 per year.

57. 35% of $2300 = (0.35)(2300)$
$= 805$

805 miles of the trip were by air.

59. 15% of $420 = (0.15)(420)$
$= 63$

He could drive 63 extra miles.

61. Since the family spends 90% of the $2000 each month, they must save 10% of the $2000.

10% of $2000 = (0.10)(2000)$
$= \$200$

Since there are 12 months in a year, the annual savings are

$12 \cdot 200 = 2400,$ or $2400.$

APPENDIX C SETS

1. The set of all natural numbers less than 8 is
$$\{1, 2, 3, 4, 5, 6, 7\}.$$

3. The set of seasons is
$$\{\text{winter, spring, summer, fall}\}.$$
The seasons may be written in any order within the braces.

5. To date, there have been no women presidents, so this set is the empty set, written $\emptyset$, or { }.

7. The set of letters of the alphabet between K and M is {L}.

9. The set of positive even integers is
$$\{2, 4, 6, 8, 10, \dots\}.$$

11. The sets in Exercises 9 and 10 are infinite, since each contains an unlimited number of elements.

13. $5 \in \{1, 2, 5, 8\}$

    5 is an element of the set, so the statement is true.

15. $2 \in \{1, 3, 5, 7, 9\}$

    2 is not an element of the set, so the statement is false.

17. $7 \notin \{2, 4, 6, 8\}$

    7 is not an element of the set, so the statement is true.

19. $\{2, 4, 9, 12, 13\} = \{13, 12, 9, 4, 2\}$

    The two sets have exactly the same elements, so they are equal. The statement is true. (The order in which the elements are written does not matter.)

21. $A \subseteq U$

    $A = \{1, 3, 4, 5, 7, 8\}$

    $U = \{1, 2, 3, 4, 5, 6, 7, 8, 9, 10\}$

    Since all the elements of A are elements of U, the statement $A \subseteq U$ is true.

23. $\emptyset \subseteq A$

    Since the empty set contains no elements, the empty set is a subset of every set. The statement $\emptyset \subseteq A$ is true.

25. $C \subseteq A$

    $C = \{1, 3, 5, 7\}$

    $A = \{1, 3, 4, 5, 7, 8\}$

    Since all the elements of C are elements of A, the statement $C \subseteq A$ is true.

27. $D \subseteq B$

    $D = \{1, 2, 3\}$

    $B = \{2, 4, 6, 8\}$

    Since 1 and 3 are elements of D but are not elements of B, the statement $D \subseteq B$ is false.

29. $D \nsubseteq E$

    $D = \{1, 2, 3\}$

    $E = \{3, 7\}$

    Since 1 and 2 are elements of D and are not elements of E, D is not a subset of E, so the statement $D \nsubseteq E$ is true.

31. There are exactly 4 subsets of E.

    $E = \{3, 7\}$

    Since E has 2 elements, the number of subsets is $2^2 = 4$. The statement is true.

33. There are exactly 12 subsets of C.

    $C = \{1, 3, 5, 7\}$

    Since C has 4 elements, the number of subsets is $2^4 = 16$. The statement is false.

35. $\{4, 6, 8, 12\} \cap \{6, 8, 14, 17\} = \{6, 8\}$

    The symbol $\cap$ means the intersection of the two sets, which is the set of elements that belong to both sets. Since 6 and 8 are the only elements belonging to both sets, the statement is true.

37. $\{3, 1, 0\} \cap \{0, 2, 4\} = \{0\}$

    Only 0 belongs to both sets, so the statement is true.

39. $\{3, 9, 12\} \cap \emptyset = \{3, 9, 12\}$

    Since 3, 9, and 12 are not elements of the empty set, they are not in the intersection of the two sets. The intersection of any set with the empty set is the empty set. The statement is false.

41. $\{3, 5, 7, 9\} \cup \{4, 6, 8\} = \emptyset$

    The union of the two sets is the set of all elements that belong to either one of the sets or to both sets.

    $\{3, 5, 7, 9\} \cup \{4, 6, 8\}$

    $= \{3, 4, 5, 6, 7, 8, 9\} \neq \emptyset$

    The statement is false.

43. $\{4, 9, 11, 7, 3\} \cup \{1, 2, 3, 4, 5\}$

    $= \{1, 2, 3, 4, 5, 7, 9, 11\}$

    The union of the two sets is the set of all elements that belong to either one of the sets or to both sets. The statement is true.

45. A'

$U = \{a, b, c, d, e, f, g, h\}$

$A = \{a, b, c, d, e, f\}$

A' contains all elements in U that are not in A, so

$$A' = \{g, h\}.$$

47. C'

$U = \{a, b, c, d, e, f, g, h\}$

$C = \{a, f\}$

C' contains all elements in U that are not in C, so

$$C' = \{b, c, d, e, g, h\}.$$

49. $A \cap B$

$A = \{a, b, c, d, e, f\}$

$B = \{a, c, e\}$

The intersection of A and B is the set of all elements belonging to both A and B, so

$$A \cap B = \{a, c, e\} = B.$$

51. $A \cap D$

$A = \{a, b, c, d, e, f\}$

$D = \{d\}$

Since d is the only element in both A and D,

$$A \cap D = \{d\} = D.$$

53. $B \cap C$

$B = \{a, c, e\}$

$C = \{a, f\}$

Since a is the only element that belongs to both sets, the intersection is the set with a as its only element, so

$$B \cap C = \{a\}.$$

55. $B \cup D$

$B = \{a, c, e\}$

$D = \{d\}$

The union of B and D is the set of elements belonging to either B or D or both, so

$$B \cup D = \{a, c, d, e\}.$$

57. $C \cup B$

$C = \{a, f\}$

$B = \{a, c, e\}$

The union of C and B is the set of elements belonging to either C or B or both, so

$$C \cup B = \{a, c, e, f\}.$$

59. $A \cap \emptyset$

Since $\emptyset$ has no elements, there is no element that belongs to both A and $\emptyset$, so the intersection is the empty set. Thus,

$$A \cap \emptyset = \emptyset.$$

61. $A = \{a, b, c, d, e, f\}$ $C = \{a, f\}$

$B = \{a, c, e\}$ $D = \{d\}$

Disjoint sets are sets which have no elements in common.

B and D are disjoint since they have no elements in common. Also, C and D are disjoint since they have no elements in common.

APPENDIX D MEAN, MEDIAN, AND MODE

1. $\text{mean} = \dfrac{\text{sum of all values}}{\text{number of values}}$

$= \dfrac{4+9+6+4+7+10+9}{7}$

$= \dfrac{49}{7} = 7$

The mean (average) age of the infants at the child care center was 7 months.

3. $\text{mean} = \dfrac{\text{sum of all values}}{\text{number of values}}$

$= \dfrac{92+51+59+86+68+73+49+80}{8}$

$= \dfrac{558}{8} = 69.75$

The mean (average) final exam score was 69.8 (rounded).

5. $\text{mean} = \dfrac{\text{sum of all values}}{\text{number of values}}$

$= \dfrac{\begin{array}{c}(31{,}900+32{,}850+34{,}930+39{,}712\\+\,38{,}340+60{,}000)\end{array}}{6}$

$= \dfrac{237{,}732}{6} = 39{,}622$

The mean (average) annual salary was $39,622.

7. $\text{mean} = \dfrac{\text{sum of all values}}{\text{number of values}}$

$= \dfrac{\begin{array}{c}(75.52+36.15+58.24+21.86+47.68\\+\,106.57+82.72+52.14+28.60+72.92)\end{array}}{10}$

$= \dfrac{582.40}{10} = 58.24$

The mean (average) shoe sales amount was $58.24.

9.

Policy Amount ($)	Number of Policies Sold	Product ($)
10,000	6	60,000
20,000	24	480,000
25,000	12	300,000
30,000	8	240,000
50,000	5	250,000
100,000	3	300,000
250,000	2	500,000
Totals	60	2,130,000

$\text{weighted mean} = \dfrac{\text{sum of products}}{\text{total number of policies}}$

$= \dfrac{2{,}130{,}000}{60} = 35{,}500$

The mean (average) amount for the policies sold was $35,500.

11.

Quiz Scores	Frequency	Product
3	4	12
5	2	10
6	5	30
8	5	40
9	2	18
Totals	18	110

$\text{weighted mean} = \dfrac{\text{sum of products}}{\text{total number of quizzes}}$

$= \dfrac{110}{18} = 6.\overline{1}$

The mean (average) quiz score was 6.1 (rounded).

13.

Hours Worked	Frequency	Product
12	4	48
13	2	26
15	5	75
19	3	57
22	1	22
23	5	115
Totals	20	343

$\text{weighted mean} = \dfrac{\text{sum of products}}{\text{total number of workers}}$

$= \dfrac{343}{20} = 17.15$

The mean (average) number of hours worked was 17.2 (rounded).

15.

Course	Credits	Grade	Credits · Grade
Biology	4	B (= 3)	$4 \cdot 3 = 12$
Biology Lab	2	A (= 4)	$2 \cdot 4 = 8$
Mathematics	5	C (= 2)	$5 \cdot 2 = 10$
Health	1	F (= 0)	$1 \cdot 0 = 0$
Psychology	3	B (= 3)	$3 \cdot 3 = 9$
Totals	15		39

$\text{GPA} = \dfrac{\text{sum of Credits} \cdot \text{Grade}}{\text{total number of credits}}$

$= \dfrac{39}{15} = 2.60$

17. **(a)** In Exercise 15, replace $1 \cdot 0$ with $1 \cdot 3$ to get $\text{GPA} = \frac{42}{15} = 2.80$.

(b) In Exercise 15, replace $5 \cdot 2$ with $5 \cdot 3$ to get $\text{GPA} = \frac{44}{15} = 2.9\overline{3} = 2.93$ (rounded).

(c) Making both of those changes gives us $\text{GPA} = \frac{47}{15} = 3.1\overline{3} = 3.13$ (rounded).

19. Arrange the numbers in numerical order from smallest to largest (they already are).

9, 12, 14, 15, 23, 24, 28

The list has 7 numbers. The middle number is the 4th number, so the median is 15.

21. Arrange the numbers in numerical order from smallest to largest.

   328, 420, 483, 549, 592, 715

   The list has 6 numbers. The middle numbers are the 3rd and 4th numbers, so the median is

$$\frac{483 + 549}{2} = 516.$$

23. Arrange the numbers in numerical order from smallest to largest.

   23, 34, 40, 47, 48, 48, 51, 56, 95, 96

   The list has 10 numbers. The middle numbers are the 5th and 6th numbers, so the median is

$$\frac{48 + 48}{2} = 48.$$

25. 3, <u>8</u>, 5, 1, 7, 6, <u>8</u>, 4, 5, <u>8</u>

   The number 8 occurs three times, which is more often than any other number. Therefore, 8 is the mode.

27. <u>74</u>, <u>68</u>, <u>68</u>, <u>68</u>, 75, 75, <u>74</u>, <u>74</u>, 70, 77

   Because both 68 and 74 occur three times, each is a mode. This list is *bimodal*.

29. 5, 9, 17, 3, 2, 8, 19, 1, 4, 20, 10, 6

   No number occurs more than once. This list has *no mode*.

APPENDIX E THE METRIC SYSTEM AND CONVERSIONS

1. Answers will vary. The width of a person's hand is about 9 to 11 cm.

3. Answers will vary. The width of a person's thumb is about 20 to 25 mm.

5. The child was _cm_ tall.

7. Ming-Na swam in the 200 _m_ backstroke race.

9. Adriana drove 400 _km_ on her vacation.

11. An aspirin tablet is 10 _mm_ across.

13. A paper clip is about 3 _cm_ long.

15. Dave's truck is 5 _m_ long.

17. Some possible answers are: 35 mm film for cameras, track and field events, metric auto parts, and lead refills for mechanical pencils.

19. $7 \text{ m} = \dfrac{7 \text{ m̸}}{1} \cdot \dfrac{100 \text{ cm}}{1 \text{ m̸}} = \dfrac{7 \cdot 100 \text{ cm}}{1} = 700 \text{ cm}$

Or: From m to cm is two places to the right.

$7 \text{ m} = 7.00 \text{ m} = 700 \text{ cm}$

21. $40 \text{ mm} = \dfrac{40 \text{ m̸m̸}}{1} \cdot \dfrac{1 \text{ m}}{1000 \text{ m̸m̸}} = 0.04 \text{ m}$

Or: From mm to m is three places left.

$40 \text{ mm} = 0.040 \text{ m} \quad \text{or} \quad 0.04 \text{ m}$

23. $9.4 \text{ km} = \dfrac{9.4 \text{ k̸m̸}}{1} \cdot \dfrac{1000 \text{ m}}{1 \text{ k̸m̸}} = 9400 \text{ m}$

Or: From km to m is three places right.

$9.4 \text{ km} = 9.400 \text{ km} = 9400 \text{ m}$

25. $509 \text{ cm} = \dfrac{509 \text{ c̸m̸}}{1} \cdot \dfrac{1 \text{ m}}{100 \text{ c̸m̸}} = 5.09 \text{ m}$

Or: From cm to m is two places left.

$509 \text{ cm} = 5.09 \text{ m}$

27. $400 \text{ mm} = \dfrac{400 \text{ m̸m̸}}{1} \cdot \dfrac{1 \text{ cm}}{10 \text{ m̸m̸}} = 40 \text{ cm}$

Or: From mm to cm is one place left.

$400 \text{ mm} = 40 \text{ cm}$

29. $0.91 \text{ m} = \dfrac{0.91 \text{ m̸}}{1} \cdot \dfrac{1000 \text{ mm}}{1 \text{ m̸}} = 910 \text{ mm}$

Or: From m to mm is three places right.

$0.91 \text{ m} = 0.910 \text{ m} = 910 \text{ mm}$

31. From mm to cm is one place left.

$35 \text{ mm} = 3.5 \text{ cm}$

$70 \text{ mm} = 7 \text{ cm}$

33. From m to km is 3 places left.

$18 \text{ m} = 18.0 \text{ m} = 0.018 \text{ km}$

35. $5.6 \text{ mm} = \dfrac{5.6 \text{ m̸m̸}}{1} \cdot \dfrac{1 \text{ m̸}}{1000 \text{ m̸m̸}} \cdot \dfrac{1 \text{ km}}{1000 \text{ m̸}}$

$= \dfrac{5.6}{1,000,000} \text{ km} = 0.000\,005\,6 \text{ km}$

37. The glass held 250 _mL_ of water. (Liquids are measured in mL or L.)

39. Dolores can make 10 _L_ of soup in that pot. (Liquids are measured in mL or L.)

41. Our yellow Labrador dog grew up to weigh 40 _kg_. (Weight is measured in mg, g, or kg.)

43. Lori caught a small sunfish weighting 150 _g_. (Weight is measured in mg, g, or kg.)

45. Andre donated 500 _mL_ of blood today. (Blood is a liquid, and liquids are measured in mL or L.)

47. The patient received a 250 _mg_ tablet of medication each hour. (Weight is measured in mg, g, or kg.)

49. The gas can for the lawn mower holds 4 _L_. (Gasoline is a liquid, and liquids are measured in mL or L.)

51. Unreasonable; 4.1 L of Kaopectate is about 4 qt (or 1 gallon), which is too much.

53. Unreasonable; 5 kg of Epsom salts is about 11 pounds, which is too much.

55. Reasonable; 15 mL is about 3 tsp.

57. Some capacity examples are 2 L bottles of soda and shampoo bottles marked in mL; weight examples are grams of fat listed on cereal boxes and vitamin doses in milligrams.

59. $15 \text{ L} = \dfrac{15 \text{ L̸}}{1} \cdot \dfrac{1000 \text{ mL}}{1 \text{ L̸}}$

$= \dfrac{15 \cdot 1000 \text{ mL}}{1} = 15,000 \text{ mL}$

Or: From L to mL is three places right.

$15 \text{ L} = 15.000 \text{ L} = 15,000 \text{ mL}$

61. $3000 \text{ mL} = \dfrac{3000 \text{ m̸L̸}}{1} \cdot \dfrac{1 \text{ L}}{1000 \text{ m̸L̸}} = \dfrac{3000}{1000} \text{ L} = 3 \text{ L}$

Or: From mL to L is three places left.

$3000 \text{ mL} = 3000.0 \text{ mL} = 3 \text{ L}$

63. $925 \text{ mL} = \dfrac{925 \text{ mL}}{1} \cdot \dfrac{1 \text{ L}}{1000 \text{ mL}} = \dfrac{925}{1000} \text{ L} = 0.925 \text{ L}$

Or: From mL to L is three places left.

$$925 \text{ mL} = 925.0 \text{ mL} = 0.925 \text{ L}$$

65. $8 \text{ mL} = \dfrac{8 \text{ mL}}{1} \cdot \dfrac{1 \text{ L}}{1000 \text{ mL}} = \dfrac{8}{1000} \text{ L} = 0.008 \text{ L}$

Or: From mL to L is three places left.

$$8 \text{ mL} = 8.0 \text{ mL} = 0.008 \text{ L}$$

67. $4.15 \text{ L} = \dfrac{4.15 \text{ L}}{1} \cdot \dfrac{1000 \text{ mL}}{1 \text{ L}} = 4150 \text{ mL}$

Or: From L to mL is three places right.

$$4.15 \text{ L} = 4.150 \text{ L} = 4150 \text{ mL}$$

69. $8000 \text{ g} = \dfrac{8000 \text{ g}}{1} \cdot \dfrac{1 \text{ kg}}{1000 \text{ g}} = \dfrac{8000}{1000} \text{ kg} = 8 \text{ kg}$

Or: From g to kg is three places left.

$$8000 \text{ g} = 8 \text{ kg}$$

71. $5.2 \text{ kg} = \dfrac{5.2 \text{ kg}}{1} \cdot \dfrac{1000 \text{ g}}{1 \text{ kg}} = 5200 \text{ g}$

Or: From kg to g is three places right.

$$5.2 \text{ kg} = 5.200 \text{ kg} = 5200 \text{ g}$$

73. $0.85 \text{ g} = \dfrac{0.85 \text{ g}}{1} \cdot \dfrac{1000 \text{ mg}}{1 \text{ g}} = 850 \text{ mg}$

Or: From g to mg is three places right.

$$0.85 \text{ g} = 0.850 \text{ g} = 850 \text{ mg}$$

75. $30{,}000 \text{ mg} = \dfrac{30{,}000 \text{ mg}}{1} \cdot \dfrac{1 \text{ g}}{1000 \text{ mg}} = 30 \text{ g}$

Or: From mg to g is three places left.

$$30{,}000 \text{ mg} = 30 \text{ g}$$

77. $\dfrac{300 \text{ mL}}{1} \cdot \dfrac{1 \text{ L}}{1000 \text{ mL}} = 0.3 \text{ L}$

Each day, 0.3 L of sweat is released.

79. $\dfrac{1.34 \text{ kg}}{1} \cdot \dfrac{1000 \text{ g}}{1 \text{ kg}} = 1340 \text{ g}$

The average weight of a human brain is 1340 g.

81. $\dfrac{70 \text{ mL}}{1} \cdot \dfrac{1 \text{ L}}{1000 \text{ mL}} = 0.07 \text{ L}$

A healthy human heart pumps about 0.07 L of blood per beat.

83. $\dfrac{3000 \text{ g}}{1} \cdot \dfrac{1 \text{ kg}}{1000 \text{ g}} = 3 \text{ kg}$

$$\dfrac{4000 \text{ g}}{1} \cdot \dfrac{1 \text{ kg}}{1000 \text{ g}} = 4 \text{ kg}$$

A small adult cat weighs from 3 kg to 4 kg.

85. $1 \text{ kg} = \dfrac{1 \text{ kg}}{1} \cdot \dfrac{1000 \text{ g}}{1 \text{ kg}} = 1000 \text{ g}$

$\dfrac{1000 \text{ g}}{5 \text{ g}} = 200$; therefore, there are 200 nickels in 1 kg of nickels.

APPENDIX F   REVIEW OF EXPONENTS, POLYNOMIALS, AND FACTORING (Transition from Beginning to Intermediate Algebra)

1.
$$\left(a^4 b^{-3}\right)\left(a^{-6} b^2\right) = \left(a^{4+(-6)}\right)\left(b^{-3+2}\right)$$
$$= a^{-2} b^{-1}$$
$$= \frac{1}{a^2 b}$$

3.
$$\left(5x^{-2}y\right)^2 \left(2xy^4\right)^2 = \left(5^2 x^{-4} y^2\right)\left(2^2 x^2 y^8\right)$$
$$= 25 \cdot 4 x^{-4+2} y^{2+8}$$
$$= 100 x^{-2} y^{10}$$
$$= \frac{100 y^{10}}{x^2}$$

5.
$$-6^0 + (-6)^0 = -1 \cdot 6^0 + 1$$
$$= -1 \cdot 1 + 1$$
$$= -1 + 1 = 0$$

7.
$$\frac{\left(2w^{-1} x^2 y^{-1}\right)^3}{\left(4w^5 x^{-2} y\right)^2} = \frac{2^3 w^{-3} x^6 y^{-3}}{4^2 w^{10} x^{-4} y^2}$$
$$= \frac{8 x^6 x^4}{16 w^{10} y^2 w^3 y^3}$$
$$= \frac{x^{10}}{2 w^{13} y^5}$$

9.
$$\left(\frac{-4a^{-2} b^4}{a^3 b^{-1}}\right)^{-3} = \left(\frac{a^3 b^{-1}}{-4a^{-2} b^4}\right)^3$$
$$= \frac{a^9 b^{-3}}{(-4)^3 a^{-6} b^{12}}$$
$$= \frac{a^9 a^6}{-64 b^{12} b^3}$$
$$= \frac{a^{15}}{-64 b^{15}}$$

11.
$$\left(7x^{-4} y^2 z^{-2}\right)^{-2}\left(7x^4 y^{-1} z^3\right)^2$$
$$= \left(7^{-2} x^8 y^{-4} z^4\right)\left(7^2 x^8 y^{-2} z^6\right)$$
$$= \frac{7^2 x^8 z^4 x^8 z^6}{7^2 y^4 y^2}$$
$$= \frac{x^{16} z^{10}}{y^6}$$

13.
$$\left(2a^4 + 3a^3 - 6a^2 + 5a - 12\right)$$
$$+ \left(-8a^4 + 8a^3 - 14a^2 + 21a - 3\right)$$
$$= (2-8)a^4 + (3+8)a^3 + (-6-14)a^2$$
$$+ (5+21)a + (-12-3)$$
$$= -6a^4 + 11a^3 - 20a^2 + 26a - 15$$

15.
$$\left(6x^3 - 12x^2 + 3x - 4\right)$$
$$- \left(-2x^3 + 6x^2 - 3x + 12\right)$$
$$= 6x^3 - 12x^2 + 3x - 4 + 2x^3 - 6x^2 + 3x - 12$$
$$= 8x^3 - 18x^2 + 6x - 16$$

17. Add.
$$\begin{array}{r} 5x^2 y + 2xy^2 + y^3 \\ -4x^2 y - 3xy^2 + 5y^3 \\ \hline x^2 y - xy^2 + 6y^3 \end{array}$$

19.
$$3\left(5x^2 - 12x + 4\right) - 2\left(9x^2 + 13x - 10\right)$$
$$= 15x^2 - 36x + 12 - 18x^2 - 26x + 20$$
$$= -3x^2 - 62x + 32$$

21. Subtract.
$$\begin{array}{r} 6x^3 - 2x^2 + 3x - 1 \\ -4x^3 + 2x^2 - 6x + 3 \\ \hline \end{array}$$

Change the sign of each term in
$-4x^3 + 2x^2 - 6x + 3$, and add.

$$\begin{array}{r} 6x^3 - 2x^2 + 3x - 1 \\ 4x^3 - 2x^2 + 6x - 3 \\ \hline 10x^3 - 4x^2 + 9x - 4 \end{array}$$

23.
$$(3x + 1)(2x - 7)$$
$$= (3x)(2x) + (3x)(-7)$$
$$+ 1(2x) + 1(-7) \qquad FOIL$$
$$= 6x^2 - 21x + 2x - 7$$
$$= 6x^2 - 19x - 7$$

25.
$$(4x - 1)(x - 2)$$
$$= 4x^2 - 8x - x + 2$$
$$= 4x^2 - 9x + 2$$

27.
$$(4t + 3)(4t - 3) = (4t)^2 - 3^2$$
$$= 16t^2 - 9$$

29.
$$(2y^2 + 4)(2y^2 - 4) = (2y^2)^2 - 4^2$$
$$= 4y^4 - 16$$

31.
$$(4x - 3)^2 = (4x)^2 - 2(4x)(3) + 3^2$$
$$= 16x^2 - 24x + 9$$

33.
$$(6r + 5y)^2 = (6r)^2 + 2(6r)(5y) + (5y)^2$$
$$= 36r^2 + 60ry + 25y^2$$

35.
$$(c + 2d)\left(c^2 - 2cd + 4d^2\right)$$
Multiply vertically.

$$\begin{array}{r} c^2 - 2cd + 4d^2 \\ c + 2d \\ \hline 2c^2 d - 4cd^2 + 8d^3 \\ c^3 - 2c^2 d + 4cd^2 \\ \hline c^3 + 8d^3 \end{array}$$

37.
$$16x^2 + 4x + 1$$
$$\underline{4x - 1}$$
$$\underline{-16x^2 - 4x - 1}$$
$$\underline{64x^3 + 16x^2 + 4x}$$
$$64x^3 - 1$$

39.
$$2t^2 + 5st - s^2$$
$$\underline{7t + 5s}$$
$$\underline{10st^2 + 25s^2t - 5s^3}$$
$$\underline{14t^3 + 35st^2 - 7s^2t}$$
$$14t^3 + 45st^2 + 18s^2t - 5s^3$$

41. $8x^3y^4 + 12x^2y^3 + 36xy^4$
The GCF is $4xy^3$.
$$= 4xy^3\left(2x^2y + 3x + 9y\right)$$

43. $x^2 - 2x - 15 = (x+3)(x-5)$

45. $2x^2 - 9x - 18 = (2x+3)(x-6)$

47. $36t^2 - 25 = (6t)^2 - 5^2$
$$= (6t+5)(6t-5)$$

49. $16t^2 + 24t + 9 = (4t)^2 + 2(4t)(3) + 3^2$
$$= (4t+3)^2$$

51. $4m^2p - 12mnp + 9n^2p$
$$= p\left(4m^2 - 12mn + 9n^2\right)$$
$$= p\left[(2m)^2 - 2(2m)(3n) + (3n)^2\right]$$
$$= p(2m - 3n)^2$$

53. $x^3 + 1$
$$= (x)^3 + (1)^3$$
$$= (x+1)\left[(x)^2 - (x)(1) + (1)^2\right] \quad \textit{Sum of cubes}$$
$$= (x+1)\left(x^2 - x + 1\right)$$

55. $8t^3 + 125 = (2t)^3 + (5)^3$
$$= (2t+5)\left[(2t)^2 - (2t)(5) + (5)^2\right]$$
$$= (2t+5)\left(4t^2 - 10t + 25\right)$$

57. $t^6 - 125 = \left(t^2\right)^3 - (5)^3$
$$= \left(t^2 - 5\right)\left[\left(t^2\right)^2 + \left(t^2\right)(5) + (5)^2\right]$$
$$\textit{Difference of cubes}$$
$$= \left(t^2 - 5\right)\left(t^4 + 5t^2 + 25\right)$$

59. $5xt + 15xr + 2yt + 6yr$
$$= (5xt + 15xr) + (2yt + 6yr)$$
$$= 5x(t + 3r) + 2y(t + 3r)$$
$$= (t + 3r)(5x + 2y)$$

61. $6ar + 12br - 5as - 10bs$
$$= (6ar + 12br) + (-5as - 10bs)$$
$$= 6r(a + 2b) - 5s(a + 2b)$$
$$= (a + 2b)(6r - 5s)$$

63. $t^4 - 1 = \left(t^2\right)^2 - 1^2$
$$= \left(t^2 + 1\right)\left(t^2 - 1\right)$$
$$= \left(t^2 + 1\right)(t + 1)(t - 1)$$

65. $4x^2 + 12xy + 9y^2 - 1$
The first three terms form a
perfect square trinomial.
$$= \left(4x^2 + 12xy + 9y^2\right) - 1$$
$$= \left[(2x)^2 + 2(2x)(3y) + (3y)^2\right] - 1$$
$$= (2x + 3y)^2 - 1^2$$
Now factor the difference of squares.
$$= [(2x + 3y) + 1][(2x + 3y) - 1]$$
$$= (2x + 3y + 1)(2x + 3y - 1)$$

APPENDIX G SYNTHETIC DIVISION

1. Synthetic division provides a quick, easy way to divide a polynomial by a binomial of the form $x - k$.

3. $\dfrac{x^2 - 6x + 5}{x - 1}$

$1 \mid \begin{array}{ccc} 1 & -6 & 5 \end{array}$ ← *Coefficients of numerator*

$ \begin{array}{ccc} & 1 & -5 \end{array}$

$ \begin{array}{ccc} 1 & -5 & 0 \end{array}$ *Write the answer from*

$ \begin{array}{cc} \downarrow & \downarrow \end{array}$ *the bottom row.*

$ \begin{array}{cc} x & - 5 \end{array}$

Answer: $x - 5$

5. $\dfrac{4m^2 + 19m - 5}{m + 5}$

$m + 5 = m - (-5)$, so use -5.

$-5 \mid \begin{array}{ccc} 4 & 19 & -5 \end{array}$

$ \begin{array}{ccc} & -20 & 5 \end{array}$

$ \begin{array}{ccc} 4 & -1 & 0 \end{array}$

Answer: $4m - 1$

7. $\dfrac{2a^2 + 8a + 13}{a + 2}$

$a + 2 = a - (-2)$, so use -2.

$-2 \mid \begin{array}{ccc} 2 & 8 & 13 \end{array}$

$ \begin{array}{ccc} & -4 & -8 \end{array}$

$ \begin{array}{ccc} 2 & 4 & 5 \end{array}$ ← *Remainder*

The quotient polynomial is $2a + 4$ and the remainder is 5.

Answer: $2a + 4 + \dfrac{5}{a + 2}$

9. $(p^2 - 3p + 5) \div (p + 1)$

$-1 \mid \begin{array}{ccc} 1 & -3 & 5 \end{array}$

$ \begin{array}{ccc} & -1 & 4 \end{array}$

$ \begin{array}{ccc} 1 & -4 & 9 \end{array}$

Answer: $p - 4 + \dfrac{9}{p + 1}$

11. $\dfrac{4a^3 - 3a^2 + 2a - 3}{a - 1}$

$1 \mid \begin{array}{cccc} 4 & -3 & 2 & -3 \end{array}$

$ \begin{array}{cccc} & 4 & 1 & 3 \end{array}$

$ \begin{array}{cccc} 4 & 1 & 3 & 0 \end{array}$

Answer: $4a^2 + a + 3$

13. $(x^5 - 2x^3 + 3x^2 - 4x - 2) \div (x - 2)$

Insert 0 for the missing x^4-term.

$2 \mid \begin{array}{cccccc} 1 & 0 & -2 & 3 & -4 & -2 \end{array}$

$ \begin{array}{cccccc} & 2 & 4 & 4 & 14 & 20 \end{array}$

$ \begin{array}{cccccc} 1 & 2 & 2 & 7 & 10 & 18 \end{array}$ ← *Remainder*

Answer: $x^4 + 2x^3 + 2x^2 + 7x + 10 + \dfrac{18}{x - 2}$

15. $(-4r^6 - 3r^5 - 3r^4 + 5r^3 - 6r^2 + 3r + 3) \div (r - 1)$

$1 \mid \begin{array}{ccccccc} -4 & -3 & -3 & 5 & -6 & 3 & 3 \end{array}$

$ \begin{array}{ccccccc} & -4 & -7 & -10 & -5 & -11 & -8 \end{array}$

$ \begin{array}{ccccccc} -4 & -7 & -10 & -5 & -11 & -8 & -5 \end{array}$ ← *Remainder*

Answer:

$-4r^5 - 7r^4 - 10r^3 - 5r^2 - 11r - 8 + \dfrac{-5}{r - 1}$

17. $(-3y^5 + 2y^4 - 5y^3 - 6y^2 - 1) \div (y + 2)$

Insert 0 for the missing y-term.

$-2 \mid \begin{array}{cccccc} -3 & 2 & -5 & -6 & 0 & -1 \end{array}$

$ \begin{array}{cccccc} & 6 & -16 & 42 & -72 & 144 \end{array}$

$ \begin{array}{cccccc} -3 & 8 & -21 & 36 & -72 & 143 \end{array}$ ← *Remainder*

Answer:

$-3y^4 + 8y^3 - 21y^2 + 36y - 72 + \dfrac{143}{y + 2}$

19. $\dfrac{y^3 + 1}{y - 1} = \dfrac{y^3 + 0y^2 + 0y + 1}{y - 1}$

$1 \mid \begin{array}{cccc} 1 & 0 & 0 & 1 \end{array}$

$ \begin{array}{cccc} & 1 & 1 & 1 \end{array}$

$ \begin{array}{cccc} 1 & 1 & 1 & 2 \end{array}$ ← *Remainder*

Answer: $y^2 + y + 1 + \dfrac{2}{y - 1}$

21. $P(x) = 2x^3 - 4x^2 + 5x - 3; k = 2$

To find $P(2)$, divide the polynomial by $x - 2$. $P(2)$ will be the remainder.

$2 \mid \begin{array}{cccc} 2 & -4 & 5 & -3 \end{array}$

$ \begin{array}{cccc} & 4 & 0 & 10 \end{array}$

$ \begin{array}{cccc} 2 & 0 & 5 & 7 \end{array}$ ← *Remainder*

By the remainder theorem, $P(2) = 7$.

23. $P(r) = -r^3 - 5r^2 - 4r - 2; k = -4$

Divide by $r + 4$. The remainder is equal to $P(-4)$.

$-4 \mid \begin{array}{cccc} -1 & -5 & -4 & -2 \end{array}$

$ \begin{array}{cccc} & 4 & 4 & 0 \end{array}$

$ \begin{array}{cccc} -1 & -1 & 0 & -2 \end{array}$ ← *Remainder*

By the remainder theorem, $P(-4) = -2$.

25. $P(y) = 2y^3 - 4y^2 + 5y - 33; k = 3$

Divide by $y - 3$. The remainder is equal to $P(3)$.

$$3 \ \overline{\big)\ 2 \quad -4 \quad 5 \quad -33}$$
$$ \quad\quad 6 \quad\ 6 \quad\ 33$$
$$ \overline{2 \quad\ 2 \quad 11 \quad\ 0} \quad \leftarrow \quad \textit{Remainder}$$

By the remainder theorem, $P(3) = 0$.

27. By the remainder theorem, a zero remainder means that $P(k) = 0$; that is, k is a number that makes $P(x) = 0$.

29. Is $x = -2$ a solution of
$$x^3 - 2x^2 - 3x + 10 = 0?$$

To decide whether -2 is a solution to the given equation, divide the polynomial by $x + 2$.

$$-2 \ \overline{\big)\ 1 \quad -2 \quad -3 \quad 10}$$
$$ \quad\quad\ -2 \quad\ 8 \quad -10$$
$$ \overline{1 \quad -4 \quad\ 5 \quad\ 0} \quad \leftarrow \quad \textit{Remainder}$$

Since the remainder is 0, -2 is a solution of the equation.

31. Is $m = -2$ a solution of
$$m^4 + 2m^3 - 3m^2 + 8m - 8 = 0?$$

To decide whether -2 is a solution to the given equation, divide the polynomial by $m + 2$.

$$-2 \ \overline{\big)\ 1 \quad 2 \quad -3 \quad 8 \quad -8}$$
$$ \quad\quad -2 \quad\ 0 \quad\ 6 \quad -28$$
$$ \overline{1 \quad 0 \quad -3 \quad 14 \quad -36} \quad \leftarrow \quad \textit{Remainder}$$

Since the remainder is not 0, -2 is not a solution of the equation.

33. Is $a = -2$ a solution of
$$3a^3 + 2a^2 - 2a + 11 = 0?$$

$$-2 \ \overline{\big)\ 3 \quad 2 \quad -2 \quad 11}$$
$$ \quad\quad -6 \quad\ 8 \quad -12$$
$$ \overline{3 \quad -4 \quad\ 6 \quad -1} \quad \leftarrow \quad \textit{Remainder}$$

Since the remainder is not 0, -2 is not a solution of the equation.

35. Is $x = -3$ a solution of
$$2x^3 - x^2 - 13x + 24 = 0?$$

$$-3 \ \overline{\big)\ 2 \quad -1 \quad -13 \quad 24}$$
$$ \quad\quad -6 \quad\ 21 \quad -24$$
$$ \overline{2 \quad -7 \quad\ 8 \quad\ 0} \quad \leftarrow \quad \textit{Remainder}$$

Since the remainder is 0, -3 is a solution of the equation.

In Exercises 37–41,
$$P(x) = 2x^2 + 5x - 12.$$

37. Factor $P(x)$.

$$2x^2 + 5x - 12 = (2x - 3)(x + 4)$$

38. Solve $P(x) = 0$.
$$2x^2 + 5x - 12 = 0$$
$$(2x - 3)(x + 4) = 0$$

$$2x - 3 = 0 \quad \text{or} \quad x + 4 = 0$$
$$2x = 3 \quad\quad\quad\quad x = -4$$
$$x = \tfrac{3}{2}$$

The solution set is $\left\{-4, \tfrac{3}{2}\right\}$.

39. $P(-4) = 2(-4)^2 + 5(-4) - 12$
$$= 2(16) - 20 - 12$$
$$= 32 - 20 - 12 = 0$$

$P\left(\tfrac{3}{2}\right) = 2\left(\tfrac{3}{2}\right)^2 + 5\left(\tfrac{3}{2}\right) - 12$
$$= 2\left(\tfrac{9}{4}\right) + \tfrac{15}{2} - 12$$
$$= \tfrac{9}{2} + \tfrac{15}{2} - \tfrac{24}{2} = 0$$

40. If $P(a) = 0$, then $x - \underline{a}$ is a factor of $P(x)$.

41. $Q(x) = 3x^3 - 4x^2 - 17x + 6$
$Q(3) = 3(3)^3 - 4(3)^2 - 17(3) + 6$
$$= 81 - 36 - 51 + 6 = 0$$

Since $Q(3) = 0$, $x - 3$ is a factor of $Q(x)$. To check, use synthetic division to see if 3 is a solution of the equation.

$$3 \ \overline{\big)\ 3 \quad -4 \quad -17 \quad 6}$$
$$ \quad\quad\ 9 \quad\ 15 \quad -6$$
$$ \overline{3 \quad\ 5 \quad -2 \quad\ 0}$$

Therefore, $x - 3$ is a factor of the polynomial and

$$3x^3 - 4x^2 - 17x + 6 = (x - 3)\left(3x^2 + 5x - 2\right)$$
$$Q(x) = (x - 3)(3x - 1)(x + 2).$$

43. From the graph, it appears that $x = 3$ is a solution of the equation
$$x^3 - x^2 - 21x + 45 = 0.$$

Check this with synthetic division.

$$3 \ \overline{\big)\ 1 \quad -1 \quad -21 \quad 45}$$
$$ \quad\quad\ 3 \quad\ 6 \quad -45$$
$$ \overline{1 \quad\ 2 \quad -15 \quad\ 0} \quad \leftarrow \quad \textit{Remainder}$$

Since the remainder is 0, 3 is a solution of the equation.

45. From the graph, it appears that $x = -1$ is a solution of the equation
$$x^3 + 3x^2 - 13x - 15 = 0.$$

Check this with synthetic division.

$$-1 \ \overline{\big)\ 1 \quad 3 \quad -13 \quad -15}$$
$$ \quad\quad -1 \quad -2 \quad\ 15$$
$$ \overline{1 \quad 2 \quad -15 \quad\ 0} \quad \leftarrow \quad \textit{Remainder}$$

Since the remainder is 0, -1 is a solution of the equation.

APPENDIX H DETERMINANTS AND CRAMER'S RULE

1. **(a)** *True*, a matrix is an array of numbers, while a determinant is a single number.

(b) *True*, a square matrix has the same number of rows as columns.

(c) *False*, the determinant $\begin{vmatrix} a & b \\ c & d \end{vmatrix}$ is equal to $ad - bc$.

(d) The value $\begin{vmatrix} 0 & 0 \\ x & y \end{vmatrix}$ is zero for any replacements of x and y.

$$\begin{vmatrix} 0 & 0 \\ x & y \end{vmatrix} = 0(y) - 0(x) = 0$$

No matter what replacements are used for x and y, the value of the determinant is zero since both x and y are being multiplied by zero. The statement is *true*.

3. $\begin{vmatrix} -2 & 5 \\ -1 & 4 \end{vmatrix} = -2(4) - 5(-1)$
$$= -8 + 5 = -3$$

5. $\begin{vmatrix} 1 & -2 \\ 7 & 0 \end{vmatrix} = 1(0) - (-2)7$
$$= 0 + 14 = 14$$

7. $\begin{vmatrix} 0 & 4 \\ 0 & 4 \end{vmatrix} = 0(4) - 4(0)$
$$= 0 - 0 = 0$$

9. $\begin{vmatrix} -1 & 2 & 4 \\ -3 & -2 & -3 \\ 2 & -1 & 5 \end{vmatrix}$ Expand by minors about the first column.

$= -1\begin{vmatrix} -2 & -3 \\ -1 & 5 \end{vmatrix} - (-3)\begin{vmatrix} 2 & 4 \\ -1 & 5 \end{vmatrix}$

$\quad + 2\begin{vmatrix} 2 & 4 \\ -2 & -3 \end{vmatrix}$

$= -1[-2(5) - (-3)(-1)]$
$\quad + 3[2(5) - 4(-1)] + 2[2(-3) - 4(-2)]$

$= -1(-13) + 3(14) + 2(2)$

$= 13 + 42 + 4 = 59$

11. $\begin{vmatrix} 1 & 0 & -2 \\ 0 & 2 & 3 \\ 1 & 0 & 5 \end{vmatrix}$ There are two 0s in column 2. We'll expand about that column since there is only 1 minor to evaluate.

$= -0\begin{vmatrix} 0 & 3 \\ 1 & 5 \end{vmatrix} + 2\begin{vmatrix} 1 & -2 \\ 1 & 5 \end{vmatrix} - 0\begin{vmatrix} 1 & -2 \\ 0 & 3 \end{vmatrix}$

$= 0 + 2[1(5) - (-2)(1)] - 0$

$= 2[5 + 2] = 2(7) = 14$

13. Multiply the upper left and lower right entries. Then multiply the upper right and lower left entries. Subtract the second product from the first to obtain the determinant. For example,

$$\begin{vmatrix} 4 & 2 \\ 7 & 1 \end{vmatrix} = 4 \cdot 1 - 2 \cdot 7$$
$$= 4 - 14 = -10.$$

15. $\begin{vmatrix} 3 & -1 & 2 \\ 1 & 5 & -2 \\ 0 & 2 & 0 \end{vmatrix}$ Expand about row 3.

$= 0 - 2\begin{vmatrix} 3 & 2 \\ 1 & -2 \end{vmatrix} + 0$

$= -2[3(-2) - 2(1)]$

$= -2[-6 - 2] = -2[-8] = 16$

17. $\begin{vmatrix} 0 & 0 & 3 \\ 4 & 0 & -2 \\ 2 & -1 & 3 \end{vmatrix}$ Expand about row 1.

$= 0 - 0 + 3\begin{vmatrix} 4 & 0 \\ 2 & -1 \end{vmatrix}$

$= 3[4(-1) - 0(2)]$

$= 3(-4) = -12$

19. $\begin{vmatrix} 1 & 1 & 2 \\ 5 & 5 & 7 \\ 3 & 3 & 1 \end{vmatrix}$ Expand about row 1.

$= 1\begin{vmatrix} 5 & 7 \\ 3 & 1 \end{vmatrix} - 1\begin{vmatrix} 5 & 7 \\ 3 & 1 \end{vmatrix} + 2\begin{vmatrix} 5 & 5 \\ 3 & 3 \end{vmatrix}$

$= 1[5(1) - 7(3)] - 1[5(1) - 7(3)]$
$\quad + 2[5(3) - 5(3)]$

$= 1(-16) - 1(-16) + 2(0)$

$= -16 + 16 + 0 = 0$

21. $x = \dfrac{D_x}{D} = \dfrac{-43}{-43} = 1$

$y = \dfrac{D_y}{D} = \dfrac{0}{-43} = 0$

$z = \dfrac{D_z}{D} = \dfrac{43}{-43} = -1$

The solution set is $\{(1, 0, -1)\}$.

23. $5x + 2y = -3$
$4x - 3y = -30$

$D = \begin{vmatrix} 5 & 2 \\ 4 & -3 \end{vmatrix} = -15 - 8 = -23$

$D_x = \begin{vmatrix} -3 & 2 \\ -30 & -3 \end{vmatrix} = 9 + 60 = 69$

$D_y = \begin{vmatrix} 5 & -3 \\ 4 & -30 \end{vmatrix} = -150 + 12 = -138$

$x = \dfrac{D_x}{D} = \dfrac{69}{-23} = -3; y = \dfrac{D_y}{D} = \dfrac{-138}{-23} = 6$

The solution set is $\{(-3, 6)\}$.

25. $3x - y = 9$
$2x + 5y = 8$

$$D = \begin{vmatrix} 3 & -1 \\ 2 & 5 \end{vmatrix} = 15 + 2 = 17$$

$$D_x = \begin{vmatrix} 9 & -1 \\ 8 & 5 \end{vmatrix} = 45 + 8 = 53$$

$$D_y = \begin{vmatrix} 3 & 9 \\ 2 & 8 \end{vmatrix} = 24 - 18 = 6$$

$$x = \frac{D_x}{D} = \frac{53}{17}; y = \frac{D_y}{D} = \frac{6}{17}$$

The solution set is $\left\{ \left(\frac{53}{17}, \frac{6}{17} \right) \right\}$.

27. $4x + 5y = 6$
$7x + 8y = 9$

$$D = \begin{vmatrix} 4 & 5 \\ 7 & 8 \end{vmatrix} = 32 - 35 = -3$$

$$D_x = \begin{vmatrix} 6 & 5 \\ 9 & 8 \end{vmatrix} = 48 - 45 = 3$$

$$D_y = \begin{vmatrix} 4 & 6 \\ 7 & 9 \end{vmatrix} = 36 - 42 = -6$$

$$x = \frac{D_x}{D} = \frac{3}{-3} = -1; y = \frac{D_y}{D} = \frac{-6}{-3} = 2$$

The solution set is $\{(-1, 2)\}$.

29. $x - y + 6z = 19$
$3x + 3y - z = 1$
$x + 9y + 2z = -19$

$$D = \begin{vmatrix} 1 & -1 & 6 \\ 3 & 3 & -1 \\ 1 & 9 & 2 \end{vmatrix} \text{ Expand about column 1.}$$

$$= 1\begin{vmatrix} 3 & -1 \\ 9 & 2 \end{vmatrix} - 3\begin{vmatrix} -1 & 6 \\ 9 & 2 \end{vmatrix} + 1\begin{vmatrix} -1 & 6 \\ 3 & -1 \end{vmatrix}$$

$$= 1(6 + 9) - 3(-2 - 54) + 1(1 - 18)$$
$$= 15 + 168 - 17 = 166$$

$$D_x = \begin{vmatrix} 19 & -1 & 6 \\ 1 & 3 & -1 \\ -19 & 9 & 2 \end{vmatrix} \text{ Expand about column 3.}$$

$$= 6\begin{vmatrix} 1 & 3 \\ -19 & 9 \end{vmatrix} - (-1)\begin{vmatrix} 19 & -1 \\ -19 & 9 \end{vmatrix}$$

$$+ 2\begin{vmatrix} 19 & -1 \\ 1 & 3 \end{vmatrix}$$

$$= 6(9 + 57) + 1(171 - 19) + 2(57 + 1)$$
$$= 396 + 152 + 116 = 664$$

$$D_y = \begin{vmatrix} 1 & 19 & 6 \\ 3 & 1 & -1 \\ 1 & -19 & 2 \end{vmatrix} \text{ Expand about column 1.}$$

$$= 1\begin{vmatrix} 1 & -1 \\ -19 & 2 \end{vmatrix} - 3\begin{vmatrix} 19 & 6 \\ -19 & 2 \end{vmatrix} + 1\begin{vmatrix} 19 & 6 \\ 1 & -1 \end{vmatrix}$$

$$= 1(2 - 19) - 3(38 + 114) + 1(-19 - 6)$$
$$= -17 - 456 - 25 = -498$$

$$D_z = \begin{vmatrix} 1 & -1 & 19 \\ 3 & 3 & 1 \\ 1 & 9 & -19 \end{vmatrix} \text{ Expand about column 1.}$$

$$= 1\begin{vmatrix} 3 & 1 \\ 9 & -19 \end{vmatrix} - 3\begin{vmatrix} -1 & 19 \\ 9 & -19 \end{vmatrix} + 1\begin{vmatrix} -1 & 19 \\ 3 & 1 \end{vmatrix}$$

$$= 1(-57 - 9) - 3(19 - 171) + 1(-1 - 57)$$
$$= -66 + 456 - 58 = 332$$

$$x = \frac{D_x}{D} = \frac{664}{166} = 4; y = \frac{D_y}{D} = \frac{-498}{166} = -3$$

$$z = \frac{D_z}{D} = \frac{332}{166} = 2$$

The solution set is $\{(4, -3, 2)\}$.

31. $7x + y - z = 4$
$2x - 3y + z = 2$
$-6x + 9y - 3z = -6$

$$D = \begin{vmatrix} 7 & 1 & -1 \\ 2 & -3 & 1 \\ -6 & 9 & -3 \end{vmatrix} \text{ Expand about column 3.}$$

$$= -1\begin{vmatrix} 2 & -3 \\ -6 & 9 \end{vmatrix} - 1\begin{vmatrix} 7 & 1 \\ -6 & 9 \end{vmatrix} - 3\begin{vmatrix} 7 & 1 \\ 2 & -3 \end{vmatrix}$$

$$= -1(18 - 18) - 1(63 + 6) - 3(-21 - 2)$$
$$= 0 - 69 + 69 = 0$$

Because $D = 0$, Cramer's rule does not apply.

33. $-x + 2y = 4$
$3x + y = -5$
$2x + z = -1$

$$D = \begin{vmatrix} -1 & 2 & 0 \\ 3 & 1 & 0 \\ 2 & 0 & 1 \end{vmatrix} \text{ Expand about column 3.}$$

$$= 0 - 0 + 1\begin{vmatrix} -1 & 2 \\ 3 & 1 \end{vmatrix}$$

$$= 1(-1 - 6) = -7$$

$$D_x = \begin{vmatrix} 4 & 2 & 0 \\ -5 & 1 & 0 \\ -1 & 0 & 1 \end{vmatrix} \text{ Expand about column 3.}$$

$$= 0 - 0 + 1\begin{vmatrix} 4 & 2 \\ -5 & 1 \end{vmatrix}$$

$$= 1(4 + 10) = 14$$

$$D_y = \begin{vmatrix} -1 & 4 & 0 \\ 3 & -5 & 0 \\ 2 & -1 & 1 \end{vmatrix} \text{ Expand about column 3.}$$

$$= 0 - 0 + 1\begin{vmatrix} -1 & 4 \\ 3 & -5 \end{vmatrix}$$

$$= 1(5 - 12) = -7$$

$$D_z = \begin{vmatrix} -1 & 2 & 4 \\ 3 & 1 & -5 \\ 2 & 0 & -1 \end{vmatrix} \quad \text{Expand about row 3.}$$

$$= 2 \begin{vmatrix} 2 & 4 \\ 1 & -5 \end{vmatrix} - 0 - 1 \begin{vmatrix} -1 & 2 \\ 3 & 1 \end{vmatrix}$$

$$= 2(-10 - 4) - 1(-1 - 6)$$

$$= -28 + 7 = -21$$

$$x = \frac{D_x}{D} = \frac{14}{-7} = -2; \; y = \frac{D_y}{D} = \frac{-7}{-7} = 1$$

$$z = \frac{D_z}{D} = \frac{-21}{-7} = 3$$

The solution set is $\{(-2, 1, 3)\}$.

35.
$$\begin{array}{rcrcrcr} -5x & - & y & & & = & -10 \\ 3x & + & 2y & + & z & = & -3 \\ & & -y & - & 2z & = & -13 \end{array}$$

$$D = \begin{vmatrix} -5 & -1 & 0 \\ 3 & 2 & 1 \\ 0 & -1 & -2 \end{vmatrix} \quad \text{Expand about column 3.}$$

$$= 0 - 1 \begin{vmatrix} -5 & -1 \\ 0 & -1 \end{vmatrix} - 2 \begin{vmatrix} -5 & -1 \\ 3 & 2 \end{vmatrix}$$

$$= -1(5 - 0) - 2(-10 + 3)$$

$$= -5 + 14 = 9$$

$$D_x = \begin{vmatrix} -10 & -1 & 0 \\ -3 & 2 & 1 \\ -13 & -1 & -2 \end{vmatrix} \quad \text{Expand about column 3.}$$

$$= 0 - 1 \begin{vmatrix} -10 & -1 \\ -13 & -1 \end{vmatrix} - 2 \begin{vmatrix} -10 & -1 \\ -3 & 2 \end{vmatrix}$$

$$= -1(10 - 13) - 2(-20 - 3)$$

$$= 3 + 46 = 49$$

$$D_y = \begin{vmatrix} -5 & -10 & 0 \\ 3 & -3 & 1 \\ 0 & -13 & -2 \end{vmatrix} \quad \text{Expand about column 3.}$$

$$= 0 - 1 \begin{vmatrix} -5 & -10 \\ 0 & -13 \end{vmatrix} - 2 \begin{vmatrix} -5 & -10 \\ 3 & -3 \end{vmatrix}$$

$$= -1(65) - 2(15 + 30)$$

$$= -65 - 90 = -155$$

$$D_z = \begin{vmatrix} -5 & -1 & -10 \\ 3 & 2 & -3 \\ 0 & -1 & -13 \end{vmatrix} \quad \text{Expand about row 3.}$$

$$= 0 - (-1) \begin{vmatrix} -5 & -10 \\ 3 & -3 \end{vmatrix} - 13 \begin{vmatrix} -5 & -1 \\ 3 & 2 \end{vmatrix}$$

$$= 1(15 + 30) - 13(-10 + 3)$$

$$= 45 + 91 = 136$$

$$x = \frac{D_x}{D} = \frac{49}{9}; \; y = \frac{D_y}{D} = \frac{-155}{9} = -\frac{155}{9}$$

$$z = \frac{D_z}{D} = \frac{136}{9}$$

The solution set is $\left\{ \left(\frac{49}{9}, -\frac{155}{9}, \frac{136}{9} \right) \right\}$.

37.
$$\begin{vmatrix} 4 & x \\ 2 & 3 \end{vmatrix} = 8$$

Evaluate the determinant.

$$\begin{vmatrix} 4 & x \\ 2 & 3 \end{vmatrix} = 4(3) - x(2)$$

$$= 12 - 2x$$

Solve the equation.

$$12 - 2x = 8$$
$$-2x = -4$$
$$x = 2$$

The solution set is $\{2\}$.

39.
$$\begin{vmatrix} x & 4 \\ x & -3 \end{vmatrix} = 0$$

Evaluate the determinant.

$$\begin{vmatrix} x & 4 \\ x & -3 \end{vmatrix} = x(-3) - 4x$$

$$= -3x - 4x = -7x$$

Solve the equation.

$$-7x = 0$$
$$x = 0$$

The solution set is $\{0\}$.